MW00716972

Concrete Under Severe Conditions

Volume Two

BOOKS ON CONCRETE MATERIALS FROM E & FN SPON

Application of Admixtures in Concrete
Edited by A.M. Paillere
Blended Cements in Construction
Edited by R.N. Swamy
Calcium Aluminate Cements
Edited by R.G. Mangabhai
Concrete in Hot Environments
I. Soroka
Concrete in Marine Environments
P.K. Mehta
Concrete Mix Design, Quality Control and Specification
K.W. Day
Corrosion of Reinforcement in Concrete
Edited by C. Page, K. Treadaway and P. Bamforth
Disposal and Recycling of Organic and Polymeric Construction Materials
Edited by Y. Ohama
Durability of Building Materials and Components
Edited by S. Nagataki, T. Nireki and F. Tomosawa
Durability of Concrete in Cold Climates
M. Pigeon and R. Pleau
Euro-Cements: The Impact of ENV 197 on Concrete Construction
Edited by R.K. Dhir and M.R. Jones
Ferrocement
Edited by P.J. Nedwell and R.N. Swamy
Fibre Reinforced Cementitious Composites
A. Bentur and S. Mindess
Fly Ash in Concrete: Properties and Performance
Edited by K. Wesche
Hydration and Setting of Cements
Edited by A. Nonat and J-C. Mutin
Interfacial Transition Zone in Concrete
Edited by J.C. Maso
Manual of Ready-Mixed Concrete
J.D. Dewar and R. Anderson
Performance Criteria for Concrete Durability
Edited by J. Kropp and H.K. Hilsdorf
Protection of Concrete
Edited by R.K. Dhir and J.W. Green
Special Concretes: Workability and Mixing
Edited by P.J.M. Bartos
Thermal Cracking in Concrete at Early Ages
Edited by R. Springenschmid

Concrete Under Severe Conditions

Environment and loading

Proceedings of the International Conference on Concrete under Severe Conditions, CONSEC '95

Sapporo, Japan
2–4 August 1995

Volume Two

EDITED BY

K. Sakai
Civil Engineering Research Institute, Hokkaido Development Bureau, Japan

N. Banthia
University of British Columbia, Vancouver, Canada

O.E. Gjørv
Norwegian Institute of Technology, Trondheim, Norway

E & FN SPON
An Imprint of Chapman & Hall
London · Glasgow · Weinheim · New York · Tokyo · Melbourne · Madras

Published by E & FN Spon, an imprint of Chapman & Hall,
2–6 Boundary Row, London SE1 8HN, UK

Chapman & Hall, 2–6 Boundary Row, London SE1 8HN, UK

Blackie Academic & Professional, Wester Cleddens Road, Bishopbriggs, Glasgow G64 2NZ, UK

Chapman & Hall GmbH, Pappelallee 3, 69469 Weinheim, Germany

Chapman & Hall USA, 115 Fifth Avenue, New York, NY 10003, USA

Chapman & Hall Japan, ITP-Japan, Kyowa Building, 3F, 2-2-1 Hirakawacho, Chiyoda-ku, Tokyo 102, Japan

Chapman & Hall Australia, 102 Dodds Street, South Melbourne, Victoria 3205, Australia

Chapman & Hall India, R. Seshadri, 32 Second Main Road, CIT East, Madras 600 035, India

First edition 1995

Printed in Hong Kong by Wellprint Production

ISBN Volume One 0 419 19850 4
Volume Two 0 419 19860 1
2 volume set 0 419 19870 9

A catalogue record for this book is available from the British Library

Publisher's Note This book has been prepared from camera ready copy provided by the individual contributors in order to make the book available for the Conference.

Contents

VOLUME TWO

FOREWORD

Hokkaido accounts for one–fifth of Japan's land area, and it is expected to play an important role in realizing well–balanced development of the country toward the 21st century, by utilizing the prefecture's rich land resources. Therefore, the development of Hokkaido is being carried out systematically by the Hokkaido Development Agency under the Hokkaido Development Law.

The Hokkaido Development Bureau, which was founded as an organization to make development plans for Hokkaido and to execute projects, established the Civil Engineering Research Institute (CERI) in order to conduct such development efficiently and effectively in this cold and snowy region. CERI is playing the role of the only center in Japan for cold–region research related to public works.

CERI engages in various activities, which include international exchanges in science and technology. As for concrete technology, workshops on "Low Temperature Effects on Concrete" were held in both Sapporo and Ottawa in 1988 and 1991 on the basis of the Canada/Japan intergovernmental agreement on cooperation in science and technology, and fruitful results were gained.

This international conference, CONSEC '95, was planned to provide a place for exchanges of research information throughout the world.

Dr K. Sakai, the first editor of this book, is among those energetically promoting international activities as the Head of the Materials Section of CERI, and he is a prime organizer of CONSEC '95. Co–editors Prof N. Banthia and Prof O.E. Gjørv are internationally renowned scholars engaging in research and education at the University of British Columbia in Canada and the Norwegian Institute of Technology–NTH in Norway, respectively. CERI is conducting joint research with both universities under bilateral agreements between the governments.

There can be no doubt that concrete is a significant construction material that has supported industrial development in the 20th century. However, further development is required in concrete technology to lessen the burden imposed on the 21st century generation, by the efficient construction of infrastructures. In light of this circumstance, it is timely that researchers and engineers gather, exchange information and share the fruits of research to prepare for the coming century. I sincerely hope that this conference will greatly contribute to a future of mankind.

Yoshihiro Oyamada
Director General
Civil Engineering Research Institute
Hokkaido Development Bureau

PREFACE

While the origin of concrete is said to date back to the Greek and Roman periods, the modern history of concrete essentially finds its beginning in 1824, in Great Britain, when J. Aspdin invented the technology for producing cement. In 1918, the law of water–cement ratio was established by D.A. Abrams and in 1938, the effect of an air entraining agent was found. Thereafter, the use of ready–mixed concrete became prevalent. Concrete technology has progressed dramatically over the last 50 years. With the recent development of high–range water–reducing admixture, producing concrete with a strength of 100MPa is no longer as difficult as it was once considered to be. Furthermore, the development of structural analysis methods has facilitated the construction of concrete structures of exceedingly sophisticated forms, even under complex loading conditions.

Within this context, the prospects for concrete may seem bright but there are also significant problems. Concrete has been the most widely used construction material due to its low cost and solidity. However, as a result, we have been left with a vast infrastructure constructed of concrete. Costs for the maintenance, management and repair of these concrete structures have been increasing year by year. It is predicted that this will cause serious social problems in many countries in the near future.

A particularly serious issue in recent years has been the premature deterioration of concrete structures. While the causes of such deterioration are considered to be varied, the principal reason appears ultimately to stem from the failure to recognize the significance of "long–term durability", the most important feature of concrete, in spite of an extensive array of results from research and practical experience. It is no exaggeration, indeed, to say that progress in concrete technology remained entirely synonymous with improvements in construction methods. The premature deterioration could be the "side–effect" of these so–called "improvements". The gap between research and practice has also contributed greatly to the incidence of this side–effect.

In order to alter the present situation, it is necessary to reconstruct the framework of conventional concrete technology. In other words, the direction of future research must be reconsidered. This conference was planned with a keen awareness of the above issues, as may clearly be recognized in the subtitle, "Environment and Loading". The majority of the past international conferences concerning concrete was for certain specialized areas, such as structure, materials and durability. This has led to the participants' apparent lack of interest in matters outside of their specialties. However, in order to avoid bequeathing a "negative heritage" to the next generation, it is crucial to take a comprehensive look at concrete.

Consequently, this international conference aimed to provide specialists in the field with the opportunity of discussing issues transcending their own fields and identifying the interface between their specialties with the shared recognition of the need to "construct durable concrete structures". The keynote papers of Dr Hoff and Professor Shah were specifically written with this intention. In addition, we tried to devise a program so that this purpose would be realized. Although this was an ambitious attempt, we hope that it will provide impetus to the consideration of the future direction of concrete–related research.

More than 160 papers from 28 countries were submitted for presentation at the conference. The conference was one of the largest international events on the subject of concrete, the fruit of the splendid efforts of many people involved. The members of the Advisory Committee, in particular, contributed substantially to the success of the conference and we would like to express our heartfelt gratitude to them. We were delighted that more than 150 researchers and engineers cooperated with us in reviewing the papers, this being one of the most demanding tasks in the holding of an international conference, and we greatly appreciate their contribution.

It is with the greatest pleasure that the editors were able to compile the papers in two volumes. These papers represent the outstanding efforts of the authors who readily cooperated, even within the limited time. We owe the utmost respect to these authors.

We would also like to thank the chairpersons of each session, the large number of participants, co-sponsors, exhibitors, and colleagues of the Civil Engineering Research Institute, and the Hokkaido Development Bureau who brought about the success of this conference.

Furthermore, we would like to express our sincere gratitude to Professors T. Horiguchi, T. Ueda, K. Maruyama and T. Miyagawa for their tremendous efforts as members of the Working Group of the Conference Technical Committee, and finally, to Ms Werawan Manakul who did an excellent job as the secretary to the first editor.

Koji Sakai
Nemkumar Banthia
Odd E. Gjørv

Sapporo
August 1995

Organizing Committee

Chairman **Y. Oyamada**, Hokkaido Development Bureau
Vice Chairman **T. Shibata**, JCI Hokkaido Chapter
Secretary **K. Sakai**, Hokkaido Development Bureau
Members **N. Dohgakinai**, Hokkaido Development Engineering Center
Y. Sato, Hokkaido Development Association
T. Ohkoshi, Hokkaido Road Administration Engineering Center
T. Ohta, Hokkaido Ready-Mixed Concrete Technical Center
H. Ozaki, Cold Region Port and Harbor Engineering Research Center
K. Tateya, Hokkaido River Disaster Prevention Research Center
K. Tsukamoto, Agricultural Engineering Consultants Association

Executive Steering Committee

Chairman **Y. Oyamada**, Hokkaido Development Bureau
Vice Chairman **Y. Kakuta**, Hokkaido University
Secretary **K. Sakai**, Hokkaido Development Bureau
Members **K. Honda**, Hokkaido Development Bureau
K. Hoshi, Hokkaido Development Bureau
E. Kamada, Hokkaido University
Y. Mizuno, Hokkaido Development Bureau
M. Negishi, Hokkaido Development Bureau
S. Noto, Hokkaido Development Bureau
S. Ohmura, Hokkaido Development Bureau
T. Ohta, Hokkaido Ready Mixed Concrete Technical Center
H. Ohtaki, Hokkaido Development Bureau
N. Saeki, Hokkaido University
S. Shimohirao, Hokkaido Development Bureau
T. Soma, Hokkaido Development Bureau
K. Takezawa, Hokkaido Development Bureau

Technical Committee

Chairman	**K. Sakai**, Hokkaido Development Bureau
Vice Chairmen	**N. Banthia**, The University of British Columbia
	O.E. Gjørv, The Norwegian Institute of Technology-NTH
Secretary	**T. Horiguchi**, Hokkaido Institute of Technology
Members	**K. Ayuta**, Kitami Institute of Technology
	T. Fukute, Ministry of Transport
	O. Joh, Hokkaido University
	S. Kakuta, Akashi College of Technology
	T. Kawai, Shimizu Corporation
	H. Kawano, Ministry of Construction
	K. Maekawa, University of Tokyo
	T. Makizumi, Kyushu University
	K. Maruyama, Nagaoka University of Technology
	Y. Matsuoka, Taisei Corporation
	T. Miyagawa, Kyoto University
	A. Miyamoto, Kobe University
	K. Motohashi, Kajima Corporation
	H. Mutsuyoshi, Saitama University
	J. Niwa, University of Nagoya
	M. Ohtsu, Kumamoto University
	N. Ohtsuki, Tokyo Institute of Technology
	K. Rokugo, University of Gifu
	S. Sogo, Obayashi Corporation
	M. Suzuki, Ministry of Construction
	N. Takagi, Ritsumeikan University
	K. Takewaka, Kagoshima University
	T. Ueda, Hokkaido University
	T. Uomoto, University of Tokyo
	N. Vitharana, Department of Water Resources, Australia
	M. Yurugi, Kajima Corporation

Local Advisory Committee

M. Fujii, Kyoto University
Y. Kakuta, Hokkaido University
E. Kamada, Hokkaido University
K. Kohno, University of Tokushima
H. Mihashi, Tohoku University
T. Miura, Tohoku University
S. Nagataki, Tokyo Institute of Technology
T. Ohta, Hokkaido Ready Mixed Concrete Technical Center
H. Okamura, University of Tokyo
N. Saeki, Hokkaido University

International Advisory Committee

A.K. Aggarwal, University of Technology, PNG
V. Agopyan, University of São Paulo, Brazil
P.N. Balaguru, Rutgers University, USA
J.J. Beaudoin, IRC–NRC, Canada
A. Bentur, Technion, Israel
J.C. Chern, National University of Taiwan, Taiwan
G. Fagerlund, Lund Institute of Technology, Sweden
S. Gong, Shanghai Institute of Urban Construction, P.R. China
B.L. Karihaloo, University of Sydney, Australia
M.H. Kim, Chung–nam National University, Korea
V.M. Malhotra, CANMET, Canada
S. Mindess, University of British Columbia, Canada
J.P. Ollivier, INSA, France
C.L. Page, Aston University, UK
A.M. Paillere, LCPC, France
R. Park, University of Canterbury, New Zealand
V. Penttala, Helsinki University of Technology, Finland
M. Pigeon, Laval University, Canada
M.J. Setzer, University of Essen, Germany
S.P. Shah, Northwestern University, USA
S.A. Sheikh, University of Toronto, Canada

Sponsors and Co–sponsors

Sponsors
Civil Engineering Research Institute, Hokkaido Development Bureau
Japan Concrete Institute's Hokkaido Chapter
American Concrete Institute

Co–sponsors
Architectural Institute of Japan
International Union of Testing and Research Laboratories for Materials and Structures
Japan Concrete Institute
Japan Society of Civil Engineers
National Research Council Canada
The Norwegian Institute of Technology–NTH
The University of British Columbia
Agricultural Engineering Consultants Association
Cold Region Port and Harbor Engineering Research Center
Hokkaido Ready Mixed Concrete Technical Center
Hokkaido Development Association
Hokkaido Development Engineering Center
Hokkaido River Disaster Prevention Research Center
Hokkaido Road Administration Engineering Center

Reviewers

S. Ahmad
P.C. Aitcin
M. Al–Asaly
A. Al–Manaseer
S. Amasaki
C. Andrade
G. Arligue
K. Ayuta
D. Bager
P.N. Balaguru
N. Banthia
J. Baron
A. Bentur
N. Carino
R. Carrasquillo
R. Day
K. Demura
L. Elfgren
D. Feldman
F. Fouad
T. Fujii
T. Fukute
M.R. Geiker
N. Ghafoori
O.E. Gjørv
S. Goñi
R. Gray
T. Hakkinen
T. Hara
T. Hasegawa
G. Hoff
R.D. Hooton
T. Horiguchi
G. Horrigmo
M. Hossain
K. Hover
T. Ichinose
H. Imai
S. Inoue
S. Inoue
O. Joh
T. Kaku
S. Kakuta
Y. Kakuta
E. Kamada
T. Kanazu
T. Kawai
M. Kawakami
H. Kawano
R.J. Kettle
K. Khayat
N. Kishi
K. Kokubu
H. Kukko
C. Leung
G. Litvan
K.E. Løland
K. Maekawa
N. Mailvagnam
T. Makizumi
V.M. Malhotra
J. Marchand
K. Maruyama
M. Mashima
N. Matsumoto
Y. Matsuoka
H. Mihashi
R.H. Mills
S. Mindess
T. Miura
T. Miyagawa
A. Miyamoto
S. Miyazawa
H. Mizuguchi
B. Mobasher
P.J.M. Monteiro
D.R. Morgan
K. Motohashi
H. Mutsuyoshi
A.E. Naaman
T. Naik
E. Nawy
I. Nilsson
L.–O. Nilsson
J. Niwa
H. Nomachi
S. Numata
H. Ohga
Y. Ohno
M. Ohtsu
K. Ohtsuka
H. Ohuchi
T. Okamoto
J–P. Ollivier
S. Ono
T. Oshiro
N. Otsuki
S. Ozaki
C.L. Page
A.M. Paillere
V. Penttala
M. Pigeon
S. Pihlajavaara
R. Pleau
S. Popovics
J.A. Purkiss
V.S. Ramachandran
V. Ramakrishnan
D.V. Reddy
K. Rokugo
D.M. Roy
N. Saeki
S. Saiidi
E. Sakai
K. Sakai
A. Sarja
S. Sarkar
R. Sato
J. Scanlon
P. Seabrook
H. Seki
O. Senbu
M.J. Setzer
S.P. Shah
S.A. Sheikh
H. Shima
N. Short
M. Shoya
K. Shuttoh
J. Skalny
S. Sogo
E. Sørensen
M. Suzuki
N. Takagi
K. Takewaka
M. Tamai
K. Torii
T. Tsubaki
Y. Tsuji
P.J. Tumidjski
K. Tuutti
K. Uchida
T. Ueda
T. Uomoto
S. Ushijima
N. Vitharana
D. Whiting
D. Wood
Y. Yamamoto
A. Yonekura
T. Yonezawa
F. Young
M. Yurugi
P. Zia

PART FIFTEEN
CONCRETING UNDER SEVERE CONDITIONS

85 HOT WEATHER AND CONCRETE PRACTICES IN THE ARABIAN GULF COUNTRIES

H.M. ZEIN AL-ABIDEEN
Ministry for Public Works and Housing, Riyadh, Kingdom of Saudi Arabia

Abstract
A historical review of past, present and future concrete practices in the hot weather of the Arabian Gulf countries is presented in this paper. It is an attempt to answer some important questions such as: Is it true that the Gulf climate is the severest in the world in absolute terms? What is meant by the Arabian Gulf region? What is the ratio of deteriorated buildings to those which are still sound? What are the assumed ages of structures and what is the actual age? How can we get out of this dilemma? Is it by preparing building codes, by adopting modern technologies, by doing more research or by performing different types of maintenance?
There are no ready-made solutions. However, this is an attempt to shed some light on the practical approaches that engineers in this region could take. The aim is to preserve existing buildings and have confidence in future structures by improving their quality and performance and thus, lengthening their life in the hot weather and severe environment of the Gulf.
Keywords: Aggressive environment, Arabian Gulf countries, concrete practices, deterioration of structures, hot weather, repair methods, severe climate.

1 Introduction

There is insufficient information or database regarding the history of concrete practices, its background and the present and future situation in Arabian Gulf countries. One of the main objectives of this paper is to initiate the establishment of such a database to provide information to researchers solving local problems and to help development and to avoid negative aspects.

Native and foreign engineers and others concerned with concrete in Gulf Countries should assume their role to serve the building sector and, in particular, the concrete industry, which is the cornerstone for construction in the Gulf. Part of this can be achieved by attending and taking an active role in the technical committees which prepare reports about hot weather, such as ACI Committee 305 or RILEM International Committee 94, where Europeans who have no substantial hot weather in their countries prepare reports and specifications for Gulf countries, whereas natives seldom participate in such activities.

It is believed that establishing an institute in the Gulf to serve this important function is long overdue, as well as being pioneers in presenting scientific research, specifications, instructions, and suitable methods to improve the quality of concrete, protect and maintain it in order to extend its lifespan in the hot and aggressive environment of the Gulf and its severe climate.

Concrete Under Severe Conditions: Environment and loading (Volume Two) Edited by K. Sakai, N. Banthia and O.E. Gjørv. Published in 1995 by E & FN Spon. ISBN 0 419 19860 1

2 Historical background

It is difficult to generalize the experiences in Gulf countries. However, the obvious similarities between Arabian Gulf countries provide a historical background which might be different in its scale, importance and priority from one country to another, but in general, indicate the problems of concrete in our Arabian Gulf. The historical background presented here reflects the author's own experience and readings [1-8]. The following subsections summarize the history of concrete in the Gulf:

2.1 Before the 1960's until the 1970's

- Available materials were used without understanding their limitations, problems and correct proportions.
- Labourers, contractors, technicians, and engineers were poorly trained.
- Unavailability of construction standards and specifications.
- Shortage of supervision in most projects.
- Wrong designs.
- No attention was given to environmental and climatic conditions.
- No maintenance was conducted.

2.2 Between the 1970's until 1980's

Additional negative aspects appeared, and more developments took place in various proportions:

- Fast designs and quick execution due to a boom in business; owners accepted low quality in return for fast income.
- Inexperienced contractors joined the construction business.
- Partial knowledge started to accumulate regarding material, specifications, and solving problems.
- Little development in the experience of construction staff, together with the entry of foreign staff.
- Using different specifications, guidelines and technologies from the Far East, U.S.A. and Europe.
- Unqualified supervision.
- New problems appeared regarding the environment such as the rise of ground water table.
- Discovery of consequences of using foreign specifications which do not address local problems such as hot weather, chloride and sulphate effects... etc.
- Using ready mixed concrete without knowing its negative and positive effects.
- Starting maintenance and repair works.

2.3 From the 1980's until now

- Results of research to date indicate the causes of failures and deterioration of structures and which assists in providing repair methods.
- Attention is particularly given to deterioration of concrete buildings caused by corrosion of reinforcement on the Gulf coast.
- Some booklets, specifications, and general guidelines became available, covering some environmental conditions, but not all necessary aspects and not at the required level.
- The skill level of workmanship improved especially in large projects.
- Expansion in using ready mixed concrete, with the continuation of its negative and positive aspects (among the negative aspects, users took it for granted that it is always of good quality).
- Increased appearance of deterioration and expansion of remedial repair, without regular preventive maintenance.
- Negative aspects of improper repair and maintenance started to appear.

3 The definition of hot weather

According to ACI committee 305 [9], the definition of hot weather is any combination of the following conditions: a) high ambient temperature, b) high concrete temperature, c) low relative humidity, d) high wind velocity, and e) solar radiation. These conditions tend to impair the quality of freshly mixed or hardened concrete by accelerating the rate of moisture loss and rate of cement hydration. The potential problems of hot weather concreting may occur at any time of the year in warm tropical or arid climates, and generally occur during the summer season in other climates.

However, the draft report prepared by RILEM technical committee No.94 [10] details more divisions and differences, among which are: hot - dry or humid, moderate, arid, coastal, inland.. etc. The study concentrates on the evaluation of hot weather effects and environmental conditions that surrounds the structure directly (micro climate), rather than general environmental conditions of a vast area (macro climate). It is important to design structures in accordance with the climate, geography, topography ... etc. of the surrounding zone or the city where the construction will take place. This is believed to be a good approach that should be considered in any report which deals with hot weather concrete.

4 The need for accurate definition of Gulf environment for concrete industry

Environment means, for the purpose of this paper, all the external conditions surrounding concrete structures including, but not restricted to, climatic conditions, topography, geomorphology... etc. which affect raw materials, concrete production, and required serviceability.

Discussions are carried out in many researches about the Gulf environment without supplying an accurate definition of it. In some cases, it is taken as the coastal cities directly on the Gulf. In other cases it is considered as the entire Arabian Gulf region or even the entire area of the Gulf states and the Arabian peninsula, i.e. [3-8].

The deterioration of concrete and corrosion of reinforcement represents an extensive problem for coastal cities in the Gulf region. It is somewhat exaggerated through studies carried out by some of universities, many foreign companies, active researchers, and the local engineering societies working in this area. Therefore, all these together have led to the effect that the environment in this region was generalized in many researches to cover the entire Gulf states including the vast area of Saudi Arabia (Fig.1).

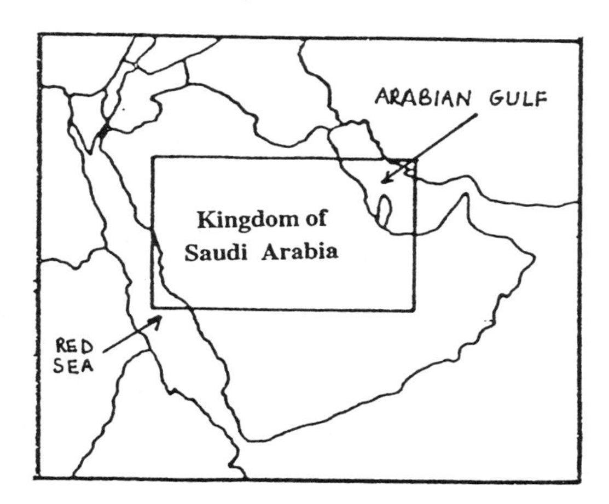

Fig. 1. The Kingdom of Saudi Arabia and the Arabian peninsula

The difference in temperatures, relative humidity, wind velocity and amount of rain and the extreme differences in geomorphological formations for aggregates, soil, and raw materials is remarkably distinct in each state, so how about the differences among east and west, north and south, and centre of the Arabian peninsula. Figure 2 shows some of these differences among the cities of Kuwait, Muscat, Jeddah and Riyadh [10,11].

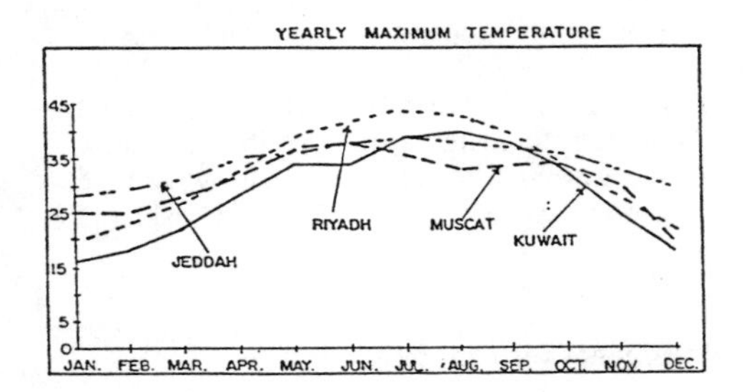

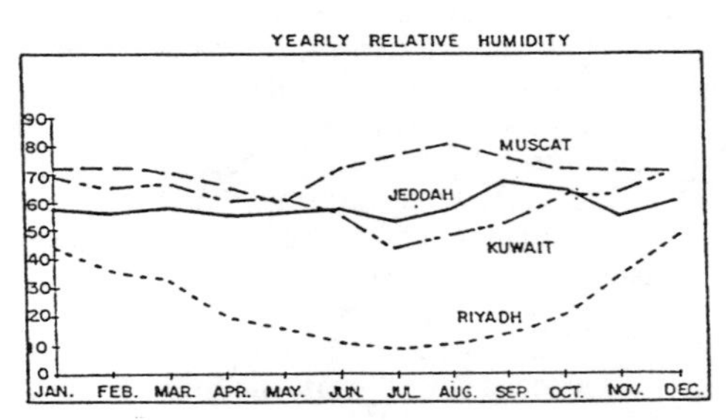

Fig. 2. Climatic variation in Gulf region [11]

Figure 3 indicates that climatic and other conditions could change between the shores and coastal regions extending few metres to about 5 km and more.

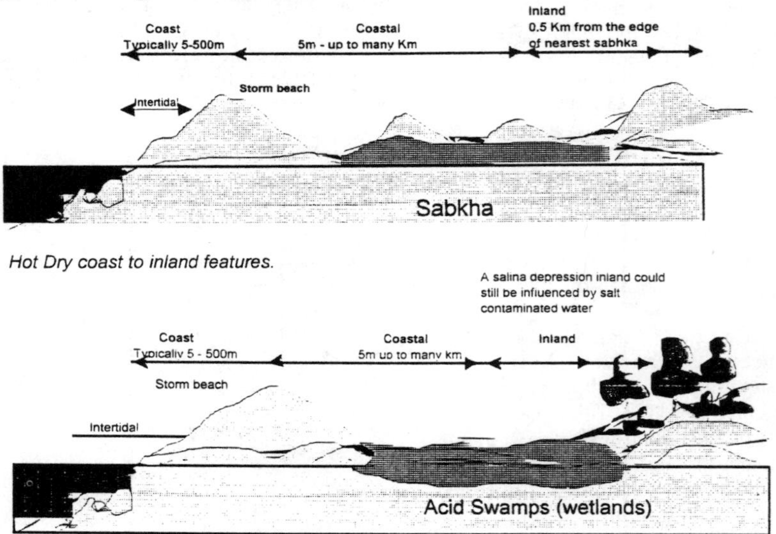

Hot wet coast to inland features.

Fig. 3. Sections through possible ground formations from coast to inland [10]

Figure 4 depicts the large differences in chloride attack caused by salt formation[10], and how they decline as we move further from the shore.

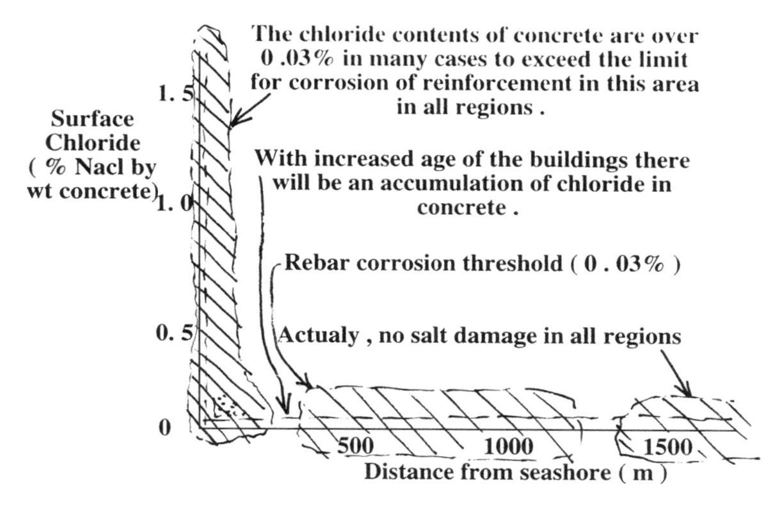

Fig. 4. Chlorides measured at concrete surface layer with distance from seashore (in Okinawa [10]).

An extensive study by the author [1,12] indicates that there is a fundamental difference between the aggregates available in the Eastern, Central and Western regions of Saudi Arabia. Therefore, the Gulf climate and environment cannot simply be considered as one. It is more appropriate to define the Gulf environment as a series of micro environments which represent topographical and geomorphological conditions in each region. In order to account for environmental differences and the availability of suitable raw materials, contamination of soil and atmosphere with salts, and other elements harmful to the concrete; it is required to divide the Gulf states into several secondary subdivisions. Draft[10]which the author participated in preparing with other researchers, has taken some of these aspects into consideration. However to obtain and complete the required guidelines, specifications ... etc., a large and coordinated effort of engineers, contractors and others who work in the concrete industry in this part of the world is needed.

5 Discussion of some important statistics and research results

Presented in this section is a summary of the current situation, as reported in the literature, in order to evaluate the size of the problem, its causes and remedies.

- Investigating 62 buildings on the Gulf coastal region mainly showed deterioration due to steel corrosion [13]. This proves that the cause was poor quality concrete with neither adequate quality control nor construction supervision.
- In another research investigating 42 concrete structures[14], it was found that inadequate concrete cover was one of the main reasons causing the deterioration of these buildings. Ref.[15] proves this by examining more than 100 specimens taken from different old buildings in the coastal Gulf area. They found that deterioration decreases or even disappears with increased cover. Secondly, the chloride limit in concrete and its constituents considerably exceeded the allowable values in the specifications [16].
- In conclusion, the shortcomings in design, construction, bad materials and exposure to the environment played a great role in the extent of damage to the concrete[17-22] .

. Some researchers categorized the Gulf environment as the worst and severest in the world [17] and that it is impossible for a concrete structure to survive its required life. Others[13,18] were less critical and referred the problem to bad mix design, not following modern technologies, lack of quality control, improper construction practices and untrained workmanship.
. There are indications that the problem is intentionally or otherwise exaggerated by some researchers working for companies that try to sell different products, material and services. The problem is not denied, but it must be put in its right prospective.
. Some researchers exaggerate the results. Figure 5-a indicates a division of problems encountered in one research, whereas one could put it in such a way as that given in Figure 5-b. This clarifies the problem with fewer divisions and puts it in a more practical and realistic prospective.

5a - As given in Ref .14

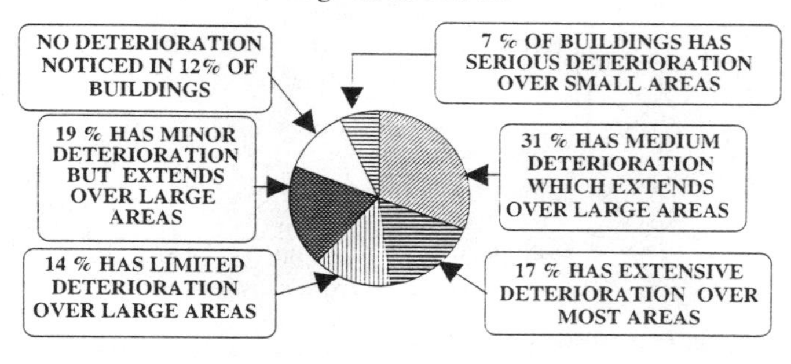

5b - Reduced subdivisions lessens the exaggeration

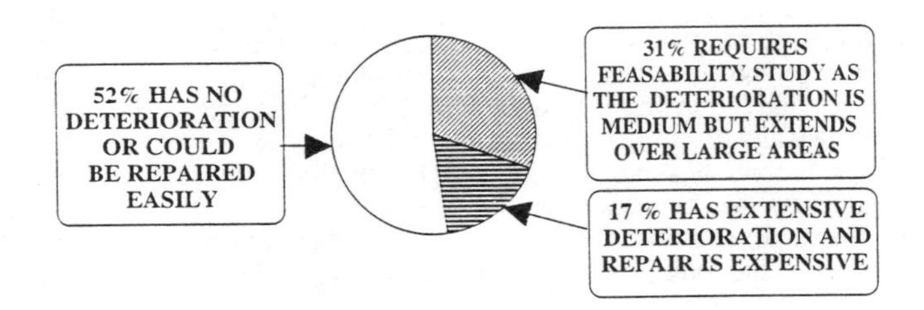

Fig. 5. Exaggeration in presenting research results

. Most of the problems could have been avoided if preventive maintenance was applied. The problems could have been less severe had the buildings been originally designed properly and constructed according to proper standards that provide impermeable concrete.
. The author conducted research [23] to investigate the quality of ready mixed concrete in the three main regions of Saudi Arabia and found that most plants do not produce good quality concrete.
. In a study investigating 70 buildings in the Mecca area[24], which is not a coastal area, the damage found in these buildings were mostly cracks and deflections due to thermal expansion or leakage in sanitary pipes. No chloride corrosion was found.
. In Saudi Arabia as a whole [25] the problem of steel corrosion comes in third place when cases of building deterioration are taken from all over the country, preceded by bad design and execution, and soil problems such as high ground water table problems. However, steel corrosion problem comes first in the Gulf coastal region.

- In the literature review, which the author has conducted, no report was found indicating the ratio of damaged structures compared to undamaged buildings in a region, city or neighbourhood. From my experience, there is a high percentage of good structures on the coastal areas of the Gulf, some of them have survived 30-40 years without problems, even with no or very little maintenance [2,26].
- Figure 6 illustrates the above point, that good concrete can withstand a severe environment. The pictures taken are for two sub-columns only a few metres apart. One is extremely deteriorated while the other is in excellent condition.
- The matter requires various studies to survey some cities representative of the different micro environments. These studies should consider buildings and structures, both damaged and undamaged, i.e. compare damaged structures with undamaged ones over the entire area. This would give a clear and true picture of the size of the problem and its severity. It should indicate the true average age of structures in the Arabian Gulf.

Fig. 6. Two subcolumns in the same building, one is badly deteriorated while the other is intact

6 How to deal with the existing situation

The problems and it's magnitude and causes were emphasized in the previous sections. Now some suggestions and recommendations are introduced on how to treat the existing situation and to take a look at the needed future work to avoid past mistakes. Accordingly, this section has been subdivided to deal mainly with existing and new structures. The new structures are considered as either ordinary or large (important) structures.

6.1 Existing structures

These cannot be left without any treatment, especially those on the coast. They require at least routine checks, regular maintenance and, if necessary repair.

It is emphasized that a great deal can be achieved through proper maintenance and effective repair which improves the performance of structures and extends their life. Excellent experiences were developed in this area which can be adopted in saving existing structures[2,27, 28]. This requires consideration of the following:

- Government Buildings should be investigated by a qualified organization that would perform routine inspections and supervise preventive maintenance or repair, if needed, every three to five years [29,30].
- Similarly, private buildings should be inspected by municipalities or qualified consulting offices. It should be done at reasonable periods according to the severity of climate and environment [30].

- Awareness about the substantial return from funds spent on proper preventive maintenance and repair should be directed to owners of buildings and those in charge of financial appropriations in the construction sector. This should be conducted through media campaigns and all other means.
- Design, supervision and construction of repairs should be carried out and approved by qualified and experienced agencies and contractors (Give the bread to its baker).

6. 2 New structures

6. 2. 1 Ordinary structures

Ordinary structures means those buildings constructed by citizens as private homes or residential and commercial buildings of limited number of floors and small budget. It is difficult to demand high technology, expensive specifications, and stringent requirements because this renders them expensive and unprofitable. Meanwhile, these projects are usually executed by contractors, technicians and labourers who know little about modern technology which requires special skills. Therefore, positive aspects desired from such a requirement will be lost, and some negative aspects may even appear. For example when using new types of cement which contain slag, flyash or silica fume without proper mixing, pouring, vibration and curing, these cements become ineffective or even of less value than ordinary Portland Cement. Similar problems could occur by using coated steel bars.

Therefore, the design for concrete elements in such structures should be in accordance with hot weather [9] and severe environment requirements to ensure impermeability and provide sufficient concrete cover to protect steel. This is sufficient until the development and approval of an alternative document more appropriate for the Gulf environment, such as the one proposed by the author[31]. Municipalities (if they have the manpower) or qualified consulting offices should act as neutral qualified bodies to ensure the design, approve it, inspect the supervision for the construction and supervise concrete quality in the plant and proper execution at the site.

6. 2. 2 Large structures (Important)

Those are projects of prestigious value and higher cost which their owners should spend an additional percentage of their value during the design and construction stages to reflect the potential reduced cost for maintenance, repairs and to improve the performance of the structure. It is recommended that the planning, design and construction, occupancy stages.. etc. should follow the approach (How to design and construct in hot weather) demonstrated in [10] and outlined in this paper. In addition, a clear and comprehensive Quality Assurance and Quality Control system should cover the entire life cycle of the structure in accordance with the systems given in [32-34]. Accordingly, useful usage of modern methods, high technology and new materials in such projects is profitable and should be evaluated and considered (such as proper protection by reinforced plastic layer, additional permanent forms, materials or facades, coated bars, cathodic protection ... etc.).

7 Future view (recommendations and suggestions)

1. Expedite preparation of relevant unified building codes taking into consideration the severe Gulf environmental conditions.
2. Prepare and approve systems for Quality Assurance and Quality Control of concrete in the plant and at the site such as that proposed by the author[32].
3. Approving codes without assigning agencies for implementing them does not give the required benefits. Therefore, we should have authorized agencies for implementing the codes, quality control and quality assurance systems.
4. Not all engineers and technicians are qualified to deal with concrete work in plant and/or on

the site, so necessary qualification and structured training program should be detailed. The responsible bodies for this training and certification should be assigned.

5. Applied research should be directed to serve the concrete industry in the Gulf through studying hot weather and aggressive environment effects on concrete, evaluating sources for appropriate raw materials, applying new technologies, developing maps for various micro environments in the Gulf.. etc.
6. Establish a Gulf organization, subordinate to the council of Gulf Ministers for Housing and Public Works, with the responsibility of unifying the efforts of serving concrete industry by developing a database and information system for concrete practices in the Gulf, including a general survey covering the points mentioned at the end of item 5 of this paper.
7. Gulf countries should coordinate efforts to actively participate in international organizations which are interested in concrete in hot weather and severe climates.
8. It should be strongly emphasized that routine protective maintenance and proper repair should never be forgotten as it is the basic factor that saves concrete from deterioration in the Gulf region. Sufficient funds should be allocated for such purposes.

8 References

1. Zein Al-Abideen, H. (1993) *Concrete Technology*, 2nd Edition, Ministry of Public Works and Housing, Riyadh, Kingdom of Saudi Arabia (In Arabic).
2. Zein Al-Abideen, H. (1993) *Safety Assessment of Concrete structures*, 2nd Edition, Ministry of Public Works and Housing , Riyadh, Kingdom of Saudi Arabia (In Arabic).
3. Zein Al-Abideen, H. (1983) *Reasons for Deterioration and Failure of Structures*, Saudi Arabian Standards Organization Journal, No. 5 Vol. 2, SASO, Riyadh, Kingdom of Saudi Arabia (In Arabic).
4. Al-Amoudi, O., Al-Kurdi, S., Rasheeduzafar, Maslehuddin, M. (1992) *Field Study on the influence of salt contaminated concrete on the type of reinforcement and depth of carbonation*, Buildings Deterioration in the Arab World and Methods of Repair, Riyadh, Ministry of Public Works and Housing, Kingdom of Saudi Arabia.
5. Al-Amoudi, O.S.B., Maslehuddin, M., and Rasheeduzafar (1993) *Permeability of Concrete: Influential Factors,* 4th International conference "Deterioration and repair of R. C. in the Arabian Gulf", Bahrain.
6. Collis , L., and Fookes, P.G. (1985) *Some Reflections on Concrete Deterioration in The Arabian Gulf,* 1st Int'l Conference "Deterioration and repair of R.C. in the Arabian Gulf", Bahrain.
7. Zein Al-Abideen, H. (1992) *Errors in the Assessment and Rehabilitation of R.C. Structures in Saudi Arabia,* Presented in CIB-92 in Montreal , Canada.
8. Al-Tayyib, A.J., Rasheeduzzafar, and Al-Mana, A.I (1985) *Deterioration of Concrete Structure in the Gulf States,* Proceedings 1st International Conference" Deterioration and repair of R.C. in the Arabian Gulf", Bahrain, pp. 27-47.
9. ACI Committee 305 (1991) *Hot Weather Concreting,* Comm. report ACI 305 R-91 American Concrete Institute. P. O. Box 19150, Detroit, Mich., U. S. A.
10. RILEM TC 94 CHC (1993) *Concrete in Hot Weather Environments,* DRAFT "Part I: Influence of the Environment on Reinforced Concrete Durability, Part II: Design approach for Durability".
11. Agency for Meteorological and Environmental protection (1993) *Report on Climatic Information for Riyadh and Jeddah*, National Center, Department of Climate, Jeddah, Kingdom of Saudi Arabia (In Arabic).
12. Zein Al-Abideen, H. (1984) *Properties of Concrete Aggregates in Saudi Arabia,* First Saudi Engineering Symposium, King Abdul Aziz Univrersity, Jeddah (In Arabic).
13. Rasheeduzafar, Dakhil, F.H., Al-Gahtani, A.S., Al-Sadoun , S.S., Bader,M andMedallah, K.Y (1987) *Proposal For a Code of Practice to Ensure Durability of Concrete Construction in the Arabian Gulf Environment*, 2nd International Conference "Deterioration and repair of R. C. in the Arabian Gulf", Bahrain, pp. 595-631.

14. Rasheeduzzafar, Dakhil, F.H., and Al-Gahtani, A.S (1984) *Deterioration of Concrete Structures in the Environment of the Middle East*, Journal, American Concrete Institute, Proceedings V. 81 , No. 1, pp. 13-20.
15. Rasheeduzzafar, Dakhil, F.H and Bader, M.A (1985) *Toward Solving the Concrete Deterioration Problem in the Gulf Region*, The Arabian Jouranal for Science and Engineering , Theme Issue on Concrete Durability.
16. Ministry of Public Works and Housing (1982) *General Specifications for Building Construction*, First Edition, Riyadh, Kingdom of Saudi Arabia.
17. Saricimen, H. (1993) *Concrete Durability Problems in the Arabian Gulf Region- A Review*, 4th Int'l Conference "Deterioration and repair of R. C. in the Arabian Gulf" Bahrain, pp. 943-959.
18. Pollock, D. J., Kay, E. A. and Fookes, P. G. (1981) *Crack mapping for investigation of Middle East Concrete*, Concrete, pp. 12-18.
19. Fookes, P.G., Collis, L. (1975) *Problems in the Middle East*, Concrete, Vol. 9, No. 7.
20. Alidi, S. H., Al-Shiha, M. M. (1989) *Condition Assessment of Concrete Structures in the Eastern Province of Saudi Arabia*, 3rd International Conference "Deterioration and repair of R. C. in the Arabian Gulf", Bahrain, pp. 123-138.
21. Al-Amoudi, O.S.B, Rasheeduzzafar, Maslehuddin, M., and Almusallam, A.A. (1993) *Improving Concrete Durability in the Arabian Gulf,* 4th International Conference "Deterioration and Repair of R. C.in the Arabian Gulf", Bahrain, pp. 927-941.
22. Matta, Z. G. (1992 , 1993) *Deterioration of Reinforced Concrete in the Arabian Gulf*, Concrete International, November, 1993, pp. 50-51, and V.14, No. 5, May 1992, pp.47- 48, andV.15, No.7, July 1993 .
23. Zein Al-Abideen, H. (1989) *Quality of Ready Mixed concrete in Saudi Arabia*, 1st Report 1988, and 2nd Report 1989, King Abdul Aziz City for Science and Technology, Kingdom of Saudi Arabia.
24. Bayazeed, A.A., Mahmood, K.(1989) *Common Structural Defects in Small - sized Reinforced Concrete Buildings*, 3rd International Conference "Deterioration and repair of R. C. in the Arabian Gulf", Bahrain, pp. 189-202.
25. Zein Al-Abideen, H. (1989) *Insight on the Structural Safety and Age of Concrete Buildings in Saudi Arabia*, International Seminar on" The life of Structures", Brighton,England.
26. Internal Reports (1980 until now) *Inspection and Evaluation of various Buildings in The Kingdom of Saudi Arabia*, Ministry of Public Works & Housing, Riyadh (In Arabic).
27. Proceedings (1992) *Buildings Deterioration in the Arab World and Methods of Repair,* Riyadh, Ministry of Public Works and Housing, Kingdom of Saudi Arabia (mostly in Arabic).
28. Proceedings of the International Conferences (October, 1985,1987,1989 and1993) *Deterioration and Repair of Reinforced Concrete in the Arabian Gulf*, Bahrain.
29. Zein Al-Abideen, H. (1993) *Remarks on Assessment & Repair of Concrete with Application to a parking Structure, in K. S. A.,* ACI - Saudi Arabia Chapter Technical Meeting, Jeddah.
30. Internal Report (1991) *Periodic Inspection of Buildings*, Joint Committee from seven different concerned civil governmental agencies, Kingdom of Saudi Arabia (In Arabic).
31. Zein Al-Abideen, H. (1992) *Problems of Unsupervized Quality Control in Ready Mixed Concrete Plants in Saudi Arabia,* Proceedings of the Third International RILEM Conference.
32. Zein Al-Abideen, H. (1994) *Field Supervision for project Construction,* published in Riyadh (In Arabic).
33. Zein Al-Abideen, H. (1990) *Status of Quality Control of Ready Mixed Concrete in Saudi Arabia* ,Third Report, King Abdul Aziz City for Science and Technology, Riyadh.
34. ISO 9000, ISO 9001, ISO 9002, ISO 9003, ISO 9004 (1987) *Quality management and quality assurance standards,* International organization for standardization, Technical committee ISO / TC 176 Quality assurance.

86 PLASTIC SHRINKAGE TESTS ON FIBER-REINFORCED CONCRETE IN HOT WEATHER

F.F. WAFA, T.A. SAMMAN and W.H. MIRZA
King Abdulaziz University, Jeddah, Saudi Arabia

Abstract
Freshly-placed concrete surfaces are prone to cracking due to plastic shrinkage under exposure conditions which are conducive to excessive evaporation rates. This paper describes a study in which the plastic shrinkage behavior of forty-eight flat panels of normal and high-strength concrete were investigated under hot weather conditions. Immediately after finishing the concrete surface, some of the panels were placed in an indoor environment or exposed to the outdoor hot weather conditions. During the test period of five hours for each panel, readings for moisture loss and rate of evaporation were taken. Also, concrete cracking was observed and whenever it occurred, the crack width, number of cracks and the crack initiation time were recorded. From the panel test results, it was observed that plain concrete (without fibers) cracked early and, in most cases, within the first hour of casting. The addition of fibers not only delayed the onset of plastic shrinkage cracking but also helped to reduce both the maximum crack width and the total cracking area. Of the two fiber types used, polypropylene fiber was found to be more effective than steel fiber, the latter requiring relatively much higher percent contents to completely eliminate plastic shrinkage cracking.
Keywords: Concrete, cracking, fiber-reinforced, hot weather, moisture loss, normal and high-strength concrete, plastic shrinkage.

1 Introduction

Plastic shrinkage of freshly placed concrete, due mainly to excessive rates of evaporation, has been thoroughly studied for normal-strength concrete by various researchers [1-5]. In its simplest form, when the evaporation loss exceeds the bleeding rate, the concrete surface shrinks and cracks. Plastic shrinkage is more common and severe in areas when wet concrete is exposed to an ambient environment of hot weather conditions, namely, high temperature, variable humidity and surface drying winds.

The authors in a previous study [6,7] compared the plastic shrinkage behavior of plain normal-strength concrete (23-40 MPa) with high-strength concrete (60-84 MPa) and observed that the concrete mixes exhibited different shrinkage characteristics. High- strength concrete surfaces showed greater cracking potential despite their lower

Concrete Under Severe Conditions: Environment and loading (Volume Two) Edited by K. Sakai, N. Banthia and O.E. Gjørv. Published in 1995 by E & FN Spon. ISBN 0 419 19860 1

rates of evaporation as compared to the normal- strength concrete. In this paper, the influence of fibers in concrete mixes is discussed.

2 Research significance

Freshly-placed concrete needs to be protected from plastic shrinkage as it not only spoils the surface but also poses serviceability and durability problems. This is particularly true for concretes exposed to hot weather conditions. This paper discusses the use of fibers in normal and high-strength concrete mixes and explains the results of flat panel test results carried out under different exposure conditions. The type of fiber and its percentage content in each mix were evaluated in terms of the cracking potential and the crack initiation time.

3 Casting and testing of flat panels

Plastic shrinkage characteristics of plain and fiber-reinforced concrete (FRC) were determined by casting flat rectangular panels measuring 900x600x30 mm, based on the procedure suggested by Kraai [5].

3.1 Materials used

The following ingredients were used in making concrete panels:

1. Cement - Ordinary Portland (Type I).
2. Valley Sand, with a fineness modulus of 2.95.
3. Crushed coarse aggregate, with a maximum size of 9.5 mm, and with a bulk unit weight of 1660 Kg/m^3.
4. Densified silica fume with a specific gravity of 2.20.
5. Mild carbon, hooked end, steel fibers with an average length of 60 mm and a nominal diameter of 0.8 mm.
6. Polypropylene fibers with an average length of 19 mm and specific gravity of 0.90.
7. Admixture - Sulphonated naphthalene synthetic plasticizer.

3.2 Mix proportions

The four concrete mixes selected after extensive laboratory trials for casting panels are given in Table 1.

Table 1. Details of concrete mixes.

Mix No.	Mix Proportion	W/C	Admixture (%)	Silica Fume (%)	28-day Comp. Strength (MPa)
1	1 : 1.5 : 3.0	0.70	0.0	0.0	23
2	1 : 2.2 : 1.6	0.40	1.0	0.0	40
3	1 : 2.2 : 1.6	0.33	5.0	0.0	60
4	1 : 1.0 : 2.0	0.25	5.0	10.0	84

3.3 Panel casting

Flat wooden moulds with an impervious inner lining were used to cast concrete panels. For each concrete mix, individual ingredients were accurately batched and then mixed in a rotating drum-type laboratory mixer. The freshly-mixed concrete was transferred directly from the drum to the panel. The levelling and screeding operations was carried out by holding the straight edge of rectangular wooden bars parallel to the 600 mm edge and always moving along the length in one direction only.

Each finished panel was placed quickly on top of the pan of a weighing balance. Wherever required, an electric blower fan was placed along the 600 mm edge and switched on to simulate wind. During initial testing, it was observed that concrete started to separate from the edges. Therefore, a nominal steel-mesh peripheral reinforcement was provided in subsequent tests. This arrangement prevented shrinkage along the edges, and the concrete, then, became more prone to cracking away from the mould boundaries.

3.4 Panel designation

The testing program of concrete panel specimens involved a number of variables and it was therefore, decided to label each specimen in such a manner as to reflect most of its test conditions. The panels were either plain (P), with steel fibers (SF) or with polypropylene fibers (PF). Four concrete strengths were used (23, 40, 60 and 84 MPa). Some of the panels were restrained (R) and the rest were unrestrained (U). The panels were exposed to either indoor (I) or outdoor (O) environmental condition. The panels were subjected to wind (W) or no wind (0). Finally, the fiber contents were varied for both steel fibers (0, 0.5, 1.0 or 1.5%) and the polypropylene fibers (0, 0.2, 0.25, 0.35 and 0.5 %). Thus, a panel designated as P-23-U-I-0-0 would mean a plain concrete panel, made with 23 MPa mix, having no restraining mesh, placed indoors without wind and containing no fibers. Similarly, SF-84-R-O-W-0.5 means a steel fiber reinforced concrete panel, made with 84 MPa mix, containing restraining mesh, placed outdoors with fan-blown wind and having 0.5 % steel fibers.

4 Discussion of panel test results

A total of forty-eight concrete panels, cast and designated as per the description above, were subjected to either indoor (ID) environment of constant air temperature and relative humidity or to outdoor (OD) exposure conditions where the air temperature and relative humidity (R.H.) varied over the plastic shrinkage period of five hours every test day. While conducting the plastic shrinkage tests, data were recorded about the climatic conditions (air temperature and relative humidity), wind speed, moisture loss, crack initiation time and the total cracking area.

Based on the data thus obtained, a summary of test conditions and results is presented in Table 2. Concrete panels made with the high-strength mix of 84 MPa were the most susceptible to plastic shrinkage cracking as ten panels out of a total of twenty showed varying degrees of cracking potential. On the other hand, only two panels of the 23 MPa mix cracked. The maximum cracking areas, up to 1230 mm^2, were also observed in 84 MPa concrete panels. Thus, a low aggregate-cement ratio of 3.0, coupled with a water-cement ratio of 0.25 and a 10% silica fume content, appeared to be quite conducive to plastic shrinkage cracking. The 23 MPa mix, though with a water-cement ratio of 0.70, showed lesser cracking, chiefly because of higher aggregate content and a greater ability to supply bleed water. The 40 and 60 MPa mixes showed intermediate behavior as their crack density and crack areas were lower than the 84 MPa concrete mixtures.

The climatic conditions of air temperature, relative humidity and wind speed had a direct bearing on the maximum rate of evaporation, the total evaporation loss and the cracking of panels. When placed indoors (at a constant air temperature of 26 ^{o}C and 50% R.H.), the panels showed far lower rates of evaporation (0.11 to 0.67 Kg/m^2-h) than those placed outdoors at variable temperatures (36 to 50 ^{o}C) and relative humidities (75 to 20% R.H). In the latter case, the rates were six to ten times greater in

Table 2 Summary of concrete panel test conditions and results

Panel No.	Panel Designation	28 Day Strength	A/C	W/C	Adm.	Silica Fume	Fiber Content (%)		Air Temp	Rel. Humidity	Wind	Max. Rate of Evapn.	Total H_20 Loss	Total Cracking Area	Crack Initiation Time
		(MPa)			(%)	(%)	SF	PF	(^oC)	(%)	(km/h)	(kg/ m^2-h)	(%)	(mm^2)	(min.)
1.	P-23-U-I-0-0	23	4.5	0.70	0.0	0.0	0.0	0.0	26	50	0	0.11	2.69	0	-
2.	P-23-R-I-W-0	23	4.5	0.70	0.0	0.0	0.0	0.0	26	50	15	0.55	17.46	0	-
3.	P-23-U-O-0-0	23	4.5	0.70	0.0	0.0	0.0	0.0	36-48	70-25	0	1.11	38.41	0	-
4.	P-23-R-O-W-0	23	4.5	0.70	0.0	0.0	0.0	0.0	38-45	65.40	15	1.48	44.44	130	42
5.	SF-23-R-O-W-0.5	23	4.5	0.70	0.0	0.0	0.5	0.0	36-45	70-45	15	2.22	56.43	35	100
6.	SF-23-R-O-W-1.0	23	4.5	0.70	0.0	0.0	1.0	0.0	37-43	55-40	15	1.93	47.32	0	-
7.	PF-23-R-O-W-0.2	23	4.5	0.70	0.0	0.0	0.0	0.2	40-45	55-50	15	2.22	45.75	0	-
8.	P-40-U-I-0-0	40	3.8	0.40	1.0	0.0	0.0	0.0	26	50	0	0.11	5.28	0	-
9.	P-40-R-I-W-0	40	3.8	0.40	1.0	0.0	0.0	0.0	26	50	15	0.44	23.06	54	115
10.	SF-40-R-I-W-0.5	40	3.8	0.40	1.0	0.0	0.5	0.0	26	50	15	0.55	22.00	10	140
11.	SF-40-R-I-W-1.0	40	3.8	0.40	1.0	0.0	1.0	0.0	26	50	15	0.55	22.00	0	-
12.	P-40-U-O-0-0	40	3.8	0.40	1.0	0.0	0.0	0.0	43-50	55-20	0	1.11	37.50	110	90
13.	P-40-R-O-W-0	40	3.8	0.40	1.0	0.0	0.0	0.0	40-46	55-35	15	1.22	44.00	465	45
14.	SF-40-R-O-W-0.5	40	3.8	0.40	1.0	0.0	0.5	0.0	39-42	50-40	15	1.22	44.96	85	120
15.	SF-40-R-O-W-1.0	40	3.8	0.40	1.0	0.0	1.0	0.0	39-42	50-40	15	1.33	45.27	0	-
16.	PF-40-R-O-W-0.25	40	3.8	0.40	1.0	0.0	0.0	0.25	43-47	50-30	15	1.33	42.5	0	-
17.	PF-40-R-O-W-0.5	40	3.8	0.40	1.0	0.0	0.0	0.5	43-47	50-30	15	1.55	45.28	0	-
18.	P-60-U-I-0-0	60	3.8	0.33	5.0	0.0	0.0	0.0	26	50	0	0.22	3.70	0	-
19.	P-60-R-I-W-0	60	3.8	0.33	5.0	0.0	0.0	0.0	26	50	15	0.44	20.54	30	95
20.	SF-60-R-I-W-0.5	60	3.8	0.33	5.0	0.0	0.5	0.0	26	50	15	0.67	19.87	0	-
21.	PF-60-R-I-W-0.2	60	3.8	0.33	5.0	0.0	1.0	0.2	26	50	15	0.55	21.88	0	-
22.	P-60-U-O-0-0	60	3.8	0.33	5.0	0.0	0.0	0.0	40-47	50-35	0	1.11	43.03	280	55
23.	P-60-R-O-W-0	60	3.8	0.33	5.0	0.0	0.0	0.0	40-48	45-30	15	1.33	46.97	410	30
24.	SF-60-R-O-W-0.5	60	3.8	0.33	5.0	0.0	0.5	0.0	40-45	30-45	15	1.11	46.47	660	20

A/C : Aggregate/Cement Ratio W/C : Water/Cement Ratio SF : Steel Fiber PF : Polypropylene Fiber Adm : Admixture

Table 2 (Cont'd) Summary of concrete panel test conditions and results

Panel No.	Panel Designation	28 Day Strength	A/C	W/C	Adm.	Silica Fume	Fiber Content (%)		Air Temp	Rel. Humidity	Wind	Max. Rate of Evapn.	Total H_2O Loss	Total Cracking Area	Crack Initiation Time
		(MPa)			(%)	(%)	SF	PF	(°C)	(%)	(km/h)	(kg/ m^2-h)	(%)	(mm^2)	(min.)
25.	SF-60-R-O-W-1.0	60	3.8	0.33	5.0	0.0	1.0	0.0	40-45	40-45	15	1.33	51.85	143	30
26.	SF-60-R-O-W-1.5	60	3.8	0.33	5.0	0.0	1.5	0.0	40-45	40-45	15	1.11	47.48	-	-
27.	PF-60-R-O-W-0.5	60	3.8	0.33	5.0	0.0	0.0	0.50	39-42	40-50	15	1.00	48.15	71	40
28.	PF-60-R-O-W-0.35	60	3.8	0.33	5.0	0.0	0.0	0.35	44-47	40-50	15	1.22	43.55	0	-
29.	P-84-U-I-0-0	84	3.0	0.25	5.0	10.0	0.0	0.0	26	50	0	0.11	5.33	0	-
30.	P-84-R-I-0-0	84	3.0	0.25	5.0	10.0	0.0	0.0	26	50	0	0.11	4.82	0	-
31	P-84-R-I-W-0	84	3.0	0.25	5.0	10.0	0.0	0.0	26	50	15	0.33	14.48	965	25
32.	SF-84-R-I-W-0.5	84	3.0	0.25	5.0	10.0	0.5	0.0	26	50	15	0.44	21.80	540	40
33.	SF-84-R-I-W-1.0	84	3.0	0.25	5.0	10.0	1.0	0.0	26	50	15	0.44	25.10	0	-
34.	SF-84-R-I-W-1.5	84	3.0	0.25	5.0	10.0	1.5	0.0	26	50	15	0.44	22.90	0	-
35.	PF-84-R-I-W-0.2	84	3.0	0.25	5.0	10.0	0.0	0.2	26	50	15	0.33	24.10	0	-
36.	PF-84-R-I-W-1.0	84	3.0	0.25	5.0	10.0	0.0	1.0	26	50	15	0.44	21.82	0	-
37.	PF-84-R-I-W-1.5	84	3.0	0.25	5.0	10.0	0.0	1.5	26	50	15	0.44	20.36	0	-
38.	P-84-U-O-0-0	84	3.0	0.25	5.0	10.0	0.0	0.0	43-45	50-40	0	1.04	32.72	45	50
39.	P-84-R-O-W-0	84	3.0	0.25	5.0	10.0	0.0	0.0	39-47	60-30	15	1.11	31.27	1230	32
40.	SF-84-R-O-0-0.5	84	3.0	0.25	5.0	10.0	0.5	0.0	38-49	60-20	0	0.44	36.36	204	68
41.	SF-84-R-O-0-1.0	84	3.0	0.25	5.0	10.0	1.0	0.0	40-45	75-40	0	0.44	34.90	60	75
42.	SF-84-R-O-0-1.5	84	3.0	0.25	5.0	10.0	1.5	0.0	43-46	50-35	0	0.89	37.09	0	-
43.	SF-84-R-O-W-0.5	84	3.0	0.25	5.0	10.0	0.5	0.0	38-45	60-35	15	1.11	36.36	250	8
44.	SF-84-R-O-W-1.0	84	3.0	0.25	5.0	10.0	1.0	0.0	40-45	50-35	15	1.11	37.09	20	115
45.	SF-84-R-O-W-1.5	84	3.0	0.25	5.0	10.0	1.5	0.0	38-44	55-30	15	1.11	45.10	0	-
46.	PF-84-R-O-W-0.2	84	3.0	0.25	5.0	10.0	0.0	0.2	41-45	50-30	15	1.44	44.73	208	20
47.	PF-84-R-O-W-0.35	84	3.0	0.25	5.0	10.0	0.0	0.35	41-50	50-30	15	1.00	40.33	47	30
48.	PF-84-R-O-W-0.5	84	3.0	0.25	5.0	10.0	0.0	0.5	38-44	50-30	15	0.44	27.00	0	-

A/C - Aggregate/Cement Ratio   W/C : Water/Cement Ratio   SF : Steel Fiber   PF : Polypropylene Fiber   Adm: Admixture some panels. Introduction of a 15 km/h wind on the panel surface aggravated the situation further, and the maximum rates of evaporation jumped up to 2.22 Kg/m^2-h in 23 MPa concrete.

From the results in Table 2, it is observed that plain concrete panels (without fibers) showed plastic shrinkage cracks for mixes of all four strength ranges. The 23 MPa concrete cracked only under an OD environment with the surface wind of 15 km/h (panel # 4), while the 40 MPa concrete cracked even without the surface wind (panel # 12). The high-strength concrete mixes of 60 and 84 MPa showed cracks even in an ID environment (panels 19 and 31, respectively).

The efficiency of the steel fibers (SF) and polypropylene fibers (PF) was then evaluated by including these fibers separately in the four concrete mixes. The SF content ranged between 0.5 and 1.5% and the PF content was between 0.2 and 0.5%. The two fiber types had a marked influence on the plastic shrinkage behavior of both normal and high-strength concrete.

In plain concrete panels, the crack initiation time was less than one hour after casting. When the 15 km/h wind was blown on the surface, the first cracks appeared within the first 45 minutes. Inclusion of fibers prolonged the crack initiation period well beyond one hour of casting.

It was observed that the total cracking area in plain concrete panels was a function of concrete strength. Thus, normal-strength (23 and 40 MPa) concrete panels showed less cracking potential than the high-strength (60 and 84 MPa) concrete panels (see Table 2). The introduction of fibers (either SF or PF) into the four concrete mixes helped to either reduce or eliminate the cracks. This property was dependent on the fiber type, content, and the concrete strength. The SF content of up to 1.0% was found sufficient to eliminate plastic shrinkage cracks in 23 and 40 MPa concrete mixes. For high-strength mixes, this content had to be increased to 1.5% to avoid cracking. The PF contents were relatively smaller for the same purpose, and 0.2, 0.25, 0.35 and 0.5% contents were found adequate for 23, 40, 60 and 84 MPa concretes, respectively.

5 Concluding remarks

Hot weather elements, i.e., high air temperature, fluctuating relative humidity and drying winds, cause a considerable increase in the plastic shrinkage cracking of concrete, particularly in mixtures which are designed for high-strength (60 and 84 MPa) and contain silica fume. The addition of fibers helps to reduce the cracking potential and delays the onset of cracking. A higher content of steel fibers is required as compared to the polypropylene fibers for eliminating plastic shrinkage cracks. It is observed that the higher the concrete design strength, the greater is the fiber content necessary to eliminate cracking.

6 Acknowledgements

The financial support provided by King Abdulaziz City for Science and Technology (KACST) under Research Grant No. AR-12-41 is gratefully acknowledged.

7 References

1. ACI Committee 305. (1989) Hot Weather Concreting (ACI 305-89), American Concrete Institute, Detroit, 17p.
2. Blakey, B.E., and Beresford, F.D. (1959) Cracking of Concrete, Constructional Review (Sydney), Vol. 32, No. 2.

3. Ravina, D., and Shalon, R. (1968) Plastic Shrinkage Cracking. ACI Journal, Proceedings Vol. 65, No. 4, pp. 282-292.
4. Shaeles, C.A., and Hover, K.C. (1988) Influence of Mix Proportions and Construction Operation on Plastic Shrinkage Cracking in Thin Slabs. ACI Materials Journal, Proceedings Vol. 85, No. 6, pp. 495-504.
5. Kraai, P.P. (1985) A Proposed Test to Determine the Cracking Potential Due to Drying Shrinkage of Concrete. Concrete Construction, Vol. 30, No. 9, pp. 775-778.
6. Wafa, F.F., Mirza, W.H., and Samman, T.A. (1993) Shrinkage Cracking of Fiber-Reinforced Concrete under Hot Weather Conditions, KACST Project No. AR-12-41, Second Progress Report, Riyadh, Saudi Arabia.
7. Samman, T.A., Mirza, W. H., Wafa, F. F. Shrinkage Cracking of Normal and High-strength Concrete: A Comparative study, Accepted for publication by the ACI Material Journal, U.S.A.

87 EFFICIENCY OF CONCRETE CURING IN RIYADH AREA

A.M. ARAFAH, M.S. AL-HADDAD, R.Z. AL-ZAID
and G.H. SIDDIQI
King Saud University, Riyadh, Saudi Arabia

Abstract

This paper presents the efficiency of two concrete curing methods commonly employed in Riyadh as a case study for evaluation of curing efficiency under severe hot and dry weather conditions. The efficiency of curing methods was measured in terms of concrete compressive strength at 28 days. The methods were water sprinkling two times a day for seven days, which is designated as SWC; and water sprinkling two times a day for seven days with a burlap cover, which is designated as SBC. The specimens were tested at 28 days and compared with standard cured specimens, which were designated as STD. A total of 56 cube specimens were collected from construction sites in Riyadh during the sampling period and cured under STD-, SWC-, and SBC- conditions. The ratios of SWC and SBC to STD strengths were plotted on normal probability forms. The mean values of these ratios were 0.84 and 0.93, respectively. The probability of these ratios being less than the ACI-318 general requirement for concrete curing of 0.85 was 54 and 20 percent, respectively. These high probabilities indicate a deficiency in the curing methods, specially for SWC curing. Recommendations for improving the curing methods in Riyadh or, for that matter, in any other region of similar weather conditions are made.

Keywords: Concrete curing, hot weather concreting, concrete durability.

Concrete Under Severe Conditions: Environment and loading (Volume Two) Edited by K. Sakai, N. Banthia and O.E. Gjørv. Published in 1995 by E & FN Spon. ISBN 0 419 19860 1

1 Introduction

The temperature of the concrete is affected by the surrounding air, absorption of solar heat, heat of hydration of cement and initial temperature of materials. An undesirable reduction in moisture content of the cement paste at this stage tends to reduce hydration and results in drying shrinkage and development of cracks in the paste. In the parlance of concrete technology, hot weather is defined as any combination of high air temperature, low relative humidity, and wind velocity [1]. The effects of hot weather are most critical during periods of rising temperature or falling relative humidity, or both. Undesirable hot weather effects on concrete in the plastic state may include: (a) increased water demand, (b) increased rate of slump loss, (c) increased rate of setting and (d) increased tendency for plastic cracking [1]. Thus, a continuous curing, particularly during the first few hours, is acutely needed.

ACI 318 [2] and ACI 308 [3] recommend that concrete be maintained in a moist condition for at least the first 7 days after placement. Alternate cycles of wetting and drying promote the development of pattern cracking and should be avoided. ACI 318 [2] specifies that the procedure for curing concrete shall be improved when the strength ratio of field cured specimens to the companion laboratory cured specimens is less than 0.85 unless the field-cured strength exceeds the specified strength by more than 3.5 MPa.

Spears [4] indicated that proper curing maintains relative humidity above 80 percent and, thereby, advances hydration to the maximum attainable limit. Proper curing decreases concrete permeability, surface dusting, thermal-shock effects, scaling tendency and cracking. It increases strength development, abrasion resistance, durability, pozzolanic activity and weatherability. Haque [5] investigated the strength development of concrete under the conditions of fog, temperate dry, warm-wet and warm-dry weather conditions. He found that the lack of any moist curing adversely affects the compressive strength of plain concrete at all ages.

Martin [6] demonstrated that rising placing temperatures do not, as a rule, lead to lower strengths. With favorable combinations of cementitious materials and admixtures, the strength performance of concrete can remain unaffected by higher placing temperatures, or it can even improve over that at lower temperatures. Malvin and Odd [7] conducted a large-scale field investigation of high-strength light-weight concrete and concluded that maximum curing temperatures of up to 85 °C (153F) did not adversely affect the mechanical properties of the concrete. On the contrary, they observed a slight increase in compressive strength.

Khan [8] quantified the effect of interrupted curing. He found that the losses in strength of concrete due to an interruption in moist curing can be regained significantly by recuring the concrete.

Carrier [9] indicated that a short period of drying early in the curing life of concrete specimens prevents water molecules from reaching unhydrated cement particles and prevents concrete from gaining full strength. He also indicated that much of the concrete deterioration that takes place each year should be blamed on inadequate curing. Early and rapid drying can lead to failure such as shrinkage cracks, crazing, wear, dusting, scaling, and spalling. Once a surface has cracked, dusted, scaled or spalled, the entire member is more susceptible to other types of deterioration.

2 Research significance

Twice a day sprinkling of water with or without burlap cover for seven days are the curing practices used on the majority of construction jobs in Saudi Arabia [10]. The average annual maximum temperature of 46 °C and the average annual minimum humidity of 4.0 % in summer in the central region (arid zone of the Kingdom) and the prevalent methods of curing which are below the required standard practice call for studying their effects on the strength of concrete.

3 Objective and scope

This paper presents results of an experimental program designed to investigate the influence of the prevalent curing practice on the strength of concrete cast during ten month period in an arid area.

4 Experimental program

The experimental program was designed to evaluate the influence of the prevalent curing practices on the compressive strength of the concrete. The Three curing methods which were employed are described and designated as in Table 1.

Concrete samples were collected from randomly selected construction sites in Riyadh (in the central province) of Saudi Arabia. The sampling was done during ten

Table 1. Curing methods used and their designation

Designation	Curing method
SWC	Twice a day sprinkling without cover for seven days
SBC	Twice a day sprinkling with burlap cover for seven days
STD	Twenty eight day immersion in water, considered standard curing

months (September to June). The annual minimum relative humidity and maximum temperature data of the central region for 25 years between 1965 and 1989 as recorded by MEPA [11] are presented in Figs. 1 and 2, respectively.

A total of 56 concrete samples were collected at construction sites in Riyadh during the sampling period and cast into standard cubes of 150x150x150 mm. Each sample consisted of three pairs of cubes and each cube in a pair was collected from a separate truck. The cubes were left at the site for about 24 hours and then transferred to the laboratory. A pair in a sample was cured by SWC, the second by SBC and the third by STD method. The cubes were tested for strength at age of 28 days.

5 Results, analysis and discussion

The ratios R_1 and R_2 of the compressive strength of the SWC and SBC cured cubes, respectively, to the STD cured cubes were subjected to analysis by order statistics. The results are presented in Table 2 and plotted on a normal probability paper along with the best fit by linear regression in Fig. 3. The mean values of R_1 and R_2 are 0.84 and 0.93, respectively. This clearly indicates the beneficial effect of curing with burlap cover in dry-hot weather. The maximum values of the two ratios are 1.09 and 1.14, respectively, and their minima are 0.63 and 0.75.

Table 2. Basic statistics of the strength ratios R_1 and R_2 in the Riyadh Area.

R_i	Min.	5 Percentile	Mean	Max	COV%	P(R<0.85)
R_1	0.63	0.67	0.84	1.09	12.60	0.54
R_2	0.75	0.78	0.93	1.14	9.80	0.20

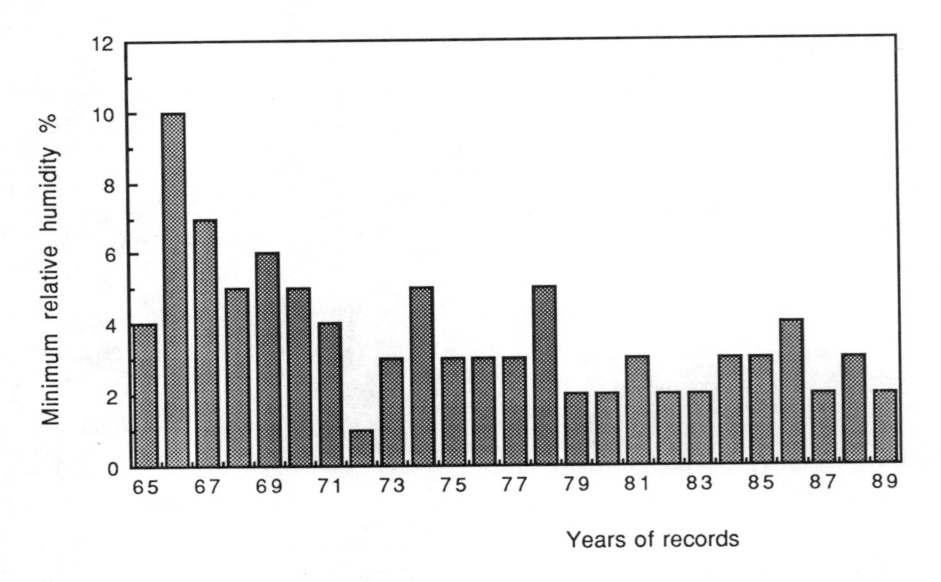

Fig. 1 Annual minimum relative humidity in the central region.

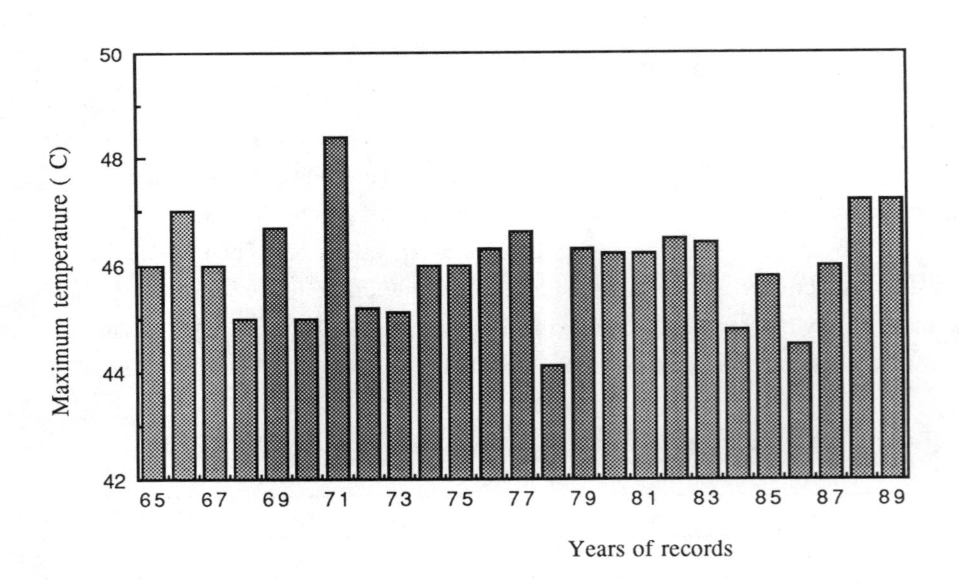

Fig. 2. Annual maximum temperature in the central region.

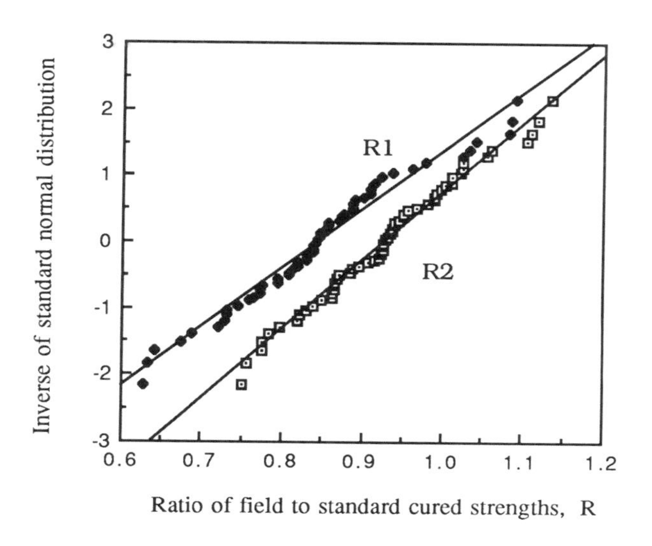

Fig. 3 Influence of curing methods on concrete strength in the central province

It is interesting to note that the variability in compressive strength and curing process can cause these ratios to be higher than unity. The five percentiles (the values with probability of 5 percent of being not exceeded) of the two ratios are 0.67 and 0.78, respectively.

ACI 318 [2] specifies that "procedures for protecting and curing concrete shall be improved when strength of field cured cylinders at test age designated for determination of f'_c is less than 85 percent of that of companion laboratory-cured cylinders." In the presence of strength variability, there is a possibility of having this ratio less than 0.85 which is very small with good curing practices, however, this probability will increase when poor practices are employed.

Results indicate that the mean values of R_1 are less than those for R_2 while the coefficients of variation of R_1 are higher than that of R_2. As a result, the probability of being R_1 less than 0.85 is 54%, which is very high. This indicates that the SWC curing method does not meet the ACI-318 requirement in arid areas. The probability of having R_2 less than 0.85 in Riyadh area is about 20%, which is also relatively high.

The authors suggest that the efficiency of curing methods should be based on the 5 percentile of the distribution function of the ratio R which is affected by both its mean and COV. The results indicate that the 5 percentiles for R_1 and R_2 are 0.67 and 0.78, respectively. The curing methods should be improved to bring these values to 0.85.

The ACI-308 recommends that concrete be maintained in a moist condition for at least the first 7 days after placement. In arid areas, it is impossible to meet this recommendation using SWC where the available water for concrete curing is very limited. The SBC curing method is more efficient; however, the frequency of water sprinkling per day should be increased so that the 5 percentile of the distribution of R_2 is at least equal to 0.85. The efficiency of water sprinkling three times a day with a burlap cover is under investigation.

6 Conclusions

The effect of curing practice was evaluated by statistical analysis of strength ratio, R, of field cured to standard cured cubes. The field curing methods used were sprinkling without cover and with burlap cover. The results in the Riyadh area showed that the mean values of R_1 and R_2 are 0.84 and 0.93, respectively. The higher value of R_2 indicates the effectiveness of burlap cover in improving the curing process. The probability of having R_1 less than 0.85, specified by ACI-318, is 54 percent, indicating that curing by water sprinkling without cover twice a day for 7 days will not satisfy the ACI-318 requirement. The SBC curing method is more efficient; however, the frequency of water sprinkling per day should be increased so that the 5 percentile of the distribution of R_2 is at least equal to 0.85. The efficiency of water sprinkling three times a day with a burlap cover is under investigation.

7 Acknowledgment

The study was sponsored by King Abdul-Aziz City for Science and Technology. The authors would like to express their thanks for this support.

8 References

1. ACI Committee 305. (1989) *Hot Weather Concreting*. ACI-305, American Concrete Institute, Detroit.

2. ACI Committee 318. (1989) *Building Code Requirements for Reinforced Concrete.* ACI-318, American Concrete Institute, Detroit.
3. ACI Committee 308. (1989) *Recommended practice for curing concrete.* ACI-308, American Concrete Institute, Detroit.
4. Spears, R. E. (1983) *The 80 Percent Solution to Inadequate Curing Problems.* Concrete International. pp.15-18.
5. Haque, M.N. (1990) *Some Concretes Need 7 Days Initial Curing.* Concrete International. Vol. 12, No. 2, pp. 42-46.
6. Mittelaeher, M. (1992) *Compressive Strength and the Rising Temperature of Field Concrete.* Concrete International. pp. 29-33.
7. Sandvik, M. and Gjorv, O. E. (1992) *High Curing Temperatures in Light Weight High-Strength Concrete.* Concrete International. pp. 40-42.
8. Khan, M., S. and Ayers, M., E., (1993) *Interrupted Concrete Curing.* Concrete International. pp. 25-28.
9. Carrier, R. E. (1983) *Concrete Curing Tests.* Concrete International. pp. 23-27.
10. Arafah, A. M., et al.(1990) *Development of a Solid Foundation for a Local Reinforced Concrete Building Code.* Final Report on KACST Project No. AR-9-34, Riyadh.
11. Meteorology and Environmental Protection Administration. (1990) *Records from 1965 to 1990*, Riyadh, Saudi Arabia.

88 ENERGETICALLY MODIFIED CEMENT (EMC) FOR HIGH PERFORMANCE CONCRETE IN WINTER CONCRETING

V. RONIN and J.-E. JONASSON
Luleå University of Technology, Sweden

Abstract
The present paper discusses experimental results concerning one important direction of modern winter concrete technology - the development and usage of Portland cement based binders with high rate of hardening. Such a binder material, an energetically modified cement(EMC), has recently been developed at Luleå University of Technology.

The EMC binder is produced by intensive vibration of Portland cement and mainly silica fume (SF). The new binder has improved properties compared with non-modified cement in several ways. For instance, higher reactivity, especially at early ages, in a broad range of curing conditions including negative temperatures; high strength concrete with low binder content; effective hardening at low dosage of antifreezing admixtures; low rate and low ultimate self-desiccation shrinkage; high durability potential in salt-freeze-testing. This more effective EMC binder will have a number of interesting applications in future concrete technology.
Keywords: Energetically modified cement, high performance concrete, high strength concrete, winter concreting.

1 Background

As it has been shown in our previous work [1,2,3,4,5,6] the usage of the energetically modified cement (EMC) with improved and regulated properties leads to new classes of high strength and high performance concretes with moderate levels of cement content. The improved properties of the EMC concretes (especially the early age maturity) create more effective technological solutions, specifically, in the field of winter concreting [3,4,5]. This is of a great importance for the Scandinavian countries, USA, Canada, Russia and Japan, where concreting at negative temperatures takes place at a very big scale.

Concrete Under Severe Conditions: Environment and loading (Volume Two) Edited by K. Sakai, N. Banthia and O.E. Gjørv. Published in 1995 by E & FN Spon. ISBN 0 419 19860 1

The experimental results represented in this paper should be considered as a continuation of the research in the above mentioned area.

2 Methodology

The moderate heat liberating Portland cement (Std P Anl, or Anl for simplicity) produced by Cementa AB, Sweden; silica fume (SiO_2 content = 96.0 %) produced by Elkem A/S, Norway; and the Betec = chloride-free antifreezing admixture produced by Finja-Betec AB, Sweden, were used in this study. A naphthalene type superplasticizer, Mighty 100, was introduced in different amounts (up to 3.5 %) in pastes and concretes with non-modified cement to keep workability at low w/B.

The energetically modification of the Portland cement included a special mechanochemical treatment [6]. The composition of the modified cement was kept constant for all cement pastes and concretes used in these experiments. In this paper, the above mentioned treatment is referred to by the abbreviation "EMC", which stands for Energetically Modified Cement, e.g., EM(Anl) means the energetically modification of cement type Anl.

The following concrete mixtures were used in the experiments: B480 (binder content 480 kg/m3 +10% SF),B340 and B270, the ratios by weight of cement of type Anl : fine aggregate (0-8 mm) : coarse aggregate (12-16 mm) were 1 : 1.58 : 2.18, 1 : 3.38 : 2.24 and 1 : 4.63 : 2.57, respectively.

Both the cement paste and concrete samples were cast at room temperature (about +20 °C) in 20x20x20 mm and 100x100x100 mm steel moulds, respectively. After casting, the samples were put in water. The samples cured at negative temperatures were cast at room temperature and then sealed with plastic foil and put into a freezer within 30 minutes after casting. The samples cured at negative temperatures were heated up to + 15 °C in water before testing.

The relative humidity (RH) and shrinkage measurements were made with the use of TESTOTERM and STAEGER instruments, respectively, on 100x100x250 mm concrete samples cured at room temperature.

The durability test cycle, test No 1, was conducted according to [7] and consisted of the following: 2 hours storage of concrete slices (thickness 10 mm, diameter 70 mm) in a 7.5% solution of NaCl, followed by 4 hours of heating at 105 °C, with storage again in the same solution for another 2 hours, and after that, storage in a freezer at -20 °C for 16 hours. Durability test cycle of test No 2 replaced the heating of the slices with freezing them for 4 hours at -20 °C. The EMC concrete was subjected to tests No 1 and 2, however, the reference concrete only to test No 1. The relative mass loss of the samples as a controlling factor was measured. More detailed information concerning experimental procedures is presented in [5].

3 Results and discussion

3.1 Strength development

The combinations of the pre- and postcuring conditions in the temperature interval from +20 °C down to -10 °C of the EMC and reference cement pastes and concretes are represented by Figs. 1, 3, 5, and 7. The measured strength developments in accordance with these curing conditions are presented in Figs. 2, 4, 6, and 8, respectively. Analysis of experimental results showed that the EM(Anl) pastes were characterized by more intensive at negative temperatures in comparison with non-modified pastes, see Fig. 2, regime C.

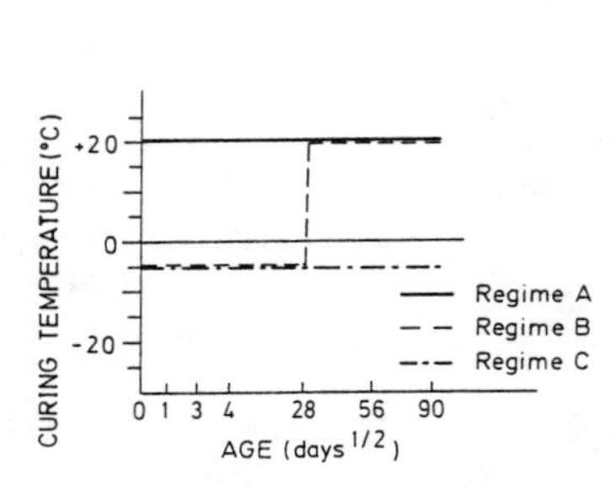

Fig. 1. Curing regimes of the cement pastes in Fig. 2.

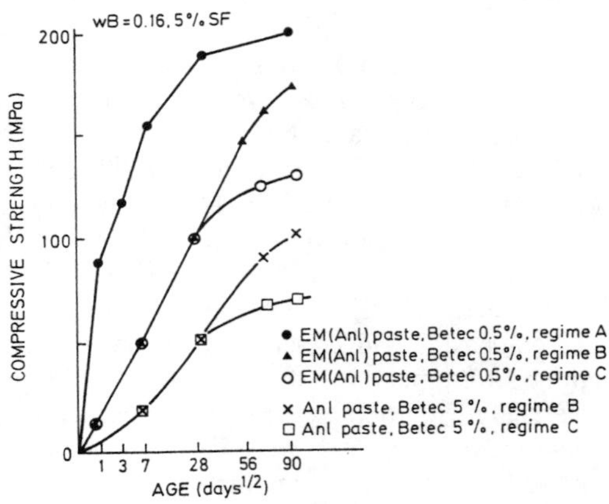

Fig. 2. Strength developments of the cement pastes cured according to Fig. 1.

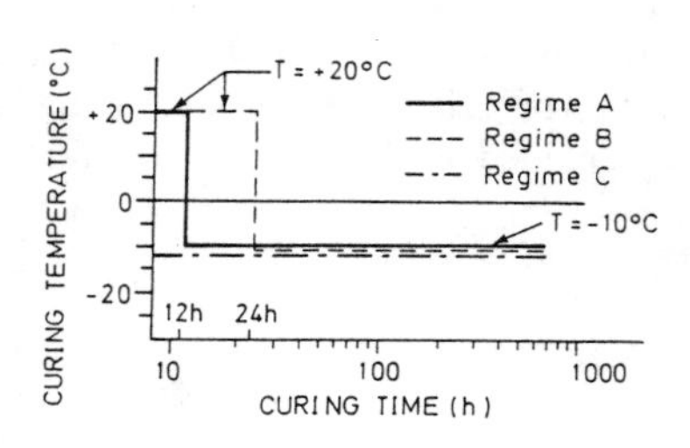

Fig. 3. Curing regimes of the cement pastes in Fig. 4.

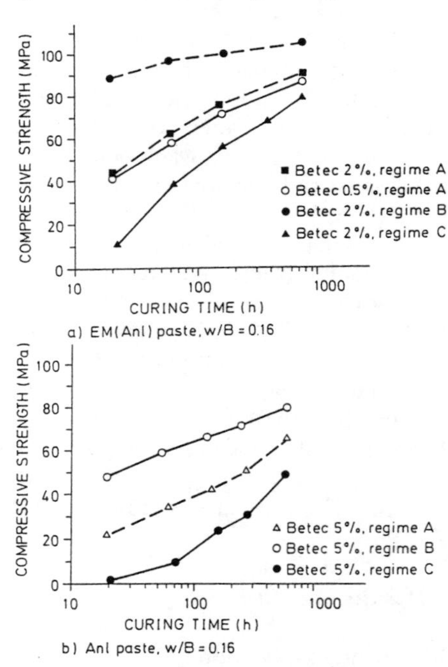

Fig. 4. Strength developments of the cement pastes cured according to Fig. 3.

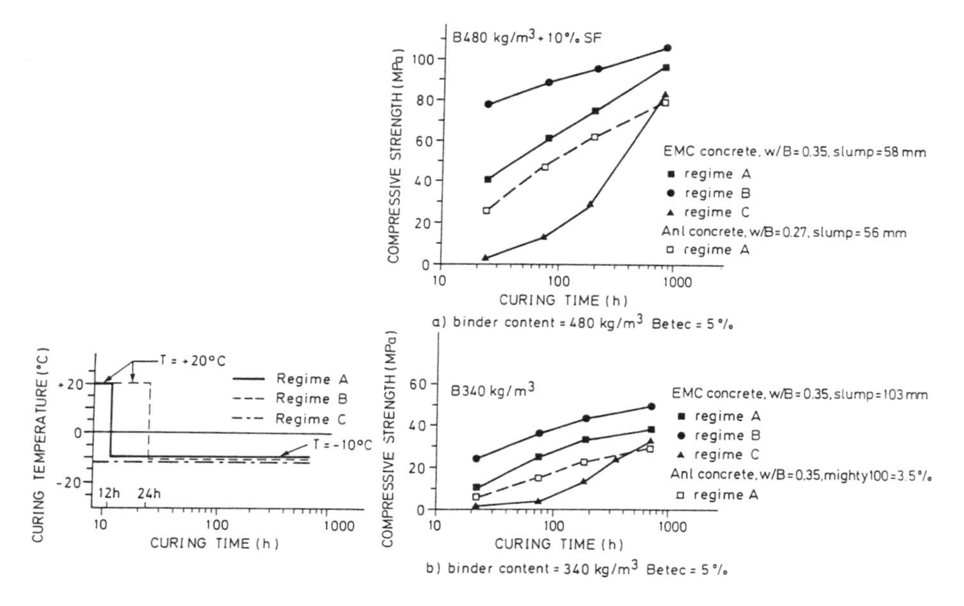

Fig. 5. Curing regimes of the concretes in Fig. 6.

Fig. 6. Strength developments of the concretes cured according to Fig. 5.

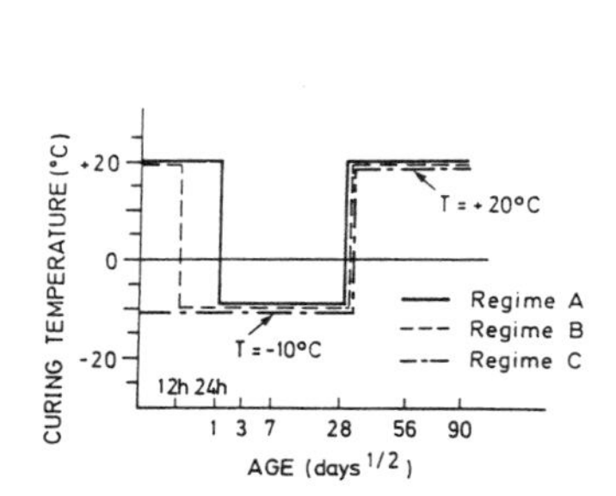

Fig. 7. Curing regimes of the concretes in Fig. 8.

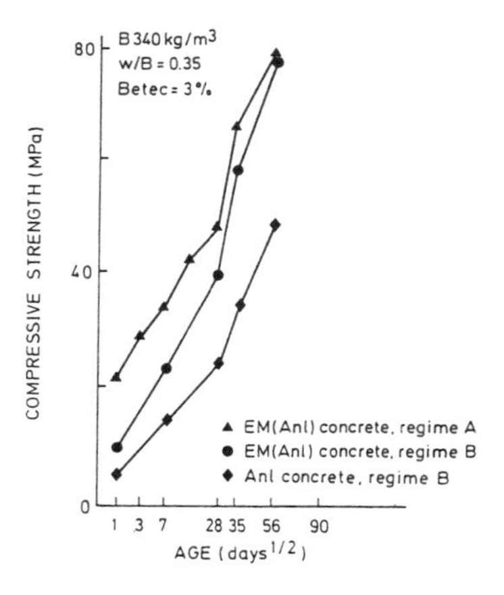

Fig. 8. Strength developments of the concretes (B = 340 kg/m^3) cured according to Fig. 7.

The compressive strength difference between the EMC pastes and non-modified pastes also existed after the postcuring period and equaled about 75-80 MPa for regime B in Fig. 1 after total 90 days of hardening, see Fig. 2. This figure also shows that at the curing time of 90 days, the postcured EMC paste reached approximately 80% of the same paste cured all the time at +20 °C.

Precuring was performed in two regimes, a starting period at +20 °C of either 12 or 24 hours, see Fig. 3. The subsequent curing was at -10 °C; it can be seen in Fig. 4 that the EMC pastes after 3 days of curing had compressive strengths from 65 to 100 MPa, respectively. The corresponding strengths of the non-modified pastes were 35 and 65 MPa. We have to emphasize that in the tests with EMC pastes, the amount of antifreezing admixture Betec was 2.5 to 10 times lower than in pastes with non-modified cement, i. e. from 0.5 to 2.0% by cement weight compared with 5% by cement weight.

A tendency similar to that for pastes can be observed for the precuring of EMC concretes with different cement content, see Figs. 6 and 8. Due to the high chemical reactivity of the EMC binder, precuring of the B480 concretes for 12 and 24 hours at +20 °C followed by curing at -10 °C resulted in levels of compressive strength of high strength concrete: from 60 to 85 MPa 3 days after casting, see Fig. 6.

As can be seen from Fig. 8, the same curing regime led to one week compressive strengths from 22 to 34 MPa in the case of the EMC B340 concrete. The corresponding strength of the concrete with non-modified cement was about 15 MPa. During the postcuring period at +20 °C, one month after the start of postcuring , i. e. the total curing time 56 days, the EMC concretes reached the strength level of about 75-80 MPa in comparison with 45-48 MPa for the reference concretes.

Different classes of the EMC concrete with binder contents of 270 kg/m^3, 340 kg/m^3 and 480 kg/m^3+10% SF were characterized by a continuous increase of strength, see Fig. 9. The duration of hardening was 11 months for EMC concretes B270 and B340, water cured, and 15 months for concrete B480+10% SF, both water and air cured. During this period of testing, no significant variations of strength were observed. In general, the air cured EMC concrete (curing temperature about +20 °C and the samples were wrapped with polyethylene film and stored at a relative humidity about 40%) showed slightly lower (up to 15%) values of compressive strength in comparison with the water-cured samples.

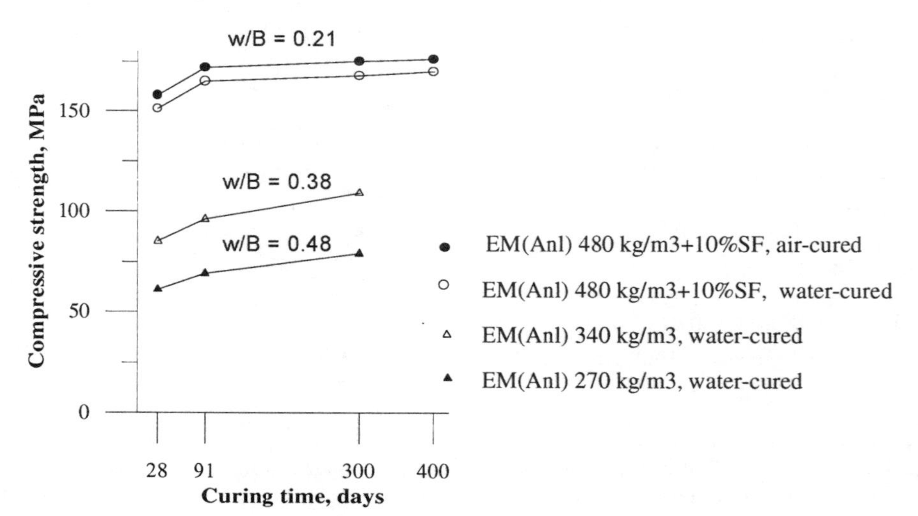

Fig. 9. Long-term strength development of EMC concretes. w/B means water-to-binder ratio.

3.2 Relative humidity (RH) and shrinkage development

Figures 10a and 10b illustrate RH-development curves in the centre of the samples for B480 concrete with 10% SF with non-modified and modified cement, respectively. The EMC concrete exhibited a higher rate of RH- reduction for all types of curing conditions, especially during the first 14-20 days after casting. In this case, RH levels of about 90%, 80% and 70% were reached for the EMC concrete after approximately 2, 5 and 14 days of curing, respectively.

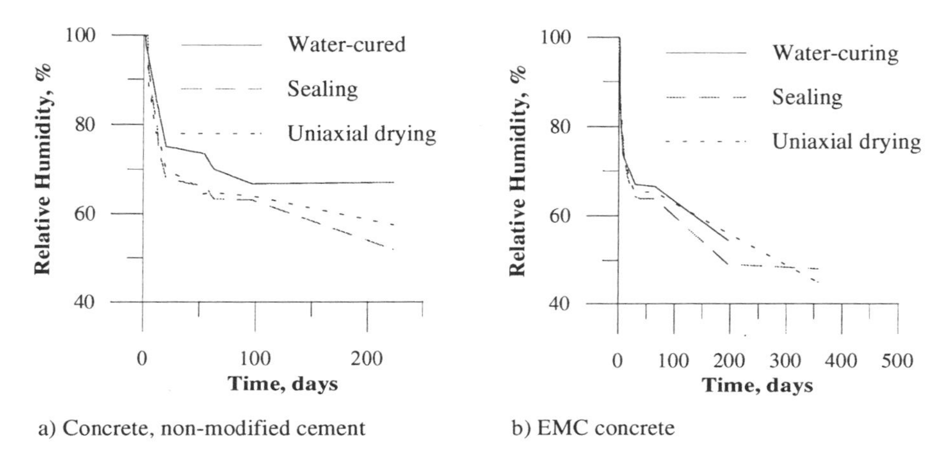

a) Concrete, non-modified cement b) EMC concrete

Fig. 10. RH development for concrete with non-modified cement and with EMC, respectively.

An important peculiarity of the EMC concrete is its behaviour in water curing conditions during the first 20-25 days of hardening; i. e. when due to high density and self-desiccation of the material, the curing conditions had a rather slight influence on RH-development.

The facts mentioned above demonstrated the high self-desiccation potential of high performance concretes produced with the energetically modified cement. In this case, the hydration process took place and concrete strength development continued at rather low values of internal relative humidities, see Figs. 9 and 10.

Figure 11 shows the results of the concrete strain measurements. As can be seen, the rate and absolute values of the self-desiccation shrinkage were higher for the concrete with non-modified cement throughout the hardening, especially during the early age curing period. Approximately 80-85% of the ultimate self-desiccation shrinkage strain was reached in 40-45 days after casting for the EMC and reference concretes, and after this time, the rate of the self-desiccation shrinkage development became practically negligible.

The development of the drying shrinkage (double-sided drying out for a thickness of 100 mm) at a room temperature of about 20 °C and a RH of about 30% is also presented in Fig. 11. The drying shrinkage curves are similar in character for both studied concretes, and also, both concretes had approximately the same level of strains after 200-220 days of hardening.

The self-desiccation shrinkage (sealed samples) seemed to be correlated in time with the decrease of internal concrete relative humidity, see Fig. 12. The EMC concrete showed significantly lower self-desiccation shrinkage strains at the same values of relative humidities in comparison with the reference concrete.

The curing in water of the EMC concrete showed practically no swelling after 3-5 days of curing. Due to high density and well pronounced self-desiccation, even a slight shrinkage of the material was observed.

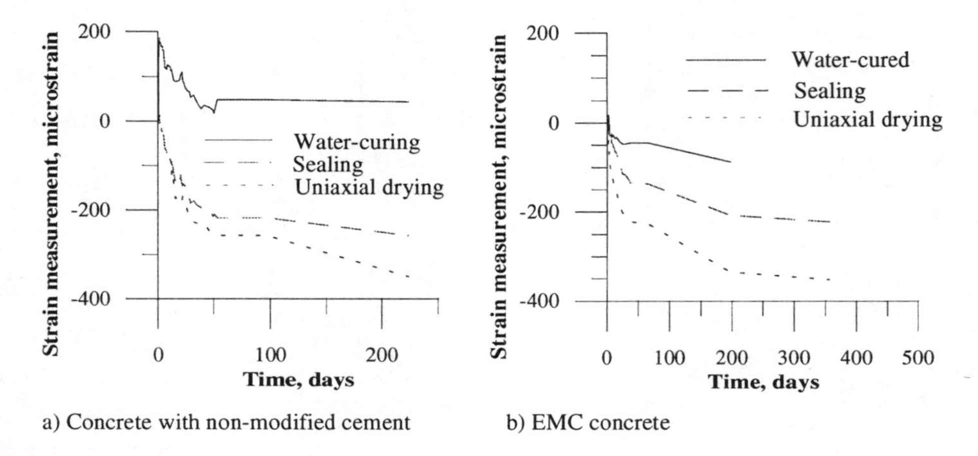

a) Concrete with non-modified cement

b) EMC concrete

Fig. 11. Strain development for concrete with non-modified cement and EMC, respectively.

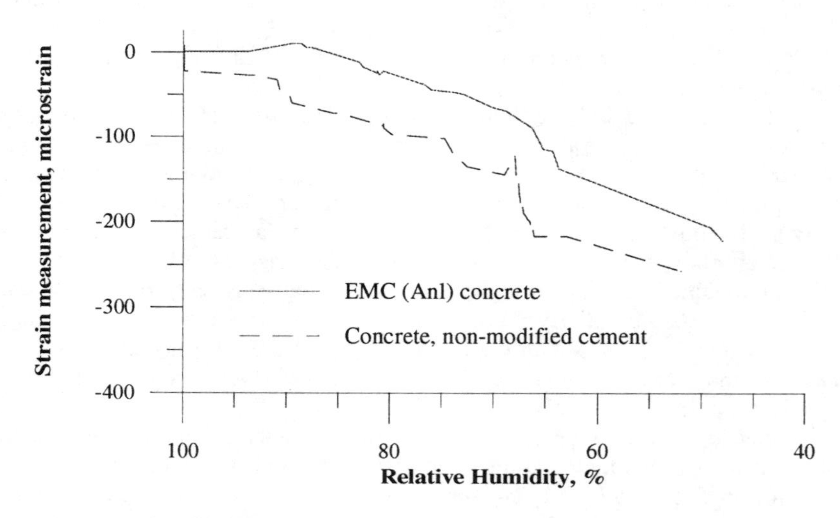

Fig. 12. Self-desiccation deformation versus relative humidity.

3.3 Durability tests

The evaluation of the durability of the EMC concrete in comparison with the reference concrete was done under rather severe test conditions described in Fig. 13. Samples of B480 concrete with 10% of SF cured 2 months in water at about 20 °C were used. The compressive strength of the EMC and reference concrete before testing were 180.3 MPa and 128.4 MPa, respectively.

The EMC concrete showed a rather high level of durability during the period of testing ,see Fig. 14. After a slight increase of the weight during the first 10-15 cycles of testing due to precipitation and crystallisation of NaCl in porous media, the system stabilized and showed practically no weight changes and no signs of deterioration.

In the case of reference concrete after approximately 10-12 cycles, a gradual deterioration took place. This process was accompanied by the development of crack networks and mass losses, see Fig. 14. The test with reference concrete was stopped after 25 cycles when mass losses reached about 9 %. The same rate and character of the deterioration in similar test conditions of high strength silica fume concrete were reported by Bache [7].

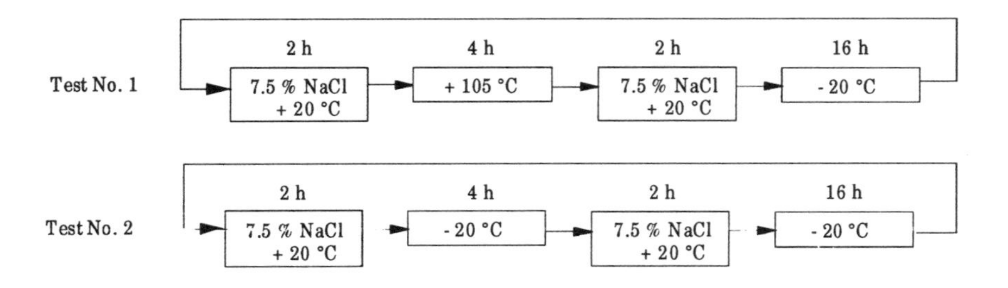

Fig. 13. The scheme of durability tests (see Fig. 14).

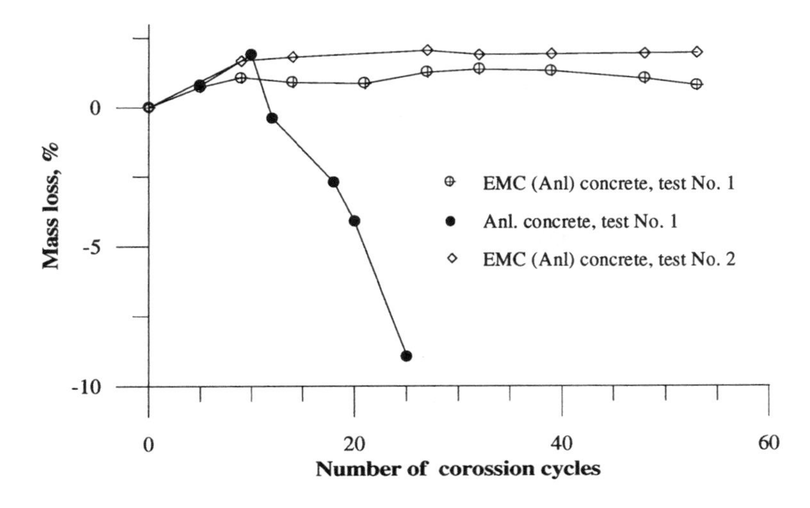

Fig. 14. Mass losses of the concrete versus number of corrosion cycles.

4 Conclusions

The EMC binders in combination with antifreezing admixtures provide different options for cement paste and concrete strength development at negative temperatures, depending on pre- and postcuring conditions.

The long-term strength development (test duration up to 15 months) of the EMC concretes showed an increase of compressive strength without any registration of strength drops.

The EMC concrete exhibit a high rate of internal relative humidity decrease at different curing conditions. The EMC concrete was characterized by lower levels of self-desiccation shrinkage development and ultimate values of shrinkage strains in comparison with reference concrete.

The EMC concrete showed higher durability potential in comparison with reference concrete: no mass losses during 2 months of the very severe durability tests.

5 References

1. Jonasson J-E. and Ronin V. (1993b) *Energeticall y modified cement (EMC).* Contribution at the Third International Conference on High-Strength Concrete, 20-24 June 1993 in Lillehammer, Norway (edited by I. Holand and E. Sellevold), Proceedings, pp 752-759.
2. Jonasson J-E. and Ronin V. (1994) *Winter concreting with the use of energetically modified cement (EMC).* Contribution at Polar Tech Conference, March 22-24, 1994 in Luleå, Sweden.
3. Ronin V. and Jonasson J-E. (1993) *New concrete technology with the use of energetically modified cement (EMC).* Contribution at the Nordic Concrete Research Meeting, August 1993 in Göteborg, Sweden, Proceedings, pp 53-55.
4. Ronin V. and Jonasson J-E. (1994a) *Concreting at low temperatures - improved methods of using inorganic modifiers for high strength concrete - Part II.* Skrift 1994:01. Department of Civil Engineering, Luleå University of Technology, Luleå, January 1994.
5. Jonasson J-E. and Ronin V. (1993b) *Concreting at low temperatures- improved methods of using inorganic modifiers for high strength silica fume concrete.* Skrift 1993:02. Department of Civil Engineering, Luleå University of Technology, Luleå, January 1993.
6. *Swedish Patent application* N 9301493-4, April 1993.
7. Bache M. (1983) *Densified cement/ultra fine particle-based materials.* The Second International Conference on Superplasticizers in Concrete, 1983.

Acknowledgement

This work is supported by the Swedish Council for Building Research and Cold Region Technology Center (COLDTECH) in Sweden. The work is supervised by Professor Lennart Elfgren, head of the Division of Structural Engineering at Luleå University of Technology.

89 STUDY ON THERMOS CURING OF CONCRETE UNDER WINTER CONDITIONS IN BULGARIA

V. VALEV and P. STANEVA
University of Architecture and Civil Engineering, Sofia, Bulgaria

Abstract
The paper presents a multi-factorial experimental study of the thermos method application for winter concreting. The parameters investigated were: temperature of concrete (t_c), average temperature of concrete (t_{ca}), duration of thermos period (T), relative compressive strength of concrete (f_{ct}) and necessary initial temperature of concrete (t_{ip}) during placing. Multi-factorial empirical relationships, based on the analysis of the experimental results, were formulated for these parameters: t_c, t_{ca}, T, f_{ct}, t_{ip} and then a design method for thermos curing of concrete under winter conditions was developed.
Keywords: Design, multi-factorial experiment, multi-factorial relationship, relative compressive strength, temperature of concrete, thermos method, thermos curing, winter conditions.

1 Introduction

Bulgaria is a country with a relatively small area (111000 km^2) and geographically located in the zone of moderate continental climate. Owning to the complicated morphology of land, the climatic diversity has wide boundaries. The winter lasts in the different areas for 1.5 to 6 months with average temperature from 3°C to -10°C.

Concreting takes place uninterruptedly all the year in areas with up to 4 month winter and average winter temperature down to -5°C. Because of the high costs involved for winter concreting, the winter concreting methods, even thermos method, are applied after giving special reasons.

The requirements for the effective winter concreting determined the objective of the present study: to investigate the basic parameters of thermos process in order to develop a more precise and effective design method for winter concreting by means of thermos method.

Concrete Under Severe Conditions: Environment and loading (Volume Two) Edited by K. Sakai, N. Banthia and O.E. Gjørv. Published in 1995 by E & FN Spon. ISBN 0 419 19860 1

2 Experimental investigation on cooling of concrete placed during winter

2.1 Theoretical conception

Thermos-hardened concrete structures are subjected to the influence of normal and exothermal cooling. The temperature of normal cooled concrete (t_c , °C) depends on the duration of thermos period (age T, h), initial temperature of concrete (t_{ip} , °C), ambient temperature (t_a , °C), or on $\Delta t = t_{ip} - t_a$, thermal resistance of heat insulation (r, m^2.°C/W), surface modulus (ratio of cooled surface of concrete and volume of the structure, M_s, m^{-1}) and on specific heat of concrete (q_c, kJ/kg.°C) in the following relationship:

$$t_c = f_{ip}(T, \Delta t, M_s, r, q_c). \tag{1}$$

The temperature of exothermal cooled concrete (t_e , °C) depends on the factors involved in (1), as well as on the cement activity (R_c, MPa), unit cement amount (C_a, kg/m^3), water-cement ratio (W/C_a) and technological effect of applied admixtures (δ), or

$$t_e = f_{1e}(T, \Delta t, M_s, r, q_c, R_c, C_a, W/C_a, \delta). \tag{2}$$

Therefore, the temperature of complex cooled concrete can be expressed as the sum of t_c and t_e , i.e.

$$t_p = t_c + t_e, \tag{3}$$

which is presented graphically in Fig. 1.

2.2 Determination of t_c

An experimental study was carried out to derive a function for (1). The cooling material was water, placed in steel boxes with different M_s (42.5 ÷ 14 m^{-1}). Three different types of heat insulation with r (0.13; 0.55; 1.05 m^2.°C/W) and Δt (40; 65 °C) were used. Based on the results, some of them presented in Fig. 2, the function was found:

$$t_c = \frac{a}{T^2 + bT - c} + t_a , \tag{4}$$

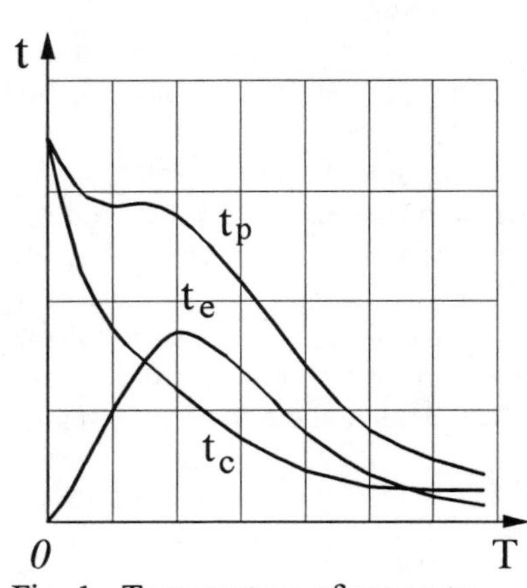

Fig. 1. Temperature of concrete.

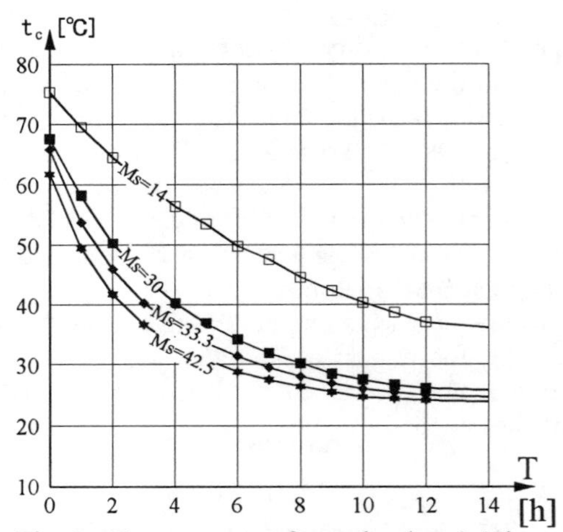

Fig. 2. Temperature of samples (r = 0.13).

where a, b, c are coefficients describing the influence of t_c, M_s, r and q_c. By means of the method of the averages, the relationships for a, b and c were determined:

$$a = -\Delta t_c, b = \frac{2}{e^{\mu}-1}, c = -\frac{2}{(e^{\mu}-1)^2}, \mu = \frac{M_s}{rq_c}. \tag{5}$$

The discrepancy of the temperatures, calculated according to (4) and experimentally recorded one, is within (-0.7÷+0.6) °C, which is precise enough for the practical purposes.

2.3 Determination of t_e

The factors, included in (2), could be divided into 2 groups: heat insulation factors (M_s, r, Δt, q_c, T) and exothermal factors (C_a, R_c, W/C_a, t_{ip}, δ, T). The common factor in both groups was T and the influence of the former factors has been already explained by (4) and (5). Therefore, the influence of latter factors was described under adiabatic conditions of cooling. The radiated exothermal heat E_T (kJ/kg) is

$$E_T = f_T(T, C_a, R_c, W/C_a, t_{ip}, \delta). \tag{6}$$

The same function expressed by means of t_e under adiabatic conditions (t_{ea}) is

$$E_T = \frac{1}{\tau} t_{ea}, \tau = \frac{C_a}{q_c \gamma}. \tag{7}$$

Based on mathematical and graphical analyses of a considerable amount of data, presented in [1], [2], [3] and [4], the exothermal heat was determined and t_{ea} can be expressed as

$$t_{ea} = \tau\alpha\left[T_P^{\frac{1}{3}} + (T - T_P)^{\frac{1}{3}}\right], \tag{8}$$

where τ is a heat constant of concrete, γ is specific concrete weight, and

$$\alpha = 1.92 + 0.22R_c + 0.64W/C_a + 0.024t_{ip} + 0.051\delta + 0.095R_cW/C_a, \tag{9}$$

$$T_p = (18.8 - 0.2R_c)(0.13 + 1.46W/C_a)(1.24 - 0.027t_{ip})(1 - 0.21\delta). \tag{10}$$

2.4 Determination of t_p

Using the relationships (4) and (5), and assuming that $\Delta t = t_{ip}+t_{ea}-t_a$, t_p can be expressed as

$$t_p = \frac{-c\left(t_{ip}+t_{ea}-t_a\right)}{T^2+bT-c} + t_a. \tag{11}$$

The precision of (11) was checked experimentally. The experimental conditions included four molds with M_s (14÷53 m^{-1}), heat insulation with r (0.25; 0.90 $m^2.°C/W$), C_a (322 kg/m^3), R_c (39.1 MPa), W/C_a (0.56), t_{ip} (30 °C) and δ=0. The experimental results are shown in Fig. 3. The discrepancy (t_p-t_{pr}) is less than ±0.5 °C, excluding some cases at very early age T (with discrepancy from 1 °C to 3.5 °C).

3 Average temperature of thermos cooled concrete

3.1 Basic idea

The average temperature of concrete cured under winter conditions and when the thermos method is applied, t_{ca} can be initially determined by integration of (11) under T, i.e.

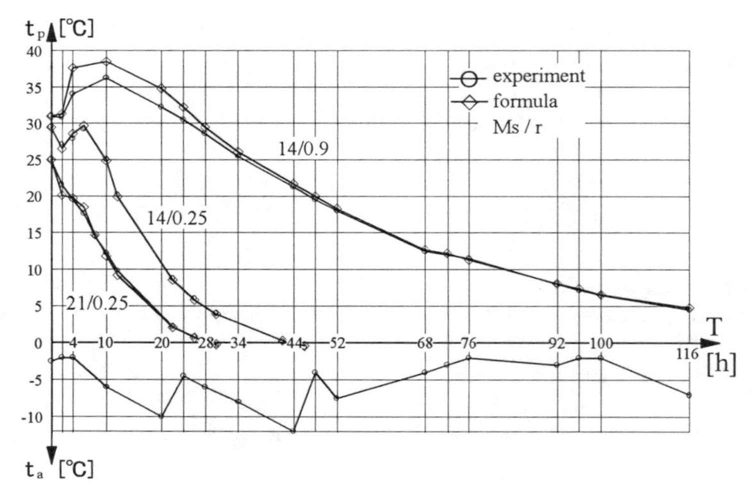

Fig. 3. Comparative temperatures of concrete.

$$t_{ca} = \frac{1}{T}\int_0^T t_p \,, \tag{12}$$

and this mathematical operation is relatively complicated. Additionally, t_{ca} is influenced by such factors of accidental character, as variation of t_a, the wind velocity, the shape of structure, formwork equipment etc. In the known design methods for thermos curing of concrete, t_{ca} is determined according to the relationship $t_{ca} = f(t_{ip}, M_s)$. Because of this, the discrepancy is too high (from -4 °C to +7.5 °C). In order to decrease it and for higher precision, additionally three basic factors of cooling: r, Δt and δ are considered. Table 1 presents data about the technological effect of different type admixtures.

Table 1. Details for different Bulgarian admixtures for concrete

Admixture	Type	Doze	Effects			Technological effect, δ
			water reduction (%)	hardening acceleration (%, 1st day)	ambient temperature (°C)	
Tekozim-H	antifreezer, plasticizer	1-2	4-7	-	-10	1.06-1.1
T-3	antifreezer, plasticizer	1.6-2.5	6-9	-	-10	1.08-1.14
Lutol KH	antifreezer, plasticizer	1.6-2.2	3-5	-	-10	1.05-1.09
Uskoplast-H	accelerator, plasticizer	0.8-1.5	5-8	20-35	normal	1.07-1.11
KK-T	accelerator, plasticizer	0.5-1	6-9	25-35	normal	1.08-1.13
BP-3	accelerator, plasticizer	1.5-2	6-12	25-40	normal	1.11-1.16
BBP-1	superplasticizer	0.5-1.5	10-15	20-35	normal	1.07-1.15
Tekoplast	superplasticizer	0.4-0.8	10-15	30-40	normal	1.08-1.14
Skleroment	superplasticizer	0.5-1	8-13	15-30	normal	1.06-1.12

3.2 Experimental determination of t_{ca}

Experimental study was conducted with four types and four different sizes of specimens in order to determine a function $f(t_{ip}, M_s, r, \Delta t, \delta)$ for t_{ca}. Based on the results obtained, the following relationship was derived;

$$t_{ca} = \frac{(1+Nr+P\delta)t_{ip}}{A} + BM_s - \frac{t_{ip}}{\Delta t}, \tag{13}$$

where A, B, N and P are constants. They were found by the least squares method while (13) is applied in the form

$At_{ca} + BM_s t_{ca} - Nrt_{ip} = t_{ip} + \frac{t_{ip}}{\Delta t} t_{ca}$, and A=3.15, B=0.03, N=0.11 and P=0.07. Finally, the relationship (13) was reformed as

$$t_{ca} = \frac{(1+0.11r+0.07\delta)t_{ip}}{3.15+0.03M_s - \frac{t_{ip}}{\Delta t}}. \tag{14}$$

Table 2. Characteristics of the series and results obtained

No series	M_s (m^{-1})	t_{ip} (°C)	r (m^2°C/W)	t_a (°C)	Δt (°C)	T (h)	$t_{ca,r}$ (°C)	$t_{ca,r}$-t_{ca}(°C)
I	53	26	0.25	-3.5	29.5	14	7.4	0.4
	33	27	0.25	-4	31	21	7.8	-0.6
	21	25	0.25	-4.5	29.5	30	8.9	0.2
	15	29.5	0.25	-5.5	35	45	11.2	0.2
II	53	25	0.75	-3.5	28.5	14	7.2	0.1
	33	26.5	1.05	-3.5	30	22	8.3	-0.7
	21	31.5	1.49	-0.5	32	67	11.2	-1.6
	15	30	1.98	-2.5	32.5	102	15.2	1.7
III	53	24	1.12	-4.5	28.5	22	7.4	0.4
	33	27	1.35	-4.5	31.5	32	9.1	-0.2
	21	31	1.68	-6.1	37.1	54	12.4	0.1
	15	31	2.05	-3.6	34.6	171	12.8	-0.9
IV	53	22	1.5	-3	25	40	6.1	-0.4
	33	25.5	1.65	-3	28.5	48	8.9	-0.2
	21	26.5	1.87	-2.5	29	102	10.8	-0.1
	15	27	2.11	-3.3	30.3	214	13.4	1.2
[5]	15.3	29	0.52	-2	31	-	10	-1.4
	9.3	31	0.96	-7	38	-	11.6	0.4
	6	18.5	0.28	-1	19.5	-	6.6	-1.4
	3.7	23	1.89	-4	27	-	12.1	0.7
	3.3	20.5	1.98	-12	32.5	-	9.5	-0.1
	1.06	7	6.7	-5	12	-	4.8	0.3
[6]	20	10	0.38	-13	23	-	3.4	0.2

The precision of (14) was described through data presented in Table 2 and Fig. 4, including experimental results obtained by other authors [5], [6]. It was seen that the discrepancy $\Delta t_{ca} = t_{ca,r} - t_{ca}$ was around ±0.5 ℃ and in few cases it reached ±1.7 ℃. Therefore, equation (14) meets the practical requirements.

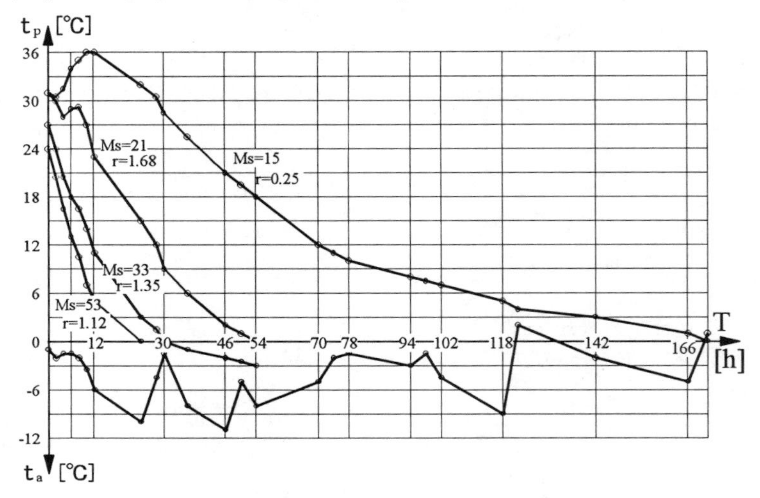

Fig. 4. Cooling of concrete test specimens (series No. 3).

4 Relative compressive strength of thermos cured concrete

4.1 Basic idea

Absolute compressive strength of conventional concrete f_c was studied comprehensively and different formulas were derived for its prediction depending on W/C_a and R_c.

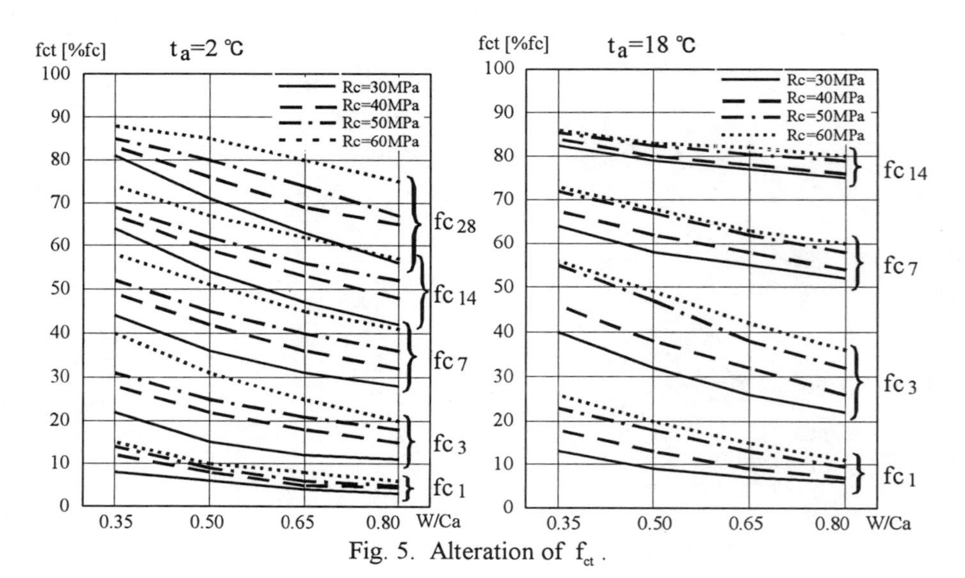

Fig. 5. Alteration of f_{ct}.

Additionally, the design and practice for concreting under winter conditions required the determination of relative compressive strength f_{ct} (% of f_c), which depends on W/C_a, R_c and t_{ca} as well. The influence of W/C_a and R_c on f_{ct} has not been studied well enough. An experimental program was carried out in order to determine $f_{ct} = f(R_c, C_a, W/C_a, t_{ca}, T)$. The program included eight series with factors as follows: R_c (30;40;50;60 MPa), C_a (250;350;450 kg/m^3), W/C_a (0.35-0.80), $t_a = t_{ca}$(0-4; 15-18; 38-40 ℃) , T (1;3;7;14;28 days). The results are presented in Fig. 5 and Fig. 6.

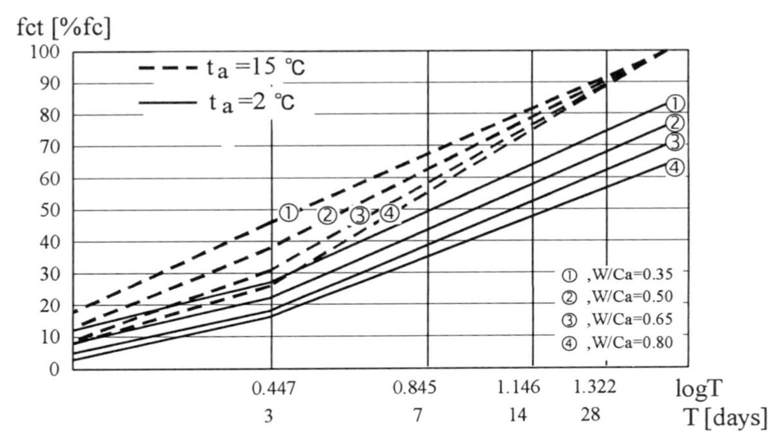

Fig. 6. Alteration of f_{ct} as f(logT, t_a, W/C_a).

4.2 Analytical model for determination of f_{ct}

Based on the Fig. 6, the function for f_{ct} could be described as two-part linear function of logT for any R_c, t_{ca} and W/C_a:

$$f_{ct_1} = A_1 + K_1 \log T, \text{ when } T \le 3; \quad (15)$$

$$f_{ct_2} = A_2 + K_2(\log T - 0.477), \text{ when } 3 < T \le 28, \quad (16)$$

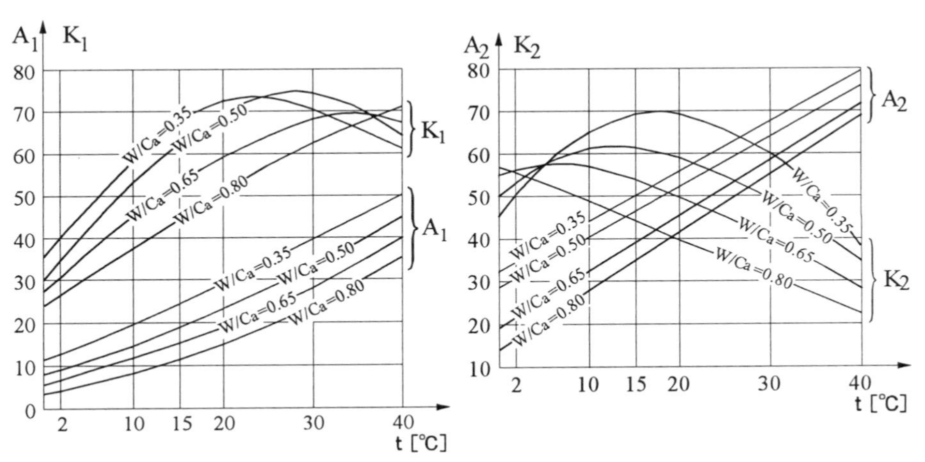

Fig. 7. Charts for coefficients A_1, K_1 and A_2, K_2 (R_c=50 MPa).

where A_1, A_2, K_1 and K_2 are coefficients, which described the influence of R_c, t_{ca} and W/C_a on f_{ct}. They could be determined graphically using Fig. 7 and analytically using the next relationships:

$$A_1 = \alpha_1 + \alpha_2 t_{ca}^2, K_1 = \beta_1 + \beta_2 t_{ca} - \beta_3 t_{ca}^2; \quad (17)$$
$$A_2 = \gamma_1 + \gamma_2 t_{ca}, K_2 = \varphi_1 + \varphi_2 t_{ca} - \varphi_3 t_{ca}^2. \quad (18)$$

After replacing A_1, A_2, K_1 and K_2 in (15) and (16) with (17) and (18) respectively, f_{ct} could be determined as follows:

$$f_{ct_1} = \alpha_1 + \alpha_2 t_{ca}^2 + (\beta_1 + \beta_2 t_{ca} - \beta_3 t_{ca}^2)\log T, \text{ when } T \leq 3; \quad (19)$$
$$f_{ct_2} = \psi_1 + \psi_2 t_{ca} + \psi_3 t_{ca}^2 + (\varphi_1 + \varphi_2 t_{ca} - \varphi_3 t_{ca}^2)\log T, \text{ when } 3 < T \leq 28, \quad (20)$$

and $\psi_1 = \gamma_1 - 0.477\varphi_1, \psi_2 = \gamma_2 - 0.477\varphi_2, \psi_3 = 0.477\varphi_3$.

The values of $\alpha_i, \beta_i, \psi_i$ and φ_i are presented in Table 3 and for different R_c and W/C_a. By means of linear interpolation, the application of chart (Fig. 7) and Table 3 was fast and precisely enough.

Table 3. Coefficients α_i, β_i, ψ_i and φ_i

Coefficients	$W/C_a = 0.50$				$W/C_a = 0.65$			
	R_c (MPa)				R_c (MPa)			
	30	40	50	60	30	40	50	60
α_1	5.2	6.4	8	10	3.9	4.5	5.9	7.4
α_2	0.02	0.02	0.02	0.03	0.02	0.02	0.02	0.02
β_1	17	28	28.5	41	15.7	25	27	31
β_2	2.47	2.01	3.02	2.02	1.68	1.69	2.22	2.7
β_3	0.03	0.02	0.05	0.04	0	0.01	0.03	0.05
ψ_1	-13.2	-5.9	0.8	4.9	-12.5	-8.9	-4.3	-3.2
ψ_2	0.48	0.57	1.01	0.82	-0.22	0.13	0.61	0.81
ψ_3	0.02	0.02	0.01	0.02	0.04	0.03	0.02	0.02
φ_1	55	53	55	51.5	46	49	50	54
φ_2	1.68	1.5	0.39	0.78	3.15	2.48	1.53	1.05
φ_3	0.05	0.05	0.03	0.04	0.08	0.07	0.05	0.04

5 Design of thermos curing of concrete under winter conditions

5.1 Basic idea

Design for thermos cured concrete under winter conditions was developed based on the theory, experiments and results described above. The initial known parameters of given production conditions were: reinforced concrete structure with its M_s, expected average air temperature t_a, absolute compressive strength at 28 day f_c, W/C_a, R_c and C_a, as well as required relative compressive strength of concrete f_{ct} at the end of thermos period. Corresponding to (13) and the paragraph above, it was derived that

$$f_{ct} \geq F\left(r, \delta, t_{ip}, M_s, t_a, W/C_a, R_c, T, C_a\right). \tag{21}$$

Based on these assumptions, therefore, it was needed to determine other factors influenced on f_{ct} as r, T, t_{ip} and δ. In order to intensify thermos hardening of concrete, it was preferable that the known factors R_c, W/C_a and C_a be changed and made more precise in the process of concrete design. That meant an increase of R_c and C_a, and also a decrease of W/C_a. The other way was to use an admixture. Such a change in the mix proportion of concrete led to an increase of exothermic heat (i.e. t_{ca}), an acceleration of hardening and an improvement of concrete strength f_c. In such a case, the thermos process was reinsured and the requirements about f_{ct} (combined with known f_c) could be changed. The design of thermos curing was considered in two cases: I. Known thermos duration T; II. Unknown thermos duration T.

5.2 Case I

In this case, first the factors t_{ca} and t_{ip} were determined while the other unknown factors were initially accepted and thereafter made more precise.

5.2.1 Determination of t_{ca}

Depending on T, t_{ca} could be found by considering (19) when $T \leq 3$,or (20) when $3 < T \leq 28$ days. The necessary t_{ca} was found as a solution of full quadratic equation (19) or (20), while $\alpha_i, \beta_i, \varphi_i, \psi_i$ were taken from Table 3.

5.2.2 Determination of t_{ip}

Relationship (14) was used and transformed in modified (full quadratic) equation:

$$(1+0.11r+0.07\delta)t_{ip}^2 - [(1+0.11r+0.07\delta)+(3.15+0.03M_s)t_{ca}]t_{ip}+ (3.15+0.03M_s)t_{ca}t_a = 0. \tag{22}$$

In order to solve (22), r and δ were determined in advance.

5.3 Case II

Relationships (14), (19) and (20) were used in order to find t_{ca} and T, while the other unknown factors were initially accepted.

5.3.1 Determination of t_{ca}

The factor Δt in (14) was replaced with $(t_{ip}-t_a)$ and t_{ca} was obtained as:

$$t_{ca} = \frac{(1+0.11r+0.07\delta)t_{ip}}{3.15+0.03M_s - \frac{t_{ip}}{t_{ip}-t_a}}. \tag{23}$$

In (23), the factors r, δ, t_{ip} were chosen in advance.

5.3.2 Determination of T

Here, (19) and (20) were presented as follows:

$$\log T \leq \frac{f_{ct}-\left(\alpha_1+\alpha_2 t_{ca}^2\right)}{\beta_1+\beta_2 t_{ca}-\beta_3 t_{ca}^2}, \text{ when } 3 \leq T, \tag{24}$$

$$\log T \leq \frac{f_{ct}-\left(\psi_1+\psi_2 t_{ca}-\psi_3 t_{ca}^2\right)}{\varphi_1+\varphi_2 t_{ca}-\varphi_3 t_{ca}^2}, \text{ when } 3 < T \leq 28. \tag{25}$$

If (24) or (25) have been not satisfied for received in (23) t_{ca}, then r, δ and t_{ip} (or only t_{ip}) were changed in order to find larger t_{ca}. The procedure finished when (24) and (25) were satisfied and T was determined.

6 Conclusions

The basic factors and parameters of thermos method for winter concreting were examined in the current study. As a result of comprehensive experimental work, the next important suggestions were made for more effective application of thermos method.

- Average temperature of thermos cooled concrete t_{ca} was determined with regard of technological effect of applied admixture δ, (14).
- Relative compressive strength of concrete f_{ct} was estimated through thermos duration T and tabulated coefficients $\alpha_i, \beta_i, \varphi_i, \psi_i$.
- Fast design procedure was proposed for concreting by means of thermos method and it was mainly based on the average temperature of cooling t_{ca} and required relative compressive strength f_{ct}.

7 References

1. Zasedatelev, I.B. (1960) Exothermal heat of cement in the concrete under high temperatures, *Proceedings,* Goststroyizdat, Moscow, (in Russian).
2. Miagkov, A.T. (1963) Determination of exothermal heat of portland cement in concrete, where accelerated hardening admixtures are applied, *Concrete and Reinforced Concrete,* No12.
3. Simeonov, J.T. (1959) Exothermal heat and shrinkage of cement and concrete in the mass dam walls, *Proceedings,* BAN, Sofia, (in Bulgarian).
4. Valev, V.N. and Chaiverova, R. (1992) Experimental study of exothermal process by hardening of concrete with admixture Retament, *Proceedings*, UACEG, Sofia, Vol. XXXVI, (in Bulgarian).
5. Mironov, S.A. (1974) *Theory and Methods of Winter Concreting,* Goststroyizdat, Moscow, (in Russian).
6. Svenson, E.G. (1956) Climatic conditions and their influence on winter concrete works, *Proceedings,* Moscow, (in Russian).

90 TEMPERATURE INFLUENCE ON CONCRETE STRUCTURES AND ITS HARDENING

B.A. KRYLOV and A.I. ZVEZDOV
NIIZHB, Moscow, Russia

Abstract
Frost attacks on freshly laid concrete, its structure and its properties from the position of structural changes and incorporated water are considered. Early freezing of concrete prior to achieving a specified maturity leads to weakening of a concrete structure, influencing its integrity and considerably aggravating concrete properties. Deteriorating factors resulting from the freezing of concrete are an expansion in volume of water being transformed into ice and internal mass exchange. The methods of concrete curing under sub-zero ambient temperatures are briefly discussed.
Keywords: Concrete structure, aggregation of molecules, antifreeze admixtures, thermos, electrothermocuring, electrode heating.

1 Introduction

Execution of construction works in winter always requires extra labour and expense. In some countries, construction works are stopped in cold periods and resumed after the return of a favourable season. But in Russia, Canada, Scandinavian countries, northern China and Japan, the cold period with low temperatures is quite prolonged, and to stop construction for several months is not reasonable. Methods of winter construction have been developed that enable construction to continue at a reasonable cost increase in comparison with warm periods without harmful effects on quality and durability.

Most difficulties arise at execution of concreting in cold weather. At present various methods of winter concreting have been developed which enable high quality reinforced concrete structures to be built at minimum expense by the acceleration of concrete

Concrete Under Severe Conditions: Environment and loading (Volume Two) Edited by K. Sakai, N. Banthia and O.E. Gjørv. Published in 1995 by E & FN Spon. ISBN 0 419 19860 1

hardening. But it should be pointed out that there is no universal method; each method has its own most effective field of application in accordance with the ambient air temperature, type and solidity of structure, type of reinforcement and maintenance conditions.

The correct choice of the curing method in cold weather will guarantee minimal extra expense for concreting works in comparison with a more favourable period and with the specified finished quality.

Before starting to analyze methods of curing concrete in cast–in–situ structures in cold weather, let us consider the phenomena of frost action on concrete.

2 Frost influence on concrete

Frost acts on all solid concrete components — cement and aggregates expand or contract in accordance with the temperature variation and thermal expansion coefficients. But water behaves differently: depending on the temperature, definite structural modifications take place. Water at a normal temperature (+20°C) is comprised of an association of molecules, which interact with cement grains at their surfaces. At elevated water temperatures its viscosity and surface tension is decreased, the bond between associated molecules weakens and they divide into smaller particles; at 60°C, many separate molecules are only weekly connected to each others, and the process of evaporation, or molecules freely escaping from the system, starts becoming easily visible. At higher temperatures up to 100°C, the association between separate molecules are disintegrating, part of them into oxygen and hydrogen ions which are exceptionally active. At boiling temperature, the avalanche type escape of water molecules into adjacent space is taking place; because of the disruption of bonds between them their movement is sharply intensified and interaction with cement particles occurs very actively. It is obvious because the rate of contacts of water particles in the form of separate and small groups of molecules and ions with the surface of cement grains happens more frequently than at lower temperatures with higher molecules borders.

Thus, at elevated temperatures the chemical activity of water increases and the cement hydration process accelerates approximately 2–3 times for each 10°C gradient. This phenomena forms the foundation of the method of concrete hardening acceleration by means of thermal curing.

Temperature decrease leads to the opposite situation — the chemical activity of water decreases, and its interaction with cement is lowered. At temperatures approaching 0°C, interaction is so lowered that the setting time prolonged 2–4 times. At 0°C, free or mechanically bound water is transforming into solid state — ice. Physically bound water, depending on the form of bond, is transformed into a solid state gradually under sub–zero temperatures up to −100°C and lower. Chemically bound water is included in the composition of new–formations, which react on alteration of temperature by expansion or contraction in accordance with the ratio of linear expansion for the given material.

It is known that substances in a solid state are the most dense due to the drawing closer of their molecules and because of stronger ties in comparison with liquid and gaseous states. An analogous picture — the drawing closer of water molecules and the strengthening of their ties — takes place in ice as well. But its density is about 9% lower than for the water at 4°C temperature. How can such a paradox be explained? Anomalous properties of water are the answer. It is known that the molecules of water are characterized by the distance between the ions of oxygen and hydrogen (ℓ) and the angle (α) between them (Fig. 1). At the transition to a solid state, ℓ and α values increase. As a consequence, in spite the fact that molecules of water come closer to each other, each molecule occupies a bigger volume, so the density of the ice becomes lower than the density of water. Because transformation is realized on the molecular level, the 9.07% water expansion at freezing generates immense forces reaching values up to 250 MPa. Such forces are sufficient to destroy a vessel made of any material if it is filled fully by water and hermetically sealed. Thus the conclusion — the strongest concrete could not resist the rupture action of freezing water contained inside. As a consequence, it is necessary to prevent rupture by decreasing the quantity of free water inside at the moment of freezing and to arrange in its structure a sufficient quantity of closed spherical pores filled by air.

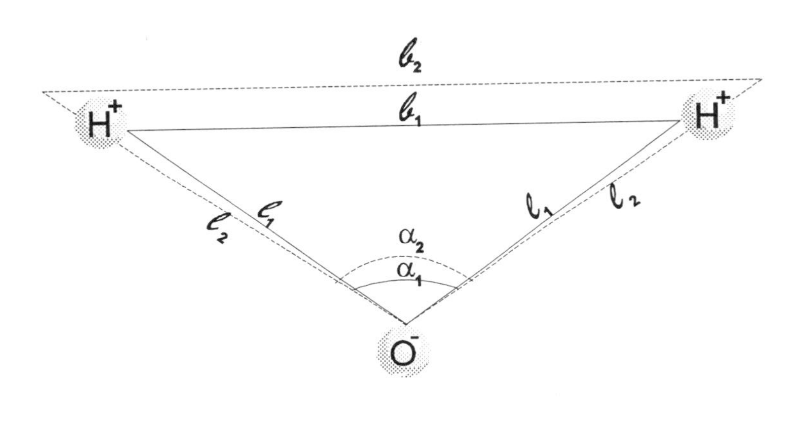

$\ell 1$ = 0.9568Å $\qquad$ $\ell 1$ = 0.99Å
α1 = 105°03 $\qquad$ α1 = 109°28
b1 = 1.54 Å $\qquad$ b1 = 1.62 Å
Intermolecular distance in water — 2.90Å, in ice — 2.76Å

Fig. 1. Scheme of deformation of water molecule at transition into ice.

Many works of different scientists have been devoted to frost action on concrete. Theoretical conceptions of I.A. Kireenko, Powers, Collins and many others from different points of view analyze the mechanism of rupture of freshly laid concrete by frost action, but a solution to the problem is complex. Completed studies show that there exists two main reasons for rupture of freshly laid concrete by frost.

The first is the high internal pressure induced in concrete at the transition of free water into ice, forcing apart concrete components and loosening material. The second is internal mass exchange facilitating formation of ice prisms, loosening the concrete structure and disintegrating material solidity due to formation of ice films in the contacts area between coarse aggregate and cement paste.

After thawing the films to above-zero temperatures, air sublayers are formed on the contact surfaces which destroy fully, or reduce dramatically, the bond between aggregate particles and cement paste.

The freezing of freshly laid concrete can lead to reduction of concrete strength up to 40%, frost resistance in tenth times, water tightness — in several times, bond with ordinary round steel reinforcement — on 80%, with rebars — on 20%. Evidently, the degree of structural damage depends on the quantity of free water in concrete and kinetics of its freezing, but in any case, structural defects will always take place, and their negative influence on concrete properties and durability of structures will be considerable.

The quantity of free water in concrete in the process of freezing decreases due to chemical and physical bonds in the new formations. This leads to a reduction of the disintegration effect in concrete during prolonged freezing (Table 1). This is expressed in terms of the freezing of concrete of different maturity (Table 2). Taking all this into account, Russian Building Codes require that at the moment of freezing concrete of class up to 12.5 shall attain not less than 50% of R_{28} strength; class 22.5, 40% of R_{28}, and class 30 and others, 30% of R_{28}. If an increased requirement for frost resistance is specified, the concrete strength at the moment of freezing, independent of its class, shall be not less than 80% of R_{28}.

Table 1. Amount of mixing water transforming into ice with w/c = 0.72 (in %% of initially added water)

Concrete strength prior to freezing point %% of R_{28}	Freezing temperature, °C				
	−3	−10	−15	−20	−45
0	90	92	93	94	95–96
15	50	65	72	75	80
50	20	45	55	60	68
70	15	35	45	50	55

In practice to achieve a specified concrete strength before freezing, various methods of hardening acceleration are used, which require a certain amount of energy, material and labour consumption. Existing methods of concrete curing in cold weather allow the construction of any structure and provide for a specified level of quality. Everything depends on the chosen method and the extra expense. Therefore, the choice of method becomes not only a technical but an economic problem as well.

Table 2. Loss of concrete strength due to freezing at different levels of maturity.

Concrete strength at the point of freezing, in %% of R_{28}	Loss of concrete strength, in %% of R_{28}
15	30
30	10–15
50	5–10
70	0

3 Method of curing

All existing methods belong to two groups: requiring and not requiring heating of structure. In the first group are the concreting methods with antifreeze admixtures and the thermos method. In the second group, all methods use heating.

In some countries, the curing of concrete in cold weather is not unusual. And anti-freeze admixtures, sodium nitrite ($NaNO_2$), potassium carbonate (K_2CO_3), calcium chloride combined with sodium chloride or with sodium nitrite and calcium nitrite combined with carbamide are used; they are not causing steel reinforcement corrosion. There exist inexpensive antifreeze admixtures that enable concrete to harden at temperatures from −15 to −25°C. The amount of introduced admixture depends upon the ambient air temperature (Table 3).

Table 3. The amount of antifreeze admixtures added to concrete mix in % of mix weight

Ambient air temperature, °C	$NaNO_2$	K_2CO_3	Composite admixtures+carbamide
0...-5	4—6	5—6	2.0+1.0...4.0+1.0
-6...-10	6—8	6—8	4.5+1.5...7.0+2.5
-11...-15	8—10	8—10	6.0+2.0...8.0+3.0
-16...-20	—	10—12	7.0+2.0...9.0+4.0
-21...-25	—	12—15	8.0+3.0...10.0+4.0

Because the chemical activity of water in salt solutions under sub-zero temperatures is very low, concrete strength development is realized very slowly; some 2—3 months are needed to attain a design strength by curing under a calculated temperature curve. If the temperature is raised or lowered relative to the calculated curve, the strength gain can be accelerated or slowed down practically to a complete stop. Even in the event of freezing of the concrete however, with antifreeze admixtures it will not be affected adversely.

In summing up the properties of the mentioned method it is to be noted that in comparison with heating methods, it is cost effective, but the slow rise only of concrete strength limits its application only to structures that will be loaded after 2 or 3 months.

One of the most efficient is the thermos curing method which is widely used for concreting massive structures. Basically, with this method, the concrete mix is delivered with the highest possible temperature, placed, compacted and covered. Due to the applied heat and the heat evolving during hydration of the cement acting in combination with reliable thermal insulation the structure is cured under the favourable conditions for quite a prolonged time, and thus, ensuring a sufficient rate of concrete hardening.

Studies show that delivery of concrete mix with a temperature above 30°C is not reasonable, because it could lose the required workability during transportation. All necessary measures should be taken to keep heat applied as long as possible using closed insulated bunkers or mobile mixers. After placing and compacting the concrete mi should be adequately thermo- and vaporinsulated. Normally, internal layers of concr in the structure experience temperature rises due to cement exothermy (1kg of cer releases during hydration up to 80 kcal = 335 KJ). In external layers, the temper can gradually decrease, so normally the monitoring for the external temperature i stantly necessary. It is desirable to control temperature in the most exposed place corners and at open surfaces. In case of a rapid fall of the concrete temperature sudden fall of ambient air temperature, extra precautions should be developed — of the most frozen areas of the whole structure.

The thermos method features the greatest cost saving — extra expenses in con with warm period concreting do not exceed 8-12%. But this method is not ap for all structures. The most applicable areas are massive concrete and reinfor crete structures with a modulus of surface (ratio of the open surface to the volu structure) about 4 to 6. This method is used widely in Russia and has proved reliable. Structures executed by the thermos curing method are norm outstanding quality; no cracks are formed due to a favourable distribution stresses.

The curing methods utilizing applied heating are the most effective from view of the construction rate of cast-in-situ structures. As a rule, at the en process, the concrete attains up to 70-75% of the design strength value anc duration does not exceed 8-12 hours, in dependent on cement type and temperature.

Heating methods posses their own peculiarities, and in Russia, four met in general: preheating of concrete mix, electrode heating, electroheating induction heating. Preheating of concrete mix before its placing into the structure was proposed in the 60 s by Prof. A.S. Arbenjev and used at present on construction sites in Russia. The essence of the method is as following: the concrete mix is delivered from the mixing plant, heated quickly at the placing site, poured into design position, compacted, covered and cured further by the thermos method until reaching the specified concrete strength.

For preheating of the concrete mix, original installations have been developed, enabling the preheating of concrete mixes up to any desired temperature in several minutes using 220–380 V current. The most rational preheating temperature is 60°C and the duration of heating, 5–15 minutes. Because the heated mix loses quickly its plasticity, it is important to place and compact the concrete within 15 minutes.

This method is effective and cost saving, and the consumption of electroenergy comprises about 40–50 kW·hr per 1m^3 of concrete depending to ambient air temperature. It is applicable to massive structures with a modulus of surface up to 8–10. Extra expenses in comparison with the warm weather concreting amount up to 20%. Placing preheated concrete mix into a massive structure contributes to a favourable distribution of internal temperature, as happens in the thermos method.

The electrode heating method was proposed in the 1930 s by Swedish engineers A. Brund and H. Bohlin, and is one of the most frequently used at construction sites. The method is economical (electroenergy consumption comprise up to 80–100 kW·hr per m^3 of concrete) and effective; extra expenses in comparison to warm period concreting do not exceed 25%.

However, it should be noted that this method requires very thorough execution (selection of electrodes, their placement, heating regime, character of reinforcement of structure, safety of works, etc.). The voltage should not exceed 120V, the heating regime must be very mild: not over 20°C temperature rise per hour, curing temperature in the range between 60 and 80°C, dependent on the cement type used. It is applicable to non-reinforced and low reinforced concrete structures.

The heating of concrete by various heaters allows curing any type of structure inde–pendently of their reinforcement and size. The heat delivered to the concrete surface is istributed into internal layers, but the effective depth of heating normally does not ceed 20 cm. Therefore, when concrete structures are over 20cm in depth, heating is ommended from both sides. Heating regimes must be mild as in any other method, open surfaces should be protected from moisture loss. This method is used exten–ly in many countries, but electroenergy consumption is higher in comparison with lectrode heating method and reaches 100–120 kW·hr per m^3 of concrete. Extra ses in comparison with warm period comprise about 25–35 percent.

induction heating of concrete was developed in Russia by Prof. A. Netoushil and d on the principle of warming up the ferro–containing elements (rebars, inserted etc.) with an applied electromagnetic field. This method is normally used for reinforced concrete structures with heavy and evenly distributed reinforcement columns, etc.). The method is suitable for heating joints of precast reinforced structures with outlet connecting ends of reinforcement and steel inserts. Con– of electroenergy in this method could reach 120–150 kW·hr per m^3 of con– sulting increase in cost of 30–50% as compared to ordinary curing.

t possible to analyze here in detail all methods of intensified curing of concrete res built in winter. Several other methods exist as well, for instance, curing in s, combined curing methods, etc. But to achieve best results as economically e, the correct technical methods should be selected carefully in accordance pe of structure and ambient conditions.

4 References

1. RILEM Recommendations for concreting in cold weather (1988), VTT, ESPOO.
2. Finnish Construction (1991). Technology of construction in severe climate, A/O Rachentajan Kustannus.
3. Third International RILEM Symposium on Winter Concreting (1985), ESPOO.
4. B.A, Krylov. and A.I, Li. (1975) Accelerated Electric Heating of Concrete, Mos–cow, Stroyizdat

PART SIXTEEN
COATING

91 EFFECTIVENESS OF COATINGS ON CONCRETE UNDER FREEZING AND THAWING CONDITIONS

T. SATO
Dai Nippon Toryo Co. Ltd, Kamakura, Japan
K. SAKAI and M. KUMAGAI
Hokkaido Development Bureau, Sapporo, Japan

Abstract
There are several types of coatings used to protect concrete from the damage which is caused by various environmental conditions. In this study, the protection properties and durability of these coatings are evaluated under different conditions. The conditions considered were freezing and thawing as well as freezing and thawing followed by drying at about 60°C. The latter test method was developed to simulate the drying condition in a real environment. The effect of defects in concrete on the performance of coatings was also examined.

From test results, it was found that the degree of deterioration of coatings and concrete under freezing, thawing and drying depends on: (a) the type of coating, (b) the coating method and (c) the test methods. Furthermore, the test method proposed by the authors was proved to be suitable for predicting the performance of concrete under real environmental conditions.
Keywords: Coating, concrete, drying, freezing and thawing, mortar.

1 Introduction

Despite the general belief that it is possible to protect concrete from frost damage if an air-entraining agent is used and a proper water/cement ratio is chosen in consideration of the environmental conditions, concrete structures are still subject to frost damage owing to a variety of factors. One of the countermeasures against frost damage is to apply a coating material to the surface of the concrete to prevent the infiltration of water through the surface of the concrete. At present, several kinds of coating materials

Concrete Under Severe Conditions: Environment and loading (Volume Two) Edited by K. Sakai, N. Banthia and O.E. Gjørv. Published in 1995 by E & FN Spon. ISBN 0 419 19860 1

for concrete are being used for aesthetic purposes and to protect concrete from damage by chloride ion, carbonation, and alkali-aggregate reaction. However, only a few studies have been made regarding the use of surface coating materials to prevent frost damage, and the results of such studies have been found unsatisfactory [1] .

In order to investigate the effect of surface coating materials in protecting concrete from frost damage, experimental studies were carried out with different kinds of coating materials and coating methods.

2 Outline of the experiment

2.1 Specimens

Mortar specimens with dimension of 40 x 40 x 160 mm, were prepared from ordinary Portland cement and Toyoura standard sand, with a water/cement ratio of 0.5 and a target flow value of 190. No air-entraining agent was used.

Following demolding one day after casting, the specimens were cured in water at 20 °C for 6 days and then dried in air (at 20°C and 60% RH) for 21 days. Each specimen was coated with any one of the eight coating materials shown in Table 1 and by any one of the six coating methods shown in Fig. 1. The coated specimens were dried again in air for 28 days. The surface coating materials and the method of coating them are summarized in Table 2. Note that 3 specimens each are produced.

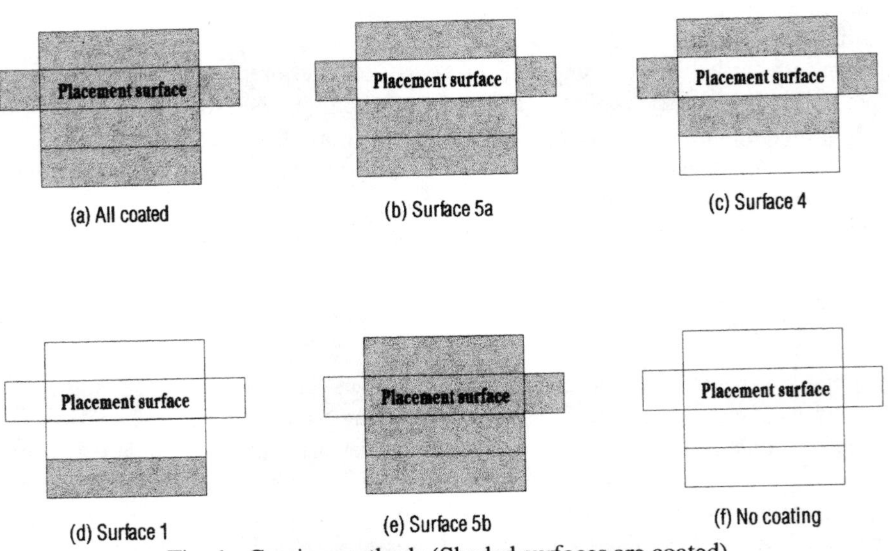

Fig. 1. Coating methods (Shaded surfaces are coated)

Table 1. Types of materials and specifications of coatings

Code No.	①	②	③	④
Specifications / Coating process	Type A coating system	Type B coating system	High−build and flexible type epoxy resin coating system	Polybutadiene rubber coating system
Code name	Type−A	Type−B	Flexible epoxy type	Rubber−based type
Primary coat	Epoxy resin primer			
Putty	Epoxy resin putty			
Intermediate coat	Epoxy resin coating	Flexible polyurethane resin coating	High−build and flexible type epoxy resin coating	Polybutadiene rubber coating
Top coat	Polyurethane resin coating	Flexible polyurethane resin coating		
Total film thickness	90μm	90μm	430μm	530μm

Code No.	⑤	⑥	⑦	⑧
Specifications / Coating process	Wet surface cureing type epoxy resin coating system	Polymer cement coating system	Oil paint coating system	Modified polyester resin type
Code name	Wet−cure type	Polymer cement type	Oil−based type	Permeation type
Primary coat	Wet surface cureing type epoxy resin primer	Polymer cement primer	Long oil phthalic acid resin coati	One package permeation type
Putty	−	Polymer cement putty	−	
Intermediate coat	Wet surface cureing type epoxy resin coating	Polymer cement coating	Long oil phthalic acid resin coati	
Top coat	Wet surface cureing type epoxy resin coating	Polymer cement coating	Long oil phthalic acid resin coati	
Total film thickness	1000μm	400μm	90μm	−

Table 2. Coating types and coated surfaces

Coating types	Coated surfaces				
	All coated	Surface 5a	Surface 4	Surface 1	Surface 5b
Type-A	x	x			
Type-B	x				
Flexible-epoxy type	x	x	x	x	x
Rubber-based type	x				
Wet-cure type	x				
Polymer-cement type	x	x			
Oil-based type	x				
Permeation type	x				
No coating					

2.2 Test method

2.2.1 Test I
Three hundred cycles of freezing and thawing were performed in accordance with the test method of the Japan Society of Civil Engineers so as to measure the rate of the weight change of the specimens and to observe the change of their external appearance. Freezing and thawing were carried out in a rubber vessel containing four specimens separated by rubber plates such that each specimen is entirely covered with a layer of water about 3 mm thick.

2.2.2 Test II
In this test the freezing and thawing (as in Test I) were combined with drying in order to simulate the actual environment to which concrete structures are exposed. This test consisted of 250 cycles of freezing in water at -18°C for 2 hours followed by thawing in water at 5°C for 2 hours, drying in the air at 30°C and 60% RH for 2 hours and additional 50 cycles of freezing in water at -18°C for 2 hours, thawing in water at 5°C for 2 hours and drying in the air at 40°C and 30% RH for 2 hours. The weights of specimens were measured, and the appearances were observed.

3 Test results and discussion

3.1 Test I

3.1.1 Observation of external appearance
The specimens having a coating on all of their surfaces remained intact, except those coated with the permeation-type coating material or the oil-based coating material. The specimens coated with the permeation-type coating material showed severe scaling and had poorer appearance than the specimen without a coating. Some of the specimens coated with the oil-based coating material exhibited cracking of the coating film. The specimens with a coating on five surfaces deteriorated in appearance to almost the same extent regardless of the type of coating materials used. The type-A coating specimens suffered cracking of the coating film. The flexible-epoxy-type-coating specimens material had an intact coating film. The polymer-cement-type-coating specimens had stretches in the coating film. In addition, many small cracks were observed. The surface 1 coating specimen deteriorated in appearance most remarkably. The surface 5a and surface 4 coating specimens deteriorated in appearance to the same degree as the specimen without a coating.

3.1.2 Rate of weight change
In Fig. 2 the rate of weight change of the specimens after each cycle are shown. The specimen entirely coated with the permeation type coating material increased in weight due to water absorption to the same extent as the specimen without a coating; however, its subsequent decrease in weight due to scaling was more than the specimen without a coating. The coating film of the specimens coated with the oil-based coating material was damaged due to rapid water absorption. One of the three specimens coated with the oil-based coating material deteriorated more remarkably than the other two; it deteriorated faster than the specimen without a coating, although it was slow in water absorption. The surface 5 coating specimens were all slower in water absorption than the specimens without a coating but they rapidly deteriorated after water absorption. The surface 1 coating specimen deteriorated remarkably after water absorption. The surface 4 and 5a coating specimens deteriorated to the same extent as the specimen without a coating. The surface 5a coating specimens or all coated specimens did not show a decrease in weight due to deterioration. The foregoing results suggests that partially-coated specimens occasionally deteriorate more severely than uncoated specimens. In other words, coating may produce adverse effects depending on its type and method of coating. The thick, soft-type coating material gave rise to a coating film which did not crack or stretch even after peeling. These results suggest that a surface coating material subject to freezing and thawing should be of the flexible epoxy type rather than the hard, thin type (such as those of the permeation type and oil-based type).

3.2 Test II

3.2.1 Observation of external appearance
The coating films of the all coated specimens were damaged in distinctly different manners. The Type-A and Type-B coating films had small blisters. The oil-based-type coating film suffered blistering, cracking, and peeling. The permeation type coating material suffered slight surface scaling and permitted the specimen to swell. The flexible-epoxy type, rubber-based type, wet-surface-cure type, and polymer-cement-type-coating films showed no change in appearance, and they were considered to be satisfactory. Namely, the thick-film-type-coating materials produced good results, presumably due to the water-proofing performance of the coating film. The specimens with the surface 5 coatings experienced almost the same mortar deterioration regardless of the type of the coating material used. However, they differed in the mode of film deterioration. That is, the coating films of almost all of the coating materials suffered blistering, whereas the coating film of the polymer-cement- type-casting material did not suffer blistering. This is noteworthy. A probable reason for this is that the polymer-cement-type-coating film permits the permeation of moisture from the specimen. By contrast, Type-A coating film suffered cracking, presumably because it is thin and hard. The surface 1 and 5a coating specimens deteriorated most rapidly. They were followed by the specimens which did not have a coating. The surface 4 and 5b coating specimens retained a good appearance without suffering damage or scaling. The surface 1 coating specimen eventually suffered severe scaling, whereas the surface 5b coating specimens suffered scaling to a lesser extent although scaling started almost at the same time as the former. However, the latter suffered severe blistering of the coating film.

3.2.2 Rate of weight change
In Fig. 3 the rate of weight change of the specimens after each cycle are shown. In the specimen entirely coated with the permeation-type coating material, the increase in weight due to high water absorption was larger than that of the specimen without the coating; however, its subsequent decrease in weight due to scaling was less than that in the specimen without the coating. The oil-based coating specimen absorbed a large amount of water, probably due to the damage to the coating film. The weight change in

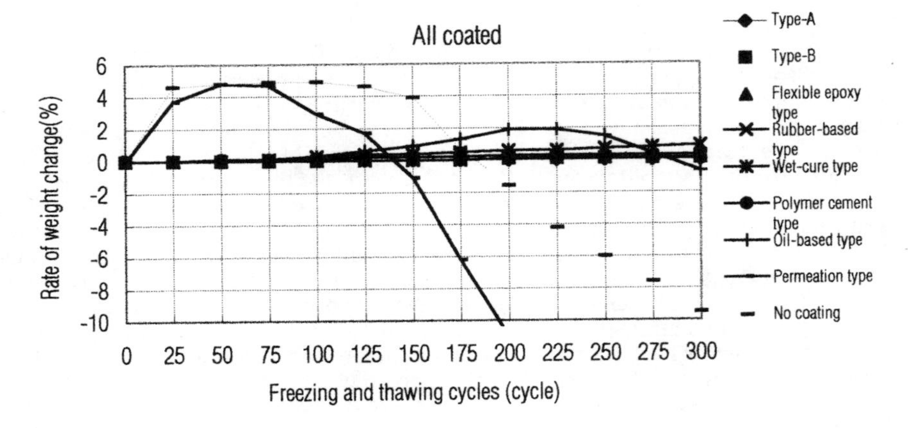

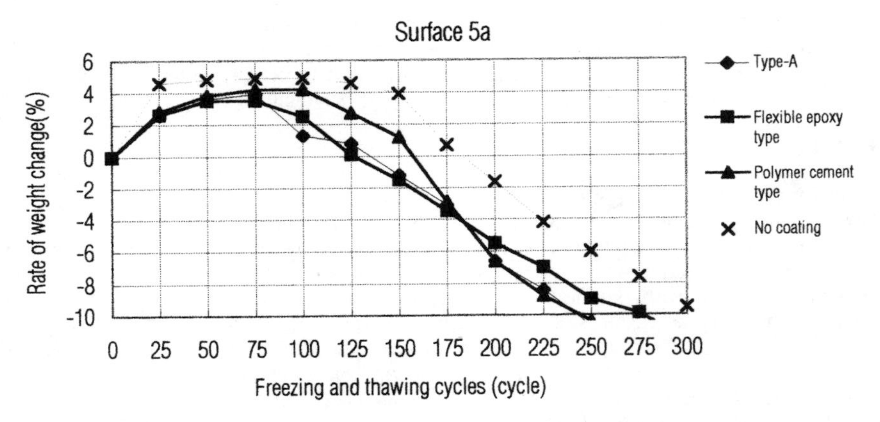

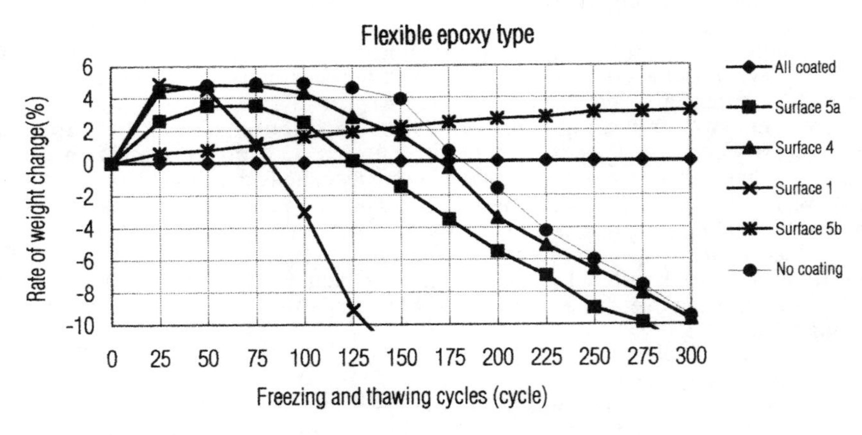

Fig. 2. Rate of weight change in Test I

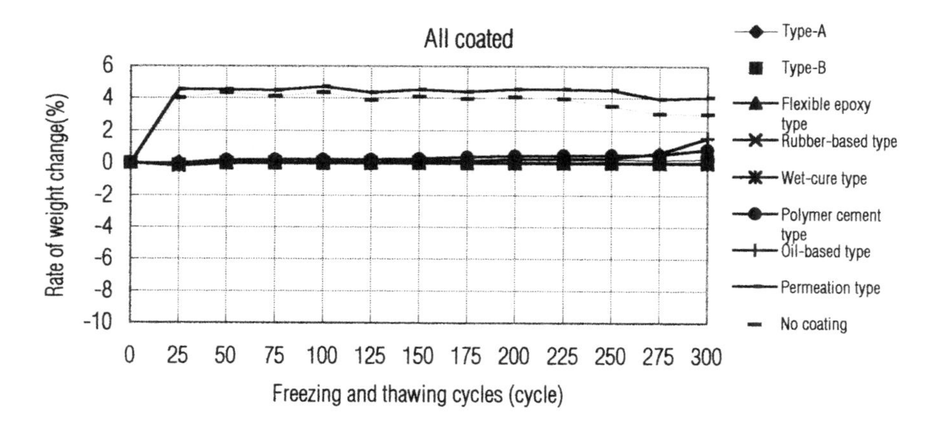

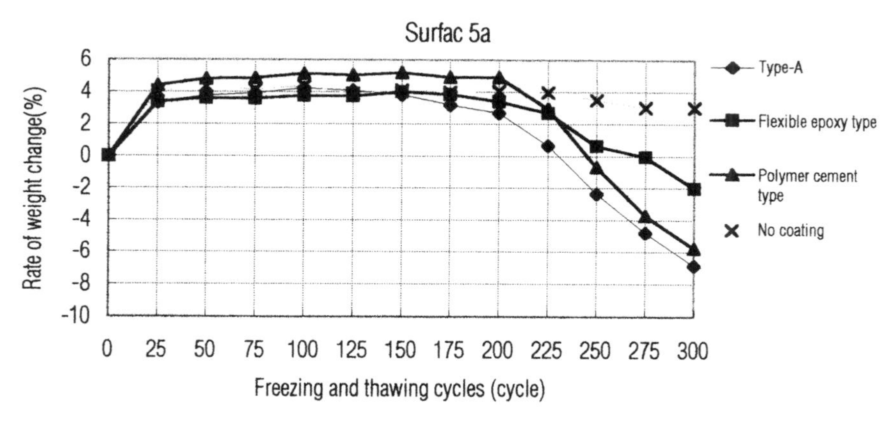

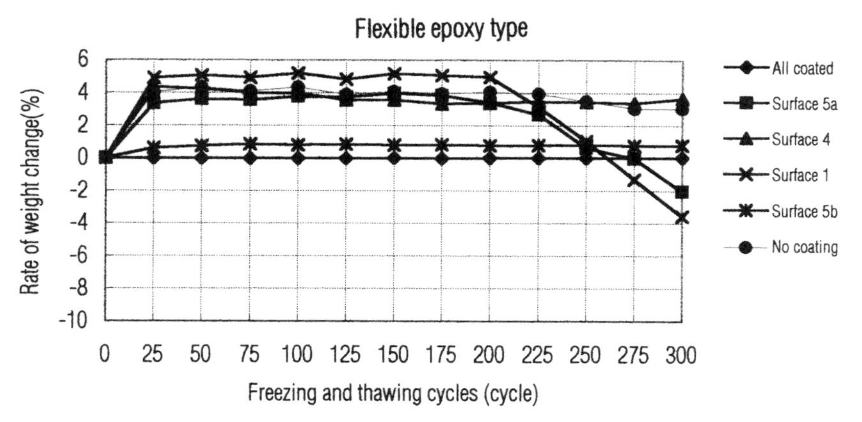

Fig. 3. Rate of weight change in Test II

the other specimens was less than 1%, on average. This is satisfactory. The surface 5 coating specimens showed almost the same weight change regardless of the type of coating material used. After water absorption, all the coated specimens decreased in weight more rapidly than the uncoated specimens. The surface 1 and 4 coating specimens absorbed water rapidly. The surface 1 coating specimen absorbed water as rapidly as the uncoated specimen. They are followed by the surface 5a coating specimen, the surface 5 coating specimen, and the entirely coated specimen. Although it is generally considered that a specimen showing a quicker weight increase due to water absorption shows a quicker weight decrease due to scaling and the like, the extent of weight loss does not necessarily coincide with the extent of weight increase.

4 Conclusions

1. A surface coating on mortar is effective in protecting concrete from frost damage if the waterproofness of the coating material is high.
2. Some coating materials should be avoided such as those of the permeation type, which partially lacks waterproofness, and those of Type-A and oil-based type, which provide hard or thin films. The permeation-type specimens are occasionally more liable to deteriorate than the uncoated ones.
3. Mortar specimens with some uncoated surfaces are occasionally more liable to deteriorate than those without any coating.
4. The new freezing-and-thawing test developed in this study causes the coating on mortar to suffer deterioration such as blistering and cracking. Therefore, it permits the prediction of the performance of concrete coating in actual environments.

5. References

1. Sato, Y. and Yoshida, S. (1970) Studies on paints to prevent the destruction of concrete on account of freezing and thawing. *Railway Technical Research Report,* No. 716
2. Litvan, G.G. (1992) The effect of sealers on the freeze thaw resistance of mortar. *Cement and Concrete Research 22*, pp. 1141-1147.

92 CHLORIDE PENETRATION INTO CONCRETE INCORPORATING MINERAL ADMIXTURES OR PROTECTED WITH SURFACE COATING MATERIAL UNDER CHLORIDE ENVIRONMENTS

R.N. SWAMY
University of Sheffield, UK
H. HAMADA and T. FUKUTE
Port and Harbour Research Institute, Ministry of Transport, Japan
S. TANIKAWA
Toa-gosei Co. Ltd, Nagoya, Japan
J.C. LAIW
Ministry of Transportation and Communications, Taiwan, ROC

Abstract
The effects of mineral admixtures and concrete surface coatings on the prevention of chloride intrusion into concrete is reported. Concretes of high, low and medium water to binder ratio, with and without mineral admixtures, or with and without a protected surface coating were designed. Reinforced concrete slab specimens were subjected up to 70 cycles of wetting and drying by ponding 4.0% sodium chloride solution and drying in atmospheric air. The chloride content in concrete was evaluated after 10, 20, 60 and 70 cycles of wetting and drying. The results show that the incorporation of mineral admixtures has a significant effect in reducing the chloride intrusion into concrete. On the other hand, a good durable surface coating can more or less completely prevent the penetration of chlorides into concrete. As a result, it can be said that the service life of concrete structures incorporating mineral admixtures or protected with surface coating materials becomes significantly longer than that of the structures made of plain concretes when exposed to chloride environments, but, a good surface coating is far more effective and superior in enhancing service life than the incorporation of mineral admixtures.
Keywords : Chloride environments, chloride ion penetration, concrete surface coating, durability, mineral admixture, service life, steel corrosion, wetting and drying.

1 Introduction

Chloride-induced corrosion is now recognized as a major cause of distress in many on-shore and off-shore structures. It is known that concrete provides good protection to embedded reinforcing steel against corrosion : the highly alkaline cement paste forms a tightly adhering film on the steel which passivates it and protects it from corrosion. However, the presence of an adequate amount of chloride ions will permit steel

Concrete Under Severe Conditions: Environment and loading (Volume Two) Edited by K. Sakai, N. Banthia and O.E. Gjørv. Published in 1995 by E & FN Spon. ISBN 0 419 19860 1

corrosion to occur even in a highly alkaline concrete. Chloride ions can exist as a result of aggregate or water contamination within the concrete itself. However, chloride ions can also penetrate into concrete from the outer environment if the concrete is in contact with seawater, saline water, deicing salts or salt spray [1]. The intrusion of chloride ions into reinforced concrete can destroy the inherent protection given to steel by the cement paste, and allow the reinforcement to corrode if oxygen and moisture are also present. This study is concerned with two methods of prevention of the intrusion of external chlorides into concrete. This experimental study provides quantitative data on the effectiveness of mineral admixtures and surface coatings in concrete structures exposed to chloride environments.

2 Experimental procedure

2.1 General

The experimental program consisted of the exposure of ten reinforced concrete slabs to 4.0% sodium chloride solution cyclic ponding on the top surface. The main variables studied are : water to binder ratios (0.45, 0.60 and 0.75), mineral admixtures (ground granulated blastfurnace slag, fly ash and silica fume) and a highly elastic surface coating. The slabs were subjected up to 70 cycles of exposure, and chloride profiles were tested after 10, 20, 60 and 70 cycles.

2.2 Details of slab specimens

Fig. 1 gives detail of the slab specimens and the location of steel in the slabs. The size is 1000 x 500 x 150 mm. All the slabs were reinforced with 20 mm diameter deformed bars at the cover depth of 115mm from the upper surface. Each slab was provided with a rectangular acrylic frame on the top surface to enable the NaCl solution to be ponded on the upper surface. Table 1 lists all the slab specimens tested in this study. Specimen No.1, 2 and 3 were made with normal concrete without any surface coating or mineral admixture. Specimen No.4, 5 and 6 were made with normal concrete without mineral admixtures, but were protected with a coating on the upper surface. Specimen No.7, 8, 9 and 10 were made with concrete incorporating mineral admixtures; however, their surfaces were left uncoated.

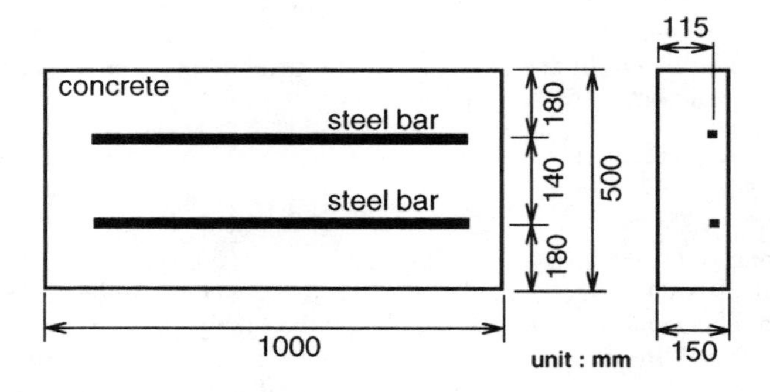

Fig. 1 Detail of the slab specimen

Table 1 List of specimens

Specimen No.	W/B ratio	Mineral admixture replacement	Surface coating application
1	0.45	-------	-------
2	0.60	-------	-------
3	0.75	-------	-------
4	0.45	-------	Surface coated
5	0.60	-------	Surface coated
6	0.75	-------	Surface coated
7	0.60	65% GGBFS	-------
8	0.60	30% FA	-------
9	0.60	10% SF	-------
10	0.75	65% GGBFS	-------

2.3 Concrete mix details

Ordinary portland cement (ASTM Type 1) was used in all the slabs in the test. The specific surface area of the cement was 345 m^2/kg, and its chemical composition is presented in Table 2. The fine and coarse aggregates used were washed natural aggregates. The sand had a water absorption coefficient of 1.37%, and a bulk specific gravity of 2.60 in the saturated surface dry condition. The coarse aggregate consisted of a mixture of rounded and crushed gravel with 10mm maximum particle size. Its water absorption was 0.80% and bulk specific gravity was 2.62 in the saturated surface dry condition. The total cementious content was 350kg/m^3 in all the slab specimens. The mineral admixtures were used to replace the cement directly by weight. The aggregate-cementious ratio was 5.26/1 and the percentage of sand in the total aggregate was 32% for all slab specimens. The mix proportions for all the slab specimen are presented in Table 3. The compressive strength of the concrete determined on 100mm cubes at various ages up to 556 days are presented in Table 4.

Table 2 Percentage chemical composition of cement, GGBFS, FA and SF

	SiO_2 (%)	Al_2O_3 (%)	Fe_2O_3 (%)	CaO (%)	MgO (%)	L.O.I .(%)
Cement	21.0	5.3	3.1	64.4	2.6	----
GGBFS	34.2	11.3	1.2	41.6	8.2	----
FA	51.4	28.1	11.1	1.4	1.6	3.5
SF	97.0	----	----	----	----	1.1

Table 3 Mix proportions of concrete

Specimen No.	W/B ratio	Cement (kg/m^3)	Water (kg/m^3)	Coarse aggregate (kg/m^3)	Fine aggregate (kg/m^3)	Mineral admixture (kg/m^3)
1 , 4	0.45	350	158	1250	590	-----
2 , 5	0.60	350	210	1250	590	-----
3 , 6	0.75	350	263	1250	590	-----
7	0.60	122	210	1250	590	228 (GGBFS)
8	0.60	245	210	1250	590	105 (FA)
9	0.60	315	210	1250	590	35 (SF)
10	0.75	122	263	1250	590	228 (GGBFS)

Table 4 Compressive strength of concrete used

Specimen No.	Compressive strength (MPa)				
	7 days	28 days	128 days	242 days	556 days
1 , 4	54	72	70	71	81
2 , 5	32	46	45	46	50
3 , 6	22	31	33	34	37
7	17	33	35	47	53
8	17	26	33	33	42
9	35	50	55	61	65
10	15	23	29	35	43

2.4 Mineral admixtures

The chemical composition of the admixtures used in this study are presented in Table 2. The fineness of ground granulated blast furnace slag (GGBFS) was $417m^2/kg$ which satisfies BS (British Standard) 6699:1986 which restricts particle size to not less than 275 m^2/kg. The MgO content of the fly ash (FA) was 1.62% which satisfies BS 3892 : part 1 : 1982 which restricts the MgO content to less than 4%. The specific gravity of silica fume (SF) used in the tests was 2.2.

2.5 Casting and curing of concrete

The mixed concrete was poured into the slab molds in three layers, and each layer was vibrated. A rectangular acrylic frame of 50 mm height was placed on the top surface of the concrete which formed an embankment on the top surface for later chloride solution ponding. After casting, the slabs were covered with polythene sheets for 24 hours. After 24 hours, the slabs were cured by water ponding on the top surface for 6 days, then demoulded and exposed to ambient conditions for 21 days further air curing.

2.6 Surface coating materials

The coating was built up from three layers of material, these being the primer, base coat and top coat. The total thickness was about 1mm. This is a rubber coating made by polymerization of acrylic ester with other raw materials. The typical characteristics of these materials are presented in Table 5. The total coating period was about 14 days.

Table 5 Specification of coating material

	Appearance and property	Viscosity (cps)	Quantity (kg/m2)
Primer	Synthetic resin, Organic solvent Flammable ; Hazard Class 4	30	0.3
Base coat	Acrylic rubber emulsion Viscous slurry material Non-flammable	25000 to 35000	1.7
Top coat	Acrylic-urethane enamel Main agent / Hardner / Thinner = 4 / 1 / 2.5 Organic solvent based Frammable ; Hazard Class 4	100 to 200	0.3

2.7 Exposure regime
The slabs were exposed to cyclic ponding with 4% sodium chloride solution on the top surface and subsequent drying. Each cycle was composed of ponding for 7 days, then removal of the solution and the surface dried at ambient condition for 3 days.

2.8 Chloride content test method
Samples for chloride content analysis were taken by rotary hammer drill with a 20mm diameter bit. The samples were taken from six different positions of the slab. Each sampling position was divided into five different depths from the top surface of the slab, the sampling depths being 5-25, 25-45, 45-65, 65-85 and 85-105mm. The chloride analysis was carried out according to BS 1881 : part 124 : 1988. Hot nitric acid solution was used as a solvent, and a standard silver nitrate solution was used for titration. The chloride content is then expressed as acid-soluble chloride content by weight of cement, and each chloride content presented is the average of two analysis results.

3 Experimental results

Restricted to chloride penetration from the outer surface of concrete, and within the time period of this study, it can be said that the chloride content in concrete reached almost a saturation level after 60 cycles of wetting and drying. The chloride profiles in slab specimens after 20 and 60 cycles of exposure are shown in Figs. 2 to 4.

Fig. 2 shows chloride profiles of plain concrete with W/B of 0.45, 0.60 and 0.75. This figure shows typical profiles of decreasing chloride content with depth and increasing chloride content with increasing exposure cycles. It can also be seen that a relationship exists between chloride content and W/B ratio, the higher chloride content profiles occurring with higher W/B ratio.

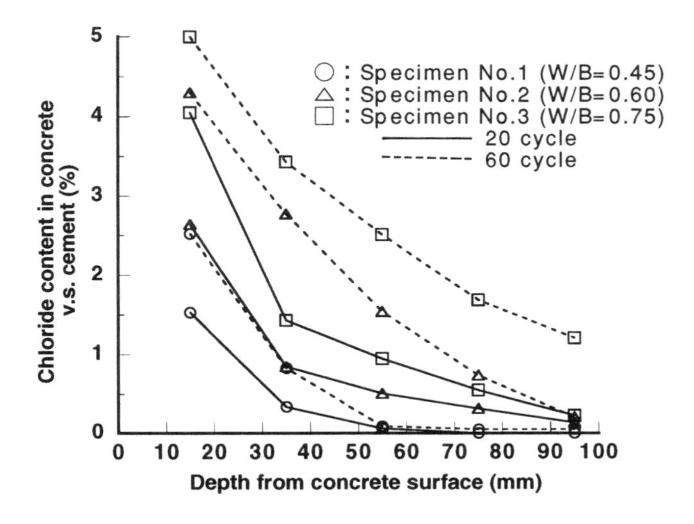

Fig. 2 Chloride profile in plain concrete (Specimen No.1, 2 and 3)

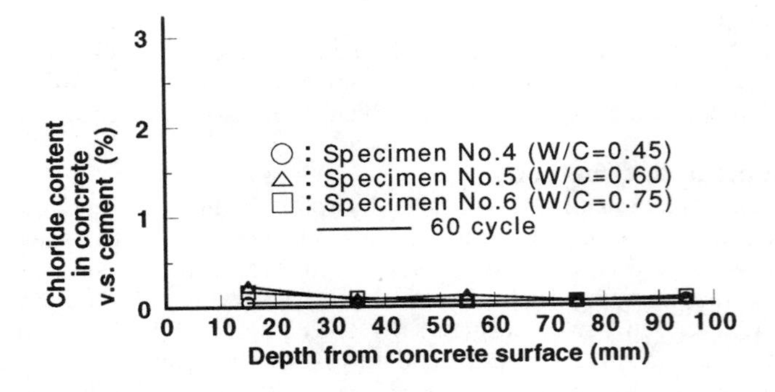

Fig. 3 Chloride profile in surface-coated concrete (Specimen No.4, 5 and 6)

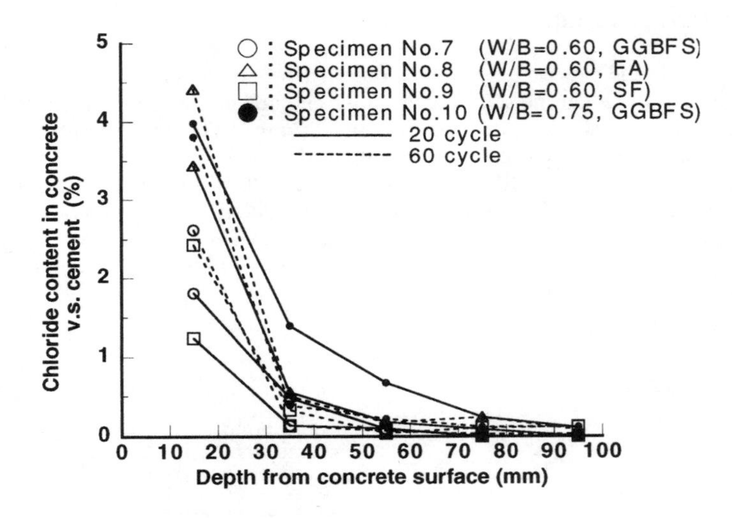

Fig. 4 Chloride profile in concrete incorporating mineral admixtures (Specimen No.7, 8, 9, and 10)

Fig. 3 shows chloride profiles of surface-coated concrete slabs with W/B ratios 0.45, 0.60 and 0.75 after 60 cycles of wetting and drying. From this figure, it is clearly seen that the chloride content in surface-coated concrete is almost zero, that is, chloride penetration is almost completely prevented by the surface coating layers. The effect of the coating in preventing chloride intrusion is not affected by the water to binder ratio of the concrete. The acrylic rubber coating used in these tests has been shown to maintain its integrity and effectiveness after a long period of time even when exposed to severe aggressive environments [2,3].

Fig. 4 shows chloride profiles of concrete containing mineral admixtures. The results show typical profiles of decreasing chloride content with depth and increasing chloride penetration with increasing exposure cycles. By comparing this figure with Fig. 2, it is seen that mineral admixtures reduce the intrusion of chloride, especially beyond the depth larger than 30mm. The specimen No.9 containing 10% SF exhibits the lowest chloride content compared to other specimens with GGBFS and FA. This figure also suggests that the water to binder ratio is an important factor which affects the chloride intrusion into concrete even when it contains mineral admixture, in this case GGBFS.

In summary, it can be said that the incorporation of mineral admixtures in concrete has a significant effect in reducing the chloride intrusion into concrete from the surrounding saline water. On the other hand, the coating more or less completely prevents the penetration of chlorides into concrete. One can thus envisage two distinct roles for mineral admixtures and surface coatings in chloride-containing environments. For new construction exposed to moderate chloride environments, the use of mineral admixtures can substantially reduce chloride penetration. In more severe chloride exposure conditions the provision of good quality surface coating can almost totally prevent chloride penetration. For existing, damaged or deteriorating structure, a surface coating can prevent or substantially reduce further chloride penetration.

4 Discussion of results

4.1 General

Three of the authors have reported a literature review on chloride penetration into concrete when it is exposed to marine environments [4]. They have also clarified the critical chloride level in concrete under various zones in marine environments [4]. It will be interesting to compare the results obtained from this laboratory study with the analysis of data extracted from literature survey which relates to a wider range of chloride conditions of concrete in the field. In this section, the results of this comparison and some related discussions are presented.

4.2 Concept of analysis

According to Tuutti [5] and some other researchers, the deterioration process of RC structures by chloride attack can be envisaged to consist of two stages, namely, the "Initiation stage" and the "Propagation stage". The boundary of these two stages is the initiation of steel corrosion : in other words, corrosion occurs when the chloride content

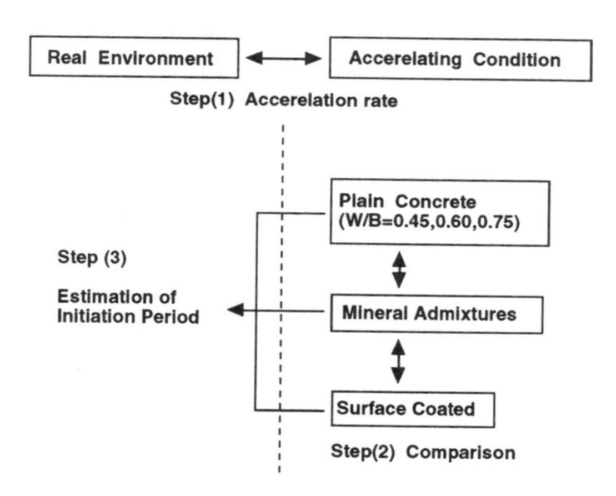

Fig. 5 Concept of analysis

around the steel bar reaches the threshold value for steel corrosion. Many research data on the threshold chloride content for steel corrosion exist, and several researchers suggest that the threshold chloride level occurs around 0.4% Cl^- v.s. cement weight [6]. In this analysis, this value of 0.4% is used as the threshold value. Fig. 5 shows the concept of the analysis. The first step is the correlation between the acceleration condition and the real environment to decide the "accerelation rate". The second step is a comparison of the various kinds of concretes used in the acceleration tests to decide the required number of wet and dry cycles for the chloride content around the steel bar to exceed the 0.4%. The final step is to estimate the initiation period (the period of the initiation stage) for each concrete based on the accerelation rate.

4.3 Accerelation rate of wet and dry cycles

Here, the term "acceleration rate" is defined as the ratio of the magnitude of chloride penetration into plain concrete under the accelerating condition (i. e. wetting and drying cycles) and the critical (maximum level) chloride penetration into concrete under real marine environment, which is taken here to be the splash zone which is one of the most severe exposure states amongst various marine conditions. Fig. 6 shows a comparison of the critical chloride profiles under splash zone obtained from the literature survey [4] and the chloride content in the Specimen No.3 (W/B=0.75) after various accerelation cycles of wetting and drying obtained in this laboratory study. By using the chloride profile of the weakest and most permeable concrete of W/B=0.75, this chloride profile can be considered to be almost the "critical level" under this accerelating condition. Thus, by comparing the two critical chloride profiles under real environment and the accerelating condition, the "accerelation rate" can be obtained. The comparison is carried out by using the chloride profile for 60 cycles, because the chloride content has

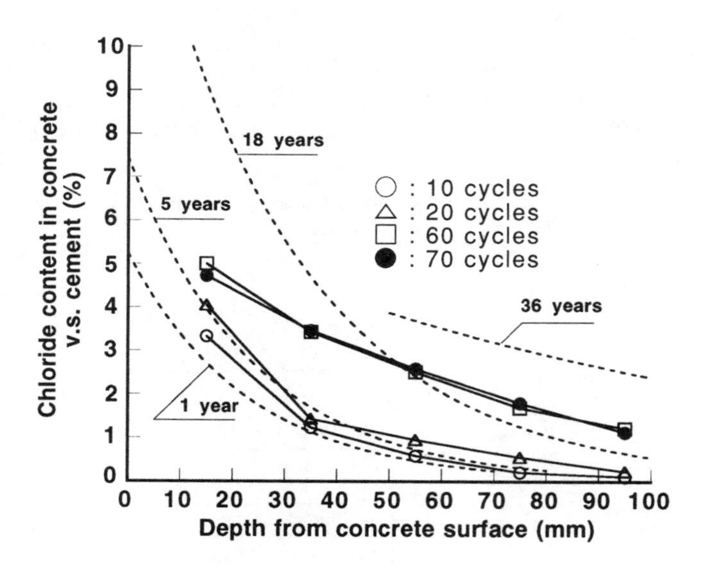

Fig. 6 Comparison of real environment and accerelation condition

Table 6 Accerelation rate

	Accerelation rate	
	Depth = 35mm	Depth = 55mm
Specimen No.3 (w/c = 0.75)	6.7	11.0

almost reached a saturation level after 60 cycles of wetting and drying. The accerelation rates so obtained are presented in Table 6. The rate takes different values for different depths, and for a depth of 35mm the value is 6.7, and for the depth of 55mm, it is 11.0. The 60 cycles of wetting and drying takes about 1.65 years, the value of 6.7 is obtained from 11years/1.65years and the 11.0 is from 18years/1.65 years.

4.4 Estimation of the initiation period

Table 7 presents the "Estimated Initiation Period" for each concrete mix. The calculation is carried out for two different cover depths of 35mm and 55mm. From this table, the protection effects provided by the surface coating on the prevention of chloride ingress into concrete is obvious. Even after 60 cycles of wetting and drying, the chloride contents in concretes protected with surface layer are all below 0.4% (threshold value for corrosion initiation), therefore it is impossible to estimate the duration of initiation period. In other words, the initiation period of surface-coated concrete becomes very long. As for the concretes incorporating mineral admixtures, it becomes clear that the effects of mineral admixtures are different for the outer surface area (where cover depth is less than about 50 mm) or for the inner area (where cover depth is more than about 50mm). The effects of mineral admixtures become predominant and significant in the inner area, therefore the cover depth of steel bars in concretes incorporating mineral admixtures should be more than 50 mm. In summary, the service life of concrete structures incorporating mineral admixtures or protected with surface coating materials can be expected to become significantly longer than that of the structures made of plain concretes under chloride environments.

Table 7 Estimated initiation period

Specimen No.	Estimated initiation period (years)	
	Cover depth = 35 mm	Cover depth = 55 mm
1 (plain, w/c=0.45)	3.9 year (n=21)	more than 21 years
2 (Plain, w/c=0.60)	1.8 year (n=10)	5.1 year (n=17)
3 (Plain, w/c=0.75)	0.6 year (n=3)	2.4 year (n=8)
4 (Surface-coated, w/c=0.45)	more than 12 years	more than 21 years
5 (Surface-coated, w/c=0.60)	more than 12 years	more than 21 years
6 (Surface-coated, w/c=0.75)	more than 12 years	more than 21 years
7 (GGBFS, w/b=0.60)	more than 12 years	more than 21 years
8 (FA, w/b=0.60)	3.7 year (n=20)	more than 21 years
9 (SF, w/b=0.60)	more than 12 years	more than 21 years
10 (GGBFS, w/b=0.75)	2.4 year (n=13)	19.6 year (n=65)

Estimated initiation period (year) = n × 10 (days) / 365 (days) × r

n : the number of cycles when chloride content reach the threshold level of 0.4 %

r : accerelating rate (6.7 for 35mm of cover depth, 11.0 for 55mm)

5 Conclusions

The main conclusions obtained from this study are as follows :

1. Within the range of plain concrete tested, chloride content in concrete decreases as the water to binder ratio decreases. In particular, a water to binder ratio below 0.50 is recommended to reduce chloride ingress into concrete.
2. Within the range of tested specimens, mineral admixtures reduce substantially chloride intrusion into concrete compared to plain concrete having the same water to binder ratio. That effectiveness and protection become much more significant especially in the inner core of the concrete.
3. Even when concretes incorporating mineral admixtures are used, low water to binder ratios are recommended to reduce chloride ingress into concrete.
4. Surface coating layers used in this study almost fully prevent the chloride intrusion into concrete.
5. With surface coating layers, even low quality concrete of high W/B=0.75 displayed good behavior in preventing chloride ingress into concrete.
6. Within the range of parameters, chloride content in concrete investigated is seem to reach a saturated level after about 60 cycles of wetting and drying.
7. The service life of concrete structure is substantially enhanced by incorporating mineral admixtures such as GGBFS, FA and SF in concrete. If protected by a good quality surface coating, the service life becomes significantly longer when compared to unprotected plain concrete.

6 Acknowledgement

The authors wish to express their thanks to Mr. N. Morton and Mr. A. Smart of The University of Sheffield for their kind assistance in the experimental work.

7 References

1. R. N. Swamy (1994) Design - The Key to Concrete Material Durability and Structural Integrity, Reinforced Concrete Materials in Hot Climate, Vol.1, pp. 1-36.
2. S. Tanikawa and R. N. Swamy (1994) Protection of Steel in Chloride Contaminated Concrete Using an Acrylic Rubber Coating, Proc. of Int'l Conf. on Corrosion and Corrosion Protection of Steel in Concrete, Editor R.N.Swamy, Vol.2, pp. 1055 - 1068.
3. S. Tanikawa and R.N.Swamy (1994) Unprotected and Protected Concrete On-site Chloride Penetration with Time in an Aggressive Environment, Proc. of Int'l Conf. on Corrosion and Corrosion Protection of Steel in Concrete, Editor R.N.Swamy, Vol.2, pp. 1069 - 1080.
4. R. N. Swamy, H. Hamada and J. C. Laiw (1994) A Critical Evaluation of Chloride Penetration into Concrete in Marine Environments, Proc. of Int'l Conf. on Corrosion and Corrosion Protection of Steel in Concrete, Editor R.N.Swamy, Vol.1, pp. 404 - 419.
5. K. Tuutti (1980) Service Life of Structures with Regard to Corrosion of Embedded Steel, ACI Special Publication No.65, pp. 223 - 236.
6. D. W. Pfeifer, W. F. Perenchio and W. G. Hime (1992) A Critique of the ACI 318 Chloride Limits, PCI Journal, March - April, pp. 68 - 71.

93 PERFORMANCE OF CONCRETE COATED WITH FLEXIBLE POLYMER CEMENT MORTAR

H. KAMIMOTO
Sumitomo Osaka Cement Co. Ltd, Funabashi, Japan
M. WAKASUGI
Sumitomo Osaka Cement Co. Ltd, Osaka, Japan
T. MIYAGAWA
Kyoto University, Kyoto, Japan

Abstract
Flexible polymer cement mortar (PCM) coatings, which consist of a polyacrylic latex, portland cement and some admixtures, have been used instead of organic resin coatings to repair damaged and deteriorated concrete structure due to alkali silica reaction (ASR) in Japan.

In this study, physical properties of flexible PCM such as elongation, water and water vapor permeability, adhesive strength to concrete, and resistance to chloride penetration are measured at various polymer cement ratios (P/C). Water and water vapor permeability of flexible PCM are reduced with an increase in P/C because of its dense structure in which the larger pores are filled by polymers and sealed by continuous polymer films. Elongation of flexible PCM tends to increase with increasing P/C, whereas adhesive strength have an optimum P/C.

Furthermore, a field exposure test is performed using concrete specimens damaged by ASR. The specimens are coated with either epoxy resin or flexible PCM with various P/C. As a result of the test it is found that coating of flexible PCM inhibits ASR, but epoxy resin coating promotes ASR. The expansion due to ASR is reduced with increase in water vapor permeability.
Keywords: Adhesive strength, alkali silica reaction, chloride penetration, coating, elongation, polymer cement mortar, polymer cement ratio.

1 Introduction

Concrete structures damaged by alkali silica reaction (ASR) have poor resistance to the neutralization of concrete and the corrosion of reinforcement since rain water and

Concrete Under Severe Conditions: Environment and loading (Volume Two) Edited by K. Sakai, N. Banthia and O.E. Gjørv. Published in 1995 by E & FN Spon. ISBN 0 419 19860 1

carbon dioxide in the air can penetrate into the concrete along cracks caused by ASR. In general, ASR occurs when all the following conditions coexist,
(1) a critical amount of reactive silica in the aggregate,
(2) a sufficient amount of alkali in the concrete,
(3) a sufficient amount of water in the concrete.
If one of these conditions is eliminated, ASR will not occur.

Control of the water content in the concrete is the most practical method to inhibit ASR. It was reported, however, that the coating materials such as polyurethane, polybutadiene and epoxy resin accelerated ASR since they had no water vapor permeability. The water remaining in the concrete could slowly accelerate ASR.

Coating materials for the repair, therefore, are required to have water vapor permeability for the water remaining in the concrete, as well as waterproofness. Furthermore, strong adhesion to the concrete and sufficient elongation to accommodate the remaining expansion of ASR are also required for the materials. A flexible polymer cement mortar (PCM), which consists of polyacrylic latex, portland cement and admixtures, has developed with such performance.

This paper presents the physical properties of flexible PCM with various polymer cement ratios (P/C) and the result of field exposure test using concrete specimens damaged by ASR and coated with either an epoxy resin or flexible PCM.

2 Experiments

2.1 Materials

Normal portland cement with the specific surface area of 3,300cm^2 /g was used as a cement. Calcium carbonate with the specific surface area of 10,500cm^2/g was used as a filler. Polymer latex was used to make mortar flexible. The properties of flexible polymer latex are shown in Table 1. The main component is polyacrylic ester (PAE).

Table 1. Properties of polymer latex

Type of polymer latex	Appearance	Total solids (%)	pH (20℃)	Specific gravity (20℃)	Viscosity (cP,20℃)	Tg (℃)
PAE	Milk-white	50	8.5	1.02	<100	-50

2.2 Physical testings

2.2.1 Preparation of specimens

Table 2 shows mix proportions of flexible PCM. The materials of each mix proportion was mixed for a minute by a fast stirrer with a speed of 1,100 rpm. The amount of mixing water was determined to have a viscosity of about 10,000 cP. Free films of flexible PCM were used to measure elongation. Asbestos-cement sheets

Table 2 Mix proportion of flexible PCM

Normal portland	Calcium carbonate	Polymer latex	Water	P/C
50	50	85	0	0.85
50	50	75	0	0.75
50	50	65	1.5	0.65
50	50	55	4.5	0.55
50	50	45	10.0	0.45
50	50	35	20.0	0.35

coated with flexible PCM were used for measurement of water and water vapor permeability, and chloride ion permeability. Flexible PCM was coated on concrete plates for adhesive strength test. The free film was 1.2mm in thickness and the quantity applied was 2.1 kg/m^2. In the preparation of flexible PCM coated specimens, flexible PCM was coated on the sheets or plates by brushing. Free films and flexible PMC coated specimens were cured at 20℃ and 60% R.H. for 28 days.

2.2.2 Water permeability test

Water permeability was determined in accordance with JIS A 6910 (Multi-Layer Wall Coating for Glossy Textured Finishes), using the apparatus shown in Fig. 1. The height of waterhead was measured after 24 hours.

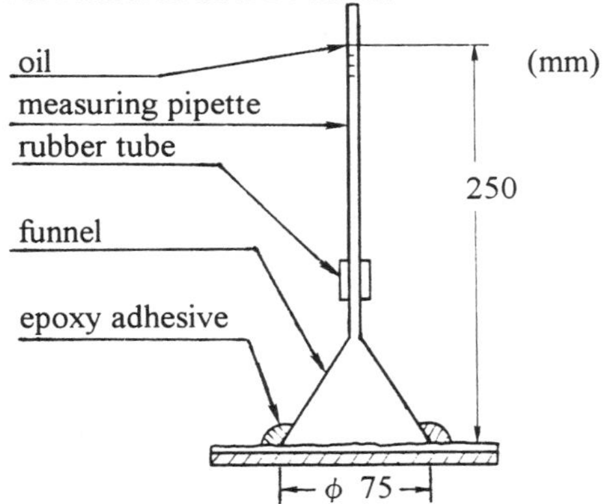

Fig. 1. Specimen for water permeability test.

2.2.3 Water vapor permeability test

Water vapor permeability was determined in accordance with JIS Z 0208 〔Testing Method for Determination of the Water Vapor Transmission Rate of Moisture-Proof Packaging Materials (Dish Method)〕, using the apparatus shown in Fig. 2. The specimens were put in the chamber of 40℃ and 90% R.H., and their weight were measured once every 24 hours.

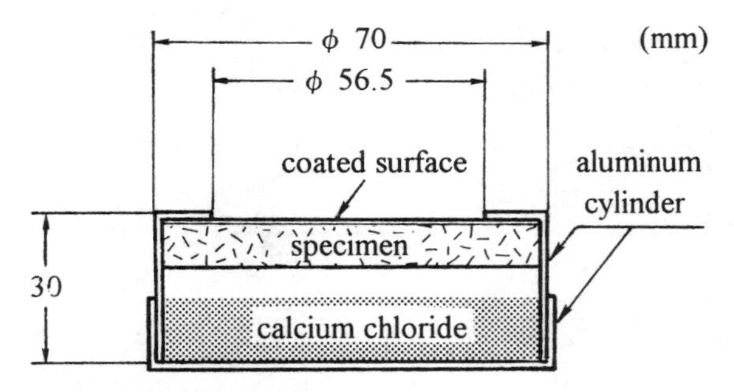

Fig. 2. Specimen for water vapor permeability test.

2.2.4 Elongation test for free film

Elongation of free film was determined in accordance with JIS A 6910, using the specimens shown in Fig. 3. The tensile speed was 5mm/min.. The distance between standard lines was measured as shown in Fig.4.

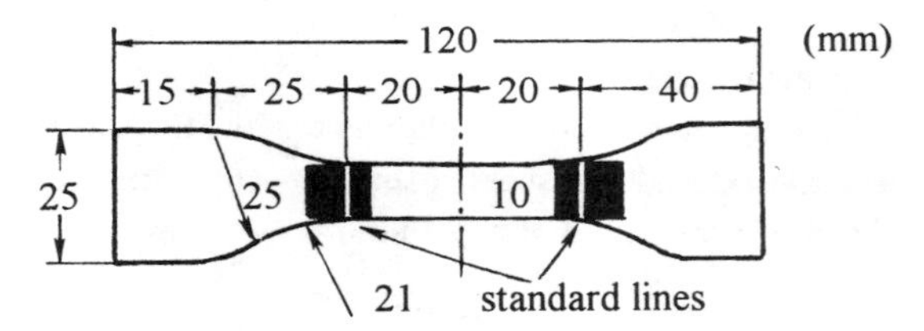

Fig. 3. Specimen for Elongation test.

Fig. 4. Elongation test.

2.2.5 Adhesive strength test

A 40mm square was cut into the specimens until reaching the concrete plate, then an attachment was fixed at 24 hours before test by epoxy resin adhesive as shown in Fig. 5. Adhesive strength test was performed using the hydraulically operated testing macine.

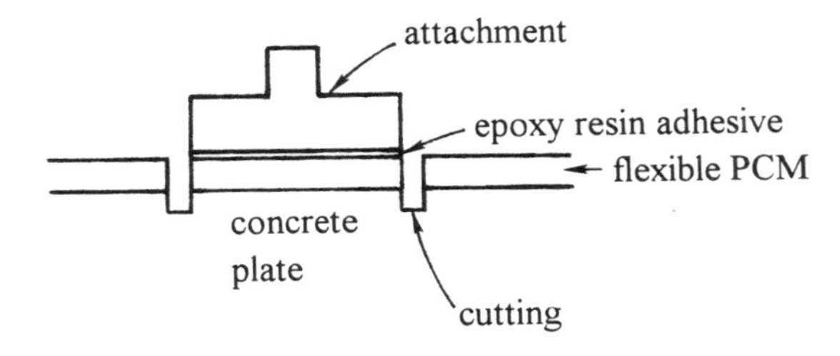

Fig. 5. Adhesive strength test.

2.2.6 Chloride ion permeability test

The test apparatus is shown in Fig. 6. The specimen was held between two cells. A 3% NaCl solution was poured into the cell on the coated surface side and distilled water into the other cell. The apparatus was kept at 20℃. After 30 days, the chloride ion content in the distilled water was measured by the potentiometric titration method.

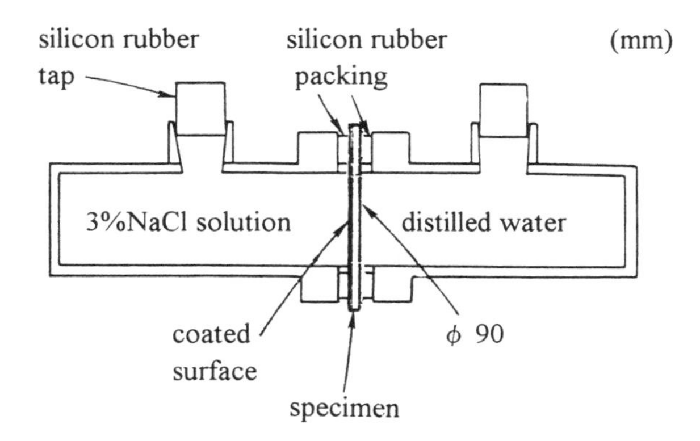

Fig. 6. Apparatus for chloride ion permeability test.

2.3 Field exposure test

2.3.1 Materials and mix proportion of concrete

Normal portland cement (alkali content 0.62%) was used. Sand from the River Ibi in Japan was used as the fine aggregate. It was judged innocuous by ASTM C 289 Standard Method of Test for Potential Reactivity of Aggregates (Chemical Method). Reactive andesite with pessimum effect was used as the coarse aggregate. Distilled water was used as the mixing water. AE water reducing agent was used as the chemical admixture. The unit alkali content of the concrete was adjusted by sodium hydroxide (NaOH), sodium chloride (NaCl) and soudium nitrite ($NaNO_2$) equivalent to an alkali level of 8kg/m^3. Table 3 presents the mix proportions of concrete.

Table 3 Mix proportions of the concrete

W/C (%)	S/a (%)	Air content (%)	Slump (cm)	Unit content (kg/m) Water	Cement	Fine aggregate	Coarse aggregate Reactive	Non-reactive	AE water reducing ag.	Total alkali
54.3	43.9	4.0	18	190	350	749	494	472	0.0105	8.0

2.3.2 Procedure for field exposure test

Concrete specimens measuring 7.5×7.5×40cm were removed from molds after one day and stored in a room at 40 and 90% R.H. for approximately one month. When the expansion reached a level of approximately 0.05%, the specimens were placed in a room at 20 and 85% R.H. until the surface moisture content decreased to 9-11%. They were then coated with either epoxy resin or flexible PCM. The specimens were subjected to the outdoor exposure in Osaka Japan. Length and weight change were measured in accordance with JIS A 1129 Method of Test for Length Change of Mortar and Concrete. Initial length was measured one day after the surface coating.

3 Results and discussion

3.1 Physical testings

3.1.1 Water and water vapor permeability

Fig.7 and 8 show the relation between P/C and water and water vapor permeability respectively. Water and water vapor permeability of flexible PCM were reduced with an increase in P/C because of its dense structure in which the larger pores were filled by polymers and sealed by continuous polymer films.

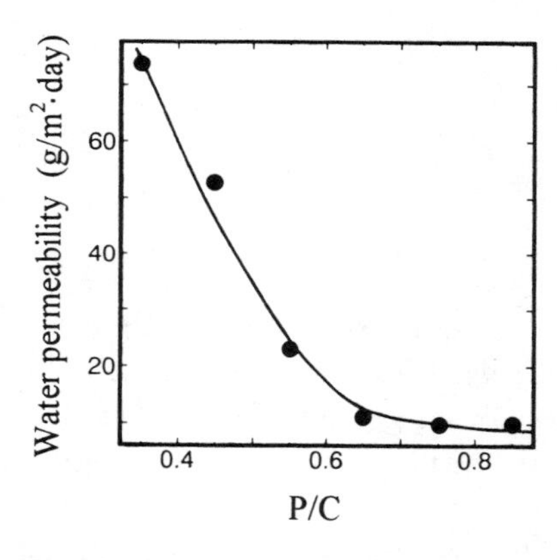

Fig.7 Relation between P/C and water permeability.

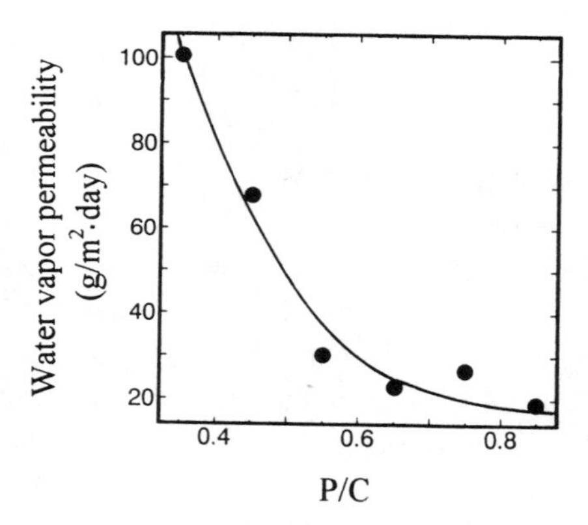

Fig.8 Relation between P/C and water vapor permeability.

3.1.2 Elongation

Fig.9 shows the relation between P/C and elongation. Elongation of flexible PCM tends to increase with increasing P/C. Especially, the elongation was markedly increased at levels of P/C higher than 0.5.

3.1.3 Adhesive strength

Fig.10 shows the relation between P/C and adhesive strength. Adhesive strength had an optimum P/C. At levels of P/C lower than 0.55, pores among the hydrated and unhydrated cement particles increased with reduction of P/C. On the other hand, at levels of P/C higher than 0.55, unhydrated cement particles exist within a polymer film as shown in Fig. 11, and the adhesive strength depend on the strength of the polymer film.

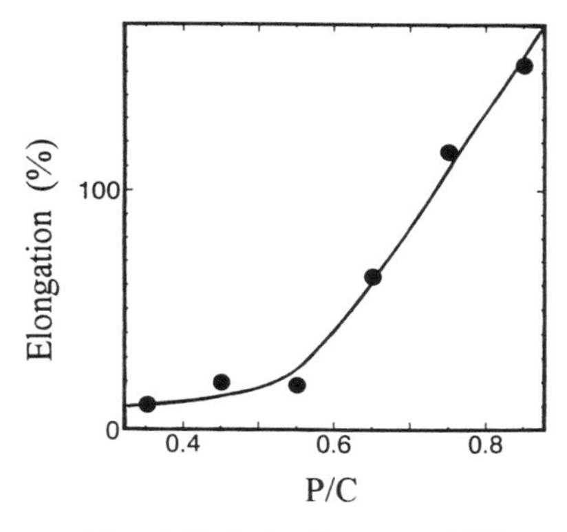

Fig. 9 Relation between P/C and elongation.

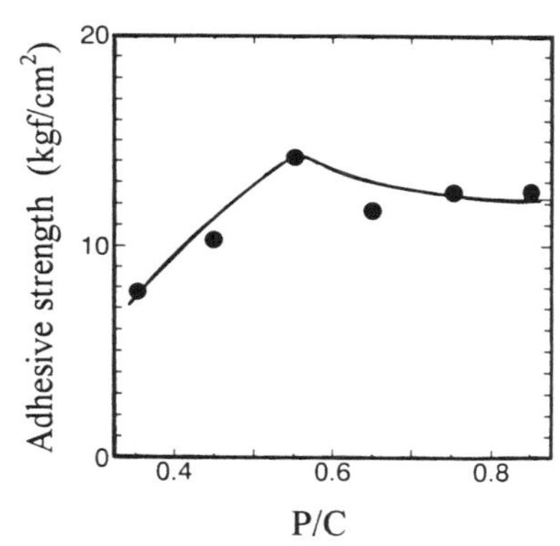

Fig. 10 Relation between P/C and adhesive strength.

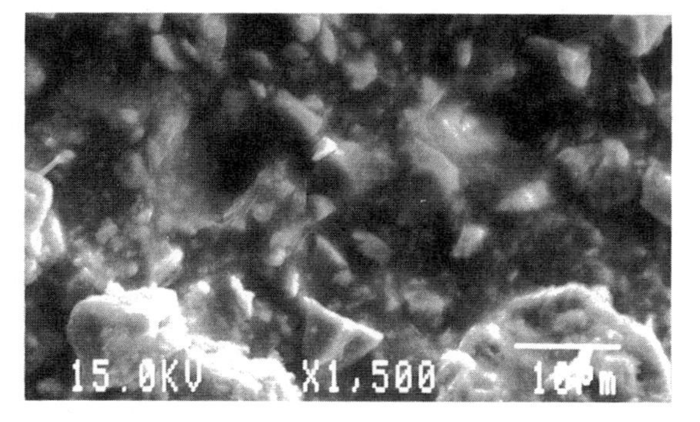

Fig. 11. Scanning electron microscope of flexible PCM. (P/C$>$0.55)

3.1.4 Chloride ion permeability

Fig.12 shows the relation between P/C and chloride ion permeability. Chloride ion permeability of flexible PCM was reduced with increasing P/C, the same tendency as in the results of water and water vapor permeability tests.

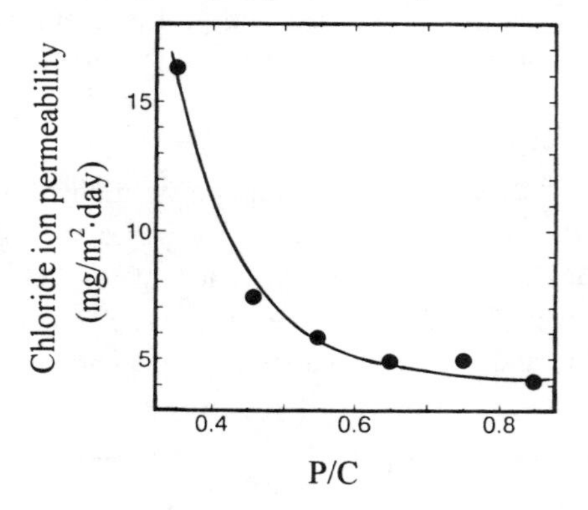

Fig. 12. Relation between P/C and chloride ion permeability.

3.2 Field exposure test

Fig.13 shows the length change and Fig.14 shows the weight change in the field exposure test. The weight of uncoated specimen increased due to rain water absorption and it decreased when it was fine. The uncoated specimens show a tendency to expand, but the specimens coated with flexible PCM demonstrate an ability to inhibit ASR. On the other hand, expansion of the specimens coated with epoxy resin was smaller than that of the uncoated specimen until about 48 weeks. However, it will become larger than that of uncoated specimen after 50 weeks. The water remaining in the concrete may have slowly promoted ASR since the epoxy resin has no water vapor permeability [2] .As shown in Fig. 15, there exists an obvious quantitative relation between water vapor permeability and ASR expansion. The expansion due to ASR is reduced with increasing water vapor permeability.

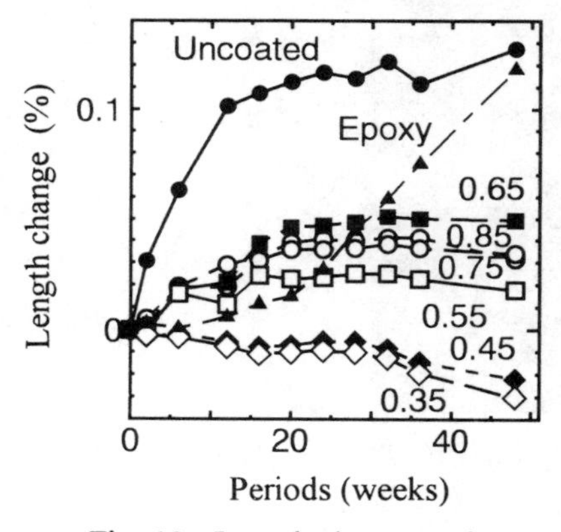

Fig. 13. Length change under the field exposure test.

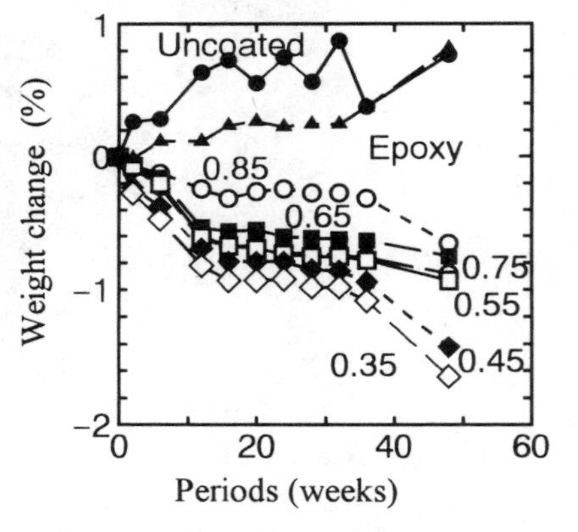

Fig. 14. Weight change under the field exposure test.

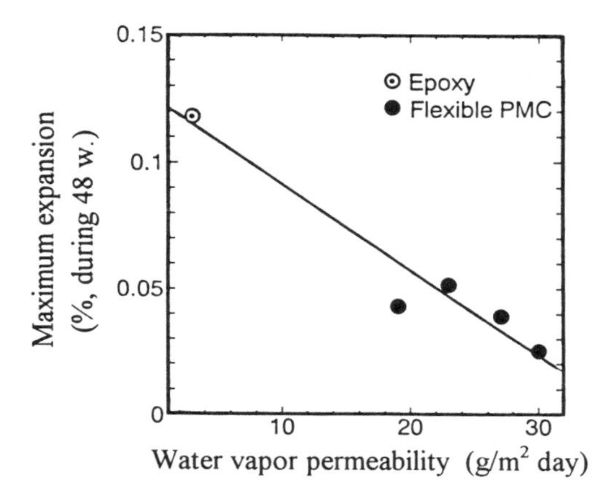

Fig. 15. Relation between water vapor permeability and ASR expansion. (48 weeks)

4 Conclusion

As results of the physical testings of flexible PCM which consist of a polyacrylic latex, portland cement and some admixtures, it is found that water and water vapor permeability are reduced with an increase in P/C. Elongation tends to increase with increasing P/C, whereas adhesive strength has an optimum P/C. As results of the field exposure test using concrete specimens damaged by ASR and coated with either epoxy resin or flexible PCM with various P/C, it is found that epoxy resin coating promotes ASR, but coating of flexible PCM inhibits ASR and the expansion due to ASR is reduced with increase in water vapor permeability of flexible PCM.

5 References

1. Ono, K. (1986) Examples of Deterioration of Concrete Structure due to Alkali Aggregate Reaction. Concrete Journal of Japan Concrete Institute, Vol. 24, No.11, pp.50-56.
2. Okada, K., Kobayashi, K., Miyagawa, T. and Sugashima A. (1987) Repair of Concrete Structure Damaged by Alkali Aggregate Expansion. Proceeding of the Japan Concrete Institute, Vol.9, No.1, pp.621-626.

94 EVALUATION SCHEME FOR PENETRATING SEALERS TO PREVENT CONCRETE DETERIORATION IN SAUDI ARABIA

E.A. AL-JURAIFANI
Saudi Aramco, Dhahran, Saudi Arabia

Abstract
In this paper, an evaluation scheme for penetrating sealers is proposed and the optimum generic type of penetrating sealers that is suitable for the Arabian Gulf region environment is promoted.
Keywords: Absorption, breathabilty, evaluation scheme, organosilicon, penetrating sealers, thermal resistant.

1 Introduction

The increasing problems of concrete deterioration observed throughout the Arabian Gulf area highlight the need to examine and evaluate additional means and techniques to overcome this phenomenom. There are many techniques and alternatives available to combat these problems, one of which is the application of penetrating sealers.

Many studies have been carried out in the last two decades to study the effect of penetrating sealers in minimizing or preventing the ingress of deleterious material into concrete. These studies, however, do not consider the severe environment of the Gulf region.

The main objectives of this study are to:
1. Propose an evaluation scheme that establishes the basis of a national standard for evaluating penetrating sealers.
2. Promote an optimum generic type of penetrating sealer that will resist the harsh environment of the Gulf area.

The final evaluation scheme was achieved based on pre-established criteria that were modified based on the results of the following four test phases:-

Concrete Under Severe Conditions: Environment and loading (Volume Two) Edited by K. Sakai, N. Banthia and O.E. Gjørv. Published in 1995 by E & FN Spon. ISBN 0 419 19860 1

Phase one :Sealers effectiveness in preventing intrusion of harmful materials into concrete.
Phase two :Sealers effectiveness against abrasion.(Depth of penetration).
Phase three :Sealers effectiveness in resisting thermal cycling.
Phase four :Sealers effectiveness in allowing the concrete to release the trapped moisture " Breathability".

Eleven sealers representing four generic types that are available in the local market were tested to accomplish the evaluation scheme.
The organosilicon group revealed excellent performance among the four generic types used in this study. Silane with 40% solids by weight was found to have the best performance among the organosilicon group.

2 Research significance

In the Arabian Gulf region, numerous concrete structures are undergoing accelerated deterioration caused by the corrosion of reinforcement steel. Corrosion is mainly caused by the ingress of deleterious material to the body of concrete. Chloride ion intrusion into the concrete from different sources (i.e. salt water, contaminated aggregate, brackish water used for curing, etc.) is one of the major causes of deterioration in the Gulf region . Other factors such as sulfate, carbon dioxide, inadequate cover and poor mix design promote the corrosion mechanism[1]. The corrosion rate of reinforcement steel depends on the moisture and chloride content of concrete[2].

Preventing deleterious material from propagating into concrete by applying sealers could extend the life of the structures significantly. Such sealers material could be used on new structures as well as older ones that are not already contaminated with chloride beyond tolerance limits[1].

The main objective of this study is to investigate the effectiveness of the eleven penetrating sealers when subjected to severe environmental conditions and propose an evaluation scheme that can be utilized to evaluate new types of sealers. The effectiveness of the sealers will be accomplished based on the ability of the sealers to prevent or reduce the absorption of water and other deleterious material by concrete samples.

3 Objectives

1. Propose an evaluation scheme for the acceptance of penetrating sealers based on the absorption limits specified by BS1881 Part 122 and ASTM C 642 [3,4]. Table 1 illustrate the pre-established criteria for the preliminary evaluation scheme. The final evaluation scheme will be established by modifying the preliminary scheme based on the results of the following test phases:

   Phase one : Sealers effectiveness in preventing intrusion of harmful materials into concrete.

Phase two : Sealers effectiveness against abrasion.
Phase three : Sealers effectiveness in resisting thermal cycling.
Phase four : Sealers effectiveness in allowing the concrete to release the trapped moisture, " Breathability".

2. Identify the generic type of sealer that will result in an optimum overall performance in the Gulf environment.

Table 1. Preliminary evaluation scheme

No.	Characteristic	Value
1	Absorption	Absorption 3 % and lower after 21-day soaking in salt solution
2	Penetration of sealer	Absorption 3 to 5 % and lower after removing 2 to 3 mm of treated surface then 21-day soaking in salt solution
3	Thermal Resistant	Absorption 3 to 5 % and lower after subjecting to 120 thermal cycles then 21-day soaking in salt solution
4	Breathability	Any reduction in the original weight of concrete sample
5	Handling	Direct application, no mixing is required
6	Hazard	Non toxic, non flammable

4 Test phases

The test phases were designed to simulate the actual exposure in the Arabian Gulf region. Two types of cement (Type I & V) were used. The eleven sealers used in this test lay under four generic types even though they have different brand names and are listed as follows:

Type	Test code
1- Acrylic	EJ4
2- Organosilicon	
a) Siloxane	EJ1, EJ10, EJ11
b) Silane	EJ5, EJ8
c) Silane-Siloxane	EJ2
3- Epoxy	EJ6, EJ7, EJ9
4- Silicates	EJ3

4.1 Phase one (absorption)

This phase was performed to evaluate the efficiency of the sealers in preventing the ingress of water and other deleterious material into concrete. Sealers were applied on concrete cube samples, then the cubes were immersed in 15% ($NaCl$+ Na_2So_4) salt solution for 21 days. The weight of the cubes were taken before and after immersion to obtain the absorption percentage with reference to uncoated control samples which were immersed in the same solution. Concrete cubes which show no or minor increase in their weight after immersion indicate that the applied sealers were able to prevent the water and salts from propagating into the concrete.

4.2 Phase two (abrasion)

This test phase was designed to examine the ability of sealers to penetrate into concrete surface. 2 mm of the outer surface of the cube samples were removed to simulate the abrasion of concrete surfaces in the field. The abraded samples were then placed in a salt solution for 21 days. Samples showing satisfactory water repelling characteristic indicate that the sealer has penetrated beyond the 2 mm depth.

4.3 Phase three (thermal resistance)

This test phase was designed to examine the ability of the sealers to resist the thermal cycling. Treated samples were placed in an oven for 120 cycles at temperature range of 20 - 65 C. These temperature cycles simulate 120 days (4 months) of the hottest months of the year.

After the thermal cycles were completed, the samples were placed in a salt solution for 21 days to measure absorption percentage. Samples showing satisfactory water repelling indicate that sealers were not affected by thermal cycling and thus were most likely to resist the outdoor environment.

4.4 Phase four (breathability)

This test phase is an extension of phase one. The same samples used in **phase one** were left to dry for 40 days and then were tested for "Breathability" of the sealers. The percentage of loss in weight due to the evaporation of trapped water in concrete sample were measured. Samples which lose all or most of the trapped water indicate that the sealer film allows breathing. Figures 1a,b,c and d show a comparison between test phases.

5 Discussion

Several studies have proven that the application of sealers would increase the life span of concrete structures. However, most of those studies were not designed for the Gulf environment. In this study, the test methodology was modified to simulate the severe exposure conditions of the Gulf region.

Out of the four test phases, the thermal cycling test was the most severe test. Except for the organo-silicons group, all the other generic types were affected significantly during this test.

Sealers applied on Type I cement exhibited better performance than the same sealers applied on Type V cement. This observation was made from the data collected in this study. The reasons behind such different performances is beyond the scope of this study.

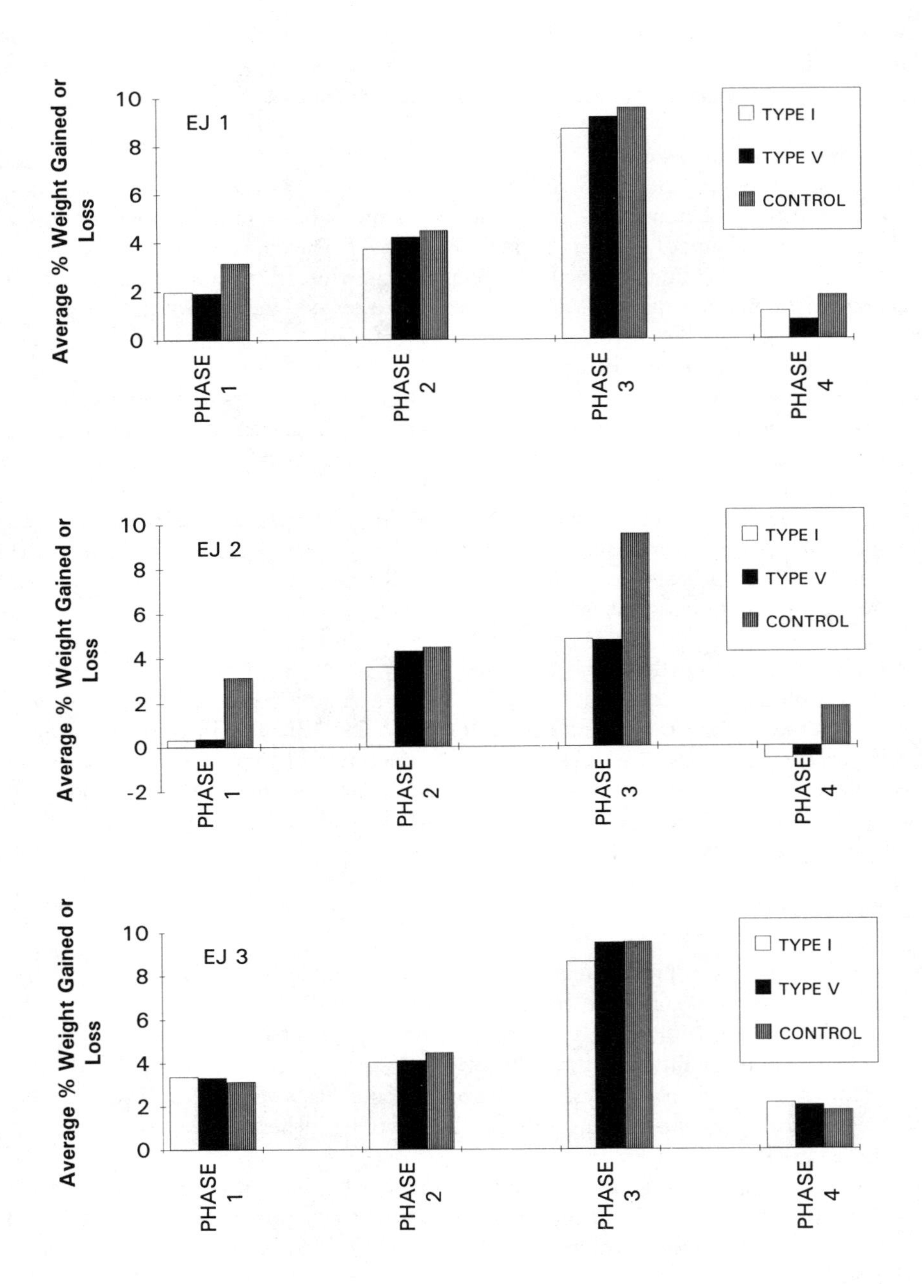

Figure 1.a : Comparison Between Test Phases

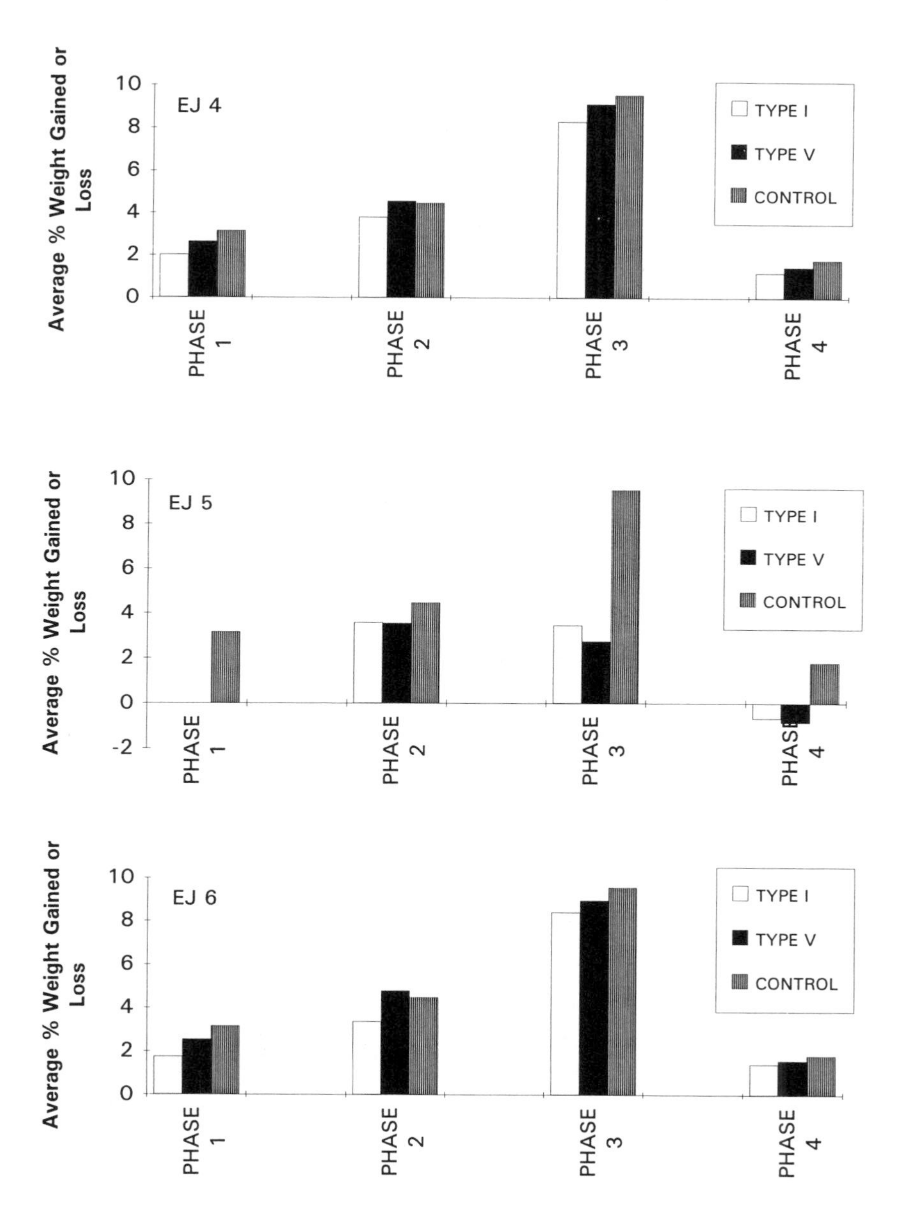

Figure 1.b : Comparison Between Test Phases

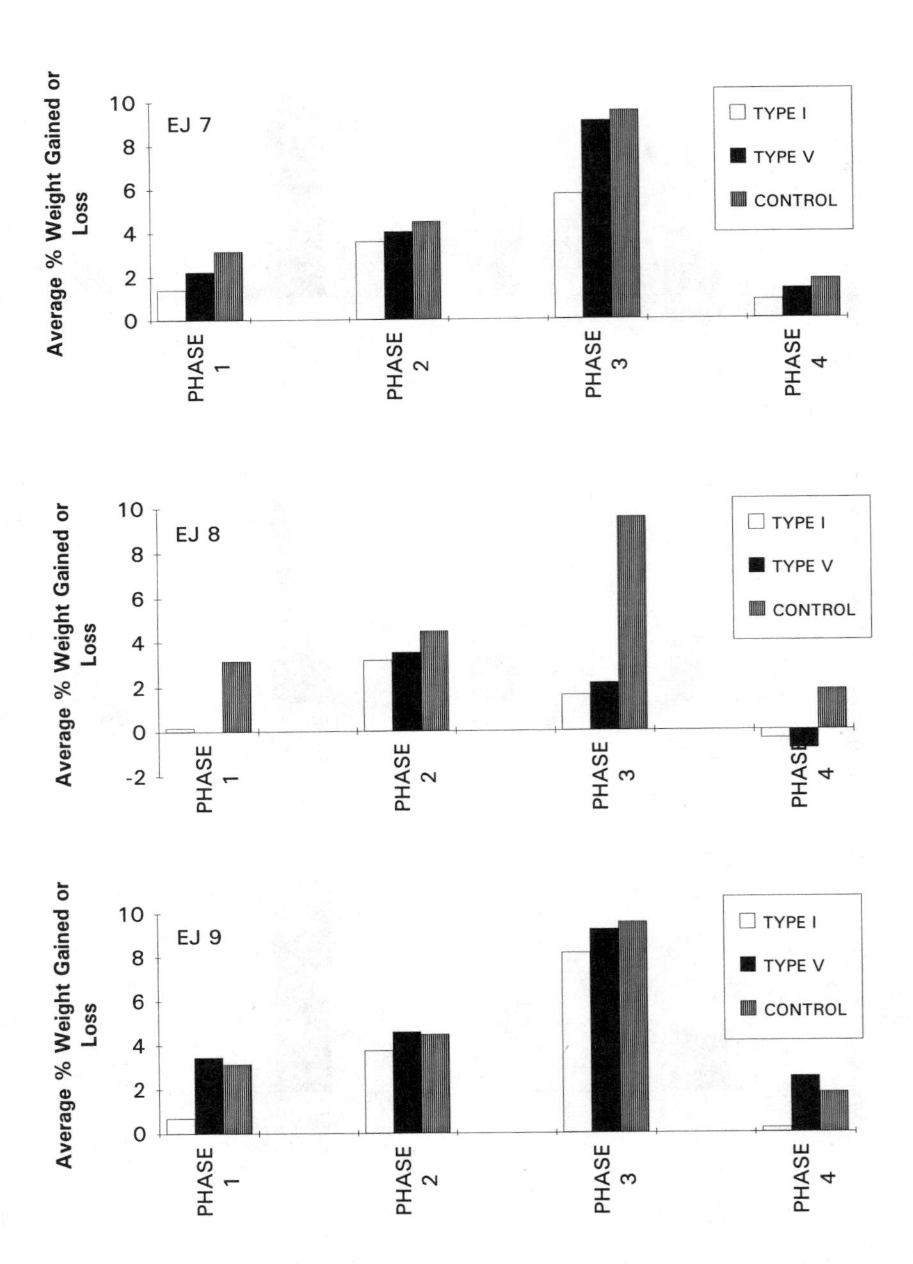

Figure 1.c : Comparison Between Test Phases

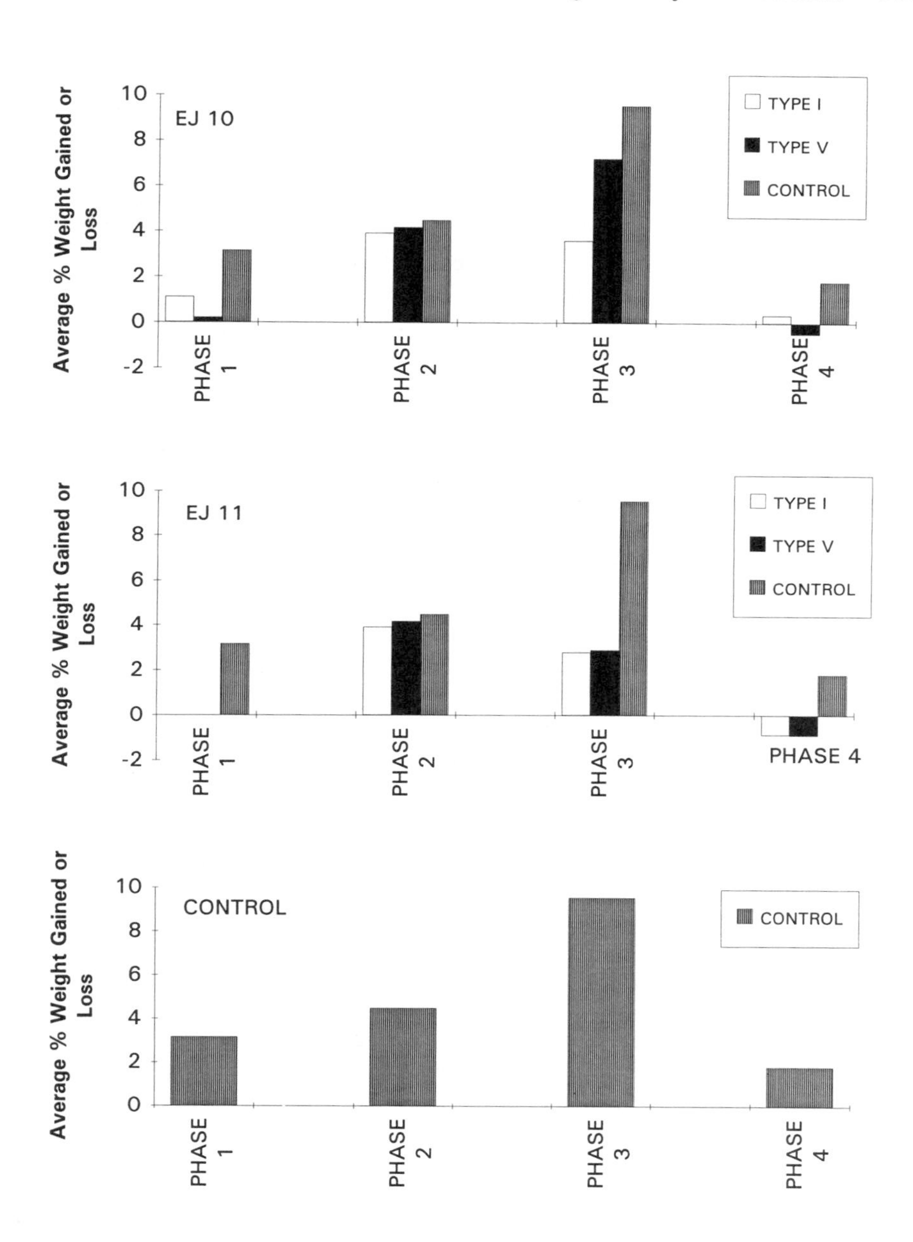

Figure 1.d : Comparison Between Test Phases

5.1 Evaluation scheme

One of the main objectives of this study was to propose a scheme that can be used for the evaluation of any penetrating sealer system. This objective was met by pre-establishing a preliminary scheme then modifying it based on the test phases carried out during this study.

The scheme was based on six main characteristics, four of them were from the laboratory tests and the remaining were from manufacturer data sheets. See Table 2.

The first characteristic specifies the maximum allowable percent of weight gained after immersing a treated concrete sample in a salt solution for 21 days. The second one specifies the maximum allowable percent of weight gained after removing 2 to 3 mm of the concrete sample surface then immersing it in a salt solution for 21 days. The third characteristic specifies the maximum allowable percent of weight gained after subjecting a treated concrete sample to 120 thermal cycles with temperatures ranging between 25 to 65 degree Centigrade then immersing it in a salt solution for 21 days. The fourth one specifies the minimum percent of weight loss from the original weight of a treated sample after immersing it in a salt solution for 21 days then allowing it to dry for 40 days. The fifth and sixth characteristics were from the actual application of the sealers in the laboratory and the available manufacturer data sheet.

Table 2. Proposed Evaluation Scheme

No.	Characteristic	Value
1	Absorption	Absorption 1 % and lower after 21-day soaking in salt solution.
2	Penetration of sealer	Absorption 4% and lower after removing 2 to 3 mm of treated surface then 21-day soaking in salt solution.
3	Thermal Resistant	Absorption 4% and lower after subjecting to 120 thermal cycles then 21-day soaking in salt solution.
4	Breathability	>0.3 % reduction in the original weight of concrete sample.
5	Handling	Direct application, no mixing is required.
6	Hazard	Non toxic, non flammable.

5.2 Optimum generic type

Among the four generic types of sealers used in this study, the organo-silicon compound revealed relatively better performance. This generic type has proven its good performance in previous studies. However, this study tested this generic type to exposures that are more relevant to the Gulf region environment.

The organosilicon group consists of many types that are distinguished based on the carrier (organic solvent) used in preparing them as a final product. The silane penetrating sealer (monomeric alkyl alkoxysiloxane) indicated relatively better performance compared with the other types in the same group. The percent solids by weight of this type was 40%.

The organosilicon group performance was the best among the other generic groups. Silane with 40% solids by weight was the best among the organosilicon group.

6 Summary

1. The evaluation scheme was based on the ability of the sealers to prevent or reduce the absorption of water.
2. The organo-silicones group performance was the best among the generic groups. Silane with 40% solids by weight was the best among the organo-silicones.
3. The organo-silicones group has better thermal resistance than other generic groups.
4. Sealers applied on Type I cement exibit better performance than similar sealers applied on Type V cement.

7 Suggested studies

Further investigations may produce more, significant results on this topic. The following are suggested studies :

1. Investigate in detail the reasons for the different performances with the two types of cement.
2. Investigate the effect of hydrostatic pressure on the performance of the penetrating sealers.
3. Investigate the effect of field exposure on the performance of the penetrating sealers and verify them with laboratory test results.
4. Investigate the effect of surface preparation on the performance of penetrating sealers.
5. Investigate the life cycle of penetrating sealers when subjected to combined exposure conditions.

8 References

1. Pfeifer , D. W. and Scali, M.J. (1981) Concrete Sealing For Protection Of Bridge Structures, *National Cooperative Highway Research Program* (NSHRP 244). Transportation Research Board, Washington, D.C. .
2. Mailvaganam, N.P, J. J. Deans, and K. Cleary (1992) *Sealing and Waterproofing Materials Repair and Protection of Concrete Structures,* CRC Press .
3. Maslehuddin, M. , H. Saricimen and A.I. Al Mana (1993) Performance Evaluation of Concrete in the Arabian Gulf, *4th International Conference Deterioration and repair of reinforced concrete in the Arabian Gulf ,Bahran,.*
4. Maslehuddin, M. (1981) Optimization of Concrete Mix Design for Durability in the Eastren Province of Saudi Arabia, *M.S. Thesis, King Fahd University* of *Petroleum and Minerals, Dhahran,* .

95 MONITORING OF THE DEPTH-DEPENDENT MOISTURE CONTENT OF CONCRETE USING MULTI-RING-ELECTRODES

P. SCHIESSL and W. BREIT
RWTH University of Technology, Aachen, Germany

Abstract
To determine the moisture distribution within the concrete between the reinforcement and the concrete surface, a so-called multi-ring-electrode has been developed. The time- and depth-dependent distribution of the electrolytic resistance, which is influenced mainly by the distribution of water content in the concrete, can be determined by AC current measurements. Basic results for drying behaviour, behaviour after wetting with water and the effect of surface protection systems on water penetration into concrete are demonstrated. It is shown that the multi-ring-electrode is suitable for laboratory tests for examining the response of changing concrete water content on concrete technology parameters, for investigations into the influence of environmental conditions, and finally, for checking the influence and efficiency of concrete coatings. In practice, the sensor can be placed in high-risk areas of new or existing structures.
Keywords: Coatings, concrete, hydration, moisture, moisture-distribution, monitoring, surface protection systems.

1 Introduction

Moisture distributions in concrete structures are relevant to many problems of construction technology since a number of important material parameters are moisture dependent. Concrete moisture has a decisive influence on, for example, the corrosion rate of steel in concrete, the frost resistance of concrete and its resistance to aggressive agents like sulphates, and the alkali-silica reaction rate. Moisture distribution in the surface zone is of special interest in regard to measures to reduce water content by surface coating of the concrete.

Concrete Under Severe Conditions: Environment and loading (Volume Two) Edited by K. Sakai, N. Banthia and O.E. Gjørv. Published in 1995 by E & FN Spon. ISBN 0 419 19860 1

A precise determination of concrete moisture is possible only by the direct method of weighing the amount of water in a sample before and after drying. A variety of indirect testing methods are available as an alternative to direct moisture determination; these infer concrete moisture from moisture-dependent concrete parameters, without extracting concrete samples. The large group of indirect methods includes the method described in this paper, in which electrical resistance is measured as an indicator of moisture distribution in the concrete surface zone.

2 Electrolytic resistance of concrete

The electrolytic resistance of concrete depends on both the chemical composition of the pore solution and the pore structure (concrete mixture, curing and compacting) and on the concrete moisture and temperature. Concrete exhibits a special electrical behaviour, since concrete acts either as an insulator or as a good conductor depending on its moisture content. Conductive behaviour corresponds to the hydrated cement-matrix [1]. Changes in ion concentration can occur due to penetration of ions. The most important parameters which influence the electrolytic resistance of concrete are described in [2] and [3].

The time- and depth-dependent distribution of electrolytic resistance, which in a specific concrete member is influenced mainly by the distribution of moisture content in the concrete, can be determined using multi-ring-electrodes developed at the Institute for Building Materials Research, Aachen.

3 Multi-ring-electrode

The multi-ring-electrode consists of a series of alternate noble metal and insulating rings, arranged above one another. The design is modular, allowing the user to vary the number of rings according to his requirements. The geometrically optimized version of the sensor has a diameter of 20 mm and normally consists of nine noble metal rings, providing eight measuring points. The thickness of each ring is 2.5 mm and they are 2.5 mm apart, yielding a profile 50 mm in depth with values every five mm. The rings are connected to cables arranged inside the electrode in such a way that they do not influence the surrounding concrete (measuring area). The remaining interspaces of the electrode are filled with epoxy resin. The arrangement of the multi-ring-electrode is shown in Fig. 1.

The multi-ring-electrode is embedded in concrete so that the uppermost ring is at a distance of 2.5 mm from the concrete surface, the thickness of the layer of non-conducting material. This defines the positions of the other rings. The electrolytic resistance between neighbouring noble metal rings can be measured externally using a suitable AC ohmmeter. A profile is then inferred from the data. Figure 1 indicates the qualitative results.

In principle, electrolytic resistance can be determined by AC or by DC current measurements. In this case, AC measurements were used to determine electrolytic resistance, since this virtually eliminates the effect of test electrode polarisation which could occur with DC current. The movable ions between the test electrodes are made

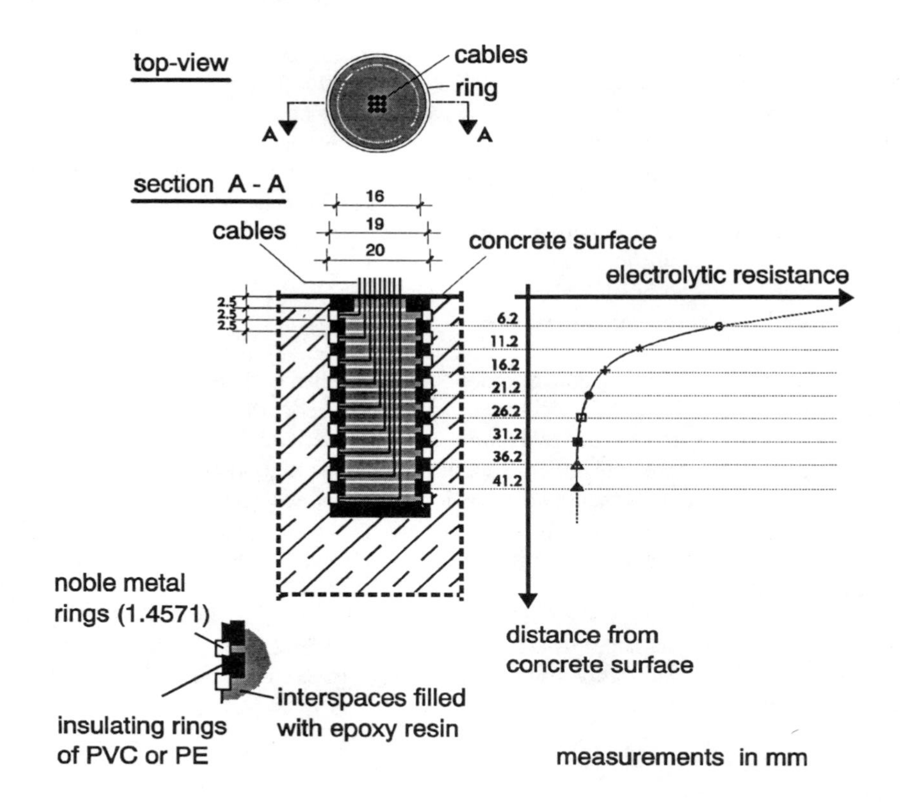

Fig. 1. Schematic representation of the multi-ring-electrode and a qualitative diagram of measured values.

to oscillate, preventing chemical reactions or polarisation on the test electrode (noble metal ring) surface [4]. A unit designed specially for measurements with the multi-ring-electrode has been developed at the ibac. This battery-powered hand-held unit has a selector switch for up to 12 measuring points. An alternative is computer-controlled data logging and analysis.

4 Investigations and results

After preliminary tests in earlier research projects intended to optimize the geometry and the manufacturing process for the multi-ring-electrodes, the present experiments are aimed at calibrating the sensor for concrete with varying properties (type of cement, water/cement ratio, grading curve, curing and condition of storage). Fifty test specimens with multi-ring-electrodes were made, together with reference specimens, in order to determine electrolytic resistance [3]. The test specimens measure 350·270·230 mm³ and were sealed on five sides. Therefore, water exchange with the environment was possible only via a defined area of 350·270 mm², the simulated surface of the structure. The multi-ring-electrode was centred on the unsealed surface and concreted in. The depth of carbonation was also tested on concrete beams made for the purpose. Simultaneously, various types of "environmental actions" (wetting, chloride attack and rapid carbonation) were simulated, to investigate the time- and depth-dependent response of the concrete surface area.

The following section indicates basic results for drying behaviour, behaviour after wetting with water and the effect of surface protection systems on water penetration into concrete to illustrate potential fields of application for the multi-ring-electrode method.

4.1 Results of drying and wetting studies

Figure 2 shows the results of resistance measurements using the multi-ring-electrode on a test specimen made with ordinary portland cement (c = 300 kg/m³, w/c = 0.6) at ages between 200 and 400 days after concreting. The test specimen was stored two days in a humidity chamber and subsequently at 20°C and 80% relative humidity. It shows clearly that the electrolytic resistance decreased with the distance from the concrete surface. This depended mainly on depth-dependent drying and on hydration. The difference in resistance at 6.2 and 11.2 mm distance from the concrete surface was about one order of magnitude. The difference in deeper areas was not so great. This means that moisture changes in the concrete cover, e.g. due to drying, can be determined clearly with this kind of measurement.

At an age of roughly 260 days, the test specimen surface was wetted with water over a period of three days. Wetting caused resistance to decrease at different rates and by different amounts. Immediately after wetting, there was a pronounced reduction in

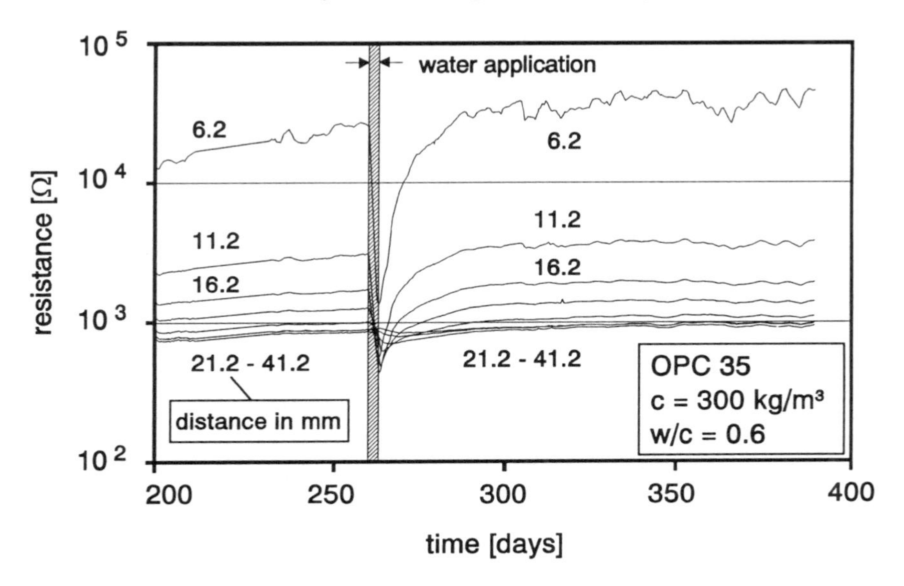

Fig. 2. Depth-dependent effect of drying and wetting on the electrolytic resistance of concrete.

electrolytic resistance at a distance of 6.2 mm from the concrete surface, while the resistance in deeper areas changed more slowly and less markedly. After wetting ceased, drying of the concrete surface zone led to an immediate increase in resistance, while in deeper areas (deeper than 21 mm) the minimum resistance value was not yet reached. This may be attributed to the time-delayed movement of the water within the concrete.

After strong initial capillary suction, the water therefore penetrated the concrete only at a slow rate, leading to a minimum electrolytic resistance value in deeper areas of the reinforcement with the usual practical concrete cover after a lengthy time-lag of some hours or days.

Figure 2 also shows that the original distribution of resistance is restored as early as about one month after wetting and that a further increase in resistance takes place in the surface zone. This behaviour is attributed to a change in pore structure due to carbonation, as explained in the following section.

4.2 Results of investigations using surface protection systems

Three among the twelve existing surface protection systems, according to guideline [5], were chosen for lab experiments [6]. The systems examined had already passed the basic tests required according to the guideline and had been admitted to the catalogue of licensed coatings in Germany.

Some 600 days after preparation, the test specimens were coated with the chosen surface protection system. To check the efficiency of the coating, the test specimens were wetted with 30 mm water applied to the (horizontal) concrete surface after about one month. A percentage distribution relative to the resistance value before wetting is the most useful means of analyzing changes in the distribution of resistance due to such wetting.

Figure 3 shows the changes in resistance distribution of an uncoated reference specimen. It clearly shows the rapid decrease in resistance in the near-surface areas,

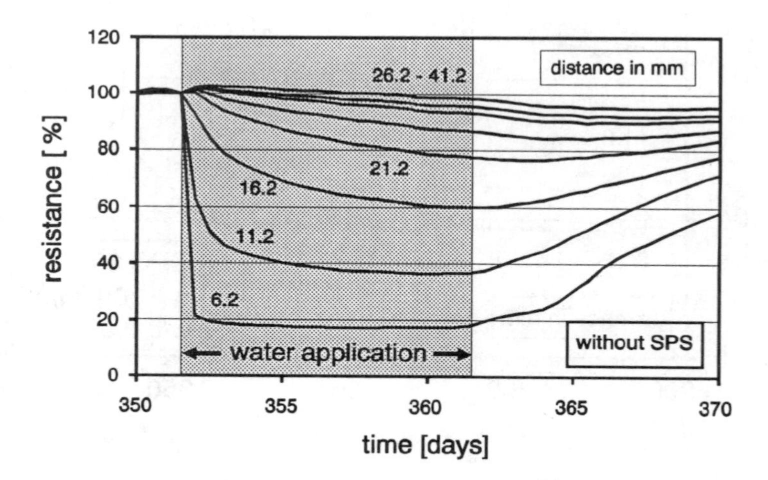

Fig. 3. Influence of wetting on the electrolytic resistance of uncoated concrete specimen (without SPS - Surface Protection System)

which is in contrast with deeper areas where the minimum resistance value had not yet been reached when wetting ceased. At a depth of 6.2 mm, electrolytic resistance was reduced to less than 20 % of the starting value.

Figure 4 shows the percentage distribution of resistances during wetting for the surface protection systems (SPS) 2, 5 and 11 that were investigated.

SPS 2 is a sealing for non-trafficable surfaces. The field of application is the preventive protection of outdoor-weathered concrete surfaces in the new construction sector, for vertical surfaces and undersides [5]. The structure of SPS 2 is shown in Fig.5. After wetting, significant changes in resistance were measured on the carbonated specimen (depth of carbonation 23 mm) subsequently coated with SPS 2, as compared to the reference specimen in Fig. 3. The main difference is the rate of decrease in electrolytic resistance, which was not so rapid with SPS 2. Due to carbonation (23 mm), the electrolytic resistance down to a depth of 21.2 mm was significantly reduced. The water reached the deeper areas after a time-lag. Behaviour was estimated in relation to the field of applications, which does not entail direct wetting.

SPS 5 is defined as a coating for non-trafficable surfaces with low crack-bridging capability. The field of application are facades, engineering structures and other non-trafficable, outdoor weathered concrete surfaces not subjected to mechanical loading. The system may be used in case of de-icing salt attack, but its suitability must be approved. It is a standard measure in case of repair according to corrosion principle W (reduction of the water content in the concrete) and C (application of a protective coating on the steel surface), provided low crack-bridging capacity is acceptable [5]. Schematically, the two possible structures of SPS 5 are shown in Fig. 5. The investigated system is classified as SPS 5, structure c1. A distinctly improved protective effect was achieved with this SPS. A slow decrease in resistance was, however, evident at a depth of 6.2 mm. Clear blistering and sporadic cracking occurred in the surface layer, and remain to be investigated.

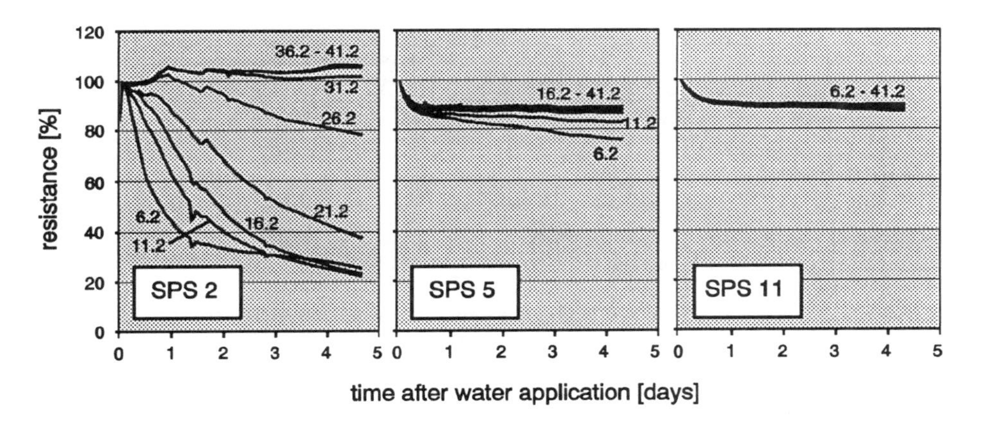

Fig. 4. Influence of wetting on the electrolytic resistance of concrete specimens coated with different types of surface protection system (SPS 2, 5, 11).

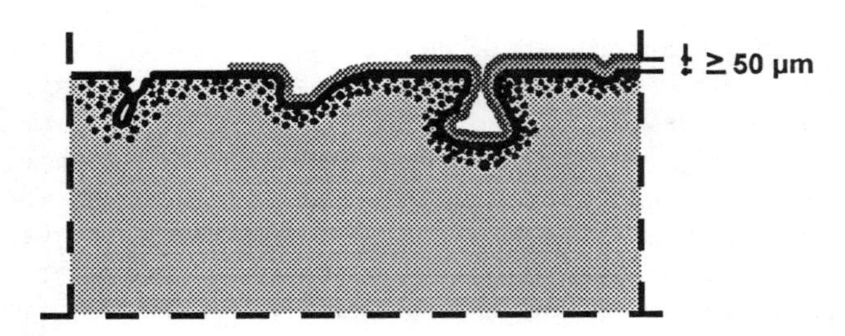

SPS 2

Structure:
- water-repellent impregnation
- application of non-pigmented, colourless primer
- 2 colourless, glazing or pigmented top coats

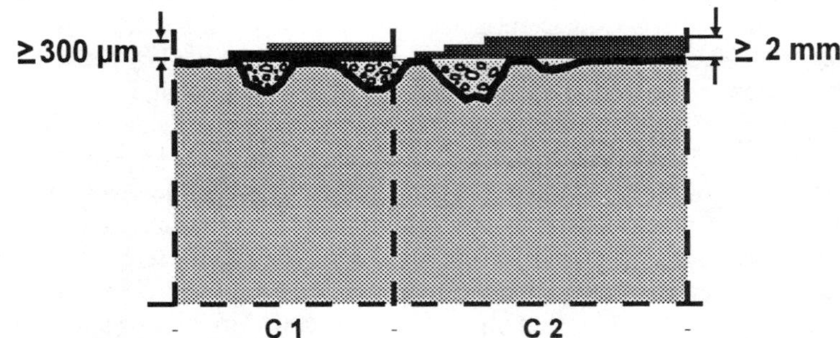

SPS 5

Structure:
- c1/c2: stopping to fill flaws, pores and cavities
- c1: non-pigmented, consolidating and water-repellent primer
- c1: 2-4 top coats
- c2: sealing slurry

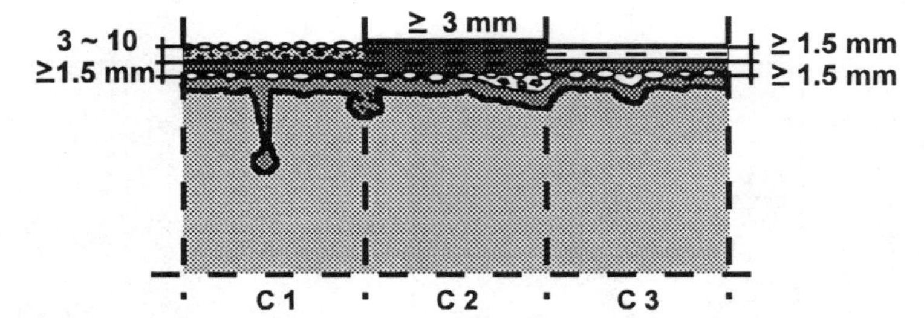

SPS 11

Structure:
- c1/c2/c3: non-pigmented, low-viscosity, solvent-free cold-curing-resin-based primer, surfaced with dry quartz sand
- c1/c2/c3: stopping with a fine cold-curing resin stopper to fill flaws, pores and cavities
- c1/c3: elastic coating as crack-bridging intermediate layer
- c1: elastic, filled bedding mortar as wear course, consisting of a mineral-filled coating compound, mineral surfaced to increase grip
- c2: 3 coatings of a high-strength cold-curing resin binding material, first and second coats containing fibre-glass tissue or mat (500 g/m²)
- c3: top coat of a high-strength cold-curing resin binder fibre-glass tissue or mat (100g/m²)

Fig. 5. Schematic representation of the structures of surface protection systems [5].

SPS 11 is classified as a coating for trafficable surfaces with at least increased crack-bridging capability. The field of application is crack-endangered concrete surfaces such as curbs and bridge caps and surfaces subjected to heavy mechanical loads such as parking or bridge decks [5]. In Fig. 5 the three possible structures are represented. In this case, structure c1 was investigated. With SPS 11, no significant changes in resistance for the different areas due to wetting were found after an initial decrease of 10 %, which means that no water penetrated the coating.

The initial decrease in the resistance curve, also found with SPS 5, can be explained by a warming effect due to the temperature difference of about 3°C between the specimen (~18°C) and water (~21°C) applied to the concrete surface. The Arrhenius equation can be used to formulate the relationship between temperature and electrolytic resistance. In accordance with this equation, a temperature difference of 3 °C leads to a decrease in resistance of about 10 %. Both test specimens (SPS 5 and 11) were stored under the same conditions. The water and concrete temperatures were identical to the case of SPS 2.

5 Subsequent installation of the multi-ring-electrode

Apart from its use for fundamental laboratory studies and direct embedding in new concrete components, the multi-ring-electrode has also been designed as a sensor for moisture distribution monitoring in the surface zone of restored components. Therefore, a subsequent installation of the sensor in existing structures must be possible, which is mainly a problem of coupling it to the concrete.

5.1 Coupling to existing concrete

The material used for subsequent installations must be a cementitious substance which connects the electrode to the surrounding concrete without negatively affecting or falsifying the resistance measurements. It must also be workable enough to allow injection into the narrowest possible interspace between the sensor and the surrounding concrete.

Cement pastes and fine mortars, including ready-mixed products, were tested for suitability. Results indicated the following requirements:

- maximum grain size ≤ 1 mm,
- fluid consistency and
- conductivity in the hardened state roughly equivalent to that of the existing concrete.

A cement paste made up with the same cement type and the same or a lower cement strength class as those used for the surrounding component provides an optimum connection between the sensor and the structure.

5.2 Coupling method

Figure 6 shows an application of the multi-ring-electrode in a vertical structure. The diameter of the drilled fitting hole should be 26 mm and its depth 80 mm to ensure that the electrode can be moved laterally in the fitting hole prior to coupling. The sensor is

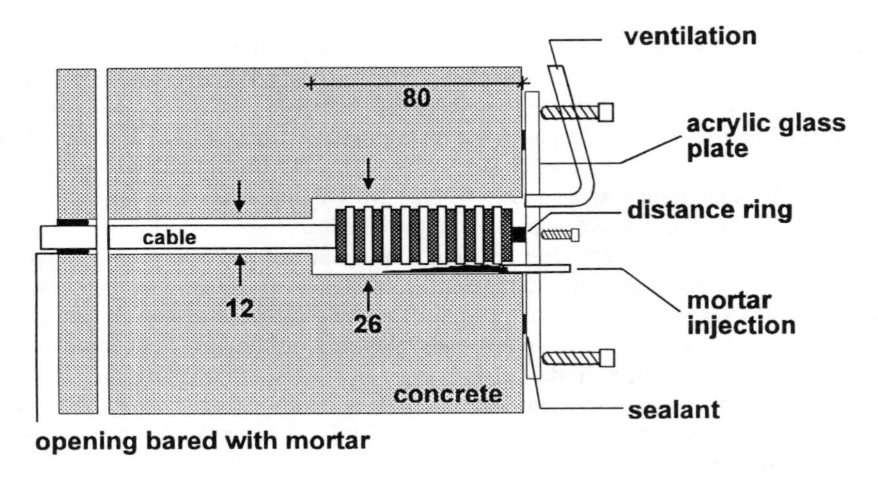

Fig. 6. Installation of multi-ring-electrode into vertical part of the building.

secured to an acrylic glass plate mounted on the concrete surface, with an interspace of 4 mm so that a layer of mortar covers the electrode after fitting. The cables are led to the rear opening and sealed with sealing compound.

Cement paste is injected into the fitting hole via an injection tube. The air from the fitting hole escapes through a ventilation tube until the cement paste enters the tube. The injection tube is finally sealed with a plug. The acrylic glass plate may be removed after about 24 h.

6 Conclusions

To determine the water distribution within the concrete near the reinforcement and the concrete surface, a multi-ring-electrode was developed. Studies to date show that both the measuring procedure and the multi-ring-electrode works without any problems and respond clearly to moisture changes.

Moisture is the dominant parameter influencing the electrolytic resistance of concrete. Resistance differences between water saturated and dry concrete in the concrete surface zone are in the order of some decades. It has not been possible to indicate all the influencing parameters for the electrolytic resistance of concrete in this paper. For the sake of completeness, other observations from laboratory tests [2] [3] [6] [7] and literature are summarised below. An increase in electrolytic resistance was found under the following conditions:

- Increasing hydration,
- decreasing water/cement ratio when concrete is moist,
- increasing water/cement ratio when concrete is dry,

- decreasing content of cement matrix,
- increasing content of slag (PBFC, BFSC)
- increasing content of fly ash and silica fume,
- increasing curing,
- decreasing chloride content and
- decreasing temperature.

It was shown that the multi-ring-electrode is suitable for laboratory tests to examine concrete and coating parameters, for investigations into the influence of environmental conditions and, finally, for checking the influence and efficiency of concrete coatings.

The sensor can be used for both lab examinations and in situ installations. For instance, the multi-ring-electrode can be placed in high-risk areas of new or existing structures (e.g. bridges). Especially for repair work, where final coatings are usually applied, the multi-ring-electrode performs adequately as a sensor monitoring the efficiency of the coating during use. The signals give an early warning should the efficiency of the coating decrease so that interventions are possible before damage, e.g. by reinforcement corrosion, occurs.

7 References

1. Monfore, G.E. (1968) The Electrical Resistivity of Concrete. *Journal of the PCA Research and Development Laboratories*, Vol. 10, Nr. 2, pp. 35-48.
2. Raupach, M. (1991) Zur chloridinduzierten Makroelementkorrosion von Stahl in Beton. *Heft 433 des DAfStb*, Beuth Verlag Berlin.
3. Schießl, P., Breit, W. and Souchon, T. (1994) Überwachung der Korrosionsgefahr für die Bewehrung bei Trägern mit geringer Betonüberdeckung mittels Einbausensoren. *Forschungsbericht Nr. F 389*, Institut für Bauforschung, Aachen.
4. Forker, W. (1989) *Elektrochemische Kinetik*. 2. Auflage, Akademie Verlag Berlin.
5. German Committee on Reinforced Concrete *Guidelines for Protection and Repair of Concrete Components.* (1990) *Part 1: General Regulations and Planning Principles,* (1990) *Part 2: Planning and Execution of Works,* (1991) *Part 3: Quality Assurance in Execution of the Works,* (1992) *Part 4: Terms of Delivery and Standardized Principles for Directions on the Execution of Construction Works,* Beuth Verlag Berlin
6. Schießl, P., Breit, W. and Raupach, M. (1993) Investigations into the Effect of Coatings on Water Distribution in Concrete Using Multi-Ring-Electrodes. *International Symposium on the Condition Assessment, Protection, Repair and Rehabilitation of Concrete Bridges Exposed to Aggressive Environments,* Minneapolis, November 7-12
7. Schießl, P., Breit, W. and Raupach, M. (1994) Durability of Local Repair Measures on Concrete Structures Damaged by Reinforcement Corrosion. *Durability of Concrete*, Detroit: ACI SP-145, Third CANMET/ACI International Conference on Durability of Concrete, May, 22-28, 1994, Nice, pp. 1195-1215

96 INTERNAL PROTECTION OF CONCRETE PIPELINE FOR THE CONVEYANCE OF AGGRESSIVE WATER (CO_2)

O.A. ABUAZZA and A. A. IBRAHIM
Great Man-Made River Project, Libya

Abstract
Hundreds of thousands of prestressed concrete cylinder pipes (PCCP) are required for the Great Man-Made River Project to convey water from southern desert aquifers to the northern coast of Libya. The water from the wellfields is considered to be aggressive to the internal concrete surface of the pipes so it was decided to provide treatment to the water at a central water treatment facility and to provide an internal lining to the wellfield pipes upstream of this treatment facility. It is necessary for the lining to be durable for 50 years in line with the design life of the project and PCCP.

Four different coating systems were chosen and subjected to trials in Libya to demonstrate their suitability. This paper describes the laboratory tests which were performed on weathered and unweathered samples on the lining system adopted.
Keywords: Adhesion, carbonation, disinfectant, resistance, electrical continuity, hardness, permeability, polyurea, weathering.

1 Introduction

The Great Man-Made River Project (GMRP) was conceived in 1980 to exploit the vast aquifers underlying the southern deserts of Libya, which were identified during the exploration for oil in the 1960s; and convey the water to the populated coastal strip for the benefit of agriculture, industry and the urban population.

Early planning envisaged two major sub-systems of the project to supply the Benghazi and Tripoli urban and agricultural regions from eastern and western wellfields respectively. Each of these sub-systems was planned to be developed in Phases; Phase I of which comprises the Sarir-Sirt and Tazerbo-Benghazi (SS/TB) components of the eastern sub-system. This will later be extended as Phase III

Concrete Under Severe Conditions: Environment and loading (Volume Two) Edited by K. Sakai, N. Banthia and O.E. Gjørv. Published in 1995 by E & FN Spon. ISBN 0 419 19860 1

to provide additional water to Tobruk in the east and Tripoli in the west from a new wellfield near Kufra. Phase II is known as the Western Jamahiriya System (WJS) and will provide water to the Tripoli region from the Jebel Hasouna wellfields south of Tripoli.

Phase I is approaching completion with water already being supplied to Benghazi, Sirt and Ajdabiya municipalities and to agricultural developments between Benghazi and Sirt. Phase II at present is being concurrently designed and constructed, and Phase III is at the planning and preliminary design stage. The three phases conveyance pipe line net work, consisting of approximately 3,380 km of prestressed concrete cylinder pipe sized from 1.6m to 4m diameter and approximately 980 wells varying in depth from 450 to 750 meters, produce approximately 5.68 million m^3 of water per day. To prevent the upstream conveyance pipe line net work from the aggressive water, an experimental program was performed to investigate and find the most suitable coating or lining system.

2 Problems

The main degradation mechanisms for the PCCP can be summarized as follows:

(1) Acid attack due to free CO_2
(2) Sulphate attack of the concrete
(3) Erosion damage
(4) Microbiologically induced corrosion
(5) Corrosion damage to the embedded cylinder

From the above items the single most aggressive species in the water as far as the concrete core is concerned is the dissolved carbon dioxide. So it was decided to concentrate the study on the main degradation mechanism for the PCCP which is the acid attack due to free CO_2.

The acid produced by excess carbon dioxide in water reacts with alkalies in the cement paste and causes leaching of the cement binder with a consequential loss in both mechanical strength and section.

The hydration reaction for ordinary portland cement is complex but one of the main reactions can be represented :

$$2[(CaO)_3\ SiO_2] + 6H_2O \rightleftharpoons (CaO)_3(SiO_2)_2\ 3H_2O + 3Ca(OH)_2 \quad (1)$$

An equilibrium is established between unreacted tricalcium, hydrated silicate compounds and calcium hydroxide.

The carbonate/bicarbonate/carbon dioxide equilibrium system for the water can be represented:

$$Ca(HCO_3)_2 \rightleftharpoons H_2O + CO_2 + CaCO_3 \quad (2)$$

Increasing amount of carbon dioxide in the water increases the amount of calcium bicarbonate in the water. Conversely, decreasing the amount of carbon dioxide results in the formation of calcium carbonate.

The free carbon dioxide in the water is in excess of that required to form calcium carbonate. Some carbon dioxide in the water is necessary to maintain the above equilibrium and this concentration does not affect the calcium carbonate concentration. With increasing amounts of bicarbonate in solution the amount of carbon dioxide which can be present can also be increased. Only free dioxide present in excess of this equilibrium amount can be considered as agressive carbon dioxide.

When the calcium and carbon dioxide levels in the water reach a certain concentration, leaching of free lime from the concrete matrix will occur causing precipitation carbonate.

The level of carbon dioxide in the well water is such that, PCCP will not achieve its design life without water treatment to remove CO_2 and the lining of the PCCP upstream of such treatment. The protective lining must have extremely low permeability to dissolved CO_2 gas.

3 Experimental program and testing results (Polyurea coat)

Four different coating/lining systems were subjected to trials, the result of the trials demonstrate that 100% solid, two component polyurea coating/lining is the most suitable material which can be used to prevent the internal core of PCCP from CO_2 attack, the other three coating/lining systems can be listed as follows:

- 100% solid, two component polyethylene coat.
- polyethylene sheet liner system.
- P.V.C. sheet liner system.

Samples of coated/lined concrete were removed from the internal cores of the polyurea coated pipe and subjected to the tests which are listed in table 1 before and after subjected to accelerated weathering for a period of 250 and 500 hours, accelerated weathering samples are prepared by placing the core on the galley of an enclosed carbon arc weatherometer programmed to run in accordance with the cycle given in BS 3900, part F3 and is tabulated on Table 2. The results are presented and discussed as follows:

Table 1. Samples required for testing

Test number	Test	Type of sample required	Approximately size of sample required (mm)	Number of samples required for testing	
				Coated	Uncoated
1	Dry film thickness and Electrical continuity	Core	100 (diameter)	4	2
2	Coating hardness and adhesion	Panel	300 x 300	9	0
3	Shear test	Panel	50 x 40	8	0
4	Mandrel flexibility	Aluminum panel	100 x 50 x 0.3	8	0
5	Free film	Plastic sheet	200 x 15 x 1	15	0
6	Chloride permeability	Core	100 (diameter)	4	2
7	Oxygen permeability	Core	100 (diameter)	3	1
8	CO_2 Permeability	Core	100)diameter)	3	1
9	Water permeability	Core	100 (diameter)	4	2
10	Weathering	Core	50 x 50	4	0

Table 2. Samples accelerated on weathered core

Duration	UV	Water Spray	Extractor Fan
4 hours	On	On	Off
2 hours	On	Off	On
10 hours	On	On	Off
2 hours	On	Off	On
5 hours	On	On	Off
1 hour	On	Off	Off

3.1 Film thickness and electrical continuity of polyurea coat

The last two samples as presented on Table 3 are control samples. The dry film thickness was determined at a magnification of 32 times. 100 graticule unit is equal to 2.5mm. Thickness is variable due to some control mistakes during the spraying application.

The electrical resistance of the polyurea coat was acceptable because it is a dielectrical coat and it can cover all the small voids, airpockets which normally appear on the susbstrate. Also thickness of the coat is sufficient to get acceptable electrical resistance. Weathering does not make any different on the electrical resistance.

Table 3. Film thickness and electrical continuity of polyurea coat

Dry film thickness (D.F.T.) and electrical continuity (E.C.) of the polyurea coat					
Dry film thickness			Electrical continuity		
Test number	D.F.T. (Graticule measurement)	D.F.T. (mm)	Average specimen thickness (mm)	Tested area (cm^2)	Resistance reading of cell (ohm)
1 Unweathered	88	2.20	16.19	75.12	>20 M ohm
2 Unweathered	80	2.00	15.34	75.12	>20 M ohm
3 Weathered 500 hrs	70	1.75	15.58	75.12	>20 M ohm
4 Weathered 500 hrs	92	2.30	16.53	75.12	>20 M ohm
5	-	-	15.61	75.12	0.52 k ohm
6	-	-	14.92	75.12	0.71 k ohm

3.2 Coating hardness and adhesion

Weathered sealed and unsealed samples were subjected to accelerated weathering for 500 hours. Hardness was measured by Buchholz indentation. No indentation occured, confirming the hardness to be adequate.

There is also no evidence to show that hardness is affected by weathering as no increase in hardness occured.

Table 4. Coating hardness and adhesion

Coating hardness			Coating adhesion	
Sample number	Unweathered/ weathered	Indentation length	Load (kN)	Stress N/mm^2
1	Unweathered	No indentation	5.95	3.03
2	Unweathered	No indentation	4.37	2.23
3	Unweathered	No indentation	4.59	2.85
4	Weathered-sealed	No indentation	4.24	2.16
5	Weathered-sealed	No indentation	5.10	2.60
6	Weathered-sealed	No indentation	4.62	2.35
7	Weathered-sealed	No indentation	3.27	1.67
8	Weathered-sealed	No indentation	2.60	1.32
9	Weathered-sealed	No indentation	4.08	2.08

3.3 Shear strength

Non-weathered test results see Table 5, show that the coating to concrete bond strength was much greater than the cohesive strength of concrete, also the weathered sealed test results have not shown significant changes from the initial tests but the weathered unsealed test show a little bit decreasing of adhesion strength. But this result is still satisfactory and the bond strength is still greater than the cohesive strength of the substrate.

This bonding is due to using Epoxy primer to cure dry concrete before dry area application and not due to penetration into concrete.

Table 5. Shear strength

Sample number	Weathered/ unweathered	Load at Break (kN)	Tested Area (mm^2)	Shear Strength (N/mm^2)	Mode of failure
1	Unweathered	8.53	2×10^3	4.27	100% concrete
2	Unweathered	5.39	2×10^3	2.70	80% concrete 20% coating adhesion
3	Unweathered	8.53	2×10^3	4.27	90% concrete 10% coating adhesion
4	Unweathered	6.40	2×10^3	3.20	90% concrete 10% coating adhesion
5	Unweathered	1.72	2×10^3	0.86	75% concrete 25% coating adhesion
6	Weathered	3.60	2×10^3	1.80	60% concrete 40% coating adhesion
7	Weathered	2.57	2×10^3	1.29	70% concrete 30% coating adhesion
8	Weathered	2.05	2×10^3	1.03	60% concrete 40% coating adhesion

Unweathered samples were subjected to accelerated weathering for 500 hours. The weathered samples showed significant decrease in shear and bond strengths. The mode of failure is mostly in concrete. Result of adhesion and shear indicate that the coating can withstand the handling transportation.

3.4 Mandrel flexibility

The weathered samples were subjected to accelerated weathering for 500 hours. The above mandrel flexibility of the coated aluminum samples were determined in accordance with BS 3900 part H4: 1983, using normal corrected vision and 10 times magnification.

A description of the extent of any cracking and/or detachment of the coating from the substrate is reported for each test panel, against each mandrel size. If failure of this type did not occur, the test result is recorded as a 'pass'.

Unweathered panel showed crack like stretch marks but no actual cracking was observed in 10 X magnification. But the weathered (500 hours) panels displayed cracking with one preferential direction, along the bend.

Table 6. Mandrel flexibility tests and results

Sample number	Weathered/ Unweathered	Mandrel size	Thickness range coating (mm)	Degree of cracking	Detachment of coating from substrate	Test Results
1	Unweathered	3	1.80 - 2.02	none	none	pass
2	Unweathered	3	1.78 - 2.22	none	none	pass
3	Unweathered	8	1.79 - 2.10	none	none	pass
4	Unweathered	8	1.79 - 1.97	none	none	pass
5	Weathered	3	1.82 - 1.97	4(S1)9	none	fail
6	Weathered	3	1.78 - 1.96	4(S1)9	none	fail
7	Weathered	8	1.79 - 1.90	4(S1)9	none	fail
8	Weathered	8	1.78 - 1.95	4(S1)9	none	fail

3.5 Free film and tensile test

The tensile polythene sheets were removed from the steel panels with the aid of a sharp knife and the straight edge. The free film samples were then removed from the flexible polythene sheet using a double bladed precision cutter. The tensile stress/strain and elongation at break of the free film samples were then determined in accordance with ASTMD 2370-82. The resulting measurements indicate the polyurea coat flexibility decreased after weathering and the elongation reduced from 58% of unweathered to 37% after 500 hours weathering, see Table 7.

Table 7. Free film and tensile test results

Free film and tensile, stress/strain and elongation at break result					
Sample number	Hours Weathered	Average dry film thickness (mm)	Tensile strength (N/mm^2)	Elongation at break (%)	Modulus of electricity (N/mm^2)
1	0	1.670	6.32	55.0	2.32
2	0	1.881	6.11	78.0	1.77
3	0	1.806	6.06	38.0	2.66
4	0	1.714	6.03	64.0	2.88
5	0	1.762	5.77	56.0	2.65
6	250	1.957	6.34	77.0	2.25
7	250	1.790	5.96	46.0	0.89
8	250	1.816	5.91	21.0	0.81
9	250	1.816	5.63	78.5	1.62
10	250	1.934	5.22	37.0	1.65
11	500	2.187	5.90	25.0	3.17
12	500	1.924	5.60	25.0	3.17
13	500	1.707	5.58	55.0	2.34
14	500	2.034	5.53	47.5	2.36
15	500	2.004	5.52	36.0	2.86

3.6 Chloride-oxygen-carbon dioxide and water permeability results. The samples which were used to determine permeability of polyurea coat were removed from the coated and uncoated concrete panels with the aid of a 100 mm diamond coring tool. The uncoated core were used as control specimens. The pressure of all permeability tests was 10 bar.

Chloride permeability is negligible. These results must be considered as providing reassurance that the coating will be able to perform the fundamental barrier functions and would continue to do so provided they remain continuous and intact. No transmission within 30 days. Oxygen gas permeability coeffecient is presented in Table 8, water analysis of test solution and carbon dioxide permeability are presented in tables 9 & 10 respectively.

Table 8. Oxygen gas permeability

Sample number	Hours weathered	Average thickness (mm)	Test area	Oxygen gas permeability coeffecient (m^2)
1	0	45.1	75.12	9.21×10^{-22}
2	500	39.9	75.12	7.12×10^{-20}
3 (control)	0	36.2	75.12	2.02×10^{-16}

3rd sample is control uncoated unweathered specimen.

Table 9. Water analysis of test solution

Temperature (^oC)	23
$NaHCO_3$ (mg/l)	309
H_2SO_4 (mg/l)	50
$CaCl_2$ (mg/l)	50
Total dissolved solids (mg/l)	409
Measured pH	6.5
CCPP (as mg/l equiv. $CaCO_3$)	- 86.4
Hardness Ca as mg/l $CaCO_3$)	43.5
Free CO_2 (mg/l)	102.00

CCPP - Calcium carbonate precipitate potential

Table 10. Carbon dioxide permeability

Sample type	Unweathered		Weathered (500 hours)	
	Coated	Uncoated (control sample)	Coated	Coated
Sample no.	1	2	3	4
Sample thickness (mm)	0.050	0.039	0.044	0.045
Test area (m^2)	7.5×10^{-3} 7.	7.5×10^{-3}	7.5×10^{-3}	7.5×10^{-3}
Penetration (hrs)	n/a	0.75	n/a	n/a
Flow rate (m^3/sec.)	n/a	1.71×10^{-9}	n/a	n/a
Depth of carbonation (mm)	<0.1	<0.1	<0.1	<0.1

*depth of carbonation measured on dry sample and found to be minimum measurable depth.

Samples 1,3 & 4 after 21 days in the carbon dioxide permeability ring no water appeared on the surface of the coated sample. The measured depth of carbonation was found to be <0.1 mm. No change in the coating was observed. No solution penetrated the coating.

But water appeared on the surface of the uncoated sample after 45 minutes in the carbon dioxide permeability ring. The measured depth of carbonation was found to be <0.1 mm.

Water permeability after 14 days in the water permeability apparatus no water appeared on the surface of the unweathered and weathered coated samples. No cracks were detected when the coated samples were investigated under a microscope. Water appeared within 30 minutes on the surface of uncoated samples.

The results are presented in Table 11, weathered samples results are exactly the same as unweathered one so they are not presented.

Table 11. Water permeability results

Sample type	Coated (unweathered)		Uncoated (control samples)	
Sample number	1	2	3	4
Coefficient of permeability by flow (m/s)	n/a	n/a	6.79×10^{-11}	6.18×10^{-11}
Coeffecient of permeability by penetration (m/s)	n/a	n/a	3.58×10^{-10}	2.27×10^{-10}

3.8 Weathering test

The coated samples were subjected to accelerated weathering for 500 hours then the coating is inspected and the results as follows:

Other than a change in colour and a loss of gloss from the surface of the weathered samples, there were no visible differences or deterioration. The change in colour and loss of gloss to the coating is rated as 5 (severe i.e. intense change) in accordance with BS3900: H, 1983. There was no blistering, cracking, chalking or flaking of the sample according to the schemes given in BS3900: 1983.

4 Findings

From the experimental program and site trial applications and laboratory test results on polyurea coat the following findings could be drawn:

1. Thickness was sufficient to cover all surface defects and to get good electrical resistivity, but the thickness was very variable, this variation was due to mistakes of application.
2. The adhesion of coating with concrete was sufficient to withstand a back pressure of 10 bar during all permeabilities test, and the bond strength was always greater than the cohesive strength of substrate. This is due to using of epoxy primer.
3. The shear strength is satisfactory, even in weathered 500 hours tests.
4. The hardness and impact resistance were satisfactory.
5. The coating was an effective barrier to the passage of dissolved carbon dioxide, chloride ions, to attack by acidic water.
6. Weathering of 500 hours (representing 7 months in Libya) does not make a significant difference in the above mentioned properties.

7. It is indicated that the coat remains less flexible after weathering, a loss in flexibility was shown in a flexibility mandrel test. Also the elongation reduces from 58% of unweathered to 37% after 500 hours of weathering.
8. 58% of elongation may not be sufficient to stand pipe movement without allowing leakage behind the joint seal between every two pipes.

5 Conclusion and recommendation

From the above findings the conclusion that can be drawn is that the polyurea coating satisfies the acceptance criteria for internal lining application of PCCP to prevent acid attack due to free carbon dioxide. Further tests have been recommended on the use of this material at joint positions because of its lack of flexibility in this formulation.

6 References

1. Management & Implementation Authority of the Great Man-Made River Project (1992) *River of Life*, the Revolution Printing and Publishing House, Benghazi, Libya.
2. Englinton, M.S. (1975) *Review of Concrete Behaviour in Acidic Soils and Groundwaters*, Ciria Publication 69.
3. Werner, Giertz-Hedstrom, D. (1934) *Physical and Chemical Properties of Cement and Concrete*, The Engineer, pp. 235-1182.
4. Flentje, M.E. and Sweitzer, R.J. (1955) *Solution Effects of Water on Cement and Concrete in Pipe*, Journ AWWA, pp. 1173-1182.

97 TESTING CONCRETE COATING RESISTANCE TO CO_2 PERMEABILITY

C.S. KAZMIERCZAK
UNISINOS, São Leopoldo, Brazil
P.R. HELENE
EPUSP, São Paulo, Brazil

Abstract
Concrete carbonation in reinforced concrete leads to depassivation of the reinforcing bars, and therefore to initiation of corrosion. This deterioration, caused by the penetration of carbon dioxide, can be prevented by the use of coatings.

In this paper, a testing methodology using accelerated carbonation to evaluate coating efficiency is presented. The study was carried out on pozzolanic cement mortar specimens through analysis of their characteristics and the influence of high CO_2 concentration on the accelerated tests for evaluation of anti-carbonation coating efficiency. Microscopy and chemical analysis were used.
Keywords: Accelerated tests, carbonation, coating performance, concrete protection, CO_2 permeability.

1 Introduction

In the last years, serious pathological evidences in reinforced concrete structures have been observed, and the most frequent one was the corrosion of the reinforcing bars caused by the action of aggressive atmospheres and harmful gases and salts.

Reinforced concrete offers two kinds of protection to the reinforcing bars: physical, through the concrete cover, and chemical, through a passive oxide film formed on the steel surface and the high alkalinity (pH ≥ 13) of the environment.

When the concrete structure is exposed to the atmosphere, the CO_2 penetrates the interior of the concrete through the pore network formed during cement hydration and reacts with the alkaline interstitial fluid, reducing the pH of the concrete and destroying the oxide film. The main reaction occurs with the carbon dioxide (Equation 1).

Concrete Under Severe Conditions: Environment and loading (Volume Two) Edited by K. Sakai, N. Banthia and O.E. Gjørv. Published in 1995 by E & FN Spon. ISBN 0 419 19860 1

$$Ca(OH)_2 + CO_2 = CaCO_3 + H_2O \quad (1)$$

The main resources used in the maintenance of the integrity of passive oxide film are low porosity concretes, switable concrete covers and surface protection systems (coatings).

The utilization of coatings with the purpose of protecting concrete structures is a recent technique, and so far no recognized testing methodology to evaluate the efficiency of coatings to CO_2 penetration exists. A method used by chemical industries to determine coating permeability to carbon dioxide consists of determining the coating CO_2 diffusion coefficient and specifying a cover thickness with a higher performance than the one obtained by a 125 mm concrete layer possessing 30 MPa compression strength, according to Kopler's method (or the same as a $5 g/m^2/day$ flow, according to Leeming [1]). Several authors [2], [3], [4] show comparative results for coating performance based on this principle.

However, there are some limitations when these results are transposed to practice. The application of a coat in the concrete requires previous surface preparation, with the application of primer and of one or two coatings. The substratum characteristics (such as surface roughness and porosity) and the quality of the application will influence the performance of the coating, which should be continuous and applied in the thickness specified by the manufacturer.

Considering these factors, recent research has sought to analyse the properties of coatings when applied on concrete or mortar specimens in order to obtain results closer to reality. The efficiency of the protection system tends to be determined by accelerated carbonation tests, indicating carbonation depth in the concrete or mortar specimens on which the coating is applied.

Andrade [5] proposes a methodology for accelerated carbonation tests in which the coating is applied on specimens made from cement and sand mortar. The proposed method consists in determining the increase on weight caused by carbonation in mortar specimens exposed to high CO_2 contents, and monitoring the depassivation in "corrosion sensors" inside the specimens. In further research [6] [7], the efficacy of tests on specimens exposed to several CO_2 concentrations was analysed, with the conclusion that the accelerated carbonation tests (in atmospheres with CO_2 concentrations up to 100%) proved to be fast and adequate for classifying coating protection capacity against carbon dioxide. Also in Building Research Establishment - BRE [8], a series of tests was developed using cement, lime and sand specimens exposed to a 15% CO_2 atmosphere. In these tests, carbonation was determined by the weight increase and carbonation depth was determined by aspersion with phenolphthalein.

Research has emphasized the importance of tests being carried out with the coating applied on concrete or mortar.

2 Methodology

The methodology proposed here consists of applying coatings on standardized mortar specimens and then determining the relationship between carbonation observed in accelerated tests and carbonation in specimens submitted to natural exposure in a laboratory with constant temperature and relative humidity.

With the purpose of obtaining substratum characteristics similar to those of reality, specimens were made using mortar in the same proportions of cement/sand and water/cement ratio as specified for a conventional concrete. For a compression strength of 26 MPa (proportion of mixture: 1 : 2.2 : 2.9 (cement : sand : aggregate) and water/cement ratio 0.6), a mortar of 1 : 2.2 (w/c = 0.6) was used. Portland Pozzolanic cement (fly ash content 30%) and standardized Brazilian sand were used.

The specimens measure 55 x 80 x 20 mm, and only the two rectangular faces of 55 x 80 mm are considered in the test. The moulds corresponding to these faces are made of resin-coated plywood in an attempt to reproduce the characteristics of the concrete cover in the specimens. The specimens dimensions are small because of the carbonation measuring system adopted (controlled by weight increase).

The specimens are removed from their moulds after 24 hours, and remain in submerse cure for 7 days, and then are kept in a CO_2-free environment at the temperature of 20±2°C until they are tested.

Due to the great variability of the results when using specimens presenting initial carbonation, a curing chamber was developed to allow specimen curing and stabilization of internal humidity in a CO_2-free environment. The air which circulates in this chamber is constantly filtered through a KOH solution that retains the CO_2.

The carbonating chamber (Fig. 1) has a small internal volume to help control relative humidity and CO_2 content.

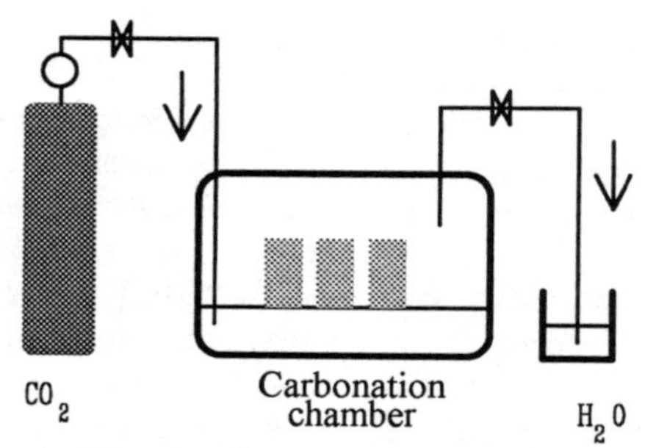

Fig. 1. Chamber for accelerated carbonation

The chamber allows the accomplishment of accelerated tests by keeping relative humidity between 95% and 100% in a 100% CO_2 atmosphere (this rate has been chosen in order to allow the maintenance of relative humidity during the experiment, when water vapour is released from the specimens). Specimen weight increases are measured periodically. At the end of the test, the increase in weight is related with the carbonation depth.

With the purpose of comparing the performance of several substrata, accelerated tests were carried out on non-painted specimens, mixed with three different compression strengths (w/c = 0.6 - reference; w/c = 0.4; w/c = 0.8) in a 100% CO_2 atmosphere. The mix proportions had the same mortar/concrete ratio (0.52) and the same proportion of water to aggregates (9.98%).

Figure 2 shows the results of these tests.

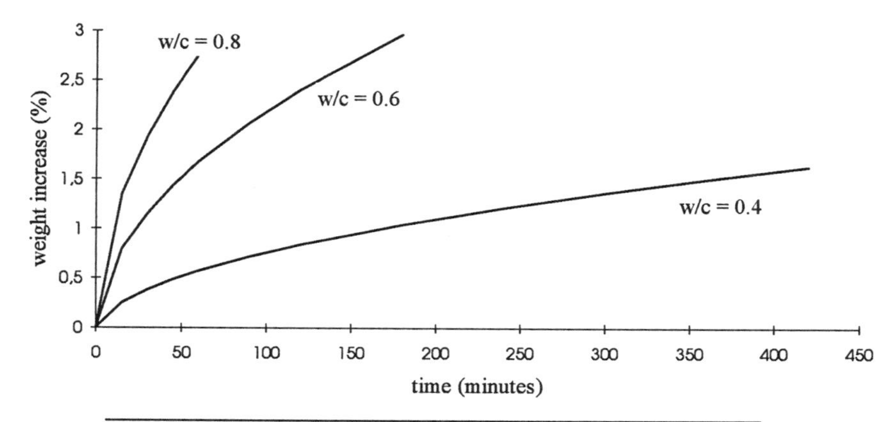

Mix proportion	w/c	Regression	r
1 : 1.10	0.4	w% = – 0.0743 + 0.0833 * t ½	0.97
1 : 2.17	0.6	w% = – 0.0056 + 0.2459 * t ½	0.97
1 : 3.16	0.8	w% = – 0.0705 + 0.3661 * t ½	0.95

w% = percentage of weight change
t = test duration (in minutes)

Fig. 2. Accelerated tests with non-painted specimens of different w/c ratios

Equation 2 relates the length of exposure (in the laboratory, at a temperature of 20±2°C and relative humidity between 70 and 80%) to the carbonation depth observed in the reference samples (w/c=0.60), after an eighteen-month test. These specimens were not painted, and carbonation depth was measured with phenolphthalein.

$$p = -2.692 + 0.5357 * t^{1/2} \quad (2)$$

p = carbonation depth (in mm)
t = length of test (in days)

Equation 3 presents the correlation between the weight increase observed in the non-painted specimens and the carbonation depth found by phenolphthalein aspersion in specimens prepared under the same conditions, but subjected to natural carbonation, in laboratory (valid to $p \geq 2$ mm).

$$p = 2.034\, w\% + 0.821 \quad (3)$$

w% = percentage of weight change
p = carbonation depth (in mm)

The comparison between the carbonation in real time and in accelerated tests was made using Equations 2 and 3 and those presented in Fig. 2. The use of phenolphthalein was validated through the analysis of carbonated specimens with a polarized light microscope. In the specimens exposed to natural carbonation, the carbonation front exhibited a transition zone between the carbonated and the non-carbonated area, which sometimes reached 2.5 mm thickness. In the specimens exposed to accelerated carbonation, the highest thickness observed was 1.5 mm. In both cases, the depth observed by the phenolphthalein color change corresponded to the beginning of this transition zone.

It was also observed that, due to the high rate of the accelerated tests, the indication of carbonation depth with phenolphthalein in non-painted specimens presented a greater dispersion in the results during the first hour of testing.

The evaluation tests on the permeability of coatings to CO_2 were carried out with four types of resins, as specified in Table 1. At least ten samples were tested for each system. The test length was thirty hours, with six specimens being tested at a time. The specimens were tested in sequences of aleatory groups.

Table 1. Coatings tested using the proposed methodology

Nature of Coating	Coat	Number of Coats
Acrylic emulsion	AE1	primer + 2 coats
	AE2	2 coats
	AE3	2 coats
Acrylic dispersion	AD1	primer + 2 coats
	AD2	2 coats
	AD3	2 coats
Methylmethacrylate dispersion	MM	1 coat
Polyurethane (1 component)	PM	2 coats
Polyurethane (2 components)	PB	1 coat
Silane/Siloxane + Acrylic dispersion	SSA	1 coat of silane/siloxane + 1 coat of acrylic dispersion

The experiment was developed as follows: (1) Measurement of weight increase during the experiment and a final verification of carbonation depth after thirty hours; (2) Calculation of the equation corresponding to the change in weight of each specimen. In this analysis, weight increase values over 3% obtained in the experiment were excluded, because at greater values the two carbonation fronts became too close, causing interference in each other; (3) Estimation of the equation corresponding to each type of coating. In this analysis, a prediction limit of 90% and a 5% significance level (α) were considered as confidence limits. The specimens presenting values in excess of these limits were excluded; (4) Analysis of weight increase and carbonation depth values obtained for each painting system; (5) Classification of each system formed by concrete and coating according to their performance.

3 Results and discussion

Table 2 presents the equations corresponding to the weight increases obtained in the systems tested.

Table 2. Weight increases of protection systems tested

Coat	Estimated weight increase (w%)	Correlation coefficient (r)
AE1	$-0.455 + 0.1088 * t^{1/2}$	0.93
AE2	$-0.517 + 0.1139 * t^{1/2}$	0.96
AE3	$-0.448 + 0.1108 * t^{1/2}$	0.97
AD1	$-0.449 + 0.0878 * t^{1/2}$	0.97
AD2	$-0.504 + 0.2210 * t^{1/2}$	0.96
AD3	$-0.314 + 0.0800 * t^{1/2}$	0.93
MM	$-0.458 + 0.1412 * t^{1/2}$	0.98
PM	$-0.089 + 0.0264 * t^{1/2}$	0.97
PB	$-0.481 + 0.1323 * t^{1/2}$	0.99
SSA	$-0.624 + 0.1701 * t^{1/2}$	0.99

The equations obtained correspond to the performance of each system formed by concrete and coating for protection against carbonation, and are then used for comparison between their efficiencies, as further explained.

Considering that the carbonation depth in a structure should not reach the reinforcement, and that a usual covering in Brazil corresponds to a 20 mm thickness, an evaluation of the performance of the various systems is proposed, aiming to fulfill this requirement.

From the results presented in Figure 2, it can be noticed that the reference specimens (non-painted) do not offer an efficient protection for a service life estimated at fifty years. For this age, the carbonation depth is expected to be 68.4 mm. Considering the interest in ensuring structure durability, the depths reached by the various systems were estimated. The same procedure was adopted for a ten-year-service life.

The decrease in coating performance that occurs naturally due to the action of the environment and deficient maintenance was not considered for estimation of carbonation depth (it is assumed that periodic repainting and good maintenance procedures are performed throughout the concrete service life). The systems that exhibit carbonation depths lower than 20 mm at the age of fifty years are classified as high performance systems; those presenting carbonation depths over 20 mm at the age of ten years are considered as low performance ones; and the others are considered as medium performance.

Figures 3 and 4 show the estimation of carbonation depths for the various systems at these reference ages.

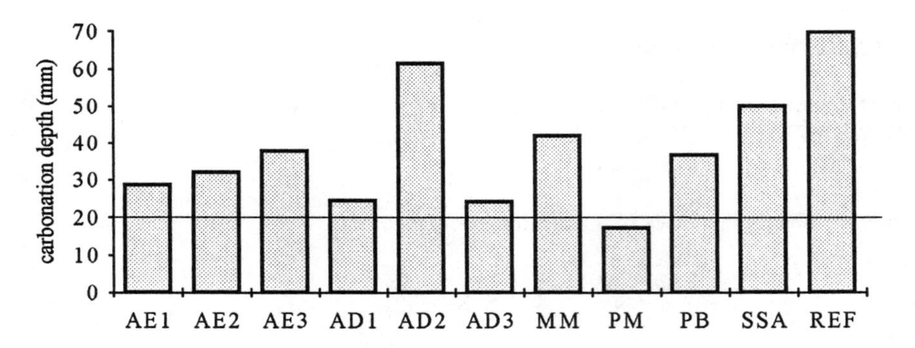

Fig. 3. Estimation of carbonation depth after 50 years of exposure at the temperature of 20±2°C and relative humidity between 70 and 80% (18.934 minutes of accelerated test)

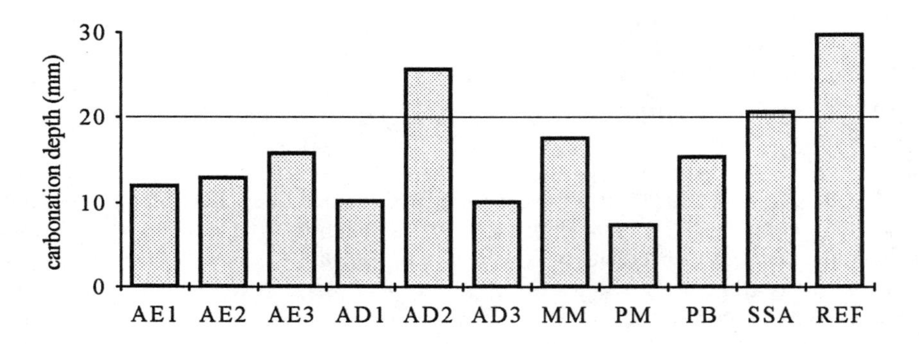

Fig. 4. Estimation of carbonation depth after 10 years of exposure at a temperature of 20±2°C and relative humidity between 70 and 80% (3.323 minutes of accelerated test)

According to the criterion proposed, coating PM was considered efficient (high performance coating), and the coatings AD2 and SSA did not offer enough protection to the concrete, being classified as low performance coatings.

By analyzing the results obtained, a clear distinction could be observed in the performance of the several types of coatings, where some kinds of painting almost did not offer any protection to the concrete (such as the AD2 coating).

The influence of coating thickness is evidenced in the case of the PB coating, in which an excellent performance was expected because of the kind of the resin used (polyurethane). It was not achieved probably due to the application of only one coat. Another example is the AE1 coating, applied in three coats, which showed a superior performance in relation to the other emulsion acrylic resisns.

A tendency for better performance was observed in dispersion coatings in comparison with emulsion coatings, just as expected.

4 Considerations on the methodology

The methodology presented proposes the use of specimens with surface roughness, porosity and physic-chemical characteristics similar to those of the concrete on which the painting will be applied. As this is a very accelerated test, it is possible to determine the efficiency of the systems tested in a very short time.

Monitoring weight increase is a simple and efficient technique, and has an excellent correlation to the results observed in natural carbonation. A proper number of specimens of each type of coating is necessary for regression estimation, due to the dispersion found in results.

The painting procedure results in different coating thicknesses, which will result in different carbonation depths in the same specimen, sometimes detectable with phenolphthalein. This confirms that the coatings must be evaluated when applied on the substratum in which they will be used.

Despite not taking into account the effects of coating degeneration over time and not considering that the carbonation rate in concrete structures is lower than the one observed in the reference specimens due to the humidity variations to which concrete is exposed, this methodology allows an initial evaluation of the behavior of the system formed by concrete and coating to a satisfactory level.

5 Acknowledgment

The authors wish to thank *FAPERGS - Fundação de Amparo à Pesquisa do Estado do Rio Grande do Sul* and *FAPESP* for providing financial support for this research.

6 References

1. Leeming, M. (1990) Surface treatments for the protection of concrete, in *Protection of Concrete*, (ed. R.K. Dhir and J.W. Green), E & FN Spon, Dundee, pp. 135-148.
2. Browne, R.D. and Robery, P.C. (1987) Practical Experience in the testing of surface coatings for reinforced concrete, in *Fourth International Conference on Durability of Building Materials & Components*, Vol.1, Singapore, pp. 325-333.
3. Lindberg, B. (1987) Protection of concrete against aggressive atmospheric deterioration by use of surface treatment (painting), in *Fourth International Conference on Durability of Building Materials & Components*, Vol.1, Singapore, pp. 309-316.
4. McGill, L.P. and Humpage, M. (1990) Prolonging the life of reinforced concrete structures by surface treatment, in *Protection of Concrete*, (ed. R.K. Dhir and J.W.Green), E & FN Spon, Dundee, pp. 191-200.
5. Andrade, C. and others (1988) Accelerated testing methodology for evaluating carbonation and chloride resistance of concrete coatings, in *FIP Symposium*, Jerusalem, pp. 61-67.
6. Garcia, A.M., Alonso, C. and Andrade, C. (1990) Evaluation of the resistance of concrete coatings against carbonation and water penetration, in

Protection of Concrete, (ed. R.K. Dhir and J.W. Green), E & FN Spon, Dundee, pp. 233-243.

7. Guillot, A.M.G., Andrade, C. and Alonso, C. (1990) Metodologia de ensayo evaluadora de la capacidad protectora de pinturas para hormigón frente a la carbonatación, in *Hormigón y Acero*, No. 199. pp. 135-139.
8. Wei, W., Rothwell G.W. and Davies, H. (1990) Investigation of the gas and vapor resistance of surface coatings on concrete and effects of weathering on their carbonation protective performance, in *Corrosion of reinforcement in concrete,* (ed. C.L. Page and others), Elsevier, Essex, pp. 409-419.
9. Sergi,G., Lattey S.E. and Page, C.L. (1990) Influence of surface treatments on corrosion rates of steel in carbonated concrete, in *Corrosion of reinforcement in concrete,* (ed. C.L. Page and others), Elsevier, Essex, pp. 409-419.

PART SEVENTEEN
MIX DESIGN

98 VISCOSITY AGENT AND MINERAL ADMIXTURES FOR HIGHLY FLUIDIZED CONCRETE

M. YURUGI and G. SAKAI
Kajima Technical Research Institute, Tokyo, Japan
N. SAKATA
Kajima Corporation, Tokyo, Japan

Abstract
In using highly fluidized concrete that does not require to be vibrated and consolidates under its own weight at sites, the variation of flowability has a significant influence on the quality of the structures because such concrete can be charged into the formwork without mechanical vibration. In addition, the flowability of the concrete is markedly influenced by the type of powdered materials because a large amount of powdered materials is used for maintaining the viscosity of the concrete.

The usage of a polysaccharide-based viscosity agent as one method to restrict the variation in flowability and the effects of powdered materials on the flowability of the mortar were investigated. From these studies, it was found that addition of polysaccharide-based viscosity agent could reduce the variation of flowability and that flowability of the mortar was extremely improved by replacing a portion of the cement with stone dust, fly ash or blast-furnace slag.
Keywords: Blast-furnace slag, flowability, fly ash, highly fluidized concrete, stone dust, variation, viscosity agent

1 Introduction

In recent years, there has been increased construction of large concrete structures such as prestressed concrete bridges with long spans or high rise offshore towers for oil excavation platforms and so on, and many new types of concrete have been developed in accordance with these needs.

Highly fluidized concrete that would not require to be vibrated but would consolidate under its own weight is one of the new types of concrete. The concept of such a fluidized concrete was proposed several years ago in Japan[1], and many research studies on both fundamental characteristics and practical applications for construction have

Concrete Under Severe Conditions: Environment and loading (Volume Two) Edited by K. Sakai, N. Banthia and O.E. Gjørv. Published in 1995 by E & FN Spon. ISBN 0 419 19860 1

been carried out extensively. From the results of this research, it has been shown that a relatively high dosage of high range water reducing agent (HRWR), 10 – 15 l/m^3, and a large amount of finely powdered cementitious materials were used to obtain high flowability and high segregation resistance of the concrete. However, in manufacturing the concrete, the use of HRWR at high dosages sometimes resulted in large variations of the flowability due to the so–called cement–HRWR incompatibility[2]. Variation of flowability of high fluidized concrete has a significant influence on the quality of the structures because such concrete can be sometimes charged into the formwork without consolidation using vibrators. In addition, the flowability of the concrete is markedly influenced by the type of powdered materials used for maintaining the viscosity of the concrete.

The authors are carrying out investigations on the usage of a polysaccharide–based viscosity agent to reduce the variation in flowability and also on the effects of powdered materials on the flowability of mortars with the viscosity agent. The results of the experimental study are reported here.

2 Polysaccharide–based viscosity agent

The polysaccharide–based viscosity agent is a natural water–soluble gum produced by carefully controlled aerobic fermentation processes. Its structure is shown in Fig.1.

Rheological properties of the alkali solution with 1 % polysaccharide were measured using a rheometer. The results are shown in Fig.2 together with the results using three reference viscosity agents (methyl cellulose derivatives, hydroxyethle based derivatives and acrylic polymers) which have already been used as concrete materials. Compared to the reference agents, the polysaccacharide solution showed a larger yielding stress and a smaller viscosity coefficient at high strain rates. This means that the concrete with polysaccharide possessed higher deformability of the fresh concrete in a flowing state and poorer deformability in a state of rest. This is very favorable for making a flowing concrete with high segregation resistance.

Investigation was also carried out on the suspension stabilizing characteristics of pastes with these viscosity agents. Stone dust that has the same order of fineness as

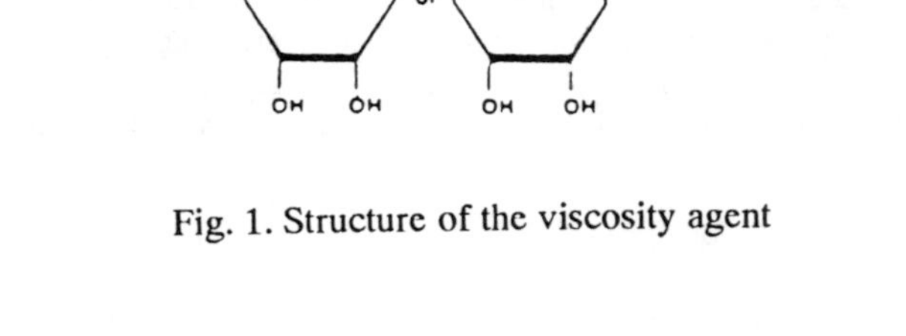

Fig. 1. Structure of the viscosity agent

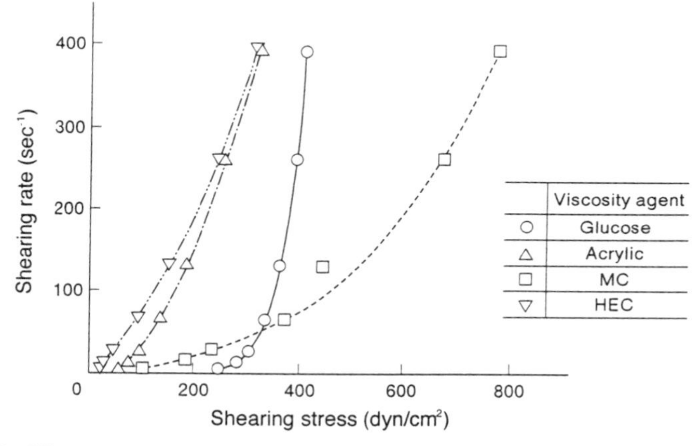

Fig.2. Effect of the viscosity agent on the rheological properties of the solution

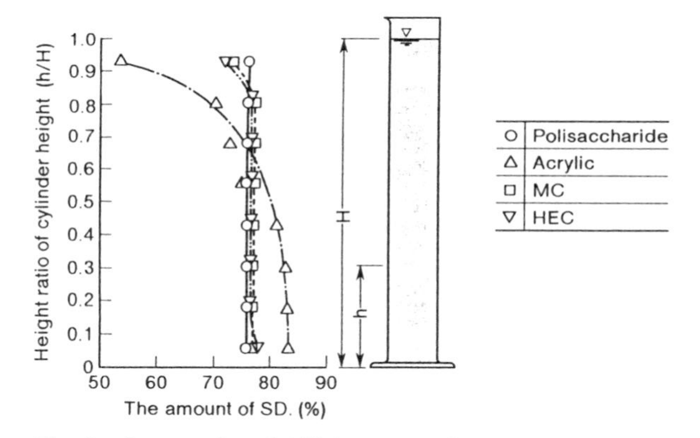

Fig. 3. Suspension stabilizing property

ordinary cement was used for the pastes. A water to stone dust ratio of 30%, HRWR of 0.5% of stone dust and viscosity agent of 1 % of water were used for the tests. A fixed quantity of each constituent was vigorously mixed. The suspensions were transferred and stored in 300 ml cylinders for 3 hours. Small quantities of the suspensions were then pipeted off from the same portion at different heights of each cylinder, and the amount of stone dust in each suspension was measured by drying out.

When using the polysaccharide-based viscosity agent, almost the same amount of stone dust particles were obtained in each height of the cylinder as shown in Fig.3. On the other hand, some distribution of the stone dust amount was observed for other viscosity agents.

Although these results were obtained from a limited number of tests using alkaline solution, the polysaccharide-based viscosity agent showed the most suitable properties for self-consolidating fluidized concrete.

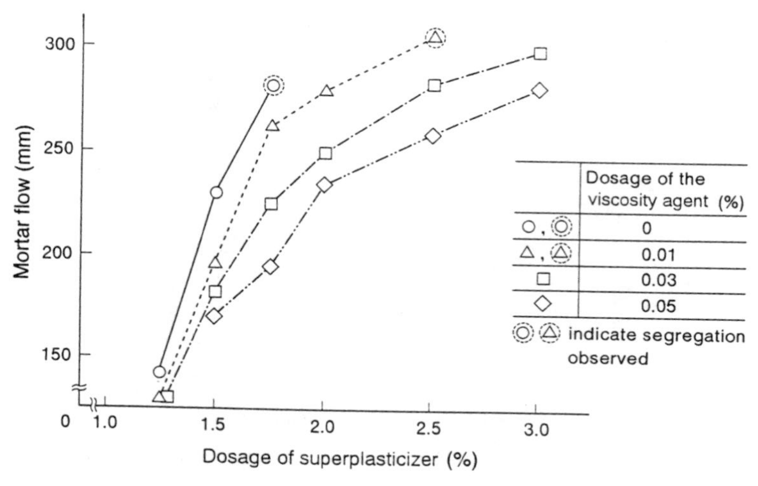

Fig.4. Effect of viscosity agent on mortar flow

3 Flowability of mortars

The use of the viscosity agent tended to make the mortar flowability less sensitive to the addition of HRWR. Using mortars, the effect of the addition of small quantities of the polysaccharide – based viscosity agent together with a high dosage of HRWR on the flow of mortar was examined. The water to powdered material ratio by volume (W/Pd) was 95 %, sand to paste ratio by volume (S/Pt) was 75% and the proportion of stone dust in total powdered materials was 43 %.

Flowability of mortars was tested by the flow table method described in the ASTM C109. Flowability of mortars was so high that the diameter of the mortar flow was measured immediately after removing the mold without any dropping of the table. Noticeable bleeding was also examined.

The results shown in Fig.4 indicate the followings : (1) the use of polysaccharide-based viscosity agent made mortar flow stable even with a high dosage of HRWR and (2) although bleeding was observed in mortar having a high flow value (> 270mm) without a viscosity agent, no bleeding was observed when the dosage of the viscosity agent exceeded 0.03 %.

4. Flowability of concrete

4.1 Outline of experiments

Three series of experiments were carried out to investigate the effect of concrete temperature, cement quality and grading of fine aggregate on flowability and determine whether the variation in the flowability can be controlled by using a polysaccharide-based viscosity agent. As shown in Table 1, experiments in Series I were carried out at concrete temperatures of 10, 20 and 30 °C. In Series II, tests were carried out using six sets of ordinary portland cement having almost the same fineness and manu–

Table 1 Experimental program for mortar flow

Series	Mix	VA*	Variable	Level
I	1	Used	Concrete Temperature	10, 20, 30 ℃
	2	Not used		
II	1	Used	Cement**	A (3,300)
				B (3,250)
				C (3,420)
	2	Not used		D (3,180)
				E (3,310)
				F (3,250)
III	1	Used	Gradation of sand	FM=2.08, 2.43, 3.06
	2	Not used		

* Viscosity agent

** Obtained from 6 Companies, Blaine finess value given in ( )

Table 2. Mix proportion of concrete

Mix No.	W/C (%)	s/a (%)	Slump flow (cm)	Air (%)	Unit content (Kg/m^3) W	C	SD	S	G	HRWR[1] (%)	V.A[2] (Kg/m^3)
1	53.0	45.1	65	4	175	331	216	703	861	2.5	0.35
2	53.0	45.1	65	4	175	331	216	703	861	1.8	—

1 Calculated on the basis of (C+SD) content
β-Naphthalene sulfonate type HRWR

2 Polysaccharide-based viscosity agent

factured by six different companies. Furthermore, assuming that variation in sand grading in an actual concrete plant, three types of sand of different fineness modulus were used in Series III.

The experimental program is summarized in Table 1, and the mix proportion of concrete used is shown in Table 2. As powdered materials ordinary portland cement and limestone dust were used.

The proportioning was carried out so as to obtain a highly fluidized concrete with a slump flow of 65 cm at 20 °C using cement A and sand of FM 2.43.

A forced mixer with dual axis was used, and crushed coarse aggregate, fine aggregate, cement, stone dust and water with admixtures were charged into the mixer sequentially and mixed for 120 sec. Slump flow tests were carried out to evaluate the flowability of the concrete. Also, V–shaped funnel tests were carried out[3]. In this test, the time required for all of the concrete placed into the funnel to flow down through the opening was measured; this V funnel value is related to the viscosity of the concrete.

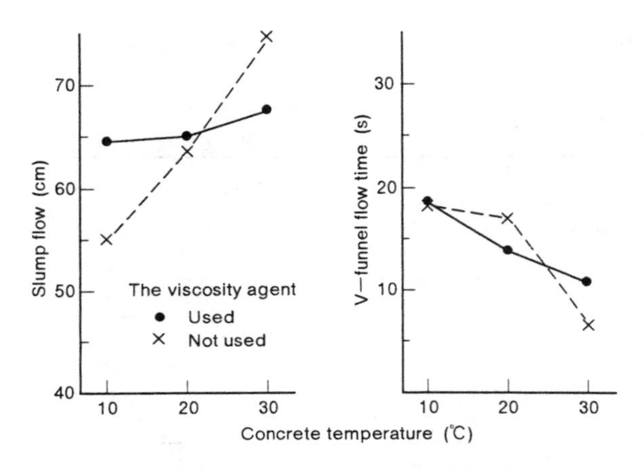

Fig. 5. Effect of temperature on flowability

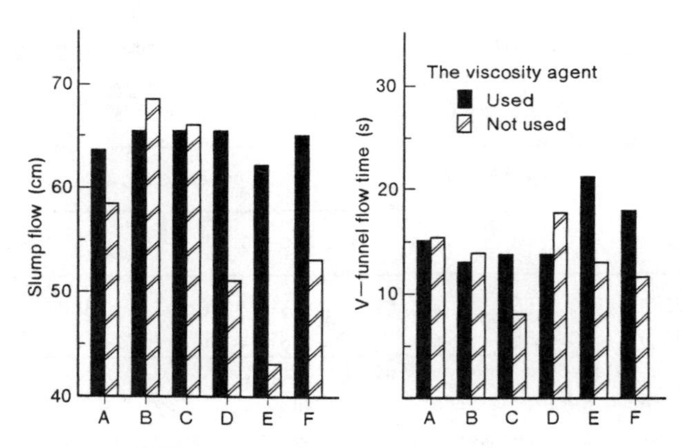

Fig. 6. Effect of cement quality on flowability

4.2 Experimental results and discussion

Figure 5 shows the results of experiments in Series I, where the concrete temperature was varied in the range of 10 to 30 °C. Regardless of the presence or absence of the viscosity agent, the higher the temperature, the larger was the slump flow and the smaller was the V funnel value. This is explained by the fact that the high dispersing action of HRWR is observed at a higher temperature. The results plotted in Fig.5 show that the variation in the slump flow with temperature was much smaller for mix 1 with the viscosity agent relative to the mix without the agent. For temperatures from 10 to 30 °C, slump flows were almost unchanged when using the viscosity agent. On the other hand, a large increase of 55 to 75 cm in slump flow was observed without the agent.

The results of experiments in Series II, using cements from six companies, are shown in Fig.6. It can be seen that even though the fineness of the cement was almost

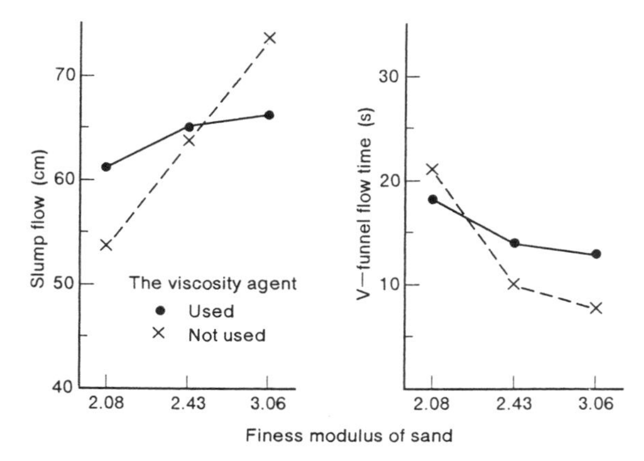

Fig. 7. Effect of sand gradation on flowability

identical, there was a large variation in the slump flow of concrete when the viscosity agent was not used. This can be attributed to differences in the particle size distribution of cements, HRWR absorption to cement particles and chemical composition of cements. On the other hand, the addition of the viscosity agent resulted in stable flows in the range of 62 to 66 cm for all cements.

Figure 7 shows the results in experiments in Series III, in which the main parameter was variation of sand grading. The coarser the gradation of sand, the larger was the slump flow and the smaller was the V–funnel value. Both the variation of slump flow and V–funnel value were much smaller for the concrete with the viscosity agent relative to the concrete without it. For large scale concreting, it is desirable that the variation in the sand grade be controlled within the range of +/– 0.2 of FM[4]. When the variation of FM is maintained in this range, the variation of slump flow would be restricted to a range of 62 to 66cm with use of the viscosity agent. In contrast, a wide variation of slump flow from 60 to 70cm would occur when using no viscosity agent.

It was found that the flow properties of fresh, highly fluidized concrete are extremely sensitive to change in concrete temperature, quality of cement and grading of sand, and the addition of a polysaccharide– based viscosity agent could reduce the degree of sensitivity.

5 Effects of finely powdered materials

In proportioning highly fluidized concrete, a large amount of finely powdered materials is apt to be used to maintain high segregation resistance. As a result, the powdered materials would markedly affect the flowability of the concrete. Experiments were carried out using different kinds of cements and mineral admixtures to investigate the effects of powdered materials on the flowability of concrete.

5.1 Outline of experiments

Table 3 Experimental program

Composition of powdered materials	OPC, OPC/SD, OPC/FA, OPC/BFS RHPC, RHPC/SD			
W/Pd (Volume %)	90	95	100	105
W/C (Weight %)	28.5	30.1	31.6	33.2
	40.0	40.0	40.0	40.0
	50.0	50.0	50.0	50.0
	60.0	60.0	60.0	60.0

Table 4 Physical properties of powdered materials

Type of Cement	Specific gravity	Particle* size (μm)	Fineness** (m²/g)
OPC	3.16	11.8	0.82
RHPC	3.19	7.8	1.17
SD	2.71	16.8	0.85
FA	2.22	15.0	1.27
BFS	2.90	9.7	1.24

* ; Average

** ; Measured by BET method

The experimental program is summarized in Table 3. Limestone dust(SD) conforming to JIS A 5008, fly ash(FA) conforming to JIS A 6201 and granulated blast–furnace slag(BFS) conforming to JSCE–1986 were used together with Type I ordinary portland cement (OPC). In the case of using Type III rapid–hardening portland cement(RHPC), limestone dust was only added as a powdered material.

As shown in Table 3, water to powered material ratio by volume was in the range of 90 to 105 %, and water to cement ratio by weight was varied from 30 to 60 %. Volume fraction of sand to paste ratio was fixed at 75 %. The dosage of HRWR was 2.5 % of powdered materials, and the amount of polysaccharide–based viscosity agent used was 0.05 % of water for all mixes. The flow table test previously mentioned and the J_{14} cone test, described in KODAN 304, were carried out immediately after mixing the mortar. In the J_{14} cone test, the time required for all the mortar poured into the 640ml cone to flow out the opening was measured. The characteristics of the materials used in the test are shown in Table 4.

5.2 Experimental results and discussion

Figures 8 to 10 show the results using stone dust, fly ash and blast –furnace slag respectively with OPC. Regardless of the type of mineral admixture, the larger the amount of mineral admixture replacement, the larger was the table flow value and the smaller was the J_{14} cone flow time. This was observed more clearly when the mortar

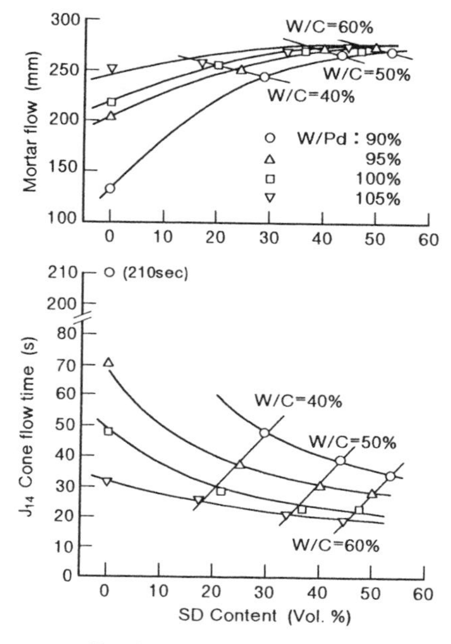

Fig. 8. Effect of SD with OPC

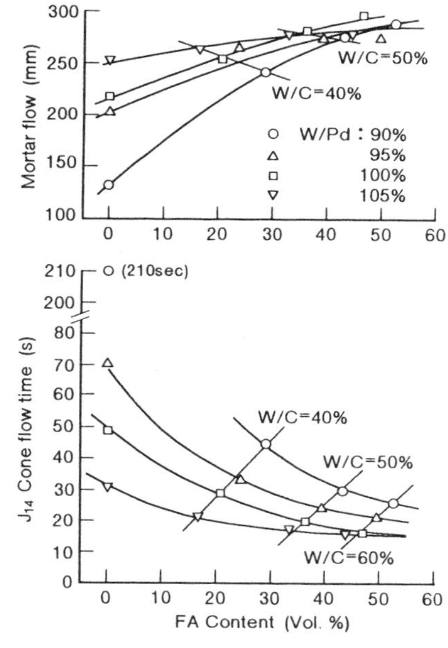

Fig. 9. Effect of fly ash.

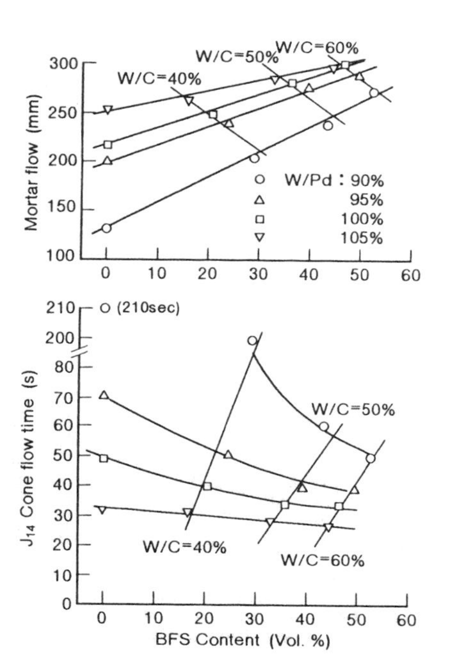

Fig. 10. Effect of blast-furnace slag

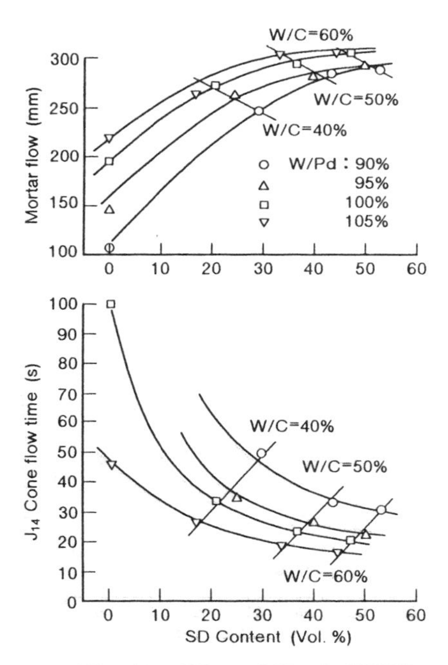

Fig. 11. Effect of SD with RHPC

had a smaller water to powder ratio. The flowability improvement of the mortar with stone dust and fly ash was better than using blast–furnace slag. This was because the slag had finer particles and higher activity in the first stage of hydration than other materials.

Figure 11 shows the results using stone dust with RHPC. Flowability of the mortar using only RHPC as powdered materials was very poor. However, flowability could be remarkably improved by adding the stone dust. RHPC with the stone dust replacement by mass of 30 % showed the same order of flowability as using OPC. When the water to cement ratio is constant, there seems to be a linear relationship between table flow value or J_{14} flow time and replacement ratio of these admixtures, regardless of the type of admixture. It was assumed that, for highly fluidized concrete to be self–consolidating, mortar should have a table flow value of 250 – 280mm and J_{14} flow time of 20 – 50 seconds. Using Figs. 8 to 10, the optimum combination of W/Pd and W/C (i.e. the replacement ratio of the admixture) for each powdered materials could be obtained.

6 Concluding remarks

The investigation was carried out using a polysaccharide–based viscosity agent to control or eliminate the variation in flowability of highly fluidized concrete, and the effects of mineral admixtures on the flowability of mortar with the viscosity agent were also studied. Within the limited range of the experiments carried out, it was found that;

1. Addition of a polysaccharide–based viscosity agent could reduce the variation of flowability, and thus enable production of more uniform quality concrete at actual batching plants.
2. Flowability of mortar was extremely improved by replacing a portion of the cement with stone dust, fly ash and/or blast – furnace slag, especially when combining stone dust and fly ash.
3. In order to have the required flowability of highly fluidized concrete, the optimum combination of W/Pd and W/C for each powdered material could be obtained.

7 References

1. Ozawa,K. Maekawa,K. Kunishima,M. Okamura,H. (1989) "High–Performance Concrete based on the durability design of concrete structures", Proc. of the Second East Asia–Pacific Conference on Structural Engineering and Construction.
2. Aitcin,P. Jolicoeur,C. MacGregor,J.G. (1989) "Superplasticizer; how they work and why they sometimes don't", V.16,No.5, Concrete international, ACI, May.
3. Ozawa,K. Sakata,N. Okamura,H. (1994) "Evaluation of self–compactability of fresh concrete using the funnel test" Proc. of JSCE, No.490, V–23.
4. "Concrete Manual"(1975) U.S.Department of the interior, Bureau of Reclamation.

99 FUNDAMENTAL STUDY ON THE PRACTICAL USE OF HIGH PERFORMANCE CONCRETE

S. USHIJIMA, K. HARADA and H. TANIGUCHI
Technical Research Institute of AOKI Corporation, Tsukuba, Japan

Abstract

High performance concrete(HPC) has gained attention as concrete with high self–compactability, appropriate stability and high durability. Application of the HPC to structures with complex shape and dense bar arrangement is increased every year because of those special features. Test methods for evaluating the quality of HPC prepared using various binders and viscosity improver and measures for various problems have, however, never been established. In order to put the HPC to practical use, therefore, it is necessary to investigate the properties of the HPC by experiments in advance.

This paper presents the following items aiming at obtaining the basic data to put HPC to practical use;

- Effects of the type of binder and the existence of viscosity improver on the properties of fresh and hardened concrete, and the durability.
- Measures for segregation caused by the variation of the surface moisture ratio of fine aggregate.

Valuable data on the properties of HPC with various proportions were obtained to allow its use for future concrete projects.

Keywords: Binder, fluidity, high performance concrete, self–compactability, stability, surface moisture.

1 Introduction

Research and development of high performance concrete(HPC) have recently been made to reduce the labor for concrete works and improve the reliability of structures[1]. For HPC, while having high fluidity, it have moderate stability. In addition it has self–compactability. So, it can flow into the spaces between the reinforcing bars and forms without vibration.

Concrete Under Severe Conditions: Environment and loading (Volume Two) Edited by K. Sakai, N. Banthia and O.E. Gjørv. Published in 1995 by E & FN Spon. ISBN 0 419 19860 1

Previous references[2] report that HPC is generally prepared using a superplasticizer to provide the concrete with fluidity and large quantities of powder, for example blast−furnace slag, fly ash and so on, to improve stability. Since such a method cause thermal cracking by heat of hydration, various viscosity improvers as the subsutitute for powder have been developed. The viscosity improver can not only give the stability to concrete but also reduce the variation of the stability accompanied with the variation of the surface moisture ratio of fine aggregate[3] in a concrete manufacturing plant. Recently, therefore, HPC is prepared by controlling the additions of the binders and viscosity improver.

The selection method of the materials, the design method for proportion and the evaluation method of fluidity have, however, not been completely established yet. Few papers present the effects of the binders and viscosity improvers for HPC on the properties of hardened concrete. Also, the measures for problems, for example, quantity control method of HPC, arising at the construction have never been established.

The authors carried out the experiments aiming at elucidating the effects of the type of binder, water/binder ratio and the existence of viscosity improver on the fluidity of concrete, the properties of hardened concretes and the durability.

2 Experimental outline

2.1 Materials

The materials used for the experiment are listed in Table 1. Ordinary portland cement, blast−furnace slag and fly ash were used as the binder to understand the fluidity of HPC. Since it is expected that HPC containing high amount of binder will be applied to construction of mass concrete, belite type portland cement and low heat three−component cement were used to control the heat of hydration.

The viscosity improver used for the experiment was a natural bacterial composite which is hardly soluble in water, alcohol and acetone but is in an alkaline aqueous solution. And it is able to be stably kept in a polyethylene bag or bottle for a long time.

2.2 Proportion

The fluidity of mortar prepared at water binder ratios of 35 and 45% and fine aggregate binder ratios (S/P) of 1.4 to 2.0 was examined by a preliminary experiment.

The mix proportions of the concrete are listed in Table 2. Generally, HPC of slump flow to approximately 600 mm is excellent in fresh property[4]. So the amount of superplasticizer and viscosity improver were determined from the mortar experiments and the proportion of concrete was determined so as to adjust the slump flow to approximately 600 mm. The water binder ratio and water content were fixed at 33 and 43% and 165 and 170 kg/m^3, respectively. The adiabatic temperature rise in case of using belite type portland cement and low heat three−component cement were examined to obtain the heat generating properties of concrete. The test items are shown in table 3.

Table 1. Materials

Matarials	Detail	Specific gravity	Chiper
Binder	Ordinary portland cement Belite type portland cement Low heat three−component cement (Moedrate heat poltland cement : Blast furnace slag : Flyash = 2:2:1)	3.16 3.24 2.85	C LP MBF
	Blast furnace slag (Blain value:6000cm ^{2} /g) Flyash	2.90 2.30	B F
Aggregate	Fine aggregate :Land sand(F.M=2.71) Coarce aggregate :Crushed stone(Max.size=20mm)	2.60 2.70	S G
Chemical admixtrue	Super plasticizer Air entraining agent Viscosity improver	− − −	SP AE V

Table 2. Mix proportions of concrete

Chiper	W/P (%)	s/a (%)	Unit weight (kg/m ^{3})								
			W	P			S	G	* SP	* AE	** V
				C	B	F					
C33−1 C33−2	33	48	165	500	0	0	788	888	1.50 1.70	0.000 0.000	0.7 0.7
C43−1	43	50	170	400	0	0	855	888	1.20	0.003	1.0
CB33−1 CB33−2 CB33−3 CB33−4	33	48	165	250	250	0	780	878	1.10 1.20 1.30 1.40	0.003 0.006 0.008 0.001	0.0 0.0 0.7 0.7
CB43−1 CB43−2 CB43−3	43	50	170	200	200	0	842	888	1.00 1.20 1.40	0.004 0.003 0.005	0.0 1.0 3.0
CBF33−1 CBF33−2 CBF33−3 CBF33−4	33	48	165	200	200	100	759	856	1.50 1.50 1.50 1.50	0.010 0.010 0.010 0.010	0.5 1.0 1.5 0.0
LP32−1 LP32−2	32.4	48	165	509	0	0	788	888	1.10 1.10	0.000 0.020	0.6 0.9
MBF32−1 MBF32−2 MBF32−3	32.4	48	165	509	0	0	762	859	1.20 1.30 1.40	0.020 0.020 0.020	1.0 1.0 1.0
MBF37−4	36.9	48	165	447	0	0	788	888	1.20	0.010	1.0

*:P × wt.%, **:W × wt.%

3 Effect of type of binder and viscosity improver on the fluidity

Figure 1 illustrates the relationship between fluidity and amount of superplasticizer for HPC using various binders and viscosity improvers. To obtain a specified slump flow, the addition to ordinary portland cement is the largest and that to belite type portland cement was the smallest. This result was caused by the differences of shape of binder. Since the addition of the

Table 3. Test item and test method

Aim	Test item	Test method
Fluidity	Slump flow	According to JSCE* standard
	O−funnel [6]	Falling time was measured (see Fig.10)
Properties of hardened concrete	Adiabatic temperature rise	Adiabatic temp. rise was measured for 10 days
	Compressive strength	JIS** A 1108 (Age=3,7,14,28days)
Durability	Drying shrinkage	JIS A 1129
	Freezing and thawing	JIS A 6204
	Carbonation	According to AIJ***
	Chloride ion penetration into concrete	Chloride ion content were measured after brine spray test (Test condition:brine temp. =40℃, concentration=3%), (Test cycle:3days continuous spraying followed by 4days forced drying at 50℃)
	Pore size distribution	Measured by porosimeter
Practical use	Valuation of stability against the change of surface moisture	Change the surface moisture ratio (−1.0〜+1.5%) and slump flow and O−funnel test were carried out
	Valuation of stability, a case of viscosity improver was charged after mixing	Charge the viscosity improver to segregated concrete after mixed up, and atability was observed by slump flow and O−funnel test after remixing

*Japan Society of Civil Engineering, **Architectural Institute of Japan, ***Japanese Industrial Standard

superplasticizer to the cement using two− and three−component binders was larger than that to the cement content in the binder, it is assumed that the addition of superplasticizer is required also for the binder other than cement. Although the slump flow is generally decreased by adding a viscosity improver, the addition of the present viscosity improver, was resulted in relatively less decrease[5].

The O−funnel falling time[6] was affected by the additions of the superplasticizer and viscosity improver. For CBF33 (the additions of superplasticizer was fixed to 1.5%, while additions of viscosity improver varied from 0.5 to 1.5 wt.%), the slump flow were almost same, while the O−funnel falling time varied from 13 to 18 seconds.

In the case of concrete without viscosity improver which W/P was 33%, the segregation was observed in concrete showing the slump flow exceeding 700 mm. And the case of W/P was 43%(without viscosity improver), even if slump flow was approximately 600mm, the tendency of segregation was observed. This suggests that the viscosity improver will be needed from the point of view of practical use.

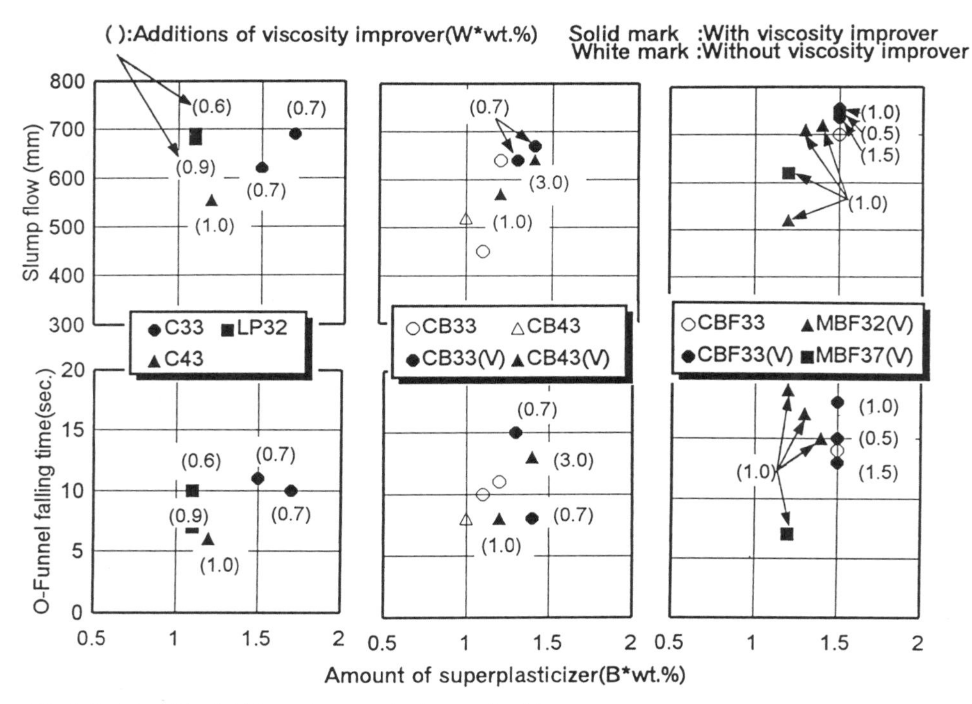

Fig.1. Relationship fluidity and amonunt of superplasticizer

4 Effect of type of binder and viscosity improver on hardened concrete

4.1 Adiabatic temperature rise test

Figure 2 shows the result of adiabatic temperature rise tests. This figure reveals that the heat−generating rates of two types of low heat cement were much lower than that of ordinary portland cement. The adiabatic temperature rise curves of belite type portland cement was not as steep as that of low heat three−component cement.

4.2 Compressive strength

The test results of compressive strength are illustrated in Fig. 3. Although the compressive strength depended upon the type of binder, the result of two−component cement(CB33 and 43) and three−component cement(CBF33) indicated that the viscosity improver has no effect on the compressive strength. The compressive strengths of concrete samples using belite type portland cement (LP32−1) and low heat three−component cement(MBF32−2) at the same water binder ratio were almost same each other. The strength development of these concrete samples at and beyond the age of 28 days were higher than that of concrete using other binders.

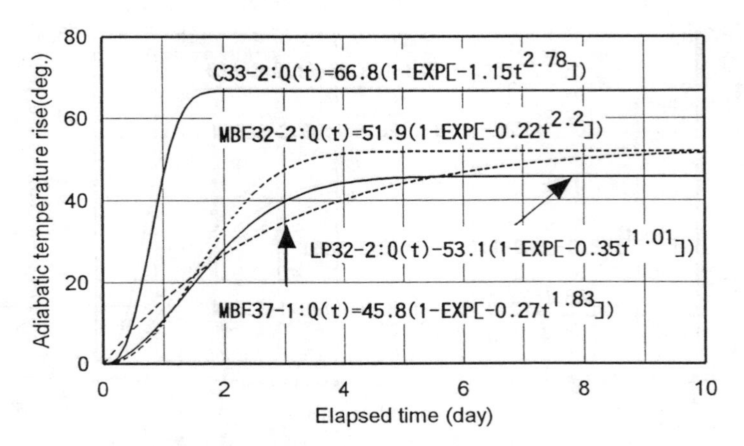

Fig.2. Adiabatic temperature rise test

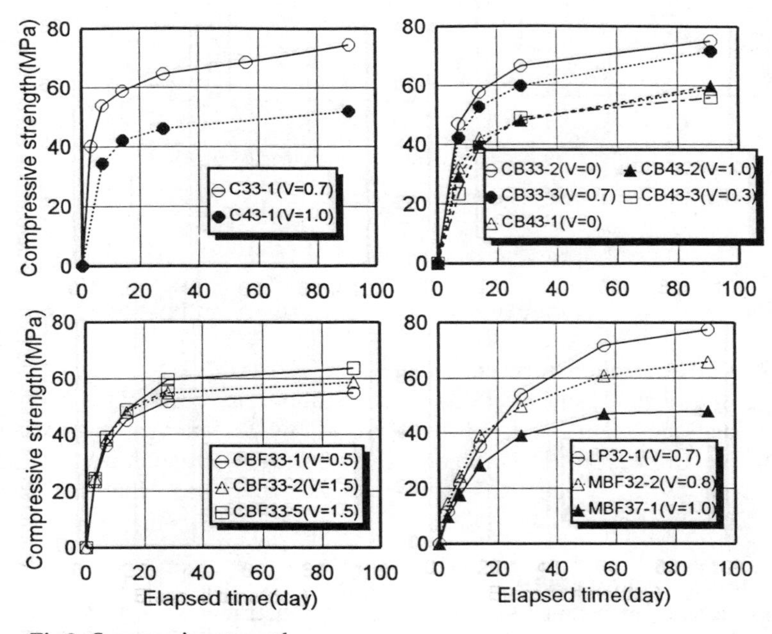

Fig.3. Compressive strength

5 Durability

5.1 Length change

Figure 4 illustrates the length changes of HPC. The length change tests were started after standard curing for 7 days. In the case of concrete samples prepared at the W/B of 32%, the length change of C33−1(using one component cement) was the largest and CB33−2, MBF32−2 (using tree or two components cement) were the smallest. While, concrete samples prepared at the W/B of 43% using these cements, had opposite results. These results indicate that the length changes of concrete samples were affected by not only drying shrinkage but also autogeneous shrinkage.

5.2 Frost resistance

Figure 5 illustrates the results of freezing and thawing tests. This figure reveals that the high performance concrete samples have superior frost resistance regardless of the water binder ratio, type of binder and existence of viscosity improver. The curves of the rate of weight change indicate that some scaling was observed on the surface of concrete prepared using low heat cement including moderate heat three−component cement(MBF32−2) and belite type portland cement (LP32−1). This may result in insufficient development at the begining of test.

5.3 Carbonated depth

Figure 6 illustrates the carbonated depth until the age of 8 weeks. It is considered that the carbonation proceeds more slowly than that of general

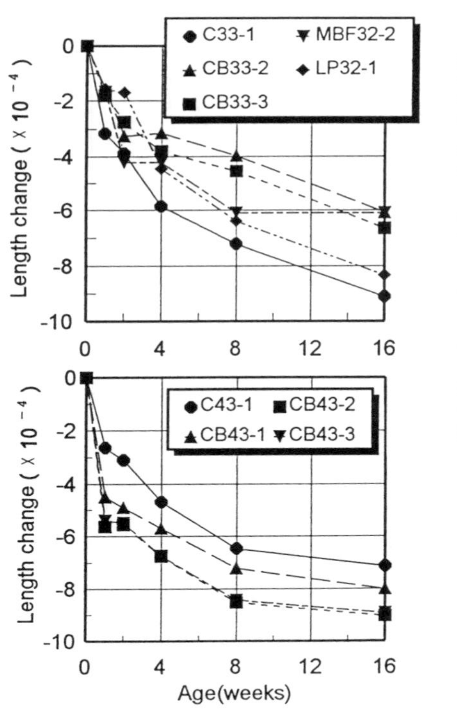

Fig.4. Length change test

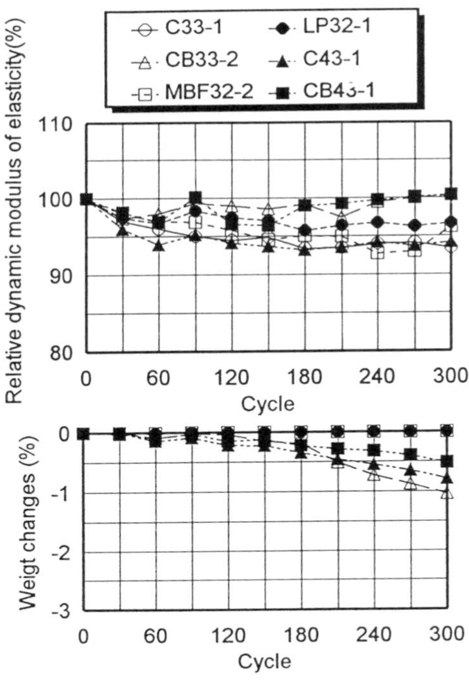

Fig.5. Freezing and thawing test

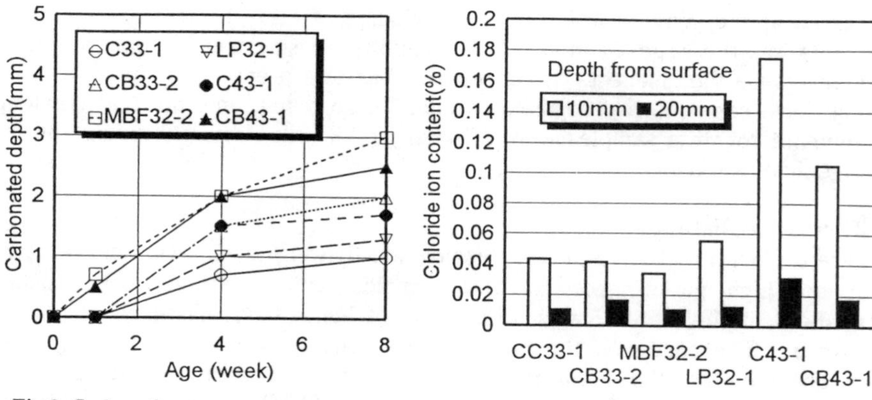

Fig.6. Carbonation test

Fig.7. Chloride ion penetration test

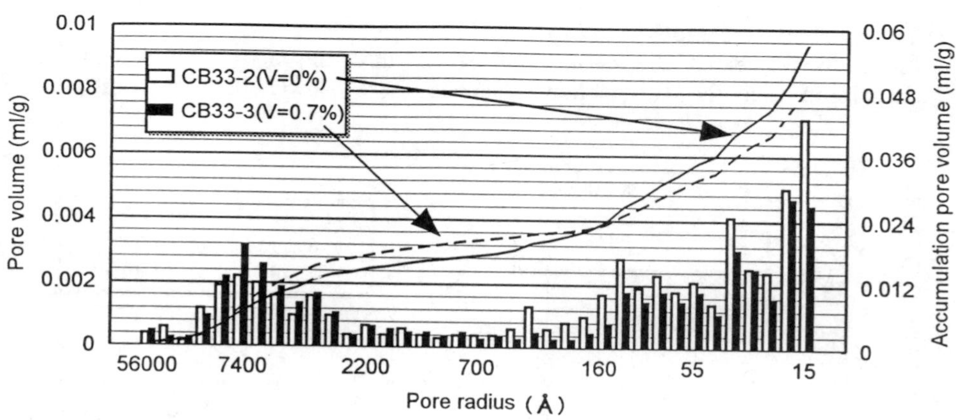

Fig.8. Pore size distribution

concrete because the water binder ratio of the HPC samples was low and the strength of them was high. The carbonated depth of concrete with blast−furnace slag and fly ash was slightly higher.

5.4 Chloride ion penetration

The results of brine spray test after 8 cycles are shown in Fig.7. The chloride ion content of the samples with water binder ratio of 33% is smaller than those with water binder ratio of 43%. And the chloride ion content of belite type portland cement is slightly higher than other samples with water binder ratio of 33%.

5.5 Pore size distribution

Pore size distribution are shown in Fig.8. The figure reveals that the total pore volume in concrete prepared using the viscosity improver (CB33−3) is slightly less than in concrete prepared without using it (CB33−2).

6 Measures to count the segregation

6.1 Effect of the variation of surface moisture on stability

The fluidity of concrete of CB33−3 using fine aggregate with a surface moisture ratio of −1.0 to 1.5% was measured by the slump flow test and O−funnel test. The results are illustrated in Fig.9. The slump flow of concrete was increased by approximately 100mm by an increase of 1.5% in the surface moisture ratio, while it is remarkably decreased by decreasing the surface moisture ratio. Although the falling time was not changed very much by variations in the surface moisture ratio, it was increased by the arching (see Fig.10) of aggregate accompanied with the segregation at the surface moisture ratio of 1.5%.

6.2 Improvements of segregation

The concrete sample without viscosity improver(CBF33−4) showing segregation was returned to the mixer and the viscosity improver was added (post−addition).

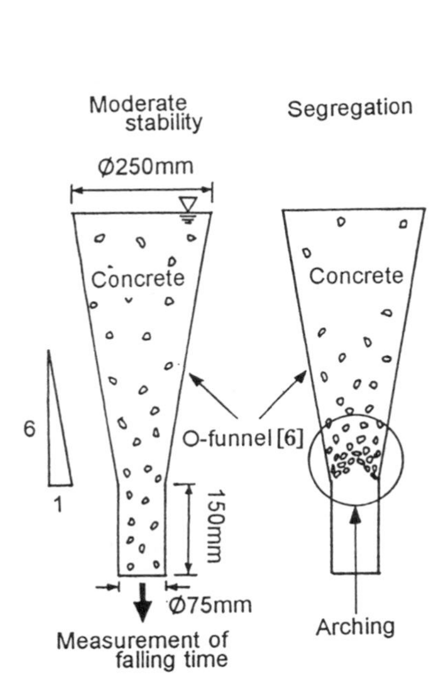

Fig.10. Conception of arching of aggregate

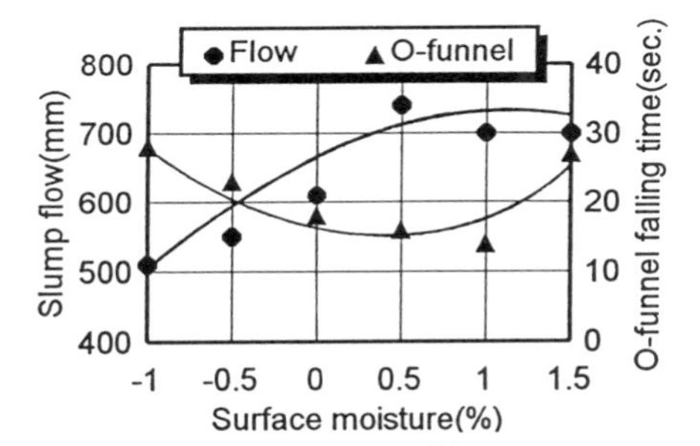

Fig.9. Stability of concrete with change in surface moisture

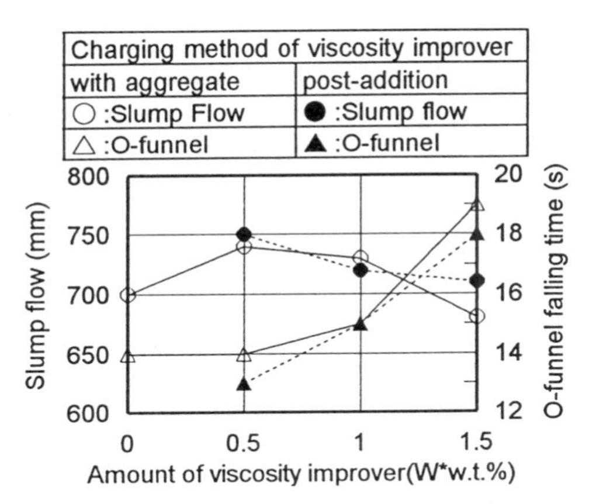

Fig.11. Effect of improvements of segregation

The change of segregation in concrete was visually observed and evaluated by the slump flow and O−funnel tests. The results are illustrated in Fig. 11. The figure reveals that the test results were almost same as that in the ordinary mixing method (charge with aggregate). This indicates that the post−addition of viscosity improver can easily improve the segregating concrete. Namely, it will be possible that the segregating concrete in the truck agitator will be improved by charging the viscosity improver to the truck agitator.

7 Conclusion

The following results were obtained by the experiment:

1. Segregation can be eliminated, even at a slump flow of 700 mm or more, by adding 0.5 to 1.0 wt% of the viscosity improver to the water content.
2. Although the effect of the viscosity improver on the properties of hardened and durability of high performance concretes is small, they depends upon the type of binder and the water binder ratio.
3. Segregation of concrete caused by the variation of surface moisture ratio in the construction can be reduced by the addition of viscosity improver. Also the segregation occurring after mixing can be improved by the post−addition of viscosity improver.

8 References

1. Ozawa,K., Maekawa,K. and Okamura,H. (1989) Development of high performance concrete. *Proceedings of JCI*, Vol.11, No.1, pp.699−704, (in Japanese)
2. Japan Concrete Institute. (1993) *Report of Super Workable Concrete Committee*. JCI (in Japanese)
3. Shindo,T., Matuoka,Y., S.Tangtermsirikul and Sakamoto,J. (1992) Effect of quality variation of materials on properties of super workable concrete. *Proceedings of JCI*, Vol.14, No.1, pp.75−78 (in Japanese)
4. Taniguchi,H., Harada,K. and Ushijima,S. (1994) Study on properties of fluidity of mortar and concerte for highly workable concrete. *Extended Abstracts; The 48th Annual Meeting of JCA*, pp.838−843 (in Japanese)
5. Ushijima,S., Taniguchi,H., Goami,Y. and Tateishi,A. (1994) Study on the property of super−workable−concrete with various binders and special viscosity improver. *Proceedings of JCI 2nd Symposium on Super Flowable Concrete*, pp.95−102 (in Japanese)
6. Miura,R., Chikamatu,R., Aoki,S. and Sogo,S. (1991) Fundamental study on super workable concrete. *Proceedings of JCI* vol.13, No.1, pp.185−190 (in Japanese)

100 HIGH REACTIVITY METAKAOLIN – A MINERAL ADMIXTURE FOR HIGH-PERFORMANCE CONCRETE

M.A. CALDARONE
Dolese Brothers, Oklahoma City, Oklahoma, USA
K.A. GRUBER
Engelhard Corporation, Iselin, New Jersey, USA

Abstract

The development of high-performance concrete (HPC) has brought about the need for additives, both chemical and mineral, to improve the performance of concrete. High Reactivity Metakaolin (HRM) is one such material. HRM is a manufactured pozzolanic mineral admixture that can significantly improve the performance of portland-cement-based products.

This study investigates the improvements in compressive strength and the reduction of chloride permeability, based upon the AASHTO T-277 rapid chloride test, associated with the addition of HRM. HRM was added to the concrete mixtures alone or in conjunction with two less reactive admixtures, Class F fly ash and ground granulated blast-furnace slag.

In summary, HRM was found in this study to be a highly effective mineral admixture for high-performance concrete with properties similar to silica fume. Ternary blends of concrete containing HRM in conjunction with Class F fly ash or blast-furnace slag improved the performance over the original formulations.
Keywords: AASHTO T-277 rapid chloride permeability, blast-furnace slag, Class F fly ash, compressive strength, high-strength concrete, high-performance concrete, High Reactivity Metakaolin, silica fume

1 Introduction

The development of high-performance concrete (HPC) has brought forth the need for additives, both chemical and mineral, to improve the performance of the concrete. High Reactivity Metakaolin (HRM) is one such material. HRM is a

Concrete Under Severe Conditions: Environment and loading (Volume Two) Edited by K. Sakai, N. Banthia and O.E. Gjørv. Published in 1995 by E & FN Spon. ISBN 0 419 19860 1

manufactured pozzolanic mineral admixture that can significantly improve the performance of portland cement-based products.

Significant research has been performed on natural pozzolans, predominantly mixed clays or volcanic ashes that contain some quantity of thermally activated kaolinitic clay [1]. These unpurified materials have often been termed "metakaolin". The designation High Reactivity Metakaolin (HRM) has been used to differentiate this white, purified, manufactured, thermally activated kaolinite from the less active calcined mixed clays which contain some undefined concentration of kaolinitic clay. HRM is a highly reactive pozzolan that has been shown to exhibit properties similar to silica fume without some of the typical drawbacks, such as color, workability and finishability [2].

The focus of this investigation is to study the performance of HRM in three typical formulations common to the state of Illinois, USA. First, a comparison of HRM versus silica fume in a high-performance formulation was studied. Secondly, this study explores ternary systems using HRM in conjunction with other less reactive admixtures, a Class F fly ash and a ground granulated blast-furnace slag.

Results of the evaluations indicate that HRM is an effective pozzolan with properties similar to silica fume. Not only did HRM improve the performance of concrete containing no other pozzolanic material, HRM also enhanced concretes containing other admixtures.

2 Materials

2.1 Portland cement

ASTM C-150 Type I portland cement was used in all of the concrete mixtures.

2.2 High Reactivity Metakaolin (HRM)

The HRM used in each of the studies is a white, highly purified, manufactured product that meets or exceeds all of the specifications of ASTM C-618 Class N pozzolans. The material was a commercial product that is manufactured in the USA. Table 1 shows a typical analysis of the HRM.

2.3 Silica fume (SF)

The silica fume used in the first concrete mixture formulation was a densified powder. The SF was a commercially available product from the USA. Table 1 includes a typical analysis of the SF.

2.4 Fly ash (FA)

The fly ash added to the second concrete formulation was a Class F fly ash available in the Midwest of the United States. The typical properties of the Class F fly ash are in Table 1.

2.5 Ground granulated blast-furnace slag (GGBS)

The slag used in the third concrete formulation was a commercially available product from the USA. The typical properties of the GGBS appear in Table 1.

2.6 Aggregates

The fine and coarse aggregate both originate in the Midwest Region of the United States. The fine aggregate used in all of the concrete mixtures was an ASTM C-33 manufactured sand, a blend of natural sand and limestone from crusher screenings. The coarse aggregate used in all of the concrete mixtures was a #57 38.1 mm (1½-inch) nominal diameter stone.

2.7 Air-entraining admixture

When employed, a commercially available neutralized vinsol resin (ASTM C-260) was used as the air-entraining agent.

2.8 Water-reducing admixtures

A standard ASTM C-494 Type A water-reducer was added to the first two concrete mixtures to initially disperse the cement.

An ASTM C-494 Type F high-range water-reducer (HRWR), superplasticizer, was added to the first two formulations. The superplasticizer was a commercial naphthalene-based product. The HRWR was added to achieve an equal slump.

Table 1. Typical properties of admixtures

Property	HRM	SF	FA	GGBS
Specific Gravity	2.5	2.2	2.6	2.8
$SiO_2 + Al_2O_3 + Fe_2O_3$, %	>95	>95	>70	39-59
Color	White	Grey	Grey/Tan	White

3 Results and Discussion

3.1 Comparison of pozzolanic activity of high reactivity metakaolin with silica fume

The concrete mixture used to compare the activity of high reactivity metakaolin (HRM) and silica fume (SF) are listed in Table 2. A weight replacement of the silica fume with the HRM was performed. The formulation contained both ASTM C-494 Type A and Type F water reducers. No air-entraining admixture was added so that a more exact assessment of the effect of the materials on strength could be determined.

Table 2. Mixture proportions for HRM and SF concretes

Material	HRM concrete	SF concrete
Type I cement (kg)	390	390
High Reactivity Metakaolin (kg)	33	0
Silica fume (kg)	0	33
Water (kg)	154	154
Coarse aggregate, SSD (kg)	1,032	1,032
Fine aggregate, SSD (kg)	837	837
Type A water-reducer (ml)	755	755
Type F HRWR (l)	6.4	8.2

The properties of the fresh concretes are in Table 3. The concrete containing HRM required significantly less superplasticizer than the silica fume concrete to achieve and equal workability as measured by the slump cone.

Table 3. Fresh concrete properties of HRM and SF concretes

Property	HRM concrete	SF concrete
Slump (mm)	267	254
Concrete temperature (°C)	22	22
Air content (%)	3.0	3.1
Unit weight (kg/m^3)	2,440	2,410

The compressive strength development of these two concretes are in Table 4. The strength measurements were performed on continuously moist-cured 102 *x* 203 mm cylinders at 22°C. The measurements listed are an average of two cylinders.

Table 4. Compressive strength development

Testing age (days)	Compressive strength (MPa)	
	HRM concrete	SF concrete
7	79.8	70.3
28	94.6	88.6
56	104.9	96.2
90	113.8	108.9
180	119.6	114.8
365	127.5	124.0

Rapid chloride permeability measurements were performed on these two concretes after 56 days of moist-curing at 22°C. The AASHTO T-277 test method was followed. The rating scale was:

> 4,000	high
2,000-4,000	moderate
1,000-2,000	low
100-1,000	very low
<100	negligible

The top sections of two cylinders were tested and averaged. The average coulombs passed after a six hour period was 900 for the HRM concrete and 600 for the SF concrete. Both concretes had a very low rating according to AASHTO and similar impermeability.

3.2 Ternary system using Class F fly ash

The concrete formulations containing Class F fly ash appear in Table 5. The formulations compare a typical fly ash mixture with a concrete that contains an addition of a small amount of HRM. As a second comparison, an addition of fly ash was included as a second "control" of equal cementitious content. Both the fly ash and the HRM were added in powder form with the cement.

An ASTM C-494 Type A water-reducer was added to all of the mixtures at a constant concentration of approximately 195 ml/100 kg (3 oz./100 lb.) of cement. Adjustments were made in the ASTM C-494 Type F high-range water-reducer (superplasticizer) to achieve a slump of 152.4 ± 12.7 mm (6 ± 0.5 inches). An ASTM C-260 air-entraining agent was added to reach a target air content of 6-8%. The water-to-cement ratio was maintained at 0.50.

Table 5. Mixture proportions of FA containing concretes

Material	Original FA concrete	Addition of HRM	Addition of FA
Type I cement (kg)	300	300	300
Class F fly ash (kg)	60	60	90
High Reactivity Metakaolin (kg)	0	30	0
Fine aggregate, SSD (kg)	800	800	800
Coarse aggregate, SSD (kg)	1,030	1,030	1,030
Water (kg)	150	150	150
Type A water-reducer (ml)	580	580	580
Air-entraining agent (ml)	190	350	350
Type F HRWR (l)	1.3	3.7	2.4

Table 6 shows the properties of the fresh concretes. The HRWR was added to achieve an equal slump. The HRM required more HRWR than the concrete which contained the fly ash addition. The air contents were approximately the same, although additional air-entraining agent was required.

Table 6. Properties of fresh concretes containing FA

Property	Original FA concrete	Addition of HRM	Addition of FA
Slump (mm)	160	160	180
Air content (%)	7.4	7.0	7.4
Unit weight (kg/m^3)	2,260	2,280	2,290

Table 7 shows the compressive strength development of the concretes containing fly ash. The concrete containing an addition of HRM exhibited higher strengths than either of the concretes containing only fly ash. This indicates that the HRM is compatible with fly ash systems and possesses higher pozzolanic activity than Class F fly ash.

Table 7. Compressive strength development of FA concretes

Testing age (days)	Compressive strength (MPa)		
	Original FA concrete	Addition of HRM	Addition of FA
3	22.9	31.4	25.9
7	31.0	47.0	34.7
28	40.8	62.1	47.6
56	47.4	65.3	52.4
90	48.0	65.8	53.2
180	51.3	68.0	56.3

Rapid chloride permeability testing was performed in accordance with AASHTO T-277 after the concrete had been moist-cured at 22°C for 56 days. Two measurements were recorded for each concrete. The average charge passed after six hours was 2,500 for the original fly ash concrete, 800 for the concrete containing an addition of HRM and 1,500 for the concrete containing an addition of fly ash. According to the AASHTO rating scale, the concrete containing HRM showed a very low permeability. Although the additional fly ash reduced the permeability from a moderate rating (original formulation) to a low rating, the additional fly ash did not reduce the permeability as significantly as the HRM.

3.3 Ternary system using blast-furnace slag

The formulation of the ground granulated blast-furnace slag (GGBS) concretes are in Table 8. The HRM was used in addition to the cementitious material. The slag and HRM were added to the cement during batching. No water-reducers or air-entraining admixtures were added to the concrete. Two different HRM additions were made to the original slag concrete. Previous studies of the slag formulation indicated that the optimum amount of slag for the highest 28 day compressive strength in this mixture was approximately 35%. Two different HRM additions were made to the original slag concrete.

Table 8. Mixture proportions of concretes containing GGBS

Material	Original GGBS concrete	4.4% HRM addition	6.2% HRM addition
Type I cement (kg)	217	217	217
Ground granulated blast-furnace slag (kg)	116	116	116
High Reactivity Metakaolin (kg)	0	15.5	22
Fine aggregate, SSD (kg)	862	862	862
Coarse aggregate, SSD (kg)	1,040	1,040	1,040
Water (kg)	217	217	217

Table 9 shows the results of the fresh concrete testing. Because of the addition of the extra cementitious material, the slump of the concretes containing HRM was slightly less than the original slag formulation. The set time of the concrete was approximately 30 minutes less when the 6.2% HRM addition was made.

Table 9. Properties of fresh concretes containing GGBS

Property	Original GGBS concrete	4.4% HRM addition	6.2% HRM addition
Slump (mm)	171	146	127
Air content (%)	1.1	1.2	1.5
Initial set (hr)	5.4	5.4	5.1
Final set (hr)	7.7	7.5	7.1
Unit weight (kg/m^3)	2,380	2,380	2,376

Table 10 shows the compressive strength development of the concretes containing blast-furnace slag. The compressive strengths were measured on two 102 *x* 203 mm cylinders that had been continuously moist-cured at 22°C until testing. The addition of small amounts of HRM increased the strength of the concrete. These results indicate that HRM is compatible with slag and that small additions of HRM can improve the performance of the slag concrete.

Table 10. Compressive strength development of GGBS concretes

Testing age (days)	Compressive strength (MPa)		
	Original GGBS concrete	4.4% HRM addition	6.2% HRM addition
3	17.0	19.0	20.3
7	23.7	29.4	32.6
28	33.5	38.7	41.4
56	36.6	42.7	42.7
90	39.0	44.7	45.6
180	41.2	45.8	48.6

Rapid chloride permeability testing was performed in accordance with AASHTO T-277 after the concrete cylinders had been moist-cured at 22°C for 56 days. The average charge passed after six hours was 1,500 for the original blast-furnace slag concrete, 800 for the concrete containing an addition of 4.4% HRM and 700 for the concrete containing an addition of 6.2% HRM. Both concretes containing HRM exhibited very low permeability. The original slag concrete had a low permeability rating.

4 Conclusions

High Reactivity Metakaolin (HRM) is a white, purified product that has been found to exhibit pozzolanic activity similar to silica fume. According to this study, HRM produced compressive strength and rapid chloride permeability ratings equivalent to silica fume. However, the HRM concrete required significantly less superplasticizer, approximately 28%, than the silica fume concrete to achieve equal workability as measured by the slump cone.

HRM was also found to be effective in ternary systems containing Class F fly ash or ground granulated blast-furnace slag. In both ternary systems, the compressive strength was improved when small quantities of HRM were added. The rapid chloride permeability ratings were reduced more significantly than the addition of the same quantity of the less reactive cementitious/pozzolanic material.

The need for high-performance concrete has produced a need for admixtures that can reduce the permeability of the concrete while providing the necessary strengths. In addition, environmental concerns have brought about the need for higher usages of by-products which do not always produce the desired final properties of the concrete. This study indicates that HRM may allow for the use of increased levels of less active by-product additions while maintaining the necessary performance level of the concrete.

5 References

1. Lea, F.M. (1970) *The Chemistry of Cement and Concrete*, 3rd Edition, Chemical Publishing Company, Inc., New York, pp. 414-453.
2. Caldarone, M.A., Gruber, K.A. and Burg, R.G. (1994) High-Reactivity Metakaolin: A New Generation Mineral Admixture, *Concrete International: Formwork*, Vol. 16, No. 11, pp. 37-40.

101 MIX DESIGN OF AIR-ENTRAINED, HIGH-PERFORMANCE CONCRETE

M. LESSARD, M. BAALBAKI and P.-C. AÏTCIN
Université de Sherbrooke, Sherbrooke, Canada

Abstract
Designing high-performance concrete without air entrainment is a difficult and complex task. It comes down to combining the various components that go into the concrete to obtain the required compressive strength while ensuring that the mixture maintains adequate workability until final placement. Balancing these two requirements is often problematic. Consequently, designing high-performance concrete consists in weighing advantages against disadvantages.

The situation is rendered even more complex when there is a requirement to create air-void systems with specific volume and spacing factors. In fact, in some cases, the combination of design requirements and constituent compatibility make it difficult to produce a specific air-entrained high-performance concrete.

Based on our experience in designing the concrete used in the abutment and deck of the Portneuf Bridge in Quebec, Canada, it was found that working step-by-step is more effective in determining the correct formulation than making a lot of trial batches. The different concrete ingredients and their proportions are progressively optimized in order to meet the design requirements. The following summarises the steps involved:

(1) Test the cement/superplasticizer compatibility on grouts.
(2) Ensure that a non-air-entrained concrete of adequate compressive strength can be produced.
(3) Carry out trials to create a stable air-void system that meets the requirements.
(4) Test the freeze/thaw and scaling resistance of the air-entrained high-performance concrete.

Keywords: Air–entrained, cement/superplasticizer compatibility, freeze/thaw resistance, high–performance concrete, mix design, scaling resistance, superplasticizer.

Concrete Under Severe Conditions: Environment and loading (Volume Two) Edited by K. Sakai, N. Banthia and O.E. Gjørv. Published in 1995 by E & FN Spon. ISBN 0 419 19860 1

1 Introduction

As long as Procedure A (freezing and thawing in water) of ASTM C666 Standard will be used to determine the freeze-thaw durability of a high-performance concrete, it will be found that it is mandatory for high-performance concrete having a compressive strength of between 50 and 100 MPa to contain entrained air in order to successfully pass the 300 freezing and thawing cycles [1, 2, 3]. Although this test has been found to be too severe in comparison to natural conditions [4, 5], it is the test that is and will be used as reference by designers to specify frost durable high-performance concrete.

The mix design of a non-air-entrained, high-performance concrete is not an easy task [6, 7]. It is necessary to find a combination of concrete ingredients to achieve the level of strength that is required provided that the concrete could be easily transported and placed at a competitive price. The simultaneous satisfaction of these requirements is very often conflictual, so that the mix proportions of a high-performance concrete is, for most of the time, the result of a compromise [8].

It is necessary to develop an adequate system of microscopic bubbles having adequate diameter and spacing factor to protect concrete against freezing and thawing. Moreover, the bubble system developed by the tensioactive agent introduced during the mixing must be stable long enough so that the concrete keeps its bubble system until its final placement. It is foreseen that this task is not always easy for the mix designer.

The experience gained during the construction of the Portneuf Bridge in Quebec has shown that instead of making numerous trial batches to find an adequate formulation, it is better to proceed step-by-step, so that the different ingredients of the concrete and their proportions are progressively optimized to fulfil the design requirements.

2 Portneuf Bridge

Portneuf Bridge has a span of 24.8 m and a width of 10.25 m, it was built in 1992 on Portneuf River, on the North shore of St. Lawrence River. The bridge deck has been built on top of five prestressed girders. Concrete was cast in place and post tensioned transversally and longitudinally. The girders and the deck were designed with a characteristic strength of 60 MPa.

The design requirements for the concrete used to build the deck were as follows:

(1) The concrete should contain silica fume in order to insure adequate protection against corrosion for the reinforcing steels;

(2) The average bubble spacing should be lower than 230 μm without any individual value being greater than 260 μm. This is the spacing factor required by Canadian Standards CSA A23.1 in order to make conventional concrete durable to freezing and thawing in the most severe conditions;

(3) The concrete temperature during placement should be between 15 and 25 °C;

(4) The slump should be equal to 180 ± 40 mm.

2.1 Final composition

The composition of 26 concrete deliveries on 29th October 1992 is presented in Table 1.

Table 1. Composition of fresh concrete delivered at job site

Materials (kg/m^3)					Admixtures (L/m^3)		
W/B	Water	Binder	Aggregates		Superplasticizer[a]	A.E.A.	Retarder
			Coarse	Fine			
0.30	134[b]	450[c]	1150	710	5	0.125	0.450

[a] Polynaphthalene sulfonate base
[b] Including the water of the superplasticizer
[c] The blended cement includes approximately 7.5% by mass of silica fume

2.2 Field results

The abutment and the bridge deck were cast by an average ambiant temperature of +3°C. It was necessary to heat the mixing water in order to bring concrete temperature within the specified range.

It was decided that two testings involving a greater number of specimens and tests was to be performed on the 26 concrete deliveries in order to evaluate compressive and flexural strengths, abrasion resistance, elastic modulus, shrinkage, spacing factor, freeze-thaw durability, scaling resistance, and chloride ion permeability. For the 24 other deliveries the testing program was reduced to compressive strength at 28 days, and spacing factor on the hardened concrete. The properties of the fresh concrete (temperature, slump, air content and unit mass) were systematically evaluated at the unloading point for all the 26 deliveries.

All of the specified requirements were achieved: the temperature of the delivered concrete was between 15 and 25°C, the slump was between 140 and 220 mm except for 3 deliveries. The control of the quality was excellent, the average compressive strength was 75.3 MPa with a standard deviation of 3.4 MPa corresponding to a coefficient of variation of 4.5 percent, which is excellent, according to the recommendation of ACI Committee 363 on high-strength concrete [9]. The quality of the testing was also excellent with a within-test coefficient of variation of 2.3 percent.

Taking into account the acceptance criteria selected, it was found that the characteristic strength of the delivered concrete was 70.7 MPa which was well above the specified 60 MPa. Finally, the average total air content of 6.2 percent resulted in an average spacing factor of 190 μm.

It is interesting to note that compressive strength was found to vary according to the total air content, according to the two following relationships:

$f_c' = 94 - 3A_f$ (fresh concrete) and $f_c' = 92 - 3A_h$ (hardened concrete)

where f_c' is expressed in MPa and A_f and A_h represent the total percentage of air contents in the fresh and hardened concretes, respectively.

In other words, an increase of one percent in the air content resulted in a strength reduction in the order of four percent.

3 Proposed mix design method for air-entrained high-performance concrete

Based on the experience gained, it seems appropriate to propose the following four-step mix-design method for air-entrained, high-performance concrete.
(1) Evaluation of the cement/superplasticizer compatibility on grouts;
(2) Achieving compressive strength on a non-air-entrained concrete;
(3) Developing an adequate bubble system in the preceding concrete;
(4) Verifying the freeze thaw and scaling durability of the air-entrained concrete.

This method consists of designing a non-air-entrained concrete (steps 1 and 2), introducing an adequate bubble system into it (step 3) and verifying that the obtained concrete complies with the freeze thaw and scaling durability requirements (step 4).

For the first step, a cement/superplasticizer combination that does not present any critical fluidity loss during 60-90 minutes is selected, corresponding to the usual delivery time of concrete in most field applications. The next step consists of selecting the aggregate and the water/binder ratio in order to achieve the required specific strength. During this step the mix design is carried out on a non-air-entrained concrete with compressive strength higher than that of the air-entrained concrete that is to be designed.

It is only when this strength objective has been reached that an adequate bubble system is introduced into the mix so that the concrete be freeze thaw resistant after 300 cycles of freezing and thawing in water if the ASTM C666 Standard has been selected as a criteria for freeze thaw durability. As entrained air improves the fluidity of concrete, it is appropriate to slightly lower the superplasticizer dosage in order to obtain an air-entrained concrete having the same slump of concrete without air.

4 The cement superplasticizer compatibility

As previously mentioned, this study can be made on grouts. Different methods based on Marsh cone flow time have been proposed by de Larrard et al. [10], Hanna et al [11], and Rollet et al [12]. Alternatively, a minislump type test like the one proposed by Rollet et al [12] can be used. It has been found that there is a good correlation between the results of these two tests and the slump retention of the concrete [13].

The Marsh cone test consists of measuring the flow time of 1 L of grout placed in a cone having a lower diameter varying between 0.5 to 10 mm, while the minislump test consists of measuring the spread of a small cone of grout. In both cases, the grout is vigorously mixed in order to deflocculate cement particles. The water/cement ratio of the grout is selected within the range of 0.30 to 0.35 corresponding to the usual water/cement ratio for high-performance concrete. The amount of superplasticizer is expressed in solid content with respect to the mass of cement, and is varied by 0.2 percent steps around an average value of 1 percent.

The fluidity of the grout is measured just after mixing and one hour later. During this period of time, the grout can be slightly agitated (60 rpm) to simulate transportation conditions. The results of the tests are transformed into a graph giving the fluidity or the spread of the cement/superplasticizer system as a function of the dosage of superplasticizer (Fig. 1). According to the type of curves obtained, three essential values governing the rheological behaviour of the cement/superplasticizer combination are determined [10, 11, 13, 14]:

- the critical dosage of superplasticizer corresponding to the saturation point where there is a break in the fluidity curve and beyond which the fluidity of the system is not changed as more superplasticizer is added,
- the flow time or spread obtained for that critical dosage,
- the variation with time of the fluidity or of the spread of the grout shown by the relative position of the curve at 5 and 60 min.

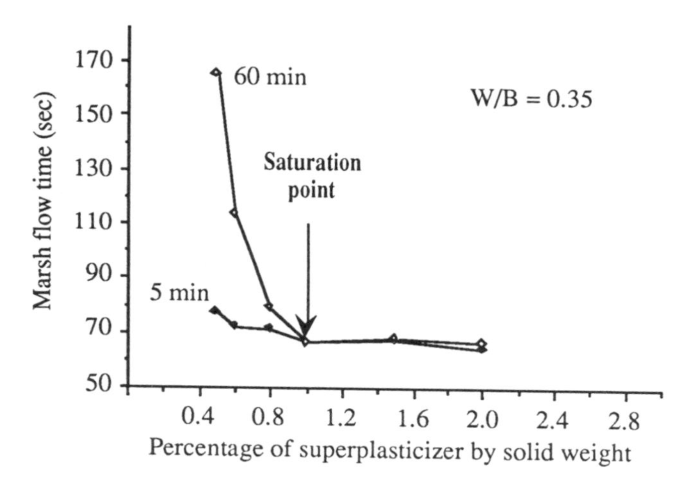

Fig. 1. Variation of grout fluidity in terms of dosage of Superplasticizer.

When a superplasticizer is added beyond the saturation point, the fluidity of the system is not improved but the risk of sedimentation or set retardation can increase.

When studying compatibility between the cement and the superplasticizer the four different cases, schematically represented on Fig. 2, can be found when the test is made with a Marsh cone [15].

The rheological properties of cement grouts containing superplasticizer are significantly affected by cement fineness, chemical composition, etc. The effect of such properties on fluidity, point of saturation of superplasticizer, and the loss of fluidity with time of the grouts shown in Fig. 2 have been explained in References 16-18.

Fig. 2a represents the case of a perfectly compatible cement/superplasticizer combination: the dosage corresponding to the saturation point is not very high, usually in the order of 1 percent, the flow time is quite low even for low W/C ratio values and does not increase or only slightly with time during the first hour.

Fig. 2b on the contrary represents the case of an incompatibility between the cement and the superplasticizer: the dosage corresponding to the saturation point is very high, (usually greater than 2 percent), the flow time of the system is high and increases inexorably with time so that in some cases it is impossible to measure any flow time after 10 minutes. Of course, in such a case, it is better to look for another combination of cement/superplasticizer. When it is difficult to find another cement at a competitive price, it is suggested to try another more efficient superplasticizer, or see if the addition of a retarder to the superplasticizer improves the behaviour of the grout so, that the behaviour of the ternary combination is close to the behaviour shown in Fig. 2a. Whenever possible, it is preferable to find a cement/superplasticizer combination that is easier to master.

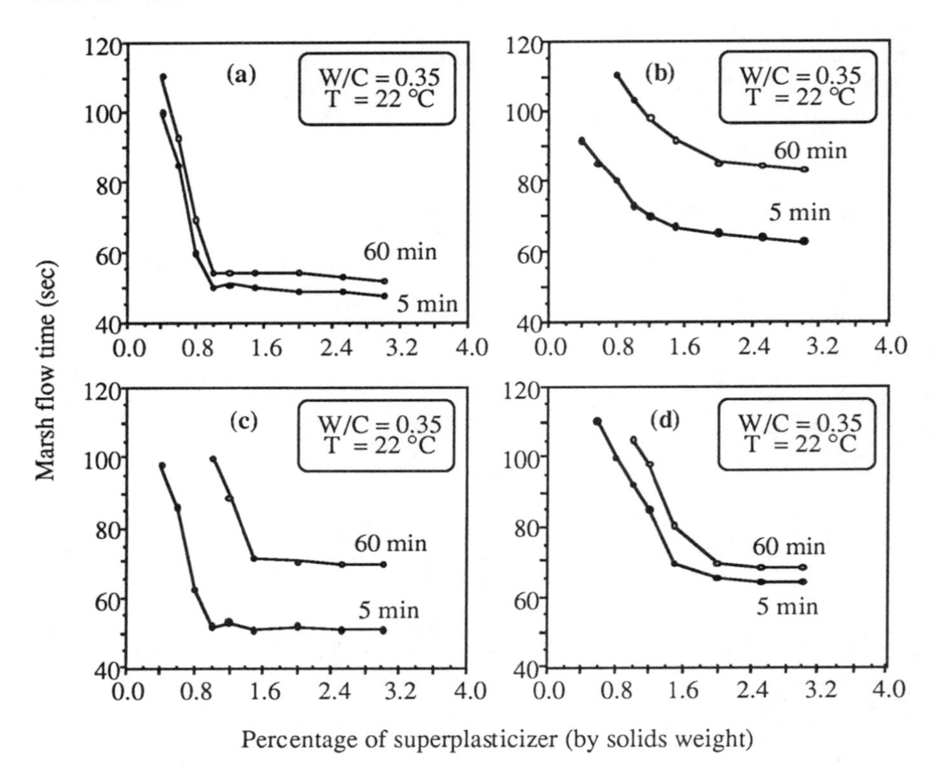

Fig. 2. Different rheological properties of cement grouts.

Fig. 2c and 2d represent intermediate cases. Fig. 2c corresponds to the case of a cement/superplasticizer combination that presents an excellent initial fluidity, but which deteriorates with time. In this particular case, the situation can be corrected by introducing an adequate amount of retarding agent so that the flow time curves as a function of the superplasticizer dosage looks like that of Fig. 2a.

Fig. 2d represents the case of a cement/superplasticizer combination that does not present a good initial rheology (the saturation point corresponds to a high dosage of superplasticizer) and the flow time is high, but the rheological property in this case does not deteriorate with time. In this case too, the right amount of retarding agent can be found to correct the initial situation.

At the end of this step, according to circumstances one or several cement/superplasticizer combinations that will not result in any premature slump loss that could interfere with the placing of the high-performance concrete has been selected, the next step will consist of looking if these different combinations can be used to achieve the aimed compressive strength.

As an example, Fig. 3 present the curves giving the fluidity of grouts, having a 0.35 water/binder ratio as a function of the superplasticizer dosage for the two cements that could be used for the construction of Portneuf Bridge. This figure shows that the Type 10 Canadian cement similar to the Type I ASTM cement was perfectly compatible with the superplasticizer selected, and that the blended silica fume cement was still acceptable if a retarding agent was used. In fact, the use of a retarding agent had a secondary economical impact since the saturation point decreased from 1.2 to 1 percent.

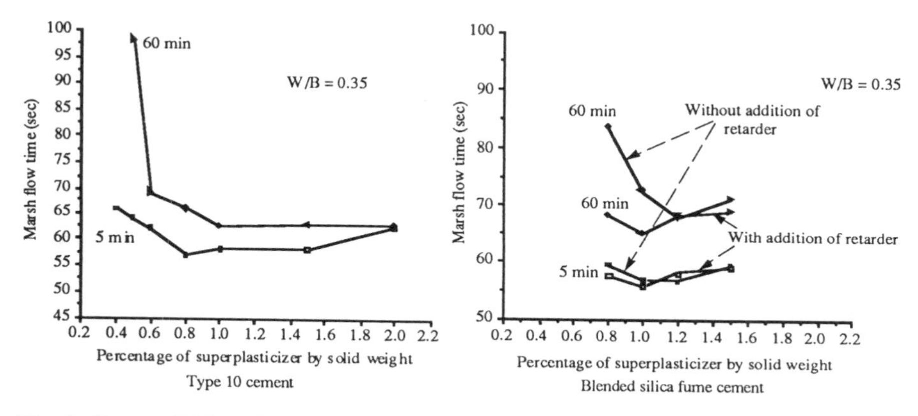

Fig. 3. Compatibility of type 10 and blended silica fume cement with polynaphthalene sulfonate based superplasticizer.

5 Reaching the strength requirement

This step is carried out on a non-air-entrained concrete. Of course, it is necessary to achieve a higher compressive strength than that of the air-entrained concrete. Based on the experience obtained during the construction of Portneuf Bridge and on the rule of thumb (1 percent increase in air content results in a drop of 5 percent in compressive strength). Therefore the compressive strength of the concrete should be increased by a factor equal to 5 (a - p) percent, where a represents the average air content necessary to achieve the right spacing factor, and p represents the amount of entrapped air found in the non-air-entrained high-performance concrete. The required compressive strength in the non-air-entrained concrete is then expressed as:

$$f'_{na} = f_a \left[1 + \frac{5}{100}(a - p)\right] \qquad (1)$$

As an example, we can suppose that the necessary air content is a = 6 percent and that p = 2 percent. In this case, it will be necessary to make a non-air-entrained concrete having a compressive strength 20 percent higher than that of the air-entrained high performance concrete that has to be made.

Of course, it will be also necessary to take into account the variability of the concrete production in order to define the average compressive strength that the concrete producer will have to obtain. In North America, the most used criteria for bridge construction is that the average of three consecutive tests should be higher than the specified strength 99 percent of the times. Therefore, the average compressive strength of the concrete that has to be delivered can be calculated as:

$$f_{cr} = f'_{na} + 2.33 \frac{\sigma}{\sqrt{3}} \tag{2}$$

where σ represents concrete standard deviation, f_{cr} the average strength, and f'_{na} the compressive strength of the non-air-entrained concrete given by equation (1).

If there are no statistical data for the production of high-performance concrete, a coefficient of variation of 10 percent could be taken. According to ACI Committee 363 [9] this value corresponds to an average quality control for high-strength concrete production. Of course, the designer could select any other value to his discretion.

In order to achieve a high-performance concrete having a given compressive strength, different approaches can be used. For example, a theoretical formula derived from Féret's Law [6] could be used, or an experimental approach can be taken using three trial batches having three different W/C or W/B ratios depending on the average aimed strength (Fig. 4). When it is decided to use this last method, the slump retention of the tested concrete should be monitored for one hour.

During this step of the optimisation process, the influence of the different coarse aggregates available at an economical price can be studied in order to see their influence on compressive strength of the concrete [19, 20, 21]. It will be sufficient for a given W/C or W/B ratio to use different aggregates and compare the obtained strength.

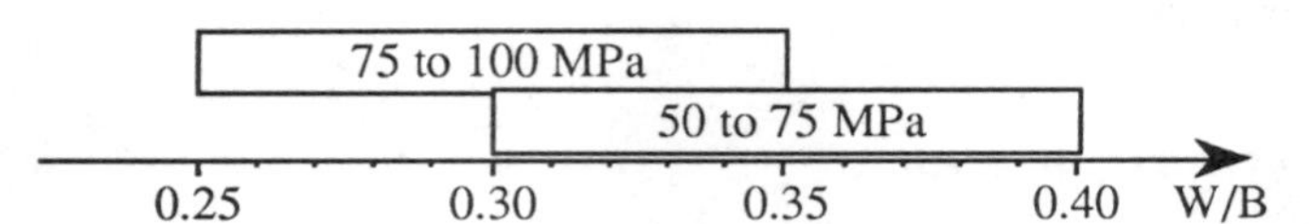

Fig. 4. Range of W/B necessary to achieve targeted compressive strength of air-entrained concrete.

6 The development of an adequate bubble system

When the formulation of the mix design of a non-air-entrained concrete has been optimised, it is necessary to introduce an air-entraining agent during the mixing. Three different dosages of the air-entraining agent can be used in order to bracket the required air volume. One of these three dosages can be based on the recommendation of the manufacturer. If the specifications do not require any given total air content but rather a spacing factor, it is then suggested to start with a total air content of 6 percent, which in general, corresponds to a spacing factor lower than 230 μm when an efficient air-entraining agent is used.

When doing this adjustment of the air content the slump loss during one hour can be measured. During this step it is appropriate to measure compressive strength variation as a function of air content to verify if the 5 percent law is valid in this particular case.

7 Freeze thaw durability

In order to ensure that the selected concrete is resistant to freezing and thawing and to deicer salt scaling, two of the following tests could be carried out:
(1) ASTM C666 "Test Method for Resistance of Concrete to Rapid Freezing and Thawing";
(2) ASTM C672 "Test Method for Scaling Resistance of Concrete Surfaces Exposed to Deicing Chemicals".

8 Conclusion

The construction of Portneuf Bridge has shown that it is possible to deliver an air-entrained, high-performance concrete in difficult field conditions (low temperature) with high quality standards: average compressive strength of 75.3 MPa, average air content of 6.2% with an average spacing factor of 190 μm.

A four step method is proposed to design an air-entrained, high-performance concrete. The first step, based on a grout study, consists of selecting a cement/superplasticizer combination that is compatible, the second one is determining the W/C and W/B ratios that must be achieved in order to comply with compressive strength requirements. This study is made on a non-air-entrained concrete. The third step consists of finding the right amount of air entraining agent to achieve the necessary total air volume content having the specified spacing factor. The last step consists of verifying the freeze thaw durability and the scaling resistance of the optimised concrete.

9 References

1. Pigeon, M., Gagné, R., Foy, C. (1987) Critical air-void spacing factors for low water/cement ratio concretes, *Cement and Concrete Research*, vol. 17, No. 7, pp. 896-906.
2. Gagné, R., Pigeon, M., Aïtcin, P.-C., (1992) The frost durability of high-performance concrete, in *High-performance Concrete*, Chapman & Hall, London, pp. 239-51.
3. Gagné, R., Pigeon, M., Aïtcin, P.-C. (1989) Deicer salt scaling resistance of high performance concrete, *Paul Klieger Symposium on Performance of Concrete*, San Diego, California, SP-122-3, pp. 29-44.
4. Philéo, R.E. (1986) Freezing and thawing resistance of high-strength concrete, *National Cooperative Highway Research Program Synthesis of Highway Practice 129*, Transportation Research Board, Washington, D.C., 31 p.
5. Hammer, T.A., Sellevold, E.J. (1990) Frost resistance of high-strength concrete, *2nd International Symposium on Utilization of High-Strength Concrete*, University of California, Berkeley, SP121-14, pp. 457-487.
6. de Larrard, F. (1990) Prévision des résistances mécaniques des bétons à hautes performances aux fumées de silice, ou une nouvelle jeunesse pour la loi de Féret, *Annales de l'Institut Technique du Bâtiment et des Travaux Publics*, 483, pp. 92-98.

7. Mehta, P.K., Aïtcin, P.-C. (1990) Microstructural basis of selection of materials and mix proportions for high-strength concrete, *2nd International Symposium on Utilization of High-Strength Concrete*, University of California, Berkeley, SP121-14, pp. 265-86.
8. Aïtcin, P.-C., Neville, A.M. (1993) High-performance concrete demystified, *Concrete International*, Vol. 15, 1 pp, 21-26.
9. ACI Committee 363 (1993) *Guide to inspection and testing of high-strength concrete*, 26 p.
10. de Larrard, F., Puch, C. (1989) Formulation des bétons à hautes performances: la méthode des coulis, *Bulletin de liaison du Laboratoire des Ponts et Chaussées*, 161, May-June, pp. 75-83.
11. Hanna, E., Luke, K., Perraton, D., Aïtcin, P.-C. (1990) Rheological behaviour of Portland cement in the presence of a superplasticizer, *3rd Canmet/ACI International Conference on Superplasticizer and Other Chemical Admixtures in Concrete*, Ottawa, Canada, SP119-9, pp. 171-188.
12. Rollet, M., Lévy, C., Cavailles, R. (1992) Evaluation of compatible-superplasticizer for the production of high-strength concrete, *9th International Congress on the Chemistry of Cement,* New Delhi, Proceedings.
13. Baalbaki, M. (1990) Façon pratique d'évaluer le dosage en superplastifiant: la détermination du point de saturation, *Séminaire sur les superplastifiants, Centre de recherche interuniversitaire sur le béton*, Université de Sherbrooke, Sherbrooke, pp.69-79.
14. Khalifé, M. (1991) *Contribution à l'étude de la compatibilité ciment/superplastifiant*, Mémoire de maîtrise, Département de génie civil, Université de Sherbrooke, Sherbrooke, 125 p.
15. Jolicoeur, C., Simard, M.-A, Aïtcin, P.-C., Baalbaki, M. (1992) Cement-superplasticizer compatibility in high-performance concretes: the role of sulfates, *Progrès dans le Domaine du Béton*, Montréal, Canada, 29 p.
16. Ranc, R. (1990) Interaction entre les réducteurs d'eau plastifiants et les ciments, *Ciments, Bétons, Plâtres, Chaux*, 782-1/90, pp. 19-21.
17. Uchikawa, H., Haneharas, S., Shirasaka, T., Sawaki, D. (1992) Effect of admixture on hydration of cement, absorptive behaviour of admixture and fluidity and setting of fresh cement paste, *Cement and Concrete Research*, Vol. 22, 6, pp. 1115-1129.
18. Tagnit-Hamou, A., Baalbaki, M., Aïtcin, P.-C. (1992) Calcium sulphonate optimization in low water/cement ratio concrete for rheological purposes, *9th International Congress on the Chemistry of Cement*, New Delhi, Proceedings, Vol. 5, pp. 21-25.
19. Sarkar, S.L., Lessard, M., Aïtcin, P.-C. (1990) Correlation of petrographic characteristics of aggregates with type of failures in high strength concretes, *Canadian Aggregates*, Vol. 4, No. 6, pp. 21-25.
20. Baalbaki, W., Benmokrane, B., Chaallal, O., Aïtcin, P.-C. (1991) Influence of coarse aggregate on elastic properties of high-performance concrete, *ACI Materials Journal*, Vol. 88, No. 5, pp. 499-503.
21. Aïtcin, P.-C., Mehta, P.K. (1990) Effect of coarse aggregates characteristics on mechanical properties of high-strength concrete, *ACI Materials Journal,* Vol. 87, No. 2, pp. 103-107.

102 CONCRETE MIXTURE PROPORTIONING TECHNIQUES FOR REMOTE ALASKAN LOCATIONS

M.R. NICHOLS and G.S. CHRISTENSEN
Alaska Testlab, Anchorage, Alaska, USA

Abstract
Alaska is huge, representing twenty percent of the area of the United States while spanning roughly the latitudinal and longitudinal equivalent of the continental US. There is almost no road access to this enormous land mass. This size and lack of transportation infrastructure present unusual challenges for both the design and the construction of remote projects. Alaska's environmental extremes are equally challenging. Temperatures of 100°F (38°C) in the summer to -70°F (-57°C) in the winter occur along with high winds, frequent earthquakes, marine environments, and some of the wettest and driest climates in the world.

These conditions present extreme and unique problems for designing and producing concrete. The standard specifications and recommendations of the American Society for Testing Materials (ASTM) and the American Concrete Institute (ACI) for concrete aggregates and concrete mixture proportion are often not practical and must be modified to assure that high quality concrete can be economically produced and placed. Over thirty years, Alaska Testlab (ATL) has developed solutions to these problems.
Keywords: Alaska, Arctic, concrete mixture proportioning, concrete production, sub-Arctic.

1 Introduction

The most common US standards and procedures covering the production of concrete aggregates and the design of concrete mixes are ASTM C 33, "Standard Specification for Concrete Aggregates," ACI 318, "Building Code Requirements for Reinforced Concrete" and ACI 211, "Standard Practice for Selecting Proportions for Normal, Heavyweight, and Mass Concrete." Inherent in each is the assumption that the unit cost of the mixture ingredients in ascending order are: aggregates, cement, and admixtures.

Aggregate grading limits do not specify the combination of coarse and fine which results in a workable mix. This combination is found by following broad empirical rules and validation through trial batches. Cement is used to provide strength and finishability. The assumption is always that since cement is far more expensive than aggregates, the amount of cement in the mixture should be minimized, while meeting the requirements of the application, to reduce overall cost. Admixtures are used to improve concrete performance without adding additional cement; to give the concrete

Concrete Under Severe Conditions: Environment and loading (Volume Two) Edited by K. Sakai, N. Banthia and O.E. Gjørv. Published in 1995 by E & FN Spon. ISBN 0 419 19860 1

characteristics that could not be imparted by the cement alone; or to reduce the amount of cement in the mix.

These assumptions work well in the continental US where commercial batch plant operations are nearby and where there is ready road or rail access to commercial aggregate sources. However, they do not work in most of Alaska.

The only urban areas in the state are Anchorage (225,000 population) and Fairbanks (75,000). After these two cities, populations drop off quickly. There are only another 10 to 15 communities large enough to have an established commercial aggregate source and some sort of fixed concrete batch plant. Even then most of these aggregate sources yield materials that cannot economically meet ASTM C 33 and have batch plants that do not meet current standards.

Each year there are many military installations, public buildings, communications facilities, and mining and petroleum projects scattered throughout remote locations of the state that require concrete. Additionally, there are over 200 Eskimo villages, all of which periodically have schools, utility, or other construction projects that require concrete. None of these locations have road access. All have an airport that provides access by air, but the length of the runway, the condition of the runway surface, and the navigational equipment available varies widely which often limits the size and type of aircraft that can service these locations. Almost all have some form of water access for two to four months in the summer, generally by barge.

Aggregates must be either mined locally or barged in at tremendous cost. Cement is generally shipped in by barge. Sometimes it is flown in, but this is cost prohibitive unless the quantities are small. Admixtures are either barged in or flown in. This reverses the cost scale so that the cheapest material in the concrete is admixtures followed by cement while the aggregates are the most expensive. Generally concrete quantities are small which makes it very expensive to bring in, set up, and operate crushing, screening, and washing facilities to produce aggregates that meet the grading requirements of ASTM C 33.

These economics require a significantly different approach to mixture proportioning. The aggregates must be accepted as they are and a way found to make quality concrete using them. ATL has developed procedures, modifying the ASTM and ACI standards and methods, that address these problems and result in economic, quality concrete. These same procedures can be used in other remote locations of the world where similar conditions exist.

2 Concrete mixture proportioning

There are a series of crucial questions the client must answer before trial proportioning can begin with confidence.

Where is the project? The project location immediately provides invaluable information. It reveals if there is barge access and how soon in the spring it will be possible to get a barge to the site -- many Alaskan locations are ice bound for much of the year. It also defines aircraft access. For example, if it is a military installation air access is probably unlimited. If it is an Eskimo village on an interior river there may be no more than 2,500 feet (762 m) of gravel airstrip available.

What is the volume of concrete to be placed? Since producing, forming, placing, and finishing remote site concrete often costs \$500 US to \$1,000 US per cubic yard (0.76 m^3), the amount of time spent developing and processing aggregates and refining the concrete mixture proportioning will vary considerably based on the amount of concrete required. If there is only a 100 cubic yards (76 m^3) it is cost effective to rely more on admixtures and cement and to minimize aggregate processing. However, with the high unit cost of the concrete, a small increase in volume quickly makes it economical to spend more effort on the mixture proportioning.

What will the concrete be used for? In the US it is normal to use standard specifications in the preparation of project specifications. Unless the designer has experience with designs in remote Alaska, usually he will do little to modify the standard language used in his technical specifications. By knowing what the concrete will be used for we can review the project specifications for practicality and applicability at this location.

What are the project specifications for the concrete? Obviously we must know what is required of the concrete before we can properly proportion the mixture. Our first review of the specifications determines if they can be met, given what we know of the project location and the available materials. Often it is necessary to recommend changes to the project specifications so the concrete can be produced or so the concrete can better meet the environmental conditions to which it will be subjected.

For example, it is not unusual for specifications to discourage or preclude the use of a high range water reducing admixture (HRWRA). Often with marginal aggregates in remote locations, an HRWRA is mandatory to produce the concrete required. Another example is specifying a low water cement ratio. With marginal, or out of normal specification aggregates, generally water demand is much higher. Good, high strength concrete can still be produced, but unrealistically low water cement ratios can not be meet. Other changes frequently recommended include maximum aggregate size [3/4-inch (19.0 mm) maximum is preferred], the type of cement used, or the use of accelerating admixtures or other techniques for cold weather concrete placement and curing.

Will the mixture be pumped? This question is critical because it is difficult to proportion a mixture that can be pumped using aggregates that fall outside the grading limits of ASTM C 33. Also the pump has a tremendous influence on the ability to pump a particular mix. In remote locations a contractor is stuck with whatever pump he takes to the job. He cannot get a new pump on the job in a few hours or even in a few days. The combination of abnormal aggregate grading and shape, and lack of flexibility in pump selection almost always results in failure.

Has an aggregate source been identified, and if so what is it? After 30 years proportioning mixtures throughout Alaska, often we will have done other mixtures using the same source. If so, we have a considerable advantage as we consult with our client on how to approach the mixture design. If we do not have such experience, we ask if they know if concrete has been produced from that source before, and what they know, or can find out, about how well it performed.

Have the materials for the mixture already been shipped, and if so what are they? Normally the answer is "Yes". Because of the long lead time needed to order materials, reserve barge space, and catch the earliest possible barge to allow the contractor maximum advantage of a short construction season (sometimes mid-June through mid-September), materials have been selected and shipped before we have been contacted. This presents additional problems in the design. For example, an adequate amount, or the right type, of cement may not have been ordered. On the other hand, it can also be an advantage. If excess cement was ordered, there is little reason to economize its use because it will never leave the job site. It has already been paid for and shipped. The expense has been incurred, therefore it might as well be used.

How will the aggregates be processed and the concrete batched? All aggregate processing and concrete batching equipment will be shipped to the site by barge. Usually it is already on a barge before the mixture proportion is requested. There generally are no alternatives to what the contractor has already decided to use. This quickly determines how much the aggregates can be processed; if the concrete can be batched using two piles of aggregate (coarse or fine); or whether it can be batched by weight or must be batched by volume.

When must the mixture proportion be completed by? This is often a problem. Since concrete is used in footings, the contractor wants to pour concrete soon after he begins work on the project. This is right after the ground thaws and just as the barge arrives. However, aggregate samples are not available because none have been produced; or if produced, they are coming from a frozen stockpile or beach. Since there is little time between when the aggregates can be sampled and when the contractor wants to place concrete, the time to proportion the mixture is usually compressed and this also affects the approach to the design.

2.1 Aggregates

Normally the aggregates will be mined from a beach, or produced by crushing and screening from a quarry. Many Alaskan villages are on the coast where beach deposits are common. Most other villages are on major, navigable rivers where the usual source is quarried rock. Alluvial deposits are unusual.

Despite concerns about potential corrosion of reinforcing steel from salt in the aggregates, beach deposits usually work well. The concern over salinity is often overstated. By following a few simple procedures, usually aggregates with low salinity adequate for normal applications can be found. However, selecting the location from which the aggregates are mined makes a good deal of difference.

There is a considerable variation in salinity along a beach. It is not uncommon to find small fresh water streams that periodically flow across the beach. Usually beach aggregates mined from the channels of such streams have much lower salinity than beach deposits not regularly washed by fresh water.

There is also a considerable difference in salinity of aggregates mined at different locations along any section perpendicular to the water. Surprisingly, the further up the beach the aggregates are, generally the higher the salinity, while the closer to the mean low low water line the aggregates are mined, the lower the salinity. Along the Alaskan coast, the wind blows across the water on shore almost constantly, frequently at high velocity. The wind picks up salt spray and deposits that spray against the upper parts of the beach at the spray line. This concentrates the salts and results in a greater salinity the higher up the beach the aggregates lay. The lowest salt contents are right at the mean low low water line where they have been least affected by salt spray.

Quarried aggregates are used about as frequently as beach deposits. Most locations, particularly those inland or those without accessible sand and gravel beaches, have established quarries. From these quarries, shot rock is mined, crushed, and screened to produce angular aggregates. They have a high water demand and their gradings usually fall below the coarse limit of the ASTM C 33 recommended grading for coarse and fine aggregates. The cost of producing aggregates at even this low standard is high. To process them more, to reduce the amount of fines present by washing, or to produce greater quantities of sand sizes is cost prohibitive even with large projects.

Occasionally fine, uniformly graded sands are available to blend with the fine aggregates to improve the grading. Adding five to ten percent blend sand to the fine aggregate can frequently improve the characteristics of the mixture significantly.

Well graded alluvial deposits are rarely available in remote Alaskan locations. When they are, developing a high quality concrete mixture is easier.

2.1.1 Aggregate gradings

In the mainland US by far the most widely used specification for concrete aggregate grading and quality is ASTM C 33 which establishes a single fine aggregate grading and a variety of coarse aggregate gradings, the most frequently used of which is the No. 67 grading -- a 3/4-inch (19.0 mm) maximum aggregate size (MAS).

A significant failing of this standard is that it does not address the proportion in which the fine and coarse aggregate are to be combined. The total grading of the aggregate in the mixture is much more important than the grading of the coarse or fine aggregate alone. The total grading of the mixture, including aggregate, cement, and water is even more important in determining workability, finishability, and pumpability of a particular mixture. It is essential to understand and take advantage of this principal when designing concrete mixture proportions in remote locations.

By varying the proportion of sand to coarse, sometimes taking the sand fraction to as much as 70% of the total aggregate weight (as opposed to the more normal fine fraction of 40%), by adding cement, and by the liberal use of admixtures high quality concrete mixes can be produced even with aggregate gradings considerably outside the ASTM C 33 standards.

A typical mixture produced at a remote mine location, 125 miles (200 km) north of the Arctic Circle and 50 miles (80 km) inland from the Bering Sea, was a 5,000 psi (34.5 MPa), air entrained concrete mixture with a 4-inch (10.2 cm) slump. The aggregates were produced by blasting, crushing, and screening rock from a quarry. The coarse and fine aggregate gradings are tabulated below.

Table 1. Coarse aggregate

		ASTM C 33, No. 67	
Sieve	Sample	Minimum	Maximum
1-inch (25.0 mm)	100.0	100	100
3/4-inch (19.0 mm)	100.0	90	100
1/2-inch (12.5 mm)	81.0	...	...
3/8-inch (9.5 mm)	45.0	20	55
#4 (4.75 mm)	8.2	0	10
#8 (2.36 mm)	0	0	5

Table 2. Fine aggregate

		ASTM C 33	
Sieve	Sample	Minimum	Maximum
3/8-inch (9.5 mm)	100.0	100	100
#4 (4.75 mm)	90.0	95	100
#8 (2.36 mm)	64.0	80	100
#16 (1.18 mm)	41.0	50	85
#30 (600 μM)	26.0	25	60
#50 (300 μM)	15.0	10	30
#100 (150 μM)	9.0	2	10
#200 (75 μM)	6.4	0	3

Even though the fine aggregate was far outside the coarse limit of the ASTM C 33 fine aggregate grading, it produced a good, workable mixture by increasing the fine aggregate proportion to 69% of the total mix, using 7.5 sacks (320 kg) of cement, and high dosages of a normal range water reducing admixture and an HRWRA in combination. Over 10,000 cubic yards (7,646 m^3) of concrete was successfully placed and finished. Average strength was over 6,000 psi (41.4 MPa). It continues to perform well. This approach proved far less costly than additional processing of the aggregates.

ASTM C 33 limits the percentage of material finer than a No. 200 (75-μm) sieve to 3% unless the aggregate is produced by crushing, and then the fines can be as much as 5%. We have successfully produced good quality concrete with fine aggregates having as much as 13% passing the No. 200 (75-μm) sieve, if the aggregates are produced by crushing and the fines are a product of the crushing process and not from silts and clays in the aggregates. The presence of these "crusher fines" increases water demand, but that can usually be overcome by the increase use of cement or admixtures which is cheaper than further processing the fine aggregates.

A typical concrete mixture proportion using crushed quarry rock with a high fines content is a mixture developed for use at the Adak Naval Air Station on Adak Island in the Aleutian Chain. Adak is located at 177°W longitude in the North Pacific. It has some of the worst weather in the world. The wind blows almost constantly and is salt laden. In the winter the temperature constantly fluctuates around freezing.

Three aggregates were used -- coarse and fine aggregates produced by crushing and screening quarry rock and a uniform, fine, native sand. The fines fraction for the crushed sand was 11.3%, and 1.8% for the blend sand. These values are well outside the ASTM C 33 recommended ranges.

Table 3. Adak coarse aggregate

		ASTM C 33, No. 67	
Sieve	Sample	Minimum	Maximum
1-inch (25.0 mm)	100.0	100	100
3/4-inch (19.0 mm)	98.0	90	100
1/2-inch (12.5 mm)	61.0	...	...
3/8-inch (9.5 mm)	34.0	20	55
#4 (4.75 mm)	4.8	0	10
#8 (2.36 mm)	0	0	5

Table 4. Adak fine aggregate

		ASTM C 33	
Sieve	Sample	Minimum	Maximum
3/8-inch (9.5 mm)	100.0	100	100
#4 (4.75 mm)	100.0	95	100
#8 (2.36 mm)	76.0	80	100
#16 (1.18 mm)	49.0	50	85
#30 (600 μM)	33.0	25	60
#50 (300 μM)	22.0	10	30
#100 (150 μM)	16.0	2	10
#200 (75 μM)	11.3	0	3

Table 5. Adak blend sand

Sieve	Sample
#16 (1.18 mm)	100.0
#30 (600 μM)	99.0
#50 (300 μM)	76.0
#100 (150 μM)	8.0
#200 (75 μM)	1.8

Table 6. Adak concrete mixture proportion aggregate weights per cubic yard (0.76 m^3)

	Coarse	Fine	Blend Sand
Weight, lb. (kg)	1,768 (802)	925 (420)	386 (175)
Percentage	56.4%	30.6%	13.1%

The cement content was 564 pounds (256 kg) per cubic yard (0.76 m^3) and the water content to produce a 3.0 inch (7.6 cm) slump was 319 pounds (145 kg) per cubic yard (0.76 m^3) with an air content of 5.5% ±1.5%. The average strength for this mixture was 3,800 psi (26.2 MPa) and it has adequately withstood Adak's severe weather.

It is near impossible to make pumpable concrete from aggregates outside the middle of the grading limits of ASTM C 33. Consequently, it is almost always impractical and uneconomical to produce pumpable mixes in remote locations. The extra cost of producing aggregates that will result in a pumpable mixture universally exceeds the cost of the extra labor required to place the concrete without pumping.

2.1.2 Other aggregate characteristics
The most significant aggregate characteristics after grading, in order of importance are: organic content, salinity (if beach deposits are used), reactivity, resistance to degrading from freeze thaw, and abrasion resistance.

The presence of even small amounts of organics in the aggregates drives up water demand dramatically. If there is any indication organics may be present, aggregates should be tested (ASTM C 40, "Standard Test Method for Organic Impurities in Fine Aggregate for Concrete"). It is unusual that good concrete can be produced using aggregates with organic contamination.

The salinity of both the coarse and fine aggregates should be checked if beach deposits are used. Using the chloride content of the aggregates, the weight of chlorides in the mixture can be estimated and expressed as a percentage of the cement content of the mix. The results should be compared against the recommendations of Table 4.4.1, "Maximum Chloride Ion Content for Corrosion Protection" in ACI 318. This simple calculation is conservative, as all chlorides in the aggregates are not soluble ions. However, if the chloride content is 500 ppm or less, generally this calculation will reveal that for normal applications the chlorides in the aggregates are not of concern. If the calculation results in higher values, or if the applications for the concrete require exposure to extreme corrosion conditions, further testing of the soluble chloride ion content of hardened concrete specimens from trial batches should be conducted.

Aggregates from throughout Alaska are rarely found to be reactive when tested [ASTM C 289, "Standard Test Method for Potential Reactivity of Aggregates (Chemical Method)"]. However, since the cost of shipping cement to the site exceeds the cost of the cement, it costs little to use low alkali cements to avoid problems. However, since materials have to be shipped well in advance of need, or even in advance of availability of the aggregates for testing, contractors and owners often order the cement and have it on a barge bound for the site before they request the production of a mixture proportion. This precludes specifying a low alkali cement if that is not what was shipped. On the rare occasions when this has been a problem it can be overcome by the addition of a small dosage of silica fume which is considerably more economical to purchase and ship in by air than the cost to replace all the cement.

The degrading of concrete because of freezing and thawing is rarely a function of aggregate quality, but is far more a function of the quality of the paste surrounding the aggregates. With proper air entrainment and low water-cement ratios, the majority of concrete under a wide range of normal exposures, even in Alaska, will function well. ASTM C 88, "Standard Test Method for Soundness of Aggregates by Use of Sodium or Magnesium Sulfate," normally used for estimating the freeze thaw resistance of aggregates is of little use unless test results show high losses, which is unusual.

It is simple to run abrasion tests on aggregates using standard test methods (ASTM C 131, "Standard Test Method for Resistance to Degradation of Small-Size Coarse Aggregate by Abrasion and Impact in the Los Angeles Machine"). However, much like testing aggregates for freeze thaw resistance, aggregates rarely fail to comply with the abrasion recommendations of ASTM C 33.

2.2 Cement

The two biggest problems with cement are: the poorest quality cement is shipped by suppliers to the most remote locations (they rightly figure it will not be rejected and shipped back); and it is not unusual for the owner or contractor to order and ship the cement before contacting a consultant to produce the mixture proportion. This means you must use what you have. Whenever possible, trial batches should be performed using the actual cement to be used on the job.

2.3 Admixtures

The use of normal range water reducing admixtures is the single most important and simplest procedure that can be used to improve the quality of concrete mixes in remote locations. They should always be used, and generally used at or near the maximum dosages. They add little to cost because they are used in relatively small volumes and therefore do not cost much to ship. Use newer products which do not significantly retard the mixture when higher than recommended dosages are used. This is important because the ability to use higher dosages gives added flexibility in adjusting the mix, and because it is not unusual with remote site concrete production for mistakes to be made with admixture dosages.

The use of high range water reducing admixtures can be very effective when using marginal aggregates. They are often mandatory to meet the project specifications because high water demand inevitably is part of out-of-normal specification aggregates. They should be used in combination with normal range water reducing admixtures. The normal range water reducing admixture is introduced with the batch water and provides enough workability that the HRWRA will be effective. If the mixture does not have at least 1.5 inches (3.8 cm) of slump, and preferably 2.0 inches (5.1 cm), the HRWRA is generally ineffective.

2.4 Trial batches

Laboratory versus field trial batches. Whenever possible, concrete mixture proportions should be done with trial field batches utilizing the aggregates, cement, admixtures, and equipment that will actually produce the mixture. This makes for better mixtures and often is more economical in remote locations. The cost of bagging and shipping aggregates from the site into the laboratory can easily run $2.00 US to $3.00 US per pound (0.45 kg). Since it takes 600 (272 kg) to 1,000 pounds (454 kg) of aggregate to do an effective mixture proportion, these are significant costs. It is usually cheaper to send an engineer or technician to the site to do the trial batches. This also produces more consistent concrete, as that same person can check equipment calibration and train on-site contractor personnel to produce the mixture.

3 Mixture production

3.1 Getting started right

The key to producing good concrete is to get started right. As the batching operation starts up, the problems begin. If those problems are not recognized and corrected, they will be carried through the entire production. Since volumes are generally low, it is likely all the concrete will be produced and placed before the problems are recognized in the hardened concrete.

The single most important thing that can be done to avoid problems is to put an engineer or senior technician, experienced in concrete mixture proportion and production and knowledgeable about the mixture, on the site to check out the plant, calibrate it, and be certain the personnel who will be producing the mixture understand it and know how to produce it with the available equipment. These two or three days on-site, prior to the start up of actual production, are very cost effective.

3.2 Typical problems

There are several typical problems that occur in remote site concrete batching operations. It is common that the personnel assigned to make the concrete are not knowledgeable of the equipment being used, or of concrete production.

Typographical or calculation errors are common. If there are no scales available, the mixture must be produced volumetrically. There are often problems converting batch weights to volumes. Typically the mixture will be produced in several batch sizes. It is not uncommon for the personnel to make a table of weights or volumes for several size batches and for those tables to have typographical errors.

The single biggest problem is the failure to understand the importance of, or how to go about making, aggregate moisture corrections from saturated surface dry (SSD) moistures to actual moisture content at the time of production. Include a chart in the mixture proportion that allows for easy conversion of batch weights, based on typical combinations of, and variations in, moisture content.

4 Quality control

4.1 Slump, air content, and unit weight

It is important that a quality control program be set up to measure slump, entrained air content, and unit weight of each batch of concrete. This is particularly important with volumetric batching. If volumes are small, to reduce costs one of the production personnel should be trained in performing these tests. This training is best done on site by the engineer or technician at the start of production.

4.2 Making, curing, and testing cylinders

Generally, strength is not of great concern for remote site concrete production. Normally the concrete required is unsophisticated 3,000 psi (20.7 MPa) or 4,000 psi (27.6 MPa) concrete. Because the aggregates are never ideal, and require the use of considerable cement and admixtures to provide adequate workability and finishability, once the mixture proportion has addressed those concerns, strength is easily attained. However, there are too many things that could go wrong, to not make test cylinders and break them.

It is difficult to handle and ship standard 6-inch (15.2 cm) by 12-inch (30.5 cm) cylinders. It is also expensive since they weigh almost 30 pounds (13.6 kg) uncrated and shipping costs are approaching $2.00 US to $3.00 US per pound (0.45 kg). The best solution, assuming a maximum aggregate size of 3/4-inch (19.0 mm), is to use smaller 4-inch (10.2 cm) by 8-inch (20.3 cm) cylinders. Once made on site they are stored in a lime bath, in an insulated tank. The lime batch temperature is held steady with a thermostatically controlled electric heater. When transportation is available, the cylinders are crated, and shipped to a testing laboratory for further curing and testing. Since these smaller cylinders weigh about eight pounds (3.6 kg), the amount of storage area required on-site, the crating needed for shipment, and the cost of handling and shipping is far less. Considerable parallel testing of various mixtures using the standard cylinders and the smaller cylinders have shown insignificant differences for a 3/4-inch (19.0 mm) maximum (or smaller) aggregate size mixture.

5 Conclusion

In remote locations, far removed from normal transportation systems, good concrete can still be consistently produced. However, to do so requires modifying the standard specifications for aggregates and the standard methods for concrete mixture proportions to accommodate the considerably different conditions that prevail in such locations.

6 References

1. American Society for Testing Materials, (1994) *Standard Specification for Concrete Aggregates*, Sec. C 33.
2. American Concrete Institute, (1994) *Building Code Requirements for Reinforced Concrete* Sec. 318.
3. American Concrete Institute, (1994) *Standard Practice for Selecting Proportions for Normal, Heavyweight, and Mass Concrete* , Sec. 211.
4. American Society for Testing Materials, (1994) *Standard Test Method for Organic Impurities in Fine Aggregate for Concrete*, Sec. C 40.
5. American Society for Testing Materials, (1994) *Standard Test Method for Potential Reactivity of Aggregates (Chemical Method)*, Sec. C 289.
6. American Society for Testing Materials, (1994) *Standard Test Method for Soundness of Aggregates by Use of Sodium or Magnesium Sulfate*, Sec. C 88.
7. American Society for Testing Materials, (1994) *Standard Test Method for Resistance to Degradation of Small-Size Coarse Aggregate by Abrasion and Impact in the Los Angeles Machine* (Sec. C 131).

PART EIGHTEEN

TESTING AND EVALUATION

103 DESTRUCTIVE AND NONDESTRUCTIVE TESTS FOR EVALUATING THE STRENGTH PROPERTIES OF CONCRETE CONTAINING SILICA FUME

F.H. AL-SUGAIR
King Saud University, Riyadh, Saudi Arabia

Abstract
The use of silica fume (SF) in concrete as an additive for improving durability is gaining acceptance. Concrete containing SF has displayed excellent properties and resistance to aggressive chemicals in many test programs. It has also been shown to produce high quality concrete with high compressive strength. The performance of concrete containing SF, in terms of strength properties, is investigated in this work. Three strength properties were studied using destructive tests: compressive strength, tensile strength and abrasion resistance. In addition, two common nondestructive tests were used to measure the ultrasonic pulse velocity and the rebound number (using the Schmidt hammer). Specimens were cast containing 4 different amounts of SF as an additive. From each mix, specimens were placed in three different drying conditions simulating optimum, average and severe conditions. The results indicated that the performance of concrete containing SF was consistently better with increasing amounts of SF. In addition, SF concrete performed just as well in severe environmental conditions as it did in average and optimum conditions of concrete drying.
Keywords: Concrete strength, durability, high-quality concrete, nondestructive testing, silica fume.

1 Introduction

Silica fume (SF) is a byproduct resulting from the reduction of high-purity quartz with coal in electric arc furnaces in the production of silicon and ferrosilicon alloys [1]. Silica fume (SF) has been called different names in the literature, including microsilica, ferrosilicon dust, and arc furnace silica. The main characteristics of SF are [1,2]:

1. It contains 85-98% silicon dioxide.

Concrete Under Severe Conditions: Environment and loading (Volume Two) Edited by K. Sakai, N. Banthia and O.E. Gjørv. Published in 1995 by E & FN Spon. ISBN 0 419 19860 1

2. Its mean particle size is in the range 0.1 to 0.2 micron.
3 .Its particles are spherical in shape.

Many investigations of the performance of SF in concrete were conducted in the 1970s and 1980s, and promising results were shown for the use of SF to produce high-performance concrete. For example, one of the earlier studies by Malhotra and Carette [3] presented SF concrete properties, applications and limitations. They reported that SF is a highly efficient pozzolanic material and has a considerable potential for use in concrete to improve its mechanical properties and durability [3]. A summary of the literature available on the use of SF in concrete can be found in two important documents which were published by the American Concrete Institute [1] and the FIP Commission on Concrete [4].

The mechanisms by which SF improves concrete performance are both physical and chemical. The physical effect is due to SF being much finer than cement; hence it acts as a filler of the spaces between the cement grains [5]. The chemical effect is the secondary pozzolanic reaction occuring between silicon dioxide and calcium hydroxide which results from the cement hardening process [6]. This pozzolanic effect takes place mainly between the ages of 3 and 28 days for curing at 20 degrees centigrade [7]. A third effect of SF on the compressive strength of concrete is to improve the interfacial bond between aggregates and cement paste [5]. The relative contribution of the three different effects described above to the increase in concrete compressive strength depends on the properties of the other mix constituents and is still a subject of research, for example,[5,8]. There are two ways to use SF in producing concrete: as a replacement for cement or as an additive [3]. The first way helps reduce the extra costs incurred from adding SF by reducing the cement content. Different percentages of SF have been used in concrete, ranging from 5% to 30%. It has been reported that 24% SF as a replacement for cement is optimum in the sense that it consumes all the calcium hydroxide that is produced by the cement hydration process [2].

One of the main problems associated with the use of SF is the loss of workability of fresh concrete. To correct this problem, superplasticizers are usually used in amounts increasing with the increase in SF used.

In this study, the strength properties of concrete containing SF were investigated using both destructive and nondestructive tests. Concrete mixes with four different amounts of SF added to the concrete mix were cast and tested. Specimens were cast containing 0, 5, 10, and 15% added SF. In addition, specimens were dried in three different conditions to simulate optimum, average and severe conditions.

The investigation included three destructive tests: compressive strength tensile strength and abrasion resistance tests; it also included two nondestructive tests: the ultrasonic pulse velocity and the Schmidt hammer tests. For all mixes, a superplasticizer was used to obtain a slump of 80-100 mm.

2 Experimental information

The material properties were the same for all mixes. A portland cement type I satisfying ASTM C150 [9] was used. Crushed limestone aggregate and sand meeting ASTM C33 [9] grading requirements was used. The maximum aggregate size was 10 mm, and the fineness modulus of the sand was 2.7. The mixing water was potable from the city distribution network. The superplasticizer used was based on sulphonated naphthalene.

As previously mentioned, one basic concrete mix was designed The mix had a water/cement (w/c) ratio of 0.5, and its mixture proportions are given in Table 1. Four mixes from this basic mix were cast with 0, 5, 10, and 15% of the cement weight as added SF. The amount of superplasticizer (SP) added was to maintain a slump of 80-100 mm. The mixture proportions for mixes designated M1 (0% SF), M2 (5% SF), M3 (10% SF) and M4 (15% SF) are given in Table 1.

It is important to note that only M1 in Table 1 had a volume of 1 cubic meter. Mixes M2, M3 and M4 had volumes a little more than 1 cubic meter due to the amount of added SF. The issue of objective comparison between mixes is a subject of discussion among researchers. Whether to design mixes based on a constant volume, constant compressive strength or any other criterion is a subject of controversy. It is the opinion of the author that the mix design approach used in this paper reveals the effect of adding SF in a clear manner.

Standard concrete cylinders for measuring compressive strength at 28 days in accordance with ASTM C39 [9] were cast. Two identical concrete cylinders were cast from each mix for each drying condition and were used to measure the compressive strength in accordance with ASTM C39 [9]. Two extra cylinders were cast and used to measure the pulse velocity in accordance with ASTM C597[9] and the rebound number in accordance with ASTM C805 [9] throughout a testing period of 154 days; they were then tested for compressive strength.

Table 1. Mix design proportions in kilograms per cubic meter

Material	M1	M2	M3	M4
Cement	380	380	380	380
Water(free)	190	190	190	190
Coarse Agg.	1033	1033	1033	1033
Sand	669	669	669	669
SF	0	19	38	57
SP(ml/kg cement)	3.88	12.4	13.7	23.6
Slump(mm)	97	95	92	100

Two standard concrete briquettes were cast from each mix for each drying condition to measure the tensile strength in accordance with ASTM C190 [9]. In addition, small slabs were cast and tested for abrasion resistance in accordance with ASTM C944 [9]. Special care was taken to avoid plastic shrinkage cracking, which is usually a problem with concrete containing SF.

All specimens were covered with wet burlap for 24 hours immediately after casting . After that, the specimens were transferred to stainless steel curing tanks containing regular water. The specimens were kept in the tank for 6 more days; subsequently, some were transferred to two environmental conditions. Condition B was a controlled room set at a temperature of 30 degrees and relative humidity (RH) of 50%, and condition C was a controlled room which had a temperature of 55 degrees and RH of 5%. The rest of the specimens were kept in the water tank until the different tests were conducted. This was denoted condition A.

3 Results and discussion

3.1 Compressive strength

The compressive strength results are given in Table 2. The results include results at ages of 28 days and 154 days. As expected, the compressive strength increased with increasing SF content. The specimens for this test were continuously cured under water in condition A. Generally there was not a great improvement in strength when the amount of added SF is increased from 10 to 15%.

3.2 Tensile strength

The 28-day tensile strength for all mixes is given in Table 3. Generally, the tensile strength increased with increasing amounts of SF. The tensile strength for specimens in condition A was higher than in other conditions. This was mainly due to better curing in condition A. On the other hand, condition C had the lowest tensile strength due to it being a relatively severe drying condition. The results for M2 in condition A are not given in Table 3 since an error occurred during testing of the specimens.

Table 2 . Compressive strength results (MPa)

Age(days)	M1	M2	M3	M4
28	27	35	40	37.8
154	29.1	37.2	43.5	45

Table 3. Tensile strength results (MPa)

Condition	M1	M2	M3	M4
A	3.45	--	3.31	3.79
B	2.76	2.96	2.96	3.1
C	2.45	2.55	2.62	2.86

3.3 Abrasion resistance

The abrasion resistance test in accordance with ASTM C944 involves subjecting specimens to a wearing test for a fixed amount of time and measuring the percentage of weight loss. The comparison is valid as long as the specimens are of identical geometry and are subjected to the wearing action for the same time period. This was 5 minutes in this test program. The results are given in Table 4. The percentage of weight loss for specimens in condition A at two ages of 7 and 154 days are given, and that for specimens in conditions B and C at 154 days only. Generally, specimens with more SF experienced a smaller loss due to abrasion. It is also seen that specimens in condition A were more resistance to abrasion, again due to better curing than conditions B and C. An obvious observation is that the abrasion resistance was better with time because of better hydration.

3.4 Pulse velocity

In general, the pulse velocity (PV) increased with increasing moisture content and with concrete strength [10]. Therefore, it is expected that at the early stages of drying, the PV will increase since the strength gain rate is larger than the moisture loss rate. At some time during drying, the PV should start decreasing when the rate of moisture loss exceeds that of strength gain. It is expected that the PV values should eventually stabilize at a constant value after some time when both rates of strength gain and moisture loss become very small. Generally, the PV reflects the soundness of the concrete. The larger the PV, the sounder is the concrete.

Fig. 1 shows the change in PV with time for specimens from all mixes in condition A. The PV was larger for specimens with larger percentages of SF. This is consistent with the previous findings that the strength of specimens increased with the percentage of SF. The PV increased for the first 50 days then starts to decrease. This is expected as previously explained. The results for specimens in conditions B and C are shown in Figs 2 and 3 and provide the same conclusions as those in Fig. 1.

Table 4. Abrasion resistance as percentage loss of weight

Age(Days)	Condition	M1	M2	M3	M4
7	A	0.52	0.50	0.47	0.43
154	A	0.24	0.23	0.2	0.18
154	B	0.29	0.27	0.24	0.23
154	C	0.32	0.29	0.27	0.26

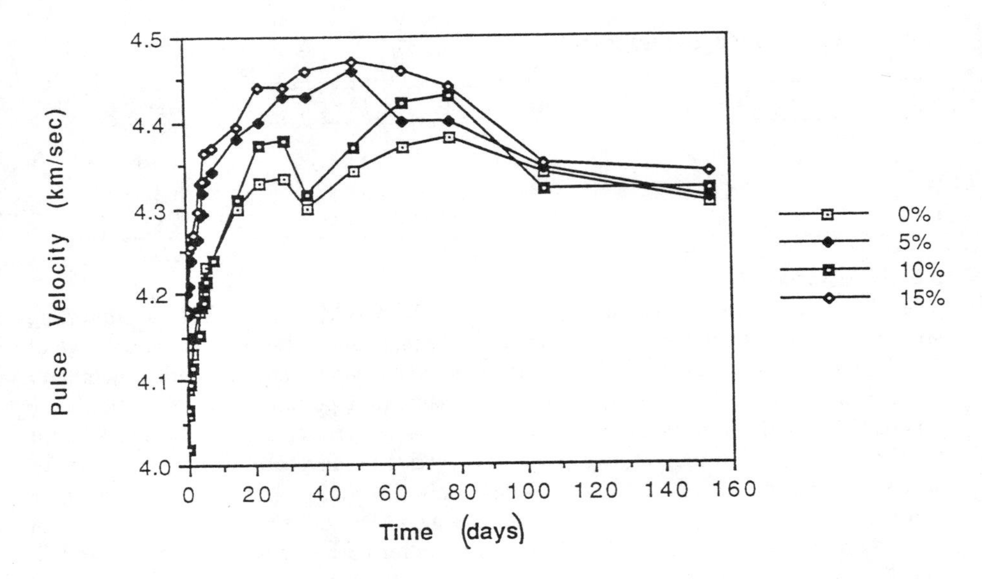

Fig. 1. Pulse velocity of specimens in condition A.

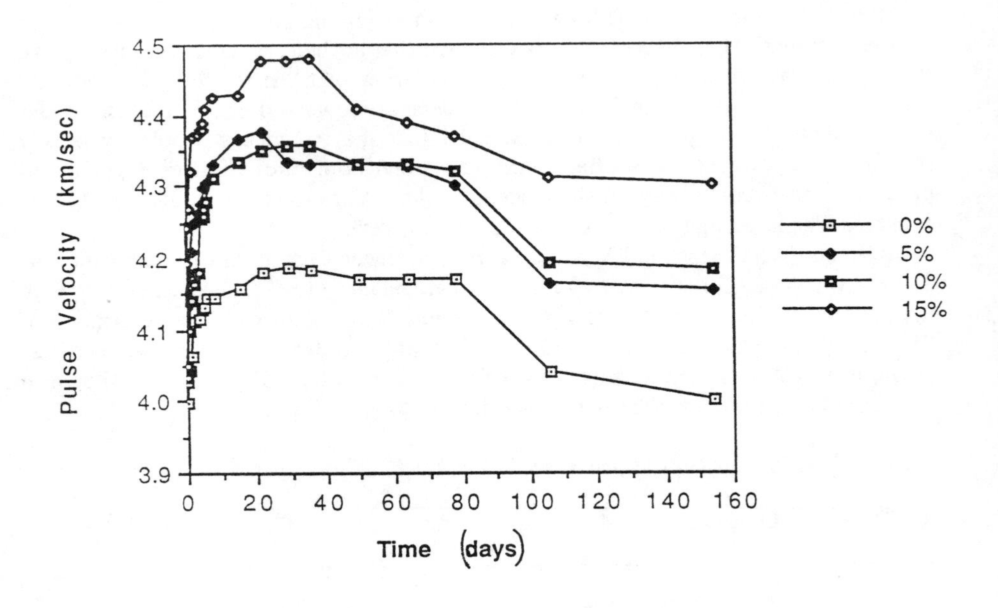

Fig. 2. Pulse velocity of specimens in condition B.

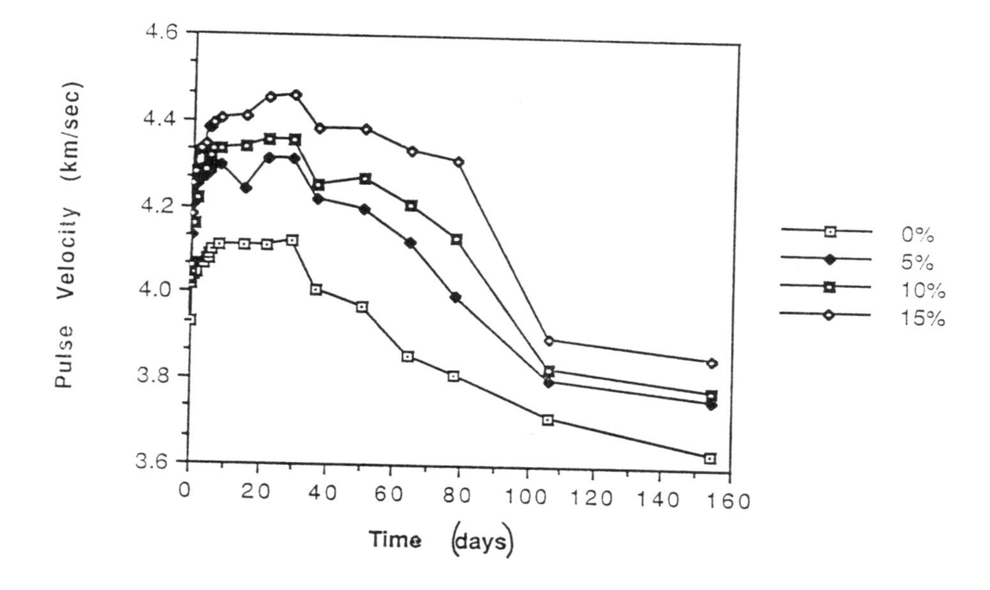

Fig. 3. Pulse velocity of specimens in condition C.

3.5 Rebound number

In general, the physical processes described above in relation to the PV can be used to explain the results of the Schmidt hammer rebound number (RN). One important note, however, is that the rebound number is usually less accurate than the PV and is sensitive to any changes in the environment; thus it exhibits a large variability in its results [11]. The rebound number reflects the hardness of the concrete. The larger the RN, the harder is the concrete.

Figs. 4,5 and 6 show the results of the RN with time for specimens in conditions A,B and C. In general, the increase in SF percent increased the RN and, thus, the strength.

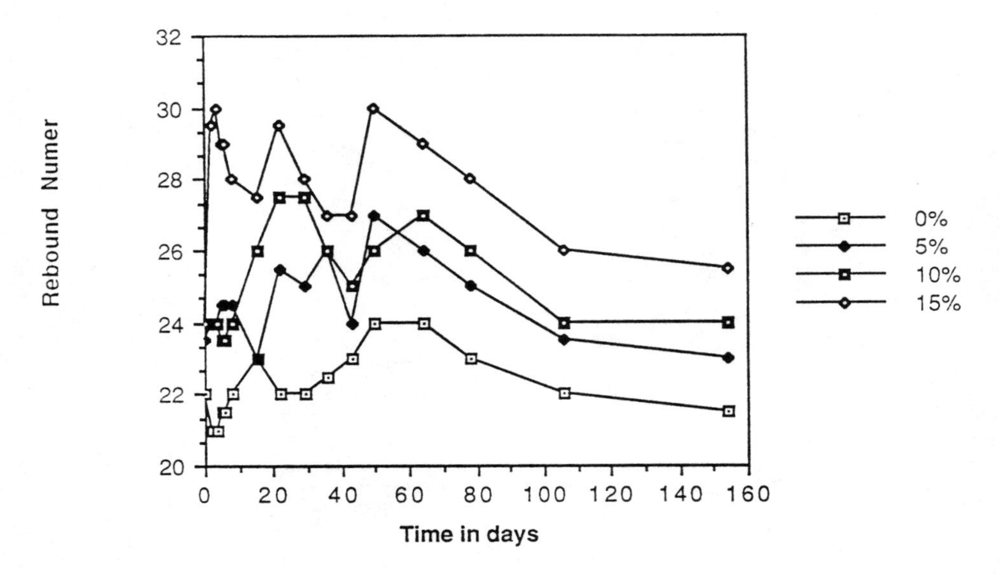

Fig. 4. Rebound number of specimens in condition A.

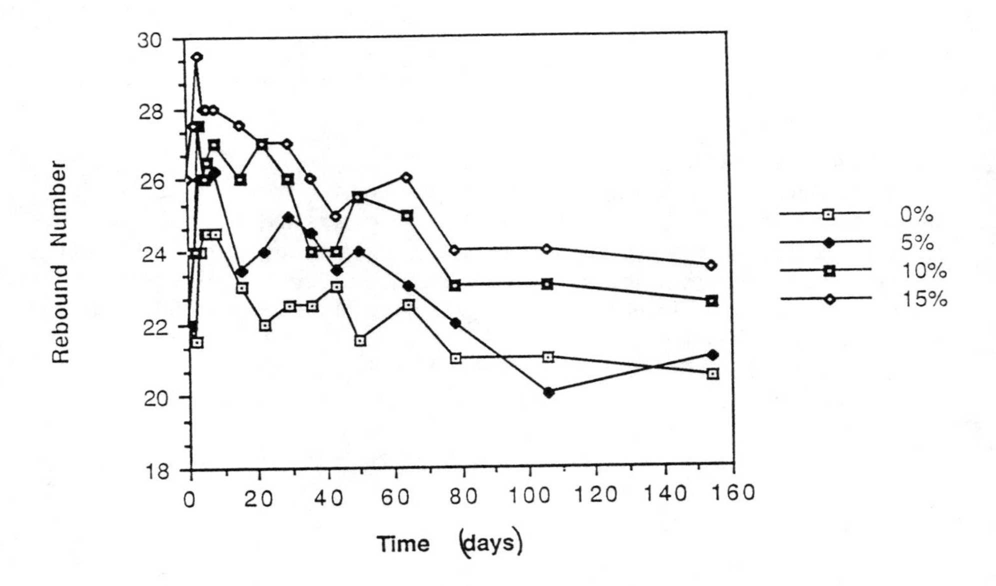

Fig. 5. Rebound number of specimens in condition B.

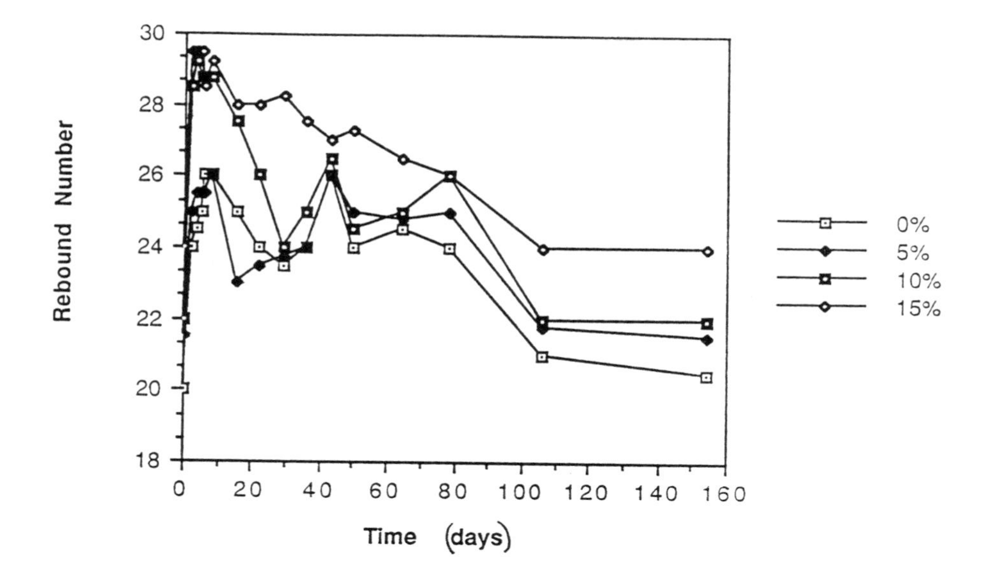

Fig. 6. Rebound number of specimens in condition C.

4 Conclusions

The quantitative data presented herein lead to the general conclusion that increased amounts of SF in the concrete mixes increase their compressive and tensile strengths, resistance to abrasion, soundness, and hardness. The results are consistent and lead to the same general conclusion for all of the test types and drying conditions; silica fume improves concrete strength properties even in severe drying conditions.

5 References

1. ACI Committee 226 (1987) Silica fume in concrete. *ACI Materials Journal*, Vol. 84, No. 2, pp. 158-166.
2. Sellevold, E.J. et al. (1981) Silica fume-cement pastes: hydration and pore structure, in *Nordisk Miniseminar on Silica in Concrete*, Cement and Concrete Research Institute, Trondheim, Norway.

3. Malhotra, V.M. and Carette, G.G. (1983) Silica fume concrete-properties, applications and limitations. *Concrete International*, Vol. 5, No. 5. pp. 40-46.
4. FIP Commision on Concrete (1988) *Condensed silica fume in concrete*, Thomas Telford Ltd., London, England.
5. Detwiler, R.J. and Mehta, P.K. (1989) Chemical and physical effects of silica fume on the mechanical behavior of concrete. *ACI Materials Journal,* Vol. 86, No. 6. pp. 609-614.
6. Sellevold, E.J. (1987) Condensed silica fume in concrete: a world review, in *International Workshop on Condensed Silica Fume in Concrete*, Montreal Canada.
7. Sellevold, E.J. and Radjy, F.F. (1983) Condensed silica fume in concrete: water demand and strength development, in *SP-79; Fly Ash, SilicaFume, Slag and Other Mineral By-Products in Concrete*, pages 677-694, ACI Detroit.
8. Cong, X., Gong, S., Darwin, D. and McCabe, S.L. (1992) Role of silica fume in compressive strength of cement paste, mortar, and concrete. *ACI Materials Journal*, Vol. 89, No. 4. pp. 375-387.
9. American Society for Testing and Materials (1992) S*tandards. Section 4-Construction,* ASTM, Philadelphia, USA.
10. Neville, A.M. (1983) *Properties of Concrete*, Pitman Publishing, London, Third Edition.
11. Troxell, G.E., Davis, H.E. and Kelly, J.W. (1968) *Composition and Properties of Concrete*, McGraw-Hill, NewYork.

104 EVALUATION OF DETERIORATED CONCRETE STRUCTURES BY QUANTITATIVE IMPACT TEST AND SPECTROSCOPY OF TRANSVERSE ELASTIC WAVE

M. OHTSU
Kumamoto University, Kumamoto, Japan
S. UESUGI
Kumamoto Institute of Technology, Kumamoto, Japan

Abstract
To ensure that concrete structures function properly under severe environments, inspection methods for extent of deterioration need to be established. With this concern, impact tests and spectral analysis of transverse elastic waves are quantitatively investigated for nondestructive evaluation of concrete.

In the impact loading tests, resonance vibrations of reinforced concrete beams are analyzed by the boundary element method (BEM). Based on analytical results, the quantitative relation between the dynamic behavior of the beams and the deterioration are discussed in respect to spectral responses. It is suggested that peak frequencies of given resonance modes decrease with an increase of deterioration.

In the spectroscopy study of transverse elastic waves, the effect of a deteriorated layer is studied by a coupling analysis of the BEM and the finite element method (FEM). In concrete beams subjected to freezing and thawing, the peak frequencies of the spectral response shift to lower frequencies with an increase in thickness of the deteriorated layer. These relations were analytically confirmed and could be applied to evaluate the thickness of a deteriorated layer.
Keywords: Impact test, nondestructive evaluation, spectroscopy of transverse elastic waves.

1 Introduction

Some concrete structures deteriorate seriously under severe environments. Therefore, monitoring and estimating the deterioration have been major concerns in concrete

Concrete Under Severe Conditions: Environment and loading (Volume Two) Edited by K. Sakai, N. Banthia and O.E. Gjørv. Published in 1995 by E & FN Spon. ISBN 0 419 19860 1

engineering for a long time. A number of inspection techniques have been investigated and some are developed as nondestructive evaluation (NDE) techniques. Recently, NDE techniques for infrastructure were extensively studied [1], because it was realized that a large number of civil structures are approaching their service-life limit. To evaluate deteriorated concrete structures quantitatively, one promising NDE is the measurement of spectral responses to dynamic behaviors. In the low frequency range, global vibration modes are measurable by employing an impact test [2]. Ultrasonic testings are available for the high frequency range. These are applied to detect defects [3] and to estimate the depth of a crack visible at concrete surface [4]. Concerning the impact test, analytical studies based on the finite element method (FEM) [5] and the boundary element method (BEM) [6] have been reported. In ultrasonic testing, the spectroscopy of transverse elastic waves is applied to the thickness determination of the deteriorated layer due to freezing and thawing [7]. On the basis of these results, quantitative NDE of deterioration in concrete members is reported here, by the impact test and by the spectroscopy of transverse elastic waves.

2 Elastodynamic theory and spectral responses

In elastodynamics, a governing equation of elastic displacements is given, as follows;

$$(\lambda+\mu)u_{j,ji}(\mathbf{x},t) + \mu u_{i,jj}(\mathbf{x},t) = \rho\ddot{u}_i(\mathbf{x},t), \quad \text{for } i,j = 1,2,3 \tag{1}$$

where λ and μ are Lame constants and ρ is the density of the material. The term $u_i(\mathbf{x},t)$ is the displacement vector in 3-D space, which could be assumed as $u_i(\mathbf{x})\exp(i2\pi ft)$ in steady-state motion. Taking into account the plane-strain state in 2-D space, Eq. (1) is converted:

$$[(k_T/k_L)^2-1]u_{j,ji}(\mathbf{x}) + u_{i,jj}(\mathbf{x}) + k_T^2u(\mathbf{x}) = 0, \quad \text{for } i,j = 1,2 \tag{2}$$

$$u_{3,jj}(\mathbf{x}) + k_T^2u_3(\mathbf{x}) = 0, \tag{3}$$

where k_L and k_T are the wave numbers for P wave and S wave, and are given as $k_L = 2\pi f/v_p$ and $k_T = 2\pi f/v_s$. v_p and v_s are velocities of P wave and S wave, respectively. For a 2-D problem, in-plane motion of a concrete beam is represented by Eq. (2). The propagation of transverse elastic waves is governed by Eq. (3) as out-of-plane motion. Both equations hold for a particular frequency, f. Thus, spectral responses of concrete members are obtained by solving Eqs. (2) and (3) for sequential frequencies in the numerical analysis of BEM and FEM.

3 Experiment and analysis

3.1 Specimens

Mixture proportions employed are summarized in Table 1. In all mixture, the maximum size of gravel was 10 mm. For the impact test, reinforced concrete beams were made of three concrete mixes. These are mix A, mix B, and mix C in the table. The water to cement ratios were 30%, 45%, and 60%, resulting in compressive strengths estimated as 86.4 MPa, 46.8 MPa, and 23.5 MPa at 28 days under moisture curing. Three types of reinforced concrete beams were made of each concrete mix. These were specimen H6 with dimensions 1200 mm x 50 mm x 60 mm, specimen H8 with dimensions 1200 mm x 50 mm x 80 mm, and specimen H12 with dimensions 1200 mm x 50 mm x 120 mm. For each specimen, a reinforcing steel bar of 10 mm diameter was embedded with 15 mm cover thickness, although 15 mm cover thickness is relatively small because of scale-down models.

For plain concrete beams subjected to freezing and thawing, concrete mix D was employed. Specimens with dimensions 400 mm x 100 mm x 100 mm were cast. After 28-day moisture curing, temperature under a freezing-thawing cycle was varied from 4°C to -18°C during three hours.

3.2 Impact test of reinforced concrete beam

A steel ball of 20 mm diameter was dropped from 50 mm height on the specimen. The impacted position on the top surface was 100 mm apart from the center of the specimen. Displacement and acceleration were measured at the center of the bottom surface. Corresponding analytical models are shown in Fig. 1. In this figure, locations of strain gauge, displacement-meter, and accelerometer are given. To estimate the impact stress, the strain gauge was attached to the specimen. Numerals in the figures denote the number of representative boundary elements. Displacements measured were digitally recorded at 400 μs sampling and accelerations were at 5 μs sampling, because displacement-meters measure frequency response lower than 200 Hz and accelerometers could measure up to 10 kHz.

3.3 Spectroscopy of transverse elastic wave

By employing ultrasonic sensors in the transverse mode, spectral responses of concrete beams deteriorated by freezing and thawing were measured. A sketch of the

Table 1 Mixture proportions and concrete properties

mix	unit weight (kg/m^2)* W	C	S	G	air (%)	slump (cm)	compressive strength(MPa)
A	178	593	695	793	2.4	5.1	86.4
B	180	400	816	827	8.1	9.3	46.8
C	178	297	913	820	11.8	15.2	23.5
D	158	306	875	950	5.7	6.0	33.8

*W = water, C = cement, S = sand, and G = gravel

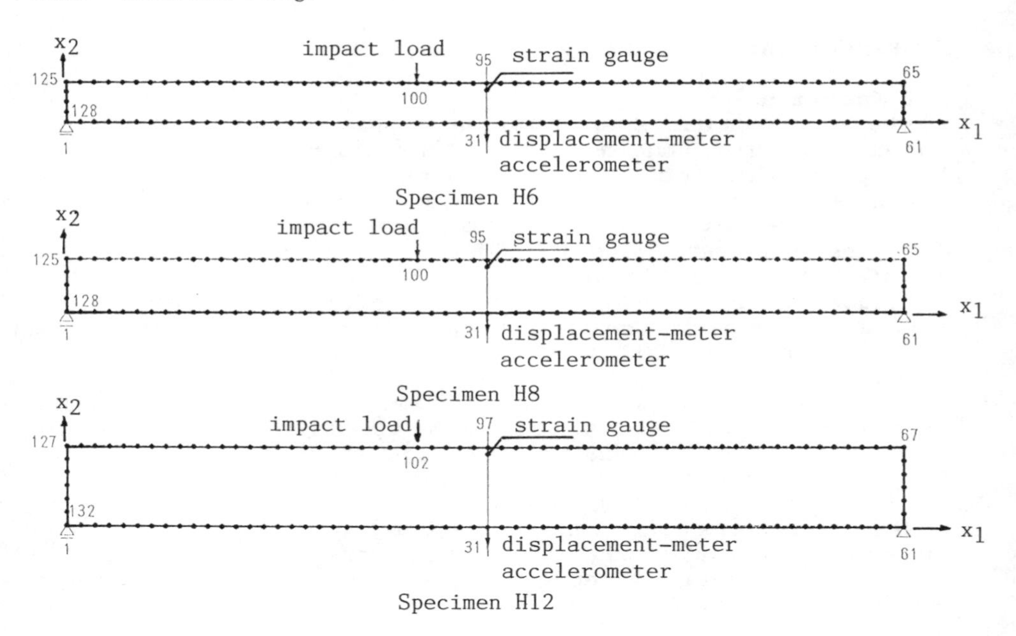

Fig. 1. BEM models for impact tests of reinforced concrete beams.

measurement procedure is shown in Fig. 2. A function synthesizer was used to generate sinusoidal signals of constant amplitudes with sweep-mode frequency modulation. At 100 mm apart from the input sensor location on the top surface, propagated waves were detected. Average amplitudes were recorded against the driving frequency. Thus, spectral responses were directly measured. Due to the freezing-thawing action, the concrete surface deteriorates. In the analysis, the deteriorated layer was modeled by FEM and the base concrete was modeled by BEM.

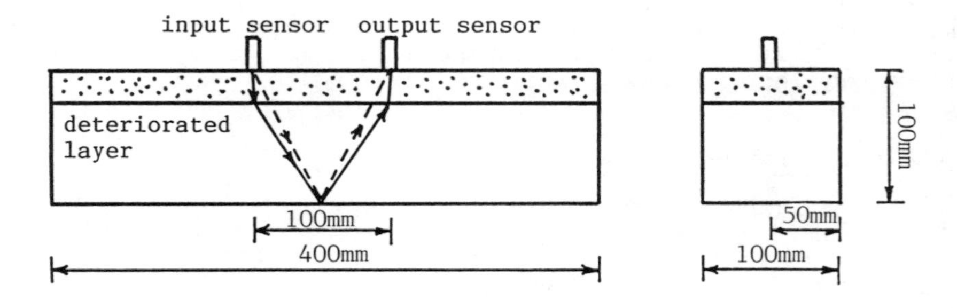

Fig. 2. A sketch of the spectroscopy measurement, and an analytical model for the deteriorated layer.

4 Results and discussion

4.1 Impact response of deteriorated RC beam

To analyze Eq. (2) by BEM, Young's modulus should be known because relations between Young's modulus, E, and velocities of elastic waves are given as,

$$v_p = [(1-\nu)E/\{\rho(1+\nu)(1-2\nu)\}]^{0.5}, \tag{4}$$

$$v_s = [E/\{2\rho(1+\nu)\}]^{0.5}. \tag{5}$$

Here, ν is Poisson's ratio. In the analysis, to take into account the effect of the reinforcing steel bar, the fundamental resonant frequency of a simply supported beam is compared with the spectrum of the displacement. The resonant frequencies of the simple beam are given by:

$$f_n = \pi n^2/(2L^2)[EIg/A\rho]^{0.5}, \tag{6}$$

where n is the number of the vibration mode, L is the span length, I is the moment of inertia, g is the acceleration of gravity, and A is the cross-section of the beam. Because only one peak frequency was observed in the frequency spectrum of the displacement record, this peak frequency was substituted into the first resonance f_1 determined from Eq. (6) (the case n=1). Thus, equivalent Young's modulus of each specimen was determined. These are summarized in Table 2. These values are approximately 30% greater than those of plain concrete. It is observed that Young's moduli vary, depending not only on the mixture but also on the specimen size. Small moduli of specimen H12 may result from poor compaction during the casting procedure.

By utilizing these Young's moduli, theoretical spectral responses at the center of the bottom of the beam were computed in the BEM models, sequentially giving frequencies. The case of specimen H12 is given in Fig. 3. Agreement between frequency spectra transformed from experimental records of acceleration and spectral responses by the BEM analysis is excellent. From the results of concrete with mix A, four dominant peak frequencies are identified. In the experiments, it was observed that these peak frequencies shift to lower frequencies with the decrease of Young's modulus (also compressive strengths) among mixes A, B, and C. In the analysis, these peaks KS

Table 2 Equivalent Young's moduli for BEM analysis

mix	specimen H6 (GPa)	specimen H8 (GPa)	specimen H12 (GPa)
A	49.8	46.3	37.2
B	40.8	39.3	32.4
C	25.7	30.7	28.3

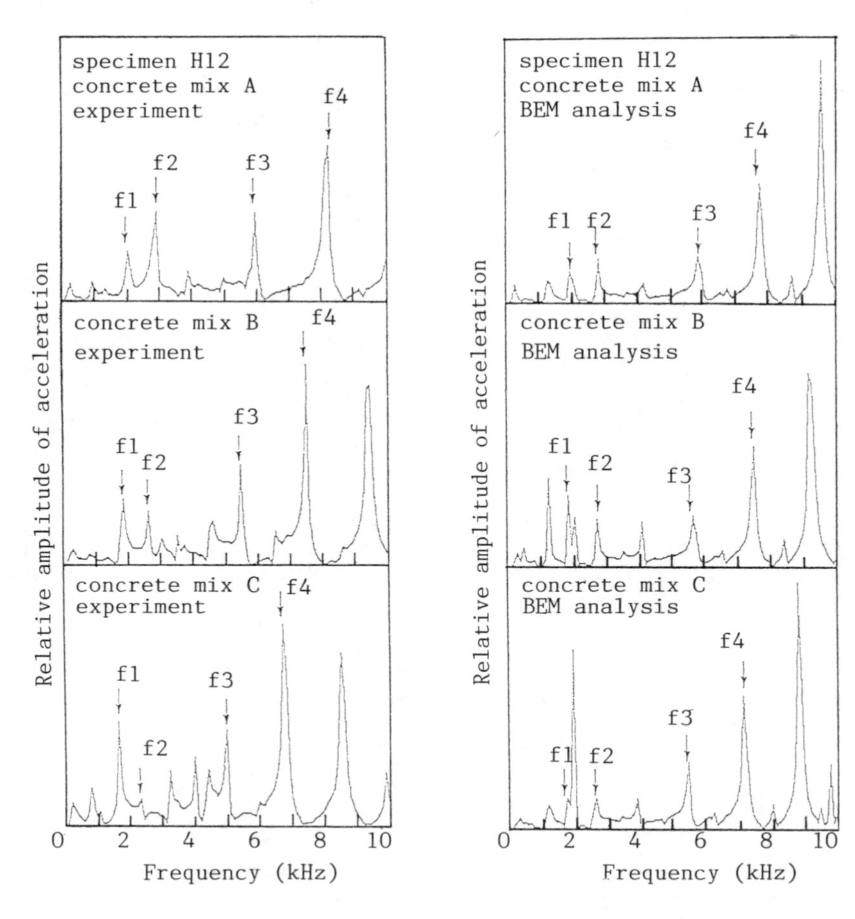

Fig. 3. Spectral responses of RC beams for specimen H12.

were carefully identified from the vibration mode. Resonance modes of the peak frequency f2 are shown in Fig. 4. The peak frequencies decrease with the decrease of Young's modulus, whereas the resonance modes are identical. Thus, identification of the corresponding peak frequency is readily performed. These results imply that the deterioration could be evaluated from the decrease of the resonance frequency of the identical vibration mode.

The significance of the resonance mode for the peak frequency identification is emphasized from comparison with Eq. (6). Resonance frequencies of the simple beam were calculated, and then the resonance modes of the corresponding frequencies in the analysis were obtained. Results are given in Fig. 5. Except for the first resonance mode, resonance frequencies were determined as close as those of the simple beam in Eq. (6). It is realized

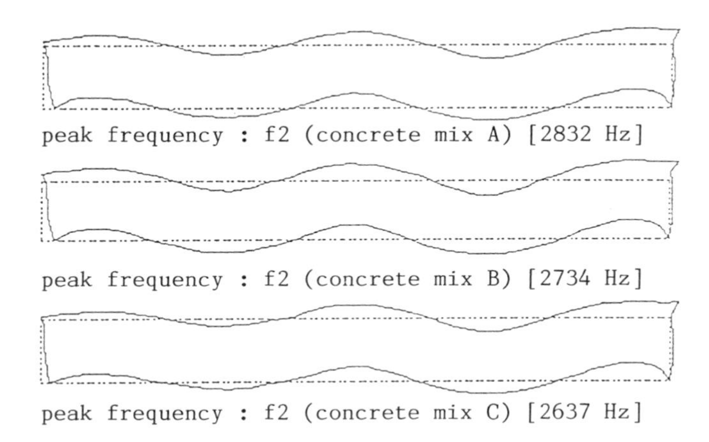

Fig. 4. Peak frequency, f2, for different concrete mix.

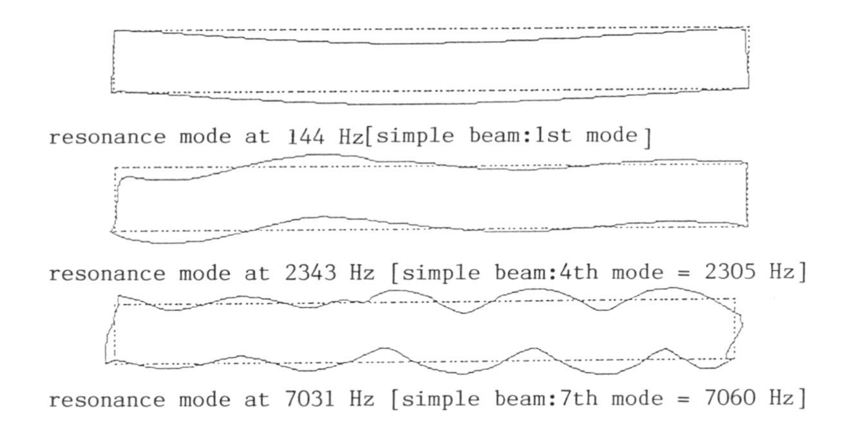

Fig. 5. Resonance modes for specimen H12 of concrete mix A and the corresponding modes of the simple beam.

the resonance vibration modes are not so simple as predicted from the resonance vibration number of the simple beam, although resonance frequencies are comparable.

The shifts of the peak frequencies are summarized in Fig. 6 for the case of the specimen H12. In this figure, relations between peak frequency ratio to mix A and the decrease of peak frequencies are given. The decrease of peak frequencies in the experiments is in remarkable agreement with those of the BEM analysis. It is observed that frequency shift due to the mixture variation is quite similar for all peak frequencies. This implies that the evaluation of deterioration could be performed from only

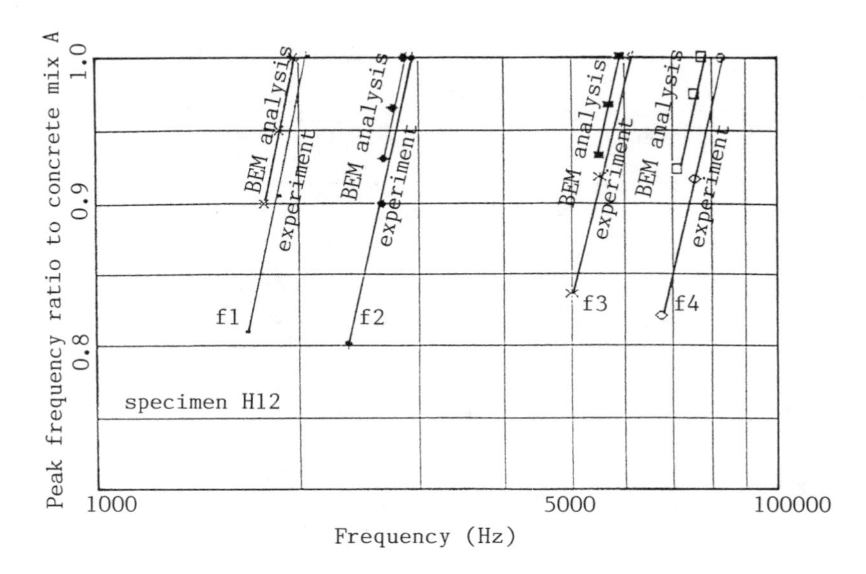

Fig. 6. Relations between the peak frequency ratio to concrete mix A and the decrease of the peak frequencies.

one particular peak frequency, when the vibration mode is identified.

4.2 Deteriorated layer due to freezing and thawing

Variation of spectral responses during freezing-thawing cycles is given in Fig. 7. There exist two dominant peak frequencies, f_o and f_n. It was realized that the peak frequency f_o is independent of the cycles, while the peak frequency f_n shifts to the lower frequencies with an increase of the freeze-thaw cycles.

For the comparison, relative dynamic modulus of elasticity (R.D.M.E.) was also measured. Under the axial vibration, the resonance frequency f_r is given by,

$$f_r = v_p/(2L). \quad (7)$$

Substituting Eq. (4) into Eq. (7), the resonance frequency f_r is related to Young's modulus E. Thus, the decrease of relative Young's modulus E_1/E_0 is given from the corresponding resonance frequencies f_{r1} and f_{r0},

$$E_1/E_0 = [f_{r1}/f_{r0}]^2. \quad (8)$$

In Fig. 8, it is observed that R.D.M.E. decreased from 1.0 to about 0.7 during 145 freezing-thawing cycles.

In BEM, Young's modulus of the deteriorated layer was assumed to be a half of the base concrete, and the depth of the layer was varied from 0 mm to 10 mm, based on the visual observation. The normalized frequency (f_n/f_o) is compared with the analytical results as shown in Fig. 8.

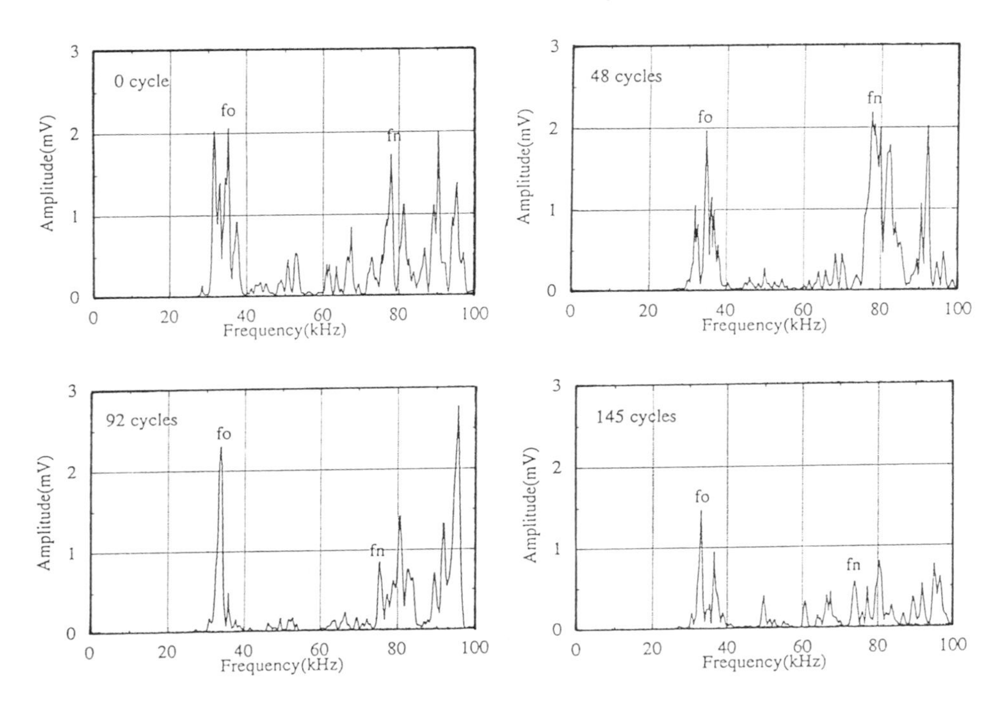

Fig. 7. Spectral responses of deteriorated concrete beams due to freezing-thawing cycles.

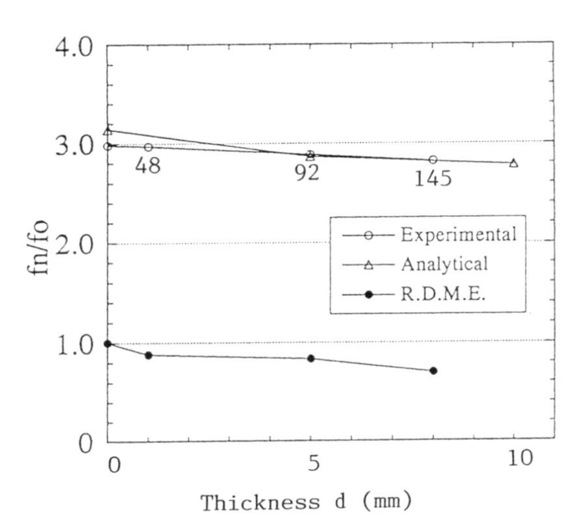

Fig. 8. The decrease of the normalized peak frequencies with the increase of the deteriorated layer.

Agreement between experimental and analytical results is quite good. The decrease of R.D.M.E. is also similar to that of the normalized frequency. This implies that the depth of the deteriorated layer and the decrease of R.D.M.E. could be evaluated from the decrease of the normalized frequency.

5 Conclusion

For NDE of deteriorated concrete structures under severe environments, the impact test is investigated for the evaluation of the dynamic behavior of reinforced concrete beams, and the spectroscopy of transverse elastic waves is applied to evaluate the thickness of a deteriorated layer.

In the impact loading tests, the dynamic behavior is analyzed by BEM. Based on the analytical results, a quantitative impact test is suggested from spectral information. By identifying the resonance mode, the deterioration could be evaluated from the decrease of the resonance frequency.

In the spectroscopy study, the effect of a deteriorated layer is studied by a coupling analysis of BEM and FEM. Peak frequencies in the spectral response are shifted to lower frequencies with the increase of the deteriorated layer due to freezing and thawing.

6 References

1. Suprenant, B.A. et al. eds. (1992) *Proc. Nondestructive Evaluation of Civil Structures and Materials,* University of Colorado, Boulder.
2. Alampalli, S., Fu, G. and Aziz, I.A. (1992) Nondestructive evaluation of highway bridges by dynamic monitoring, *Proc. Nondestructive Evaluation of Civil Structures and Materials,* University of Colorado, Boulder, pp. 211-26.
3. Sansalone, M. and Carino, N.J. (1988) Laboratory and field studies of the impact-echo method for flaw detection in concrete, *Nondestructive Testing,* ACI, SP-112, pp. 1-20.
4. Ohtsu, M. and Sakata, Y. (1992) Nondestructive crack identification by acoustic emission analysis and ultrasonic frequency response, *Nondestructive Testing of Concrete Elements and Structures,* ASCE, pp. 162-70.
5. Shiartori, M., Higai, T. and Okamura, Y. (1992) Analytical study on the response of concrete members for light impact load, *Proc. of the JCI,* Vol. 14, No. 1, pp. 679-84.
6. Kaneda, K. and Ohtsu, M. (1993) Study on impact response of deteriorated concrete members, *Proc. of the JCI,* Vol. 15, No. 1, pp. 631-36.

105 ASSESSMENT OF FROST DAMAGE OF CONCRETE AND ITS REPAIR TECHNIQUE

Y. TSUKINAGA, M. SHOYA and S. SUGITA
Hachinohe Institute of Technology, Hachinohe, Japan
K. DOMON
Sho-Bond Corporation, Tokyo, Japan

Abstract
In this study, experimental investigations were made on the technique of assessing the thickness of damaged sections, on the ability of repair materials to bond tightly to the concrete surface and to protect it from frost damage.

The degree of damage at different of depths was assessed fairly well by applying the improved pull-off test method. The rebound number and depth of pin penetration could be also used for this measurement. The rapid air permeability and the water absorption tests were also useful for getting information on the changes of air and water tightness caused by the deterioration due to freeze-thaw action.

Furthermore, by using two kinds of repair materials in nine different repair procedures, the performance after repair was examined to evaluate the protecting effect from suffering subsequent freeze-thaw damage.
Keywords: Depth of deteriorated portion, depth of pin penetration, frost damage, pull-off method, rapid air permeability, rebound number, repair material, water absorption.

1 Introduction

The deterioration caused by freeze-thaw action seems to develop gradually from the surface to the inner part. When the frost damage is limited in the cover, the weakened portion is chipped off and the recovering is accomplished by using some repair material. In this case, it is very important to totally remove the weakened portion, but this often depends on personal experiences and the conditions at the job site; for example, the thickness is usually determined empirically by the chipping force and so on. Economical and reliable repair can be attained if a diagnostic technique for assessing the thickness of the weakened portion

Concrete Under Severe Conditions: Environment and loading (Volume Two) Edited by K. Sakai, N. Banthia and O.E. Gjørv. Published in 1995 by E & FN Spon. ISBN 0 419 19860 1

and for clearly deciding the chipping depth could be developed.

Many coating and repair materials of high performance have been developed. However, it has not been long since these materials were developed; therefore, their fundamental properties have been made clear, but there is still much left to be studied about the interface behavior between these materials and concrete.

This study describes the results of assessments of the thickness of weakened portions caused by freeze-thaw action and evaluations of the interface behavior after repairing the damaged concrete; the pull-off tensile strength, rebound number, depth of pin penetration, rapid air permeability and water absorption methods which were used for these assessments and evaluations were, themselves, analyzed.

2 Experimental

2.1 Test procedure

The test procedure is shown in Figure 1. In this experiment, the degree of frost damage of concrete was evaluated; subsequently, the performance of the repaired portion under freeze-thaw action was evaluated.

2.2 Materials

The cement was high-early-strength portland cement. The coarse aggregate was crushed stone with a maximum size of 20㎜, fineness modulus of 6.62, and specific gravity of 2.70. The fine aggregate was land sand with a fineness modulus of 2.73 and a specific gravity of 2.59. An air entraining admixture, Vinsol-resin, was used in the concrete. For this study, two repairing materials and coating materials, shown in Tables 1 and 2, were used.

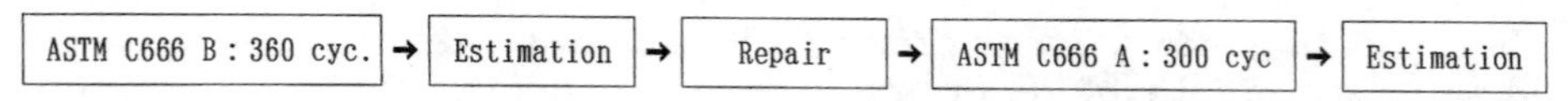

Fig.1. Test procedures.

Table 1. Repairing materials

Sym.	Repairing materials	Composition
RA	Non repair	—
RB	Polymer cement mortar	polymer emulsion glass fiber high early strength portland cement
RC	Prepacked polymer cement concrete	epoxy emulsion coarse aggregate with maximum size of 10mm high early strength portland cement

Table 2. Coating materials

Primer	Putty	Under	Second	Finish	Thickness(μm)
Epoxy resin	Epoxy resin	polyurethane resin	polyurethane resin	polyurethane resin	150

Table 3. Mix proportion of concrete

W/C (%)	Target slump (cm)	Target air (%)	s/a (%)	Unit weight (kg/m^3)				
				W	C	S	G	AE
65	8.0	2.5	47.5	188	396	857	986	0.040

Table 4. Strength and air-void characteristics of concrete

Compressive strength 28 days (MPa)	Splitting tensile strength 28 days (MPa)	Air content		Mean of air void diameter (μm)	Spacing factor (μm)
		Fresh (%)	Hardened (%)		
28.1	2.8	2.4	1.9	72.5	493

2.3 Specimen

The concrete mixture used in tests is shown in Table 3. The specimens, with dimensions 10x10x40cm, were cast and cured for 14 days in water at 20°C, and then continuously cured in air at 20°C until an age of 16 days. Then, the coating shown in Table 2 was applied to five surfaces; one 10x40 cm side of the specimen was left uncoated. The coated specimens were cured with vinyl wrapping in air at 20°C until an age of 28 days. Strength and air-void characteristics are shown in Table 4.

2.4 Freezing-thawing test

Freeze-thaw resistance was tested according to ASTM C 666 procedure B and A, as shown in Figure 1. In the tests, the relative dynamic modulus of elasticity and weight change were measured at predetermined cycles.

2.5 Test assessing the damage

(1) Pull-off tensile strength [1][2]: The test apparatus is shown in Figure 2. The test shown in Figure 2 (a) is the revised pull-off method, which enables the measurement of strength at an arbitrary depth using a circular steel probe with a hollow cylinder. For the test, circular grooves with a diameter of 75 mm were cut to depths of 5, 15, 30 and 50 mm into the surface of concrete using a dry coring bit. A circular steel probe with a hollow cylinder equal to the depth of the groove in concrete was bonded to the surface of the concrete core with epoxy resin adhesives. After being bonded, the pull-off strength tester was applied to pull off the concrete core. The pull-off tensile strength was calculated by dividing the recorded maximum load by the circular concrete area surrounded by the groove.

$$\sigma_t = P/A \qquad (1)$$

where σ_t=Pull-off tensile strength (Pa), P=Maximum load (N) and A=Area of fractured section ($10^{-4}m^2$).

The strength after repairing was measured as shown in Figure 2 (b); after making a circular groove to a depth of a 10 mm from the concrete surface, the pull-off rest was accomplished using a flat circular steel probe without a hollow cylinder.

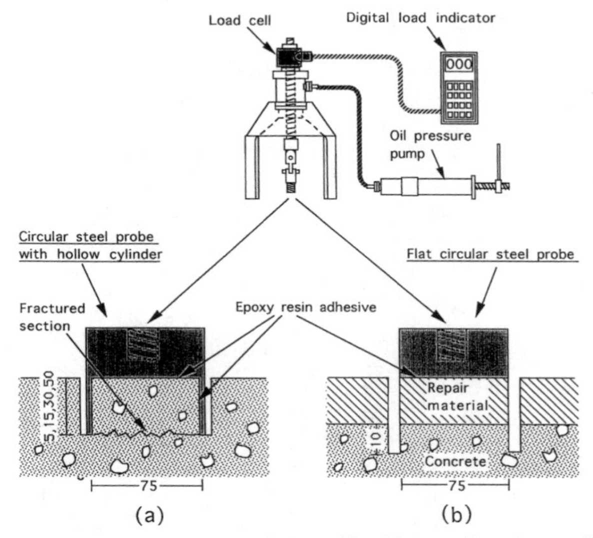

Fig.2. Test apparatus for pull-off tensile strength.

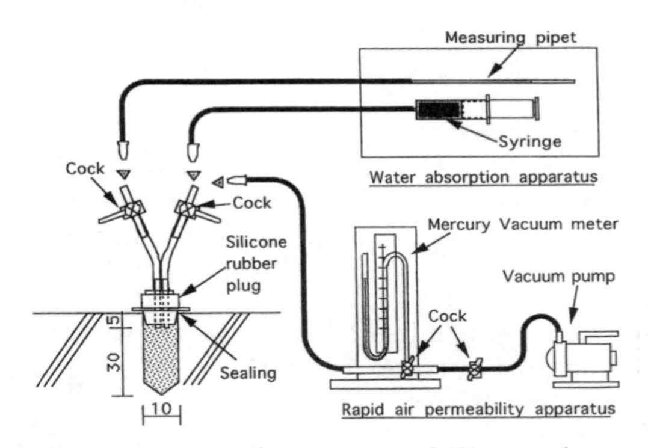

Fig.3. Test apparatus for rapid air permeability and water absorption.

(2) Rebound number: The rebound number was measured with a N type Schmidt concrete hammer under a pressure of 2.5 MPa.
(3) Depth of pin penetration[3]: The steel pin, whose diameter, length and point angle were 3.6 mm, 30.5 mm and 22.5 degrees, respectively, was driven into the concrete surface by 108 N·m of spring repulsion force energy, and the depth of the crater developed by pin penetration was measured.
(4) Rapid air permeability[2][4]: The test apparatus is shown in Figure 3. For the test, a test hole, 35 mm in depth and 10 mm in diameter, was drilled into the surface of the concrete. The recovering speed of the test hole was calculated by dividing the decrease in the gauge reading, 2.7kPa decrease from 8.0kPa to 10.7kPa, by the elapsed recovery time.

$$S=2.7/T \tag{2}$$

where S=Recovering speed (kPa/sec) and T=Recovery time (sec).
(5) Water absorption (sorptivity)[2][4]: The test apparatus is also shown

in Figure 3. After the rapid air permeability test, the hole was reused for the water absorption test. The syringe and measuring pipet were connected to separate pipes. Water was forced into the hole with the syringe water; after the hole was filled up, water then flowed out into the measuring pipet. The coefficient of water absorption was calculated by the time taken for the water meniscus to travel through the pipet.

$$W/A = \alpha \cdot \sqrt{T} \quad (3)$$

Where α=Coefficient of water absorption (10^{-2}m/sec$^{1/2}$), W=Amount of water absorption (10^{-6}m^3), A=Inside surface area of test hole (10^{-4}m^2) and T=Time of water absorption (sec).

3 Results and consideration

3.1 Variation in test data

Number of samples and the coefficient of variation in each test are shown in Table 5. The coefficient of variation in recovering speed and water absorption were 24 % and 22 %, respectively, showing larger values than those in other three tests.

3.2 Assessment of the damage

Figure 4 shows the relative dynamic modulus of elasticity and weight change of concrete specimens as functions of the number of freeze-thaw cycles. The relative dynamic modulus of elasticity showed a decrease to a little more than 80 % as a result of 360 cycles, but the weight showed little

Table 5. Number of samples and the coefficient of variation in each test

Measured characteristics	Number of samples	coefficient of variation (%)
Pull-off tensile strength	6	16
Rebound number	3 (twenty measuring points)	8
Depth of pin penetration	3 (ten measuring points)	11
Recovering speed	6	24
Coefficient water absorption	6	22

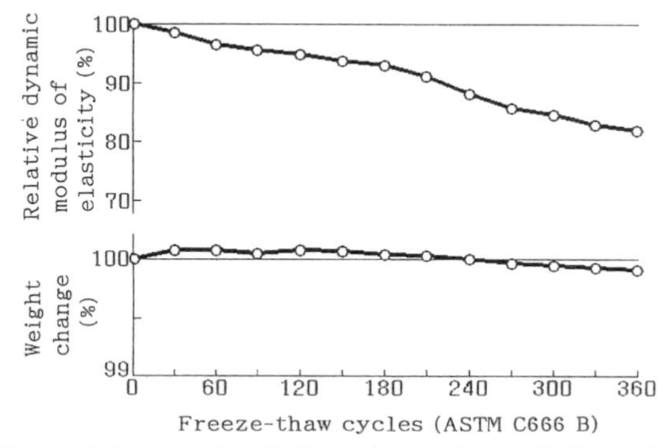

Fig.4. The relations of relative dynamic modulus of elasticity and weight change versus freeze-thaw cycle.

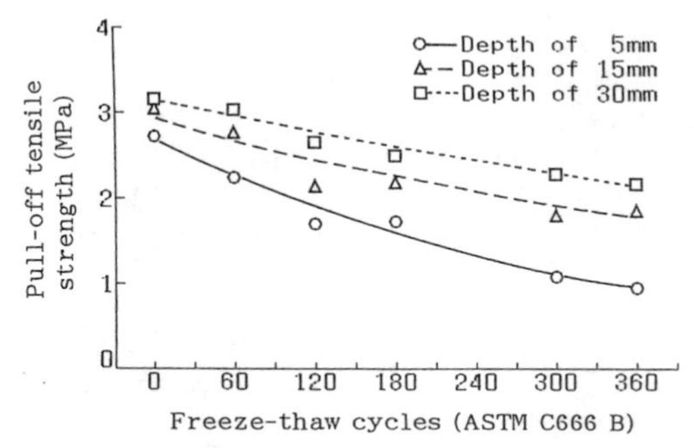

Fig.5. Pull-off tensile strength versus freeze-thaw cycle.

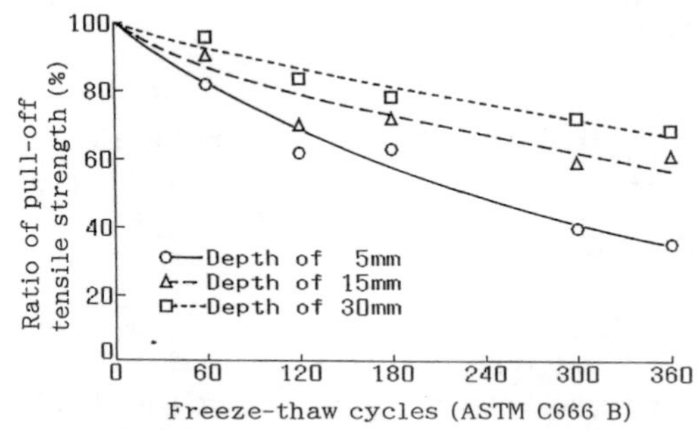

Fig.6. Ratio of pull-off tensile strength versus freeze-thaw cycle.

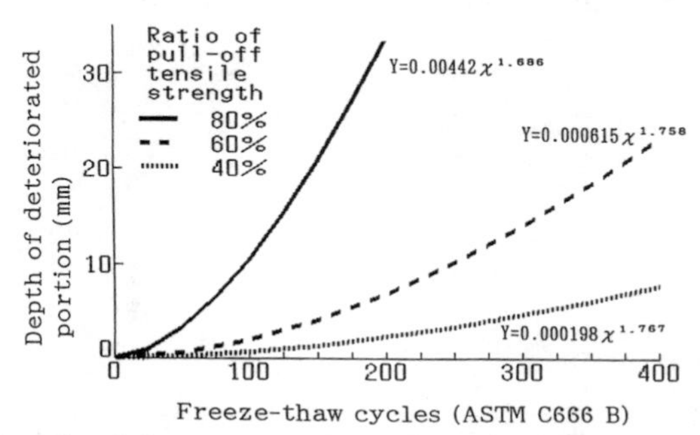

Fig.7. Depth of deteriorated portion versus freeze-thaw cycle.

change. This is considered to be partly caused by having only one face exposured to the alternative freezing and thawing action with the other five faces coated.

Figure 5 shows the pull-off tensile strength in different depths in the specimens. The pull-off tensile strength gradually decreased with freeze -thaw cycle; the strength of the portion at a depth of 5 mm showed the largest decrease. Figure 6 shows the ratio of pull-off tensile strength as a functions of the freeze-thaw cycles, where the strength is expressed as a percentage of the 0 cycle strength. Figure 7 shows the estimated depth of portions whose ratio of pull-off tensile strength was lowered to 80 %, 60 % and 40 % as functions of the freeze-thaw cycles derived from Figure 6.

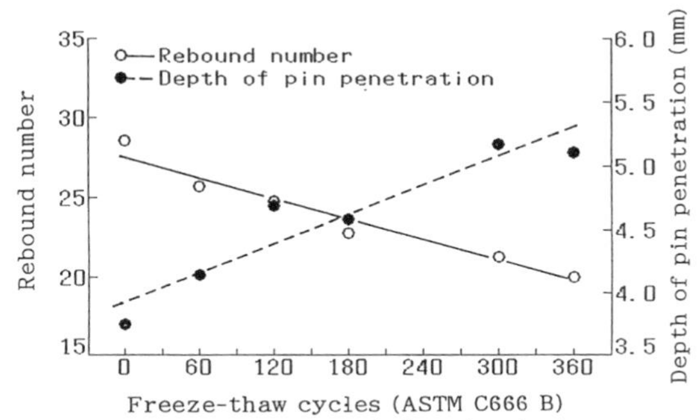

Fig.8. Rebound number and depth of pin penetration versus freeze-thaw cycle.

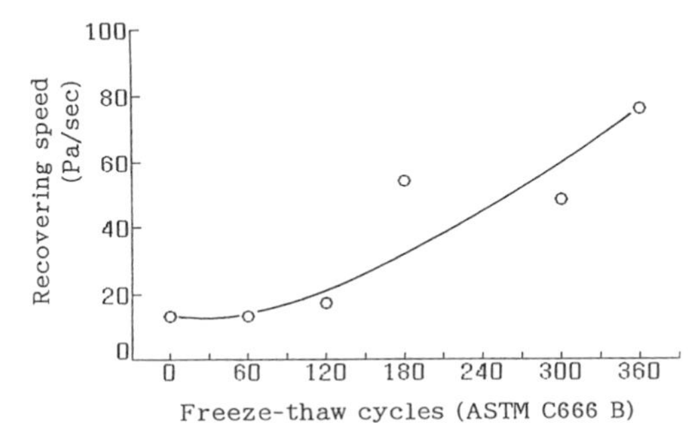

Fig.9. Recovering speed versus freeze-thaw cycle.

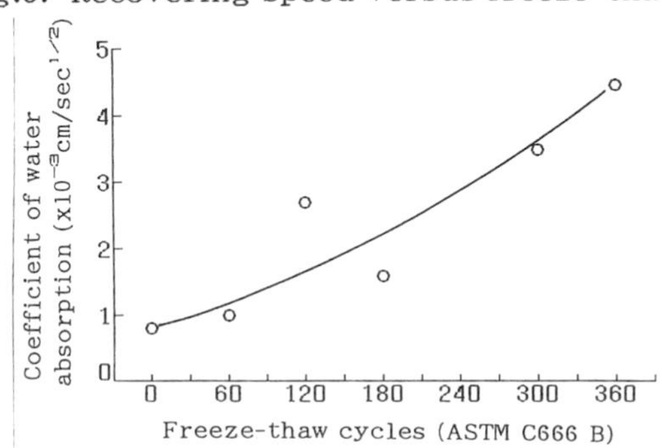

Fig.10 Coefficient of water absorption versus freeze-thaw cycle.

It was recognized that the relation between depth of deteriorated portion and freeze-thaw cycles in above lowering levels; 80 %, 60 % and 40 %, could be expressed by a power expression, and the relation suggests the possibility that the depth of the deteriorated portion can be estimated at the required cycle from a view of pull-off strength.

Figure 8 shows the rebound number and depth of pin penetration as functions of the freeze-thaw cycles. The rebound number decreased with increasing cycles: 70 % of the 0 cycle value at 360 cycles. On the other hand, the depth of pin penetration increased to 136 %, relative to the initial value, by 360 cycles. The rebound number and depth of pin penetration were also found to be useful as parameters to assess the surface damage due to freeze–thaw action.

When using pull-off strength to assess the damage due to freeze–thaw by first determining the damaged portion from the rebound number or action, a high accuracy diagnosis can be possible depth of pin penetration, and then determining the damaged depth from the pull-off tensile strength.

Figures 9 and 10 show the recovering speed and coefficient of water absorption as functions of the freeze–thaw cycles. These two values increased with freeze–thaw cycles; these increases were at higher values than those found in the pull-off strength, rebound number and depth of pin penetration tests. These two parameters were closely related to the porosity [2], and the increases of these values were considered to reflect sensitively the weakening of the surface layer structure in the concrete.

The recovering speed and the coefficient of water absorption were found to be useful as parameters to assess frost damage since they can be simply measured in situ.

Table 6. Type of repair

Repair time	RA※1	RB※1	RC※1
After 180 cycles σt= 1. 7 ※2	Sym. 180RA (5-0)※3 Cut of 5mm (5 / 95) ↓ (95)	Sym. 180RB (5-5)※3 Cut of 5mm (5 / 95) ↓ Repairing material (5 / 95)	Sym. 180RC (5-15)※3 Cut of 5mm Cut of 10mm (5 / 85 / 10) ↓ Repairing material (15 / 85)
After 300 cycles σt= 1. 1 ※2	Sym. 300RA (5-0)※3 Cut of 5mm (5 / 95) ↓ (95)	Sym. 300RB (5-5)※3 Cut of 5mm (5 / 95) ↓ Repairing material (5 / 95)	Sym. 300RC (5-15)※3 Cut of 5mm Cut of 10mm (5 / 85 / 10) ↓ Repairing material (15 / 85)
After 360 cycles σt= 0. 9 ※2	Sym. 360RA (0-0)※3 Brush with a wire brush (100) ↓ (100)	Sym. 360RB (0-5)※3 Brush with a wire brush Cut of 5mm (95 / 5) ↓ Repairing material (5 / 95)	Sym. 360RC (0-15)※3 Brush with a wire brush Cut of 15mm (85 / 15) ↓ Repairing material (15 / 85)

※1 : Refer to Table 1.
※2 : Pull-off tensile strength of 5mm depth. Unit in "MPa".
※3 (x1-x2) : x1 : Depth of cutting. x2 : Depth of filling.

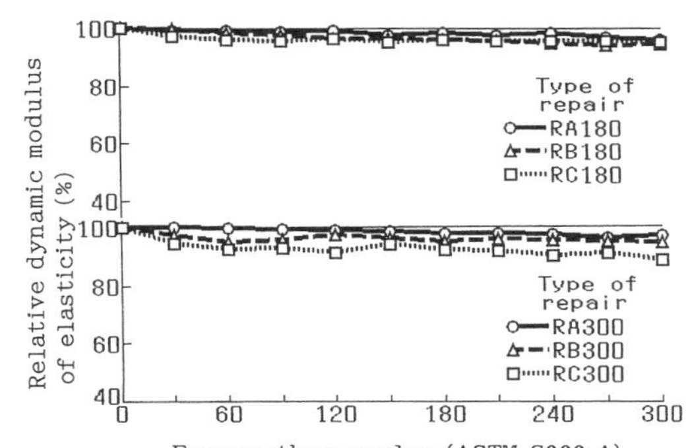

Fig.11. Relative dynamic modulus of elasticity versus freeze-thaw cycle of repaired concrete.

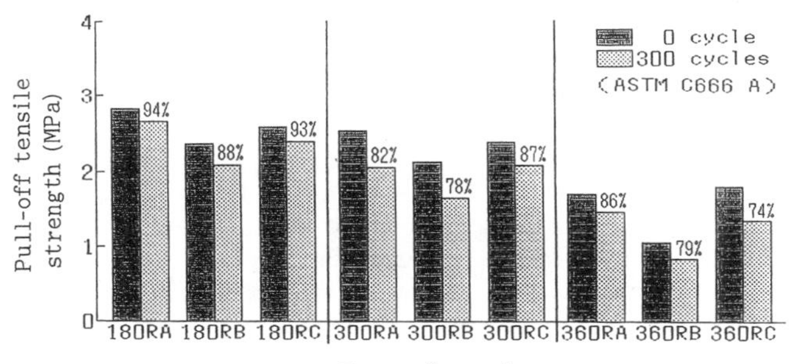

Fig.12. Pull-off tensile strength of repaired concrete.

3.3 Repairing

Table 6 shows the outline of nine types of repair. These repairs were completed, provided that; ①the sound state corresponded to 1.7 MPa in pull-off strength in the depth of 5 mm after 180 freeze-thaw cycles, ②intermediate state corresponded to 1.1 MPa in the depth of 5 mm after 300 cycles, and ③deteriorated state did to 0.9 MPa in the depth of 5 mm after 360 cycles. After 180 and 300 cycles, concrete was removed to the depth of 5 mm with a diamond wheel cutter. After 360 cycles, the exposed surface of concrete was only brushed with a wire brush. Continuously, concrete specimens were repaired through repairing procedures of RA, RB and RC shown in Table 1. The repair was conducted one week after the finish of 360 freeze-thaw cycles; specimens were left with vinyl wrapping in the conditioning chamber at 20°C until repairing time.

Figure 11 shows an example of the relative dynamic modulus of elasticity of repaired concrete. The relative dynamic modulus of elasticity showed little decrease regardless of the repair type. Figure 12 shows the pull-off tensile strength of repaired concrete. In the tests, fractures were mostly observed in the base concrete, except that some of the fractures happened near the interface of the repair materials and base concrete. On the whole, the pull-off tensile strength of repaired concrete ranked as follows; sound state after 180 freeze-thaw cycles > intermediate state after 300 cycles > deteriorated state after 360 cycles. However, all of the pull-off tensile strengths of repaired concrete exhibited higher values than those in the sound state, intermediate state and deteriorated state. These tendencies are considered to be influenced by the effects of self-healing due to leaving the specimens in a sealed state until the repair time and the possible increase of strength due to penetration of repair materials. From the figure, it is shown that an economical repair with minimal chipping thickness is possible if the repair was made in an early stage of deterioration, and the full effectiveness of the repair depended upon the weakened portion being chipped off completely. Comparing the differences in repair type, the RA type repair resulted in the highest pull-off tensile strength. The pull-off tensile strength was a little higher with RC type repair than with RB type repair. The repair performance of prepacked polymer cement seems better than that of polymer cement mixed glass fiber.

4 Conclusions

The conclusions obtained in this study are summarized as follows;

1. The pull-off tensile strength was found to be an excellent measure of both the concrete strength in the weakened portion due to freeze-thaw action and of the required chipping depth. It is also available to evaluate the interface behavior such as bond strength after repairing.
2. The rebound number and depth of pin penetration were found to be useful parameters to assess the surface damage.
3. The recovering speed and the coefficient of water absorption were found to be useful parameters to assess the frost damage due to the ease of estimating air tightness and water tightness on site.
4. It will be possible to make a synthetic diagnosis of damaged concrete due to freeze-thaw action from a side view of both strength and air tightness or water tightness, as shown in conclusions 1, 2, 3.
5. Economical repairs with minimal chipping depth is possible if the repair occurs in the early stage of the deterioration. Effective repair can't be expected unless the weakened portion is chipped off completely.
6. The repair performance of prepacked polymer cement seems better than that of polymer cement mixed glass fiber.

5 Acknowledgement

The authors would like to express our thanks to Professor Kasai, Nihon University, for his instructive advice and support to this study.

6 References

1. Long,A.E. and Murray,A.McC.(1984) The Pull-off Partially Destructive Test for Concrete, *ACI SP82-17,* pp.327-350.
2. Tsukinaga,Y., Shoya M. and Kasai Y.(1992) In-situ test methods assessing the quality of the surface layer of concrete, *Proc. of International Symposium on NDT & SSM, FENDT'92,* Vol.1, pp.477-484.
3. Nasser,K.W. and Al-Manaseer,A.A.(1987) New Nondestructive Test, *Concrete International:Design & Construction,* Vol.9, No.1, pp.41-44.
4. Kasai,Y., Matui,I and Nagano,M.(1984) On Site Rapid Air Permeability Test for Concrete, *ACI SP82-26,* pp.525-541.

106 MEASUREMENT OF CRACKING IN MORTAR DUE TO FREEZING AND THAWING BY IMAGE PROCESSING METHOD

T. NARITA
Tohoku Electric Power Co. Inc., Sendai, Japan
H. MIHASHI and K. HIRAI
Tokohu University, Sendai, Japan
T. UMEOKA
Shimizu Co. Inc., Nagoya, Japan

Abstract
Cracking properties of damaged concrete is very complicated and many details are still far from well understood because of its heterogeneous structure. Recently, a technique to evaluate picture using an image analyzing system has been developed.

In this paper, cracking properties of mortar gradually damaged by freezing and thawing were analyzed by using the image processing technique. The damage area ratio was calculated as well as the maximum size of the damaged domain and the length of cracking. These results were discussed in relation to the number of freezing and thawing cycles. The relationship of the damaged area ratio and crack length were strongly correlated with the number.
Keywords: cement paste, crack angle, crack length, freezing and thawing action, Image processing method, mortar, relative dynamic modulus of elasticity.

1 Introduction

It is important to evaluate the deterioration of concrete for determining the durability of concrete structures. Deterioration of concrete due to freezing and thawing is caused by the accumulation of microcracking which is resulted by internal stress and relaxation of the system. It is expected the degree of deterioration can be evaluated by distribution of cracking.

The most conventional method to observe cracking in a section of concrete specimen is optical microscope examination. Although it is possible to measure the width and the length of cracking with this technique, much labor and time would be needed. On the other hand, prompt image processing and quantitative evaluation of the internal microstructure of concrete have become possible because of the

Concrete Under Severe Conditions: Environment and loading (Volume Two) Edited by K. Sakai, N. Banthia and O.E. Gjørv. Published in 1995 by E & FN Spon. ISBN 0 419 19860 1

development of the computer science. Although many studies on the air-void system and the internal constitution of concrete have been performed by image processing method[1,2,3,4], researches on cracking seem to be few in number[5,6].

The purpose of this study is to extract cracking data from mortar and hardened cement paste specimens that were damaged by freezing and thawing by means of an image processing method.

2 Method of image processing

In order to calculate the geometrical parameters of cracking with an image processing technique, the objective image must be reduced to binary data. A cut surface of damaged specimens were first put in a red polymer and then the surface was polished. Photographs of cracking on cut surfaces of a specimen were input to the computer as a color image: a specimen area of 4x4 cm corresponds to 800x800 pixels. The image was input as binary data depending on the presence of red or not. Hence, cracks and voids were distinguished from solid areas.

Density histograms were produced from the red shaded images obtained. A threshold value was calculated by a distinguished Automatic Threshold Selection Method Based on Discriminant[7], and binary images were made.

Since air voids exist in the obtained binary images, their removal was attempted. By this procedure, we obtained the following three types of images for the analysis:

(1) R-Image : only cracking
(2) M-Image : cracking plus lacking
(3) T-Image : thinned R-Image

Thus, cracking and lacking caused by the frost damage can be recognized, and cracking can be quantified.

3 Experiment of mortar specimens

3.1 Materials and mix proportion of mortar

High early strength Portland cement and river sand from Miyagi prefecture's Shiroishi River were used. The mix proportion of mortar used in this experiment is shown in Table 1. The water-cement ratio (w/c) was 65%, and the sand-cement ratio was 3.5. The target flow was 210±10.

3.2 Specimen

Mortar specimens of 4x4x16 cm were used; eighteen specimens were made in steel

Table 1. Mix proportion of mortar.

Water–Cement Ratio	Sand–Cement Ratio	Water (g)	Cement (g)	Sand (g)	Flow
65%	3.5	280	430	1,505	207

molds, according to Japanese Industrial Standard (JIS) R 5201. The molds were stored in a moisture room for 24 hours after casting, and specimens were cured in water at 20 ℃ for 7 days after form removal.

3.3 Freezing and thawing test

The specimens were tested for resistance to frost damage by rapid freezing in air(-18℃) and thawing in water(+5℃) in an automatic test machine. One cycle of freezing and thawing took about two hours. The dynamic modulus of elasticity was measured according to JIS A 1127. Three specimens were sampled randomly for the tests at 0, 20, 60, 70, 80 freezing and thawing cycles to reduce error from deviation among molding boxes. The mean value was adopted to the analysis. Cracks were impregnated with red color polymer.

3.4 Evaluation of cracking by image processing

The following three parameters were evaluated by the image processing technique. Damage area ratio is a parameter for evaluating deterioration degree, which is the ratio of damaged area to specimen's section area. The maximum size of damaged area is also a parameter for evaluating deterioration degree. Crack length is obtained by counting the number of pixels in the T-Image.

3.5 Experimental result and discussion

An example of photographs showing a section of one mortar specimen, and R, M, T-Images are shown in Photo 1 and Fig.1, respectively.

In Photo 1, observation by optical microscope showed the cracks impregnated with the red polymer. It might be sufficient for our purpose when 50x50 μm is the minimum size that can be input into the computer as a pixel.

In Fig.1, it was found that many cracks were generated parallel to the placement surface. As for the binary method attempted in this study, the threshold value evaluation

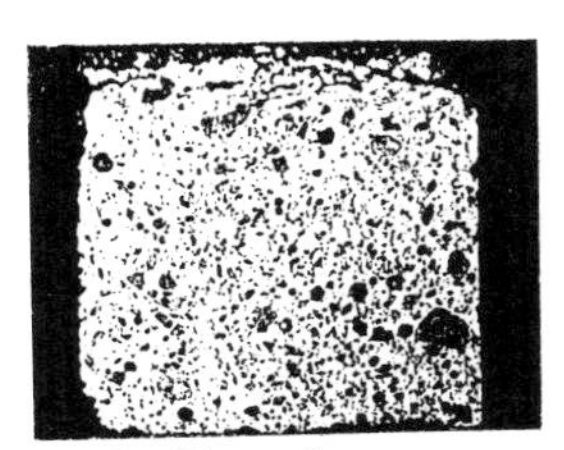

Photo.1. Section of mortar specimen.

R-Image M-Image T-Image

Fig.1. Obtained binary image.

Table 2. Test results.

Specimen Series	Relative Dynamic Modulus of Elasticity	Freezing and Thawing Cycles	Damage Ratio(%) R-Image*	Area M-Image*	Maximum Damage R-Image*	Size of (mm2) M-Image*	CrackLength (mm) T-Image*
A	100.0	0	2.41/0.18	4.02/0.24	5.56/ 1.80	15.37/ 1.39	113.0/12.03
B	99.5	20	3.34/1.49	4.31/1.91	17.44/21.38	22.93/27.74	198.3/70.83
C	91.9	20	3.49/0.73	4.22/1.02	8.79/ 3.14	11.65/ 6.17	203.8/84.38
D	83.2	80	5.53/0.53	7.38/0.73	19.10/ 5.46	31.22/12.14	392.3/112.57
E	80.1	70	6.42/1.11	8.41/1.60	33.78/ 9.98	58.25/21.44	345.3/57.62
F	72.5	60	4.53/0.40	8.25/1.22	16.74/ 4.68	67.90/40.37	283.0/22.69

*Mean / Standard Deviation

was automated. This simplified the method of sampling that was conventionally done by try and error, and the accuracy of the sampling could be much improved.

The maximum size and the crack length are shown in Table 2; the mean is an arithmetical mean. As shown in Table 2, it can be seen that the damaged area ratio, the maximum size and the crack length in the various sections of the same specimen are scattered. This might be resulted from the following reasons: (1) the microstructure of specimen is usually random; (2) damage distribution is not uniform; (3) cracking and lacking might have developed during the cutting and grinding process.

The relationship between the damaged area ratio and the number of freezing and thawing cycles obtained by this image processing method is shown in Fig.2, and that between the damaged area ratio and the relative dynamic modulus of elasticity is shown in Fig.3. As for the R-Image, which is only of cracking, and the M-Image, which is the cracking plus lacking part, it is recognized that the damaged area ratio calculated from each sample has a good correlation with the number of freezing and thawing cycles. As for Fig.3, it seems that the correlation of the damaged area calculated from the M-Image with the relative dynamic modulus of elasticity is strong.

The relationship between the maximum size of damage and the relative dynamic modulus of elasticity is shown in Fig.4. It can be seen that the correlation of the maximum size calculated from the R-Image or the M-Image with the relative dynamic modulus of elasticity is not strong. The relationship between crack length and number of freezing and thawing cycles is likewise shown in Fig.5, where the relation between the cracking length obtained from the T-Image and the number of freezing and thawing cycles is remarkably strong.

The decline in the relative dynamic modulus of elasticity of damaged mortar is explained by the occurrence of relaxation in the mortar system due to the development of internal cracking. The correspondence of the relative dynamic modulus of elasticity to the damaged area ratio and the crack length is inferior to that which exists between the latter two parameters and the number of freezing and thawing cycles. The reason is that there were many cracks which did not appear in the observed surface. It is important to observe more than two planes crossing each other as the next step for future research and development.

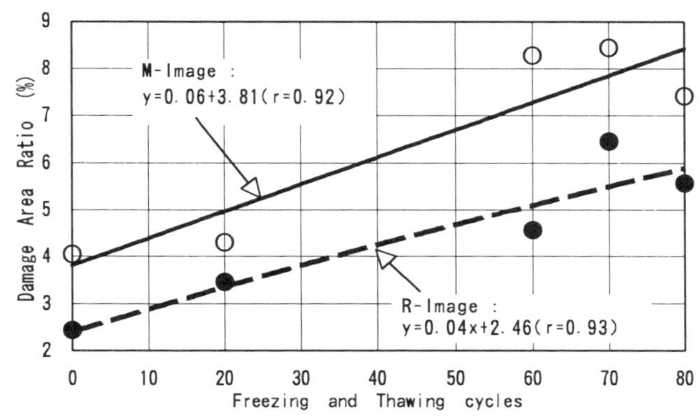

Fig.2. Relationship between damaged area ratio and number of freezing and thawing cycles.

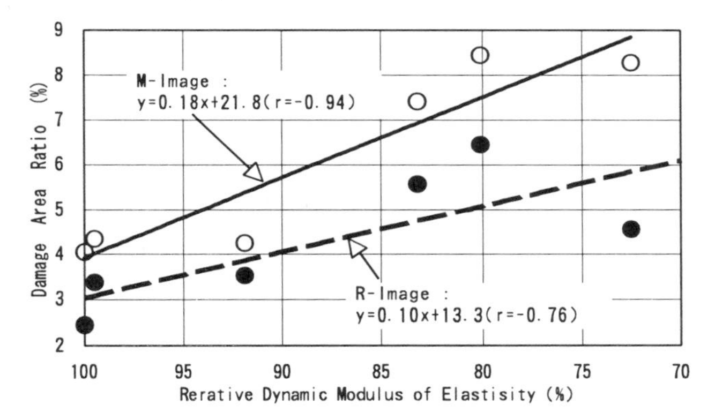

Fig.3. Relationship between damaged area ratio and relative dynamic modulus of elasticity.

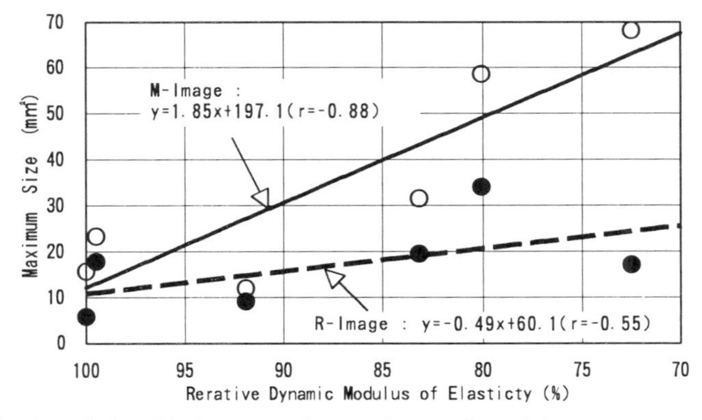

Fig.4. Relationship between the maximum size of damage and relative dynamic modulus of elasticity.

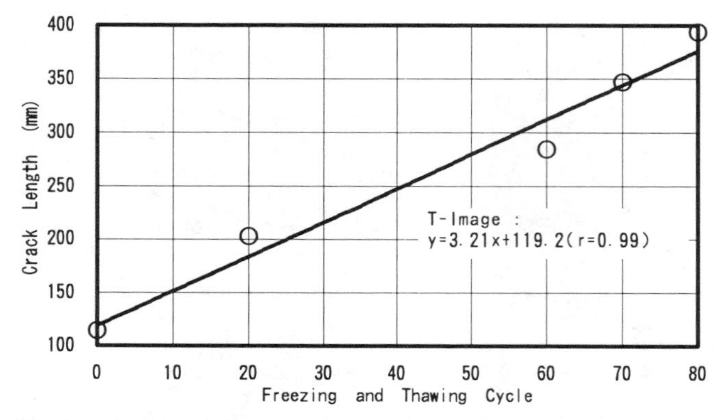

Fig.5. Relationship between crack length and number of freezing and thawing cycles.

4 Experiment with hardened cement paste

4.1 Test procedure

Hardened cement paste specimens with water-cement ratios of 65% and 45% using ordinary Portland cement were tested. Eleven prism specimens of 4x4x16 cm were made for each water-cement ratio. Moisture curing was conducted for 24 hours after casting. Test pieces were cured in water at 20℃ for 28 days and the freezing and thawing tests were conducted. Sampling is carried out every several cycles, and the dynamic modulus of elasticity was measured. Test pieces damaged to a certain level were taken out from the chamber. Image processing analysis was carried out on sections of the specimens cut with a concrete cutter. The eleven test pieces shown in Table 3 were used for the image processing analysis.

4.2 Measurement

Crack length was measured by two methods: the length was determined by counting number of pixels in the fine crack image; and the geometrical length was determined by the vector description of the fine crack image. Crack patterns were further finely divided

Table 3. Test pieces.

Specimen	Water-Cement Ratio(%)	Relative Dynamic Modulus of Elasticity(%)	Freezing and Thawing Cycle
1	65	100.0	0
2	65	88.7	8
3	65	70.0	10
4	65	57.9	20
5	65	53.5	12
6	65	27.9	30
7	45	100.0	0
8	45	92.2	50
9	45	83.4	140
10	45	73.7	200
11	45	22.1	100

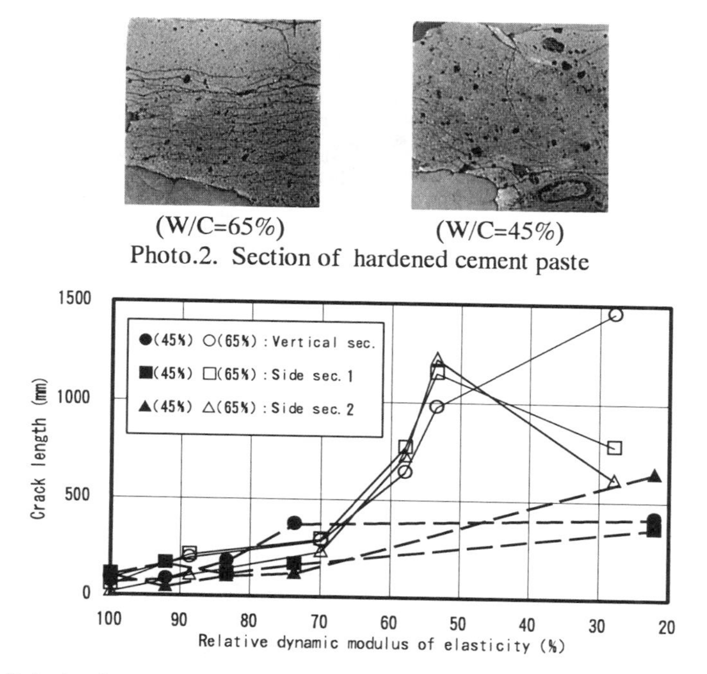

Photo.2. Section of hardened cement paste

Fig.6. Relation between relative dynamic modulus of elasticity and crack length.

and local crack angles from the casting side were measured for each crack.

Fractal dimension is given by the following equation.

$$N(r) = N_0 r^{-D} \tag{1}$$

where N(r) is the number of boxes which contain at least one crack in the area of r x r; N_0 is a constant; r is the size of the grid; D is the fractal dimension which describes the complexity of cracking and is obtained as the slope of the semilogarithmic line. This method to calculate the fractal dimension is called box count method.

4.3 Results and discussion

The relationship between the relative dynamic modulus of elasticity and crack length is shown in Fig. 6. Crack length shown here is the length measured by the number of pixel. According to these results, crack length increases with a reduction of the relative dynamic modulus of elasticity in the case of two different water-cement ratio. When the relative dynamic modulus of elasticity becomes less than 70%, crack length changes considerably depending on the water-cement ratio. After a certain extent, the crack length is reduced in the case of water-cement ratio 65%. This is because the cracks in the cement paste linked together and the cracked parts were lacked.

In Fig. 7, the relation between the crack length shown in Photo 2 and the local crack angle is shown. The crack length shown here is the geometrical length. IP1 - IP4 are the data resulting from one input image divided into four parts, and IPA is the sum. Many

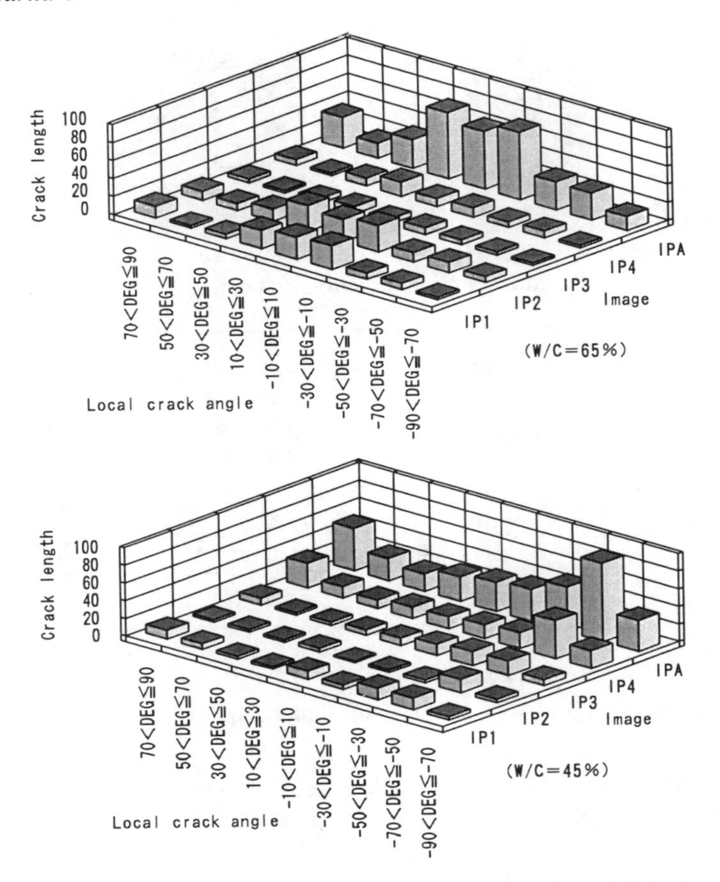

Fig.7. Relationship between local crack angle and crack length.

local angles are distributed in the range of 0 - 30 degrees. This result means that cracks occurred parallel to the casting side in the case of W/C=65%. On the other hand, cracks occurred in the range of 70 ~ 90 degrees and -50 ~ -70 degrees in the case of W/C=45%, which means that cracks occurred perpendicular to the casting surface.

The relations of crack length measured by the two methods and fractal dimension are shown in Figs. 8 and 9. A good correlation is observed between the crack length and the fractal dimension in Fig.8. In this figure, the slope of the linear regression line in W/C=45% is slightly less than that in W/C=65%. This tendency is obvious in Fig. 9, in which the crack length was measured as a geometrical length. It is generally accepted that the fractal dimension becomes larger as the complexity of the figure increases. Accordingly, the following remarks result from Figs. 8 and 9. In the case of hardened cement paste with W/C=65%, cracking becomes more tortuous and more complex with increase of crack length. On the other hand, such a tendency is less in the case of W/C=45%.

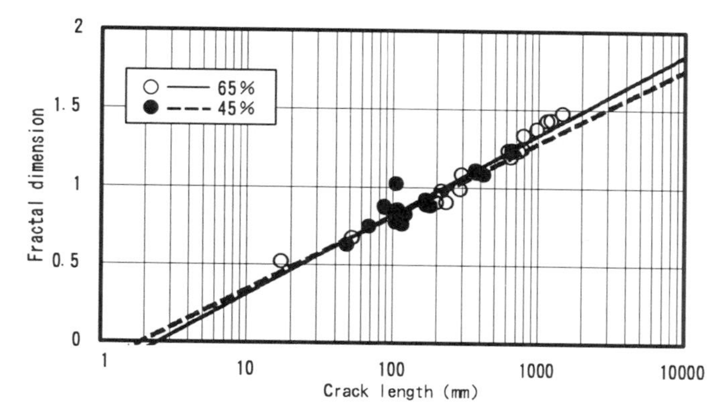

Fig.8. Relation between crack length(number of pixels) and fractal dimension.

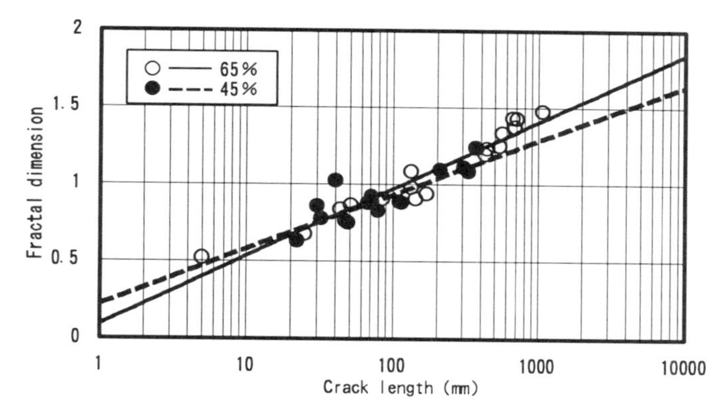

Fig.9. Relation between crack length(geometrical length) and fractal dimension.

5. Conclusions

From the results of the image analysis of mortar and hardened cement paste specimens damaged by freezing and thawing action, the following conclusions were obtained:

1. Damaged area ratios, the maximum size of damaged area and crack length in mortar were quantitatively obtained by image processing method, which were related to changes in the dynamic modulus of elasticity.
2. The relationship between damaged area ratio and crack length were highly correlated with the number of freezing and thawing cycles was strong.
3. In the case of hardened cement paste, cracking became more complex for a higher water-cement ratio. Measuring local crack angle distribution is a useful method to describe a cracking pattern.

References

1. Hamada,K.,et al. (1986) Method of measurement of air-void system in hardened concrete using an image analysis. *Cement Concrete*, No.471, pp.22-28.
2. Nishiyama,T.,et al. (1988) Coloring and observation of air-void system in hard ened concrete by cyanoacrylate. *CAJ Review of The 42nd General Meeting*. pp.212-214
3. Hirai, K., Mihashi, H. and Kurita, H. (1988) Fundamental sturdy on evaluation of structures of concrete using image processing method. *Proceedings of AIJ TOHOKU Branch*, No.51, pp.223-226.
4. Ayuta, K.,et al. (1990) Image analysis of an air-void system in hardened concrete. *Proceedings of JSCE*, No.420, pp.81-86.
5. Tanimichi, T.,et al. (1985) Study of image analysis of cracking using personal computer. *Summaries of Technical Papers of Annual Meeting AIJ (Tokai) A*, pp.765-766.
6. Kurita, H., Mihashi, H. and Hirai, K. (1990) Estimation of durability property of concrete using image processing. *Summaries of Technical Papers of Annual Meeting AIJ (Chugoku) A*, pp.63-64.
7. Ohtsu, N. (1980) An automatic threshold selection method based on discriminant and least squares criteria. *The Transactions of Institute of Electronics and Communication Engineers of Japan*, Vol.J63-D No.4.

107 THERMOPOROMETRIC APPROACH IN CHARACTERIZATION OF PORE STRUCTURE IN CONCRETE

S.P. MATALA
Helsinki University of Technology, Concrete Laboratory, Finland

Abstract
A calculation method to analyze the pore volume and pore size distribution in concrete based on thermoporometry was developed. A hypothetical pore size distribution typical for concretes was used in the determination of the correction factor of the entropy changes due to the superficial phase transformations. Comparison of the results of pore analysis by different methods showed that the pore structural changes which were only slightly detectable by the mercury intrusion method were clearly noticeable by the pore analysis with low-temperature calorimetry. The freezing of the major part of the gel pore water that occurred near the pore radius of 2 nm was in good agreement with the results by the N_2-gas adsorption method. Fusion thermograms showed a very clear onset temperature below which the thermogram behaved almost linearly showing the smallest pore where nucleation can take place. The slope of the curve below this temperature provided information on the change in the energy state of the non-freezable water layer on the pore walls; Therefore, it is a measure of the specific surface area of concrete. The hysteresis between the solidification and fusion thermograms reflected the cylindrical or layered pore shape of concrete.
Keywords: Concrete, low-temperature calorimetry, mercury intrusion porosimetry, pore size distribution, pore structure, thermoporometry

1 Introduction

In studies of the pore structure, the selection of the testing procedure is of great importance. Each method used in the studies had its own limitations. The conventional mercury intrusion (MIP) method used mostly in the characterization of pore structure

Concrete Under Severe Conditions: Environment and loading (Volume Two) Edited by K. Sakai, N. Banthia and O.E. Gjørv. Published in 1995 by E & FN Spon. ISBN 0 419 19860 1

is able to detect pore sizes over 2 to 4 nm in radius. This method, however, is quite sensitive to the preparation of samples, and the need to dry the sample beforehand obviously changes the pore structure of the finest pore range. In order to avoid the preparational problems, a new test method, thermoporometry, was selected to confirm the results observed by the MIP-method. During the last twenty years, this method has been used in the characterization of pore structure of cement based materials, e.g., by Fagerlund [1], Sellevold with his coworkers [2], Bager [3] and Beddoe and Setzer [4]. These methods are based on the utilization of Kelvin's equation, and in the entropy of phase transformation, the effect of the superficial phase transformations on the entropy changes were neglected. The group of French researchers leaded by Brun [5] included these factors in the method. These researchers have done the pioneer work that made the further development of the calculation method possible. In the following sections, only the theoretical background and the final results and some applications are presented. A full description of the method will be presented elsewhere.

2 Thermoporometric approach to determine porosity

2.1 Theoretical basis

The measurable solidification energy of pore water can be evaluated theoretically if the change of entropy during the phase transformation can be formulated. In a fully water saturated porous system, the solid-gas interface is plane and the change in the entropy of the water-ice system of pores for the change in temperature, $T_0 \rightarrow T$, and for the corresponding changes in the pressure of liquid phase, $P_0 \rightarrow P_l$, and the solid phase, $P_0 \rightarrow P_s$, can be expressed by the following equation:

$$\Delta S_f = \Delta S_{f0} + \int_{P_l}^{P_0} \left(-\frac{\partial v_l}{\partial T} \right)_P dP + \int_{T}^{T_0} \left(\frac{c_l}{T} \right) dT + \int_{T_0}^{T} \left(\frac{c_s}{T} \right) dT + \int_{P_0}^{P_s} \left(-\frac{\partial v_s}{\partial T} \right)_P dP \qquad (1)$$

The first term in Equation (1) is the basic entropy and the second and the fifth term represent the effects of the compression heat of phases due to the change in pressure, assuming that the phases are incompressible. The third and the fourth term take into account the changes in the specific heat of the liquid and solid due to the depression of temperature. Equation (1) can be rearranged into the following form :

$$\Delta S_f = \Delta S_{f0} + \int_{T_0}^{T} \left(\frac{c_s - c_l}{T} \right) dT + \left[\left(\frac{\partial v_l}{\partial T} \right)_P - \left(\frac{\partial v_s}{\partial T} \right)_P \right]_T \cdot (P_s - P_0) + \left[\left(\frac{\partial v_l}{\partial T} \right)_P \right]_T \cdot (P_l - P_s) \qquad (2)$$

In equation (2) P_s is the vapor pressure of ice at temperature T, and P_l-P_s can be expressed by Laplace's equation:

$$P_l - P_s = -\frac{\gamma_{ls}}{R_n} = -\int_{T_0}^{T} \left(\frac{\Delta S_f}{v_l} \right) dT \qquad (3)$$

Brun has stated that Equation (2) is not fully accurate when the solidification of ice is considered. It does not take into account the effect of the entropy of superficial phases. During the solidification of ice there occur changes of interphase between the layers that do not freeze and the adjacent phase (water before solidification and ice afterwards). Brun derived the correction factor of entropy due to this additional entropy variation by considering the equality of chemical potentials and the equilibrium of three interphases at triple point [5].

We derived a new relationship between the temperature depression and the enthalpy in the solidification of pore water. The comprehensive derivation of the solidification enthalpy is not presented here. It is based on the numeric formulation of Equation (2) derived by Brun, but the effect of superficial phase transformations on the total entropy has been considered in a way that is more convenient for the pore structure of concrete. Since this increase in entropy is a function of the pore geometry, the correction factor was derived in our study by utilizing a hypothetical pore size distribution corresponding to the finest capillary pore structure of cement pastes and mortars. In thermoporometric approaches of the first generation, the effect of the superficial phase transformations on the total entropy has been neglected although its influence may be within the limits of 20% to 40% of the total entropy, depending on the pore structure. The finer the pore structure and the lower the temperature, the higher this effect is.

Using the linearized correcting factor due to the entropy changes of the superficial phase transformations and multiplying this by the equation derived by Brun, the numeric expression for the entropy at the nucleation point was developed into the following form:

$$\Delta S_{f_{np}} = \frac{(1+0.004\cdot\theta)\cdot\left[-1.2227-4.889\cdot\ln\left(1+\frac{\theta}{T_0}\right)+10.124\cdot10^{-3}\cdot\theta+1.265\cdot10^{-5}\cdot\theta^2\right]}{1-4.556\cdot10^{-5}\cdot\left(\theta-0.227\cdot\theta^2\right)}, \quad (4)$$

where θ is the freezing point temperature in °C. The enthalpy at nucleation point $W_a=(273.15+\theta)\Delta S_{f,np}$ has been presented in Fig. 1 together with the enthalpy values proposed by other investigators. Numerically, the solidification energy W_a based on Equation 4 can be expressed within the temperature range $0 > T > -60°$ C by a quadratic equation:

$$W_a = 334 + 4.83\cdot\theta + 0.0125\cdot\theta^2 , \quad (5)$$

where W_a is expressed in J/g and θ in °C. Equation 5 is valid in the temperature range $0 > T > -60°$ C. The enthalpy proposed by Fagerlund in Fig. 1 differs from the values given by Equation 5 and proposed by Brun since Fagerlund's expression neglects the effects of superficial phase transformations and compression heat of phases. The difference between Equation 5 and Brun's expression is due to different assumptions in the determination of the entropy of superficial phase transformations. Brun's correcting factor is a function of surface tension of liquid-solid interface and temperature depression. At a specified nucleation point, it corrects the entropy values of the superficial phases in previously frozen pores by an amount dependent on the surface tension of the nucleating pore. Obviously, this overestimates the effect of superficial phase transformations on the entropy values for the wide range pore structure normally observed in cement pastes. Obviously, Brun's equation gives correct values with the pore structure in the vicinity of the nucleation point. The hypothetical pore structures used in the testing of Brun's equation were very fine for ordinary cement pastes and mortars observed by different testing methods. A coarser pore size distribution resulted in an increase of 20% at -50° C for the entropy at the nucleation point, and the curve was approximately linear. Theoretically, the way by which the correction factor was determined is not absolutely correct since it is a function of pore size. However, this approximation must be used for the calculations. Otherwise, the calculations consist of a very complicated iterative process, where the correct pore size distribution must be derived step by step and the entropy of the previous pore system should be taken as a starting value for the next calculation step.

2.2 Pore size distribution in thermodynamic approach

The relation between an ice sphere R_n and freezing point temperature θ can be calculated using Laplace's equation (3), which gives the relationship between the difference in the vapor pressures of the liquid and solid and the surface tension of liquid-solid interface and its curvature. In the literature, there are many contradictory proposals for the values of surface tension between water and ice. We collected experimental values from 0 to -30° C presented in the literature and assumed γ_{ls} to vary linearly with temperature. Those values gave the following linear regression line:

$$\gamma_{ls} = (36 + 0.25 \cdot \theta) \cdot 10^{-3}, \tag{6}$$

where γ_{ls} is expressed in N/m and θ in °C. The relation between the ice sphere R_n and freezing point temperature θ can be calculated directly from Equation 3 by applying the relationships derived before and inserting the equations of γ_{ls} (Eq. 6), $\Delta S_{f,np}$ (Eq. 4) and v_l (Eq. 7) into Equation 8 and setting $T\text{-}T_0 = \theta$.

$$v_l = 1.000132 \cdot \left(1 - 9.1 \cdot 10^{-5} \cdot \theta + 1.035 \cdot 10^{-5} \cdot \theta^2\right) \tag{7}$$

$$R_n = \frac{2 \cdot \gamma_{ls}}{\int_{T_0}^{T} \frac{\Delta S_{f_{np}}}{v_l} dT} \tag{8}$$

Since all the parameters in Equation 8 are expressed as a function of temperature, the numeric integration is possible. The relationship between ice sphere radius and temperature according to Equation 8 is presented in Fig. 4, and it can be expressed by quadratic Equation 9 with good accuracy.

$$R_n = \frac{-2 \cdot (36 + 0.25 \cdot \theta)}{1.222 \cdot \theta + 0.0068 \cdot \theta^2 - 8.67 \cdot 10^{-7} \cdot \theta^3} \tag{9}$$

The ice sphere radius as a function of nucleation temperature is strongly dependent on the assumptions taken in the evaluation of the surface tension of the water-ice interphase. It gives clearly smaller values for the ice sphere than the equations based on the surface tension of the liquid-vapor interphase, e.g., Fagerlund's expression (Fig. 2).

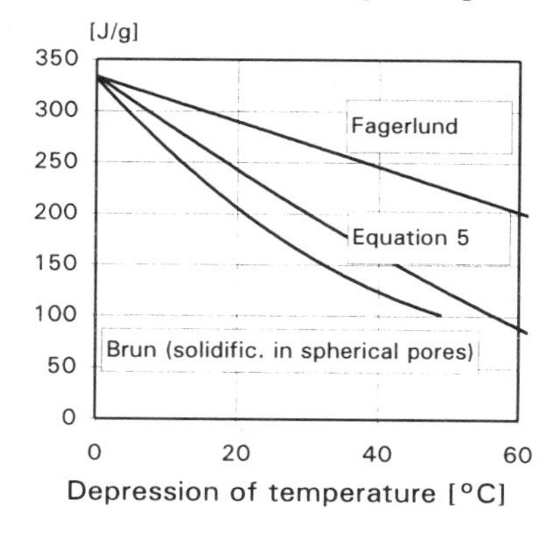

Fig. 1. Enthalpy at nucleation point for spherical pore structure.

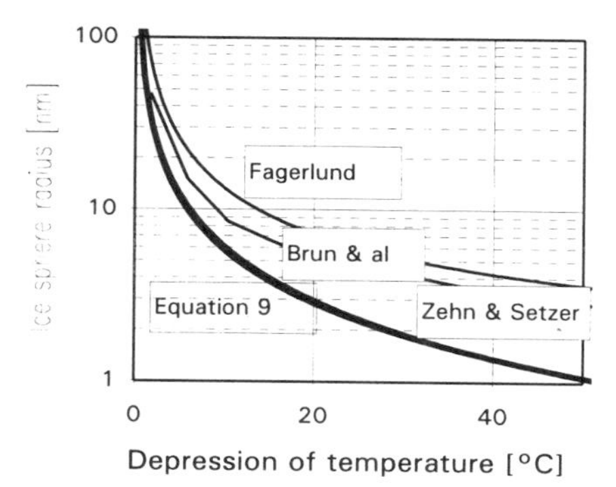

Fig. 2. Different expressions for the size of ice sphere radius

Fig. 2 shows that the curves calculated by Equation 9 and by the method presented by Brun deviates only somewhat from each other. This is due to the fact that, although the entropy in Brun's equation differs markedly from the entropy value used in Equation 9, this effect is compensated by the higher value of surface tension .

When the ice sphere radius R_n has been formulated as a function of temperature depression, the pore size R_p can be calculated if we know the thicknesses of the adsorbed water layer and the liquid-like layer. Here we utilized the thickness of the adsorbed water layer derived by Fagerlund:

$$t_{ad} = 1.97 \cdot \sqrt[3]{1/(-\theta)} \, , \tag{10}$$

where t_{ad} is expressed in nm and θ in °C. A new formula for the thickness of the liquid-like layer was derived starting from Takagi's quintic equation [6]. A general approximation for the thickness of the liquid-like layer δ [nm] as a function of the radius of ice sphere r_{ice} [nm] and temperature depression θ [°C] is as follows:

$$\delta = \frac{0.9 + 0.1 \cdot \log r_{ice}}{\dfrac{3}{(\log r_{ice})^{0.938}} + \theta \cdot \left(0.07 \cdot \log r_{ice} - 0.0365 + \ln \sqrt[3]{\dfrac{\log r_{ice} + 0.12}{\log r_{ice} - 0.12}} \right)} \tag{11}$$

The total thickness of the adsorbed water and the liquid-like layer of ice sphere can be calculated by summing Equations 10 and 11. Fig. 3 shows the magnitude of this layer as a function of temperature depression at the nucleation point.

Now, the pore size R_p can be calculated by adding the values of the adsorbed water layer and the thickness of the liquid-like layer to the value of ice sphere radius (Eq. 9).

For a spherical pore shape and always during nucleation, R_n is taken directly from Equation 9. In fusion, for a cylindrical pore, the multiplier 2 in the numerator of Equation 9 vanishes according to Laplace's equation. The curves for the spherical and cylindrical pore radii in nucleation and for the cylindrical pore radius in fusion are presented in Fig. 4. The direct relationship between the temperature depression and pore radius in nucleation and in fusion is given by Equations 12 and 13.

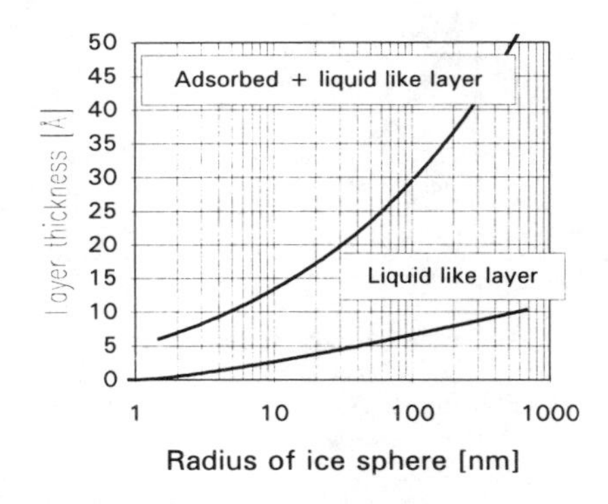

Fig. 3. Thickness of the non frozen layers at nucleation point.

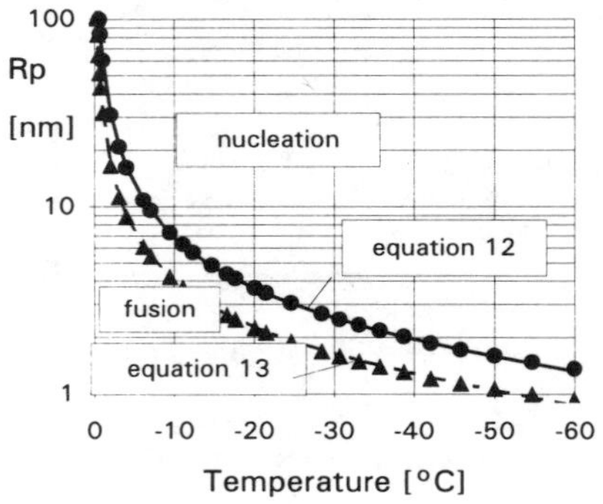

Fig. 4. Pore radius R_p in nucleation and in fusion for a cylindrical pore.

$$R_{p_n} = 0.584 + 0.0052 \cdot \theta - \frac{63.46}{\theta} \tag{12}$$

$$R_{p_f} = 0.757 + 0.0074 \cdot \theta - \frac{33.45}{\theta} \tag{13}$$

In Equations 12 and 13, $R_{p,n}$ and $R_{p,f}$ are the values of pore radius in nucleation and in fusion in nm and θ is temperature in °C.

2.3 Calculation method

The energy evolved during phase transformations of the pore water of concrete can be measured calorimetrically. Fig. 5 presents a typical heat capacity curve of the frozen sample.

The total amount of ice formed during freezing can be calculated without making any assumptions regarding the temperature dependence of the heat of fusion or of the heat of solidification and assuming that all ice is formed at 0° C. When considering the heat capacity curve in Figure 5, we see that the curve is linear when the temperature is below -55° C. This lets us to assume that no phase transformations take place below this temperature. The linear decrease in the specific heat of the specimen is due to the changes in the entropy of bulk ice and the entropy changes of the surface energy between the ice and the non-freezable water layer of the pore wall. Also, the entropy changes in the non-freezable water and in the paste and aggregates of the specimen affect this decrease in the specific heat of the specimen. However, we can assume that the change in the entropy of all these parameters behaves linearly along the linear part of the heat capacity curve. Now, if we extend the straight line a-b in Figure 5 up to the higher temperatures, this line crosses the y-axis at point c. If the assumptions taken above are correct, the straight line a-b-c represents the lower limit for the specific heat of the specimen and the real specific heat curve must lie above this line at all the temperatures during phase transformations.

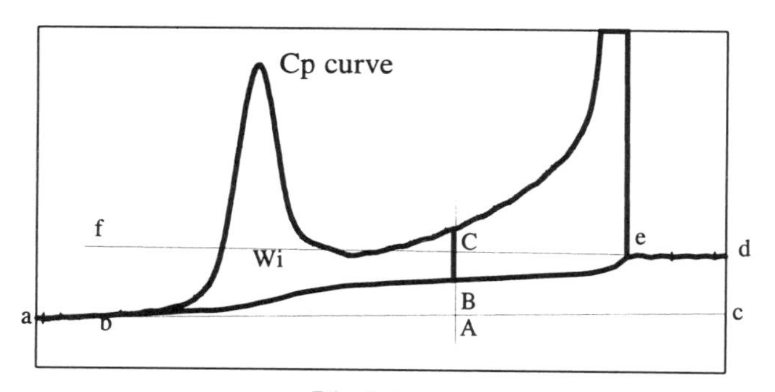

Fig. 5. Freezing thermogram in LTC-test and the construction of the integration baseline.

The assumptions taken in defining the heat capacity of supercooled water have only a slight effect when analyzing the freezing thermogram. Naturally, when regarding the fusion thermogram, the supercooling has no effect. Since the literature does not give any reliable data for the heat capacity of supercooled water and since its effect is of minor importance, we assume here the linear relationship measured every time separately on the basis of test data. Therefore, in the construction of the upper limit for the specific heat of the specimen, it was assumed that the change in the specific heat of supercooled water is linear and the line f-e-d can be drawn by extending the regression line determined at the temperature range between the values e and d. In Fig. 5 point c represents the heat capacity of the frozen sample and point d the heat capacity of the sample before freezing at temperature of 0° C. The difference in the heat capacities from d to c, Δc_p^0, is the amount of ice, w_{ice}^0, multiplied by the heat capacity of ice, $c_{p_{ice}}^0$, at 0° C. Since both the quantities Δc_p^0 and $c_{p_{ice}}^0$ are known, the total amount of ice formed during the freezing process can be calculated directly from Equation 14:

$$w_{ice} = \frac{\int \left(c_p - c_p^{bl}\right) dT}{W_0}, \tag{14}$$

where c_p^{bl} is the specific heat of baseline bc and W_0 is the heat of fusion of ice at 0° C.

When considering Fig. 5, we can conclude that all water freezes between the points e and b. Hence, the lines bc and ef are the upper and lower bounds between which the zero line (baseline) for freezing thermogram lies. The baseline between the points e and b can be evaluated by an iterative method; the difference of the ordinates of lines bc and ef represents the measure for the total amount of freezable water at a specified temperature. In Fig. 5, w_i is the amount of ice formed during the temperature decrease from point A to point b. The line AC represents the total amount of freezable water, w_{tot}, and the parts AB and BC represent the freezable water between the temperature ranges A to b (w_i) and e to A (w_{tot}-w_i). When point B has been measured at each measuring point of heat flow thermogram by iterative calculation process, the difference between the curve bBe and the line bc represents the distribution of freezable water over the block temperature of calorimeter. Since the upper and lower baselines (line ef in Fig. 5) represent the thermodynamic states where no solidification (line ef in Fig. 5) and no fusion (line bc in Fig. 5) takes place, line AC corresponds to the total freezable water measured by the method described before and line AB corresponds to the freezing of water taking place between the points A and b. Freezable water between any temperature range can be calculated by numeric integration with Equation 15

$$w = \int_{T_1}^{T_2} \frac{c_p(T) - c_p^{bl}(T)}{W_a(T)} dT, \qquad (15)$$

where $c_p(T)$ is measured by the calorimeter, $c_p^{bl}(T)$ is determined by an iteration process and $W_a(T)$ is calculated by Equation 5 from which the enthalpy effect of superfacial phase transformations must be subtracted.

3 Application of the method

The method presented before was utilized in order to determine the effects of aging on the pore structure of ordinary portland cement and granulated blast furnace cement mortars. In Fig. 6 the pore size distribution measured by MIP and by themoporometry (LTC) for non-aged, carbonated and cycled (repeated drying and wetting) specimens are presented. The pore size distribution by the LTC method using layered pore model was calculated from a freezing thermogram, assuming the first freezing for fully water saturated samples to take place at 0° C. The freezing thermogram was moved correspondingly. This assumption introduced an error for small pore sizes at lower temperatures since, there, the supercooling was not probably of the same magnitude. The sample size of the prism, 12 by 12 by 70 mm, caused a delay in temperature distribution although a very slow rate of temperature decrease of 3°C/h was used. These facts mean that the LTC method could not be very accurate for pore size distributions over 50 to 100 nm. These uncertainties did not affect the observed pore volumes at pore ranges from 1 to 50 nm. Carbonation affected the coarsening of the pore structure for granulated blast furnace slag concrete that was not clearly visible with the MIP test, as Figure 6 shows. Carbonation products were blocking the pore necks preventing the intrusion of mercury into pores larger than 4 nm. The pore volume became underestimated in the MIP test of carbonated blast furnace slag concrete where the carbonated products mainly originated from CSH gel.

Another example shows the use of the method in the determination of the gel pore volume of high strength paste. In Fig. 7a, the gel porosity is presented by the LTC method for pastes prepared from extra rapid hardening cement and hydrated for 1 and 28 days. A gel porosity from 1.5 to 2.5 nm was increased from 8.8% to 10% of the paste volume when the hydration proceeded. The same pastes were also tested by the N_2-gas adsorption method with samples dried in vacuum for six weeks after the initial water curing periods of 1 and 28 days. The results in Fig. 7b show that the gel porosity from 1.5 to 2.5 nm was exceptional low, being only 1.1 and 0.9 vol.-% at the ages of 1 and 28 days. This illustrated the detrimental effect of the vacuum drying on the pore structure. There was, however, one surprising similarity in the results. The main peak of the derivative curve of both methods exists at the same pore size from 2.1 to 2.2 nm. This pore range seems to be dominant for the gel pore range in cement pastes.

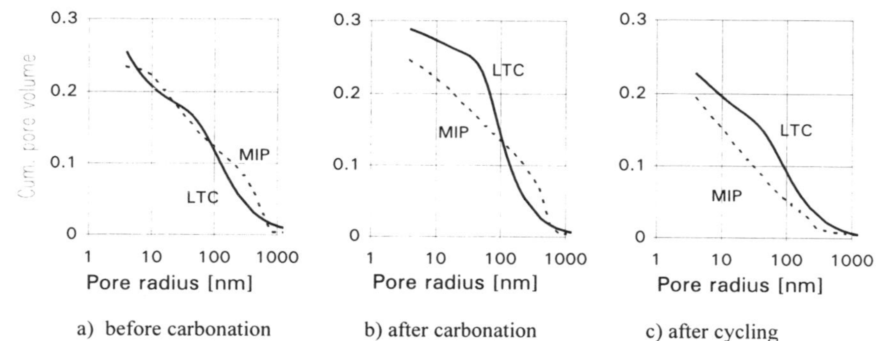

Fig. 6. Cumulative pore volumes for non-aged and aged mortar specimens measured by mercury intrusion porosimetry (MIP) and by themoporometry (LTC).

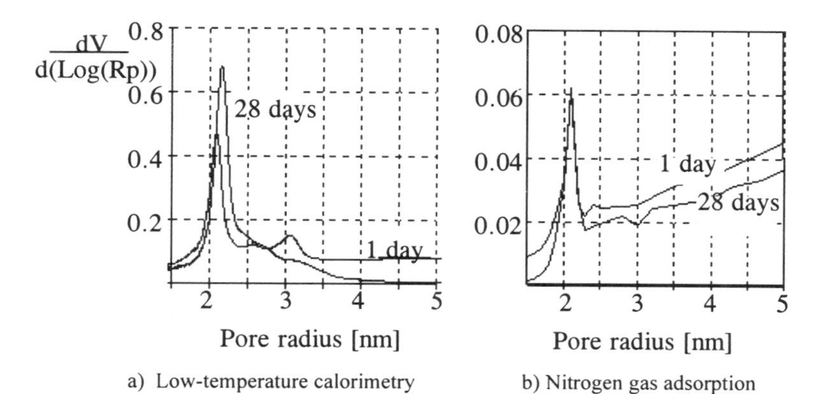

Fig. 7. Derivative pore size distribution for high strength pastes after 1 and 28 days water curing.

Freezing and thawing thermograms showed strong hysteresis. This can be explained by the different solidification and fusion energies of the cylindrical and layered pore structure (Fig. 4) and by the phenomena, where freezing is controlled by the necks of pores and thawing by the cavities, proposed by Sellevold [2]. The analysis of fusion thermograms showed a clear change in the derivative of the fusion curve at the block temperature of -37° C, which corresponds to the first fusion in the pores of 1.35 nm. Temperature is in line with the minimum freezing temperature of -63° C that gives the smallest freezable pore radius of 1.26 nm according to Equation 12. This observation proves the gel pore shape to be somewhere between layered or spherical. In the author's opinion, this also certifies that the main peak at -37 to -44° C in the freezing thermogram represents the freezing of the gel pore water and that the previous speculations about the anomaly of the entropy values of supercooled water at this temperature range are not necessarily correct.

4 Conclusions

The method developed enables the calculation of the amount of freezable water of porous material and the porosity and the pore size distribution from the data collected in calorimetric measurements. The defined pore ranges over 50 nm are not very accurate due to the supercooling of water and due to the size effects of specimens. The cooling rate of the calorimeter must be very low if the accurate pore size distribution is studied. The background of the method is very theoretical. The uncertainties lay in the values of surface tension of the water-ice interphase. There is also some uncertainty regarding the entropy variation of supercooled water at lower temperatures and the assumption that the nucleation follows homogenous nucleation theory. The method proved to be efficient when analyzing the pore structural changes in aged concretes. Changes were clearly noticeable by the LTC method but only slightly detectable by the MIP method, where the intrusion of mercury was prevented by the carbonation products closing the entrances of the wider pores. The superiority of the method lies in the fact that no preparation of the sample is needed; Thus no changes of the original pore structure occurs. The test results of the N_2-gas adsorption method showed that the reliability of the gas adsorption method suffers due to the vacuum drying of samples.

5 References

1. Fagerlund, G. (1973) Determination of pore-size distribution from freezing-point depression. *Matériaux et Constructions*, Vol.6-No 33. pp. 215-25.
2. Sellevold, E.J., and Bager D.H. (1980) *Low temperature calorimetry as a pore structure probe*. Proceedings of 7th Int. cong. on the Chemistry of Cement. Vol.IV. Paris 1980. pp. 394-99.
3. Bager D.H. (1984) *Ice formation in hardened cement paste*; Technical University of Denmark. Ph.D. Thesis. Technical Report 141/84. 66 p.
4. Beddoe, R E., and Setzer, M.J. (1990) Phase transformations of water in hardened cement paste, a low-temperature DSC investigation. *Cement and Concrete Research*, Vol 20. pp. 236-42.
5. Brun, M., Lallemand, A., Quinson. J-F. and Euraud. C. (1977) A new method for the simultaneous determination of the size and the shape of pores: The thermoporometry. *Thermochimica Acta*, 21. pp. 59-88.
6. Takagi, S. (1990) Approximate Thermodynamics of the Liquid-like Layer on an Ice Sphere Based on an Interpretation of the Wetting Parameter. *Journal of Colloid and Interface Science*, Vol. 137, No. 2. pp. 446-55.

108 METHODS FOR INSPECTING GROUTING IN PRESTRESSED CONCRETE

S. KAKUTA
Akashi College of Technology, Akashi, Japan
K. KUZUME and T. OHNISHI
Kokusai Structural Engineering, Osaka, Japan

Abstract

This paper discusses the characteristics of nondestructive testing methods based on model and field tests conducted on a post-tensioned prestressed girder bridge constructed about 25 years ago. Ultrasonic and radiographic tests were conducted for inspecting grouting defects in the web. A new technique of spectrum analysis was applied to the ultrasonic test. Results from this technique showed that the linear spectrum of the recorded waveform gave useful information on grouting defects. We also discovered through spectral analysis that a difference in spectra was produced before and after fresh grout material was injected into the voids. A computer aided radiographic system with a filter to prevent the scattering of X-ray was used in this study. Tests provided us with X-ray photographs clear enough for detecting insufficient grouting in the sheath. In addition, a miniature video camera was used to confirm the results of nondestructive testing.
Keywords: Charged coupled device camera, grouting, nondestructive test, post tensioned prestressed concrete, spectrum analysis, ultrasonic, X-ray.

1 Introduction

The post-tensioned prestressed concrete (PC) structure is a reliable system that has been used in many railway and road bridges. Recently, damage such as longitudinal cracks and water leakage, have been found on the lower flange of post-tensioned PC girders. It is assumed that this damage is caused by insufficient grouting in the sheath. This system requires that the sheath space be fully filled with grout material. Especially, PC ducts where penetrated by

Concrete Under Severe Conditions: Environment and loading (Volume Two) Edited by K. Sakai, N. Banthia and O.E. Gjørv. Published in 1995 by E & FN Spon. ISBN 0 419 19860 1

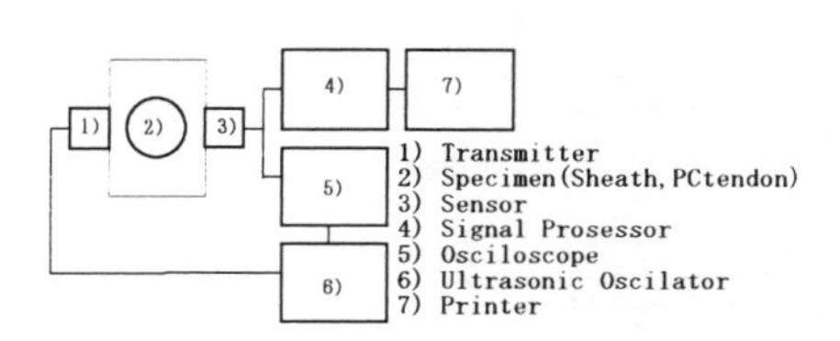

Fig. 1 Ultrasonic test and spectral analysis system

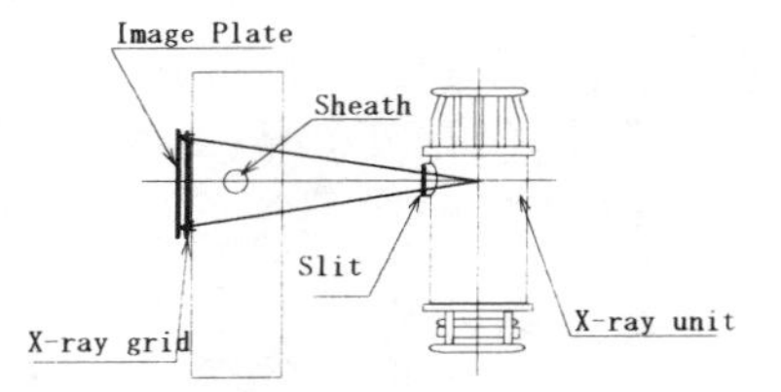

Fig. 2 Set up of X-ray radiograph unit (IP unit)

water will be leading frost and salt penetration in cold region. In the maintenance of post-tensioned prestressed concrete bridges, grouting inspection is becoming one of the most important items to check. In spite of its importance, there are few satisfactory inspecting techniques for this problem.

This paper discusses the characteristics and limitations of some nondestructive testing methods for defecting insufficient grouting in PC ducts. Ultrasonic, radiographic tests and the use of a miniature camera test directed at insufficient grouting at the web were conducted on a model test specimen, and field tests were carried out on a post-tensioned PC girder bridge constructed about 25 years ago.

Data drawn from tests conducted according to the conventional ultrasonic pulse velocity method were no good. We applied a new technique of spectrum analysis 1)2)3), similar to ultrasonic spectroscopy 4), to the recorded wave passing through the concrete structure. Results from this technique showed us that the amplitude spectrum of the recorded waveform gave us more useful information on grouting defects in the duct.

Radiographic inspection provides a permanent record of the state of the inspected area in the developed film. X-rays are commonly used for radiographic inspection in Japan because of legal regulations. The maximum thickness attainable is up to 40 ~ 50 cm when using a sensitive film and portable equipment. A computer aided fluoroscopic system with a slit for preventing X-ray scattering was used in this study. Tests provided us with photographs clear enough for detecting insufficient grouting in the sheath.

2 Test Procedure

2.1 Test Equipment

2.1.1 Ultrasonic Test System

The block diagram of the ultrasonic test system employed in this study is given in Figure 1. A transmitter and a receiver made of PZT with a 50 kHz resonant frequency for the longitudinal wave, were attached to both sides of a concrete member with grease. The velocity through the member was calculated from the travel distance and the propagation time measured reading from the waveform on an oscilloscope. Frequency spectra of detected waves passing through the concrete were calculated by a signal processor composed of a Fast Fourier Transform analyzer-equipped micro-computer.

2.1.2 X-ray I.P. (Image Plate) Processing
In this technique, a portable X-ray system for measuring the state of cracking on-site was used. X-ray are projected from the system which is installed on one side of a concrete structure. The radiant ray is captured and accumulated by a sensitized material after going through the structure. The image is processed and then recorded on computed graphic film. For the case of thick concrete, it becomes difficult to get a clear picture because the X-rays tend to scatter. By projecting through a grid, we were able to reduce the scattering effect (See Figure 2).

3 Small-scale model test

3.1 Test Procedure

A model test specimen of the PC girder was manufactured. Details of specimen are illustrated in Figure 3. We based the specimen on the web portion of a PC girder taken near the support and controlled the grouting charge when filling the sheath. The web thickness varied from 30 to 40 cm. In the middle part of the specimen, 4.5 cm diameter sheaths and 12 - 7 mm diameter PC cables were placed at 20 cm intervals. Cables were not prestressed. From the upper part of the specimen, the sheaths were charged with grout up to the 100%, 50% (indeed, by I.P. picture analyzing it was cleared they varied from 53 to 73 %) and 0% levels, respectively. For X-ray tests, two web thickness of 32 cm and 38 cm, with a 50% grouting charge were used in order to test the effect of thickness on image quality.

Ultrasonic tests were conducted at each measuring point where thickness was 32 cm and grouting charges were evaluated in areas filled to 100%, 50% and 0% where the thickness was 32 cm. Also, X-ray tests were conducted at 100% and 0% points where the thickness was 38 cm.

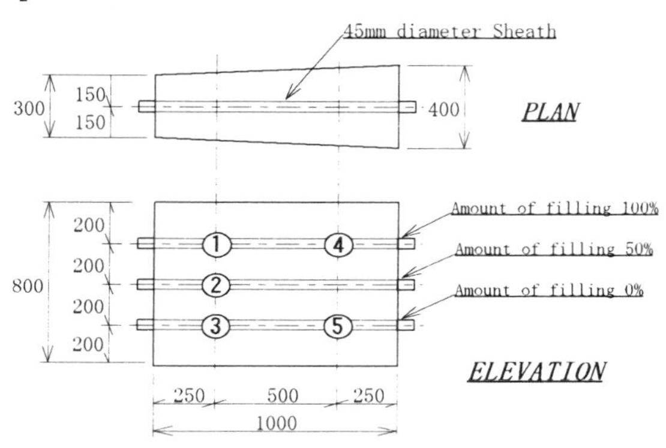

Fig. 3 Dimensions of model specimen. (in mm)

Table 1 . Longitudinal Wave Velocities through model specimens

Measured point	Web width (mm)	Amount of filling (%)	Velocity (km/s)
①	325	100	4.58
②	325	50	4.65
③	325	0	4.58
④	375	100	4.47
⑤	375	0	4.58

3.2 Results and Discussions

3.2.1 Ultrasonic Test

Table 1 indicates the velocity at each measuring point. It shows that the speed was slower for the lower the grouting charge in comparison with the same thickness point, like ①,② and ③, or④ and ⑤. The travel distance became longer because of voids, which in turn reduced the travel speed.

Frequency spectra of the recorded signals from the model specimen are given in Figure 4. These spectra exhibit an amplitude maxima at the respective carrier frequencies. The transducers represent an essential link between the specimen and the electronic equipment used for generating and processing signals. There is a maximum intensity in the spectrum at a transducer resonant frequency of 41.25 kHz (marked A). and the second small peak represents the frequency due to the analyzing system. Frequencies higher than 50 kHz were deliberately omitted in this experiment. The frequency-dependency of the response is seen by comparison of the maximum peak A for the 100% grouting charge (Points ① and ④), 50% grouting charge (Point ②) and 0% grouting charge (Points ③ and ⑤). We thought that a round empty sheath might generate a high-frequency spectrum by diffraction. The peak B around 10 kHz represents the resonant frequency corresponding to specimen thickness. The resonant frequency range derived from the measured velocity and travel distance was estimated as 5 ~ 8 kHz. An increase in traveling distance necessarily meant a decrease in resonant frequency. Therefore, it was assumed that the resonant frequency of concrete with void should be moved to low. Figure 4 shows this estimation. Peak B was an important point in this test.

3.2.2 X-ray Fluoroscopy

Figure 5 are a pictures of the film output by an image plate (I.P.) processing system. It was taken with a 35 mm lens camera. A radiant ray which passes through the matter sensitizes the I.P. upon contact. Also, the stornger the radiant ray, the darker is the image due to the sensitivity of the IP. The intensity of the radiant ray is influenced by the thickness and quality of the specimen. For example, a PC tendon cannot be easily penetrated by a radiant ray, leaving the image lighter in shade than surroundings areas. On the other hand, a ray easily passes through void, rendering darker image than other parts. Consquently, pictures taken show voids of any shading difference across the concrete as insufficient grouting.

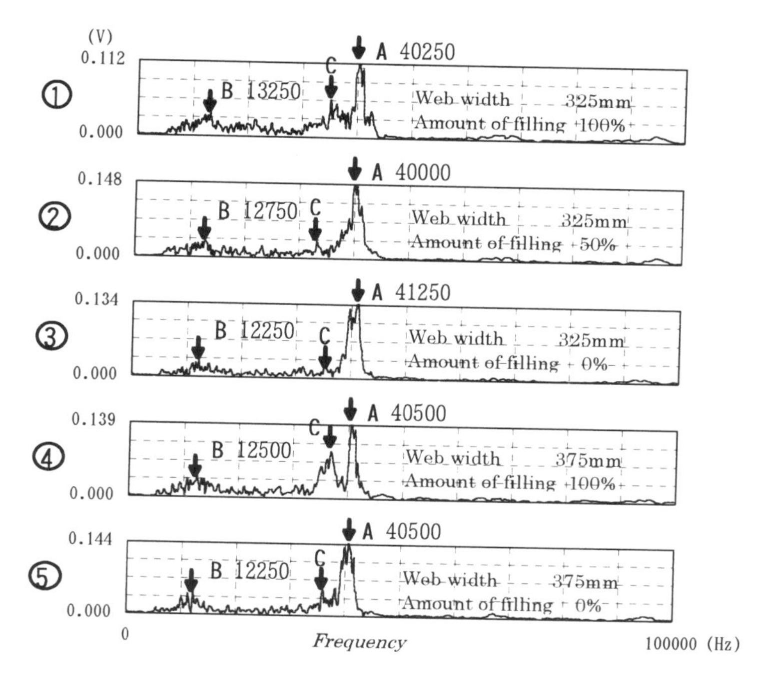

Fig. 4 Amplitude spectra of ultrasonic through model specimen

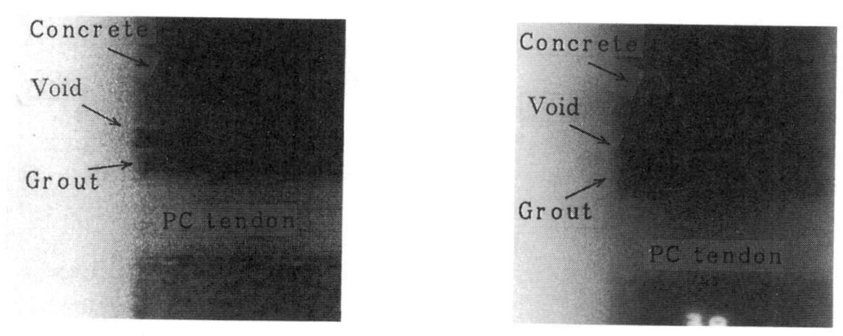

a) Insufficient grouting (w=32cm) b) Insufficient grouting (w=38cm)
Fig. 5 I.P. pictures of insufficient grouting in model specimen

4 Field test

4.1 Test procedure

Experiment were conducted on a post-tentioned PC gireder bridge completed about 25 years ago. A PC strand of 12-7 mm was composed inside of a 4.5 cm diameter sheath. Fig 6 shows the profile of the inspectedPC girders. At test points, the web thickness varied from 18 to 40 cm. Cable position were were

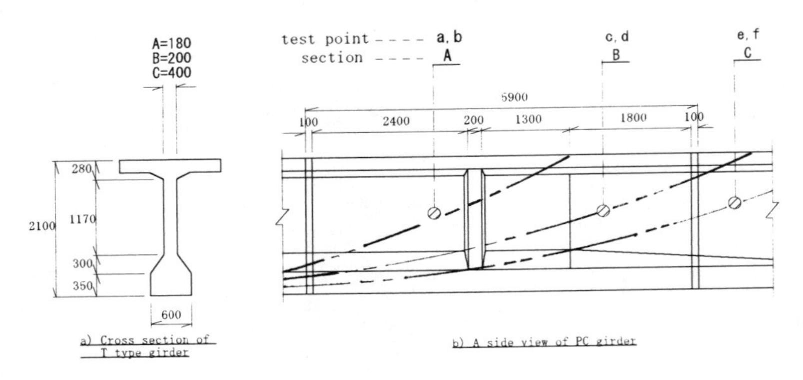

Fig.6 View of inspected post-tensioned PC girders

determined by microwave method (ground penetrating radar was used) and construction photograghs. The surface of concrete at the test points were ground flat.

Ultrasonic tests were carried out in the same manner as with the model test. A hole was drilled at suspected points and then a C.C.D. (Charge Coupled Device) camera was introduced into the hole. A video tape recording was taken to make sure grouting was in fact insufficient. Points where insufficient grouting was detected were injected with a fresh charge of grout and retested, using ultrasonic and X-ray fluoroscopic techniques, to confirm grouting quality.

4.2 Results and Observations

4.2.1 Ultrasonic Test

Table 2 shows the velocities of the inspected PC girders. In comparing the velocities with the state of grouting charge, we observed no relationship between charge and velocity as in the model tests.

The primary reason for this was that the difference in travel distance between an empty and a grouted sheath never exceed some 0.3 cm.

In concrete structures, it is difficult to defect a differential distance on the order of one centimeter or less using existing ultrasonic wave techniques. Also, the test measured surface was in a good polished state and the quality of the concrete was uniform with the model specimen. In contrast to this, the

Table 2. Longitudinal wave velocities through a web of PC girder

Measured point	Web width (mm)	Amount of filling (%)	Velocity (km/s)
a	180	100	4.06
b	180	0	4.06
c	200	100	4.08
d	200	50	4.35
e	400	100	4.60
f	400	0	4.22

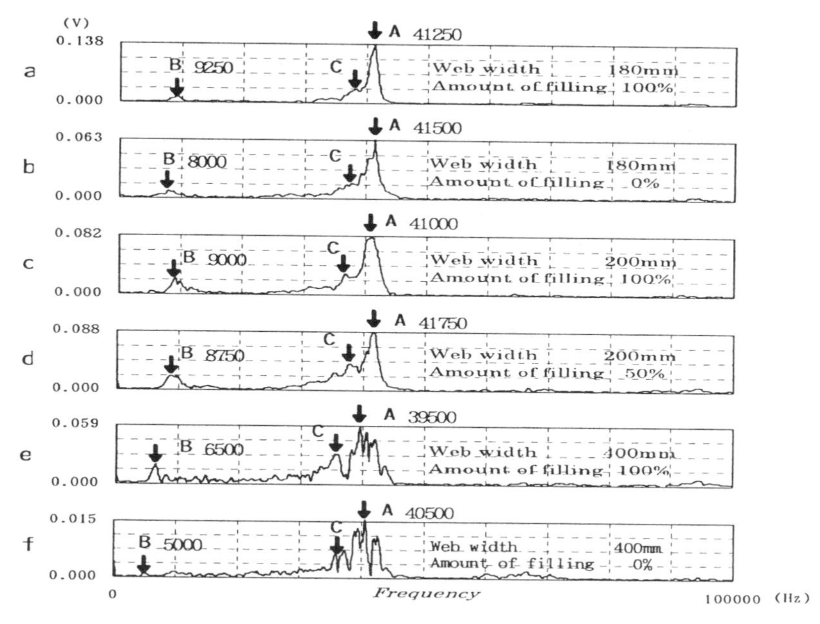

Fig. 7. Amplitude spectra of inspected PC girder

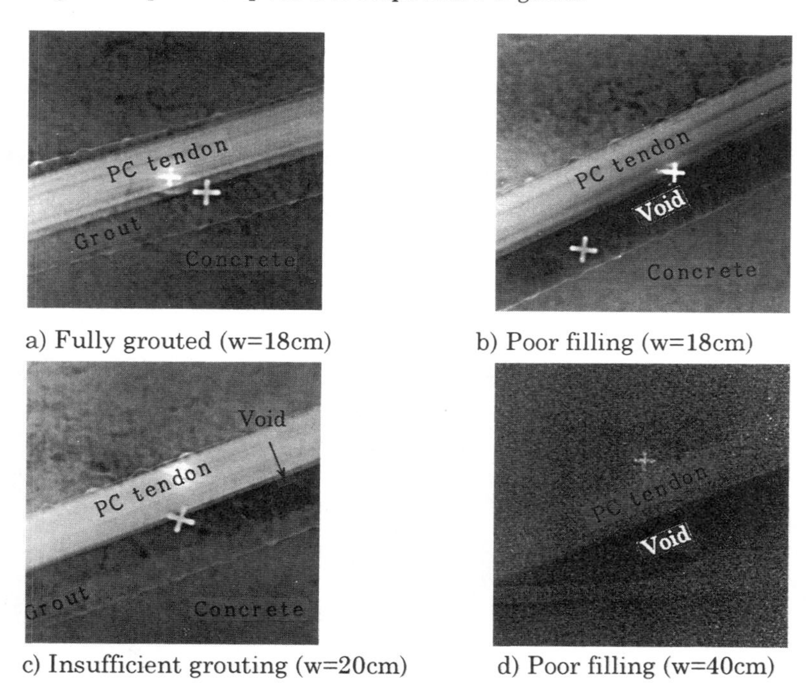

a) Fully grouted (w=18cm)

b) Poor filling (w=18cm)

c) Insufficient grouting (w=20cm)

d) Poor filling (w=40cm)

Fig. 8. I.P. pictures of insufficient grouting in PC girders concrete surface had deteriorated in the actual structure, and it was highly unlikely that the quality of the concrete was uniform in the separate segments. This suggests that velocity testing alone is insufficient in detecting sheath cavities in actual structures.

The amplitude spectra are given in Figure 7. and amplification in the test system were kept constant in the tests.

From comparisons of the amplitude spectra obtained at the same web width - points **a** and **b**, **c** and **d**, and **e** and **f** - we observed that peak **A** moved towards a higher frequency whereas peak **B** moved towards a lower frequency when voids were present. These results suggest the possibility of detecting defects for the same reasons deduced in model tests.

4.2.2 X-ray Fluoroscopy

The pictures shown in Figure 8 were taken at suspected points detected by ultrasonic tests using spectrum analysis. These I.P. pictures clearly indicate the grout, sheath and PC cable. However, this method can be applied only to concrete members of no more than 50 cm thickness. Although there are many legal regulations in Japan, the image plate processing method employed in this study is a useful and accurate method for inspecting middle and short span PC bridges, because the thickness of most members is less than 50 cm.

A C.C.D. camera was used to confirm the above results (Figure 9). The diameter of the camera used was 11 mm and diameter of holes drilled for viewing was about 3 cm. These pictures shows voids in ducts. Insufficient grouting points detected by the ultrasonic test and X-ray fluoroscopy were confirmed with the C.C.D. camera.

4.2.3 Post-Grouting Measurement

After grouting, we confirmed the state of grouting by ultrasonic testing and X-ray fluoroscopy, at the same points measured before grouting was performed. Results from the ultrasonic tests and X-ray fluoroscopy at point **b** are shown in Figure 10 and Figure 11, respectively.

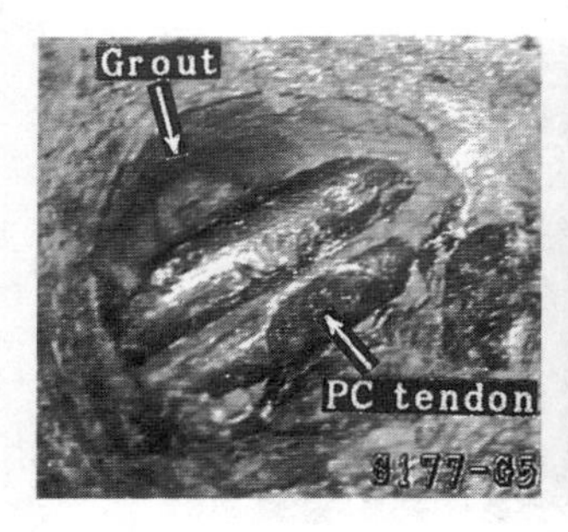

a) Fully grouted

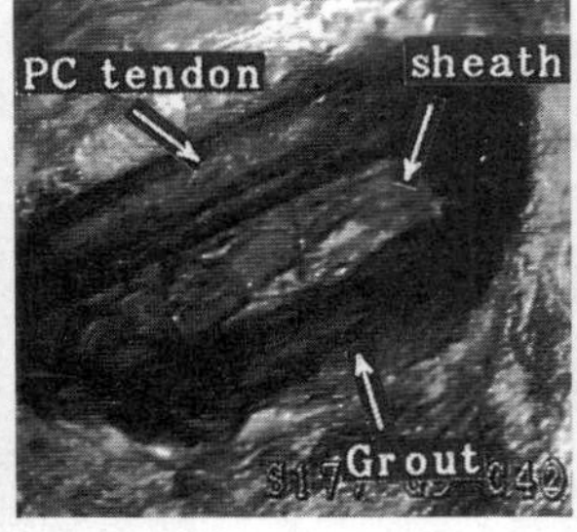

b)Insufficient grouting

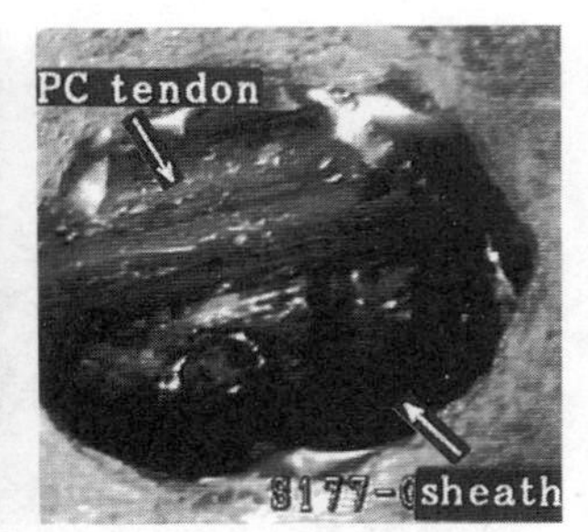

c) Poor filling

Fig. 9. Pictures from C.C.D. camera inspection

When looking at the recorded waveform, we discovered amplitude was always larger after grouting than before. And, peak A of the amplitude spectrum moved towards a lower frequency while peak B moved towards a higher frequency.

A solid circle in the fluoroscopic picture is the cross section of the area restored with grout, as viewed by the C.C.D. camera through the prepared hole. The contrast between the void and PC members seen before grouting has decreased in the post-grouting picture, in comparison with other potions of concrete.

These results were confirmed in other locations as well, suggesting that insufficiently grouted parts were filled with the fresh grout.

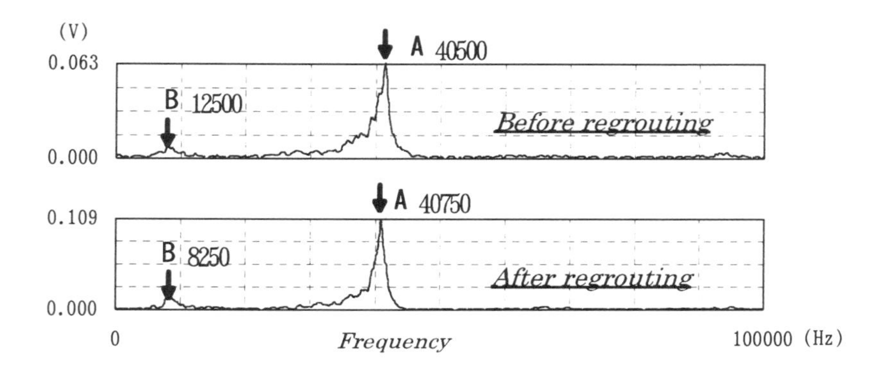

Fig. 10. A typical recorded waveform and the amplitude spectra at point b after post-grouting

Fig. 11. An I.P. picture after post-grouting

5 Conclusions

The conclusions obtained from the above test results are summarized as follows:

(1) Frequency analysis of the ultrasonic waveforms gives us clearer informationon any insufficient grouting in the duct than we can obtain by calculating velocity alone.

(2) Image plate processing of X-ray fluoroscopy gives a clearer picture of insufficient grouting when inspecting post-tensioned PC girders, than conventional X-ray fluoroscopy does.

(3) The ultrasonic frequency analysis method is recommended for initial inspections due to the ease of testing and the portable equipment. After a suspected location is detected by ultrasonic tests, X-ray I.P. fluoroscopy or the C.C.D. camera method must be employed as a more accurate method in next step.

(4) The ultrasonic frequency analysis is an effective technique for verifying the amount of grout charged in to a deficient region after regrouting has been performed.

6 References

1. Sakata, Y. and Ohtsu, M. (1990) Crack evaluation in concrete by ultrasonic spectroscopy, Proc. of JSCE, No. 414/V-12, pp. 69-78.
2. Carino, N.J. et al. (1986) Flaw defection in concrete by frequency spectrum analysis of impact echo waveforms, International Advances in Nondestructive Testing, pp. 1-30.
3. Kakuta,S. et al.(1994) Nondestructive testing methods for inspection of grouting defect, Developments in Short and Medium Span Bridge Engineering '94, CSCE. Montreal Canada, pp. 1007-1018
4. Brown, A.F. (1982) Ultrasonic spectroscopy, Ultrasonic Testing, Chap. 5, edited by J. Szilard, John & Wiley & Sons, pp. 167-215.

109 ON THE MEASUREMENT OF WATER-CEMENT RATIO OF FRESH CONCRETE WITH THE KANSAS WATER-CEMENT RATIO METER

M. HOSSAIN, J. KOELLIKER and H. IBRAHIM
Kansas State University, Manhattan, Kansas, USA
J. WOJAKOWSKI
Kansas Department of Transportation, Topeka, Kansas, USA

Abstract
The water-cement ratio of fresh concrete is recognized as being the one factor that affects the strength and permeability of an adequately compacted concrete mix. Although water-cement ratio is the predominant factor affecting strength of hardened concrete, currently, no widely-used, reliable method is available for measuring water-cement ratio in the field. A prototype device has been developed to measure the water-cement ratio of a plastic concrete mix. The method is based on the measurement of turbidity of water-cement slurry separated out of a concrete mixture by "pressure sieving." Consistent results were obtained for both air-entrained and non-air-entrained concrete. Statistical analyses of the test results have shown that this water-cement ratio meter can predict the water-cement ratio of fresh concrete with an accuracy of ±0.01 on the water-cement ratio scale for a single test at 90% confidence interval. If the method works as well in the field as it does in the laboratory, accurate determination of water-cement ratio could dramatically improve the ability of the concrete industry to assure the quality of concrete construction.
Keywords: Concrete, confidence interval, construction, quality, water-cement ratio.

1 Background and problem statement

In engineering practice, the strength and durability of Portland cement concrete that has a given age and is cured at a prescribed temperature condition is assumed to depend primarily on two factors: the water-cement ratio and the degree of compaction [1]. When a standard level of compaction is achieved, the only variable affecting the strength and durability of any concrete mix is the water-cement ratio. For a well compacted concrete, the strength is taken to be inversely proportional to the water-cement ratio. This relation is preceded by a "law" established by Duff

Concrete Under Severe Conditions: Environment and loading (Volume Two) Edited by K. Sakai, N. Banthia and O.E. Gjørv. Published in 1995 by E & FN Spon. ISBN 0 419 19860 1

Abrams in 1919 who found strength to be equal to

$$f_c = K_1/K_2^{\,w/c} \tag{1}$$

where w/c represents the water-cement ratio of the mix (by volume), and K_1 and K_2 are empirical constants.

Although water-cement ratio is the predominant factor affecting strength of hardened concrete, currently, no widely-used, reliable method is available for measuring water-cement ratio in the field or at the job site. The water-cement ratio of a concrete mix is usually evaluated indirectly through a "consistency" or "workability" evaluation with the slump test according to ASTM standard method C143: Slump of Portland Cement Concrete [2]. It is recognized, however, that slump has no unique relation with the workability. The simplicity and ability of this test to detect variations in the uniformity of a particular mix of nominal proportion is responsible for its widespread use [1].

The end-product of a recent research project entitled "Development of a Water-Cement Ratio Meter" sponsored by the Kansas Department of Transportation (KDOT) was the Kansas water-cement ratio meter, a test instrument, for measuring water-cement ratio of fresh (plastic) concrete [3]. This meter was envisioned by the researchers at the Department of Civil Engineering of Kansas State University and a prototype was fabricated in that research project. The meter will be used to find the water-cement ratio of a fresh concrete mixture in the field.

2 Kansas water-cement ratio meter

Kansas water-cement ratio meter is an electro-mechanical system consisting of the following components:

1. A state-of-the art turbidimeter (Model No. DRT-100B) manufactured by the HF Scientific, Inc. of Fort Myers, FL. This turbidimeter provides linear turbidity measurements over four switch-selectable ranges: 0 to 1, 10, 100 and 1000 NTU (nephelometric turbidity unit). It is sensitive to changes as small as 0.01 NTU, even in colored liquids.
2. A nest of 305-mm diameter U.S. #4, #10, #50 and #100 brass sieves.
3. A water-distribution system at the top of the sieves.
4. A splitter at the bottom of the cone below the sieves to prevent whirlpooling.
5. A 373-watt, 90-liter/min capacity pump.
6. Flexible tubing of 16 mm diameter for fluid passage and associated control valves.

Figure 1 shows the schematic diagram of the system.

In usual operation, 454 g of fresh concrete is introduced onto the top of the sieves and approximately 9 liter of water are added. The pump is then turned on and it recirculates the water through the system. After two minutes, the valve controlling the flow through the turbidimeter cuvette is turned on and turbidimeter readings are

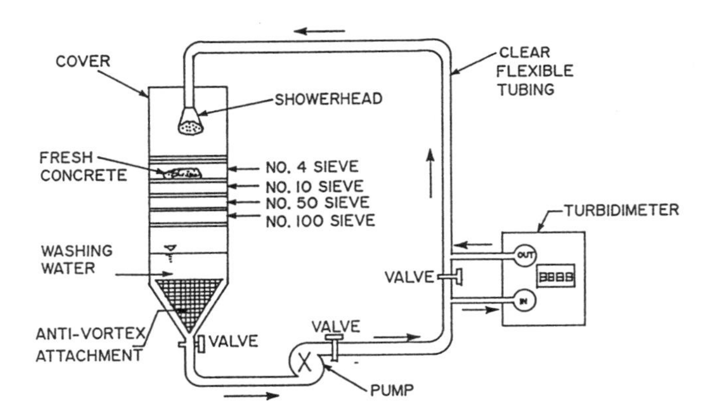

Fig. 1. Schematic representation of Kansas water-cement ratio meter.

taken at specific time intervals. The readings usually stabilize after a few minutes.

The measurement of water-cement ratio with this equipment is based on the principle of measurement of turbidity and "pressure sieving" of concrete to separate the cement-water slurry phase from the aggregates. Turbidity is a measure of extent to which light is either absorbed or scattered by the suspended particles in water. Although, turbidity is not a direct quantitative measurement of suspended solids, turbidity changes when there are more suspended particles in water. Thus, as the amount of cement in a water-cement slurry changes, the turbidity reading changes, too. This resulting turbidity reading of a water-cement slurry separated out of the concrete mix can be correlated with the water-cement ratio of the mix to produce a direct reading of water-cement ratio.

3 Results and discussions

3.1 Tests on concrete mix

Tests with the Kansas water-cement ratio meter were carried out on three types of concrete mixtures: mix without air-entraining admixture, mix with air-entraining admixtures, and mix with air-detraining admixture. Details of the mixtures are described below:

3.2 Mix characteristics

3.2.1 Constituent Specifications

Type of Cement: Monarch Type I
Coarse Aggregate: KDOT CA-4
Fine Aggregate: KDOT FA-A

3.2.2 Mix Proportions

1:2:3 (by weight)

Cement content:	379 kg/m^3
Water-cement ratio (by weight):	0.40, 0.44, 0.48.
Air-entraining admixture:	Vinsol Resin
Air-detraining admixture:	Polypropylene Glycol

To ensure consistency, only two samples were taken from a single batch of concrete. A third sample could not be taken because the concrete tended to dry up by the time the system was ready for the third run (a single test and cleaning takes around 10 minutes). This is especially critical when the water-cement ratio is low.

In every test, once the pump was turned on, turbidimeter readings were taken at 60-second intervals for the first 2 minutes, then at 30-second intervals for the following 90 seconds, and finally at 15-second intervals until the end of the test (after 6 minutes). After about four minutes from the beginning of the test, the turbidimeter readings tended to stabilize.

3.3 Tests on mix without air-entraining admixture

The readings from these tests were considered to form the "control" set because the concrete was free of admixtures. Table 1 shows the turbidimeter readings obtained for a concrete mixture with a water-cement ratio of 0.44, and Figure 2 illustrates the two trials for this mix. It is evident that after around 4 minutes of "pressure" sieving the system gives consistent readings for a finite period of time (about 2 minutes). Figure 3 illustrates the average turbidity readings versus the elapsed time (from the start of the test) for all three water-cement ratios. It is apparent that after about 5 or 6 minutes the readings are most consistent, and the turbidimeter readings and water-cement ratios are approximately linearly related with a difference in readings of about 20 NTU near or at the end of the tests. Since only 454 g of concrete were used in each of the tests, by using proportioning, there was a difference of a few grams of water between mixes with different water-cement ratios (the amount of cement, fine aggregate, and coarse aggregate are constant in that 454-gm sample). Thus, there was a considerable difference in the turbidimeter readings for the different water-cement ratios tested. This indicates how sensitive the system is in its ability to expand the small difference in water content to a rather considerable difference in turbidimeter readings.

3.4 Tests on mix with air-entraining admixture

Concrete used in Kansas, however, is usually air–entrained. In this part of the test program, air-entraining agent was used at a dosage so that it would produce about

Table 1. Turbidimeter readings of mix without air-entraining admixture with water-cement ratio of 0.44.

Elapsed Time (sec)	Turbidimeter Readings (NTU)			
	Trial 1	Trial 2	Avg.	Std. Dev.
0	5	4	5	1
60	348	342	345	4
120	245	255	250	7
150	235	228	232	5
180	221	220	221	1
210	217	205	211	8
225	215	205	210	7
240	219	202	211	12
255	205	198	202	5
270	203	203	203	0
285	202	205	204	2
300	208	210	209	1
315	208	209	209	1
330	202	211	207	6
345	205	194	200	8
360	207	197	202	7

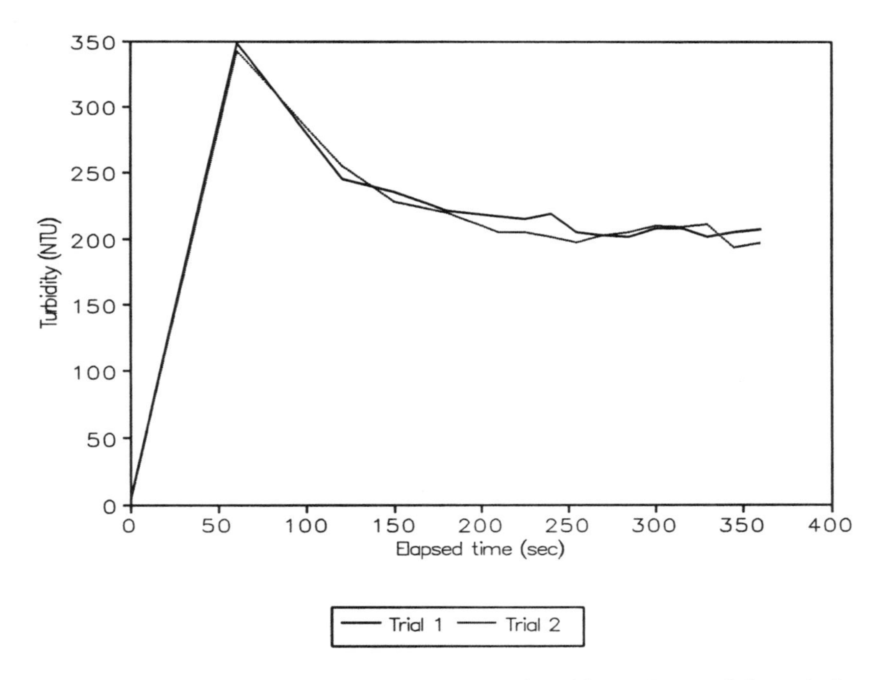

Fig. 2. Turbidity vs. elapsed time for concrete mix without air-entraining admixture (w/c = 0.44).

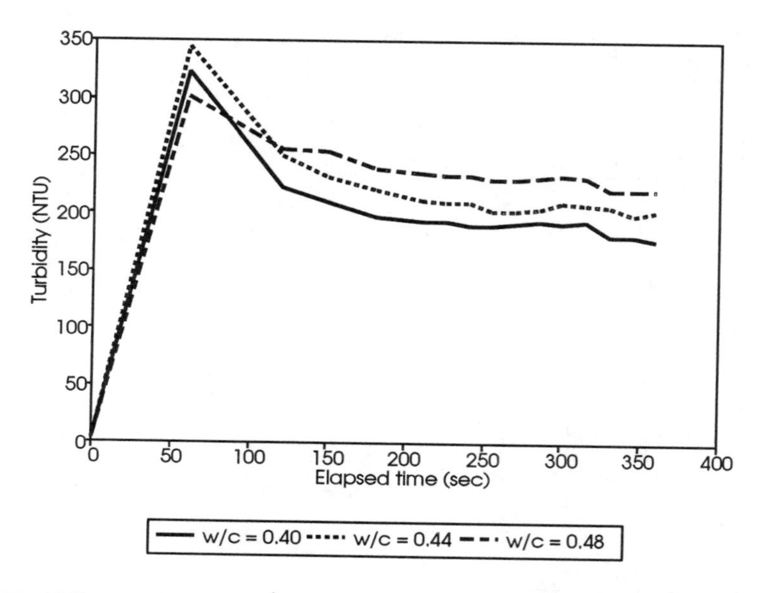

Fig. 3. Turbidity vs. elapsed time for concrete mix without air-entraining admixture.

5 to 7 percent of air in the mix. It was expected that the readings would be different from those of mix without air-entrainment because entrained air would cause air bubbles to interfere with the turbidimeter readings. Table 2 shows the turbidimeter readings for an air-entrained concrete mix with a water-cement ratio of 0.44, and Figure 4 illustrates the average turbidity readings versus elapsed time (from the start of the test) for air-entrained concrete mixes with water-cement ratios of 0.40, 0.44 and 0.48.

As evident, the results of mix with air-entraining admixture are different from those without air-entrainment. Also, there is no longer linear correlation between the turbidimeter readings and the water-cement ratios. This was due to the presence of air bubbles in the system; bubbles caused the readings to be higher and to fluctuate rapidly even after 5 minutes of testing. This problem with air-entrained mix necessitated the use of an air-detraining agent.

3.5 Tests on mix with air-detraining admixture

An air-detraining agent was used to eliminate air bubbles in the fresh concrete due to the air-entraining admixture. Two air-detraining agents, Tributyl Phosphate (TBP) and Polypropylene Glycol (PPG), supplied by Fritz Chemical Co, Dallas, Tex., were used. These chemicals were added to the wash water, not into the materials for batching the concrete. Several trials were performed to determine the optimum dosage of each air-detrainer to counteract the presence of air in the system. By comparing the results obtained from mixes without air-entraining admixture, it was determined that 20 ml of TBP or 2 ml of PPG can neutralize all air bubbles in the mixes. Table 3 shows the turbidimeter readings obtained with air-detrainment (using

Table 2. Turbidimeter readings of mix with air-entraining admixture with water-cement ratio of 0.44.

Elapsed Time (sec)	Turbidimeter Readings (NTU)			
	Trial 1	Trial 2	Avg.	Std. Dev.
0	4	5	5	1
60	301	337	319	25
120	222	216	219	4
150	206	203	205	2
180	194	197	196	2
210	189	190	190	1
225	190	187	189	2
240	192	186	189	4
255	188	186	187	1
270	188	185	187	2
285	187	191	189	3
300	183	189	186	4
315	181	190	186	6
330	184	190	187	4
345	181	192	187	8
360	186	198	192	8

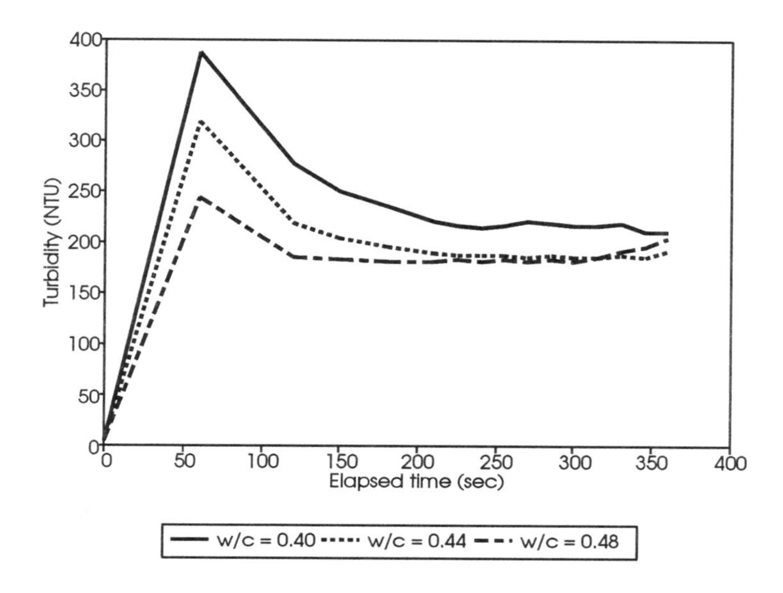

Fig. 4. Turbidity vs. elapsed time for concrete mix with air-entraining admixture.

Table 3. Turbidimeter readings of air-detrained mix with water-cement ratio of 0.44.

Elapsed Time (sec)	Turbidimeter Readings (NTU)			
	Trial 1	Trial 2	Avg.	Std. Dev.
0	5	4	5	1
60	366	325	346	29
120	267	257	262	7
150	239	244	242	4
180	232	233	233	1
210	216	226	221	7
225	216	222	219	4
240	218	219	219	1
255	217	218	218	1
270	204	214	209	7
285	207	217	212	7
300	209	213	211	3
315	208	212	210	3
330	203	209	206	4
345	206	208	207	1
360	198	208	203	7

PPG) of concrete with air-entaining admixture with a water-cement ratio of 0.44. It was found that air-detrainment neutralized the effect of air-entrainment on the turbidimeter readings for all mixes.

3.7 Tests on non air-entrained mix with water-reducing admixture

Tests were also conducted on a non air-entrained mix with a water reducer. The mix had a water-cement ratio of 0.40, and the water reducer was added at a rate of 118 ml per 45.4 kg of cement. The objective of adding the water reducer was not to reduce the amount of mixing water but to increase the slump up to about 25 mm as compared to the "zero" slump in the mix without any air-entraining agent. The results of turbidimeter tests on this mix showed that the addition of water-reducer did not have any appreciable effect on the results of turbidimeter tests and after 5 minutes, the readings tend to converge.

4 Statistical analysis

A key question for the system will be: "How confident are we in determining the water-cement ratio from a single NTU reading and at what time" ? To answer this question on a preliminary basis, the average turbidimeter readings for air-detrained mixes at 300, 330 and 360 seconds were chosen. Regression analysis was performed by taking the water-cement ratio as a dependent variable and the average turbidimeter reading as an independent variable. The data set used the three water-cement ratios– 0.40, 0.44, and 0.48 for each time period. A curve–fitting package (Tablecurve) by Jandel Scientific, was used in the analysis. A curvilinear equation gave almost identical fits for all three time periods. Based on this analysis, and in order to select a shorter time period, the 300-second time period was chosen as the best test period.

An expanded analysis was done to determine the precision of measuring water-cement ratio by the Kansas water-cement ratio meter using the observations for the 300-second time period. Again, regression analysis was performed using Tablecurve by taking the water-cement ratio as a dependent variable and the turbidimeter reading as an independent variable. The data set used the two replications from the three water-cement ratios- 0.40, 0.44, and 0.48, a total of six values.

A simple linear fit was chosen, although curvilinear equations might give somewhat better fits at the expense of more degrees of freedom. The equation is as below:

$$\text{Water-Cement Ratio} = 0.097 + 0.0016 * \text{Turbidimeter Reading (in NTU)} \quad (2)$$
$$(R^2 = 0.988,\ n = 6)$$

The R-squared value of the linear fit for the 300-second data is 0.988. For a single test on a particular concrete mix that gave an NTU value of 210 at 300 seconds, this relationship indicates that one could be 90% confident that the water-cement ratio for the sample lies between 0.433 and 0.454, and the average value would be 0.444. It is clear that the band-width is about ±0.01 on the water-cement ratio scale for the true value for a single test [3]. However, this needs to be established by testing more samples with varying water-cement ratios.

5 Conclusions

Although somewhat empirical, the proposed method appears to be very promising for measuring water-cement ratio of fresh concrete. The method is simple and unique. Consistent results were obtained for mixes with and without air-entraining admixtures. Statistical analyses of the test results have shown that this equipment can predict the water-cement ratio of fresh concrete with an accuracy of ±0.01 on the water-cement ratio scale for a single test at 90% confidence interval. If the method works as well in the field as it does in the laboratory, accurate determination of water-cement ratio could dramatically improve the ability of the concrete industry to assure the quality of concrete construction.

6 Acknowledgements

The authors wish to acknowledge the financial support for this study provided by the Kansas Department of Transportation and the Advanced Manufacturing Institute of Kansas State University. Cooperation and advice of Mr. Richard McReynolds, P.E. of KDOT and Mr. Dan Montgomery of Fritz Chemical Co. are gratefully acknowledged. All admixtures used in this study were supplied by Fritz Chemical Co. of Dallas, Tex.

7 References

1. Neville, A.M. (1981) *Properties of Concrete.* 3rd ed. London: Longman Scientific and Technical.
2. American Society for Testing and Materials (1992) *Annual Book of ASTM Standards.* Section 4, Vol. 04.02.
3. Koelliker, J., Hossain, M. and Ibrahim, H. (1994) *Development of a Water–Cement Ratio Meter.* Report No. KSU–265, Kansas Department of Transportation, Topeka, Kansas.

PART NINETEEN
HIGH TEMPERATURE ENVIRONMENT

110 STUDY ON THE DEGRADATION OF CONCRETE CHARACTERISTICS IN HIGH TEMPERATURE ENVIRONMENTS

T. SUZUKI and M. TABUCHI
Mitsubishi Heavy Industries Ltd, Yokohama, Japan
K. NAGAO
Obayashi Corporation Technical Research Institute, Tokyo, Japan

Abstract
Research was carried out on the influences of temperature and moisture migration on the mechanical properties of concrete subjected to temperatures up to 175 ℃. In the experiment, tests measured compressive strength and elastic modulus as mechanical properties by using test samples. Major experimental factors were exposure conditions during heating, such as whether test samples were in a sealed or an unsealed condition.

This paper describes the results of a series of such heating tests, involving up to 3.5 years of high temperature, and discusses the mechanical properties of concrete under the influence of heat.

As a result, it was confirmed that the mechanical properties of concrete subjected to high temperature are affected by the moisture migration and content in concrete during heating.
Keywords: Compressive strength, concrete, elastic modulus, high temperature, moisture migration, weight loss.

1 Introduction

At nuclear power plant facilities, concrete used for the foundations of equipment may be subjected to elevated temperatures for a long time, due to the heat conduction from the equipment.

Consequently, in order to evaluate the integrity of the foundations, it is necessary to investigate the changes in the physical properties of the concrete itself when subjected to elevated temperatures.

Concrete Under Severe Conditions: Environment and loading (Volume Two) Edited by K. Sakai, N. Banthia and O.E. Gjørv. Published in 1995 by E & FN Spon. ISBN 0 419 19860 1

However, there have been only a few examples of research concerned with the physical properties of concrete subjected to a long–term exposure to high temperatures. Especially,the moisture content and evaporation in concrete during long–term heating have not been taken into account in past studies.

For this study, research was carried out on the influences of temperature and moisture migration on the mechanical properties of concrete subjected to temperatures of up to 175 ℃. In the experiment, tests measured compressive strength and elastic modulus as mechanical properties by using test specimens. Major experimental factors included in the test were exposure conditions during heating, such as a sealed condition (evaporation of moisture is prevented) or an unsealed condition (evaporation of moisture is allowed).

This paper describes the results of a series of such heating tests over a period of 3.5 years and discusses the mechanical properties of concrete subjected to high temperature.

2 Test procedures

2.1 Materials and concrete mixes

Materials and mixtures of concrete used for the tests are listed in Table 1,based on specifications of mass concrete foundation of eqipment at nuclear power plant facilities.

2.2 Test items

Measurements taken before and after heating are compressive strength, elastic modulus, and weight loss.

Table.1 Materials and mix proportion of concrete

Type of Mix	Fine Aggregate	Coarse Aggregate	Type of Cement	Water-Cement Ratio* (%)	Sand Coarse Aggregate Ratio* (%)	Unit Water Content (kg/m³)	Cement (kg/m³)	Fine Aggregate (kg/m³)	Coarse Aggregate (kg/m³)	Fly Ash (kg/m³)	AE Water Reducing Agent (kg/m³)	Auxiliary AE Agent (kg/m³)	Super-plasticizer (kg/m³)
A	Crushed-sand of basalt + Land sand	Crushed-stone of basalt	Moderate heat cement + fly ash cemeht	50	46.4	157	256	872	1058	58	3.45	0.047	1.00
B			fly ash cement (class B)	50	46.4	159	318	866	1050	0	3.50	0.041	1.00
C			Normal portland cement	50	46.4	163	326	866	1053	0	3.59	0.008	1.00
D		Hard sand stone	Moderate heat cement + fly ash cement	50	46.4	157	256	872	991	58	3.45	0.047	1.00

(Note) *The design strength of Mix A is F_c=300 kgf/cm²(13 weeks). The water-cement ratio for Mix A was as used for the remaining mixes

2.3 Heating conditions

A programmable electric heating furnace with a temperature controller was used for the heating tests. Heating condition are shown in Fig.1.

For the long–term heating tests,specimens were exposed to sustained temperatures of 65 ℃ (temperature limitation except for local areas ,such as around penetration,of nuclear power plant facilities for normal operation or any other long–term period[4]),90℃(temperature limitation for local areas of nuclear power plant facilities for normal operation or any other long– term period[4])and 110℃(temperature at which water evaporates rapidly)in constant temperature up to 3.5 years.

For the short–term heating tests,specimens were exposed to sustained temperature of 175 ℃ (temperature limitation except for local areas of nuclear power plant facilities for accident or any other short–term period[4])in constant temperature up to 91 days.

For the thermal cycle heating tests,specimens were exposed to cycled heating temperature of 20 ℃ to 110 ℃(simulated temperature variation during operation periods of nuclear power plant facilities)up to 120 cycles.

2.4 Curing conditions before heating and age when heated

Since the drying process of concrete actually used in a power plant, which is mass concrete, is expected to be slow, curing before heating was performed by the sealed curing method and its age when heated was 91 days.

2.5 Exposure conditions

For evaluation with respect to exposure, specimens were put under either sealed conditions, where evaporation of water was prevented, or unsealed conditions, where evaporation of water was allowed. The method for the sealing of specimens is illustrated in Fig.2.

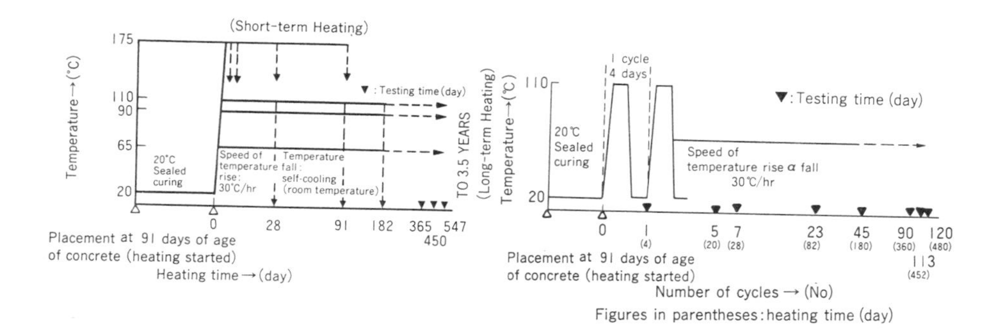

(a) Constant heating (b) Thermal cycle heating

Fig.1. Heating condition.

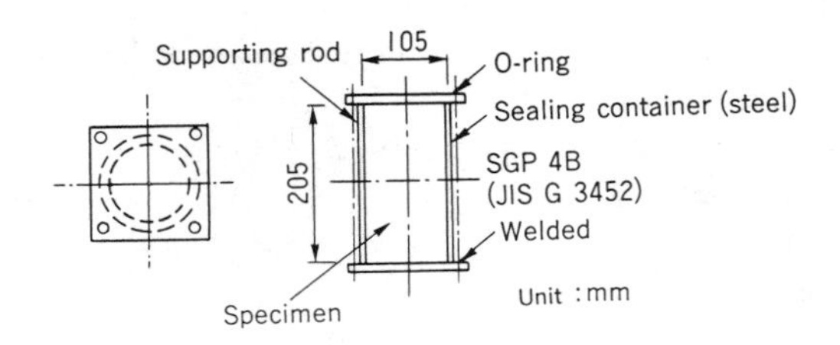

Fig.2. Sealing method.

2.6 Specimens

Three cylindrical specimens(ϕ 10 × 20cm) were prepared for each testing period. The types of specimens are given in Table 2.

Table 2. Types of specimens

	Heating	Type of Mix	Curing before Heating	Heating Temperature (°C)	Exposure during Heating
Constant Heating	Long-Term Heating	A	sealed	65	sealed/unsealed
		A	sealed	90	sealed/unsealed
		A	sealed	110	sealed/unsealed
		B	sealed	110	sealed/unsealed
		C	sealed	110	sealed/unsealed
		D	sealed	110	sealed/unsealed
	Short-Term Heating	A	sealed	175	unsealed
Themal cycle Heating		A	sealed	20-110cycles	sealed/unsealed
		B	sealed	20-110cycles	sealed/unsealed

3 Consideration

Results of the heating tests are summarized in the form of ratios(%) against corresponding values before heating (hereafter called "residual ratio").

3.1 Results of the constant heating test

3.1.1 Long–term heating test

(1) Compressive strength

The test results for mixture A are shown in Fig.3.

Under both sealed and unsealed conditions during heating, the strength (or residual ratio) after heating was higher than before heating, regardless of heating temperature. Under unsealed conditions during heating, the residual ratio became greater as the temperature rose.

Especially under sealed conditions, the compressive strength increased for 1.5 years and the residual ratio reached approximately 200 percent. These results are considered to be caused by the effect of auto–clave curing.

Under unsealed conditions, it is considered that micro–cracking in concrete, which was caused by moisture migration and evaporation under high temperatures, caused a degradation of compressive strength. However, the acceleration of hydration by high temperatures in any non–hydrated sections of concrete caused increased the compressive strength more than any degradation caused by micro–cracking. The results after 3.5 years were relatively unchanged from the measured after 1 year. Therefore, it was found that the compressive strength ceased to fluctuate at an early stage.

(2) Elastic modulus

The tests results for mixture A are shown in Fig.4.

Under sealed conditions during heating, the increase in the elastic modulus due to heating was

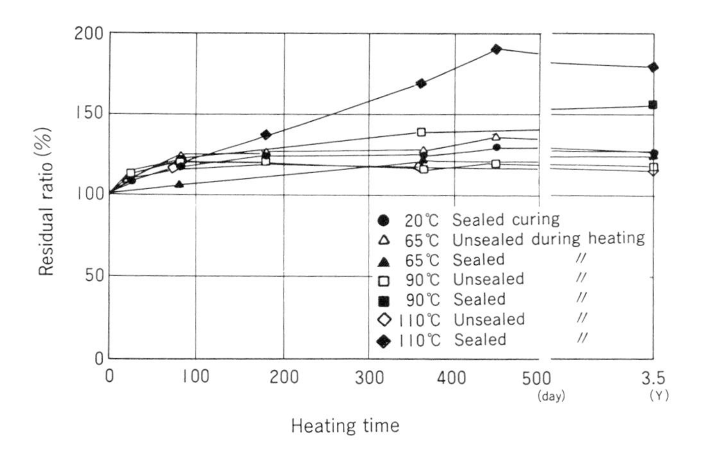

Fig.3. Result of compressive strength (Long–term heating test).

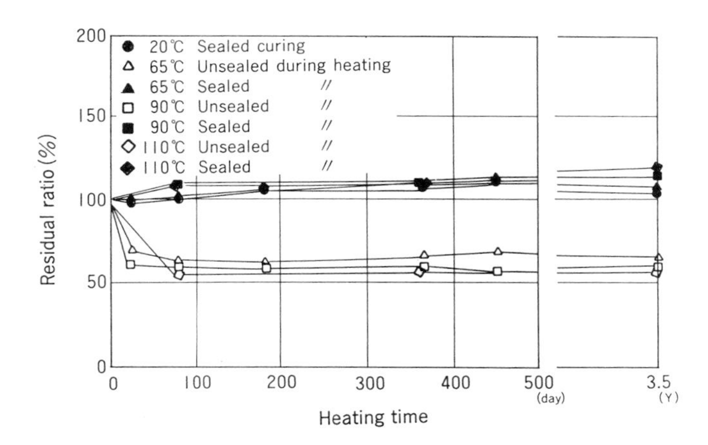

Fig.4. Result of elastic modulus (Long–term heating test).

small, and the elastic modulus had a tendency to rise, though not as much as the compressive strength. The elastic modulus under sealed remained relatively unchanged under heating even at 3.5 years.

Under unsealed conditions during heating, the residual ratio was considerably reduced. The reduction of the elastic modulus under unsealed conditions can be explained by the closing of micro–cracks at an early stage of stress. These results can be explained by the $\sigma - \varepsilon$ relation which indicated a smooth S–type curve of unsealed specimens(Fig.5).

Thus, the elastic modulus of concrete heated to high temperatures, with moisture migration and evaporation present, was markedly reduced. In addition, it was found that the elastic modulus became stabilized at an early stage, not changing much after 91 days through 3.5 years, even under heating.

(3) Influences of mixtures

The residual ratio of the compressive strength and elastic modulus of various concrete mixtures heated to 110 ℃ for 1.5 years and 3.5 years are shown in Fig.6.

There were few differences observed in the residual ratio of the elastic modulus between different mixtures types under either sealed or unsealed conditions. On the other hand, there was no difference observed in residual ratio of compressive strength under unsealed conditions, but differences were observed under sealed conditions after heating for 1.5 years. Thus, the residual ratio of compressive strength using normal portland cement (mixture C) was smaller than for mixtures A, B,and D. Otherwise,differences in the residual ratio of mixtures (mixed cement and normal portland cement) were not significant after heating for long periods as demonstrated by our test results after 3.5 years of heating.

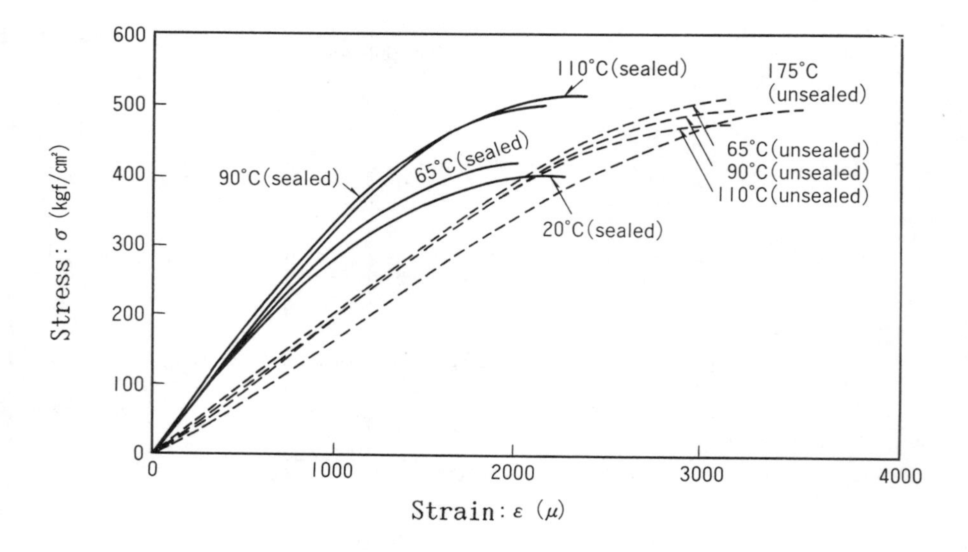

Fig.5 Stress–Strain relationships for concrete.

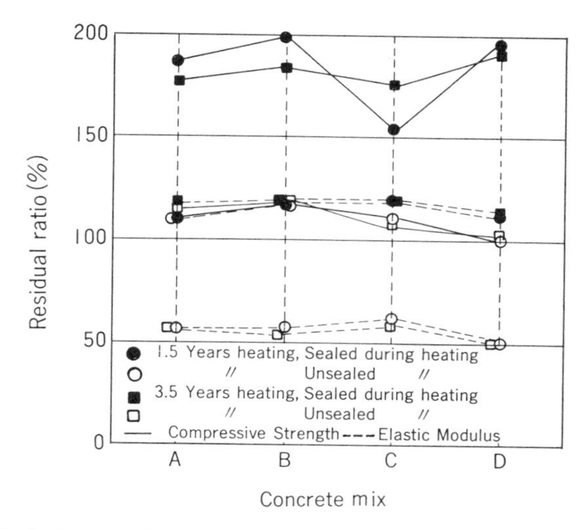

Fig.6 Compressive strength and elastic modulus by mix types.

3.1.2 Short–term heating test

Results of the short–term heating test are shown in Fig.7.

(1) Compressive strength

Compressive strength decreased slightly in the early stages of heating, but soon began to increase. No significant changes were observed during the period from day 1 to 91, and the final residual ratio was about 120 percent.

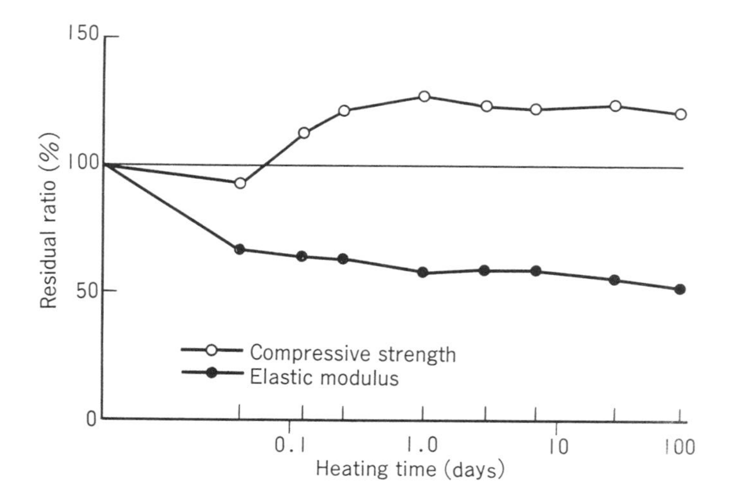

Fig.7. Result of short–term heating test.

(2) Elastic modulus
The decrease in the elastic modulus due to heating was larger than the decrease in compressive strength. In particular, the modulus fell considerably in the early stages of heating, and thereafter, decreased gradually until day 91 when the residual ratio was 50 percent or so.

3.2 Results of the thermal–cycle heating test

(1) Compressive strength
Results of compressive strength are shown in Fig.8.

Under sealed conditions during heating, compressive strength fell slightly as the number of cycles increased, while the residual ratio rose during and after the fifth cycle. Under unsealed conditions, the residual ratio rose again, but not as much an under sealed conditions. It leveled off during and after the fifth cycle.

(2) Elastic modulus
Results of elastic modulus are shown in Fig.9.

Under sealed conditions during heating, the elastic modulus showed a tendency similar to that of compressive strength, but the residual ratio did not increase as much as the compressive strength. Under unsealed conditions, the residual ratio fell as much as in the constant heating test, and was almost reduced by half. A major part of the reduction occurred in the early stages of thermal–cycle heating. No significant differences between mixtureA and B were observed.

3.3 Result of the weight loss

Figure 10 shows the relationship between changes in weight of mixure A and the compressive strength and elastic modulus obtained from long–term, thermal cycle, and short–term heating tests.

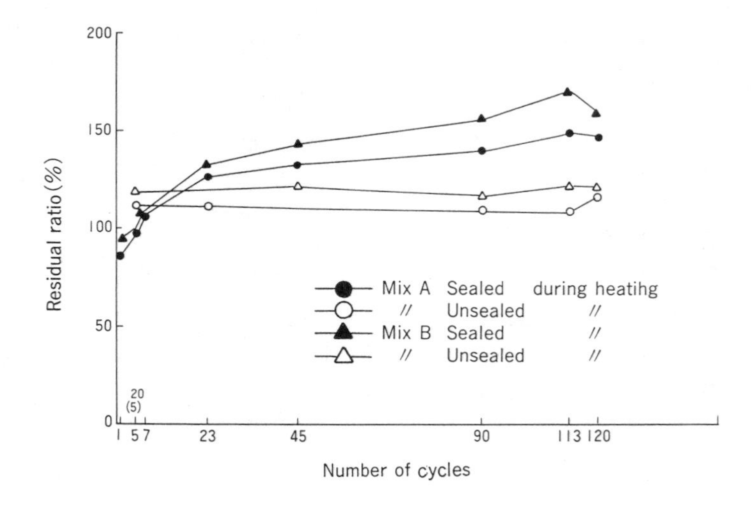

Fig.8. Result of compressive strength (Thermal cycle heating test).

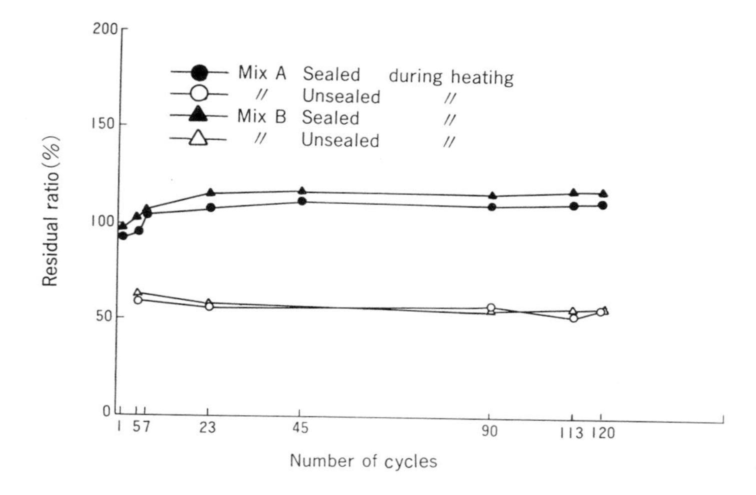

Fig.9. Result of elastic modulus (Thermal cycle heating test).

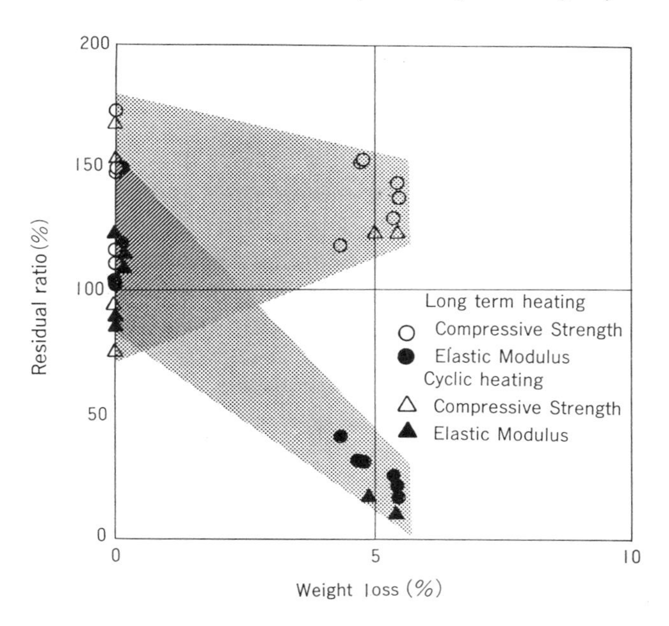

Fig.10. Relationship between weight loss and compressive strength/elastic modulus (Mix A).

This does not necessarily reveal any distinct relationship between the compressive strength and weight reduction ratio. However, it shows that the greater the weight reduction became, the greater the decrease in the elastic modulus tended to be.

4 Conclusions

Based on the tests result, the following conclusions may be drawn.

1. Under long–term heating at 65 ℃, 90℃, and 110 ℃,compressive strength after heating was greater than before heating, under both sealed and unsealed conditions. The elastic modulus varied slightly under sealed conditions. In addition the residual ratio decreased under unsealed conditions.
2. Compressive strength increased under short–term heating at 175℃, and the residual ratio of the elastic modulus decreased about 50 percent.
3. During the thermal–cycle heating test, compressive strength after heating was greater than before heating under both sealed and unsealed conditions. However the ratio of increase was smaller than under constant heating, suggesting the influence of thermal–cycle heating.Under unsealed conditions during heating, the residual ratio of the elastic modulus fell considerably; greater decreases were observed in the early stages of thermal–cycle heating.
4. The greater the weight reduction became, the greater the decrease in the elastic modulus tended to be, which indicated that moisture migration and evaporation during heating affected the reduction of elastic modulus.

Therefore, in order to estimate the properties of massive concrete structure subjected to high temperatures as accurately, it is necessary to study moisture migration in mass concrete members which are subjected to high temperatures further the over long periods of time.

5 References

1. Shneider, U. (1982) Behaviour of Concrete at High Temperatures. Deutscher Ausschuss für Stahlbeton, pp.50–69
2. Nassar, K.W. and Lohtia, R.P.(1971) Mass Concrete Properties at High Temperature. ACI Journal, No.3, pp.180–186
3. Tabuchi,M., Iriyama,M., Suzuki,T. and Nagao,K. (1991) Study on degradation of Concrete Characteristics in the High Temperature Environment. 11th International Conference on Structural Mechanics in Reactor Technology, Vol.H, pp.43–48
4. ASME (1983) Boiler and Pressure Vessel Code, Sept.Ⅲ, Div2, pp.214

111 HIGH PERFORMANCE CONCRETE FOR SEVERE THERMAL CONDITIONS

A.N. NOUMOWE, P. CLASTRES and G. DEBICKI
INSA LYON, Villeurbanne, France
J-L. COSTAZ
EDF/SEPTEN, Villeurbanne, France

Abstract
This study is part of an intensive experimental investigation carried out in France on High Performance Concrete under high temperature. Cylindrical and I-specimens were heated to different temperatures: 150, 300, 450 and 500°C. Measurements of temperature distributions in the concrete during heating above 300°C showed high temperature differences between the center and the surface of the specimens. At some points, the thermal gradient was greater than 40°C/cm, suggesting high thermal stresses in the concrete. Weight measurements during heating up to 450°C were suitable for gathering information on microstructural changes. This permitted quantifying the specimen water loss. After a heating/cooling cycle at each target temperature, decreases were recorded in the residual compressive strength, tensile strength and modulus of elasticity. A comparative study was made with the residual properties of an ordinary concrete tested in the same conditions. Furthermore, the tests showed main parameters for the understanding of the violent explosion which took place in some specimens: the thermal gradient, the moisture content and the tensile strength.
Keywords: High performance concrete, modulus of elasticity, spalling, strength, temperature, thermal gradient, water, weight.

Concrete Under Severe Conditions: Environment and loading (Volume Two) Edited by K. Sakai, N. Banthia and O.E. Gjørv. Published in 1995 by E & FN Spon. ISBN 0 419 19860 1

1 Introduction

Although high performance concrete is often considered a relatively new material, its development has been gradual over many years. The applications of high performance concrete have increased, and this material has been used in many parts of the world. The economy of the structure in energy technology can be enhanced considerably by the use of high performance concretes [1].

In designing building elements, the conditions arising not only in normal service but also under unusual circumstances, such as fire, nuclear or offshore accidents, should be taken into account. The success of designing high performance concrete elements for a required fire endurance depends on how accurately the concrete properties are known.

Considerable data are available on the effect of high temperatures on the properties of ordinary strength concrete [2]. Since the use of high performance concrete is a recent development, there is little information on the effects of high temperature on its properties.

This paper presents part of an experimental investigation on high performance concrete after exposure to elevated temperatures. It includes a comparative study of the properties of an ordinary concrete (OC) and high performance concrete (HPC) tested in the same conditions. Both types of concrete were cast with the same cement and with aggregates having the same mineralogical origin.

Measurements of temperature distributions in the concrete during heating were conducted in order to quantify the thermal gradient in the concrete at high temperature. Weight loss measurements during heating were suitable for gathering information on microstructural changes that occurred in the cement paste. Uniaxial compression tests and direct tensile tests were conducted to determine residual strengths and residual modulus of elasticity. The obtained results were useful for understanding the behaviour of high performance concrete under severe thermal conditions.

2 Experimental program

2.1 Materials

The mix proportions and some concrete data are given in Table 1. Two concretes were investigated experimentaly. The types of aggregates were the same in both concretes, and the grading of the aggregates was similar. Because the concrete is used for thick structures, high performance concrete was densified by silica fume and calcareous fillers in order to reduce the hydration heat [3] [4]. Its workability was improved with water reducer and retarding admixture. Specimens were cast as 16x32 cm cylinders for compression tests and I-specimens (useful section: 100x100 cm) for direct tensile tests (see Fig. 1). Compressive strength was measured at 2 months. It was 38.1 MPa for OC and 61.1 MPa for HPC.

Table 1. Mix proportions for 1 m^3.

Constituents	Ordinary Concrete (kg)	H P Concrete (kg)
calcareous sand	772	782
calcareous gravel	316	318
calcareous gravel	784	815
calcareous fillers	0	57
silica fume	0	40.3
water	195	161
cement CPJ 55 PM	350	266
plasticiser	0.35 %	0
water reducer	0	9.08
retarding admixture	0	0.931

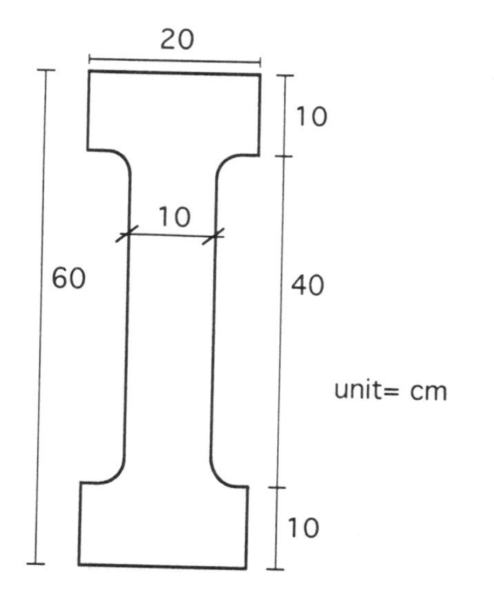

Fig. 1. Testing of I-specimens.

2.2 Test methods

The specimens were stored in impermeable packages until the age of two months. Then, the specimens were heated without load at 1°C/min to the desired temperature (150, 300, 450 and 500°C) and maintained there until dimensional stability was achieved (about 1 hour). After that, they were cooled back to room temperature at a cooling rate less than 1°C/min before loading.

Thermocouples were installed at points 1, 2, 3 and 4 (Fig. 2) during the cylindrical specimen moulding in order to record the temperature distributions between the surface (point 5) and the center.

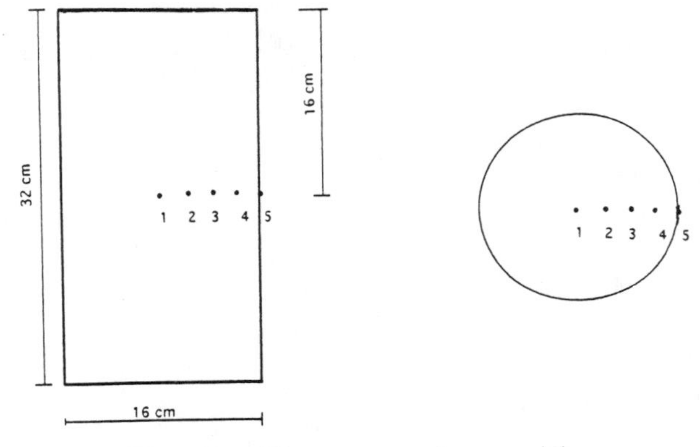

Fig. 2. Thermocouples positions.

Weight measurements were carried out during heating by hanging the specimen on a load cell (Fig. 3). After the heating/cooling cycle, compression and direct tensile tests were carried out to obtain residual mechanical properties.

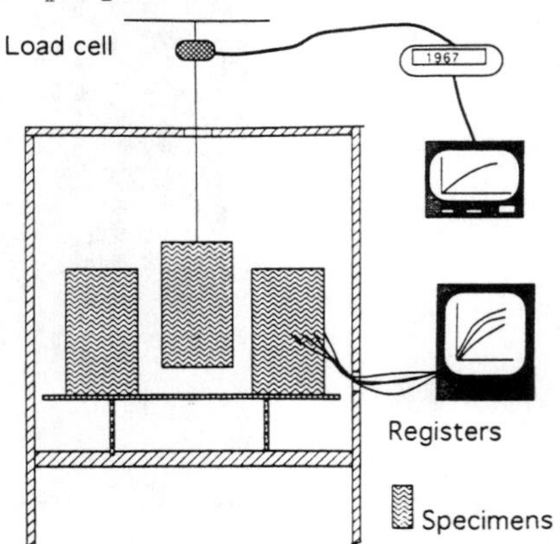

Fig. 3. Electric oven and registers.

3 Results and discussion

3.1 Temperature distributions

There were great temperature differences between the surface and the center of the specimen. In Figs. 4 and 5, the temperature differences $\Delta\theta$, in which $\Delta\theta_i = \theta_5 - \theta_i$, between the surface and the interior points (1, 2, 3, 4) are plotted.

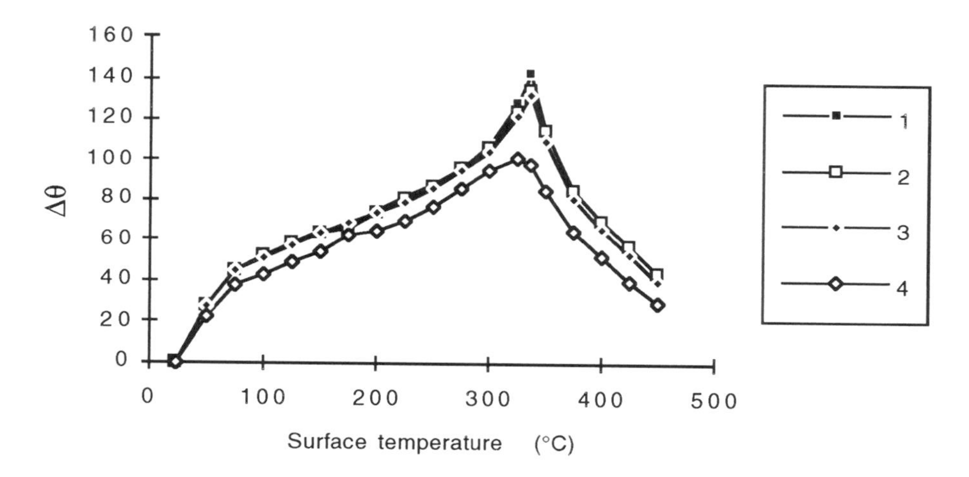

Fig. 4. Temperature differences in OC during heating. $\Delta\theta_i = \theta_5 - \theta_i$

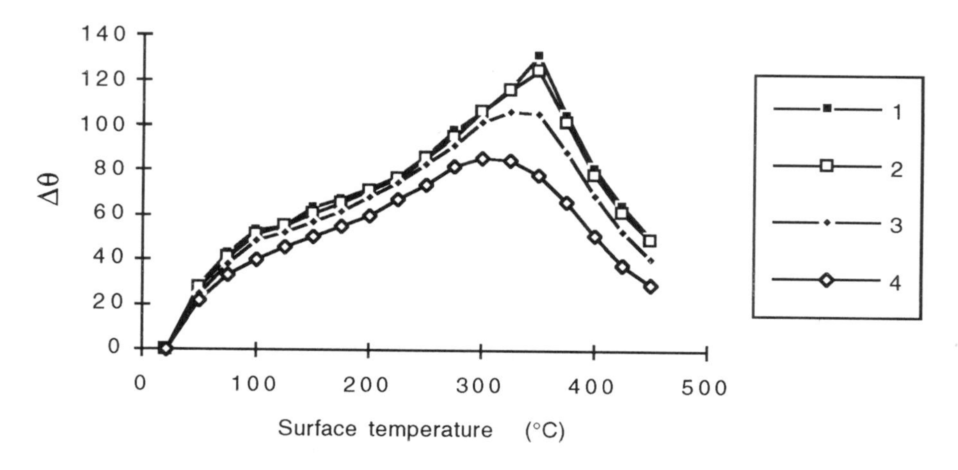

Fig. 5. Temperature differences in HPC during heating. $\Delta\theta_i = \theta_5 - \theta_i$

The maximum temperature difference was 143°C for OC and 132°C for HPC. This means an average gradient of more than 16°C/cm. The temperature gradients can induce high thermal stresses in the concrete. It is notable that the thermal gradients increased from the center of the specimen to the surface. A very sharp temperature gradient developed in the region from the surface to a depth of 2 cm (43 to 50°C/cm of concrete).

3.2 Weight losses

Water was free to escape from the concrete during heating. Assuming that weight loss during first heating is due essentially to water escaping from the concrete, we measured the specimen weight during heating. Weight variations are plotted in Fig. 6.

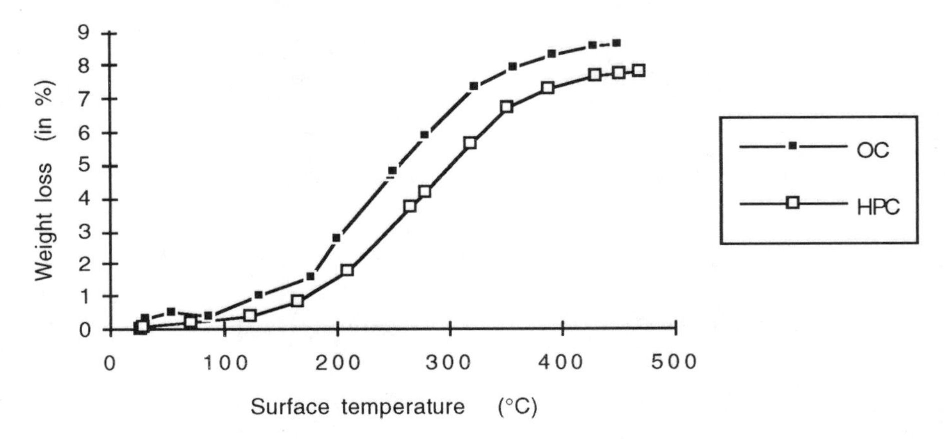

Fig. 6. Weight variations of OC and HPC (in % of the initial weight).

Wet concrete includes capillary water, physically adsorbed water and interlayer and chemically bound water in C-S-H and $Ca(OH)_2$. During heating, the cement paste dried. When heated at 450°C, OC specimens lost 8.6% of their initial weight. At the same temperature, HPC specimens lost 7.7% of their initial weight.

Three regions were notable on the weight variation curves:

- From ambient temperature to 150°C, there was only a little weight loss (1% for OC and 0.5% for HPC). 150°C at the surface corresponded to about 100°C at the center of the specimen. Weight loss is attributed to free water escaping. More than 83% of the total water was still retained in the concrete when the temperature in the specimen center achieved 100°C

- Between 150 and 350°C, both curves had similar slopes. The rates of water escaping were likely the same. The OC weight loss was greater than that of HPC. In this region, weight loss is generally attributed to the escape of the water contained in hydrated cement. In both tested concretes, the main hydrates were C-S-H. Chemically bound water became free water (vapor) and escaped from the concrete if the porosity and permeability allowed that. Bazant et al [5] reported that when passing from 95 to 150° C, there was an upward jump of permeability by two orders of magnitude (about 200 times). After 105° C, the moisture transfer must be governed chiefly by the viscosity of the steam instead of the adsorbed water migration at normal temperature. This can explain why according to our experimental results, above 150° C, concrete dried much faster than below 150°C.
- The temperature region above 350° C is characterized by the decomposition of $Ca(OH)_2$ above 400°C. The weight loss was 8.6% for OC and 7.8% for HPC. $Ca(OH)_2$ dissociation was not clearly evident in the measurements although we noticed whitish marks on the specimens after exposure to 450 and 500° C. These marks could be released lime from $Ca(OH)_2$. One should keep in mind that the equation is $Ca(OH)_2 \Longrightarrow CaO + H_2O$. At 500° C, both OC and HPC specimens were almost completely dehydrated.

3.3 Residual compressive strength

Fig. 7 shows the measured residual compressive strengths of both concretes. Each point in the Fig. represents an average of the compressive strengths of at least three specimens normalized with respect to the average compressive strength at room temperature.

Similar evolutions were observed for both concretes. As the temperature increased to 300°C, the strength did not decrease very much compared to the strength at room temperature. The residual compressive strength was about 85% of the initial strength. With further increases in temperature, between 300 and 500°C, the loss of strength became more significant. The residual compressive strength dropped from 85 to 60% of its initial value. Formation of microcracks due to thermal incompatibilities of cement paste and aggregates led to a considerable reduction of strength in concrete after 300°C. At 500°C, the residual strengths of OC and HPC were, respectively, 23.2 and 38.7 MPa instead of 38.1 and 61.1 MPa.

3.4 Residual modulus of elasticity

Stresses and strains were measured during compression tests on 16x32 cm cylinders. The residual modulus of elasticity was calculated from the $\sigma-\varepsilon$ curves.

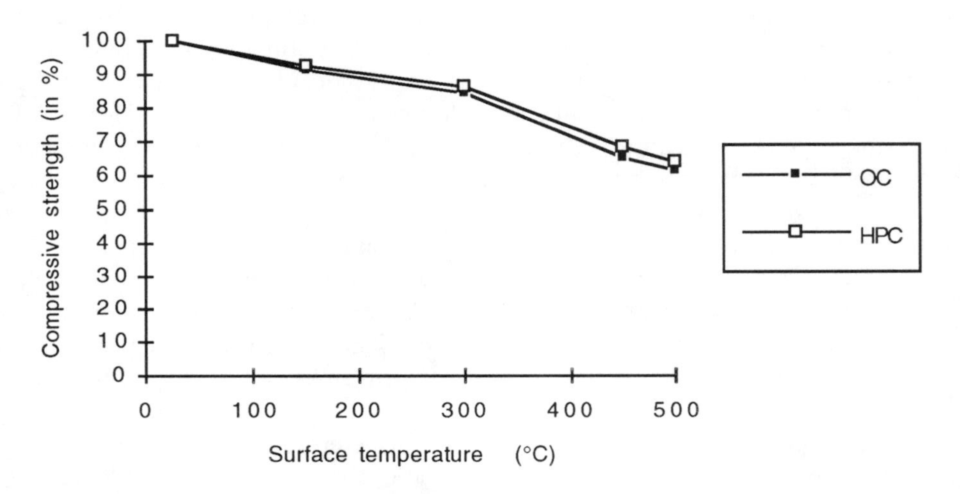

Fig. 7. Residual compressive strength/Initial strength.

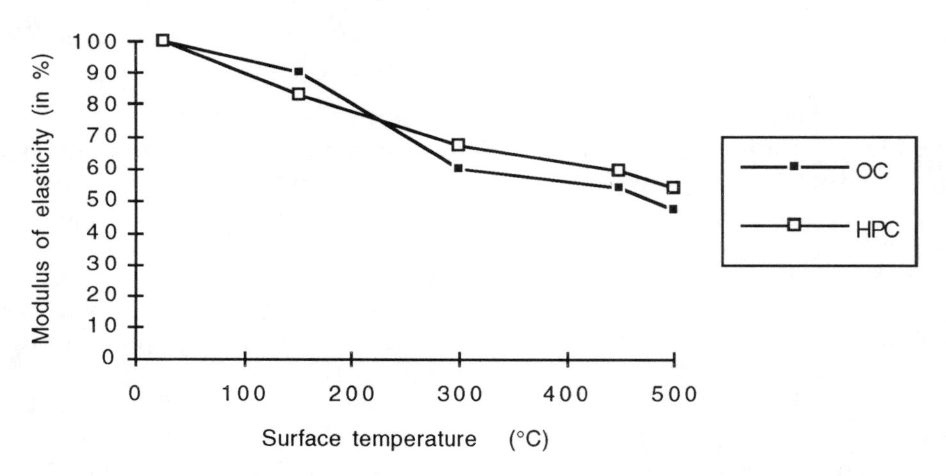

Fig. 8. Residual modulus of elasticity/ Initial modulus of elasticity.

The modulus of elasticity taken into account was the slope of the $\sigma-\varepsilon$ curve in the portion encompassing the range from 0 to 25% of the compressive stength. Fig. 8 reports the residual modulus of elasticity normalized with respect to the initial modulus of elasticity.

At ambient temperature, the compressive modulus of elasticity was 29.1 GPa for OC and 34.6 GPa for HPC. Between 22 and 500°C, the modulus of elasticity dropped to 13.8 for OC and 18.7 for HPC. That means, after exposure to 500°C, both concretes retained only about 50% of their initial modulus of elasticity. It is also notable that, at 150°C, the relative value of the residual modulus of elasticity of OC (90%) was greater than that of HPC (83%). It seems that in the region near 150°C, the stiffness of HPC was more affected than that of OC. During the compression tests, it was clearly noticeable that when increasing the temperature test, the specimen failures were less brittle and more gradual.

3.5 Residual direct tensile strength

Direct tensile tests permitted obtaining the evolution of the residual direct tensile strength of both concretes. Fig. 9 presents the residual direct tensile strength normalized with respect to the initial direct tensile strength.

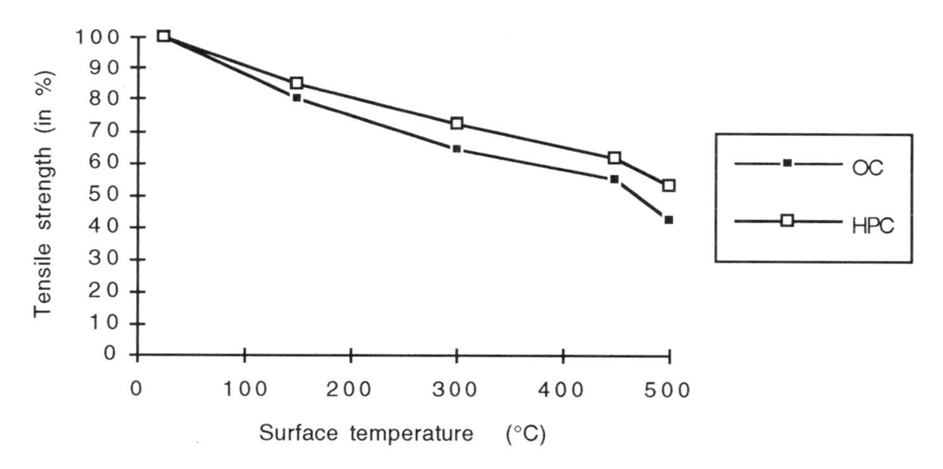

Fig. 9. Residual tensile strength/ Initial strength.

At ambient temperature, the tensile strength was 3.0 MPa for OC and 3.5 MPa for HPC. Direct tensile tests are very difficult to carry out. The problem is to avoid flexural effects in the specimen. The problem was solved by putting double knee-joints at the upper and lower ends of the specimen (see Fig. 1). The boundary conditions were then pinned supports. The results were more realistic than those of tensile splitting tests.

Residual tensile strength decreased with temperature. The decrease was greater than that of residual compressive strength for each temperature and for both concretes. At 150°C, tensile strength decrease was 15 to 20% of the initial strength instead of 8 to 9% for compressive strength. After exposure to 500°C, both concretes retained 42 to 53% of their initial tensile strength instead of 60 to 63% of their initial compressive strength.

3.6 Thermal stability

Two of the eight tested HPC specimens failed explosively at 330 and 340°C during the heating process. However, the other specimens sustained 500°C. Mercury porosimetry measurements showed that the total pore volume of the exploded samples was greater than that of unexploded samples [6]. Explosive spalling is attributed to high thermal gradients and high vapour pressure in low permeability concretes. The phenomenon was previously reported by several research workers [7] [8] [9] [10].

Four specimens which were predried at 105°C for 10 hours did not explode when heating up to 450°C. During heating, the temperature difference between the center and the surface of the specimens increased. For unexploded specimens, the temperature difference presented a peak between 300 and 350°C (see Fig. 5). However, it was notable that for specimens which exploded, the temperature increased until the violent failure took place. The temperature gradient was then higher than that of unexploded specimens. It was greater than 43°C/cm in the region between the surface and a point located at 2 cm. High thermal gradients in concrete cause high thermal stresses, while high temperatures cause tensile strength decreases. From the tests, it appeared that explosions are closely related to the moisture content. None of the specimens predried at 105°C did explode. The heating of HPC (low porosity and small capillaries) may produce a significant build-up of pore pressure.

In silica fume concrete, the matrix is homogenous. Silica fume particles are uniformly distributed between cement grains and constitute nucleation sites for hydrates. Capillary porosity is reduced and discontinuous. As fillers and pozzolana, silica fume particles densify the cement paste/aggregate interface. Water vapour can then be confined in discontinuous small pores.

Weight measurements showed that water vapour alone can not explain the HPC specimen explosions above 300°C. At this temperature, a large part of water has already escaped from the specimens. There may still be water vapour confined in closed small pores, but to this pore pressure, one has to add thermal

stresses. It can be said that explosive spalling of HPC at high temperatures proceeds by the combined action of the following:
-High thermal stresses in the dry region,
-Spontaneous development of unstable cracks,
-Tension stresses due to pore-water pressure.

Brittle mortar characteristics (low tensile strength and poor extensibility) before breakage may be improved to varying degrees by incorporating fibres [9]. It has also been shown that the explosion of high performance concrete can be dramatically reduced by adding certain low melting point fibres [10]. Theoretical analysis remains necessary to determine the evolution of some parameters: porosity/permeability, moisture content, thermal diffusivity [11], tensile strength and to estimate the stresses in the concrete during heating.

4 Conclusion

Any technology has limitations that must be determined before the technology can be applied appropriately. This study contributes to the improvement of the knowledge of the behaviour of high performance concrete and to the safety of construction.

The following conclusions were derived from this experimental investigation:

- High thermal gradients were recorded in both OC and HPC. At some points, the gradients were higher than 43°C/cm. This may cause high thermal stresses in concrete and development of unstable cracks.
- Up to 150°C, there was only a minor decrease of weight for both OC and HPC. Free water escaped from the concrete. Between 150 and 350°C, the rate of water loss increased. Concrete dried much faster than before 150°C. At 500°C, the specimens were almost completely dehydrated.
- High temperature exposure led to a reduction of strength of both OC and HPC. The loss of compressive strength was significant after 300°C. From 22 to 500°C, the tensile strength losses were greater than that of compressive strength.
- Furthermore, the residual modulus of elasticity decreased. At 500°C, it was about 50% of its initial value.
- There was a high risk of explosive spalling in HPC between 300 and 340°C due to the special porous structure (wet concrete with small capillaries, low permeability). Analytical studies remains necessary to determine the stresses in the concrete during heating.

5 References

1. Costaz, J.L. (1992) The High Performance Concrete. *Revue Générale Nucléaire*, N° 4. pp. 314-317.
2. Schneider, U. (1988) Concrete at high temperature. A general review. *Fire Safety Journal*, Vol. 13, 1988. pp. 55-68.
3. Larrard, F. de, Ithurralde, G., Acker, P. and Chauvel, D. (1990) High-Performance Concrete for a Nuclear Containment. *2nd International Symposium*, ACI, SP-121. pp. 549-576
4. Ithurralde, G.J.B. and Olivier, J. (1993) High Performance Concretes for French Nuclear Reactor Containment Vessels. *Utilization of High Strength Concrete*, Proceedings Vol. 1, Symposium in Lillehammer, Norway. pp. 217-224.
5. Bazant, Z.P. and Thonguthai, W. (1979) Pore Pressure in Heated Concrete Walls: Theoretical prediction. *Magazine of Concrete Research*, Vol. 31, N° 107. pp. 67-76.
6. Nourmowé, A.N., Clastres, P., Debicki, G. and Bolvin, M. (1994) High Temperature Effect on High Performance Concrete (70-600 °C) Strength and Porosity. *Third CANMET/ACI International Conference on Durability of Concrete*, Nice. pp. 157-172.
7. Diederichs, U., Jumppanen, U. and Pentalla, V. (1989) Behaviour of high strength concrete at high temperatures. *Helsinki University of Technology*, Report 92, Espoo.
8. Jahren, P.A. Fire resistance of high strength/dense concrete with particular reference to the use of condensed silica fume. A review, *ACI SP-114, 50 Trondheim Conference*, 1989
9. Sarvaranta, L., Elomaa, M. and Järvelä, E. (1993) A Study of Spalling Behaviour of PAN Fibre-reinforced Concrete by Thermal Analysis. *Fire and Materials*, Vol. 17. pp. 227-230.
10. Diederichs, U., Jumppanen, U.-M., Morita, T., Nause, P. and Schneider, U. (1994) (in Germany) Zum Abplatzverhalten von Stützen aus hochfestem Normalbeton unter Brandbeanspruchung. Concerning Spalling Behaviour of High Strength Concrete Columns under Fire Exposure. Report, TU Braunschweig, p.12
11. Khoury, G.A., Sullivan, P.J.E. and Grainger, B.N. (1984) Radial temperature distributions within solid concrete cylinders under thermal states. *Magazine of concrete research*, Vol. 36, N° 128, Sept.

112 STUDY ON THE PROPERTIES OF HIGH STRENGTH CONCRETE WITH SHORT POLYPROPYLENE FIBER FOR SPALLING RESISTANCE

A. NISHIDA, N. YAMAZAKI and H. INOUE
Shimizu Corporation, Tokyo, Japan
U. SCHNEIDER
Technical University of Vienna, Vienna, Austria
U. DIEDERICHS
Technical University of Braunschweig, Braunschweig, Germany

Abstract
This paper presents the effect of polypropylene fibers on the spalling resistance of high strength concrete.

The compressive strength and the elastic modulus of fiber concrete were basically the same as in concrete without fiber. Cylinder specimens (ø=100mm) were subjected to a fire test. It was observed that while most of the concrete specimens without fiber showed deep spalling and rupture after the fire test, fiber concrete specimens, on the contrary, showed only partial or non-spalling behavior. In the fire tests carried out on concrete column specimens, those with a water-cementitious binder ratio=0.25 and without fibers showed severe spalling of their concrete covers. However, in the case of column specimens with short polypropylene fibers, very slight spalling was observed. Hence from these test results, the significant contribution of polypropylene fiber to the spalling resistance of high strength concrete could be verified. The spalling properties could be estimated by fire tests with cylinder specimen under no loading, because similar results were obtained for fire tests of both column specimens and cylinder specimens.
Keywords: Compressive strength, elastic modulus, fire test, high strength concrete, polypropylene fiber, spalling.

1 Introduction

High strength concrete has recently been utilized for structures such as ultra-high-rise buildings and prestressed concrete bridges [1]. In the future, the utilization of high strength concrete will spread widely.

Concrete Under Severe Conditions: Environment and loading (Volume Two) Edited by K. Sakai, N. Banthia and O.E. Gjørv. Published in 1995 by E & FN Spon. ISBN 0 419 19860 1

It is generally known that high strength concrete can easily spall during a fire [2]. However, some researchers disagree with this view and reported no spalling of the concrete during a fire [3]. The spalling mechanism of concrete is not clear yet. There are few effective methods available to improve the spalling resistance [4][5].

In this context, this paper reports on the effect of short polypropylene fibers on the spalling resistance of high strength concrete and on the strength properties of the fiber concrete.

2 Outline of tests

The concrete used in the test specimens had a compressive strength of 41 to 102 MPa. The effect on spalling resistance, compressive strength and elastic modulus results from the introduction of polypropylene fiber into concrete were experimentally investigated to compare with concrete without polypropylene fibers.

As the first step in the current study, the compressive strength and elastic modulus were measured and fire tests using cylinder specimens(ø=100mm) were carried out.

Secondly, fire tests of concrete column specimens were carried out to estimate the effect of polypropylene fiber, simultaneous loading on the column specimen, and the allowance of compressive strength of concrete on the spalling resistance of structural members.

3 Materials and mix proportion

3.1 Materials

The materials used for the experiments are given in Table 1. A normal portland cement, crushed sandstone coarse aggregate and river sand were used. A type of superplasticizer was employed as a chemical admixture. Silica fume was used for the concrete to produce a compressive strength of 102 MPa.

The polypropylene fiber was 19 mm long and the melting point of 160 - 170 ℃.

3.2 Mix proportion

The mix proportions were designed to result in compressive strengths between 41 and

Table 1. Materials.

Materials	Kind	Characteristics
Cement	Normal Portland	Specific gravity 3.15, Specific surface area=0.32m^2/g
Coarse aggregate	Crushed Sandstone	Specific gravity 2.63, absorption ratio 0.85%
Fine aggregate	Land Sand	Specific gravity 2.59, absorption ratio 1.74%
Mineral admixture	Silica fume	Specific gravity 2.20, Specific surface area=20m^2/g
Chemical admixture	Superplasticizer	Principal ingredient - Sulphonated naphthalene formaldehyde condensates
Water	Potable	———
Spalling protection material	Polypropylene	———

102 MPa. The water-cementitious binder ratios were set to 0.25, 0.30, 0.45 and 0.55. The unit water content was taken as 165 kg/m³ or 175 kg/m³. The slumps of concretes without polypropylene fiber were set to 23 cm for water-cementitious binder ratios=0.25 and 0.30, 21 cm for 0.45, 18 cm for 0.55. The slumps of fiber concrete specimens were not controlled.

The polypropylene fiber contents, taking into consideration the distribution of polypropylene fiber in the cement matrix, were set to 1.5 or 3.0 kg/m³ for water-cementitious binder ratios=0.25 and 0.30 and 1.5 kg/m³ for 0.45 and 0.55. The concretes were mixed by a forced mixing type mixer for 2 minutes after all materials with the polypropylene fiber were charged. The mix proportion of the concrete specimens is shown in Table 2.

Table 2. Mix proportion of concrete specimens.

Name	W/(C+SF) (%)	W (kg/m³)	C (kg/m³)	SF (kg/m³)	s/a (%)	Fiber (kg/m³)	Ad content (%)
25-0	25.0	165	594	66	38.0	0	2.8
25-1.5	25.0	165	594	66	38.0	1.5	2.8
25-3	25.0	165	594	66	38.0	3.0	2.8
30-0	30.0	165	550	——	40.0	0	2.0
30-1.5	30.0	165	550	——	40.0	1.5	2.0
30-3	30.0	165	550	——	40.0	3.0	2.0
45-0	45.0	175	389	——	45.0	0	1.2
45-1.5	45.0	175	389	——	45.0	1.5	1.2
55-0	55.0	175	318	——	48.0	0	0.9
55-1.5	55.0	175	318	——	48.0	1.5	0.9

4 Strength properties

4.1 Test items

Compressive strength and elastic modulus tests were performed according to JIS A 1108 and ASTM C 469 (secant modulus), respectively.

The cylindrical specimen size for the compressive strength and the elastic modulus was ø=100x200 mm. The specimens were cured in water at 20 ℃ or in air (20 ℃, 60% R.H.) or sealed in test room. Testing ages were 28 and 91days.

4.2 Test results

Specimen compressive strengths ranged from 26.5 to 114.7 MPa with water-cementitious binder ratios between 0.25 and 0.55.

The relationship between the compressive strength of fiber concretes and compressive strength of concretes without fibers is given in Fig. 1. The compressive strength of polypropylene fiber concrete was basically the same as concrete without polypropylene fibers. Irrespective of whether the concrete was water cured, sealed or air-cured, the same results could be seen.

The relationship between the elastic modulus of fiber concretes and the elastic modulus of concretes without fibers is given in Fig. 2. The elastic modulus of polypropylene fiber concrete was basically the same or decreased slightly compared with concrete without polypropylene fibers.

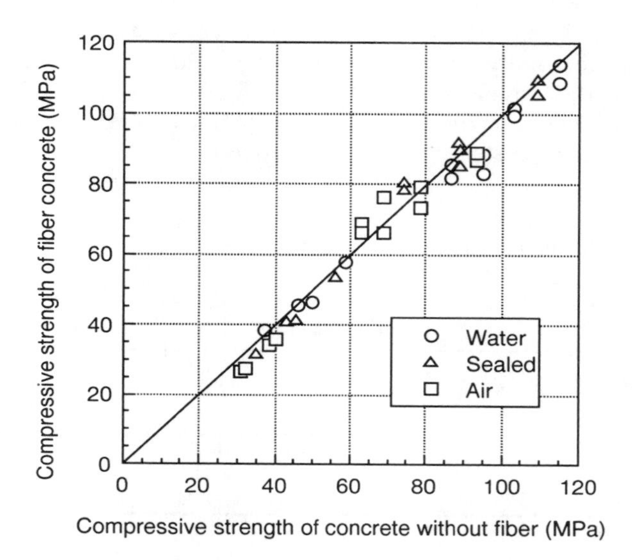

Fig. 1. Comparison of compressive strength.

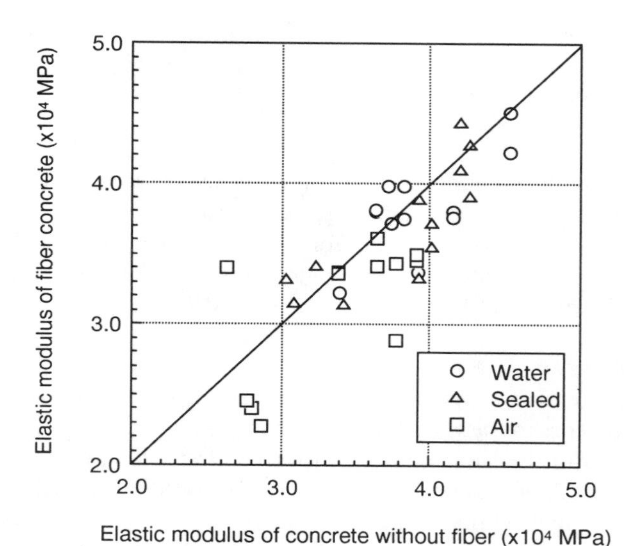

Fig. 2. Comparison of elastic modulus.

5 Fire test of cylinder specimens

5.1 Test items

The fire tests were performed according to the heating temperature curve specified in JIS A 1304, "Method of Fire Resistance Tests for Structural Parts of Buildings." The heat curve of JIS A 1304 is shown in Fig. 3. The cylindrical specimen size for the fire test was ø=100x200mm. Testing ages were 28 and 91days.

5.2 Test results

The degree of spalling is shown in Fig. 4. The test results are given in Fig. 5 and Table 3.

The general trend of results showed that the spalling of specimens was more severe in the case of lower water-cementitious binder ratios and an age of 28 days compared to

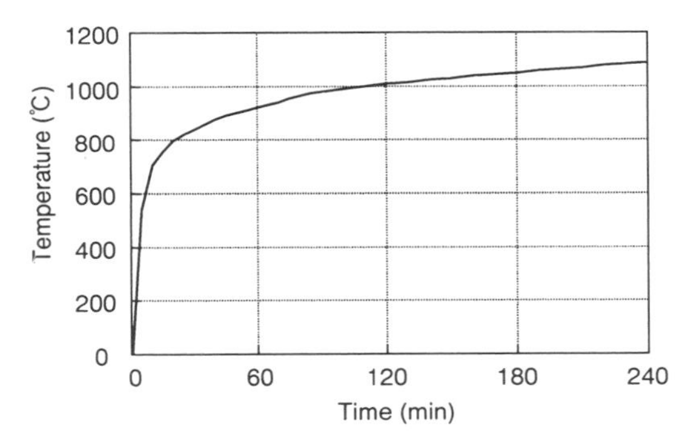

Fig. 3. JIS standard heat-curve.

Fig. 4. Degree of spalling.

Table 3. Fire test results of cylinder specimens.

Series	28days Water	28days Sealed	28days Air	91days Water	91days Sealed	91days Air
25-0	×	×	×	■	□	○
30-0	■	○	○	□	○	○
45-0	△	○	○	□	○	○
55-0	□	○	○	△	○	○
25-1.5	△	△	○	○	○	○
30-1.5	○	○	○	○	○	○
45-1.5	○	○	○	○	○	○
55-1.5	○	○	○	○	○	○
25-3	△	○	○	○	○	○
30-3	○	○	○	○	○	○

Degree of spalling	
○	None spalling
△	Shallow spalling
□	Deep spalling
■	Partial rupture
×	Complete rupture

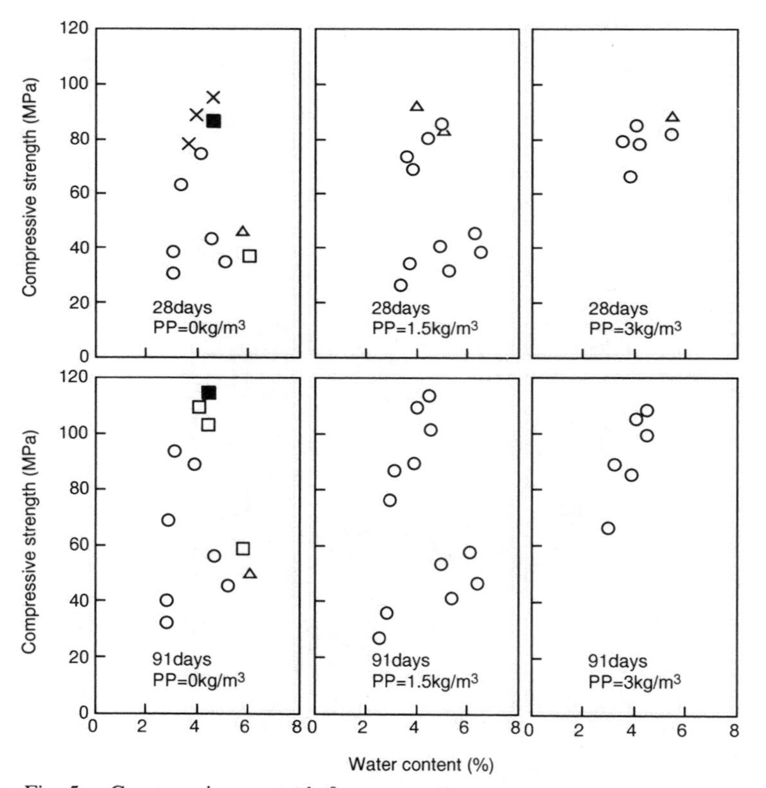

Fig. 5. Compressive strength & water content.

91days.

The degree of spalling at the same water-cementitious binder ratio and the same age increased in severity in the following order : water, sealed and air curing. However, no clear relation was observed between water content of concretes and spalling behavior (Fig.5).

It was observed that most of the concrete specimens without polypropylene fiber showed deep spalling and rupture. Polypropylene fiber concrete specimens, on the other hand, showed partial or non-spalling behavior. After the fire tests, the polypropylene fiber was not found in the specimens because of destruction by fire. It was considered that the polypropylene fiber melted and the traces performed the part to reduce the vapor pressure inside the specimens. These test results verified the significant contribution of polypropylene fiber to the spalling resistance of concrete.

6 Fire test of concrete column specimens

6.1 Test items

In this experiment, the water-cementitious binder ratio of the concretes was set to 0.25 and 0.55. Concrete column specimens without polypropylene fiber, and with water-cementitious binder ratios=0.25 with 3.0 kg/m^3 fiber and 0.55 with 1.5 kg/m^3 fiber were subjected to a fire test. The materials used and the mix proportion of the concretes are given in Tables 1 and 2.

With each water-cementitious binder ratio, loaded and without loading specimens were subjected to a fire test. The details of the specimens are given in Table 4.

Concrete specimens were wet-cured for 7 days in the test room. They were then allowed to dry in air for approximately 4 weeks after the forms were removed. Specimens were heated with a two-hour heating curve (Fig. 3). The internal temperature of the specimens were measured at 12 points.

6.2 Column specimen and apparatus

The reinforced concrete specimens for the column heating tests were 300x300x1,200 mm, Fig. 6. The concrete cover for embedded bars was 40 mm. The specimens had a centre hole of 100 mm diameter for loading.

The test apparatus is shown in Fig. 7. The apparatus consisted of centre hole jack, load cell, tension rod and support assemblies. The load on a column specimen was produced by the centre hole jack and transmitted to the specimen with the tension rod. The applied load was equaled to 1/4 of the compressive strength of a cylindrical

Table 4. Axial load applied to column specimen.

Series	Compressive strength(MPa)	Sectional area(cm^2)	Ultimate load(t)	Load ratio	Applied Load(t)
C25-0-1	100.2		807	1/4	202
C25-0-2				0	0
C25-3-1	90.5		729	1/4	182
C25-3-2		821.5		0	0
C55-0-1	39.2		315	1/4	79
C55-0-2				0	0
C55-1.5-1	35.8		288	1/4	72
C55-1.5-2				0	0

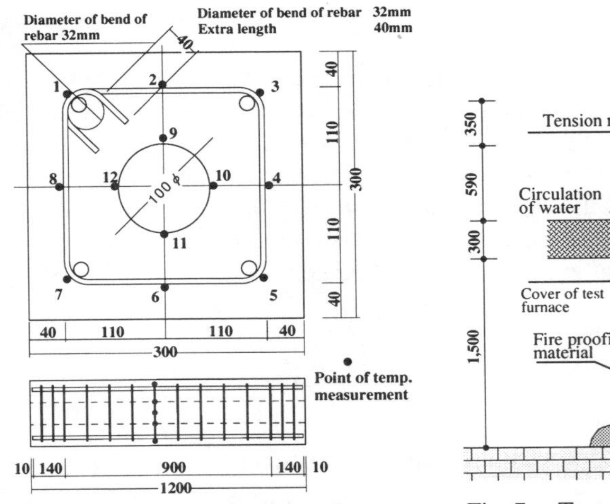

Fig. 6. Geometry of specimen.

Fig. 7. Test apparatus.

specimen cured under the same condition as that of the column specimen.

The loading on the specimens was started just before the test and maintained for 2 hours after the fire exposure [6].

6.3 Test results

Figure 8 shows heated C25 specimens after subjecting to the fire test. In this test, the concrete column specimens with a water-cementitious binder ratio=0.25 without polypropylene fibers showed severe spalling of their concrete cover. This can be seen from the figure where the reinforcing bars of the column specimen become visible.

However, for the column specimens with polypropylene fiber, very little spalling could be observed. Comparison of the figures (with and without polypropylene fiber) confirms the significant effect of polypropylene fiber on the spalling resistance of

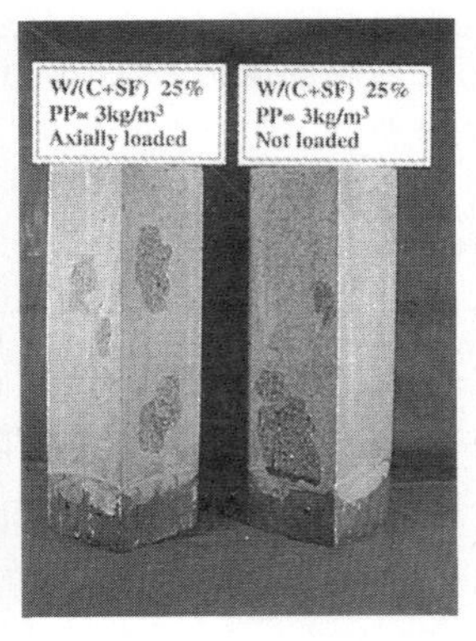

Fig. 8. Test specimen after the fire test.

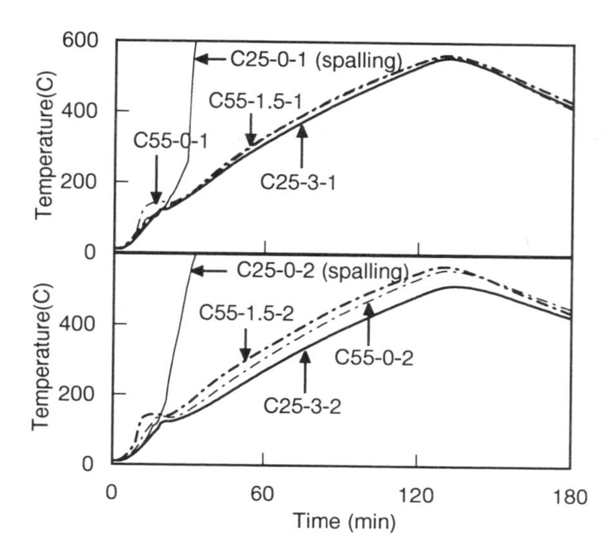

Fig. 9. Temperature vs elapsed time.

concrete column specimen.

In the case of column specimens with a water-cementitious binder ratio=0.55, no spalling occurred whether or not they contained polypropylene fiber.

As for the effect of loading on column specimens having either a water-cementitious binder ratio=0.25 or 0.55, it could be seen that the effect of loading during the fire test appeared not to significantly affect the spalling resistance of column specimen.

The spalling properties could be estimated by fire tests with cylinder specimens under no loading, because similar results were obtained from fire tests on column specimens and cylinder specimens.

Figure 9 presents the relationship between the temperature of the hoop reinforcement and elapsed time. The C25-0 specimens indicated over 500 ℃ at 30 minutes because of spalling. The C25-3 and C55 specimens indicated almost the same hoop reinforcement temperatures on because of little spalling.

7 Conclusions

The effect of polypropylene fibers on the spalling resistance of high strength concrete was investigated experimentally. The results obtained within the scope of this study can be summarized as follows.

The compressive strength and the elastic modulus of fiber concrete were basically the same as in concrete without fibers.

Cylinder specimens (ø=100mm) were subjected to fire tests. It was observed that most of the concrete specimens without fiber showed deep spalling and rupture after the tests. While fiber concrete specimens, on the other hand, showed partial or non-spalling behavior.

In the fire tests of concrete column specimens, concrete column specimens with a water-cementitious binder ratio=0.25 and without fibers showed severe spalling of their concrete cover. However, for column specimens with polypropylene fiber, very little spalling could be observed. In the case of column specimens with a water-cementitious binder ratio=0.55, it was observed that no spalling occurred whether or not they contained polypropylene fibers.

From these results, the significant contribution of polypropylene fiber to the spalling resistance of concrete could be verified. The spalling properties could be estimated by fire tests with cylinder specimens under no loading, because similar results were obtained from fire tests on column specimens and cylinder specimens.

8 References

1. Russel, H.G. (1987) High strength concrete in North America. *FIP Notes,* April. pp. 103-115.
2. Castillo, C. and Durrani, A.J. (1990) Effect of transient high temperature on high-strength concrete. *ACI Materials Journal,* Vol. 87, No. 1. pp. 47-53.
3. Scott, T.S., Ronald, G.B. and Anthony, E.F. (1988) Fire endurance of high-strength concrete slabs. *ACI Materials Journal,* Vol. 85, No. 2. pp. 102-108.
4. Sarvaranta, L., Elomaa, M. and Järvelä, E. (1993) A study of spalling behavior of PAN fibre-reinforced concrete by thermal analysis. *FIRE AND MATERIALS,* Vol. 17. pp. 225-230
5. Diederichs, U., Spitzner, J., Sandvik, M., Kepp, B. and Gillen, M. (1993) The behavior of high-strength lightweight aggregate concrete at elevated temperatures. *Proceedings of UTILIZATION OF HIGH STRENGTH CONCRETE* Symposium in Lillehammer, Norway, Vol. 2, pp.1046-1053.
6. Morita, T., Schneider, U. and Franssen, J.M. (1994) A concrete model considering the load history applied to centrally loaded columns under fire attack. *Fourth International Symposium on Fire Safety Science,* Canada.

113 PORE PRESSURE IN SEALED CONCRETE AT SUSTAINED HIGH TEMPERATURES

O. KONTANI
Kajima Corporation, Tokyo, Japan
S.P. SHAH
Northwestern University, Evanston, Illinois, USA

Abstract
In the present study, pore pressure was measured in concrete under sealed conditions and at a sustained high temperature of 171°C. The maximum pressures measured were 15 to 45% higher than the sum of the saturated vapor pressure (SVP) and partial air pressure (PAP). The volume of water and non-condensable gas released from concrete was measured. At 171°C, 50 to 70% of all the water in concrete was released. The released gas volume was found to be two to seven times greater than the volume of water and ten to forty times that of the entrapped air.

A theoretical model is presented in which the pore pressures in sealed concrete at high temperatures were predicted. The Powers-Brownyard model was adopted to calculate the volume of water to be released from a given concrete mix. The calculation of released gas volume from the specimens at high temperatures could not be achieved since the sources and mechanisms of gas generation were not well understood.

The tests demonstrated that a significant amount of water could be released at a temperature above 105°C to cause SVP. Hence, all possible mixes of concrete used for structural purposes will generate pore pressures greater than the sum of SVP and PAP (SVP+PAP) when concrete is heated.
Keywords: Evaporable water, non-condensable gas (gas), non-evaporable water, partial air pressure (PAP), pore pressure, saturated vapor pressure (SVP), sealed concrete

1 Introduction

A new type of nuclear power reactor, called the Simplified Boiling Water Reactor (SBWR), is under development by General Electric Corporation and other companies. The design idea is quite different from the conventional Boiling Water Reactor (BWR) designs. The reactor is enclosed in a reinforced concrete containment vessel, as shown in Fig. 1, which is a massive structure with a thin interior steel liner for leak tightness to prevent the escape of radioactive fission products to the outside environment. The wall thickness is 2.0 m and the liner thickness is 6.4 mm. In the case of an accident, the temperature inside the containment vessel may suddenly increase. Because of the passive nature of SBWR, the temperature inside the containment vessel may reach as high as 171°C following a loss-of coolant accident (LOCA).

Concrete Under Severe Conditions: Environment and loading (Volume Two) Edited by K. Sakai, N. Banthia and O.E. Gjørv. Published in 1995 by E & FN Spon. ISBN 0 419 19860 1

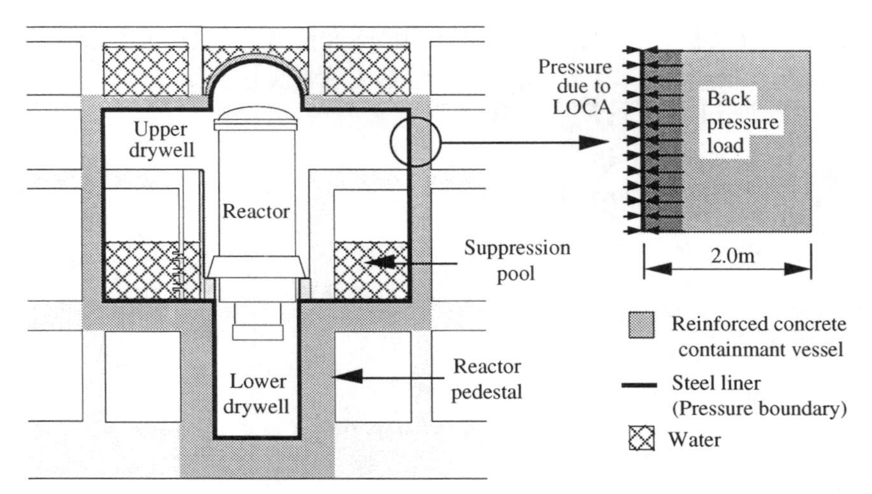

Fig. 1. Reinforced concrete containment vessel.

If the concrete vessel is heated from inside due to a LOCA, the capillary water and gel water in the concrete will vaporize. The pore vapor pressure that is developed may impose a back pressure load on the thin liner. If the pressure difference across the liner is sufficiently higher than the ASME Code allowables, the liner integrity may be challenged and the purpose of leak-tightness may be defeated. Therefore, it is very important to determine the pore vapor pressure in concrete at sustained high temperatures. If the pore vapor pressure is large enough to cause the liner back-pressurization problem, a possible solution is to install a venting system between the liner and concrete to relieve the vapor pressure towards the inside. Therefore, the volume and release rates of water and non-condensable gas from heated concrete are key parameters for designing the venting system.

2 Objectives

The objectives of this study are to determine the following parameters at 171°C: (1) pore pressure in concrete, (2) volume of water and gas released from concrete, and (3) release rates of water and gas from concrete. In order to obtain the above information, pressure tests and water/gas release tests were performed on concrete specimens having three different mix proportions.

3 Test details

3.1 Concrete specimens

Mix proportions and materials for concrete are shown in Table 1. Four specimens were made of concrete having w/c = 0.25 and 0.30. These specimens were first used for pressure tests, and then for water/gas release tests. Eight specimens were made of concrete having w/c = 0.45. Four of them were for pressure tests, and the other four were for water/gas release tests. The specimens were cast in a steel container. The steel container and locations of devices are shown in Fig. 2. The container was completely sealed immediately following the concrete casting to prevent moisture loss. The specimens were cured for more than six months.

Table 1. Materials specifications and mix proportions.

w/c ratio (1)	0.25	0.30	0.45	0.45	Specifications
Batch ID	Mix25	Mix30	Mix45-B1	Mix45-B2	
Water	137	160	175	175	
Cement	595	535	375	376	ASTM C150, Type II
Sand (2)	668	655	698	701	ASTM C33, Grade 2
Gravel (3)	1015	1023	1086	1090	ASTM C33, Size 8, Max 9.5mm
Silica Fume	27.7	26.8	18.7	18.8	ACI 212.2R, SiO2 85% min.
Superplasticizer	15.0	8.03	-	-	ASTM C494, Type G
Retarder	-	-	0.75	0.75	ASTM C494, Type D
Air Content (vol. %)	1.5	1.9	2.5	2.1	ASTM C197
Strength (MPa)	81.3	70.7	54.1	57.5	ASTM C39

Unit: kg/m3, (1): w/c = (Water+Water in Admixtures)/(Cement+Silica Fume)
(2) & (3): water absorption is 2.3% and 1.3% by weight respectively.

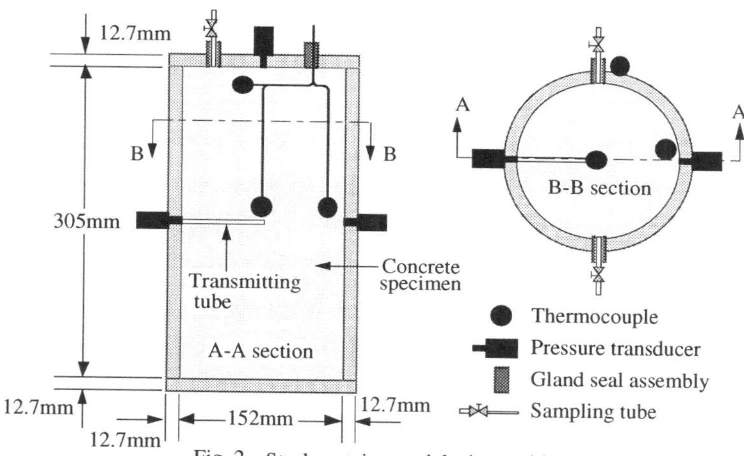

Fig. 2. Steel container and device positions.

3.2 Instrumentation

Fig. 3 shows the instrumentation for the pressure tests and the water gas release tests. Pressure transducers were installed to measure pore pressures in the concrete at three different locations. Thermocouples were used to measure temperature of the concrete specimens where pore pressures were measured. Data gathered from the sensing devices were continuously monitored and recorded by a data acquisition system. Water and gas were released only through sampling tubes and were collected by equipment consisting of a vapor condenser and a water/gas accumulator.

3.3 Pressure transducer calibration

In order to calibrate the pressure transducers, steam tests were performed before and after the pressure tests. All pressure transducers were mounted on empty steel containers. A stainless steel tube, with an in-line valve, was connected to the container in order to release all the air inside. After the air was expelled, the in-line valve was closed and the oven temperature was increased stepwise up to 190°C. Fig. 4 shows one of the test results. The data points at 100, 130, 150, 170, 180, and 190°C were taken at a steady state condition. The remaining data were taken in a transient state. Since the test results were very consistent, the pressure readings are considered reliable with an error of 70kPa.

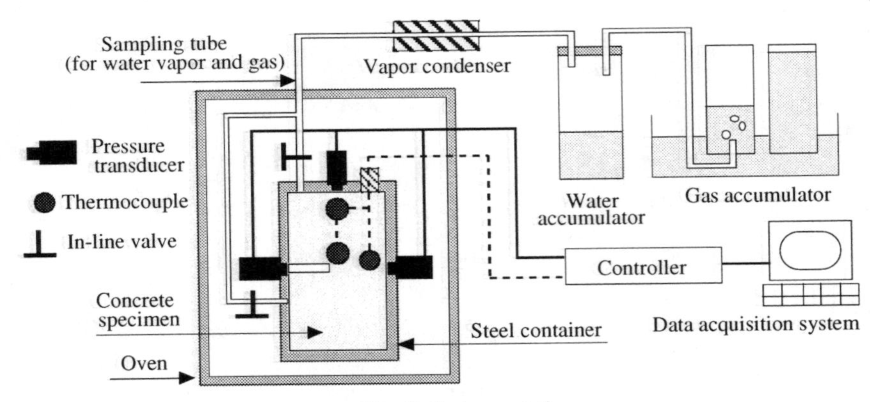

Fig. 3. Instrumentation.

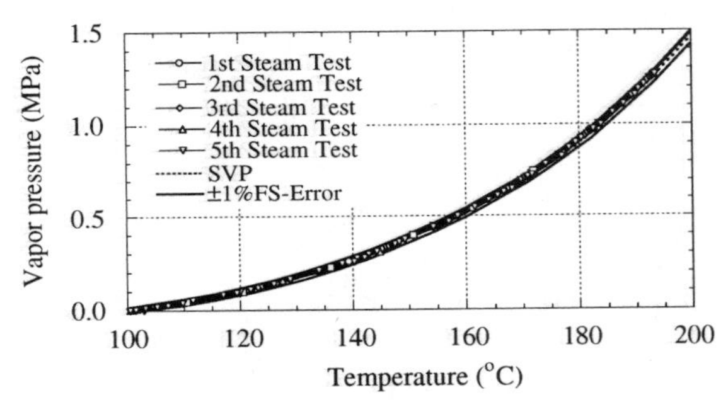

Fig. 4. Steam test results.

4 Test results and evaluations

4.1 Pressure tests

The sum of the saturated vapor pressure and the partial air pressure (SVP+PAP) is used as a basic scale to discuss the magnitude of the pore pressures measured, as described in Section 6.1. Fig. 5 represents one of the typical results showing the relationship between pore pressure and time. The specimens never reached a steady state in terms of pressure although the temperature was maintained at 171°C. The maximum pore pressures measured were much higher than SVP+PAP. Fig. 6 shows one of the typical relationships between pore pressure and temperature. Since pore pressure and temperature were measured at the same locations, they most likely represent some local equilibrium. The figure indicates that pore pressures increased, while the oven temperature was held at 171°C.

England et al. [1] reported pore pressures measured in terms of the weight loss of a specimen as shown in Fig. 7. Pore pressures were measured at a constant temperature of 350°C with controlled moisture release. The maximum pressure measured was much higher than SVP+PAP, which is similar to what was observed in the present study. After releasing all of the air and some water, the pore pressures dropped to SVP. Pore pressures stayed at the SVP plateau until a certain amount of water was further expelled. This indicates that the air contained in concrete may be one reason for the additional pressure over the SVP.

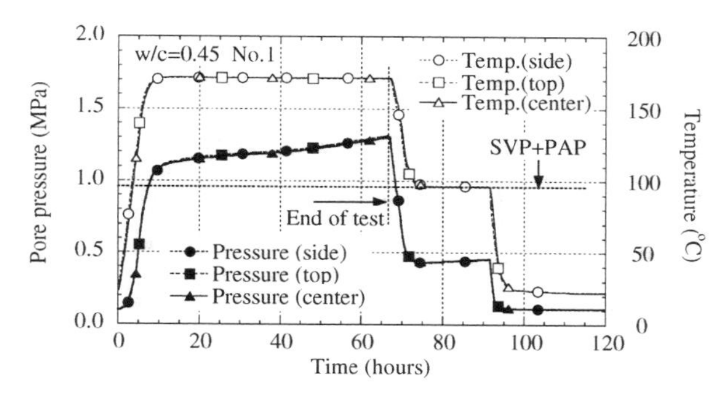

Fig. 5. Pore pressure and temperature vs. time.

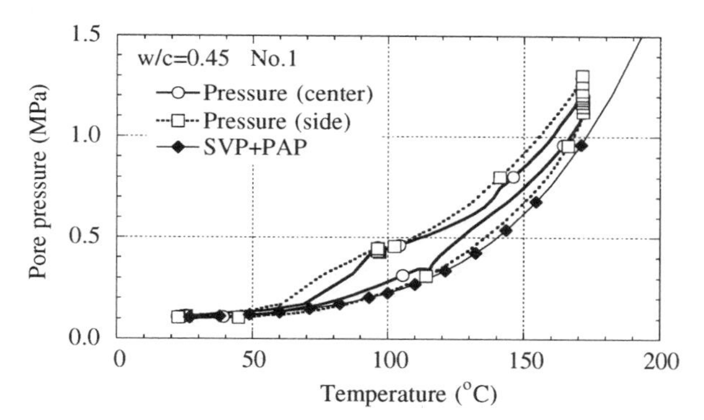

Fig. 6. Pore pressure and temperature.

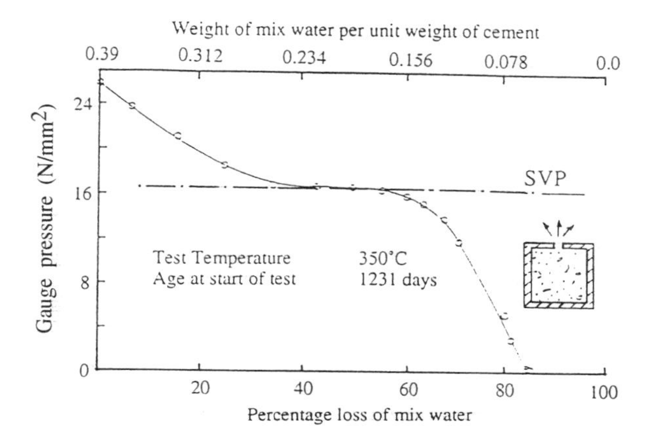

Fig. 7. Variation of pore pressure with moisture loss from sealed specimen [1].

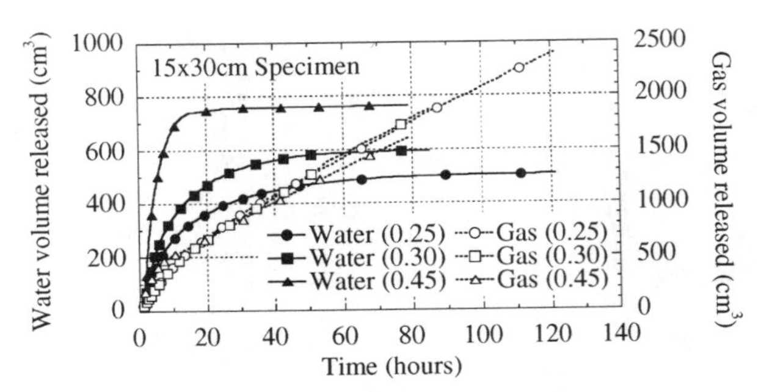

Fig. 8. Average volume of water and gas.

4.2 Water/gas release tests

Fig. 8 shows a comparison between the average volume of water and gas released from the tested specimens. About 50 to 70% of the total water in concrete was expelled when concrete was heated up to 171°C. As can be seen in the graph, the specimens made of higher w/c concrete released water faster than that of lower w/c due, most likely, to a higher permeability. The release rates of water and non-condensable gas were calculated from this relationship.

4.3 Chemical analysis of condensed water

Chemical analysis of condensed water released from concrete by Inductively Coupled Plasma was performed. It was deduced that admixtures decomposed chemically at 171°C because a very high concentration of carbon was found in condensed water from specimens.

5 Model for estimating pore pressure

A model should incorporate the following observations: (1) The presence of a sufficient amount of water in concrete to maintain the SVP; (2) The presence of air in concrete; (3) The volume expansion of liquid water is considerably larger than that of concrete at high temperature; and (4) Superplasticizer and retarder, added to the mixes, probably decompose chemically at high temperatures.

5.1 Expanding liquid water and compressed air - Basic concept

Fig. 9 shows a conceptualization for the model which is based on (1) to (3) of the above observations. Since it is very difficult to determine how the individual components of the admixtures decompose chemically, observation (4) is not accounted for in the present model. At room temperature, there is no liquid water in the capillary pores and the air voids in concrete having a lower w/c. As the temperature is increased, water expands and capillary pores and air voids are partially filled with liquid water. This compresses the air. Thus, the partial air pressure is increased by raising the temperature as well as reducing air volume.

5.2 Evaporable and non-evaporable water

There are two classifications of water in cement paste based on the drying process [2]. Evaporable water is released from cement paste upon drying up to 105°C. This contains capillary water, gel water, and water absorbed by aggregates. Non-evaporable chemically bound water is retained in cement paste even after drying at 105°C and only released upon further heating at a higher temperature.

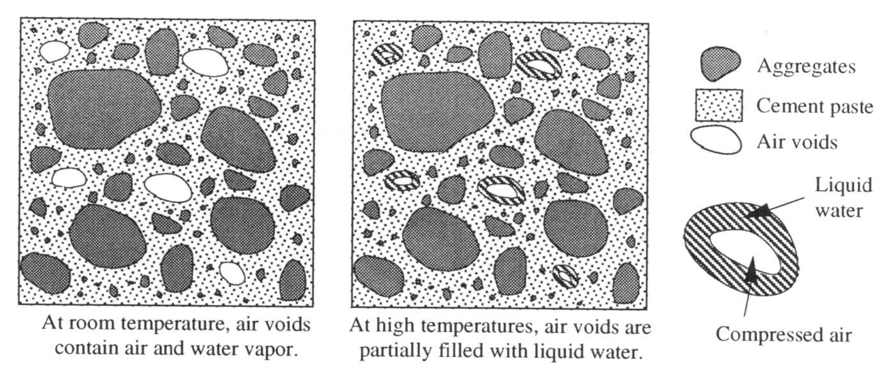

Fig. 9. Expansion of liquid water and compressed air.

5.3 Powers and Brownyard's model for cement paste

In order to determine how air is compressed by the expanding liquid water, it is very important to evaluate the total volume of pores and voids, and the volume fraction of liquid water in the pores and voids. To do so, Powers and Brownyard's Model was adopted in order to physically describe the cement paste of concrete.

The minimum w/c ratio of 0.438 is required for complete hydration [3, 4]. The non-evaporable water to cement ratio (wn/c) and gel water to cement ratio (wg/c) are 0.227 and 0.211, respectively. The total volume of cement gel is slightly less than the original volume of unhydrated cement and water. This difference is regarded as capillary pores. Once the volume of capillary pores is known, the volume of cement gel is easily obtained. Fig. 10 shows complete hydration under sealed curing. Cement gel is considered water-saturated if gel pores are filled with water. Cement gel has a fixed amount of gel water. Each component of cement paste is calculated using the following equations:

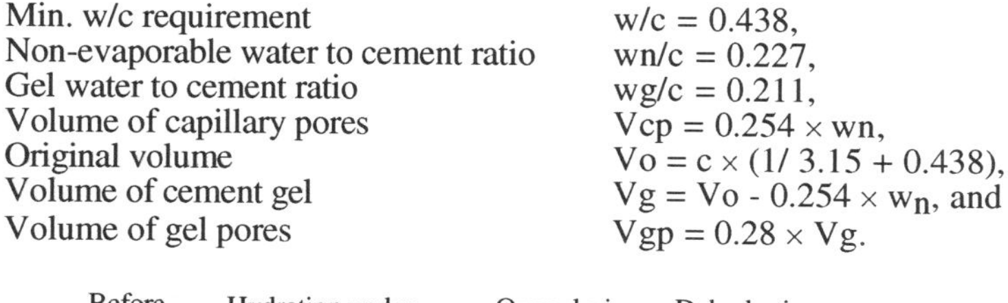

Min. w/c requirement	$w/c = 0.438$,
Non-evaporable water to cement ratio	$wn/c = 0.227$,
Gel water to cement ratio	$wg/c = 0.211$,
Volume of capillary pores	$Vcp = 0.254 \times wn$,
Original volume	$Vo = c \times (1/3.15 + 0.438)$,
Volume of cement gel	$Vg = Vo - 0.254 \times w_n$, and
Volume of gel pores	$Vgp = 0.28 \times Vg$.

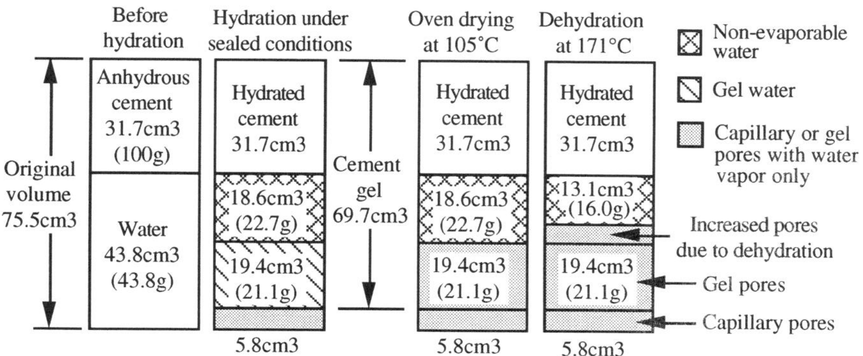

Fig. 10. Hydration under sealed conditions and dehydration at 171°C.

Original volume means the volume of unhydrated cement and water. A hundred g of cement and 43.8g of water produces 69.7 cm^3 of cement gel. Gel pores are completely filled with water, and capillary pores are empty.

5.4 Water released due to dehydration
Fig. 10 also shows the dehydration of cement paste having a w/c = 0.438 cured under sealed conditions. On oven drying, all gel water is released. Based on available data, about 30% of non-evaporable water will be released at 171°C due to the process of dehydration. The volume of cement gel is reduced by 5.5 cm^3 because the non-evaporable water is part of the cement gel. The volume of porosity is increased.

5.5 Physical model for concrete
In order to calculate components in concrete which are needed for evaluating pore pressure, it is necessary to make the following assumptions because sufficient information is not available:

(1) Total porosity in the cement paste of concrete
It is assumed that aggregates do not change the total porosity of concrete. Silica fume is replaced with cement. Therefore, Powers and Brownyard's model can be used for evaluating volume of pores and hydration products in the cement paste of concrete.
(2) Total porosity in aggregates
For fine aggregate, it is assumes that the total porosity is equal to the volume of absorbed water. For coarse aggregate, however, this may not yield a correct porosity due to the presence of isolated and air-filled voids and pores which water cannot reach. Since the total porosity of course aggregates may be more than the volume of absorbed water, in the proposed model, the following two cases are considered.

Case G1: Total porosity = Volume of absorbed water

Case G2: Total porosity = 2 × Volume of absorbed water

(3) Air content in the cement paste of concrete
Air content is measured on fresh concrete. It is assumed that air content does not change after setting.

6 Calculation of pore pressure in concrete

6.1 Mixture of air and saturated water vapor
Air and water are present in concrete. If air and water vapor are assumed to behave as ideal gases and if the volume expansion of liquid water is neglected, the total pressure is equal to the sum of the saturated vapor pressure (SVP) and the partial pressure of the entrapped air (PAP). SVP can be found in a standard steam table. PAP can be calculated using Equation (1). The total pressure is the sum of SVP and PAP, which will be denoted by SVP+PAP. It is assumed that the total pressure in the sealed container is 0.1013 MPa at 21.1°C. Thus, SVP = 0.0025 MPa and PAP = 0.0988 MPa.

$$PAP(T) = 0.0988 \times (273.15+T)/(273.15+T_o) \quad (1)$$

where T is the temperature of air and T_o is 21.1°C. At 171°C, SVP = 0.814 MPa and PAP = 0.149 MPa, then SVP+PAP = 0.963 MPa.

6.2 Increase in specific volume of liquid water
Once the total volume of pores and voids and the volume of water are known, the pore pressure at high temperatures can be calculated. However, there is one more assumption to be made because of insufficient information. That is the physical state in

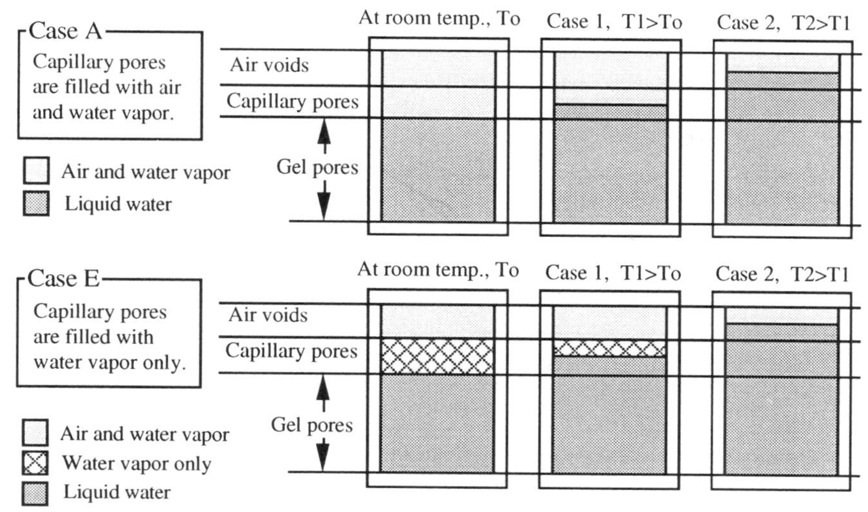

Fig. 11. Proportions of pores and conditions in pores.

the capillary pores. It is not clear what is in the capillary pores. Water can diffuse into capillary pores through cement paste without applying any pressure. It is not clear, however, if air can diffuse into the capillary pores in cement paste. For modeling, two extreme cases are considered. The real situation lies between those two cases.

6.2.1 Capillary pores with air and water vapor (Case A)

In the first case, it is assumed that air can diffuse into the capillary pores of cement paste under conditions of ambient pressure and room temperature. Before heating, the capillary pores are completely filled with air and water vapor. With an increase in temperature, water expands. Upon expansion of the water, the air in the capillary pores begins to compress. This is shown in the upper drawing of Fig. 11. Since air and water can diffuse in the cement paste, the pore pressure is uniform throughout the specimen.

To find the volume of liquid water, only the equilibrium between the liquid water and the water vapor, at a given temperature, is considered. The partial pressure of air can be calculated separately because air has no effect on the equilibrium of water. The total mass of liquid water and water vapor is constant at any temperature. Let V_W and V_V be volume fractions of liquid water and water vapor in the container.

$$V_W + V_V = 1 \tag{2}$$

$$\rho_W \times V_W + \rho_V \times V_V = \text{mass (constant)} \tag{3}$$

where ρ_W and ρ_V are the mass density of liquid water and vapor at given temperature. The constant in Equation (3) can be evaluated at any temperature because water is contained in the sealed container. Then V_W and V_V will be obtained by solving Equations (2) and (3) at a given temperature. Once V_V is known, the partial pressure of the entrapped air is calculated from the following equation since V_V is also the volume of vapor.

$$\text{PAP2}(T) = 0.0988 \times V_V(T_O)/V_V(T) \times (273.15+T)/(273.15+T_O) \tag{4}$$

where T_O is 21.1°C. $V_V(T_O)$ and $V_V(T)$ are the volume fractions of vapor (or air) at temperatures of T_O and T, respectively. The partial air pressure will be denoted by PAP2 to distinguish it from PAP. Equation (4) is based on the ideal gas equation and similar to Equation (1) except for the effect of the volume change. The pressure of the mixture of dry air and saturated vapor is equal to SVP+PAP2.

6.2.2 Capillary pores with Water Vapor Only (Case E)
In the second case, it is assumed that air cannot diffuse into the capillary pores. In other words, air is confined within the air voids. Before heating, the capillary pores contain only saturated water vapor. With an increase in temperature, water begins to expand, but air is not compressed until the capillary pores are completely filled with liquid water. This is shown in the lower drawing of Fig. 11. In this case, pore pressure in the capillary pores is different from pressure in the air voids and the difference is held by the hydrated solid of cement gel.

In case 1, the pressure in the air voids is equal to SVP+PAP and the pressure in the gel pores is equal to SVP only because there is no air in gel pores. The PAP is calculated using Equation (1). In case 2, the air voids begin to fill with water and air is compressed due to water expansion. The pressure in the air voids is equal to SVP+PAP2. The PAP2 is calculated using Equation (4).

6.3 Predicted Pore Pressures
There are two unknown variables due to lack of information. One is the total porosity of coarse aggregates (refer to Section 5.5). The other is the physical state of capillary pores (refer to Section 6.2). Since these variables are independent of each other, the following combinations will be considered:

A&G1 (capillary pores with air and water vapor and 1 × porosity of gravel)
A&G2 (capillary pores with air and water vapor and 2 × porosity of gravel)
E&G1 (capillary pores with water vapor only and 1 × porosity of gravel)
E&G2 (capillary pores with water vapor only and 2 × porosity of gravel)

Fig. 12 shows the comparison between predicted pore pressures and SVP+PAP for concrete of w/c = 0.25. Case A always yields a higher pressure than case E because air is more compressed in case A than in case E. Case G1 yields a higher pressure than case G2 because case G2 introduces additional air. Therefore, case A&G1 yields the highest pressure and case E&G2 yields the lowest pressure of the four cases.

In Fig. 13, the predicted pore pressures are compared with experimentally measured pressure. Pore pressures measured at the side of a specimen are compared

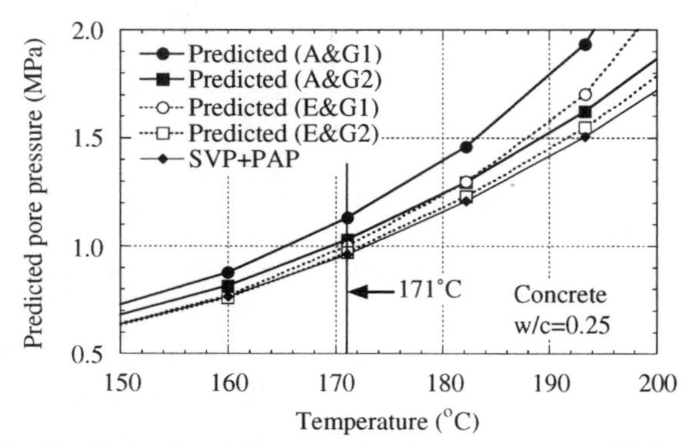

Fig. 12. Comparison between predicted pore pressures and SVP+PAP.

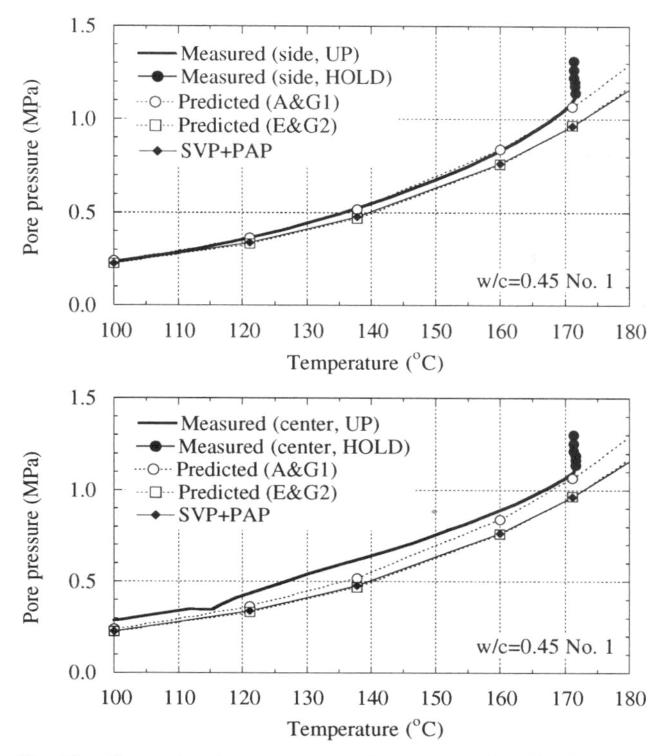

Fig. 13. Comparison between measured pressures and predicted pressures.

with the predicted pore pressure curves. UP and HOLD indicate pore pressures measured when specimen temperature was going up (UP) and while the temperature was held (HOLD) at 171°C, respectively. When the temperature was increased, the pore pressures were within the predicted range. But, when the temperature was held at 171°C, pore pressures continued to increase.

Pore pressures at the center of the same specimen are also compared with the predicted pressures. When the temperature was increased, the measured pressures were always higher than the predicted values. The pore pressure continued to increase although the temperature was held at 171°C.

The increase in pore pressure when the temperature was held at 171°C may be attributed to chemical decomposition of the admixtures.

7 Conclusions

7.1 Pore pressure in concrete at 171°C

(1) The pressure tests were performed on sealed concrete at a sustained high temperature of 171°C. The measured pore pressures were considerably higher than the sum of SVP and PAP.

(2) The pore pressures were found to exceed the design tolerances of the thin liner. Instead of dealing with these pore pressures, a venting system will be installed between the liner and concrete to relieve the pore pressures towards the inside.

(3) A model for predicting pore pressure of sealed concrete at high temperatures was proposed.

7.2 Water and gas released from concrete at 171°C

(1) In the water/gas release tests performed, 50 to 70% of the total water in concrete was released at 171°C. The released gas volume was found to be two to seven times the released water volume, and ten to forty times the entrapped air volume.
(2) The volume of water can be predicted using Powers and Brownyard's model, however, allowance needs to be made for the release of non-evaporable water due to dehydration at temperatures above 105°C because not enough information is available. On the other hand, it is very difficult to estimate the volume of non-condensable gas because the sources and mechanisms of gas generation are not well understood.
(3) The water volume data can be used in the design of the venting system in the SBWR, but gas volume data is of little value.

7.3 Release rates of water and gas from concrete at 171°C

(1) Release rates of water and gas were calculated based on the relationship between released volume of water and gas and time.
(2) The data may not be used for the design of the venting system of SBWR since the release rates have a very strong dependency on size and geometry of specimens.
(3) It is very difficult to predict release rates of water and gas because an appropriate diffusion model is not available.

7.4 Concrete mix

(1) The tests demonstrated that gel water, chemically bound water and water absorbed by aggregates can be released in significant quantities at temperatures above 105°C and cause high pressures.
(2) All possible mixes of concrete used for structural purposes will generate pore pressures greater than SVP+PAP when the concrete is heated.

8 Acknowledgments

The experimental research was done at NSF Center for Advanced Cement-Based Materials (ACBM), Northwestern University. The authors appreciate the financial support provided by Burns and Roe, a subcontract management company, for experimental work from a Department of Energy contract through General Electric Corporation.

9 References

1. England, G. L., Greathead, R. J. and Khan, S. A.(1991) Influence of High Temperature on Water Content, Permeability and Pore Pressure in Concrete. *Transactions of the 10th International Conference on Structural Mechanics in Reactor Technology (SMIRT)*, Vol. H, pp. 31-36.
2. Mindess, S. and Young, J. F.(1981) *Concrete*, Prentice-Hall, Inc.
3. Powers, T. C. and Brownyard, T. L.(1946-1947) Studies of the Physical Properties of Hardened Portland Cement Paste. *Proceedings, ACI*, Vol. 41.
4. Powers, T. C.(1960) Physical Properties of Cement Paste. *4th International Symposium on Chemistry of Cement,* Washington D. C., Vol. II, Session V, Paper V-1, pp. 557-609.

PART TWENTY
CARBONATION AND ASR

114 ALTERATION OF C-S-H AND ASR GELS IN DETERIORATED CONCRETES, NEWFOUNDLAND, CANADA

T. KATAYAMA
Sumitomo Osaka Cement Co. Ltd, Osaka, Japan
D.J. BRAGG
Government of Newfoundland and Labrador, St. John's, Canada

Abstract
A petrographic study was made of 50-70 year old concrete structures in Newfoundland that have undergone combined freeze/thaw weathering and alkali-silica reaction (ASR). Electron-probe microanalysis (EPMA) confirmed the general presence of alkali-aluminates in these old concretes with up to 3.5% of Na_2O_{eq}, suggesting that high-alkali cements were commonly used. It was also found that calcium silicates, when weathered, hydrate to C-S-H gels, leaching calcium down to a Ca/Si atomic ratio 1.2. In contrast, ASR gels during migration lose alkalies and take up calcium from the cement paste, transforming into or replacing C-S-H gels at a Ca/Si ratio of up to 1.0-1.2. This ratio may be an "equilibrium ratio" between the two gels in these concretes, at which the leaching of calcium from the C-S-H gels apparently equilibrates with the absorption of liberated calcium in the ASR gels. This ratio also corresponds to products of a pozzolanic reaction, which represents a missing-link between the products of cement hydration and ASR.
Keywords: Aggregate, alkali-silica reaction, ASR gel, cement hydration, EPMA, concrete deterioration, C-S-H gel, freeze/thaw weathering, pozzolanic reaction.

1 Introduction

A study of alkali-aggregate reaction was started in Newfoundland, Canada, after signs of alkali-aggregate reactivity in concrete structures near St. John's were first noticed in 1983 by the second author [1]. Because the climate conditions are very severe in this province, freeze/thaw weathering is invariably found in old concrete structures, particularly those constructed more than 50 years ago before the air-entraining technique was introduced.

A field survey was made of the damaged structures on the Avalon Peninsula by the authors in 1992, and concrete cores were taken in 1993 [2]. This paper describes the results of petrographic examinations of these old concretes that have undergone combined, long-term deterioration of freeze/thaw weathering and alkali-aggregate reaction, focusing on the nature of chemical reactions as detected by EPMA measurements. A process of chemical alterations found in concretes, resembling the pozzolanic reaction, will be discussed in detail.

Concrete Under Severe Conditions: Environment and loading (Volume Two) Edited by K. Sakai, N. Banthia and O.E. Gjørv. Published in 1995 by E & FN Spon. ISBN 0 419 19860 1

2 Petrography of concrete

The damaged structures examined were constructed 50 to 70 years ago, and included three abandoned US Army installations, two restored bridges and one reservor dam, and a retaining wall (Table 1). They were in various degrees of deterioration caused by alkali-aggregate reaction and freeze/thaw weathering, as evidenced by map cracking, reaction rims, exudation, stalactite formation, spalling and stress cracking, mostly indicative of poor quality control [2].

2.1 Concrete

White gel deposits are visible on the reacted aggregate and surrounding cement paste in the fracture surface of concretes. They are, however, not evident on the sawed surface unless a gel fluorescence test is made. Carbonation and leaching prevails in the weathered portions near the concrete surface, but are also noticeable inside the structures, along open cracks and open interspaces between the cement paste and aggregate. In thin sections, unhydrated clusters of cement minerals are seen. Cement hydrates such as portlandite, ettringite and hydrocalumite, as well as ASR gels, fill air-voids and dissolution vugs in the cement paste, but portlandite is depleted in the carbonated surface.

2.2 Aggregate

Aggregates used in the structures came from the Avalon Peninsula, whose local geology is similar to Nova Scotia and Southern New Brunswick which consists of Late Precambrian to Early Paleozoic sedimentary rocks with minor volcanic suites [3]. The reacted aggregates are Late Precambrian coarse siltstone and sandstone aggregates with a crypto- to microcrystalline matrix of authigenic quartz (Table 2), having a lithology common to those found in a late-expansive alkali-silica reaction, formerly called the alkali-silicate reaction in these provinces [4]. Rhyolitic tuff in the sand also reacted in concretes.

Table 1. Deterioration of concrete structures near St.John's, Newfoundland

Location	Structure	Construction	Deterioration with diagnosis
Red Cliff	Radar base	1941	Freeze/thaw (S), ASR (M,R,E), Stress (L)
Water Street	Retaining wall and gate	1923	ASR (M,R,E), Freeze/thaw (S)
Fort Amherst	Gun bunkers	1941	ASR (M,R,E), Freeze/thaw (S), Stress (L)
Cape Spear	Gun bunkers	1941	ASR (M,R,E), Freeze/thaw (S)
Petty Harbor	Dam	1923	ASR (M,R), Stress (L)
Tors Cove	Bridge (old)	1935	Freeze/thaw (S), ASR (R), Stress (L)
Cape Broyle	Bridge (old)	1935	ASR (M,R), Stress (L)

M:map cracking, R:reaction rims, E:exudation, S:spalling, L:longitudinal crack

Table 2. Type of reacted coarse aggregates in the deteriorated NF structures

Location	Aggregate	Rock type	Reactive phase	Geologic origin
Red Cliff	Gravel	Siltstone (red/green)	CQ	Conception Group
Water Street	Crushed stone	Sandstone (green)	MQ	Signal Hill Group
Fort Amherst	Crushed stone	Sandstone (green)	MQ	Signal Hill Group
	and gravel	and siltstone (green)	CQ	Signal Hill Group
Cape Spear	Crushed stone	Sandstone (brownish green)	MQ	Signal Hill Group
Petty Harbor	Crushed stone	Sandstone (red)	MQ	Signal Hill Group
Tors Cove	Gravel	Siltstone (green)	CQ	Conception Group
Cape Broyle	Gravel	Siltstone (green)	CQ	Conception Group
		and sandstone (green)	MQ	Conception Group

CQ:cryptocrystalline quartz, MQ:microcrystalline quartz

3 Electron-probe microanalysis of concrete

Thin sections were subjected to quantitative electron-probe microanalysis in Japan using a WDS-type EPMA (JEOL JCMA 733: 15KV, 12 nA, measuring time 5s x 2 for each element, beam diameter 1 micrometer), employing Bence/Albee correction mostly. Selected data from about 500 measurements are given below for the clinker minerals (Tables 3, 4), C-S-H gels (Tables 5, 6) and ASR gels (Tables 7-9), representing the first comprehensive data of concretes in Newfoundland.

EPMA measurements of hardened concretes revealed the general presence of alkali-bearing cement minerals in the structures. The greater part of the alkalies came from aluminate and belite (total Na_2O_{eq} 0.6%) in clinkers, with a slight contribution from alite (0.1% Na_2O_{eq}). Thus the cements used were high-alkali cements containing more than 0.7% of Na_2O_{eq}, because water-soluble alkali-sulfates were also likely contained in the clinkers.

3.1 Unhydrated cement minerals in concrete

Alite is occasionaly found unhydrated in large clusters of clinker minerals in concretes. Unhydrated alites are colorless and transparent without inclusions of dark interstitial materials and show uniform birefringence, lacking distinctive optical and compositional zoning within one crystal. This suggests that SO_3 was not especially high in the original clinkers [5]. Appreciable amounts of MgO, up to 1.7% (Table 3), are found entering the alite solid solution, which indicates that the clinkers generally contained much more MgO in their bulk compositions [6].

Table 3. Unhydrated alite and belite in the clusters in Newfoundland concrete

	Alite					Belite				
	Water Street Wb265	Fort Amherst Fc9	Cape Spear Cb219	Tors Cove Tb91	Cape Broyle Cc102	Water Street Wb259	Fort Amherst Fb178	Cape Spear Cb228	Tors Cove Tb86	Cape Broyle Cc113
SiO_2	23.07	24.15	24.32	24.65	25.01	28.94	31.09	30.45	31.53	30.92
TiO_2	0.31	0.19	0.40	0.27	0.16	0.36	0.13	0.44	0.13	0.20
Al_2O_3	1.38	0.97	1.14	0.98	0.88	2.14	1.64	1.37	1.37	1.50
Fe_2O_3	0.65	0.61	0.72	0.48	0.52	1.22	0.94	0.93	0.69	0.97
Mn_2O_3	0.05		0.10	0.09		0.08	0.12	0.12	0.17	0.01
MgO	1.64	0.88	1.63	1.73	1.48	0.33	0.17	0.53	0.38	0.52
CaO	67.04	70.82	69.50	70.33	69.77	58.43	63.81	61.80	62.82	60.99
Na_2O	0.18	0.14	0.23	0.28	0.12	0.60	0.45	0.63	0.65	0.71
K_2O	0.05	0.03	0.15	0.02	0.06	0.50	0.61	0.49	0.34	0.70
SO_3	0.04	0.05	0.01	0.01		0.30	0.37	0.16	0.14	0.06
P_2O_5	0.35	0.26	0.17	0.14	0.08	0.61	0.33	0.34	0.21	0.41
Total	94.70	98.03	98.29	98.94	98.04	93.52	99.84	97.25	98.42	97.02
Si	0.92	0.94	0.94	0.95	0.97	0.90	0.91	0.91	0.93	0.92
Ti	0.01	0.01	0.01	0.01	0.01	0.01	0.01	0.01		
Al	0.07	0.04	0.05	0.04	0.04	0.08	0.06	0.05	0.05	0.05
Fe	0.02	0.02	0.02	0.01	0.02	0.03	0.02	0.02	0.02	0.02
Mg	0.10	0.05	0.09	0.10	0.09	0.02	0.01	0.02	0.02	0.02
Ca	2.88	2.95	2.88	2.89	2.89	1.94	1.99	1.98	1.98	1.95
Na	0.01	0.01	0.02	0.02	0.01	0.04	0.03	0.04	0.04	0.04
K			0.01			0.02	0.02	0.02	0.01	0.03
P	0.01	0.01	0.01	0.01		0.01	0.01	0.01	0.01	0.01
Total	4.02	4.02	4.03	4.03	4.01	3.05	3.06	3.06	3.06	3.06
O	5.00	5.00	5.00	5.00	5.00	4.00	4.00	4.00	4.00	4.00
Ca/Si	3.11	3.14	3.06	3.06	2.99	2.16	2.20	2.17	2.13	2.11

Alite: $(Ca, Mg, Na, K)_3(Si, Al, Fe, Ti, P)_1O_5$, Belite: $(Ca, Na, K, Mg)_2(Si, Al, Fe, Ti, P)_1O_4$

Belite is more often unhydrated than alite in concretes, accompanied by alite or forming entire clusters. The optical properties are variable within one thin section. It is almost colorless and transparent with cross-lamellae in some clusters, while in others it is brownish without lamellae, reflecting variations in the thermal histories during clinker formation. Belite contains much more alkali than alite, up to 1% (Table 3), corresponding to 0.25% Na_2O_{eq} in the original clinkers. The Ca/Si atomic ratio always exceeds 2.1, but this is usual for belites containing minor elements in the solid solution [7].

Aluminate is found from all structures as optically anisotropic, as sharply elongated crystals with a colorless to pale-brownish tint, and intergrown with dark brownish ferrite in a coarsely grained, well-differentiated matrix within the clusters. The optical property indicates that this is the "alkali-aluminate" variety. EPMA data confirmed that the alkali-aluminates contain up to 3.5% alkalies, corresponding 0.35% Na_2O_{eq} in the clinkers (Table 4). Such a high level of alkali in the alkali-aluminate, up to 4.4% Na_2O_{eq}, has also been detected by the first author from an old concrete mortar made 95 years ago in Japan. Hydration of alkali-aluminates in concretes is generally slower than alite and belite, but even faster than ferrite.

Ferrite varies widely in the grain-size, color and chemical compositions within the same thin sections, reflecting a different cooling rate and oxidation state in clinkers. This mineral shows distinctive pleochroism of greenish to orange brown in the well-differentiated matrix, but it may be less pleochroic with an orange brownish tint, where the Fe/Al ratio or oxidation state is low (Table 4). Ferrite is the least hydrated phase in concretes.

Table 4. Unhydrated aluminate and ferrite in the clusters in NF concrete

	Aluminate							Ferrite		
	Water Street		Fort Amherst			Cape Broyle		Fort Amherst		Cape Spear
	Wb279	Wb280	Fc15	Fc13	Fb205	Cc142	Cc90	Fb214	Fc6	Cb225
SiO_2	3.82	3.90	4.51	3.34	3.56	4.85	3.52	1.52	3.95	3.60
TiO_2	0.11	0.36	0.21	0.28	0.22	0.03	0.25	1.67	1.17	1.67
Al_2O_3	29.90	29.91	26.82	29.23	30.09	29.27	29.57	21.72	21.16	23.66
Fe_2O_3	4.68	4.93	6.05	5.77	4.98	4.40	5.13	22.91	19.92	18.24
Mn_2O_3	0.04		0.07		0.29	0.03		0.51	0.41	0.46
MgO	0.52	0.56	0.59	0.59	0.52	0.92	0.49	2.37	1.88	1.98
CaO	50.70	50.30	53.30	53.73	55.39	53.74	54.82	46.77	48.86	49.56
Na_2O	3.02	2.81	2.81	2.54	2.13	2.53	2.02	0.07	0.01	0.88
K_2O	0.90	0.91	1.15	1.25	0.96	1.39	1.03	0.05	0.20	0.27
SO_3			0.08		0.02	0.05	0.01	0.02	0.04	
P_2O_5	0.02	0.01	0.16	0.08	0.01	0.14	0.07		0.13	0.01
Total	93.71	93.69	95.75	96.81	98.17	97.35	96.91	97.61	97.73	100.33
Si	0.18	0.18	0.22	0.16	0.16	0.23	0.17	0.06	0.16	0.14
Ti		0.01	0.01	0.01	0.01		0.01	0.05	0.04	0.05
Al	1.70	1.69	1.51	1.63	1.64	1.62	1.64	1.03	0.99	1.07
Fe	0.17	0.18	0.22	0.20	0.17	0.16	0.18	0.69	0.60	0.53
Mn								0.02	0.01	0.01
Mg	0.04	0.04	0.04	0.04	0.04	0.06	0.04	0.14	0.11	0.11
Ca	2.62	2.59	2.73	2.72	2.75	2.70	2.76	2.02	2.08	2.05
Na	0.28	0.26	0.26	0.23	0.19	0.23	0.18	0.01		0.07
K	0.06	0.06	0.07	0.08	0.06	0.08	0.06		0.01	0.01
P			0.01			0.01				
Total	5.06	5.01	5.07	5.07	5.02	5.09	5.04	4.02	4.00	4.04
O	6.00	6.00	6.00	6.00	6.00	6.00	6.00	5.00	5.00	5.00
Fe/Al	0.10	0.11	0.13	0.13	0.11	0.10	0.11	0.67	0.60	0.49

Aluminate: $(Ca,Na,K)_3(Al,Fe,Mg,Si,Ti)_2O_6$, Ferrite: $Ca_2(Al,Fe,Mg,Si,Ti,Mn)_2O_5$

3.2 Hydration products in concrete

There are strong simlarities between hydration products of alite and belite in concretes. C-S-H gels of these calcium silicates have differing Ca/Si ratios according to their location within the cement grains and clusters, but their range is remarkably similar (Tables 5, 6). In the tables, the difference between 100% and totals, ranging 20-30%, can be H_2O in these gels, since carbonated portions were excluded from analysis by petrographic checking. Where the total is extremely low, it may be a poor analysis caused by a high porosity of gels.

3.2.1 Calcium silicate hydrates

Both alite and belite are mostly hydrated to form very similar, optically amorphous C-S-H gels, leaving "pseudomorphs" of these minerals. The gel hydrates are colorless and transparent in thin sections, with a low refractive index and low reflectivity, forming rims of "inner hydration products" around the unhydrated core. Alite of isolated small grains, or, less commonly, alite in a cluster (Fig. 1), may show a cyclic zoning of C-S-H gels, suggestive of the formation of annual-rings due to periodical hydration in concretes.

The Ca/Si atomic ratio of C-S-H gels in the pseudomorphs of calcium silicates differs greatly, reflecting the degree of hydration, leaching and weathering in concretes (Tables 5, 6). However, the range of variation is similar among products with advanced hydration of alite and belite, as reported [8]. Both Al and Mg tend to be concentrated in the C-S-H gels during hydration, which were probably derived from hydrating aluminates and ferrites intimately associated.

Table 5. C-S-H gels forming alite "pseudomorphs" in Newfoundland concrete

	Water Street (block)				Fort Amherst (block)				Fort Amherst (core)			
	Wb254	Wb266	Wb220	Wb261	Fb185	Fb211	Fb176	Fb175	Fc36	Fc10	Fc38	Fc11
SiO_2	24.39	26.92	25.57	25.56	22.06	27.02	29.17	30.64	25.78	28.10	30.47	31.48
TiO_2	0.26	0.20	0.19	0.20	0.44	0.36	0.39	0.51	0.22	0.10	0.33	0.33
Al_2O_3	2.17	3.36	3.88	3.88	4.11	2.12	1.68	2.08	1.44	2.02	1.89	2.30
Fe_2O_3	0.93	1.04	0.82	0.57	2.51	0.88	0.85	0.80	0.88	1.02	0.77	0.70
Mn_2O_3	0.03		0.14	0.08		0.03	0.08	0.04				
MgO	2.34	2.69	4.71	5.16	2.48	1.06	0.92	0.83	1.07	1.04	1.12	1.33
CaO	37.15	35.21	32.38	29.63	40.36	39.98	38.17	37.36	48.39	48.17	46.60	44.89
Na_2O		0.01	0.04			0.04	0.01		0.07			
K_2O		0.04	0.02		0.01	0.02	0.01	0.01	0.08	0.12		
SO_3	0.19	0.12	0.23	0.21	0.48	0.31	0.29	0.24	0.14	0.07	0.08	0.12
P_2O_5	0.15	0.29	0.36	0.23	0.20	0.30	0.27	0.40	0.33	0.25	0.25	0.38
Total	67.60	69.96	68.33	65.46	72.63	72.12	71.82	72.90	78.38	80.88	81.51	81.53
Ca/Si	1.63	1.40	1.36	1.24	1.96	1.59	1.40	1.31	2.01	1.84	1.64	1.53

	Cape Spear (block)				Tors Cove (block)				Cape Broyle (core)			
	Cb233	Cb224	Cb230	Cb220	Tb121	Tb101	Tb99	Tb120	Cc92	Cc108	Cc112	Cc103
SiO_2	23.28	24.80	21.73	25.57	26.15	27.53	28.42	28.33	25.19	25.92	28.61	29.12
TiO_2	0.50	0.40	0.71	0.19	0.23	0.25	0.22	0.32	0.61	0.22	0.19	0.24
Al_2O_3	2.70	3.04	3.47	3.88	1.80	3.80	3.89	3.56	3.19	2.72	4.54	3.60
Fe_2O_3	0.92	0.81	1.05	0.82	0.52	0.87	0.79	1.02	1.07	0.60	0.87	1.04
Mn_2O_3		0.01	0.04	0.16	0.02	0.09		0.64		0.03		0.02
MgO	3.44	3.15	5.98	4.71	1.86	1.34	1.66	2.72	1.83	2.43	2.97	2.48
CaO	36.03	34.02	27.90	32.38	38.79	38.36	37.44	34.17	43.04	40.37	41.11	40.99
Na_2O	0.03			0.04								
K_2O	0.08		0.04	0.02	0.02	0.02	0.06	0.05				
SO_3	0.69	0.20	0.44	0.23	0.15	0.76	0.67	0.08	0.16	0.10	0.02	0.14
P_2O_5	0.10	0.29	0.02	0.36	0.29	0.12	0.13	0.27	0.15	0.12	0.28	0.18
Total	67.76	66.71	61.37	68.33	69.83	73.12	73.28	70.23	75.23	72.50	78.60	77.79
Ca/Si	1.66	1.50	1.38	1.36	1.59	1.49	1.41	1.29	1.83	1.67	1.54	1.51

In an unweathered section of the concretes (the interior of core samples), calcium is leached from calcium silicates through normal hydration processes, down to Ca/Si 1.5-1.6 for alite hydrates (Table 5) and 1.6-1.7 for belite hydrates (Table 6), which are equivalent to the "final hydration products" of these silicates under Ca-saturated conditions [8]. By contrast, in the weathered concretes, i.e. block samples from the structure surface, the Ca/Si ratio of C-S-H gels continuously decreases to 1.2 or less, irrespective of whether they were originally alite or belite. This is the result of excessive leaching of calcium by percolating water near the surface during long-term weathering, such as freeze/thaw and rain-fall. The calcium leached out from the structures forms stalactites and exudations mixed with ASR products. At a ratio of Ca/Si 1.0-1.2, inter-replacement takes place between the C-S-H and ASR gels.

3.2.2 Other hydrates

Hydrocalumite, $3CaO{\cdot}Al_2O_3{\cdot}Ca(OH)_2{\cdot}12H_2O$, occurs as an aggregation of platy to bladed crystals filling air-voids in concretes (Fig. 2), or replacing tufts of needle-like ettringite leaving pseudomorphs of this mineral. EPMA analysis of a core sample from Cape Broyle gave a composition of CaO 36.37, Al_2O_3 19.78, SO_3 3.23, P_2O_5 0.16, MgO 0.05, with a total 59.59%, which suggests this hydrate forms a solid solution with monosulfate, $3CaO{\cdot}Al_2O_3{\cdot}CaSO_4{\cdot}12H_2O$. Portlandite, $Ca(OH)_2$, occurs interstitially within fresh, non-carbonated cement pastes in concretes (Fig. 2). A sample from the same concrete gave analysis of CaO 74.68, SiO_2 0.34, P_2O_5 0.09, MgO 0.08, Fe_2O_3 0.04, Al_2O_3 0.02, and a total 75.25%.

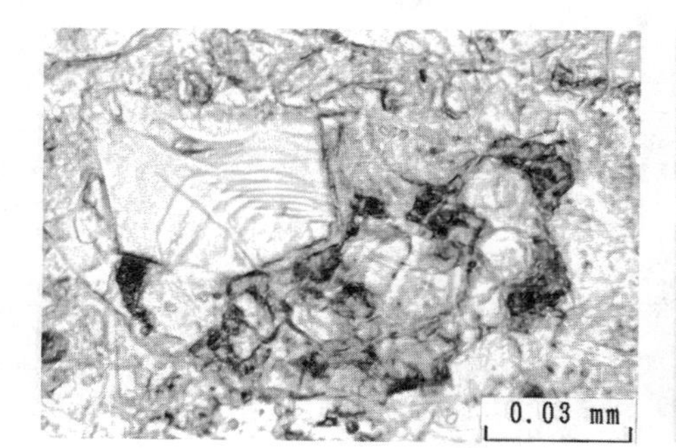

Fig. 1 Pseudomorph of alite in cluster

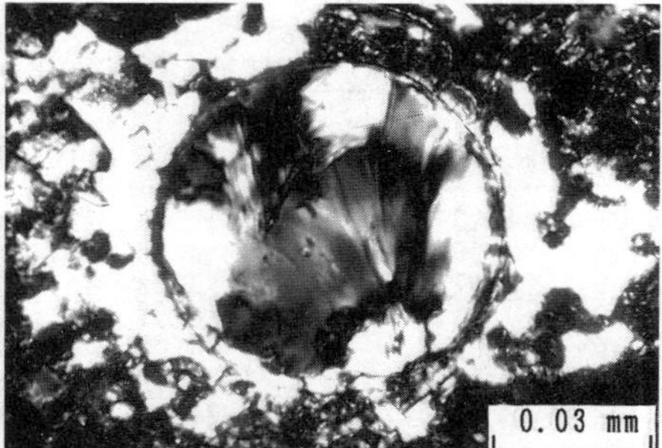

Fig. 2 Hydrocalumite filling air-void

Table 6. C-S-H gels forming belite "pseudomorphs" in Newfoundland concrete

	Water Street			Fort Amherst		Tors Cove (block)				Cape Broyle	
	Wb262	Wb244	Wb245*	Fb168	Fc16	Tb146	Tb87	Tb130	Tb131*	Cc115	Cc119
SiO_2	27.60	20.95	22.45	24.50	27.85	26.79	27.21	28.47	32.38	26.52	25.97
TiO_2	0.33			0.30	0.42	0.19	0.30	0.40		0.15	0.50
Al_2O_3	2.62	1.45	1.47	1.48	2.94	2.53	2.95	4.35	3.11	3.09	4.23
Fe_2O_3	1.53	0.17	0.14	0.62	1.02	0.97	0.53	0.72	0.08	0.62	1.28
Mn_2O_3	0.03	0.02	0.03	0.03		0.20		0.14	0.04		
MgO	0.42			0.81	1.61	1.99	2.44	3.53	4.33	1.00	2.95
CaO	47.39	31.75	25.91	43.43	44.26	41.29	36.87	31.70	29.68	42.48	40.36
Na_2O	0.26			0.14		0.16	0.02		0.18		
K_2O	0.11		0.02	0.10		0.15	0.02		0.24		
SO_3	0.21	0.04	0.07	0.12	0.13	0.06	0.19	0.12	0.06	0.08	0.43
P_2O_5	0.30		0.03	0.13	0.19	0.19	0.15	0.37	0.22	0.15	0.27
Total	80.81	54.39	50.14	71.65	78.51	74.51	70.66	69.79	70.32	74.08	75.99
Ca/Si	1.84	1.62	1.24	1.90	1.70	1.65	1.45	1.19	0.98	1.71	1.67

* Replaced by ASR gel.

3.3 Alteration of ASR gels in concrete

Alkali-silica reaction in Newfoundland concretes has produced ASR gels, which 1) occur in the cracks in the reacted aggregates extending into the cement paste, 2) fill air-voids in the cement paste and cement/aggregate interspaces, 3) replace the cement paste and reaction rims. They represent different stages of formation altering into C-S-H gels. Alkalies are depleted from ASR gels near the concrete surface, due to freeze/thaw weathering and/or leaching.

3.3.1 Crack-filling ASR gels

The early stage of gel alteration is represented by "younger gels" filling cracks in the reacted aggregates, with highest alkali and lowest Ca/Si ratios of 0.3-0.5 (Table 7). They are, however, much lower in alkalies and higher in calcium, than typical ASR gels [9] and those in concretes undergoing similar ASR in New Brunswick [10]. This shows that leaching of alkalies and moving of calcium are intense in Newfoundland, even at 10 cm, below the structure surface, suggesting that concrete weathering is most severe in this province.

The middle stage of gel alteration is characterized by the ASR gels filling cracks in the cement paste. During migration from reacted aggregates, these gels lose alkaies and gain calcium in a very short distance with a Ca/Si ratio increasing from 0.6 to 1.2 (Table 7). This means that they chemically approach the C-S-H gels, resembling products in the pozzolanic reaction [9], taking up calcium from the cement paste. Cracks in some aggregates may be filled with composit gel-veins with varying Ca/Si ratios, formed at different stages.

Table 7. ASR gels filling cracks from aggregate to the cement paste in concrete

	Fort Amherst (core) In sandstone				Cape Broyle (core) In sandstone				Cape Broyle (block) In siltstone			
	Fc22	Fc31	Fc21	Fc29	Cc73	Cc78	Cc76	Cc77	Cb136	Cb139	Cb141	Cb140
	A	A	A	A	A	A	C	C	A	A	A	A
SiO_2	56.51	59.20	59.26	58.10	51.84	53.50	34.76	34.19	50.55	59.49	51.20	54.43
Al_2O_3	0.05		0.05	0.02	2.86	0.02	1.83	1.81	0.03	0.48	0.84	0.03
Fe_2O_3					0.68	0.15		0.05		1.01	0.32	0.02
MgO				0.02	0.21					0.11	0.02	
CaO	22.89	22.35	22.77	27.83	26.12	27.26	40.94	42.32	21.15	18.35	20.04	22.50
Na_2O	1.54	1.42	0.70	0.05	0.25	0.15			1.23	0.18	0.28	0.09
K_2O	3.76	3.68	3.23	1.07	2.12	1.12	0.01		6.64	2.75	2.56	1.19
SO_3					0.01	0.07	0.04	0.03	0.03	0.12		
P_2O_5	0.78	0.12	0.20	0.07	0.04		0.09	0.06	-	-	-	-
Total	85.52	86.77	86.15	87.16	84.13	82.26	77.67	78.46	79.64	82.41	75.40	78.37
Ca/Si	0.43	0.40	0.41	0.51	0.54	0.55	1.26	1.33	0.45	0.33	0.42	0.44

	Water Street (block)									
	In sandstone			Boundary	In cement paste (outside)					
	-0.08	-0.06	-0.01(mm)	0.00	+0.09	+0.18	+0.22	+0.23	+0.54	+0.54(mm)
	Wb46	Wb42	Wb45	Wb43	Wb47	Wb49	Wb51	Wb50	Wb52	Wb53
	A	A	A	A	A	A	W	W	W	W
SiO_2	48.98	52.65	55.91	48.58	48.06	45.06	38.90	39.25	30.64	33.35
Al_2O_3	0.05		0.03	0.03	0.01	0.03	0.11	0.22	1.92	1.96
Fe_2O_3	0.34	0.24	0.06	0.24	0.13	0.12	0.23	0.26	0.26	0.10
MgO			0.02	0.05	0.03	0.05		0.03	0.04	0.05
CaO	26.90	29.14	29.69	30.24	26.45	29.05	28.20	33.14	35.03	35.10
Na_2O	0.18	0.23	0.26	0.10	0.02	0.05	0.05	0.03		0.04
K_2O	0.14	0.13	0.25	0.34	0.15	0.09	0.02	0.03		0.03
SO_3					0.03	0.12	0.06			
Total	76.59	82.39	86.21	72.61	74.88	74.56	67.57	72.95	67.89	70.64
Ca/Si	0.59	0.59	0.57	0.67	0.59	0.69	0.78	0.90	1.23	1.13

A: Optically amorphous, W: poorly crystalline, C: crystalline

3.3.2 Air void-filling ASR gels

The late stage of gel alteration is represented by the ASR gels, almost chemically equivalent to the C-S-H gels, having modified compositions with higher Ca/Si ratios of 0.8-1.2 and negligible alkalies. These gels migrate from cracks or porous textures in the cement paste, into the air-voids and "dissolution vugs", filling these openings or lining their inner walls.

ASR gels that line large air-voids, generally consist of alternate layers of different crystallinity from amorphous to crystalline, but they show no marked compositional differences (Table 8). In the table, analysis points are, to the right, arranged outwardly or from center to periphery in the gels. It is noteworthy that ASR gels filling cracks in the reacted aggregates (Table 7), with much alkali and lower Ca/Si ratios of less than 0.7, are optically amorphous, whereas those occurring near the cement paste, with little alkali and higher Ca/Si ratios up to 1.2, tend to be more crystalline.

The dissolution vugs are the irregular-shaped secondary texture, developed at the sites of air-voids between the closely arranged coarse aggregate grains. They are formed through the leaching and the dissolution of hydrated cement pastes during freeze/thaw weathering, in close association with alkali-silica reaction. ASR gels line the inner wall of the dissolution vugs (Fig.3) and/or fill them entirely, replacing adjacent C-S-H gels (Table 8), which suggests that alkali-silica reaction became prominent in the concretes after freeze/thaw had caused intense leaching and dissolution of the cement paste.

Table 8. ASR gels filling/lining air-voids in the cement paste of NF concrete

	Water Street(block) Air void-lining			Fort Amherst(block) Air void-filling			Fort Amherst (core) Dissolution vug/pore-filling				
	Wb76	Wb77	Wb78	Fb85	Fb86	Fb87	Fc46	Fc45	Fc44	Fc42	Fc41*
	A	A	A	A	A	A	A	A	A	A	A
SiO_2	31.46	32.88	32.65	34.11	35.28	34.60	38.62	39.78	40.95	43.55	32.63
TiO_2	-	-	-	-	-	-		0.15	0.04	0.01	0.04
Al_2O_3	2.26	1.79	2.47	1.94	2.09	1.97	0.02	0.05	0.06	0.19	2.22
Fe_2O_3		0.29	0.10	0.24				0.01			0.67
MgO	0.06	0.01	0.05	0.05	0.04	0.04				0.01	0.39
CaO	31.53	32.25	33.05	39.04	37.43	38.03	36.21	36.02	36.77	33.68	30.53
Na_2O			0.04							0.01	
K_2O			0.05	0.04				0.14	0.15	0.19	0.18
SO_3	0.07	0.09	0.14	0.13	0.10	0.10			0.02		0.12
P_2O_5	-	-	-	-	-	-		0.04			0.05
Total	65.38	67.31	68.54	75.54	74.94	74.75	74.92	76.18	77.98	77.65	66.82
Ca/Si	1.07	1.05	1.08	1.23	1.14	1.18	1.00	0.97	0.96	0.83	1.00

	Tors Cove (block) Large air void-lining				Tors Cove (block) Large air void-lining						
	Tb39	Tb37	Tb36	Tb41	Tb14	Tb16	Tb19	Tb20	Tb22	Tb23	Tb24
	W	A	A	A	C	W	C	W	C	A	A
SiO_2	42.20	39.83	42.70	32.81	37.95	36.28	36.85	37.05	34.02	28.18	35.23
Al_2O_3	0.15	0.16	0.23	0.64	0.96	0.65	1.01	1.25	1.87	1.74	1.84
Fe_2O_3	0.32		0.13			0.13	0.13		0.07	0.19	0.24
MgO	0.01	0.04	0.01	0.02	0.02	0.02	0.05	0.02	0.04	0.05	0.04
CaO	34.33	33.84	33.82	32.20	34.89	34.26	33.98	34.02	35.62	31.48	34.50
Na_2O	0.01	0.02	0.02		0.01	0.02	0.01		0.02	0.02	0.02
K_2O	0.05		0.01		0.05	0.05	0.08	0.03	0.03	0.08	
SO_3	0.41	0.26	0.39	0.43	0.06	0.19	0.14	0.20	0.30	0.48	0.37
Total	77.49	74.15	77.32	66.11	73.94	71.59	72.31	72.58	71.97	62.21	72.25
Ca/Si	0.87	0.91	0.85	1.05	0.99	1.01	0.99	0.98	1.12	1.20	1.05

A: Optically amorphous, W: poorly crystalline, C: crystalline

*: Replacing C-S-H gel in the cement paste

3.3.3 ASR gels inter-replacing C-S-H gels

The final stage of alteration of ASR gels is characterized by the replacement of C-S-H gels at highest Ca/Si ratios up to 1.2. Such alteration has been scarcely documented in the deteriorated field concretes, except for a laboratory test to alter the ASR gels into C-S-H gels [11].

ASR gels invade the cement paste and texturally replace the C-S-H gels at a Ca/Si ratio 1.0-1.2, leaving pseudomorphs of hydrated alite and belite (Table 9, Fig.4). This process is an inter-replacement between the two gels, because ASR gels are chemically replaced by C-S-H gels, transforming into C-S-H gels. ASR gels also replace the cement paste filling dissolution vugs, which present a characteristic texture whose original outlines are obscured by the gels.

The Ca/Si ratio of 1.0-1.2 is important because it possibly represents an "equilibrium ratio", at which leaching of calcium from C-S-H gels and takeup of calcium by ASR gels are apparently balanced. That is, C-S-H gels leach calcium down to a Ca/Si ratio of 1.2 during hydration and weathering (Tables 5,6), while ASR gels absorb calcium up to this ratio during alteration into C-S-H gels (Tables 7-9). This ratio is also significant because it represents a "missing-link" between the typical ASR gels and the C-S-H gels in concretes, corresponding to products of a pozzolanic reaction[9].This means the formation of harmless, pozzolanic C-S-H gels occurred in weathered deteriated concretes.

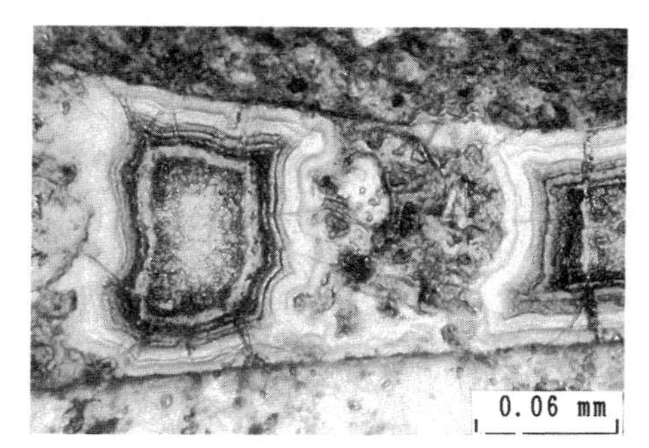

Fig.3 ASR gel lining dissolution vugs

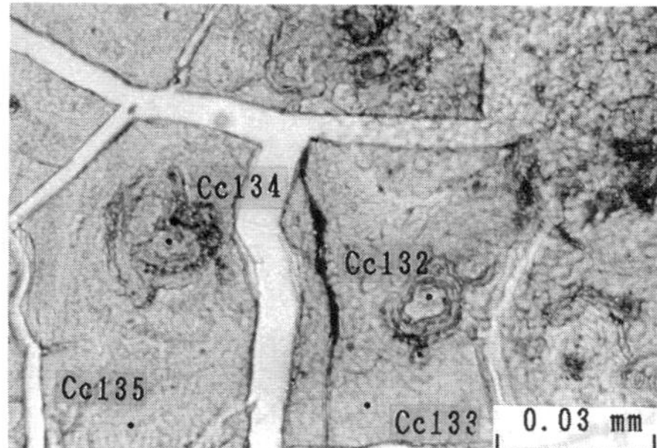

Fig.4 ASR gel replacing C-S-H gels

Table 9. ASR gels replacing C-S-H gels of alite "pseudomorphs" in NF concrete

	Water Street (block)				Tors Cove (block)				Cape Broyle (core)			
	"alite"		ASR gel		"alite"		ASR gel		"alite"		ASR gel	
	Fb247	Fb249	Fb253	Fb243	Tb152	Tb151	Tb110	Tb114	Cc132	Cc134	Cc133	Cc135
	A	A	A	A	A	A	A	W	W	W	C	C
SiO_2	24.07	25.31	30.35	30.58	30.82	34.58	33.94	34.17	35.35	38.06	37.45	38.75
TiO_2	0.20	0.29		0.01	0.64	0.65	0.02				0.05	0.05
Al_2O_3	1.97	2.44	2.44	2.38			1.08	0.49	0.58	0.28	0.08	0.05
Cr_2O_3							0.04	0.05	0.02			0.06
Fe_2O_3	0.70	0.88	0.09	0.03					0.02	0.02		
Mn_2O_3				0.09			0.08		0.05			
MgO	1.92	3.28			0.17	0.09	0.07	0.04		0.01	0.01	0.01
CaO	30.53	28.02	33.52	30.50	32.27	31.08	35.27	35.01	39.99	41.25	38.74	36.72
Na_2O	0.02				0.08	0.15	0.06	0.11				
K_2O		0.01	0.01	0.03	0.10	0.12	0.09	0.12	0.05	0.11	0.13	0.30
SO_3	0.02	0.03	0.02	0.05			0.03	0.07			0.02	
P_2O_5	0.05	0.15	0.01		0.05	0.01	0.06		0.24	0.26	0.13	0.15
Total	59.39	60.40	66.44	63.66	64.13	66.67	70.73	70.07	76.29	79.99	76.62	76.08
Ca/Si	1.36	1.19	1.18	1.07	1.12	0.96	1.11	1.10	1.21	1.16	1.11	1.02

A: Optically amorphous, W: poorly crystalline, C: crystalline

4 Conclusion

Petrographic study coupled with EPMA analysis, revealed that a full spectrum of gel products in the Ca/Si ratio, ranging from C-S-H gels to ASR gels, are present in the deteriorated Newfoundland concretes. Intermediate gel products, corresponding to a pozzolanic reaction, were formed at an apparent equilibrium between the C-S-H and ASR gels, the former leaching calcium and the latter absorbing such liberated calcium in these concretes. All these products were formed through the combined alteration of concretes, due to alkali-silica reaction and freeze/thaw weathering which is most severe in Newfoundland.

5 Acknowledgements

Mr.Keith Foster and Mr.Tom Ring of the Department of Works, Services and Transportation, Government of Newfoundland and Labrador, are acknowledged for providing the authors with convenience for core sampling. Mr.Colin Keane and Mr.Mike Blyde helped in the collection of core samples. Dr.Eungi Min Haga of the Sumitomo Cement Co.,Ltd. prepared a Bence/Albee program for EPMA analysis.

6 References

1. Bragg,D.J.and Foster,K.(1992) Relationship between petrography and alkali-reactivity testing, samples from Newfoundland, Canada. Proc. 9th intern. conf. on alkali-aggregate reactions in concrete, London, pp.127-135.
2. Bragg, D.J.and Katayama,T.(1993) Alkali-aggregate reactivity of concrete structures on the Avalon Peninsula. Field trip guidebook, prepared for CSA subcommittee meeting on cement-aggregate reactivity, Newfoundland.
3. King,A.F.(1990) Geology of the St.John's Area. Geological Survey Branch, Department of Mines and Energy, Government of Newfoundland and Labrador, Report 90-2.
4. Lewczuk,L.,DeMerchant,D.and MacNeill,G.S.(1990) Identification and characterization of alkali silicate reactive argillites and greywackes in Canada Proc. Canadian developments in testing concrete aggregates for alkali-aggregate reactivity. Ministry of Transportation, Ontario, Report EM-92, pp.60-80.
5. Maki,I.and Goto,K.(1982) Factors influencing the phase constitution of alite in Portland cement clinker.Cement and Concr.Res., Vol.12,pp.301-308.
6. Kristmann, M.(1978) Portland cement clinker. Mineralogical and chemical investigations, Part 2: Electron microprobe analysis. Cement and Concrete Research, Vol.8,pp.93-102.
7. Boikova,A.I.,Gristchenko,L.,V.,Nilova,G.P.and Domansky,A.I.(1980) Belites of complex composition. Tsement, 1980, no.7, pp.9-11 (in Russian).
8. Brunauer,S.and Kantro,D.L.(1964) The hydration of tricalcium silicate and beta-dicalcium silicate from 5°C to 50°C. in Chemistry of Cement (ed. H.F.W.Taylor). Academic Press. Vol.1,pp.287-309.
9. Urhan,S.(1987) Alkali-silica and pozzolanic reactions in concrete. Part 1: Interpretation of published results and an hypothesis concerning the mechanism. Cement and Concrete Research, Vol.17,pp.141-152.
10. Katayama,T. and Futagawa,T. (1989) Alkali-aggregate reaction in New Brunswick, Eastern Canada - Petrographic diagnosis of the deterioration. Proc. 8th Intern. conf. on alkali-aggrgate reaction, Kyoto, pp.531-536.
11. Dent Glasser,L.S.and Kataoka,N.(1982) On the role of calcium in the alkali-aggregate reaction. Cement and Concrete Research,Vol.12,pp.321-331.

115 STUDY OF AUTOCLAVE TEST METHOD FOR DETERMINING ALKALI SILICA REACTION OF CONCRETE

T. WANG
Chuken Consultant Co. Ltd, Osaka, Japan
S. NISHIBAYASHI
Tottori University, Tottori, Japan
K. NAKANO
Sumitomo Osaka Cement Co. Ltd, Osaka, Japan
Q. BIAN
Nanjing Institute of Chemical Technology, Nanjing, China

Abstract
The optimum experiment conditions for the autoclave was discussed which influences obviously expansion of concrete due to alkali silica reaction (ASR). The effects of the type and content of reactive aggregate, alkali content, water-cement ratio and storage temperature on ASR expansion were investigated. The expansion characteristics of concrete for the autoclave were also compared with that of exposure to the natural environment. Purpose of this study was to develop a useful rapid method for determining the expansion characteristics and degree of alkali aggregate reaction of concrete.
Keywords: Alkali content, alkali silica reaction, autoclave test, reactive aggregate, steam pressure, storage temperature, treatment time, water–cement ratio

1 Introduction

In order to examine the expansion characteristics and the degree of cracking of concrete subjected to alkali silica reaction, it was necessary to study a rapid test method for various factors and under several conditions. Some mortar bar methods with rapid test were proposed by some authors such as Chatterji (saturated NaCl solution, 50℃)[1], Nishibayashi (added NaOH in bar, autoclaved at 125 ℃ for 4 hours)[2] and Tang (steam cured, then autoclaved in KOH at 150℃ for 6 hours)[3] etc. in the literature. These methods have been widely adopted to determine the experiment conditions and the degree of alkali silica reaction with several factors. However, proposed test methods using an autoclave on concrete were fewer. To take into consideration the different mixture conditions from mortar and concrete which influence results of testing, there is an urgent need to develop a method with rapid tests, in order to directly determine the experiment conditions and the degree of alkali silica reaction of concrete.

This paper presents the results of tests to determine the optimum experiment conditions for the autoclave of concrete and the evaluation of the effects of this rapid test method.

Concrete Under Severe Conditions: Environment and loading (Volume Two) Edited by K. Sakai, N. Banthia and O.E. Gjørv. Published in 1995 by E & FN Spon. ISBN 0 419 19860 1

2 Experiments

2.1 Concrete materials

A ordinary portland cement (made in Japan) with an alkali content of 0.5% (Na_2O equiv.) of the mass of cement was used.

The aggregates used in the testing program included three types of reactive pyroxene andesite (Types T1, T2 and O), a non-reactive natural sand (Type NS) and a crushed non reactive coarse aggregate (Type NT). The reactive fine aggregate was from crushed reactive aggregate T2. These aggregates are characterized in Table 1.

NaOH (reagent, JIS), dissolved in the mixing water was used as an additive.

Table 1 Physical and Chemical Properties of Aggregate (JCI AAR-1)

Kind and symbols of aggregate	Type	F.M	Density	Water absorption (%)	JCI chemical mehtod (m mol / L)		Sc/Rc
					Rc	Sc	
Reactive T1	Pyroxene andesite	6.52	2.60	1.93	101.0	558	5.53
Reactive T2	Pyroxene andesite	6.53	2.64	1.59	67.5	301	4.46
Reactive O	Pyroxene andesite	6.54	2.25	1.81	177.0	732	4.14
Non reactive NT	Sandstone	6.64	2.70	0.65	21.5	30	1.39
Non reactive NS	Crushed sand	2.79	2.67	1.40	----	----	----

2.2 Testing plan

The alkali at rates of 1.0, 1.5, 2.0, 2.5, 3.0 and 3.5% of the mass of cement were added to the cement. Reactive coarse and fine aggregates of 100% were used in the concrete mixtures. The concrete specimens were 75 by 75 by 400-mm prisms. A 350 kg/m^3 cement content and water-cement ratios of 0.54, 0.58 and 0.62 were cast in several mixture compositions, and three concrete prisms were made from each mixture. In the autoclave, steam pressures of 0.10, 0.15, 0.20, 0.25, 0.30 MPa and treatment times of 1, 2, 4, 6, 8, 12 hours were used in the study.

For all series, the mixtures were proportioned to have a slump of 12 to 15 cm without the addition of NaOH, and adjustments for variations in slump due to the addition of excessive alkali were not carried out. A marking code of reactive aggregate proportion was used, for example, 100/100, which specified the ratio of reactive coarse to fine aggregate.

2.3 Test procedures

The concrete prims were made according to JCI AAR-3. To measure length, plugs for measuring length variation were arranged in the concrete specimens. After molding, the specimens were placed in a moist cabinet for 24 hours (20℃, R.H.60%). Thereafter, the initial reading of length was taken, and the specimens were autoclaved at several pressures and treatment times (at about 130 ℃). The equipment developed for the autoclave test is shown schematically in Fig.1. Immediately after the autoclave, the change in length was measured, and then the specimens were stored in curing tanks at 20, 40 and 60 ℃ and R.H.100 %. The concrete specimens not subjected to the autoclave test were placed outdoors (Tottori, Japan).

The specimens were measured for length variation after curing for 24 hours and

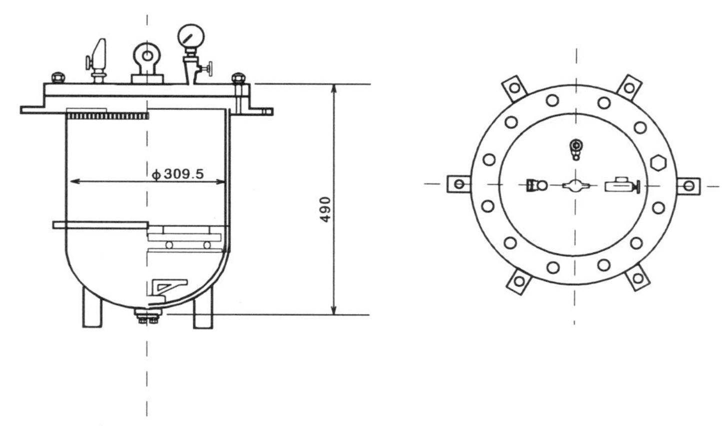

Fig. 1. Autoclave equipment in ASR test of concrete.

Table 2 Testing Plan and Conditions

Reactive aggregates	T1, T2, O	Size of specimen (mm)	75×75×400
Non reactive aggregates	NT, NS	Kind of added alkali	NaOH
Max.size of aggregate (cm)	20	Added alkali contents (Na_2O eq. %)	1.0, 1.5, 2.0, 2.5, 3.0, 3.5
Water cement ratios (%)	0.54, 0.58, 0.62, 0.68	Pressures in the autoclave (MPa)	0.10, 0.15, 0.20, 0.25, 0.30
Unit weight of cement (kg/m³)	350	Treating times (h)	1, 2, 4, 6, 8, 12
Alkali content in cement (%)	0.5	Storage conditions	20, 40, 60℃ R,H.100% after autoclave, outdoors
Reactive aggregate contents(%)	0, 100	Items of measurement	Length
Slump (cm)	12~15		

at the age of 0.5, 1.0, 1.5, 2.0 and 2.5 months and once every month thereafter. The testing plan and conditions are given in Table 2.

3 Discussion

3.1 Effect of reactive aggregate

Figure.2 shows the relationship between expansion and treatment time for different ratios of reactive aggregate. For a pressure of 0.2 MPa and an alkali content of 2.5%, the expansion increased with the increase in the treatment time. When a 100 % reactive fine aggregate was used, the expansion was larger than that obtained with other ratios.

Figure.3 shows the relationship between expansion and alkali content at these reactive aggregate contents. For 4 hours of treatment in the autoclave, the expansion was the greatest when 100% reactive fine aggregate was used, and it increased with the increase in the alkali content. Immediately after the autoclave, the concrete prisms expanded more than 0.10%, which is typically deemed harmful for 0/100 and 100/100 concrete specimens, and the largest expansion exceeded 0.40 %. The expansion did not increase

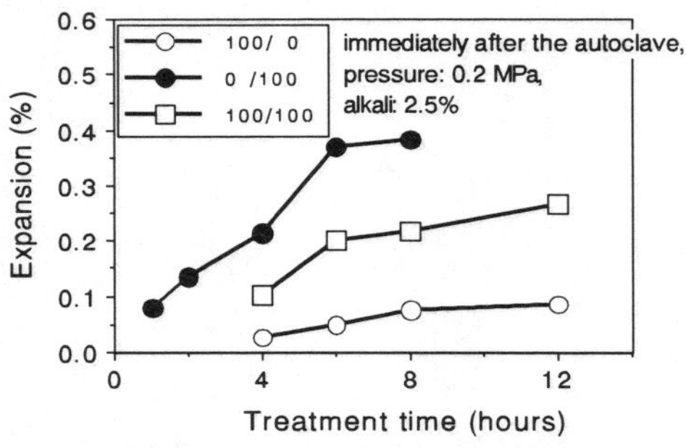

Fig. 2. Relationship between expansion and treatment time with ratios of reactive aggregate.

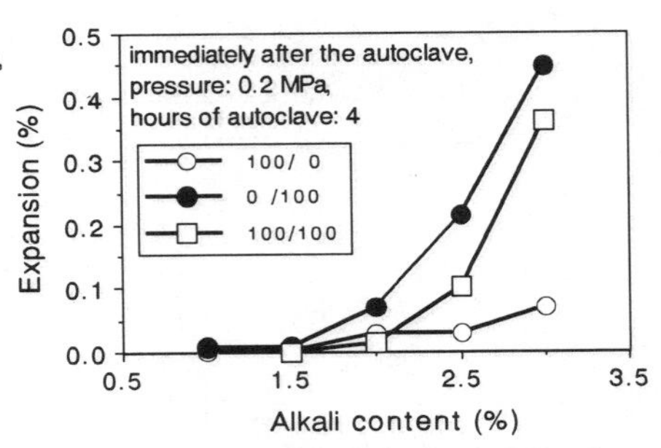

Fig. 3. Relationship betweeen expansion and alkali content with ratios of reactive aggregate.

with the increase in the reactive aggregate content and there is a pessimum content on expansion for reactive aggregate. This may have been due to specific alkali / reactive silica ratios with the aggregate and alkali used for the mixture proportion in this test.

The relationship among expansion, treatment time and alkali content in concrete for three kinds of reactive aggregates is given in Figs.4 and 5. The expansion of specimens increased with increasing treatment time and alkali content. As to the effect of the kind of reactive aggregate, the expansion differed with the kind of reactive aggregate. The concrete specimens with type O aggregate had a marked effect in the autoclaving, and a higher expansion occurred than with other kinds of aggregate. The concrete specimen with type T1 aggregate had a small expansion after 6 hours treatment and did not expand below 3.0 % alkali.

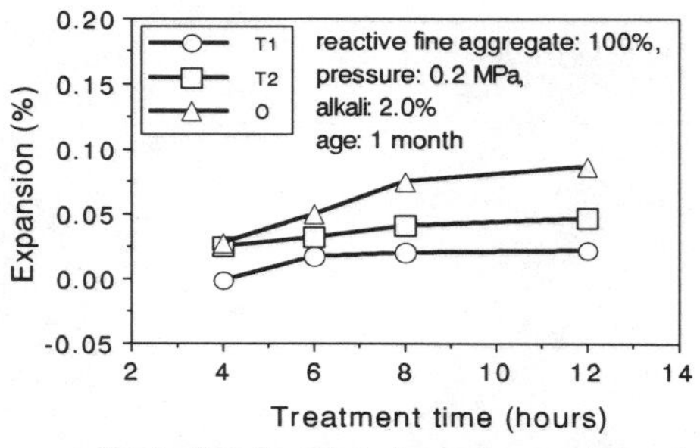

Fig. 4. Relationship between expansion and treatment time with three aggregates.

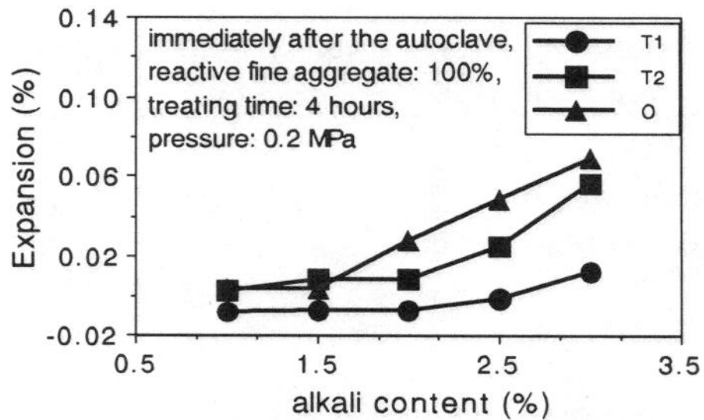

Fig. 5. Relationship between expansion and alkali content with three aggregates.

3.2 Effect of treatment time and pressure

The autoclave test is a method which rapidly determines the results of alkali silica reaction. Treatment time in the autoclave is a critical factor which influences expansion. The effect of treatment time on expansion is shown in Figs.6 and 7. Figure.6 suggests that the curves of expansion have approximately the same tendency. The expansion did not continue to increase when the treatment time was raised to 8 hours, but it reached a maximum at 4 hours. It was concluded that a rather long period of treatment in the autoclave was not effective for achieving maximum expansion. For a pressure of 0.2 MPa and an alkali content of 2.0 %, the concrete specimen treated for 6 hours experienced the greatest expansion, followed by 1 hour and 4 hours. Figure.7 suggests that the effect of treatment time on expansion differed with the alkali content, and the expansion increased with the increase in the alkali content, however, it did not increase with an increasing period of treatment in the autoclave. The results indicated that the applicable treatment times are about 2 hours below 1.5% alkali, 6 hours at 2.0 or 2.5% alkali and 4 hours at 3.0% alkali, at which point, the expansion reached a maximum value. The effect of accelerating expansion in the autoclave was not marked below 1.5% alkali.

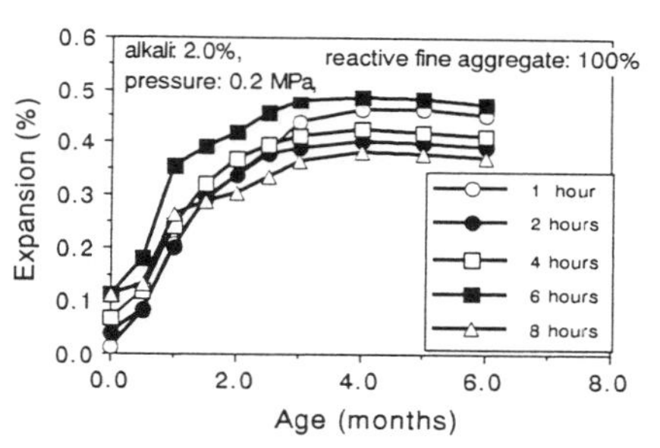

Fig. 6. Effect of treatment time on expansion.

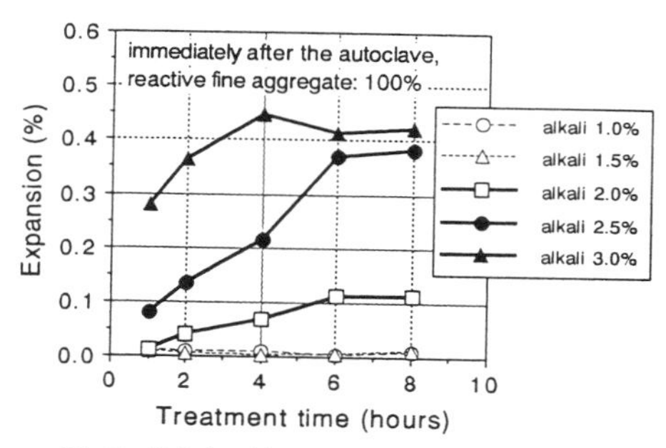

Fig. 7. Relationship between treatment time and expansion.

Figure.8 shows the expansion versus time development for different pressures (4 hours of treatment, 3.0% alkali). The expansion of concrete without autoclave treatment was smaller than that obtained by the autoclave. The expansion reached up to 0.3% immediately after the autoclave. However, the expansion does not increase with the rise in the pressure. The expansion did not reach to the highest value, even though a pressure as high as 0.3 MPa was used. The concrete specimen autoclaved at a pressure of 0.2 MPa experienced a higher expansion than at other pressures. Thus, there is a pessimum expansion at about 0.2 MPa when 4 hours treatment was carried out. This may have been due to a rapid alkali silica reaction caused by this pressure, and the pessimum behavior on expansion is applicable to steam pressure. Figure.7 and 8 suggest that the pressure of 0.2 MPa and 4 to 6 hours of treatment are appropriate testing conditions for this autoclave.

3.3 Effect of alkali content

Figure.9 shows the expansion versus time development for several alkali contents. The expansion increased with the increase in the alkali content. With the exception of 3.0% alkali, the expansion reached a maximum value after 4 months. The concrete specimen with 3.0% alkali had an after-expansion after 6 months. These concrete specimens experienced a rapid early expansion up to 2 months at all alkali contents.

The effect of alkali content on expansion with several treatment times is given in Fig.10. The expansion increased with increasing alkali content. However, the alkali contents at which expansion reaches a maximum value differed with the treatment

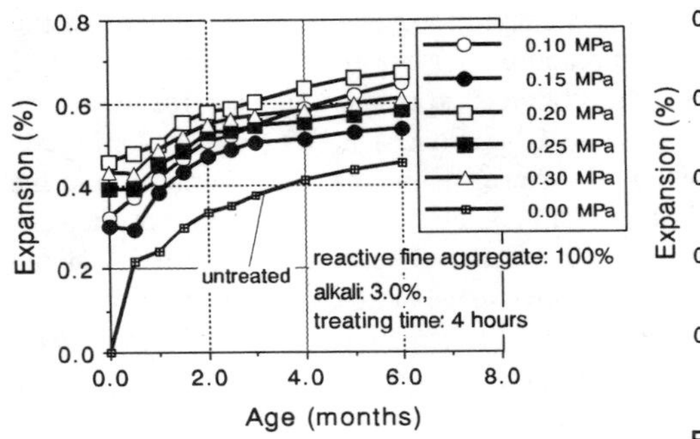

Fig. 8. Effect of pressure in the autoclave on expansion.

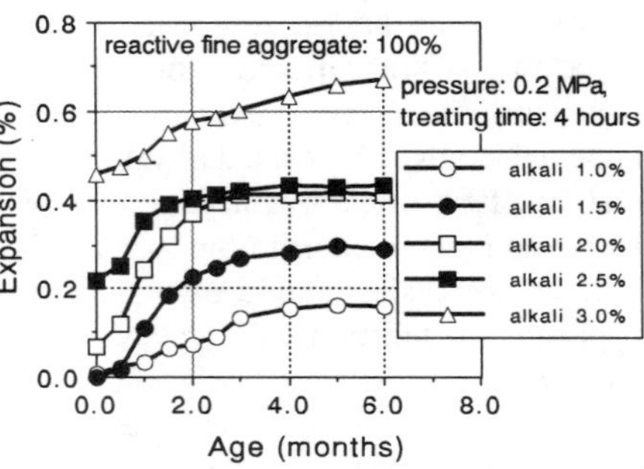

Fig. 9. Effect of alkali content on expansion.

periods. The pessimum value on expansion for alkali content was not marked here. It is suggested that the alkali contents at which expansion reached a maximum value were 1.0% for 2 hours of treatment, 1.5% for 1 hour, 2.0 and 2.5% for 6 hours and 3.0% for 4 hours.

It has been reported that the pessimum value on expansion for alkali content is about 2.0% (100% reactive fine aggregate, 40℃ storage and without autoclave)[4]. The expansions of concretes from autoclave treatment and 40℃ storage had approximately the same change tendency. However, the results from the autoclave indicated that the concept of the pessimum limit for alkali content is not applicable here. It is understood that the pessimum behavior may not be applicable to the autoclave treatment of concrete with alkali below 3.0% .

3.4 Effect of water-cement ratio and storage temperature

Figure.11 shows the relationship between expansion and water-cement ratio. The expansion increased with the decrease in the water-cement ratio and with increasing alkali content. It is deemed that the large amount of holes formed in hardened cement paste when increasing the water-cement ratio, and the gel (reactive compound from ASR) fills into these holes. Thus, the pressure from expansion was decreased. In addition, the expansion had a pessimum value at 3.0% alkali when a water-cement ratio of 0.54 was used.

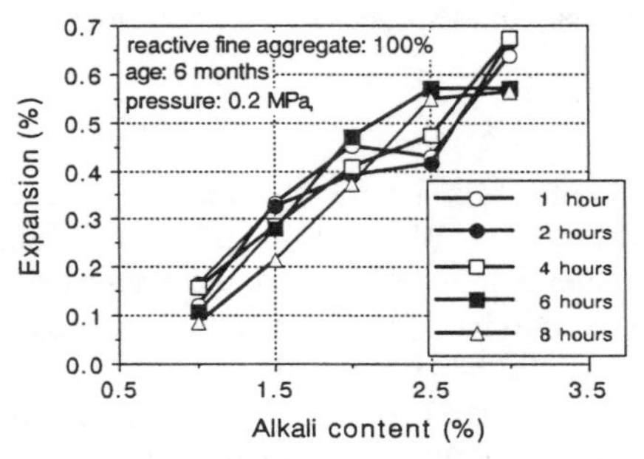

Fig. 10. Relationship between expansion and alkali content with several treatment times.

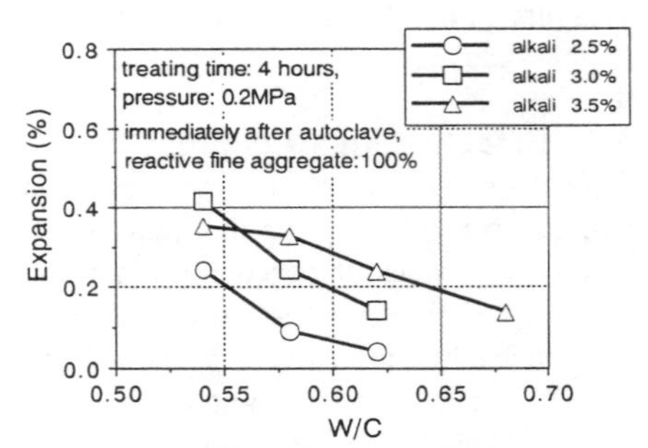

Fig. 11. Effect of water cement ratio on expansion with several alkali contents.

The relationship between expansion and storage temperature after the autoclave is given in Fig.12. The expansion increased with the rise in the temperature and with increasing alkali content. The expansion of concrete stored at 60 ℃ was larger than that obtained at 40 and 20 ℃. The concrete achieved expansion of 0.35% at 6 months. The alkali silica reaction was accelerated with rising temperature. However, this result from concrete at higher alkali content was different to that obtained from a mortar test. It is reported that predominantly raised temperatures caused increased and accelerated expansion; in the case of a mortar bar, the 40℃ cured specimen achieved a higher value[5].

3.5 Acceleration of expansion effect in the autoclave

The expansion of concrete due to ASR was accelerated with the autoclave test. Figure.13 shows a comparison between the expansions in the autoclaved and in a natural environment. The concrete in the natural environment had a tendency of long term and slower expansion. In addition, the concrete with the autoclave had a rapid expansion. The expansion of concrete at 5 months after the autoclave was almost the same as that at 12 months outdoors. From these curves of expansion, it is concluded that elevated temperatures and pressures in the autoclave increased the initial rates of concrete expansion, but they decreased the total expansion.

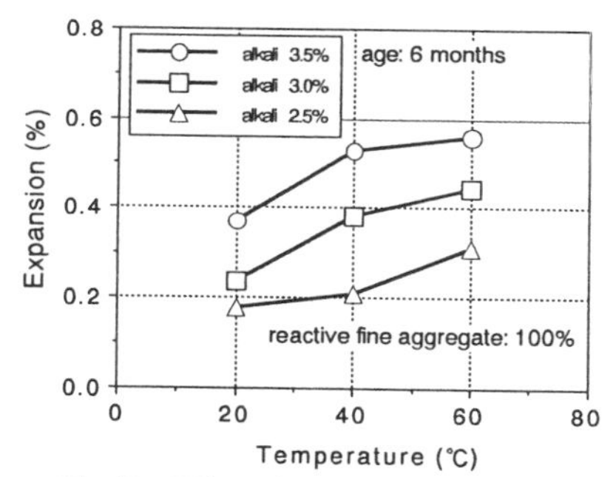

Fig. 12. Effect of storage temperature on expansion after the autoclave

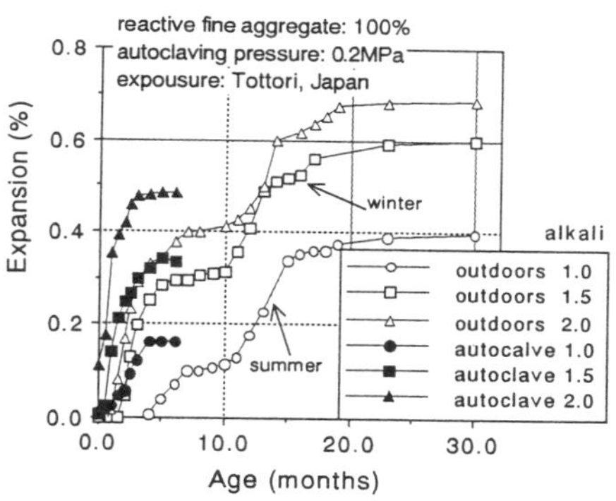

Fig. 13. Comparison between the expansions under the two conditions.

3.6 Reliability analysis of data

Figure.14 provides coefficients of variation which show the dispersion of the measured value immediately after the autoclave. The coefficient of variation is defined as that the

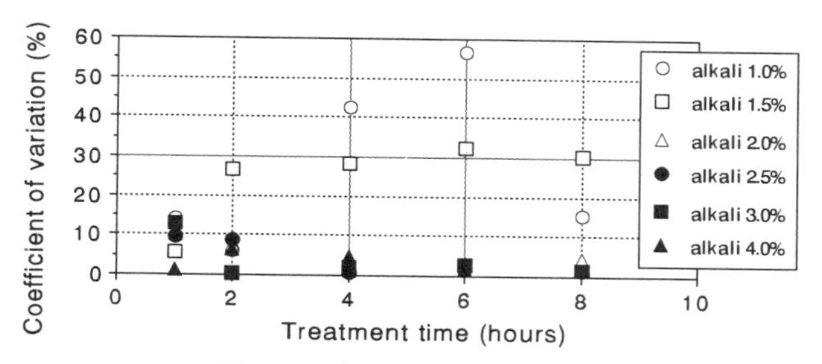

Fig. 14. Coefficient of variation.

standard deviation value divides by the average value for expansion of three specimens. The coefficient of variation was below 0.15, and it has a high reliability with alkali over 1.5%. In the case of alkali below 1.5%, it is understood that the coefficient of variation became greater, because the value of expansion measured was lower (Fig.7). However, the standard deviation was almost the same as that of the higher alkali content, that is to say, the measurements were also reliable.

4 Conclusion

The followings can be concluded from the results of the autoclave tests on concrete made with various alkali contents, water-cement ratios at various temperatures, pressures and treatment times:

1) The expansion differed with the kind and ratio of reactive aggregate. There were pessimum expansion for reactive fine aggregate at 100% and a for a treatment pressure of 0.2 MPa.
2) An applicable treatment pressure in the autoclave was about 0.2 MPa. Applicable treatment times were 4 hours at 3.0% alkali, 6 hours at 2.0 and 2.5% alkali and about 2 hours at 1.0 and 1.5% alkali.
3) The expansion had a pessimum value for the treatment pressure of about 0.2 MPa. The expansion increased with the decrease in the water–cement ratio and with the rise in the temperature. The concrete stored at 60°C expanded more than that at 40 and 20°C.
4) The expansion was accelerated with autoclaving. The expansion of concrete at 5 months after autoclave treatment was about the same as that obtained in specimen left 12 months in a natural environment.

This autoclave test method was developed at Tottori University in Japan from 1992 to 1993. The authors would like to appreciate the research fellows for their enthusiastic cooperation in the study.

5 References

1. Chatterji,S. (1978) An Accelerated Method for the Detection of Alkali-Aggregate Reactivity of Aggregate, Cement and Concrete Research, Vol. 8, pp. 647-649
2. Nishibayashi,S.,Yamura,K. and Matsushita,H. (1987) A Rapid Method of Determining the Alkali-Aggregate Reaction in Concrete by Autoclave. In: Alkali Aggregate Reaction. (Proc. 7th Int. Conf. on AAR, Ottawa, Canada). Noyes, Park Ridge, NJ,USA, pp.299-303
3. Tang,M., Han,S.F. and Zhen,S.H. (1983) A Rapid Method for Identification of Alkali Reactivity of Aggregate. Cement and Concrete Research, Vol.13, pp.417-422
4. T.Wang, S.Nishibayashi, K.Nakano and A.Yoshino (1994) Effect of Alkali Content on Expansion due to ASR in Concrete. Proceedings of the Japan Concrete Institute, Vol.16, No.1, pp.1067-1072
5. K.Nakano, S.Kobayashi, A.Nakaue and H.Ishibashi (1986): Effect of Alkali Content and Curing Conditions on Expansion of Mortar. Cement Technology Annual (Japanese). Vol. 40, No.73 pp.316-319

116 STRENGTH AND COMPOSITION CHANGES IN CARBONATED GBFS AND OPC MORTARS DURING EXTENDED WATER CURING

V.E. PENTTALA and S.P. MATALA
Helsinki University of Technology, Helsinki, Finland

Abstract
The secondary hydration of the binders after carbonation of granulated blast furnace slag concrete transforms the binder matrix into a carbonated concrete where capillary porosity diminishes significantly. After an extended water curing of 5 months, the compressive strength of the mortar increased by 26% to 44% when OPC was used as binder; with GBFS mortars the strength increase was from 43% to 120%, depending on the composition and water binder ratio. The products of the secondary hydration fill up the previously (before carbonation) capsulated layered pore structure of slag mortars and is the main cause for the strength increase. New hydration products, such as magnesium aluminate silicate hydrate, mangnesium aluminate hydrate and strätlingite may also contribute to the improved properties, which was studied by low temperature calorimetry, mercury intrusion porosimetry, thermogravimetry, infra-red spectrometry, X-ray diffractiometry, and environmental scanning electron microscopy together with EDX analysis in which different concrete phases were determined by a statistical principal component analysis. The healing effect of water curing after carbonation can be applied in renovation of concrete and reinforced concrete structures, and when the method is applicable, it will provide a new and cost efficient repairing procedure.
Keywords: Carbonation, compositional phase analysis, compressive strength, curing, porosity, portland cement, slag, thermoporosimetry.

1 Introduction

Carbonation has been observed to coarsen and open the pore structure of slag concrete. This phenomenon is harmful for the design of durable reinforced concrete

Concrete Under Severe Conditions: Environment and loading (Volume Two) Edited by K. Sakai, N. Banthia and O.E. Gjørv. Published in 1995 by E & FN Spon. ISBN 0 419 19860 1

structures in which granulated blast furnace slag offers in many cases an excellent binder choice due to its various good properties, cost savings, chemical resistance, especially the sulphate resistance. Due to the low heat of hydration, it is an appropriate alternative in massive structures. In order to improve the properties of carbonated granulated blast furnace slag concretes, it was decided to study the effects of extended water curing after carbonation on compressive strength, porosity, and composition of slag and OPC mortars.

2 Materials and sample preparation

The paste and mortar samples were produced at three different water binder ratios, 0.35, 0.45, and 0.6, and the aggregate binder ratios of the mortar test specimens were 1.22, 1.35, and 1.45, respectively. The slag contents of the test samples were 0, 50, and 70% of the total binder amount. The ordinary portland cement had a prism compressive strength of 51.0 MPa at the age of 28 days. The specific surface area of both binder types was 400 m^2/kg (Blaine). The chemical composition of the binders is presented in Table 1. The maximum aggregate size in the mortar specimens was 2 mm and the mineralogical composition was quartz.

Table 1. Chemical and mineralogical composition of the binders

	Chemical composition [%]		Mineralogical composition of the cement [%]	
	Cement (OPC)	Granulated blast furnace slag		
SiO_2	20.4	36.1	C_3S	46.07
Al_2O_3	4.7	9.9	C_2S	23.74
Fe_2O_3	2.7	0.49	C_3A	7.88
MgO	2.4	10.9	C_4AF	8.21
CaO	62.0	38.5		
Na_2O	0.77	0.4		
K_2O	0.99	0.5		
S	1.1	1.6		
LOI	2.8			

Mixing time of the paste and mortar batches was 4.5 minutes. A sulfonated naphthalene condensate was used as a superplasticizer, and it was added into the mixer after a mixing time of 1 minute. After mixing, the pastes were poured into plastic bottles which were rotated for 8 hours to avoid segregation. At the age of 24 hours, the bottles were demoulded, and the paste specimens were placed in a curing chamber (20° C and over 95% relative humidity) for six days. Mortar samples were prepared according to Finnish standard SFS-EN196-1 using prism moulds 40 x 40 x 160 mm. Demoulding time and precuring conditions up to seven days were the same as with pastes.

At the age of seven days, ochtaedhral specimens were cut from the center line of the paste cylinders and mortar prisms in order to obtain as homogeneous specimens as possible. The size of the specimens was 12 x 12 x 70 mm after having cut off corners of 1.5 mm. Specimens were stored for 15 months at 45% RH and in a normal CO_2-

content atmosphere. Temperature during the carbonation process was 20° C. The first test series was performed after this treatment. For the second test series, similar samples were treated in vacuum for two days and, thereafter, saturated in distilled water and stored in water for 11 months. The specimen size for the low temperature calorimetry (LTC), thermogravimetry (TG), and infra-red spectrometry (FTIR) tests was 12 x 12 x 70 mm after having cut off corners of 1.5 mm. The specimens for the environmental scanning electron microscope (ESEM) and the EDX phase map analysis were cut and polished from the before mentioned test specimens. The sample size for the mercury intrusion porosimetry tests was 12 x 12 x 35 mm. For compression strength tests, 40 mm cubes were cut from the mortar prisms stored for 34 months at 45% RH and in a normal CO_2-content atmosphere. The first test series was tested immediately and the second series after the water immersion time of 5 months.

3 Test methods

For mercury intrusion porosimetry (MIP) tests, three samples for each curing procedure and for each mix were prepared. After the curing regimes, the samples were dried in vacuum for six weeks and then tested in a porosimeter where the maximum mercury pressure was 200 MPa. The contact angle and surface tension was 141.2° and 0.48 N/m, respectively. The minimum measurable pore radius was 3.7 nm.

For another porosity test method, low temperature calorimetry was used. After the immersion time of 5 days or 11 months in water, the samples were weighed in both water and air in a saturated surface dry state. For each mix proportion, one sample was tested in a calorimeter. The microcalorimeter, model CALVET BT 2.15, was used for calorimetric measurements. The test procedure in the calorimetric measurements is presented in detail by Matala in [1]. After the test, the sample was weighed and cut into two parts. One half of the sample was ground immediately for TG+EGA and FTIR tests. The other half of the sample was dried in a ventilated oven at 105° C to a constant weight in order to measure the total porosity of the sample; thereafter, the sample was ignited at 1000°C for two hours to determine the total amount of non-evaporable water and gases evolved during ignition.

The degree of hydration and of carbonation were studied by a thermogravimeter which was used in combination with a gas analyser. The gases evolving from the samples, H_2O, CO_2 and SO_2, were analysed and measured simultaneously during the TG run by a Fourier transform infra-red spectrometer. The powdered 120 mg sample was first dried in the TG for 20 min at 50° C. Thereafter, the temperature was increased linearly at the rate of 20° C/min up to 110° C, where a 5 min pause took place. Subsequently, the temperature was increased up to 1100° C at the rate of 20° C/min. Helium was used as a protective gas in the TG+EGA test. The collection of IR scans during the TG run, transforming and converting the scans into an absorbance spectrum, and the spectrum manipulation are described in detail by Penttala [2]. The detection limits for H_2O and CO_2 were 0.2-0.4 µg/s and 0.04 µg/s, respectively.

The composition of the pastes was studied by Fourier transform infra-red spectrometry (FTIR) using diffuse reflectance analysis and by X-ray diffraction

(XRD) analysis. In the diffuse reflectance analysis, a powdered sample was mixed with KBr powder, and the spectra were gathered by a liquid nitrogen cooled wide-band mercury cadmium tellurium detector with a resolution of 4 cm^{-1} using 200 scans. In the XRD tests, a Philips PW1710 diffractometer and CuKα radiation were used, the voltage being 40 kV and the current 40 mA.

The SEM images were gathered by an environmental scanning electron microscope. The phase maps were produced from the EDX element maps of the micrographs using statistical principal component analysis. The EDX element micrograph constituted 256 by 256 pixels image. EDX element map technique is a well known investigation method while the phase map technique is a new method where the accuracy of the results is mainly dependent on the sample preparation. It is difficult to polish the samples to adequate evenness and, especially, pores can give rise to erroneous results. With statistical principal component analysis part of the errors caused by the unevenness of the sample surface can be minimized. During the polishing procedure sample surface becomes coated with a carbon layer. In the analysis, the carbon content on the micrographs was calibrated so that the carbon content on the aggregate particle was set to zero.

4 Compressive strength

The compressive strength results after carbonation and after the subsequent water curing for 5 months are presented in Table 2. The results show that the extended water curing of carbonated samples increased the compression strength remarkably. The strength increase correlates both with increasing slag content and with increasing water binder ratio. The hydration degree of the carbonated GBFS mortars where the slag content was 70% and the water binder ratio was 0.45 increased by 80%, and when the w/b ratio was 0.35 the increase was 45%. The corresponding strength increase was 91% and 44%, respectively. The hydration degrees after carbonation of the 70% slag mortars having w/b ratios 0.45 and 0.35 were 0.49 and 0.56, respectively. After the extended water curing, hydration degrees increased to 0.88 and 0.81, respectively. The hydration degree of the carbonated OPC mortar having a w/b ratio 0.45 increased by 42% from 0.61 to 0.87 during the extended water curing, and the compressive strength increased by 30%.

The strength values and the hydration degrees can not be directly compared because the immersion time in water for the TG+EGA samples was 11 months and for the compression only 5 months. In order to study if the observed strength increase due to the extended water curing after carbonation corresponds to the ordinary strength gain in standard water curing for 5 months, we calculated these compressive strength values by comparing the corresponding hydration degrees. These values are presented in parentheses in Table 2. The OPC mortars gain approximately same strength increase in both cases, in standard curing for 5 months and in the case where the carbonated samples were cured in water for same time. However, the strength gain of GBFS mortars in standard curing conditions is clearly smaller than in the situation of extended water curing after carbonation.

Table 2. Compressive strength and carbonated volume of the 40-mm mortar test cubes

Mix composition	Water binder ratio	Carbonated volume [%] after carbon-ation/1100 d	Carbonated volume [%] after ext. water curing/1250 d	Compressive strength [MPa] after carbon-ation/1100 d	Compressive strength [MPa] after ext. water curing/1250 d[1)]	Strength increase [%]
OPC 100%	0.60	82.1	87.1	35.6	51.5 (48.5)	44.7
OPC 95%+Silica 5%	0.60	90.4	71.6	37.7	47.6	26.3
OPC 50%+Slag 50%	0.60	91.6	93.2	32.4	50.9 (42.7)	57.1
OPC 30%+Slag 70%	0.60	98.3	98.0	23.8	52.5 (47.6)	120.6
OPC 95%+Silica 5%	0.45	69.9	57.0	50.5	66.3 (64.3)	31.3
OPC 50%+Slag 50%	0.45	67.0	67.3	49.9	73.0 (56.7)	46.0
OPC 30%+Slag 70%	0.45	72.4	85.7	40.7	77.6 (66.4)	90.7
OPC 30%+Slag 70%	0.35	68.2	72.5	57.7	83.0 (74.9)	43.8

[1)] The values in parentheses correspond to the calculated compressive strength values of standard water cured specimens at the age of 5 months. Calculations are based on hydration degree measurements.

5 Porosity

The effects of extended water curing on the porosity of carbonated test mortars was measured by mercury intrusion porosimetry (MIP) and by thermoporosimetry. With MIP, it is possible to study the changes in capillary porosity, but the method does not give a completely correct pore size distribution. Actually, the measured pore sizes correspond to the neck dimensions between the pores, and therefore, the pore size distribution exaggerates the volume amounts of smaller pore sizes and the distribution needs to be shifted towards larger pore sizes. Pore necks wider than a radius of 4 nm are detectable with this apparatus. Thermoporosimetry gives information about the pore volumes wider than 1.4 nm. The test method is presented in another paper at this conference [3].

The freezing thermograms in Fig. 1 show that carbonation has increased the freezable water amount above -20° C in all mortars. At the same time, gel pore area at -40° C has clearly diminished, especially with the GBFS mortars. After the extended water curing, the first peak above -20° C, which represents the capillary pore range has disappeared. Similarly, this peak has also drastically decreased with OPC mortars. At the same time, the second peak shows that the gel pore range is increasing. The same conclusion can be drawn from Table 3. There, the capillary porosity ($R_p > 4$ nm) in the thermoporosimetry results has almost diasappeared with GBFS mortars, and it has diminished even more with OPC mortars due to the extended water curing. MIP results show the same tendency, but the decrease in capillary porosity of OPC mortars is only from 30 to 50 % and of GBFS mortars from 70 to 80 %.

The changes in gel porosity ($R_p < 4$ nm) were studied by calorimetric measurements. The results of Table 3 show that gel porosities of OPC mortars having w/b ratios of 0.45 and 0.35 have increased from 0.021 cm^3/g_{ssd} to 0.064 cm^3/g_{ssd} and from 0.038 cm^3/g_{ssd} to 0.063 cm^3/g_{ssd}, respectively. Notation g_{ssd} denotes sample weight in grams in the saturated surface dry state. The changes in the gel porosities of GBFS mortars are clearly larger, being from 0.018 cm^3/g_{ssd} to 0.090 cm^3/g_{ssd}, and from 0.014 cm^3/g_{ssd} to 0.051 cm^3/g_{ssd}, respectively. Porosity changes, the increase in

Table 3. The porosity results of the carbonated test mortars before and after the extended water curing

Mix	Water-binder ratio	Vp [cm3/g(ssd)] Thermoporometry, layered pore model						Vp [cm3/g(ssd)] MIP	
		Total porosity		Rp > 1.4 nm		Rp > 4 nm		Rp > 4 nm	
		carb.	carb.+wet	carb.	carb.+wet	carb.	carb.+wet	carb.	carb.+wet
OPC 100%	0.45	0.074	0.079	0.069	0.053	0.053	0.015	0.055	0.038
OPC 30%+Slag 70%	0.45	0.100	0.091	0.090	0.067	0.082	0.001	0.085	0.018
OPC 100%	0.35	0.072	0.069	0.048	0.037	0.034	0.006	0.049	0.023
OPC 30%+Slag 70%	0.35	0.054	0.056	0.047	0.038	0.040	0.003	0.047	0.015

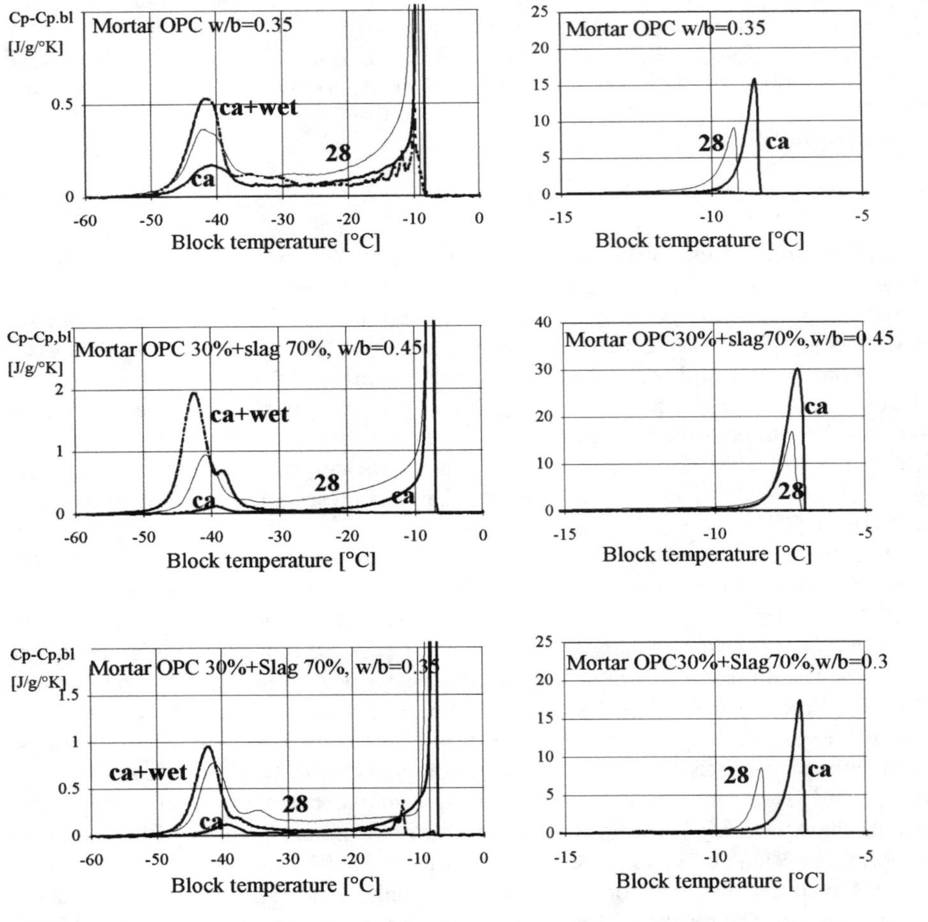

Fig. 1. The freezing thermograms of different mortar samples at the age of 28 days, after carbonation, and after the extended water curing.

gel porosity, and the decrease in capillary porosity explain the observed strength increases in the test mortars. The carbonation of GBFS mortars coarsens and modifies the pore structure, creating new entrances for water to ingress into the vicinity of the non-hydrated parts of matrix and the new hydration products densify the binder matrix. This is the main reason for the remarkable strength increase in carbonated GBFS mortars during extended water curing.

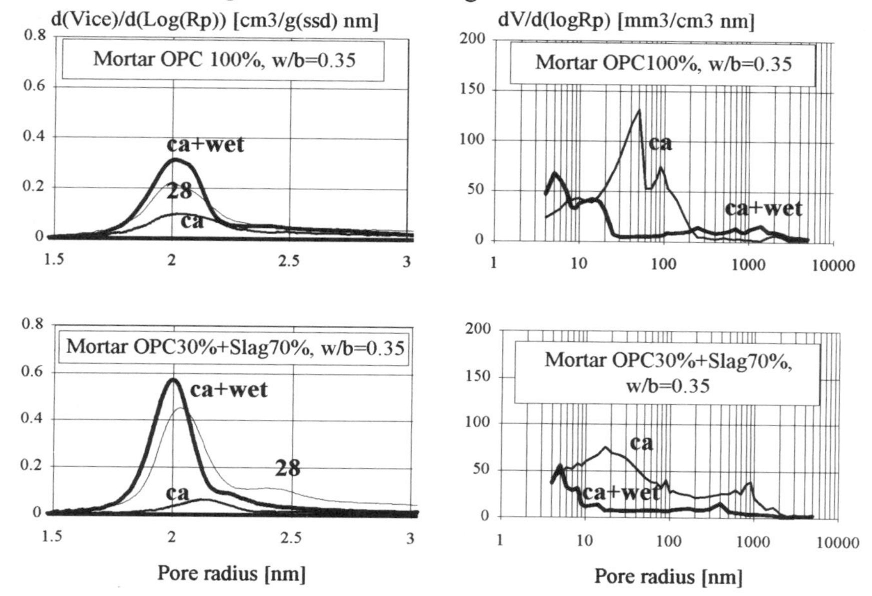

Fig. 2. The pore size distributions of test mortars determined by thermoporosimetry (pore range1.5-3 nm) and by mercury intrusion porosimetry (pore range 4-5000 nm) at the age of 28 days, after carbonation, and after extended water curing.

6 Composition

The changes in the composition of pastes and mortars were analyzed by using thermo-gravimetry together with an evolved gas analyzer (TG+EGA), infra-red spectroscopy (FTIR), and X-ray diffractiometry (XRD). According to the XRD results, some changes in the relative proportions of different carbonates have taken place. In the OPC pastes, vaterite seems to crystallize further forming calcite, but with slag pastes the relative proportion of vaterite increases according to Fig 3. Since this phenomenon was observed also with other samples, the reason for it must lie in the different hydration products of OPC and GBFS pastes and mortars. It can be noticed that the hydrotalcite detectable in GBFS pastes before carbonation has disappeared during the carbonation. It is analogous with the carbonation of carboaluminates. This may indicate that in water storage the CO_3^{2-}-group of hydrotalcite $M_6A\bar{C}H_{12}$ is exchanged with the silicon group of the silicate hydrate formed due to the carbonation of CSH and magnesium aluminate silicate hydrate (e.g. M_6ASH_x), and carbonates precipitate as vaterite. Vaterite precipitates in the narrow pores where the further precipitation of calcite is impossible due to the lack of space and due to the deviating ionic radius of Mg^{2+} ions incorporated in the formed carbonates [6].

These magnesium aluminate silicate hydrates, or possibly also pure magnesium aluminate hydrates, may have the binding properties that are partly responsible for the increased compressive strength properties in GBFS mortars. The secondary hydration together with the new hydration compounds densifies the matrix and increases the strength properties due to the pore filling effect. The results show that carbonation decomposes tri- and monosulphates. Aluminates released from these products or due to the carbonation of the cubic or the hexagonal forms of calcium aluminate hydrates may react with the silicate hydrates formed from the carbonation of CSH gel. They may also react with calcium possibly originating from calcium sulphates, and as a result, hexagonal strätlingite is formed. In the XRD patterns of the slag binder, there exists peaks at 21.2° and 31.1° which may belong to strätlingite C_2ASH_8 and the latter peak to a hydrogarnet with a composition of C_3ASH_4. According to the literature, C_2ASH_8 rarely exists in pure form, and it is relatively unstable in saturated lime solutions but stable in saturated sulfate solutions [4,5]. C_2ASH_8 can be identified by TG since it decomposes at 250° C where the decomposition into hydrogarnet takes place and almost all water is lost between 350-400°C. In the studied cases, the existence of strätlingite is possible in slag concretes after carbonation. In XRD patterns, no sulfate compounds were detected after carbonation. In carbonated pastes and mortars, no lime or CH were found by XRD or by TG. The coexistence of $C_4A\overline{C}H_{11}$ and C_4AH_{13} phases were clear in XRD patterns. Further, the water loss curves in TG runs support the existence of this compound.

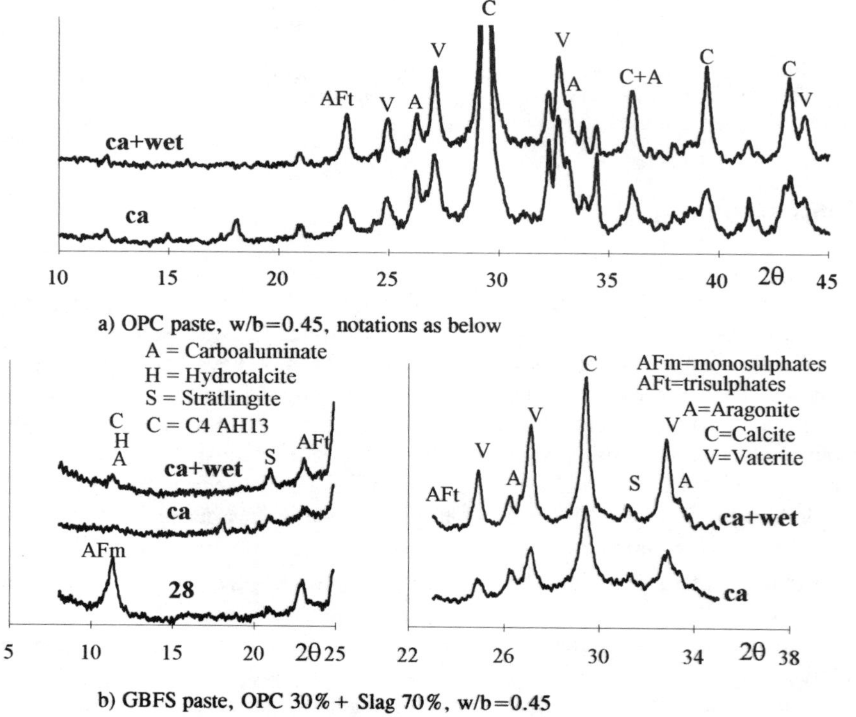

Fig. 3. XRD patterns of the pastes having a water binder ratio 0.45.

Infra-red spectroscopy was used in order to verify qualitatively the changes in the CSH phases during carbonation. The diffusion reflectance analysis by FTIR spectroscopy was performed using powdered samples having the water binder ratios of 0.60 and 0.45 and slag contents 0 and 70%. Figure 4 shows the absorbance bands of OPC and GBFS pastes (water binder ratio 0.45) at a wavenumber range from 600 to 1600 1/cm. The results show that CSH gel is decomposed to SH gel during carbonation. An increase in slag content enhances the decomposition. When carbonated samples were stored in water for 11 months, a new formation of CSH begins and the absorbance bands move back to the typical position of CSH-gel, Fig. 4. The FTIR results also reveal the decomposition of tri- and monosulphates during carbonation. A higher proportion of vaterite was observed in the XRD tests. Indirectly, the same conclusion can be drawn also from FTIR tests by which the relative proportions of calcite+vaterite (band at 879 1/cm) to aragonite (band at 855 1/cm) can be compared during the extended water curing.

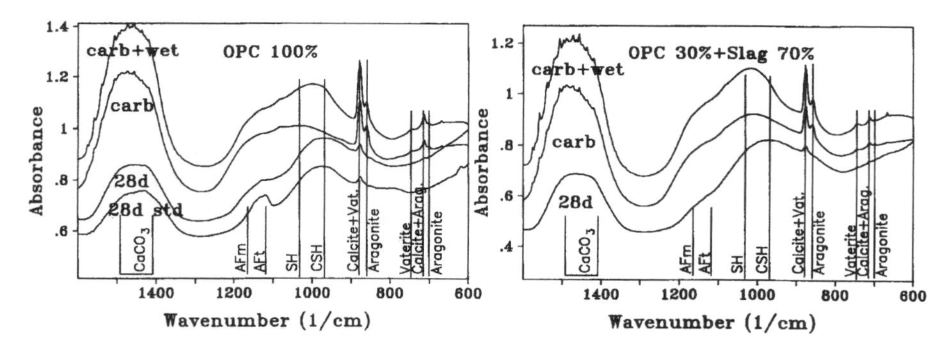

Fig. 4. The FTIR spectra of pastes having a water binder ratio of 0.45.

Table 4. Element weight percents in the different phases of the EDX phase images presented in Fig. 5

Mix composition water binder ratio carb/ext.wet curing	Phase color	Ca/Si	Element weight percent Ca	Si	C	O	Al	Fe	S	K	Mg
OPC 100%	Raster	0.17	9.31	53.3	-	33.0	1.8	0.7	0.7	-	1.2
w/b=0.35	White	1.19	31.3	26.2	3.0	29.5	3.0	1.6	1.6	-	3.8
carbonated	Gray	1.64	38.0	23.1	2.6	28.5	2.7	1.5	1.5	-	2.1
	Black	2.16	43.6	20.2	2.1	27.6	2.2	1.8	1.3	-	1.7
OPC 100%	Raster	0.21	10.9	52.1	-	32.7	1.9	1.0	0.9	0.4	-
w/b=0.35	White	1.67	38.9	23.3	2.0	28.2	2.8	2.2	1.9	0.8	-
extend. wet curing	Gray	2.58	46.0	17.8	1.8	27.1	2.6	2.1	1.9	0.8	-
	Black	3.36	50.6	15.1	1.6	26.3	2.3	1.9	1.6	0.7	-
OPC 30%+	Raster	0.23	11.2	50.0	-	32.6	2.2	0.9	1.4	0.6	1.8
GBFS 70%	White	1.12	29.5	26.3	2.3	29.5	3.6	1.1	3.1	1.2	3.5
w/b=0.45	Gray	1.52	35.5	23.3	1.2	28.5	3.2	1.0	2.9	1.1	3.2
carbonated	Black	1.96	40.9	20.9	0.4	27.6	2.8	0.9	2.6	1.0	2.8
OPC 30%+	Raster	0.25	11.8	47.9	-	32.3	3.0	0.5	1.0	0.7	2.9
GBFS 70%	White	1.59	32.1	20.4	2.9	28.8	4.9	0.9	2.0	1.3	6.8
w/b=0.45	Gray	2.11	37.6	17.8	2.6	28.0	4.2	0.8	1.8	1.1	6.0
extend. wet curing	Black	2.77	43.3	15.7	1.6	27.1	3.7	0.7	1.7	1.0	5.2

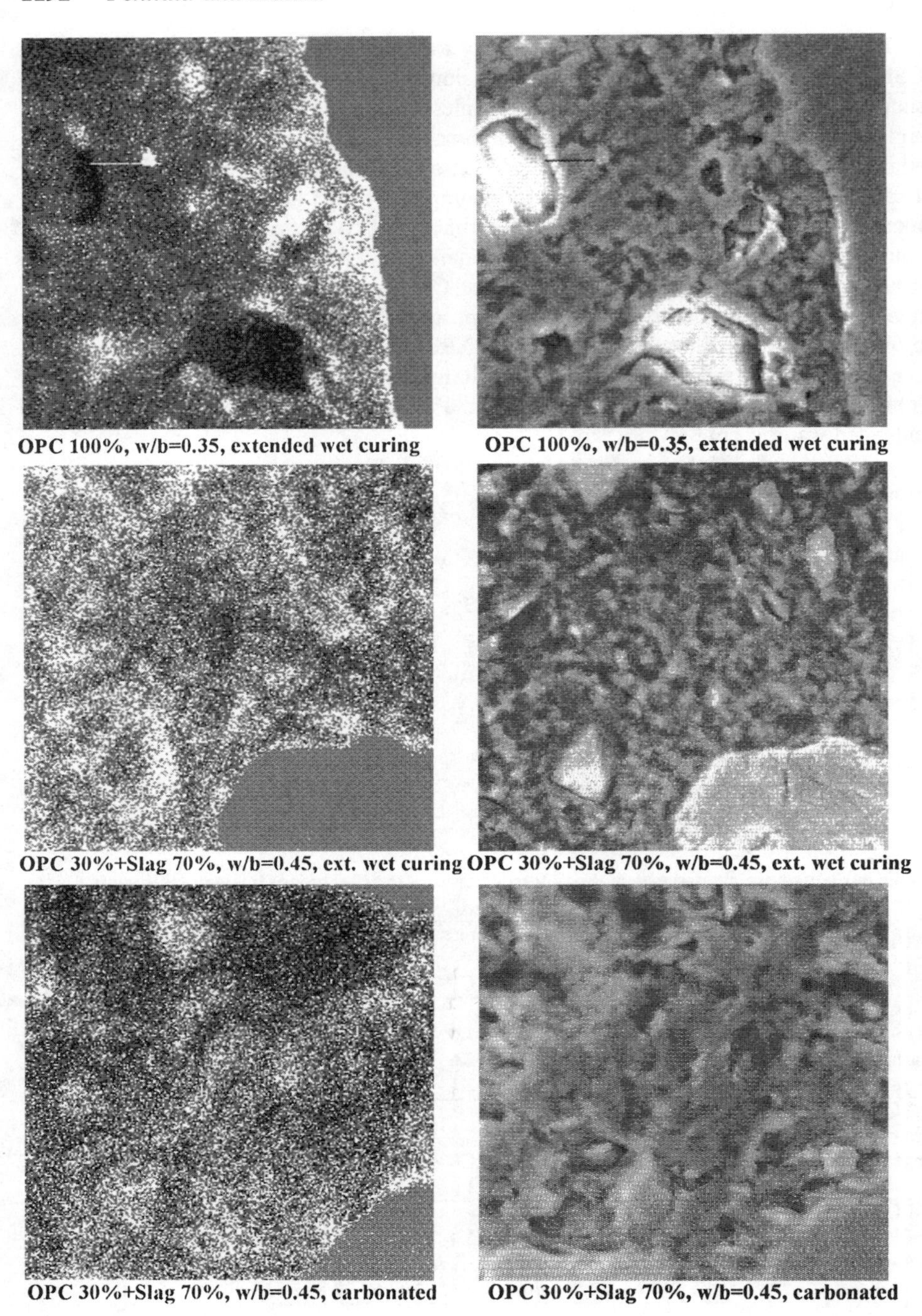

Fig. 5. EDX phase images and ESEM micrographs of test mortars. Element composition of the phases presented on the left-side images is presented in Table 4. The edge dimension of the micrographs is 60 mm.

The ESEM micrographs in Fig. 5 clearly show the porosity differences before and after wet curing. After the wet curing the binder matrix capillary porosity is clearly smaller and more homogeneous than after the carbonation. Keeping in mind the many error possibilities in the EDX phase analysis, especially with light element quantification, the overall element composition of the test mortars presented in Table 4 seems to be reasonable. The element weight percent accuracy is certainly not 0.1% even though the results are presented with one decimal. The results of the different mortars should be compared with each other in a broad sense, and they should not be considered too representative especially because they present a more or less randomly selected area of 60 by 60 μm^2.

In the phase maps of Fig. 5, aggregate particle phases have been presented with a raster shading. The binder paste has been divided into three phases according to the results of the statistical principal component analysis, so that the Ca/Si ratio is increasing in the order white, gray, and black. Comparing the results in Table 4 and the phase maps, one can notice that the distribution of carbon is logical. In all test mortars, the carbon weight percent is highest in the white phase, which has the lowest Ca/Si ratio of the binder matrix phases. The lowest carbon weight percentages are found in the black phases, where Ca/Si ratio is the highest and the hydration degree the lowest. In the phase map of the OPC mortar, the white phase represents a transition zone around the aggregate particle, where the porosity is higher and carbonate minerals have space to form. After extended water curing, the carbon content in the binder matrix phases is more evenly distributed compared to the carbon distribution after carbonation. This supports the assumption that the vaterite amount increases due to the decomposition of hydrotalcite.

Comparing the two lowest phase maps in Fig. 5 of the slag mortars it can be noticed that the hydration degree is much higher after the extended water curing; the area of the black phase is much smaller compared to the lowest phase map analysed after carbonation. Similarly, the binder matrix after wet curing is more homogeneous than after carbonation. In the phase map of the slag mortar after extended water curing, the slag particles are surrounded by the white phase, which denotes the inner slag hydrate, and the skeletal hydrate from where calcium has migrated. The white rim around these slag particles represents the area of the original unhydrated slag particle where Mg has been retained and concentrated. In the literature, not only calcium but also aluminum is reported to migrate from this area [7]. However, the results of this investigation do not confirm this supposition. Possibly the long carbonation period prior to the secondary hydration has liberated the above mentioned elements which have been distributed in a different manner.

7 Conclusions

Extended water curing of carbonated GBFS mortars increases remarkably the compressive strength. The strength increase is smaller with OPC mortars. The explanations for the phenomenon are:

1. Carbonation coarsens and opens the binder matrix pore structure, especially in slag mortars. Water curing after carbonation initiates hydration again, and it is able to

proceed further inside the binder matrix. Also new hydration products form and fill up the more porous carbonated layered pore structure of slag mortars.
2. The relative proportions of the different carbonates in slag pastes and mortars change during the extended water curing. This is probably due to the fact that hydrotalcite may transform into magnesium aluminate silicate hydrate or into mangnesium aluminate hydrate. Carbon group of hydrotalcite will form vaterite that precipitates on hydration products and in narrow pores. As a consequense carbonates and new hydration products close the previously porous SH-gel structure of carbonated slag mortars.
3. It is also possible that due to the decomposition of tri- and monosulphates the conditions to form stable strätlingite in slag matrix are favourable. Perhaps these two hydration products, magnesium aluminate silicate hydrate and strätlingite, may also contribute to the strength increase in slag concretes.

Results show that the harmful property of carbonated slag concrete, coarsened and open pore structure, may also be beneficial. The extended water curing of deteriorated slag concrete strengthens the matrix and makes it less porous than the original non-carbonated well cured slag concrete. OPC concretes show similar behaviour, but less extensively, mainly due to the secondary CSH formation. New types of hydration products were not detected in OPC concretes after extended water curing. The healing effect of water curing after carbonation can be applied in renovation of concrete and reinforced concrete structures. When the method is applicable, it will provide a new and cost efficient repairing procedure.

8 References

1. Matala S. (1994) The changes in pore structure of carbonated granulated blast furnace slag and OPC concretes. The Second Joint Finland-Japan Workshop on Service Life and Maintenance of Buildings. Technical Research Centre of Finland. 17p.
2. Penttala, V.E. (1992) Nature of compression strength in concrete. *Magazine of Concrete Research,* Vol. 44, No. 159. pp. 87-106.
3. Matala S. (1995) Thermoporometric Approach in Characterization of Pore structure in Concrete. Proceedings of Int. Conf. on Concrete under Severe Conditions. Sapporo. Japan 1995. 10p.
4. Lea, M. (1980) *The Chemistry of Cement and Concrete*, The Gresham Press, Surrey, 727p.
5. Taylor, H.F.W. (1990) *Cement Chemistry*, Academic Press, London, 475p.
6. Sawada, K., Ogino, T. and Suzuki, T. (1990) The distribution coefficient of Mg^{2+} ion between $CaCO_3$ polymorphs and solution and the effects on the formation and transformation of $CaCO_3$ in water. *Journal of Crystal Growth*, Vol. 106, pp. 393-399.
7. Tanaka, H., Totani, Y. and Saito, Y. (1983) Structure of hydrated slag in concrete, in *Fly Ash, Silica Fume, Slag & Other Mineral By-Products in Concrete*, (ed. V.M. Malhotra), Detroit, American Concrete Institute, ACI Publication SP-79. Vol. II, pp. 963-977.

117 MEASUREMENT OF THE DEGREE OF CARBONATION IN CEMENT BASED MATERIALS

J.S. MÖLLER
Chalmers University of Technology, Gothenburg, Sweden

Abstract
The primary object of the study upon which this article is based has been to find and develop a practical and precise method to measure the degree of carbonation in cement based materials, particularly in concrete. The article briefly summarises some of the findings in the study, and some examples of the results so far obtained are shown.

Methods in use at present for the investigation of carbonation in concrete are briefly discussed.

The method being developed is based on acid digestion in a closed compartment (pressure) and chemical analysis of the acid solution. By applying the new method it is easy to quantify the amount of material that has carbonated and to compare this with the carbonatable material at hand. This gives a better and broader base than previously for improved carbonation models.
Keywords: Carbonation, concrete, measurements.

1 Introduction

Carbonation of concrete is generally defined as an ongoing chemical reaction between carbon dioxide (CO_2) and some of the calcareous components of the cementitious matrix. Carbonation is also generally considered to be a transport or diffusion governed process. To reach to the point of reaction in the core of the cementitious material the carbon dioxide must normally travel inwards from the surface and the external layers. Considering this transport phenomenon, the carbonation process has been described as a diffusion process accompanied by chemical reactions. The transport part of the carbonation process is generally considered to be a gas in gas diffusion, but other transport processes may also be of some importance. In time, the concrete becomes stratified into distinct zones with different chemical and physical characteristics.

Concrete Under Severe Conditions: Environment and loading (Volume Two) Edited by K. Sakai, N. Banthia and O.E. Gjørv. Published in 1995 by E & FN Spon. ISBN 0 419 19860 1

Carbonation affects some important properties of the concrete, but the best known and probably also the most important effect is the change in alkalinity or the decreased pH level of the pore solution. The decrease in the pH level (from between 13.5 and 12.5 to 9.0 or below.) depassivates the concrete steel reinforcement, resulting in corrosion if oxygen and humidity conditions are adequate. If the pH value of the pore fluid next to the reinforcement is lower than 11, the concrete cannot protect the reinforcement from corrosion [1]. The consequence of reinforcement corrosion is the spalling of the concrete cover and a decrease in the reinforcement cross-sectional area, in some cases leading to irreversible deterioration or mechanical failure of the structure.

Carbonation has been investigated by a considerable number of researchers during the last few decades. Hamada [2] described a comprehensive study of the relationship between the depth of carbonation and various influencing factors, both in test specimens and existing structures. Extensive studies, with the main emphasis placed on different aspects, have also been published by, for example, Richardson [3], Tuutti [4], Parrott [5] and Hergenröder [6], who studied and compared different models for the propagation of carbonation.

The majority of the work published is, however, concentrated around the measurement and modelling of carbonation depth or the progression of a carbonation front. Usually resulting in a $\sqrt{t}$ law, the models can be summarised in equations similar to Equation (1).

$$x_c = A \cdot \sqrt{t} \tag{1}$$

The ability of these models to predict the advance of carbonation, while being good under some special conditions, is not general. In many cases, the studies on which the models are based have not attempted to clarify the underlying mechanisms of carbonation.

Some researchers, for example, Reardon [7,8,9] and Papadakis et al. [10,11], have contributed to the chemical modelling of the processes. Relatively few, however, have quantitatively considered the extent or degree of carbonation in concrete.

The relationship between the pH value in the pore solution, on the one hand, and the amount of carbonated material, on the other is a parameter of primary importance when modelling of the carbonation process is concerned. Surprisingly few successful attempts have, however, been made in the past to establish and explain this relationship. One of the main reasons for this is probably the lack of effective methods to measure the extent of carbonation, that is, the degree of carbonation in concrete samples. The method presented in this paper provides a simple and cost effective solution to this problem. This article is, to a large extent, based upon the author's licentiate thesis [12] on the subject of measurement of carbonation in cement based materials.

2 Measurement of carbonation

2.1 Traditional approach

The traditional approach of measurement, when carbonation is concerned, is to measure the carbonation depth. The term carbonation depth generally refers to the depth from the concrete surface to which the carbonation reaction has progressed to some, often undefined, stage. As the parameter of principal interest where carbonation is concerned is the pH level of the pore fluid, or more exactly of the pore solution, the carbonation depth is often referred to as the depth from the concrete surface where the pH level is above some predefined level. Generally, the "measurement" of carbonation depth does not include any quantitative evaluation of carbonation, either in the concrete considered as being carbonated or in the concrete zone where carbonation is considered to be progressing.

The method most generally applied is the so-called phenolphthalein method. It is generally concluded that the phenolphthalein indicator provides an approximate and incomplete picture of carbonation since it cannot show the width of the carbonating zone [13]. On the other hand, the main advantage of the phenolphthalein method is its simplicity, low cost and production of immediate results.

Another method that has been applied by a number of researchers is XRD. However, Richardson [3] points out that the method might be too detailed for use as a routine test method for the simple identification of carbonated and non-carbonated areas. Kropp [14] as well as Bier [15] have used XRD for qualitative identification of the crystal phases, but they make no attempt to quantify the phases on the basis of XRD patterns.

A method for determining the quantity of carbon dioxide in carbonated concrete in relation to the content of CaO has been described by Tuutti [16] and Pettersson [17]. The quantity of carbon dioxide is determined by heating the samples together with iron and tin powder, whereupon carbon dioxide is released. The liberated carbon dioxide is recorded by means of an absorption and weighing procedure. The quantity of CaO is determined by means of titration.

Thermogravimetry, TGA, has been applied by some researchers in order to quantitatively determine both the hydration products and the carbonation phases in cement based materials [18]. However, the interpretation of the results seems to be rather difficult and uncertain.

2.2 Degree of carbonation

Another approach is to define and measure the degree of carbonation as a function of the depth from the concrete surface. The degree of carbonation, χ, can be defined as the ratio between the calcium carbonate existing in the concrete sample and the calcium containing components of the cement matrix, presented as CaO. The ratio χ is a dimensionless factor, but as Equation (2) suggests, it represents mole ratios (mole/mole) [12].

$$\chi = \frac{[CO_2]_{released}}{[CaO]_{measured}} \quad (2)$$

The use or presentation of the degree of carbonation is rarely seen in literature, but for example Kropp [15] applied a slightly different form for expressing the degree of carbonation:

$$k = \frac{CO_2 \text{ - Bound}}{\text{max. } CO_2 \text{ - Binding capacity}} \quad (3)$$

The definition in Equation (3) has a major disadvantage in that the evaluation of the maximum carbon dioxide binding capacity is difficult, and it is generally necessary to estimate this value.

Measuring the degree of carbonation as a function of the depth from the concrete surface brings a few advantages:

- A carbonation profile can be drawn up and compared with the phenolphthalein measurement and the cover to the reinforcement.
- The total amount of carbon dioxide and matrix phases, involved in the carbonation reactions, can be determined.
- The data obtained can be applied in a more accurate and detailed modelling than before.

3 The new method

3.1 Background

Assuming that the carbonation of cementitious materials is basically a formation of calcium carbonate within the cementitious matrix, the chemistry of carbonation can be summarised in the following two reaction equations:

$$Ca(OH)_2 + CO_2 \Rightarrow CaCO_3 + H_2O \tag{4}$$

and

$$3CaO \cdot 2SiO_2 \cdot 3H_2O + 3CO_2 \Rightarrow 3CaCO_3 + 2SiO_2 + 3H_2O \tag{5}$$

The change in the quantity of calcium carbonate is, thus, a measure of the degree of carbonation. The initial content of calcium carbonate, which may, for example, originate in the aggregates or even in the cement, can be estimated from an analysis of samples from the inner, non-carbonated, parts of the concrete. The second quantity, the denominator, in the equation for the degree of carbonation is the calcium content or, more commonly, the calcium oxide content. The calcium oxide referred to in this text is the calcium oxide detectable by acid digestion and titration, as described in the text below. It must, however, be pointed out here that it is not very likely that all the calcium oxide which is thus detectable is available for carbonation, and that the maximum degree of carbonation probably lies between 0.8 and 0.9 [12].

In this study, a major emphasis has been placed on determining both the bound carbon dioxide and the calcium content from the same sample rather than from parallel samples. The main and most logical reason for this is the fact that the samples used are generally small and homogeneous component representation between samples can hardly be obtained with reasonable certainty. Two factors of uncertainty are the trend towards cement paste concentration near the concrete surface and the risk of variations in the aggregate content between small samples. The method proposed is, therefore, divided into the following three main stages:

- Preparation of samples for analysis by cutting powder from concrete cores with a lathe and a rotating diamond cutting blade.
- Acid digestion in a closed compartment for the determination of bound carbon dioxide by means of pressure measurement.
- Titration for determination of digested calcium oxide.

3.2 Theoretical background

The idea behind acid digestion in a closed compartment is that, prior to mixing the solvent (acid) and the sample, the system should be closed and the pre-digestion temperature and pressure should be measured. The volume of the solvent and the air space within the compartment are known, and the gas evolution from the digestion can, therefore, be easily determined by measuring the changes in pressure and temperature after complete digestion. During digestion, the system will establish an equilibrium condition. Even when the dissolved material is pure calcium carbonate, the equilibrium condition is very complex and controlled by a number of different parameters. As the system is constituted from three phases, the solution, the solid (powder) and the air space above, we also have three equilibrium systems.

3.2.1 The gas equilibrium

Assuming a closed gaseous system, where only energy but not matter can escape or intrude, the thermodynamic laws for gases apply. A good approximation of the relationship between the pressure P, the molar volume V and the temperature T is the ideal gas equation:

$$P \approx \frac{R \cdot T}{V} = \frac{n \cdot R \cdot T}{V_t} \Rightarrow n(CO_2) \approx \frac{P(CO_2) \cdot V_t}{R \cdot T} \qquad (6)$$

where R is the universal gas constant, V_t is the actual volume and n is the number of moles. A description of the theories of gases and gas behaviour can be found in almost any textbook on physical chemistry or engineering thermodynamics [19].

3.2.2 Gas-liquid equilibrium

The equilibrium between the gas phase and the liquid can be approximated by Henry's law[20,21]:

$$n(CO_2) \approx \frac{P(CO_2) \cdot \kappa \cdot n(H_2O)}{H(CO_2)} \qquad (7)$$

where $n(CO_2)$ is the number of moles of carbon dioxide within the liquid phase, represented as $H_2CO_3^0$, $P(CO_2)$ is the partial pressure of the carbon dioxide, $n(H_2O)$ is the number of moles of water in the solution, $H(CO_2)$ is Henry's constant for carbon dioxide in water at a given temperature, and κ is a correction factor depending on the composition of the liquid phase, for example the acid strength.

3.2.3 The solution equilibrium

If the acid strength is sufficient for maintaining strong acidity after digestion (pH<1), the $H_2CO_3^0$ ion is the predominant carbonate ion and other carbonate ions can therefore be ignored. Calcite is a salt of a weak acid and will, therefore, dissolve in any stronger acid. The chemical reaction is explained by the following reaction:

$$CaCO_3 + 2H^+ \Rightarrow Ca^{++} + H_2CO_3^0 \qquad (8)$$

The digestion also involves the dissolution of the cement phases, with both hydrated and unhydrated cement contributing calcium ions to the solution.

3.3 Experimental setup

3.3.1 Acid digestion

The powder sample is dried and weighed in an acid proof crucible, 5-10 gramme for each digestion. The digesting acid, 8.75% HCl, is poured into the compartment, the crucible placed in the sample holder, the lid tightened and the pressure compartment tilted. The pressure rise due to release of carbon dioxide (carbonate ions) is measured at predefined time intervals after tilting. After six minutes, the pressure compartment is returned to level position and the final pressure reading taken. The acid digestion apparatus is shown in the figure below.

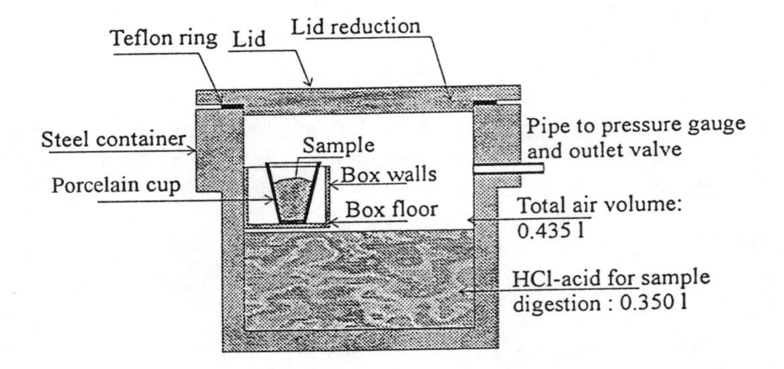

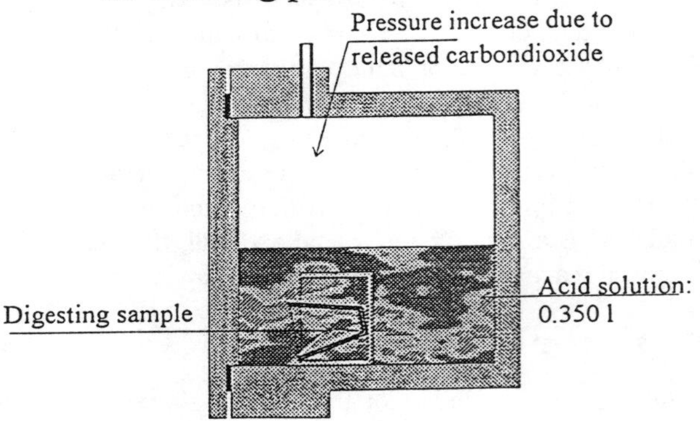

Figure 1 Acid digestion apparatus.

3.3.2 Calcium titration

After digestion in the pressure compartment, the solution is filtered and diluted with water. The calcium, CaO, content is determined by using potentiometric titration of a sample from the digested solution. The apparatus for the titration is a Metroohm 702 SET/MET Titrino. A calcium selective electrode and a calomel reference electrode are used. The volume increment of titrant (0.1 N EDTA) is set at 0.1 ml. Before titration, 5 ml of triethanolamine is added to the sample (solution) and the pH value of the solution is adjusted to >12 by adding NaOH solution while constantly stirring the solution.

3.3.3 Data interpretation

The pressure rise during digestion has an approximately linear relationship with the number of moles of carbon dioxide released.

$$n = n_1 + n_2 = P \cdot (\frac{V_{air}}{RT} + \kappa_1 \cdot V_{sol} \cdot f(T, A)) \tag{9}$$

4 Carbonation profiles

4.1 General

A number of concrete cores have been analysed by the above method. Three interesting examples are presented here, one analysis of a laboratory specimen along with the analyses of two field specimens. The samples were in all cases taken by applying the lathe and a rotating cutting blade as mentioned above.

4.2 Evaluation

The field specimens are marked F-391 and F-459 and come from a thirty year old apartment building. The specimen cores were taken from balcony slabs on two different sides of the building. The carbonation depth was found by applying the phenolphthalein solution to the outside of the cores. The depth from surface is measured from the sheltered side of the slabs.

The laboratory specimen, marked L-001, was in this case an 8 year old cube of normal concrete. The concrete had been kept in normal indoor conditions (20°C, 30-60%RH and 400-600ppm carbon dioxide). Two sides of the cube had been exposed, while the other four had been sealed. First a 70 mm core was drilled out and stored, exposed, in the laboratory for a few hours. The rest of the cube was split to measure the carbonation depth using the phenolphthalein method. The carbonation depth was measured as accurately as possible; the difference between the largest and the smallest carbonation depths was only 2 mm.

Between 10 and 15 powder samples were taken from the cores for acid digestion and titration. All the samples were treated in exactly the same way, from cutting through drying, acid digestion and titration. In all three cases, it could be seen quite clearly how the profile of released carbon dioxide, representing the calcium carbonate, profile changes with the depth from the surface.

The results from the analysis are shown in Figure 2. It is interesting to note that there is a very considerable difference in the carbonation depths measured by the phenolphtahlein, while the degree of carbonation profiles are more similar. In fact, the only difference seems to be at or near the surface. The background carbonation, that is, the degree of carbonation in the inner core of the concretes is almost identical and the sloping parts of the curves lie much nearer to each other than the phenolphthalein fronts indicated.

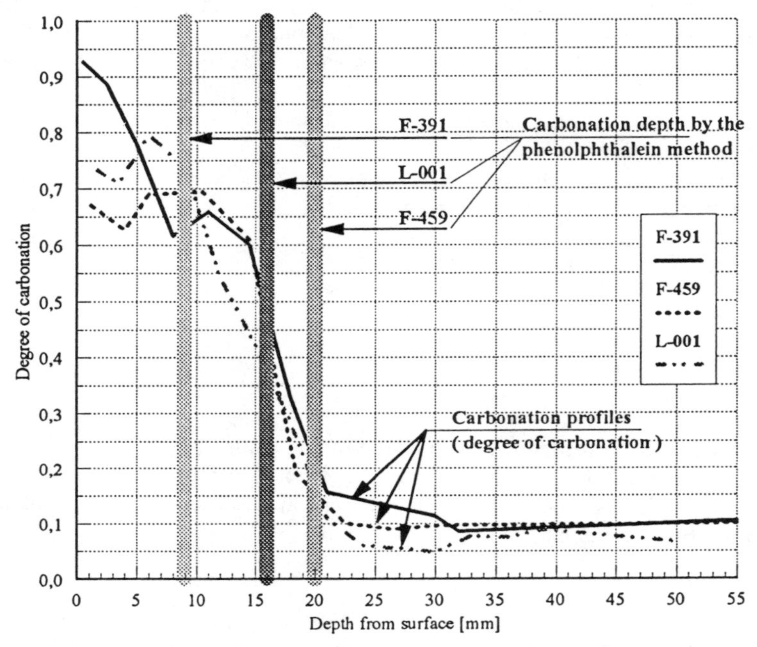

Figure 2 Carbonation profiles and carbonation depths.

5 Applications and discussion

The primary object of the study presented has been to find and develop a practical and precise method to measure the degree of carbonation in cement based materials, particularly in concrete.

Methods used at present for investigation of carbonation have been studied and some comparisons made with the new method. In methods such as the phenolphthalein method, only the depth of carbonation is investigated. No quantitative assessment is generally made of how much of the cementitious material has carbonated at different depths.

Here, only one aspect of the problems involved in measuring the carbonation depth by the phenolphthalein method has been examined. This is the question: How good is the relationship between the depth measurements and the actual degree of carbonation? From the diagram in Figure 2, we can see that the point where the phenolphthalein front cuts the degree of carbonation profile does not seem to follow a very narrow pattern. An error in the estimation of the carbonation depth can have serious consequences in evaluating the risk of reinforcement corrosion. Some 10 mm in concrete cover is a large deviation when the total concrete cover is generally between 15 and 40 mm. Degree of carbonation measurement should, therefore, replace, or at least be performed parallel to, the phenolphthalein tests when it is considered that there is a high risk of carbonation induced corrosion and expected repairs are expensive.

It is very important to study and map the carbonation process as a combined system, rather than to look at either the chemistry or the transport processes in isolation. A more complete description of the complex mechanisms of the combined transport and chemical binding of carbon dioxide in concrete would be of great value for the evaluation and prediction of the service life of a reinforced concrete structure. This is however somewhat meaningless if good methods and equipment to monitor the physicochemical processes involved are not available. The work so briefly presented here is an endeavour to improve this situation.

6 References

1. RILEM Technical Committee 12-CRC, (1976) Corrosion of reinforcement and prestressing tendons. A. State of the art report; *Materiaux et Constructions*, Vol. 9, No. 51, pp. 187-206
2. Hamada, M. (1969) Neutralisation of concrete and corrosion of reinforcing steel., *5ICCC*, ,III-3, pp. 343-368
3. Richardson, M.G. (1988) *Carbonation of reinforced concrete: Its causes and management*, CITIS Ltd.,Dublin.
4. Tuutti, K.(1982) *Corrosion of Steel in Concrete*, CBI fo 4•82.
5. Parrott, L.J. (1987) *A review of carbonation in reinforced concrete.*, Cement and Concrete Association / Building Research Establishment.
6. Hergenröder, M. (1992) *Zur statistischen Instandhaltungsplanung für bestehende Betonbauwerke bei Karbonatiserung des Betons und möglicher Korrosion der Bewherung*, Technische Universität München.
7. Reardon, E.J.(1992) Problems and approaches to the prediction of the chemical composition in cement/water systems. Waste Management, V12, 1992, pp. 221-239
8. Reardon, E.J. and Dewaele, P. (1990) Chemical Model for the Carbonation, *Journal of the American Ceramic Society*, Vol. 73, No. 6, pp. 1681-1690
9. Reardon, E.J.(1990) An Ion Interaction Model for the Determination of Chemical Equilibria in Cement/Water Systems., *Cement and Concrete Research*, Vol. 20, pp. 175-192
10. Papadakis, V., Vayenas, C., and Fardis, M.1989 A Reaction engineering approach to the problem of concrete carbonation, *Journ Amer Inst Chem Eng (AIChE Journ)*, Vol.35, No.10, pp. 1639-1650
11. Papadakis, V., Vayenas, C., and Fardis, M. (1992) Experimental investigation and mathematical modelling of the concrete carbonation problem, *Chemical Engineering Science*, Vol.46, No.5/6, pp. 1333-1338
12. Möller, J.S.. (1994), *Measurement of Carbonation in Cement Based Materials*, Licentiate Thesis, CTH.
13. RILEM Concrete Permanent Committee 18-CPC, (1984) Measurement of hardened concrete carbonation depth, *Materiaux et Constructions*, Vol. 17, No. 102, pp. 435-440

14. Bier, T.A.(1988) *Karbonatiserung und Realkaliserung von Zementstein und Betong*, Dissertation, Karlsruhe.
15. Kropp, J.(1983) *Karbonatiserung und transportvorgänge in Zementstein*, Dissertation, Universität Karlsruhe.
16. Tuutti, K. (1979) *Service life of concrete structures - Corrosion test methods.*, Swedish Cement and Concrete Research Institute.
17. Pettersson, O. (1979) *Sätt att följa betongens karbonatisering som funktion av djupet under ytan.*, Cementa R&D Section. (In Swedish)
18. Atlassi, E. (1993), *A Quantitative Thermogravimetric Study on the Nonevaporable water in Silica Fume Concrete*, Dissertation, CTH.
19. Zemansky, M.W., Abbott, M.M. and van ESS, H.C. (1975) *Basic Engineering Thermodynamics*, McGraw-Hill.
20. Atkins, P.W.(1978) *Physical Chemistry*, Oxford University Press.
21. Fogg, P.T.G and Gerrard, W. (1991) Solubility of gases in liquids, John Wiley & Sons.

118 CONTROLLED PERMEABILITY FORMWORK: INFLUENCE ON CARBONATION AND CHLORIDE INGRESS IN CONCRETE

P.A.M. BASHEER, A.A-H. SHA'AT
and A.E. LONG
Queen's University, Belfast, UK

Abstract
Ideally, from the viewpoint of concrete structures, the penetration of chloride ions and carbon dioxide can be minimised by producing a high quality cover concrete (covercrete). Efforts have been made in the recent past to develop new techniques by which the water-cement ratio in the near surface region can be lowered and a dense matrix achieved. One way of achieving this is to use a controlled permeability formwork system, (CPF), in which the surplus mixing water and entrapped air are removed from the fresh concrete via a fibre liner. Relatively little information is available at present on the efficiency of this material in improving the protection of the concrete against carbonation and chloride ingress, and how it compares with other techniques such as the application of silane. This paper discusses results of an experimental programme carried out in an attempt to evaluate critically the use of CPF for normal concrete. Specimens were subjected to the two types of environmental attack stated above and the results indicate that concrete made with CPF can provide a very effective defence against mechanisms which cause the corrosion.
Keywords: Carbonation, chloride ingress, durability of concrete, formwork.

1 Introduction

Steel embedded in concrete is protected both chemically and physically by the concrete. Various corrosion theories, laboratory experiments and field investigations have shown that in general steel does not corrode immediately but only as a result of:

- Neutralisation of the environment surrounding the metal e.g. carbonation. In this case the corrosion rate is generally very low.
- The action of corrosion inducing ions, e.g. chlorides. In this case the corrosion rate will be normally higher than that resulting from carbonation.

Concrete Under Severe Conditions: Environment and loading (Volume Two) Edited by K. Sakai, N. Banthia and O.E. Gjørv. Published in 1995 by E & FN Spon. ISBN 0 419 19860 1

- Combined action, e.g. carbonation and chlorides. In this case the corrosion rate will be very high relative to that indicated by the separate mechanisms.

Non- carbonated concrete affords embedded steel a high degree of protection against corrosion owing to the high alkalinity of the environment which inhibits corrosion. However, the protective capacity will be diminished when the alkalinity is reduced by carbonation [1]. If, for some reason, such as:
(i) inadequate depth of cover
(ii) cracking of the concrete
or (iii) poor quality concrete of high permeability
the concrete adjacent to embedded steel becomes carbonated, then the normal protective action of the concrete against corrosion of steel will be impaired. In this situation corrosion can take place, but it will progress more rapidly if both water and oxygen are present. Thus, carbonation is not necessarily accompanied by severe corrosion of the steel and associated cracking or spalling of the concrete cover. For example, carbonated concrete kept indoors in dry conditions will have minimum risk of corrosion whereas the risk will be increased considerably when carbonated concrete is exposed to wet conditions either indoors or outdoors [2]. The presence of significant amounts of chloride further accelerate the corrosive action on embedded steel [3].

The resistance of concrete to the diffusion of carbon dioxide and chloride ions is of direct relevance to its durability as it provides corrosion protection to steel reinforcing bars. Improved protection to reinforcement can be achieved by better curing [4], application of surface treatments such as silane [5] and reducing the w/c ratio in the cover concrete region [6, 7]. (This can be achieved by using a Controlled Permeability Formwork (CPF) system.) CPF liner is a fibrous material made from a polypropylene fabric which is attached to the inside of the formwork. The principle of operation is that it drains away the surplus mixing water and entrapped air from the surface of the wet concrete during casting as illustrated in Figure 1. As a result of this it produces concrete with a high strength surface layer [7], lower permeability and greater durability.

The aim of this paper is to compare the performance of CPF concrete and silane treated concrete in relation to carbonation resistance and chloride ingress at two water-cement ratios.

2 Experimental Programme

2.1 Variables

Variables in the investigation were two w/c (0.45 and 0.65) and the following types of concretes:

(i) Normal concrete (concrete made with timber formwork), (N)
(ii) Concrete made with controlled permeability formwork, (CPF)
(iii) Normal concrete coated with silane, (NS)
(iv) Concrete as in (ii) coated with Silane, (CPFS).

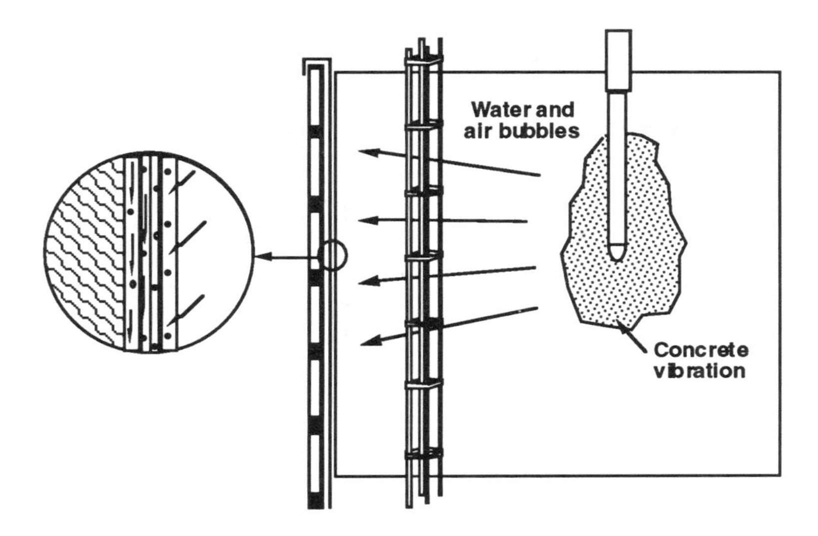

Fig. 1. Action of controlled permeability formwork.

2.2 Test Specimens

Samples of size 950x300x150 mm were cast vertically (Figure 2) with one 950x300 mm face against CPF and the opposite face against conventional timber formwork. Slump tests were carried out and three cubes were cast in order to determine the compressive strength of concrete after 28 days of curing in water at 20 °C. Table 1 shows the mix details. The formwork was removed 24 hours after casting and the specimens were stored at 20 °C and 55 % relative humidity until they were tested. At an age of 75 days, cores of 75 mm diameter and 150 mm length were taken from the hardened concrete. These cores were placed at 40 °C and 22% relative humidity for 1 week and then allowed to cool for 3 days. At an age of 85 days, the relevant specimens were sprayed evenly with silane in two coats on both faces at a rate of 0.3 l/m^2, with two hours between each coats. Figure 2 shows the locations where cores were taken and surface treatment applied. On the basis of results obtained at Queen's, it is considered that the effect of the depth of test locations on measured properties of concrete is insignificant in comparison with the variability of concrete. At an age of 90 days, the testing started.

2.3 Carbonation

The curved surfaces of the cores removed were coated with an epoxy emulsion in two coats and allowed to dry leaving the test surfaces uncoated. The specimens were then placed in a carbonation chamber at 16 (± 2) °C and 70 % relative humidity, and at a carbon dioxide concentration of 15%. After 10 days of exposure to this environment, the specimens were split into two halves and the carbonation depth measured by spraying a phenolphthalein indicator solution to the broken surfaces. A selected number of cores, after the completion of the carbonation test, was not split after carbonation as these were subsequently used to study the effect of carbonation on chloride ingress into the concrete. Section 2.5 reports the procedure used for testing these samples.

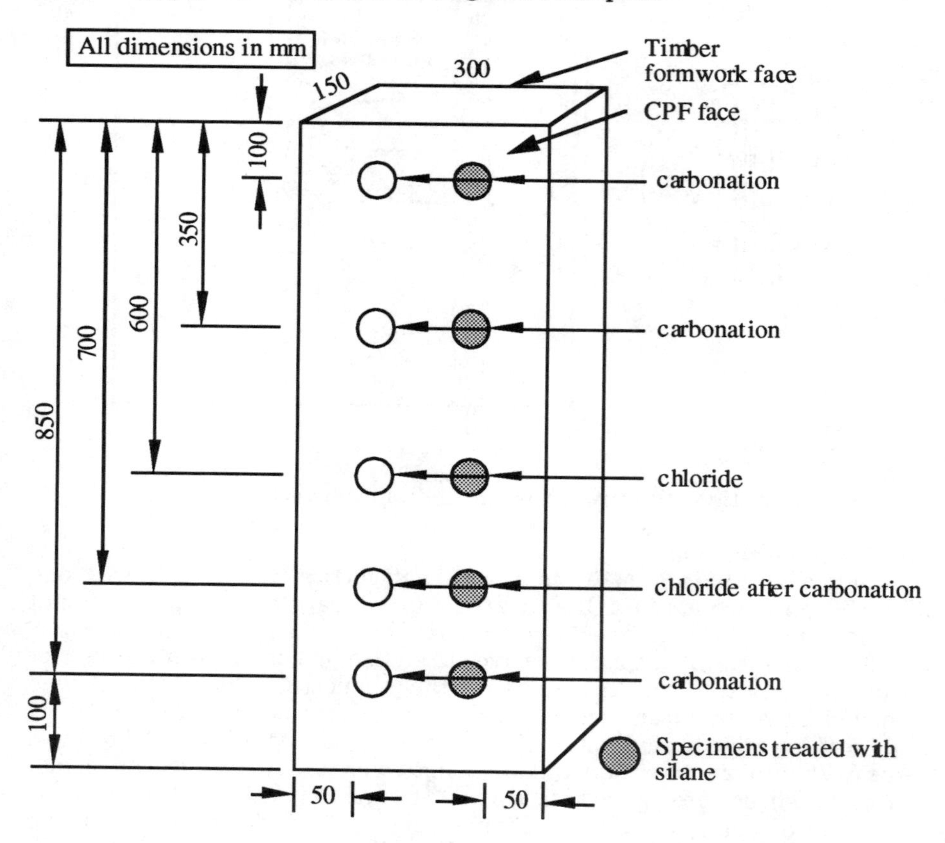

Fig. 2. Details of specimens and locations of tests.

2.4 Chloride Ingress

The locations of the cores used for the chloride tests are shown in Figure 2. These were cut into 2 halves of 75 mm diameter and 73 mm length (approximately) and the curved sides were coated with epoxy emulsion in two coats, leaving the test surface and the cut surface uncoated. The test surfaces, as stated before, were N, CPF, NS and CPFS. The cores were placed on a stainless steel wire mesh in a tray containing Sodium Chloride solution (0.55 m/l), immersing the test surface to a depth of 15 mm and leaving the cut surface exposed to the air. The specimens were tested for chloride penetration at the completion of 24 days of exposure.

After exposure to chlorides, the cores were sliced into four separate 3 mm portions at the following depths:

depth from the test face(mm)	average depth (mm)
2-3.5	2.5
5-8	6.5
15-18	16.5
25-28	26.5
35-38	36.5

During the slicing isopropane alcohol was used as a coolant to ensure that any chloride present was not washed away by the coolant. Each 3 mm slice was ground into a fine powder. Chemical analysis [8, 9] was carried out to determine the acid soluble chloride ions and the results were expressed as chloride content by percentage weight of cement.

Table 1. Properties of concrete mix

Cement :		Normal portland cement	
Aggregate (coarse)		10 mm crushed basalt	
Aggregate (fine)		Zone 3 Sand (BS 882 ; 1983)	
Mix ratio		1 : 1.65 : 3 cement sand aggregate (coarse)	
W/C	Slump (mm)	Compressive strength (KN/m^2)	Density (Kg/m^3)
0.45	5-10	61.8	2507
0.65	210	31.2	2448

2.5 Carbonation Followed by Chloride Ingress

The carbonated cores of 75 mm diameter and 150 mm length were broken into two approximately at the middle, leaving two cores of 75 mm diameter and 73 mm length for the chloride test. These were tested as described in section 2.4.

3 Results and Discussion

3.1 Carbonation

Figure 3 presents an average of three results of the carbonation test for both the water-cement ratios. It can be seen that CPF resulted in a significant reduction in carbonation relative to the normal concrete at the two w/c levels. Also CPF concrete at the higher w/c concrete had a better resistance to carbonation than that of 0.45 w/c normal concrete. This is because when CPF is used with the 0.65 w/c concrete it presumably reduced the w/c near to the cast surface to a level lower than that of the 0.45 w/c concrete [6]. As expected, the application of silane did not have any effect on the resistance to carbonation [5].

3.2 Chloride ingress (Before and after carbonation)

Figures 4 to 7 show the chloride profiles for 0.45 and 0.65 water-cement ratios before and after the carbonation. In these diagrams, the depth of carbonation also is presented. The chloride content by % weight of cement before exposure to chlorides for specimens taken from the hardened concrete was 0.028. The following comparisons are possible with data presented in Figures 4 to 7:

- (i) the effect of water-cement ratio on the chloride penetration resistance;
- (ii) the influence of CPF and silane on the chloride ingress;
- (iii) the effect of carbonation on the chloride profile.

It may be noted that, although there is a decrease in concentration of chlorides with depth, the chloride levels at the first depth were higher in some cases and lower in some other cases than the values beyond 5 mm depth from the surface. It is presumed that these variations are due to the way the specimens were lifted from the chloride tanks. The surface carbonation cannot be considered to be the reason for this variation because none of the samples carbonated before they were tested.

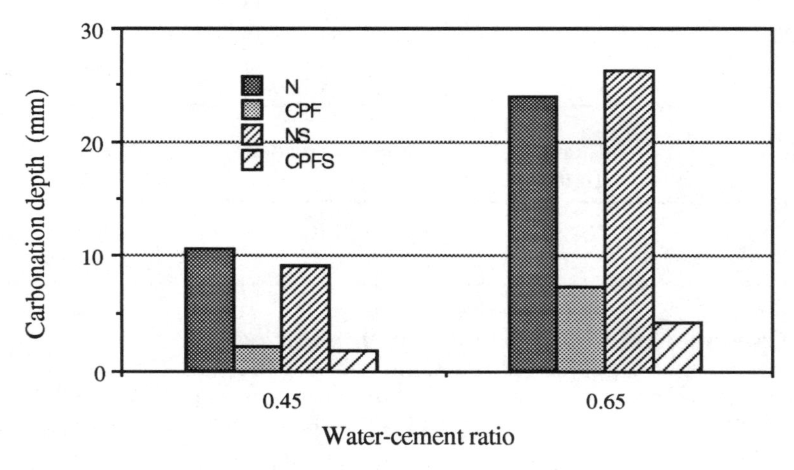

Fig. 3. Comparison of depth of carbonation.

A comparison of two sets of data (a's and b's) in each Figure clearly illustrates that the chloride ingress can be reduced greatly by reducing the water-cement ratio of the mix. This was true even when other methods such as the use of CPF and silane were employed. However, the chloride concentrations didnot differ much at a depth of 37.5 mm from the surface after 25 weeks in chloride solution. Obviously, this supports the general perception that a low water-cement ratio concrete with adequate cover to the reinforcing steel will provide long lasting protection against corrosion.

In Figure 4, the carbonation prior to the chloride exposure increased the concentrations of chloride at various depths from the surface for both water-cement ratio concretes. However, this effect was very small for the

0.65 w/c concrete. In the case of 0.45 water-cement ratio concrete, the carbonation might have induced microcracks in the cover zone which might have been the reason for the increase in chloride concentrations. However, the effect of carbonation in 0.65 w/c concrete might have been to block some of the capillary pores with calcium carbonate formed during the carbonation in addition to the introduction of microcracks. The net effect was not to have a significant change in the chloride ingress.

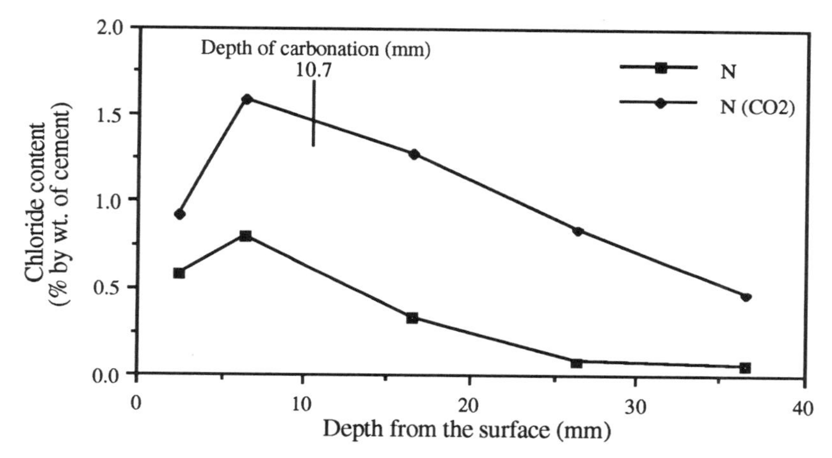

Fig. 4a. Chloride profile of 0.45 w/c normal concrete with and without carbonation.

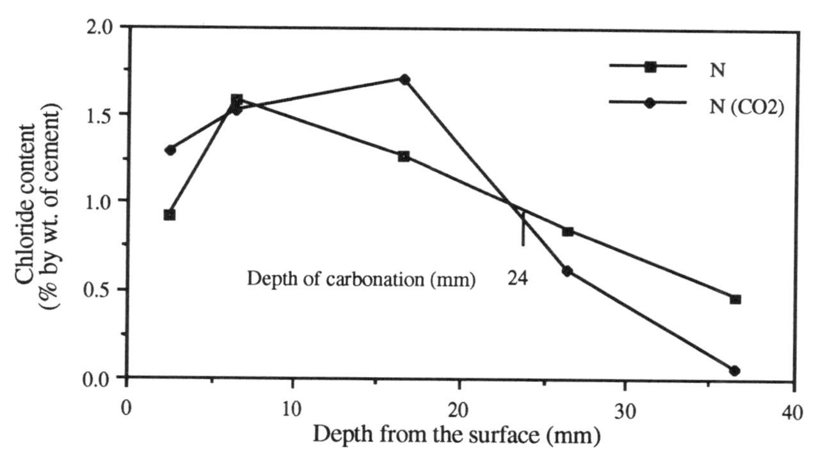

Fig. 4b. Chloride profile of 0.65 w/c normal concrete with and without carbonation.

The effect of CPF on chloride ingress can be obtained by comparing the data in Figures 4 and 5. Both for the carbonated and noncarbonated concretes, there was a substantial reduction in chloride concentrations at various depths from the surface. The chloride concentrations at 16.5 mm for noncarbonated CPF concretes were similar in magnitude to the values at 36.5 mm for the corresponding normal concretes. This would mean that relatively smaller cover to reinforcement could be used when CPF is used at the time of casting concrete.

Apart from a small change in the surface concentration levels, the chloride content did not change very significantly as a result of carbonation for the CPF concrete.

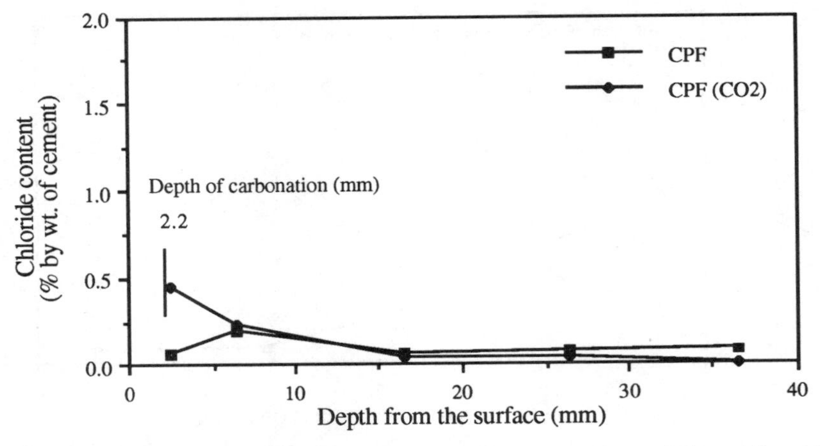

Fig. 5a. Chloride profile of 0.45 w/c CPF concrete with and without carbonation.

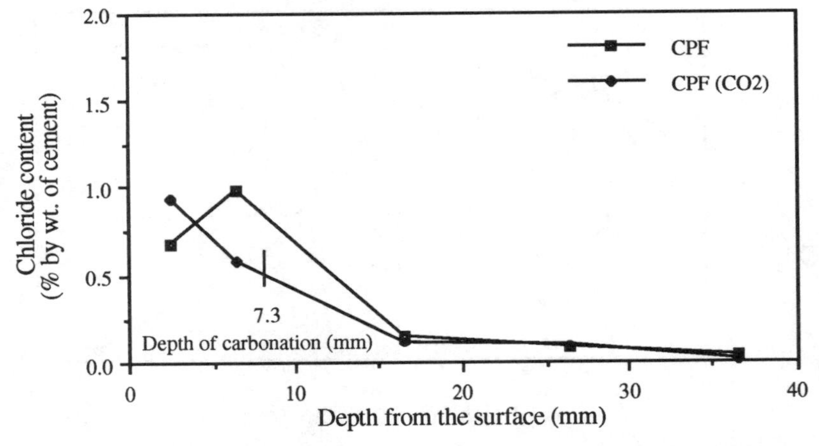

Fig. 5b. Chloride profile of 0.65 w/c CPF concrete with and without carbonation.

The data in Figures 4 and 6 can be used to arrive at the effect of silane on the chloride ingress. Very low chloride concentrations were obtained for both water-cement ratios in Figure 6 for the silane treated concretes. Similar to the values obtained with the CPF concrete, the chloride concentrations were very low beyond 16.5 mm from the surface for the silane treated concretes. Once again, the effect of carbonation prior to the exposure to chlorides on the chloride profile was confined to the surface.

A comparison between Figures 5, 6 and 7 would indicate that there is no additional benefit in applying silane onto concrete made with CPF as far as the chloride ingress is concerned. However, if protection is required against both carbonation and chloride induced corrosion, CPF was found to better. Although results in Figure 7a follow a pattern similar to those in Figure 5 and 6, Figure 7b deviated from them. As the pattern in chloride profile in Figure 7b was the same for both carbonated and noncarbonated concretes, the data in this Figure cannot be considered to be spurious.

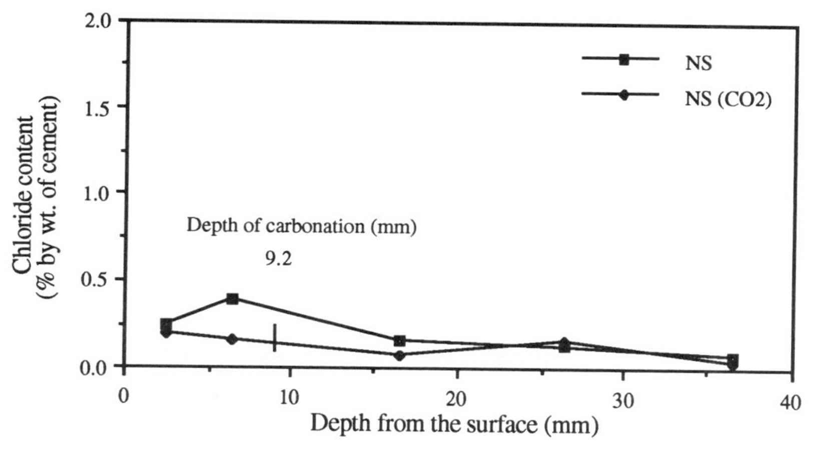

Fig. 6a. Chloride profile of 0.45 w/c normal concrete treated with silane, with and without carbonation.

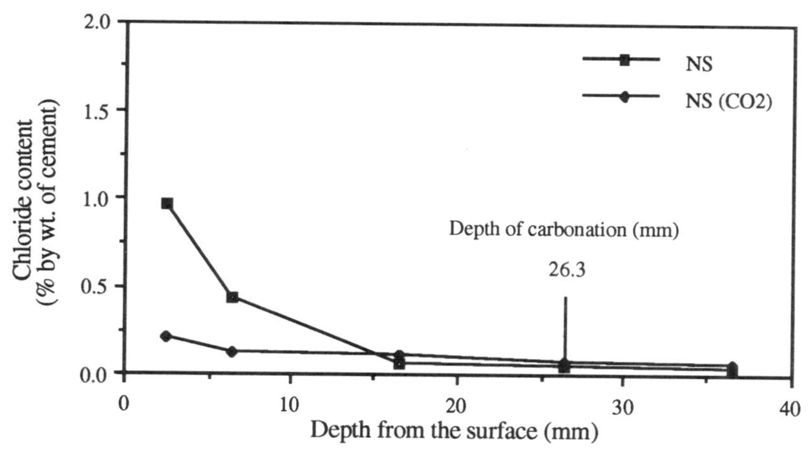

Fig. 6b. Chloride profile of 0.65 w/c normal concrete treated with silane, with and without carbonation.

One possible explanation is that chlorides might have penetrated through some defects in the silane treatment, which were pushed farther away from the surface due to the hydrophobic effect of silane.

On the basis of research at Queen's and elsewhere [6, 7], it is now understood that the depth of concrete which the CPF can remove the surplus mixing water and entrapped air from the fresh concrete is in the range of 5 to 10 mm. This depth depends on the w/c of the concrete. As the near surface layer provides the resistance to the transport of aggressive substances, it is considered that the results discussed above are not depended on the variation of the above depth.

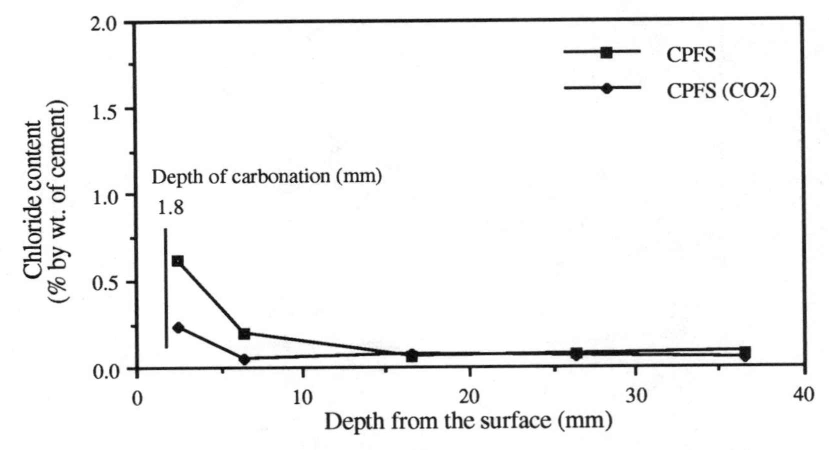

Fig. 7a. Chloride profile of 0.45 w/c CPF concrete treated with silane, with and without carbonation.

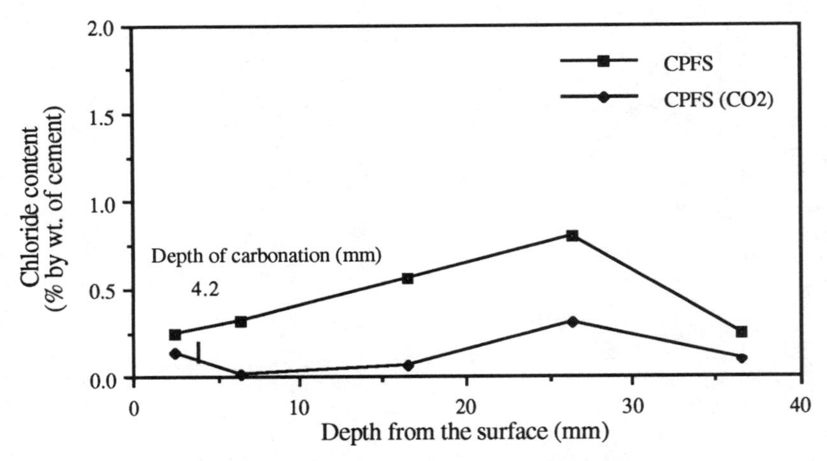

Fig. 7b. Chloride profile of 0.65 w/c CPF concrete treated with silane, with and without carbonation.

4 Conclusions

Based on the results reported in this paper, the following conclusions have been drawn.

The ingress of chloride into concrete can be reduced significantly with the use of CPF and silane. Apart from some surface effects, the subsequent carbonation does not affect their protective effects.

Carbonated concrete was found to be less resistant to the penetration of chlorides, however, this effect could be reduced by applying silane onto the concrete prior to carbonation.

Although silane was effective in reducing the chloride ingress, it did not have any effect on carbonation. Therefore, if resistance against carbonation induced corrosion is required, silane is not effective.

In these studies it was found that there is no additional benefit in applying silane onto a concrete made with CPF; sufficient protection is provided by CPF alone.

5 Acknowledgements

The investigation included in this paper was carried out with a grant from the Engineering and Physical Sciences Research Council (U.K).

6 References

1. Lawrence C. D. (1981) Durability of concrete : molecular transport process and test methods, *Cement and Concrete Association*, Slough, Technical report 455, 25 pages.
2. Roberts, M. H. (1981) Carbonation of concrete made with dense natural aggregate, *Building Research Establishment*, IP 6/81, April.
3. Brown J.H. (1982) Factors effecting real life performance of concrete in bridge substructures in the U.K., *British Cement Association*, PP/527, July.
4. Potter R. and Ho, D. (1987) Quality of cover concrete and its influence on durability, *American Concrete Institute*, SP 100-25, pp 423-445.
5. Robinson, H.L. (1987) An evaluation of silane treated concrete, *Taywood Engineering Limited*, Report, Department of Transport, U.K.
6. Price W. F. (1992) The use of zemdrain controlled permeability formwork with concretes containing blended cements, *Taywood Engineering Limited*, R & D division, London, report 1303/92/6157, July.
7. Sha'at A. A-H., Long A.E., Montgomery F.R. and Basheer P.A.M. (1993) The influence of controlled permeability formwork liner on the quality of cover concrete, *ACI*, SP 139-6, Durable concrete in hot climates.
8. Building Research Station (1977) *Simplified method for detection and determination of chloride in hardened concrete*, IS 12/77, July.
9. Goto S. and Diamond M. (1986) Ions diffusion in cement paste, *8th international congress in chemistry of cement*, Vol. VI, pp 405-409.

119 EFFECT OF MACRO CLIMATE ON DURABILITY OF CONCRETE STRUCTURE

X.T. DEE and Y. ZHOU
China Academy of Building Research, Beijing, China

Abstract
A comprehensive study of the test data from fast tests, long—term observation tests and investigations—on—site was carried out. The effect of macro climate on concrete carbonation and reinforcement rust rate are reflected in the Equations contained in this paper. Durability designs with different ultimate states are indicated.
Keywords: Carbonation, durability design, macro climate, rust, ultimate state.

1 Introduction

Durability design is important for concrete structures. For a durable structure, there are at least five problems that must be treated: the classification of exposure environment, design service life of structure, ultimate state of durability, the effect of the environment and resistance of the structure to this effect. The atmosphere, soil, natural water, ocean, chemical and special environments were classified according to the degradation mechanism and model. Service life can be designed as 100 years, 50 years or 25 years as required by the owner. The start of reinforcement rust and the time required for reinforcement rust to attain a certain level represent the ultimate state of durability in an atmosphere environment required. When a structure is designed to be durable, Equa-

Concrete Under Severe Conditions: Environment and loading (Volume Two) Edited by K. Sakai, N. Banthia and O.E. Gjørv. Published in 1995 by E & FN Spon. ISBN 0 419 19860 1

tion (1) must be satisfied.

$$R \geqslant F \tag{1}$$

where R=resistance of the structure to the environmental action, F =effect of the environment.

This paper mainly discusses R and F in an atmosphere environment: that is, the effect of macro climate on the carbonation rate of concrete and the rust rate of reinforcement. A comprehensive study of the test data from fast tests, long—term observation tests and investigations—on—site was carried out, and the equations developed. The statistical character of the measured carbonation depth and the ultimate state of reinforcement rust are discussed.

2 Carbonation and rust

2.1 Statistical character of the measured carbonation depth

The statistical data of carbonation depth collected from a long—term observation on 128 specimen—beams laid in Beijing city are listed in Table 1. Figure 1 is a typical carbonation depth histogram which approximates a normal probability. There is a linear relation between statistic parameter Sx and concrete standard strength $f_{cu,k}$ (see Fig. 2).

Table 1. Statistical character of carbonation depth

Type of cement	Environ—ment	$f_{cu,k}$ (MPa)	Number of data	Statistic parameter of carbonation depth		$D_{c,2}$ (mm)
				D_c (mm)	S_x	
SC	outdoor	10.6	215	12.9	3.82	12.6
	outdoor	20.0	123	4.0	1.34	5.41
	indoor	10.6	91	13.6	3.03	14.86
	indoor	20.0	73	6.4	1.59	6.37
PC	outdoor	13.1	263	7.2	3.41	7.46
	outdoor	18.3	120	5.0	1.76	4.47
	indoor	13.1	184	8.2	2.56	8.44
	indoor	18.3	72	5.5	1.22	5.58

Note: PC—Portland cement; SC—blast furnace slag cement.

2.2 Carbonation regularity

The carbonation depth of concrete affected by macro climate may be describde by Equation (2).

$$D_c=\alpha_1 \cdot \alpha_2 \cdot \alpha_3\left(\frac{60.0}{f_{cu,k}}-1.0\right)\sqrt{t} \quad (mm) \tag{2}$$

where t=the period of carbonation (y), $f_{cu,k}$=standard strength of concrete (MPa), $f_{cu,k} \leqslant 40$MPa, α_1=cure coefficient, α_2=coefficient of cement type, Portland cement $\alpha_2=1.0$, blast furnace slag cement $\alpha_2=1.3$, α_3=coefficient of environment effect.

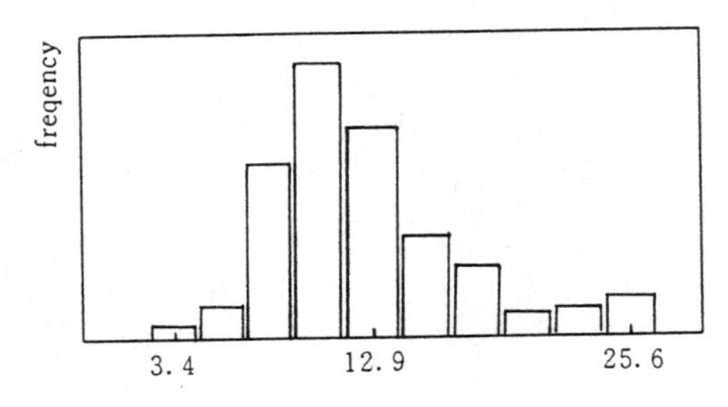

Fig. 1. Carbonation depth histogram.

Fig. 2. Relation between Sx and $f_{cu,k}$.

2.3 Carbonation period t

In Equation (2), the exponent of carbonation period t is 0.5, even though long-term observation tests show that the exponent is approximately 0.2 for blast furnace slag cement and 0.45 for Portland cement (see Table 2).

Table 2. Exponent of t in long-term tests

Type of cement	City	$f_{cu,k}$ (MPa)	t (y)	D_c (mm)	Exponent	Mean	$D_{c,2}$ (mm)	$D_c/D_{c,2}$
SC	Ji'nan	13.4	5	8.5	0.316	0.187	7.58	1.121
			13	11.5			11.5	1.00
		17.4	5	7.0	0.127		5.32	1.315
			13	7.9			8.58	0.921
		20.8	5	5.0	0.119		4.11	1.217
			13	5.6			3.63	0.845
PC	Ji'nan	16.8	5	4.0	0.389	0.453	4.56	0.877
			13	5.8			7.35	0.289
		20.1	5	3.2	0.446		3.31	0.967
			13	4.9			5.36	0.914
		23.5	5	2.0	0.670		2.60	0.769
			13	3.8			4.20	0.905

2.4 Effect of $f_{cu,k}$

There is a linear relation between the depth of carbonation D_c and the reciprocal of $f_{cu,k}$ that was verified by long-term tests and investigations-on-site (see Figs. 3 and 4).

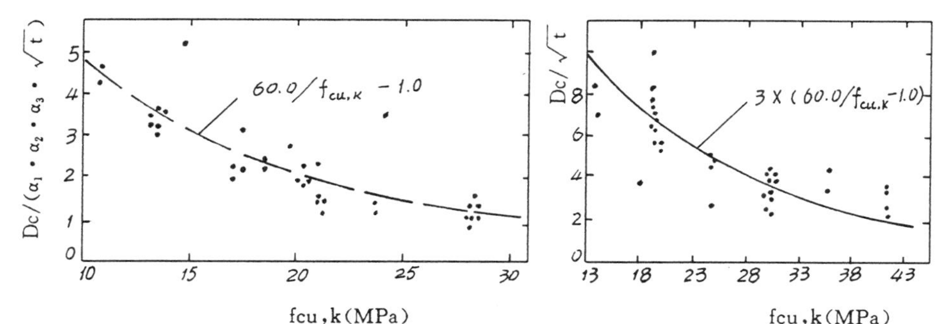

Fig. 3. The relation between Dc and $f_{cu,k}$ by long—term test.

Fig. 4. The relation between Dc and $f_{cu,k}$ from investigations—on—site.

2.5 Cure coefficient α_1

Because of the obvious effect of cure on the concrete carbonation rate, α_1 is needed in Equation (2). The carbonation ratios are listed in Table 3 to indicate the effect.

Table 3. Standard cure period and carbonation ratio

Cure period (day)		28	14	7	3	1
α_1	SC	1.0	1.29	1.43	1.81	2.34
	PC	1.0	1.19	1.43	1.56	2.34

Table 4. Long—term test for α_3

$f_{cu,k}$ (MPa)	Type of cement	t (y)	City	D_c (mm)	$D_{c,2}$	City	D_c (mm)	$D_{c,2}$ (mm)
28.2	PC	11	Lanzhou	4.2	3.21	Wuhan	2.5	2.21
27.9	PC	11	"	3.4	3.36	"	2.4	2.24
20.8	SC	11	"	5.0	6.96	"	3.6	3.68
13.4	SC	11	"	12.1	12.87	"	9.1	8.87

2.6 Effect of the environment α_3

After the relationships of α_1, α_2 and $f_{cu,k}$ with D_c were determined, environmental effect coefficient α_3 could be determined through the data derived from long—term tests and investigations—on—site. Table 1 lists the data from our long—term tests. The specimens were laid up at a civic building in Beijing. The test period was six years. In the Table, $D_{c,2}$ is the calculated carbonation depth. Long—term test data of other cities are shown in Tables 2 and 4. The investigation data in Figure 4 were derived from investigations—on—site.

2.7 Mechanic characteristics of rusted reinforcement

The ultimate tensile strength and ductility of rusted reinforcement were determined with a special test in which the reinforcement was rusted using a fast method to form an obviously pitted area on part of it. The diameters of the reinforcement are 12mm and 16mm. The regularities produced in the test are shown in Equation (3).

$$\left.\begin{array}{ll}\sigma_b^* = (1-0.695\lambda)\sigma_b & \\ \delta_b^* = (1-0.625\lambda)\delta_b & (\lambda \leqslant 8\%) \\ \delta_b^* = (0.628-1.03\lambda)\delta_b & (\lambda > 8\%)\end{array}\right\} \quad (3)$$

Where σ_b, δ_b = the strength and ultimate specific elongation, λ = section loss percentage of rusted reinforcement, δ_b^* = specific elongation of rusted reinforcement, σ_b^* = the tensile strength of rusted reinforcement.

2.8 Permitted rust state of reinforcement

When concrete cover is cracked by rust production to a certain width such as 0.15~0.25mm — a permitted rust state of reinforcement — the rust percentage of the reinforcement section is related to the diameter of reinforcement, the thickness of cover and the strength of the concrete. The test data are listed in Table 5, and the regularity of the test is described by Equation (4). The parameter λ_2 in Table 5 is the calculated value of Equation (4).

Table 5. Relation between λ and other factors

Diameter D (mm)	25		16			6.5		
Thickness of cover (mm)	30	20	30	20	20	20	10	30
$f_{cu,k}$ (MPa)	23.1	28.9	15.9	15.9	23.1	15.9	23.1	28.9
λ_1 (%)	1.81	1.11	3.02	2.13	2.26	12.1	6.52	7.11
λ_2 (%)	1.65	1.51	2.54	2.07	2.50	2.95	2.07	9.82
λ_1/λ_2	1.10	0.74	1.19	1.03	1.05	2.04	1.29	0.72

$$\lambda = \frac{\sqrt{f_{cu,k}} \cdot C}{24} \left[\frac{57.1}{D} - 0.78\right] \quad (4)$$

where λ = rust percentage of reinforcement section, $f_{cu,k}$ = standard strength of concrete (MPa), $f_{cu,k} \leqslant 40$MPa, C = thickness of concrete cover (mm), D = diameter of reinforcement (mm).

2.9 Regularity of reinforcement rust

The rust rate of reinforcement may be described by Equation (5).

$$\lambda = \beta_1 \cdot \beta_2 \cdot \beta_3 \left[\frac{4.18}{f_{cu,k}} - 0.073\right](1.85 - 0.04C)\left[\frac{5.18}{D} + 0.13\right] t \quad (5)$$

where t=the period of reinforcement rust (y), C, D, $f_{cu,k}$ and λ= see Equation (4), β_1=cure coefficient of concrete, when cure is standard $\beta_1 = 0.4 \sim 0.5$, when cure is nonstandard $\beta_1 = 1.0 \sim 1.2$, β_2=coefficient of cement type, for Portland cement $\beta_2 = 1.0$, for blast furnace slag cement $\beta_2 = 1.7$, β_3=effect coefficient of environmenal action.

2.10 Rust rate and concrete strength $f_{cu,k}$

Figure 5 shows the relation between concrete standard strength $f_{cu,k}$ and the rust rate of reinforcement. The relation was obtained from fast tests.

2.11 Diameter of reinforcement D

A two—year observation test, specimens of which were laid up in a thermostastic and humidistatic chamber, showed the effect of reinforcement diameter on rust rate (see Fig. 6).

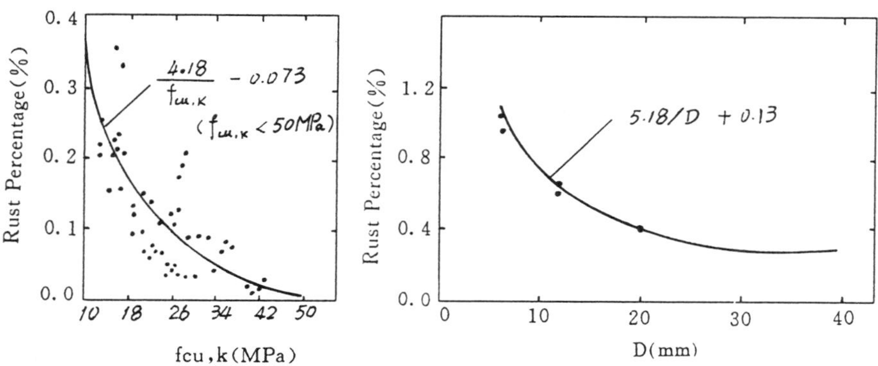

Fig. 5. The relation beween $f_{cu,k}$ and rust loss percentage.

Fig. 6. The relation between reinforcement diameter and rust loss percentage.

2.12 Thickness of concrete cover C

A two—year observation test, outdoor specimens of which were laid up in Ji'nan City, Shandong Province, showed that the ratio of rust rate were 1.25, 1.00 and 0.75, when the thicknesses of concrete cover were 15mm, 20mm and 25mm, respectively.

2.13 Rust period t

There is a linear relation between rust loss percentage of reinforcement section λ and the square root of rust period t. This relationship was proved by a two—year and a six—year observation test of outdoor specimens laid up in Ji'nan City, Shandong Province. The test data are listed in Table 6 in which λ_2 is the calculated value. A linear relation between λ and t is used in Equation (5) because of the faster rust rate of the investigations—on—site data.

Table 6. Two—year and six—year observation test

Rust period t (y)	6		2	
Type of cement	SC	PC	SC	PC
λ_1(%)	0.262	0.167	0.160	0.090
λ_2(%)	0.260	0.153	0.150	0.088
λ_1/λ_2	1.008	1.092	1.067	1.023

Note: $f_{cu,k}$=23.4MPa, C=20mm D=12mm, $F(t)=\sqrt{t}$

2.14 Coefficient of cement type β_2

The rust rate of reinforcement when protected by concrete cover composed of blast furnace slag cement is 1.7～2.0 times that when Portland cement was used. The certification data are listed in Table 6 and Table 7.

Table 7. Six—year observation test in six cities

City	Beijing		Changchun		Lanzhou		Wuhan	
Cement	SC	PC	SC	PC	SC	PC	SC	PC
λ_1(%)	0.273	0.163	0.175	0.108	0.388	0.188	0.381	0.183
λ_2(%)	0.636	0.374	0.635	0.374	0.635	0.374	0.635	0.374
λ_1/λ_2	0.430	0.436	0.276	0.289	0.611	0.503	0.600	0.499
β_3	0.993	1.000	0.637	0.667	1.411	1.163	1.386	1.129
β_1	0.433							

Note: outdoor specimen of civic building, $f_{cu,k}$ = 23.4MPa, D = 12mm, C = 20mm, t = 6 years.

2.15 Coefficient of cure β_1 and environment action effect β_3

Because the specimens were formed and cured in the same condition, the information in Table 7 might reflect the effect of macro climate on the rust rate of reinforcement. If the coefficient of cure is determined as β_1=0.433 for a standard cure, β_3 reflects the effect of macro climate. The rust rate of indoor specimens was 20%～60% of the rates listed in Table 7.

The investigations — on — site were carried out in Beijing, Hangzhou, Guiyang and Xi'ning cities. Most of the objects investigated were elements in industrial buildings, the concrete of which was composed of blast furnace slag cement. The other features of the elements are natural carbonation and nonstandard cure of the concrete. In Table 8, investigation data are listed.

Table 8. Investigation data analysis

City	Environ—ment	t (y)	D_c (mm)	$f_{cu,k}$ (MPa)	D (mm)	C (mm)	λ_1	λ_2
Guiyang	indoor	30	35. 2	17. 4	10	16	0. 17	0. 222
〃	〃	〃	35. 2	〃	〃	〃	0. 55	0. 222
〃	〃	〃	15. 6	28. 7	16	10	0. 50	0. 115
Hangzhou	indoor	35	50. 9	14. 0	25	49	0. 16	0. 11
〃	〃	〃	48. 5	14. 5	8. 5	28	0. 49	0. 20
〃	〃	〃	48. 5	14. 5	6. 0	20	0. 55	0. 35
Hangzhou	outdoor	〃	25. 5	20. 2	16	21	0. 55	0. 10
〃	〃	〃	25. 5	20. 2	16	21	0. 22	0. 10
〃	〃	10	18. 5	16. 3	6	5. 9	0. 51	0. 50

3 Durability design

There are at least two ultimate states for durability design of concrete structures in atmosphere environment: 1) rust is not permitted and 2) finite rust is permitted. Equation (6) must be satisfied for the first ultimate state.

$$C(1+\gamma)/\alpha_1 \cdot \alpha_2 \cdot \left[\frac{60.0}{f_{cu,k}} - 1.0\right] \geqslant \alpha_3 \sqrt{T_d} \qquad (6)$$

where α_1, α_2, α_3 and $f_{cu,k}$ = see Equation (2), T_d = designed service life of the structure (y), γ = modification coefficient of coating, C = thickness of concrete cover (mm).

For the other, Equation(7) must be satisfied.

$$C(1+\gamma)/\alpha_1 \cdot \alpha_2 \cdot \left[\frac{60.0}{f_{cu,k}} - 1.0\right] \geqslant \alpha_3 \sqrt{T_d - T_r} \qquad (7)$$

where T_r = the perminent rust period of the reinforcement.

In Equation (6) and (7), the left side items represent the resistance of the structure to the action of the environment, the right side items represent the effect of the environment.

4 Conclusions

1. Durability of concrete structure is obviously affected by macro climate.
2. Equation (2) describes the carbonation regularity of concrete at

some cities in China and Equation (5) describes the reinforcement rust regularity of the concrete structures in some cities in China.

5 References

1. Dee, X.T. and Zhou, Y. (1994) Durability Design of Concrete Structures. *Report of Building Research*, China Academy of Building Research.
2. Dee, X.T. and Zhou, Y. (1994) Regularity of Concrete Carbonation. *Report of Building Research*, China Academy of Building Research.
3. Dee, X.T. and Zhou, Y. (1994) Regularity of Reinforcement Rust. *Report of Building Research*, China Academy of Building Research.

PART TWENTY-ONE
CHEMICAL RESISTANCE

120 EFFECT OF SILICA FUME ON CONCRETES AGAINST SULFATE ATTACK

L.C.P. SILVA FILHO
Federal University of Rio Grande do Sul, Porto Alegre, Brazil
V. AGOPYAN
University of São Paulo, São Paulo, Brazil

Abstract
Resistance against sulfate attack of concretes and mortars made with two types of portland cements (ordinary and pozzolanic), w/c ratios of 0.28, 0.43, and 0.67, with or without silica fume addition of 5% and 10% was investigated. The concrete and mortars specimens were immersed in a 5% sodium sulfate solution. As expected, the main conclusion of this research is that the low permeability of the concrete is the major factor increasing its resistance to sulfate attack and that as long as the silica fume addition can reduce the permeability of the concrete, it is useful for increasing the durability of the material against this sort of attack. However, for pozzolanic cement specimens, probably due to a competition between the silica fume and the fly-ash for the available lime, this addition cannot be considered adequate for this purpose. Therefore, the contribution of the addition of silica fume to increasing the sulfate attack resistance must be better investigated because the effects of the addition of silica fume and those of the use of a superplasticizer are intermingled, the superplasticizer playing a major role in reducing the permeability of the concretes.
Keywords: Chemical attack, concrete, durability, fly-ash, pozzolanic cement, silica fume sulfate attack.

1 Introduction

The rapid deterioration of some concrete structures in hostile environments, such as foundations and seashore structures, is one of the major concerns of the construction industry. Sulfate attack is a recognized and relevant process of chemical attack, playing an important role in the degradation of concrete in some cases. This type of

Concrete Under Severe Conditions: Environment and loading (Volume Two) Edited by K. Sakai, N. Banthia and O.E. Gjørv. Published in 1995 by E & FN Spon. ISBN 0 419 19860 1

aggression has been widely studied since the early 20's [1] but, in spite of that, the mechanism of attack has not been fully understood yet. The additions like silica fume in concretes nowadays increase the doubts about the behaviour of this material in a sulfate contaminated environment [2]. The paper describes experimental work on the chemical resistance of some types of concretes and mortars to sulfate attack.

2 Sulfate attack processes

Sulfate attack was traditionally accepted as an aggressive process originated by the reaction of the sulfate ions from the environment with the calcium aluminate hydrate present in the structure of portland cement-based materials. This reaction generates calcium-sulfoaluminate (ettringite), which occupies more space because of the adsorption of water and results in an expansion process that, with time, causes in the total disruption of the material. Lawrence [2] pointed out that the sulfate could react with other hydrated constituents, like monosulfate and even with C-S-H, when the cation of the salt is magnesium.

There are some uncertainties about how the reactions with sulfate ions cause the disruption of the matrix. While the formation of crystals of sulfoaluminate and gypsum is well established, many experimental observations throw doubt on the assumption that physical forces of crystallisation are the primary cause of expansion. Some observations of Thorvaldson, cited by Mehta [3], suggest other mechanisms, for example: volume changes caused by osmotic forces associated with swelling and shrinkage of gels or chemical reactions conditioning the gel systems and leading to the destruction of the cementing substances. The existence of two types of ettringite, one expansive and the other non expansive, is also proposed [4].

In abstract, concrete degradation by sulfate attack involves three stages [5]:

- diffusion of sulfate ions into matrix;
- reactions of sulfate ions with some hydrate constituents (primarily calcium hydroxide, calcium aluminate and monosulfoaluminate) to form expansive compounds, as ettringite and gypsum;
- cracking of matrix leading to strength loss and disintegration.

The first stage is basically controlled by the diffusion characteristics of the concrete, related to porosity and permeability. The second stage is influenced by the chemical aspects, as the constituents of the cement and the solubility conditions of the solution of the pores. In the last stage, the disruption of the concrete structure occurs and the degradation process accelerates and is controlled mainly by the maintenance of the aggressive ions inflow.

To control the degradation, two major approaches are considered: reduction of sulfate ions penetration or modification of the chemistry of the concrete. These two solutions are concerned, respectively, with the first two stages of the degradation process, because we intend to avoid the third stage, where the structure of the material was already damaged. The use of additions is generally considered beneficial in regard to sulfate resistance. The most widely used additions are pulverised fly-ash (PFA), blast furnace slag (BFS) and silica fume (SF). According to previous studies, the beneficial effects of PFA against the sulfate attack have been attributed to the following:

- reduction of pore size and, therefore, slower penetration of sulfate (the pore size can be reduced from 240 to 166 Angstroms [6];
- lower amount of $Ca(OH)_2$ available for the formation of gypsum;
- smaller volume of the ettringite formed;
- lower quantity of C_3A susceptible to attack, since part has already reacted with the SO_3 content of PFA.

The silica fume has similar effects to the fly ash, but its greater fineness and high content of reactive silica give it a greater reactivity, usually being considered as suitable to improve the performance of concrete. The addition of BFS is generally considered beneficial, specially at rates of 40% to 65%, showing high stability to sulfate attack [7]. However, in some cases, this good performance was not achieved [2]; this fact is generally attributed to a unsuitable mixture of clinker and low hydraulic slags.

3 Experimental investigation

The experimental study developed in this work consists of an external sulfate attack on mortar and concrete prisms, with the consequent deterioration controlled by means of the compression and, mainly, the bending strength variations with time. Three water/cement ratios were used, 0.28; 0.43 and 0.67, in order to obtain a regular spacing of the points in the Abrams curve of strength versus w/c. Additionally, these ratios represent three very different types of mixtures, from a high water mixture similar to the ones employed normally on site to a very low water content mixture, typically used for high-strength concretes. Three amounts of SF addition were studied: 0%, 5% and 10% by the weight of the cement. The effect of the fly ash addition was evaluated through the use of a pozzolanic portland cement.

The combinations of these variables resulted in eighteen different samples presented in Table 1. A superplasticizer additive must be used to obtain a workable mixture for the lowest w/c ratio, so the same additive was employed in all of the samples, but in different amounts. Due to the use of the additions and the additive, the w/c ratio was calculated as the relation between all cementitious materials (OPC+PFA+SF) and the total water content (mix water plus 60% of the additive).

3.1 Materials

The aggregates were a normal sand from Guaíba river (State of Rio Grande do Sul), with a fineness module of 4.17, a maximum size of 2.4 mm and classified as fine sand according to the Brazilian standard and a crushed basalt coarse aggregate, with a maximum size of 25 mm and a shape factor of 3.64.

The samples were prepared with ordinary portland cement (OPC) or pozzolanic portland cement (PPC), with 40% of fly ash addition, because the latter is the most common type of cement on the market in the State of Rio Grande do Sul, with up to 93% of the sales; the ordinary portland cement was used for comparison reasons. The chemical analyses of the cements are presented in Table 2.

Table 1. Concrete specimens series.

Code	Cement Type	w/c ratio	Silica Fume Addition (%)	Proportions* (cem:sand:agg.)	Super-plasticiz. (%)
O20	OPC	0.28	0	1:0.72:2.27	4.0
O40	OPC	0.43	0	1:0.72:2.27	3.0
O60	OPC	0.67	0	1:0.72:2.27	3.0
O25	OPC	0.28	5	1:1.58:3.41	1.1
O45	OPC	0.43	5	1:1.58:3.41	1.5
O65	OPC	0.67	5	1:1.58:3.41	1.5
O210	OPC	0.28	10	1:3.12:5.45	1.5
O410	OPC	0.43	10	1:3.12:5.45	1.2
O610	OPC	0.67	10	1:3.12:5.45	0.8
P20	Pozzolanic	0.28	0	1:0.72:2.27	2.5
P40	Pozzolanic	0.43	0	1:0.72:2.27	2.0
P60	Pozzolanic	0.67	0	1:0.72:2.27	2.0
P25	Pozzolanic	0.28	5	1:1.58:3.41	1.0
P45	Pozzolanic	0.43	5	1:1.58:3.41	1.5
P65	Pozzolanic	0.67	5	1:1.58:3.41	0.8
P210	Pozzolanic	0.28	10	1:3.12:5.45	1.5
P410	Pozzolanic	0.43	10	1:3.12:5.45	1.1
P610	Pozzolanic	0.67	10	1:3.12:5.45	0.8

*by weight

Table 2. Chemical analysis of the cements.

	Proportion of constituents (%)					
Type of cement	Al_2O_3	SiO_2	CaO	MgO	SO_3	Ignition Loss
OPC	6.89	19.90	53.84	8.84	3.02	6.51
PPC	6.10	33.30	47.30	5.31	1.80	6.81

The silica fume employed had a specific surface area (BET) of 15 m^2/g and a mean diameter of 0.20 micrometers, with a silica content of 94%. The superplasticizer was a commercial compound of condensed sulfonate naphthalene formaldehyde, whose properties were studied by Raabe [8].

3.2 Procedures

Concrete specimens, cylinders of 100 mm diameter and 40x40x160 mm^3 prisms, were stored in 5% sodium sulfate solution to evaluate the effects of the attack on the compression and bending strength, respectively. The tests were carried out at 28 days, when the specimens were immersed in the aggressive solution, at 148 days (120 days of immersion), 500 days and 3 years (≈1000 days).

For the mortars, the K-S test was used to control the variation in the bending strength of 1x1x6 cm^3 specimens immersed in sodium and magnesium sulfate solutions up to 120 days. This method is recommended by Calleja [9] as one of the best methods to evaluate the performance of blended cements against sulfate deterioration. The method is based in the procedure developed by Koch-Steinegger [10]. The original procedure uses a w/c ratio of 0.6 and a 1:3 cement:sand relation.

In the present work, other values of w/c ratio were employed to evaluate the influence of this parameter on the resistance to sulfate attack. The material proportions for the mortars were similar to the ones used for the concrete, without the coarse aggregate, see Table 1. The mortar specimens were exposed to the sulfate environment after 21 days of curing, according to the original K-S test procedure, and the tests were performed after 7, 56 and 99 days of immersion.

The workability of the concretes was controlled with the slump test and that of the mortars, with the flow-test and the mini-slump test developed by Kantro, cited by Raabe [8], which has been widely used to evaluate the characteristics of cement pastes. After the casting, the specimens were stored in a wet room with relative humidity U > 85% and temperature range of 23-25°C.

The aggressive solution was prepared by the dilution of sodium sulfate salt in tap water. During the exposure time, the pH value was controlled by the replacement of the solution or by the use of a dissolved sulfuric acid solution.

3.3 Results

As expected, the aggressive process in both the concrete and the mortar began with an increase of the bending strength, which can be associated with a densification of the matrix [11] due to the filling of the pores with the products of the chemical reactions.

Mortar - bending strength

In the case of mortars, the analysis of the results is based on the evaluation of parameter R (corrosion factor) with time. The R-factor is the ratio of the bending strength of the specimens immersed in the sulfate solution (f_s) to the bending strength of the control specimens stored in tap water (f_w). The criteria adopted to evaluate the resistance to the sulfate attack, according to the K-S test recommendations, states that the cement could be considered potentially sulfate resistant when the R-factor is greater than 0.7 at 77 days (56 days of exposure). Nevertheless, for blended cements, Irassar [12] recommends an extension of the test duration for up to 120 days, at least.

Figures 1 to 6 present the curves of R-factor versus time for all the samples. Each point is an average of four observations, statistically analysed to determine the presence of spurious values.

All samples showed an R-factor greater than 0.77 at 77 days, except at 120 days for the O60 and the O65 samples, suggesting that the OPC concrete with a w/c ratio of 0.67 was moderately susceptible to sulfate attack. In this case, the addition of 10% of SF significantly improved the performance, because sample O610 had an R-factor greater than 1 at 120 days (Fig. 5).

For the OPC cement mortars, the samples with w/c ratios of 0.28 and 0.43, Figs. 1 and 3, had similar behaviour. The mixes without SF had higher R-factors, and at 120 days, the degradation was increasing, being in the descending part of the curve. The 10% addition had a positive effect on both w/c ratios, showing little densification and

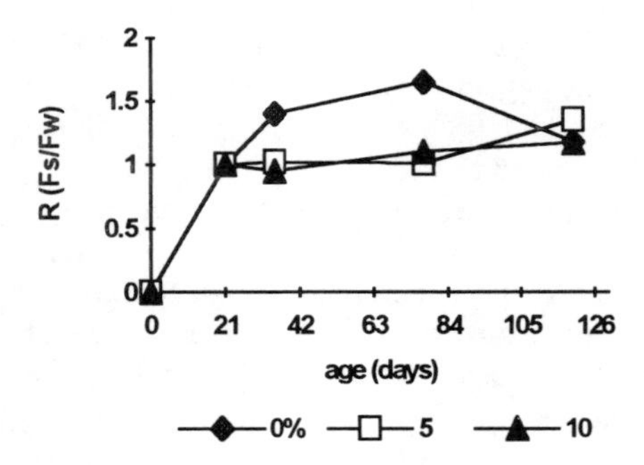

Fig. 1. Evolution of R-factor with time. OPC - w/c of 0.28.

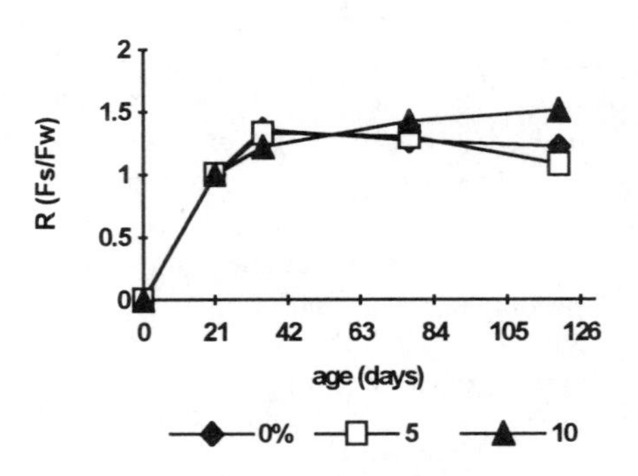

Fig. 2. Evolution of R-factor with time. PPC - w/c of 0.28.

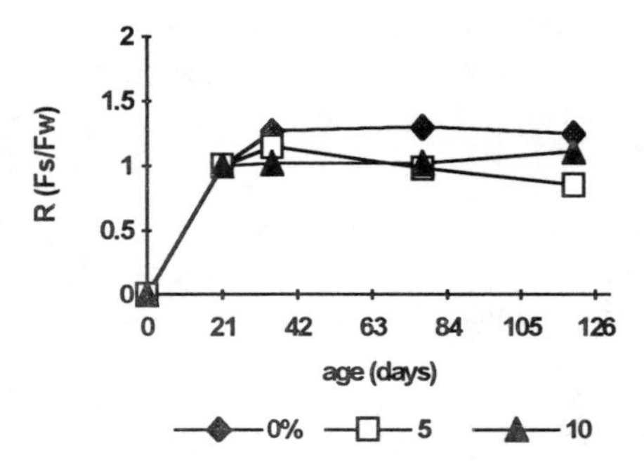

Fig. 3. Evolution of R-factor with time. OPC - w/c of 0.43.

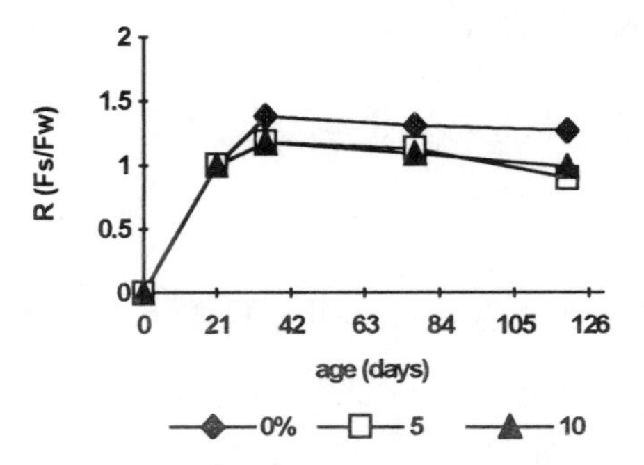

Fig. 4. Evolution of R-factor with time. PPC - w/c of 0.43.

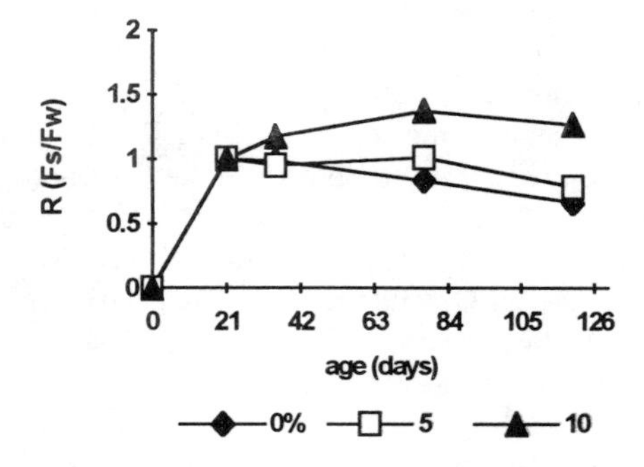

Fig. 5. Evolution of R-factor with time. OPC - w/c of 0.67.

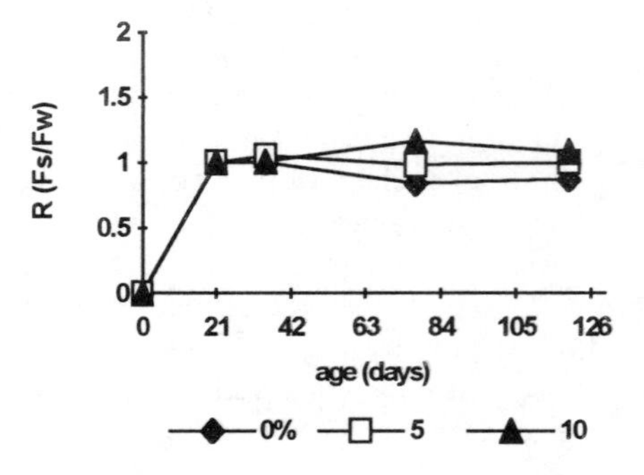

Fig. 6. Evolution of R-factor with time. PPC - w/c of 0.67.

no signs of reduction of the R-factor until 120 days. The 5% addition of SF resulted in a good performance for the 0.28 w/c mix, but for the O45 sample, it showed a premature diminution of the R-factor, going into a descending curve after the 35 day observation (14 days of immersion).

In the 0.67 mixes (Fig. 5), all mixes showed signs of degradation at 120 days, demonstrating the importance of the w/c ratio in durability. The O60 sample had a descending curve from the beginning of the ageing, and the O65 has a similar behaviour. The 10% addition of SF retarded the drop in the R-factor, and gave an R > 1 at 120 days, but the amount of densification was very large, showing that a lot of sulfate ions penetrated the mortar structure.

For the PPC mortars (Figs. 2, 4 and 6) all samples showed signs of degradation at 120 days, expressed as a tendency of R-factor to decrease, except the P210 mix. In general, the SF addition improved the sulfate resistance of the mortars, but this improvement was not statistically significant.

In Fig. 6, it can be observed that the R-factor of the P60 and P65 samples increased in the period from 77 to 120 days of immersion. This fact is attributed to a later manifestation of the pozzolanic effect, which was compensated for the drop in strength caused by the sulfate attack.

Concrete - compressive strength

The analysis of the compressive strengths at 120 days of immersion shows that all the PPC samples with w/c ratio 0.67 and 0.43 had an R-factor higher than 1, indicating the occurrence of densification, but the magnitude of the phenomena was unimportant because the high surface-volume relation of the specimen restrained the degradation to the superficial zone, minimizing the effect on the strength.

The addition of SF was slightly beneficial in this case, because it diminished the densification. Nevertheless, at 500 days, the specimens with SF addition showed a tendency to lose resistance (R < 1), but the O60 and the P60 samples still had an R-factor above 1, suggesting that a high quantity of voids could retard the start of the deterioration.

In respect to the type of cement, for the w/c ratio of 0.28, at 120 days, samples of PPC concrete reached, when stored in water, strengths similar to those of the OPC concrete and lower strengths at the other w/c ratios. When exposed to sulfates, there was no statistically significant difference in the performances, but for PPC specimens, lower increases due to densification were observed.

At 500 days, the general tendency was for the samples to begin to show signs of matrix disruption. The pozzolanic mixes seemed to have major drops in the strength. At 3 years, all the samples were seriously deteriorated, with great fissuration, serious spalling and softening of the concrete matrix.

Concrete - bending strength

The specimens for the bending tests were maintained in the same solution as the ones for the compressive tests and tested at the same ages. Figures 7 to 12 show the comparison between the bending strengths of the prisms stored in water and in sulfate solutions for 120 days. The points are the mean values of three observations, statistically analyzed for determining the spurious values.

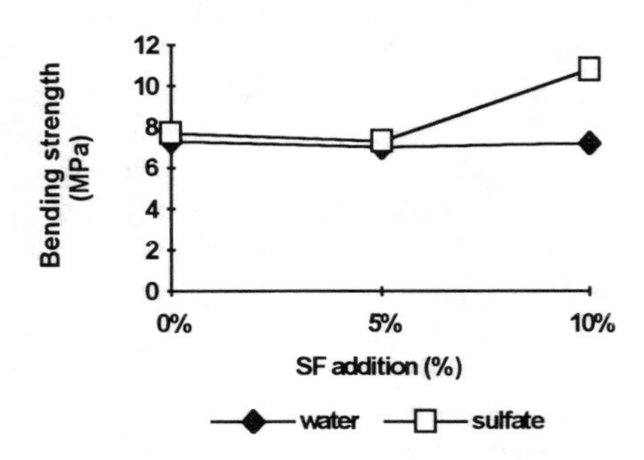

Fig. 7. Comparison between fs and fw at 120 days. OPC - w/c of 0.28.

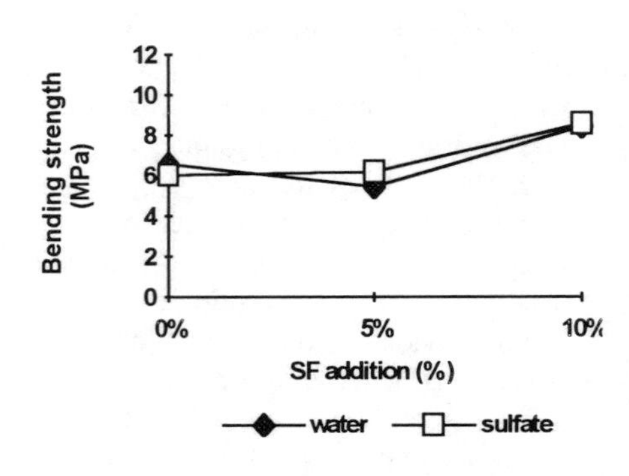

Fig. 8. Comparison between fs and fw at 120 days. PPC - w/c of 0.28.

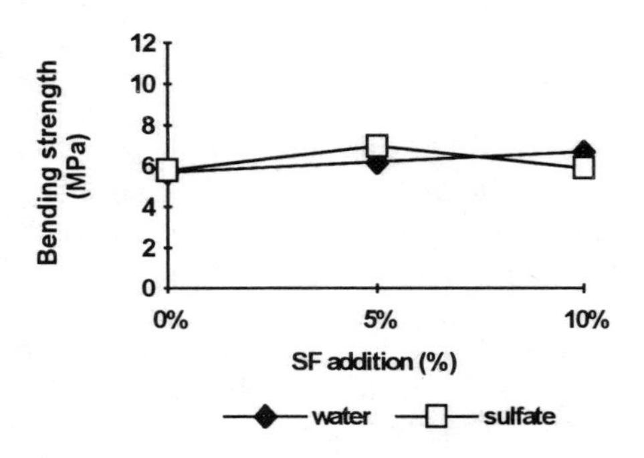

Fig. 9. Comparison between fs and fw at 120 days. OPC - w/c of 0.43.

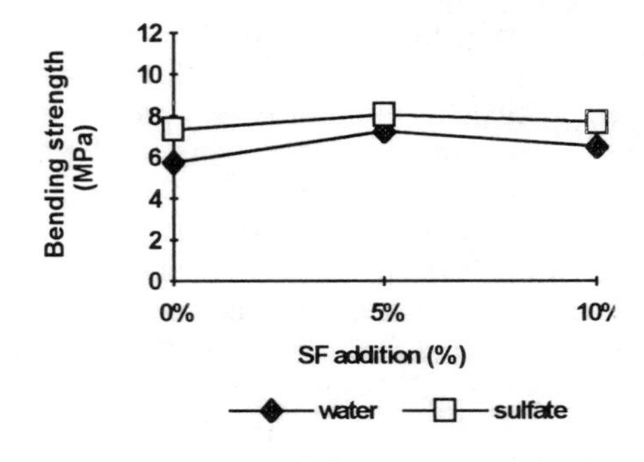

Fig. 10. Comparison between fs and fw at 120 days. PPC - w/c of 0.43.

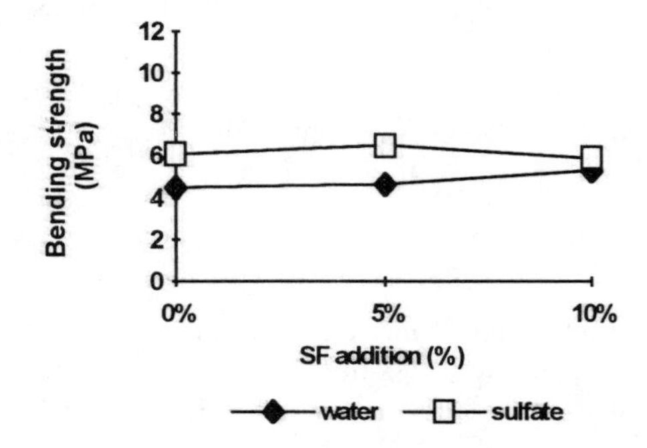

Fig. 11. Comparison between fs and fw at 120 days. OPC - w/c of 0.67.

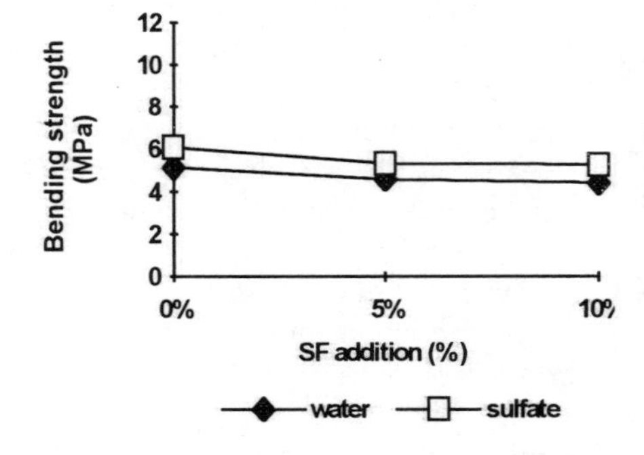

Fig. 12. Comparison between fs and fw at 120 days. PPC - w/c of 0.67.

For the OPC samples with an w/c ratio of 0.67 (Fig. 11), the sulfate attack caused a great densification, with significant increases in the bending strength. The sample with the 10% SF addition gave better results, showing a lower increase in strength, but the addition of 5% SF did not cause any improvement in the performance.

For samples with a w/c ratio of 0.43 (Fig. 9), the tendency was the same; the 10% SF addition gave the best results, with no signs of densification, and the 5% addition had the worst performance. However, for samples with a w/c ratio of 0.28 (Fig. 7), the 10% SF addition did not seem to give good results, showing a great densification, while the O20 and O25 samples presented small improvements of the strength. At 500 days, the general tendencies were the same, and at 3 years, the specimens were totally deteriorated, with transversal ruptures caused by the expansion.

For PPC specimens, the SF addition did not significantly change their performance when the w/c ratios were 0.67 and 0.43 (Figs. 12 and 13). Moreover, the addition seemed even to have a negative effect on the strength of the 0.67 samples; for specimens immersed in tap water, the addition caused a slight reduction in strength. This problem must be linked to the competition between the PFA of this blended cement and the SF because of the small amount of CH present in the mixture. For specimens with a w/c ratio of 0.28 (Fig. 8), the presence of SF caused a reduction in the densification process, indicating that the addition produced a beneficial effect to the permeation characteristics of the concrete, reducing the ingress of the sulfate ions.

4 Conclusions

The results of the experimental program show that sulfate attack is a very aggressive process causing concrete and mortar deterioration, as shown by the wholly degraded specimens of concrete after 3 years of exposition, and must, therefore, be well studied, especially when additions and additives are used in the mixture.

Considering the test procedures, the K-S test seemed to be very suitable to evaluate the sulfate resistance. It gave more consistent results and provided a large amount of information with a smaller amount of raw materials in a shorter period of time than the tests performed with concrete. The coefficients of variation of the K-S tests were generally lower than 10%, the maximum being 15%.

In regard to the effects of SF addition, the available data pointed out that the addition could be advantageous in some cases and without any effect in some others. In reality, in the case of PPC specimens, the addition of silica fume can even be deleterious because of the strong competition between the PFA of the blended cement and the SF for the small amount of CH available in the mixture. The main conclusion is that the effects of SF in increasing the sulfate attack resistance must be investigated further. This is emphasised by the fact that a superposition of effects could be happening between that of the addition itself and the effects derived from the use of a superplasticizer, with the latter being primarily responsible for the observed improvements obtained with the use of silica fume.

The authors have enlarged the scope of the research, studying the effects of the magnesium sulfate and the effects of different specimens conditioning (in soil, water in movement, etc.), and those concerned with surface protection. These data can be requested from the authors. The next phase is designed to try to identify the differences between the effects of the silica fume and those of the superplasticizer.

5 References

1. Tuthill, L.H. (1988) Lasting concrete in a sulfate environment. *Concrete International*, Vol.10, No.12. pp. 85-6.
2. Lawrence, C.D. (1990) Sulphate attack on concrete. *Magazine of Concrete Research*, No. 153. pp. 249-64.
3. Mehta, P.K. (1993) Durability of concrete - Fifty years of progress? *Seminário sobre qualidade e durabilidade das estruturas de concreto armado*, Porto Alegre. (Appendix).
4. Mehta, P.K. (1983) Mechanism of sulfate attack on portland cement concrete - Another look. *Cement and Concrete Research,* Vol. 13. pp. 401-6.
5. Roy, D.M. (1986) Mechanisms of cement paste degradation due to chemical and physical factors, in *Int. Con. on Chemistry of Cements, 8th,* Rio de Janeiro, Vol. 4.2. pp. 1-19.
6. Hussain, S. E., Rasheeduzzafar (1994) Corrosion resistance performance of fly ash cement concrete. *ACI Materials Journal*, Vol. 91. pp. 364-72.
7. Malhotra, V.M. (1993) Fly ash, slag, silica fume and rice-husk ash in concrete: a review. *Concrete International,* Vol. 15, No. 4. pp. 23-8.
8. Raabe, A.L. (1991) *Aditivos Superplastificantes em Concretos de Cimento Portland Pozolânico*, UFRGS, Porto Alegre (Master of Engineering Thesis).
9. Calleja, M. (1980) Durability of concrete, in *Int. Con. on Chemistry of Cements, 7th,* Paris, Vol. VII-2. pp. 1-48.
10. Koch, A., Steinegger, H. (1960) Ein schnellprüfverfahren für zemente auf ihr verhalten bei sulfatangriff. *Zement-Kalk-Gips*, No. 7. pp. 317-24.
11. Moukwa, M. (1990) Characteristics of the attack of cement paste by $MgSO_4$ and $MgCl_2$ from the pore structure measurements. *Cement and Concrete Research,* Vol. 20. pp. 148-58.
12. Irassar, F. (1990) Sulfate resistance of blended cement: prediction and relation with flexural strength. *Cement and Concrete Research,* Vol. 20. pp. 209-18.

121 HIGH PERFORMANCE CONCRETES IN CONCENTRATED MAGNESIUM SULFATES, MAGNESIUM CHLORIDES AND SODIUM SULFATES SOLUTION

A. BENTUR, A. GOLDMAN and M. BEN-BASSAT
National Building Research Institute, Technion – Israel Institute of Technology, Haifa, Israel

Abstract
The durability of high strength concretes in highly concentrated magnesium and sulfate solutions was evaluated to develop optimal compositions that might be applied in structures exposed to such environments, as is the case for example in the Dead Sea plants in Israel. The compositional variables evaluated were the water/cement ratio, presence of silica fume and the composition of the portland cement (ordinary vs. sulfate resistant). The present paper describes the results of six years study. It was resolved that silica fume was extremely effective in reducing sulfate attack, providing performance which is equal to or better than that obtained by sulfate resistant cement. However, in magnesium containing solutions, decomposition typical of magnesium attack could not be eliminated. There were indications that under such conditions the durability performance was somewhat better in the low water/cement concrete without silica fume.
Key words: durability, high performance concrete, magnesium chloride, magnesium sulfate, sodium sulfate.

1 Introduction

High strength concretes, with silica fume, have been used extensively for production of components which should resist aggressive environmental conditions [1]. The improved sulfate resistance and durability in various acid solutions are well documented by Cohen and Bentur [2], Mather [3] and Mehta [4]. This improved performance is the result of the physical nature of these concretes, which are much more impermeable than conventional concretes, due to the pozzolanic reactivity of the silica fume.

Concrete Under Severe Conditions: Environment and loading (Volume Two) Edited by K. Sakai, N. Banthia and O.E. Gjørv. Published in 1995 by E & FN Spon. ISBN 0 419 19860 1

The improved durability achieved with such concretes has made them attractive for use in industrial plants exposed to a particularly aggressive environment. In Israel, there was a special interest in evaluating such concretes for use in the Dead Sea Works, where the structures are exposed to a highly concentrated magnesium chloride brine, in combination with an extremely harsh climate, in which the temperature readily reaches 40 °C. In these climatic conditions the chemical attack by the brine can be aggravated by wetting and drying cycles to which some parts of the structures are exposed, as they are being partially submerged in the brine.

A previous study of the durability of high strength silica fume concretes in highly concentrated sodium sulfate and magnesium sulfate solutions indicated their superior performance in the sodium sulfate solution, Cohen and Bentur[2] (being comparable or exceeding that of sulfate resistant cement). However, the results of the magnesium sulfate immersion tests did not indicate superior behaviour of the silica fume concrete. These tendencies implied that additional evaluation of such concretes is required when considering their use in magnesium rich brines.

In a previous paper [5], results of exposure of various types of high strength concretes to Dead Sea brines was described, providing data of up to 2½ years exposure. The present paper describes the results obtained after additional monitoring to 4 years exposure, as well as evaluation of similar concretes under laboratory conditions after six years exposure to magnesium sulfate and sodium sulfate solutions. The parameters evaluated were the water/cement ratio, the type of silica fume and ordinary vs. sulfate resistant cement.

2 Experimental Procedures

Two series of tests were carried out in the present work. The first one was already described in a previous paper and it was intended to evaluate the influence of the composition of the concrete on its durability performance under conditions representing actual exposure to the Dead Sea brines which is a highly concentrated magnesium chloride solution [5] (25% concentrated solution in which the salt consists primarily of 22% $MgC\ell_2$, 59% $CaC\ell_2$, 15% $CK\ell$, and 6% $NaC\ell$). The variables studied were the water/cement ratio, ranging from 0.72 to 0.31 and the influence of silica fume of different sources. Their influence was evaluated by testing the performance of concretes with water/binder ratio of 0.31. The silica fume content was 15% by cement weight in one case (microsilica of Elkem, Norway) which is considered the upper recommended limit by that manufacturer, and 20% in another case (silica fume, labelled Corrocem), which is the dosage recommended by its producer. The Corrocem product is based on silica fume, but the producer specifies that it was particularly developed and formulated to provide superior durability performance. The composition of the concretes is provided in Table 1. The concretes were cast as 100 mm cubes. Another composition parameter was a special commercial mortar formulation (VGM - Germany) intended for protective coating of concrete. It is essentially a low water/binder ratio mix containing special admixtures. The specimens in the case consisted of a 70 mm core of 0.72 water/cement ratio portland cement concrete covered with 25 mm layer of VGM product.

Table 1. Composition and characteristics of the concretes of Series I

Mix Notation		B 75	B 40	B 31	MS 31	CO 31
Coarse aggregates,	kg/m^3	1324	1370	1342	1349	1363
Fine aggregates,	kg/m^3	606	523	410	463	442
Portland cement,	kg/m^3	263	392	512	386	390
Silica fume,	kg/m^3	--	--	--	58(1)	78(2)
Water,	kg/m^3	190	157	159	142	146
Superplasticizer,	kg/m^3	--	3.9	5.8	8.2	--
Water/binder ratio		0.72	0.40	0.31	0.31	0.31
Unit Weight,	kg/m^3	2384	2446	2429	2406	2419
Slump,	mm	150	60	90	60	100
28-day compressive strength,	MPa	25.0	69.0	78.3	85.3	94.2

(1)Microsilica of Elkem, Norway
(2)Corrocem, Norway

The evaluation of the concretes in the first series might be considered as a site study, as the cube specimens were immersed to half height in the brine solution, and exposed to the natural environment of the Dead Sea, in a bath placed on the roof of one of the buildings there. The performance was evaluated by means of visual inspection and determination of chloride content at a depth of 5 to 10 mm.

The second series of experiments consisted of a more conventional laboratory evaluation of concrete performance, based on immersion of concrete specimens in a "synthetic" solution and measuring length and weight changes, as well a visual appearance. The variables studied were the composition of the cementing material in the concretes: types of portland cement (ordinary and sulfate resistant) and presence of silica fume. All the concretes were of a low water binder ratio, in the range of 0.31 to 0.34 and their composition is given in Table 2. The specimens were prepared

Table 2. Composition and strength of concretes of Series II

Type of Cement		Portland Cement			Sulfate Resistant Cement	
Silica Fume Content (%) Weight of Cement		0	5%	15%	0	15%
Coarse aggregate,	kg/m^3	870	886	883	848	882
Medium aggregate,	kg/m^3	485	494	492	473	491
Sand,	kg/m^3	453	522	447	492	476
Cement,	kg/m^3	493	409	407	522	406
Silica fume,	kg/m^3	--	20	62	--	61
Water,	kg/m^3	156	138	144	154	144
Superplasticizer,	kg/m^3	10.0	12.4	15.5	10.5	15.5
Water/binder ratio		0.33	0.34	0.33	0.31	0.33
28 day compressive strength,	MPa	76.6	79.7	107.6	80.0	88.8

as 70×70×280 mm beams and their length was measured periodically. They were immersed 28 days after casing in 10% concentrated sodium sulfate and magnesium sulfate solutions.

3 Results and discussion

The present paper presents four and six year results of exposure of Series I and II, respectively. The 2½ year results of Series I were presented previously [5].

Series I

The visual observations after four years are presented in Fig. 1. The classification of the durability performance based on these observations is summarized in Table 3. It is similar to that of the previous classification, after 2½ years, except that the extent of deterioration has become greater. Also, at 2½ years there was a difference between the extent of deterioration of the two silica fume concretes, with one of them being at a lower state of deterioration. At 4 years, they seem to be equally deteriorated. It is of interest to note that the 4 and 2½ years visual observations indicate a better performance of portland cement concrete with the low water/cement ratio (0.31) compared to both silica fume concretes of similar water/binder ratio.

In the 2½ years results the visual classification of the durability performance was found to be in agreement with the chloride content at a depth of 5 to 10 mm. In the 4 years' results this was not the case (Fig. 2), the reason being that in some of the specimens, corrosion was so severe that the external concrete layers were peeled off, and the 5 to 10 mm depth exposed at 4 years was practically a greater depth than that of the original concrete specimen.

Series II

Visual observation of the specimens in the second series of tests after six years of exposure is shown in Fig. 3. Results of length measurements are shown in Table 4. The behaviour in sodium sulfate solution is in agreement with numerous publications: No deterioration for the concretes with sulfate resistant cements or with ordinary portland cement + silica fume, but considerable corrosion in the ordinary portland cement. The visual observations are in agreement with the length measurements: addition of silica fume and use of sulfate resistant cement are highly effective in reducing expansion.

The situation is quite different in the concretes exposed to magnesium sulfate solution: Here, all the concretes seem to have been damaged, and the nature of corrosion is quite different than that in the sodium sulfate solution: External layers seemed to be disintegrating and peeling off (Fig. 3) without necessarily having excessive expansion (Table 4) which is similar to observations in Series I. In the concretes in sodium sulfate solution, however, cracking seemed to have occurred. In the exposure to magnesium sulfate, the presence of silica fume did not bring about any advantage, and there are signs to suggest that it even aggravated the situation. There

Fig. 1. Visual characterization of corrosion after 4 years of exposure in Dead Sea brines. (a) B 75; (b) B 40; (c) B 31; (d) MS 31; (e) CO 31; (f) VGM coating.

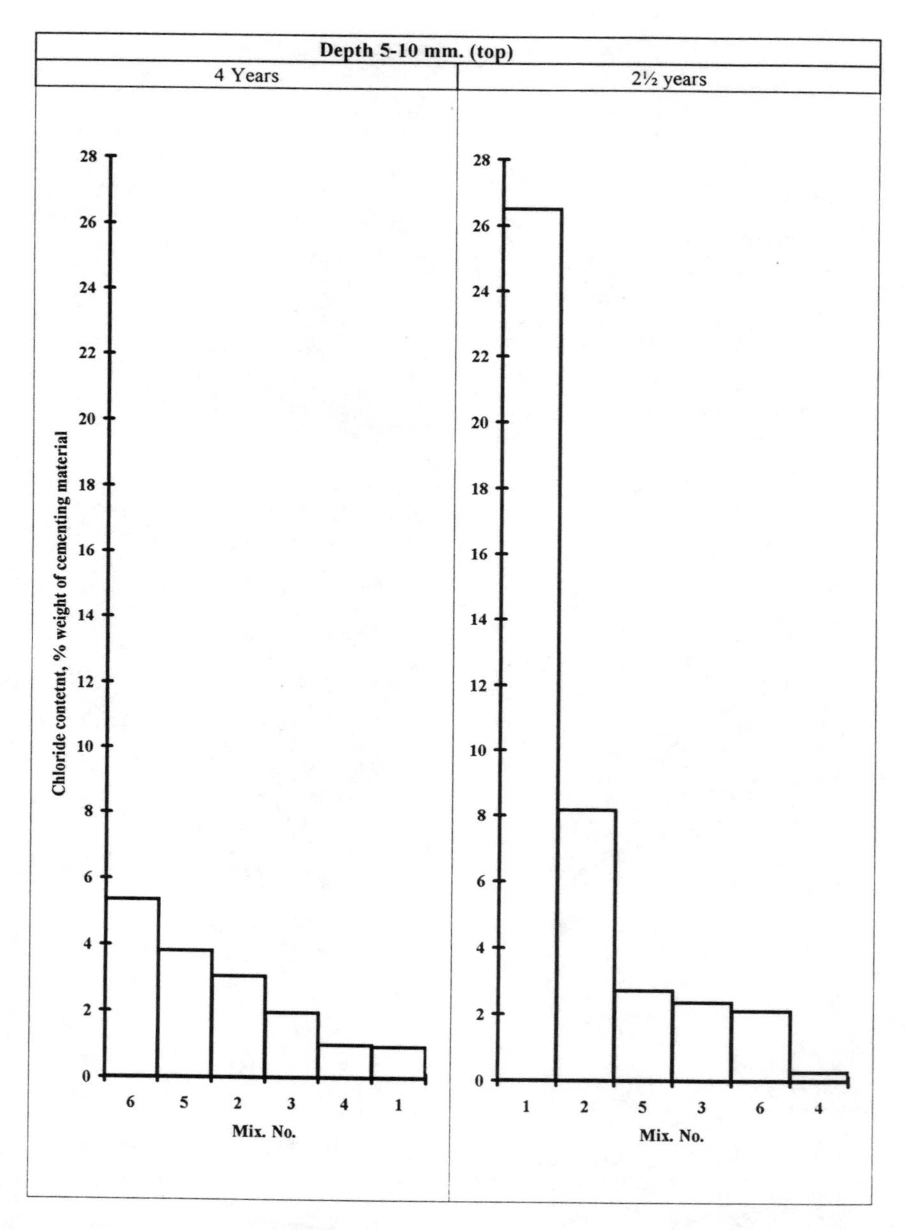

Fig. 2. Classification of concrete durability performance according to chloride content at a depth of 5-10 mm from the surface after exposure for 2½ years and 4 years in Dead Sea brines: (1) B 75; (2) B 40; (3) B 31; (4) VGM coating; (5) MS 31; (6) CO 31.

Table 3. Classification of mixes according to their durability performance in Dead Sea brines.

	Mix No.	Mix Composition: Cement content kg/m³	Mix Composition: Silica fume kg/m³	Mix Composition: Water/binder ratio	Characterization of Deterioration: 8 months & 2½ years	Characterization of Deterioration: 4 years
Increasing Durability	Coated (VGM)	Coated conventional concrete of 0.72 water/cement ratio			No signs of deterioration	No signs of deterioration
	B 31	500	--	0.31	Early signs of deterioration only after 2½ years	Partial deterioration on the surface
	CO 31	400	80; Corrocem	0.31	Early signs of deterioration after 8 months: advanced deterioration after 2½ years	All the surface shows deterioration and aggregates are slightly exposed
	MS 31	400	60; Elkem Microsilica	0.31	Mild corrosion after 8 months, heavily corroded after 2½ years	All the surface is disintegrating
	B 40	400	--	0.40	Heavy deterioration after 8 months	All the surface is disintegrating
	B 72	200	--	0.72	Very heavily deteriorated after 8 months; disintegrating after 2½ years	The bulk concrete is disintegrating

was no benefit to having the ordinary portland cement replaced by sulfate resistant cement.

One may expect that, in the magnesium sulfate solution, both types of degradation processes will occur, softening and disintegration induced by magnesium and expansion induced by sulfate. However, it seems that the magnesium attack dominates as seen by the visual observations of the portland cement concrete, as well as its expansion which is lower than that obtained in the sodium sulfate solution. It might be suggested that aggressiveness of magnesium attack is higher, and as the concrete softens there is no sufficient rigid skeleton against which expansion may occur.

Fig. 3. Visual characterization of corrosion after 6 years of exposure to concentrated magnesium sulfate ($M\overline{S}$) and sodium sulfate ($N\overline{S}$) solutions of concretes of ordinary portland cement (OPC) and sulfate resistant (SR) cement without and with 15% silica fume (SF): (a) OPC in $N\overline{S}$; (b) OPC+SF in $N\overline{S}$; (c) SR in $N\overline{S}$; (d) SR+SF in $N\overline{S}$; (e) OPC in $M\overline{S}$; (f) OPC+SF in $M\overline{S}$; (g) SR in $M\overline{S}$; (h) SR+SF in $M\overline{S}$.

Table 4. Linear expansion of high performance concretes (0.33 water/binder ratio) in various aggressive solutions

Type of Cement		Linear Expansion (%) Ordinary Portland Cement			Sulfate Resistant Cement	
Silica Fume Content		0	5%	15%	0	15%
Exposure in H_2O	9 months	0.047	0.027	0.000	0.047	0.027
	3½ years	0.092	0.040	0.016	0.073	0.084
	6 years	0.107	0.050	0.050	0.096	0.089
Exposure in $N\overline{S}$	9 months	0.042	0.029	0.000	0.039	0.000
	3½ years	0.089	0.104	0.022	0.092	0.044
	6 years	0.233	0.156	0.051	0.104	0.040
Exposure in $M\overline{S}$	9 months	0.071	0.047	0.000	0.044	0.042
	3½ years	0.121	0.071	0.058	0.110	0.100
	6 years	0.122	0.107	0.061	0.140	0.147

4. Conclusions

1. Sulfate resistant cement and ordinary portland cement with silica fume added are highly effective in reducing deterioration which may occur in exposure to sodium sulfate solution. Their presence eliminates cracking and reduces expansion considerably.
2. Silica fume and sulfate resistant cements do not seem to be effective in reducing the aggressiveness of magnesium solutions. There are signs to suggest that a very low water/cement ratio concrete of ordinary portland cement will be more effective than a composition containing silica fume. This is in agreement with studies in paste [2] and a possible explanation is the lower C/S ratio of the pozzolanic CSH, making it more susceptible to attack by magnesium solutions [6].
3. In conditions where both sulfate and magnesium attack may occur, like in magnesium sulfate solution, it seems that the magnesium attack will dominate.

5 References

1. FP State of the art report (1988) *Condensed Silica Fume in Concrete*, Thomas Telford, U.K.
2. Cohen, M.D. and Bentur, A. (1988) Durability of Portland Cement-Silica Fume Pastes in Magnesium and Sodium Sulfate Solutions, *ACI Materials Journal* 85(3), pp.148–157.
3. Mather, K. (1982) Current Research in Sulfate Resistance at the Waterways Experiment Station, *George Verbeck Symp. on Sulfate Resistance of Concrete*, ACI SP-77, Amer. Concr. Inst., Detroit.

4. Mehta, P.K. (1985) Chemical Attack of Low Water Cement Ratio Concretes Containing Latex or Silica Fume as Admixtures, Proc., *RILEM/ACI Symp. On Technology of Concrete when Pozzolans, Slags and Chemical Admixtures are Used*, Monterey, Mexico.
5. Bentur, A. and Ben-Bassat, M. (1992) Durability of High Performance Concretes in Highly Concentrated Magnesium Solutions, pp. 26-29, in *Durability of Building Materials and Components 6*, S. Nagataki, T. Nireki and F. Tomosawa (Eds.), Proc. Int. Conf. Japan, E&FN Spon.
6. Bonen, D. and Cohen, M.D. (1992) Magnesium Sulfate Attack on Portland Cement Paste II. Chemical and Mineralogical Analyses, *Cement and Concrete Research*, 22(4), pp.707–718.

122 COMBINED EFFECT OF ACID AND DEICER SOLUTIONS ON FIBER-REINFORCED HARDENED CEMENT PASTES

T. FUJII
Tokyo University of Agriculture and Technology, Tokyo, Japan

Abstract

Fiber-reinforced composite materials were immersed in combined solutions of sulfuric acid and deicing chemicals such as NaCl and calcium magnesium acetate (CMA). Time-dependent change in pH and concentration of Ca^{2+} ions in the solutions was measured during the immersion. Residual amount of calcium hydroxide and microhardness on the specimens before and after the immersions were also measured. The residual amounts of calcium hydroxide decreased with increasing in immersion period in the combined solutions, especially in those of acid and CMA. The lowest value in microhardness was observed on any specimen immersed in the mixed solution of acid and CMA among the combined solutions. It is found that the combined action of the sulfuric acid and the deicer solution is more deleterious to the cement-based composite materials than the action of solely sulfuric acid solution of pH3.

Keywords: Aramid fiber, calcium hydroxide, calcium ion, carbon fiber, deicing chemicals, hardened cement paste, microhardness, sulfuric acid.

1 Introduction

Deicing chemicals such as NaCl and $CaCl_2$ have been used to prevent freezing or remove snow and ice from pavements for the purpose of traffic safety in cold and snowy regions. Such chlorides not only deteriorate concrete and steel bars in reinforced concrete structures, but also significantly influence the environment due to penetration of Cl^- ions into the soil, rivers and lakes. Calcium magnesium acetate, so called CMA deicer, has been developed in the United States as a harmless deicing substitute for those harmful chlorides. It is reported that CMA neutralizes acid rain water which has penetrated into the soil, rivers and lakes [1]. It could become one of the most ideal

Concrete Under Severe Conditions: Environment and loading (Volume Two) Edited by K. Sakai, N. Banthia and O.E. Gjørv. Published in 1995 by E & FN Spon. ISBN 0 419 19860 1

deicing chemicals from the view point of environmental preservation. However, it has become clear that CMA is as deleterious as the chlorides to concrete [2].

On the other hand, acid rain has become relatively widespread over much of Japan as in Europe and North America [3]. Its influence on the environment is becoming a serious problem. As a consequence, concrete structures are now subjected to an aggressive environmental condition which was not expected at the time of construction. Research concerned with effects of acid solutions on concrete has been carried out to investigate the deterioration on concrete structures exposed to acid soil or acid ground water [4], while study on the influence of acid rain on deterioration in concrete is a very recent subject [5][6][7][8]. The behavior of fiber-reinforced hardened cement pastes exposed to acid solutions has been already discussed for the purpose of developing a surface treatment material for protecting concrete against acid rain [9].

The deterioration in concrete structures due to the combined chemical action of acid rain and the above mentioned deicers should be investigated, especially in cold and snowy regions. However, a proper countermeasure for the deterioration of concrete structures, which are chemically attacked by the combined action of acid rain and the deicers, has not been investigated. In this sudy, the fiber-reinforced hardened cement pastes, which are composed of portland cement paste matrix and the inclusion of short fibers such as carbon fibers or aramid fibers, were immersed in acid solutions, deicer solutions, or mixed solutions of acid and deicers. The influence of the kinds of fiber and solution, and immersing orders of the solutions on the properties of cement-based composite materials was discussed on the basis of the measurements of the changes in pH value and concentration of Ca^{2+} ions in the solutions during the immersions, and of the residual amount of calcium hydroxide and the microhardness on the specimens before and after the immersions.

2 Experimental procedure

The mix proportions for specimens are given in Table 1. Ordinary portland cement was used, and anionic super-plasticizer was added at 2% of cement by weight. Composite hardened cement pastes were prepared, i.e., carbon or aramid fibers was added to plain cement paste. The properties of the fibers are shown in Table 2. The specimen size was 8 mm in diameter and 30 mm in length, and Teflon tube was used as a mold.

Table 1. Mix proportions for specimens

Symbol	Additive	W/C (%)	Plasticizer (%)	V_f (vol.%)
NC	Non	30	2.0	0
CF	Carbon fiber	30	2.0	1.0
AF	Aramid fiber	30	2.0	1.0

Table 2. Properties of fibers

Kind of fiber	L (mm)	ϕ (μm)	f_t (MPa)	E_t (GPa)	ε (%)	γ (g/cm^3)
Carbon fiber	24	7	3,600	240	1.5	1.76
Aramid fiber	13	13	2,800	63	4.0	1.44

The specimens were set on a swing-roller at 20 ℃ for 24 hours to maintain homogeneity of the materials. Then they were cured in saturated calcium hydroxide solution at 20 ℃ for 6 days to avoid leaching from crystallized calcium hydroxide. The immersing processes after the curing are as follows:

(1) Continuous immersion in H_2SO_4 of pH1, 3 or 5 for 56 days [HS-pH1, 3 or 5]
(2) Subsequent immersion in 3% NaCl solution for 28 days after immersion in H_2SO_4 of pH3 for 28 days [HS>NaCl]
(3) Continuous immersion in a mixed solution of H_2SO_4 of pH3 and 3% NaCl solution in the ratio of 1:1 (weight) for 56 days [HS+NaCl]
(4) Subsequent immersion in 3% CMA solution for 28 days after immersion in H_2SO_4 of pH3 for 28 days [HS>CMA]
(5) Continuous immersion in a mixed solution of H_2SO_4 of pH3 and 3% CMA solution in the ratio of 1:1 (weight) for 56 days [HS+CMA]
(6) Continuous immersion in saturated $Ca(OH)_2$ solution for 56 days for control [CH56]

CMA in the mole ratio $Ca(CH_3COO)_2/Mg(CH_3COO)_2$ of 1:1 was selected. Volume of the solutions for immersion was ten times that of the specimens, i.e., about 50 milliliter. All solutions were replaced after every seven days, and changes in pH and concentration of Ca^{2+} ions in the solutions were measured during the immersions. An ion meter was used for measurements of pH and concentration of Ca^{2+} ions. Samples of 3 mm thickness were cut from the specimens of ϕ 8x3 mm before and after the immersions, and they were doused into acetone to cease hydration. Then the surfaces of the samples for measuring microhardness were mirror-finished using emery paper. A dynamic ultra-microhardness tester was used with the applied load of 500 mN. Microhardness was measured at 20 points which were at intervals of 200 μm from the starting point of 50 μm from the circular edge of each sample. The samples for a differential thermal analysis (DTA) were pulverized in acetone and dried in a vacuum. Then residual amounts of calcium hydroxide were analyzed on each sample of 20 mg, which was heated up to a temperature of 650 ℃ at a heating rate of 20 ℃/min. in nitrogen gas.

3 Results and considerations

3.1 Time-dependent pH change in solutions

Fig.1 shows, as a typical example, the time-dependent change in pH of the solutions in which the carbon fiber composites were immersed. Each solution shows pH value of about 12.5 after one week of the immersion. The pH of the solutions maintains a value of about 12 after eight weeks in the immersing processes (1) [HS-pH3, -5], (2) [HS>NaCl], and (3) [HS+NaCl]. In the immersing process (5) [HS+CMA], the pH decreases gradually after two weeks and reaches about 10 after eight weeks, almost the same as in the process (4)[HS>CMA], in which the pH of the solution decreases after five weeks and reaches about 10 after eight weeks. On the contrary, in the immersing process (1) [HS-pH1], the pH of the solution decreases abruptly and reaches about 2 after three weeks. Differences in the change in pH were scarcely observed among the carbon fiber composite, the aramid fiber composite, and the plain paste.

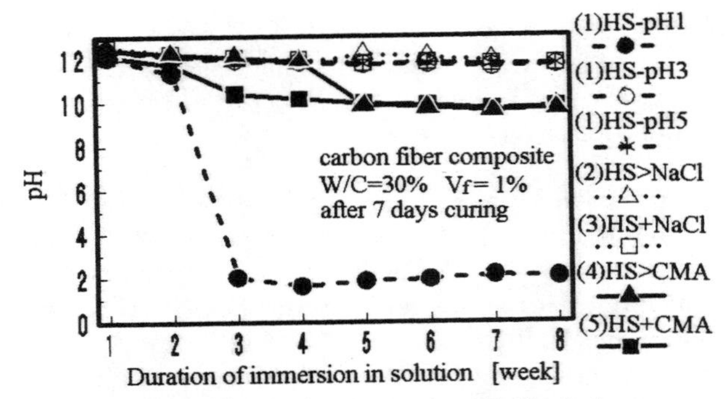

Fig. 1. Typical example of change in pH of solutions.

As a result of the above investigation, it is found that the pH values of the solutions, in which the specimens were immersed, are almost the same in the combined solutions of acid and NaCl as in the solely acidic solution of pH3, and that the pH value decreases by about 2 in the combined solution of acid and CMA.

3.2 Time-dependent change in Ca^{2+} ion concentration in solutions

Fig.2 shows the time-dependent change in concentrations of Ca^{2+} ions in the solutions in the case of carbon fiber composites as a typical example. The Ca^{2+} ions in the initial CMA solutions have been subtracted in the Ca^{2+} ion concentrations shown in Fig.2 . The concentrations of Ca^{2+} ions in the immersing processes (2) [HS>NaCl] and (3) [HS+NaCl] show almost the same behavior as that in the processes (1) [HS-pH3] and [HS-pH5] during the period from three to eight weeks. The concentrations of Ca^{2+} ions are greater in the immersing processes (4) [HS>CMA] and (5) [HS+CMA] than in the other processes. The concentrations of Ca^{2+} ions for the immersing process (4) [HS>CMA] show similar behavior as in the process (5) [HS+CMA] at the stage after seven weeks of the immersion. It is probably explained by a following reaction formula that the concentrations of Ca^{2+} ions increases in the combined solution of CMA.

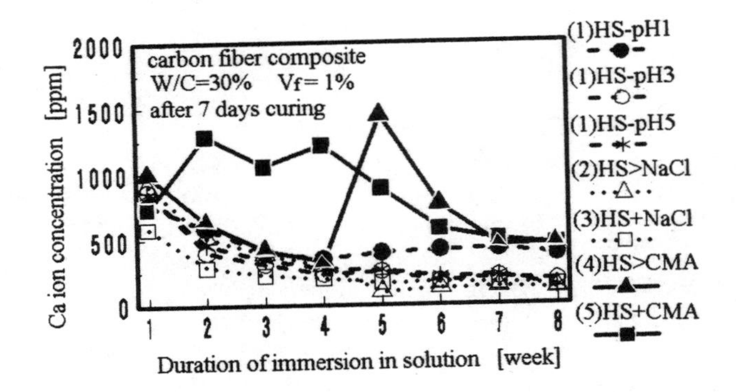

Fig. 2. Typical example of change in concentration of Ca^{2+} ions in solutions.

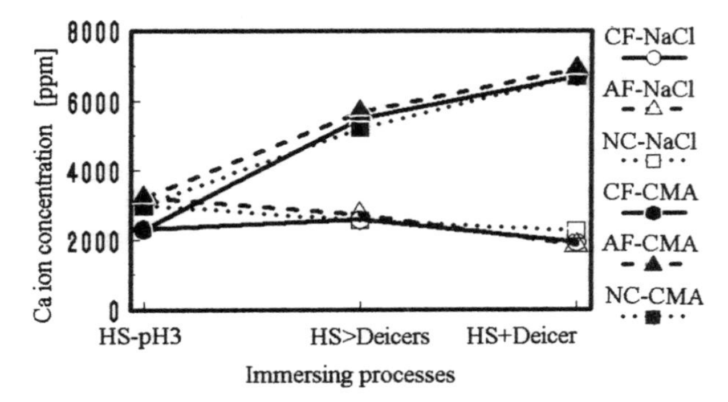

Fig. 3. Effect of combinations of deicers and acid solution on leach of Ca^{2+} ions.

$$Ca(OH)_2 + Mg(CH_3COO)_2 \rightarrow Mg(OH)_2 + Ca^{2+} + 2(CH_3COO)^-$$

It should be noted that the total amount of Ca^{2+} ions leached in the mixed solution is greater than in the individual solution of acid and CMA. Decreasing behavior in the concentrations of Ca^{2+} ions in each solution corresponds to the changes in pH as shown in Fig.1. The pH and the concentration of Ca^{2+} ions in the solutions for the hardened paste without fibers were slightly lower than those for the fiber-reinforced pastes at the first stage of immersion. However, there were ultimately no significant differences among them.

Fig.3 shows the effect of combinations of the deicers and the sulfuric acid solution on the accumulated concentrations of Ca^{2+} ions during eight weeks of immersions, i.e., the sum of eight measures. The accumulated concentrations of Ca^{2+} ions in the combined solutions of acid and NaCl (the processes (2) [HS>NaCl] and (3) [HS+NaCl]) are slightly lower than those in the solely sulfuric acid solution of pH3 (the process (1) [HS-pH3]). On the other hand, the concentrations of Ca^{2+} ions are greater in the combined solutions of acid and CMA (the processes (4) [HS>CMA] and (5) [HS+CMA]) than in the solely acidic solution. This is caused by the leach of Ca^{2+} ions due to the chemical reaction between calcium hydroxide and magnesium acetate as shown previously. There was an adequate correlation between the concentrations of Ca^{2+} ions and pH values in the acid solution of pH3 in which the specimens were immersed [9]. This suggests that the change in concentration of Ca^{2+} ions is attributed to leaching from calcium hydroxide, although it is pointed out that the supply of Ca^{2+} ions is caused by the decomposition of cement hydrates such as calcium silicate hydrate and ettringite [6]. As a result, it can be said that the influence of fiber-reinforcing and the kinds of fiber is not significant on the difference in concentrations of Ca^{2+} ions in the solutions, although the effect of CMA on the leach of Ca^{2+} ions is greater than that of other solutions.

3.3 Residual amount of calcium hydroxide

Fig.4 shows the effect of the initial pH values of the acid solutions on the residual amount of calcium hydroxide in the specimens after eight weeks of the immersions. The residual amount of calcium hydroxide decreases with decreasing initial pH value of the

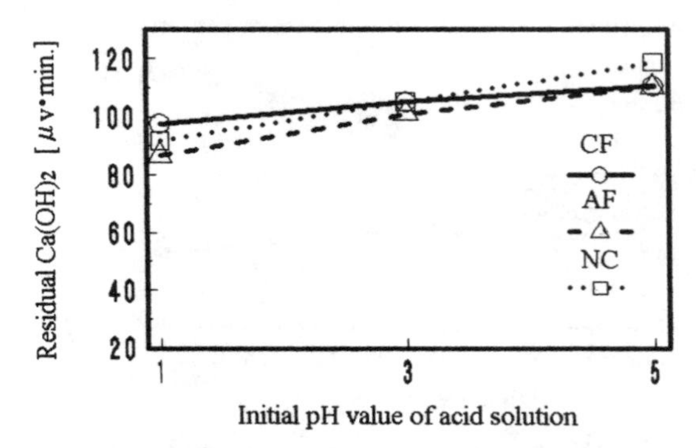

Fig. 4. Effect of pH of acid solutions on residual $Ca(OH)_2$ amount.

acid solution. There is not much difference in the amount of calcium hydroxide among the specimens. It is indicated that leaching from calcium hydroxide increases, not depending on the type of composite, with decreasing initial pH value of the acid solution.

Fig.5 shows the effect of the combinations of deicers and sulfuric acid solution of pH3 on the residual amount of calcium hydroxide. The residual amount of calcium hydroxide decreases much more in the processes (3) [HS+NaCl], (4) [HS>CMA], and (5) [HS+CMA] than in the solely acidic solution (the process (1) [HS-pH3]) in any case. It is also less in the mixed solutions (the processes (3) [HS+NaCl] and (5) [HS+CMA]) than in the subsequent immersions into the deicer solutions (the processes (2) [HS>NaCl] and (4) [HS>CMA]). Moreover, the residual amount of calcium hydroxide is significantly less in the CMA solution than in the NaCl solution. In these experiments, there is not much difference in the residual amount of calcium hydroxide between carbon fiber composites and aramid fiber composites. However, it is pointed out that the difference in the residual amount of calcium hydroxide between the carbon fiber composites and the aramid fiber composites was attributed to the deposition of calcium hydroxide at the interfacial zones between the fibers and the matrix [10]. Leaching from calcium hydroxide is slightly greater in the combined solutions of NaCl than in the solely sulfuric acid solution for any composite, and the corresponding cumulative concentration of Ca^{2+} ions decreases as shown in Fig.3. Therefore, it is indicated that

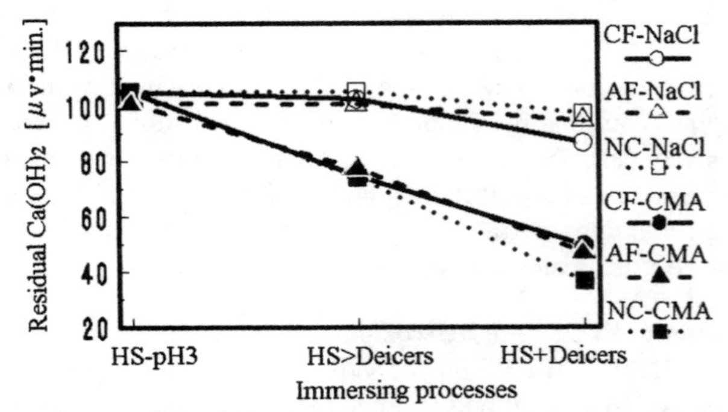

Fig. 5. Effect of combinations of deicers and acid solution on residual $Ca(OH)_2$ amount.

the leached Ca^{2+} ions were consumed by the formation of compound having lower solubility. Leaching from calcium hydroxide is significantly greater in the combined solutions of acid and CMA than in the solely sulfuric acid solution for any composite, and the corresponding cumulative concentration of Ca^{2+} ions increases as shown in Fig.3. It suggests that the effect of CMA, above all magnesium acetate, on leaching from calcium hydroxide is greater than that of the solely sulfuric acid solution of pH3.

3.4 Distributions of microhardness

According to the order statistics, probabilities can be calculated by the equation, $P_n = (n - 1/2)/N$, in which n is ordinal numbers in ascending ordered data, and N is a total number of data. A graph of data versus probabilities is called Hazen plots which is useful to analyze a distribution of data. Fig.6 shows Hazen plots for distributions of microhardness on the carbon fiber composites before and after the immersions for an example. In the case of the mixed solution of acid and NaCl (the immersing process (2) [HS>NaCl]), the hardness shows a slightly increasing tendency among the solutions of acid and deicers. Increasing rates in hardness are greater at the range of higher values than at that of lower values in hardness. However, the hardnesses decrease in any immersion of the CMA solution, particularly in the mixed solution of acid and CMA (the process (5) [HS+CMA]). Among the solutions of pH3 acid, NaCl and CMA, an order of the hardnesses at 50% probability, which correspond to average values, indicates that the hardness for the mixed solution of acid and CMA is lower than that for the solely acidic solution for the carbon fiber (Fig.6) and aramid fiber composites. On the contrary, the hardness value for the mixed solution of acid and CMA was greater than that for the solely acidic solution for the plain pastes.

When hardness increases after immersion in the solutions, the distribution of hardness shifts generally toward the higher value, and has a tendency that an increasing rate in hardness is greater at the range of higher values. Consequently the variations in hardnesses become greater, i.e., the slope becomes gentler. On the other hand, when hardness decreases after immersion in the solutions, the distribution of hardness has a tendency that a decreasing rate in hardness is greater at the range of lower values. This tendency on the distribution of hardness was observed on hardened cement pastes subjected to freeze-thaw cycles [11], and on cement-based fiber composites

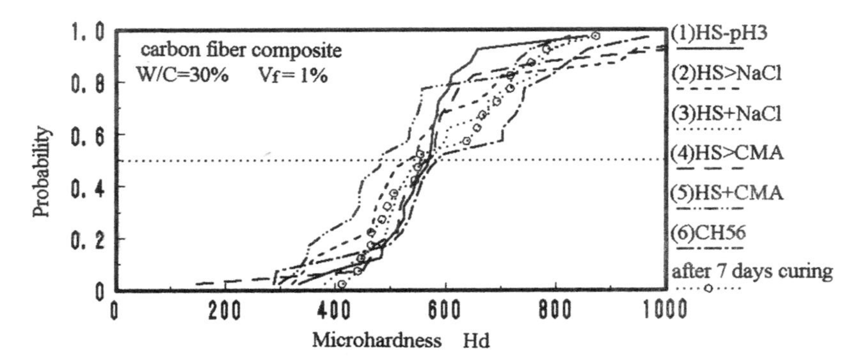

Fig. 6. Change in microhardness distributions.

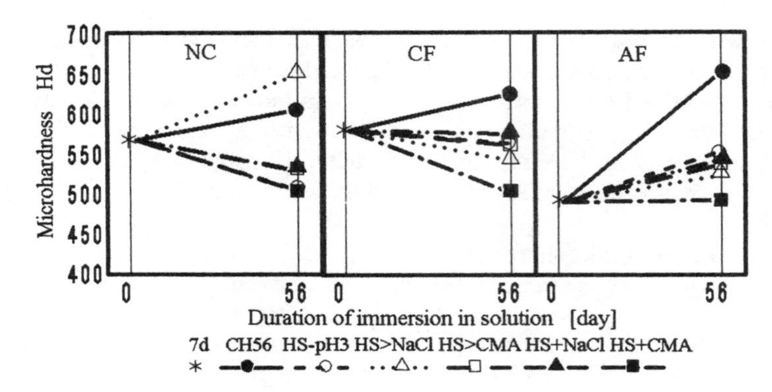

Fig. 7. Change in microhardness after immersions.

exposed to deicing chemicals [10]. This suggests that the structural system of the fiber-reinforced hardened cement pastes is composed of several kinds of particle which mechanical properties are distributed over a wide range, depending on hydrated compounds, degree of hydration, their structural arrangement, and interaction between fibers and matrix.

3.5 Change in microhardness

Fig.7 shows comparison in the hardnesses before and after the immersions, which are the average values of 20 measures for each sample. The hardness values both on the plain paste and the carbon fiber composite are similar after seven days of curing, while the aramid fiber composite indicates lower value than other two. The highest hardness values are obtained on each composite after immersion in the saturated calcium hydroxide solution (the process (6)[CH56]). In contrast, the highest hardness value is observed on the plain paste in the process (2)[HS>NaCl]. On the other hand, the lowest hardness is shown on any specimen in the mixed solution of acid and CMA (the process (5) [HS+CMA]).

Fig.8 shows the combined effects of acid and deicer solutions on the microhardness. The hardness has a tendency to decrease much more in the combined solution of acid

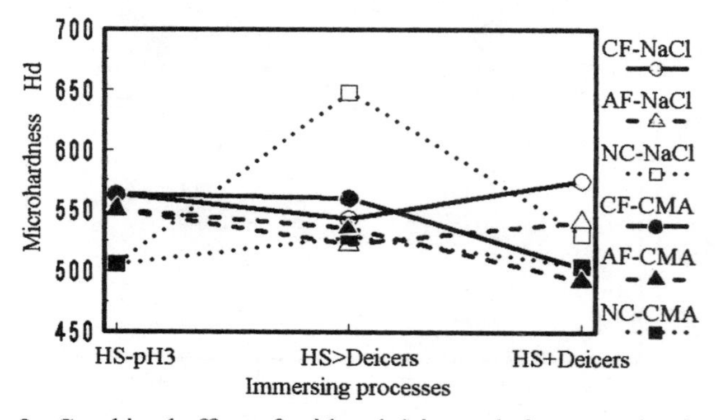

Fig. 8. Combined effect of acid and deicer solutions on microhardness.

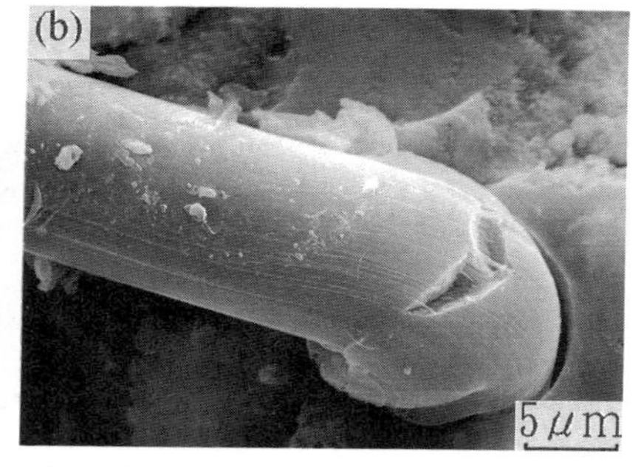

Fig. 9. (a) Collapse of $Ca(OH)_2$ at interfacial zone of carbon fiber and (b) bond slip failure between aramid fiber and matrix.

and the deicers than in the solely sulfuric acid solution of pH3 on the fiber-reinforced composites. Particularly the hardness decreases in the combined solutions of acid and CMA (the processes (4)[HS>CMA] and (5)[HS+CMA]). However, the hardness increases slightly on the fiber-reinforced composites immersed in the mixed solution of acid and NaCl (the process (3)[HS+NaCl]). The decreasing tendency in hardness corresponds to the reduction in residual amount of calcium hydroxide in the combined solution of acid and CMA as shown in Fig.5.

It should be noted that there is not much difference in hardnesses among the specimens in the combined solution of acid and CMA (the processes (4)[HS>CMA] and (5)[HS+CMA]), although the fiber-reinforced composites are greater in hardness than the plain paste in the solely acidic solution (the process (1)[HS-pH3]). This suggests that fiber-reinforcing is not effective in the combined solution of acid and CMA. It can be pointed out that the variation in hardness is caused by whether the matrix structure is dense or porous in the system depending on crystallization of calcium hydroxide or leaching from it, and on decomposition of C-S-H. Furthermore, there is a number of factors affecting the behavior of hardness such as interfacial zones between fibers and paste matrix, and the difference in compounds deposited in the systems [7]. Calcium hydroxide increases interfacial bond strength between fibers and matrix due to crystallization, although leaching from calcium hydroxide causes reduction in it (Fig.9) [12].

4 Summary and conclusions

From the results of this investigation, following summary and conclusions can be drawn.

1. The pH values decrease from 12 to 10 after eight weeks of immersion in the combined solutions of pH3 sulfuric acid and CMA. On the other hand, the pH values are kept almost constant at 12 in other solutions, except the acid solution of pH1.
2. The time-dependent change in concentrations of Ca^{2+} ions is higher in the combined solutions of acid and CMA than in the solely sulfuric acid solution.
3. The residual amount of calcium hydroxide in the specimens after the immersion is generally lower in the mixed solutions of acid and the deicers than in the solely

acidic solution.

4. The hardnesses on the fiber-reinforced composites are lower in the mixed solutions of acid and CMA than in the solely acidic solution.
5. The factors affecting on the behavior of these composites are addition of fibers, the kinds of solution, and the immersing orders into the solutions. However, the kinds of fiber do not much affect on the behavior.
6. The combined action of the acid and the deicer solutions is more deleterious to the cement-based composite materials than the action of the solely sulfuric acid solution of pH3.

5 Acknowledgment

The author extends his appreciation to the Ministry of Education, Science and Culture for supporting this research by the Grant-in-aid for general scientific research.

6 References

1. Schenk, R.U. (1986) Ice-melting characteristics of calcium magnesium acetate, PB Report, PB-86-142742.
2. Fujii, T. and Fujita, Y. (1987) Influence of CMA on scaling deterioration of hardened cement pastes, *Proc. of Japan Concrete Institute*, Vol.9, No.1. pp.543-548.
3. Nomura, K. (1993) *Acid Rain and Acid Fog*, Shokabo Co. Ltd., Tokyo.
4. Moum J. and Rosenqvst, I.T. (1959) Sulfate attack on concrete in the Oslo region, *J. of American Concrete Institute*, Vol.56, No.9, pp.257-265.
5. Kong, L.H. and Orbison, J.G. (1987) Concrete deterioration due to acid precipitation, *ACI Materials Journal*, Vol.84, No.2, pp.110-116.
6. Kobayashi, K. and Uno, Y. (1991) Influence of acid rain on deterioration of concrete structures, *Proc. of Japan Concrete Institute*, Vol.13, No.1. pp.615-620.
7. Nakano, T., Noda, S., Urakami, Y. and Mori, T. (1992) Analysis of secondary products from mortar specimens corroded by sulfuric acid, *Trans. of the Japanese Society of Irrigation, Drainage and Reclamation Engineering*, No.161, pp.25-30.
8. Yoshida, C. and Higashiyama, I. (1993) Acid-proof properties of concrete and behavior of Ca ion, *Trans. of the Japanese Society of Irrigation, Drainage and Reclamation Engineering*, No.165, pp.129-137.
9. Fujii, T. and Nakamura, A. (1994) Behavior of fiber-reinforced hardened cement pastes immersed in acid solutions, *Proc. of Japan Concrete Institute*, Vol.16, No.1, pp.907-912.
10. Fujii, T. (1993) Freeze-thaw resistance of composite hardened cement pastes subjected to deicing chemicals, *Proc. of Japan Concrete Institute Symp. on Performance of Concrete under Natural Weathering Conditions,* pp.93-100.
11. Fujii, T. and Fujita, Y. (1988) Microhardness of hardened cement paste subjected to deicer and freeze-thaw action, *Japan Cement Association Review 1988*, pp.198-201.
12. Fujii, T. (1993) Microhardness of composite hardened cement pastes, *Japan Cement Association Proc. of Cement & Concrete*, No.47, pp.372-377.

123 SEAWATER RESISTANCE OF ZERO-SLUMP POROUS CONCRETE USING FOUR TYPES OF CEMENT

K. AMO
Anan College of Technology, Tokushima, Japan
K. KOHNO
University of Tokushima, Tokushima, Japan
N. KINOSHITA
Hanshin Expressway Public Corporation, Osaka, Japan

Abstract

Porous concrete having continuous voids is expected to be used for an artificial fish reef because it may lead to the attachment of seaweed and the growth of marine ecological resources. This paper presents an investigation relating to the seawater resistance of zero-slump porous concretes using four types of cement. The void ratio of each specimen made was about 15 percent. One group of specimens was repeatedly soaked and dried. The others were soaked in still seawater, flowing seawater or a solution of 10% $MgSO_4$. The durability of the porous concrete in seawater is discussed from the results obtained.
Keywords: Compressive strength, porous concrete, relative dynamic modulus of elasticity, seawater resistance, types of cement, x–ray analysis.

1 Introduction

Recently, it has been reported that porous concrete having many continuous voids may contribute to the preservation of the environment by cleaning up seawater, seeping rainwater into the ground,etc.[1],[2]. The use of porous concrete for an artificial fish reef [3] is also expected to solve several problems. In general, concrete for an artificial fish reef needs to have high water–tightness and good durability; conventional concrete has these qualities but is not considered best for growing seaweed. Porous concrete, although good for marine ecology, has lower compressive strength and durability than conventional concrete because of the increase interior continuous voids.

Therefore, in this paper the seawater resistance of zero-slump porous concrete using four types of cement, ordinaly portland cement, fly ash cement, blast–furnace slag cement and sulfate resisting portland cement, are discussed on the basis of

Concrete Under Severe Conditions: Environment and loading (Volume Two) Edited by K. Sakai, N. Banthia and O.E. Gjørv. Published in 1995 by E & FN Spon. ISBN 0 419 19860 1

experimental results in order to obtain basic data for future application as an artificial fish reef. The specimens having the void ratio of about 15 percent for compressive strength, dynamic modulus of elasticity and x–ray analysis were repeated wet–and–dry treatment, and were soaked in still seawater, flowing seawater and a solution of 10% $MgSO_4$, respectively.

2 Experimental procedure

2.1 Materials

Ordinary portland(OP) cement, B– type fly ash(FA) cement, B– type blast– furnace slag(BS) cement and sulfate resisting portland(SR) cement were used. The aggregate used was crushed limestone, No.7(2.5 ~ 5mm) from Kochi prefecture.

The main physical properties of materials used are shown in Tables 1 and 2.

Table 1. Physical properties of cement

Type of cement	Specific gravity	Blaine's value (cm^2/g)	Compressive strength (MPa)	
			7d	28d
OP	3.15	3420	25.6	40.6
FA	2.96	3500	22.0	36.8
BS	3.05	3890	22.1	42.1
SR	3.20	3390	22.7	36.4

Table 2. Physical properties of aggregate

Type of aggregate	Specific gravity	Water absorption (%)
Limestone	2.68	0.62

2.2 Mixture proportion and fabrication of specimens

The mixture proportions of concrete are shown in Table 3. The target of for the void percentage was 15%. Concrete was mixed by a pan–type forced circulating mixer of 50 liter. The batching of materials,mixing method and time are shown in Fig.1.

The Void percentage was measured according to JIS A 1116 immediately after mixing the concrete. The concrete was placed in to a cylindrical mold, ϕ of 7.5 and length of 15cm, and then compacted by a table vibrator (frequency : 4000rpm, amplitude : 1.0mm) for 40 seconds.

Specimens were cured in a water tank at 20 ℃ until the age of 7 days or 28 days, and then were used for each tests.

Table 3. Mixture proportion of concrete

Type of mix	W/C (%)	Weight per unit volume(kg/m³)			Target of void percentage (%)
		W	C	A	
OC	26	142	547	1431	15 ±1
FC	26	135	521	1445	15 ±1
BC	26	138	531	1440	15 ±1
SC	26	139	537	1456	15 ±1
CC	26	240	400	1574	4 ±1

Note)OC : Ordinay portland cement
FC : Fly ash cement
BC : Blast–furnace slag cement
SC : Sulfate resisting portland cement
CC : Conventional concrete (s/a=50%, river sand was used)

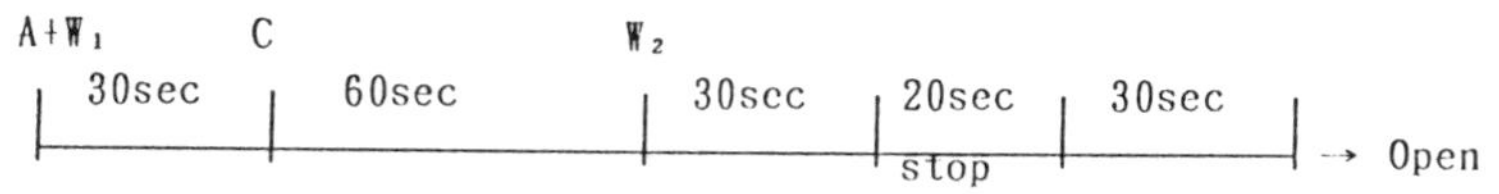

A : aggregate, C : cement, W_1 : primaly water 25% for total water content, W_2 : second water

Fig.1. Batching materials, mixing method and time.

3 Result and discussion

3.1 Results of cycles of wet–and–dry treatment

Figures 2 and 3 show the relationship between the number of wet–and–dry treatment cycles and the relative dynamic modulus of elasticity(RE_D) of concrete using various types of cement. One cycle of treatment involved soaking the test piece in still seawater at 20 ℃ for 24 hours and drying in 65 ℃ air for 24 hours. As can be seen from these figures, the RE_D of the porous concrete made of BS cement of BS dropped sharply after 60 cycles of wet-and-dry treatment in the case of the 7–day standard curing and after 40 cycles in the case of the 28–day standard curing. With each of the other types of cement, however, the RE_D increased slightly throughout the treatment. The drops in the BS cement concrete data were probably due to the fact that the drying shrinkage was much greater than that of the other types of cement [4].

According to Nishibayashi et al.[5], their experimentation with the repetition of wet-and-dry treatments as a means of accelerated testing shows that one cycle of treatment corresponds to 50 to 60 days of continuous soaking. Therefore, the exception of BS concrete, it can be estimated that concrete soaked in seawater will not deteriorate before 13 to 16 years. Similar assumption might be made as to the seawater resistance of porous concrete, if the proper type of cement to be used has been decided.

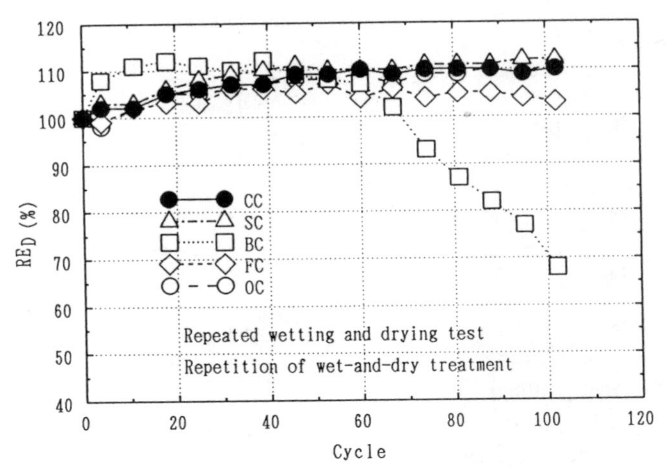

Fig.2. Relationship between cycles of treatment and RE_D.
(after 7 days in water curing)

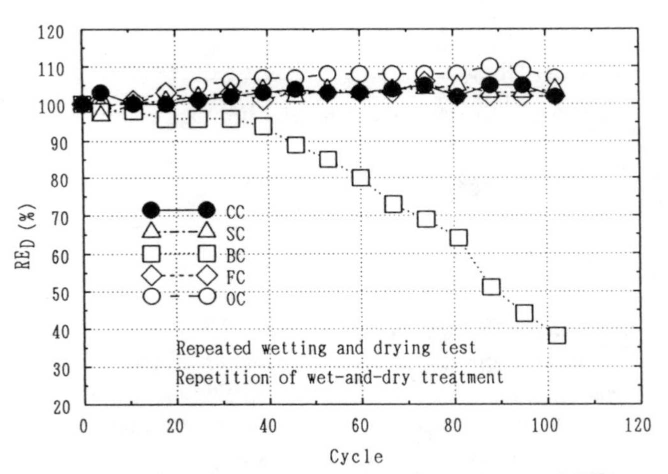

Fig.3. Relationship between cycles of treatment and RE_D.
(after 28 days in water curing)

The effect of concrete type on compressive strength after repeated wet– and– dry treatments is shown in Fig.4.

It is obvious that the compressive strength of porous concrete after 50– cycles of treatment were generally greater than those just after standard curing. After 100 cycles of wet–and–dry treatment were performed after 7–day and 28–day standard curing, the compressive strength of concrete using BS cement was the lowest relative to the other concrete.

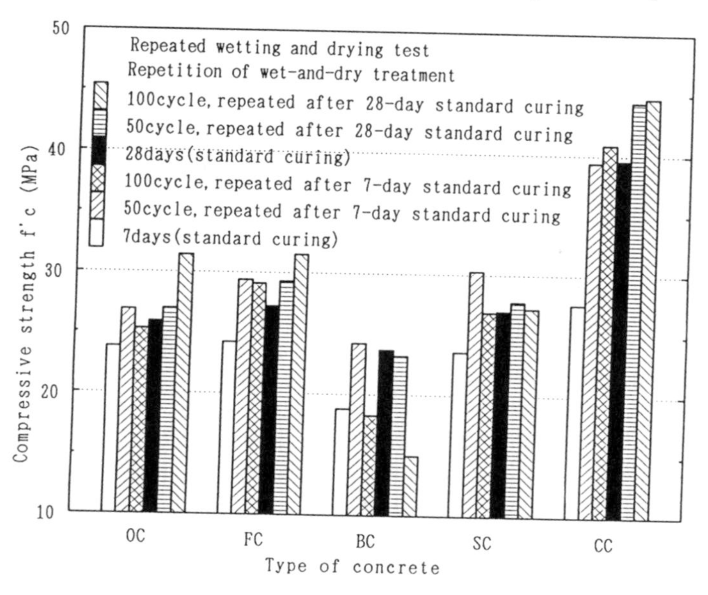

Fig.4. Effect of types of concrete on compressive strength.

3.2 Results of tests of soaking in still water, still seawater and flowing seawater

As can be seen in Fig.5, the rate of strength development with age of porous concrete was small in the case of standard curing. In addition, the effect of concrete type on compressive strength was also found to be small. It should be noted that early strength development of porous concrete was promoted by the presence of continous voids, which increased the area of cement paste in contact with water[6]. Therefore, the addition of such substances as fly ash or blast–furnace slag, which shows slow hydration,

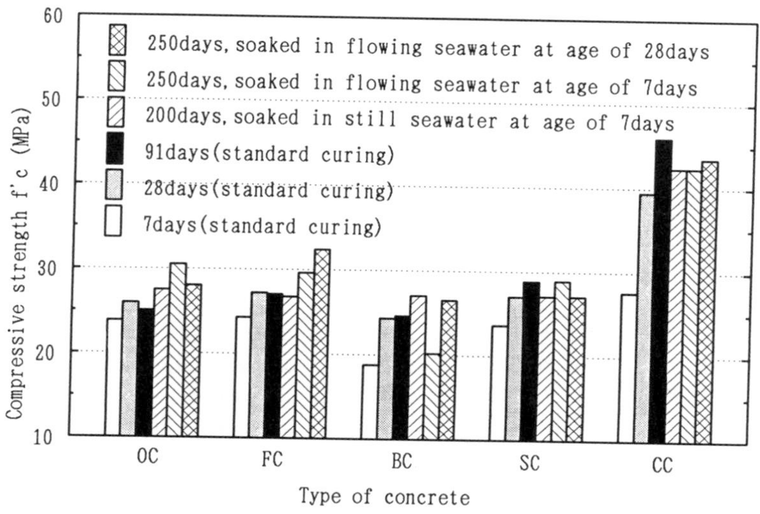

Fig.5. Effect of types of concrete on compressive strength.

can be effective for making porous concrete.

With the soaking test in flowing water(at a flowing rate of 30 to 40cm per second), the specimensshowed no decrease in strength after 250 days. $Ca(OH)_2$ products resulting from hydration easily flow out of the concrete. This tendency is intensified by an increase of interior continuous voids. Though it may seem that the increase of voids affects the strength of concrete, there was little influence in the case of porous concrete with about 15% voids, and it made no difference in the strength of porous concrete whether the specimens were soaked in still seawater or flowing seawater.

The effects of the soaking method and the types of concrete soaked after the 7−day standard curing, excepting BS cement concrete, on compressive strength were small.

3.3 Results of soaking test in a solution of 10% MgSO4

It is generally stated that concrete collapses in seawater because of ettringite produced in a reaction between $3CaO{\cdot}Al_2O_3$ and sulfate existing in seawater. For that reason, the specimens were soaked in a solution of 10% $MgSO_4$ to estimate their resistance to sulfate attack. Subsequently, the amount of deterioration was evaluated by the measurements of dynamic modulus of elasticity and the compressive strength as a function of age. Figure 6 shows the relationship between the soaking period in a solution of 10% $MgSO_4$ and RE_D.

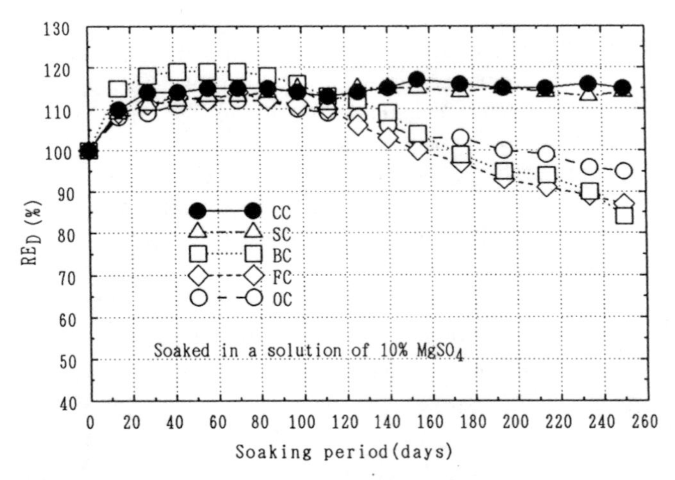

Fig.6. Relationship between soaking period and RE_D.

The specimens of conventional concrete and porous concrete using SR cement did not show any decrease in the RE_D after 250 soaking days. This occurred because the permeability of conventional concrete is extremely low as relative to types of porous concrete and the amount of $3Ca{\cdot}Al_2O_3$ in SR cement is less in other cement. Therefore, the use of SR cement is effective for improving the resistance of porous concrete to sulfate attack.

From the tests of compressive strength, conventional concrete and SR cement concrete were also showed high strength at a soaking age of 250 days, and the resistances to $MgSO_4$ were higher relative to the other types of concrete.

3.4 Results of x–ray analysis

Figure 7 and 8 show examples of the results of the x–ray analysis. As can be seen in Figures 7 and 8, $Ca(OH)_2$ products were recognized in all types of concrete in the case of standard curing in water, but not in the case of porous concrete after a 100–cycle treatment. In addition, products such as ettringite($3CaO{\cdot}Al_2O_3{\cdot}3CaSO_4{\cdot}32H_2O$), Friedel's salts($3CaO{\cdot}Al_2O_3{\cdot}CaCl_2{\cdot}10H_2O{\cdot}3CaO{\cdot}Al_2O_3{\cdot}Ca(SO_4Cl){\cdot}12H_2O$) could not be observed.

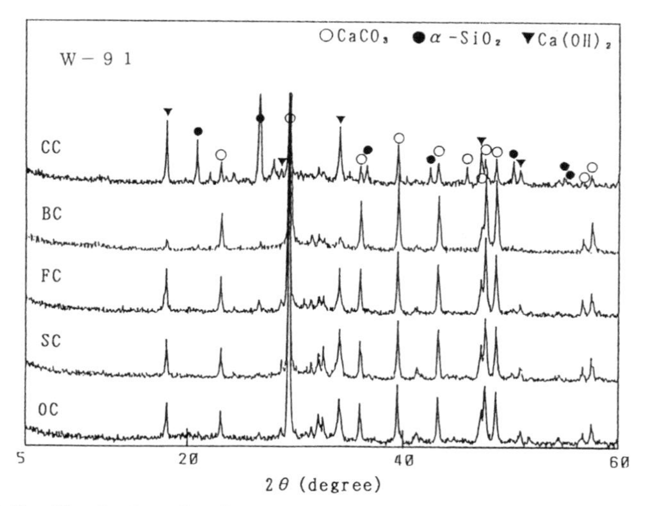

note) W : Standard curing in water, OC, FC, et al : Types of concrete
91 : The nuber of day of curing period,

Fig.7. Result of x–ray analysis of the specimen under standerd curing.

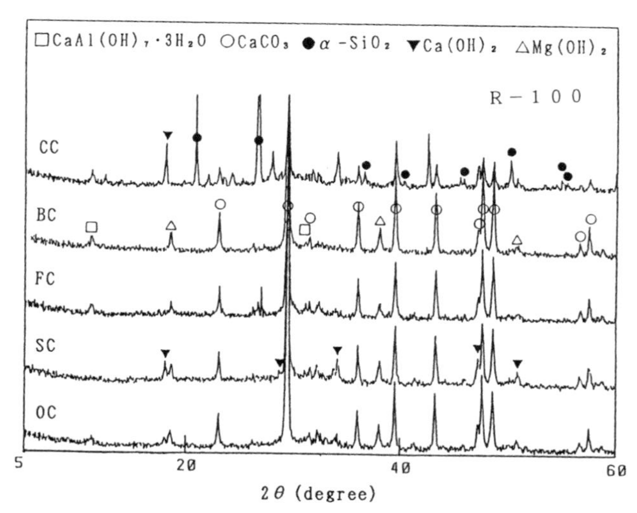

note) R : Repetition of treatment, 100 : The number of repetition
OC, FC, et al : Types of concrete

Fig.8. Results of x–ray analysis of the specimen under repetition of treatment.

4 Conclusion

Investigations on the durability of porous concrete in seawater resulted in the following conclusions:

(1) The dynamic modulus of elasticity of porous concrete made with B–type blast–furnace slag cement dropped after repetitive wet–and–dry treatment.

(2) Apart from concrete using B–type blast–furnace slag cement, the conpressive strength of porous concrete after 100 cycles of wet–and–dry treatment were slightly higher than those after 7–day or 28–day curing in water.

(3) The compressive strength development of porous concrete became lower after 7 days of standard curing in water.

(4) There was no difference in the strength of porous concrete having 15% voids after the soaking in still or flowing seawater.

(5) The use of sulfate resistant portland cement was effective in improving the sulfate resistance of porous concrete.

(6) $Ca(OH)_2$ products were found in all types of concrete after a 91–day standard curing. However, it was not found in any type of porous concrete after 100 cycles of wet–and–dry treatment.

5 Reference

1. Tamai,M., Kawai,A. (1992) Changes of marine epilithic organisms on the concrete with continuous voids. Proceedings of JSCE, No.452, pp.81–90.
2. Meiniger,R.C. (1988) No–fines pervious concrete for paving. Concrete international, Vol.10, No.8, pp.20–27.
3. Kohno,K., Amo,K., Kinoshita,N., and Kanazawa, E. (1992) Effect of mixture proportion on zero–slump concrete for artificial fish reef. JCA Proceeding of Cement and Concrete, No.46, pp.446–451.
4. Nagataki,S., Shiire,T. (1974) Cracks in concrete. Important points of RC technique, Japan Concrete Institute, pp.121–149.
5. Nishibayashi,S.(1981) Durability of concrete in seawater. Cement and Concrete, No.410, pp.2–9.
6. Tamai.M.(1989) Dynamic modulus of elasticity and freeze–thaw resistance of no–fines concrete. The 45th annual meeting of JCI, pp.524–527.

124 CHEMICAL ATTACK OF DE-ICING SALTS ON PORTLAND CEMENT CONCRETE

T. SASATANI
Kokudo Kaihatsu Center Co. Ltd, Nonoichi, Japan
K. TORII and M. KAWAMURA
Kanazawa University, Kanazawa, Japan
E. DOKYU
Fukui Prefecture, Fukui, Japan

Abstract
It has recently been pointed out that a high-concentration solution of $CaCl_2$, which is widely used along with NaCl as de-icing salts in snowy and cold regions in Japan, may cause serious damage to concrete. The damage to concrete is attributed to a detrimental effect of the chemical and physical processes of the leaching of calcium hydroxide and the formation of a complex salt. Portland cement concrete specimens were exposed to a 30 % $CaCl_2$ solution for 1 year, during which changes in length, compressive strength and dynamic modulus of elasticity were periodically measured. The effects of the water : cement ratio and entrained air on the deterioration of portland cement concrete due to the chemical attack caused by a high-concentration solution of $CaCl_2$ were evaluated based on the measurements on the physical and chemical properties of concrete. From the results, it was confirmed that the entrained air in concrete improved the resistance to the calcium chloride attack to some extent, but the reduced water : cement ratio had little influence at later stages of exposure time. Furthermore, a further deterioration accompanied by cracking was found to be caused by repetitive drying and wetting into 30% $CaCl_2$ solution.
Keywords : calcium chloride attack, chloride ion penetration, compressive strength, de-icing salt, dynamic modulus of elasticity, fluorescence microscopic observation, mechanism of deterioration

1 Introduction

In Japan, the use of de-icing salts has rapidly been increasing to keep main roads free of ice and snow during the winter season since the use of studded tires was prohibited throughout the country in 1991 [1]. The most popular de-icing salts used on roads or bridges for traffic safety in Japan are sodium chloride (NaCl) and calcium chloride ($CaCl_2$) ; calcium chloride is especially used in very cold or mountainous regions. It is well known that de-icing salts, NaCl and $CaCl_2$, can penetrate into concrete structures and accelerate the corrosion of the steel reinforcement. In some cases, deicing-salts promote the expansion of concrete due to the alkali-silica reaction and the surface scaling of concrete due to the freezing and thawing action [2],[3]. In addition, it has recently been pointed out by several researchers that a high-concentration solution of

Concrete Under Severe Conditions: Environment and loading (Volume Two) Edited by K. Sakai, N. Banthia and O.E. Gjørv. Published in 1995 by E & FN Spon. ISBN 0 419 19860 1

$CaCl_2$ is particularly aggressive even towards the concrete itself [4],[5]. The authors have also revealed that severe deterioration occurs in mortars immersed in a $CaCl_2$ solution if the temperature is below 20 °C and the concentration of the solution is 30 %, but that no sign of deterioration is observed in mortars immersed in a NaCl solution regardless of the temperature and concentration of the solution [6]. It has been found that the deterioration of mortar caused by a high-concentration solution of $CaCl_2$ is mainly attributed to the dissolution of $Ca(OH)_2$ and the formation of a complex salt of calcium oxychloride hydrate ($3CaO{\bullet}CaCl_2{\bullet}15H_2O$) [7],[8]. This peculiar type of deterioration may possibly occur when the calcium chloride that was distributed on roads or bridges is condensed on the surface layer of concrete structures in the presence of repeated wetting and drying. However, there have been very few studies concerning the deterioration of concrete caused by the calcium chloride attack and measures for its prevention.

The objective of this study is to clarify the effects of mix proportions on the damage of concrete due to the calcium chloride attack. Portland cement concrete specimens were exposed to a 30 % $CaCl_2$ solution at 20 °C for 1 year. The effects of the water : cement ratio and entrained air content on the deterioration of portland cement concrete due to the chemical attack caused by a high-concentration solution of $CaCl_2$ were evaluated based on the measurements of the mechanical properties of concrete such as compressive strength, dynamic modulus of elasticity and changes in length and weight.

2 Experimental

2.1 Materials and mix proportions of concrete

An ordinary portland cement (OPC) with a specific gravity of 3.16 and a Blaine fineness of 3300 cm^2/g was used. Its chemical composition is presented in Table 1.

Table 1. Chemical composition of the tested ordinary portland cement .

Ig.loss	Insol	SiO_2	Al_2O_3	Fe_2O_3	CaO	MgO	SO_3	Na_2O	K_2O	TiO_2	P_2O_5	MnO
1.3	0.1	21.8	5.1	2.8	63.7	1.9	2.0	0.16	0.52	0.32	0.13	0.11

The fine and coarse aggregates used were river sand (specific gravity : 2.61, water absorption : 1.3 %, fineness modulus : 2.46) and crushed gravel (specific gravity : 2.60, water absorption : 1.4 %, maximum size of aggregate : 20 mm), respectively, which were supplied from the Hayatsuki river in Toyama prefecture in Japan. All concrete mixtures with water : cement ratios of 0.45, 0.55 and 0.65 were made so as to have a slump of 80 ± 20 mm. AE admixture-free concretes with an air content of 1 ± 0.5 % and AE agent-bearing ones with an air content of 4 ± 1 %, using a resin-type air-entraining admixture, were prepared. Mix proportions of non-AE and AE OPC mixtures are presented in Table 2.

2.2 Experimental procedures

Cylindrical specimens, 75 mm in diameter and 150 mm in height, were prepared for the compressive strength test, and prismatic specimens, 75 mm by 75 mm by 400 mm, for the measurements of changes in weight, length and dynamic modulus of elasticity. The chemical resistance test against calcium chloride attack was performed in a solution with a concentration of 30 % at a temperature of 20 °C, in which the solution was renewed every month. Prior to the exposure to the 30 % $CaCl_2$ solution, the specimens were cured in water at 20 °C for 7 days. Two types of exposure conditions

Table 2. Mix proportions and properties of non-AE and AE concretes.

	W/C	s/a	Unit content (kg/m^3)		Slump	Air content
		(%)	Water	Cement	(mm)	(%)
OPC45 non-AE	0.45	43	190	422	80±20	1±0.5
OPC55 non-AE	0.55	45	190	345	80±20	1±0.5
OPC65 non-AE	0.65	47	190	292	80±20	1±0.5
OPC45 AE	0.45	39	165	367	80±20	4±1
OPC55 AE	0.55	41	165	300	80±20	4±1
OPC65 AE	0.65	43	165	254	80±20	4±1

were selected : one series of specimens continuous immersed in a 30 % $CaCl_2$ solution, labeled CIC and the other series of specimens intermittently immersed, that is, the repetitions of 7-day immersion in a 30 % $CaCl_2$ solution and subsequent 7-day drying at 20 °C and 60 % R.H., labeled IIC.

Changes in compressive strength (JIS A 1132), length (JIS A 1129), weight and dynamic modulus of elasticity (JIS A 1127) of the specimens were measured at prescribed intervals up to 1 year. The depths of chloride ion penetration into concrete were measured by spraying a 0.1 N $AgNO_3$ solution on fracture surfaces of the cylindrical specimens. The occurrence of micro-cracks and scaling in the surface layers of concrete was examined by means of the fluorescence microscopy. Furthermore, in order to investigate changes in the microstructural features and the reaction products in the deteriorated portion of the concrete, X-ray diffraction analysis, DSC-TG analysis and SEM observations were carried out. Control specimens were also stored in water at 20 °C and measured by the same methods, labeled CIW.

3 Results and discussion

3.1 Chloride ion penetration and compressive strength

The depths of chloride ion penetration and the deterioration of cylindrical specimens at 3 months exposure are presented in Table 3. As shown in Table 3, the depths of chloride ion penetration into concrete proportionally increased with increasing the water : cement ratio. It was observed that at 3 months immersion in a 30 % $CaCl_2$ solution, concrete specimens with a water : cement ratio of 0.65 showed a thorough penetration of chloride ion in the cross section of the cylindrical specimens, 75 mm in diameter and 150 mm in height, while the depths of chloride ion penetration into concrete specimens with a water : cement ratio of 0.45 were half as much. However, the depths of chloride ion penetration into concrete were almost the same for both non-AE and AE mixtures. Furthermore, continuous immersion in a 30 % $CaCl_2$ solution (CIC) showed a slightly higher depth of chloride ion penetration in comparison with an intermittent one (IIC) at the early stages of exposure time, that is, up to 3 months.

Figures 1 and 2 show fluorescence microscopic observations of cylindrical specimens immersed in a 30 % $CaCl_2$ solution for 3 months. At the early ages of immersion in a 30 % $CaCl_2$ solution, the deterioration of concrete due to the calcium chloride attack is accompanied by the occurrence of cracks and scaling on the surfaces of the specimens. However, it is apparent from Figures 1 and 2 that the formation of connected cracks and voids parallel to along the edge of specimens, distinguished by the lights areas and the areas of various shades of yellow color in a fluorescence

Table 3. Depths of chloride ion penetration and deterioration for cylindrical specimens.

	Depth of chloride ion penetration (mm)			Depth of deterioration (mm) ※
	1 month	2 months	3 months	3 months
Continuous immersion in 30% $CaCl_2$ solution (CIC)				
OPC45 non-AE	11	15	15	5.3
OPC45 AE	13	16	16	6.4
OPC55 non-AE	17	20	24	8.2
OPC55 AE	20	20	25	7.4
OPC65 no-AE	18	28	32	8.9
OPC65 AE	21	28	35	12.8
Intermittent immersion in 30% $CaCl_2$ solution (IIC)				
OPC45 non-AE	10	12	15	8.8
OPC45 AE	10	12	14	8.2
OPC55 non-AE	14	18	22	8.7
OPC55 AE	14	20	21	8.7
OPC65 non-AE	18	26	33	7.6
OPC65 AE	20	26	29	10.4

※ determined for cylindrical specimens by means of fluorescence microscope

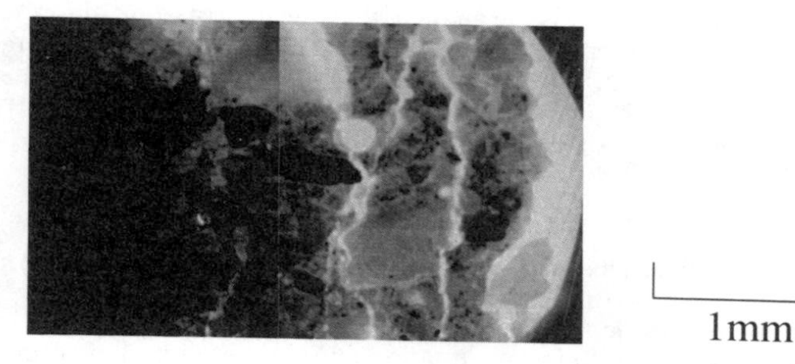

Fig. 1 Fluorescence microscopic observation for cylindrical specimens at 3 months. (OPC 55 AE, CIC).

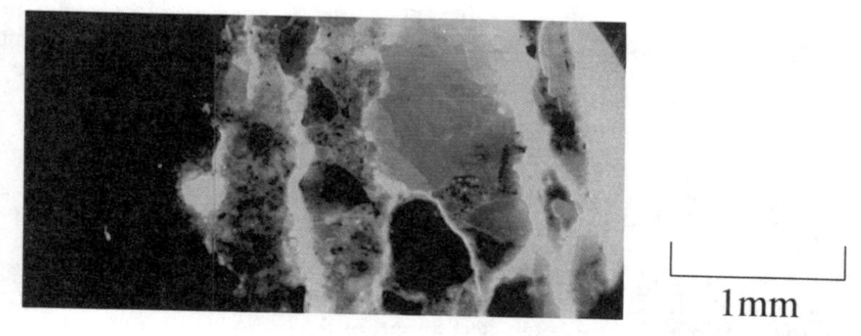

Fig. 2 Fluorescence microscopic observation for cylindrical specimens at 3 months. (OPC 55 AE, IIC)

microscopic observation, occurs only in the surface layers, and that in the interior beyond the surface, there are no changes in fluorescence color. The depths of deterioration measured by the fluorescence microscope were about 13 mm at 3 months exposure time, which was less than half the depths of chloride ion penetration observed in the concretes. This suggests that the portion deteriorated by the calcium chloride attack is limited to only the regions into which the chloride ion has already penetrated to a considerable extent.

Figure 3 shows changes in compressive strength of cylindrical specimens as a function of the exposure time. In the case of continuous immersion in a 30 % $CaCl_2$ solution (CIC), a large reduction in compressive strength was observed only for non-AE mixtures with a water : cement ratio of 0.65; other mixtures kept their initial compressive strength, although a small reduction was observed for non-AE mixtures with water : cement ratios of 0.45 and 0.55. On the other hand, in the case of an intermittent immersion in a 30 % $CaCl_2$ solution (IIC), all non-AE mixtures showed a decrease of more than 20 % in compressive strength, which was independent of the water : cement ratio. As shown by a comparison between non-AE and AE mixtures, the entrained air in concrete effectively improved the resistance to the calcium chloride attack, but the effect of a reduced water : cement ratio was not as significant. This result is in agreement with the recent data obtaind by Collepardi et al. [9]. Furthermore, when the concrete was exposed to repetitions of drying and wetting into a 30% $CaCl_2$ solution, the compressive strength due to the calcium chloride attack was reduced more greatly.

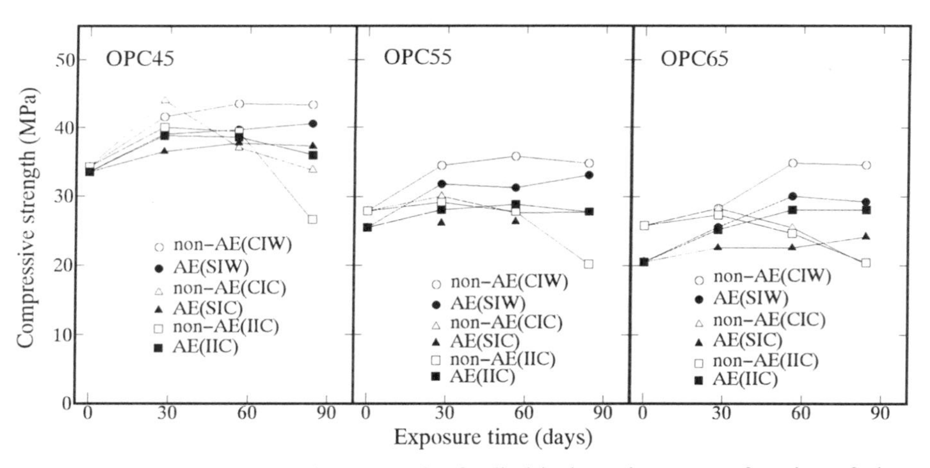

Fig. 3 Changes in compressive strength of cylindrical specimens as a function of the exposure time.

3.2 Overall view of deteriorated specimens

Figures 4 and 5 show overall views of deteriorated prismatic specimens after immersion in a 30 % $CaCl_2$ solution. In the case of continuous immersion in a 30 % $CaCl_2$ solution (CIC), crumbling at first occurred at the edges and corners of specimens around 1 month exposure time, followed by swelling and spalling on the surfaces of specimens. At 3 months exposure time, this type of deterioration became more serious for non-AE mixtures with water : cement ratio of 0.55 and 0.65. A reduction in water : cement ratio mitigated the crumbling at the corners and the scaling on the surfaces at early stages of immersion in a 30 % $CaCl_2$ solution, but at 6

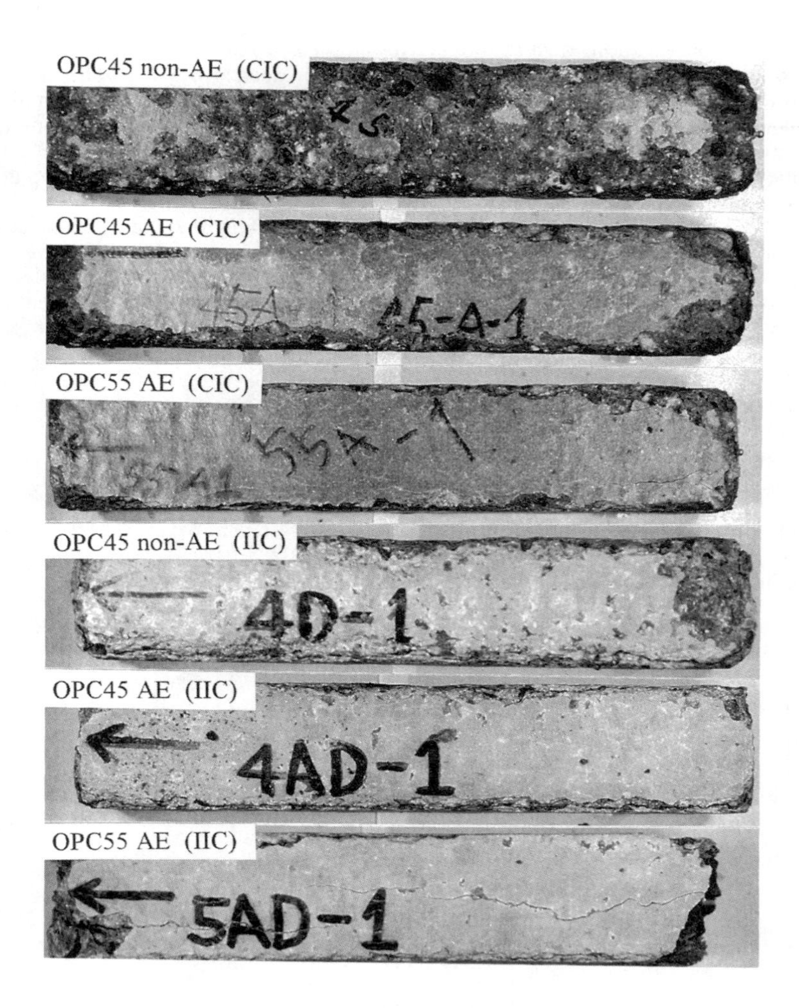

Fig. 4 Overall view of deteriorated prismatic specimens after immersion in a 30 % $CaCl_2$ solution (OPC 45,55).

Fig. 5 Overall view of severely cracked prismatic specimens after immersion in a 30 % $CaCl_2$ solution (OPC 65).

months of exposure time, its effect was not as significant as expected. However, an air-entrainment of 4 % in concrete was effective in improving the crumbling and scaling on the surfaces of specimens independently of water : cement ratio, as shown in Figure 4. On the other hand, the process of deterioration in the case of an intermittent immersion in a 30 % $CaCl_2$ solution was different from that in the case of continuous immersion, which was accompanied by severe cracking on the surfaces of the specimens. That is, in the case of an intermittent immersion (IIC), around 3 months exposure time, small longitudinal cracks occurred along the edges of specimens with water : cement ratios of 0.55 and 0.65. The cracks developed into pattern-cracking over the whole surface after 6 months exposure time, leading to the failure in the center of the specimens at 8 months of immersion as shown in Figure 5. However, concrete specimens with a water : cement ratio of 0.45 showed little cracking, although relatively large amounts of crumbling or scaling portions were found at the later stages of the exposure time. It has been found that damage to concrete specimens with a water : cement ratio of 0.45 generally consists of scaling or crumbling at corners and edges, while in concrete specimens with a water : cement ratios of 0.55 and 0.65, damage is accompanied by a formation of severe pattern-cracking [10].

3.3 Change in length and weight

Figures 6 and 7 show changes in length and weight of prismatic specimens as functions of the exposure time. It can be observed from Figure 6 that a large expansion of the specimens did not occur during the early stages of exposure and in some cases, the specimens shrank considerably in a 30 % $CaCl_2$ solution. The shrinkage strain of concrete specimens attained approximately 0.035 %, which was especially marked for an intermittent immersion in a 30 % $CaCl_2$ solution. It is considered that the unusual specimen shrinkage in a 30 % $CaCl_2$ solution may have resulted from either thermal shock or osmotic pressure, but a further investigation on the mechanism of shrinkage in a 30 % $CaCl_2$ solution is needed. Concrete specimens with a water : cement ratio of 0.45 did not expand at all during the exposure in a 30 % $CaCl_2$ solution; their changes in length being always in the range of ± 0.02 %. An overall expansion in excess of 0.1 % was found only for concrete specimens with water : cement ratios of 0.55 and 0.65 in which pattern-cracking had developed. This

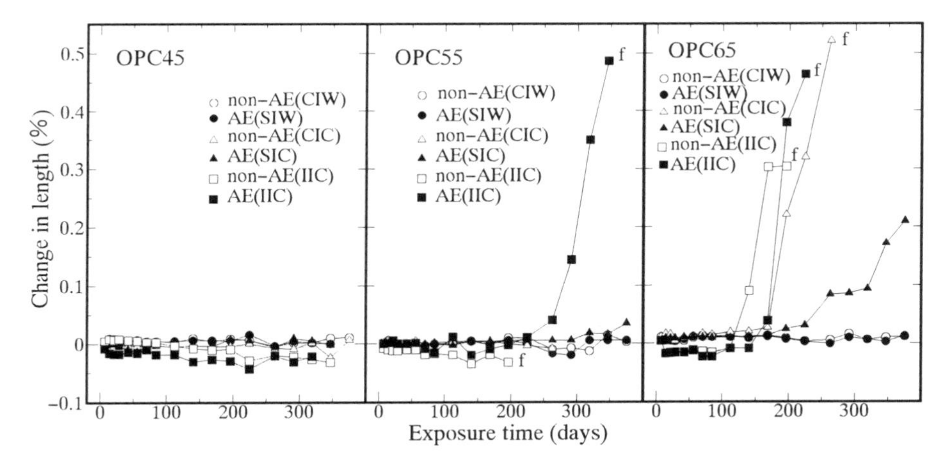

Fig. 6 Changes in length of prismatic specimens with the exposure time.

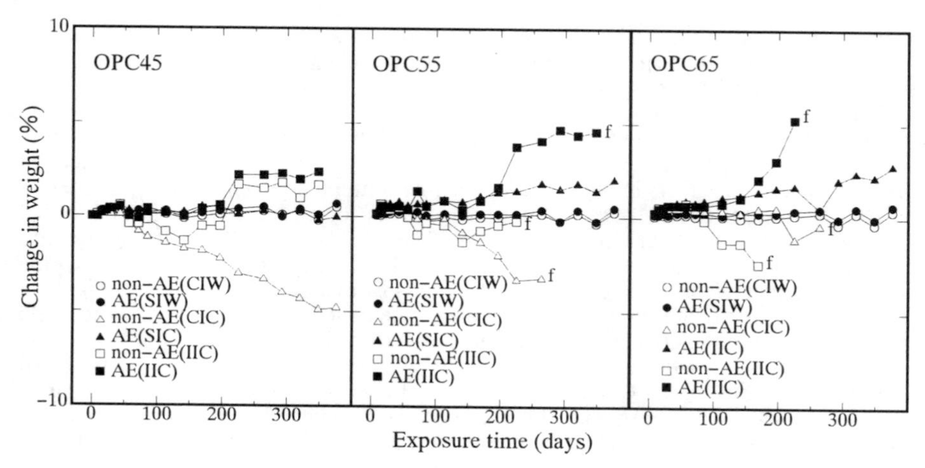

Fig. 7 Changes in weight of prismatic specimens with the exposure time.

was just a couple of months before concrete specimens failed. The trends in dimensional change are in good agreement with the results observed for the mortar specimens [6]. It is also evident that the calcium chloride attack is quite different from the alkali-silica reaction or sulfate attack in its expansive behavior of the concrete. On the other hand, as shown in Figure 7, non-AE mixtures showed a relatively larger amount of weight loss caused by surface scaling than AE mixtures when they were continuous immersed in a 30 % $CaCl_2$ solution , but the weight losses were 4 to 5 % at the maximum. Also the case of an intermittent immersion in a 30 % $CaCl_2$ solution, the weight of cracked specimens gradually increased during the exposure to the $CaCl_2$ solution, resulting in the final failure.

3.4 Dynamic modulus of elasticity

Figure 8 shows changes in dynamic modulus of elasticity of prismatic specimens as a function of exposure time. From the results, it is apparent that the change in dynamic modulus of elasticity with the exposure time was not very significant during the early stages of exposure when the deterioration of concrete was limited to the corners or edges of the specimens, but the dynamic modulus of elasticity rapidly decreases at 6 months of exposure when longitudinal cracks have developed on the surface of the specimens, in many cases resulting in failure at 1 year. The measurements of dynamic modulus of elasticity seems to very accurately reflect the degree of deterioration of concrete due to the calcium chloride attack. From the results of the dynamic modulus of elasticity, it is confirmed that AE mixtures with a low water : cement ratio are superior in their resistance to the calcium chloride attack relative to non-AE mixtures with a high water : cement ratio. In addition, a rapid decrease in dynamic modulus of elasticity is found in non-AE and AE mixtures with a high water : cement ratio, especially when they are exposed to the repetitions of wetting and drying in a 30 % $CaCl_2$ solution. The accelerated deterioration observed in the intermittent immersion in a 30 % $CaCl_2$ solution may be attributed not only to the condensation of $CaCl_2$ solution in the process of drying at room temperature, but to the precipitation of calcium chloride salt ($CaCl_2 \cdot 2H_2O$). The re-crystallization of calcium oxychloride hydrate is also responsible for deterioration of concretes due to the calcium chloride attack. Especially, the crystallization pressure is generated in the cracks or voids in concrete.

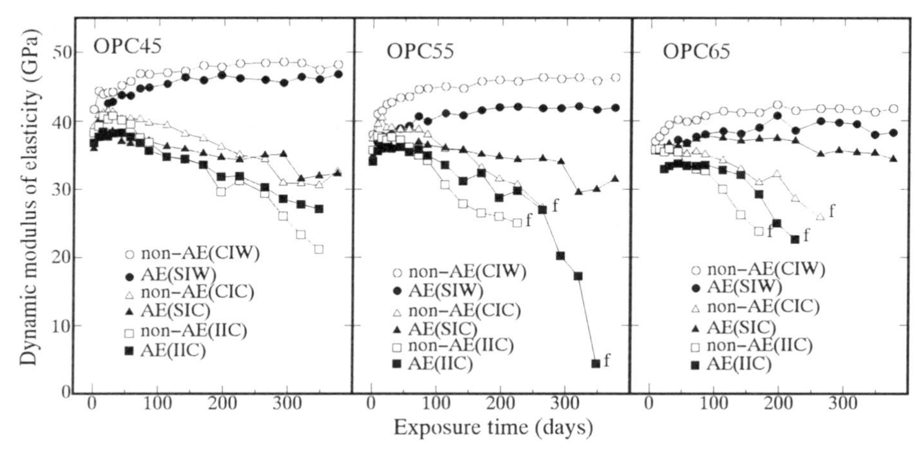

Fig. 8 Changes in dynamic modulus of elasticity of prismatic specimens as a function of the exposure time.

4 Conclusions

The laboratory test data showed that a 30 % $CaCl_2$ solution acted very aggressively on portland cement concrete at 20 C°, in which the damage to concrete was associated with cracking and scaling on the surfaces of specimens. From the results, it was found that the entrained air in concrete improved the resistance to the calcium chloride attack to some extent, but the reduced water : cement ratio had little influence at later stages of exposure. Furthermore, a further deterioration accompanied by a severe cracking was found to be caused by repeated drying and wetting in a 30% $CaCl_2$ solution. Thus, it can be concluded that both a low : water cement ratio and an entrained air can control the calcium chloride attack on portland cement concretes to some extent, but the joint use of other measures, such as blended cements or water-proof surface treatment, is essential to effectively prevent damage to the concrete.

Under the present conditions, it is preferable to avoid the use of a high-concentration solution of $CaCl_2$ as a de-icing salt, especially on concrete bridges or concrete pavement. Hereafter, it is clearly necessary to search for a harmless and inexpensive de-icing agent to replace chloride-bearing materials such as NaCl and $CaCl_2$.

5 Acknowledgments

The authors would like to thank Dr. S. Chatterji, Danish Technological Institute, for his useful guidance in carrying out the experiment, and also Prof. Y. Kajikawa of Kanazawa University for his advice and encouragement.

6 References

1. Working Group on Deicing Salt, Committee on Concrete Research Activities, Committee on Concrete (1994) Influence of deicing salt on concrete structures in Japan, *Journal of Materials, Concrete Structures and Pavements*, Japan Society of Civil Engineers, No.490, pp.15-9 (in Japanese).
2. Kawamura, M. and Ichise, M. (1990) Characteristics of alkali-silica reaction in the presence of sodium and calcium chloride, *Cement and Concrete Research*, Vol.20, pp.757-66.
3. Fujii, T. and Fujita, Y. (1984) Influence of chloride on freeze-thaw deterioration of hardened cement pastes, Proc. of Japan Society of Civil Engineering, No.343, pp.209-17 (in Japanese).
4. Chatterji, S. and Jensen, A.D. (1975) Studies of the mechanism of calcium chloride attack on portland cement , *Nordisk Beton*, Vol.5, pp.5-6.
5. Chatterji, S. (1978) Mechanism of the $CaCl_2$ attack on portland cement concrete, *Cement and Concrete Research*, Vol.8, pp.461-68.
6. Torii, K., Kawamura, M., Yamada, M. and Chatterji, S. (1992) Deterioration of cement mortars in NaCl and $CaCl_2$ solution, JCA Proceedings of Cement & Concrete, No.42, pp.504-9 (in Japanese).
7. Monosi, S. and Collepardi, M. (1989) Chemical attack of calcium chloride on the portland cement paste, *IL Cement*, Vol.86, pp.97-104.
8. Torii, K., Sasatani, T. and Kawamura, M. (1995) Effects of fly ash, blast-furnace slag and silica fume on resistance of mortar to calcium chloride attack, Proc. of 5th Int'l Conf. on Fly Ash, Silica Fume, Slag and Natural Pozzolans in Concrete, Milwakee (in press).
9. Collepardi, M., Coppola, L. and Pistolesi, C., (1994) Durability of concrete structures exposed to $CaCl_2$ based deicing salts. Proc. of 3rd Int'l Conf. on Durability of Concrete. ACI SP-145, Nice, pp.107-20.
10. Moriconi, S., Pauri, M., Alvera, I. and Collepardi, M. (1989) Damage of concrete by exposure to calcium chloride, *Materials Engineering*, Vol.1, No.2, pp.491-96.

125 A PROTECTION METHOD FOR CONCRETE UNDER SEVERE ACIDIC CONDITIONS

S. KURIHARA
Sho-Bond Corporation, Osaka, Japan
H. MATSUSHITA
Kyushu-Kyoritsu University, Fukuoka, Japan
S. MATSUI
Osaka University, Osaka, Japan

Abstract

Oita Expressway, a branch of the Kyushu Traverse Expressway, crosses a well-known Beppu spa on the way to Oita. Near Beppu spa, there is a long beautiful arched bridge called Beppu Myoban bridge on the route. The bridge substructures were covered with epoxy resin mortar to protect the concrete from deterioration under acidic environment.

In 1976, the Japan Highway Public Corporation conducted exposure test for five years in which many kinds of protective systems were investigated to find a better protection method for concrete structures under severe acidic and thermal environments. Concrete specimens coated with 11 sorts of protective materials, uncoated concrete specimens mixed in various compositions, and reinforced concrete specimens were compared from the view of deterioration. The protective materials were as follows:

1. Coatings of asphalt, epoxy and polyester strengthened with glass fiber.
2. Mortar linings of asphalt, epoxy, latex polymer and polyester.
3. Impregnation of acrylic resin on concrete surface.
4. Sheet linings of butyl gum and polypropylene.
5. Autoclave cured concrete.

After three years of exposure, only two protection systems that were the polyester mortar and the epoxy mortar had endured, but soon after, some cracks were found in the polyester mortar. The epoxy mortar covered specimen remained sound without any cracking and it was kept at the experimental yard for another four years.

After a total of nine years, the authors investigated the specimens in detail on chemical and physical properties. The results were compared with the results of another test and the endurance limits were estimated.

Keywords: Epoxy mortar, exposure test, hot spa, modified epoxy, overlay coating, protection system, soaking test, sulfuric acid, vinyl ester, weight change

Concrete Under Severe Conditions: Environment and loading (Volume Two) Edited by K. Sakai, N. Banthia and O.E. Gjørv. Published in 1995 by E & FN Spon. ISBN 0 419 19860 1

1 Introduction

In Japan, there are many hot spring spas which often contain acidic ingredients, e.g., sulfuric acid, hydrochloric acid, carbonic acid and so on. These acids may cause serious damage to concrete and steel structures when exposed to them. It is necessary, therefore, to protect such structures from acid attack.

Equations (1) and (2) show reactions between sulfuric acid and concrete. Calcium hydroxide and hydrates of cement, the chief components of concrete, disintegrate entirely into gypsum and silica gel.

$$Ca(OH)_2 + H_2SO_4 \rightarrow CaSO_4 \cdot 2H_2O \text{ (gypsum)} \quad (1)$$

$$3CaO \cdot 2SiO_2 \cdot 3H_2O + 3H_2SO_4 + 4H_2O \rightarrow 3CaSO_4 \cdot 2H_2O + 2Si(OH)_4 \text{ (silica gel)} \quad (2)$$

If the above mentioned reactions occur in the concrete under coating, the chemical and physical properties of the concrete such as alkalinity, weight and strength will change. Therefore, the specimens were cut into slices to investigate those properties.

Then, the property of epoxy mortar itself was investigated by a soaking test of the cylindrical specimens to evaluate the deterioration mechanism of the nine years exposed specimen. Furthermore, to increase the endurance limit of the epoxy mortar, the effect of overlay coated specimens of epoxy mortar were investigated by a three years of soaking test at another hot spa.

2 Damage evaluation after nine years of exposure in acidic soil

The specimen was a long prism concrete of 100x100x800mm covered with 10mm thick epoxy mortar layer. The mixture ratio of epoxy to sand was 1:5 by weight and fine dry silica sand (the maximum grain size was 1.2mm) was used.

The lower 500mm of the length of the specimens was buried in an artificial clay soil bed as shown in Fig.1. Under the soil bed, steam pipes were arranged to blow up the hot spa steam. Under the soil, severe thermal and acidic conditions were created as seen in Table 1.

Table 1. Environment in the artificial soil [1].

Point of depth(mm)	pH	Temperature(℃)
0	1.7～1.9	–
-100	–	45～60
-250	2.0～3.3	92～98
-500	3.0～3.7	98～99

The specimen was compared to the specimen kept in the laboratory as a control for the same period of nine years. Judging from the appearance of the exposed specimen, the degree of damage of covering material could assess easily. Then, the specimen was cut

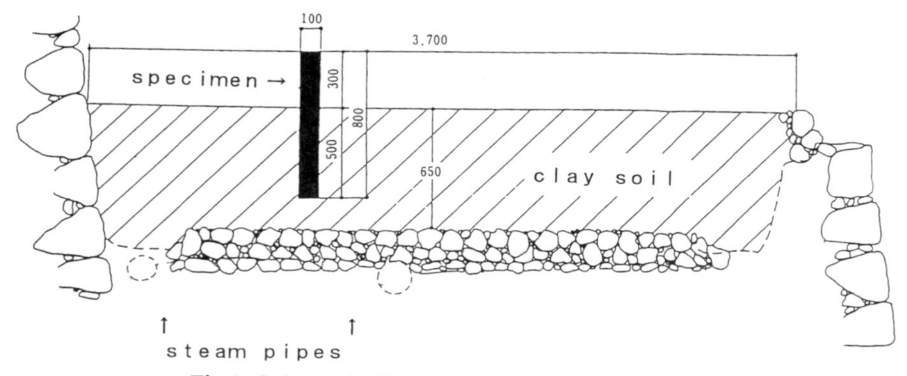

Fig.1 Schematic diagram of exposure test(section).

into three zones of the exposed zone, buried zone and boundary zone.

Three properties of the specimen were examined: the degree of neutralization of the concrete, weight changes of the epoxy mortar coating and adhesion between concrete and the epoxy mortar.

Table 2 shows the results of the examination. The neutralization of concrete did not occur in the specimen. Therefore, it can be said that the inner concrete was never attacked by the acid and the epoxy mortar was quite effective for stopping penetration of acid. Weight changes were calculated on the assumption that all changes occurred within the epoxy mortar layer. Comparing to the control specimen's weight, a little bit of increase occurred due to penetration of hot acid. By the detailed observation of the cut section of the epoxy mortar layer, the color change seems to occur about 5mm from the outer surface. Adhesive strength between concrete and the epoxy mortar of the control specimen was over 45kgf/cm^2 which is almost the same as original state. The strength of the exposed specimen has dropped about 40%. The drop seems to be due to temperature effect of hot steam.

Table 2. Results of inspection.

Zone	Neutralization of concrete (mm)	Weight changes (%)	Strength ratio of adhesion (%)
1 exposed	0	2.78	59.5
2 Boundary	0	5.42	55.7
3 Buried	0	6.85	65.9

From the above results, it was concluded that the concrete had been perfectly protected against corrosion by the epoxy mortar of 10mm thickness. But, in the actual construction of Beppu Myoban bridge the modified epoxy mortar was used with the overlay coating of vinyl ester based on experimental results as shown in Table 3.

The authors have carried out a soaking test to prove the durability of the modified epoxy mortar and the vinyl ester overlay. The major difference between the epoxy and the modified epoxy was the curing agent which was exchanged from polyamide to aromatic amine. The overlay was a thin layer coating without sand on the surface.

Table 3. Recommended protection system.

Layer	Materials	Thickness(mm)
1st	Epoxy primer	–
2nd	Modified epoxy mortar	5.0
3rd	Modified epoxy mortar	5.0
4th	Vinyl ester overlay	0.25
5th	Vinyl ester overlay	0.25

3 Soaking examination in sulfuric acid solution

In the soaking test, the epoxy mortar, the modified epoxy mortar and the vinyl ester mortar were compared on durability against acid. Small cylindrical specimens of 25x50mm were set in a hot bath of tap water and in a hot sulfuric acid solution. The solutions were set to a temperature of 70℃ and the concentration of sulfuric acid was 5% by weight (the pH range between 3 and 4) in order to accelerate the test. The test was conducted for five months. The mixture ratio of resin to sand of each specimen was 1:5 by weight. After soaking, the specimens were investigated for change in weight, compressive strength and Young's modulus.

Table 4 shows the results of the soaking test. The values in parentheses were obtained from the specimens set in a hot bath of tap water. As seen in the table, the damage in a hot water bath was not so serious. However the epoxy mortar which was fully saturated by hot sulfuric acid for one month, the weight change of the modified epoxy mortar was under half even after 5 months. Furthermore, the weight changes of the vinyl ester mortar specimen were very little. The test results of compressive strength and Young's modulus are summarized in Table 4. The compressive strength of the epoxy mortar decreased 41% in a hot bath of tap water after a month. In the case of hot sulfuric acid, it decreased 72%, so the difference of about 30% is considered as acid attack. But the modified epoxy mortar was deteriorated to the same degree of 28% in both solutions after five months. It means that the decrease is due to the effect of hot water and is not due to acid. The same tendency can be seen in the Young's modulus. The vinyl ester mortar showed excellent property of resistance against thermal and acidic condition. From all the results, it can be said that the vinyl ester mortar will not be deteriorated easily by sulfuric acid.

Table 4. Results of the soaking test.

Materials	Soaking period (Month)	Weight changes (%)	Change in compressive strength (%)	Change in Young's modulus (%)
1 Epoxy mortar	1	6.23(0.39)*	-72(-41)*	-80(-20)*
	3	6.76(0.43)	-78(-50)	-87(-36)
2 Modified epoxy mortar	1	1.01(0.51)	-28(-19)	-50(-57)
	5	2.79(0.57)	-24(-28)	-69(-54)
3 Vinyl ester mortar	1	0 (0)	0(-1)	-21(-21)
	5	0.10(0.10)	0(-4)	-35(-31)

*(Hot bath of tap water)

Though the epoxy mortar seemed to reach the same deterioration state as the specimen of above mentioned exposure test by soaking in a hot sulfuric acid solution for only three months, the modified epoxy mortar and the vinyl ester mortar were still durable and quite sound even after five months of soaking, respectively. Therefore, the authors decided to use actually the modified epoxy mortar as the base protective layer and the vinyl ester as the overlay coating because the vinyl ester is easy to crack when used for thicker layer such as mortar.

4 Verification of soaking test of the modified epoxy at another spa

This test was a joint research with the Ministry of Construction at Tamagawa spa in Akita prefecture. The test objective was the same as that of the Beppu spa tests.

Concrete prism specimens of 70x100x400mm covered with various protective materials were soaked in a hot spa drainage. The drainage is also a quite severe environment of the pH range between 1 and 1.5 and the temperature about 60℃. The authors investigated four types of specimens based on the epoxy mortar and the modified epoxy mortar as shown in Table 5. The latter two specimens are having overlay coating which gave them a higher density surface than mortar. The thickness of the resin mortars in the specimens of No.1 to 3 was 5mm and that of overlay coatings was 0.5mm. The specimen of No.4 had 10mm thick mortar and 0.5mm thick overlay.

Table 5. Weight changes after the immersion test.

Covered materials	Weight changes of specimens(%)
1 Epoxy mortar	2.18 (10.5)
2 Modified epoxy mortar	0.93 (4.5)
3 Epoxy mortar overlaid with modified epoxy	0.77 (3.6)
4 Modified epoxy mortar over-laid with modified epoxy	0.34 (0.9)

The weight changes of the specimens were measured after about three years of soaking. The results are shown in Table 5 with the calculated value which indicate that the all changes occurred within the covered materials. From these results, the followings can be said that the modified epoxy mortar showed almost twice durability comparing to the epoxy mortar. Also the overlaying of the materials will give almost twice durability. Therefore, No.4 specimen showed over four times' durability than No.1 specimen by the superposition of effects of the modified epoxy mortar and the overlay coating.

5 Recommendation of protection system

From the three series of exposure and soaking tests, the authors have proposed the actual protection system for Beppu Myoban bridge as shown in Table 3. Each material was decided to be applied by half thickness for securing a better workability and uniform density. On the basis of the experimental results, the recommended system will have at least a 20 years of guarantee for efficient working, because the 10mm thick modified epoxy mortar can be expected to endure more than 18 years. From the experimental results, the epoxy mortar deteriorated only 5mm thick during 9 years and the durability of the modified epoxy mortar will be twice than the epoxy mortar, simply 36 years durability will be expected, also the vinyl ester overlay coating will resist for an over 3 years (overlay coating effect is expected more than 3 years in the modified epoxy coating and the vinyl ester is more durable than the modified epoxy).

For the abutments and piers of the Beppu Myoban bridge, the total concrete surface area of over 3000m^2 was covered with the above recommended protection system. After the finishing of the overlay coating work, some cracks in the protection system were found just above some construction joints of concrete. It seemed to be due to insufficient curing period for the massive concrete. The cracks were repaired, and follow-up investigations were carried out continuously.

6 Conclusions

The paper described the essential environmental tests to recommend the protection system for foundation concrete structure against thermal and acidic environments. The series of tests seemed to be not systematic and the data are not enough to say rigorously the durability lives of the protection system. The authors are intending to continue that research and would like to get the deteriorating rate of the materials.

7 References

1. Japan-Highway-Corp., Naigai engineering corporation. (1977) *Report on environm-ental exmination for corrosion protection system in Kyusyu-Traverse-Expressway*, pp. 14 and pp. 229
2. Ministry of Construction, Sho-Bond corporation. (1985) *A survey of concrete corro-sion against acidic conditions*, pp. 103

PART TWENTY-TWO
SPECIAL CONCRETE

126 EFFECTS OF CEMENTS AND ADMIXTURES ON THE PROPERTIES OF VACUUM EXPOSED CONCRETE

H. KANAMORI
Shimizu Corporation, Tokyo, Japan

Abstract

Concrete specimens made with various cements and admixtures were exposed to vacuum conditions. Changes in weight, strain, and strength of these specimens were measured in order to see how these properties were affected by the vacuum and to find suitable cements and admixtures to be used for lunar concrete. One of the interesting results was that the vacuum exposed specimen made with alumina cement showed the least shrinkage strain and the largest increase in compressive strength among all specimens tested.
Keywords: Admixture, cement, compressive strength, concrete, shrinkage, vacuum, weight loss.

1 Introduction

Soon after we obtain more detailed and accurate information about the moon through initial surface explorations, construction of permanent lunar structures will be started. It has been proposed that concrete could be one of the most suitable construction materials for these structures, and various studies are being conducted.

The author has been performing studies on the properties of concrete in a lunar environment, especially focusing on the effect of vacuum. The harsh conditions of the moon include a vacuum effect, low gravity, extreme temperatures, radiation and meteorite impact with the vacuum effect having the greatest influence on the properties of concrete. The other conditions, except for low gravity, can be eliminated by covering concrete with lunar sand called 'regolith'. Regarding gravity, experimental study has shown that the strength of concrete produced under lunar gravity is estimated to be

Concrete Under Severe Conditions: Environment and loading (Volume Two) Edited by K. Sakai, N. Banthia and O.E. Gjørv. Published in 1995 by E & FN Spon. ISBN 0 419 19860 1

about 90 % of the strength of concrete produced on earth [1]. Thus, it seems that the influence of low gravity on the moon will not present a big problem with regard to the properties of concrete. On the contrary, however, the influence of vacuum conditions remains the major issue to be investigated.

Even though concrete is manufactured in pressurized air, the concrete products will be eventually used under a vacuum condition.

In a previous study, a series of vacuum exposure tests was performed to see how mortar behaves under a vacuum condition. According to the results of this test, a significant water loss and shrinkage strain were observed, even though the well pre-cured mortar showed an increase in strength upon exposure to a vacuum [2].

Following this test, a new series of vacuum exposure experiments was conducted using concrete specimens made with various cements and admixtures. This paper will describe the results of the experiments.

2 Experimental

2.1 Materials

2.1.1 Cements

Cement for lunar concrete is expected to be produced from lunar basalt or anorthite. Ordinary portland cement and blast-furnace slag cement will be produced by sintering and grinding a mixture of calcium oxide (CaO) and anorthite, while alumina cement will be extracted from lunar basalt by means of melting and separating methods [3][4].

Since it is feasible to use these cements on the moon, ordinary portland cement (specific gravity = 3.15, Blaine fineness = 3280 cm^2/g), blast-furnace slag cement (specific gravity = 3.08, Blaine fineness = 3770 cm^2/g), and alumina cement (specific gravity = 2.96, Blaine fineness = 4800 cm^2/g) were selected and used in these experiments.

2.1.2 Admixtures

Several admixtures were selected which were thought to reduce shrinkage in vacuum exposed concrete. Superplasticizer (specific gravity = 1.05, dosage = C x 1.1 wt %), silica fume (specific gravity = 2.20, specific surface = 200000 cm^2/g, silica fume/(cement + silica fume) = 10 wt %), and expansive admixture (specific gravity = 3.00, Blaine fineness = 3200 cm^2/g, dosage = 30 kg/m^3) were used.

Superplasticizer was used to reduce the water content of the concrete and hence the shrinkage strain. Silica fume was used to give a resistant capability to the concrete against shrinkage cracks by increasing the strength of cement matrix. The expansive admixture was expected to reduce shrinkage strain.

2.1.3 Water

Water can be produced on the moon by reducing lunar oxides with hydrogen [5]. Tap water was used in this test.

Table 1. Concrete mix proportions (kg/m^3).

Mix No.	Water	Cement			Admixture			Sand	Gravel
		Ordinary Portland	Blast furnace	Alumina	Super-plasticizer	Silica fume	Expansive admix.		
1	208	520						645	982
2	210		525					638	972
3	180			450				686	1045
4	160	400			4.40			702	1069
5	163		408		4.48			702	1070
6	160			400	4.40			723	1101
7	191	430			5.26	47.8		661	1007
8	193		434		5.30	48.3		657	1001
9	178			401	4.90	44.5		689	1049
10	209	493					30	637	970
11	212		500				30	632	964
12	197			463			30	652	993

2.1.4 Aggregates

A past study showed that lunar rock and soil could be used as concrete aggregates [6].

Fine aggregate (specific gravity = 2.60, Fineness Modulus (F.M.) = 2.88, water absorption = 1.13%) and coarse aggregate (maximum size = 20 mm, specific gravity = 2.62, F.M. = 6.85, water absorption = 0.71%) were used.

2.2 Specimens

Concrete specimens were cylindrical (ϕ10 x 20 cm), and tailored following the standard methods specified in JIS A 1132. Specimens were made for 12 mixes, shown in Table 1, with various combinations of cements and admixtures. The water/cement (including silica fume and expansive admixture) ratio was kept at 0.4 for all mixes, and the water content of each mix was determined by preliminary tests to obtain a slump of 12 cm. Three specimens were tested for each test condition.

2.3 Measurements

The following properties of the concrete specimens were measured before and after the vacuum exposure.

- Weight
- Strain
- Compressive strength

2.4 Testing apparatus

The major tools and testing devices used in the experiments were as follows:

- A vertical-shaft mixer (100 l)
- A dial gauge (mechanical) strain meter with ball-point tags
- A vacuum exposure apparatus
- A vacuum desiccator

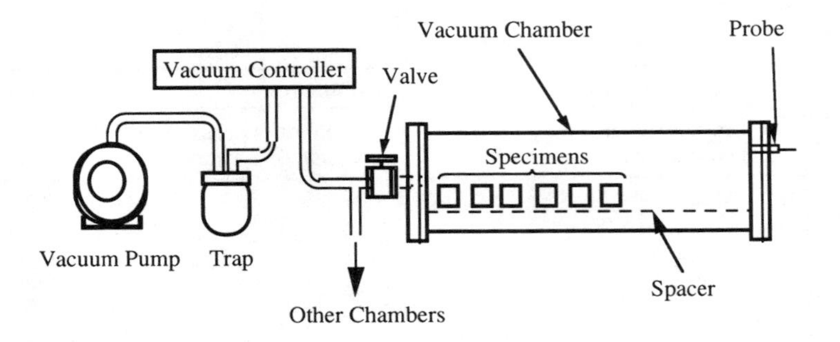

Fig. 1. Vacuum exposure apparatus.

The vacuum exposure apparatus consists of five vacuum chambers, a vacuum controller, a trap, a pump, and valves with complete pipe system as shown in Fig. 1. The steel vacuum chamber is a cylinder, 400 mm in inside diameter and 700 mm in length. Both ends of the cylinder were covered with circular steel plates. One end plate was equipped with a gas outlet and a valve, and a probe for the vacuum meter was attached to the other plate. The vacuum controller measures the degree of vacuum in the chambers and keeps it constant by controlling electromagnetic valves. Maximum gas suction rate and minimum achievable pressure of the oil-sealed rotary vacuum pump were 200 liter/min and 6.7×10^{-2} Pa, respectively.

2.5 Test procedure

Concrete specimens were produced in air at 20 °C and R.H. 80 %. The specimens were demolded at the age of 1 day, and a set of ball-point tags (base distance = 100 mm) was immediately attached to each specimen to provide reference points for measuring length change through the use of a mechanical strain meter. After measuring the weight of the specimens, they were placed in water at 20 °C.

The specimens for vacuum exposure were removed from the water at 28 days. After the weight and the tag distance were measured, they were placed in the vacuum chamber. The degree of vacuum and the temperature inside the chambers were kept at 10 Pa and 20 °C, respectively.

At the age of 91 days and 182 days, specimens were taken from the chamber, and the weight and the distance between the tags were measured again. After that, the compressive strength of each specimen was measured. During this sequence, all vacuum exposed specimens were stored in a vacuum desiccator in order to prevent moisture adsorption by the specimens.

Measurements of weight, tag distance, and compressive strength for the water cured specimens were conducted in the same manner.

3 Results and discussion

3.1 Weight change

Changes in the weight of the concrete specimens with respect to age are shown in Fig. 2. The weight change is expressed as a ratio of the changes in sample weight from the 1-day sample to the weight of paste (cement + water + admixtures) of the 1-day sample, because the amount of water initially contained in the mix was different among the mixes.

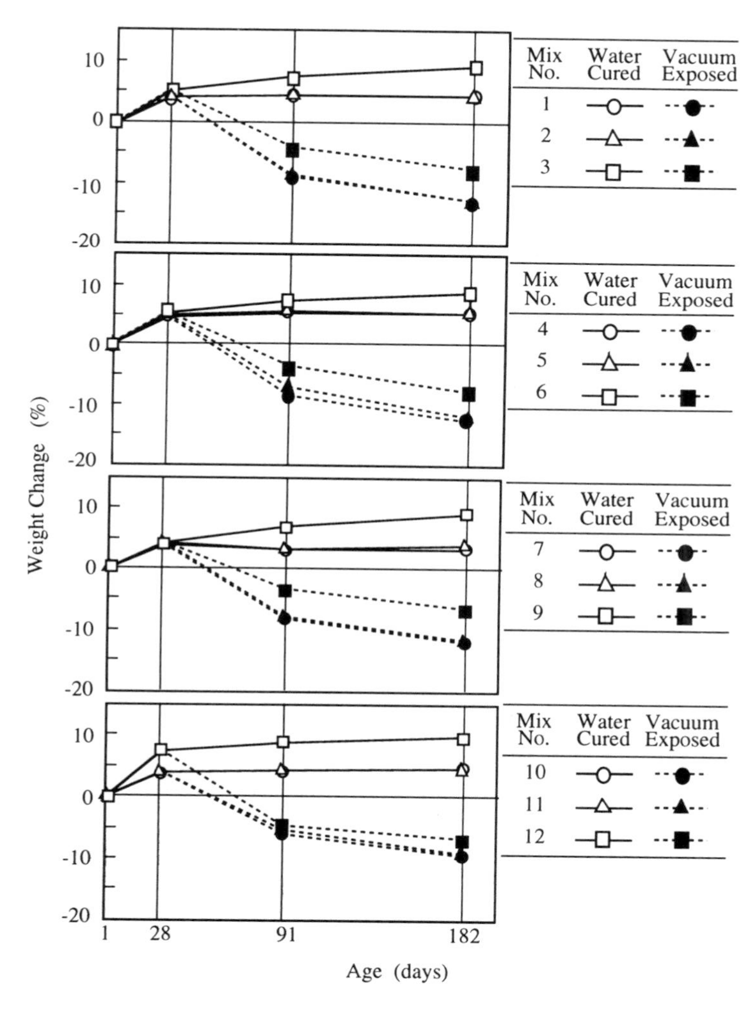

Fig. 2. Changes in weight of concrete specimens.

Regarding the cement effects, both water-cured and vacuum-exposed specimens made with ordinary portland cement showed similar weight changes to the specimens made with blast-furnace slag cement. On the other hand, vacuum-exposed specimens containing alumina cement (mixes 3, 6, 9 and 12) showed smaller weight losses than the specimens containing other cements.

Since the initial W/C ratio for every mix was 0.4, the water content in the initial paste was calculated as 0.29 (0.4/(1+0.4)). The weight changes of the non-admixture mixes (mixes 1 to 3) at the age of 182 days were approximately 7 % for alumina cement, and 14 % for other cements. From these figures, W/C ratios of vacuum exposed specimens at the age of 182 days were estimated as 0.28 for alumina cement and 0.17 for other cements. Since the past study showed that the W/C ratio for the complete hydration of cement was approximately 0.23, we can conclude that most of the water, other than bound water, evaporated from the concrete through long-term vacuum exposure[7].

Regarding the effects of admixture, specimens made with silica fume (mixes 7 to 9) or expansive admixture (mixes 10 to 12) showed slightly smaller weight loss than other specimens. These admixtures seem to reduce the water evaporation to some extent.

3.2 Strain change

Changes in strains of vacuum exposed concrete specimens and water cured specimens are shown in Fig. 3. Strains were calculated from the changes in tag distance from the 1-day sample.

The strain of water cured specimens containing an expansive admixture or alumina cement showed a slight expansion, while other mixes did not show any particular changes. In the case of vacuum exposed specimens, specimens containing alumina cement showed the least shrinkage strain among all specimens tested. However, the effect of admixtures on the shrinkage strain could not be identified clearly from this experiment.

3.3 Compressive strength change

Changes in compressive strength of vacuum exposed and water cured concrete specimens with respect to age are shown in Fig. 4.

As a general tendency, vacuum exposed specimens showed higher or the same compressive strength compared with water cured specimens. The mixes containing alumina cement (mixes 3, 6, 9 and 12) particularly showed increases in strength caused by the vacuum exposure. In the past study, the vacuum exposed mortar specimen containing various cement showed increase in the strength [8]. In the case of concrete, however, the compressive strength of vacuum exposed specimen containing cement other than alumina cement did not clearly show the increase. This result suggested that the shrinkage stresses caused by the presence of coarse aggregates could generate some defects in the cement matrix of vacuum exposed concrete.

The effect of admixtures on the compressive strength could not be identified clearly from this experiment. This means that these admixtures are not very effective for reducing shrinkage of vacuum exposed concrete.

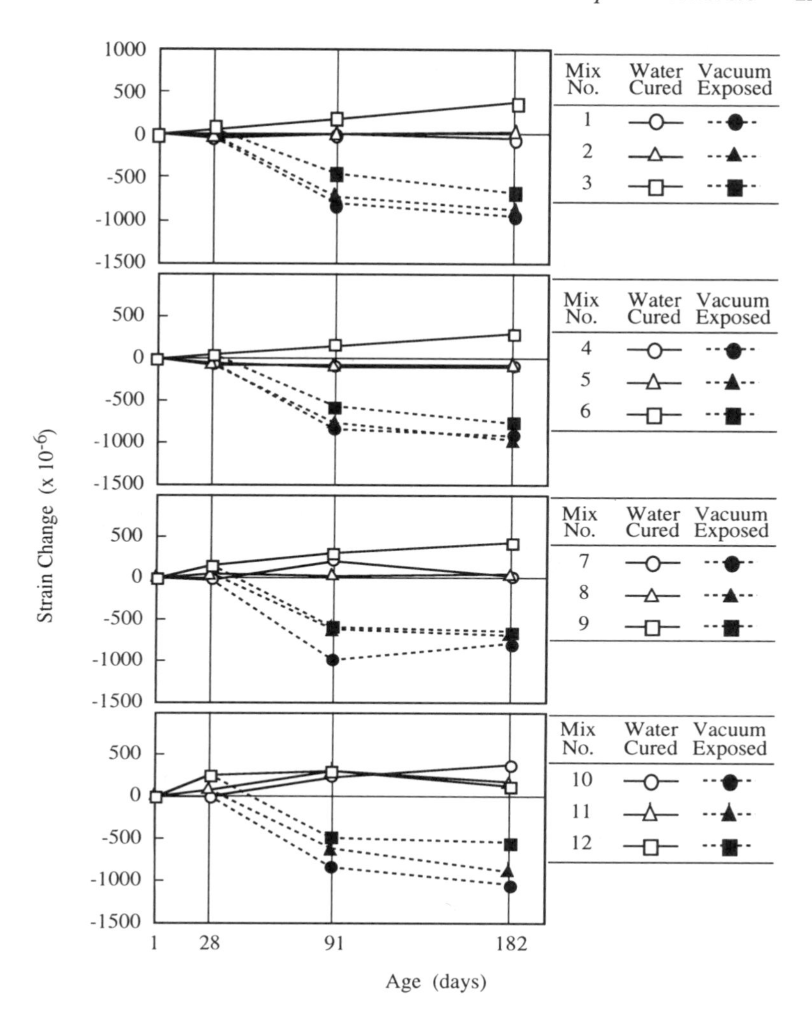

Fig. 3. Changes in strain of concrete specimens

4 Conclusions

The effect of cements and admixtures on the properties of vacuum exposed concrete was studied in order to determine suitable materials for lunar concrete. Results of the experimental study are summarized as follows:

1. Vacuum exposed concrete specimens made with alumina cement showed less weight loss than those made with other cements.

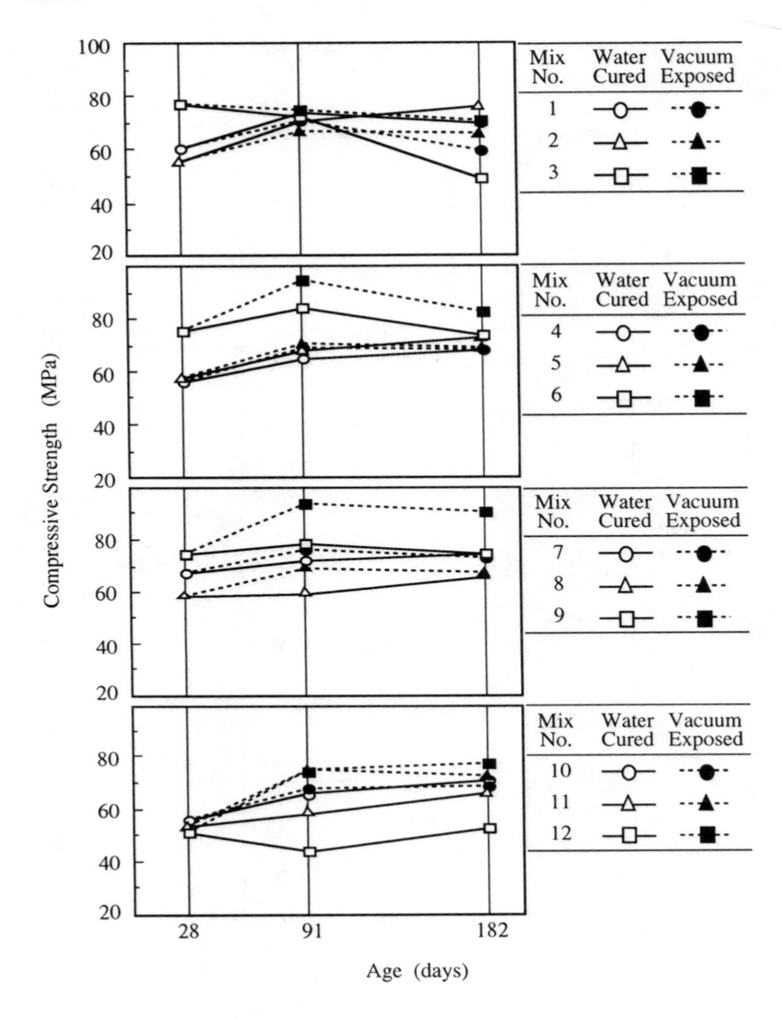

Fig. 4. Changes in compressive strength of concrete specimens

2. Vacuum exposed concrete specimens containing silica fume or an expansive admixture showed slightly less weight losses than those not containing these admixtures.
3. Vacuum exposed concrete specimens made with alumina cement showed the least shrinkage strain among all specimens tested.
4. Vacuum exposed concrete specimens made with alumina cement showed the largest increase in compressive strength among all specimens tested.
5. The effects of admixtures on the shrinkage and compressive strength of vacuum exposed concrete specimens could not be identified from this experiment.

It is concluded from this study that alumina cement could be one of the most realistic materials for lunar concrete. These studies will, however, continue in order that our lunar concrete production goal an become a reality and technologies derived from lunar concrete studies can contribute to existing terrestrial concrete work.

5 References

1. Ishikawa, N., Kanamori, H., and Okada, T. (1988) Possibility of concrete production on the moon, in *Lunar Bases and Space Activities in the 21st Century,* LPI, Houston.
2. Kanamori, H., Matsumoto, S., and Ishikawa, N. (1991) Long–term properties of mortar exposed to a vacuum. *ACI SP–125*, pp 57–69.
3. Burt, D.M., (1991) Lime production from lunar anorthite. *ACI SP–125*, pp 237–244.
4. Mishulovich, A., Lin, T.D., Tresouthick. S.W., and West, P.B. (1991) Lunar cement formulation. *ACI SP–125*, 255–263.
5. Knudsen, C.W., Gibson, M.A., Brueneman, D.J., Suzuki, S., Yoshida, T., and Kanamori, H. (1992) Recent development of the carbotek process for production of lunar oxygen, in *Space 92/Engineering, Construction, and Operations in Space*, ASCE, Denver.
6. Lin, T.D., Love, H., and Stark, D. (1987) Physical properties of concrete made with Apollo 16 lunar soil. *Space Manufacturing 6/Nonterrestrial Resources, Biosciences,and Space Engineering/Proceedings of the Eighth Princeton/AIAA/SSI Conference,* 361 pp.
7. Neville, A.M. (1973) *Properties of Concrete*, Pitman Publishing Ltd., London.
8. Kanamori, H. (1994) For the realization of lunar concrete, in *Space 94/Engineering, Construction, and Operations in Space IV,* ASCE, Albuquerque, pp. 942–51.

127 APPLICATION OF RCD CONCRETE WITH BLENDED CEMENT CONTAINING SLAG TO COLD REGIONS: EXECUTION AT THE SATSUNAIGAWA DAM

S. YASUNAKA, H. NAKA and Y. IDE
Hokkaido Development Bureau, Sapporo, Japan

Abstract

Satsunaigawa Dam is a 114-m-high concrete gravity dam having a crest length of 300m and a total volume of 770,000 m^3. The dam is being constructed in an upper reach of the Satsunai River in the Tokachi River system in Hokkaido, Japan. The purposes of the dam are flood control, maintenance of normal functions of river water, irrigation and drinking water supply, and hydroelectric power generation. The construction of the Satsunaigawa Dam began in 1985, and the placement of the dam concrete began in May 1991. The dam will be completed by 1997 (as of June 1994, 77% of concrete has been placed). Characteristics of the Satsunaigawa Dam include the following:

(1) This is the first dam in Japan to be constructed by the RCD method using moderate-heat portland cement as base cement and powdered blast-furnace slag as admixture.

(2) Since the dam site is located in the Hidaka mountain range in east Hokkaido, a typical cold heavy-snow district, the RCD method is being executed under severe environmental conditions.

This paper discusses the characteristics of concrete containing blast-furnace slag, describes newly developed construction and quality control techniques specifically designed for use in cold regions, and reports on satisfactory results achieved by the new techniques.

Keywords: Blast-furnace-slag-mixed cement, cold region, construction technique, gravity dam, quality control, RCD (Roller-Compacted concrete for Dam).

1 Introduction

Since Satsunaigawa Dam was to become the first RCD concrete dam using blast-furnace slag, it was necessary to make sure that construction data reflected the

Concrete Under Severe Conditions: Environment and loading (Volume Two) Edited by K. Sakai, N. Banthia and O.E. Gjørv. Published in 1995 by E & FN Spon. ISBN 0 419 19860 1

characteristics of blast-furnace slag, and so, to produce high-quality RCD concrete and minimize quality variations. Therefore, it was decided to establish a strict standard for the quality of moderate-heat portland cement and blast-furnace slag used as binders.

Although blast-furnace slag is as effective as or more effective than fly ash in reducing the heat of hydration and increasing the long-term strength of concrete, blast-furnace slag concrete is more temperature-dependent in strength and other properties than ordinary concrete. Since the development of initial strength had been expected to be delayed in a cold region like the Satsunaigawa Dam, curing was planned carefully.

2 Quality standard of cement for RCD concrete

The JIS specifications and JSCE's proposed specifications deal either with the upper or lower limits. They cannot be applied directly to construction control for Satsunaigawa Dam because they do not fully take account of the purpose of cement for use in dams. It was decided, therefore, to make more stringent specifications for important quality items while conforming to JIS and JSCE specifications (Fig. 2). A total of about 30 items including specific gravity, fineness, chemical components, and the slag replacement ratio were studied paying particular attention to the following:

(1) Which quality items are more important to ensure the required quality of dam cement?
(2) What is the significance of the upper and lower limit specifications and tolerance specifications?
(3) To what extent can variations due to production plants be permitted?
(4) How should variations in concrete quality due to variations of raw materials be treated?

Different slag ratios for interior concrete and exterior concrete were used. Hence, quality specifications for base cement, slag and their mixtures were determined separately (Table 1). These specifications were close to the upper limits of the current cement production technology and quality control standard. Production targets were also set to minimize quality variations in production. Strict quality specifications and production targets are shown in Table 2.

Fig. 1. Location of Satsunaigawa Dam.

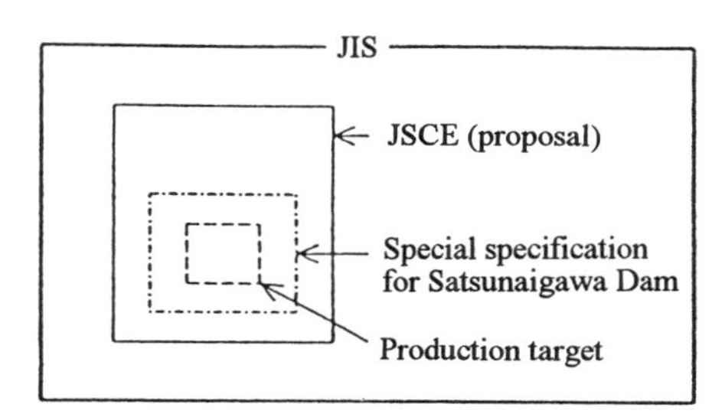

Fig. 2. Concept of quality specification.

Table 1. Quality specifications for blended cement for Satsunaigawa Dam.

Item		Moderate-heat portland cement	Blast-furnace slag powder	Blended cement (55% slag)	Blended cement (65% slag)
Slag ratio (%)		—	—	55±2	65±2
Specific gravity		3.21±0.02	2.91±0.03	3.04±0.02	3.01±0.02
Blaine (cm^2/g)		3 450±200	4 000±250	3 750±250	3 800±250
Chemical composition (%)	Ignition loss	0.3-2.0	2.0 or less	0.2-2.0	0.2-2.0
	MgO	0.5-5.0	4.0-8.0	2.0-7.0	2.0-7.0
	SO_3	1.5-3.0	2.0 or less	2.5 or less	2.5 or less
	S	—	0.3-2.0		
	R_2O	0.30-0.60	0.30-0.60	0.30-0.60	0.30-0.60
	Moisture content	—	0.5 or less	—	—
Mineral composition (%)	C_3S	40-48	—	—	—
	C_3A	6 or less	—	—	—
Setting (h)	Initial	1.5-3.0	—	2.5-4.5	3.0-5.0
	Final	3.0-5.0	—	4.5-7.0	5.0-8.0
Stability		High	—	High	High
Flow ratio (%)		—	95 or more	—	—
Compressive strength (kgf/cm^2)	3-day	120±-20	—	60±20	50±20
	7-day	160±20	—	145±25	135±25
	28-day	330±40	—	340±50	330±50
Activity index SAI (%)	7-day	—	55 or more	—	—
	28-day	—	75 or more	—	—
	91-day	—	95 or more	—	—
Heat of hydration (cal/g)	7-day	60-70	—	50-60	47-57
	28-day	72-82	—	63-73	60-70
Basicity			1.75-1.95	—	—

Table 2. Production target.

	Quality specification (%)	Production target (%)
Slag replacement ratio	±2	±1.5
Ignition loss	0.2-2.0	0.2-1.8
Magnesia	2.0-7.0	0.2-6.5
R_2O	0.3-0.6	0.35-0.55
Initial setting	3.0-5.0	3.0-4.5
Final setting	5.0-8.0	5.5-8.0

Table 3. Cement quality data.

Classification		Item	Number of Specimens	Max	Min	Mean	Tolerance
Slag ratio: 55%	Blended cement	(a)	7	56.0	54.3	55.1	55±2
		(b)	7	3.04	3.04	3.04	3.04±0.02
		(c)	7	3 800	3 720	3 763	3 750±250
	Moderate-Heat portland cement	(b)	7	3.21	3.21	3.21	3.21±0.02
		(c)	7	3 410	3 300	3 340	3 450±200
	Blast-furnace slag powder	(b)	7	2.93	2.90	2.91	2.91±0.03
		(c)	7	4 210	4 080	4 133	4 000±250
Slag ratio: 65%	Blended cement	(a)	5	65.4	64.5	64.8	65.2
		(b)	5	3.00	3.00	3.00	3.01±0.02
		(c)	5	3 940	3 810	3 876	3 800±250
	Moderate-Heat portland cement	(b)	5	3.21	3.21	3.21	3.21±0.02
		(c)	5	3 560	3 430	3 504	3 450±200
	Blast-furnace slag powder	(b)	5	2.91	2.91	2.91	2.91±0.03
		(c)	5	4 110	4 010	4 066	4 000±250

(a) Slag replacement ratio (%)

(b) Specific gravity

(c) Blaine's value (cm^2/g)

Table 3 shows records of the cement quality which has been achieved. As seen from the table, the values obtained are well within the limits of the specified limits and production targets.

3 Placement and curing

3.1 Curing Temperatures for Green Cutting*

Green cutting must not be started until the proper time when freshly placed concrete is sufficiently strong. In order to determine the proper times at which green cutting was to be started and forms were to be removed, a series of tests at the dam site including mortar strength tests (Fig. 3) and green cutting tests were conducted. When to start green cutting was determined according to the test results (Fig. 4); on the other hand, target placement intervals between lifts were determined according to the annual placement schedule (Tables 4,5). As a result of the two above-mentioned, curing temperatures were determined (Table 6).

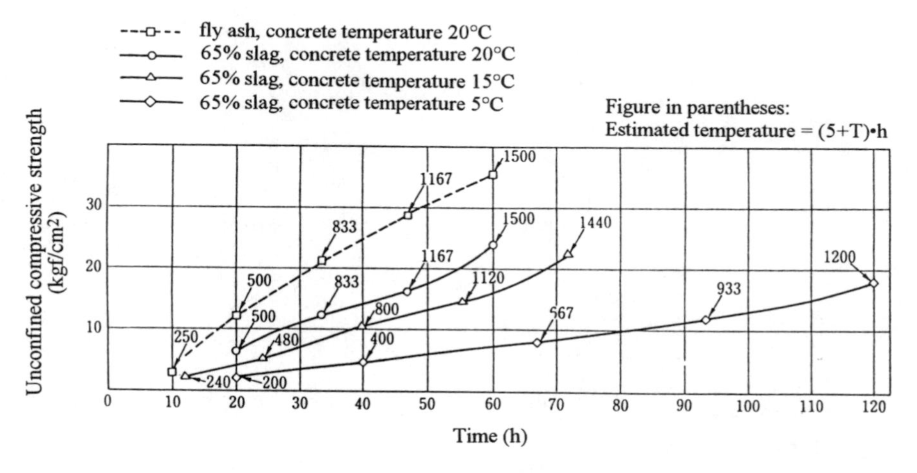

Fig. 3. Unconfined compressive strength of mortar.

* Removing laitance from the upper surface of dam concrete in early age.

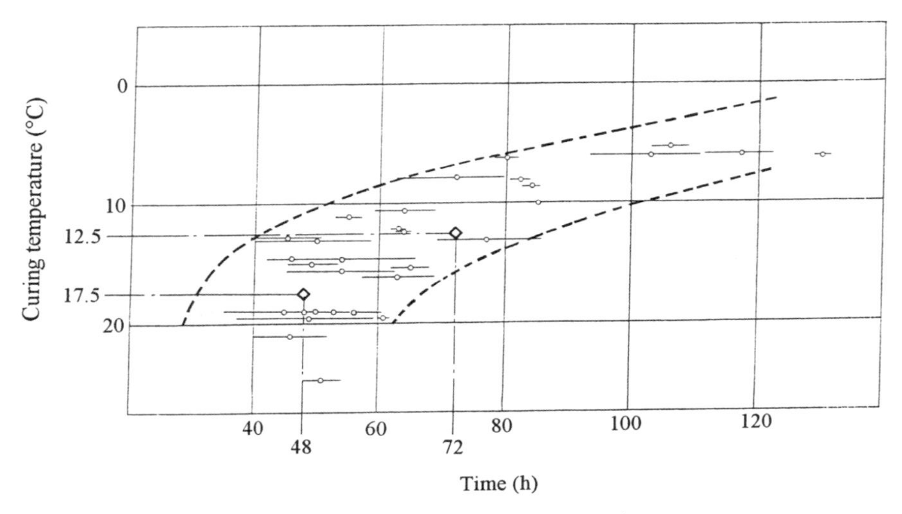

Fig. 4.(a) Timing of green cutting (interior concrete).

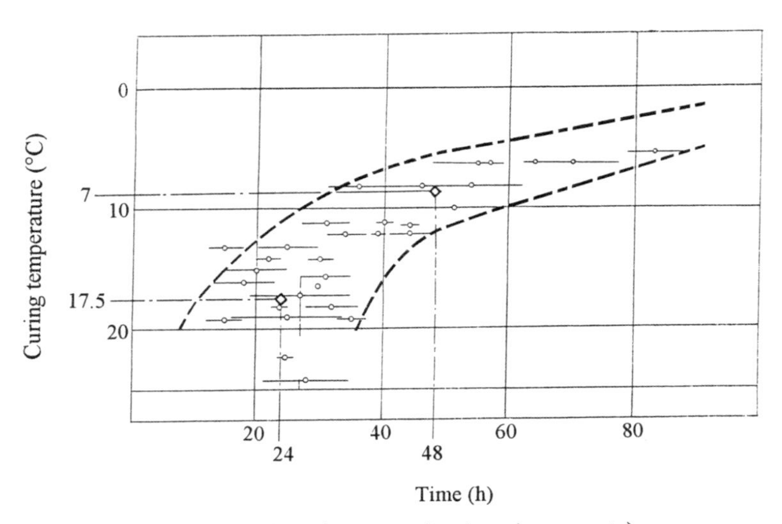

Fig. 4.(b) Timing of green cutting (exterior concrete).

Table 4. Target placing rate.

Classification	Normal temperature period (day)	Low temperature period (day)
Ordinary Lifts	2	3
Lifts involving from sliding work	3	4
Lifts involving removal of mortar from core-foundation contacts	4	5

Table 5. Target curing time.

Classification	Normal temperature period (h)	Low temperature period (h)
Before green cutting of exterior concrete	24 or less	48 or less
Before green cutting of RCD concrete	48 or less	72 or less
Before removal of forms from upstream and downstream faces	60 or less	72 or less

Table 6. Curing temperature requirements.

Classification	Normal temperature period (°C)	Low temperature period (°C)
Before green cutting of exterior concrete	17.5 or more*	7.0 or more
Before green cutting of RCD concrete	17.5 or more*	12.5 or more*
Before removal of forms from upstream and downstream faces	15.0 or more	10.0 or more

3.2 Measures for concrete placement during cold seasons

The curing and other conditions mentioned in the preceding section and the geography (cold region) of the dam site require special measures to be taken during the cold months in spring and autumn. From among many possible measures, three methods were considered effective:

(1) Heat curing using heating mats
(2) Insulated curing using high initial concrete temperature
(3) Use of different chemical admixtures (retarder No.8 -> standard type No.70*)

Of the three methods, method 1 was not used, despite its effectiveness, because of the vast surface area of concrete to be cured and anticipated light vehicle traffic between the placements of concrete. Thus, methods 2 and 3, which were considered to be

* See 3.2.2 Chemical admixtures

applicable to the dam site, were adopted.

3.2.1 Insulated curing with high concrete temperature
Initial concrete temperature can be raised by using warm mixing water. By insulating the upper surface of concrete that has just been placed, as well as the upstream and downstream surfaces, curing temperature can be raised so that the development of the early strength of concrete can be accelerated. Raising the initial concrete temperature increases the difference between concrete temperature and air temperature. Insulated curing is therefore necessary partly because of the need to prevent adverse effects of rapid changes in temperature from affecting the surface layers of concrete.

(1) Warm water: Warm mixing water was prepared by heating river water in a boiler installed at the side of the batcher plant and supplying it to the batcher plant. Water temperatures that produce an the initial temperature of RCD concrete of about 12°C are about 50°C when the air temperature is 7°C and about 30°C when the air temperature is 10°C. During actual placement, records of initial concrete temperature and the air temperature on that day were used to estimate initial concrete temperatures and adjust water temperature accordingly.

(2) Insulated curing: Freshly placed concrete was insulated with polyethylene curing mats (10mm thick, thermal conductivity K=0.03kcal/m·h·°C). Measured values of concrete temperature obtained in 1993 were used to analytically determine concrete temperatures (curing temperatures) resulting from the placing temperature of 10°C. The surface temperatures thus calculated of concrete 5 days after placement were 11°C at an air temperature of 5°C and 14.8°C at an air temperature of 10°C. These results indicate that the insulation was effective.

3.2.2 Chemical admixtures
The water-reducing agent used for Satsunaigawa Dam is Type 1 AE plasticizer (retarder type, Pozzolith No.8), which has proven performance. In view of the characteristics of the blast-furnace slag, a standard water reducer (Pozzolith No.70) was used during the cold months.

3.3 Period
Figure 5 shows air temperatures at the dam site. The placement period was divided into "low temperature period", "intermediate period" and "normal temperature period" (Table 7) according to the 10-day mean temperatures shown in Figure 5.

3.4 Protection against low air temperatures
Since temperatures at the dam site may reach -20°C or below, it was necessary to protect the concrete surface from the cold.

Freshly placed concrete to be exposed to winter was covered with two 10-mm-thick curing mats (50m long). After all forms on the upstream and downstream faces were removed, those faces were covered with a 30-mm-thick curing mat.

The latter curing mat has another benefit. Annual variations of air temperature can create differences between the exterior and interior concrete temperatures, and thus, an injurious tensile strain in the exterior concrete may result.

As a result of FEM analysis, the use of the curing mats effectively planned (Table 8).

Table 7. Countermeasures taken during each of the periods.

Period	Measures
(a) Low temperature period	Since insulation is not enough to maintain temperatures sufficiently high for strength development, it is necessary to raise initial concrete temperature by heating mixing water. • Chemical admixture : standard type • Air temperature (10-Day Average) : 5-10°C • Period : 4/1-5/30, 10/1-10/31
(b) Intermediate temperature period	Concrete temperatures required for strength development can be maintained by insulated curing using PVC mats. • Chemical admixture : standard type • Air temperature (10-Day Average) : 10-15°C • Period : 6/1-6/10, 9/20-9/30
(c) Normal temperature period	Concrete temperatures required for strength development can be maintained without taking special measures. • Chemical admixture : retarder type • Air temperature (10-Day Average) : 15°C or above • Period : 6/11-9/19

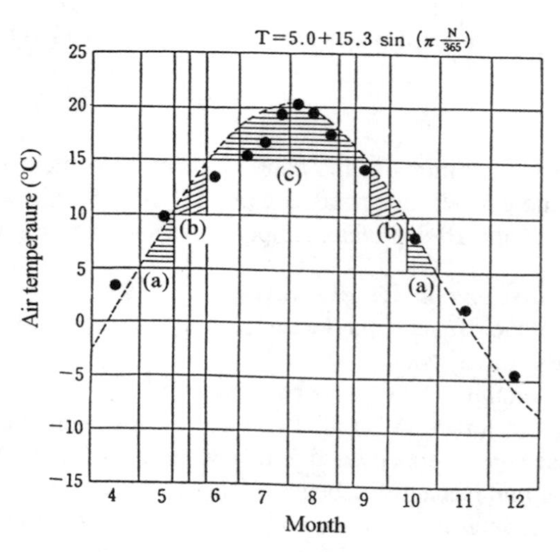

Fig. 5. Air temperature at dam site.

Table 8. Term of insulated curing.

Lift	Term (year)	Date	Lift	Term (year)	Date
17	1	10/1	35	1	6/17
18	1	10/7			
19	1	10/14			
20	1	10/19			
21	1	10/24			
22	1	10/29	44	1	7/28
23	2	11/6			
24	infinit	11/12			
25	2	4/23			
26	2	4/27			
27	1	5/6	51	1	9/15

4 Conclusions

The following conclusions can be drawn from this paper:

(1) Cement quality which has been achieved is good enough to use for the fundamental data.

(2) Use of hot mixing water and insulated curing are effective countermeasures against the delay of initial strength development of the dam concrete.

(3) Using the curing mat is also an effective countermeasure for preventing an injurious tensile strain in the exterior concrete.

Construction of Satsunaigawa Dam, which began in 1991, is now in full swing in its third year of construction. So far, the progress has been on schedule, and concrete placement during the cold months, which was one of the early concerns, has been successful. Efforts will be continued to further improve construction techniques.

5 References

1. H.Watanabe, H.Kouno, Y.Sugiyama. (1994) Study on the Simple Testing Method for Thermal Conductive Properties of Concrete. *Proceedings of the Japan Concrete Institute*, Vol. 16, No. 1. pp. 1341-1346.
2. K.Matsui, N.Nishida, Y.Dobashi, K.Shiota. (1994) Thermal Parameters Estimation of Massive Concrete Based on Inverse Analysis. *Proceedings of the Japan Concrete Institute*, Vol. 16, No. 1. pp. 1347-1352.
3. H.Chikahisa, J.Tsuzaki, T.Arai, S.Sakurai. (1992) Evaluation on Thermal Conductive Parameters of Massive Concrete Based on Inverse Analysis. *Proceedings of JSCE*, No. 451 / Vol. 17. pp. 39-47.
4. JSCE. (1994) *Standard Specifications of Concrete.*

128 PERFORMANCE OF HIGH ALUMINA CEMENT/ BLASTFURNACE SLAG CONCRETES IN AGGRESSIVE ENVIRONMENTS

G.J. OSBORNE
Building Research Establishment, Watford, UK

Abstract
The Building Research Establishment in the UK has developed a blended cement based on high alumina cement with ground granulated blastfurnace slag. The new cement has the trademark 'BRECEM' and the chemical and physical properties of concretes made with it have been compared with those of plain high alumina cement control concretes of similar mix design. This paper describes the results of concrete durability studies carried out using 100 mm cubes. Specimens were initially cured for up to 28 days at 5°, 20° and 38°C in water and in air at 20°C prior to their storage in aggressive marine and acidic water environments. HAC and BRECEM concretes have both performed well after 2 years storage. However, BRECEM concretes which are far less prone to the effects of water content, through the formation of more stable phase assemblages, also showed greater tolerance to differences in early curing regime than did equivalent HAC concretes.
Keywords: Acid attack, blastfurnace slag, concrete durability, compressive strength; freeze-thaw, high alumina cement, marine environment, seawater resistance.

1 Introduction

The Building Research Establishment (BRE) in the United Kingdom has developed a blended cement based on high alumina cement (HAC) and ground granulated blastfurnace slag (GGBS). This cement has been given the trademark 'BRECEM' [1] and the properties of concrete made from the new cement are being studied at BRE in collaboration with the industry. The reasons for the development of BRECEM and how it differs from HAC, in terms of the hydrate formation and stability, have been fully described in a recent paper [2].

Concrete Under Severe Conditions: Environment and loading (Volume Two) Edited by K. Sakai, N. Banthia and O.E. Gjørv. Published in 1995 by E & FN Spon. ISBN 0 419 19860 1

Calcium aluminate or high alumina cement is not used in structural applications in the UK at the present time, as concrete made from such cement shows reduction in strength with time under hot and humid conditions [3]. This reduction in strength is associated with a set of chemical reactions, often referred to as "conversion", that produce an increase in porosity of concrete [4], as a number of metastable hydrates which form with water convert rapidly at higher than ambient temperature to form stable hydrates. It should be noted though, that converted Fondu was found to be no more porous than ordinary Portland cement (OPC), of equal water/cement ratio [5]. The addition of slag, typically 40-60% alters the course of the hydration reactions in HAC. A chemical compound, C_2ASH_8 or $2CaO.Al_2O_3.SiO_2.8H_2O$, gehlenite hydrate (stratlingite), not seen in plain HAC in significant amounts, forms readily and becomes the major hydrate constituent in due course, and is thought to provide a more stable phase assemblage [6-10].

Studies to determine the chemical and physical properties of HAC and BRECEM pastes, mortars and concretes over a range of temperatures and storage conditions have been set up at BRE and previously reported [2, 6, 8-10]. Earlier durability results and other BRE data on the temperature rises produced in near adiabatic conditions for a series of metre cubes using similar concretes, with cores being subsequently taken to assess strength development and hydrate formation have also been reported with the results of practical trials carried out on precast concrete blocks for application purposes [2, 10]. This paper gives detailed results of tests carried out on concrete durability specimens made with BRECEM and HAC concretes of similar mixture proportions and different early curing regimes, following storage for 2 years in aggressive marine and soft acid water environments.

2 Concrete materials

The high alumina cement (HAC) was supplied by Lafarge Special Cements in the UK and the two ground granulated blastfurnace slags (GGBS), one British and the other French, originated from the Appleby Frodingham Cement and Civil and Marine Slag companies respectively. These three companies were industrial sponsors and contributed to the funding of this research. The chemical analysis of these cementitious materials and their surface areas are given in Table 1. Thames Valley gravel coarse and fine aggregates (sands) were used for all the concrete mixtures which are described in Table 2.

2.1 Chemical analysis

The powdered HAC and slags were analysed by X-ray fluorescence (main oxides), flame emission (alkalis) and by using the Leco apparatus for total sulphur. These commercially available materials had fairly low loss on ignition values and were approximately of the same surface area at about 400 m^2/kg. There were minor differences in the chemical analyses of the two slags, with the UK sample having a slightly lower CaO/SiO_2 ratio and higher Al_2O_3, than the French slag.

Table 1. Chemical analysis of HAC and GGBS (% by weight).

Oxides	HAC	Civil and Marine GGBS	Frodingham GGBS
SiO_2	3.64	34.20	35.56
Al_2O_3	39.62	11.30	13.08
Fe_2O_3	16.12	1.17	0.64
CaO	37.66	41.60	39.39
MgO	0.67	8.21	9.37
Na_2O	0.01	0.26	0.27
K_2O	0.08	0.40	0.46
Mn_3O_4	0.05	0.25	0.48
TiO_2	1.94	0.77	0.71
ZrO_2	-	0.03	-
BaO	0.04	0.06	-
SO_3	0.08	-	-
Total **S**	-	0.89	0.89
P_2O_5	0.03	-	-
LOI	0.64	-	-
Fineness (m^2/kg)	320	417	389

Table 2. Concrete mixture proportions and wet concrete properties.

Concrete Mix Proportions					Fresh Concrete Properties (range of values)		
Thames Valley Aggregates			Cement	'Water'			
20-10 mm (67%)	10-5 mm (33%)	<5 mm	(Slag+ HAC)	Total w/c* (free w/c)	Cement content (kg/m^3)	Wet density (kg/m^3)	Slump (mm)
2.91		1.75	1.0	0.5 (0.45)	380	2330-2370	60-100

* Total w/c is the water to "cementitious materials" ratio (ie HAC + GGBS)

3 Concrete durability studies

A comprehensive programme of durability studies was set up at BRE to investigate the robustness of BRECEM, in comparison with HAC concretes, to variations in mixture proportions and its performance in a variety of aggressive environments. 100 mm concrete cubes cured in water at different temperatures for up to 28 days, prior to storage in a number of sulphate [10], marine and acidic water environments, were assessed after 1 and 2 years and specimens remain for testing at 5 and 10 years. The visual appearance, attack ratings (where necessary) and retention of compressive strength of the concrete cubes were determined by an established BRE method [11]. Results presented earlier [10] have been updated and assessed in more detail in this paper.

3.1 Concrete mixtures

There are presently no recommended concrete mixes in the UK for the use of calcium aluminate cements in structural concrete. Concretes were therefore prepared to the appropriate mix proportions which satisfied; the minimum requirements of BS 5328, Part 1, 1990 for the "Guide to specifying concrete", when placed in exposure conditions classified as most severe [12]. It should be stressed that the HAC and slag cement concretes were proportioned to provide equal cement content and workability rather than equal 28 day strength.

Concretes in which one or other of the two blastfurnace slags replaced HAC in the proportions 0, 40, 50 and 60% were prepared in BRE's concrete laboratory, as 100 mm cubes. The concrete mixture proportions and fresh properties are given in Table 2. These showed that the marine and acid water specimens, of 380 kg/m^3 total cementitious content, were prepared from moderate slump concretes (60 - 100 mm) with compacting factors between 0.90-0.96. The workability and 28 day strength data are given in Table 3 and the concretes were deemed to have reasonably good workability. No problems were experienced in their preparation and placing in moulds, although the GGBS originating from the UK gave slightly more workable concretes, than did the French slag when blended with the same HAC, and had lower 28 day strengths for most of the water cured specimens.

3.2 Curing regimes prior to site storage

The concretes were vibrated into 100 mm cube moulds and stored below a cover of damp hessian and polythene sheet to maintain an initial curing condition close to 100% relative humidity. The concrete cubes were then demoulded at 24 hours, numbered and pre-cured for a further 27 days in water at 5°C, 20°C or 38°C or in air at 20°C and 65% RH to attain their "characteristic" 28 day strength, prior to placing randomly in the different aggressive storage environments for testing at 1, 2, 5 and 10 years.

Table 3. Workability and strength at 28 days of concrete stored in aggressive environments .

Cements	Workability			28 Day Compressive Strength (MPa)			
	Slump (mm)	Vebe time (sec)	CF	Water 20°C	Water 38°C	Water 5°C	Air 20°C, 65% rh
HAC	80	3	0.95	83	26	61	65
(A) Civil & Marine GGBS							
60/40 HAC+ GGBS	60	4	0.91	50	56	48	51
50/50 HAC+ GGBS	75	4	0.94	52	59	39	44
40/60 HAC+ GGBS	55	4	0.90	39	51	29	34
(B) Frodingham GGBS							
60/40 HAC+ GGBS	90	2	0.96	50	46	45	51
50/50 HAC+ GGBS	100	2	0.96	44	48	34	43
40/60 HAC+ GGBS	70	2	0.93	31	42	27	34

3.3 Aggressive storage environments

The aggressive storage environments, namely, seawater and soft acid water have been described in detail previously [13, 14], however a brief description is given below:

(1) The BRE marine exposure site, with spray, tidal and full immersion zones, is situated at Shoeburyness, 4 miles east of Southend on the River Thames estuary in the UK [14]. It is one of the major field sites worldwide, listed by the F.I.P. Commission 6 on "Concrete Sea Structures" [15], and the chemical analysis of the seawater is given in Table 4. This shows that it is typical of that which is found around the coasts of Britain. The pH is between 7.9 and 8.4.

(2) The BRE soft acid-water site is at Butterley reservoir in Yorkshire in the UK [13]. The pH of the water from the peaty soil varies between 3.5 and 4.5 in the winter and summer periods respectively, mainly due to the changes in humic acid content of the water.

Table 4. Chemical analysis of sea-water at Shoeburyness.

Ions analysed	Amount present		Atlantic sea-water (g/l)*
	(%)	(g/l)	
SO_4^{2-}	0.26	2.60	2.54
Cl^-	1.82	18.20	17.83
Ca^{2+}	0.04	0.40	0.41
Mg^{2+}	0.12	1.20	1.50
Na^+	0.97	9.74	9.95
K^+	0.04	0.40	0.33

* Lea, 1970 [16]

3.4 Methods of test and assessment

Concrete cubes (100 mm) stored in water at 20°C, (as controls), and those from the seawater and soft acid-water environments were collected in batches of three at the different ages of test for assessment in the laboratory. The physical appearance was determined visually and by measuring the "attack rating"[11]. One of the cubes from each environment was photographed prior to the determination of the compressive strength in accordance with BS 1881 : Part 116 (BSI 1983) [17].

4 Results and discussion

The use of 100 mm concrete cubes as a research facility for durability studies has sometimes been criticised, as not providing realistic data, as the specimens are quite small compared with concrete of larger dimensions in real structures. It is also difficult to assess the risk of cracking,due to shrinkage or swelling, and its influence on durability with small test specimens. However, the outer 50-100 mm of reinforced concrete is that which protects the steel reinforcement and it may not behave too differently from the concrete cubes used in the present study. The main object of this study was to compare the effects of the early curing regime on the performance of the range of concretes after storage in the different outdoor and aggressive marine and acidic environments. A secondary aim was to assess the effect of blending different GGBS materials with the same HAC on the properties of the concretes made with neat HAC and 60/40, 50/50 and 40/60 HAC/GGBS composite cements. The results for the Civil and Marine Slag and Frodingham cement slag/HAC concretes at 2 years are given in Tables 5 and 6 respectively.

4.1 Effect of early curing regime on strength development

The plain HAC concretes had developed 2 year strengths in the range 33 to 45 MPa. The strengths of HAC concretes cured in water at 38°C (33 MPa), in water at 5°C (45 MPa) or in air at 20°C (42.5 MPa) were considerably lower than the 64 MPa strength attained by the same concrete following water storage at 20°C. The reduced strength of these HAC concretes was attributed to the effects of "conversion" at 38°C, low temperature curing at 5°C in water and slower hydration at 20°C in air at 65% RH. It is noteworthy that most of these HAC concretes have lower strength at 2 years as compared to their respective 28 day strength.

However at the same age, the HAC/GGBS cement control concretes had compressive strengths greater than 50 MPa and were generally in the range 55-65 MPa. This indicated that the effects of "conversion" via the HAC component were minimal or even non existent. It is also relevant that all these HAC/GGBS concretes have increased in strength from 28 days to 2 years, unlike the neat HAC control concretes which had decreased generally.

4.2 Resistance to marine environments

The results in Tables 5 and 6 showed that generally all the HAC and HAC/GGBS blended cement concretes had good resistance to seawater attack after 2 years of exposure in the spray, tidal and full immersion zones at the BRE marine exposure site. The performance and durability of these concretes were assessed by comparing concrete cube strength data (loads and failure) and the percentage strength retained of the water stored controls at the same age.

4.2.1 HAC concretes

Good strength development was achieved with cubes pre-cured in water at 20°C, or at 5°C and in air at 20°C, 65% RH followed by storage in the full immersion, tidal and spray zones. The concrete cubes pre-cured in water at 38°C and with the lower 28 day strength showed signs of attack when stored in full immersion and tidal zones but not in the spray zone. There was slight spalling of concrete at the corners and edges with cracks appearing at the edges of some of the cubes. The lower control strengths and associated increased porosity of the "converted" HAC concretes, pre-cured at 38°C, are considered to be the primary causes for the slight damage showing on these specimens although the percentage strength retained data are very good.

4.2.2 BRECEM concretes

Concrete made with different blends of HAC/GGBS (French slag, Table 5) and (UK slag, Table 6) showed no signs of attack in any of the marine aggressive environments and were in excellent condition at two years. Those stored at the BRE Garston exposure site and the marine site spray zone gave the best durability results. The concretes with 50/50 and 40/60 blends of HAC and GGBS, pre-cured in air at 20°C, 65% RH had increased strengths from 1 to 2 years in these aggressive environments, although were still somewhat less than their water stored controls at 20°C. There were slight differences in strength development depending upon whether the concretes contained the UK or French slags and these are discussed below.

Table 5. Compressive strength data (MPa) for 100 mm concrete cubes stored in aggressive environments at 2 years (French GGBS).

Cementitious Mixture and Pre-cure	Storage Environments						
	Control		Marine Zone			Soft Acid Water	Outdoor Exposure Garston
	Water 20°C						
	28 days	2 years	Full	Tidal	Spray		
HAC							
Water 20°C	83.0	64.0	80.5	79.0	70.0	78.0	66.0
Water 38°C	26.0	33.0	34.0	34.0	40.0	33.0	38.0
Water 5°C	61.0	45.0	71.0	71.0	52.0	76.5	45.0
Air 20°C	65.0	42.5	67.5	70.0	47.0	72.0	44.0
60/40 HAC/GGBS							
Water 20°C	50.0	66.5	65.0	69.5	74.0	65.0	71.0
Water 38°C	56.0	64.0	61.0	64.0	74.0	61.0	67.0
Water 5°C	48.0	71.0	68.0	69.5	75.0	64.0	74.0
Air 20°C	51.0	64.5	67.0	67.0	69.0	60.0	63.0
50/50 HAC/GGBS							
Water 20°C	52.0	64.5	67.0	65.0	73.5	63.5	73.0
Water 38°C	59.0	52.5	62.5	61.0	69.0	56.0	68.5
Water 5°C	39.0	68.0	64.0	65.0	69.0	64.0	68.0
Air 20°C	44.0	65.0	59.0	57.0	61.0	54.0	61.0
40/60 HAC/GGBS							
Water 20°C	39.0	63.0	59.0	58.0	63.0	54.0	61.0
Water 38°C	51.0	64.0	59.0	57.0	65.0	53.0	65.0
Water 5°C	29.0	62.0	57.0	57.0	59.0	56.0	59.0
Air 20°C	34.0	56.0	51.0	51.0	51.0	47.0	50.0

Table 6. Compressive strength data (MPa) for 100 mm concrete cubes stored in aggressive environments at 2 years (UK GGBS).

Cementitious Mixture and Pre-cure	Control Water 20°C		Marine Zone			Soft Acid Water	Outdoor Exposure Garston
	28 days	2 years	Full	Tidal	Spray		
60/40 HAC/GGBS							
Water 20°C	50.0	67.0	67.5	68.5	74.0	64.5	74.5
Water 38°C	46.0	67.0	58.5	63.0	72.5	57.5	68.5
Water 5°C	45.0	68.0	65.5	65.5	72.5	59.5	69.0
Air 20°C	51.0	64.0	61.0	63.0	62.5	55.5	60.5
50/50 HAC/GGBS							
Water 20°C	44.0	64.0	61.0	62.0	69.5	57.0	69.0
Water 38°C	48.0	58.0	55.0	55.0	68.0	50.0	66.0
Water 5°C	34.0	67.0	64.0	64.0	72.0	56.0	69.0
Air 20°C	43.0	59.0	52.0	58.0	62.0	52.0	56.0
40/60 HAC/GGBS							
Water 20°C	31.0	51.0	49.0	46.0	54.0	40.0	51.0
Water 38°C	42.0	56.0	54.0	56.0	61.0	49.0	58.5
Water 5°C	27.0	55.5	43.0	42.5	52.0	43.5	50.0
Air 20°C	34.0	46.0	40.0	42.0	45.0	41.0	43.0

4.2.3 Effect of different GGBS used with same HAC

The compressive strength development of the two year old 100 mm concrete cubes made from 60/40, 50/50 and 40/60 mixtures of HAC and the UK GGBS was generally slower than that of the equivalent concrete specimens containing the French slag. These concrete cubes had no signs of attack and the trends shown in Table 6 were very similar to those obtained using the French slag (see Table 5).

The differences are partly attributed to the fact that the higher workability concretes containing the slightly coarser UK slag originally had lower 28 day strengths and developed less strength at 2 years particularly where higher levels of slag replacement for the same HAC had occurred. However factors such as; mode of quenching, glass content and oxide ratio, which determine the activity of the blastfurnace slag may also have affected the strength development.

These results show that although the HAC concretes have generally performed well, the durability of BRECEM concrete cubes was somewhat better than those of HAC cubes when stored in the same aggressive environments.

4.2.4 Effect of marine environment
Only minor differences in behaviour have been exhibited at this relatively early stage of 2 year's exposure to the three marine environments. Spray zone concretes, as might be expected, have performed the best with some slight differences showing up between those HAC concretes which having been pre-cured at 38°C and "converted", have started to show slight signs of attack in the tidal and full immersion zones. However HAC concrete strengths have continued improve, compared with their water-stored controls, whereas the 40/60 HAC/GGBS blends have exhibited some slight reductions in percentage strength retained. Either way there is at present no evidence of attack in the BRECEM concretes. The results at 5 years should provide the necessary longer term performance data, when compressive strength, wear ratings and chloride ingress will be determined and assessed.

4.3 Freeze thaw resistance in the tidal zone

None of the concrete specimens containing HAC or HAC/GGBS blends, in the tidal zone had suffered from frost damage in the form of "pop outs" or surface spalling as was witnessed with Portland cement/GGBS concretes during periods of extremely cold weather experienced in the UK winters of 1985 and 1986 [14]. The regular diurnal tidal movements, particularly when associated with extremes of temperatures, combine with other weather conditions to produce a rather unsettled, changeable environment, with wetting and drying, freezing and thawing, compared with the more evenly controlled full immersion zone. These observations are covered in the relevant British Standard for maritime structures [18], which recommend air entrainment if concrete is likely to be subjected to freezing while wet.

4.4 Resistance to acid water

The performance of the different concretes was generally good after 2 years of storage in the soft acid, moorland waters at Butterley reservoir in Yorkshire. This water has a pH of around 4.0, but has little dissolved carbon dioxide.

(1) HAC concretes with strengths of 33 to 78 MPa, depending on pre-cure showed signs of attack with the lower strength concretes as evidenced by some slight softening of edges and corners. However the higher strength, good quality concretes have performed very well, in this aggressive environment.

(2) HAC/GGBS concretes with strengths in the ranges 60-65, 54-64 and 47-56 MPa (French slag) and 56-65, 50-57 and 40-49 MPa (UK slag), for the 60/40, 50/50 and 40/60, HAC/GGBS cements have shown no signs of softening or attack. At this relatively early stage there appears to be some evidence of slight benefits from the use of GGBS.

These early data with all HAC/GGBS concrete strengths in excess of 40 MPa (cf 33 MPa for HAC) lend support to the view, or expectation that, the quality of the concrete is of greater importance than the type of cement used in such aggressive conditions [13]. However, the reduced calcium hydroxide levels and lower porosity resulting from the well-cured HAC and HAC/GGBS cement concretes, could be responsible for reducing the rate of attack. Large quantities of gehlenite hydrate were found by Xray Diffraction analysis to be present in all BRECEM concretes following their storage in the aggressive soft acid water (and marine zone) environments for 2 years.

5. Conclusions

1. BRECEM control concretes with water to cement ratios greater than 0.40, the maximum formerly specified for HAC concretes, showed no signs of strength loss following pre-curing at temperatures from 5° to 38°C, whereas higher w/c HAC concretes pre-cured at 38°C, showed considerably reduced strength due to the effects of "conversion". HAC concretes were therefore less tolerant to the effects of higher water contents than BRECEM concretes.
2. The continuing strength gain of the BRECEM concretes from 28 days to 2 years is considered to be partly due to the formation of the stable gehlenite hydrate which replaces the metastable hydrates associated with HAC which showed strength decreases for control specimens over the same period of time.
3. Two year old BRECEM concrete specimens contained large quantities of gehlenite hydrate, when pre-cured over the temperature range of 5° to 38°C and this was again responsible for their more consistent performance and improved strength characteristics when stored in water.
4. All BRECEM concretes, with total water to cement ratios of 0.5 (free 0.45), showed excellent durability in both marine and soft acid water environments with no signs of attack at 2 years. Equivalent HAC concretes also performed very well with high strength retention and reasonable strength development up to 2 years in all aggressive environments, even though there was slight softening and spalling at edges of some of the lower strength HAC concrete cubes, pre-cured at 38°C.
5. There were no signs of frost attack in any of the HAC or HAC/GGBS cement concretes after storage for 2 years in the tidal seawater zone.
6. Both GGBS materials gave similar performances when used in the BRECEM formulation, although concretes made using the UK slag developed strength at a slower rate.

The findings, at this relatively early stage, are very encouraging for all concretes and longer term tests including chloride ingress and rebar corrosion studies will be carried out at 5 and 10 years. Conversion reactions and increased porosity of HAC concretes are reflected by lower compressive strength. However, microstructural and more detailed mechanical studies are required to provide information about the reasons for the strength losses. These studies should help to elucidate whether conversion and the associated increase in porosity, or cracking due to the effects of frost or sulphates, have the predominant effects on concrete durability.

BRECEM concretes have shown a greater tolerance of high water to cement ratio mix designs in forming stable assemblages with reduced temperature rises, enhanced durability in terms of their excellent sulphate, seawater and soft acid water resistance and there are cost savings compared with HAC concretes [2, 10]. There is a great potential for the prospective use of this new cementitious material and a number of practical applications are advocated [2, 19].

6 Acknowledgement

The work described has been carried out as part of the strategic research programme of the Building Research Establishment of the Department of the Environment and this paper is published by permission of the Chief Executive. The author thanks Dr A J Majumdar, Mr B Singh and Dr K Quillin for their involvement with this project and Mrs J L Hardcastle and the Concrete Laboratory staff for carrying out the experimental work.

7 References

1. Majumdar, A.J. and Singh, B. (1991) *Cement compositions, UK.* Patent GB 2211 182 B and European Patent (1994), 0312 323 B1.
2. Osborne, G.J. (1994) *BRECEM: A rapid hardening cement based on high alumina cement.* Proceedings of Institution of Civil Engineers, Structures and Buildings, 104, pp 93-100.
3. British Standards Institution (1972) *Code for Structural Concrete.* AMD 1553 in CP 110. BSI, London.
4. Midgley, H.G. and Midgley, A. (1975) *The conversion of high alumina cement.* Magazine of Concrete Research, 27, pp 59-77.
5. George, C.M. (Mangabhai, R.J. (ED)) (1990) *Manufacture and performance of aluminous cement; a new perspective.* Calcium Aluminate Cements, E and F.N. Spon, London, pp 181-207.
6. Majumdar, et al (1991) *Blended high alumina cements,* Proceedings of International Conference on Advances in Cementitious Materials, Gaitherburg. Ceramic Transactions, 16, pp 661-678, July.
7. Hirose, S. and Yamazaki, Y. (1991) *Hydration of high alumina cement with blastfurnace slag.* Gypsum and Lime, No 233, pp 8-12.
8. Quillin, K.C. and Majumdar, A.J. (1994) *Phase equilibria in the* CaO-Al_2O_3 - SiO_2 - H_2O *system at 5°C, 20°C and 38°C.* Advances in Cement Research, 6 No 22, pp 47-56.

9. Majumdar, A.J. and Singh, B. (1992) *Properties of some blended high-alumina cements.* Cement and Concrete Research, 22, pp 101-114, .
10. Osborne, G.J. and Singh, B. (1995) *The durability of concretes made with 'BRECEM' cement comprised of blends of high alumina cement and ground granulated blastfurnace slag.* Proceedings of the Fifth CANMET/ACI International Conference on Fly Ash, Silica Fume, Slag and Natural Pozzolans in Concrete, Milwaukee, WI, USA. (to be published).
11. Osborne, G.J. (1991) *The sulphate resistance of Portland and blastfurnace slag cement concretes.* Proceedings of Second CANMET/ACI International Conference on the Durability of Concrete, Montreal, II, SP-126, pp 1047-1071.
12. British Standards Institution (1990). *Concrete. Part 1 : Guide to Specifying Concrete*, BSI, London. BS 5328 : Part 1.
13. Matthews, J.D. (1992) *The resistance of PFA concrete in acid ground waters.* Proceedings of the Ninth International Congress on the Chemistry of Cement, New Delhi, India, Vol V, pp 355-362.
14. Osborne, G.J. (1986) *Concrete durability studies at the Building Research Establishment Marine Exposure Site, Shoeburyness.* Proceedings of Marine Concrete 1986. The Concrete Society, London, pp 157-164.
15. Materials of F.I.P. Commission 6, (1989) *Concrete Sea Structures Canada. Major field sites for concrete exposure.*
16. Lea, F.M. (1970). *The chemistry of cement and concrete.* Edward Arnold, London, Third edition, 625 pp.
17. British Standards Institution, (1983) *Method for determination of compressive strength of concrete cubes*, BSI, London, BS 1881, Part 116.
18. British Standards Institution (1989) *Code of Practice for Maritime Structures.* Amendment No 4. AMD 6159. BSI, London, BS 6349, Part 1.
19. Montgomery, R.G.J., Rashid, S., Capmas, A. and Woolley, W.B. (1993) *Calcium aluminates and their derivatives: new applications and technologies.* Proceedings of Concrete 2000, E. and F.N. Spon, Dundee, Vol 2, pp 1857-1879.

129 DURABILITY OF HIGH-STRENGTH MORTARS INCORPORATING BLAST-FURNACE SLAGS OF DIFFERENT FINENESS

J. MADEJ
Beton VUIS Co. Ltd, Bratislava, Slovak Republic
Y. OHAMA and K. DEMURA
Nihon University, Koriyama, Japan

Abstract
The addition of finely ground granulated blast-furnace slags with different fineness (surface areas of 4000, 8000, and 15000 cm^2/g) was investigated in relation to the fundamental physico-mechanical properties and durability of high-strength mortars. Besides a positive effect of the slag fineness on the strength development, a decisive increase in density, an extremely low difusivity of CO_2 or chloride ions and excellent freezing and thawing resistance were achieved for water-cured mortars. Autoclave-cured high-strength mortars, though having higher strength and lower diffusivity of CO_2, proved to have slightly lower resistance to chloride ion penetration than water-cured mortars, depending also on the slag fineness.
Keywords: Blast-furnace slags, compressive strength, curing conditions, durability, fineness, high-strength mortars.

1 Introduction

It is well known that fine mineral powder such as silica fume influences the fundamental mechanical properties of cement-based materials. Due to a positive effect of ultrafine SiO_2 particles on the cement hydration and pore structure formation, improvement in the mechanical properties and durability of concrete were ascertained. Thus, the use of the silica fume combined with an admixture with dispersing properties is assumed to be the most important factor for an achievement of high-strength concrete [1].

Concrete Under Severe Conditions: Environment and loading (Volume Two) Edited by K. Sakai, N. Banthia and O.E. Gjørv. Published in 1995 by E & FN Spon. ISBN 0 419 19860 1

Besides the mix proportions of fresh mixture, the curing conditions seem to be of great importance in relation to the mechanical and other properties of cement composites. It was proved that "super" high-strength mortars having compressive strengths of the level of 200 MPa or higher can be obtained using ordinary portland cement, fine aggregates such as silica sands and stainless steel particles, high-purity silica, silica fume and various high-range water-reducing agents [2].

An interest of concrete producers all over the world has also been focused on finely ground granulated blast-furnace slags as concrete admixture. It was found that the finely ground granulated blast-furnace slags can be used as highly efficient supplementary materials which improve the properties of fresh and hardened concretes with a possibility of cement saving , higher resistance to sulphate attack, elimination of alkali-aggregate reaction, etc. [3] [4]. The application of finely ground blast-furnace slag as a component for high-strength concrete was reported [5].

The present paper deals with the investigation of granulated blast-furnace slags ground to different fineness with regard to their use in high-strength mortars. The properties such as strength development, carbonation, chloride ion penetration and freezing and thawing resistance of the high-strength mortars are examined.

2 Materials

Ordinary Portland cement (OPC) as specified by JIS R 5210 (Portland Cement) and silica sand (1:1 silica sand No.4-No.7 mixture) were used for the preparation of mortars. The blast-furnace slags (BFS) of Japanese origin, "Himent", supplied by Hitachi Cement Co., Ltd., Japan, was ground to different fineness: 4000, 8000 and 15000 cm^2/g in surface area. A commercial high-range water-reducing agent (HRWRA), Mighty 2000 WH (Kao Corporation, Japan), was used for mortars. The chemical compositions and physical properties of the finely ground granulated BFS are given in Table 1.

Table 1. Chemical compositions and physical properties of ground granulated BFS

Chemical compositions (% by mass)						
Loss of ignition	SiO_2	Al_2O_3	MgO	CaO	Fe_2O_3	SO_3
0.0	33.9	15.5	5.9	42.5	0.8	0.0

Physical properties			
Sign of slag	Specific gravity	Surface area (m^2/kg)	Average particle size (μm)
S(4)	2.92	406	10.85
S(8)	2.92	805	4.96
S(15)	2.92	1500	2.48

3 Testing procedures

3.1 Preparation of mortar specimens

The plain cement mortar (100% OPC) considered as a control and the mortars with a partial cement replacement with BFS (60% OPC + 40% BFS) were prepared with a cementitious material to sand ratio of 1:1 by mass by using a laboratory Omni mixer. The proportions of the mortars is summarized in Table 2.

The slags were added with cement and sand, and homogenized in the mixer for 3 minutes. The high-range water-reducing agent was added to the mixing water. The total mixing time was 6 minutes.

Mortar specimens, prisms 40x40x160 mm, were cast in two layers, and compacted on a vibration table with a vibration of 25 Hz. The mortar specimens in the molds were cured in a moist room, maintained at 20° C and 80% R.H. for 48 hours. After demolding, the mortar specimens were subjected to the following different curing conditions:

a) *Water cure*, cure in water at 20° C for 7, 28 and 90 days;

b) *Autoclave cure*, cure in an autoclave at a maximum temperature of 180° C under a pressure of 1.0 MPa for 3 hours.

3.2 Strength test

Mortar specimens were tested for compressive strength according to JIS 5201 (Physical Testing Methods for Cement).

3.3 Water absorption test

Mortar specimens cured in water for 28 days and in an autoclave were immersed in tap water at 20° C for 48 hours, and their water absorption was determined.

3.4 Accelerated carbonation test

Two different procedures were applied to the mortar specimens cured in water for 28 days and in an autoclave:

Table 2. Mix proportions of fresh mortars

Mix No.	Cementitious binder : sand ratio (by mass)	Water - cementitious ratio, w/c	HRWRA (% of cement by mass)	Compositions of cementitious binder (% by mass)			Unit weight (kg/m^3)	Flow
				Cement	Slag			
					Sign	% by mass		
1				100	-	-	2394	141
2	1:1	0.20	3.0	60	S(4)	40	2334	170
3				60	S(8)	40	2330	192
4				60	S(15)	40	2256	224

a) *non-pressurizing carbonation test*: the mortar specimens were coated with an epoxy resin paint, and placed in a non-pressurized carbonation test chamber for 2 years, in which temperature, humidity and CO_2 gas concentration were controlled to be 30° C, 60% R.H. and 5.0%, respectively.

b) *pressurized carbonation test*: the mortar specimens were coated with an epoxy resin paint and placed in a sealed pressurized vessel, evacuated to 667 Pa or less at ambient temperature for 30 minutes, and then exposed to CO_2 gas with a pressure of 1.0 MPa for 14 days.

After accelerated carbonation, the mortar specimens were split, and the split crosssections were sprayed with a 1% phenolphthalein alcoholic solution. The depth of rim of each crosssection without color change after spraying the phenolphthalein solution was measured with slide calipers as a carbonation depth as shown in Figure 1.

3.5 Chloride ion penetration test

Mortar specimens cured in water for 28 days and in an autoclave were coated with an epoxy resin paint by using the same procedure as in the case of the mortar specimens for carbonation test, and immersed in a 2.5% NaCl solution at 20° C for 28 and 56 days. After immersion, the mortar specimens were split, and the split crosssections were sprayed with a 0.1% sodium fluorescein solution and a 0.1N silver nitrate solution as prescribed in UNI 7928 (Concrete- Determination of the Chloride Ion Penetration). The depth of the rim of each crosssection changed to white color was measured with slide calipers as a chloride ion penetration depth as shown in Fig.1.

3.6 Freezing and thawing resistance test

Mortar specimens cured in water for 28 days were tested for freezing and thawing resistance in a temperature range of -17.8°C to +4.4°C in conformity with ASTM C 666 (Standard Test Method for Resistance of Concrete to Rapid Freezing and Thawing), Procedure B. The fundamental transverse frequency of the mortar specimens after each

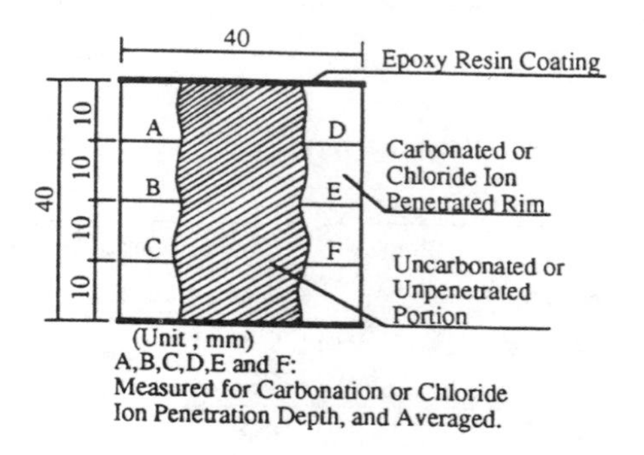

Fig.1. Crosssection of mortar specimens after carbonation or chloride ion penetration test

30 cycles of freezing and thawing was determined according to JIS A 1127 (Methods of Test for Dynamic Modulus of Elasticity, Rigidity and Dynamic Poisson's Ratio of Concrete Specimens by Resonance Vibration), and the relative modulus of elasticity was calculated. The weight change of the mortar specimens was also determined.

4 Test results and discussion

4.1 Compressive strength

Figure 2 represents the compressive strength of the mortars cured under different conditions. The compressive strength of water-cured mortars depends on the curing time and the compositions of the cementitious binder. The early-age (i.e., 2- and 7- day) compressive strengths of the mortars with BFS are either lower than or comparable with those of plain mortars. An increase in the curing period proves to have a positive effect on the strength development, positively influenced by the fineness of BFS. Even at a relatively high BFS content (40 % of cement was replaced with BFS), the higher compressive strengths than the plain mortar are obtained in longer water cure.

The compressive strength of autoclave- cured mortars is higher than that of 28-day or 90-day water-cured mortars. The finer the BFS used as the partial cement replacement, the higher the compressive strength achieved as seen in Fig. 2.

It is proved that under appropriate mix proportions and curing conditions very finely ground BFS is a highly efficient material which might be used for high-strength cement-based materials. This is related mainly to the well known influence of the fine BFS on the phase composition and pore structure formation, which are responsible for the mechanical and other characteristics of the hydrated cement system [6]. In fact, the observations of the main characteristics of the pore structure formation of the high-strength mortars under discussion showed a decrease in the volume of micropores with hydration time. The relationships between the compressive strength and the volume of micropores for the high-strength mortars prepared in a comprehensive research program were confirmed [7].

4.2 Water absorption

Table 3 shows the water absorption of mortars cured under different conditions. Regardless of curing conditions, the water absorption of the mortars with BFS is markedly smaller than that of plain mortars, and decreases with an increase in the surface area of BFS. From such results, a positive effect of the fineness of BFS on the water absorption is well visible. Except for the plain mortars, the water absorption of autoclave-cured mortars is smaller than that of 28-day water-cured mortars. This trend becomes marked with an increase in the surface area of BFS. This is attributed to the formation of cement hydrates with different porosity due to different curing conditions and BFS fineness.

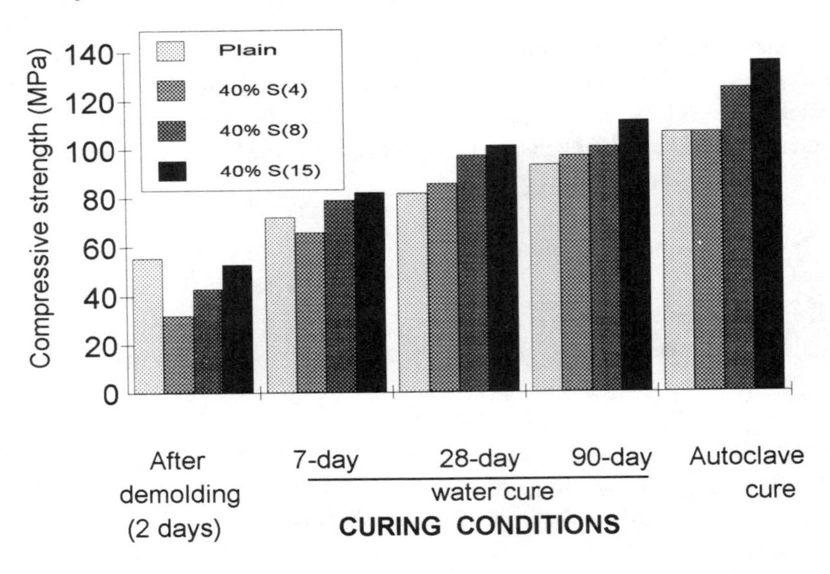

Fig.2. Compressive strength of mortars cured under different conditions

4.3 Carbonation

Table 4 gives the carbonation depth of the mortars cured under different conditions The carbonation depths obtained by two independent testing methods, which were developed for testing polymer-modified concrete [8], show an extremely small diffusivity of CO_2 to the mortars. Even in the case that the mortars had been exposed to CO_2 for a relatively long exposure period (14 days) by the pressurized method or for 2 years by the non-pressurized method, the carbonation rims in all the mortars did not reach a level of 1.5 mm. The high density of the mortars with a relatively small amount of calcium hydroxide formed seems to be responsible for the extremely small carbonation depth.

Table 3. Water absorption of mortars cured under different conditions

Type of	48-hour water absorption (% by mass)	
mortar	28-day water cure	Autoclave cure
Plain	2.12	3.47
40% S(4)	1.37	1.17
40% S(8)	1.27	0.97
40% S(15)	1.06	0.52

Table 4. Carbonation depth of mortars cured under different conditions

Type of mortar	Carbonation depth (mm)			
	Non-pressurized method 2-year depth		Pressurized method 14-day depth	
	28-day water cure	Autoclave cure	28-day water cure	Autoclave cure
Plain	0.0	1.3	0.4	0.0
40% S(4)	0.0	0.0	0.0	0.0
40% S(8)	0.0	0.0	0.0	0.0
40% S(15)	0.0	0.0	0.0	0.0

4.4 Chloride ion penetration

Figure 3 represents the chloride ion penetration of the mortars cured under different conditions and immersed in a 2.5% NaCl solution. The weight change of the mortars through immersion in the solution is illustrated in Fig. 4.

The chloride ion penetration depth of **28-day water-cured mortars** is extremely small even at a longer immersion time period (56 days). The chloride ion penetration depth of the mortars with BFS is smaller than that of plain mortars. This fact corresponds with the similar results obtained on cement pastes and mortars [9]. Due to the very small values obtained, differences in the chloride ion penetration depth among the mortars with different cementitious binder compositions are not visible.

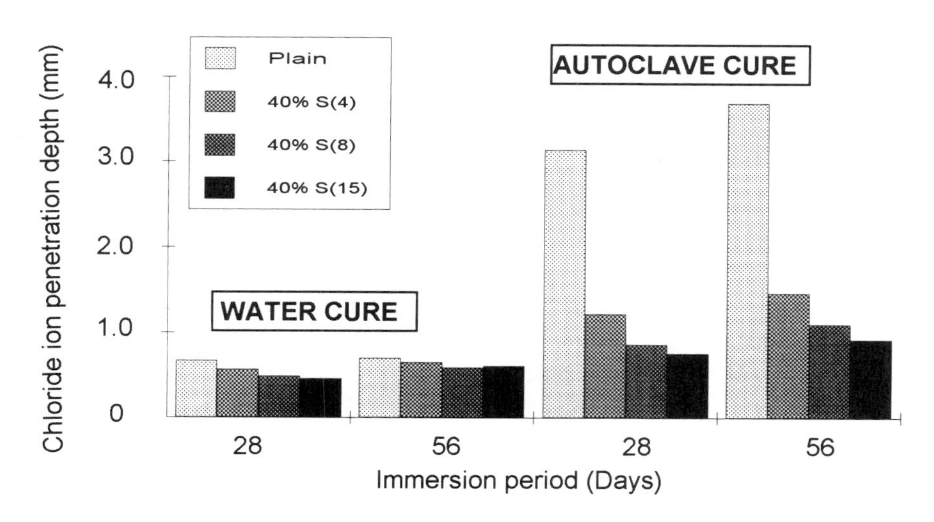

Fig.3. Chloride ion penetration depth of mortars cured under different conditions, immersed in 2.5% NaCl solution.

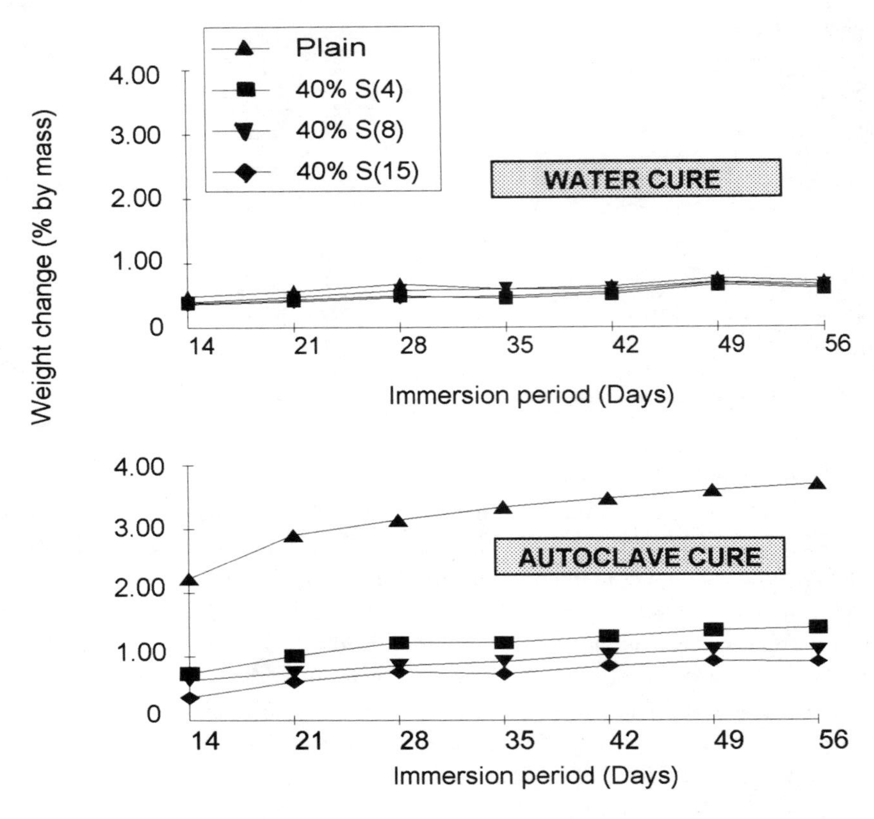

Fig.4. Weight change of mortars cured under different conditions and immersed in 2.5% NaCl solution

The chloride ion penetration depth of **autoclave-cured mortars** is relatively small compared to others mortars which had been tested in a comprehensive research program under similar laboratory conditions as the plain (control) mortars for polymer-modified mortars. The chloride ion penetration depth of the autoclave-cured mortars with BFS is remarkably smaller than that of plain mortars, and reduced with an increase in the surface area of BFS. In general, the chloride ion penetration depth of the autoclave-cured mortars is larger than that of 28-day water-cured mortars irrespective of the fineness of BFS and the immersion period.

The larger chloride ion penetration depth is obtained for the plain mortars. The above-mentioned trend corresponds with similar results obtained on concretes cured under elevated temperatures [10]. It is assumed that the results obtained are related to the formation of hydration products with different porosity due to the different curing conditions and binder compositions. A positive effect of BFS fineness on a decrease in the chloride ion penetration depth is evident as shown in Figs.3 and 4.

4.5 Freezing and thawing resistance

Table 5 exhibits the freezing and thawing resistance of the mortars cured for 28 days in water. All the mortars show very high high freezing and thawing resistance at 210 cycles. There is no visible difference in the freezing and thawing resistance among the mortars.

Table 5. Freezing and thawing resistance of 28-day water-cured mortars.

Type of mortar	Weight change at 210 cycles (% by mass)	Relative dynamic modulus of elasticity at 210 cycles (%)
Plain	0.31	101
40% S(4)	0.12	101
40% S(8)	0.10	102
40% S(15)	0.10	101

5 Conclusions

The conclusions obtained from the laboratory investigation of the fundamental mechanical properties and durability of high-strength mortars incorporating finely ground granulated blast-furnace slags (BFS) are summarized as follows:

1. Finely ground granulated BFS has a positive effect on the strength development of mortars. The finer the BFS used as a partial cement replacement in mortars with a very low water-cementitious material ratio of 0.20, the higher the strength achieved.
2. High-strength mortars have an extremely low diffusivity of CO_2. Regardless of the type of the carbonation test methods, the carbonation depth of the mortars does not reach a level of 1.5 mm.
3. High-strength mortars cured in water have a very high density and low diffusivity to chloride ions compared to autoclave-cured high-strength mortars. A positive effect of BFS fineness on a decrease in the chloride ion penatration is confirmed.
4. The incorporation of BFS in the high-strength mortars cured in water does not affect the freezing and thawing resistance of the mortars.

6 Acknowledgments

The Authors gratefully acknowledge the support of the Matsumae International Foundation in Tokyo under MIF fellowship 1992-3. Special thanks are given to Mr. Masanao Kimura from Hitachi Cement Co., Ltd., Japan, for his cooperation and assistance in the preparation of the slags used for the experiments.

7 References

1. Berntsson, L., Chandra,S. and Kutti, T. (1990) Principles and factors influencing high-strength concrete production. *Concrete International*, Vol.12, No.12, pp.59-62.
2. Ohama, Y., Demura, K. and Lin, Z. (1990) Development of super high-strength mortars using silica fume, in *Proceedings of the Concrete for the Nineties, International Conference on the Use of Fly Ash, Silica Fume and Other Siliceous Materials in Concrete* (ed. W.B.Butler and I. Hinczak), Concrete Institute of Australia, Sydney, Australia, pp. Ohama 1- Ohama 12.
3. ACI Committee 226. (1987) Ground granulated blast-furnace slag as a cementitious constituent in concrete (ACI 226.1R-87), *ACI Materials Journal*, Vol. 84, No. 4, pp.327-342.
4. Nagataki, S., Machida, A., Yamamoto, Y. and Uomoto, T. (1989) Japanese recommendation for the use of ground granulated blast-furnace slag in concrete as an admixture, in *Fly Ash, Silica Fume, Slag and Natural Pozzolans in Concrete, Proceedings of the Third International Conference* (Trondheim, Norway), SP-114, Vol.2, American Concrete Institute, Detroit, pp.1567-1576.
5. Yamamoto,Y. and Kobayashi, M. (1982) Use of mineral fines in high-strength concrete-water requirement and strength. *Concrete International*, Vol. 4, No.7, pp.33-40.
6. Uchikawa, H. (1986) Effect of blending component on hydration and structure formation, in *Proceedings of the 8th International Congress on the Chemistry of Cement*, Vol.1, Secretaria Geral do 8- CIQC, Rio de Janeiro, Brazil, pp.249-280.
7. Madej, J.(1992) *Investigation of Industrial Slags Suitable for Use in High-Strength Concrete* (*Technical Report*), College of Engineering, Nihon University, Koriyama, Japan, 78pp.
8. Ohama, Y. and Miyake, T. (1980) Accelerated carbonation of polymer-modified concrete. *Transactions of the Japan Concrete Institute*, Vol.2., pp.61-68.
9. Roy, D. M. (1989) Hydration, microstructure and chloride diffusion of slag-cement pastes and mortars, in *Fly Ash, Silica Fume, Slag and Natural Pozzolans in Concrete, Proceedings of the Third International Conference* (Trondheim, Norway), SP-114, Vol.2, American Concrete Institute, Detroit, pp.1273-1281.
10. Detwiler, R. J., Kjellsen, K.O. and Gjorv, O.E. (1991) Resistance of chloride intrusion of concrete cured at different temperatures. *ACI Materials Journal*, Vol.88, No.1, pp.19-24.

130 HIGH PERFORMANCE SLAG ASH CONCRETE

S.I. PAVLENKO
Siberian Mining & Metallurgical Academy, Novokuznetsk, Russia

Abstract
The Siberian Mining & Metallurgical Academy has developed fine-grained slag ash concrete having high strength (50 to 68 MPa), frost resistance (F 1000 to F 1200 cycles), crack resistance, water proofness (W 6 to W 12) and resistance to corrosion of reinforcement. The concrete does not contain any natural aggregates such as gravel, crushed stone and sand. Slag sand and fly ash (as an admixture) which contributes much to the environmental protection are incorporated. This concrete is 300-400 kg/m^3 lighter than that of ordinary heavy concrete containing crushed stone aggregate. The optimum composition of concrete mixtures includes 500 to 650 kg/m^3 cement, 900 to 1100 kg/m^3 slag sand, 150 to 250 kg/m^3 fly ash, 220 to 260 kg/m^3 water and 0,3% (by weight of cement) of plasticizing admixture (technical grade lignosulfonate). The strength of concrete increased by 10 MPa; investigations were made into crack resistance, deformation and protection of reinforcement from corrosion; and the results obtained allow a broader application of the concrete investigated (structures of roads, tunnels, canals, roofs, bridges ets.).
Keywords: Corrosion of reinforcement, crack resistance, fine-grained slag ash concrete, frost resistance, strength.

Concrete Under Severe Conditions: Environment and loading (Volume Two) Edited by K. Sakai, N. Banthia and O.E. Gjørv. Published in 1995 by E & FN Spon. ISBN 0 419 19860 1

1 Introduction

The present paper continues the studies of fine-grained slag ash concrete having high frost resistance (1000 to 1500 cycles) and waterproofness (W 6 to W 12), namely, its phisico-mechanical and deformation properties as well as protective properties of concrete against corrosion of reinforcement. The studies were carried out for one year beginning at the age of 28 days so that the reliability of the previously obtained data [1] could be checked in accelerated tests and the sphere of the application of concrete investigated could be expended.

Out of 18 compositions used in the previous study [1], 2 optimum compositions were selected and are described.

2 Materials and the optimum mixture proportions of concrete

2.1 Granulated slag sand from the Tom-Usinskaya Power Plant

Slag sand was obtained by water granulation and grinding. It has a particle size distribution of 0 to 5 mm and fineness modulus of 2.9. The technology of its productions was presented previously [2]. The alumina-silicate glass was mostly in the radioamorphous state. It did not contain any unburnt organic particles. It was used as a main aggregate, excluding the use of natural coarse aggregates (crushed stone or gravel). Its bulk and true densities are 1400 and 2250 kg/m^3, respectively.

2.2 Fly ash

Fly ash from the Tom-Usinskaya PP is acid, low-calcium (5 to 7%) ash and its chemical composition is similar to that of slag sand. It contained less than 5% unburnt particles. Its bulk, true densities and specific surface were 900, 2000 kg/m^3 and 2900 cm^2/g, respectively.

2.3 Technical grade lignosulfonate (TGL)

TGL is a by-product of the pulp and paper industry. It is a plasticizing partially air-entraining substance of a hydrophilic character.

2.4 Optimum compositions of concrete

Out of 18 compositions of the concrete mixtures, 2 compositions using two different cement brands (M400 and M500 or 40 and 50 MPa) were chosen to perform strength and durability studies of concrete. The compositions are given in Table 1.

The lower density of a fine-grained slag ash concrete compared wich that of an ordinary heavy concrete is explained by the presence of slag sand and fly ash particles of closed porosity and by air-entrainment during the introduction of a plasticizing air-entraining admixture of TGL (5%).

Table 1. Mixture proportions and main properties of fine-grained concrete

Mixture, No		1	2
Cement, kg/m^3	M 400 PC	640	–
	M 500 PC	–	580
Slag sand, kg/m^3	0 - 5 mm	925	975
Fly ash, kg/m^3		185	195
Water, kg/m^3		266	258
Admixture, % by cement weight (TGL)		0.3	0.3
Slump, cm		4 - 6	4 - 6
Average density, kg/m^3		2016	2008
Compressive strength at 28 days, MPa		48	54
Prism strength at 28 days, MPa		37	42

3 Physico-mechanical and deformation properties of concrete over the 1-year period

10 x 10 x 10 cm cubes and 10 x 10 x 40 cm prisms were cast. The specimens were made by using the MC-142 laboratory positive mixer and then vibrated on a vibrating table. After the vibration, 50% of the specimens were cured at + 18 - 22^{0}C and 80 to 85% relative humidity. The other 50% of the specimens were cured at room temperature (+20^{0}C) for 4 hours and then in a laboratory steam-curing chamber using a 3+8+3 hour cycle (3 hours-rise of temperature, 8 hours-curing at a constant temperature and 3 hours-cooling).

3.1 Physico-mechanical properties

The compressive strength of the concrete was determined on the 10 cm cubes and the prism strength test was performed on the 10 x 10 x 40 cm prisms after the flexural test. Flexural strength, axial tension, extensibility and compressibility were also determined. Specimens were tested according to the method developed by NIIZhB [3]; the measurements were taken at 28, 60, 90, 180 and 360 days. The test results are given in Table 2.

As can be seen from Table 2, the cube strength of the slag ash concrete was 10 to 15% higher than the SNIP requirements [4] for ordinary fine-grained sand concretes. The maximum cube strength of fine-grained concrete according to SNIP is 50 MPa (B 40 class), while the concrete investigated attained the strength of 60 to 68 MPa (B 50 to 55 classes) at the age of a year. Slag sand concrete had better indices of the relationship between the prism strength and the cube strength as compared to those of ordinary heavy concrete, these relationships being 0.74 to 0.77 and 0.65 to 0.70 for slag ash concrete and ordinary concrete, respectively. It increases the reliability of structures made of slag ash concrete subjected to loads during service.

It should be noted that heat treatment of slag ash concrete increased its early strength due to the more intensive activity of ash during heating. For comparison, ordinary concrete, hardened in natural conditions, attains the equivalent strength only at the age of 60 to 90 days.

Table 2. Strength characteristics of fine-grained slag ash concrete

Characteristics	Heat treatment	Age, days				
		28	60	90	180	360
Mixture No 1						
Compressive strength of a cube, MPa	without	31	47	51	56	59
	with	48	52	54	58	60
Prism strength, MPa	without	23	35	38	42	44
	with	37	40	42	45	46
Flexural strength,	without	3.0	3.8	3.9	4.0	4.2
	with	3.9	4.0	4.0	4.2	4.4
Axial tension, MPa	without	1.8	2.2	2.3	2.4	2.5
	with	2.3	2.4	2.4	2.5	2.6
Extensibility, mm/m	without	0.10	0.13	0.14	0.17	0.18
	with	0.12	0.14	0.15	0.18	0.19
Compressibility,	without	0.86	0.90	0.94	1.02	1.08
	with	0.91	0.93	0.97	1.03	1.08
Mixture No 2						
Compressive strength of a cube, MPa	without	37	51	58	62	66
	with	54	58	61	65	68
Prism strength, MPa	without	28	38	43	46	49
	with	42	44	47	50	52
Flexural strength, MPa	without	3.4	3.9	4.3	4.4	4.6
	with	4.0	4.3	4.4	4.6	4.8
Axial tension, MPa	without	2.0	2.3	2.5	2.6	2.7
	with	2.4	2.5	2.6	2.7	2.8
Extensibility, mm/m	without	0.12	0.14	0.17	0.19	0.20
	with	0.15	0.17	0.18	0.20	0.22
Compressibility, mm/m	without	0.92	0.98	1.05	1.08	1.11
	with	0.96	1.03	1.07	1.09	1.12

3.2 Deformation characteristics

3.2.1 Drying shrinkage

Drying shrinkage was determined on 10 x 10 x 40-cm prisms by means of a dial guage accoding to the method developed by NIIZhB. Measurements were made in the following periods: after dismantling and cooling to room temperature (18-20°C); then 30 min later; 2, 4, 12, 24 hours later; once a day between 2 and 6 days; once in 2 days, between 7 and 15 days; and once a week at ages above 15 days.

The data on the drying shrinkage of slag ash concrete and fine grained sand concrete are given in Fig.1. As can be seen from these data, the drying shrinkage of slag ash concrete developed more slowly and was less than that of an ordinary sand concrete.

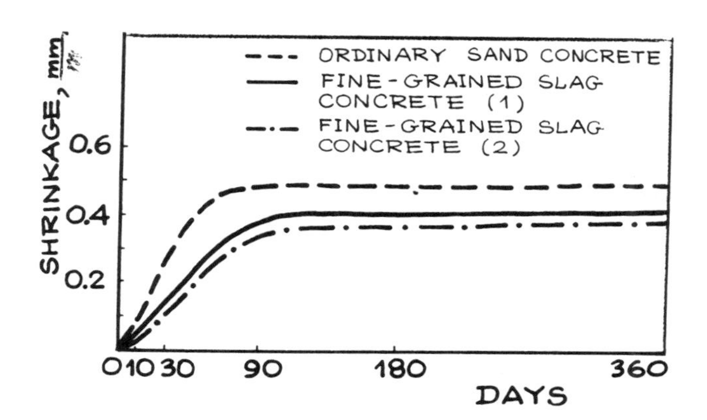

Fig. 1. Drying shrinkage of fine-grained slag ash concrete and sand concrete.

3.2.2 Creep

Creep was determined on 28 days concrete prisms. They were subjected to a long-term loading in spring devices according to the method developed by NIIZhB. The applied stresses 50% of the prism strength ($\sigma = 0.5\ R_{prism}$). Measurements were taken in the following periods: 30 min later; 1, 3, 6 and 12 hours later; once a day, between 2 and 6 days; once in a 2-day period, between 7 and 15 days; and twice a week beyond this period.

The data on creep of concrete are given in Fig. 2. It can be seen that the creep of the concrete investigated developed more slowly and the magnitude of the creep was smaller as compared with that of sand concrete.

3.2.3 Initial modulus of elasticity

The initial modulus of elasticity of the concrete was determined on 10 x 10 x 40-cm prisms by means of dial guages fixed to the sides of the prisms using the technique developed by NIIZhB [3]. The test results are given in Table 3. It is evident from the data that the initial modulus of elasticity of the concrete investigated was higher than that of a sand concrete, especially after heat treatment.

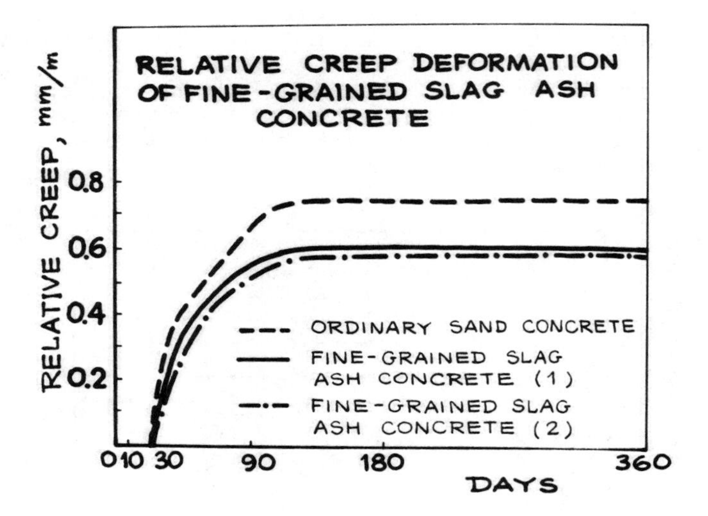

Fig. 2. Creep of fine-grained slag ash concrete and sand concrete.

Table 3. Initial modulus of elasticity of fine-grained slag ash concrete ($E_c x10^{-3}$)

Concrete mixtures	Heat treatment	Age, days				
		28	60	90	180	360
Mixture No 1, MPa	without	27.8	28.2	28.7	29.4	30.3
	with	29.6	30.1	30.6	31.8	32.4
Mixture No 2, MPa	without	28.9	29.7	32.4	34.8	35.6
	with	32.0	34.6	36.7	37.8	38.5
Fine-grained sand concrete (SNIP requrements) at 28 days, MPa	without	29.0	-	-	-	-
	with	28.0	-	-	-	-

Table 4. Summary of test results for frost resistence

Heat treatment of concrete	Coefficient of frost resistence after freezing cycles													
	100	200	300	400	500	600	700	800	900	1000	1100	1200	1300	1400
Mixture No 1. Freezing in water														
Without	1.05	1.02	1.06	1.06	1.07	1.10	1.11	1.09	1.02	0.92	0.84	-	-	-
With	1.01	1.01	1.02	1.05	1.06	1.09	1.10	1.07	0.93	0.83	-	-	-	-
Mixture No 2. Freezing in water														
Without	1.01	1.03	1.06	1.08	1.12	1.08	1.05	1.04	1.04	1.00	0.98	0.95	0.89	0.83
With	1.02	1.02	1.04	1.06	1.10	1.09	1.08	1.06	1.03	0.98	0.91	0.89	0.84	-
Mixture No 1. Freezing in a 5% solution of NaCl														
Without	1.01	1.02	1.03	1.06	1.09	1.08	1.09	1.08	0.98	0.89	0.84	-	-	-
With	1.02	1.03	1.05	1.05	1.07	1.09	1.03	1.00	0.95	0.83	-	-	-	-
Mixture No 2. Freezing in a 5% solution of NaCl														
Without	1.00	1.03	1.04	1.04	1.06	1.09	1.10	1.08	1.05	0.93	0.88	0.84	-	-
With	1.02	1.02	1.03	1.04	1.07	1.07	1.08	1.06	1.04	0.90	0.83	-	-	-

4 Frost resistance

4.1 Method

In previous investigation of the fine-grained slag ash concrete with high frost resistance (over 1000 cycles), the accelerated method (freezing at -50°C according to GOST 10060-76) was used. In order to check the reliability of the data obtained, in this work, the frost resistance of concrete with two optimum mix proportions was determined by the principal method according to the same GOST. The criterion of frost resistance was the number of cycles of alternate freezing and thawing sustained by concrete cubes without a decrease compressive strength of more than 15% and without a weight loss of more than 5% for pavement concrete [5].

4.2 Tests

A total of 35 cubes, 10 x 10 x 10 cm, were cast from each mixture; 19 of them were cast for reference purposes. Frost resistance was determined using the principal method in which F is equal to 75 cycles or more.

Specimens were tested at the age of 7 days after heat treatment and at the age 28 days after curing in natural conditions. The main specimens were saturated with water at 15 to 20°C (96h) prior to freezing and thawing. At the same time, the companion specimens were frozen in a 5% NaCL solution. This salt solution was used in irder to study the possibility of applying the concrete to pavements and airfield coverings. The temperature in the refrigerating chamber was -15 to -20°C.

The test cycle consisted of freezing the cubes for 4 hours and thawing in a water bath at +15 to +20°C for about 4 hours. The test results are given in Table 4 and Figs. 3 and 4. The data in Table 4 are characterized by the coefficient of frost resistance (CFR), i.e. the ratio of the compressive strength of concrete at every stage after freezing to its compressive strength prior to freezing. This relationship was determined every 100 cycles. Figs. 3 and 4 show the change in the compressive strength of concrete specimens with and without heat treatment, being frozen in water and in a 5% NaCL solution.

As can be seen from Table 4 and Fig. 3, the concrete investigated showed a similarity to the data obtained earlier [1] by the accelerated method of testing (at - 50°C) and the data obtained by usual method of testing (at -15° to 20°C) for 500 days. A slight decrease in the number of cycles in concrete frozen by the usual method as compared to that frozen by the accelerated method is explained by the greater number of cycles (by 10 times) of freezing and thawing. This shows that temperature not only influences the concrete durability but also the number of cycles.

The testing in salt water (Table 4 and Fig.4) also demonstrated high durability of the concrete investigated. The number of cycles exceeded 1000, so concrete can be used for pavements and airfield coverings.

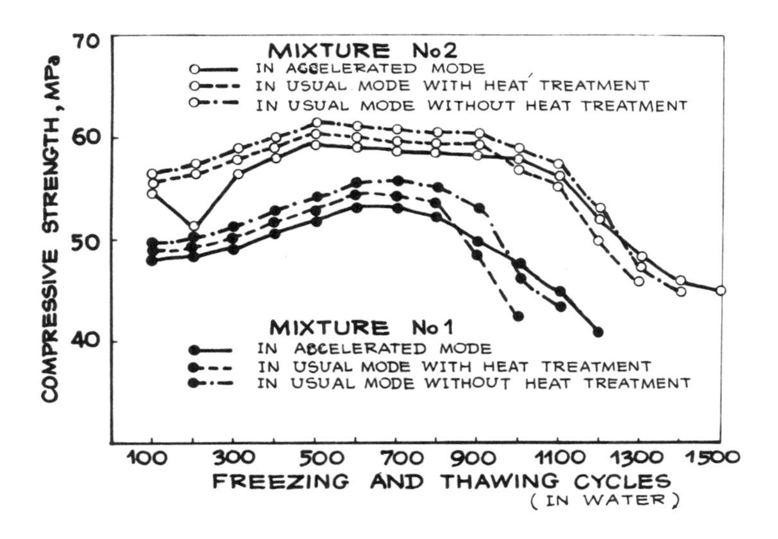

Fig. 3. Strength of concrete versus cycles of freezing and thawing (in water).

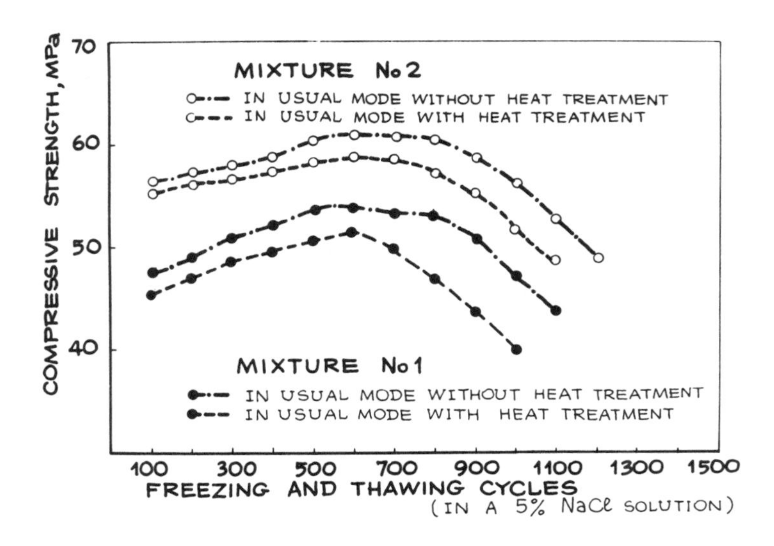

Fig. 4. Strength of concrete versus cycles of freezing and thawing (in a 5% solution of NaCL).

5 Protection of reinforcement from corrosion

In order to determine the ability of concrete to resist the corrosion of reinforcement, the following tests were performed: investigation of the electrochemical behaviour of steel in concrete, defining the pH value of the liquid phase of concrete and diffusion permeability.

5.1 Electrochemical tests

5.1.1 Method
The electrochemical investigation were carried out on specimens, 70 x 70 x 140-mm, reinforced with ground and skimmed steel wire, 6 mm in diameter and 120 mm long in accordance with the NIIZhB technique. Prior to taking polarization curves, the specimens were saturated with water in a vacuum until complete saturation resulted. The end of the specimen was cut off, and the exposured wires were insulated by coating with paint. The specimen immersed in the electrolyte was connected with the electric circuit, and the potential value of steel in concrete as well as the density of the electric current in the circuit were measured.

5.1.2 Testing
The electrochemical tests were performed in the initial state and after 3 and 6 months of storage in the condition of alternating moistening and drying. The anodic polarization curves of steel in the slag ash concrete are shown in Figs. 5 and 6. On the completion of the tests, the reinforcing rods were extracted from the concrete in order to evaluate corrosion affected areas. A visual examination of the rods showed that the reinforcement was not affected by corrosion. An analysis of the anodic polarization curves of steel also indicates that steel was passive with respect to corrosion during all periods of testing.

5.1.3 The pH value of the liquid phase of concrete
The pH value was determined according to the method developed by NIIZhB [6]. The pH values for the two concrete compositions are given in Table 5. They range between 11.8 and 12.8, which is in accordance with requirements and gives an indication of safe protection of reinforcement in concrete.

Table 5. The pH value of the liquid phase of slag-ash concrete

Mixture No	pH value of the liquid phase of concrete	
	at 28 days	at 6 months
1	12.44	12.32
2	12.29	12.18

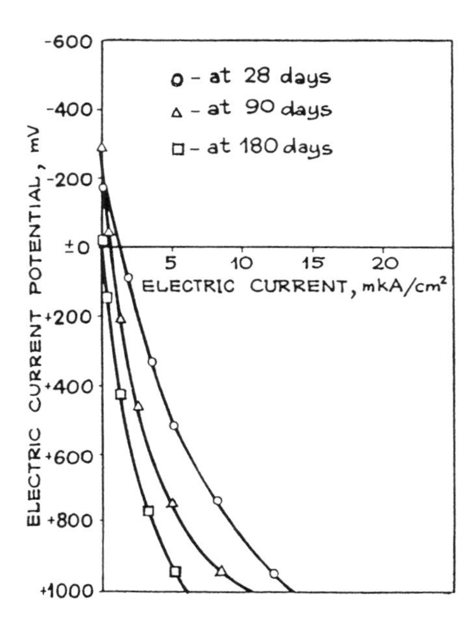

Fig. 5. Anodic polarization curves of steel in fine-grained slag ash concrete made of mixture No. 1.

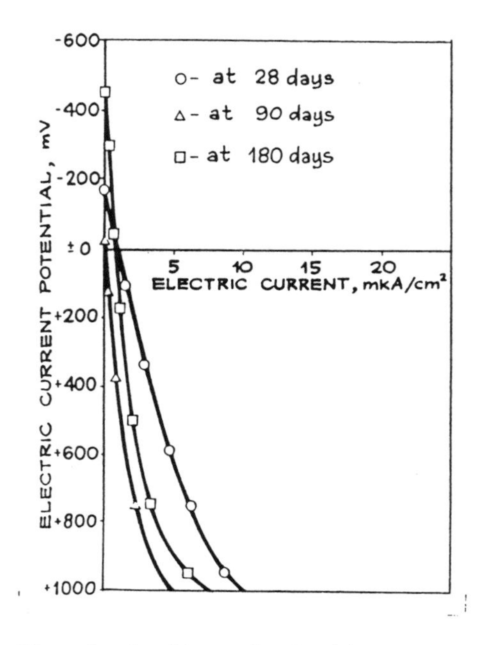

Fig. 6. Anodic polarization curves of steel in fine-grained slag ash concrete made of mixture No. 2.

5.1.4 Diffusion permeability
The diffusion permeability was defined on specimens 10 x 10 x 10 cm in size, in accordance with the NIIZhB technique [7]. Prior to testing, the specimens were cured in a chamber at 20^{0}C, 75% relative humidity and 10% carbonic acid concentration. Then the specimens were split and their surfaces were covered with a 0.1% phenolphthalein spirit solution to measure the neutralized layer of concrete. The data on CO_2 diffusion permeability in concrete are shown in Table 6. They show the safe protection of reinforcement from corrosion.

Table 6. Permeability of slag ash concrete

Mix-ture No	Quantities, kg/m^3		CO_2 diffusion coefficient $*10^{-4}$, cm^2/s	Period of neutralization of protecive concrete layer		Water absorption, %
	Cement M400 M500	Fly ash		20mm	30mm	
1	640	185	5.18	17	34	6.1
2	580	195	4.63	19	39	5.4

6 Conclusions

1. Fine-grained concrete containing slag sand and fly ash had the following quality indices:
   - compressive strength is 50 to 70 MPa;
   - average density is 2000 kg/m^3, which is 400 to 500 kg/m^3 less than that of ordinary heavy concrete;
   - high resilience and crack resistance;
   - frost resistance and resistance to gas permeability are 2-3 times higher than those of ordinary concrete;
   - provides safe protection of reinforcement from corrosion.
2. The concrete investigated may be used for pavements and airfield coverings.
3. It saves natural resources by utilizing only industrial waste products as aggregate.

7 Acknowledgment

Acknowledgment is made to the laboratory of Concrete Hardening Acceleration (Lagoida A.B.) and the laboratory of Corrosion (Stepanova V.F.), the Research Institute of Concrete and Reinforced Concrete (NIIZhB, Moscow), for their participation in this work.

8 References

1. Pavlenko, S.I. and Rekhtin, I.V. (1991) Fine-grained slag sand and fly ash concrete with higher frost resistance and waterproofness. Evaluation and rehabilitation of concrete structures and innovation in design, Proceedings ACI International Conference, Hong Kong, Vol. 1, pp. 559-575.
2. Pavlenko, S.I. (1989) Concrete of basis of the steam electric station ash and crushed slag. Third CANMET/ACI International Conference on Fly Ash, Silica Fume, Slag and Natural Pozzolans in Concrete. Supplementary Papers Compiled by Mohamad Alasali, Trodheim, Norway, pp. 727-738.
3. NIIZhB. (1976) Recommendations for Determining Strength and Structural Characteristics of Concrete during Short-Term and Long-Term Loading, Moscow, pp. 10-76.
4. SNIP 2. 03. 01-84. (1985) Concrete and Reinforced Concrete Structures Gosstroy USSR, Moscow.
5. Leshchinsky, M.Y. (1980) Test on Concrete, Guide, Stroyizdat, Moscow, 269 p.
6. Guide for providing safety of reinforcement in structures made of concrete containing porous aggregates in aggressive environment (1979), NIIZhB, Moscow, pp. 30-32.
7. Guide for determining carbonic acid diffusion permeability in concrete (1974), NIIZhB, Moscow, 19 p.

131 SOME BEHAVIOUR OF FROST RESISTANCE OF WATER PERMEABLE CONCRETE

H. TOKUSHIGE, N. SAEKI, T. MIKAMI
and K. SHIMURA
Hokkaido University, Sapporo, Japan

Abstract
The purpose in this study was to investigate the behaviour of frost deterioration and to improve the resistance of water permeable concrete. Under freezing and thawing action, permeable concrete has more severe condition than ordinary concrete; steel fiber is used to promote toughness and thermal conductivity, and water–repellent agent is used in order to make flow of water in continuous voids smooth and protect cement paste from water. Two types of freezing and thawing tests were performed to simulate natural conditions. One type showered water on specimens when they were thawing, which look like a pavement condition. In the other, the lower half of specimens were soaked in water, similar to the condition of riverside blocks. The strains, dynamic Young's modulus and weight loss of specimens were measured. The following was obtained from results of these experiments. Steel fiber in the specimens was apt to promote the frost resistance of porous concrete. The frost durability of specimens with water–repellent agent was found to be affected by the atmospheric conditions of the tests.
Keywords: Dynamic Young's modulus, frost damage, frost resistance, steel fiber, strain, water permeable concrete, water–repellent agent, weight loss

1 Introduction

Recently, concrete has required permeable performance in order to drain water from pavement, and also especially where natural environment must be strictly protected, it

Concrete Under Severe Conditions: Environment and loading (Volume Two) Edited by K. Sakai, N. Banthia and O.E. Gjørv. Published in 1995 by E & FN Spon. ISBN 0 419 19860 1

could keep underground water flow. In cold region, it is important that water permeable concrete is protected from damage due to freezing and thawing action. Study on frost deterioration of ordinary concrete was carried out; it is represented by T. C. Powers [1]. On the contrary, there are more remarkable sizes and numbers of continuous voids on permeable concrete than on ordinary concrete. Therefore, certain ice pressure just as frost heaving [2][3] has a strong effect on continuous voids instead of water pressure explained by Powers [1]. To increase toughness, and to drain smoothly through continuous voids and protect cement paste from water, steel fiber and water–repellent agent were used.

2 Experiment procedure

2.1 Freezing and thawing test

Two natural conditions in winter were simulated to follow the behaviours of water permeable concrete. Test 1 as shown by Fig. 1 is to supply water like shower when specimens are at thawing time. That is the case of smooth drainage; e.g. water permeable pavement. Test 2 is that the lower half of specimens are soaked in water as shown by Fig. 2. That condition does not have enough drainage; e.g. riverside blocks. On both experiments, temperature in testing room is from −22 to 13℃ and 24 hours is one cycle, as shown by Fig. 3. Air entrained concrete specimens using high early strength portland cement were carried into the testing room at 7 days age and the tests were started. Other specimens were carried at 14 days age.

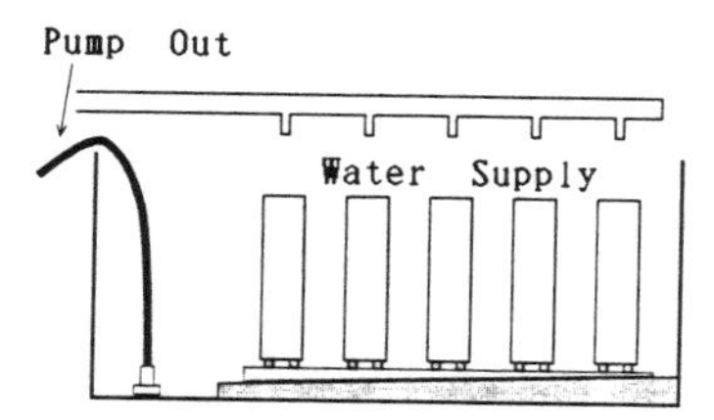

Fig. 1. Equipment for Test 1.

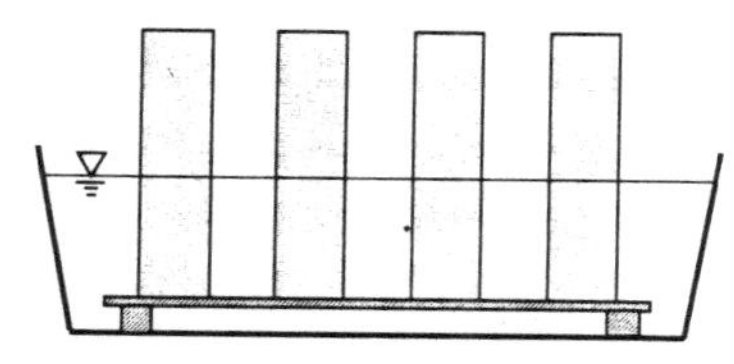

Fig. 2. Equipment for Test 2.

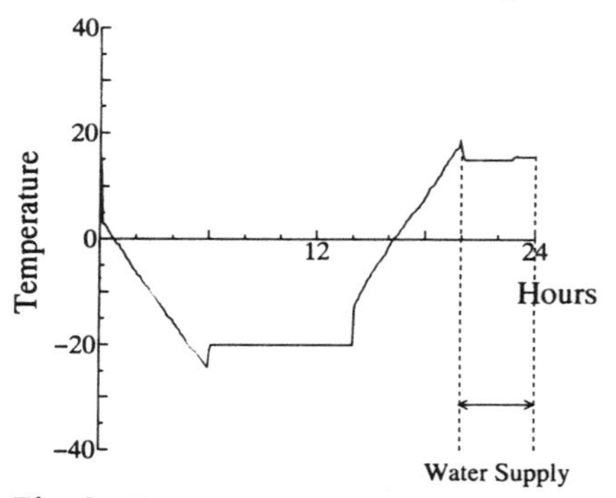

Fig. 3. Temperature in testing room.

2.2 Specimens

Two types in making water permeable concrete specimens were chosen. One is cellular concrete using foaming agent and air–entraining and high–rainge water–reducing agent. The other is porous concrete which is cast with vibrating and pressing compaction; it is stiff consistency concrete. They have remarkable continuous voids. Their water cement ratio is 0.3. Air–entraining and water–reducing agent was used. Additionally, steel fiber and water–repellent agent were used for the porous concrete in order to promote the frost resistance. Five types of specimens were used in this experiment as shown in Table 1. They are cellular concrete, porous concrete, porous concrete with steel fiber, porous concrete with water–repellent agent, and air entrained concrete whose water cement ratio is 0.5 and air content is 4.2%. The specimens are size of 10x10x40 cm rectangular. Table 2 shows the mixture proportions of specimens.

Table 1. Outline of experiments

Specimen	Type	Permeability (cm/s)	Strain gauge	Measuring method
AE	Air entrained concrete	------	Embedded[1)]	Strain and temperature in Test 1
F	Cellular concrete	10^{-7}	Embedded[1)]	
PA	Porous concrete	10^{-2}	Outside[2)]	Strain and temperature in Test 1
PC		10^{-1}		
PSA	Porous concrete with steel fiber	10^{-2}		Weight loss and dynamic Young's modulus in Test 1 & 2
PSC		10^{-1}		
PCA	Porous concrete with water–repe–llent agent	10^{-2}		
PCC		10^{-1}		

1) shown in Fig. 4
2) shown in Fig. 5

Table 2. Mixture proportions

Specimen	Maximum size of coarse aggregate (mm)	Sand percen–tage (%)	Void ratio (%)	W/C	W (kg/m³)	Chemical admixture (cc/m³)	Foaming agent (cc/m³)	Steel fiber (kg/m³)
AE	25	45	4.2	0.5	175	1750 [1)]	------	-----
F	25	40	29	0.35	123	3489 [2)]	14000	-----
PA, PCA	7	25	15	0.3	87	1500 [3)]	------	-----
PSA	7	25	15	0.3	87	1500 [3)]	------	32
PC,PCC	7	5	21	0.3	90	1500 [3)]	------	-----
PSC	7	5	21	0.3	90	1500 [3)]	------	30

1) Air entraining agent
2) Air–entraining and high–rainge water–reducing agent
3) Air–entraining and water–reducing agent

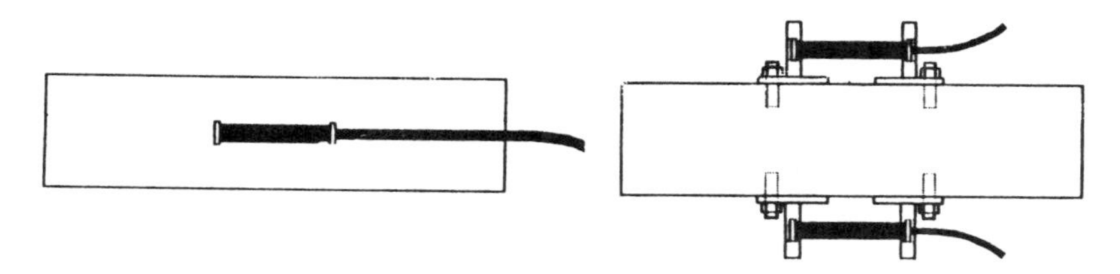

Fig. 4. Embedded gauge.

Fig. 5. Outside gauge.

Two methods were used to measure strain due to the freezing and thawing cycles. One was 'embedded gauge' method and the other was 'outside gauge' method. In the former, strain at the core of specimen was measured; one Carlson gauge was embedded in cellular concrete and air entrained concrete specimens as shown in Fig. 4. In the latter, strain at the surface of specimen was measured; two gauges were set on both sides of specimen as shown in Fig. 5. The outside gauge was used for porous concrete series.

2.3 Measuring

As shown in Table 1, on both tests weight loss and dynamic Young's modulus were measured on porous concrete, porous concrete with steel fiber and porous concrete with water-repellent agent specimens every 10 cycles. Strain and temperature of specimens were measured every 10 minutes up to 50 or 60 cycles on Test 1.

2.4 Materials

2.4.1. Steel fiber

Steel fiber is used to increase toughness of matrix. In structure of water permeable concrete, paste binder around aggregate is thin. Therefore, the type of steel fiber shown in Table 3 was selected in order to improve the toughness of matrix. Steel fiber content has been defined in advance by trial tests for mixing and wakability. According to the trial tests, 0.5 vol.% was appropriate as an average value. In order to investigate an effect of steel fiber reinforcement on porous concrete specimens, the average value was used.

2.4.2 Water-repellent agent

In order to run water-repellent agent into the continuous voids well, infiltration type using silane was selected as shown in Table 3. To compare specimen with water-repellent agent to without, only one type of water-repellent agent was used.

Table 3. Type of steel fiber and water-repellent agent

Steel Fiber	With hook and banded type(ϕ 0.5, 30mm)
Water-repellent Agent	Infiltration type(Silane)

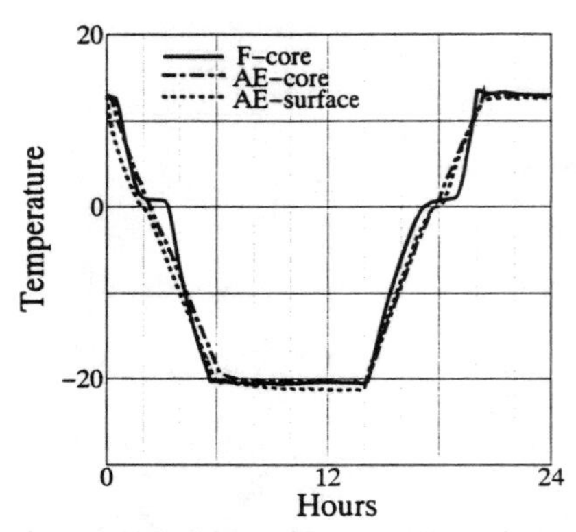

Fig. 6. Behaviour of temperature in air entrained and cellular concrete.

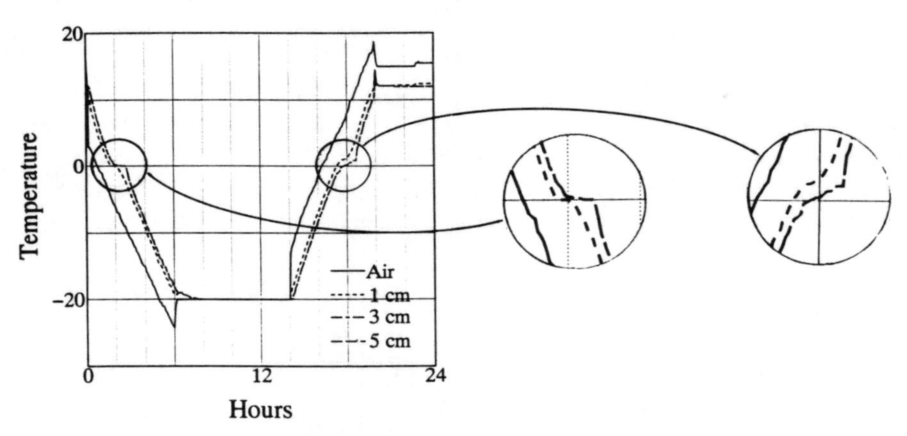

Fig. 7. Behaviour of temperature in porous concrete.

3 Result and discussion

3.1 Behaviour of temperature in specimens

Fig. 6 shows behaviours of temperature at the core of cellular concrete (F) and air entrained concrete specimens (AE). In air entrained concrete, the temperature alters almost constantly. In cellular concrete, whenever the temperature rises and falls, it remains around 0 ℃ for a few hours. It seems that the transition due to ice formation occurs. The behaviour of temperature in porous concrete are shown in Fig. 7. The measuring points of temperature were 1, 3 and 5 cm from surface. The deeper the point was, the longer the temperature stayed around 0 ℃. The above statement is characteristic of the water permeable concrete. Since there are more continuous voids of water permeable concrete than ordinary concrete, it is probably that latent heat of water for ice formation in voids has a strong effect on the behaviours. It affects the behaviours of strain described later on 3.2

3.2 Behaviour of strain due to freezing and thawing in Test 1

The effect of coefficient of thermal expansion was cleared up by subtracting from each value of strain. From Fig. 8 to Fig. 12 show the progress of strain during Test 1. Fig. 13 shows expansion on each specimen; it plots the value of residual strain at thawing period (13 ℃) in each cycle.

3.2.1 Strain in the core of ordinary air entrained and cellular specimens

On ordinary air entrained concrete as shown in Fig. 8, a little shrinkage occurred when temperature fell. As temperature raise, the routes of its strain returned close to the starting point. As freezing and thawing continued, the concrete was slightly apt to accumulate shrinkage. In the air entrained concrete, there was scarcely deterioration as shown in Fig. 13. On the contrary, as shown in Fig. 9, the line for cellular concrete suddenly bends around 0 ℃ . That behaviour appears at both of expansion and shrinkage. The behaviour is thought to have an effect on latent heat of water. As cycles continue, its residual strain due to expansion gradually increases as shown in Fig. 13. It seems that deterioration occurs. This is noticeably different from that of ordinary air entrained concrete. It suggests that theory for frost deterioration of water permeable concrete is probably different from that of ordinary air entrained concrete [1] and water permeable concrete has a tendency to be affected by a certain ice formation of continuous voids beyond more than 10^{-7} cm/s in permeability.

3.2.2 Strain at the surface of porous concrete specimens

Names of specimen (PA, PC,..., etc.) are shown in Table 1. The behaviour of strain of porous concrete with coefficient of permeability 10^{-2} cm/s was slightly different from that with 10^{-1} cm/s, while the shapes of route at each cycle were similar. Therefore,

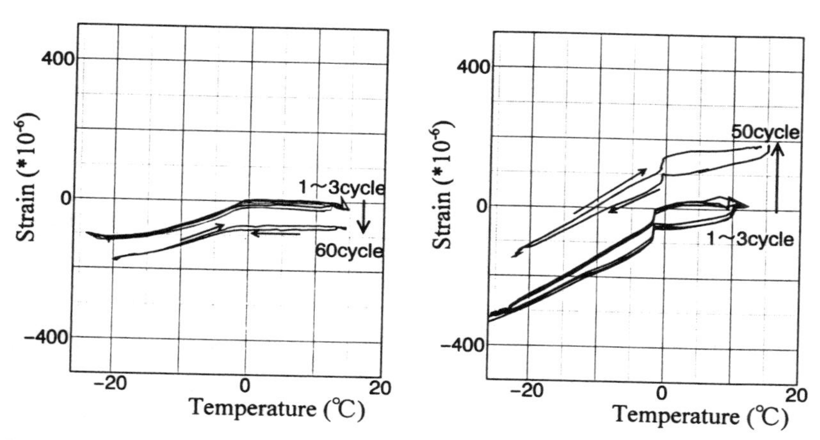

Fig. 8. Strain of air entrained concrete. Fig. 9. Strain of cellular concrete.

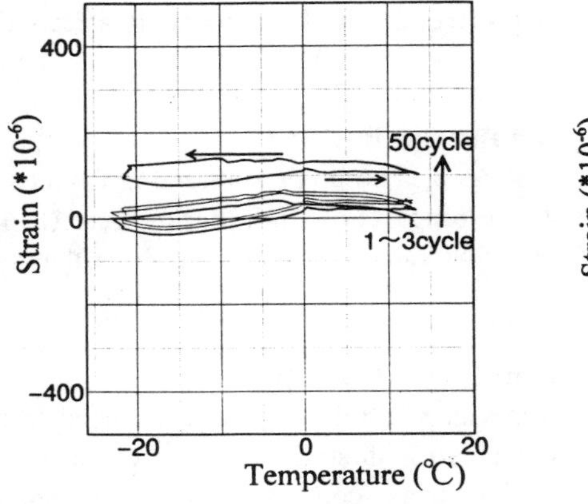

Fig. 10. Strain of porous concrete (PC).

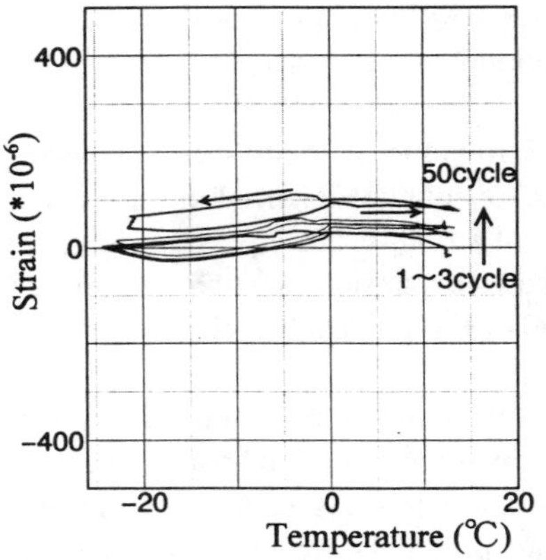

Fig. 11. Strain of porous concrete with steel fiber (PSC).

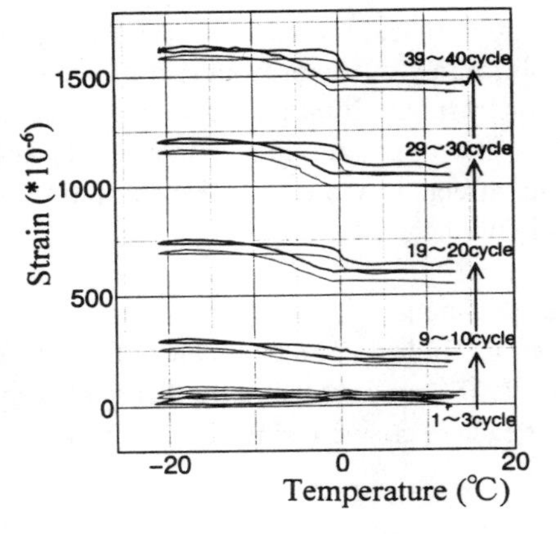

Fig. 12. Strain of porous concrete with water–repellent agent (PCC).

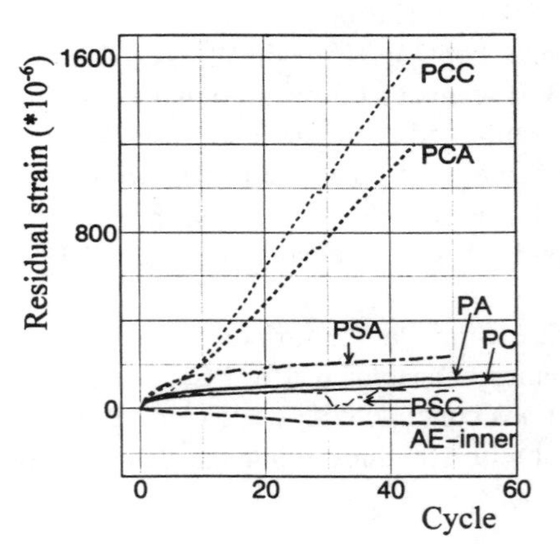

Fig. 13. Residual strain.

figures about PC, PSC and PCC at each cycle are taken as a typification in the following.

Fig. 10 shows one of porous concrete specimen (PC). The behaviour of strain at each cycle had a loop unlike that of ordinary air entrained concrete according to temperature. When it took many cycles, the expansive residual strain occurred as shown in Fig. 13.

The behaviour of specimens with steel fiber (PSC) was similar to that of without (PC) as shown in Fig. 11. The residual strain was smaller as shown in Fig. 13. Then, steel fiber had a slightly good effect on frost damage of water permeable concrete.

PCC as shown in Fig. 12, water permeable concrete with water–repellent agent expanded noticeably at each cycle. As the result of Test 1, PCC did not have stable durability for frost damage. According to a previous study [4], it is probable that voids of a certain size in water permeable concrete with water–repellent agent are tend to keep more water, because water–repellent agent makes dull contact angle between water and material surface. As shown in Fig. 13, PCC and PCA had much deterioration in Test 1.

3.3 Relative dynamic Young's modulus and weight loss

In Test 1 and 2, dynamic Young's modulus and weight loss on three types of porous concrete were measured at every 10 cycles as shown from Fig. 14 to Fig. 17 and types are follows; porous concrete with steel fiber, with water–repellent agent and without either. The steel fiber types had a good frost resistance. The water–repellent agent type showed also good results in Test 2. Deterioration of this type in Test 1 was found in dynamic Young's modulus as well as found in strain shown in Fig. 12. The deterioration depended on the condition of water in specimens. Their weight was slightly reduced at both tests shown by Fig. 16 and Fig. 17.

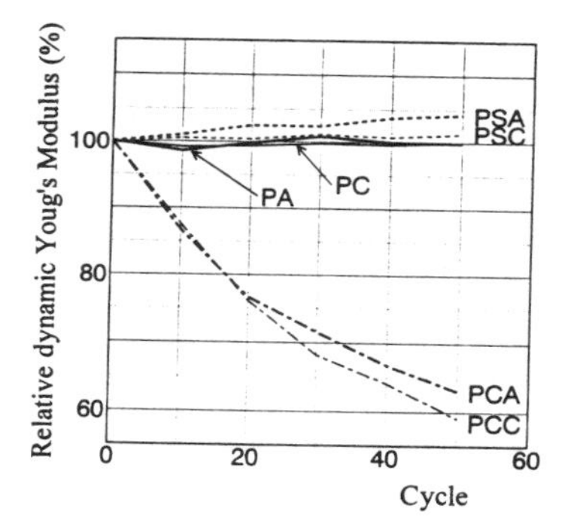

Fig. 14. Relative dynamic Young 's modulus of porous concrete series in Test 1.

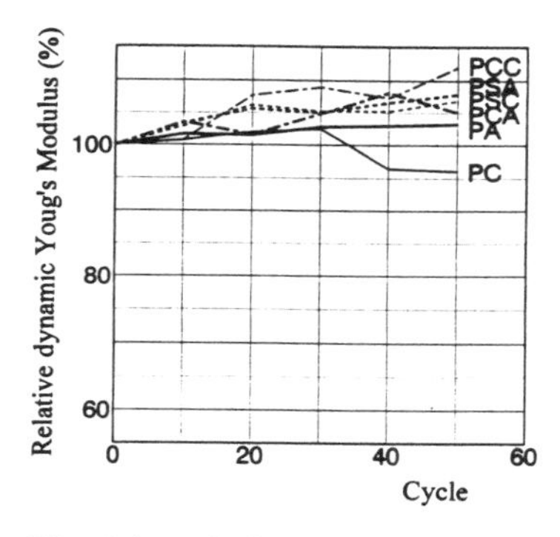

Fig. 15. Relative dynamic Young 's modulus of porous concrete series in Test 2.

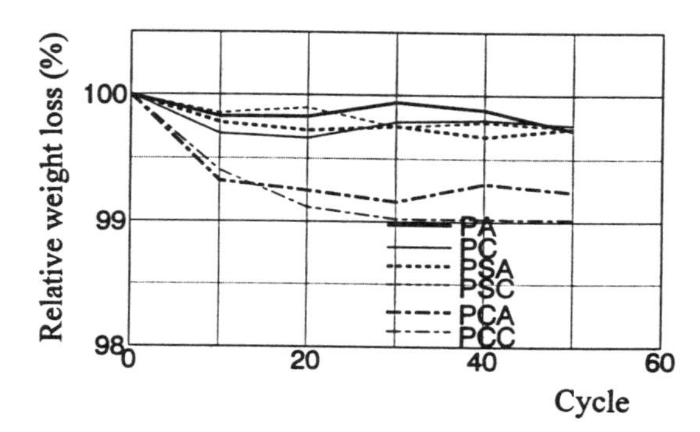

Fig. 16. Weight loss of porous concrete series in Test 1.

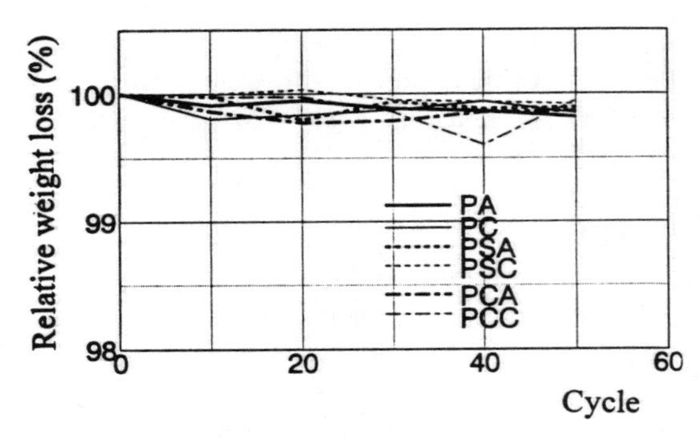

Fig. 17. Weight loss of porous concrete series in Test 2.

3.4 Consideration of frost damage of water permeable concrete

If frost deterioration occurs in non air entrained concrete, generally weight loss is related to the decrease of dynamic Young's modulus. In the water permeable concrete weight loss was scarce, although in certain types relative dynamic Young's modulus was decreased as proven in Test 1. It was noticed that the behaviour of frost damage of water permeable concrete was influenced by the condition of test. In addition, there were two interesting points in Test 2. One was a capillary phenomenon of few centimeters observed at the surface of the specimens. The other was that some specimens were broken in two around the surface of water or ice. It seems to be useful to apply other theories such as some frost heaving theory ([2] and [3]) to frost damage of water permeable concrete than general theory [1]. In water permeable concrete, there are many continuous voids and they are remarkably large. Ice formation seems to be easy there, and therefore water permeable concrete could have a strong effect on latent heat of water.

To conclude, It seems that the frost damage of water permeable concrete occurs due to an effect of some direct pressure by ice formation in each void rather than deterioration of cement paste by hydraulic pressure [1] in capillary pores. Some frost heaving theory could be used to investigate frost damage of water permeable concrete, and the study on mechanism of frost damage to water permeable concrete is proceed with it.

4 Conclusions

Conclusions in this study are as follows :

1. The temperature in water permeable concrete remained for a relatively long time around 0 ℃. The deeper the measuring point was, the longer the temperature retained around 0 ℃.

2. Steel fiber reinforcement was apt to proceed the frost resistance of water permeable concrete.
3. Water permeable concrete with water–repellent agent was subjected to strong effect on the condition of water supply. Frost resistance of it was better if the specimen was soaked in water.
4. The deterioration of water permeable concrete subjected to freezing and thawing action was different from that of ordinary concrete. While the relative dynamic Young's modulus decreased, the weight loss was scarce.

5 Acknowledgments

The authors would like to thank Mr. S. Yamada (HOKKAIDO PORACON, CO. ,LTD.), Mr. H. Fujisawa (Pozzolith Bussan Ltd.), Mr. M. Kimura (Public Works Research Institute, Ministry of Construction) and Mr. K. Kurita (Graduated School, Hokkaido University) for their help in carrying out the experiments.

6 References

1. Powers, T.C. (1949) The Air–Requirement of Frost Resistant Concrete, *Proceedings, Highway Research Board*, Vol. 29, pp. 184–211.
2. Collins, A.R. (1944) The Destruction of Concrete by Frost., *Institution of Civil Engineering, London,* pp. 29–41.
3. Grübl, V.P. (1981) Über die Roll des Eises im Gefüge Zemetgebundener Baustoffe, *Beton*, 31, H.2, pp. 54–8. (*In German*)
4. Senbu, O. (1989) *Dissertation Submitted to Hokkaido University* (*In Japanese*)

132 A STUDY ON EFFECTS OF NON-CHLORIDE AND NON-ALKALI TYPE ANTIFREEZERS ON FROST RESISTANCE OF CONCRETE AT EARLY AGES

Y. HAMA and E. KAMADA
Hokkaido University, Sapporo, Japan
Y. OKUDERA
Shimizu Corporation, Tokyo, Japan

Abstract
In this paper, the basic properties of concrete containing non-chloride and non-alkali type antifreeze admixtures (hereafter called antifreeze concrete), such as freezing point, setting, hardening and frost resistance at early ages were studied, and the utilization method of using these admixtures to protect fresh concrete against frost damage in cold weather concreting was investigated. It was found that the freezing point of antifreeze concrete was -2 to -3℃. The setting time at low temperature and the strength gains at early ages were accelerated by using antifreezers. The benefit of using antifreezers was not only the protection of fresh concrete against frost damage by lowering the freezing point but also the accelerative effect on the setting and hardening of the concrete. The relationship between the concentration of antifreezers and the allowable ambient temperature was derived from the results of the experimental and analytical studies. This was useful for assessment of the effectiveness level of the antifreezers when used in cold weather concreting.
Keywords : Allowable ambient temperature, antifreeze admixture, cold weather concreting, frost damage at early ages, non-chloride and non-alkali, setting, hardening, history of temperature.

1 Introduction

It is well known that the rate of hydration of cement in concrete depends on the ambient temperature. When fresh concrete is exposed to low temperature, the concrete may suffer frost damage due to freezing at early ages and strength development may be delayed. The use of antifreeze admixtures in concrete is one method of protecting fresh concrete from frost damage. Formerly, antifreeze admixtures contained chloride or so-

Concrete Under Severe Conditions: Environment and loading (Volume Two) Edited by K. Sakai, N. Banthia and O.E. Gjørv. Published in 1995 by E & FN Spon. ISBN 0 419 19860 1

dium ions, these ions may cause the corrosion of reinforcing steel or alkali-aggregate-reactions. Some new types of antifreeze admixture that are free of chloride and alkali have been recently developed and are on the market in Japan. However, the level of effectiveness of these admixtures on the frost resistance of concrete at early ages is not quite clear yet.

In this experimental work, an attempt has been made to study the effects of non-chloride and non-alkali type antifreezers on the frost resistance of concrete at early ages. The maturity of the concrete that gain from the time of placing up to the freezing point was calculated by analyzing the temperature history using the Finite Element Method (F.E.M.). The utilization method of these admixtures, for the protection of fresh concrete from frost damage on cold weather concreting, was obtained from considering the analytical and the experimental results.

2 Outline of the experiment

2.1 Materials

The materials used were ordinary portland cement, tap water, Mukawa sand and Tokiwa crushed stone. As chemical admixtures, non-chloride and non-alkali type antifreezers (AS, BS, CS, D, E) and accelerative type AE water reducer (H) were used. The antifreezers were free of chloride and alkali and classified as two types. One was a water reducing type (AS, BS, CS), and the other was only an antifreeze composition (D, E). The quantity of the AE agent was adjusted from one mix to another in order to obtain the target air content.

2.2 Mix proportions

Mix proportions were determined by trial mixing. Water cement ratio, slump, air content and temperature of fresh concrete were kept constant in all series of specimens. The water cement ratio was 50%. Slump, air content and temperature of fresh concrete were 18cm, 3.5% and 15℃ respectively. Mix proportions and the properties of fresh concrete are shown in Table.1.

2.3 Experimental method

A series of penetration tests and freezing and thawing tests for evaluation of the setting

Table. 1 Mix proportion and properties of fresh concret

Type of concrete	Antifreezer (ℓ /m³)	S/A (%)	Water (kg/m³)	Volume(ℓ /m³)			Properties of fresh concrete		
				C	S	G	Slump(cm)	Temp(℃)	Air (%)
P	—	48.2	184	116	320	345	18.7	14.0	5.3
H	3.5 (C×1.0%)	49.3	174	110	336	345	22.0	12.0	4.3
AS	13.8 (4 ℓ /C=100kg)	49.6	172	109	339	345	18.0	13.0	2.7
BS	20.6 (6 ℓ /C=100kg)	49.6	172	109	339	345	19.2	16.0	3.7
CS	13.8 (4 ℓ /C=100kg)	49.6	172	109	339	345	19.5	14.8	3.5
D	15.4 (4 ℓ /C=100kg)	46.9	192	122	305	345	21.1	14.0	3.6
E	23.1 (6 ℓ /C=100kg)	46.9	192	122	305	345	20.5	14.5	3.9

Table. 2 Outline of experiments

Setting time		Damage due to freezing at early ages			
Curing temp.	Testing method	Curing method	Curing time before freezing	Freezing method	Type of measurement
20℃			6hours	Freeze and Thaw (4cycles/day)	Compressive strength (at Freeze starting point and 28 days)
	Penetration resistance (JIS A6204)	10℃ (Sield curing)	12hours		
10℃			18hours		
			24hours	Keep freezing (for 24hours)	
5℃			48hours		Freezing temp.

time and frost resistance of fresh concrete was carried out. The outline of these tests is shown in Table.2. The penetration test was carried out in accordance with JIS A 6204 (appendix 1). The specimens for the penetration test were cured at three different temperatures (20, 10 and 5℃). For the tests evaluating the frost resistance of fresh concrete, the specimens were cured at 10℃ before and after freezing and were frozen under two different conditions. The resistance to frost damage at early ages was evaluated by comparing the compressive strength at 4 weeks of concrete specimens which had experienced frost action at early ages with that of concrete specimens which had been cured at 10℃ for 4 weeks. The compressive strength at freeze starting point and the freezing point of concrete and the admixture solution were also measured. The concentration of the solution was 0 to 100%.

3 Discussions of experimental results

3.1 Freezing point

The freezing point of the admixture solution and the concrete specimens are given in

Table. 3 Freezing point of antifreezer solution and concrete (℃)

Type of antifreezer	Concentration of antifreezer solution (%vol) 0	20	40	60	80	100	Concrete
P	-0.2	—	—	—	—	—	-1.5
H	—	-1.8	-1.6	-1.7	-1.5	-1.7	-1.2
AS	—	-4.3	-8.8	-13.1	-19.9	-28.5	-2.1
BS	—	-5.0	-10.6	-17.9	-28.4	<-40	-2.6
CS	—	-4.3	-9.3	-14.3	-20.7	-29.1	-3.0
D	—	-5.6	-12.3	-19.5	-29.8	-40.1	-3.6
E	—	-4.7	-9.3	-15.9	-24.7	-39.2	-3.6

Table.3. The results show that the higher the concentration of the solution, the larger the drop of th freezing point. The freezing point of the concentrated solution was very low (i.e., -30 to -40℃) but the standard concentration of antifreezers in concrete is about 8 to 10 %vol. The freezing point of antifreeze concrete was about -2 to -3℃. The depression of the freezing point was not big enough to pevent the frost damage at early ages by using antifreezers. However, it is considered that more non-freezable water exists in antifreeze concrete than in plain concrete.

3.2 Setting and hardening

Figure.1 shows that the relationship between the setting time and the curing temperature. It was recognized that the retardation of the concrete due to low temperature effect

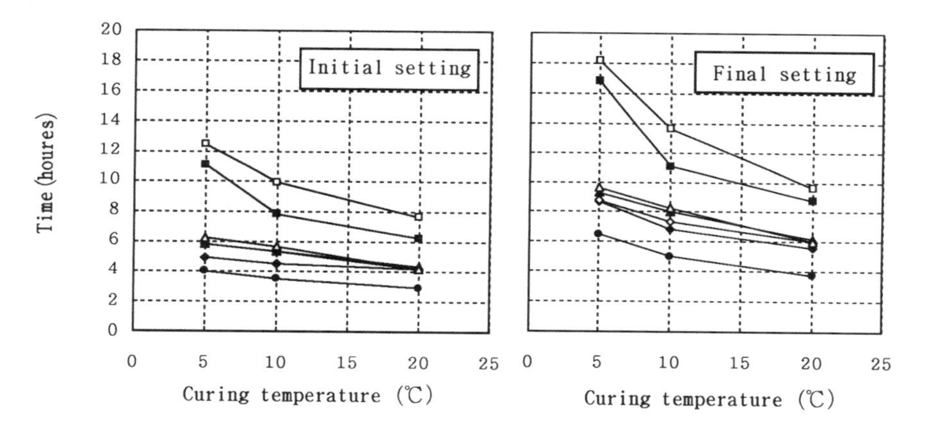

Figure. 1 Relationship between setting time and curing temperature

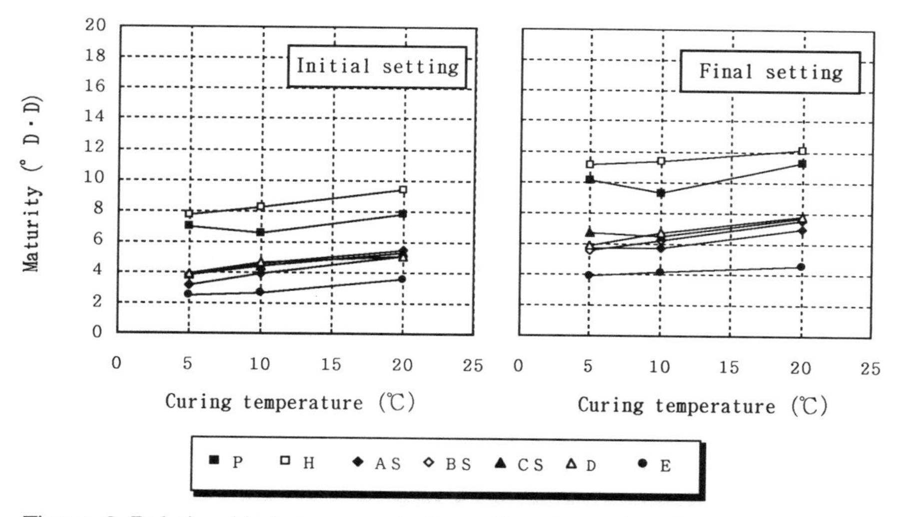

Figure. 2 Relationship between maturity until setting and curing temperature

improved by using antifreezers. The final setting time of the antifreeze concrete was about 6 to 9 hours at 5℃ which was similar to that of plain concrete cured at 20℃.

The maturity of the concrete at the initial and final setting against curing temperature is shown in Figure.2. From this figure, it can be seen that the maturity of plain concrete specimen cured at 5℃ is higher than that at 10℃, while in the antifreeze concrete specimens, an opposite behavior was observed. In other words, the low curing temperature resulted in a smaller value of maturity before final setting. However, the variation of the maturity at the initial and final setting was between 1 to 2 ° D · D, so it could be considered as a constant value.

The compressive strength gain as a function of the maturity for the various concrete specimens is shown in Figure.3. It was observed that the strength gain after final setting (i.e., at 10 and 15° D · D) of antifreeze concrete was higher than plain concrete. The required minimum values of maturity to develop compressive strength of 5 MPa recomended in JASS 5 as a safe value for frost damage at early ages are shown in Figure.4. From this figure, it is evident that the target strength of 5MPa in antifreeze concrete specimens could be obtained at lesser maturity than ordinary concrete specimens. For example, specimen CS, the maturity was 16.1° D · D, while this was increased to 36.7° D · D for specimen P. The requred strength was gained at lower maturity in antifreeze concrete, especially on CS, D, E, the maturity was half that of the plain concrete.

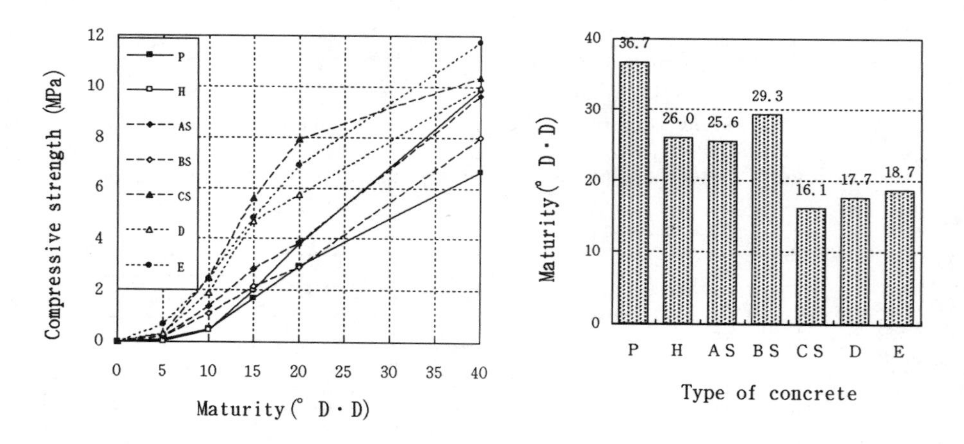

Figure. 3 Strength development at early ages

Figure. 4 Comparison of maturity at compressive strength 50MPa

---

* Maturity factor in this paper, which defined in JASS5, is caluculated in accordance with following equation.

$$M = \sum_{z} (\theta + 10) \qquad (^\circ D \cdot D)$$

where z = age(days)

θ = curing temperature of concrete (℃)

3.3 Frost resistance at early ages

In this study, the resistance to damage due to freezing at early ages was evaluated by comparing the compressive strength at 4 weeks of concrete specimens which had been under freezing and thawing action at early ages (6, 12, 18, 24 and 48 houres after placing). It was found that those specimens having a strength ration less than 90 % were deteriorated by freezing at early ages. It was observed that those specimens which had been frozen before reaching final setting deteriorated under the process of freezing and thawing or unable to gain sufficient strength by 4 weeks. However, when the compressive strength at freeze starting point was more than 2MPa, the strength ratio of all specimens, except CS, was above 90 % and they could withstand frost damage. The relationship between the compressive strength at freeze starting point and the strength ratio after 4 weeks is shown in Figure.5. Therefore, the frost resistance at early ages would be improved by air entrainment. The result of this part of the study also confirmed that the required strength of 5MPa as specified in JASS 5 is conservative and is a safe value for AE concrete.

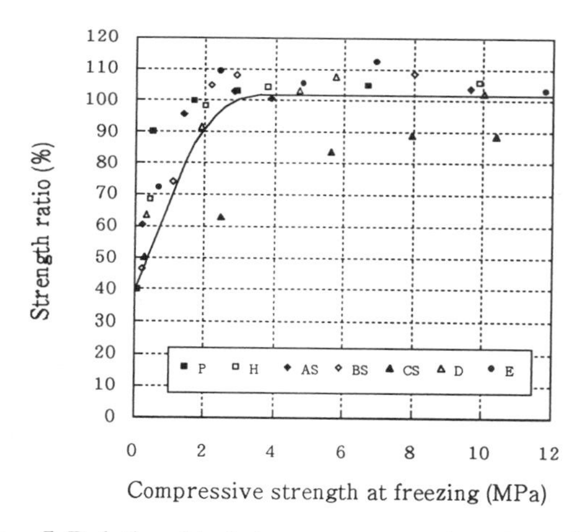

Figure. 5 Relationship between compressive strength at freezing and strength ratio at 28 days

4 Investigation of utilization method

4.1 Cooling and hardening of concrete member

From the experimental results, it was found that the use of antifreezers for protecting fresh concrete from frost damage was not only beneficial from the drop of freezing point but also the acceleration of setting and hardening. Therefore, it is important to remind that the maturity of concrete which gains from the time of placing to the freezing point of the concrete was calculated based on considering the temperature of fresh concrete, the heat of hydration, the thermal capacity and the size of concrete member to give full play the ability of the antifreezers. In this section, the maturity was calculated

Table. 4 Analytical conditions of temperature history

Cement content	Casting temp.(°C)	Ambient temp. (°C)	Type of structural member and curing condition
		0	1.Column (800×800mm)
360 (W/C=40%)		-2	2.Wall (180mm) : Plywood form
330 (W/C=50%)	15	-3	3.Wall (180mm) : Insulating form
300 (W/C=60%)		-5	4.Foundation (t150mm×h600mm) : Covered with vinyl sheet
		-10	5.Foundation (t150mm×h600mm) : Uncovered
		-18	

by analyzing the temperature history by using the Finite Element Method (F.E.M.). The analytical conditions of F.E.M. is shown in Table.4. As the results of this analysis, the maturity from the time of placing up to freezing point increased as the cement content increased for same ambient temperature. And the maturity depended on the size of concrete member and the curing condition. For wall using an insulating form and mass concrete such as column (800 x 800mm), the requred compressive strength to prevent frost damage at early ages would be gained before freezing of the concrete. And as it was shown in Figure.4, the antifreeze concrete could pass the requred compressive strength of 5MPa even at a maturity less than 20° D · D. Therefore, either suffer frost damage at early ages or not on certain condition would be evaluated from the relation the maturity with the strength development until freezing of the concrete.

4.2 Allowable temperature for making early age frost resistance concrete

The utilization method of the admixtures for the protection of fresh concrete from frost damage on cold weather concreting was investigated. The relationship between the concentration of antifreezers and the allowable ambient temperature was derived from the analytical and the experimental results. For the case of foundation covering the upper surface with vinyl sheet in order to protect it from the influence of wind and snow, as an example of severe thermal conditions, the maturity until freezing of concrete versus ambient temperature at various freezing points is shown in Fig.6. The relation between the ambient temperature and the maturity until freezing of concrete for each freezing point is expressed in Eq.(1).

$$FD = a(-AT)^{b} \tag{1}$$

Where; FD is the maturity until freezing of concrete, AT is the ambient temperature, and a,b are constants.

If the maturity, SD, which gains the requred compressive strength 5MPa is less than FD, it would be prevent frost damage of concrete at early ages. Therefore, the allowable ambient temperature, ATp, is given by Eq.(2);

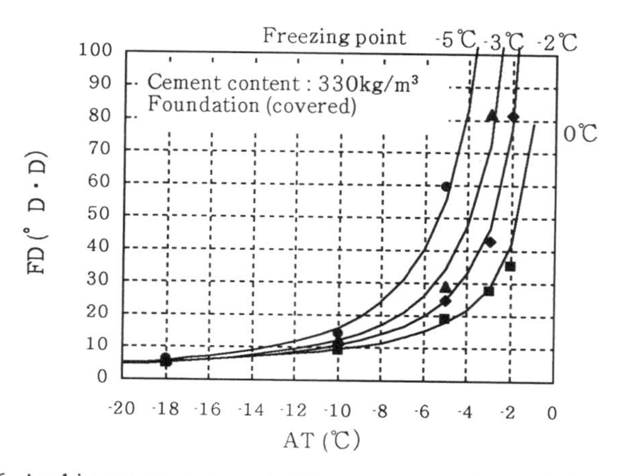

Figure. 6 Ambient temperature (AT) versus maturity until freezing (FD) at each freezing point (Eq.(1))

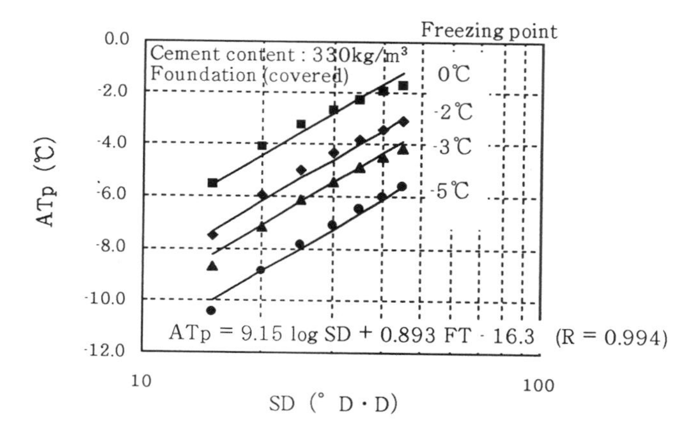

Figure. 7 Relationship between SDand ATp at each freezing point (Eq.(3))

$$ATp = -(SD / a)^{1/b} \tag{2}$$

The relationship between SD and ATp at each freezing point of concrete is shown in Figure.7. ATp is expressed by SD and FT. Equation (3) was obtained by multiple regression analysis;

$$ATp = c \log SD + d\,FT + e \tag{3}$$

From the results of experiment, SD and FT are in proportion to the concentration of antifreeze admixtures, N. SD and FT are expressed in Eq.(4) and Eq.(5);

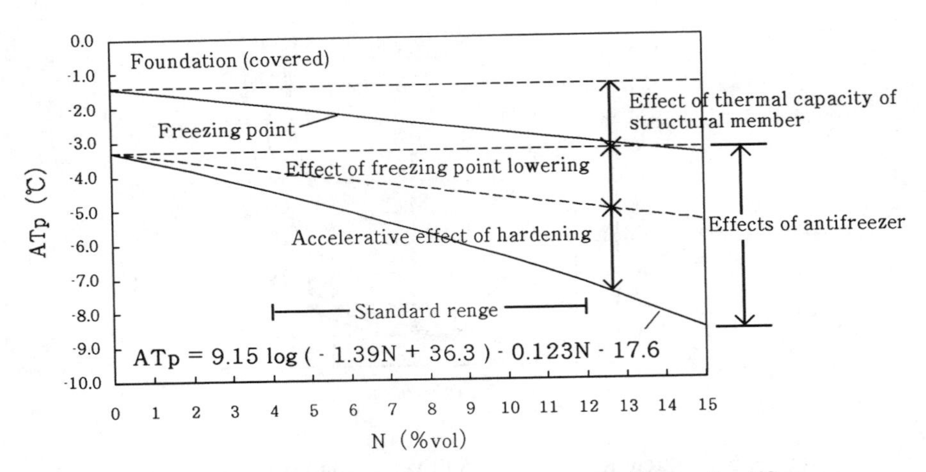

Figure. 8 Allowable ambient temperature and effects of antifreezer

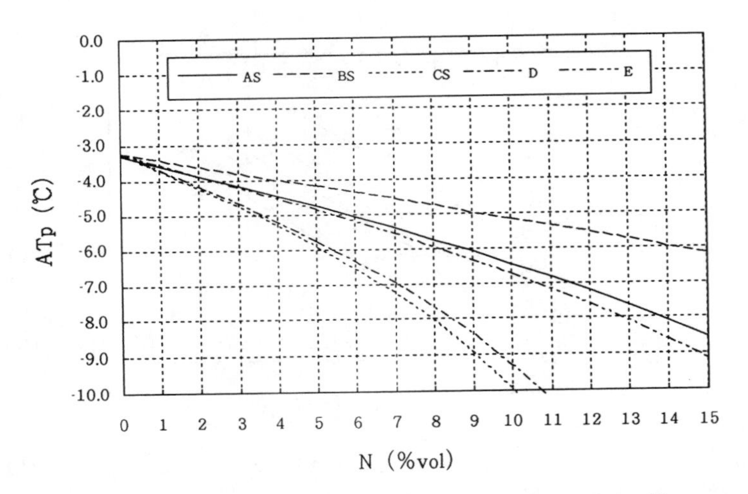

Figure. 9 Relationship between N and ATp for each antifreezers

$$SD = \alpha N + \beta \tag{4}$$

$$FT = \alpha' N + \beta' \tag{5}$$

Substituting Eq.(4) and Eq.(5) into Eq.(3), We obtain Eq.(6);

$$ATp = c \log(\alpha N + \beta) + f N + g \qquad (f = d\alpha',\ g = d\beta' + e) \tag{6}$$

The relationship between the concentration of admixture and the allowable temperature on AS is shown in Fig. 8. The effectiveness of the antifreezers results from the combination of the effects of the drop of freezing point, the acceleration of setting and hardening and the thermal capacity of concrete structural members. The standard concentration of this admixture is 4 to 12% volume. The allowable ambient temperature is −4.5 to −7.2°C corresponding to the standard concentration. The relationship between N and Apt for each admixture type is shown in Fig. 9. This is useful for evaluation of the level of effectiveness and the use of these admixtures on cold weather concreting.

5 Conclusions

1. The freezing point of concrete using antifreeze admixtures was −2 to −3°C. This depression of the freezing point was not, itself, large enough to prevent frost damage of concrete at early ages.
2. The setting and hardening process of the concrete were accelerated and the retardation of setting due to the low temperature was improved by using the antifreeze admixtures. It was found that the required compressive strength to prevent the antifreeze concrete from frost damage at early ages could be gained at lower maturity than ordinary concrete.
3. To give full play the ability of the antifreezers, it is important to remind that the maturity of the concrete, which gains from the time of placing to the freezing point of concrete, considering the temperature of fresh concrete, heat of hydration, the thermal capacity and the size of concrete member.
4. The allowable ambient temperature corresponding to the concentration of antifreeze admixture was derived by considering the analytical and experimental results. This is useful for evaluations of the level of effectiveness and the utilization of admixtures on cold weather concreting.

6 References

1. *Recommendation for practice of cold weather concreting* (1990) Architectural Institute of Japan.
2. Japanese architectural standard specification for reinforced concrete work, *JASS5* (1993) Architectural Institute of Japan.
3. K. Sakai et al (1991) Antifreeze admixture developed in Japan. *ACI Concrete International*, Vol. 13, No. 3.

133 DYNAMIC MODULUS OF ELASTICITY AND DURABILITY OF NO-FINES CONCRETE

M. TAMAI
Kinki University, Higasi-Osaka City, Japan
M. TANAKA
Daiichi Technical College, Kokubu City, Japan

Abstract

This paper presents the results of an investigation undertaken to develop additional data on the mechanical properties of No-Fines Concrete (NFC) such as the dynamic modulus of elasticity, strength, freezing and thawing resistance and the elusion of free lime in sea water.

The dynamic modulus of elasticity shows a steep rise during early periods of up to 7 days. This rise might be attributable to the fact that NFC has continuous voids and binder segment surface areas which allow contact with water so that the hydration reaction is enhanced. This tendency becomes more noticeable when the binder-void ratio decreases. The relationship between compressive strength and the dynamic modulus of elasticity is shown by a linear function.

The resistance of NFC to freezing and thawing is considered to be low, because NFC has continuous voids into which water can permeate during freezing and thawing. Increasing the binder per unit volume of NFC increases its durability. Contrary to ordinary concrete the destructive pattern of freezing and thawing is seen at the center of NFC specimen. Specimens with two layers had high densities, but early destruction occurred at the interface.

When immersed in sea water for a period of 12 months, the compressive strength of the NFC decreased about 20% as compared with standard curing, due to the elusion of free lime. The addition of silica fume could possibly prevent the elusion of free lime.
Keywords: Compressive strength, durability, dynamic modulus of elasticity, flexural strength, freezing and thawing, no-fines concrete, sea water, silica fume, superplasticizer.

Concrete Under Severe Conditions: Environment and loading (Volume Two) Edited by K. Sakai, N. Banthia and O.E. Gjørv. Published in 1995 by E & FN Spon. ISBN 0 419 19860 1

1 Introduction

No-fines concrete (NFC) is a concrete made up of coarse aggregate with viscous cement paste. Since it has continuous voids from the fresh state, as compared with ordinary concrete, various characteristics are shown in its mechanical properties including strengths and chemical properties[1].

So far, NFC has only been applicable to limited places due to the problem of strength[2]. But recently, thanks to the use of a superplasticizer and silica fume, a strength equivalent to that of ordinary concrete has been obtained, and the application of NFC could spread not only for water-permeating pavement material[3]. Attention has been paid to the fact that the permeability and the internal surface area of NFC are controllable, and that the application range to waterside structures that are friendly to living things is going to expand. Generally, the dynamic modulus of elasticity (E_d) for concrete approximated to the initial static modulus of elasticity. Since E_d can also be applied to nondestructive tests, it is utilized in evaluating the strength, durability tests and etc.. Considering the above present state, this technique was applied as one of the control methods for the NFC structures in this study.

In this research, the effect of mix design of NFC and external factors on the dynamic modulus of elasticity (E_d), resistance to freezing and thawing,and the elusion of free lime when NFC is immersed in sea water were examined.

2 Materials

Cement : Normal portland cement and blast furnace slag cement were used in these tests. Physical and chemical properties of the cement are shown in Table 1. Silica fume was obtained in Norway (ELKEM). The physical and chemical analyses are given in Table 1.

Table 1. Physical properties and chemical analysis of cement and silica fume

Physical Tests	CN*1	CB*2
Fineness-Blaine m^2/kg	329	341
Specific gravity	3.16	3.03
Setting time, min		
- Initial	160	160
- Final	259	226
Compressive strength MPa: 7-day	24.3	22.0
28-day	41.0	43.1
Flexural strength MPa: 7-day	5.3	4.7
28-day	7.3	7.8

Chemical Analysis	CN	CB	SF*3
Insoluble residue	0.20	0.1	---
Silicon dioxide(SiO_2)	22.2	26.1	95.80
Aluminum oxide (Al_2O_3)	5.5	9.1	---
Ferric oxide (Fe_2O_3)	3.1	1.9	---
Calcium oxide (CaO)	63.6	54.1	0.14
Magnesium oxide (MgO)	1.6	3.4	---
Sulphur trioxide (SO_3)	1.9	2.2	---
Sodium oxide (Na_2O)	---	0.22	0.13
Potassium oxide (K_2O)	---	0.50	0.42
Carbon (C)	---	0.58	1.60
Phosphorus oxide(P_2O_5)	---	0.11	0.16
Loss on ignition	0.7	1.1	1.80

*1 CN: Normal portland cement
*2 CB: Portland blast-furnace slag cement
*3 SF: Silica fume

Table 2. Physical properties of aggregate

Class of crushed stone (No.)	Size(mm)	Unit weight (kg/m^3)	Specific gravity	Void ratio (%)
5	13-20	1560	2.69	42.0
6	5-13	1540	2.69	42.7
7	2.5-5	1490	2.69	44.6

Table 3. Mixture proportions of NFC

W/(C+SF) (wt.%)	25					
B/V (vol.%)	30, 40 and 50					
Mixture proportion of crushed stone	No.5:6:7 = 0:0:1			No.5:6:7 = 5:1:4		
SF/(C+SF) (wt.%)	0	10	20	0	10	20
Sp/(C+SF) (wt.%)	1.0	1.5	2.0	1.0	1.5	2.0

W:Water, C:Cement, SF:Silica fume, Sp:superplasticizer
B=W+C+SF+Sp, V:Voids of aggregate

Aggregate : Three kinds of crushed stone were used. Their grading and other physical properties are shown in Table 2.
Superplasticizer : A sodium salt of naphthalene sulphonic acid was used. Its chloride content was negligible.
Air-entraining agent : A sodium salt of lignin sulphonic acid used .

3 Mixture proportion

Mixture proportions of these materials were as follows ; Water-cement ratio(W/C)=0.25, Superplasticizer-cement ratio (Sp/C)=0.01～0.02.
Silica fume-binder ratio :[SF/(C+SF)]=0, 10 and 20%. Stone weight ratio (No.5:No.6:No.7) were 0:0:1 and 5:1:4.
The binder-aggregate ratio was 0.20～0.34.
The proportioning of the NFC mixtures is summarized in Table 3. For all mixtures, the aggregate was weighed under room-saturated, surface-dry condition.

4 Test method and procedure

4.1 Mixing and fabrication

The mixing of the materials and the fabrication of the specimen were carried out according to the procedure shown in Fig. 1. NFC was placed into cylinders or cubes using a rammer and a surface vibrator and formed one or two layers. The dimensions were ϕ 10×20cm for the compression test, 10×10×40cm for the flexural, dynamic modulus of elasticity and the freezing and thawing tests.

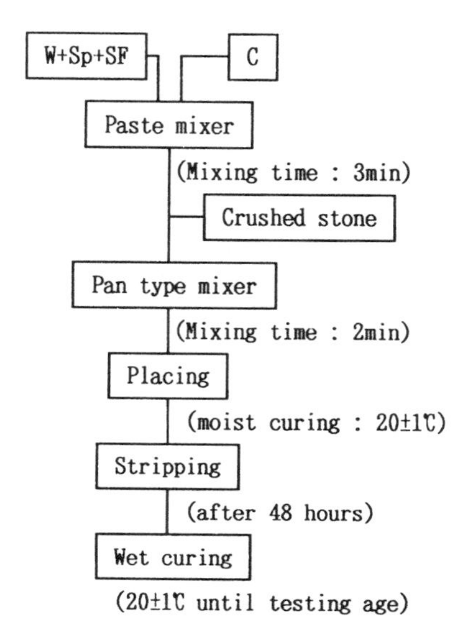

Fig. 1. Miximig and fabrication of the specimen

4.2 Tests on hardened NFC

The compressive strength and the modulus of elasticity of the NFC were measured with strain detectors attached to the specimens according to JIS A 1108. The measurement of the dynamic modulus of elasticity of the NFC during resonance was performed through an axial vibration or deflecting vibration method in accordance with JIS A 1127 regulations. In the freezing and thawing test, the cycle of freezing and thawing was performed five times each day, as required in JSCE-1986 regulations.

5 Results and discussion

5.1 Dynamic modulus of elasticity of NFC

(1) Relation of material age and E_d

The process for increasing the E_d after the form removal at an age of 2 to 56 days was investigated. Fig. 2 shows the results of the investigation of how the E_d was affected in the case of aggregate mixture proportions of No. 5:6:7=0:0:1. W/(C+SF)=25% and SF/(C+SF)=0% were kept constant, and the volume ratio of binder to voids of aggregate (B/V) was changed to 30,40 and 50%. Moreover, the difference between the cases in which the method of curing was in water or in the atmosphere was examined.

Fig. 3 shows the results of the similar experiment., having kept aggregate mixture proportion No. 5:6:7=5:1:4, W/(C+SF)=25% and SF/(C+SF)=10% constant. According to the results, the E_d which is considered to be linked to strength was high from the very beginning. And when the binder ratio (B/V) was increased, the E_d also rose. Thus, it was shown that there was a mutually proportional relation. Besides, whatever mixture proportion of these B/V, the relation of age and to E_d showed that as

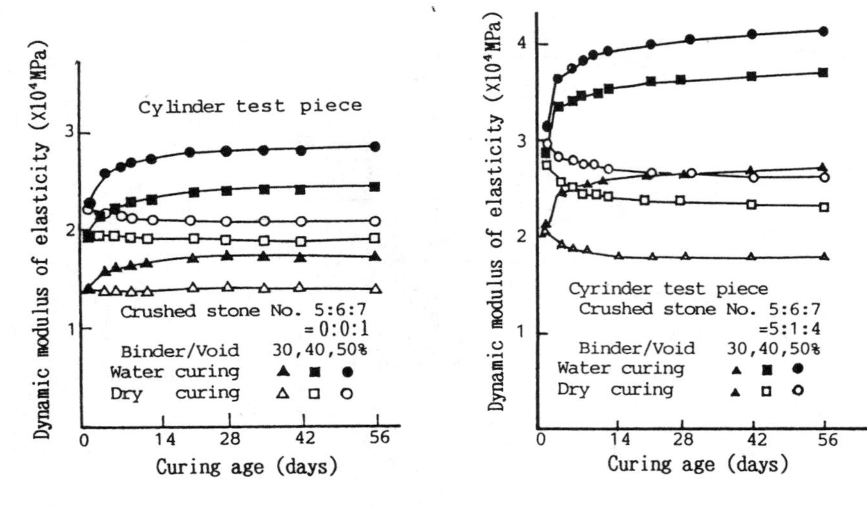

Fig. 2. Relationship between E_d and age   Fig. 3. Relationship between E_d and age

compared with that of ordinary concrete, in the case of curing in water, the Ed increased rapidly in the initial period, and reached more than 90% of the value at the age of 56 days at an age of about 14 days. Thereafter, the gradient of the rise became gentle. This is considered to be due to the fact that as NFC has continuous voids, the surface area of past which directly comes in contact with the water was large, and hydration was completed early. This tendency becomes larger as the B/V becomes smaller, namely the amount of binder was less in relation to the void ratio of aggregate.

(2) Difference in curing methods and Ed

The Ed of specimens cured in the atmosphere, at the age of 56 days was the same as that at 7 days, and a increase in strength was not observed. This seems to be caused by early drying, and the necessity to cure in water or a wet state, particularly for NFC has been shown.

(3) Effect due to mixing SF

When SF was mixed in the binder, the initial hydration reaction was slightly delayed, and Ed showed a tendency to be somewhat lower than that without mixing at quite an

Table 4. Effect of containing SF in E_d

Mixture proportion of crushed stone			No.5:6:7 = 0:0:1			No.5:6:7 = 5:1:4		
SF/(C+SF) (wt.%)			0	10	20	0	10	20
E_d ($\times10^4$) MPa	Age 28days	B/V(%) 40	2.48	2.45	2.26	3.48	3.62	3.19
		50	2.77	2.68	2.57	3.72	3.94	3.65
	56days	40	2.52	2.59	2.39	3.55	3.75	3.68
		50	2.84	2.95	2.78	3.78	4.10	3.79

early stage. But when the age surpassed 7 days or more, Ed became similar or higher at SF/(C+SF)=10%. As time elapsed, the pozzolanic reaction progressed, and it rose depicting a gentle curve.
However, when the amount of SF mixing increased, in the case of same compressive strength, the Ed showed of lowering tendency than that without mixing. Moreover, when the curing in the atmosphere was carried out, the effect of the drying shrinkage was exerted more than that without mixing, and tendency toward lowering was shown as the age progressed. Thus, it became clear that in particular the resistance to initial drying was low.
(4) Difference in mixing proportion of aggregate and Ed
In the NFC in which the mixture proportion of aggregates was set at No.5:6:7=5:1:4 (percentage of solid volume: 66.8%), not withstanding the unit amount of the binder used was less as compared with 0:0:1 (percentage of solid volume: 56.0%), the Ed rose to a high value, and it was shown that the magnitude of the percentage of solid volume of the aggregates themselves becomes one of the important factors.
(5) Relation of compressive strength to Ed
Generally, the compressive strength (fc) and the Ed of ordinary concrete are in proportional, and it has been applied to nondestructive tests.

In Fig.4, the relation of fc to the Ed determined by axial vibration method for NFC is shown. According to these results, fc and Ed showed a proportional relation, and in the case of making the mixture proportion of aggregates into 5:1:4 and 0:0:1, they become the following equations.

$$f_c = -22.4+16.1\times10^{-4}\,E_d(\mathrm{MPa}) \quad (1)$$

$$f_c = -9.2+12.6\times10^{-4}\,E_d(\mathrm{MPa}) \quad (2)$$

In addition, it was affected largely by the rate of the solid volume of aggregates.

Fig.5 shows the relation of the flexural strength (fb) and the Ed determined by deflecting vibration method. Some dispersions are observed in the result, but similarly

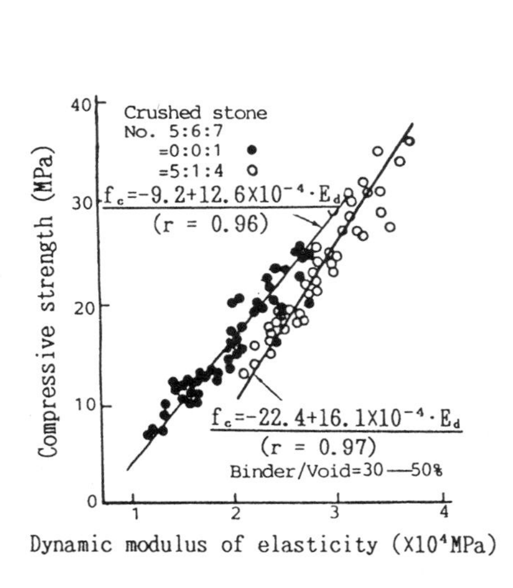

Fig.4. Relationship between f_c and E_d

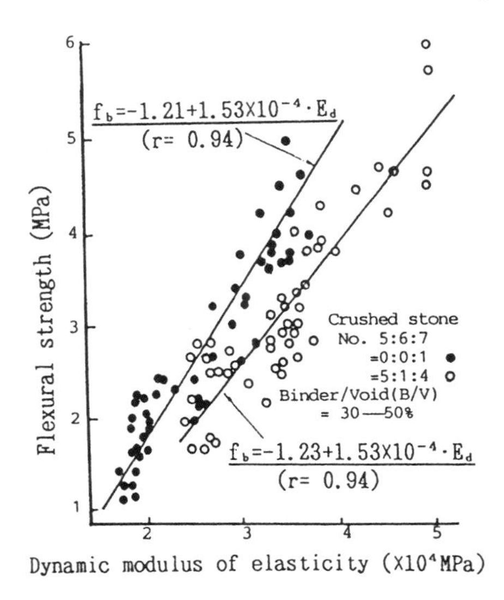

Fig.5. Relationship between f_b and E_d

to those mentioned before, they can be shown by the following equations:

$$f_b = -1.23+1.25\times10^{-4}E_d(\mathrm{MPa}) \quad (3)$$

$$f_b = -1.12+1.53\times10^{-4}E_d(\mathrm{MPa}) \quad (4)$$

5.2 Durability for freezing and thawing

Generally, the mechanism for breaking a concrete structure by freezing and thawing is attributed to the water infiltrating into the minute voids of the concrete which repeats expansion and shrinkage at the time of freezing and thawing, and it gradually breaks from surface to the internal structure.

Accordingly, the durability of NFC to freezing and thawing is predicted to be lower than that of ordinary concrete because it has continuous voids, and water infiltrates easily into them. In this study, however, the extent of the durability when Sp, SF and AE water-reducing agent were used was examined.

(1) Effect of change in the amount of binder

Fig. 6 shows the relation between the number of cycles of freezing and thawing and the relative dynamic modulus of elasticity of respective specimens, in which the mixture proportion of aggregates was set as 0:0:1 and 5:1:4, the B/V was changed to 30, 40 and 50%, and one-layer placing was carried out.

According to the results, when the mixture proportion was 0:0:1 and B/V=30%, breaking occurred after 25 cycles of freezing and thawing, and when B/V=40%, breaking occurred after 55 cycles. It is seen that when the B/V was set at 50%, the durability was noticeably higher. In this way, it was shown that as a matter of course, the resistance of NFC to freezing and thawing is proportional to the amount of binder that fills the voids of the aggregate.

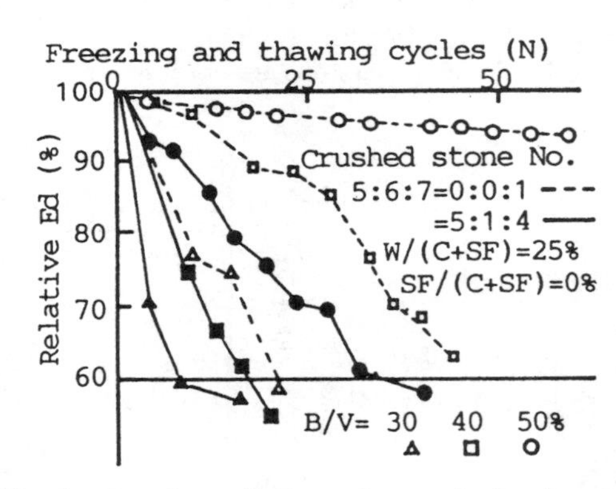

Fig. 6. Results of freezing and thawing test

(2) Effect of the mixture proportion of the aggregate on the breaking mode of the NFC

Similarly, when the mixture proportion of the aggregate was set as 5:1:4, and B/V was changed, as shown in Fig.6, the durability was proportional to B/V. However, notwithstanding the fact this mixture proportion of the aggregate is high density, and that E_d, f_b and f_c are all high, at the early stages of the freezing and thawing cycles, breaking occurred or the relative dynamic modulus of elasticity reached below 60% and the resulted in an inverse to the general prediction.

Table 5. Effect of containing SF in durability factor

Mixture proportion of crushed stone		No.5:6:7 = 0:0:1			No.5:6:7 = 5:1:4		
SF/(C+SF) (wt.%)	B/V(%)	0	10	20	0	10	20
Durability factor(%)	40	12	43	38	4	8	9
	50	65	68	49	8	8	9

In addition, as shown in Table 5, the durability factor of NFC with different mixture proportions of the aggregate had obtained similar results. These are considered to be that as compared with the former mixture proportion of the aggregate. In the mixture proportion of 5:1:4, the size of the continuous voids was fine; therefore, the infiltrated water in specimens became difficult to elude at the time of freezing[1][3]. And conditions under which it is hard to attenuate pressure of an ice expansion. It is presumed that there is some correlation between the size of continuous voids and paste strength and between these quantities and the resistance to freezing and thawing.

Regarding this point , it is known for a fact that the breaking mode of NFC due to freezing and thawing is different from that of ordinary concrete. Namely, with ordinary concrete, breaking advances gradually from the surface of the specimens toward the inside. But with NFC, the pressure caused by freezing inside becomes highest, and the pressure is distributed in an inverse state. This can be predicted from the fact that the breaking occurred from the center of the specimens.

(3) Effect of compaction method of specimens

Concrete was placed in one and two-layer packings using a surface vibrator. As a result, the two-layer packing brought about high density. But a tendency also existed for another mixture proportion to break early at the joint. (Photo 1.)

(4) Effect of air entraining(AE) agent

According to a report by V.M. Malhotra, it is said that an AE agent is effective for raising

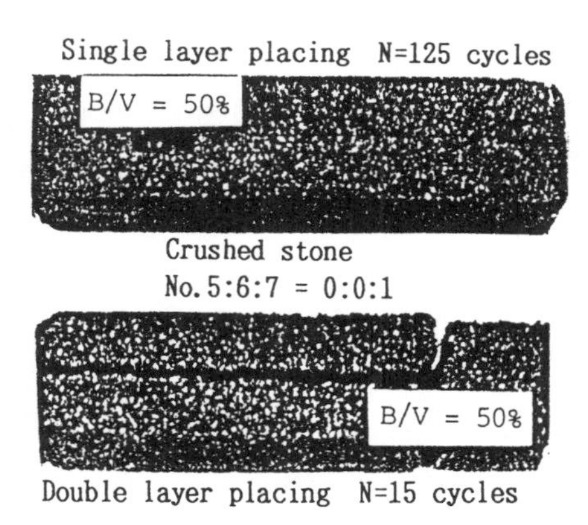

Photo.1. Destructive pattern of freezing and thawing test

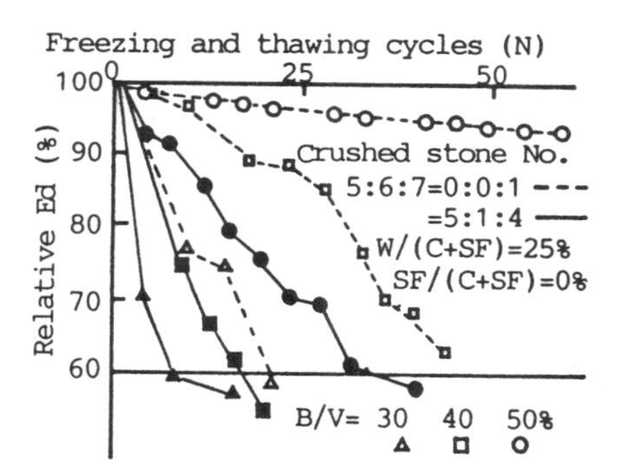

Fig.7. Results of freezing and thawing test

the durability of NFC[5]. Fig.7 shows a case in which AE water-reducing agent is used for the purpose of improving the durability and setting the amount of the AE agent in relation to the binder at 0.4%. As a result, it was found that at all values of B/V, the resistance to freezing and thawing was improved, and the AE water-reducing agent was also effective for thin cement paste films like NFC.
(5) Effect of mixing SF
When SF was mixed, although it depends on the mixture proportion of the aggregates, results showed that the durability of the NFC rose slightly as shown in Table 5 were obtained. However, when an AE water-reducing agent was used together, it was slightly lower than or became equivalent to the case without SF admixing, and a tendency to cancel both effects was shown. This is attributed to the fact that the carbon in the SF absorbs the AE water-reducing agent and decreases the AE effect.

5.3 Elusion of free lime and lowering of strength
Since the internal surface area of NFC is very large, the elusion of free lime in water is remarkable. It is not only lowers the strength, but also in a closed state in a narrow range of water area, there is sometimes a harmful effect to living things inhabiting the surrounding area. However, it is thought that in a vast water area like the sea, and in an environment with a relatively high pH, the effect to living things is rarely present.

Fig.8 shows the relation of the pH and the age of various specimens which were immersed in the natural sea area. According to the results, it was shown that NFC is neutralized earlier than ordinary concrete. It was also shown that the addition of SF, which is a kind of pozzolanic material, or the utilization of blast furnace slag promoted neutralization. Moreover, in the NFC in which the unit binder was nearly the same, in the range of good water permeability, the neutralization occurred as the particle size of the aggregate used grew smaller. This is thought to be related to the internal surface area of NFC.

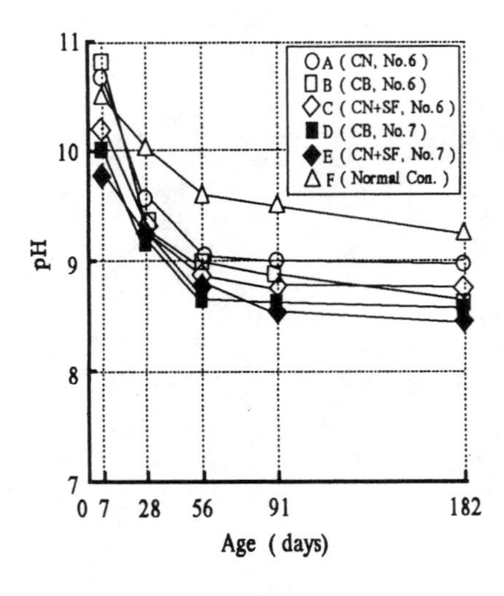

Fig.8. Relationship between pH and age (cured in natural sea water)

6 Conclusions

This research can be summarized into the following items.

1. The E_d of NFC was high from the beginning, and with curing in water, it is grew to more than 90% of the value at 56 days up to 14 days. This was because water infiltrated into continuous voids, and hydration was completed early. Besides, the E_d in the dry curing was affected largely by drying shrinkage, and sometimes it dropped below the value at the time of form removal. Also, over the long term, a rise cannot be expected.
2. When SF was mixed, E_d was subjected to the effect of pozzolanic reaction. It increased over a long period, depicting gentle curves. However, in dry curing, it was shown that a greater effect of shrinkage was exerted than samples without SF.
3. It became clear that the E_d of NFC depends largely on the rate of the solid volume of the aggregate used.
4. When the mixture proportion of the aggregates was constant, the E_d and f_c determined by the axial vibration method as well as E_d and f_b determined by the deflecting vibration method showed a linear relation. Thus, it is thought that both can be applied to nondestructive tests.
5. When the aggregates used were the same, the resistance of NFC to the freezing and thawing was proportional to the amount of binder. However, when aggregates with a high rate of solid volume were used, high values of E_d and various strengths and a low resistance to freezing and thawing were frequently observed.
6. In view of the resistance to freezing and thawing, the placing of one-layer of NFC is suitable. In the case of multiple-layer placing, it is necessary to give consideration to the homogeneity of the boundary surfaces.
7. It was found that AE water-reducing agent raised the resistance to freezing and thawing in NFC. In addition, the admixing of SF improved the resistance to freezing and thawing, but it showed a tendency of lowering the AE effect.
8. The strength of NFC in sea water is lowered by the elusion of free lime. As a method of prevention the addition of a pozzolanic material like SF is desirable. Moreover, the rate of neutralization is proportional to the internal surface area per unit volume and the amount of pozzolanic material added.

References

1. Tamai, M. (1988) Water Permeability of Hardened Materials with Continuous Voids, CAJ Review, pp.446-449.
2. Meininger, R.C. (1988) No-Fines Pervious Concrete for Paving, Concrete International, Vol.10,No.8, pp.20-27.
3. Eisenberg, D. , Kauzamann, W. (1969) The Structure and Properties of Water, Oxford at the Clarendon Press, London.
4. Tamai, M. (1989) Properties of No-Fines Concrete Containing Silica Fume, ACI, SP-144, Vol.2, pp.799-814.
5. Malhotra, V.M. (1976) No-Fines Concrete - Its Properties and Applications, Jour. of ACI, Vol.73, No.11, pp.628-644.

134 THE USE OF HIGH-STRENGTH MODIFIED NORMAL DENSITY CONCRETE IN OFFSHORE STRUCTURES

R. WALUM and J.K. WENG
Hibernia Management and Development Company, St John's, Newfoundland, Canada
G.C. HOFF
Mobil Research and Development Corporation, Dallas, Texas, USA
R.A. NUNEZ
Carolina Stalite Company, Salisbury, North Carolina, USA

Abstract
For offshore concrete structures which spend all or part of their construction and/or operational life in a floating mode, buoyancy of the structure under all conditions is essential. Contributions to this buoyancy can be obtained by reducing the density of the concrete. This paper discusses the development of high-strength concrete using a blend of normal weight coarse aggregate and structural lightweight aggregate to provide a modified normal density concrete (MNDC) which retains the desired characteristics of high-strength concrete, including durability, while having a reduced density. For the materials evaluated, a replacement of 50 percent (by volume) of the normal weight coarse aggregate with a high quality structural lightweight aggregate produced concrete with 28-day cylinder strengths of 79 MPa. Other mechanical properties are also reported. Full size production batches had very good uniformity with coefficients of variation being in the excellent category.
Keywords: Blended cement, concrete production, high-strength concrete, lightweight aggregates, lightweight concrete, mixture proportions, offshore structures, silica fume.

1 Introduction

The Hibernia offshore concrete platform [1] (Fig. 1), currently (1995) under construction in Newfoundland, Canada represents one of the largest, single uses of high strength concrete (69 MPa cylinder strength) in North America with a total concrete volume of approximately 165,000 m^3. It will operate in 80-m of water on the Grand Banks of East Coast Canada. The concrete substructure, commonly called a gravity base structure (GBS), is essentially a cylindrical concrete caisson approximately 108-m in diameter which rests on the seabed and is 85-m high. Four shafts extend above the caisson for

Concrete Under Severe Conditions: Environment and loading (Volume Two) Edited by K. Sakai, N. Banthia and O.E. Gjørv. Published in 1995 by E & FN Spon. ISBN 0 419 19860 1

Fig. 1 Hibernia Offshore Concrete Platform

another 26-m to support the topsides facilities of the platform. Details of the platform geometry can be found in [1] and [2]. The GBS is being constructed in a similar manner as many North Sea gravity base structures with both a dry-dock and wet-dock construction phase. Details of GBS construction can be found elsewhere [2,3].

The wet-dock phase of construction means that portions of the construction of the concrete base will be done while the structure is in a floating mode. This floating construction period will take several years. Upon completion of the concrete substructure, a topsides facility of approximately 38,000 tonnes will be added to the floating structure. The entire assemblage will then be towed to its final location and placed on the seabed floor. The GBS has a requirement for hydrostatic stability while floating and being towed which is influenced by both the structures' center of gravity and center of buoyancy. Both of these considerations are a function of the concrete density and its distribution, and the dimensioning and arrangement of the GBS.

For the GBS to remain hydrostatically stable, the center of gravity must always be lower than the center of buoyancy by a safe margin. To accomplish this on the Hibernia platform, the lower portions of the GBS are being constructed with a normal density concrete (NDC) while regions above the NDC are being constructed with a reduced density concrete, hereafter called modified normal density concrete (MNDC). A description of the concrete production for the NDC concrete skirts and base slab can be found in [4]. The reduced density was accomplished by replacing portions of the normal weight coarse aggregate of the NDC with structural lightweight aggregates. This paper describes the selection of materials and the evaluation of the MNDC.

2 Test Program

Previous work [5] on the use of high strength lightweight concrete in severe marine environments had indicated that not all lightweight aggregates could be used because of limiting strength of the aggregate or because of high water absorptions that adversely affected freezing and thawing durability and constructability. A screening program [6] of selected lightweight aggregates was performed in support of the of the Hibernia platform and several North American lightweight aggregates were identified as being suitable for this application. This paper describes the evaluation of only one of those aggregates which is given in further detail latter in the paper.

The criteria for the MNDC was that it have a 28-day cylinder compressive strength $\geq$ 69 MPa, a splitting tensile strength $\geq$ 5.2 MPa, a density of $\leq$ 2250 kg/m^3, a Modulus of Elasticity in the range of 30 to 36 GPa, and an air content, after pumping, of $\leq$ 4%. The submerged lower portions of the structure are not exposed to freezing so there is no need for high levels of entrained air. Further work on air-entrained MNDC for the splash zone of the structure is under development.

A series of laboratory batches were made with varying amounts (25, 50 and 70%) of lightweight aggregate being substituted (by volume) for the normal weight coarse aggregate and evaluated for the parameters noted above. Based on those results, a lightweight aggregate replacement value was selected for evaluation in full size production batches in the Hibernia project batching plant [4].

In the full size (2 m^3) batch evaluations, 32 batches of MNDC were made to evaluate the reproducibility. Each batch was evaluated for slump, air content, unit weight, temperature and compressive strength of 150- by 300-mm cylinders at 28 and 365 days age. Only the 28-day results were available at the time of preparation of this paper. Once the reliability of the MNDC production was established, some additional MNDC full size batches were made to produce specimens to evaluate other mechanical properties of the concrete. Comparison specimens were made from the normal production of normal density concrete (NDC) being used to construct the platform at that time. Some of the properties evaluated are described in Table 7 along with the test methods used for evaluation. All test specimens were 150- by 300-mm cylinders except for the 50-mm cubes.

3 Materials

The materials used in the laboratory batches and the full size production batches were identical and are described in the following paragraphs.

3.1 Cement

The cement is a blended cement produced in Canada which contains approximately 8.5 $\pm$ 1 percent silica fume. The silica fume is blended with a Type 10 portland cement at the cement mill during production. The cement had an specific gravity of 3.15 (avg.), a Blaine specific surface of 554 m^2/kg (avg.) with 93% passing 45 microns and average initial and final setting times (hr:min) of 2:20 and 4:40, respectively. The composition is shown in Table 1.

Table 1 Typical chemical composition of blended silica fume/portland cement (%)

Silica Fume	Silica in Clinker	Silica in Silica Fume	Total SiO_2
8.0	22.4	95.2	26.7

Al_2O_3	Fe_2O_3	CaO	MgO	SO_3	Na_2O (equiv.)	C_3A	LOI
3.9	2.1	59.0	2.8	2.9	0.70	6.9	1.1

3.2 Coarse aggregate

The normal weight (NW) coarse aggregate was produced from a local deposit of granitic and meta-volcanic rock. The material is screened, and oversize rock is crushed and screened with gravel to produce a mix of natural shaped and crushed material possessing excellent mechanical and physical properties [7]. The NW aggregate typically has an absorption of 0.6% and a specific density of 2.63.

The lightweight aggregate (LWA) is a rotary kiln expanded slate produced in the USA and has an apparent specific density of 1.51. The general shape is cubical resulting from a crushing operation. The general pore structure consists of relatively uniform microcells. The aggregate manufacturer provided performance data on the product which included LA abrasion (24%, AASHTO B), aggregate freezing and thawing (0.22%, AASHTO 103) and magnesium sulfate soundness (0.01%, ASTM C88).

Absorption was a primary consideration in aggregate selection for two principal reasons: 1) the offshore structure was being constructed by pumping the concrete and a high workability of the concrete (220 mm slump) was required at the end of the pumpline, and, 2) the completed structure would be exposed to a severe freezing and thawing environment. LWA's with high absorptions would defeat both of these requirements. The aggregate absorption tests indicated a typical absorption (24 hours) of 5.1%, a saturated surface dry (SSD) absorption from 6 to 9.5% at working conditions, and absorptions of 9.4 and 10.7% at pumping pressures of 1 and 2 MPa, respectively. Under consistent ambient temperature and pressure, prewetted crushed expanded slate tested at "SSD" conditions yields different values for absorbed water depending on the soaking time prior to testing. The LWA of this study showed that absorptions in the range of 6 to 9.5% can be achieved by continuous prewetting periods ranging from 48 hours to more than 20 days. The LWA producer showed evidence that adequate pumping behavior was achieved at absorption levels of at least 6%.

To meet the project requirements for moisture control, since the absorbed water tends to be stable under normal ambient conditions, the LWA was prewetted during aggregate production and shipped to the project with an absorption of approximately 8%. Field tests demonstrated that the initial moisture was retained during the transportation and handling of the aggregate.

Because the LWA needed to be of high quality, special efforts were taken in the selection and sizing of both the raw and the expanded slate. The aggregate producer utilized a selective mining process, based on careful mapping of the quarry's physical and chemical characteristics [8]. To assure consistency, the expaanded material was, in turn subjected to comprehensive testing before and after crushing, including physical and chemical analysis. The expansion temperatures during production averaged 1225°C and were consistently kept above 1150°C to minimize potential residual expansion of the aggregate if the concrete, when in service, is ever exposed to a hydrocarbon fire. The fuel for the kiln was coal.

Other considerations in aggregate selection were that the strength and elastic modulus of the aggregate did not significantly reduce the strength and modulus of the MNDC when compared to the NDC.

Gradations for both the NW aggregate and LWA (prior to shipping and from the project stockpile) are shown in Table 2. To minimize degradation of the LWA due to handling, a vessel with its own unloading conveyor was chosen for transportation directly to the project site. The gradations shown in Table 2 indicate acceptable differences as a result of the aggregate handling.

3.3 Fine aggregates

The fine aggregate is a manufactured sand from the same crushing operation that produced the coarse aggregate. The fine aggregate typically has an absorption of 0.7% and a specific density of 2.63. The gradation is shown in Table 2.

Table 2 Aggregate gradations

Sieve opening, mm	Percent Passing,				
			LW Coarse Aggregate		
	NW Fine Aggregate	NW Coarse Aggregate	Prior to Shipping	Stockpile	ASTM C 330 Requirements
20	-----	100	100	100	100
14	-----	99.2	92	98.6	90-100
10	100	72.8	53	55.9	---
5	99.6	14.8	6	3.2	10-50
2.5	86.6	4.3	2	1.3	0-15
1.25	59.4	-----	-----	-----	---
0.63	36.3	-----	-----	-----	---
0.31	17.7	-----	-----	-----	---

3.3 Chemical admixtures

A Type F (ASTM C 494) high-range water reducing admixture (HRWRA) of the sulphonated naphthalene formaldehyde condensate type was used together with a compatible Type D water reducing admixture (ASTM C 494) for both NDC and MNDC. A compatible ASTM C 260 neutralized vinsol resin air entraining admixture (AEA) was used with the MNDC to improve workability.

4 Mixture Preparation

4.1 Laboratory batching

The batching was done in a 0.2 m^3 pan mixer in the site laboratory. The LWA was presoaked to a SSD condition prior to mixing. All of the LWA and the NW aggregates along with 50% of the mixing water were added to the mixer and mixed for 30 seconds. All the cement and the remaining mixing water was then added and mixed for an additional 5 minutes. HRWRA was added during mixing to adjust to the required slump (200 ± 20 mm) without inducing aggregate segregation. The concrete was mixed for an additional 2 minutes after each addition of HRWRA. The mixture proportions for the three LWA replacement mixtures are shown in Table 3.

4.2 Field batching

The field batching was done in the permanent batch plant for the project. Mixing was done in a 2 m^3 twin-axle compulsory mixer. All the aggregates were weighed in separate bins before being discharged into a feeder belt for the mixer. The cement, water and admixtures were added directly into the mixer. A typical mixing time was 35 seconds. The LWA stockpile was pre-wetted to saturation and mixed thoroughly before loading into its designated bin in the batch plant. The mixture proportions for the MNDC field batching are shown in Table 4 along with the proportions of the NDC which was produced to provide comparison data for the MNDC.

Table 3 Laboratory mixture proportions

Mixture	Quantities, kg/m^3		
	LWA-25	LWA-50	LWA-70
Cement	439	442	440
Fine Aggregate	780	785	782
Coarse Aggregate (NW)	783	554	309
Coarse Aggregate (LWA)	156	290	431
Mix Water*	137	137	137
HRWRA, L/m^3	1.3	1.3	1.43
Water/Binder Ratio	0.31	0.31	0.31

* Water added in addition to that contained in HRWRA.

Table 4 Full size batch mixture proportions

Mixture	Quantities, kg/m^3	
	NDC	MNDC
Cement	450	450
Fine Aggregate	830	920
Coarse Aggregate (NW)	910	430
Coarse Aggregate (LWA)	0	255
Mix Water*	152	150
HRWRA, L/m^3	1.3	1.3
WRA, L/m^3	0.20	0.30
AEA, L/m^3	0	0.15
Water/Binder Ratio	0.34	0.33

* Water added in addition to that contained in chemical admixtures.

5 Test Results

The results of the laboratory batching LWA aggregate replacement tests are shown in Table 5. The mixture with 50% LWA replacement (by volume) gave results most closely approximating the project criteria for the MNDC and this percentage of replacement was adopted for the full size field batching tests.

Table 5 Effect of LWA replacement in laboratory batches

LWA Amount, percent	Air Content, percent	Hardened Density kg/m^3	Compressive Strength, MPa	Splitting Tensile Strength, MPa	Modulus of Elasticity, GPa
70	1.6	2120	72.2	5.4	29.8
50	1.8	2224	76.4	5.5	33.2
25	1.8	2313	83.6	6.2	35.2

For the full size field batch reproducibility tests, some slight adjustments were made in the laboratory mixture proportions (Table 3) to accommodate the different batching system and the fact that the final version of MNDC would have to be pumped over very long distances. The overall mortar content was increased (cement plus fine aggregate), the coarse aggregate content decreased, and the water content slightly increased as shown in Table 4. The results from the 32 batches are shown in Table 6.

Table 6 Summary of test results for full size field batches

	Average	Standard Deviation
28-day Compressive Strength, MPa	79.3	2.1
Unhardened Density, kg/m^3	2193	12
Air Content, percent	2.6	0.4
Slump, mm	216	11
Hardened Density, kg/m^3	2216	15

The results indicated that the concrete properties were very consistent and followed the good quality production that had been obtained with the NDC [4]. Proper preparation of the LWA by pre-saturating the aggregate storage piles with sprinkler systems was important as well as careful monitoring of the moisture of the aggregate during production.

For a project strength requirement of 69 MPa, the required produced concrete strength at 28-days age based on project specifications was determined as:

$$f_{cr} = f_c + (1.4 \times \text{standard deviation}) = 69 + (1.4 \times 2.1) = 72 \text{ MPa}$$

As the MNDC produced 79.3 MPa, it was deemed satisfactory for use in the structure. To support that use, other mechanical properties of the MNDC were also measured on concrete from additional full sizes batches of the mixture shown in Table 4. Some of these results are summarized in Table 7 and are compared to results from NDC concrete (Table 4).

Table 7 Comparison of MNDC and NDC mechanical properties

Mechanical Property	MNDC	NDC	Test Method
28-day Cylinder Strength, MPa	79.9	78.2	CSA A23,2-9C
28-day Cube Strength, MPa	92.6	86.0	NS 427A
Cylinder/Cube Ratio	0.86	0.91	
Splitting Tensile Strength, MPa	5.87	5.28	CSA A23.2-13C
Modulus of Elasticity, GPa	30.5	32.8	ASTM C469
Poisson's Ratio	0.22	0.21	ASTM C469
Hardened Density, kg/m^3	2216	2347	ASTM C642
Absorption, %	2.1	--	ASTM C642

As can be seen, the replacement of 50% of the normal weight coarse aggregate with a high quality structural lightweight aggregate had only minor effects, if any, on the mechanical properties. Additional tests on the MNDC such as the determination of stress-strain behavior, fracture energy, water intrusion, pull-out strength, bearing strength, creep and shrinkage are also being conducted and will be reported when the results become available. Initial tests of the freezing and thawing behavior of the MNDC in accordance with ASTM C666, Procedure A, show excellent durability after 300 cycles even with the LWA containing more moisture than the NW aggregate it replaced and having no air entrainment. Additional testing is being conducted.

6 Conclusions

The substitution of high quality structural lightweight aggregate for a portion of the normal weight coarse aggregate in high strength concrete can provide benefits from reduced density without compromising the strength and other mechanical properties of the concrete. The LWA selection is very important as the aggregate must, in itself, be strong and have relatively low water absorption if durability and constructability are also important.

For the LWA used and the replacement range (25 to 70%) examined, an average strength reduction of 0.25 MPa for each percent substitution of LWA was realized. Accompanying reductions in density averaged 4 kg/m^3 for each percent substitution. Other values may result when different LWA's are used. The splitting tensile strength appears to improve slightly when LWA's are used. This is probably the result of an improved transition zone between the cement paste and the more porous LWA which leads to reduced microcracking at the paste/aggregate interface.

Strict quality control is necessary in the LWA production to insure uniform, consistent production of both the raw materials, the aggregate bloating temperatures, and the final sizing and grading of the aggregates. As changes in the gradation of LWA's can occur during handling, caution must be exercised to insure there is no degradation of the aggregate during transportation and stockpiling.

As offshore concrete platforms containing LWA's always have the risk of being exposed to a hydrocarbon fire (peak temperatures of approximately 1100°C), the burning process in the production of LWA must be high enough so that all aggregate expansion that occurs at 1100°C is complete so that no residual expansions will occur which might disrupt the concrete.

7 References

1. Hoff, G.C., Luther, D.C., Woodhead, H.R., Abel, W. and Johnson, R.C. (1994) The Hibernia Platform, Proceedings, 4th (1994) International Offshore and Polar Engineering Conference, Osaka, Japan, April 10-15, 1994.
2. Woodhead, H.R. (1993) Hibernia Offshore Oil Platform, Concrete International, Vol. 15, No. 12, pp 23-30.

3. Hoff, G.C. (1992) Concrete for Offshore Structures, Advances in Concrete Technology,(ed. V.M. Malhotra), Canada Centre for Mineral and Energy Technology, Ottawa, Ontario, Canada, pp 79-121.
4. Hoff, G.C., Walum, R., Elimov, R. and Woodhead, H.R. (1994) Production of High-Strength Concrete for the Hibernia Offshore Concrete Platform, Proceedings, International Conference on High Performance Concrete, Singapore, November 15-18, 1994, (Available as Special Publication SP-149, American Concrete Institute, Detroit, Michigan.)
5. Hoff, G.C. (1992) High Strength Lightweight Aggregate Concrete for Arctic Applications, SP-136, Structural Lightweight Aggregate Concrete Performance,(ed. T.A. Holm and A.M. Vaysburd), American Concrete Institute, Detroit, Michigan, pp 1-245.
6. Malhottra, V.M. (1995) Evaluation of Lightweight Aggregates for Use in Offshore Concrete Structures, (in preparation), Canada Centre for Mineral and Energy Technology, Ottawa, Ontario, Canada.
7. Fournier, B., Malhotra, V.M., Langley, W.S. and Hoff, G.C. (1994) Alkali-Aggregate Reactivity (AAR) Potential of Selected Canadian Aggregates for Use in Offshore Concrete Structures, Proceedings, Third CANMET/ACI International Conference on Durabiity of Concrete, Nice, France, (Available as Special Publication SP-145, American Concrete Institute, Detroit, Michigan.)

135 OVERLAY OF THIN CONCRETE BLOCKS AND ADHESIVE LAYER

M. INUZUKA and K. SASAKI
Hokkaido Institute of Technology, Sapporo, Japan
T. YOSHIDA
Hokkon International Precast Concrete Inc., Sapporo, Japan

Abstract
This research obtained information on precast concrete block road plates which can cover existing roads, usually made of asphalt. The thinner the covers are the better. The roads surface can be subjected to various severe conditions from thermal stresses to mechanical scratching due to anti skidding devices in winter. Mechanical properties in different categories are related to protecting against such conditions. Above all, the bending strength and the abrasion resistance are key to success. Resin concrete blocks may be a solution; however, the question remains how to hold them in position. An adhesive layer is proposed. Dimensions of a block, or mechanical unit, are a main factor in the stability of the cover. Theory and experimental data are presented and discussed: case studies of the overlay in an underpath, a bridge and an ordinary road.
Keywords: Abrasion resistance, adhesive layer, asphalt membrane, bending strength, overlay, resin concrete, road plate, thin concrete block.

1 Introduction

The road surface is subjected to severe conditions such as mechanical and thermal stresses. Pavements consisting of conventional concrete blocks with sand beds underneath have proved to be much stronger than asphaltic pavements in cold regions, especially when these are subjected to the scratching of anti–skidding devices on ice. Thus, the replacement of the whole existing asphalt with conventional blocks is often carried out in the maintenance procedure even where damage is limited only to the surface and the overlay is more than adequate.

The demand for the overlay on asphaltic roads is increasing today. Even a thin

Concrete Under Severe Conditions: Environment and loading (Volume Two) Edited by K. Sakai, N. Banthia and O.E. Gjørv. Published in 1995 by E & FN Spon. ISBN 0 419 19860 1

block cover can considerably mitigate the thermal stress [1]. Stronger materials are used to enhance the durability of the overlay . To choose the material, appropriate information is necessary for the feasible design. This paper proposes an overlay method with resin blocks and adhesive layer. Road plates, or resin concrete blocks with slag aggregates from roll mills, proved in experiments to be appropriate for thin blocks in reference to mechanical properties, including abrasion resistance. The theoretical analysis centered on the relationship between the destructive stresses in materials and the block size under the various conditions to which blocks are subjected, proving that downsizing blocks can reduce the damage taking place in the overlay. Small blocks are, however, difficult to hold at given locations [2]. Therefore, an adhesive layer of bituminous materials should be employed [3]. An optimal combination between properly sized blocks and adhesive may enable the use of a very thin overlay thickness of from 20 to 40 mm. The thickness of the adhesive layer is from 3 to 10 mm depending on the geometrical conditions of the blocks and the road surface [4]. The appropriate combination between the two factors should be based not only on mechanics but work efficiency [5].

2 Outline of road plates, or thin concrete blocks, with adhesive layer

There are two objectives of road plates. One is to resist against severe conditions such as abrasion due to anti-skidding devises. Resin concrete with roll mill slag can be strong and resistive against abrasion. The brand name of an acrylic resin mixture with carbonic calcium and fine aggregates mainly of roll mill slag is Marvelrec. The cost increase of a specific material may be permitted by an enhanced performance. Thus, the choice of materials can vary. It is also preferable to perform quick and simple repairs to damaged areas. Another objective is to reduce the closing period in the maintenance procedure. Resin concrete can exhibit high adhesive strength to itself, which permits the repair work to proceed quickly and simply at the very location where damage has taken place. Precast concrete can eliminate the curing period. In order to reduce the transportation effort, the weight should be reduced by making the road plates thinner. Another point is that the road plates are applied mainly to existing asphalt roads. The alteration of the surface height may cause traffic problems even if the smoothing was successfully carried out. Thus, road plates should be made of acrylic resin and their thickness should be minimized in order to satisfy the above objectives.

The area requiring the cover protection can be limited to that subjected to the abrasion by car wheels. There are two lanes on a path such as in a railway, though these are as wide as 1000 mm. On the other hand, the length of the cover area

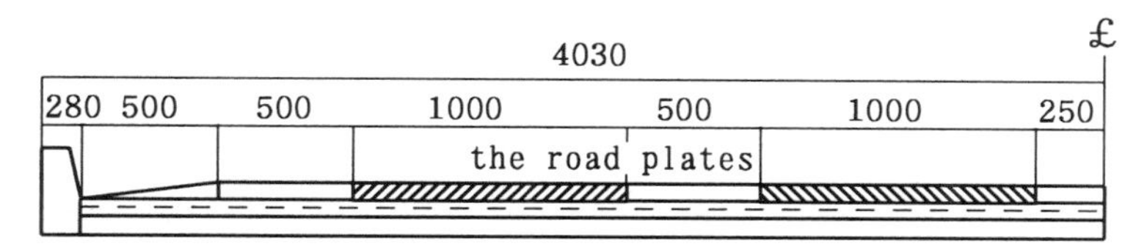

Fig.1 Cross section of road covered with road plates

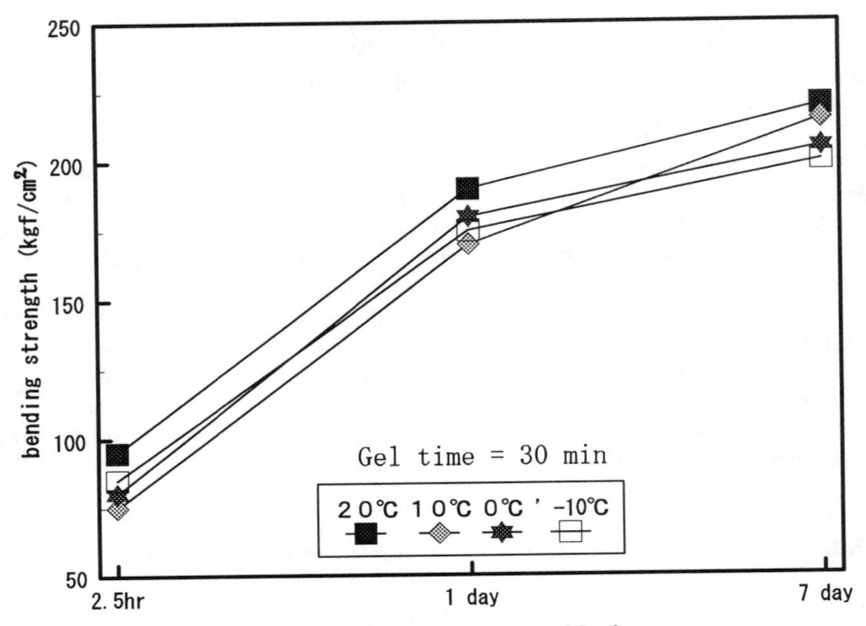

Fig.2 Bending strength of resin concrete with time.

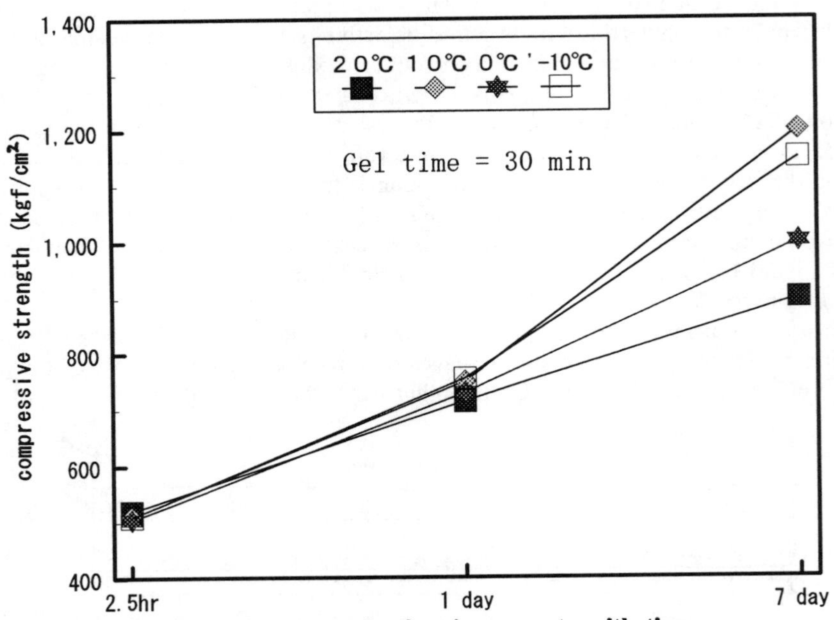

Fig.3 Compressive strength of resin concrete with time.

should be long enough to give sufficient service. It is preferable to enlarge the panel size, since large sized panels are more efficient to process a given area than smaller ones. There is, however, a limit imposed by the available transportation. The panel can be divided into smaller units which are stronger mechanically than the larger ones. The size of the panels should be smaller to prevent adverse fractures. Consequently, the panels were designed to be rectangular and 40 mm in thickness. The cross sectional diagram of a road covered with the panels is shown in Fig.1. An asphalt mixture was used as the adhesive layer for connecting the panels to the existing road. A mobile asphalt cooker was usually employed to spread the adhesive layer on which the panels were laid and pressed by a tire roller.

3 Mechanical properties of Marvelrec, or acrylic resin concrete

Fresh Marvelrec concrete has a low viscosity of 20 cps at 20 ℃ and retains good workability even at sub zero temperatures. The adhesive strength to existing concrete can quickly exceed 23 kg/sq.cm. Thus, damage such as dent or abrasion can be easily repaired by laying it directly on what is to be repaired. The surface of the damaged part ought to be processed since the adhesive strength may not be achieved on wet or dusty surfaces. It can harden within hours, even at −10 ℃. The hardening time can be controlled by adjusting the amount of admixture. A specialized mixer should be better used at the repair work site on account of its short pot life.

The strengths of Marvelrec develop with hours. The bending strength diagram is shown in Fig.2, and the compressive strength in Fig.3. The tests were carried out according to the cement concrete test. Admixtures were used in order to keep the mixture in optimum workability at given temperatures.

The abrasion resistance against studded tires proved to be from three to ten times stronger depending on locations and traffic conditions. The results of the laboratory test in which the wear depth was measured are shown in Fig.4.

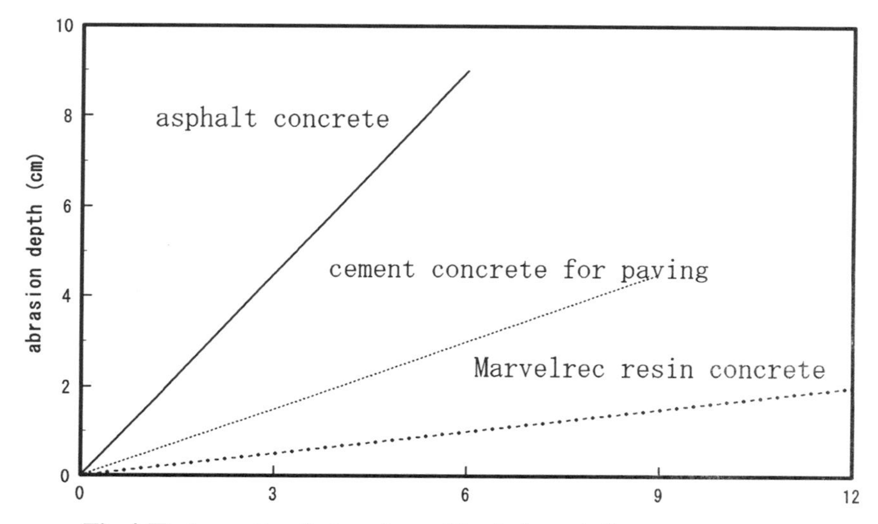

Fig.4 Test results of abrasion with chain rotations

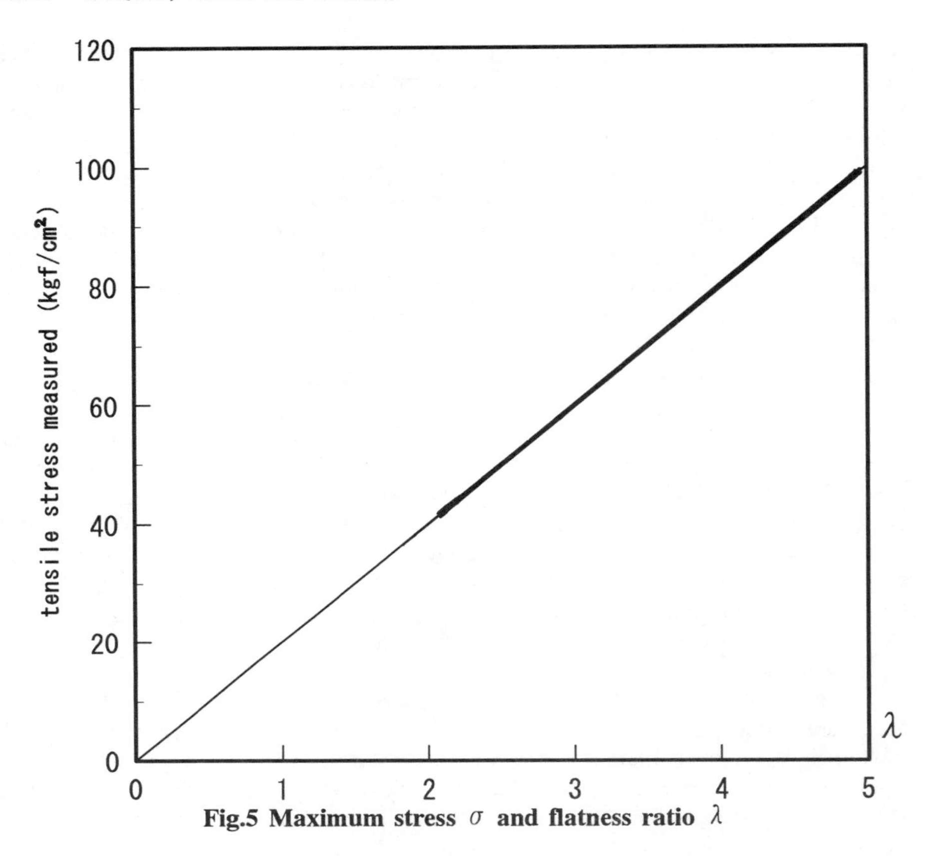

Fig.5 Maximum stress σ and flatness ratio λ

4 Dimensions of mechanical units

Although working units or panels may be enlarged from the view point of work efficiency and speed, the size of mechanical units which keep the original shape can be smaller when the panels are divided into pieces. Each piece works as an independent mechanical unit, and its dimensions are significant. Assuming that one block is supported as a simple beam on two virtual supports and that the vehicle tires will exert a constant pressure q, with the flatness ratio λ and the virtual span divided by the thickness b, the maximum tensile stress σ can be calculated.

When the thin block, adhered to the surface of the model asphalt 400 mm in depth, is subjected to the tire pressure 3 kg/sq.cm , the tensile stress measured rises with the flatness ratio λ, as shown in Fig. 5.

The first crack is easily magnified by repeated loading. Fine grains will enter the crack which is eventually closed by the alternation of loading. The stress encountered is intensified by a longer span. Therefore, the ratio λ should be minimized in order to lower the tensile stress, so far as the other conditions can allow. Although the minimum thickness may depend on the aggregate size, the tensile strength and the underneath properties should be significant factors.

5 Case studies

Characteristic examples are as follows;
5.1.1. A case of an underpath is shown in Fig.6. Because of the steep slope and the heavy traffic controlled by traffic signals, the horizontal thrust due to acceleration caused characteristic deformations in the areas where the adhesive layer was thicker than that originally designed. When some panels are repaired, addition of joint lines to reduce the width seemed to make these more resistive. Since there were relatively many panels repaired when the mechanical units were large, down sizing proved to be a practical means to considerably reduce the repair rate.
5.1.2. A case of a bridge is shown in Fig.7. This case shows the merits of downsized mechanical units. Since the surface was close to the level, the effect from horizontal thrust was not significant. The effects of thermal stress in the quasi–circadian rhythm were found to be specific. One of the cases was subjected to the highest temperatures in summer, which could exude molten asphalt onto the surface of the blocks.
5.1.3. A case of a tunnel is shown in Fig.8. Wheel paths were concentrated on a narrow path. The cover width being the same as that in other cases, the abrasion depth was found to be much more. Road plates are shaded, asphalt exudation from the adhesive layer did not take place. Membrane temperatures are kept cool. Paving speed was vital factor, since no bypass was available for the route. Thus, the use of connected thin blocks which can spare the curing procedure in the site may be advantageous.

Fig.6 Application to underpath

Fig.7 Application to bridge

Fig.8 Application to tunnel

6 Discussion and conclusion

1. The flatness ratio proved a key factor for thin blocks to remain in a satisfactory condition. Mechanical units can be small in a larger working unit by connecting smaller blocks. The repair rate could fall off considerably with the flatness ratio reduced in an original block. Thinning blocks can be achieved by connecting small blocks into a large panel. Besides small blocks enable easy adjustments to irregular shapes for paving on work site. However, when the abrasion exceeds a certain level, the increase in the rate of fractured blocks tends to be accelerated. The size of aggregates seemed to determine the level of other factors such as an underneath condition.
2. Asphaltic adhesive materials proved satisfactory from the view point of mechanics. The thicker the blocks, the longer it took for molten asphalt to ooze out of the adhesive layer in summer. This occurred because the volume ratio of asphalt was higher with the thin block covers than with the thicker ones. Thinner blocks, however, could be stabilized by a thinner adhesive layer. The required strength for the adhesive material was about 1 kg/sq. cm which is relatively low. Better adhesive means still remain to be found.
3. When each block is small and thin, repairing can be systematic and simple. The word , "Severe, " may imply in engineering, "Unpredictable" or "in a wider range of possibility." If so, the design philosophy should be different from the conventional one. Repairing at the minimum cost should be taken into account in well prepared study.

7 References

1. Inuzuka, M. and Sato, I. (1984) Thin Concrete Block Overlay, *Prc. of Paving in Cold Areas, Canada/Japan Science and Technology Consultations, Technical Memorandum of PWRI No.2136,* Ministry of Construction, pp.843–62.
2. Inuzuka, M. and Sato, I. (1984) Concrete Block Pavement on Asphalt Highways, *Prc. of Paving of 2nd Int. Conf. of Concrete Blocks*, Delft University Press, pp.322–29.
3. Inuzuka, M. and Sato, I. (1988) Adhesive Layer for Overlay with Thin Concrete Blocks, *Prc. of Paving of 3rd Int. Conf. of Concrete Blocks*, PAVITAILIA, pp.353–60
4. Inuzuka, M. and Yoshida, T. (1994) Thin Concrete Blocks with Adhesive Layer, *Prc. of 2nd Int. Workshop of Concrete Blocks*, BLF Norway, pp.299–305.
5. Yoshida, T. and Nagata, S. (1988) Precast Paving Plate, Production Process and Pavement, JP Patent 206501

PART TWENTY-THREE
INTERFACE: DURABILITY AND MECHANICS

136 CHLORIDE PERMEABILITY OF CONCRETE UNDER COMPRESSIVE STRESS

T. SUGIYAMA and Y. TSUJI
Gunma University, Kiryu, Japan
T. W. BREMNER
University of New Brunswick, Fredericton, Canada
T.A. HOLM
Solite Corporation, Richmond, USA

Abstract
The effect of compressive stress on the chloride permeability of concrete was experimentally investigated. A uniaxial compressive load was sustained on cylindrical hollow concrete at a given stress level, and the amount of chloride ions passing through the thick wall of the specimen was regularly measured until the flow rate of chloride ions became constant. No significant increase in the diffusion coefficient of chloride ions was found at stress levels below or equal to 60 percent of the ultimate strength. However, the chloride diffusion coefficient increased noticeably when the stress level reached 65 percent of the ultimate strength. In addition, the flow of electric current in concrete increased significantly when subjected to a compressive load above the static endurance level. Lightweight concrete was tested and compared with normal weight concrete. A negligible difference in the chloride permeability was found between lightweight and normal weight concrete with an equal water-to-cement ratio.
Keywords: Concrete, chloride permeability, diffusion coefficient of chloride ions, compressive loads, electrical methods, lightweight concrete.

1 Introduction

The corrosion of steel in concrete is of concern as it is the most important factor influencing the deterioration process of reinforced concrete structures in marine and other salt-laden environments. The transport mechanism of externally applied chloride ions into hardened concrete deserves extensive

Concrete Under Severe Conditions: Environment and loading (Volume Two) Edited by K. Sakai, N. Banthia and O.E. Gjørv. Published in 1995 by E & FN Spon. ISBN 0 419 19860 1

study as the mobility of chloride ions through concrete cover influences both the time to corrosion and the corrosion rate of steel reinforcement embedded in the concrete. Not all chloride ions present in concrete are responsible for causing the breakdown of the protective oxide layer on the surface of steel bars and hence causing corrosion of the steel. When chloride ions move through pores or cavities, a complex interaction takes place with the cement paste. This results in immobilizing some chloride ions. These bound chloride ions are considered not to be involved in the corrosion of steel in concrete. Only free chloride ions which can reach and accumulate in the interfacial area of reinforcing steel will destroy the passive nature of the reinforcement and consequently cause its corrosion. Of all aspects influencing chloride transport in concrete, the internal pore structure is thought to be the most important. Also, when microcracks are present in concrete, these flaws could provide a favorable passage for the flow of chloride ions [1]. Microcracks are caused by structural loads, drying shrinkage and freeze and thaw action. Among these actions, the compressive stress and microcracks caused by stress have been the subject of recent research [2, 3, 4, 5].

This study aims to clarify the effect of microcracks on chloride permeability in concrete. In order to induce microcracks in cylindrical hollow concrete under test, a sustained compressive load was applied, during which the actual amount of chloride ions passing through the concrete wall was measured. This was done at different stress levels and the stress level at which a significant increase of the chloride flow rate occurred was determined. In addition, concrete specimens subjected to one cycle compressive load to failure were also tested, and the flow of electrical current with increases in compressive load was monitored. By relating increasing compressive stress level to the occurrence and subsequent development of microcracks in concrete, the effect of compressive stress on chloride permeability was evaluated.

2 Experimental details

2.1 Specimen preparation

Mix proportions of the concrete studied and the properties of fresh and hardened concrete are shown in Table 1. Normal weight and lightweight concrete with water–to–cement ratios of 0.4, 0.5, and 0.6 were made (hereinafter abbreviated as NW04, NW05, and NW06 for normal weight concrete and LW04, LW05, and LW06 for lightweight concrete). Normal Portland cement was used (CSA A–5 Type 10). An air–entraining admixture was used for all mixing to have an air content of the fresh concrete ranging from 5 to 7 percent. Natural gravel with a specific gravity of 2.68, maximum size of 12.5 mm and absorption of 1.02 % and an expanded shale with a specific gravity of 1.52, maximum size of 12.5 mm and absorption of 12.0 % were used as coarse aggregate for normal weight concrete and lightweight

Table 1. Mix proportions and properties of fresh and hardened concrete.

	Mix proportions								Fresh concrete			Strength
Mix No.	W/C[#1]	C (kg/m³)	W	F.A.[#2]	C.A.[#3]	S[#4] (%)	AE[#5] (mL/m³)	HRWR[#6]	slump (mm)	air (%)	unit wt. (kg/m³)	28days[#7] (MPa)
NW04	0.4	460	184	745	888	46	234	534	65	7.0	2298	42.4(0.40)
LW04	0.4	460	184	745	504	46	237	855	115	6.5	1837	35.7(1.37)
NW05	0.5	368	184	823	888	48	170	690	100	7.0	2184	35.1(1.27)
LW05	0.5	368	184	823	504	48	160	705	80	5.5	1908	27.4(1.27)
NW06	0.6	305	184	876	888	50	50	0	75	7.0	2314	27.0(0.23)
LW06	0.6	305	184	876	504	50	50	0	75	6.0	1934	24.7(0.60)

#1: water-to-cement ratio by weight, #2: fine aggregate, #3: coarse aggregate, #4: fine-coarse aggregate ratio by volume, #5: air entrainment admixture, #6: Superplastisizer, #7: 28days compressive strength (standard deviation)

concrete, respectively. Natural sand with a specific gravity of 2.68, absorption of 1.14 % and fineness modulus of 2.71 was used as fine aggregate for both mixes of normal weight and lightweight concretes. A similar particle-size distribution was maintained between the gravel and lightweight aggregates. The lightweight aggregate was presoaked 24 hours prior to mixing. All water-to-cement ratios were calculated on the basis of the aggregate being saturated surface dry after 24 hours submersion in water. The volume of coarse aggregate was maintained constant for all mix designs, and thereby, the volume ratio of coarse aggregate to mortar matrix in hardened concrete was assumed to be identical. A high range water reducer (naphthalene sulphonate based superplasticizer) was added in mixing, if necessary, to achieve adequate workability.

Cylindrical hollow concrete specimens were cast with outside and inside diameters of 150 mm and 75 mm, respectively. The wall thickness was then 37.5 mm. The specimens were demoulded within 24 hours and subsequently stored in a moist curing room (humidity of over 95 % and temperature of about 25 °C) for three to four months. Then approximately 20 mm from the top of each specimen was sawed off with a diamond saw. Also the bottom was cut off so that the total height of specimens was 230 mm. For capping the bottom surface of specimens and also for sealing the center hole (see Fig. 1), an epoxy adhesive was used. This epoxy ensured no leakage of the inside solution. The top surface was capped with an ordinary sulphur capping compound. After completing the capping, all concrete specimens were again stored in the curing room until tested. All concrete specimens were tested in a moist condition.

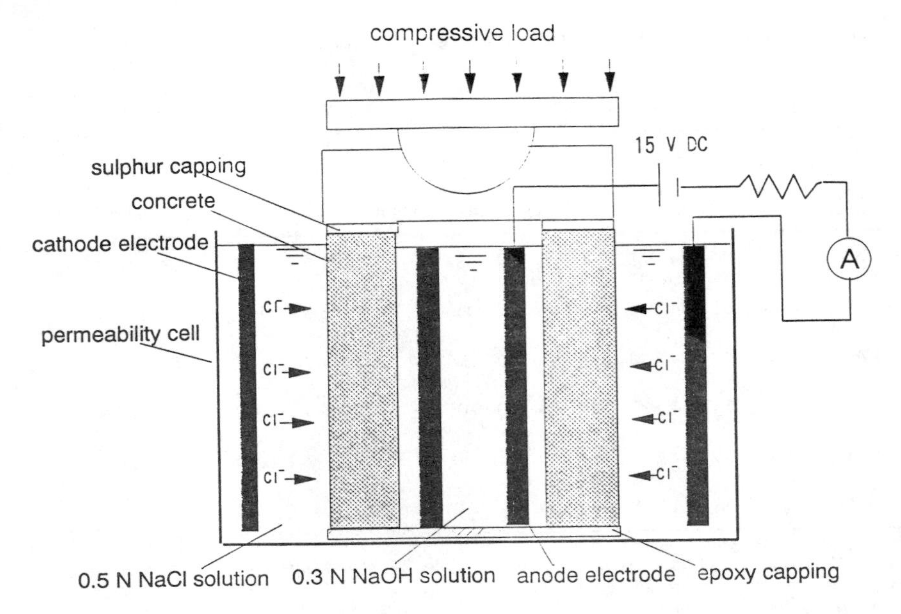

Fig. 1. Chloride permeability test arrangement.

2.2 Chloride permeability test with an electrical potential difference
The chloride permeability of concrete under stress was determined using two test methods : (1) by measuring the amount of chloride ions migrating through stressed concrete and (2) by monitoring electrical current flow in stressed concrete. The two tests were independently conducted, and the effect of compressive stress on chloride permeability was studied in each testing method.

2.2.1 Measurement of the amount of chloride ions passing through stressed concrete
The chloride permeability of concrete under compressive stress was determined by measuring the actual amount of chloride ions which passed through the full wall thickness of the cylindrical hollow concrete specimen. The arrangement for the chloride permeability test is schematically depicted in Fig. 1. An electrical potential difference was maintained at 15 volts between the outer and inner curved surface of the hollow concrete specimen. The applied voltage of 15 V was selected because the temperature increase due to the application of this voltage was found to be negligible. A stainless steel anode electrode was placed inside the hollow cylinder core. The core was filled with a 0.3 N sodium hydroxide solution. The alkali solution was used to simulate a pore solution of concrete. Also, a preliminary test showed that the NaOH solution kept the anode electrode from dissolving during the electrical test. A cathode electrode made with stainless steel mesh was wrapped around the outside surface of the cylinder, which was exposed to

a 0.5 N sodium chloride solution. Under this electrical field, chloride ions having a negative charge would migrate through the concrete wall from the outer surface towards the inner anode electrode. After a certain time had elapsed (transition time), migrating chloride ions would accumulate in the NaOH solution. The concentration of chloride ions in the neutral solution was measured at 12 hour intervals. This procedure permitted the determination of a constant flow rate of chloride ions through concrete under test. This flow rate was then used to calculate the chloride diffusion coefficient. To calculate the diffusion coefficient, the following equation was used [6].

$$D = V_2 \frac{dC_2}{dt} \frac{\ln(r_1/r_2)}{2\pi h \frac{zFV}{RT} C_1} \qquad (1)$$

where D is the diffusion coefficient of chloride ions (cm^2/s), $V_2(dC_2/dt)$ is the flow rate of chloride ions (mol/s) which is obtained from the test, z is the valency of chloride ions, F is the Faraday constant, R is the gas constant, T is the absolute temperature, V is the applied voltage (15 V), C_1 is the concentration of chloride ions in the NaCl solution (maintained constant through the test at 0.5 N), r_1 and r_2 are the outside and inside radius of the hollow cylinder specimen (75 mm and 37.5 mm, respectively), and h is the effective height of the specimen.

Chemical analysis to determine the chloride content in the NaOH solution was conducted with the mercuric nitrate titration method. It was expected that it would take some days to attain a constant flow rate of chloride ions. Thus, compressive loading at a given stress level was sustained on the specimen under test until the flow of chloride ions became constant. The applied compressive stress levels were selected at 20, 40, 60, 62.5, 65, and 70 per cent of the ultimate strength. In order to calculate each applied stress level, the ultimate strength of each concrete mix was measured by an ordinary compression test method using companion specimens. The chloride flow rate was measured at these increasing stress levels, and therefore, chloride diffusion coefficients were determined at each stress level. In this study, concrete mixes of NW04, LW04, NW06 and LW06 were tested.

2.2.2 Monitoring the electrical current flow through stressed concrete

A uniaxial compressive load was applied to a cylindrical hollow concrete specimen and increased at a loading rate of 4410 N/min. until the specimen failed. During loading, the flow of electrical current in the concrete was continuously measured. The stress level was expressed as a percentage of the maximum strength recorded in each test. In this study the testing arrangement described in the previous section (see Fig. 1) was used with an applied voltage of 15 V, and the current flowing between the outside and inside surface of the hollow specimen was measured. In this way, it was expected that an increase in the electrical current would occur if sufficient cracks in the concrete developed due to the applied compressive load. For this study, specimens of mix designs NW05 and LW05 were tested.

3 Results and discussion

3.1 Diffusion coefficient of chloride ions and stress level

Figures 2 and 3 show an increase in the chloride content in NaOH solution with elapsed time. These concrete specimens were subjected to sustained

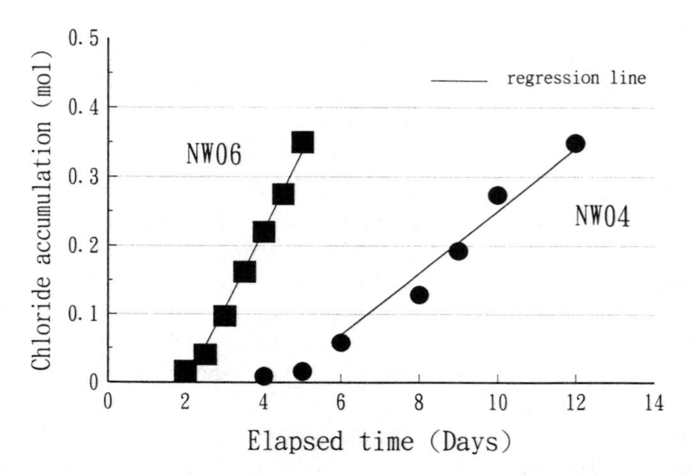

Fig. 2. Increase of chloride ions in NaOH solution at 20 % stress level : NW04 and NW06.

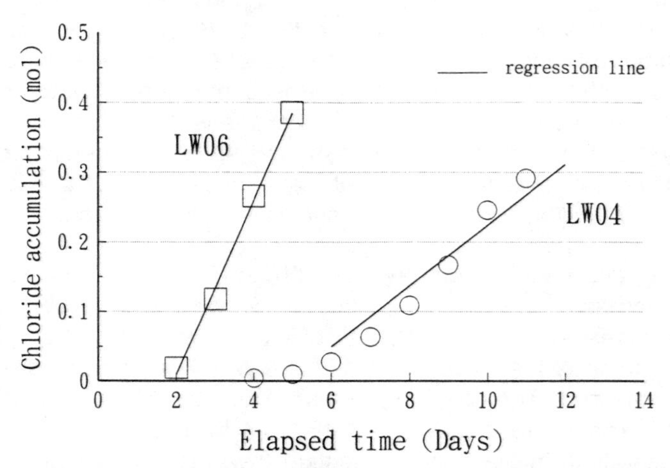

Fig. 3. Increase of chloride ions in NaOH solution at 20 % stress level : LW04 and LW06.

Table 2. The constant flow rate (J) and the diffusion coefficient of chloride ions (D) with increasing stress levels.

	NW04		LW04		NW06		LW06	
Stress level(%)	J $\times 10^{-7}$ (mol/s)	D $\times 10^{-9}$ (cm^2/s)	J $\times 10^{-7}$ (mol/s)	D $\times 10^{-9}$ (cm^2/s)	J $\times 10^{-7}$ (mol/s)	D $\times 10^{-9}$ (cm^2/s)	J $\times 10^{-7}$ (mol/s)	D $\times 10^{-9}$ (cm^2/s)
20	6.54	10.6	7.27	11.8	14.2	23.1	15.6	25.3
40	4.65	7.60	4.90	7.95	14.7	23.9	16.8	27.3
60[#1]	5.22	8.48	5.18	8.41	15.2	24.7	15.7	25.5
62.5	11.2	18.2	6.66	10.8	---	---	---	---
65[#1]	11.2	18.2	7.89	12.8	24.2	39.3	21.0	34.1
70[#1]	F[#2]		F		F		F	

#1: the entry is an average of two duplicated tests, #2: F means failure of specimen before measurable chloride accumulation was obtained.

compressive stress at 20 percent of the ultimate strength. It took 6 to 7 days (transition time) for the rate of chloride increase to become constant for NW04 and LW04. For NW06 and LW06 the transition time was about 2 days. On the basis of the relationship between the chloride accumulation and elapsed time a linear regression analysis was conducted to determine the constant flow rate of chloride ions. The slope of the linear line was taken as the constant flow rate. The chloride flow rate permitted the diffusion coefficients of chloride ions to be calculated using Eq. (1). The constant flow rates and diffusion coefficients of chloride ions determined at the stress level of 20 percent of the ultimate strength are listed in Table 2. The constant flow rates of chloride ions for NW06 and LW06 were higher than those of NW04 and LW04, respectively, by a factor of about 2.2, and therefore, so were the diffusion coefficients of the chloride ions. Also, the diffusion coefficients of chloride ions determined in this test lay within a range of comparable values to others [7]. There was little difference in the chloride diffusion coefficients of lightweight concrete and normal weight concrete with an equal water-to-cement ratio. This indicates a negligible effect of the porous aggregate on the transport of chloride ions in lightweight concrete at this stress level. It was the mortar matrix that controls the chloride permeability of concrete. The diffusion coefficients of chloride ions were calculated in the said manner for concrete specimens subjected to sustained compressive stress at stress levels of 40, 60, 62.5 and 65 percent of the ultimate strength. These results are shown in Table 2. All concrete specimens tested at 70 percent of the ultimate strength failed before measurable chloride accumulation. This

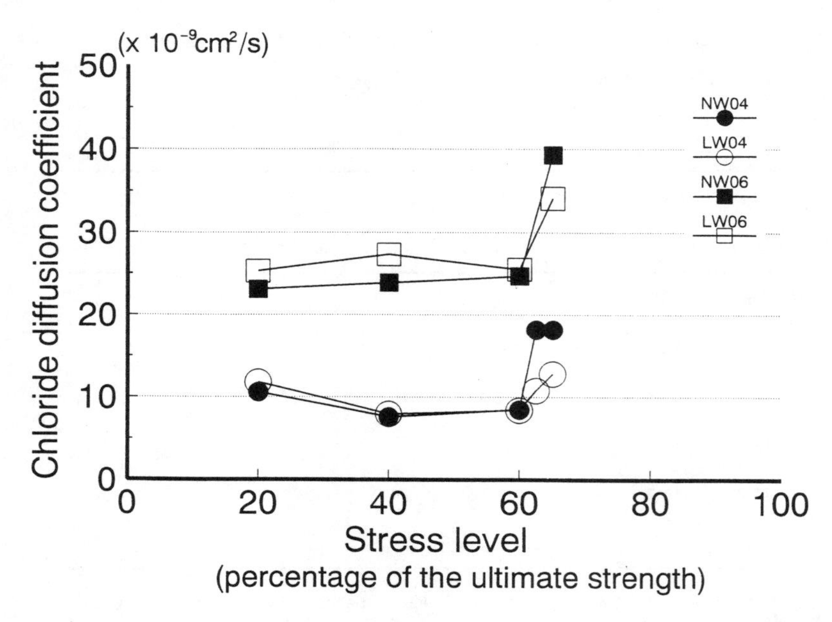

Fig. 4. Effect of compressive stress on chloride permeability.

indicates that the sustained strength of all concrete mixes is about 70 percent of the ultimate strength under this particular testing condition where the concrete was submerged in solutions. Figure 4 shows the effect of stress on chloride permeability. Chloride permeability was found to be little increased by an increase in stress below or equal to 60 percent of the ultimate strength for all types of concrete mixes. This means that the chloride permeability was insensitive to these load levels. Thus, compressive loads at these low to medium stress levels had no significant effect on chloride transport through concrete. In addition, the chloride permeability of lightweight concrete was found to be comparable to that of normal weight concrete with an equal water–to–cement ratio at stress levels below or equal to 60 percent of the ultimate strength. A significant increase in chloride permeability was found at a stress level of 65 per cent of the ultimate strength for all concrete mixes. This is believed to be due to significant development of cracks caused by the sustained compressive load.

3.2 Electrical current and stress level

The flow of electrical current with an continuous increase in the compressive load is shown in Fig. 5. It was found that the current flow through concrete subjected to one cycle uniaxial compressive load was constant until the stress level reached near 90 percent of the maximum strength. In addition, the electrical current for LW05 was slightly higher than that for NW05 within these stress levels. At stress levels in excess of about 90 percent of the ultimate strength the electrical current flow started to increase. Subsequently,

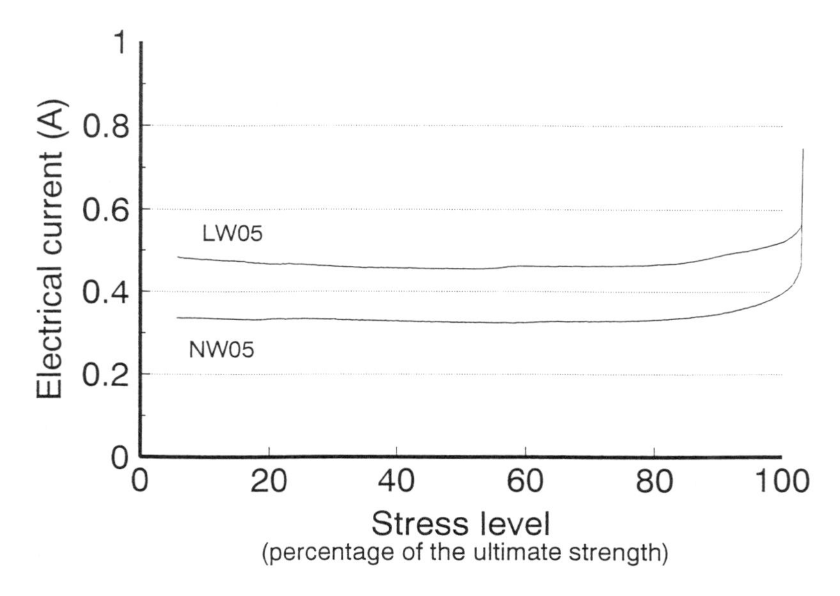

Fig. 5. The flow of electrical current with an increase in compressive stress.

rapid increase of the current occurred at the same time as the specimen was failing. According to generally accepted hypothesis, uniaxial compressive load causes bond cracks and, subsequently, mortar cracks with increases in the stress level. Thus, this study shows that these microcracks, which are assumed to develop in concrete under test, had negligible influence on the flow of current until the applied compressive load caused the collapse of the specimen. However, in the sustained loading condition, a rapid increase in the chloride diffusion coefficient of concrete occurred at 65 percent of the ultimate strength (see Fig. 4). On the basis of this, it appears that a stress level of about 90 percent of the ultimate strength in the standard method of compression testing has the same effect on chloride permeability as a stress level of 65 percent in the sustained load test. In the sustained load test carried out at 70 percent of the ultimate strength, all specimens failed before a measurable concentration of chloride ions could be obtained, indicating that at the stress level of 65 percent of the ultimate strength, failure also was imminent. Consequently, the electrical current and the diffusion coefficient of chloride ions are little influenced by the applied compressive stress unless the stress level is at or near the failure stress.

4 Conclusions

The effect of compressive stress on chloride permeability was studied by two test methods in which the diffusion coefficient of chloride ions was determined in concrete subjected to sustained compressive load on one hand and electrical current was monitored in concrete subjected to the one cycle compressive loading to failure on the other hand. Based on these investigations the following conclusions have been drawn.

1. The diffusion coefficient of chloride ions was constant at stress levels below or equal to 60 percent of the ultimate strength for all types of concrete under a sustained loading condition. However, it increased noticeably at 65 percent of the ultimate strength.
2. Change in the flow of electrical current in concrete subjected to one cycle compressive loading to failure was too small to be significant until the stress reached about 90 percent of the ultimate compressive load.
3. The chloride permeability test developed for this work was effective for calculating the diffusion coefficient of chloride ions when an electrical potential of 15 volts was used.
4. The chloride diffusion coefficients of lightweight concrete were similar to those of normal weight concrete with equal water-to-cement ratios although the electrical current flow through lightweight concrete was slightly higher than that of equivalent normal weight concrete.

5 References

1. Mehta, P.K. and Gerwick, Jr.B.C. (1982) Cracking–Corrosion interaction in concrete exposed to marine environment, *Concrete International*, Vol. 4, No. 10, pp. 45–51.
2. Samaha, H.R. and Hover, K.C. (1992) Influence of microcracking on the mass transport properties of concrete, *American Concrete Institute Materials Journal*, July–August, pp. 416–24.
3. Locoge, P., Massat, M., Olliver, J.P. and Richet, C. (1992) Ion diffusion in microcracked concrete, *Cement and Concrete Research*, Vol. 22, pp. 431–38.
4. Saito, M. and Ishimori, H. (1993) Mobility of chloride ions in concrete subjected to static and pulsating compressive loadings, *Proceedings of the 48th annual conference of the Japan Society for Civil Engineers*, V, pp. 170–71 (in Japanese).
5. Sugiyama, T., Bremner, T.W. and Holm, T.A. (1993) Effect of stress on chloride permeability of concrete, *Durability of Building Materials and Components 6*, E & FN Spon, London, pp. 239–48.
6. Sugiyama, T. (1994) *Permeability of stressed concrete*, Ph.D. thesis University of New Brunswick, Canada, pp. 201–04.
7. The Concrete Society. (1987), *Permeability testing of site concrete*, Technical Report, No. 31, pp. 75.

137 THE EFFECTS OF STRESS ON CHLORIDE PENETRATION INTO CONCRETE

H. HEIDEMAN
J & W Bygg & Anlåggning AB, Örebro, Sweden
H. SUNDSTROM
SKANSKA AB, Gothenburg, Sweden

Abstract
In this study, the penetration depth of chlorides was investigated in three concrete qualities subjected to compressive stress and flexural tension. The loading levels were chosen as 50% and 75% of the ultimate characteristic compressive strength and calculated flexural tension. Two methods were used; one method was the immersion of concrete specimens, cubes and beams, for 30 days in 3% chloride solution and the second was an electrical method, applied for the first time on loaded specimens. An electrical field was applied to provide a rapid penetration of chlorides into the concrete. From the results, it appears that the compression stress decreases the chloride penetration, whereas the flexural tension increases the chloride penetration. The method of applying the load is very important when studying the influence of load on chloride penetration in concrete structures.
Keywords: Chloride penetration, compressive stress, concrete, flexural tension, immersion method, electrical method.

1 Introduction

A bridge is planned to be built between Denmark and Sweden across the Sound. It is of interest to predict its service life, which should exceed 100 years. Therefore, one of the aims is to use a high quality concrete to slow down chloride intrusion. One of the purpose of this work was to simulate the conditions such a bridge will be exposed to.

The chloride intrusion into concrete intended for the Sound bridge was compared with two reference concretes. A bridge column has a great dead weight, and therefore, the concrete was subjected to compression stress. The columns are surrounded by sea water in and under the splashing zone, which was simulated in this experiment.

The chloride penetration depth has previously been studied on non-loaded specimens in laboratory and field tests. In this diploma work, the diffusion depth of chlorides into

Concrete Under Severe Conditions: Environment and loading (Volume Two) Edited by K. Sakai, N. Banthia and O.E. Gjørv. Published in 1995 by E & FN Spon. ISBN 0 419 19860 1

concrete under tensile and compressive stress was investigated. Micro cracks opened by the stress might influence the chloride penetration.

The scope of this analysis is to study the effects of chloride penetration in concrete under stress. Different concrete qualities as well as different stress levels were examined concerning chloride concentration. The work was carried out at the Department of Building Materials, Chalmers University of Technology, Gothenburg [1].

2 Experimental

2.1 Concrete qualities

In this diploma work, three different concrete qualities were analysed. The concretes were chosen to be representative for various types of structures. In normal building structures, concrete with a water cement ratio of 0.65 is common, whereas in bridge structures a W/C of 0.45 is used. However, for future bridges, such as in the Sound bridge, the intention is to use a high quality concrete. High strength concrete is very dense with low permeability and diffusivity. This is achieved by adding micro silica (S), reducing the water content and using superplasticisers. A water-binder ratio (water to cement plus micro silica), W/(C+S) of 0.35, was used.

2.2 Chloride solution

To simulate sea water conditions around southern Sweden, a chloride solution of 3% NaCl was used. The chloride used was NaCl mixed with standard tap water. A few grams of CaO was also added to the water to prevent CaOH leaching from the concrete. To render a more distinguished chloride penetration front, a 20% NaCl solution was used for the CTH rapid method [2].

In order to keep the chloride concentration constant, the solution was changed once a week.

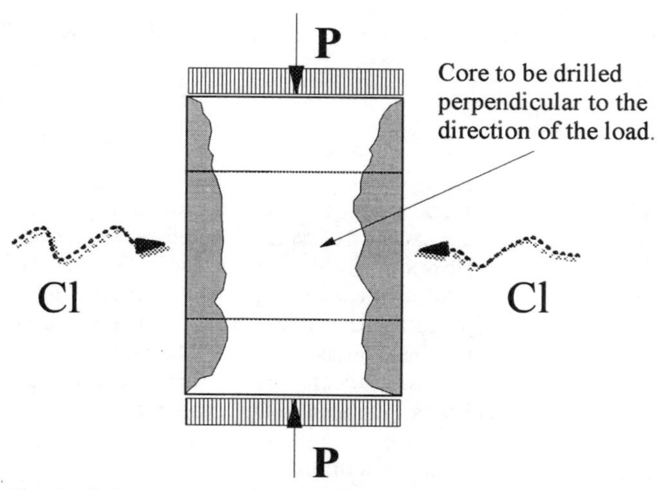

Fig. 1. Schematic 2D-view of chloride penetration to be studied. The cylindrical bore core is shown with dashed lines.

2.3 Design of test specimens

For the immersion tests in the compression analysis, 100×100×100 mm³ cubes were chosen. To avoid wall effects, it was necessary to cut off the surfaces in the penetration direction. Close to the mould surface, the proportion of aggregate is less than inside the material. This is called the wall effect. The increased concentration of cement paste in the surface layer increases the amount of chlorides.

The influence of two-dimensional chloride penetration close to the cube edges, was possible to avoid by drilling the core from the centre of the penetration surface. For the electrical method in the compression analysis, the cubes used were similar to the ones used for the immersion test, only without any surfaces cut off. It was believed unnecessary since the penetration depth was believed to be deeper than the "wall".

Swedish standard moulds for beams, 100×150×800 mm³, were used for the flexural tension capacity evaluation with the immersion method. The beams were reinforced by two 6 mm Ks 600 bars, a ribbed bar with a yield limit of 600 MPa, placed on the tension side. This was necessary to avoid any sudden collapse of the beams, which can occur if no reinforcement is used.

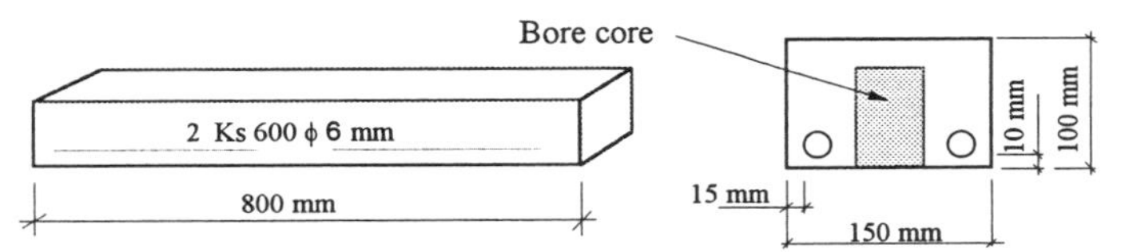

Fig. 2. Flexural tension beams, reinforced with two 6 mm Ks 600 bars. The cylindrical bore core is indicated as a shaded area with dashed borders.

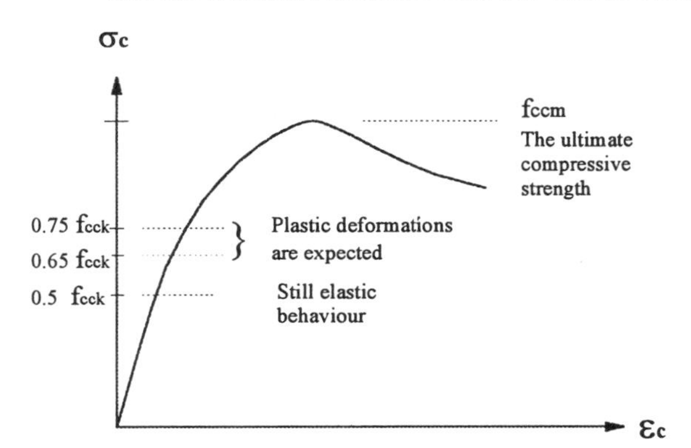

Fig. 3. Schematic stress - strain relationship for concrete. The chosen levels of stress are indicated as portions of f_{cck}.

2.4 Stress levels

2.4.1 Compression analysis

It has been shown, [3], that loads giving compressive stresses larger than 60% of the ultimate short term strength increase the chloride penetration. Hence, 50% and 75% of the ultimate *characteristic* compression strength, f_{cck}, were chosen for all three types of concrete. It was anticipated that at 0.5 f_{cck} the deformations of the concrete would be elastic, whereas at 0.75 f_{cck} some plastic deformations could be expected to have occurred.

Due to a sudden collapse during loading of a cube with W/C 0.65, it was decided to lower the stress for this concrete to 0.65 f_{cck}. This was done as it was suspected that this concrete would be more sensitive than the other two. The collapse showed a clear hourglass shape, thereby indicating that the eccentricities were small.

2.4.2 Flexural tension analysis

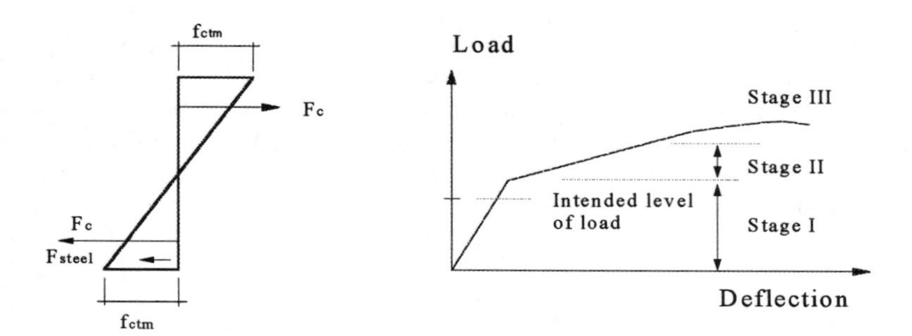

Fig. 4. Schematic strain- and load-displacement graphs for reinforced concrete beams.

It was decided to choose a tension - stress ratio in stage I, where it was assumed that the concrete had not cracked. When the concrete cracks, it can no longer carry any tensile forces. Thus, in order to study chloride penetration under tension stress, stage I was chosen.

In order to have knowledge of the actual stress level, a load-deflection graph for each concrete beam was evaluated by separate load tests. A mechanical universal test machine was used. The deflection was evaluated by measurements from three displacement gauges, one at each support and one in the centre of each beam.

2.5 Design of loading equipment

A method that compensated for the creep in the concrete had to be used to measure the load. This was done differently for compression and tension analysis.

2.5.1 Compression analysis

To apply the load, hydraulic jacks together with the concrete specimen were placed in steel rigs. The main problem when loading was the existence of eccentricities. Since the load was rather large, even very small eccentricities could have great effects. The best way to avoid this would be to have a spherical laurel placed between the jack and the plate on top of the cube. The chosen concretes had different possibilities of deforming and, therefore, also of levelling out the load. During the experiment, the pressure gauges were read and the loads adjusted every day.

2.5.2 Flexural tension analysis

By using lever beam equipment with lead weights, constancy of the load was ensured. The method is very simple, has very few moving parts and is fairly accurate.

2.6 Chloride penetration analysis

After five weeks of immersion exposure, samples were taken from different depths beneath the exposed surfaces. The chloride content of these samples would give a "chloride profile", i.e., a chloride content distribution. The profiles were expected to be deeper when the W/C ratio was higher.

From these profiles, it was possible to evaluate a diffusion coefficient, D, for the immersion method and the electrical method, assuming that Fick's second law is applicable [4].

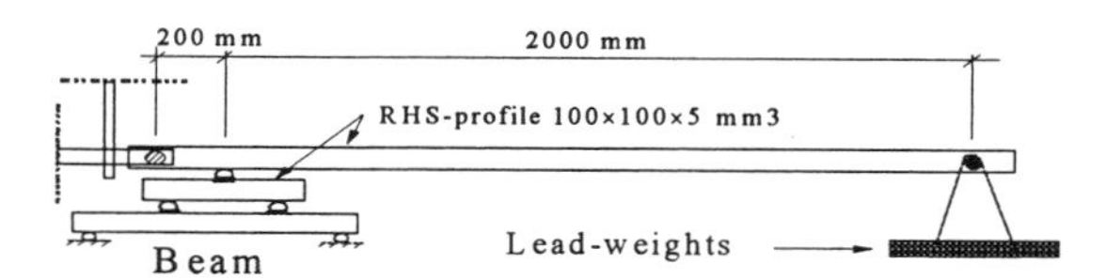

Fig. 5. Lever beam, outline of the loading equipment used for tension stress.

For the electrical method, two different solutions were put in the cups fastened on opposite sides of the cube with a thickness of 100 mm. A 20% chloride solution was used as the cathodic solution. The anodic solution was deionized water with a small addition of CaO. A copper plate, 20×30 mm^2 ,was placed in each cup as an electrode. The concrete was exposed to 60 volts potential difference for 24 hours.

To obtain a good picture of the chloride penetration, each specimen was dry sawed in the direction of penetration. The cut surfaces were sprayed with $AgNO_3$, which reacts with the chlorides. The area where chlorides were present was thereby coloured in a different grey scale, clearly displaying the penetration front. The average penetration depth was documented, and each surface was indexed and photographed. To investigate the penetration variation along the beam, each beam was dry sawed in several sections. This was of interest since the load varies along the beam.

To determine the chloride profile, a cylindrical core was drilled from each specimen. When a core is drilled, the cooling water washes away some of the chlorides from the surface layer. The magnitude of this loss is not known, thereby, rendering the surface [Cl^-] content uncertain. To obtain the chloride profile, a lathe with a diamond saw was used. The core was fixed in the lathe, and dust was collected from the layers at 1, 3, 6, 10, 15 and 20 mm. After cutting from one end of the core, the core was turned and the other end was cut. Potentiometric titration was used to determine acid soluble chlorides, which includes both the free and the bound chloride ions, since the concentrated nitric acid breaks all bonds and liberates all chloride ions.

The chloride content in each sample was calculated as follows:

$$[Cl^-] = \frac{M_{Cl} \times (V - V_0) \times N}{10 \times W} \times 100 \qquad [\%] \qquad (1)$$

Where: [Cl^-]= the concentration of total dry mass [%], M_{Cl}=35.45 is the molar weight of chlorides [g/mole], V=volume of titrant at endpoint [ml], V_0=volume of titrant consumed in blank running (V_0=0.16 [ml]), N=normality of titrant (N=0.01 N) and W=weight of the sample [g].

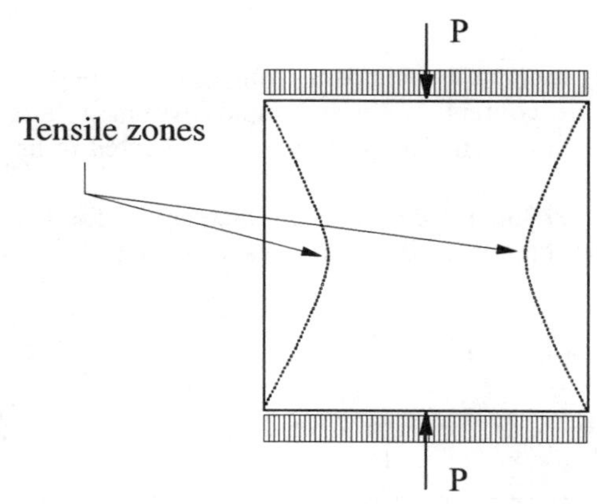

Fig. 6. Schematic view of the tensile zone, present in the cube due to the compressive load.

3 Results and discussion

3.1 Chloride penetration during compression stress

A concrete cube under compression has a tensile zone, shaped like an hour-glass in the direction of the compression. The sizes of the exposed cubes were 70×100×100 mm^3 and 100×100×100 mm^3. The three dimensional tensile zone has its narrowest part 20-30 mm from the surface of the cube. The concrete column inside and outside the tensile zone was under compression. The tensile zone should enlarge the chloride penetration depth more significantly than the compressed zone.

When deloading the W/(C+S) 0.35 cubes, it was discovered that some of the concrete surface had scaled off, although it showed no sign of not being able to carry the load during the experiment. This was probably due to the extreme stiffness of this high strength concrete and the presence of some eccentricities in the load. The chloride profiles showed, as expected, a deeper penetration in a lower concrete quality. A rough estimation of the average penetration depth, $d_{0.5}$, for the three types of concrete are given in Table 1.

The effect of the stress level was not clearly visible. For W/C 0.65, the effect was present but the small difference was hardly significant. For W/C 0.45 and W/(C+S) 0.35, the effect of the stress level was far from being significant.

During the electrical method, the currents were documented at the start of each experiment and the resistance calculated by using Ohm's law.

Table 1. Average penetration depth

Concrete	$d_{0.5}$ [mm]
W/C 0.65	7-12
W/C 0.45	3-5
W/(C+S) 0.35	1-2

Table 2. CTH Rapid Method penetration depth, cube depth 100 mm.

Concrete	Load [f_{cck}]	Depth [mm]
	0	50
W/C 0.65	0.5	44.5
	0.65	47
	0	17.5
W/C 0.45	0.5	23
	0.75	25.5
	0	1
W/(C+S) 0.35	0.5	1

When the load was increased, the magnitude of the tensile force grew. The current became larger and the initial resistance, lower.

The chloride penetration was measured by dry sawing each cube. The surfaces were sprayed with silver nitrate to show where the chloride was. The intrusion depth was measured as listed in Table 2.

The concrete with W/(C+S) 0.35 was dense, and the visually measured penetration depth was difficult to evaluate accurately. The concrete with W/C 0.45 had a penetration depth in the tensile zone. This might be a possible explanation of the deeper penetration for increased load. The concrete with W/C 0.65 had a penetration depth to the centre of the cube. This volume was under strict compression, which tended to close the pores. This is a possible explanation of the decrease in chloride penetration under loads.

3.2 Chloride penetration during flexural tension stress

The chloride penetration for the immersion test was measured by dry sawing each beam in six sections:

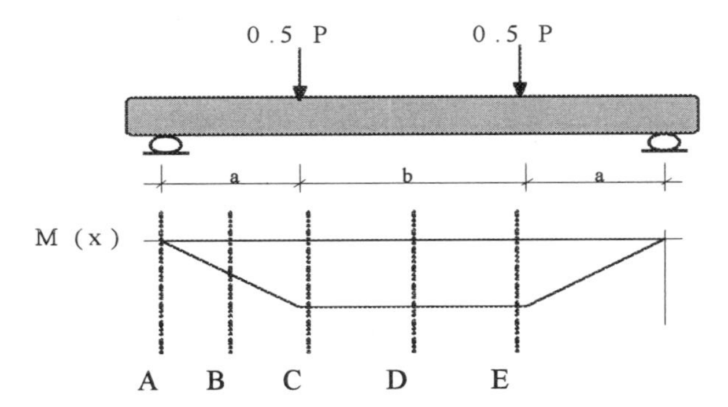

Fig.7. The beam was cut in sections A-E and analysed with $AgNO_3$. The surfaces were sprayed with silver nitrate to show where the chloride was. The average intrusion depth was measured.

Table 3. Average penetration depth, $AgNO_3$ visual indicator, beams l=800 mm.

Section	W/C 0.65	W/C 0.45	W/(C+S) 0.35
A	11.5	8.5	3.5
B	13.5	9	3
C	13	9	2
D	14	10	2
E	17	7	1

The penetration depth varies unsystematically along the beam. This might be explained by the fact that the beam was not exposed to a great stress level. Sections C, D and E had the same flexural stress. Section A had a low flexural tension and section B only half. For W/C 0.65, the penetration depth was smaller for lower stress levels. The number of measurements were, however, too small to show any significant effect. For lower W/C's, the effect of stress level was even less significant. The variation is a measure of the deviation in the method of measurement and the variation in heterogeneity of the concrete. The difference due to the stress level was not significant.

The chloride profiles from the non-loaded beams were very similar to the profiles measured in the cubes in the compressive tests. This indicates rather good reproducibility. The chloride profiles from loaded beams are shown in Fig. 8. A comparison of profiles from loaded and non-loaded beams gives some indication of an effect of stress level for W/(C+S) 0.35. For W/C 0.65, a comparison was not possible due to the lack of a reference specimen. For W/C 0.45, the effect of stress was not significant.

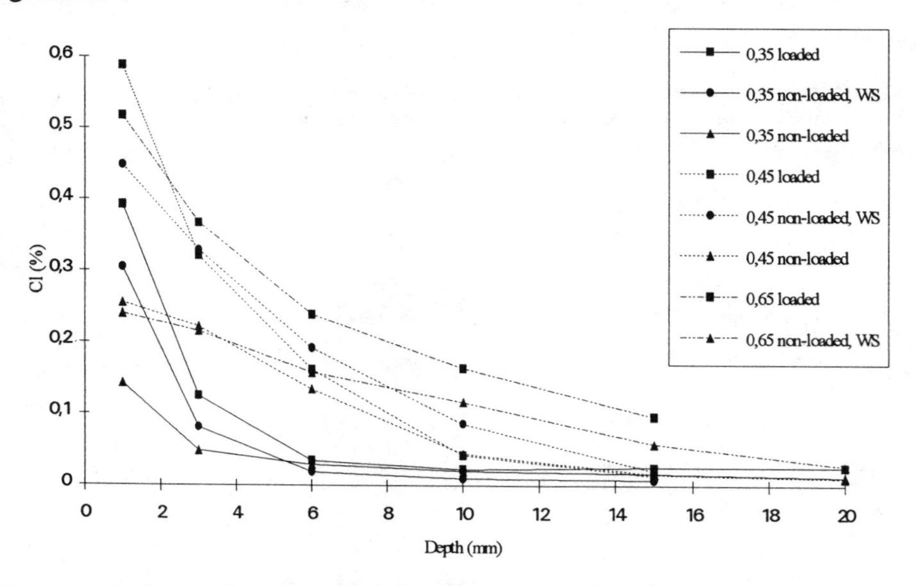

Fig. 8. Graph showing chloride content as % of total dry mass at different depth for the three concretes when subjected to flexural tension.

4 Concluding remarks

From the limited data, it can be seen that a compressive stress tended to decrease the chloride penetration (ref. the data in Table 2. W/C 0.65) and a tensile stress tended to increase the chloride penetration (ref. the data in Table 2. W/C 0.45 and Fig. 8. W/(S+C) 0.35), although the effects were not significant due to the heterogeneity of concrete materials, limited specimens, and, perhaps, incomplete experimental methods.

Further study is needed to find the quantitative influence of loading levels on chloride penetration. Particularly, the loading methods should be improved to apply a homogeneous compressive or tensile stress.

To obtain a more distinct chloride profile, it is suggested that a longer exposure period be used.

The CTH rapid method may be an effective approach to evaluate the quantitative influence of loading on chloride penetration, provided that the loading methods and the solution holders for applying an electrical field are improved.

5 References

1. Heideman, H. and Sundström, H. (1994) *Effects of stress on chloride penetration into concrete* E–94:2 Department of Building Materials, Chalmers University of Technology, Gothenburg, Sweden
2. Tang, L. and Nilsson, L–O. (1992) *Rapid determination of chloride diffusitivity of concrete by applying an electric field.* ACIMaterials Journal, Vol. 49, No. 1. pp. 49–53
3. Sugiyama, T., Bremner, T.W. and Holm, T.A. (1993) *Effects of stress on chloride permeability in concrete* 1.27 Proceedings of CIIB/RICEM Xth International Conference on Durability of Building Materials, Ohiwa, Japan
4. Tang, L. (1993) *Methods for determining chloride diffusitivity in concrete* Publication P–93:8, Department of Building Materials, Chalmers University of Technology, Göteborg

138 MICROMECHANICAL STUDY ON DETERIORATION OF CONCRETE DUE TO FREEZING AND THAWING

M. HORI
University of Tokyo, Japan
E. YANAGISAWA and H. MORIHIRO
Tohoku University, Sendai, Japan

Abstract
While deterioration due to temperature has been a major concern for concrete durability, the mechanism of deterioration is not fully understood. As concrete is expected to be used in severer environments, a more rational and quantitative understanding of the mechanism becomes essential. Based on a commonly accepted hypothesis that the deterioration is caused by damage to micro-pore structures, this paper presents a micromechanical model for deteriorating concrete; the pore structures are modeled as numerous microcracks in a brittle solid and the damaging effects of freezing and moving water on the pore structures are represented by forces which act to open each microcrack. The degree of deterioration is then analyzed by computing the overall growth and opening of microcracks and quantified in terms of the residual strain or the loss of the effective elastic moduli. Using the model, a compute simulation is performed on the deterioration process of concrete under various conditions and studies the effects of the range of cooling and heating temperatures, the number of freezing and thawing cycles, and the level of applied stresses. The results obtained from the numerical computation agreed with available experimental data, supporting the validity of the constructed model.
Keywords: Deterioration due to freezing and thawing, effective property, fracture mechanics, micromechanics

1 Introduction

In general, a porous and brittle material such as concrete is damaged when it is saturated with water and subjected to repetitive freezing and thawing in low temperature; see [1] for a concise survey of literature. Although the deterioration of concrete in cold regions has been studied [2, 3], more attentions is now being

Concrete Under Severe Conditions: Environment and loading (Volume Two) Edited by K. Sakai, N. Banthia and O.E. Gjørv. Published in 1995 by E & FN Spon. ISBN 0 419 19860 1

focused on this type of deterioration as concrete structures are to be used in more severe environments [4]. Similar deterioration of rock is studied as an underground structure in rock mass is used for the storage of very cold substances [5, 6]. A thorough understanding of the deterioration phenomena is essential in preventing or minimizing damages, and allows the prediction of the expected deterioration to ensure the safe use of structures over a long period.

It is widely accepted that the deterioration is the result of damage to the pore structure; the volume expansion of frozen pore water causes tensile stresses which may result microscopic failure, and this action can be accelerated by the movement of water within the pore structure [7, 8, 9]. While the hypothesis is supported by various experimental observations, the mechanism of the deterioration process is not examined theoretically. In addition, it has not been determined whether the hypothesis is applicable to the analytical method for predicting the degree of deterioration under various conditions. This is mainly because many factors influence the deterioration process and modeling of the phenomena is not straightforward. Indeed, the degree of deterioration depends on the material properties of concrete as well as the temperature ranges, the number of freezing and thawing cycles and the level of external loads [10, 11, 12, 13, 14].

To establish a rational method of predicting the degree of deterioration of concrete, we propose a micromechanical model of the deterioration process, which considers the local failure of a porous brittle material which has been subjected to freezing and thawing [15, 16]. The proposed model regards the deterioration as the growth of microcrack; the pore structure is modeled as a set of microcracks, and the effects of freezing and moving water as forces which apply on each crack to open it. The fracture of the microcrack is analyzed by applying the fracture mechanics, and quantities such as irreversible deformation or the loss of elasticity are computed from the micromechanical analysis. We simulate the deterioration process through the numerical computation of the proposed model, and compute effects of various factors on the deterioration. It is shown that the degree of deterioration computed from the simulation agrees with available experimental data in some quantitative manner.

2 Micromechanical modeling of deterioration process

While various mechanical and chemical processes are induced during the deterioration of a porous material, the damage is essentially caused by microfractures of the pore structure [1, 7]. As a micromechanical model of the deterioration process, we consider a two-dimensional solid containing *microcracks* with a pair of *crack-opening-forces* acting at their center. The microcracks model the micropores, and their distribution represents porosity or pore distribution. The crack-opening-displacements [16] represent those effects that may cause the deterioration (temperature history, humidity, etc.). The stress applied on the material is set as the macrostress of the solid; see Fig. 1. The deterioration process is regarded as the growth of the microcracks induced by the crack-opening-forces.

The quantities that are used in experiments as a measure of the degree of deteri-

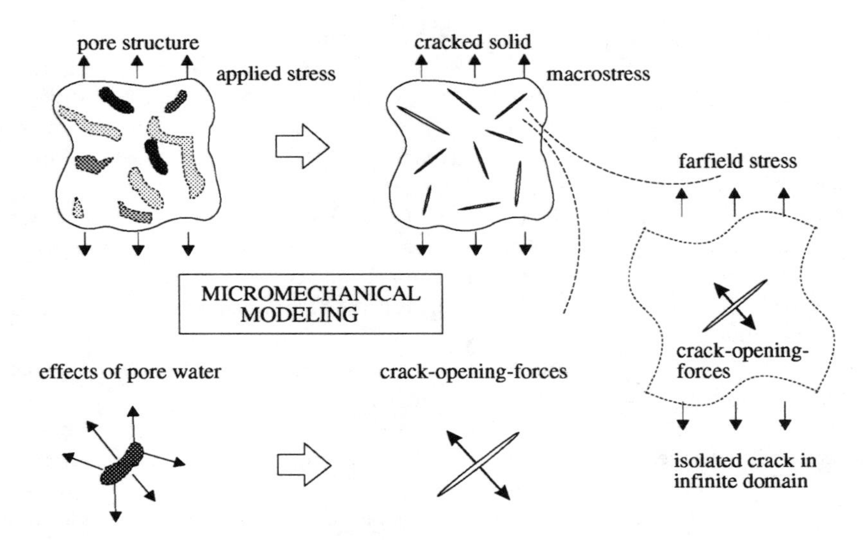

Fig. 1. Micromechanical modeling of deteriorating concrete

oration are the irreversible expansions and the loss of the dynamic Young modulus. According to the micromechanics theories, these quantities can be evaluated in the model by using a *crack strain* which is caused by the opening of the microcrack and an *effective* elasticity of the cracked solid, respectively; for more detail refer to [17]. The crack strain and the effective elasticity are analytically determined from the crack-opening-displacement and the length of all microcracks.

2.1 Approximations made in constructing model

To complete the above model, one must specify several factors, such as the distribution of the microcracks and the dependence of the crack-opening-forces on the temperature history. A fracture criterion must be prescribed as well. As more detailed and realistic specifications are given, the analysis of the model yields more accurate prediction. To verify the validity of the essential part of the model, however, we use relatively simple setting in this study. The following approximations are made:

1. Since microcracks represent micropores (mainly capillary pores) which are damaged by the freezing and movement of pore water, they are assumed to be randomly distributed and oriented in the solid. The initial length of a microcrack, a_o, is the radius of the corresponding pore, and increases to a as the pore is damaged. To determine a *crack density*, ρ, the distribution of micropores is simplified as shown in Fig. 2a. The relation between ρ and $\ln a_o$ is set as a triangle, and the triangle is specified by four parameters, a_{max}, a_{peak}, a_{min} and ρ_{peak} [16].
2. Although the crack-opening-force, P, depends on the humidity, we define $P = apw$ for a microcrack of length a assuming the pore is fully saturated, where p

is a pressure on the pore surfaces caused by frozen water, and w is a weighting factor which is given as $w_t(T,h) + w_d(h)$, such that w_t and w_d represent the dependence of the freezing process that occurs within micropores of radius a on the temperature T [18] and the damages that are accumulated by the repetition of freezing and thawing, respectively. Parameter h stands for the temperature history, e.g., $\{a_n\}$ where a_n is the crack length after the nth cycle of freezing and thawing. It is assumed that w_t and w_d are of the form of $\hat{w}_t(T,a)\Pi(1 + \alpha_t(a_m - a_{m-1}))$ and $\alpha_d(a_n - a_o)$, where α_t and α_d are constant and $\hat{w}_t(T,a)$ expresses the freezing process of water in a pore of radius a, as shown in Fig. 2b. It is assumed that T_f is function of a and ΔT is constant.

3. Although micropores are damaged in a complicated manner, it is assumed that each microcrack grows straight without kinking. The fracture criterion is common for all microcracks, and given as

$$K_c = \sqrt{K_I^2 + K_{II}^2} \qquad (K_I > 0), \tag{1}$$

where K_c is the fracture toughness and K_I and K_{II} are the stress intensity factors of Mode I and II. This criterion means that a crack grows if the strain energy accumulated at its tips attains a critical value. The condition in the parenthesis, $K_I > 0$, is to ensure that the crack must open when it grows.

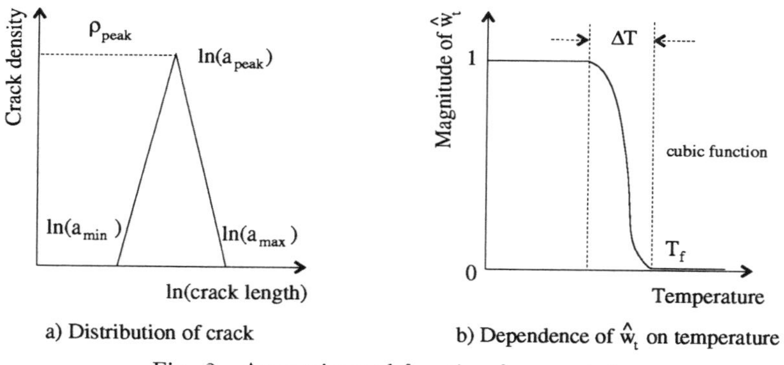

Fig. 2. Approximated function for ρ and $\hat{w}_t$

2.2 Formulation of problem

For simplicity, interaction effects among microcracks are neglected, i.e., the response of each microcrack is approximated in the analyzed by using an isolated crack in an infinite body, as shown in Fig. 1. The average stress of the solid, $\overline{\sigma}_{ij}$, is set as the farfield stress of the body, σ_{ij}^{∞}; see [17]. When the body is isotropic and linearly elastic, the stress intensity factors for a microcrack of length $2a$ and orientation θ are analytically obtained as

$$\begin{Bmatrix} K_I \\ K_{II} \end{Bmatrix} = \frac{1}{\sqrt{\pi a}} \begin{Bmatrix} P \\ 0 \end{Bmatrix} + \sqrt{\pi a} \begin{Bmatrix} \sigma_{nn}^{\infty} \\ \sigma_{ns}^{\infty} \end{Bmatrix}, \tag{2}$$

and its crack-opening-displacements at x (which is measured from the center) as

$$\frac{2G}{\kappa+1}\left\{\begin{array}{c}[u_n]\\ [u_s]\end{array}\right\}=\left\{\begin{array}{r}\frac{P}{\pi a}\left(a\ln\left|\frac{a}{x}\right|+\sqrt{\pi(a^2-x^2)}\right)+\sigma_{nn}^{\infty}\sqrt{\pi(a^2-x^2)}\\ \sigma_{ns}^{\infty}\sqrt{\pi(a^2-x^2)}\end{array}\right\}, \tag{3}$$

where subscript n and s stand for the normal component and the shear component on the crack surface. In Eq. (3), G is the shear modulus of the solid and κ is given as $3-4\nu$ for plane strain with ν being Poison's ratio.

It is shown that the effective elasticity of the cracked solid can be analytically computed from the distribution of the microcrack length [17]. Denoting by $2a(a_o,\theta)$ the increased length of a crack of initial length $2a_o$ and orientation θ, we have the asymptotic behavior of the effective Young modulus, $\overline{E}$, as

$$\frac{\overline{E}}{\overline{E}_o}\approx\frac{\int_R\int_0^{2\pi}\rho(a_o)a^2(a_o,\theta)da_od\theta}{\int_R\int_0^{2\pi}\rho(a_o)a_o^2da_od\theta}. \tag{4}$$

where $\overline{E}_o$ is the initial effective Young modulus, and R stands for the integral domain for a_o.

Similarly, it is shown that the crack strain, $\overline{\epsilon}^c_{ij}$, which can be computed from the crack-opening-displacement of each microcrack represents the loosening of the material due to the deterioration [17]. A microcrack of length $2a$ and orientation θ contributes $\overline{\epsilon}^c_{ij}$ by $1/(\pi a^2)\int_{-a}^{a} sym\{[u_i]n_j\}dx$, where $[u_i]$ is given by Eq. (3) and n_i is the unit normal of the cracked surface. If this contribution is denoted by $\epsilon^c_{ij}(a,\theta)$, the crack strain is evaluated as

$$\overline{\epsilon}^c_{ij}=\frac{1}{2\pi}\int_R\int_0^{2\pi}\rho(a_o)\;\epsilon^c_{ij}(a,\theta)da_od\theta. \tag{5}$$

Note that a is function of a_o and θ.

For an incremental change of the temperature, dT, the corresponding change of dP is determined, and the growth and opening of microcracks, if happens, is evaluated by using Eqs. (1-3). The resulting change of $\overline{E}$ and $\overline{\epsilon}^c_{ij}$ are then obtained from Eqs. (4,5). Therefore, one can analyze the deterioration process for a given temperature history in this incremental manner. This procedure is applicable even when the stress acting on the material, $\overline{\sigma}_{ij}$ or σ^{∞}_{ij}, changes according to $\overline{E}$.

3 Verification of validity of proposed model

The constructed model is based on a commonly accepted qualitative hypothesis of the deterioration process [1]. The validity of the model, however, should be verified in a quantitative manner so that it can be used to predict the degree of deterioration. The key issue in the model is the specification of P which represents various factors that influence the deterioration process. While there have been many observations reported on complicated processes, one may expect the model to reproduce them in a unified manner, if a suitable P is assumed [16].

3.1 Determination of suitable constants

In the proposed model, most of the parameters can be determined from available experimental data. The parameters used in this and the following simulations are $E = 3.3 \times 10^6$ [N/m^2], $\nu = 1/6$, and $K_c = 2.0 \times 10^4$ [Pa/$\sqrt{\text{m}}$] (the thermal expansion coefficient, $K = 9.4 \times 10^{-6}$ [/^{o}C]) for the concrete material property, $a_{min,peak,max} = 0.3, 0.6, 1.2 \times 10^{-6}$ [m] and $\rho_{peak} = 2.1$ [/m^4] for the pore structure, and $T_f = 273(1/\exp(3.3/(a-10)-1)$ [^{o}C] and $\Delta T = -70$ [^{o}C] for the function $\hat{w}_t$; see [17].

The three constants, p, α_t, and α_d, which specify P, need to be determined in the modeling. It should be recalled that p is the peak stress for the frozen water, and that α_t and α_d represent the effects of the accumulated damages. They are determined such that the following two significant effects of the temperature history on the deterioration process are reproduced:

1. The degree of deterioration due to one cycle of freezing and thawing depends on the minimum cooling temperature and the maximum heating temperature.
2. The deterioration process gradually proceeds as the freezing and thawing cycles are repeated.

See [1] for the summary of various experimental reports.

First, we determine the most suitable value of p as 1.0×10^8 [N/m^2], from a numerical simulation of the deterioration process after one cycle of thawing and heating (T decreases from 10 [^{o}C] to -80 [^{o}C] and increases to 10 [^{o}C]. Figure 3a shows a typical relation between the temperature and the strain for a stress-free concrete specimen, using this value. The uniaxial total strain, the sum of the thermal strain, $\bar{\epsilon}^t_{11} = K\Delta T$ and the crack strain, $\bar{\epsilon}^c_{11}$ ($\bar{\epsilon}^t_{11} + \bar{\epsilon}^c_{11}$), is plotted as a function of temperature T. The difference of the total strain form the thermal strain indicates the contribution of the opening of microcracks, $\bar{\epsilon}^c_{11}(=\bar{\epsilon}^c_{22})$. The relation between the residual strain and the minimum cooling temperature is summarized in Fig. 3b. Here, the strain obtained from the computation and that measured

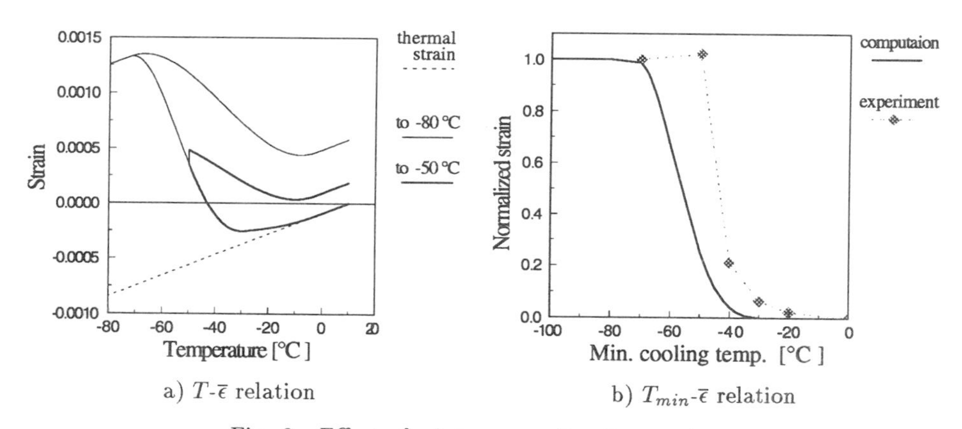

a) T-$\bar{\epsilon}$ relation

b) T_{min}-$\bar{\epsilon}$ relation

Fig. 3. Effect of minimum cooling temperature

in the experiment [4] are plotted. These strains are normalized by the maximum value when cooled to -80 [^{o}C]. In Figs. 3a and 3b, we set $\alpha_t = 0.02$ to compute an irreversible strain after one cycle. This strain corresponds to the residual strain of the specimen caused by the deterioration. As it can be seen from Fig. 3b, the model can reproduce the experimental data well, though a relatively simple dependence of P on T and h is assumed.

Next, we repeat the numerical simulation of the deterioration process during several cycles of freezing and thawing, and determine $\alpha_d = 0.05$. Figure 4a shows a typical relation between the strain $\bar{\epsilon}$ when a specimens is subjected to 12 cycles of freezing and thawing, and the maximum and residual strain during each cycle are plotted in Fig. 4b. The strain is normalized by the residual strain after the first cycle. The corresponding experimental data [10] are shown, and the agreement of the computational results seems as satisfactory as in Fig. 3b.

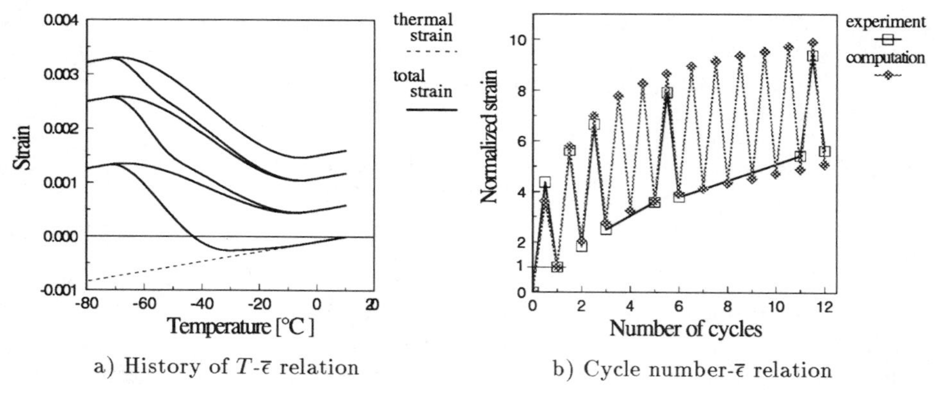

a) History of T-$\bar{\epsilon}$ relation

b) Cycle number-$\bar{\epsilon}$ relation

Fig. 4. Effect of repetition of freezing and thawing on Strain

3.2 Effect of applied stress

Since the deterioration process is modeled as the mechanical phenomenon of microfracturing, one may expect that even for the same concrete subjected to the same freezing and thawing cycles, the degree of deterioration may vary depending on the applied stress. Indeed, there are several experimental observations which report this phenomenon [12][13][14]. In general, the deterioration is suppressed when subjected to a sufficiently large confining pressure [19].

To examine the validity of the model, in particular the validity of the three constants determined above (i.e, p, α_t and α_d), we computed the deterioration process for different levels of constant macrostress, $\bar{\sigma} = \bar{\sigma}_{11}$ with $\bar{\sigma}_{22} = \bar{\sigma}_{12} = 0$. As a typical example, Fig. 5a illustrates the difference of the relation between $\bar{\epsilon}_{11}$ and T due to loading. As the confining pressure increases, the difference between the total strain and the thermal strain tends to decreases. This effect of $\bar{\sigma}$ on the residual strain is summarized in Fig. 5b, where the macrostress is normalized by the compressive strength of concrete [16]. Available experimental data [12][19] are also plotted. No additional assumptions or approximations were made in this computation and all the parameters are the same as used in the previous computation. From Fig. 5b, it is clear that the agreement of the computation results and the experiment data

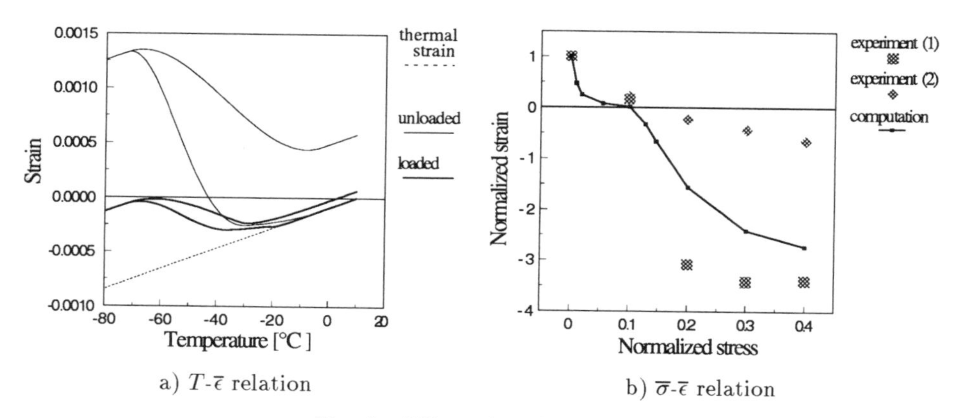

a) T-$\bar{\epsilon}$ relation

b) $\bar{\sigma}$-$\bar{\epsilon}$ relation

Fig. 5. Effect of applied stress

is worse than that shown in Fig. 3b or 4b. (The agreement becomes worse as the stress increases, because the assumed fracture criteria allows some crack to grow in a unstable manner [16].) The prediction, however, lies between the two sets of experimental data.

While relatively crude approximations are made in determining the relation between P and the temperature history, we obtain good agreement of the computational results with the experimental data. This suggests that the essential part of the constructed model, the microcracks modeling the pore structures and the crack-opening-forces representing the effects of deterioration, is reasonable. For the prediction of the deterioration, therefore, one may expect that the deterioration analysis that uses a model of this kind is applicable to members or structures which are subjected to the repeated freezing and thawing; see Appendix.

4 Conclusions

A micromechanical model is proposed to analyze the deterioration of concrete due to freezing and thawing. Based on a commonly accepted hypothesis that the deterioration is caused by damage to the pore structure, this model regards the deterioration process as the growth of microcracks (which represents the pore structure) induced by crack-opening-forces (which represent the effects of freezing and moving water). The deterioration of concrete under various conditions is simulated in this model, and the degree of the deterioration is evaluated. While approximations and assumptions are made to simplify the analysis, the results agree with experimental data in some quantitative manner, and the validity of the essential part of the model is verified. One may expect a deterioration analysis which uses a model of the proposed kind would be able to predict the deterioration of concrete members or structures.

5. References

1. Veen, Ir.C.v.d. (1987) Properties of concrete at very low temperatures: A survey of the literature, *Report of Faculty of Civil Engineering, Delft University of Technology*.
2. Powers, T.C. (1945) A working Hypothesis for further studies of frost resistance of concrete, *Proc. of ACI*, Vol. 41.
3. Hasegawa, S. and Fujiwara, T. (1988) *Frost Damage*, Gihoudou (in Japanese).
4. Miura, T. and Lee, D. (1990) Strain and deterioration of concrete under low temperatures, *Trans. JSCE*, No. 420, pp. 191-200 (in Japanese).
5. Inada, Y., Kitamura, S. and Okada, A. (1984) Stability of underground rock cavern for storage of LNG, *Trans. JSCE*, No. 345, pp. 35-44 (in Japanese).
6. Yamabe, T. and Watanabe, T. (1993) Thermal-stress-permeability-analysis of rock mass considering freezing action, *the 48the National Congress of JSCE*, pp. 1348-1349 (in Japanese).
7. Monteiro, P.J.M., Bastacky, S.J. and Hayes, T.L. (1985) Low-temperature scanning electron microscope analysis of the Portland cement paste early hydration, *Cement and Concrete Research*, Vol. 15, pp. 687–693.
8. Monteiro, P.J.M., Rashed, A.I., Bastacky, S.J. and Hayes, T.L. (1989) Ice in cement paste as analyzed in the low-temperature scanning electron microscope, *Cement and Concrete Research*, Vol. 19, pp. 306-314.
9. Rostásy, F.S., Weib, R. and Wiedemann, G. (1980) Changes of pore structures of cement mortars due to temperature, *Cement and Concrete Research*, Vol. 10, pp. 157-164.
10. Rostásy, F.S., Schneider, U. and Wiedemann, G. (1979) Behavior of mortar and concrete at extremely low temperatures, *Cement and Concrete Research*, Vol. 9, pp. 365-376.
11. Rostásy, F.S. and Wiedemann, G. (1980) Stress-strain behavior of concrete at extremely low temperature, *Cement and Concrete Research*, Vol. 10, pp. 565-572.
12. Planas, J., Corres, H. and Elices, M. (1984) Thermal deformation of loaded concrete during thermal cycles from $20C^o$ to $-165C^o$, *Cement and Concrete Research*, Vol. 14, pp. 639-644.
13. Elices, M., Planas, J. and Corres, H. (1986) Thermal deformation of loaded concrete at low temperatures 2: transverse deformation, *Cement and Concrete Research*, Vol. 16, pp. 741-748.
14. Corres, H., Elices, M. and Planas, J. (1986) Thermal deformation of loaded concrete at low temperatures 3: lightweight concrete, *Cement and Concrete Research*, Vol. 16, pp. 845-852.
15. Hori, M., Yanagisawa, E. and Morihiro, H. (1994) Micromechanical study on deterioration of porous material due to freezing and thawing, *Journal of Structural Engineering* (to appear, in Japanese).
16. Morihiro, H. (1994) Mechanical study on deterioration of rock and concrete due to freezing and thawing action, *Dissertation of Dept. Civil Eng., Tohoku University* (in Japanese).

17. Nemat-Nasser, S. and Hori, M. (1993) *Micromechanics: overall properties of heterogeneous materials*, North-Holland, New York.
18. Stockhausen, N. (1981) Die dilatation hochporöser Festkörper bei Wasseraufnahme und Eisbildung", *Dissertation Technische Universität Munchen* (in German).
19. Miura, T., Hori, M. and Matui, J. (1991) On deterioration of cooled concrete under confining pressure, *Proc. of the 45th Cement Tech.* pp. 508-513 (in Japanese).
20. Wittmann, F.H. (ed.) (1986) *Fracture toughness and fracture energy of concrete*, Elsevier, Amsterdam.

Appendix: coupled analysis of deterioration process

In the proposed model, the deterioration process is viewed as the mechanical fracture of microcracks caused by crack-opening-force which represents freezing and moving of water. The deterioration analysis, therefore, must be coupled with a mechanical analysis and a thermal conduction analysis. Indeed, the analysis made in Section 3 shows the effects of the macrostress on the degree of deterioration. As mentioned, the loss of effective elasticity may change stress distribution in a member or a structure.

The coupled analysis can be formulated in an incremental form. The strain increment is decomposed into the elastic, the thermal and the deterioration parts, $d\epsilon_{ij}^e$, $d\epsilon_{ij}^t$ and $d\epsilon_{ij}^c$ (a bar on ϵ or σ_{ij} is omitted), as follows:

$$d\epsilon_{ij} = d\epsilon_{ij}^e + d\epsilon_{ij}^t + d\epsilon_{ij}^c.$$

First, the incremental temperature, dT, is obtained from the thermal conduction analysis, and $d\epsilon_{ij}^t$ is given as $KdT\delta_{ij}$. Next, for the current state of T and σ_{ij}, the increment of the crack strain, $d\epsilon_{ij}^c$, can be evaluated from the deterioration analysis that uses the proposed model. Finally, the effective elasticity is determined from the reduced elastic constants as $\overline{C}_{ijkl}$. It follows from $d\epsilon_{ij}^e = \overline{C}_{ijkl}^{-1} d\sigma_{kl}$ that the total strain increment is related to the corresponding stress increment as

$$d\sigma_{ij} = \overline{C}_{ijkl}(d\epsilon_{kl} - \overline{K}dT\delta_{kl} - d\epsilon_{kl}^c).$$

This coupled mechanical problem can be solved if the contribution of the second and third terms in the parenthesis on the right side is treated as a body force,

The coupled analysis shown in the above is applied in [15, 16] to predict the deterioration of rock mass around a cavern where a cold substance (LNG) is stored [5, 6]. While several crude approximations and assumptions are made, the coupled analysis yields reasonable results regarding to the deformation of the cavern as well as the deterioration of the surrounding rock mass; see [16] for details.

139 DURABILITY OF E-GLASS FIBRE REINFORCED COMPOSITES WITH DIFFERENT CEMENT MATRICES

K. KOVLER and A. BENTUR
National Building Research Institute, Technion – Israel Institute of Technology, Haifa, Israel
I. ODLER
Institute for Non-Metallic Materials, Technical University–Clausthal, Claustahl, Germany

Abstract
In recent years, considerable effort has been directed at the research and development of new glass fibre reinforced cement composites intended to replace asbestos-cement as well as to generate new materials of improved properties, in particular toughness. However, due to the high surface area, the fibres are sensitive to interfacial effects with the matrix, which can be chemical or physical in nature and may have deleterious effects in regard to durability, especially when using Portland cement matrix, which is highly alkaline. Four low alkali/low lime cements were developed: glass, calcium aluminate phosphate, magnesia phosphate and ettringite cements which have the potential for providing a matrix for high durability composites. The performance of E-glass composites with these matrices was studied. The results were compared with those of ordinary Portland cement composites. It is concluded that only two E-glass cement composites can satisfy the requirements of durability: those with magnesia phosphate and ettringite cement matrices.
Keywords: Calcium aluminate phosphate cement, durability, ettringite cement, glass cement, glass fibres, load-displacement diagrams, magnesia phosphate cement, microstructure.

1 Introduction

In recent years, considerable effort has been directed into the research and development of new glass reinforced cement (GRC) composites.

Most of the fibres are made in the form of thin filaments, 10 to 20 μm in diameter. As a result of their high surface area, they are more sensitive to interfacial effects with

Concrete Under Severe Conditions: Environment and loading (Volume Two) Edited by K. Sakai, N. Banthia and O.E. Gjørv. Published in 1995 by E & FN Spon. ISBN 0 419 19860 1

the matrix, which can be chemical or physical in nature. Such effects can be beneficial, resulting in sufficiently high bond, or they could be deleterious, such as chemical attack or the development of a bond which is too strong, leading to embrittlement. Problems of this kind have been experienced with various fibres in cement matrices, and in order to overcome them, attempts were made to develop special fibre formulations to be compatible with the cement matrix [1-3]. This approach can be successful, but it is frequently costly to develop and implement in production, unless a large market is guaranteed in advance. The present study suggests a different approach, to develop special cement matrices for such composites, in which the cement composition will be adjusted to accommodate the fibres, since many of the problems cited previously are common to the different fibres incorporated in the cement matrix; they are associated with the high alkalinity of the Portland cement (PC) and its tendency to develop large, dense deposits of $Ca(OH)_2$ in the vicinity of the fibres. These two effects can be taken care of by developing low alkali/low lime cements, which could be economically feasible, since they can be made of various industrial by-products.

In the present work, different low alkali/low lime matrices were developed, and their potential for producing high durability GRC, was studied in E-glass fibre reinforced composites. This glass is particularly sensitive to ageing effects since it tends to corrode in high alkalinity matrix, and due to its brittle nature, it may break readily when microstructural densening occurs. It thus provides a good model to study the effect of these matrices.

This work deals with the study of composites with E-glass fibres at extended accelerated ageing periods of 56 days.

2 Materials

Four non-Portland cement matrices were used in the study, in addition to Portland cement which served as a reference. All of these non-Portland cements were developed and produced for this study at the Technical University-Clausthal.

2.1 Glass cement (GC)

It has been reported that some glasses in the system $CaO\text{-}Al_2O_3\text{-}SiO_2$, if ground to a high fineness, exhibit cementitious properties. The products of the hydration process are hydrogarnet $C_3AS_xH_{(6-2x)}$ and stratlingite (gehlenite hydrate) C_2ASH_6 [4]. Cements of this type may be considered candidates for GRC matrices due to both the relatively low pH value of the pore solution and the absence of calcium hydroxide in the hydration products.

For evaluation of glass fibre composites, the following composition of GC was chosen: CaO - 50%, Al_2O_3 - 30% and SiO_2 - 20%. The compressive strength of the specimens cast at a water/cement ratio of 0.3 and cured for 28 days in moist air was 83 MPa. A value of pH = 11.0 was measured on a suspension of GC in water.

2.2 Calcium aluminate phosphate cement (CAPC)

A new binder obtained by combining high alumina cement with diammonium phosphate has been reported recently [5]. This cement was also included in our study as it does not liberate $Ca(OH_2)$ (portlandite) in its hydration and exhibits a low pH value.

For work on GRC, a mix of sodium polyphosphate $(NaPO_3)_n$ and high alumina cement was chosen. It exhibited a maximum compressive strength of 48 MPa at 7 days age, and its setting was not too fast (setting time of 13 min). The pH of the liquid phase was 11.4.

The CAPC mix composition recommended was as follows: a 40% by weight solution of $(NaPO_3)_n$ in water, i.e. 40 g $(NaPO_3)_n$ in 60 g H_2O, was prepared; the solution and high alumina cement were mixed in a ratio of 40 g solution and 60 g cement.

2.3 Magnesia phosphate cement (MPC)

The blends of magnesium oxide with concentrated solutions of some phosphates, especially with diammonium phosphate, exhibit cementitious properties [5-6]. The main product of the bonding reaction is struvite $MgNH_4PO_4 \bullet 6H_2O$:

$$(NH_4)_2HPO_4MgO + 5H_2O = MgNH_4PO_4 \bullet 6H_2O + NH_3 \qquad (1)$$

Since the liquid phase in equilibrium with the reaction product has a pH value lower than that of Portland cement, and since calcium hydroxide is not among the products of reaction, this binder was the third one to be included in our study as a possible candidate for GRC composites.

Similar to the preparation of the CAPC, a 40 wt%-solution of $(NH_4)_2HPO_4$ was prepared. Then the calcined MgO (fraction < 63 μm) and the $(NH_4)_2HPO_4$-solution were thoroughly hand mixed in a bowl for 30 sec and poured. The liquid-to-solid ratio was 30:70 (e.g. 30 g $(NH_4)_2HPO_4$-solution and 70 g MgO). Because of the fineness of magnesium oxide, it was necessary to add 5% of $Na_2B_4O_7 \bullet 10H_2O$ ("Borax") as a retarding agent before mixing with MgO.

The properties of the binder prepared in this way were as follows: setting time - 10 min, 28 d compressive strength - 32 MPa, and pH of liquid phase - 11.2.

2.4 Ettringite cement (EC)

The fourth non-Portland cement to be used was an entirely new low pH, no $Ca(OH)_2$ binder. The binder is a blend of a CaO-Al_2O_3-SiO_2 glass (80%) and gypsum (20%). Its hydration product is ettringite. Usually, the ettringite phase forms partially in a topochemical and partially in a through-solution reaction. In such pastes, an expansion accompanied the hydration of non-restricted specimens. In the hydration of a paste made from a CaO-Al_2O_3-SiO_2 glass and gypsum, the ettringite formation took place exclusively in a through-solution process and was not accompanied by an expansion.

The properties of the ettringite cement chosen for this study were as follows: at a water/cement ratio of 0.4, the compressive strength at the age of 7 days was 30.3 MPa, flexural strength was 5.4 MPa, and pH of liquid phase - 11.0.

3 Specimens preparation and testing

3.1 Test specimens

The performance of the glass fibres in the cementitious matrix was evaluated by testing glass fibre-cement composite specimens in flexure. The specimens were beams, 110x20x10 mm in size, in which continuous strands of glass fibres were placed in the bottom part, 3.3 mm from the lower face. The specimens were prepared in special molds, where a 3 mm thick layer of paste was first cast. Over it, continuous strands were placed and impregnated with paste, and on top of them a second layer of cement paste was cast. The total thickness of the specimen was 10 mm, and the volume content of the glass fibres was about 1%. In each mold, six beams were prepared.

For PC and GC matrices, the water/binder ratio of 0.3 was kept constant. CAPC, MPC and EC matrices were prepared according to the compositions given in sections 2.2, 2.3 and 2.4.

3.2 Curing and ageing

The fibre reinforced beams were demoulded after one day, and then cured in water at 20°C until 28 days. The properties at this age represented those of unaged composites.

A hot water test (immersion in 60°C water) constituted the accelerated ageing. These conditions are now well accepted for the prediction of ageing of glass fibres in a cementitious matrix. For example, 1 day in 60°C water is approximately equal to 272 days (9 months) natural weathering in the UK [2]. Accelerated ageing commenced after 28 days of 20°C water curing and continued for periods of up to 56 days, with tests being carried out at 0, 14, 28, and 56 days of accelerated ageing.

3.3 Testing

The performance of the composites was evaluated by means of a flexural test, performed by four point loading at a span of 90 mm, and at a 0.5 mm/min bridge cross-head movement. The load and mid-span deflection were continuously recorded, and load-deflection curves were obtained. From the curves, the following parameters were derived: first crack stress or limit of proportionality of the diagram (LOP) representing the matrix strength, maximum flexure stress in the post-cracking zone, and the area under the curve up to a deflection of 1.6 mm.

The specific fracture energy was calculated as the area under load-deflection curve related to the cross-section. The change in the specific fracture energy or in the post-cracking stress with the period of accelerated ageing served as a means for evaluating the durability performance.

After testing, SEM observations were carried out to resolve the mechanisms involved in the ageing process, in particular those of microstructural origin (i.e. the densening of the matrix at the interfacial zone) and those of chemical origin (i.e. those which caused defects to form on the fibre surface). The SEM observations were carried out on the fractured surface exposed in the flexural test as well as the surfaces exposed by splitting of the specimen in the plane of the fibres.

4 Mechanical properties and microstructure

4.1 Mechanical properties

It was found that the PC had the highest strength matrix (LOP was 14.6 MPa before ageing), and MPC and EC had the lowest ones (5.9 and 5.3 MPa, respectively). The strengths of GC and CAPC were intermediate (7.0 and 9.8 MPa, respectively). The strengths of the investigated cements had a tendency to increase by an average of 1-3 MPa during the period of ageing, except the ettringite and Portland cement, which did not show any increase during water-saturated curing.

Only in two cases were the mechanical properties of glass fibre reinforced cement improved during ageing: MPC and EC. In the three other cases, the mechanical performance of the composite in the post-cracking zone was already weakened at the initial ageing period. For ageing longer than 14 days, the post-cracking flexural strength did not change dramatically. Thus, the initial period of 14 days of accelerated ageing can be considered as the most informative. A similar trend was observed for the fracture energy values (Fig. 1).

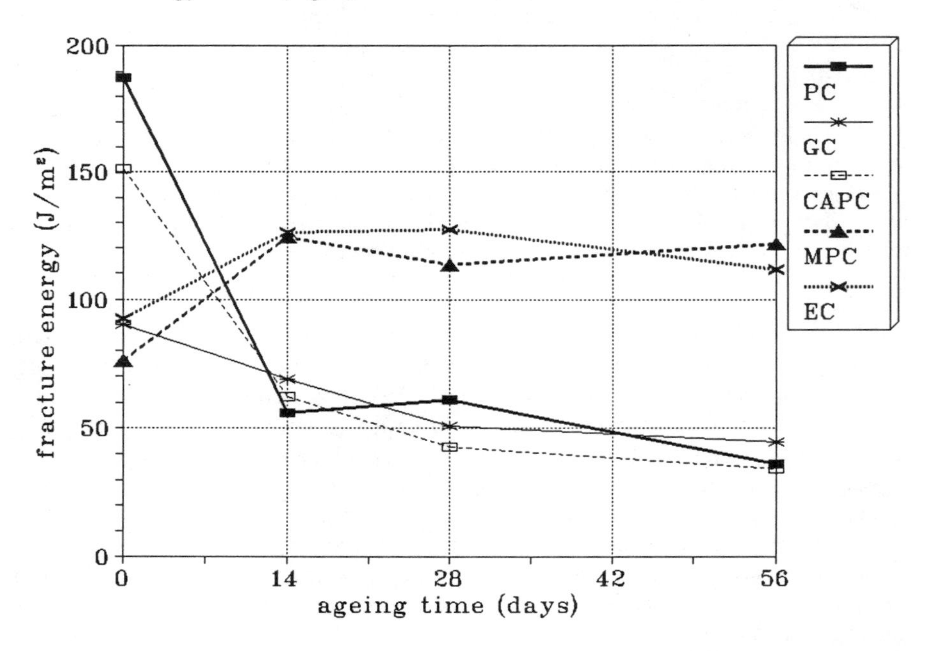

Fig. 1. Specific fracture energy of GRC composites vs. accelerated ageing time.

4.2 Microstructure

SEM observations of five matrices reinforced with E-glass fibres were carried out after 28 days of 20°C curing and after 28 days of accelerated ageing at 60°C.

Prior to ageing of GRC specimens with PC matrix, a little penetration of hydration products into the glass fibber strand was observed; the matrix in the vicinity of the strand and between the filaments was quite porous. After 28 days of accelerated ageing, there was some densening tendency at the interfaces, with more growth of hydration products observed between the filaments (Fig. 2). There was at this age, in addition, marked signs of considerable etching of the fibre surface, including peeling of external layers in the fibres, indicative of a severe chemical attack.

Fig. 2. Micrograph of GRC composite with PC matrix.

After 28 days of accelerated ageing of GRC specimens with GC matrix, the matrix around the glass filaments and between them was quite porous (Fig. 3), similar to the observations prior to ageing. However, considerable signs of chemical attack could be seen, exhibited as relatively large pits on the glass surface, as well as roughening of the surface (the right side of the glass filament in Fig. 3).

Prior to the ageing of specimens with CAPC matrix, a considerable growth of hydration products between the filaments was observed, much more than, for example, in the glass cement. After 28 days of accelerated ageing, the matrix between the filaments had grown to be very dense (Fig. 4), practically eliminating all the pores seen prior to ageing. At the same time, there were considerable signs of chemical attack in the form of densely spaced pits on the glass surface.

The superior ageing behaviour of magnesia phosphate cement matrix reinforced by E-glass fibres was consistent with microstructural observations, showing no signs of chemical attack on the fibres as well as a matrix which, also dense, was readily peeled off the fibres (Fig. 5).

As in the case of the MPC matrix, the EC matrix did not show any sign of chemical attack on the fibres (Fig. 6). The structure of the ettringite cement matrix remained friable, and the formation of ettringite needles did not lead to any damage of E-glass fibres.

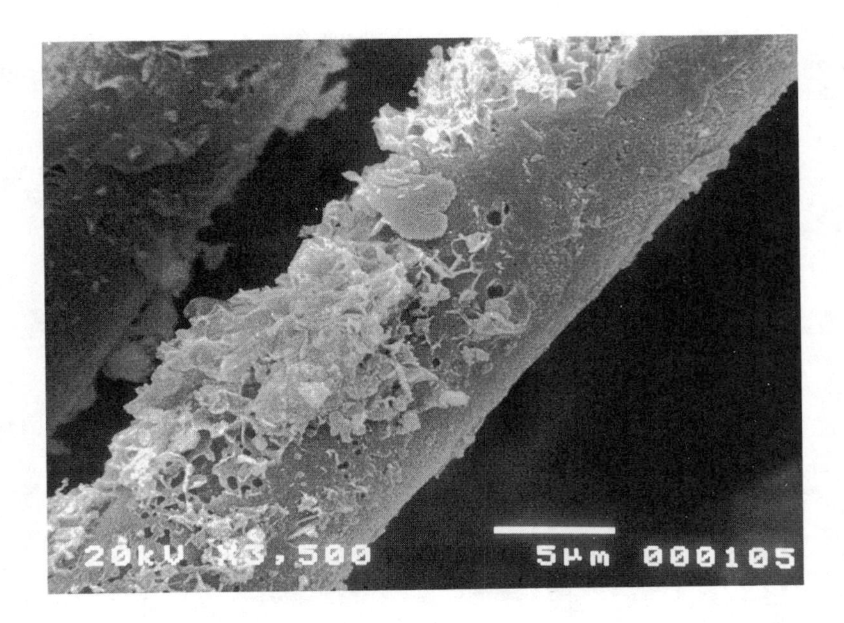

Fig. 3. Micrograph of GRC composite with GC matrix.

Fig. 4. Micrograph of GRC composite with CAPC matrix.

Fig. 5. Micrograph of GRC composite with MPC matrix.

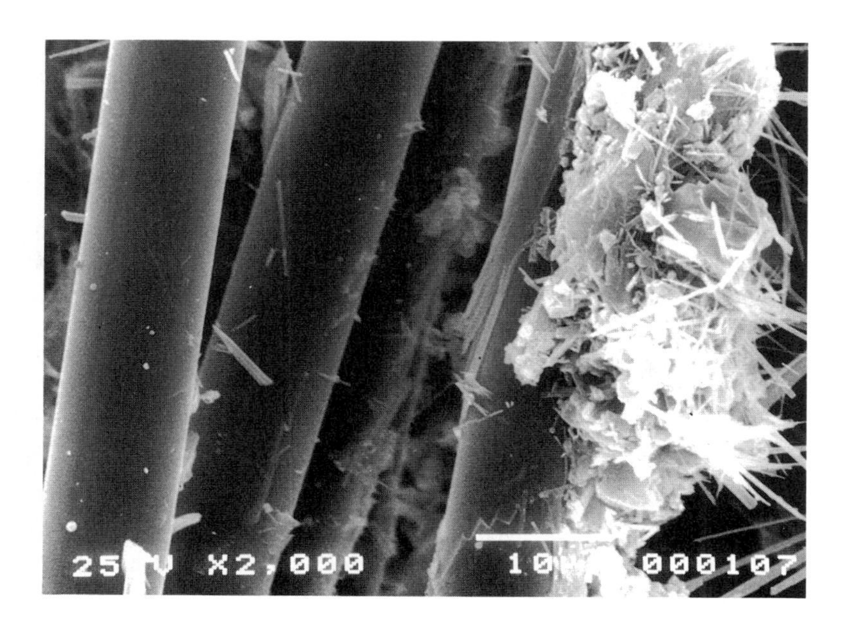

Fig. 6. Micrograph of GRC composite with EC matrix.

5 Discussion

Considerable signs of chemical attack were observed in the systems with matrices of PC, GC, and CAPC. All these systems exhibited considerable loss in post-cracking strength and toughness. The systems with MPC and EC matrices, which did not lose strength or toughness, were characterized by an absence of any signs of chemical corrosion of the fibres. Thus, in the E-glass composites, the durability performance could be readily correlated with immunity to chemical attack, and the major cause for the influence of the matrix is its influence on corrosion resistance of the fibre. However, there is still a secondary influence of the microstructure effects. In the systems which showed chemical corrosion and exhibited loss in post-cracking strength and toughness, the losses with the PC and CAPC matrices were greater (≈80%) than with the GC matrix (≈50%). This difference might be correlated with microstructural influence: the microstructure was very dense in the PC and CAPC, but much more porous in the GC composite.

The superior behavior of the MPC and EC systems was correlated with the immunity to chemical attack which they provided. Both systems exhibited a very porous interfacial microstructure, which is expected to be deficient in terms of providing sufficient bonding to mobilize the strength of the fibres. This showed up in the low first crack stress of these composites. On ageing, some additional densening was obtained with the MPC system, which may account for the increase in strength of this system with ageing. Apparently, this increase was sufficiently high to improve the strength of the composite, but it was probably still too low to cause premature fibre breakage by too high a bond or by local flexural failure.

6 Conclusions

Considerable differences in the ageing of E-glass composites with different cementitious matrices could be observed and accounted for on the basis of microstructural and chemical effects. In the case where chemical attack and microstructural densening were eliminated (magnesia phosphate and ettringite cement composites), loss in strength and toughness were also eliminated. In composites where the chemical attack was severe (Portland cement) and compounded with microstructural densening (calcium aluminate phosphate cement), the durability performance was poor. In the case of the glass cement composite where microstructural densening was eliminated and the chemical attack was rather mild (as compared to Portland cement and calcium aluminate phosphate cement), the loss in properties with ageing might be considered as tolerable, with the composite preserving 80% of its strength and 60% of its fracture energy after 56 days of accelerated ageing (compared to ≈20% preservation of properties in the Portland cement and calcium aluminate phosphate cement composites). It is concluded that only two E-glass cement composites can satisfy the requirements of durability: those with magnesia phosphate and ettringite cement matrix.

These results clearly prove the hypothesis that by adjusting the composition of the matrix, there is a potential for developing highly durable fibre-cement composites, even with E-glass, which is probably the most sensitive to corrosion of the man-made high strength fibres.

7 References

1. Bentur, A. and Mindess, S. (1990) *Fibre Reinforced Cementitious Composites*, Elsevier Science, London.
2. Majumdar, A.J. and Laws, V. (1991) *Glass Fibre Reinforced Cement*, BSP Professional Books, Oxford.
3. Bentur, A., Ben-Bassat, M. and Schneider, D. (1985) Durability of glass fiber reinforced cements with different alkali-resistant glass fibers. *J. Amer. Ceram. Soc.*, Vol. 68, pp. 203-8.
4. McDowell and Sorrentina, F. (1990) Hydration mechanism of gehlenite glass cement. *Adv. Cem. Res.*, Vol. 3, pp. 143-52.
5. Sugama, T. and Corciello, N.R. (1991). Strength development in phosphate-bonded calcium aluminate cement. *J. Amer. Ceram. Soc.*, Vol. 74, pp. 1023-30.
6. Sugama, T. and Kukacka, L.E. (1983). Magnesium mono-phosphate cements derived from diammonium phosphate solutions. *Cem. Concr. Res.*, Vol. 13, pp. 407-16.

140 STRENGTH AND TOUGHNESS OF FIBER REINFORCED CONCRETE AT LOW TEMPERATURES

N. BANTHIA and L. QU
University of British Columbia, Vancouver, Canada
K. SAKAI
Hokkaido Development Bureau, Sapporo, Japan

Abstract
Fracture toughness of fiber reinforced cement-based composites at normal and low temperatures was evaluated. Steel macro-fiber, and carbon and steel micro-fibers were investigated in mortar and concrete matrices. Composites were significantly stronger at -50°C as compared to 22°C but were also less stiff. Overall fracture energy absorption was also higher at -50°C. When compared to concrete, cement mortar appears to be more temperature sensitive.
Keywords: Fiber reinforced concrete, micro and macro-Fibers, strength, toughness, low temperature.

1 Introduction

Fracture in cement-based materials is often believed to be a thermally activated stochastic process [1-4]. This approach explains the observed sensitivity of cement-based materials to strain-rate and predicts that both strength and fracture toughness will exhibit a strong dependence on temperature. Atomic bond ruptures, which initiate the process of fracture and require finite energy, will be aided by the energy of thermal atomic vibrations such that at higher temperatures the rate of crack growth can be expected to be higher. In other words:

$$\frac{da}{dt} = \phi(K)\, e^{\frac{-U}{RT}} \qquad (1)$$

Concrete Under Severe Conditions: Environment and loading (Volume Two) Edited by K. Sakai, N. Banthia and O.E. Gjørv. Published in 1995 by E & FN Spon. ISBN 0 419 19860 1

where da/dt is the rate of crack growth, U is the activation energy of bond rupture, R is the universal gas constant, K is the stress intensity factor and ϕ (K) is an empirical function.

Limited experimental data in the literature have confirmed that concrete is stronger but more brittle at temperatures below 0°C [5]. For fiber reinforced concrete, however, the behavior must depend not only on that of the matrix itself, but also of the fiber and the fiber-matrix bond. Here again, very little information exists in the literature [6,7], and our understanding is far from being adequate.

2 Experimental

2.1 Materials, mixes and specimens

Two matrices of cement mortar (cement:water:silica-fume:sand = 1.0:0.35:0.1:2.0) and concrete (cement:water:silica-fume:sand:coarse aggregate = 1.0:0.35:0.1:2.0:2.0) were reinforced with fibers to produce the following composites:- MC1: Mortar with 1% carbon fiber; MC2: Mortar with 2% carbon fiber; MS1: Mortar with 1% steel fiber; MS2: Mortar with 2% steel fiber, CF1: Concrete with 1% Steel macro-fiber; CF1C1: Concrete with 1% steel macro-fiber and 1% carbon micro-fiber (hybrid composite) and CF1S1: Concrete with 1% steel macro-fiber and 1% steel micro-fiber (hybrid composite). The carbon micro-fiber used was pitch-based, 3 mm in length and 18 μm in diameter. The steel micro-fiber was approx. 25 μm x 5 μm in section and about 3 mm in average length. The steel macro-fiber chosen was crimped along the length with a crescent cross section. It was 25 mm in length and 2 mm in width. Six beams (50 mm x 50 mm x 450 mm) were cast from each of the above composites and cured for at least 28 days before testing. Companion sets were also cast with plain matrices without fibers and designated as M for mortar and C for the concrete matrix.

2.2 Testing and data analysis

Beams were divided into two groups. The first group was tested at a normal temperature of 22°C ± 2°C and the other at a low temperature of -50°C ± 2°C. Beams were loaded in 3-point bending at a cross-arm speed of 0.08 mm/min in a floor mounted Instron. Displacements were recorded using an LVDT mounted under the beam. The tests at -50°C were conducted in an environmental chamber which could be installed in the testing machine itself.

It is well known that the displacements measured by an LVDT mounted directly under a beam comprise not only of the true beam displacements but also of the rigid body movement of the beam as a result of the settlement of supports. The latter being an extraneous component, it had to be subtracted from the measured displacements [8]. Normally, this is done by installing a *Yoke* around the specimen. In this investigation, however, this was accomplished by first calibrating the support itself. The support was loaded vertically and its displacements were recorded. A line was then fitted through the data and all the flexural curves were corrected by subtracting the expected support settlement at a given applied load. The procedure is described in Figure 1.

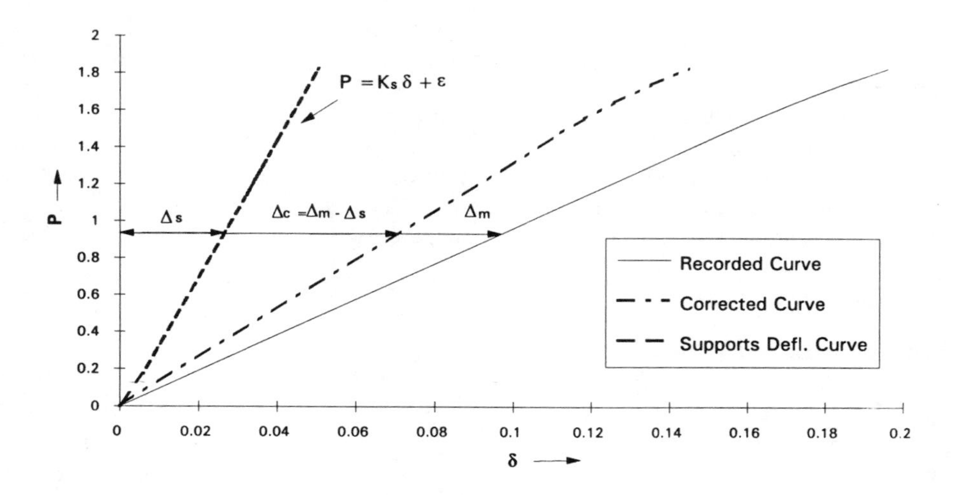

Fig. 1. Correction Applied to a Measured Curves to Account for Support Settlement

3 Results and discussion

The load-displacement curves for the various composites at normal and low temperatures were analyzed and the averaged results are given in Table 1. In Table 1, the Flexural Toughness (FT) factors calculated as per JSCE SF-4 (Method of Test for Flexural Strength and Flexural Toughness of Fiber Reinforced Concrete) are also given, Eqn. 2.

$$FT = \frac{E_{L/150}\ L}{(L/150)bh^2} \tag{2}$$

where L is the span, b and h are beam width and depth, respectively, and $E_{L/150}$ is the energy absorbed to a midspan displacement of L/150. In Figures 2-10, the load-displacement plots for the various composites at normal and low temperatures are compared.

As expected, inclusion of fibers in the matrix led to an increase in the toughness of the composite. The improvements, however, are much more significant in the case of macro-fibers than in the case of micro-fibers at equal fiber volume fractions. The strengths, however, were not altered due to fiber reinforcement. While this is well expected for macro-fiber reinforced composites, for micro-fiber reinforced composites, this is not in agreement with the previously published data [9]. It is suspected, that an adequate compaction was not achieved in the mixes reinforced with micro-fibers. When micro and macro-fibers are combined in a hybrid mix (CF1C1 and CF1S1), improvements in composite toughness occurred in the case of carbon micro-fiber but not for steel micro-fiber. This aspect needs to be explored further.

Table 1. Experimental results

	Peak load (kN)		Deflection at peak load (mm)		Energy at peak load (N-m)		Fracture energy* (N-m)		JSCE (MPa)		Ratio -50°C/22°C			
	22°C	-50°C	22°C	-50°C	22°C	-50°C	22°C	-50°C	22°C	-50°C	Peak load	Defl. @ peak load	Energy @ peak load	JSCE, FT factor
Mortar														
M**	1.60	2.80	0.14[1]	0.44[2]	0.12	0.61	0.12	0.61	0.14	0.73	1.75	3.14	5.08	5.21
MC1	1.36	2.34	0.21	0.29	0.13	0.33	0.57	0.33	0.65	0.39	1.72	1.38	2.53	0.60
MC2	0.92	2.23	0.24	0.45	0.12	0.48	0.49	0.48	0.59	0.57	2.42	1.87	4.00	0.96
MS1	1.80	2.70	0.18	0.32	0.16	0.43	0.16	0.43	0.20	0.52	1.50	1.77	2.68	2.60
MS2	1.74	2.97	0.20	0.49	0.19	0.75	0.19	0.75	0.23	0.90	1.70	2.45	3.94	3.91
Concrete														
C***	1.65	1.91	0.12[3]	0.37[4]	0.10	0.35	0.10	0.35	0.14	0.42	1.15	3.08	3.50	3.00
CF1	1.44	1.99	0.13	0.35	0.10	0.32	2.00	2.76	2.40	3.32	1.38	2.69	3.10	1.38
CF1C1	1.52	2.58	0.24	0.48	0.28	0.66	2.57	3.41	3.09	4.09	1.69	2.00	2.35	1.32
CF1S1	1.48	2.67	0.21	0.24	0.16	0.30	1.69	2.15	2.03	2.58	1.80	1.14	1.87	1.27

* Calculated to a disp. of span/150

** f'_c = 79 MPa at 22°C and 107 MPa at -50°C

*** f'_c = 76 MPa at 22°C and 90 MPa at -50°C

[1] Predicts an elastic modulus (E) of 29.2 GPa.

[2] Predicts an elastic modulus (E) of 16.2 GPa.

[3] Predicts an elastic modulus (E) of 35.8 GPa.

[4] Predicts an elastic modulus (E) of 13.4 GPa.

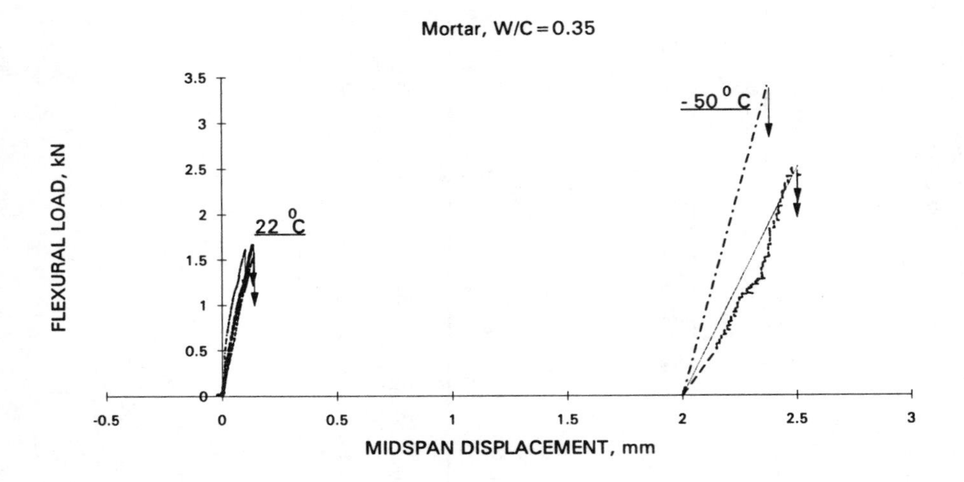

Fig. 2. Load-Displacement Plots for Plain Mortar at Normal and Low Temperatures

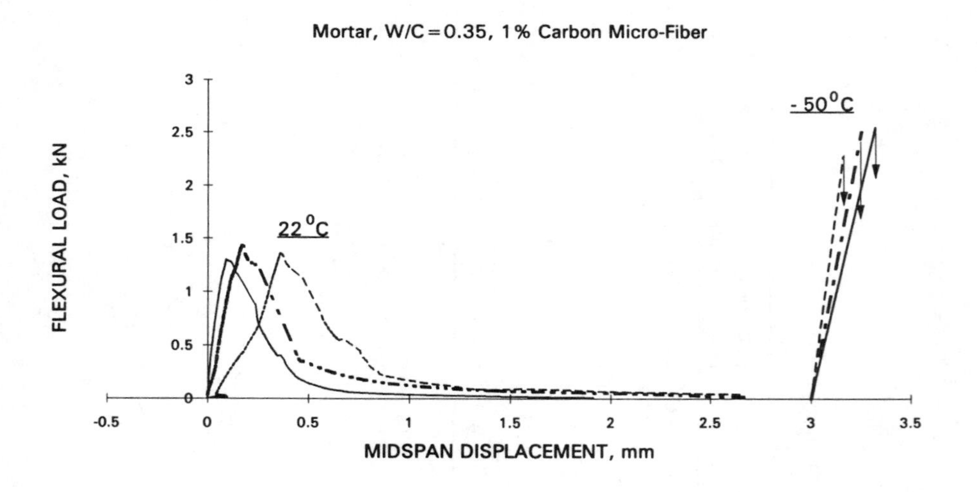

Fig. 3. Load-Displacement Plots for Mortar Reinforced with 1% Carbon Micro-Fiber at Normal and Low Temperatures

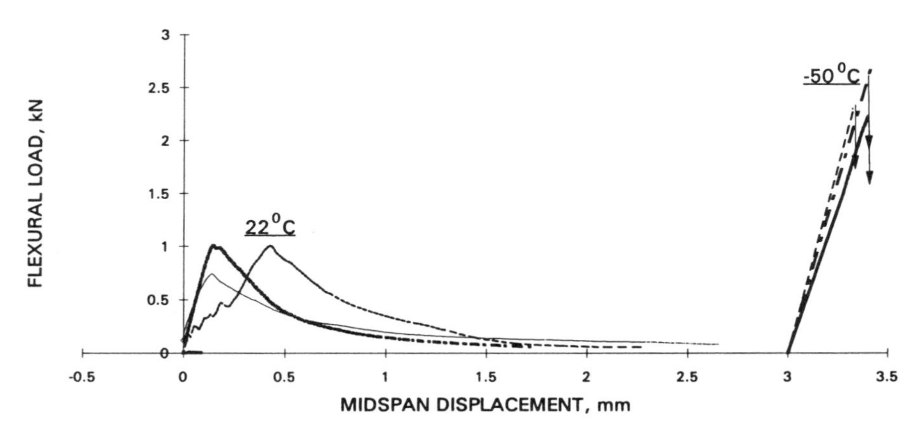

Fig. 4. Load-Displacement Plots for Mortar Reinforced with 2% Carbon Micro-Fiber at Normal and Low Temperatures

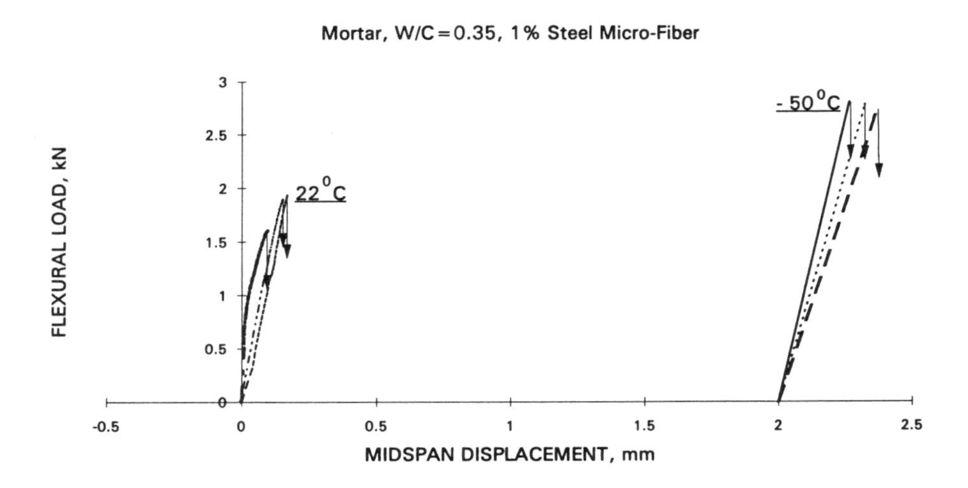

Fig. 5. Load-Displacement Plots for Mortar Reinforced with 1% Steel Micro-Fiber at Normal and Low Temperatures

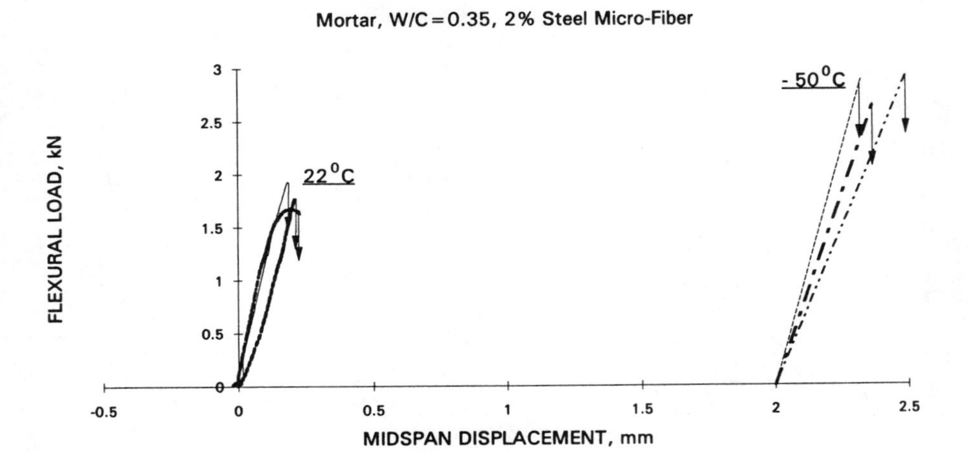

Fig. 6. Load-Displacement Plots for Mortar Reinforced with 2% Steel Micro-Fiber at Normal and Low Temperatures

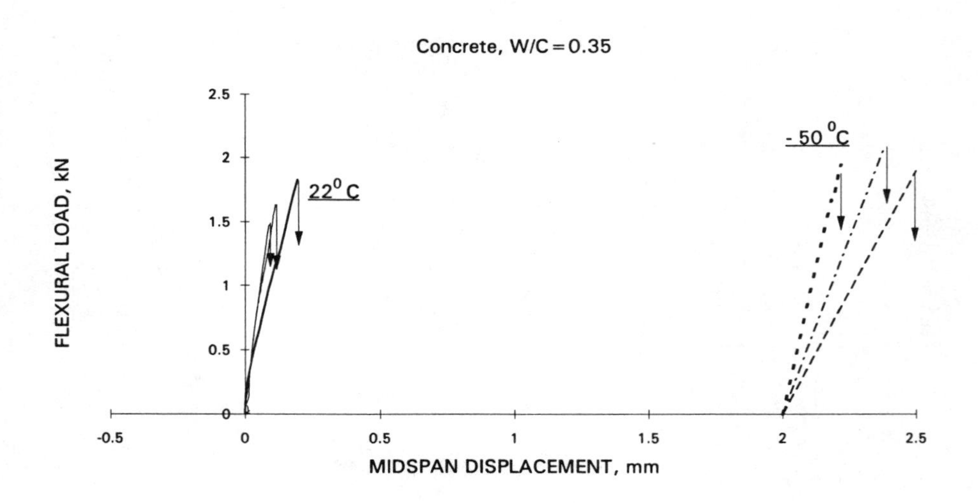

Fig. 7. Load-Displacement Plots for Plain Concrete at Normal and Low Temperatures

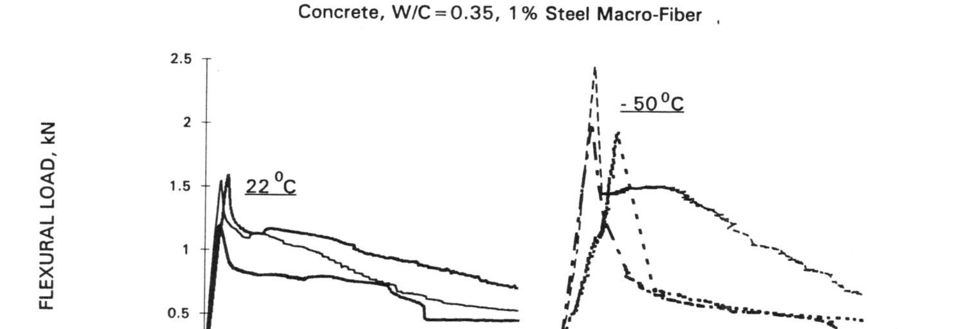

Fig. 8. Load-Displacement Plots for Concrete Reinforced with 1% Steel Macro-Fiber at Normal and Low Temperatures

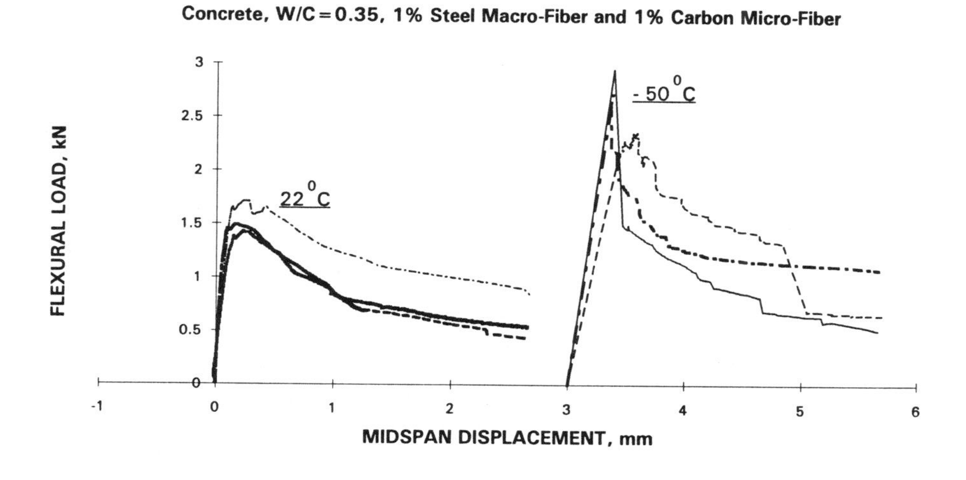

Fig. 9. Load-Displacement Plots for Concrete Reinforced with 1% Steel Macro-Fiber and 1% Carbon Micro-Fiber at Normal and Low Temperatures

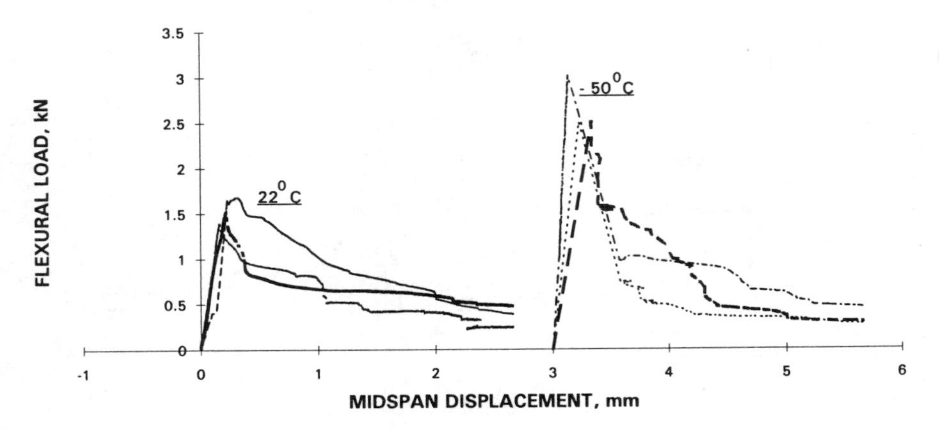

Fig. 10. Load-Displacement Plots for Concrete Reinforced with 1% Steel Macro-Fiber and 1% Steel Micro-Fiber at Normal and Low Temperatures

From the curves in Figures 2-10, an increase in the strength of the various composites at a low temperature of -50°C is clearly noticeable. Also noticeable is a greater than proportional increase in the displacements at the peak load at the low temperature. While this indicates a reduced elastic modulus (assuming an elastic response up to the peak load) at a low temperature, it may also be a result of inaccuracies stemming from the use of support calibration carried out at room temperature to correct the curves obtained at -50°C.

The specimens absorbed much greater energies up to the peak loads when tested at -50°C, and except for some micro-fiber reinforced composites, the fracture energies consumed up to a displacement of span/150 were also higher at -50°C. This is in agreement with the data reported by Staneva et al [6]. It is remarkable, however, that the increases in the energy absorption to peak load are greater than the increases in the fracture energy values which would lead to lower *toughness indices* at -50°C if the procedure outlined in ASTM C1018 is followed. Indeed, a reduction in the *toughness indices* at -50°C has been previously reported [7]. Cement mortar and its composites, appear to be more sensitive to temperature than concrete and its composites. Since the mortar phase has more freezable water per unit volume than concrete, the observed temperature sensitivity for cement-based materials apparently is linked to the extent of freezable water in the system. This however, needs to be explored further.

4 Conclusions

Some data related to the flexural behavior of fiber reinforced cement-based composites at 22°C and at a low temperature of -50°C are presented. Composites were found to be significantly stronger at -50°C. An increase in both the energy absorption up to the peak load and in the absorbed fracture energy to a displacement of span/150 were noted at -50°C. In addition to the composites being more compliant at a low temperature, the increased sensitivity of the mortar phase as compared to concrete itself were also inferred.

5 Acknowledgements

The financial support provided by the Natural Sciences and Engineering Research Council of Canada under the Japan Science and Technology Fund and the continuing cooperation of the Civil Engineering Research Institute of the Hokkaido Development Bureau, Japan, are gratefully acknowledged.

6 References

1. Bazant, Z.P. and Prat, P.C. (1988) Effect of Temperature and Humidity on Fracture Energy of Concrete, ACI Mat. Journal, July-August, pp. 262-271.
2. Mihashi, H. and Izumi, M. (1977) A Stochastic Theory for Concrete Fracture, Cement and Concrete Research, 7, pp. 411-422.
3. Reinhardt, H.W. (1986) Strain Rate Effects on the Tensile Strength of Concrete as Predicted by Thermodynamic and Fracture Mechanics Models in Cement-Based Composites: Strain Rate Effects on Fracture, MRS Symp. Proc., 64, pp. 1-14.
4. Whitmann, F.H. (1984) Influence of Time on Crack Formation and Failure of Concrete, in (preprints) Application of Frac. Mech. to Cem. Composites, NATO-ARW Workshop, pp. 443-464.
5. Tongnon, G. (1969) Behavior of Mortars and Concretes in the Temperature Range from +20°C to -196°C, Proc. Fifth Int. Symp. on Chemistry of Cements, Tokyo, The Cement Association of Japan, pp. 229-248.
6. Staneva, P., Sakai, K, Horiguchi, T and Banthia, N. (1992) Flexural Behavior of Steel Fiber Reinforced Concrete at Low Temperatures, paper presented at ACI Spring Convention, Vancouver.
7. Banthia, N. and Mani, M. (1993) Toughness Indices of Steel Fiber Reinforced Concrete at Sub-Zero Temperatures, Cement and Concrete Research, 23, pp. 863-873.
8. Banthia, N. and Trottier, J.-F. (1995) Toughness Characterization in Steel Fiber Reinforced Concrete: Some Concerns and a Proposition, ACI Materials Journal, in press.
9. Banthia, N. Chokri, K. Ohama, Y. and Mindess, S. (1994) Fiber Reinforced Cement-Based Composites under Tensile Impact, Advanced Cement Based Materials, 1, pp. 131-141.

141 EVALUATION OF STRENGTH PARAMETERS OF CEMENTITIOUS SIDING BOARDS WITH CARBONATION

H. MIHASHI and K. KIRIKOSHI
Tohoku University, Sendai, Japan
H. OKADA
Asahi Glass Co. Ltd, Yokohama, Japan

Abstract
Cementitious siding boards are widely used in Japan as facing materials for houses and other buildings. Although they are strong enough to resist usual design loads such as wind and snow, they are often cracked during transportation and construction. Furthermore, cracking occurs around nailing holes several years after the construction. Previous studies suggested that such cracking might be caused by carbonation shrinkage.

In this paper, a test method is proposed to evaluate quantitatively the brittleness by means of fracture energy G_F and strain softening diagram which are fracture mechanics parameters. The influence of carbonation on the brittleness is discussed by this approach.
Keywords: Carbonation, cracking, fracture energy, fracture mechanics of concrete, siding board, strain softening.

1 Introduction

Various kinds of cementitious siding boards, used as facing materials for houses and other buildings, have been developed in Japan. These boards are very light, easily sawn and nailed but strong enough to resist snow loads, wind loads due to typhoons, and fire. However, they are rather brittle and often fractured in the midst of handling by bending or of nailing by cracking. Although flexural strength and impact load resistance are regulated according to Japanese Industrial Standard (JIS), the brittleness is evaluated only by means of traditional test methods. Furthermore, cracking sometimes occurs around nailing holes several years after the construction. Previous investigations suggested that this kind of cracking might be caused by carbonation shrinkage restrained by the nail, though the mechanism

Concrete Under Severe Conditions: Environment and loading (Volume Two) Edited by K. Sakai, N. Banthia and O.E. Gjørv. Published in 1995 by E & FN Spon. ISBN 0 419 19860 1

Table 1. Tested specimens of siding board

Type	Full Scale (mm)	Production Process
A	455x3030x11	dip up formed, autoclave curing
B	455x3040x12	press formed, steam curing
C	455x2880x12	dip up formed, steam curing
D	455x2880x12	dip up formed (single layer), steam curing

itself has not been sufficiently clarified yet. Tashiro and his coworkers [1] studied the crack velocity of carbonated siding boards by means of the double torsion test method on the basis of linear elastic fracture mechanics. He concluded that carbonated siding boards are more brittle than uncarbonated ones.

Recently, a nonlinear fracture mechanics approach has been introduced to evaluate quasi-brittle failure properties of cementitious composite materials (for example, Li and Ward [2], Shah and Ouyang [3], and Slowik and Wittmann [4]). Fracture energy G_F was proposed by Hillerborg and his coworkers [5], and a RILEM technical committee [6] prepared a recommendation on the experimental procedure to determine G_F by means of three-point bend tests on notched beams of concrete and mortar. The parameter G_F is defined as the energy absorbed to create a unit area of fractured surface.

The purpose of the present paper is to propose the usage of the nonlinear fracture mechanics parameter G_F to quantify the brittleness and to discuss the influence of carbonation on the brittleness properties on the basis of the nonlinear fracture mechanics of concrete. Compact tension (CT) tests were carried out to determine the value of G_F which is used to discuss the brittleness together with the softening diagram.

2 Experimental procedures

2.1 Tested specimens

Four different kinds of cementitious siding boards shown in Table 1, the most typical types on the market in Japan, were tested. They were made of Portland cement or calcium silicate, perlite, fiber and some other additives. Because of the production process, most of them were anisotropic. Therefore, each specimen was tested in two directions for several properties. In most cases, the number of specimens was three. Only Izod impact tests were performed with five specimens for each series.

2.2 Traditional crack resistance properties

To evaluate the mechanical properties of cementitious siding boards, flexural strength and impact resistance by means of the Izod type test have been traditionally used. The flexural strength was evaluated by a three point bend test on beam specimens with the span length of 200 mm. The Izod sample was a notched beam that was fixed at one end and impacted on the unsupported section along the side of the beam containing the notch. Izod impact energy is usually defined by the energy absorbed per unit area of the net section. As far as crack resistance properties are concerned, nailing and handling tests have been performed in practice. The nailing test evaluates the survival probability without cracking by nailing at a corner of the board using various distances from the edge. As shown in Fig. 1, a 200x200 (mm) plate was nailed by an automatic nailing machine at points which were a distance of L from each corner to see if the plate was cracked. L was varied from 10 mm to 25 mm. The diameter and the length of stainless steel ring nail were 2.2 mm and 38 mm, respectively. The handling test was performed by two persons who vertically shook a full scale board to apply an amplitude of vibration 1.5 times larger than the deflection due to its own weight.

2.3 Evaluation of crack resistance properties by G_F

Compact tension (CT) tests were carried out by a servo-controlled hydraulic jack under the displacement control condition to determine G_F of siding boards. The size of the specimen is shown in Fig. 2, whose geometry is based on ASTM-E399 except the thickness. In addition to the load, the crack opening displacements at the notch tip (COD) and also on the loading axis were measured.

G_F was determined in conformity with the RILEM recommendation [6] as follows:

$$G_F = (W_0 + mgd)/A_{lig} \tag{1}$$

where W_0 = area under the load vs. displacement curve (N/m), m = weight of the specimen (kg), g = acceleration due to gravity, 9.81 m/s^2, d = displacement at the final failure of the specimen (m), A_{lig} = area of the ligament (m^2).

Each siding board was tested in two directions: one, notched in the longitudinal direction (Case-X specimen); and the other, notched in the lateral direction (Case-Y specimen).

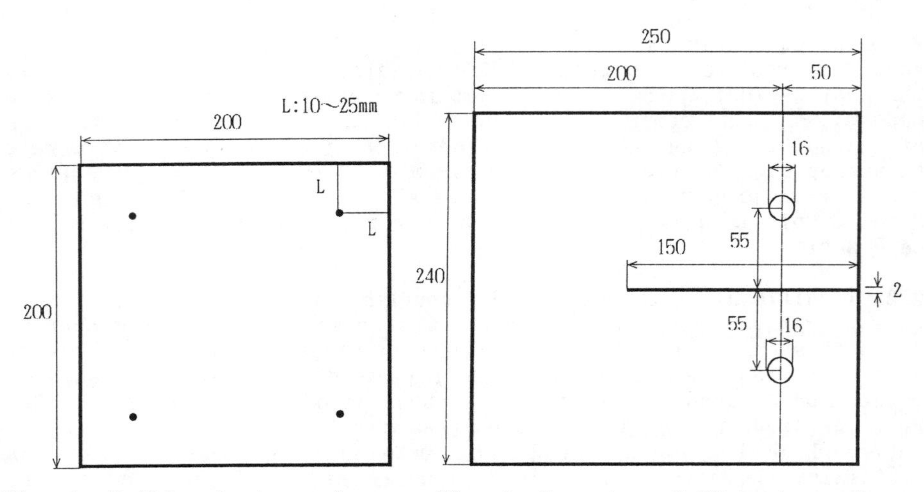

Fig. 1. Nailing test specimen. Fig. 2. Geometry of CT test specimen.

2.4 Carbonation test

Three specimens per each series were kept in a chamber of an atmosphere at a constant temperature of 40 C, a relative humidity of 80% and a CO_2 gas content of 10%. After two weeks, phenolphthalein method was used to confirm that all of the specimens were thoroughly carbonated.

3 Analytical procedure of softening diagram as a material law for cracking

If a specimen is tested in a high performance machine, the whole stress-strain curve of quasi-brittle material including the descending branch can be obtained, even in the direct tensile test. The behavior of the descending branch is usually called strain softening, and this expresses the mechanical property of the fracture process zone [5]. The strain softening diagram can be used as a constitutive law for analyzing nonlinear behavior due to cracking of quasi-brittle materials. The area below the softening curve is normally assumed to coincide with G_F. A high value of G_F is obtained by a high tensile strength and/or a large critical crack width or strain.

Usually it is not easy to obtain the softening curve by a direct tensile test. A novel technique based on the J-integral was proposed by Li and Ward [2] to experimentally determine the softening diagram in cementitious composites. This method, however, needs two types of pre-notched specimens with slightly different notch length, and the obtained results are very much influenced by the variability of each experimental curve. Uchida and his co-workers [7] modified the J-integral based test method to determine the softening diagram by a single specimen, assuming a linear distribution of the cohesive crack width over the ligament. The energy E(w) dissipated in the specimen up to the crack opening displacement w is given by eq. (2).

$$E(w) = (A_{lig}/w)\int_0^w e(w)dw \tag{2}$$

where e(w) is the energy dissipated in a unit area up to crack width w. On the other hand, the dissipated energy E(w) should correspond to the work performed by the external load P up to the crack opening displacement on the loading axis d_w. Then the softening diagram is obtained as follows:

$$\sigma(w) = \{w\ddot{E}(w) + 2\dot{E}(w)\}/A_{lig} \tag{3}$$

4 Test results

Mechanical test results are shown in Table 2. Naturally, lighter specimens (C and D) are more porous. Load vs. displacement curves of specimens without carbonation are shown in Fig. 3. It is obvious that specimens A, B and C are anisotropic materials and the toughness for cracking in the lateral direction (Case-Y) is much higher than that in the longitudinal direction (Case-X). The flexural strength and G_F

Table 2 Test results

	Before Carbonation					After Carbonation		
	Young's Modulus (GPa)	Specific Gravity	Izod Impact (J/mm^2)	Flexural Strength (MPa)	G_F (N/m)	Specific Gravity	Flexural Strength (MPa)	G_F^* (N/m)
A [Y]	3.62	1.04	21.6	10.9	1034	1.22	12.8	971
A [X]				8.7	665		10.0	602
B [Y]	2.76	1.07	33.3	13.8	1009	1.18	13.8	1146
B [X]				8.5	487		8.8	485
C [Y]	3.33	0.93	17.6	11.8	1421	1.11	11.9	1684
C [X]				10.5	740		10.3	944
D [Y]	2.62	0.92	11.8	7.8	673	1.05	6.5	836
D [X]				7.6	704		7.7	886

where G_F^* means the fracture energy after carbonated.

values quantitatively describe the anisotropic properties. Fig. 4 shows the comparison of load vs. displacement curves of uncarbonated specimens and those of carbonated ones. All of these curves are the mean of ones obtained from three specimens for each test series. An obvious difference is recognized only in Series C and D, which are more porous than A and B. According to the modified J-integral based method, softening diagrams for each specimen were obtained as shown in Fig. 5. The main difference in the softening diagram between carbonated and uncarbonated specimens is the maximum cohesive stress recognized in C-X, D-X and D-Y. Only in the case of C-Y was the maximum cohesive stress remarkably reduced after carbonation, but on the contrary, the descending branch became much more moderate, where the critical crack width in the C-Y Series was much larger than the others. These changes lead to larger values of G_F^* than G_F.

5 Discussion

According to Table 2, both G_F and the flexural strength were strongly dependent on the direction of the specimen's axis, and they are much correlative. G_F is obviously correlative to the nailing (Fig. 6) and the handling resistances (Fig. 7). These results prove that G_F is an effective parameter to quantify the crack resistance properties of cementitious siding boards.

In porous cementitious composites, CO_2 reacts with the constituents of cement paste. According to previous studies (for example, [8]), the reaction may, on the one hand, produce a finer microstructure to strengthen the material. On the other hand, however, the reaction may promote carbonation shrinkage that is not desirable in a finishing

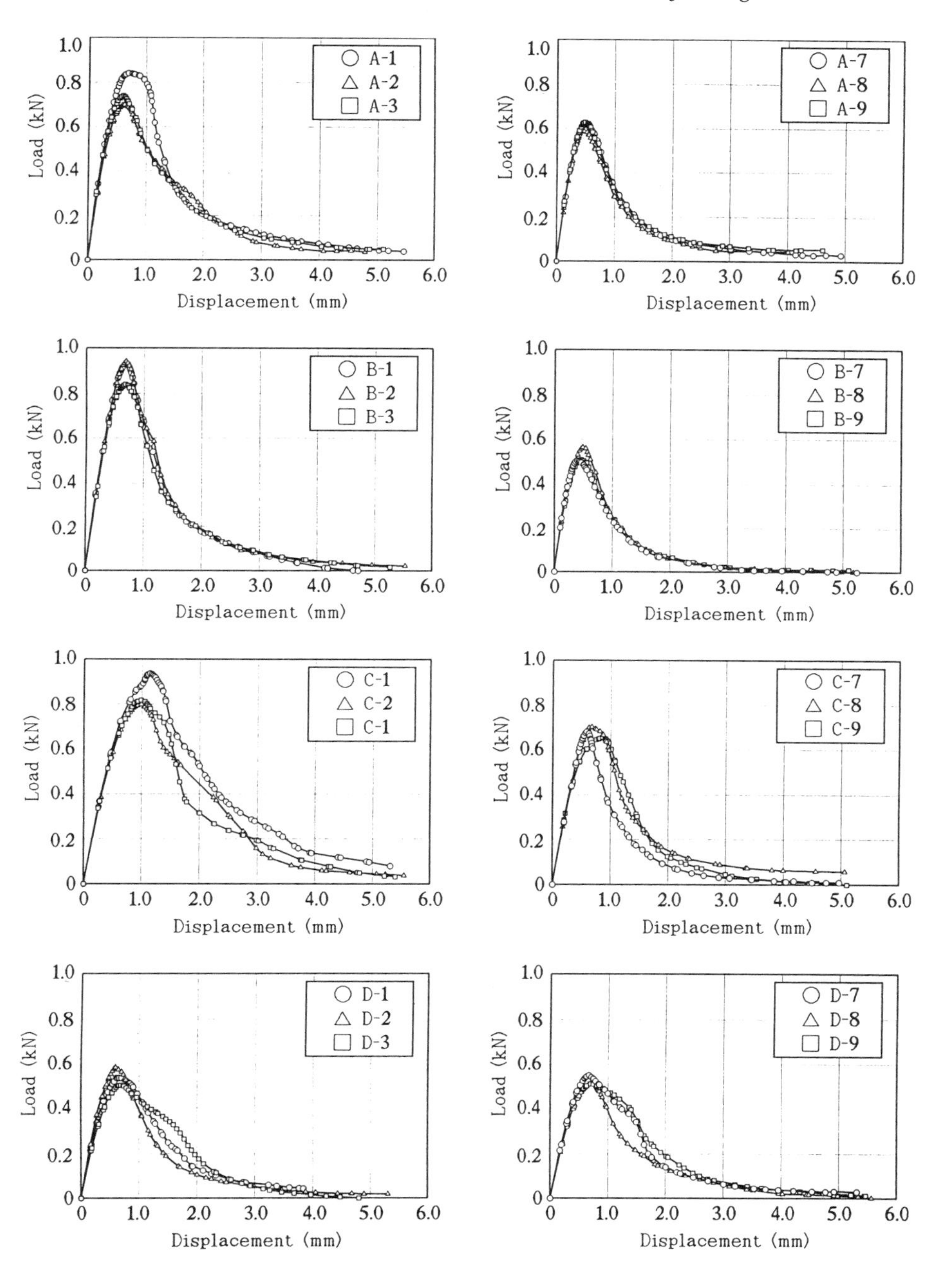

Fig. 3. Load-displacement curves before carbonation.

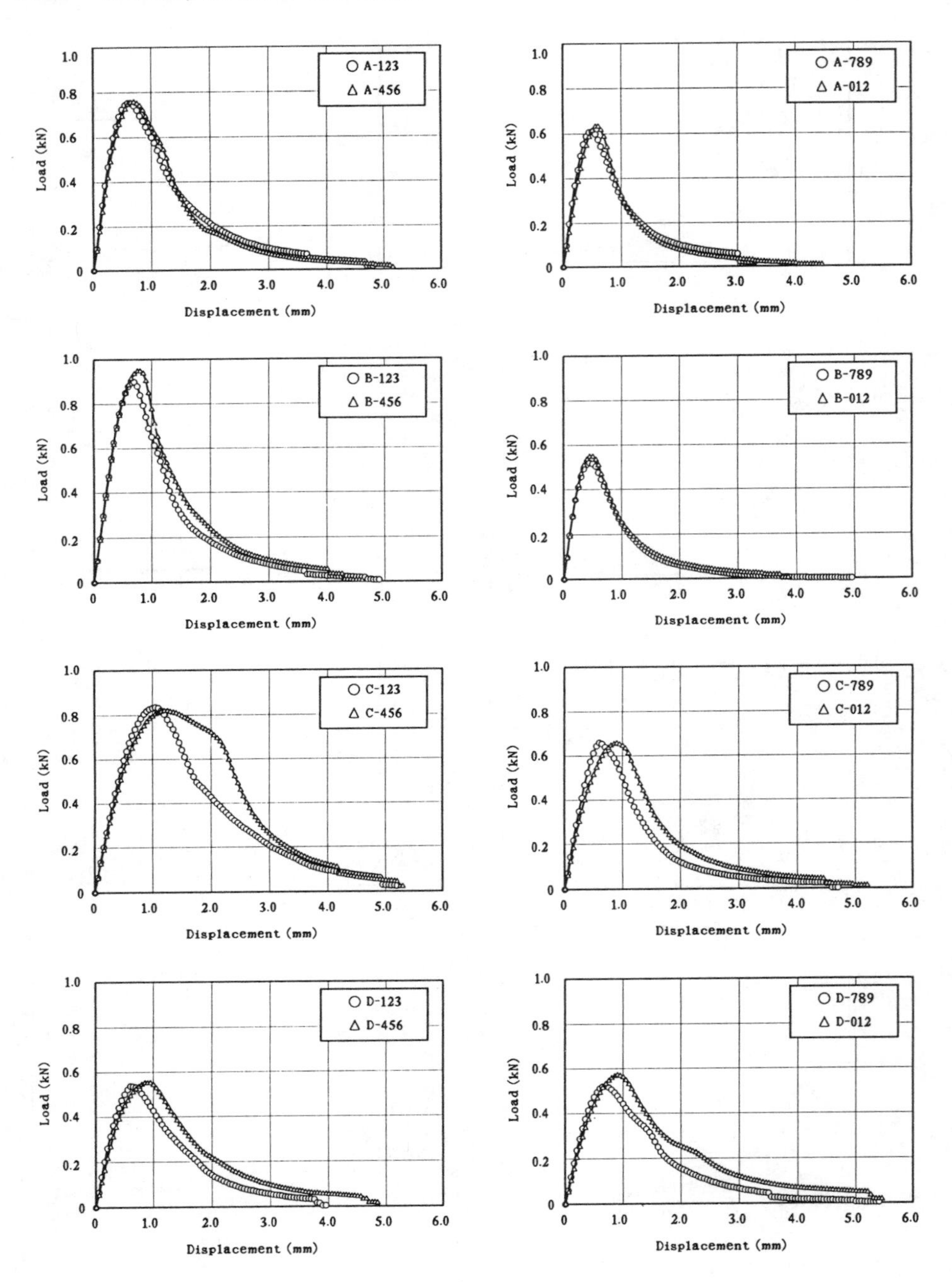

Fig. 4. Comparison of load-displacement curves before and after carbonation (○: before carbonation; △: after carbonation).

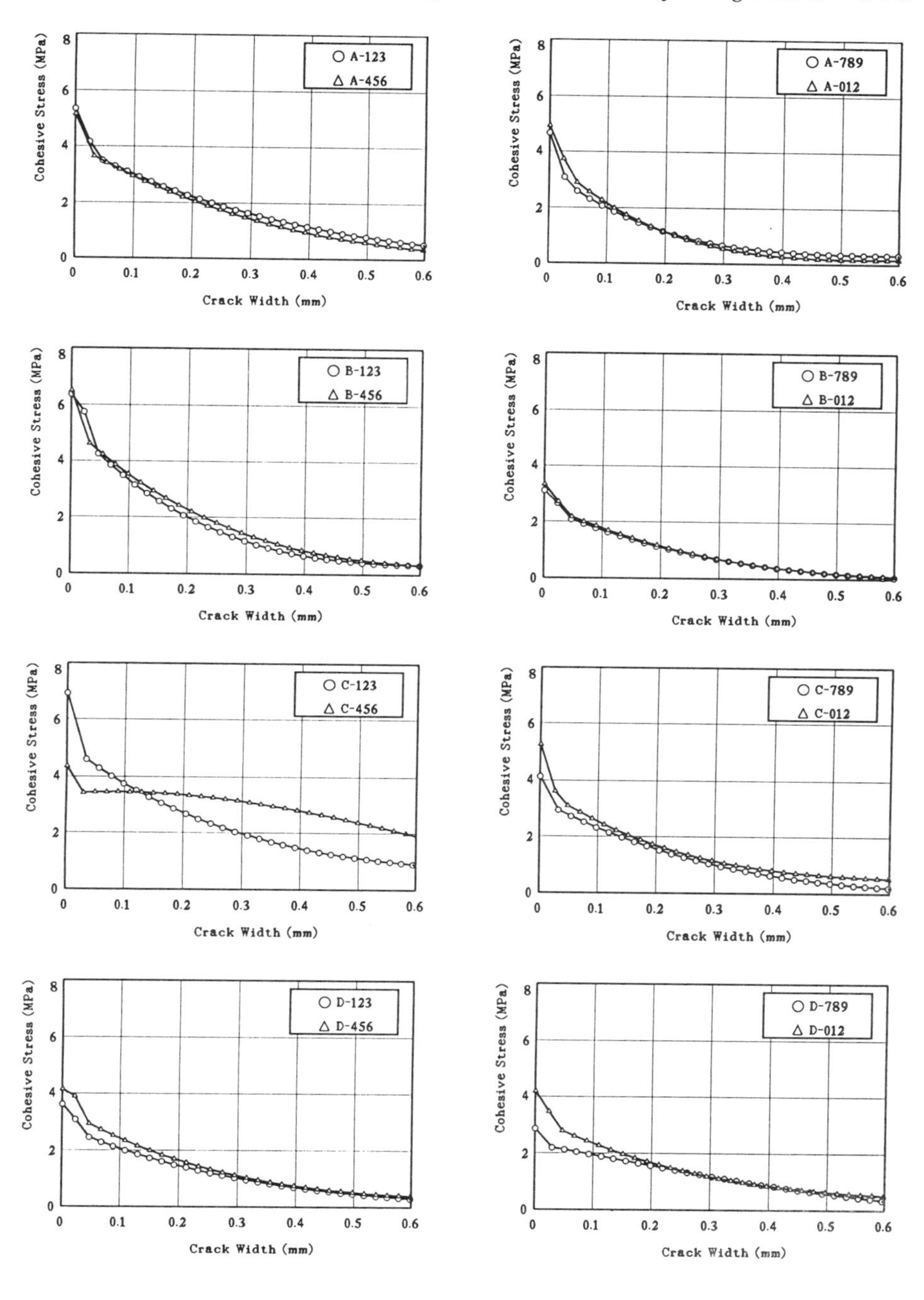

Fig. 5. Strain softening diagrams (○: before carbonation; △: after carbonation).

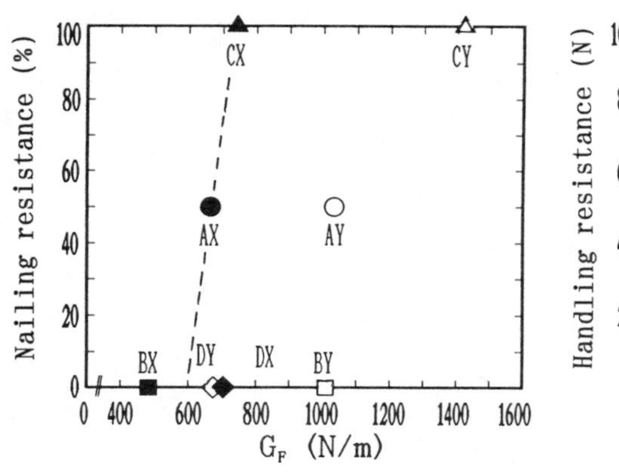

Fig. 6. Relation between G_F and nailing resistance.

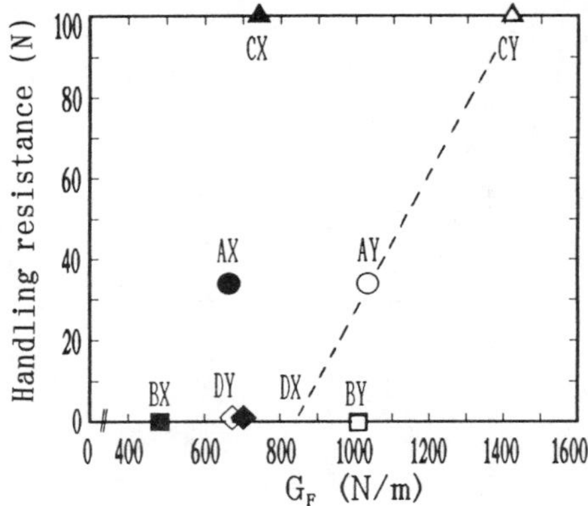

Fig. 7. Relation between G_F and handling resistance.

material.

The microstructure of C-X, D-X and D-Y might become finer after carbonation because previous studies [9, 10] suggested that the maximum cohesive stress was strongly correlated to the fineness of the microstructure. In case of C-Y, however, the carbonation reduced the maximum cohesive stress, but the fracture energy increased because of the improvement of bridging effects due to fiber. These results may suggest that cracking in cementitious siding boards which occurred several years after the construction is not simply because the boards become more brittle by carbonation. Further studies need to be done on the influence of moisture movement and changes of microstructure to prevent those kinds of cracking.

6 Conclusions

Compact tension tests were carried out to evaluate the crack resistance properties of cementitious siding boards. As a result, it was proved that G_F was clearly correlative to the nailing performance and the handling resistance. The influence of carbonation on crack resistance properties was also recognized in some cases, and it was noted that carbonation itself does not necessarily deteriorate the crack resistance properties of cementitious siding boards. It is suggested that the test method to determine G_F of cementitious siding boards needs to be standardized in order to provide a quantitative evaluation of their crack resistance properties. The softening diagram may give more comprehensive information.

7 References

1. Tashiro, C., Sano, O. and Tateishi, Y. (1990) Effect of carbonation on crack velocity of cement board. CAJ Proceedings of Cement and Concrete, No. 44, pp.460-463.

2. Li, V.C. and Ward, R.J. (1989) A novel testing technique for post-peak tensile behavior of cementitious materials. Fracture Toughness and Fracture Energy - Test Methods for Concrete and Rock (ed. H. Mihashi, H. Takahashi and F.H. Wittmann), Balkema, pp. 183-195.
3. Shah, S.P. and Ouyang, C. (1991) Mechanical behavior of fiber-reinforced cement-based composites. J. American Ceramics Society, Vol. 74, No. 11, pp. 2727-2738, 2947-2953.
4. Slowik, V. and Wittmann, F.H. (1992) Fracture energy and strain softening of AAC. Advances in Autoclaved Aerated Concrete (ed. F.H. Wittmann), Balkema, pp. 141-145.
5. Hillerborg, A., Modeer, M. and Petersson, P.E. (1976) Analysis of crack formation and crack growth in concrete by means of fracture mechanics and finite elements. Cement and Concrete Research, Vol. 6, pp.773-782.
6. RILEM Draft Recommendation (TC50-FMC) (1985) Determination of the fracture energy of mortar and concrete by means of three-point bend tests on notched beams. Materials and Structures, Vol. 18, No. 106, pp. 285-290.
7. Uchida, Y., Rokugo, K. and Koyanagi, W. (1991) Determination of tension softening diagrams of concrete by means of bending tests. Proc. of Japan Society of Civil Engineers, No. 426/V-14, pp. 203-212 (in Japanese).
8. JCI Technical Committee on Carbonation (1993) State-of-the-Art Report on Carbonation of Concrete. Japan Concrete Institute, 93p. (in Japanese).
9. Mihashi, H. (1992) Material structure and tension softening properties of concrete. Proc. of the First International Conference on Fracture Mechanics of Concrete Structures (ed. Z.P. Bazant), Elsevier Applied Science, pp. 239-250.
10. Isu, N., Teramura, S., Nomura, N., Mihashi, H. and Mitsuda, T. (1994) Influence of autoclaving process on the fracture behavior of ALC. J. of the Ceramic Society of Japan, Vol. 102, No. 8, pp. 785-789.

142 RESIDUAL CAPACITY OF CONCRETE BEAMS DAMAGED BY SALT ATTACK

A. KAWAMURA
Shimizu Corporation, Japan
K. MARUYAMA
Nagaoka University of Technology, Japan
S. YOSHIDA
Niigata Prefectural Office, Japan
T. MASUDA
Japan Highway Public Corporation, Japan

Abstract
Concrete structures which have suffered from salt attack have to be repaired or reinforced sooner or later. However it is difficult to make a decision as to when to repair or reinforce them without knowing how much has been reduced the capacity of concrete structures. Taking an account of longitudinal crack width or the amount of corrosion of steel bars as an index for damage level by salt attack, this paper deals with how the flexural capacity is reduced in RC beams.
Keywords: Bond, corrosion of steel bar, fatigue capacity, flexural capacity, longitudinal crack width, residual capacity.

1 Introduction

For maintenance and repair of existing concrete structures, it is necessary to determine the residual capacity of structures damaged by salt attack. This paper discusses the reduction of flexural capacity of RC beams when the longitudinal reinforcing bars corrode.

The test program consisted of two phases. In the first series, the fundamental behavior of damaged beams was studied using relatively small specimens (100x100x1200 mm) with a single reinforcing bar. An erosion test was conducted to corrode steel bars by compulsion, and the flexural behavior of beams was investigated under static and fatigue loadings. The main parameter was longitudinal crack widths caused by the corrosion of steel bars[1].

In the second series, relatively large specimens were used: 200 mm

Concrete Under Severe Conditions: Environment and loading (Volume Two) Edited by K. Sakai, N. Banthia and O.E. Gjørv. Published in 1995 by E & FN Spon. ISBN 0 419 19860 1

wide, 300 mm high and 2800 mm long. Two ϕ19 deformed bars were placed with a cover thickness of 40 mm. The erosion test was also conducted to corrode steel bars. The parameters were the amount of corrosion of steel bars (in other words, the longitudinal crack widths), the spacing of stirrups, and whether or not splices of longitudinal reinforcing bar existed at the mid span[2].

2 Behavior of small damaged beam

2.1 Behavior under static loading

2.1.1 Test specimens

The objective of the first series test was to study quantitatively how the longitudinal cracks caused by the corrosion of steel bars influenced the capacity of beams. Therefore, to simplify the conditions of the test, relatively small specimens (100x150x1200 mm) reinforced by a single steel bar (D13, p=1%) were used. To accelerate the corrosion of the steel bars, a chloride solution nearly equivalent to sea water (3.13% NaCl solution) was used as the mixing water. The Water-Cement ratio (W/C) was 63% and the sand-aggregate percentage (s/a) was 43% with the maximum size of coarse aggregate being 15 mm. Stirrups were arranged at an interval of 10 cm to control the longitudinal crack of the concrete surface caused by corrosion of the steel bar. The specimen is shown in Fig. 1.

2.1.2 Experimental details of static loading test

The corrosion of steel bar was induced by an erosion test. The erosion test apparatus is shown in Fig. 2. Direct current was conducted to the specimens with the main bar being the anode and copper plates placed along both sides of the specimen as the cathode. The specimens were submerged in a saline solution to accelerate corrosion. The Longitudinal crack width in each specimen was mostly controlled by measuring the amount of accumulated current. One longitudinal crack initiated along the main bar. No cracks were observed in the side of specimen.

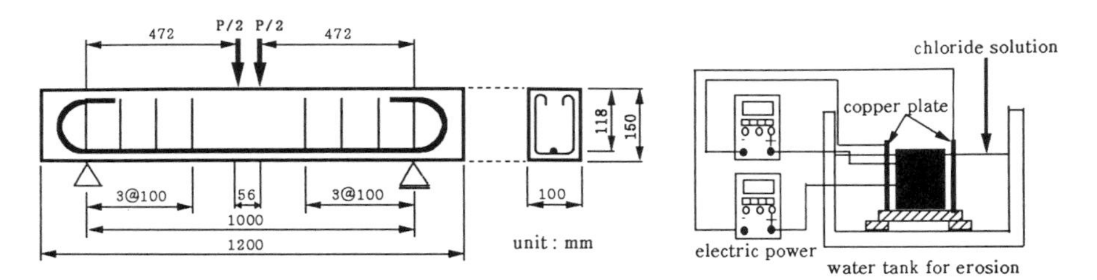

Fig. 1. Test specimen. Fig. 2. Erosion test apparatus.

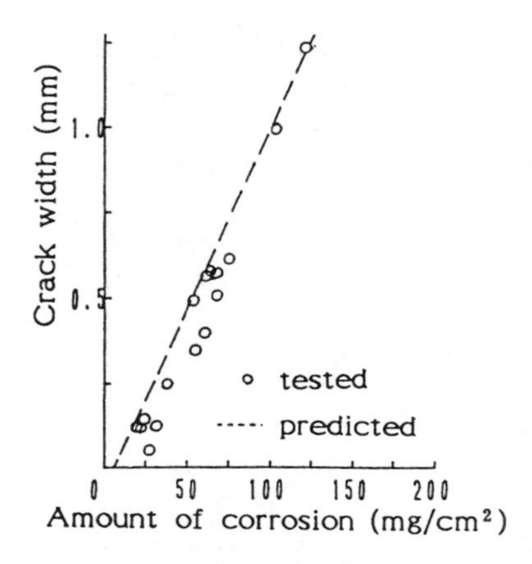

Fig. 3. Relationship between crack width and amount of corrosion.

After applying a certain amount of current to the specimen, the flexural behavior was examined under static loading. The specimens had a symmetric two point-loading. The shear span was 47.2 cm (a/d=4.0), so the specimen would fail in flexure. The load was increased incrementally with a fraction of 2 kN up to the maximum capacity. At a certain load level, measurements were conducted for load, deflection, and crack width increments of both longitudinal and flexural cracks.

2.1.3 Relationship between the amount of corrosion of steel bar and the longitudinal crack width

According to the past studies[3], the longitudinal crack width can be estimated by using concrete cover thickness, density of rust, arrangement of reinforcing bars. Based on this studies, the longitudinal crack width of these beam specimens were predicted to have some relationships with the amount of corrosion of steel bar. To examine this relationship, the test results were compared with the predictions. The compared results are shown in Fig. 3. The crack widths are proportional to the amount of corrosion.

2.1.4 Test results of static loading

The results of the maximum capacity, failure mode, initial stiffness, cracking load, and the load at the rapid increase of deflection are summarized in Table 1. The influence of the longitudinal crack width caused by corrosion of the steel bar is shown in Fig. 4 in terms of the load deflection curve.

As the longitudinal crack width grew larger, the maximum capacity tended to decrease as shown in Fig. 4 and Table 1. The capacity reduction ratio, however, was only 10%, even if the longitudinal crack width grew 1.0 mm, as shown in Fig. 5.

When the longitudinal crack width was 0.3 mm, the stiffness of the

Table 1. Specimens and result of static loading test

Specimen[1)]	f'c (MPa)	Ri[2)] (P= 5.88kN)	Pcr[3)] (kN)	Pd[4)] (kN)	ratio	Pu (kN)	ratio	Failure[5)]
BS-N	31.1	1	9.32	21.53	1	21.62	1	F
BS1-C05	31.1	0.83	9.81	21.72	1.01	22.33	1.03	F
BS2-C30	31.1	0.21	11.77	22.65	1.05	22.80	1.05	F
BS4-C50	35.1	0.19	7.85	19.65	0.91	20.24	0.94	F
BS7-C100	36.6	0.44	7.85	19.24	0.89	19.62	0.91	F
BS9-CM50	31.1	0.25	7.85	21.59	1.00	21.89	1.01	F
BS11-CM55	31.1	0.18	7.85	20.61	0.96	21.43	0.99	F
BS-Nu	31.1	1	8.83	19.81	1	20.66	1	F
BS3-C30u	37.1	0.50	9.81	20.65	1.04	20.86	1.01	F
BS5-C50u	35.1	0.10	5.88	18.57	0.94	19.06	0.92	F
BS6-C50u	34.3	0.67	11.77	17.68	0.89	18.59	0.90	D
BS8-C100u	36.6	0.39	7.85	16.65	0.84	17.65	0.85	D
BS10-CM50u	37.1	0.11	9.81	19.71	1.00	20.62	1.00	D

1) BS3-C 30 u
- u — Without stirrup confinement
- 30 — Crack width (x10^{-2} mm)
- C — N ···non corrosion; C ···re-bar corroded in all span length; CM···re-bar corroded only at the mid span 20cm

2) Initial stiffness ratio
3) Load occurring flexural crack
4) Load increasing deformation rapidly
5) F ··· flexural tensile failure
D ··· diagonal tensile failure

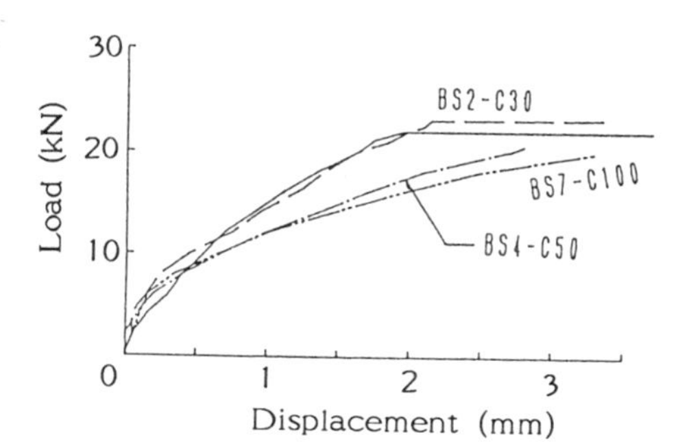

Fig. 4. Load-Displacement.

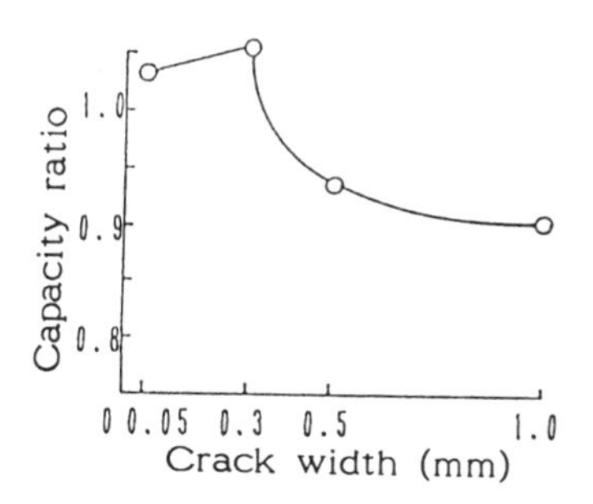

Fig. 5. The relation between capacity reduction and crack width.

beam as well as the ultimate capacity was a little improved or at least the same as the reference beam without corrosion. This indicates that a rusted bar, as long as the loss of cross-sectional area of the bar is negligible small, improves the bond between the bar and concrete, and that the rust may compensate the bond deterioration due to cracking.

On the other hand, as had been reported[4], the initial stiffness was larger in the corroded beams than in the reference beam without

Table 2. Specimens and result of cyclic loading test

Specimen	Crack width (mm)	f'c (MPa)	Maximum loading level (kN)	Maximum loading level P/Pu[1)]	Fatigue life (x10⁴)	Failure
BD-N1	non corroded	41.8	14.7	0.68	71.3	flexural failure
BD-N2	non corroded	41.7	11.8	0.55	291.0	flexural failure
BD1-C125	0.125	41.7	14.7	0.68	52.4	flexural failure
BD2-C125	0.125	41.8	11.8	0.55	133.0	flexural failure
BD3-C125	0.125	41.8	9.8	0.45	228.2	flexural failure
BD4-C125	0.125	36.8	7.8	0.36	>604.6	not failed
BD5-C50	0.50	38.8	14.7	0.68	41.2	flexural failure
BD6-C50	0.50	41.8	11.8	0.55	105.5	flexural failure
BD7-C50	0.50	38.8	9.8	0.45	176.0	flexural failure
BD8-C50	0.50	36.8	7.8	0.36	361.9	flexural failure

1) Pu is the maximum capacity of specimen BS-N at the static loading test.

longitudinal cracks. However, after flexural cracks occurred, the stiffness of the specimen without the corroded steel bar became larger when the crack width was more than 0.5 mm.

As far as the flexural behavior under static loading is concerned, the corrosion of steel bar did not cause a critical effect unless the longitudinal crack width exceeded 0.3 mm.

2.2 Behavior under fatigue loading

2.2.1 Experimental details of fatigue loading test

The dimensions of the specimens for the fatigue tests were the same as those for the static tests. The parameter was the longitudinal crack widths of 0, 0.125, 0.5 mm caused by corrosion of the steel bar.

In order to examine the fatigue life of beams, four stress ranges, P/Pu of 0.36, 0.45, 0.55, 0.68, were selected, keeping the minimum load as 1 kN. After certain numbers of loading cycles, measurements were conducted on specimens under static loading within the given stress ranges.

2.2.2 Test results of fatigue loading

The test results are summarized in Table 2, including the types of specimens, the compressive strength of concrete, and the fatigue life as well as failure mode. The relationship between the stiffness ratio and cycles number of loading are shown in Fig. 6.

The observed tendency was for the stiffness to decrease abruptly near failure. The larger the longitudinal crack width was, the more evident was the stiffness degradation with increase in number of load cycles.

The design fatigue capacity based on the fatigue strength of steel bars was calculated by the JSCE code equation[5]. Figure. 7 shows the relationship between the test and calculated results. The reference

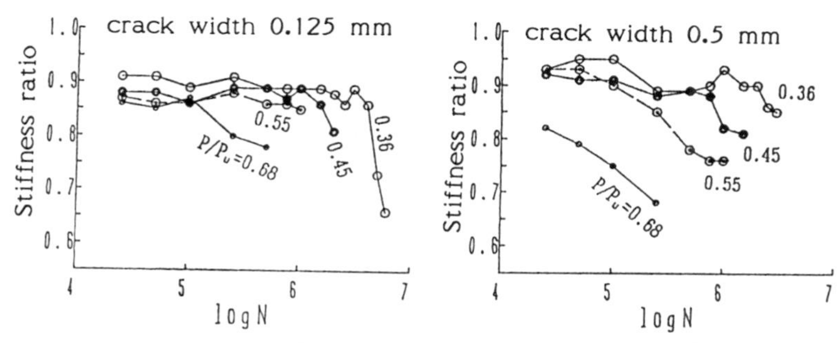

Fig. 6. Stiffness ratio.

beam without corrosion of the steel bar showed the calculated fatigue capacity in a safe side. On the contrary, the beams with longitudinal crack widths of 0.125 mm and 0.5 mm, resulted in failure prior to the predicted fatigue life.

Figure. 8 shows the relationship between the longitudinal crack width and the stress level for the predicted fatigue life of 2,000,000 cycles. Without corrosion, the beam can survive when the stress level is less than 0.55. Corrosion of steel bars reduces the fatigue life of beams or reduces the stress range for a 2,000,000 fatigue life. A beam with a 0.5 mm crack width can survive to 2,000,000 cycles only when the stress level is less than 0.45.

There are two factors which influence the reduction of the fatigue life of beams: the reduction of the fatigue life of steel bars due to corrosion and the bond deterioration due to longitudinal cracking. However, it can not be clarified which one dominates in this case.

3 Behavior of large damaged beams

3.1 Behavior under static loading

3.1.1 Test specimens

In order to extend the above mentioned findings to actual cases, it is necessary to study further the influence of the splices and stirrups. For this purpose, relatively large specimens (200x300x2800 mm) were tested under static loading as shown in Fig. 9 and Table 3.

The specified concrete strength was f'ck=300 kgf/cm^2(29.4 MPa). The specified yield strength of the main bar and stirrups was 3000 kgf/cm^2 (294 MPa) (SD295).

3.1.2 Details of static loading test

To accelerate corrosion of the main bar, the erosion test was conducted as before. Epoxy was painted on all contact points between the main bar and stirrups and plastic tape was also applied to prevent

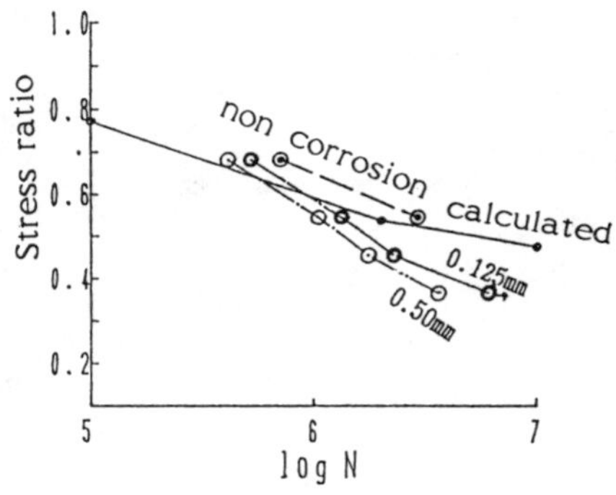

Fig. 7. Fatigue life.

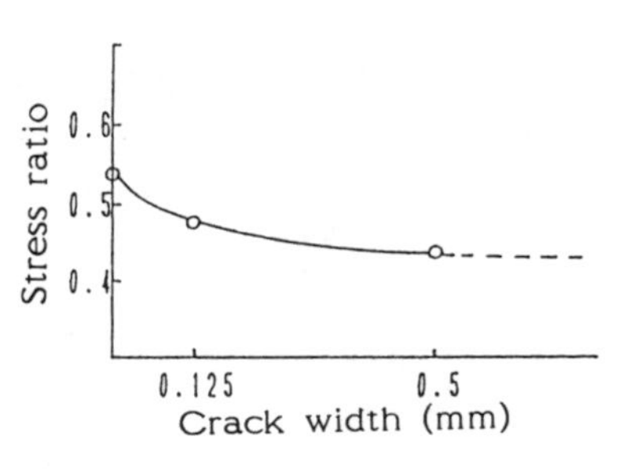

Fig. 8. Fatigue life of 2×10^6 cycles.

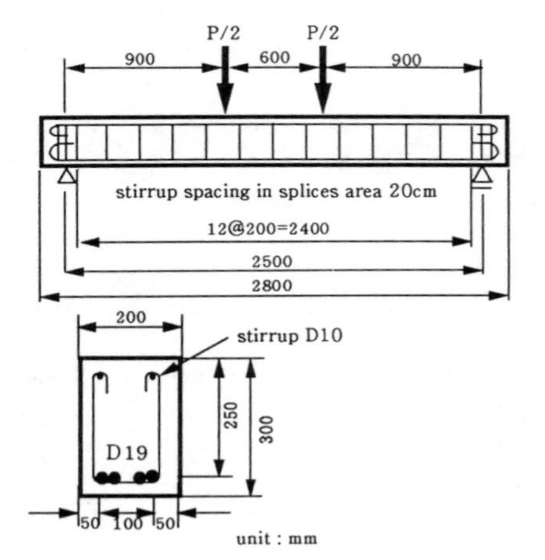

Fig. 9. Test specimens.

Table-3 Specimens

Specimen	Period of erosion (days)	Stirrup spacing in the splices area (cm)	Remarks
N1	0		non corrosion
H2	14		non splice
H2T-1	14	20	splices at mid span
H2T-2	14	10	splices at mid span
H2T-3	14	5	splices at mid span
H2TF	14	20	splices at mid span
H2T-4	14	60	splices at mid span
H3T-1	21	20	splices at mid span
H3T-2	21	10	splices at mid span
H4T-1	28	20	splices at mid span

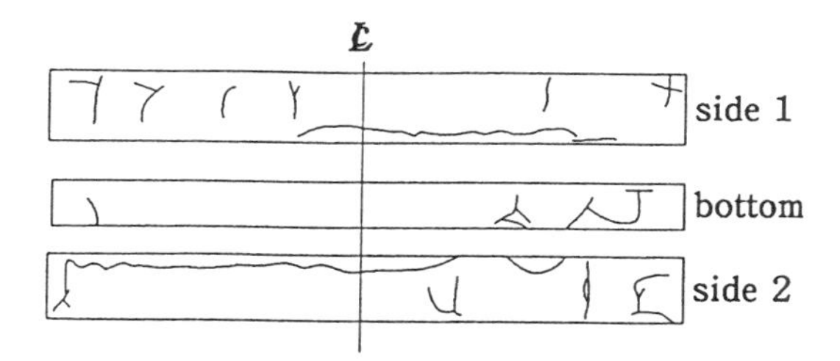

Fig. 10. Crack patterns caused by erosion.

the stirrups from being influenced by the electric corrosion. The lapping length(53 cm) was determined according to the specified lapping length by JSCE code[5].

The load was applied to specimens with a symmetric two point-loading. The shear span was 95 cm (a/d=3.8), and a splice was provided at the mid span. Displacement and crack width were measured at the loading points and a mid span.

3.1.3 Test results of static loading

The typical pattern of longitudinal cracking caused by the erosion test is shown in Fig. 10. In this series, cracking occurred on the side face of specimens rather than the bottom surface. It was observed that in the case of the specimen having splices, cracks were apt to occur through the thinner cover.

The effect of stirrups to confine and to reduce the longitudinal crack width due to corrosion of bars is shown in Fig. 11. In the case of H2T-1～3, crack widths decreased as stirrup spacings decreased. However, in the case of the H3 series, this tendency was not clearly observed. In these erosion tests of large beams, a correlation between the total electric current and the amount of corrosion or the crack width was not necessarily clear.

The load-displacement relationships are shown in Fig. 12. A reduction of the capacity was not recognized in the H2 series (except H2T-4). Specimen H2T-4 had no stirrup provided in the region of the splices; its capacity decreased suddenly after reaching a maximum. This showed that when stirrups were used, the bond between concrete and main bars could be maintained by the confinement effect of stirrups. In comparing H3T-1 with H4T-1, it can be observed that with the same stirrups spacing (20 cm), the more the bars corroded, the more the capacity was reduced. The confinement effect of stirrups might be reduced by a larger longitudinal crack width. A comparison of H3T-1 with H3T-2 indicates clearly the difference in confinement effect caused by differences in stirrup spacing.

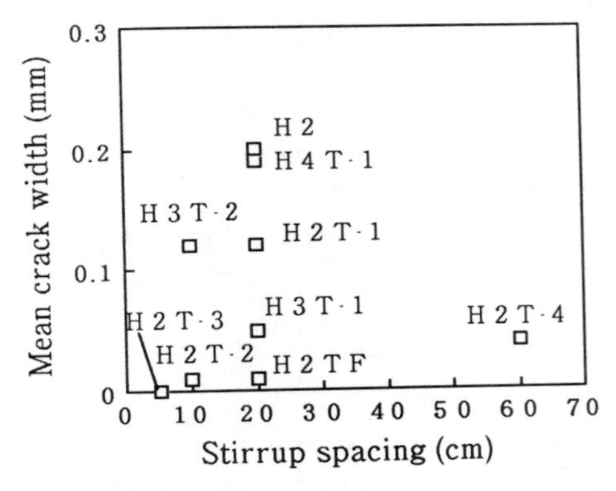

Fig. 11. Stirrup spacing — Mean crack width.

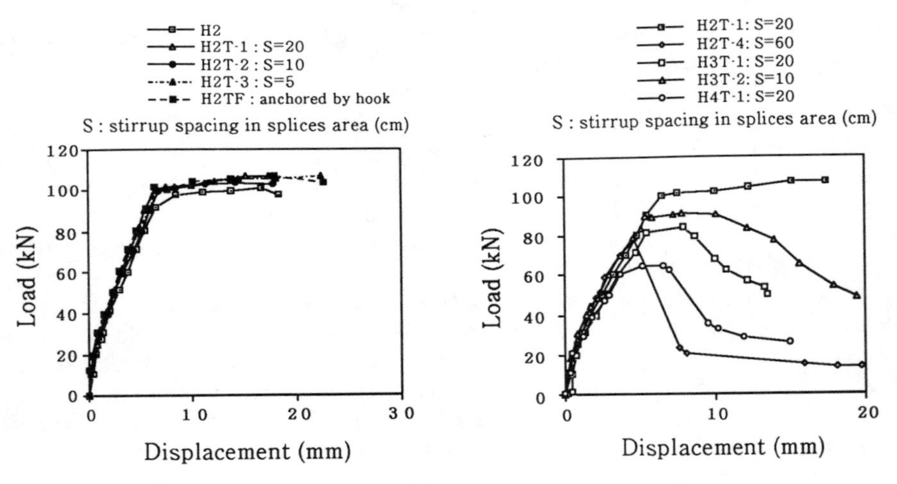

Fig. 12. Load — Displacement.

4 Conclusions

The following conclusions were drawn in this study:

(a) Small beams

1. In the static loading tests, the maximum capacity decreased as the longitudinal crack width increased due to corrosion of steel bar. However, even if the longitudinal crack width was 1 mm, the reduction of capacity was only 10%. The initial stiffness must have been affected by the surface roughness of rusted bars; thus, damaged beams showed a larger stiffness than the beams without corrosion.

2. In the fatigue behavior tests, both the fatigue life and stiffness of beams were reduced as the longitudinal crack width increased. The results indicated that the fatigue behavior of beams must be affected by the fatigue life of corroded steel bars and the bond deterioration due to cracking.

(b) Large beams
1. In the case of large beams, the crack pattern was a little different from that of small beams. The crack width was greatly influenced by splices and stirrups.
2. The decrease of capacity became large when the lapped splice was provided at the mid span. However, a proper arrangement of stirrups in the region of the lapped splice worked well to prevent the loss of capacity.

5 Acknowledgement

The experimental work was carried out by Yasuhiro Nakada and Hiroyuki Sakaya for their master's research supervised by K.Maruyama at the concrete laboratory of the Nagaoka University of Technology. The research was part of the study by the Concrete Engineer's Group belonging to the Niigata branch of JSCE.

6 References

1. Nakada, Y. et al.(1990) Influence to the loading capacity of the beam by the crack caused by corrosion of steel bar. Proceedings of JCI, Vol. 12-1. pp. 551-556. (in Japanese)
2. Katayama, S. et al.(1993) Influence to the loading capacity of the beam by the corrosion in the region of the lapped splices. Niigata society of JSCE, 11th Meeting. pp. 275-278. (in Japanese)
3. Takaoka, Y. et al.(1989) Study on the property of cracking caused by corrosion of steel bar.
   Proceedings of JCI, Vol. 11-1. pp. 591-596. (in Japanese)
4. Matsumoto, S. et al.(1983) Study on deterioration of RC structures caused by corrosion of steel bar (Part 2).
   Proceedings of JSCE, Vol. 38. pp. 261-262. (in Japanese)
5. JSCE, Standard Specification for Design and Construction of Concrete Structures-1986, Part 1[design].

143 A 3D MICROMECHANICAL MODEL FOR NUMERICAL ANALYSIS AND PREDICTION OF LONG TERM DETERIORATION IN CONCRETE

C. HUET, A GUIDOUM and P. NAVI
Swiss Federal Institute of Technology, Lausanne, Switzerland

Abstract
Three-dimensional numerical software specially developed for the purpose of overcoming the basic difficulties of long term behaviour assessment of concrete and concrete like materials is presented. It makes possible studies of the development over time of internal micro-stresses generated by environmental and/or loading effects together with internal aggression of physical or chemical origins. The coupled dissipative constitutive equations framework and specially developed meshing procedure are described. An illustrative example relating to internal stresses due to shrinkage is discussed.
Keywords: Concrete, deterioration, long term, micromechanical model, tri-dimensional analysis.

1 Introduction

Most problems of long term deterioration in concrete involve interactions between the constituents of the material at a microstructural level. For this reason the classical tools of structural analysis, based on the concept of representative volume elements to which effective properties can be assigned, cannot be used, specially for long term behaviour involving experiment duration too long for use in the design procedure. Recent developments in the micromechanical approaches of material behaviours together with the rapid increase of the power of numerical computations makes possible the development of simulation tools that may help to overcome these difficulties, see [1] [2].

This paper is devoted to a brief description of a tri-dimensional software specially developed in our laboratory for studying the time evolution of internal micro-stresses generated by environmental and/or loading effects together with internal aggression of physical or chemical origins.

Concrete Under Severe Conditions: Environment and loading (Volume Two) Edited by K. Sakai, N. Banthia and O.E. Gjørv. Published in 1995 by E & FN Spon. ISBN 0 419 19860 1

2 Bi-potential thermodynamic formalism for constitutive equations of the constituents

For numerical simulations, it is important to have available, for the constitutive equations associated with each element a formalism versatile enough to comply with the various complicated behaviours encountered in real materials without having to change the bases of the numerical procedure. This formalism should also satisfy the various restrictions of mathematical and physical origin that have to be stated about three dimensional constitutive equations. Here, the constitutive equations of the constituents are formulated in the framework of the dissipative primal internal variables approach developed in [3]. Based on a Non-linear Dissipative Continuum Thermodynamics formulation, it allows taking into account the various couplings between mechanical, physical and chemical phenomena taking place in the bulk of the material or of its constituents. This involves the kinetic behaviours of the latter and their delayed response, including cases with continuous spectra frequently met in real materials. Among several advantages, the primal approach used more easily accommodates behaviours like strain-hardening, strain-softening, micro-cracking and their couplings, see, for instance, [4]. Moreover it makes the corresponding microstructural dissipative mechanisms easier to identify.

In fact, for practical calculations, the continuous spectra of the viscoelastic or elastoviscoplastic parts of the concrete or cement response may be accurately represented by corresponding discrete spectra extending over several decades of the time scale, with two or three spectrum lines by decade. Thus, for practical calculations, the corresponding general formalism is reduced to the use of internal variables like a_j, for which the corresponding internal thermodynamic force X_j is a function of the present values of a_j and their rates $\dot{a}_j$, but not of their previous history. This means that the behaviour of the corresponding mechanism - when isolated - is of the differential type. As a result, ready for applications to concrete, this formalism may be written in the following compact and very general form :

$$\sigma = \frac{\partial \underset{\sim}{R}}{\partial \dot{\varepsilon}} + \rho \frac{\partial \underset{\sim}{\varphi}}{\partial \varepsilon} \quad ; \quad -s = \frac{\partial \underset{\sim}{R}}{\partial \dot{T}} + \frac{\partial \underset{\sim}{\varphi}}{\partial T} \quad ; \quad \mu_o = \frac{\partial \underset{\sim}{R}}{\partial \dot{w}} + \frac{\partial \underset{\sim}{\varphi}}{\partial w} \tag{1}$$

$$\frac{\partial \underset{\sim}{R}}{\partial \dot{\xi}_i} + \frac{\partial \underset{\sim}{\varphi}}{\partial \xi_i} = 0 \quad ; \quad \frac{\partial \underset{\sim}{R}}{\partial \dot{\alpha}_j} + \frac{\partial \underset{\sim}{\varphi}}{\partial \alpha_j} = 0 \tag{2}$$

In these equations, the scalar function φ denotes the free energy of the material. It must be expressed, as imposed by the Coleman theorem, in terms of the present values of the pertinent variables. Here, the latter are the density ρ in a reference configuration, the strain ε, the temperature T, the set ξ of the degrees of the chemical reactions and a set α of additional internal variables associated with the other dissipative mechanisms and, thus, leading to the following:

$$\varphi = \underset{\sim}{\varphi}(\rho\,, \varepsilon, T, \,w\,, \,\xi\,, \,\alpha\,) \tag{3}$$

On the other hand, the generalised Rayleigh dissipation function R is a scalar function of the rates defined by the following :

$$\underset{\sim}{D}\left[q;\dot{q};H^{-}(q)\right]=\frac{\partial \underset{\sim}{R}}{\partial \dot{q}_i}\dot{q}_i+\underset{\approx}{R}(q) \tag{4}$$

It may, in the non-linear case, also depend on the values taken by the state and/or the previous history H^- of the state variables, giving the following :

$$R=\underset{\sim}{R}[\ \dot{\varepsilon},\dot{T},\dot{w},\dot{\xi},\dot{\alpha};\varepsilon,T,w,\xi,\alpha;H^{-}(\varepsilon,T,w,\xi,\alpha)\ ] \tag{5}$$

The set a of internal variables may be of a scalar, vector or tensor. In the versions most often found in the literature, the internal variables a_i correspond to newtonian viscous mechanisms only. But a similar set of equations applies also to non-linear dissipative mechanisms, including plasticity, damage, cracking and degrees of chemical reactions when taking appropriate expressions for φ and R. For instance, for a plastic-strain mechanism with isotropic hardening, R is in the form :

$$\underset{\sim}{R}=k(\bar{a})\ (\dot{a}:\dot{a})^{1/2} \tag{6}$$

where $\bar{a}$ denotes the cumulated plastic deformation. From this and other similar results, the classical theories for concrete or other materials behaviours - including an extended Griffith criterion for non-linear fracture mechanics, see [4] - are easily retrieved as particular cases or extended to more general models.

Here, Eqn (1) represent three equations, one for the stress σ, one for the specific entropy s and one for the moisture potential μ_0. Eqn (2) represents a set of kinetic equations, one for each internal variable ξ_i or α_i. These equations may be applied easily to most kinds of known dissipative mechanisms and directly provide three-dimensional constitutive equations that are compatible with the general requirements of tensorial invariance, objectivity and thermodynamics. When possible, analytic elimination of the internal variables would retrieve the formulation using history functionals. But the internal variables approach appears more tractable in numerical calculations.

The first term in the R.H.S. of the first Eqn (1) cancels for behaviours with instantaneous elasticity, restoring - but only in that case - the classical property of φ to be a potential for the stress. Thus, the usual statement about this is not general enough to cope with many practical situations for which the dissipative mechanisms may have response times much shorter than the shorter time that can be considered in experiments or in the field, specially when durability problems, what may extend over several tens of years, are involved. This is true in particular for concrete or cements in quasistatic experiments, the result of which are influenced - among other things - by viscoelastic effects occurring in the acoustic or ultrasonic range, leading to a dependence of the stress upon the stretching rate. This yields, in particular, the well known experimental indeterminacy of the so-called instantaneous elasticity of concrete. When physical and/or chemical variables are introduced as internal variables, this formalism together with the positivity conditions of the dissipation imposed by the second principle provide accounts of "abnormal" environmental effects - like the Pickett or mechanosorptive effect - in a natural fashion.

It is easy to show that a Rayleigh function exists for most of the classical dissipative models already used in the literature. From Eqn (4), R may be calculated from the supposedly known dissipated power D through the following formula:

$$R = \underset{\approx}{R}(q) + \int_o^1 D(\lambda \dot{q})\, d\ln\lambda \tag{7}$$

In particular for a polynomial development of the instantaneous part of D :

$$D = \underset{\approx}{D}(q) + \sum_{k=1}^{N} D_k \tag{8}$$

with D_k a homogeneous function globally of order k in the $\dot{q}_i$, one has :

$$R = \underset{\approx}{D}(q) + \sum_{k=1}^{N} \frac{1}{k} D_k \tag{9}$$

Order one corresponds to elasto-plastic and/or damaging behaviours, order two to linear viscoelastic behaviours, orders larger than two correspond to other non-linear dissipative behaviours and, in particular, to coupled ones.

When the hydric potential μ_0 is taken as the independent variable in place of the water content w, a corresponding formalism is easily written using, in place of φ, the related Legendre transform ψ defined by the following:

$$\psi = \varphi - \mu_0 w \tag{10}$$

This is useful when μ_0 is the controlled parameter through the relative humidity of the surrounding atmosphere. The occurrence of $\dot{T}$ in R, and thus in Eqn (1), is to be expected because of the thermal counterpart of the mechanosorptive effect in concrete. This would mean that Mandel's non-duality principle, as relating to $\dot{\sigma}$, cannot be extended to $\dot{T}$. But since, in concrete, $\dot{T}$ is highly coupled to $\dot{w}$, this question needs clarification left for further research.

3 Numerical simulation and meshing of concrete-like granular composites microstructures

In the software presented here, concrete is considered as a granular composite, with random microstructure obeying a granulometric distribution prescribed at will. The shapes and orientations of the granules may also be varied. An automatic meshing procedure respecting the microstructure of the material has been elaborated.

The basic data of the microstructure generation method used are the grain size distribution and the aggregate volume content. In the examples provided below, the grain size distribution is taken according to the Füller curve. The spatial coordinates of the centre of each grain are randomly chosen. However, using this relatively simple generated method for 3D structures, one cannot obtain an aggregate content more than 50%. A typical granular structure is shown in Fig. 1 for ellipsoidal grains with random orientations. In this case, all particles have mean diameters between 7.5 and 60 mm (dam concrete). This model involves 101 grains representing 39% of the volume. The wedge length of the cubic specimen is 100 mm.

4 Finite element discretisation

In order to analyse the physical and mechanical behaviour of concrete and to determine the evolution of its internal state, computer-generated structures like the one shown in Fig. 1 are discretized with tetrahedral finite elements. A typical example of such a finite element mesh is shown in Fig. 2. First, the composite structure is subdivided into cube elements. Then, these cubes are subdivided into tetrahedrons depending on the more or less heterogeneous character of the cube. Details about the structure of the various elements may be found in [1] [5] [6].

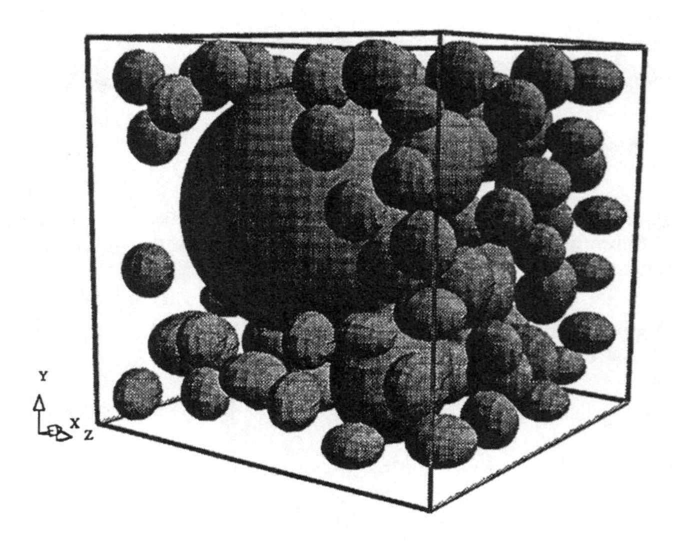

Fig. 1. Computer-generated structure .

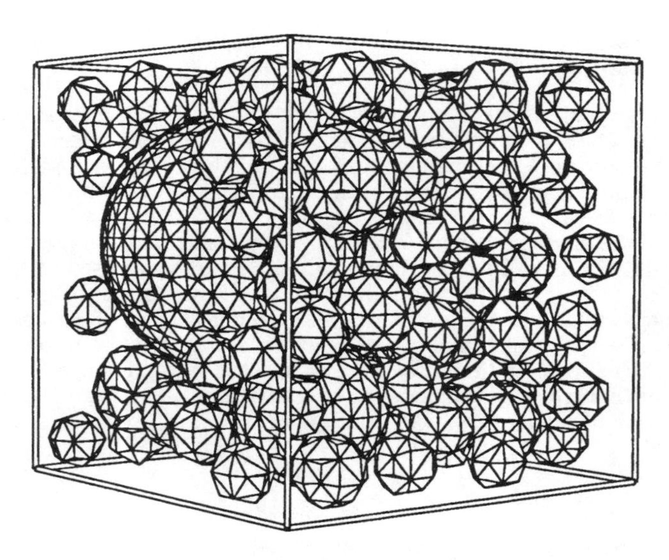

Fig. 2. Resulting finite element mesh. Only the grains are represented. There are 51 985 matrix elements, 17 135 inclusion elements and 46 875 d.o.f..

The total number of finite elements of the resulting 3D mesh in this example is 69,120 finite elements with 46,875 degrees of freedom. Hence, special attention has been paid to the storage and computational problems. The calculations are performed on a multiprocessor SG Challenge computer using a preconditioned conjugate gradient-like method for linear spare systems.

5 Finite element computation of effective shrinkage and of internal stresses due to shrinkage of the cementitious matrix

It is well known that in cement-based composite materials, the cement matrix exhibits volume changes when the moisture content changes. On the other hand, the aggregates which are embedded in the matrix are not subject to shrinkage. As a consequence, tensile stresses are created in the cement matrix. In addition to overall dimensional changes (effective shrinkage), they are responsible for micro-cracking and also macrocracking of concrete and, for this reason, highly involved in concrete durability problems, from the three viewpoints of the loading capacity of the material, its integrity and its tightness to permeation by liquids and by gas.

Temperature changes may cause similar stresses because of thermal gradients and of the differences in the coefficients of thermal expansion of the concrete constituents. In the numerical software presented here, both thermal strain and shrinkage increments are treated as initial strains in the finite element analysis.

Fig. 3 shows, for the model of Fig. 1 (grains volume content 39%), an example of calculated time evolution of the strain averages in the matrix phase and in the grains phase, respectively, together with the time evolution of the imposed free shrinkage of the matrix and the effective - or apparent - overall shrinkage of the specimen.

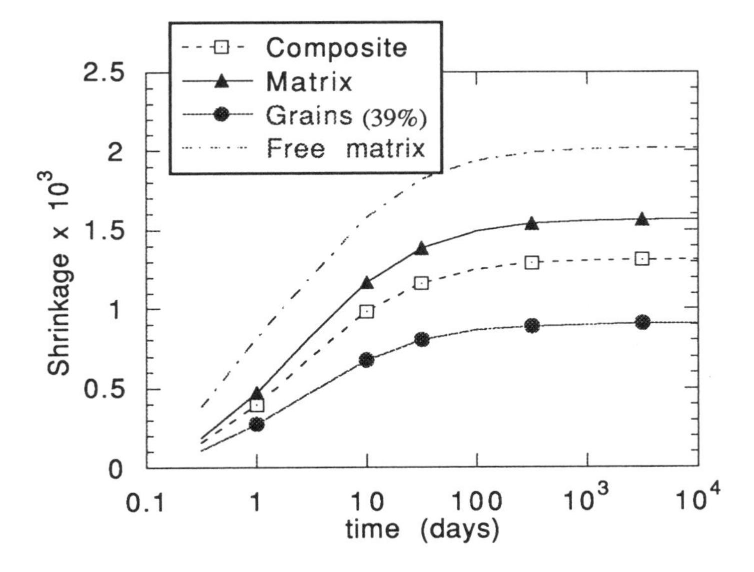

Fig. 3. Time evolution of the matrix shrinkage input and of resulting effective shrinkage and mean strains in each phase (contractions counted positive ; grains volume content 39%).

The plotted strains are the mean linear ones (one third of the relative volume change). The calculations are performed from one day up to about thirty years for the case of elastic grains, elastic matrix and thus elastic composite. The grain to matrix stiffness ratio is 1.5. No external loading nor displacement constraints are applied in this example, the specimen being free of taking its own effective shrinkage. The matrix is considered to be a mortar with maximum a grain diameter of 7.5 mm. The final free shrinkage of this mortar is taken as 2 mm/m, reached after about three year. The final effective shrinkage of the specimen is about 1.3 mm/m, also reached after the same delay as expected from the assumed elastic properties of the constituents.

Fig. 4 relates to the case of a viscoelastic matrix with wide spectrum (several decades), yielding thus a viscoelastic composite also with wide spectrum. The elastic grains volume content (13.5%) is lower than above, and the initial stiffness ratio three times higher. This figure presents the evolution for the isotropic part (one third of the first invariant of the stress tensor) of the average stress tensor in each phase.

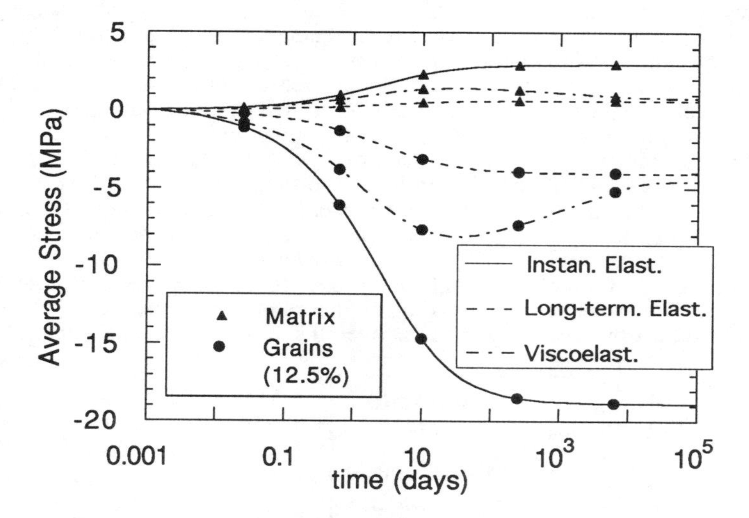

Fig. 4. Time evolution of the average stresses in matrix phase and grains phase (tensions counted positive ; grains volume content 12.5%).

Corresponding results for instantaneous elastic behaviour and long term final elastic behaviour are also monitored for the sake of comparison. As expected, compressive stresses are created in the aggregates while tensile stresses are created in the matrix in all cases. For the instantaneous elastic case with 39% of grains volume content, the maximum stress (not monitored here) was - like for the strains - reached after three years, with a magnitude of 5.9 MPa of tensile stress in the matrix. This is larger than what can be sustained by the mortar constituting the matrix. This observation confirms that elastic calculation of shrinkage stresses leads to results that are too pessimistic. But it is found here that the average shrinkage stress in the viscoelastic matrix is highly reduced by viscoelastic relaxation effects and that its evolution becomes non monotonic and exhibits a maximum. For the case with volume content 13.5% shown on Fig. 4, the maximum of the average stress in the matrix reduces from 3 MPa in the instantaneous elasticity case to 1.5 MPa in the viscoelastic one. Moreover, the maximum is reached after about one month in the viscoelastic case against three years for the elastic one. In fact, the possible existence of a maximum, its position on the time scale and its magnitude are crucial information from the viewpoint of durability prediction. It is also found in the viscoelastic case that, for the same free shrinkage evolution, the duration over which the shrinkage yields non stabilised strains and stresses may be prolonged by several orders of magnitude beyond the stabilisation time of the input. But these viscoelastic results are only preliminary ones and have still to be confirmed and extended. In fact, it is also known that shrinkage stresses in concrete yields local damage in the form of loss of adherence and microcracks, see [7] [8]. When no macroscopic localisation or unstable propagation occur, see for instance [4], this results in further reduction of the average stresses and in possible increase of the creep capacity accompanied by creep non-linearity.

Maps of some components of the stress and/or strain states at given times can also be monitored together with the evolution of the local thermal and mechanical states at a given point or of the position and magnitude of the point experiencing the largest stress. Examples of such maps for temperature, strains and stress distributions obtained by using the present software are to be found in [5] [6].

6 Concluding remarks

From the results obtained in these 3D stress and strain calculations in the concrete microstructure, some practical information may already be obtained despite the fact that strength or damage criteria are not yet implemented in the computer code. On the other hand, taking into account the occurrence and behaviour of microcracks in the 3D simulation is also presently under study. This will constitute a 3D extension of the 2D simulation code that we recently developed in our laboratory for the evolution of granular composites with a random distribution of initial microcracks, see [1] [8] and the thesis by Wang [9]. The results provided by this 2D microcracking simulation code are already very important for the qualitative understanding of damage mechanisms in concrete and the prediction of durability. But some mechanisms that are specific to the 3D features of concrete microstructure still need to be taken into account. Nevertheless, from the developments achieved and the first results obtained up to now, it may be concluded that the 3D finite element simulation code presented here turns out to be an extremely useful and powerful tool in the study of concrete durability and the evaluation of the development over time of the risk of damage. One problem to be solved in further research is to obtain realistic material properties of the constituents that, being at the limit the cement paste, may be non existent as an independent material since the presence of the granules influences the hydration process. For the long term, we plan to solve this through continuing the simulation process down to the cement grain level. Taking into consideration the whole granulometric curve still requires successive calculations at different ascending levels, for which the procedure described in [2] has to be used. On the other hand, at the level of the cement matrix, another 3D numerical code has also been developed in our laboratory in order to simulate the development over time of the cement paste microstructure, taking into account the kinetics of the main hydration reactions. It will provide some of the information that is needed to make more realistic applications of the software presented here, for instance, by taking into account the time evolution of matrix porosity, see [1] [10].

Another interesting field of applications is the study of size effects for specimens smaller than the representative volume, for which important theoretical results have recently been obtained, see [11] [12] [13]. This will constitute 3D extensions of the 2D numerical simulations already performed on this topic together with experimental testing involving successive subdivisions of initially large specimens [1] [14] [15]. The first results about the use of this 3D software in this area may be found in [1]. This topic is highly connected with durability problems since, due to wall effects, the regions of concrete structural elements that are near moulding surfaces or reinforcement locations do not have the representative volume.

In order to overcome the difficulties encountered for high grain contents with the microstructure generation used here, mainly based on the work by Wittmann, Sadouki and Roelfstra, see [16]-[19], other algorithms have been developed in our laboratory using several techniques, see, for instance, [1] [20]. A related aspect of the integrated system developed in our laboratory [2], and of which the presented software forms an important module, is the reconstruction of concrete microstructure from direct microscopic observations through the use of a confocal microscope associated with image analysis and processing, see [21]. The present software also makes possible studies of influence of the irregular grain shapes through the use of a new homogenisation technique described in [22]. The consequences for micro or macrocracks propagation may be studied using the general thermodynamic formalism that we presented in [4] with full proofs given in [23].

7 Acknowledgements

Funding support to this research from EPFL and from the Swiss National Foundation for Science under contracts n° 21-27962.89 and n° 20-32206.91 is gratefully acknowledged.

8 References

1. Huet, C. (1993) *Micromechanics of Concrete and Cementitious Composites*, Presses Polytechniques et Universitaires Romandes, Lausanne.
2. Huet, C. (1993) An integrated approach of Concrete Micromechanics, in *Micromechanics of Concrete and Cementitious Composites* , (ed. C.Huet), Presses Polytechniques et Universitaires Romandes, Lausanne, pp. 117-146.
3. Huet, C. (1993) Some basic tools and pending problems in the development of constitutive equations for the delayed behaviour of concrete, in *Creep and shrinkage of concrete* , (eds. Z. P. Bazant and I. Carol), Spon, London, pp. 189-200.
4. Huet, C. (1994) Some Aspects of the Application of Fracture Mechanics to Concrete Dams, in *Dam Fracture & Damage* , (eds E. Bourdarot, J. Mazars and V. Saouma), Balkema, Rotterdam, pp. 79-89.
5. Guidoum, A., Navi, P. and Huet, C. (1993) 3-D Numerical study of heat transfer and thermal stresses in cohesive granular composites, in *8th International Conference on Numerical Methods for Thermal Problems,* (ed. R.W. Lewis), Pineridge Press, Swansea, pp. 917-927.
6. Guidoum, A., Navi, P. and Huet, C. (1993) Numerical evaluation of microstructural effects on the shrinkage properties of concrete through a three-dimensional model of cohesive granular material, in *Creep and shrinkage of concrete* , (ed. Z.P. Bazant and I. Carol), Spon, London, pp. 127-132.
7. Huet, C. (1994) On the Concept of Adherence for Concrete, in *Adherence of Young on Old Concrete*, (ed. F.H. Wittmann), 2nd Bolomey Workshop, Sion, Switzerland, April 1-2, 1993, Balkema (in press).
8. Huet, C., Wang, J. and Navi, P. (1994) Numerical analysis of the microstructural influences on crack propagation in concrete, *Int. Symposium on Fracture and Strength of Solids*, July 4-7, 1994, Xi'an, China (in press)
9. Wang, J. (1994) *Development and application of a micromechanics based numerical approach for the study of crack propagation in concrete*, Doctoral Thesis No 1233, Swiss Federal Institute of Technology, Ecole Polytechnique Fédérale de Lausanne, Lausanne, Switzerland.

10. Navi, P. and Pignat, C. (1994) Simulation of effects of small inert grains on the cement hydration and its contact surface, *NATO/RILEM Workshop on The Modelling of microstructures and its potential for studying Transport Properties and Durability* , Saint-Rémy les Chevreuses, France (in press).
11. Huet, C. (1990) Application of variational concepts to size effects in elastic heterogeneous bodies, *J. of the Mechanics and Physics of Solids*, Vol. 38, No 6, pp. 813-841.
12. Huet, C. (1991) Hierarchies and bounds for size effects in heterogenenous bodies, in *Proceedings of the Sixth Symposium on Continuum Models and Discrete Systems*, (ed. G. Maugin), Dijon, 1989 ; Longmans, London, vol. 2, pp. 127-134.
13. Hazanov, S. and Huet, C. (1994) Order relationships for boundary conditions effects on the overall properties of elastic heterogeneous bodies smaller than the representative volume, *J. Mech. Phys. Solids* (in press).
14. Amieur, M. (1994) *Etude numérique et expérimentale des effets d'échelle et de conditions aux limites sur des éprouvettes n'ayant pas le volume représentatif*, Doctoral Thesis No 1256, Swiss Federal Institute of Technology, Ecole Polytechnique Fédérale de Lausanne, Lausanne, Switzerland.
15. Amieur, M., Hazanov, S. and Huet, C. (1994) Numerical and experimental assessment of the size and boundary conditions effects for the overall properties of granular composite bodies smaller than the representative volume, in *IUTAM & ISIMM Symp. on Anisotropy, Inhomogeneity and Nonlinearity in Solid Mechanics*, September 1994, Nottingham (in press).
16. Sadouki, H. (1987) *Simulation et analyse numérique du comportement mécanique de structures composites,* PhD. thesis 678, Swiss Federal Institute of Technology, Lausanne, Switzerland.
17. Roelfstra, P.E Sadouki, H. and Wittmann, F.H. (1985), Le Béton Numérique, *Materials and Structures*, 107, 235-248.
18. Roelfstra, P.E. (1989), *A numerical approach to investigate the properties of concrete : numerical concrete*, PhD. thesis 788, Swiss Federal Institute of Technology, Lausanne, Ecole Polytechnique Fédérale de Lausanne, Switzerland.
19. Wittmann, F.H., Sadouki, H. and Steiger, T. (1993) Experimental and numerical study of effective properties of composite materials, in *Micromechanics of Concrete and Cementitious Composites* , (ed. C. Huet), Presses Polytechniques et Universitaires Romandes, Lausanne, pp. 59-82.
20. Amieur, M., Hazanov, S. and Huet, C. (1991) Fractal modelling of concrete, *11th Int. Conference on Structural Mechanics in Reactor Technology*, Tokyo, Japan, August 18-23, HO4/2 pp. 109-114.
21. Sunderland, H., Tolou, A. and Huet, C. (1993) Multilevel numerical microscopy and tri-dimensional reconstruction of concrete microstructure, in *Micromechanics of Concrete and Cementitious Composites* , (ed. C. Huet), Presses Polytechniques et Universitaires Romandes, Lausanne, pp. 171-180.
22. Huet, C. Navi, P. and Roelfstra, P.E. (1989) A Homogenization Technique Based On Hill's Modification Theorem, in *Continuum Models and Discrete Systems* (ed. G. Maugin), Longman, Harlow, vol. 2, pp. 136-143.
23. Huet, C. (1994) A Continuum Thermodynamics Approach for Size Effects in the Failure of Concrete Type Materials and Structures, In *Size-Scale Effects in the Failure Mechanisms of Materials and Structures*, (ed. A. Carpinteri), IUTAM Symposium, Torino, October 1994, Spon, London (in press).

144 MECHANISM OF DETERIORATION OF CONCRETE UNDER SEVERE ENVIRONMENT

A.M. PODVALNY
NIIZHB, Moscow, Russia

Abstract
A mechanism of concrete deterioration under severe conditions including both environmental and loading is developed. It describes the phenomena that lead to crack formations in concrete that decrease its strength and other physical properties.

Concrete is treated as a body with hierarchical composite–in–a–composite type of structure. Scheme of the way the damage in concrete is formed includes micro– and macro–areas. The phenomena on a macro area are universal for many kinds of severe actions.

Mathematical models for estimation of the intrinsic strains and stresses are proposed. The dimensionless criterion of plain concrete deterioration summarises the influence of concrete structural and environmental conditions on concrete. It is extended on loaded and reinforced concrete.

Considering the structure of concrete and the random nature of its deterioration under a severe environment, the probability theory approach was developed. The notion of probability of concrete damage was proposed, and on its base formula for concrete strength and modulus of elasticity degradation in a single and multicycle temperature action was obtained. For the verification of developed ideas, a great amount of published data as well as the author s investigations on concrete and loaded reinforced concrete beams in the tidal zone of the Barents Sea was used.

The experimental facts concern the influence of a variety of factors on concrete durability related to the structure of concrete, properties of aggregate, resistance improving additives, schemes of loading, behaviour of concrete under different severe environments, etc.

An attempt is being made for unification on the basis of a common approach for such severe actions on concrete as freezing and thawing, cyclic high temperature influence, sulfate aggression and others.
Keywords: Concrete, deterioration mechanism, mathematical models, reinforced concrete, severe environment, strain, stress, structure.

Concrete Under Severe Conditions: Environment and loading (Volume Two) Edited by K. Sakai, N. Banthia and O.E. Gjørv. Published in 1995 by E & FN Spon. ISBN 0 419 19860 1

1 Introduction

Concrete is a porous body and its deterioration under severe environment occurs mostly due to complex processes in its capillary–porous structure. For many years, efforts were aimed at revealing the nature of those processes in pores [1]. Less attention was paid to the fact that phenomena that take place in pores, despite their extreme importance, are only the beginning of destruction. The complete picture also includes the destruction caused by mechanical forces. We can define the whole process of damage as physico–chemical mechanics of concrete deterioration under severe environment. The aim of this report is to show the close connection between such factors in concrete deterioration as the structure of material, type of construction, external loading and severe environment and suggest a way for assessment of their separate and combined action on concrete durability.

2 Model of concrete structure and two areas of deterioration process

Concrete is a heterogenous material consisting of a porous cement matrix and compact aggregates of various sizes (including clinker relics) embedded in it. Reinforced concrete can also be considered as a two–phased material. By applying principles of system the–ory, we can present their structure as a hierarchical system (Table 1).

Table 1. Model of concrete and reinforced concrete structure

Structure level	System	Components	
		Porous	Compact
I	Reinforced concrete	Concrete	Steel
II	Concrete	Cement–sand mortar	Coarse aggregate
III	Cement–sand mortar	Hard cement	Fine aggregate
IV	Hard cement	Hydrated mass	Clinker relics
V	Hydrated mass	Pores (with interporous phase)	Crystal growth

Pores are a component of concrete structure in a bit different meaning than mortar or hard cement, but it is useful to include level V in Table 1 to complete it [2].

The structure model in Table 1 has certain features. It is complete and encompasses all parts of concrete. As a porous component of a higher level, that is a system on a lower one, there is a connection between levels and continuity of transition from one level to another. All levels are formed in the same way — they consist of porous matrix and compact aggregates. According to Table 1, we can define concrete as specific material of composite–in–a–composite type. Foregoing peculiarities of structure model are important for further account.

The water or salt solutions that have penetrated into pores of concrete from the envi–ronment undergo phase transformation, temperature changes or change their location in the structure. This results in a positive or negative intrinsic pressure in the hydrated mass that leads to strains.

Since the liquid phase is contained in porous components, they are mostly affected by the strains, while compact components are unmoveable or even have temperature de–formation in the opposite direction (e.g. by freezing). Such differences of free strains of the components inevitably causes intrinsic stresses. At the points where stresses exceed the local strength of the structure (tensile as a rule), cracks appear. When aggressive action continues, maximal stresses appear at new points; thus cracks accumulate progressively.

There can be various causes of intrinsic pressure in a hydrated mass and its strains (hydraulic or osmotic pressure, crystal growth and capillary forces), but the consequence is always the same: intrinsic stresses and then internal crack formation. There is a wide range of phenomena related to influences on the durability of concrete composition, properties of aggregates, external loading, etc. whose effect (proved by experiments) are impossible to study, considering only the effects in a porous space.

3 Estimation of intrinsic strains and stresses on different structural levels

Strains that arise in a hydrated mass due to continuity of the transition from level to level appear on all other levels of structure (see Table 1). But as an effect of the presence of the compact components on every level, the magnitude of strains will regularly decrease. Assuming that strains of concrete are an additive function of the strains of its components, we can apply the simplest dependence

$$\varepsilon_S(\nu_1+\nu_2)=\varepsilon_1\nu_1+\varepsilon_2\nu_2, \tag{1}$$

where ε is the unit (relative) strain; ν is concentration by volume ($\nu_s=\nu_1+\nu_2=1$) and subscripts s, 1 and 2 denote composite (system), compact and porous components on levels II —IV, respectively.

In references [3,4], more complex formulas for concrete strain estimation can be found. However, Eq. (1) fits experimental data [5] best and gives the least valuation of strains on levels III and IV, e.g. in a range of anomalous deformations of concrete (see Fig. 2).

Transforming Eq. (1) and substituting ε for α according to equation $\varepsilon=\alpha\cdot\Delta t$ we get

$$\alpha_m=\frac{(\alpha_c-\alpha_a\nu_a)}{\nu_m}, \tag{2}$$

where α is the thermal expansivity (c.t.e.) and subscripts *c*, *m* and *a* refer to the concrete, mortar and aggregate, respectively.

Thus, if we know the composition of concrete (ν_1,ν_2) and evaluate α_c and α_a we can calculate α_m and α_{hc} and α_{hm} in a similar way for levels III and IV. As an example of an application of formulas (1) and (2), the strains of wet concrete components while freezing were calculated.

Numerous experiments show that the indication of non frost-resistant concrete is its expansion in some temperature interval (usually from −10 to −40 °C) during freezing. Average experimental values of α_c for concretes with different composition and moisture content are presented in Table 2 along with calculated values of α_m, α_{hs} and α_{hm} of porous components and referenced data of c.t.e. of compact components. For the calculation of α of porous components, it was assumed that the composition of concrete by volume is 0.16:0.09:0.025:0.5 (hydrated mass: clinker relics: fine aggregates: coarse aggregates) and the mean value for all aggregates $\alpha_a=9\cdot10^{-6}(°C)^{-1}$.

Table 2. Free strains of reinforced concrete components during freezing

Components	C.t.e. of concrete and its components $\alpha \cdot 10^{6} \cdot (°C)^{-1}$				
	Compact	Porous			
		α_c	α_m	α_{hc}	α_{hm}
1. Heavy concrete with various water contents					
the same		-10.0	-29.0	-67.0	-110.0
the same		-5.0	-19.0	-47.0	-78.0
the same		0.0	-9.0	-27.0	-47.0
the same		5.0	1.0	-7.0	-16.0
2. The same, with resistance improving additions		8.0	7.0	5.0	3.1
3. Dry heavy concrete		10.0	11.0	13.0	15.6
4. Sandstone and quartzites (as aggregates)	11.0				
5. Granites and limestones (as aggregates)	8.0				
6. Clinker relics	9.0				
7. Steel	12.0				

A minus sign of α indicates that the component expands during freezing. Components can expand by freezing while concrete as a whole is contracted (row 1d in Table 2). Concrete with a pore–forming addition (GKZH–94) have no expansion during freezing (row 2 in Table 2).

Similar to frost action, incompatibility exists when aggregates expand due to heating but the cement materials suffers from shrinkage. In a case of sulfate aggression, cement expands but the aggregates remain undeformed. Differently deformed components result in the considerable intrinsic (structural) stresses in the concrete body. As we can see from Table 1, the same model can be applied to levels II —IV. Four different elastic models were investigated:

- two single– component models: 1. solid sphere (aggregate) with a spherical shell (cement) and 2. flat disk with a ring;
- two multi– component models: 1. elastic infinite plane (cement) with circular hard disks in square order regularly embedded in it and 2. elastic infinite plane (cement) with circular hard disks in triangular order regularly embedded in it.

The models, in spite of apparent differences, give the same qualitative results and close quantitative results [6]. In the simplest analytical model (disk with a ring), the maximum tensile tangential σ_θ or radial σ_r stresses appear in the ring along its contact with the disk (we assume that the tensile strength of aggregates is much greater than that of cement materials):

$$\sigma_\theta = \frac{\Delta\varepsilon E_1 E_2\left[\left(\frac{R_2}{R_1}\right)^2 + 1\right]}{\left(\frac{R_2}{R_1}\right)^2\left[E_1 + (1-2\mu_1)\cdot E_2\right] + (1-2\mu_2)\cdot(E_1 - E_2)}, \tag{3a}$$

$$\sigma_r = \frac{-\Delta\varepsilon E_1 E_2\left[\left(\frac{R_2}{R_1}\right)^2 - 1\right]}{\left(\frac{R_2}{R_1}\right)^2\left[E_1 + (1-2\mu_1)\cdot E_2\right] + (1-2\mu_2)\cdot(E_1 - E_2)}, \tag{3b}$$

where $\Delta\varepsilon$ is the difference of unit strains of the disk and the ring; E —modulus of elasticity; μ —Poisson s ratio; R —is the disk radius and the outside radius of the ring and subscripts 1 and 2 refer to the disk and the ring, respectively.

It is noted that for the concrete components, $\mu_1 \approx \mu_2 \approx 0.2$ and, hence, $(1-2\mu_{1,2}) \approx 0.6$. Calculations show that all levels bring in a congruent contribution to a stress state of a concrete and the stresses are mostly dependent on

$$\Delta\varepsilon = \varepsilon_1 - \varepsilon_2 = (\alpha_1 - \alpha_2)\Delta t, \tag{4}$$

where $\Delta t = t - t_0$.

Depending on the character of aggressive action, radial or tangential stresses in the ring of a model may be tensile (have the + sign) and lead to cracks. There is much experimental evidence indicating two general ways in which concrete can break down under a severe environment: through cracks running along the contact between cement and aggregate (e.g. as a result of radial tensile stresses) and through cracks running normal to the aggregate surface (e.g. as a result of tangential tensile stresses).

On the basis of Eq. (4), we can compare, for instance, the stresses that arise in concrete, subjected to frost action and to sulfate aggression. As follows from Table 2 for non–resistant to frost concrete, α_1 and α_2 have different signs so that the summation is maximal. In that case, t_0 is the temperature when $\alpha_1 \approx \alpha_2$ and t is the temperature of freezing in an interval of anomalous expansion. In a case of sulfate aggression, $\varepsilon_1 = 0$; therefore, $\Delta\varepsilon$ is much less than for non frost–resistant concrete. The difference between $\Delta\varepsilon$ values in these cases can be several fold. This is one of the reasons for the much more severe action of frost in comparison with sulfate aggression.

4 Criterion of deterioration and probability assessment of concrete damage

Analyzing the concrete structure as one of a composite–inside–a–composite type, we have considered that (in terms of a mathematical model) the cement–sand mortar that forms a ring (in the case of a sphere model —a shell) around every coarse aggregate grain contains many ($\approx 1.5 \cdot 10^3$) sand grains with hard cement rings, while the hard cement ring consists of many ($\approx 1.0 \cdot 10^3$) clinker grains with rings of hydrated mass around them. Stresses appearing in the matrix (the rings at each structural level) are superimposed on one another. Thus, to a first approximation

$$\sigma_\Sigma = \sigma_{II} + \sigma_{III} + \sigma_{IV}, \tag{5}$$

where indices II, III and IV define the structural levels. The cracks form where total maximal tensile stresses $\sigma_{\Sigma,t} = \sigma_{II,t} + \sigma_{III,t} + \sigma_{IV,t}$ exceed the tensile strength of hydrated mass $R_{h,t}$ (equivalent of a strength at a point):

$$\frac{\sigma_{\Sigma,t}}{R_{h,t}} \geq 1, \tag{6}$$

The more the fraction $\frac{\sigma_{\Sigma,t}}{R_{h,t}}$ decreases below 1, the lower is the probability of cracks appearing and the higher is the durability of concrete and vice versa. The analysis shows that all the factors that influence the durability of concrete, in exactly the same way, influence the above mentioned fraction. Thus, Equation (6) is a criterion of concrete deterioration under severe environment. The number of compact ingredients on structural levels in $1dm^3$ of concrete can be characterised by the approximate ratio $0.65 \cdot 10^2$: $1.0 \cdot 10^5$: $1.0 \cdot 10^8$ (gravel : sand : clinker : grains) that justifies the statistical approach.

On the basis of Eqs. (3a) and (6), the function of concrete integrity z can be introduced [7].

$$z = 1 - \frac{\Delta\varepsilon E_1 E_2 \left[\left(\frac{R_2}{R_1}\right)^2 + 1\right]}{R_{h,t}\left(\frac{R_2}{R_1}\right)^2 \left[E_1 + 0.6 \cdot E_2\right] + 0.6 \cdot \left(E_1 - E_2\right)} > 0, \tag{7}$$

To simplify the calculations, we shall operate only with the II level of structure —concrete and its two components. If $z<0$, concrete suffers damage. The damage probability p can be evaluated from Equation [8]:

$$p(-\infty < z < 0) = 0.5 \cdot \left[1 - \Phi\left(\frac{Ez}{\sigma_z \sqrt{2}}\right)\right], \tag{8}$$

where Ez and σ_z are the mathematical expectation and mean square deviation of z and Φ the probability integral (tabulated function). A calculation shows, for instance, that using (in concrete with certain composition and certain temperature cyclic action, etc) granite with modulus of elasticity $E_2=5 \cdot 10^4$ Mpa as a coarse aggregate results in the damage probability $p(z<0)=3 \cdot 102$ ^{3}. If we replace the granite aggregate with a light one ($E_2=2.5 \cdot 10^4$ Mpa), $p(z<0)=4.5 \cdot 10$ |^4. Using steel or cast- iron as aggregate with $E_2=2.2 \cdot 10^5$MPa, we get

$$p(z<0)=11 \cdot 10 z = 1 - \frac{\Delta\varepsilon E_1 E_2 \left[\left(\frac{R_2}{R_1}\right)^2 + 1\right]}{R_{h,t}\left(\frac{R_2}{R_1}\right)^2 \left[E_1 + 0.6 \cdot E_2\right] + 0.6 \cdot \left(E_1 - E_2\right)} > 0^3.$$

Assuming that concrete breaks down after a loss of 15% of its strength, we deduce, in a rough approximation, that in the first case for heavy concrete, it breaks down after 0.15:0.003 = 50 cycles; for light weight concrete –after 333 cycles; and for radiation protective concrete —after 14 cycles. In the case when the environmental action has a cyclic nature (freezing and thawing, heating and cooling, etc) the decrease of concrete strength can be estimated by the following formula:

$$R_n = kR_0(1-p)^n, \tag{9}$$

where R_0 and R_n — initial strength of concrete and that after n cycles and k — experimental coefficient that, in a first approximation, can be taken equal to 1.

Cracks reduce the strength of concrete, and they also reduce the elastic modulus E_2. We can assume that the decrease of E_2 is directly proportional to the decrease of concrete strength and is characterised by a formula similar to Eq. (9). The decrease of E_2, according to (3a,b), reduces intrinsic stresses $\sigma_{\theta,r}$. It changes the value of z and, hence, probability $p(z<0)$ becomes a function of the n–cycles serial number. Using a recurrent process, we can obtain the formula for $p(z<0)=F(n)$:

$$R_n = k \cdot R_0(1-p_1)(1-p_2)(1-p_3)\ldots(1-p_n), \tag{10}$$

where the subscripts relate to the serial number of a cycle.

Stresses reduce step by step until they become equal to $R_{h,t}$ —the tensile strength of a hydrated mass, which remains constant for the structure. In the case when $\sigma \leq R_{h,t}$, $z \geq 0$ and destruction ceases. Theoretical curves of strength reduction for concrete with various initial $p(z<0)$ are represented in Fig. 1 along with experimental curves of changes in the prism strength in cyclic tests of heating and cooling from 10°C to 200°C [9].

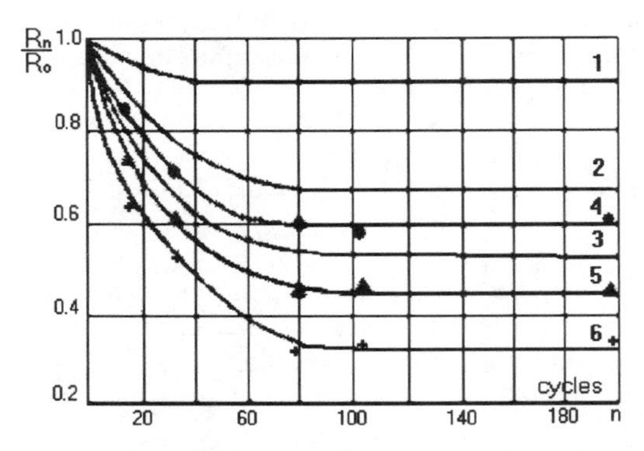

Fig. 1. Influence of cycle action on the strength of concrete in model (p=F(n)) and in experiment: 1, 2, 3 — concrete with $p_1(z<0)$=0.003;0.008;0.014 respectively; 4, 5 and 6 – concrete with coarse aggregate size 5–10 mm; 5–20 mm and 5–40 mm, respectively.

The character of experimental curves is in accordance with Eq.(10), and their disposition in accordance with Eq.(3a). In many experiments, especially in cyclic heating and cooling, it was observed that after an initial drop, the strength of concrete becomes stable as if it adapts to the environment action. The mechanism of such adaptation is revealed by formula (10).

5 Durability of concrete in construction

The stresses in concrete follow the strain discrepancies of its porous and compact components. A similar discrepancy appears between steel and concrete in reinforced concrete (see Table 2). In loaded construction, stresses that arise due to different reasons superimpose and, hence, the criterion of concrete deterioration can be expressed in the following form:

$$\frac{\Sigma\sigma_t}{R_{h,t}} = \frac{\sigma_{\Sigma,t} + \sigma_{rf} + \sigma_l + \sigma_{tel} + \sigma_{pr}}{R_{h,t}} \geq 1, \tag{11}$$

where $\sigma_{r,t}$ —stresses that arise as an effect of differences in free strains of reinforcement and concrete; σ_l —stresses due to external loading; $\sigma_{t,l}$ —thermoelastic stresses due to the temperature gradients in concrete; and σ_{pr} —stresses due to prestressing. The computation of those stresses is a special topic which can t be discussed here. As follows from (11), the stress field in actual construction is much more intensive than that in plain concrete. Also, it is quite heterogenous since stresses sum in accordance with their values and signs. The cracks appear, first of all, at the points where the total tensile stresses $\Sigma\sigma_t$ are maximal and exceed R_{ht}.

The probability approach can be applied in the case of concrete construction in the same way as was shown above. For example, if a concrete element under a severe environment is loaded, we can transform the criterion of deterioration to $\frac{(\sigma_{\Sigma,t} + \sigma_l)}{R_{h,t}} \geq 1$, the function of loaded concrete integrity as $z_l = 1 - \frac{(\sigma_{\Sigma,t} + \sigma_l)}{R_{h,t}}$,etc.

The results of experiments with water saturated samples (4x4x16cm) are presented in Fig. 2. Axial strains at various temperatures were measured, while similar samples were tested in the same chamber as beams under a central load equal to 40% of ultimate load. Deflection growth takes place when anomalous expansion begins, which can be explained assuming the summing of intrinsic and load stresses in concrete. The anomalously high creep of concrete during freezing was noted long ago [10].

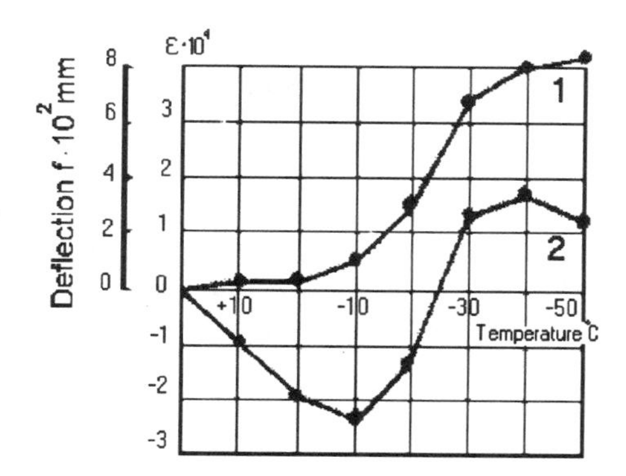

Fig. 2. Influence of freezing on deformation of concrete: 1 — loaded specimen (span length 10 cm) suffering deflection f; 2 — unloaded specimen suffering axial strains ε.

In the tidal zone of the Barents Sea, environmental conditions act upon concrete like a free-zing chamber that provides rapid freezing and thawing in mineralized water. The rate of deterioration of reinforced concrete was clearly influenced by the frost resistance of concrete itself. The deflection of loaded reinforced beams made of concrete with gas-forming and air-entrained admixtures, low w/c ratio, was much lower than that without admixtures and with high w/c ratio. Deflection of all beams in the tidal zone was much higher than those on shore (influence of $\sigma_{\Sigma,t}$). Deterioration begins at the tensile zone of beams and reaches its greatest degree in the section of maximum moment and diminishes toward the end of the beam with decreasing tensile stresses (influence of σ_l). It also begins at corners and edges of beams (influence of σ_{tel}). In all other cases, there was a close connection between the stress field in concrete and its deterioration [11].

The results of tests of mortar specimens (4x4x16 cm) partially immersed in 5% Na_2SO_4 are presented in Fig. 3. An unusual curvilinear mode of fracture was observed only if loaded specimens were tested for a long time in sulfate solution (Fig. 3c). An experiment with photo–elastic samples loaded in the same device shows that destruction follows the trajectory of main stresses in concrete (see Fig. 3a,b). The fracture follows the line of least resistance that was, beforehand, formed by micro–cracks due to the combined effect of corrosion and loading.

An analysis of the experimental results provides valuable information about the superposition of intrinsic and load stresses. Similar deterioration of the ends of reinforced beams (size 15x20x200 cm) were observed at a test station in the Barents Sea. Beams were loaded in spring devices (in a similar way) and suffered from freezing and thawing.

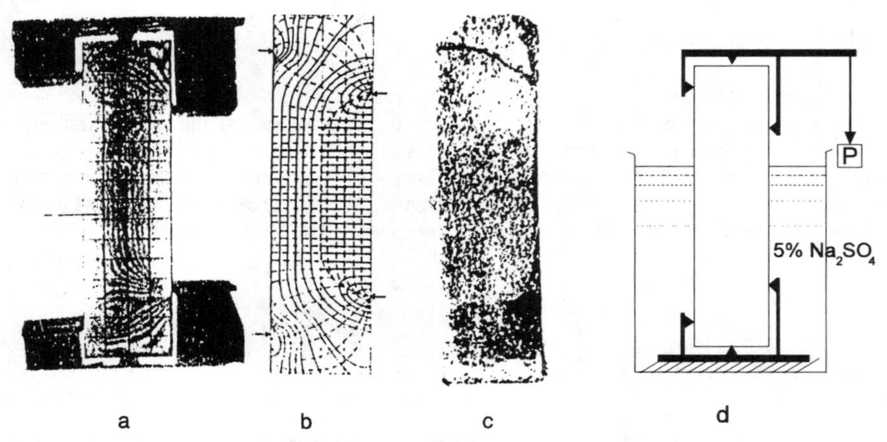

Fig. 3. Influence of combined action of corrosion and loading on fracture path in concrete: a) loaded sample of photo–elastic material; b) trajectory of main stresses in sample; c) sample of mortar after break down in sulfate corrosion test in loaded state (f–f - path of fracture); d) scheme of loading.

6 Experimental verification of mechanism

Some experimental results of mainly scientific nature were given above. It is a very limited part of the results which were analysed and taken into consideration. The data collection of experimental results published over decades of the investigation of concrete and reinforced concrete under severe environments were analyzed. Some experiments were conducted at a test station in the tidal zone of the Barents Sea, where concrete specimens and reinforced elements suffered 360–390 cycles of freezing and thawing per year. The analysed data related to the influence on concrete durability elasticity modulus and size of aggregates, volume concentration of components, strength and age of concrete, pore–forming additions, temperature and rate of freezing, schemes of loading, different kinds of reinforcing, etc. were examined as well [11]. A comparison of theoretical estimations and experimental data confirms their correspondence. Any facts that contradict the ideas stated above have still not been revealed.

7 Conclusions

The theoretical scheme of concrete deterioration under severe environment is presented in the paper without many details. A large number of explanations, consequences and conclusions are omitted. But the main idea is that in many cases (freezing and thawing, heating and cooling, wetting and drying, sulfate aggression, etc) the principal cause of concrete and reinforced concrete deterioration under severe environments lay in their structural strain heterogeneity. Concrete and its porous components (cement– sand mortar, hard cement and hydrated mass) on the one hand and steel and aggregates on the other have opposing deformations when subjected to the action of severe environment. Such discrepancy in disconnected strains of components inevitably leads to intrinsic stresses and cracks formation in the material and in the construction. This paper is devoted to tracing that theoretical scheme. It requires the introduction of some ideas related to the structure of concrete and two areas of deterioration processes. Mathematical models are suggested for calculation of the intrinsic strains and stresses. To estimate the formation of defects in concrete and the resulting decrease in strength under aggressive media, an approach is developed based on probability theory. Extensive experimental work were conducted and collections of published experimental data were gathered for verification of the proposed mechanism. Attempts were made to analyse and carry out calculations (in some cases comparative) of the effects of various factors on concrete durability that previously were determined experimentally. It has become possible to explain from a common point of view a variety of experimental results, some of which had had no logical explanations before or required special explanations in different cases.

8 References

1. Pigeon, M. (1994) Frost resistance. A critical look, in Concrete technology past, present and future. *Proceedings of V.M.Malhotra symposium*, ACI, Detroit, pp. 141–158.
2. Podvalny, A.M. and Prozenko, A.M. (1983) An investigation of the permeability of porous bodies on a mathematical model in Pore structure and properties of materials. *Proceedings of the international RILEM–JUPAC symposium*, Prague, 1973, pp. A–13 - A–31.
3. L Hermite, R.G. (1962) Volume changes of concrete, in Chemistry of cement. *Proceedings of the fourth international symposium*, Washington, pp. 639–694.
4. Hansen, T.E. (1968) Theories of multi–phase materials applied to concrete, cement, mortar and cement paste, in The structure of concrete. Cement and Concrete Association, London, pp. 68–82.
5. Walker, S., Bloem, D.L. and Mullen, W.G. (1952) Effect of temperature changes on concrete as influenced by aggregates. *Journal of the ACI,* Vol. 23, N°8, pp. 661–679.
6. Osetinsky, J.W. and Podvalny, A.M. (1982) About model choice for intrinsic stress calculation in concrete. *Mekhanica compositnykh materialov*, Riga, N°5, pp. 789–796 (In Russian).
7. Podvalny, A.M. and Osetinsky, J.W. (1982) Probability model of concrete behavior in environment action. *Stroitelnaya mechanica i raschet sorujeny*, Moscow, N°2, pp. 28–33 (In Russian).
8. Wentzel, E.S. (1969) *Teoriya veroyatnosty*. Moscow, 576 pp. (In Russian).

9. Zipenjuk, I.F. (1966) Issledovaniye svoistv betona I zhelezobetonnykh constructsi... Ph.D. Thesis, NIIZHB, Moscow, 163 pp. (In Russian).
10. Podvalny, A.M. (1963) The creep of concrete during freezing. *Docl. Acad. Nauk SSSR*, Vol. 148, N°5, pp. 1148–1151 (In Russian).
11. Podvalny, A.M. (1986) Elementy teorii stoikosty betona i zhelezobetona... Dr.Sc.(Eng.), NIIZHB, Moscow, 395 pp. (In Russian).

PART TWENTY-FOUR
ASEISMIC DESIGN

145 PERFORMANCE OF LARGE REINFORCED CONCRETE LIFELINES UNDER SEVERE EARTHQUAKES

Y. CHEN
Pennsylvania State University, Middletown, USA
T. KRAUTHAMMER
Pennsylvania State University, University Park, USA

Abstract
The earthquake resistance is studied of various reinforced concrete lifelines subjected to severe earthquakes considering lifeline embedment, structural geometry, nonlinearities and the frequency content of ground motion. The performance and behavior of the lifelines are carefully evaluated. Ground motion effects on structural responses is demonstrated. Recommendations for predicting peak structural responses and ductility factors are made, and conclusions on the overall performance of large structural concrete lifelines exposed to severe earthquake conditions are drawn.
Keywords: earthquake, embedment, frequency content, ground motion, lifeline, nonlinearity, response, structural geometry, transfer function.

1 Introduction

Gas, oil and fuel lines, water mains, highway culverts and electrical and communication utility conduits are usually classified as "lifelines", which signify the importance of such structural systems. These systems are generally susceptible to earthquake damage as evidenced by the 1940 Imperial Valley, 1964 Alaska, 1966 Parfield, 1971 San Fernando, and 1994 North Ridge Earthquakes [1-3]. Rational analysis and design of large lifeline systems has drawn increasing attention because of the impact of their survivability under severe earthquake effects on many communities.

Concrete Under Severe Conditions: Environment and loading (Volume Two) Edited by K. Sakai, N. Banthia and O.E. Gjørv. Published in 1995 by E & FN Spon. ISBN 0 419 19860 1

The available literature reveals that at least two important issues related to lifeline earthquake engineering have not been completely understood: (1) dynamic soil-structure interaction (SSI), and (2) transfer functions for predicting peak structural responses. It is noted that experimental work related to the subject is still very limited due to the complexity of the problem itself and the prohibitive cost involved. It is hopeful that the predicted results from the present study can be used for comparison with future test data.

2 Problem Description

A typical configuration of soil-straight lifeline systems is shown in Fig. 1. Typical structural dimensions are 4 m x 4 m x 0.30 m (width/height/thickness) for rectangular lifeline system, and 4.6 m x 0.33 m (outer diameter/thickness) for the circular lifeline system. The circular lifeline system is structurally equivalent to the rectangular one in that they have the same moment of inertia and cross-sectional area. The equivalency is based on the assumption that under earthquake excitations rigid body motions are dominant, while stiffness effects are less important. The 1940 El Centro California earthquake (Fig. 2) was chosen as the design earthquake. The base of the soil-lifeline systems is rigid bedrock at depth of 35 m (= H) which is based on the actual geological data measured at the earthquake site [4]. The SSI problem was solved by an effective and efficient combined finite difference (FD)-finite element (FE) with substructuring approach [5]. A two-dimensional plane strain model was employed to study the soil-lifeline systems. The seismic waves were assumed to be

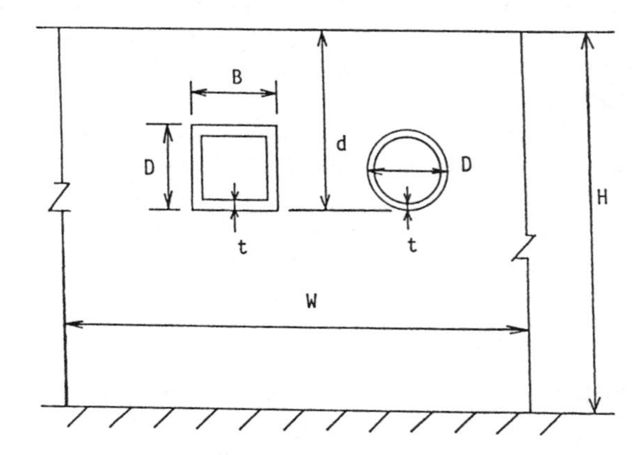

Fig. 1. Configurations of soil-lifeline systems.

the SV type. To study the effects of the frequency content of ground motion on system responses, three different earthquake motions were considered. The "standard earthquake (E1)" (Fig. 3) was simulated directly from the design ground accelerations (Fig. 2) using the free-field analysis method in which the total loading duration (T_d) of 10 seconds and time step size (Δt)of 0.02 seconds were used [6, 7]. The 10-second duration was justified to cover the predominant ground motions. The

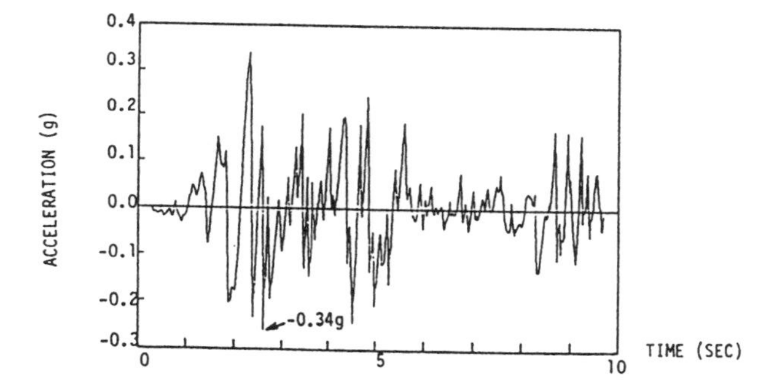

Fig. 2. The 1940 El Centro ground accelerations (NS-component).

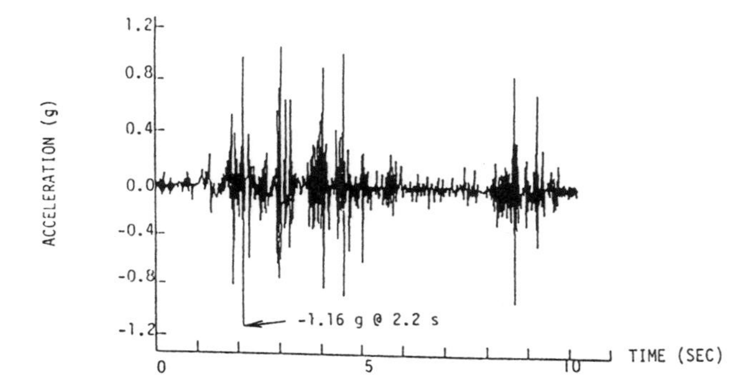

Fig. 3. The generated based accelerations, E1.

"condensed earthquake (E2)" (Fig. 4) was derived from the compressed ground accelerations which have T_d of 5 seconds, Δt of 0.01 second and the same acceleration peaks as the design earthquake (Fig. 2). The "expanded earthquake (E3)" (Fig. 5) was simulated from the stretched ground accelerations which have T_d of 15 seconds, Δt of 0.03 second and also the same acceleration peaks as the design earthquake.

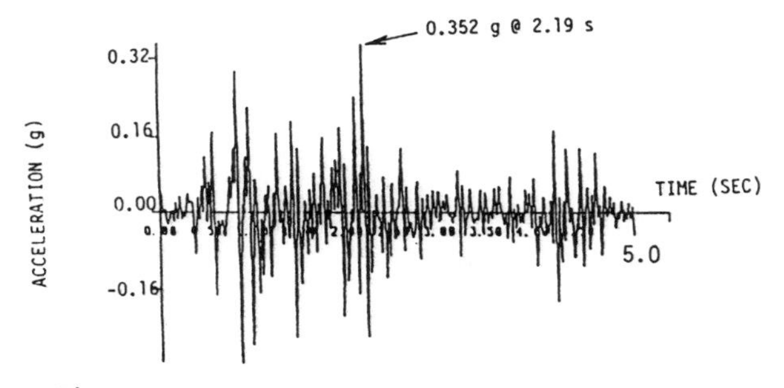

Fig. 4. The generated based accelerations, E2.

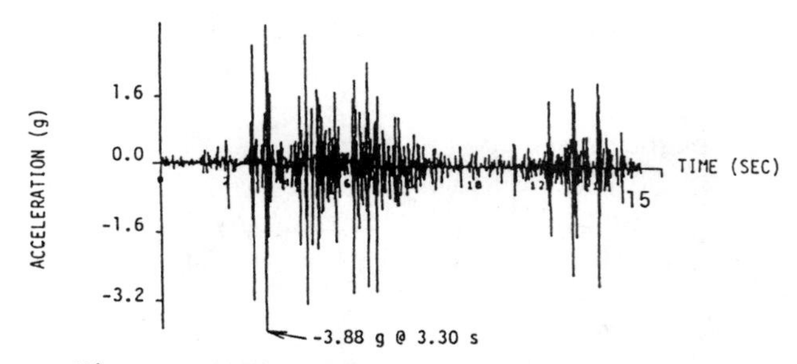

Fig. 5. The generated based accelerations, E3.

3 Governing Equations

For the FD formulation, we have

$$\rho \frac{\partial \dot{u}}{\partial t} = \frac{\partial \sigma_{ij}}{\partial x_j} - \rho \phi \, \dot{u}_i + \rho g_i, \quad (i, j = 1,2,3) \tag{1}$$

Where $\dot{u}$ are the velocities, x_j the coordinates, g_i the gravity accelerations, t the time variable, ρ the mass density, σ_{ij} the stress components, and φ is the damping constant.

For the FE formulation, we have

$$[M]\,\{\ddot{U}\}+[C]\,\{\dot{U}\}+[K]\,\{U\} = \{R(t)\} \tag{2}$$

where [M] is the total mass matrix, [C] the total damping matrix, [K] the total stiffness matrix, $\{\ddot{U}\}$ the acceleration vector, $\{\dot{U}\}$ the velocity vector, {U} the displacement vector, and {R(t)} is the loading vector.

For numerical integration, an explicit scheme was used in the FD method, while an implicit scheme was employed in the FE method. For iteration, a modified quasi-Newton scheme similar to [8] was used for the nonlinear analyses. A lumped mass approximation was assumed to form [M]. It should be noted that the combined FD-FE approach with substructuring permits the required time step size (Δt) to be determined by the softer soil material alone, which leads to significantly larger Δt, and a shorter computation time.

To avoid spurious wave reflections from the lateral boundaries of the finite model, concentrated viscous dampers with 100% damping ratio were imposed at the nodes of the boundaries. To account for both material an geometrical damping, the following damping ratios were exercised within the boundaries of the model: 7% (soil) and 5% (structure) for linear analyses, and 15% (soil) and 10% (structure) for nonlinear analyses. The overall width (W, Fig. 1) of the finite model was set to be 35 m. The criterion reported in [9] was utilized to discretize the soil medium, and the maximum aspect ratio of structural elements remained at 2 or less.

4 Nonlinear Material Models

In this study, the Mohr-Coulomb plasticity model with nonassociated flow rule, the Drucker-Prager plasticity model with cap, an the curve description model with the thin-layer concept [10] were employed, respectively, for the soil, concrete and interface.

It should be mentioned that, unlike the classical Mohr-Coulomb models, the nonlinear soil model implemented considers the effect of intermediate stresses. Also, the interface model is simpler since it does not involve any explicit yielding condition, and the aspect ratio of about 0.05 (1 to 20) for an interface element should be used when employing the interface model [10].

5 Numerical Examples and Results

The parametric studies covered: two representative structural shapes (rectangular, circular), three typical embedment ratios (α = d/D, Fig. 1) of 0.0, 0.5 and 2.0, three different simulated earthquakes (E1, E2 and E3), and ten shear strengths of clayey soils ranging from $1\text{x}10^4$ Pa to $100\text{x}10^4$ Pa with the equal increment of $11\text{x}10^4$ Pa. Totally, 180 nonlinear analyses were performed using the combined numerical approach. For correlation purposes, a frequency factor (β) of 1.0 was assigned to the E1 Earthquake, 0.5 to the E2 Earthquake, and 1.5 to the E3 Earthquake. Various concrete strengths were also considered. However, only the results corresponding to concrete strengths of 28 MPa which is mostly used in practice are presented here.
Typical FE and FD meshes are shown in Fig. 6. The computed peak normalized structural displacements of engineering significance are shown in Fig. 7-9. Variations of the peak velocities and peak accelerations were similar. Normalization of the peak structural responses was done with respect to the maximum ground motions of the design earthquake. The critical stresses in the structure and interface are shown in Tables 1 and 2, respectively.

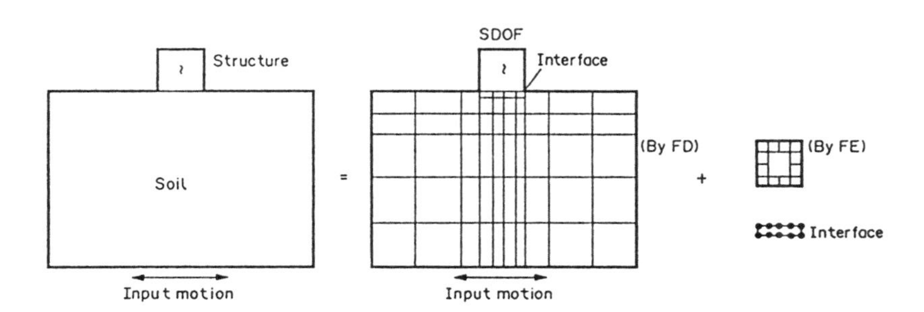

Fig. 6. Typical FD/FE meshes with substructuring scheme (SDOF: super-degree-of-freedom).

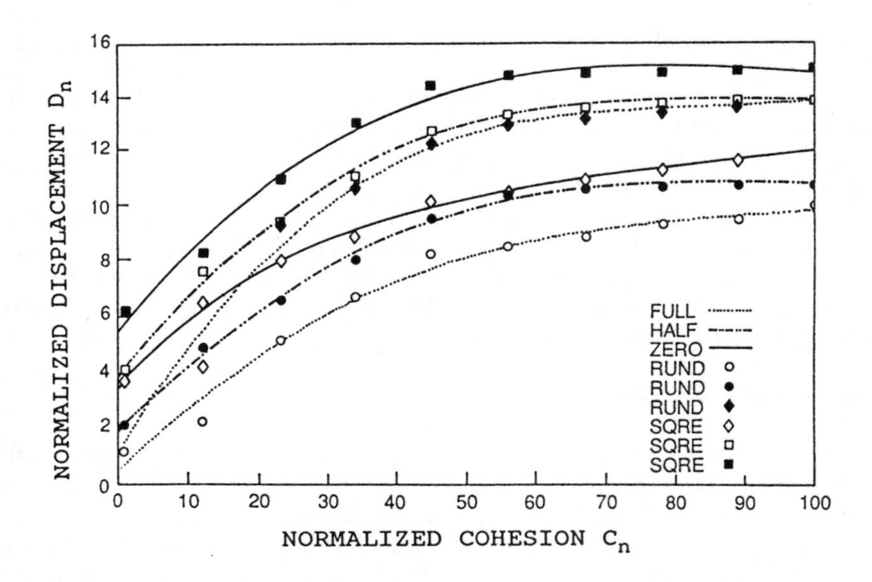

Fig. 7. The peak normalized displacements corresponding to E1 (FULL: $\alpha = 2$, HALF: $\alpha = 0.5$, ZERO: $\alpha = 0$, RUND: Round, SQRE: Square).

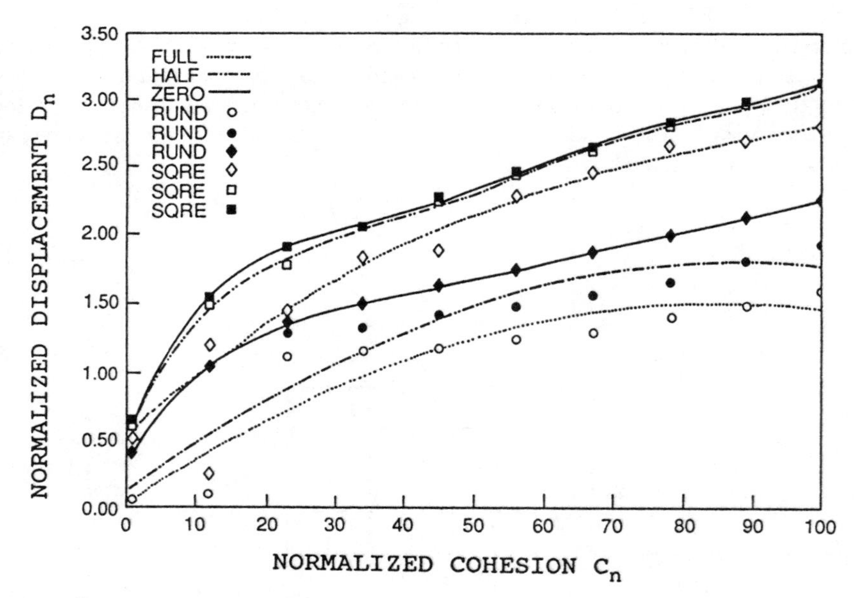

Fig. 8. The peak normalized displacements corresponding to E2.

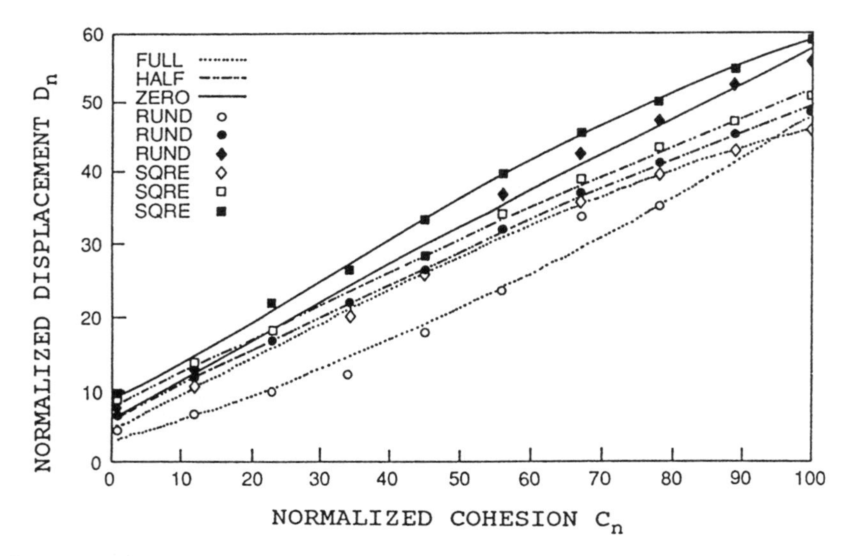

Fig. 9. The peak normalized displacements corresponding to E3.

Table 1. Peak structural stresses (for: soil shear strength = 1 MPa, and E2 earthquake)

(in MPa)

α	σ_{yy}	σ_{zz}	σ_{yz}	σ_{xx}
	For the square lifeline system			
0.0	3.63	2.84	1.00	0.88
0.5	5.41	7.26	1.93	1.80
2.0	18.42	19.39	5.83	5.38
	For the circular lifeline system			
0.0	2.11	1.65	0.58	0.51
0.5	2.97	3.99	1.06	0.99
2.0	9.38	9.87	2.97	2.74

x: longitudinal; y: vertical; z: horizontal.

Table 2. Peak interface stresses (for: soil shear strength = 1 MPa, and the square lifeline system)

(in MPa)

β	σ_{yy}	σ_{zz}	σ_{yz}	σ_{xx}
0.5	0.48	0.42	0.29	0.27
1.0	0.92	0.82	0.56	0.52
1.5	1.05	0.93	0.64	0.60

x: longitudinal; y: vertical; z: horizontal.

Transfer functions interrelating the peak lifeline responses with the maximum ground motions can be derived based on the generated data. For instance, the normalized peak displacement, $D_{n,max}$, for the rectangular and circular lifelines can be expressed by Eqns. (3) and (4), respectively, as follows:

$$\begin{aligned} D_{n,max} = {} & [(-2.49\alpha^2 + 9.72\alpha - 3.81)\beta^2 + (5.01\alpha^2 - 21.55\alpha + 15.70)\beta \\ & + (-1.87\alpha^2 + 8.33\alpha - 6.40)] + [(-0.19\alpha^2 + 0.08\alpha - 0.03)\beta^2 \\ & + (0.38\alpha^2 - 0.12\alpha + 0.41)\beta + (-0.15\alpha^2 + 0.02\alpha - 0.06)]C_n \\ & + [(0.019\alpha^2 - 0.016\alpha + 0.008)\beta^2 + (-0.040\alpha^2 + 0.023\alpha - 0.009)\beta \\ & + (0.020\alpha^2 - 0.006\alpha - 0.0025)]C_n^2, \end{aligned} \tag{3}$$

$$\begin{aligned} D_{n,max} = {} & [(-5.54\alpha^2 + 13.99\alpha - 1.57)\beta^2 + (10.16\alpha^2 - 27.51\alpha + 8.80)\beta \\ & + (-3.53\alpha^2 + 9.79\alpha - 3.70)] + [(-0.01\alpha^2 - 0.24\alpha + 0.14)\beta^2 \\ & + (-0.047\alpha^2 + 0.52\alpha + 0.17)\beta + (0.08\alpha^2 - 0.33\alpha - 0.03)]C_n \\ & + [(0.0055\alpha^2 - 0.011\alpha + 0.012)\beta^2 + (-0.007\alpha^2 + 0.014\alpha - 0.022)\beta \\ & + (-0.0005\alpha^2 + 0.025\alpha + 0.0048)]C_n^2, \end{aligned} \tag{4}$$

Ductility factors (μ) can also be derived from the elastic and inelastic design spectra by using the equal maximum deflection theory or the equal energy theory [11]. Suggested μ values for the rectangular and circular lifelines are determined by Eqns. (5) and (6), respectively, as follows:

$$\mu = 10.0 - 0.0770\ \Delta C_n \tag{5}$$

$$\mu = 8.33 - 0.0601\ \Delta C_n \tag{6}$$

where ΔC_n is the nondimensional cohesion varying from 0 to 99.

7 Discussion

Based on the results from the present study, the following emerging trends were observed:

(1) The peak structural response decreased with the increase of embedment depth, but increased with the increase of soil cohesion.
(2) The circular shape experienced lower seismic responses than the rectangular one.
(3) The expanded earthquake produced the highest maximum responses, while the standard earthquake and the condensed earthquake gave progressively lower results, respectively.
(4) The maximum stresses at the soil-structure interface derived in this study were generally comparable with those provided in [2, 12].
(5) The derived peak nodal responses can also be assessed quantitatively. For instance, one can obtain the correlation factors for the responses of the rectangular lifelines among the three typical burial conditions, as provided in Table 3.

Table 3. Correlation factors of peak structural responses for three burial conditions

	β	$D_{n,max}$	α = 0.5 $V_{n,max}$	$A_{n,max}$	$D_{n,max}$	α = 2.0 $V_{n,max}$	$A_{n,max}$
	0.5	0.99	0.84	0.58	0.89	0.45	0.43
α=0.0	1.0	0.92	0.95	0.66	0.92	0.89	0.58
	1.5	0.87	0.91	0.41	0.79	0.87	0.14

(6) Qualitatively, the peak structural responses were most influenced by the frequency content of earthquake motion, and then by the structural geometry, material nonlinearity and embedment depth in descending order.
(7) With the aid of some design charts, the transfer functions can be conveniently used to predict the maximum seismic responses of the lifelines conveniently.
(8) The effect of the mass density of soil had little effect on the maximum structural responses.
(9) The peak structural responses remained about the same for lifelines with depth-to-thickness (D/t, Fig. 1) ratio around 14.
(10) When the structural displacement becomes large, the structure tends to separate from the soil at the interface (soil tensile stress developed). Detail of interface behavior in available in the literature [10].
(11) Using the ACI standard practice [13], flexural-shear yielding was observed at the corners of the rectangular lifelines, and concrete cracking prevailed in the RC lifelines. Premateur failures were also found at the soil-lifeline interface.
(12) Depending on the system parameters, μ ranging from 2.5 to 10.0 for the rectangular lifelines, and 2.5 to 8.5 for the circular lifelines may be considered for use in design.

8 Conclusions and recommendations

The frequency content of earthquake motion affects the seismic response of RC lifeline structures significantly. Higher concrete strength in conjunction with good reinforcing steel detailing can possibly avoid the flexure-shear yielding and minimize the concrete cracking. Special means, such as using softer materials should be provided at the soil-lifeline interface to prevent premature failures.

9 References

1. Newmark, N. M. and E. Rosenblueth. (1971) *Fundamentals of Earthquake Engineering*, Prentice-Hall, Englewood Cliffs, NJ.
2. ASCE (1984) *Guidelines for the Seismic Design of Oil and Gas Pipeline Systems,* ASCE Technical Council on Lifeline Earthquake Engineering, New York.

3 Dames & Moore (1994) *The Northridge Earthquake - January 17, 1994,* A Special Report by Dames & Moore Inc., Los Angeles, CA.
4. JSEEP (1983) *NSEEP News,* No. 69.
5. Chen, Y. and Krauthammer, T. (1989) A combined ADINA-finite difference approach with substructuring for solving seismically induced nonlinear soil-structure interaction problems. *Comput. Struct.*, Vol. 32, pp. 779-785.
6. Krauthammer, T., (1987) Free field analysis considerations for dynamic soil-structure interaction. *Comput. Struct.* Vol. 26, pp. 243-251.
7. Krauthammer, T. and Chen, Y.(1988) Free field earthquake ground motions: effects of various numerical simulation approaches on soil-structure interaction results. *J. Engng. Struct.* Vol. 10, pp. 85-94.
8. Matties, H. and Strange, G. (1979) The solution of nonlinear finite element equations. *Int. j. Numer. Meth. Engng.* Vol. 14, pp. 1613-1626.
9. Kuhlemeyer, R. L. and Lysmer, J. (1973) Finite element accuracy for wave propagation problems. ASCE *J. Soil Mech. Fdns.*, Vol. 99, pp. 421-427.
10. Krauthammer, T. and Chen, Y. (1989) Soil-structure interface effects on dynamic interaction analysis of reinforced concrete lifelines. *Int. J. Soil Dyn. Earthq. Engng,* Vol. 8, pp. 32-42.
11. Housner, G. W. and Jennings, P. C. (1982) *Earthquake Design Criteria,* EERI Monograph Series, Pasadena, CA.
12. Idriss, I. M. and Seed, H. B. (1967) *Response of horizontal soil layers during earthquakes.* Research report, Soil Mechanics and Bituminous Laboratory, University of California, Berkeley, CA.
13. ACI (1989) *Building Code Requirements for Reinforced Concrete and Commentary,* ACI 318-89 and 318R-89, American Concrete Institute, Detroit, Michigan.

146 DESIGN OF LATERAL STEEL FOR CONCRETE CONFINEMENT

S.A. SHEIKH
University of Toronto, Toronto, Canada

Abstract
The current seismic design provisions of various building codes for confinement steel in reinforced concrete columns do not include any quantitative relationship between design parameters and column performance. The design based on these provisions is therefore very conservative under certain circumstances and unsafe under others. A procedure is proposed here for the design of rectilinear confining steel in columns to achieve a certain ductile performance. This procedure takes into account the effects of variables such as distribution of longitudinal and lateral steel, level of axial load and the amount of lateral steel on column behavior. The design method is calibrated against the results from an extensive experimental program involving specimens with 305 mm column sections confined with rectilinear ties and longitudinal steel. Concrete strength in these specimens varied between 27 MPa and 55 MPa.
Keywords: Columns, confined concrete, elasto-plastic behavior, earthquakes, energy dissipation.

1 Introduction

In the design of framed structures for earthquake resistance the current building design codes attempt to ensure their ductile performance by prescribing various provisions some of which require heavy confining steel in certain columns if their axial loads exceed a certain limit [1,2]. The amount of required confining steel is independent of the expected column performance. In addition, the effects of variables such as the level of axial load, distribution of longitudinal and lateral steel on the column behaviour are not considered in the design although overwhelming experimental and analytical evidences show that these variables have strong influence on the amount of

Concrete Under Severe Conditions: Environment and loading (Volume Two) Edited by K. Sakai, N. Banthia and O.E. Gjørv. Published in 1995 by E & FN Spon. ISBN 0 419 19860 1

confining steel required to achieve a certain ductile performance of a column [3,4]. As a result, the design is very conservative for columns with well-distributed steel subjected to low levels of axial load and unsafe for columns in which only four corner bars are effectively supported by tie bends and axial load is large. The New Zealand code [5], is an exception in that it includes axial load level in the design equations, although in all other respects these equations are similar to those in the North American codes. A design procedure is proposed here which considered the points mentioned above and includes the important variables in the design equations.

2 Proposed approach and limiting conditions

The proposed approach was developed using the results from an extensive experimental and analytical research program [4, 6-8]. Details from a select group of these specimens that were well-confined are shown in Table 1. Column section in each specimen was 305 mm square and the core size measured from the center of the perimeter tie was 267 mm. All the specimens with Configurations A and F (first letter in specimen designation) contained 8-19 mm (#6) diameter bars and specimens with Configuration D contained 12-16 mm (#5) diameter bars. Different steel configurations are shown in Fig. 1. Yield strength of steel varied between 462 and 517 MPa (67 and 75 ksi). Each specimen was tested with the application of axial load first followed by the application of lateral load using displacement control mode of an actuator. Moment curvature curves of failed sections from a few specimens are presented here to study a few variables. Different parameters considered in this procedure are discussed in the following.

2.1 Column performance

In evaluating the column performance and study the effects of different variables, ductility and toughness parameters defined in Fig. 2 were used. These include curvature ductility factor (μ_ϕ), cumulative ductility ratio (N_ϕ), and energy-damage indicator (E). Subscripts t and 80 wherever used indicate respectively, the value of

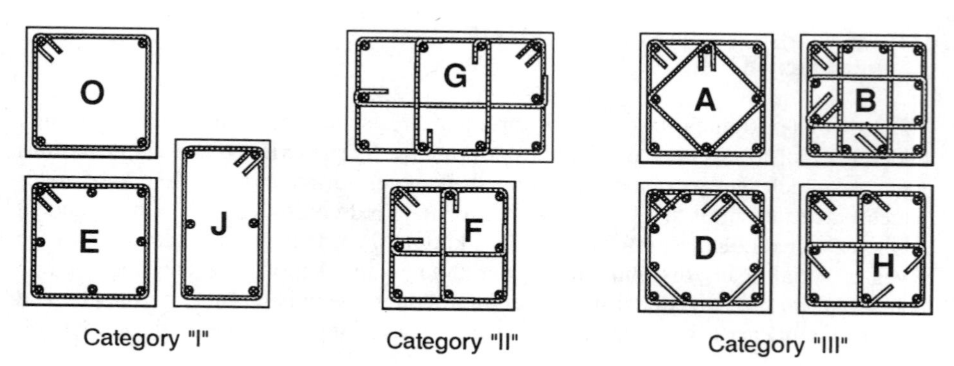

Fig. 1 Categories of steel configurations

Table 1. Specimen details and results

Spec	f'_c MPa	Lateral Steel			Axial Load		$\frac{M_{max}}{M_{ACI}}$	μ_ϕ	Ductility Ratio		Energy Indicator	
		Spacing	ρ(%)	$A_{sh}/A_{sh,c}$	$P/f''_c A_g$	P/P_o			N_{80}	N_t	E_{80}	E_t +
AS-3	33.2	108	1.68	1.43	0.60	0.50	1.25	19.0+	63	74	610	753
AS-17	31.3	108	1.68	1.52	0.77	0.63	1.43	12.0	52	58	402	443
AS-18	32.8	108	3.06	2.41	0.77	0.63	1.62	17.5	80	92	897	1156
AS-19	32.3	108	1.30	1.12	0.47	0.39	1.21	19.0	85	129	631	1230
AS-3H	54.1	108	1.68	0.88	0.62	0.59	1.20	10.5	31	35	178	204
AS-18H	54.7	108	3.06	1.44	0.64	0.61	1.30	14.0	43	59	384	458
AS-20H	53.6	76	4.30	2.10	0.64	0.61	1.49	16.5	80	98	935	1170
A-17L*	49.1	108	1.68	0.97	0.66	0.61	1.35	10.3				
A-3	31.8	108	1.68	1.44	0.61	0.49	1.23	28.5				
F-4	32.2	95	1.68	1.41	0.60	0.49	1.22	21.3				
D-5	31.2	114	1.68	1.39	0.46	0.39	1.26	20.0				
D-7	26.2	54	1.62	1.53	0.78	0.62	1.22	16.0				
A-11	27.9	108	0.77	0.72	0.74	0.58	0.97	8.3				
F-12	33.5	89	0.82	0.63	0.60	0.50	0.98	9.4				
D-14	26.9	108	0.81	0.73	0.75	0.60	1.01	7.3				
D-15	26.9	114	1.68	1.61	0.75	0.60	1.17	12.3				
A-16	33.9	108	0.77	0.70	0.60	0.50	0.95	13.2				

*Lightweight aggregate concrete specimen

+ Subscripts t and 80 refer to total and drop to 80% of the moment capacity, respectively

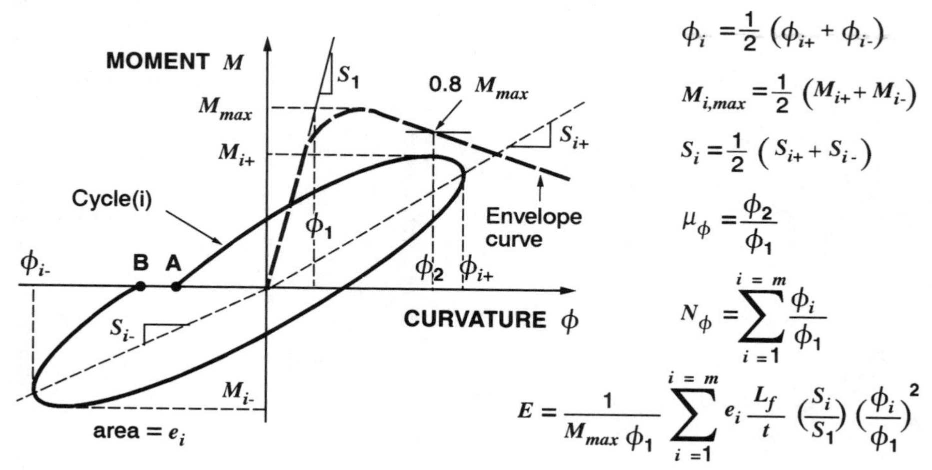

Fig. 2 Ductility parameters.

the parameter until the end of the test (total value) and the value until the end of the cycle in which the moment is dropped to 80% of the maximum value. Energy parameter e_i represents the area enclosed in the cycle i by the M-ϕ loop. All other terms are defined in Fig. 2 except L_f and t which represent the length of the most damaged region and section depth of the specimen, respectively. Parameter 'E' is similar to the one proposed by Ehsani and Wight [9] for force-deflection curves. Table 1 lists the available ductility parameters for all the specimens.

From nine specimens tested under similar conditions with constant axial load and cyclic lateral load it was observed that a reasonable correlation existed between ductility parameters N_ϕ, E and μ_ϕ. For μ_ϕ of 16 the values for $N_{\phi 80}$ and E_{80} are 64 and 575, respectively. A column section with this level of deformability is defined as highly ductile. With a μ_ϕ value of 8 to 16 the section is defined as moderately ductile and the low ductility column has $\mu_\phi < 8$. With this correlation between ductility parameters the specimens tested under monotonic flexure (e.g. last nine specimens in Table 1) could also be considered in the column performance analysis.

2.2 Axial load level

The level of axial load may be taken into consideration by introducing in the confinement equation the applied axial load (P) normalized with respect to a theoretical value for the axial capacity of the column (e.g. $f_c'A_g$ or P_o). Research [7] on columns with concrete strength varying between 27 and 55 MPa has shown that the amount of confining steel required for a certain column performance is proportional to the concrete strength as long as the axial load ratio P/P_o is constant. This is particularly important if the equations are to be applicable for a wide range of concrete strength. As an example, the moment-curvature relationships of Specimens AS-18H and AS-17 are shown in Fig. 3. The amount of lateral steel (ρ_s) is approximately in proportion to the concrete strength and P/P_o values are almost equal in both the specimens. The section behavior of the two specimens is very similar despite a large difference in the $P/f_c'A_g$ values (also see Table 1).

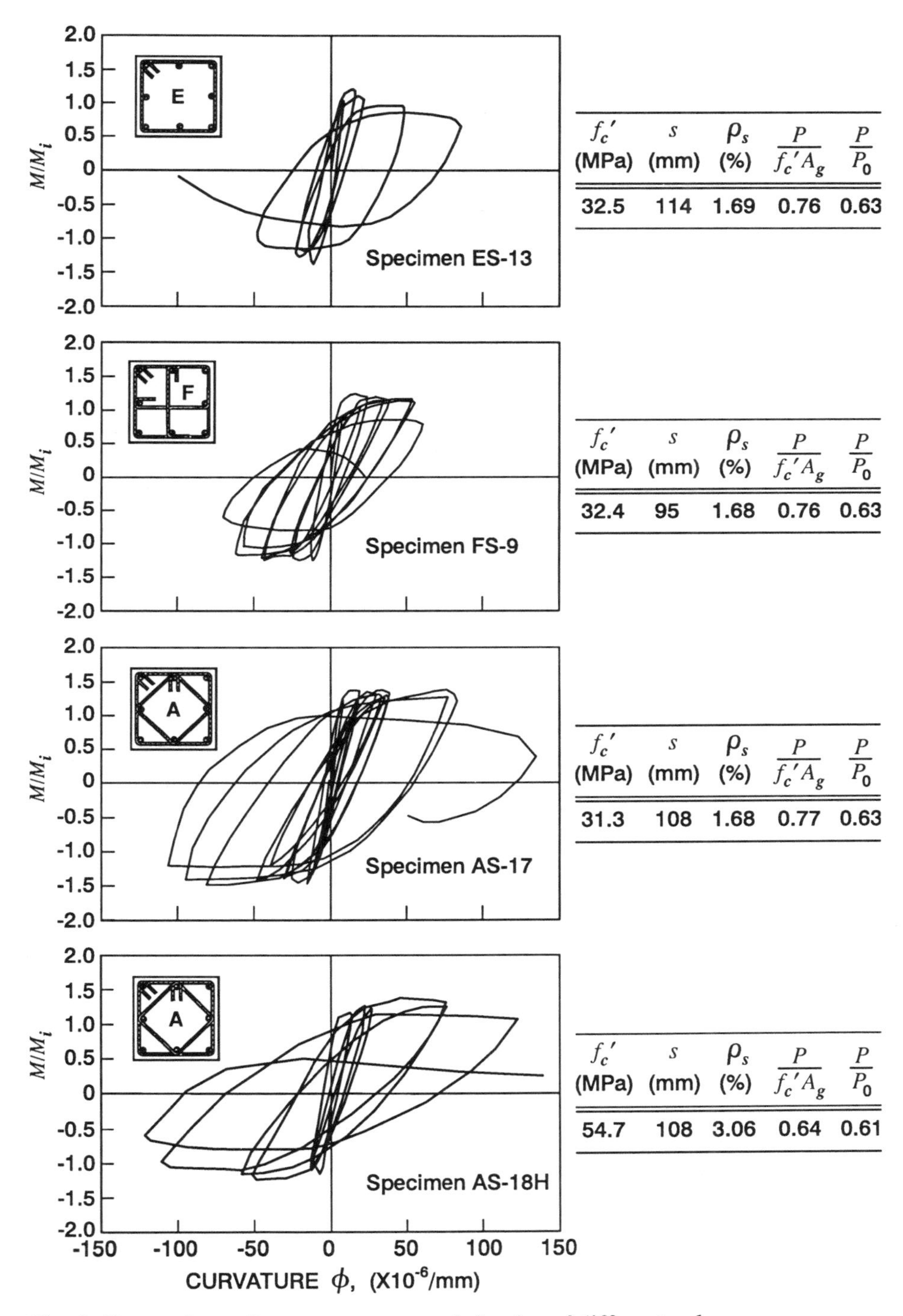

f_c' (MPa)	s (mm)	ρ_s (%)	$\frac{P}{f_c' A_g}$	$\frac{P}{P_0}$
32.5	114	1.69	0.76	0.63

f_c' (MPa)	s (mm)	ρ_s (%)	$\frac{P}{f_c' A_g}$	$\frac{P}{P_0}$
32.4	95	1.68	0.76	0.63

f_c' (MPa)	s (mm)	ρ_s (%)	$\frac{P}{f_c' A_g}$	$\frac{P}{P_0}$
31.3	108	1.68	0.77	0.63

f_c' (MPa)	s (mm)	ρ_s (%)	$\frac{P}{f_c' A_g}$	$\frac{P}{P_0}$
54.7	108	3.06	0.64	0.61

Fig. 3 Comparison of moment-curvature behavior of different columns

2.3 Steel configuration

The effectiveness of confining steel primarily depends on the area of the effectively confined concrete and distribution of confining pressure which are in turn highly affected by the distribution of longitudinal steel bars around the core and the extent of lateral restraint provided to the bars [3, 10]. On this basis steel configurations may be divided into the following three main categories (see Fig. 1).: Category "I" where only single perimeter ties are used as confining steel, Category "II" in which the middle longitudinal bars are supported at alternate points by hooks that are not anchored in the core concrete. All other hooks are anchored inside the core, and Category "III" in which a minimum of three longitudinal bars are effectively supported laterally by tie corners on each column face and all hooks are anchored into the core concrete. Experimental evidences show significant differences in the behavior of columns with different categories of steel configurations [3, 4, 6-8].

In Category I configurations, the mechanism of confinement is not very efficient [6, 10]. A large part of the core area is not effectively confined and even with a large amount of lateral steel the column may not display adequate ductility. In most specimens, it was observed that the middle unsupported bars pushed the ties outward resulting in a rapid loss of confinement and hence brittle failure. In Category II configurations, the hooks not anchored in the core may provide sufficient restraint to the bars initially, but at large deformations these hooks open out resulting in a loss of confinement and buckling of bars. Under high axial loads none of the specimens with configuration F showed satisfactory performance [4, 6] even when the lateral steel content was in excess of the ACI code [1] requirements. Specimens ES-13, FS-9 and AS-17 in Fig. 3 are almost identical in all regards except steel configuration. Specimen AS-17 (Category III) displayed more ductile behavior than Specimen FS-9 (Category II) which in turn is tougher than Specimen ES-13 (Category I).

In the light of the above discussion the limiting conditions under which the three categories of steel configurations may be reliably used for moderate and high ductility columns are outlined in Fig. 4. The loads P_o and P_b represent, respectively, the theoretical axial load capacity [1] and the balanced load for the column. It should be noted that the use of these configurations under such conditions may also require higher amount of lateral steel than that recommended by the Code [1].

2.4 Design of confining steel

The required amount of confinement steel (A_{sh}) is taken as:

$$A_{sh} = [A_{sh,c}] \, \alpha \, Y_p \, Y_\phi \qquad (1)$$

where $A_{sh,c}$ is the total cross sectional area of lateral steel within spacing "s" as defined in ACI code [1] and α, Y_p, Y_ϕ are factors to account for steel configuration, axial load level and the section performance, respectively. For Category III steel configurations, α is taken as unity. This group included all specimens of Configurations A and D. For Configuration F (Category II) specimens tested under low to moderate levels of axial load in which premature failure did not take place due to the opening of 90° hooks, α is also taken as unity. For an evaluation of Y_p and Y_ϕ

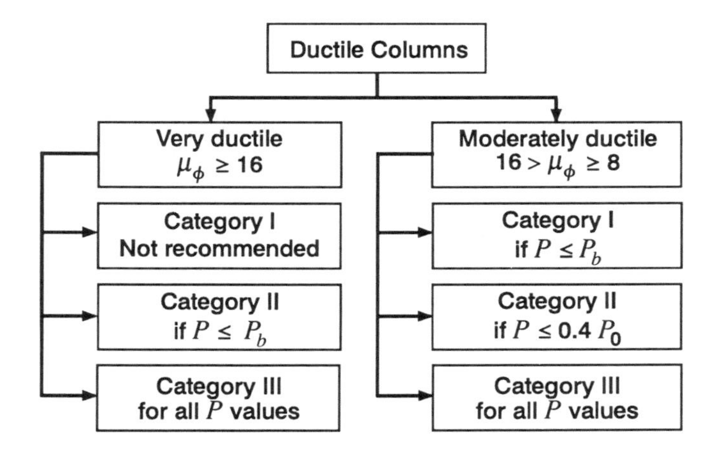

Category I	Section with only 4 corner bars laterally supported by tie bends
Category II	At least three bars on each face of the section laterally supported by tie bends including hooks not anchored in the core
Category III	Same as category II but all hooks anchored in the core

Fig. 4 Limiting conditions for steel configurations

only the specimens with $\alpha = 1.0$ were considered.

Since parameters Y_p and Y_ϕ are independent, the effect of axial load level was studied first considering those specimens which behaved in a very ductile manner ($\mu_\phi \geq 16$) and in which Y_ϕ was taken as unity. The best correlation between Y_p and P/P_o was observed using the following expression:

$$Y_p = 1 + 13\ (P/P_o)^5 \tag{2}$$

Figure 5 shows a plot of this relationship. Further simplification of this relationship also shown on the figure is given below:

$$Y_p = 6\ P/P_o - 1.4 \geq 1.0 \tag{3}$$

With the effect of axial load considered by using Eq. 2, all the specimens shown in Table 1 were considered to evaluate the effect of ductility on the required amount of confining steel. Eq. 4 resulted from this analysis with a correlation coefficient of 0.94. Eq. 5 is a simpler and relatively more conservative version of Eq. 4 as shown in Fig. 6.

$$Y_\phi = \mu_\phi^{1.15}/29 \tag{4}$$

$$Y_\phi = \mu_\phi/18 \tag{5}$$

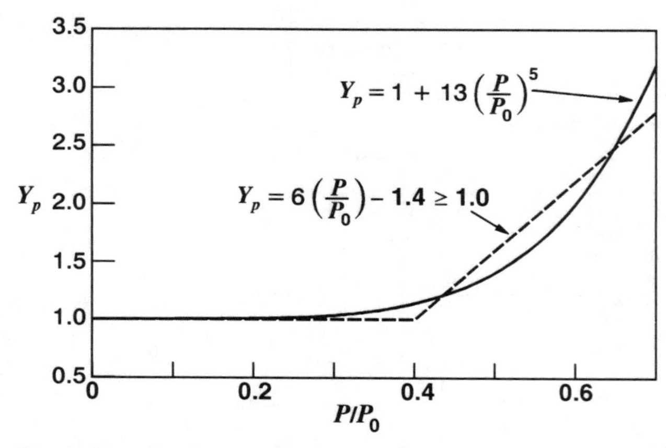

Fig. 5 Required amount of tie steel as affected by axial load level

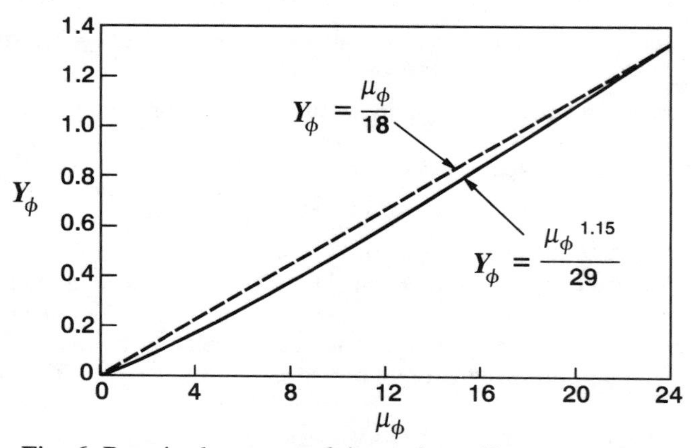

Fig. 6 Required amount of tie steel as affected by curvature ductility factor

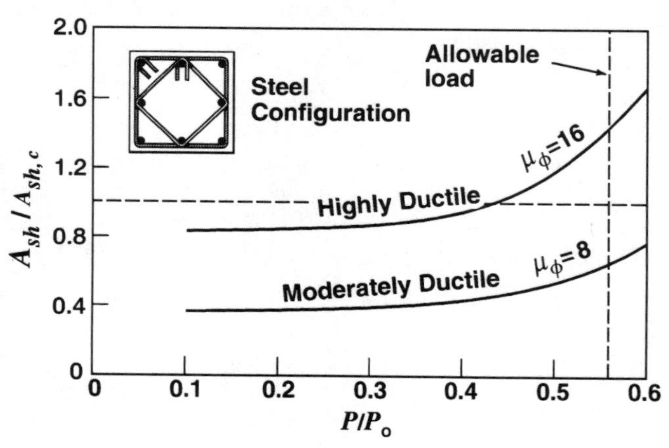

Fig. 7 Lateral steel Requirements

Figure 7 shows the required amount of confining steel as a function of axial load and section performance for a column in which steel is distributed in Configuration A or better in Category I. Eqs. 1, 2 and 4 were used for the construction of this figure. It is obvious that for low to moderate levels of axial load the amount of steel required by the code [1] results in a section which would behave in a very ductile manner. In fact, the amount of steel can be safely reduced without adversely affecting column behavior. For higher axial loads code design will result in a column which may not behave in a reasonably ductile manner. For columns with steel Configuration E which are allowed by the code, this maybe particularly serious because the code required amount of tie steel may only be about 30% of what is required for a moderately ductile behavior under high axial loads.

Application of Eqs. 1, 2 and 4 to Configuration E specimens tested under high axial load levels indicated an α value of 2 to 3. This means that for Category I configurations the amount of tie steel required for a certain ductility is about 2 to 3 times the steel required in Category III columns. The relative efficiency of confinement in these two types of configurations was found to be of the same order in an earlier analytical study [10].

As suggested before, the use of Category "I" configurations may be reliable only for moderate ductility columns under axial load level below the balance point. For this low level of axial load, the magnitude of term 13 $(P/P_o)^5$ becomes negligibly small and can be ignored in Eq. 2. Taking α = 2.5, the proposed equations for Category I configurations are as given below. Eq. 7 is a simplified form of Eq. 6.

$$A_{sh} = A_{sh,c} \{(\mu_\phi)^{1.15}/11.5] \quad (6)$$

$$A_{sh} = A_{sh,c} (\mu_\phi/7.2) \quad (7)$$

Eqs. 6 and 7 indicate that for axial load levels below the balance point, the current ACI Code steel may be sufficient to provide μ_ϕ of about 7 for sections with Category I configurations. However, for moderately ductile columns with μ_ϕ = 12, the required amount of lateral steel should be about 60% higher than that required by the Code.

Experimental and theoretical evidences show that tie spacing plays a significant role in the mechanism of confinement [3, 10]. The procedure presented here for the sake of simplicity does not include tie spacing as an active parameter. However, it should be noted that the test data based on which the equations are developed were obtained from specimens in which tie spacing varied from 0.20B to 0.43B where B is the width of confined core. In this range of spacing the confinement mechanism has high efficiency. Another important reason to limit tie spacing is to avoid premature buckling of longitudinal bars when a column is subjected to seismic excursions in the inelastic range. In the specimens considered here tie spacings varied between 3.4 d_b to 7.2 d_b where d_b is the bar diameter. In this range of s/d_b, premature buckling of the longitudinal bars can generally be avoided [11]. The proposed procedure can be used to design the confining steel for a given column performance as long as the tie spacing is less than 0.4B and 6 d_b.

3 Concluding remarks

A procedure is described which can be used for the design of rectilinear confinement steel in reinforced columns that may experience inelastic deformations during a major earthquake. The design procedure relates the performance of a column section to the design parameters such as axial load level, distribution of both longitudinal and lateral steel and the amount of confinement steel.

Results from an extensive experimental and analytical research program form the basis of this design procedure which has been calibrated again a select group of well-instrumented tests. Accuracy of the procedure has been checked against the results from a large group of tests conducted during this research program and reported by other investigators in which concrete strength varied between 27 and 55 MPa. Some of the test results are also reported in this paper which highlight the effects of variables such as axial load level and steel configuration, on column behavior.

4 References

1. Building Code Requirements for Reinforced Concrete ACI 318-89 (1989) ACI Committee 318, American Concrete Institute, Detroit, U.S.A.
2. Code for the Design of Concrete Structures for Building (CAN 3-A23.3M84) Canadian Standards Association, Rexdale, Ontario, Canada.
3. Sheikh, S.A., and Uzumeri, S.M. (1980) Strength and Ductility of Tied Concrete Columns, Journal of Structural Division, ASCE, V. 106, No. 5, pp. 1079-1102.
4. Sheikh, S.A., and Yeh, C.C. (1990) Tied Concrete Columns Under Axial Load and Flexure", Journal of Structural Division, ASCE, V. 116, No. 10, pp. 2780-2801.
5. Code of Practice for the Design of Concrete Structures, (NZS 3101:1982), Standards Association of New Zealand, Wellington, Part 1 and Part 2.
6. Sheikh, S.A., and Khoury, S. (1993) Confined Concrete Columns with Stubs, ACI Structural Journal, V. 90, No. 4, pp. 414-431.
7. Sheikh, S.A., Shah, D.V. and Khoury, S.S. (1994) Confinement of High-Strength Concrete Columns, ACI Structural Journal, V. 91, No. 1, pp. 100-111.
8. Patel, S., and Sheikh, S. (1992) Behavior of Light-Weight High-Strength Concrete Columns Under Axial Load and Cyclic Flexure, Report No. UHCEE 91-5, Department of Civil Engineering, University of Houston.
9. Ehsani, M.R., and Wight, J.K. (1990) Confinement Steel Requirements for Connections in Ductile Frames, Journal of Structural Division, ASCE, V. 116, No. 3, pp. 751-767.
10. Sheikh, S.A., and Uzumeri, S.M. (1982) Analytical Model for Concrete Confinement in Tied Columns, Journal of Structural Division, ASCE, V. 108, No. 12, pp. 2703-2722.
11. Mander, J.B., Priestley, M.J.N., and Park, R. (1988) Theoretical Stress-Strain Model for Confined Concrete, Journal of Structural Division, ASCE, V. 114, No. 8, pp. 1804-1825.

147 ANALYSIS OF CONCRETE CONFINEMENT BY CIRCULAR HOOPS

N. OGURA and T. ICHINOSE
Nagoya Institute of Technology, Nagoya, Japan

Abstract

On the confining effect of concrete circular column by hoops, experimental formulas has been proposed by many scholars, but analytical research has been rare. In this paper, the effect is analyzed with the following assumptions:

1) Mohr's theory of failure is applicable to concrete under triaxial compression.

2) The stress field is represented by functions where the principal stresses are distributed like an onion with a core.

In this analysis, "imaginary body forces" are assumed in order to satisfy the equilibrium conditions. However, these "imaginary body forces" should not exist practically. Hence, the functions of stress field are determined so that the imaginary body forces will be as small as possible. From the functions, we may choose the maximum strength as a column strength according to the lower bound theorem.

The radius of stress field decreases as hoop spacing increases. Column strength consequently decreases as hoop spacing increase, even if shear reinforcement ratio is the same. This tendency agrees with the existing experimental results quantitatively.

Keywords : Circular hoop, confining effect, imaginary body force, the lower bound theorem, triaxial compression.

1 Introduction

The confining effects of concrete are classified into two kinds: active confinement and passive.

Active confinement is applied to concrete mechanically by a hydrostatic pressure equipment, etc. This confinement can be varied at will, so that this confinement is

Concrete Under Severe Conditions: Environment and loading (Volume Two) Edited by K. Sakai, N. Banthia and O.E. Gjørv. Published in 1995 by E & FN Spon. ISBN 0 419 19860 1

used for researching the mechanical properties of concrete under triaxial compressive stress.

On the other hand, passive confinement is created in a concrete column by lateral reinforcement when axial compressive force is applied and the section is laterally expanded. This confinement is small when the strain in concrete is small and have been large when the strain is large.

Many experimental formulas have been proposed concerning this passive confinement such as Desayi's [1], Watanabe's [2] and Suzuki's [3]. However, there are few analytical formulas. In this paper, we assume the failure conditions under triaxial compression and the stress field of a concrete column. And we analyze the passive confinement using the lower bound theorem.

2 Analytical assumptions

2.1 Assumption 1 : Failure condition

The failure condition of concrete under triaxial compression, as shown in Fig. 1, is assumed as given in Eq. (1), that is, Mohr's theory of failure [4],

$$\sigma_3 = m\sigma_1 + \sigma_B \tag{1}$$

where, σ_1 : minimum principal stress
σ_3 : maximum principal stress
σ_B : unconfined uniaxial compressive strength
m : a coefficient. In this paper, m is equal to 4.1, according to the experiments of Richart. [5].

2.2 Assumption 2 : Compressive stress field

The stress field is modeled as shown in Fig. 2, where D is the diameter of the column, r_0 is the radius of the column, s is the hoop spacing, and z_0 is the half of hoop spacing. The stress field is consisted of two zones: the cylindrical central stress

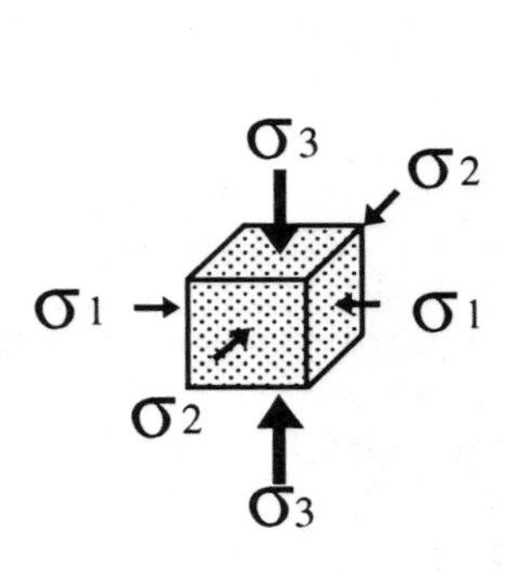

Fig. 1 Triaxial stress

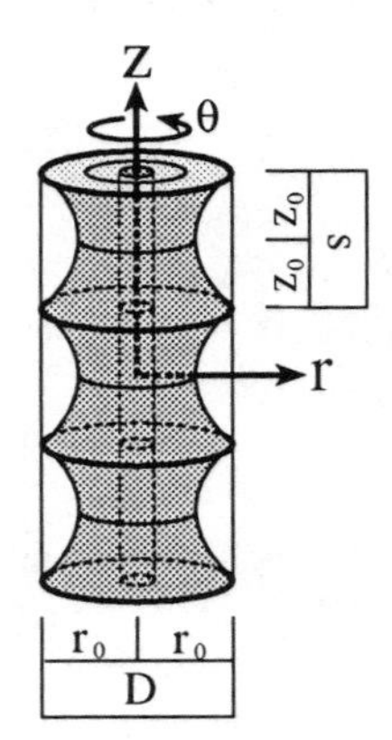

Fig.2 Compressive stress field

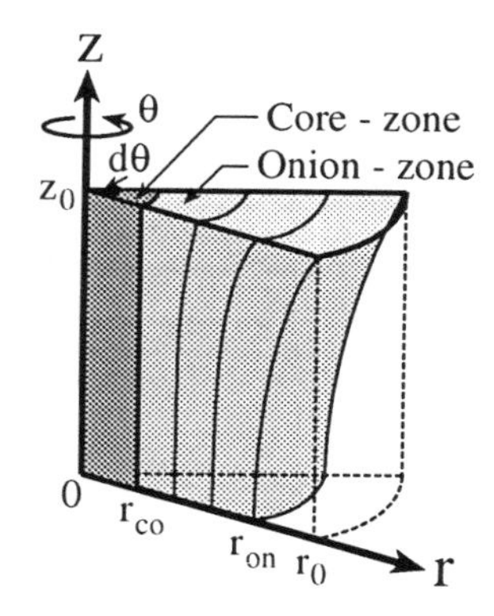

Fig.3 Core-zone and Onion-zone

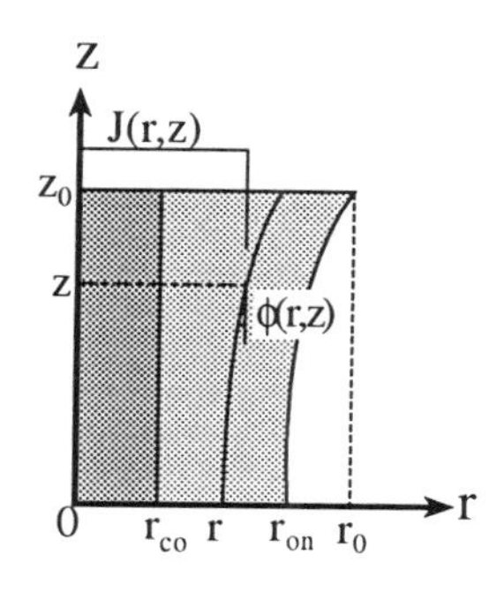

Fig.4 Trajectory of σ_3

field and the stress field around it, (hereafter, the former zone is called the core-zone and the latter is called the onion-zone,) as shown in Fig. 3. In this figure, r_{co} and r_{on} are variable, and will be determined later using the lower bound theorem.

2.2.1 Trajectory of maximum principal stress

The trajectory of maximum principal stress, J(r,z), is shown in Fig. 4, where J(r,z) is a function of r and z. In the core-zone, J(r,z) is assumed as given by Eq. (2).

$$J(r,z) = r \qquad (0 \le r \le r_{co}) \tag{2}$$

In the onion-zone, J(r,z) is assumed to be a linear function of r and a quadratic function of z as given by Eq. (3).

$$J(r,z) = a(r - r_{co})\, z^2 + r \qquad (r_{co} \le r \le r_{on}) \tag{3}$$

The direction of principal stress, $\phi(r,z)$, is given by Eq. (4).

$$\tan\phi(r,z) = \frac{\partial}{\partial z} J(r,z) \tag{4}$$

2.2.2 Principal stresses

The principal stresses are distributed as shown in Fig. 5. The minimum principal stress, $\sigma_1(r,z)$ at (J(r,z) , z), is represented by a function of r ,z and coefficients α_1 and α_2 or shown in Fig. 6. In the core-zone, $\sigma_1(r,z)$ is constant as given by Eq. (5).

$$\sigma_1(r,z) = \alpha_1(r_{co} - r_{on}) \qquad (0 \le r \le r_{co}) \tag{5}$$

In the onion-zone, $\sigma_1(r,z)$ is a quadratic function of r and a linear function of z as given by Eq. (6).

$$\sigma_1(r,z) = \left\{\alpha_1 + \alpha_2(z - z_0)(r - r_{co})\right\}(r - r_{on}) \qquad (r_{co} \le r \le r_{on}) \tag{6}$$

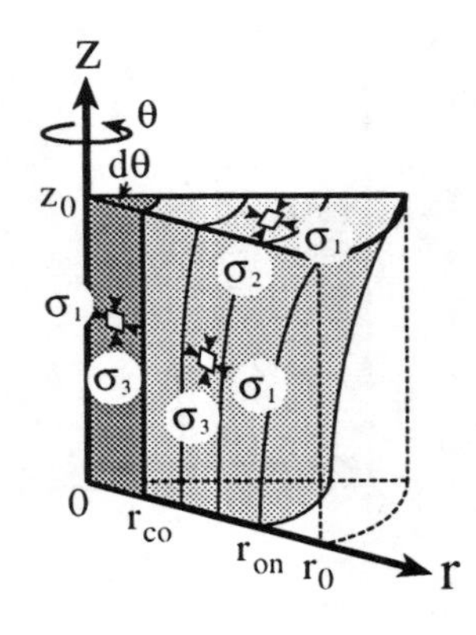

Fig. 5 Principal stress

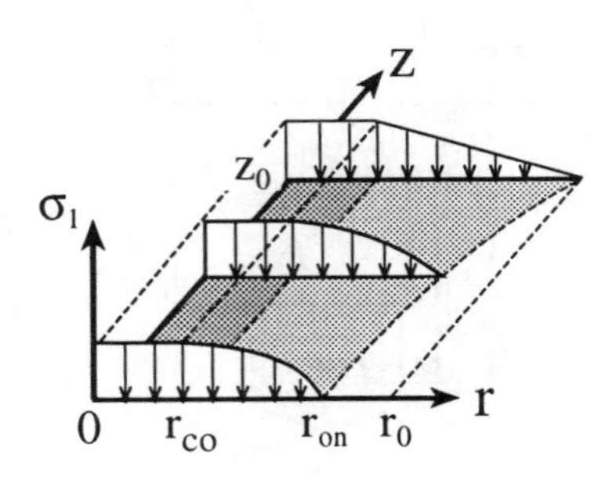

Fig. 6 Minimum principal stress $\sigma_1(r,z)$ at $(J(r,z),z)$

As the boundary condition, $\sigma_1(r,z)$ is zero on the surface of the stress field.

Next, the medium principal stress, $\sigma_2(r,z)$ is equal to the minimum principal stress, $\sigma_1(r,z)$, according to the lower bound theorem. The maximum principal stress, $\sigma_3(r,z)$, is given by assumption 1.

3 Analytical method

3.1 Equilibrium condition

In this section, we consider the equilibrium conditions which exist along the principal stress directions, σ_1, σ_2 and σ_3 axes of an element in the stress field, as shown in Fig. 7.

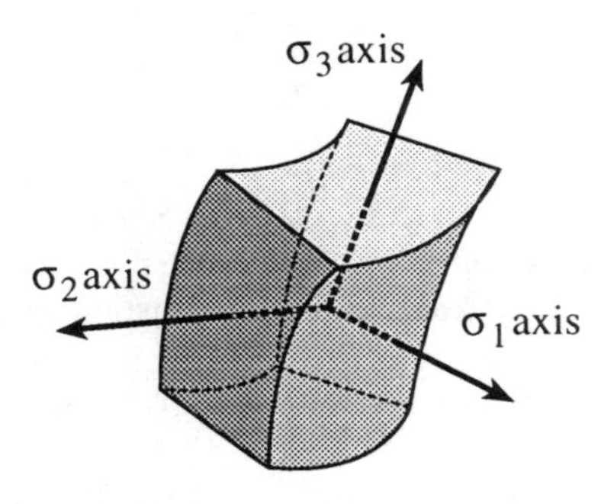

Fig. 7 Element in the stress field

3.1.3 Cross-sectional area of a element

The cross-sectional area normal to the principal axes can be obtained in the following matter. The cross-sectional area normal to the σ_1 axis, $dS_1(r,z)$ shown in Fig. 8(a), is given by Eq. (7). The cross-sectional area normal to the σ_2 axis, $dS_2(r,z)$ shown in Fig. 8(b), is given by Eq. (8). The cross-sectional area normal to the σ_3 axis, $dS_3(r,z)$ shown in Fig. 8(b), is given by Eq. (9).

$$dS_1(r,z) = 2J(r,z)\sec\phi(r,z)\,dz\,d\theta \tag{7}$$

$$dS_2(r,z) = 4\frac{\partial}{\partial r}J(r,z)\,dr\,dz \tag{8}$$

$$dS_3(r,z) = 2J(r,z)\frac{\partial}{\partial r}J(r,z)\cos\phi(r,z)\,dr\,d\theta \tag{9}$$

3.1.2 Force acting on the surface of element

From the cross-sectional area and the assumed principal stresses, the force acting on the surface of element can be obtained as follows. The force acting on the $dS_1(r \pm dr, z)$ plane, $dF_1(r \pm dr, z)$, is represented by Eq. (10). The force acting on the $dS_2(r, z)$ plane, $dF_2(r, z)$, is represented by Eq. (11). The force acting on the $dS_3(r, z \pm dz)$ plane, $dF_3(r \pm dr, z)$, is represented by Eq. (12).

$$dF_1(r \pm dr,z) = \sigma_1(r \pm dr,z)\,dS_1(r \pm dr,z) \tag{10}$$

$$dF_2(r,z) = \sigma_1(r,z)\,dS_2(r,z) \tag{11}$$

$$dF_3(r,z \pm dz) = \sigma_3(r,z \pm dz)\,dS_3(r,z \pm dz) \tag{12}$$

Moreover, these force are shown in Figs. 9(a), 9(b) and 9(c), respectively .

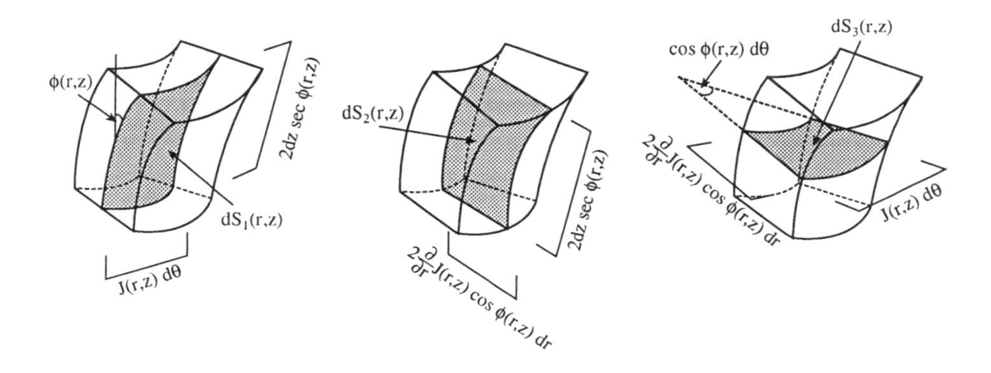

(a) Section normal to σ_1 (b) Section normal to σ_2 (c) Section normal to σ_3

Fig. 8 Cross-sections of the element

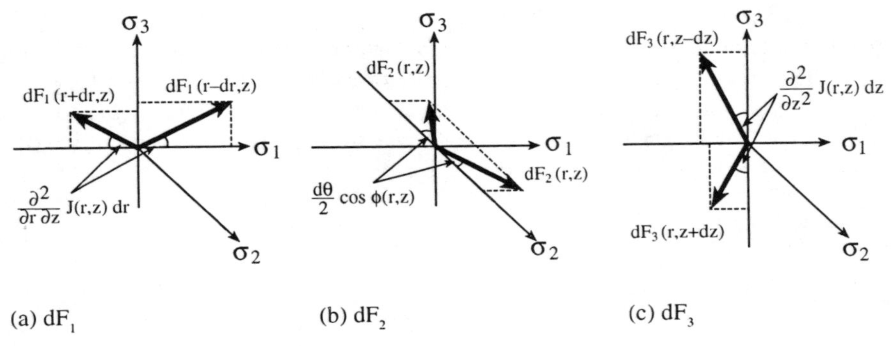

Fig. 9 Force acting on the surface of element

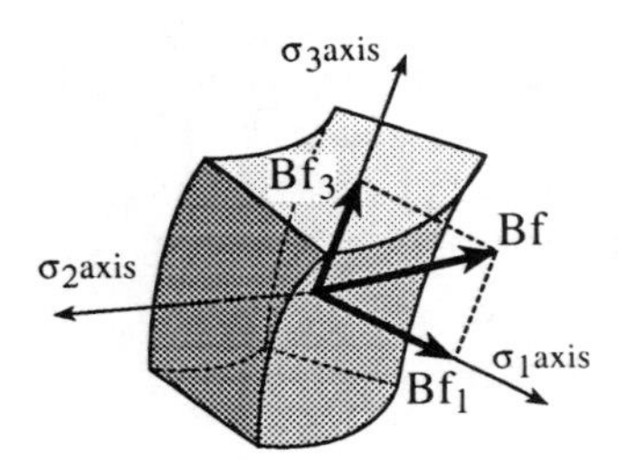

Fig. 10 Imaginary body force

3.1.3 Imaginary body force

Now we are ready to consider the equilibrium condition of the element. The equilibrium condition along the σ_2 axis is automatically satisfied throughout the stress field. However, the equilibrium conditions along the σ_1 and σ_3 axes are not satisfied tnecessarily. Thus, we must assume an imaginary body force, Bf shown in Fig. 10, in order to satisfy the equilibrium conditions. From the equilibrium conditions, the imaginary body force in the directions of σ_1 and σ_3, Bf_1 and Bf_3, are given by Eqs. (13) and (14), respectively.

$$Bf_1(r,z)\,dr\,dz\,d\theta = \{dF_1(r+dr,z) - dF_1(r-dr,z)\} - \{dF_2(r,z)\cos\phi(r,z)\,d\theta\}$$

$$+\{dF_3(r,z+dz) + dF_3(r,z-dz)\}\frac{\partial^2}{\partial z^2}J(r,z)\,dz \tag{13}$$

$$Bf_3(r,z)\,dr\,dz\,d\theta = \{dF_1(r+dr,z) + dF_1(r-dr,z)\}\frac{\partial^2}{\partial r \partial z}J(r,z)\,dr$$

$$-\{dF_3(r,z+dz) + dF_3(r,z-dz)\} \tag{14}$$

3.1.4 Determining the functions of principal stress
Because the imaginary body force should not exist practically, the imaginary body force causes an error. Here, we define this error, Er, in Eq. (15),

$$Er = \iiint \left\{ Bf_1(r,z)^2 + Bf_3(r,z)^2 \right\} dr\,dz\,d\theta \tag{15}$$

where Er is a function of the principal stresses, in other words, a function of the previously mentioned coefficients, α_1 and α_2. Solving the simultaneous linear differential equations (16) and (17) concerning α_1 and α_2, we determine the coefficients so that Er can be minimized.

$$\frac{\partial}{\partial \alpha_1} Er = 0 \tag{16}$$

$$\frac{\partial}{\partial \alpha_2} Er = 0 \tag{17}$$

3.2 Compressive strength of column

3.2.1 Determining the hoop stress, σ_s
A shear stress, $\tau(r)$, is created at $z = z_0$ as shown in Fig. 11. $\tau(r)$ is given by Eq. (18).

$$\tau(r) = \left\{ \sigma_3(r,z_0) - \sigma_1(r,z_0) \right\} \sin\phi(r,z_0) \cos\phi(r,z_0) \tag{18}$$

Considering the equilibrium condition at $z = z_0$, the hoop stress, σ_s in Fig. 11, is represented by a function of $\tau(r)$; in the other word, σ_s is represented by a function of r_{co} and r_{on} as shown in Fig. 12. On the other hand, σ_s must be smaller or equal to the yield strength of the hoop, σ_y, as in Eq. (19).

$$0 \le \sigma_s \le \sigma_y \tag{19}$$

From Eq. (19), we can determine the allowable range of r_{co} and r_{on} is represented by the shaded region as shown in Fig. 12.

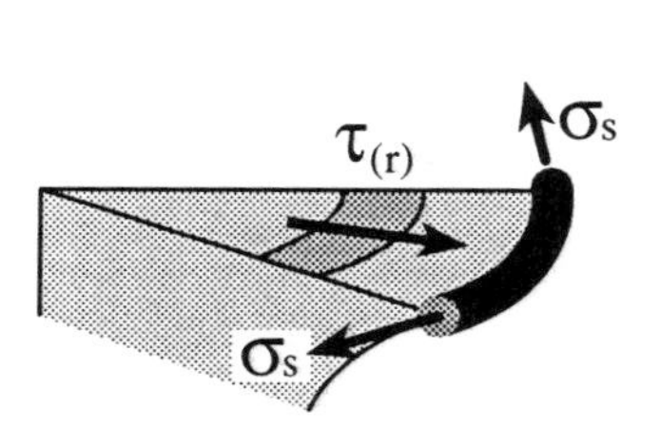

Fig. 11 Equilibrium condition between hoop stress and shear stress at $z = z_0$

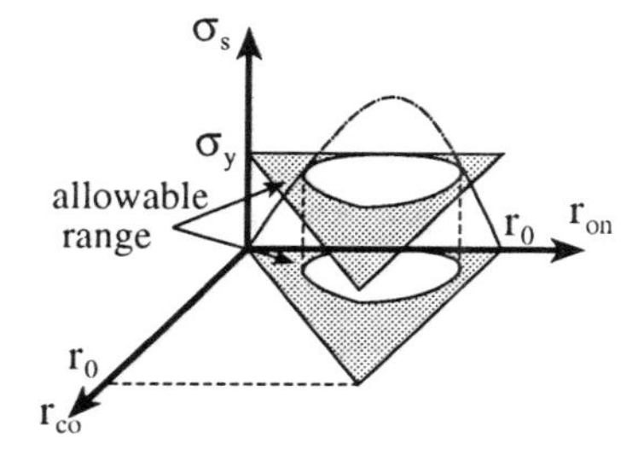

Fig. 12 σ_s versus allowable range of r_{co} and r_{on}

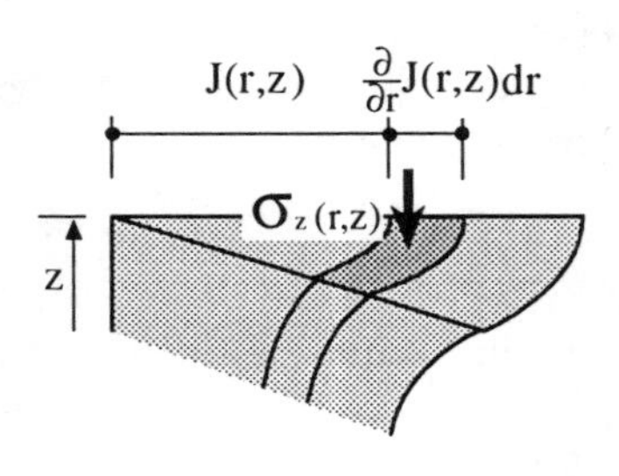

Fig. 13 Vertical stess, $\sigma_z(r,z)$

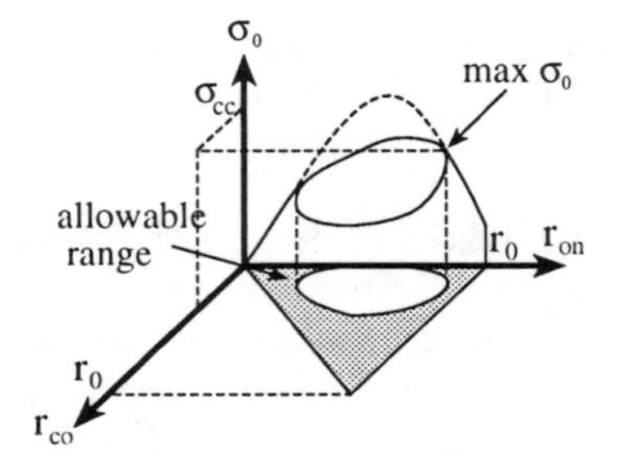

Fig. 14 σ_0 versus allowable range of r_{co} and r_{on}

3.2.2 Determining the uniaxial average strength of a column, σ_{cc}
The vertical stress, $\sigma_z(r,z)$ as shown in Fig. 13, is given by Eq. (20).

$$\sigma_z(r,z) = \sigma_1(r,z)\sin^2\phi(r,z) + \sigma_3(r,z)\,con^2\phi(r,z) \tag{20}$$

From $\sigma_z(r,z)$, the average vertical stress, σ_0 at $z = z_0$, is given by Eq. (21).

$$\sigma_0 = \frac{2}{\pi\, r_0^{\,2}}\int_0^{r_0} \sigma_z(r,z)\, J(r,z_0)\frac{\partial}{\partial r}J(r,z_0)\, dr \tag{21}$$

Because σ_1, σ_3 and σ_z are functions of r_{co} and r_{on}, this σ_0 is a function of r_{co} and r_{oo}. If the allowable range of r_{co} and r_{on} is represented by the shaded region shown in Fig. 14, the uniaxial average strength of a column, σ_{cc}, is chosen so that the value can be maximized, according to the lower bound theorem.

4 Results

4.1 Analytical conditions

The parameters are s/D (the ratio of hoop spacing to the diameter of a concrete column) and $\rho_s\,\sigma_y / \sigma_B$ (reinforcement ratio ×yield strength of hoop / unconfined uniaxial compressive strength of concrete). In this paper, we call σ_{cc} / σ_B the "Confining effect" and $\rho_s\,\sigma_y / \sigma_B$ the "Normalized hoop strength."

4.2 Analytical results

The relations between the normalized radius of the core-zone, r_{co} / r_0, and the normalized hoop strength in the cases of s/D = 0.0, 0.4, 0.8 and 1.2 are shown in Fig. 15. And the relations between the normalized radius of the onion-zone, r_{on} / r_0, and the normalized hoop strength are shown in Fig. 16. In these figures, both r_{co} and r_{on} decrease as the normalized hoop strength increases. In the case of s/D = 1.2, r_{co} decreases rapidly and becomes to zero at $\rho_s\,\sigma_y / \sigma_B = 0.12$, from where r_{on} becomes constant after $\rho_s\,\sigma_y / \sigma_B = 0.12$.

The relations between $\rho_s\,\sigma_s / \sigma_B$ (reinforcement ratio ×hoop stress / unconfined uniaxial compressive strength of concrete) and the normalized hoop strength are

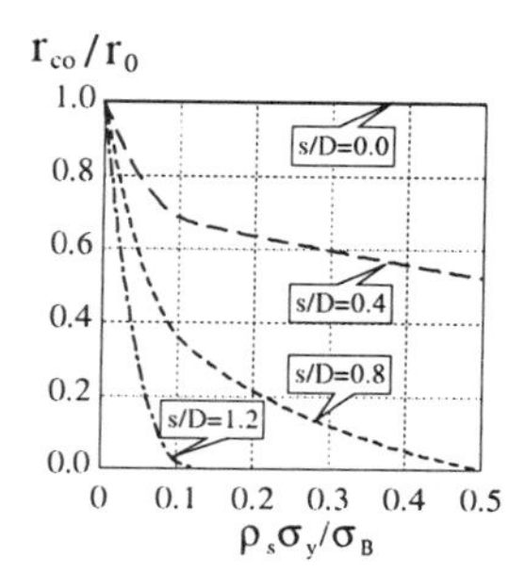

Fig. 15 The relationships between r_{co} / r_0 and $\rho_s \sigma_s / \sigma_B$

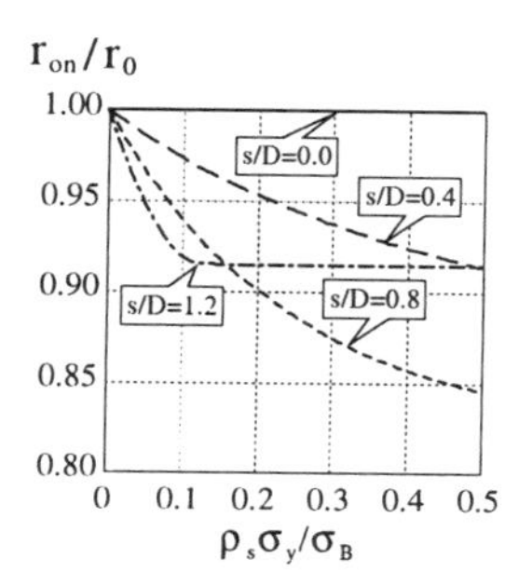

Fig. 16 The relationships between r_{on} / r_0 and $\rho_s \sigma_s / \sigma_B$

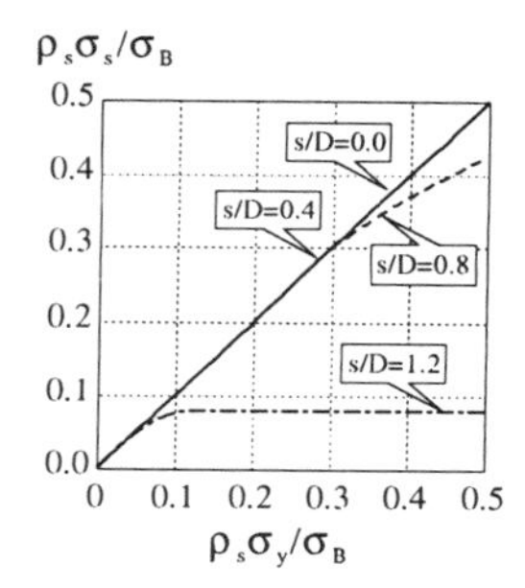

Fig. 17 The relationships between $\rho_s \sigma_s / \sigma_B$ and $\rho_s \sigma_y / \sigma_B$

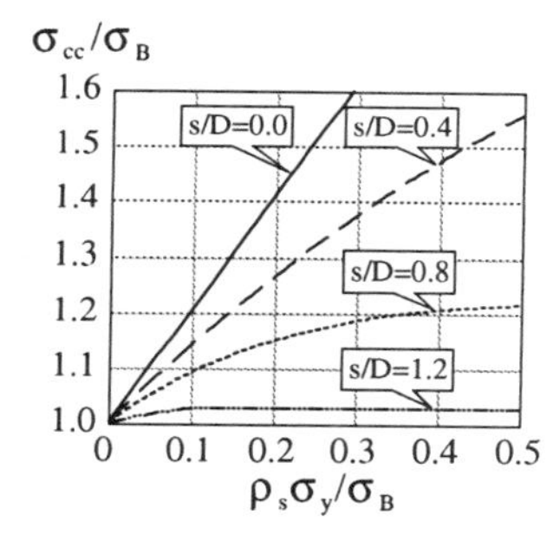

Fig. 18 The relationships between σ_{cc} / σ_B and $\rho_s \sigma_y / \sigma_B$

shown in Fig. 17. This figure signifies that the hoop does not yield in the cases of s/D = 0.8 and 1.2. Especially in the case of s/D = 1.2, the hoop stress, σ_s, is constant after $\rho_s \sigma_y / \sigma_B = 0.12$.

The relationships between the confining effect, σ_{cc} / σ_B, and the normalized hoop strength are shown in Fig. 18. This figure shows that the confining effect increases as the normalized hoop strength or s/D increases. However, in the case of s/D=1.2, the confining effect is constant after $\rho_s \sigma_y / \sigma_B = 0.12$. The reason is that the hoop stress is constant as previously mentioned.

4.3 Comparison of analytical results with experimental values

The relations between $\rho_s \sigma_s / \sigma_B$ and s/D in the cases of $\rho_s \sigma_y / \sigma_B$ =0.1, 0.3 and 0.5 are shown in Fig. 19. In this figure, the lines represent the analytical results and the shaded range represents the hoop yield range in the experiment which Suzuki et. al [3] conducted with s/D and ρ_s with constant concrete strength σ_B = 31.4 MPa. In the case of small hoop spacing, the analytical results of $\rho_s \sigma_s / \sigma_B$ are constant. This signifies that the hoop yields. In the case of large hoop spacing, the analytical results decrease and become the same value in the different normalized hoop strength (=0.1, 0.3 and 0.5) as the s/D increases. In comparison of the analytical results with the experimental values, the analytical hoop yield range almost corresponds with the

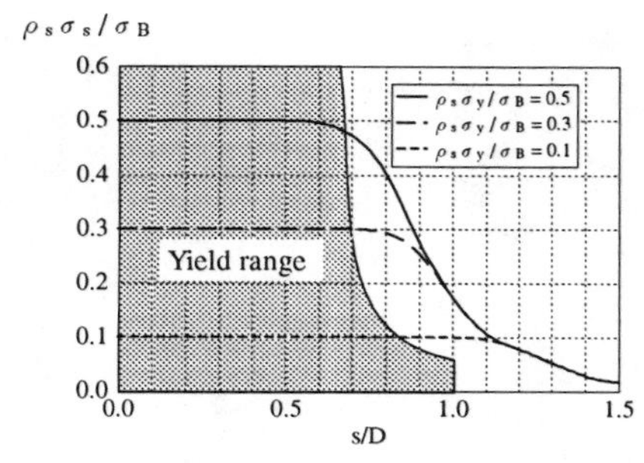

Fig. 19 The relationships between $\rho_s \sigma_s / \sigma_B$ and s/D

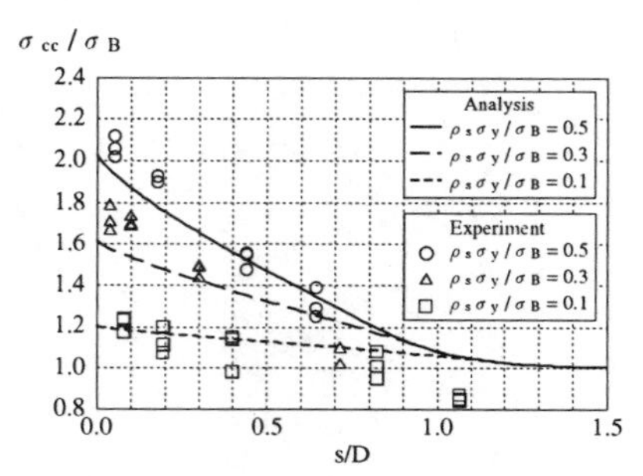

Fig. 20 The relationships between σ_{cc} / σ_B and s/D

experimental range.

The relations between the confining effect, σ_{cc} / σ_B, and s/D are shown in Fig. 20. In this figure, the lines represent the analytical results and the points represent the Suzuki's [3] experimental values. The analytical results of the confining effect decrease and become the same value in the different normalized hoop strength as s/D increases. In comparison of the analytical results with the experimental values, the analytical confining effects are smaller than the experimental confining effect in the case of small hoop spacing. However, these are similar to each other, on the whole.

5 Conclusion

1. It is possible to calculate the confining effect by circular hoops by assuming a compressive stress field and failure condition and using the lower bound theorem.
2. The analysis includes the effects of reinforcement ratio, hoop yield strength and hoop spacing.
3. The analytical results are similar to the experimental values, on the whole.

6 Reference

1. Desayi, P. (1978) Equation for stress - strain surve of concrete confined in circular steel spiral. Materiau er Constructions, Vol. 11, No. 65. pp. 339 - 345.
2. Watanabe, F. and Muguruma, H. (1980) Improving the flexural ductility of prestressed concrete beam by using the high yield strength lateral hoop reinforcement. FIP, Symposia on Partial Prestressing and Practical Construction in Reinforced Concrete, Romania, pp. 398 - 406. Part 2.
3. Suzuki, K. and Nakatuka, T. (1988) The characteristics of stress and strain of concrete confined by circular hoops and the buckling of main reinforcement distributed in the concrete, Proc. of colloquim on the ductility of reinforced concrete structures and its evaluation, JCI, pp. 21 - 31. (in Japanese)
4. Chen, W. F. (1982) Plasticity in Reinforced Concrete, McGraw - Hill International Book Company.
5. Richart. F. E. (1928) Study of the failure of concrete under combined compressive Stresses, University of Illinois Engineering Experimental Station, Bulletin, No. 185. 104 pp.

148 SPIRAL STEEL CONFINEMENT OF VERTICAL BAR CONNECTION IN PRECAST SHEAR WALLS

J.C. ADAJAR and H. IMAI
University of Tsukuba, Japan
T. YAMAGUCHI
Kabuki Construction Co., Japan

Abstract
The results of three series of experiments were used to study the spiral steel confinement of vertical bar connections for precast shear walls. An idealized model for pullout failure of the connection was developed from which an equation for the tensile capacity was proposed. The tensile capacity is calculated from the combined splitting resistance of concrete, spiral steel confinement and mesh reinforcement. The experimental results are higher by approximately 6% than the calculated ones. The splitting resistance of concrete is 41 to 64% of the tensile resistance of the connection, while spiral steel provides an additional 29 to 54% to the tensile strength.
Keywords: Connection, mesh reinforcement, precast shear wall, spiral steel, splices, splitting strength of concrete, tensile capacity, tubular steel sheath.

1 Introduction

Confining the concrete that surrounds a steel bar restricts the widening of splitting cracks that form along embedded bars and, thus, enables greater bond forces to be transmitted. Transverse reinforcements in the form of spirals or hoops can be an effective means to prevent a failure along a potential splitting crack and to enforce a shear failure associated with the maximum attainable bond strength. Such confining reinforcements cannot prevent splitting cracks, but they enable friction to be transferred along cracks and ensure that a more ductile type of bond failure occurs [1].

These phenomena were confirmed using the results of pullout tests [2][3] done on the newly developed vertical bar connection for precast shear wall (Fig. 1). These tests were essentially exploratory in nature. While there is a history of data on spiral steel confinement of concrete under external compression, there is a little or no data concerning passive confinement due to internal radial pressure caused by direct pullout of the bar. Although the primary objective in previous studies was to investigate the tensile resistance of the connection, this investigation emphasized a study of the contributions of concrete and spiral steel to the resistance.

Concrete Under Severe Conditions: Environment and loading (Volume Two) Edited by K. Sakai, N. Banthia and O.E. Gjørv. Published in 1995 by E & FN Spon. ISBN 0 419 19860 1

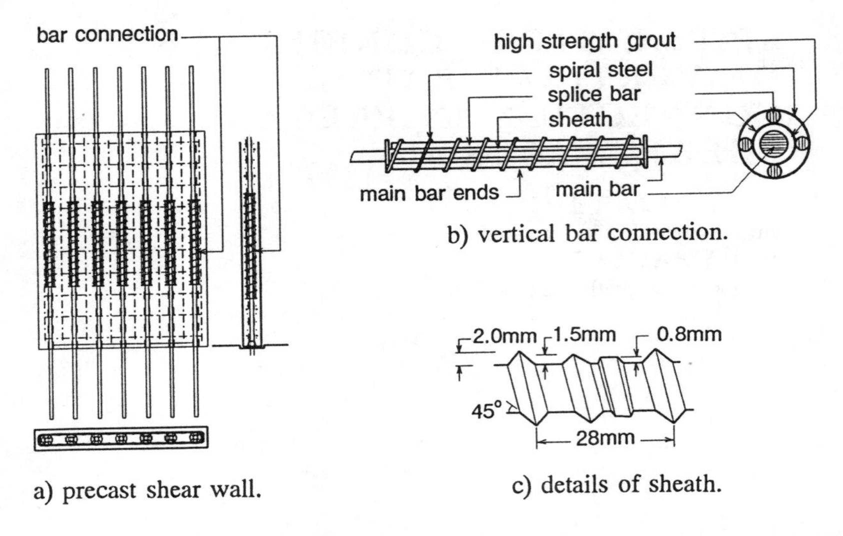

a) precast shear wall.

b) vertical bar connection.

c) details of sheath.

Fig. 1. Details of connection.

2 Details of connection

As shown in Fig. 1, two main bars are inserted at both ends of a tubular steel sheath, and the void between them is filled with high strength grout. Around the sheath are four lapped splice bars which are confined by spiral steel.

During the fabrication of precast walls, main vertical bars are not installed. Instead, tubular sheaths are set at the position of each vertical bar. Lapped bar splices confined by spiral steel are put at the location of the joint. At the construction site, the vertical bars, whose ends meet at mid–height of the wall, are inserted into the sheaths and the void between each bar and the sheath is filled with high strength grout.

A shear wall, because of its shape, is the most suitable structural member for this type of connection. Greater spacing of main bars can allow such configuration of splices. In other structural members such as beams and columns, this method may not be suitable because lapped splices may produce congestion and may interfere with proper compaction of the concrete.

3 Experimental investigation

The following discussion was based on three series of experimental tests. The first series was a direct pullout test of the connection using 45 wall specimens (mostly 600 mm x 600 mm) with 200–mm thickness. Table 1 shows the variation of parameters. Material properties are listed in Table 2. The sketch of the specimen can be seen in Fig. 2. The results of this exploratory test led to conducting the second pullout test with 48 specimens (mostly 400 mm in width and 600 mm in length). This pullout test aimed to investigate the tensile capacity at reduced thicknesses of 180 mm and 150 mm. The size and yield strength of the main bar were changed from D25(Fy = 390 MPa) in the first series to D22(Fy = 490 MPa) in the second series in order to prevent the yielding of bars before the connection collapses. Tables 3 and 4 and Fig. 3 show the details of the second experiment.

Table 1. Variation of parameters (first experiment).

Group B	W x L (in mm) t=200	splice bar no. & size	main bar spacing	lapped length	lug height	spiral pitch
B1	600x600	4-D13	200mm	20d	2.0mm	60mm
B2	600x900	2-D19	200mm	"	"	"
B3	600x750	3-D16	"	"	"	"
B4	900x600	4-D13	300mm	"	"	"
B5	800x600	"	400mm	"	"	"
B6	600x750	"	200mm	25d	"	"
B7	600x900	"	"	30d	"	"
B8	600x600	"	"	20d	1.5mm	"
B9	"	"	"	"	3.0mm	"
B10	"	"	"	"	2.0mm	30mm
B11	"	"	"	"	"	90mm
B12	"	"	"	"	"	120mm
B13						
B14		B13, B14 & B15 are similar to B1				
B15		and subjected to repeated loading.				

d = diameter of a splice bar

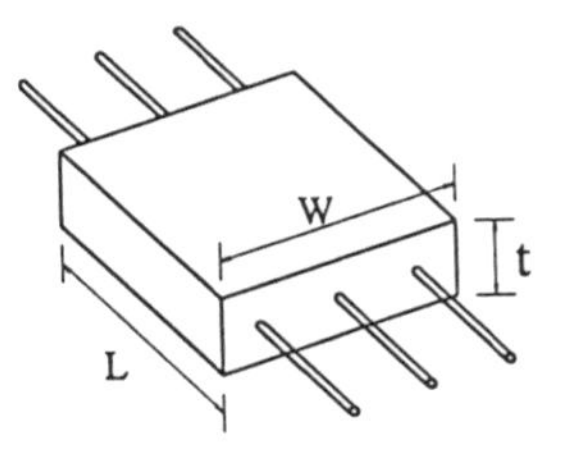

Fig. 2. Sketch of specimen.
(first experiment)

Table 2. Material properties (first experiment).

a) Concrete unit: kgf/cm^2

specimen	B1	B2	B3	B4	B5	B6	B7	B8	B9	B10	B11	B12	B13	B14	B15
compression	294	286	286	286	286	315	315	326	291	306	314	308	272	295	284
splitting	27	30	30	30	30	28	28	28	27	33	29	33	29	27	29

b) Steel unit: tonf/cm^2

size	grade	specified yield strength	actual yield strength	actual tensile stress	remarks
D25	SD390	4.0	4.33	6.04	main bar
D19	SD345	3.5	3.70	---	splice
D16	SD345	3.5	3.52	---	splice
D13	SD785	8.0	8.66	---	splice
ϕ6	---	---	4.50	7.08	spiral
D10	SD295A	3.0	3.65	3.65	mesh

c) Grout unit: kgf/cm^2

number of days	3	28	56	71	90
comp. strength	411	649	675	688	692

Specified compressive strengths:

Concrete : 300kgf/cm^2
Grout : 600kgf/cm^2

Table 3. Variation of parameters (second experiment).

Group A t=150mm	Group C t=180mm	W x L (mmxmm)	main bar	main bar spacing	lapped length	sheath diameter	spiral pitch
A1	C1	400x600	2-D22	200mm	20d	38mm	60mm
A2	C2	"	2-D25	"	"	42mm	"
A3	C3	600x600	2-D22	300mm	"	38mm	"
A4	C4	600x470	"	200mm	15d	"	"
A5	C5	600x730	"	"	25d	"	"
A6	C6	400x600	"	"	20d	42mm	"
A7	C7	"	"	"	"	38mm	30mm
A8	C8	"	"	"	"	"	90mm

d = diameter of a splice bar

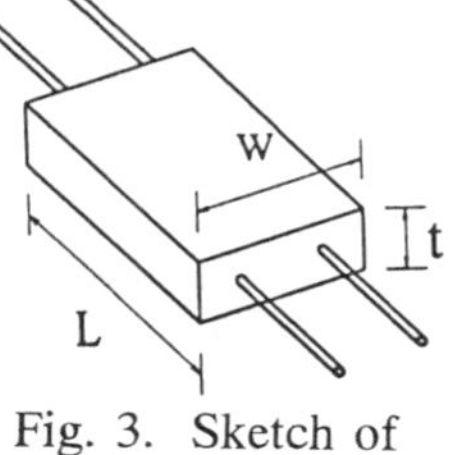

Fig. 3. Sketch of specimen.
(second experiment)

Table 4. Material properties (second experiment).

a) Steel unit: tonf/cm^2

size	grade	specified yield strength	actual yield strength	actual tensile strength	remarks
D25	SD490	5.0	4.97	5.19	main bar
D22	SD490	5.0	5.18	6.45	main bar
D13	SD780	8.0	8.76	10.60	splice
D10	SD295A	3.0	3.83	4.39	mesh
ϕ6	---	---	4.50	7.08	spiral

b) Concrete & grout unit: kgf/cm^2

	specified strength	actual compressive strength	actual splitting strength
concrete	300	415	32.5
grout	600	690	-

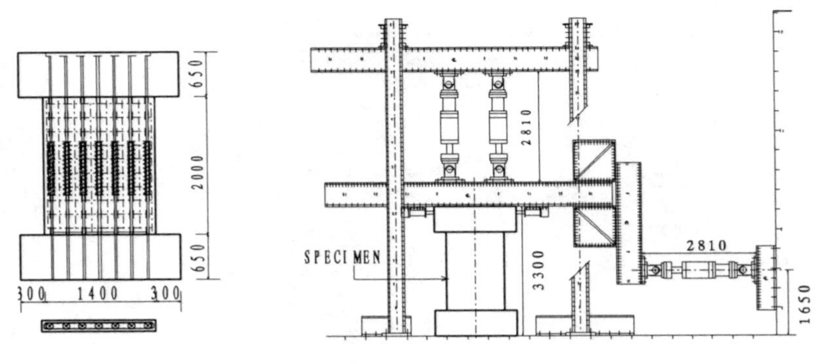

Fig. 4(a). Wall specimen. Fig. 4(b). Loading apparatus.

The third series was member testing of five precast shear walls (200–cm high x 140–cm wide x 150–mm thick) incorporating the bar connections. The sketch of a wall specimen is drawn in Fig. 4(a). One shear wall had 4–25mm vertical bars (Fy = 345 MPa) with bar connections at the lower end of the wall. The other four had 7–25mm (Fy = 390 MPa) vertical bars with the connections at mid-height of the specimen. The horizontal and vertical mesh reinforcements were 10–mm bars spaced at 200 mm. The compressive strengths of concrete and grout used were 300 kgf/cm^2 (29.4 MPa) and 600 kgf/cm^2 (58.8 MPa), respectively. Each wall was subjected to cyclic antisymmetrical bending moments and was loaded until failure by increasing the lateral displacement as shown in Fig. 4(b). Only the maximum average tensile stresses on the spiral steel were used among the results of these member tests [4]. These tensile stresses were used to calculate the confinement of spirals in the first series.

4 Idealized model for pullout failure

Four high strength (SD780) lapped bars, which splice the connection, enforce direct pullout of the main bar from the grout inside the sheath or bond failure on the surface of the sheath. Direct pullout of the main bar only happens at load levels near the maximum tensile strength of the bar. This is due to excessive cross sectional reduction of the main bar which facilitates shearing off or crushing of the grout in front of main bar lugs. In most cases, especially during the elastic stage, the connection fails at the bond on the surface of the sheath since the compressive strength of the grout is twice that of concrete. Besides, the grout is perfectly confined by the sheath against longitudinal splitting. Thus, at failure, the main bar, grout, and tubular sheath act together as one. The idealized model to be discussed herein is applicable only when the pullout failure is at the bond on the sheath.

Even a very small tensile force causes some slip on the surface of the sheath and develops a high bond stress near the loaded ends but leaves the inner part of the bars totally unstressed. As the load is increased, the slip at the loaded end increases, and both the high bond stress and slip extend deeper into the specimen [5]. The maximum bond is somewhat idealized in the sketches of Fig. 5.

When the slip first reaches the unloaded end, the maximum splitting resistance of concrete has nearly been reached. Longitudinal splitting usually occurs along a weaker plane but because of the spiral steel which starts confining the concrete, splitting collapse is prevented. After concrete splitting, additional resistance is provided by the spiral steel and mesh reinforcements across the split. This is because the concrete

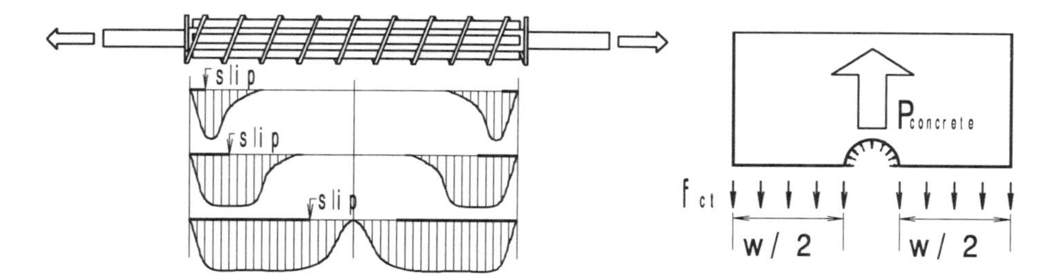

Fig. 5. Bond stress distribution and slip of main bars.

Fig. 6. Concrete subjected to splitting.

remains intact with the main bar due to confinement. It is then reasonable to hypothesize that the tensile capacity of the connection is the combined resistance due to concrete splitting strength, spiral steel confinement and transverse mesh reinforcement.

4.1 Concrete splitting strength

Splitting of concrete starts at the loaded end and progresses towards the middle of the splice as the load approaches the tensile capacity. The splitting may be on the face or on the sides of the wall. The concrete splitting stress is not uniform over the full splice length. Although concrete appears to be brittle when failing in tension, it has such a surprising ability to carry local overstress and readjust to such a change that average stress appears to be the failure criterion. The average attainable splitting strength, however, which is subject to variations for length, cover, spacing, and other factors can be idealized in Fig. 6 as follows.

The transverse tensile force $P_{concrete}$ needed to split the concrete along the weakest splitting plane (with the least concrete cover) should be equal to the average splitting strength f_{ct} of concrete multiplied by the effective width w and length l of the splitting plane which can be expressed as:

$$P_{concrete} = f_{ct}\, w\, l \tag{1}$$

4.2 Spiral steel confinement

The spiral steel which confines the splices through concrete provides a continuous confining pressure around the circumference and is in axial hoop tension as illustrated in Fig. 7. At low levels of stress in the concrete, the spiral is hardly stressed and the concrete is still nearly unconfined. A confining reaction is applied to the concrete only when the transverse strains become very high because of progressive internal cracking

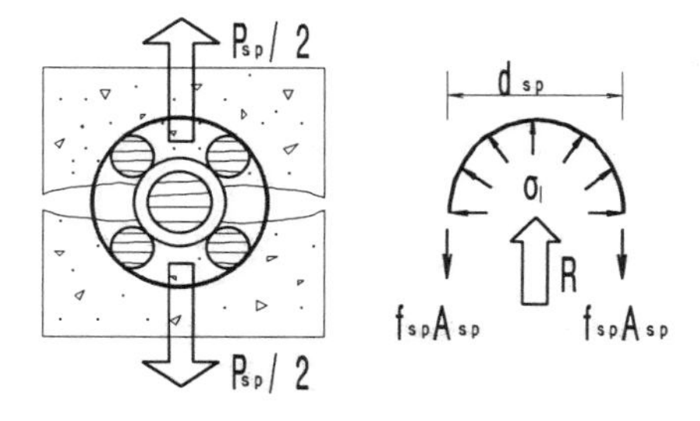

Fig. 7. Spiral steel confinement of connection.

and the concrete bearing out against the spiral steel. This passive confinement provided by the spiral steel, which approximates fluid confinement, can be calculated from the hoop tension developed by the spiral steel. The lateral pressure on the concrete σ_l reaches a maximum when the concrete between sheath lugs is crushed or sheared off causing bond failure on the sheath. The spiral reinforcement then is subjected to axial stress f_{sp}. If d_{sp} is the diameter of the spiral, A_{sp} is the area of the spiral bar, and s is the pitch of the spiral, the equilibrium of the forces [1] acting on the half turn of spiral shown in Fig. 7 requires that

$$2\, f_{sp}\, A_{sp} = \sigma_l\, d_{sp}\, s \tag{2}$$

letting $\mathbf{T_{sp}} = \mathbf{f_{sp}}\ \mathbf{A_{sp}}$,

$$\sigma_l = 2\ T_{sp}\ /\ d_{sp}\ s \tag{3}$$

In an idealized situation, in every turn of the spiral steel, the main bar is resisted by two opposite radial forces. Each force, which is denoted by N in Fig. 7, is the radial component of the bearing force exerted on the lugs of half the total surface of the sheath. This component is equal to the confining circumferential stresses σ_l multiplied by the projected area and can be expressed as:

$$N = \sigma_l\, d_{sp}\, s \tag{4}$$

Considering also the other half of the sheath, the total transverse resistance per turn of spiral is twice the radial component $2N$. When these radial forces for the whole length are summed, the total transverse resistance P_{sp} due to spiral confinement is obtained. This total resistance can be simplified to

$$P_{sp} = 4\ T_{sp}\ l\ /\ s \tag{5}$$

Due to spiral confinement, shearing–and–crushing bond failures occurred on the surface of the sheath. This was proven by checking the radial expansion of the spiral steel. Had it expanded by a displacement equivalent to the sheath lug, which was about 5.0% of the total length in one spiral turn, it would have allowed a large opening in the concrete. The sheath together with the grout and the main bar would have been released freely. But, at 5.0% increase in length, the test piece results showed that the spiral was at its ultimate load. The actual maximum stress on the spiral was less than half of its yield strength. This indicates that the spiral did not expand enough to release the sheath. There must be some crushing, bearing or shearing failure on the concrete in front of the ribs.

4.3 Transverse mesh reinforcement

Reinforcements across the longitudinal cracks shown in Fig. 8 provide additional

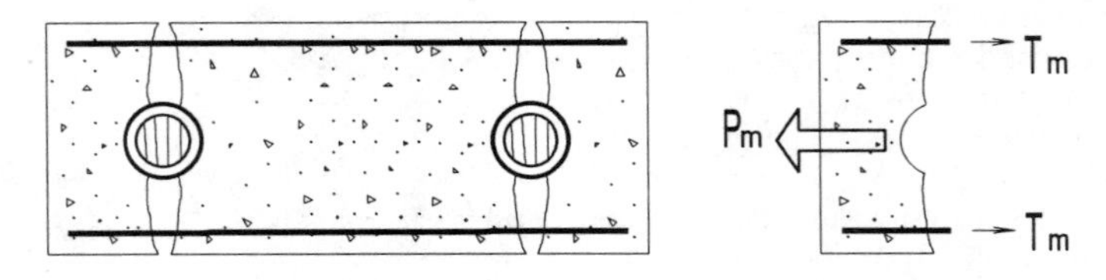

Fig. 8. Mesh reinforcement resistance against splitting of concrete.

resistance which prevent further enlargement of cracks. Although it was not evaluated quantitatively in this study, the contribution of transverse mesh reinforcements on the tensile resistance of the connection was found out to be relatively small. The mesh reinforcement resistance may correspond to the difference between the total tensile resistance and the contribution of concrete splitting strength and spiral steel confinement. A simple expression to determine the total transverse confinement P_m of mesh reinforcements can be written as

$$P_m = n\ T_m \tag{6}$$

where n is the number of bars crossing the splitting plane and T_m is the axial force on each bar.

5 Proposed equation for the tensile capacity of the connection

Assuming that the resisting tension in the concrete across the splitting section and the average bond stress over the tubular sheath is uniform at failure, a semiempirical

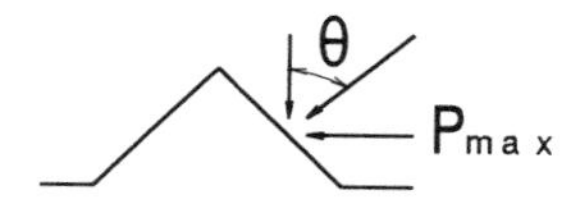

Fig. 9. Inclination of the resultant tensile resistance.

equation to calculate the tensile resistance P_{max} of the connection (see Fig. 9) is proposed as follows.

$$P_{max} = (4\ T_{sp}\ l\ /\ s\ +\ f_{ct}\ w\ l\ +\ n\ T_m)\ \tan\theta \tag{7}$$

where,

T_{sp} = $f_{sp}A_{sp}$, axial hoop tension on the spiral steel, tonf
l = lapped splice length, cm
s = pitch of spiral steel, cm
f_{ct} = splitting strength of concrete, tonf/cm^2
w = effective width of concrete along splitting plane, cm
n = number of mesh reinforcement bars across the splitting plane
T_m = axial tension on mesh reinforcement bar, tonf
θ = angle of inclination measured from the main bar axis to the resultant force exerted by the lugs of sheath on the concrete and reinforcements

The proposed equation is solely based on theories, and the values of the variables used were obtained from actual test results. It has not been verified through tests of data from other sources. The compressive strengths of concrete and grout were fixed at 29.4 MPa and 58.8 MPa, respectively. Only the material properties and parameters listed in Tables 1 to 4 were used. Beyond these limiting conditions, the authors do not claim precision for this equation.

The splitting strength of concrete f_{ct} may be approximated to be an n**th** function of the compressive strength as specified by Carino and Lew [6] and the ACI code [7].

6 Experimental and calculated tensile capacity

The values of variables in the first two terms of Eq. 7 were obtained from the results of the three series of experiments. For wall thicknesses of 150 and 180 mm, the maximum average tensile stresses on the spiral steel are plotted in Fig. 10. These were obtained from strain gauges attached to the spiral steel of the second experiment (A and C group). For 200–mm thick walls whose connections collapsed after yielding of main bar, the maximum average spiral hoop tensile stress, which is approximately 242 MPa, was based on the results of the third experiment, wherein five wall members were tested. Spiral steel stresses of wall connections with yielding main bars were the ones considered for the calculation. It was found that the stress of spiral hoops at maximum pullout load was less than half of its yield stress. The stress – strain diagram for each strand of spiral steel is plotted in Fig. 11. In Fig. 12, the results using only the first two terms inside the parenthesis of Eq. 7 are plotted against the experimental results. The third term was not realistically handled in this study, but its values were obtained after learning the value of concrete splitting resistance and spiral confinement.

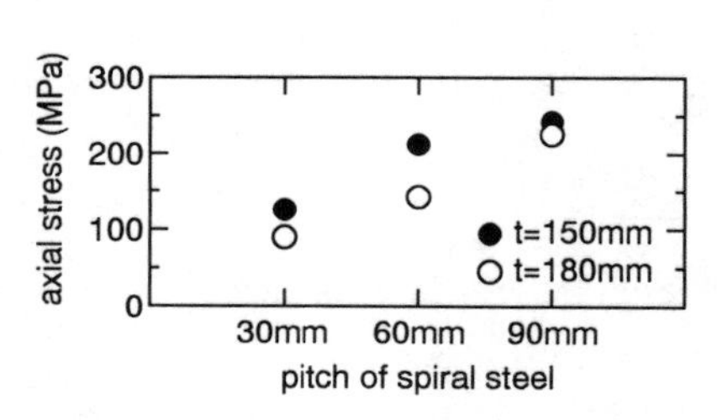

Fig. 10. Maximum average tensile stresses in the spiral steel.

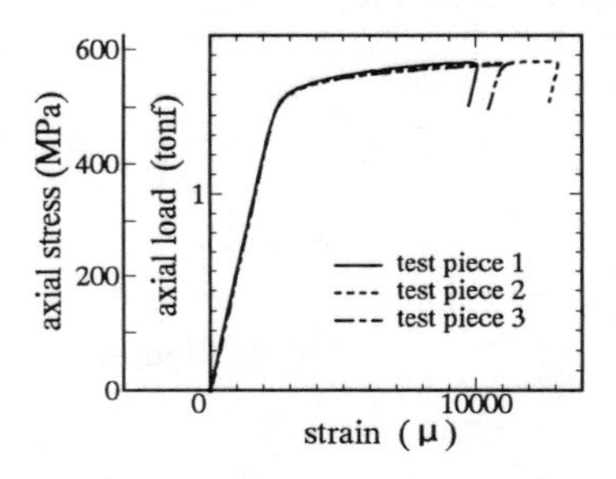

Fig. 11. Stress – strain diagram for spiral steel.

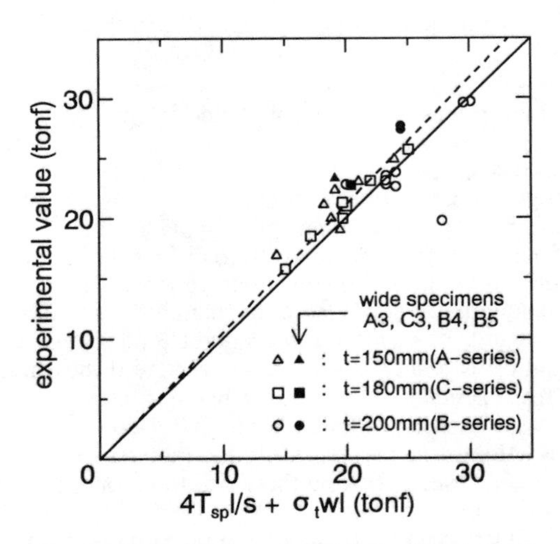

Fig. 12. Experimental and calculated results.

7 Discussion of the results

Specimens with a calculated tensile capacity more than the maximum tensile strength of main bars were not included in this study. These are groups B2, B7 and B10 (refer to Table 1) of the first series. The maximum experimental tensile strengths of these groups were almost equal to the maximum tensile strength (590 MPa) of a continuous main bar, while the calculated values were around 690 MPa. Groups B8 and B9, with sheath lug heights of 1.5 mm and 3.0 mm, respectively, were treated separately since all other specimens had lug heights of 2.0 mm. The two points in Fig. 12 for 1.5 and 3.0–mm lug heights were at (27.7, 19.8) and (23.2, 22.8), respectively. In evaluating $\tan\theta$ of Eq. 7, data must be grouped according to sheath lug height.

The results of wide specimens (groups A3, B4, B5, and C3) are plotted with shaded points in Fig. 12. In these specimens with main bar spacing of more than 200 mm, there was only face splitting of concrete along the length. Mesh reinforcements across these face splits were activated by resisting concrete separation. In such cases, the contribution of mesh reinforcements on the tensile resistance was larger. In most cases where side splitting occurs, the mesh reinforcement contribution nT_m in Eq. 7 is relatively small and can be neglected. If the spiral is stressed to less than half (less than 250 MPa, see Fig. 11) of its yield strength at maximum pullout load, the mesh bars would have a much lesser load because they are farther from the main bar than the spiral hoops. Should the mesh bars yield, which is unlikely to happen, the maximum resistance they can provide is 5.2 tonf.

Considering the results of specimens with 2.0–mm sheath lug heights, a linear regression analysis shows that $\tan\theta$ is equal to 1.06, with sum of squares of the errors equal to 1.47 and a sample correlation coefficient of 0.90. The angle θ is approximately 45° which corresponds well with the 45° inclination of sheath lugs. This suggests that the friction coefficient, which is equivalent to $\tan\theta$, on the surface of the sheath is almost 1.0 when the lug height is 2.0 mm. The maximum tensile resistance of the connection becomes

$$P_{max} = 4\, T_{sp}\, l\, /\, s \;+\; f_{ct}\, w\, l \;+\; n\, T_m \tag{8}$$

8 Contributions of spiral steel and concrete on the tensile capacity

Figure 13 shows the percentage contributions of spiral steel and concrete to the

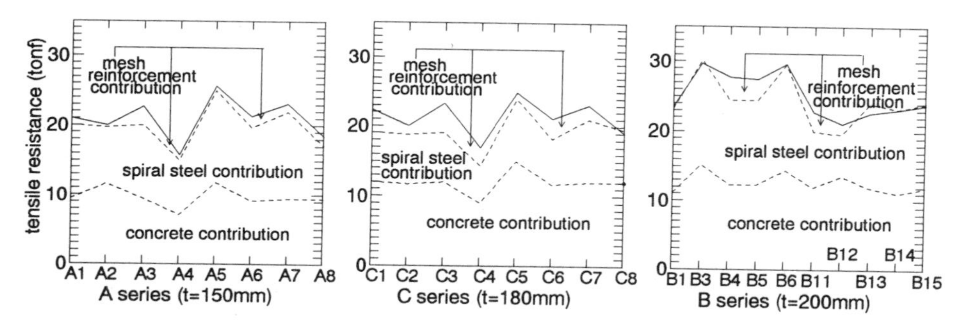

Fig. 13. Contribution of concrete, spiral steel and mesh reinforcements on the tensile capacity of the connection.

tensile capacity of the connection. The concrete resistance ranges from 41 to 64% of the total resistance, and the spiral steel adds 29 to 54%.

9 Conclusions

From the foregoing discussion on the spiral steel confinement of bar connection, the following findings were obtained.

1. Spiral steel acts only after longitudinal cracking has occurred.
2. Confining the connection with spiral steel enforces ductile failure through shearing off or crushing of concrete between sheath lugs.
3. At connection failure, the maximum tensile stress on the spiral steel is less than half its yield strength (less than 250 MPa).
4. The proposed equation can be used to calculate the tensile capacity of the connection.
5. When the sheath lug height is 2.0 mm, the angle of inclination θ of the total resistance with respect to the transverse axis is approximately 450.
6. The concrete contribution is approximately 41 to 64% of the total tensile resistance.
7. Spiral steel confinement provides an additional 29 to 54% of the total tensile resistance.
8. The mesh reinforcement contribution can be neglected when the concrete cover is less than half of the main bar spacing.
9. When the calculated tensile resistance due to concrete, spiral steel and mesh reinforcement is greater than or equal to the maximum tensile capacity of a continuous bar, the connection failure is direct pullout of the main bar inside the sheath. Otherwise, the sheath, grout and main bar act together as one and the failure is at the bond on the sheath.

10 References

1. Park, R and Paulay, T. (1975) *Reinforced concrete structures*, John Wiley & Sons, pp. 25 and 404.
2. Adajar, J.C., Yamaguchi, T and Imai H. (1993) An experimental study on the tensile capacity of vertical bar joints in a precast shear wall. *Transactions of JCI*, Vol. 15, pp. 557–564.
3. Adajar, J.C., Yamaguchi, T and Imai H. (1994) Tensile capacity of main bar splice at a reduced precast shear wall thickness. *Transactions of JCI.*
4. Adajar, J.C., Yamaguchi, T and Imai H. (1995) Seismic behavior of precast shear wall with bar splices confined to spiral steel. *Proceedings of the Fifth East Asia–Pacific Conference on Structural Engineering and Construction.*
5. Ferguson, P. M. (1979) *Reinforced concrete fundamentals*. John Wiley & Sons, fourth edition, pp. 169–212.
6. Carino N.J. and Lew H.S. (1982) Re–examination of the relation between splitting tensile and compressive strength of normal weight concrete. *ACI Structural Journal*, Vol. 79, No. 3, pp. 214–219.
7. American Concrete Institute. (1993) *Manual of Concrete Practice. Building code requirements for reinforced concrete (ACI 318)(Revised 1992)*. Chapter 11.

149 AN EXPERIMENTAL STUDY ON DETERIORATION OF ASEISMATIC BEHAVIOR OF R/C STRUCTURAL WALLS DAMAGED BY ELECTROLYTIC CORROSION TESTING METHOD

T. YAMAKAWA
University of the Ryukyus, Okinawa, Japan

Abstract
This study is a trial to investigate the influence of steel corrosion on aseismatic behavior of R/C structural walls. Six R/C structural wall specimens were provided. Two were the standard test specimens without corrosion, and the other test specimens were corroded by the electrolytic corrosion test. The experimental results of all test specimens are discussed and compared with theoretical results. As a result of this study, the lateral loading capacities for the R/C structural wall specimens were shown to be almost the same, regardless of the amount of corrosion damage. The initial stiffness of the R/C structural wall specimens which sustained damage by the electrolytic corrosion testing method were higher than normal test specimens without corrosion. However, the deterioration of the ductility was remarkable as the corrosion damages increased.
Keywords: Aseismatic behavior, corrosion, ductility, electrolytic corrosion test, exposure test, lateral loading capacity, R/C structural wall.

1 Introduction

This study is a trial to investigate the influence of steel corrosion on the aseismatic behavior of R/C structural walls under a chloride attack environment in the Ryukyu Islands, which belong to the southern islands of Japan, Okinawa Prefecture. The location has a marine environment with semitropical weather conditions of high temperature and humidity.

As the first step, R/C structural walls were adopted as the test specimens because the cover thickness, namely, the concrete protective covering for reinforcement, is thin and the corrosion area is large for steel reinforcing bars. The experimental tests were carried out in 1992 and 1993. The experimental tests consisted of two testing methods. One was an electrolytic corrosion test and an exposure test. The other was a loading test of

Concrete Under Severe Conditions: Environment and loading (Volume Two) Edited by K. Sakai, N. Banthia and O.E. Gjørv. Published in 1995 by E & FN Spon. ISBN 0 419 19860 1

corrosion damaged R/C structural walls subjected to cyclic shearing forces and a constant axial force.

Three R/C structural wall test specimens containing a sodium chloride solution (NaCl 3.3% sol.) have been exposed at the coast in Okinawa since 1992. Loading tests for the 6 test specimens were conducted under a constant gravity load and alternately repeated lateral forces. The magnitude of the constant compressive stress for the wall panel was 1.96 MPa. Two of the specimens were the standard test specimens, and the other test specimens were corroded by the electrolytic corrosion test. Experimental results and discussions of these 6 test specimens are reported in this paper as preliminary research [1].

2 Test specimen

The details of the test specimens of R/C structural walls are shown in Fig. 1. All test specimens were about one-third scale. A wall panel size was 800x950x80 mm. Rigid edge beams were attached to the top and bottom of the wall panel. The shear-span ratio M/Ql' was 1.425. The weight of a test specimen was about 1200 kg.

The steel reinforcing bars, whose diameter was 6 mm, were arranged as mesh by double layered reinforcement. The reinforcement ratio of the wall panel was 0.8%. This value is larger in Japan. Two different kinds of reinforcing arrangements were adopted in these test specimens as illustrated in Figs. 1b and 1c. Horizontal reinforcing bars were arranged as non-closed form without anchorage (see Fig. 1b) and closed form with anchorage-like hoops (see Fig. 1c).

Three concrete cylinders (ϕ100x200 mm) and a wall panel (800x500x80 mm) for corrosion monitoring were provided against each R/C structural wall test specimen. The mechanical properties of the steel reinforcing bars are shown in Table 1. Those of concrete are shown in Table 2.

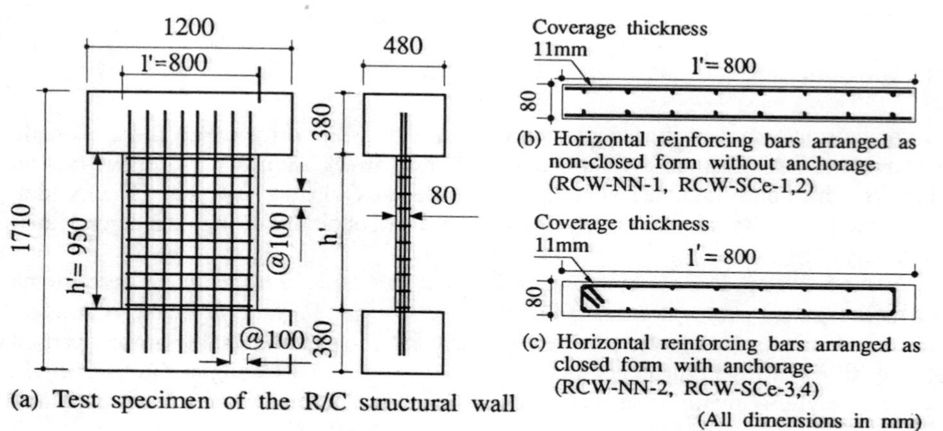

Fig. 1. Dimension of the test specimen and details of reinforcing arrangement.

Table 1. Mechanical properties of reinforcing bar

	a (cm²)	$_s\sigma_y$ (MPa)	$_s\varepsilon_y$ (%)	$_s\sigma_u$ (MPa)	$_sE$ (GPa)
Steel bar (D6-SD295A)	0.32	429.2	0.21	559.3	200.6

Note: a=sectional area, $_s\sigma_y$=yield point stress, $_s\varepsilon_y$=strain at $_s\sigma_y$, $_s\sigma_u$=tensile strength, $_sE$=Young's modulus

Table 2. Mechanical properties of concrete

	Specimen name	$_c\sigma_B$ (MPa)	$_c\varepsilon_1$ (%)	$_cE$ (GPa)	Slump (cm)
Non corrosion	RCW-NN-1	24.6	0.154	25.3	16.5
	RCW-NN-2	22.1	0.205	24.5	17.0
Electrochemical corrosion	RCW-SCe-1	24.9	0.298	23.6	18.5
	RCW-SCe-2	22.9	0.254	23.3	18.5
	RCW-SCe-3	23.9	0.230	24.7	17.0
	RCW-SCe-4	23.2	0.242	24.0	17.0

Note: $_c\sigma_B$ = concrete cylinder strength, $_c\varepsilon_1$ = strain at $_c\sigma_B$, $_cE$=Young's modulus

3 Electrolytic corrosion test

In order to accelerate corrosion of steel reinforcing bars embedded in the wall panels, an electrolytic corrosion testing method was applied to four specimens of R/C structural walls containing sodium chloride (see Table 3). As illustrated in Fig. 2, a steel mesh wrapped in gelatin was adopted as an electrode. This electrode was the same size as the wall panel and was attached on it. The electric current flowed from the steel reinforcing bars in the wall panel to the electrode. The product of the electric current flowing out per unit hour and its elapsed time is presented in Table 3.

In the same manner, the electrolytic corrosion testing method was applied to the four monitoring panels. The purpose was to investigate the corrosion loss and mechanical properties of corrosion damaged steel reinforcing bars. After the electrolytic corrosion test was concluded, these bars were taken out from the monitoring panels. The corrosion damaged steel reinforcing bars were immersed in a Diammonium Hydrogen Citrate solution (10% sol.) for a day in order to remove the corrosion product. The weight loss and mechanical properties of these steel bars are shown in Table 4. Since the nominal sectional area was used in corroded steel bars, the yield point stress was small. The strain of corrosion damaged steel bars was the mean strain in a test length of 19.93 cm.

Crack patterns due to the electrolytic corrosion test in the wall panels of R/C structural walls are illustrated in Table 5. Many cracks occurred along the reinforcing steel mesh as shown in Table 5. Maximum crack width is presented in Table 5.

Table 3. Product of current and elapsed time

Electrochemical corroded specimens			
RCW-SCe-1	RCW-SCe-2	RCW-SCe-3	RCW-SCe-4
923Ah	1595Ah	855Ah	832Ah

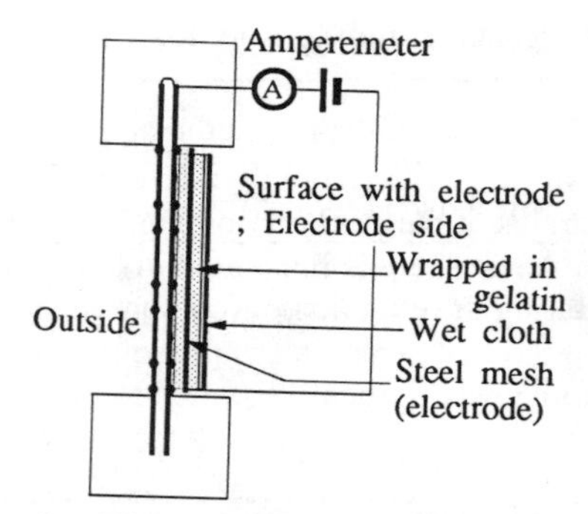

Fig. 2. Electrolytic corrosion test.

Table 4. Corrosion loss of steel reinforcing bars

Test specimens	W_0 (g)	W (g)	W/W_0 (%)	$_{sc}\sigma_y$ (MPa)	$_{sc}\sigma_y/_s\sigma_y$ (%)	$_{sc}\varepsilon_y$ (%)
RCW-SCe-1	110.1	102.1	93	365.7	85	0.185
RCW-SCe-2	111.3	95.4	86	286.7	66	0.156
RCW-SCe-3	107.7	96.6	90	341.0	80	0.172
RCW-SCe-4	107.9	97.5	91	349.5	82	0.176

Note : W_0=weight before corrosion, W=weight after corrosion, $_{sc}\sigma_y$=normal yield point stress of corrosion damaged steel bars, $_{sc}\varepsilon_y$=strain at $_{sc}\sigma_y$, $_s\sigma_y$=yield point stress of steel bars

Table 5. Crack patterns in wall panels after electrolytic corrosion test

Specimen name	RCW-SCe-1	RCW-SCe-2	RCW-SCe-3	RCW-SCe-4
Product of current and elapsed time	(923Ah)	(1595Ah)	(855Ah)	(832Ah)
Max. crack width	0.15mm	3.0mm	0.8mm	0.7mm
Outside	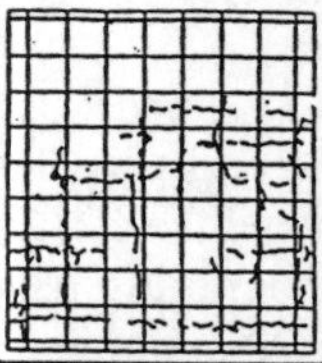	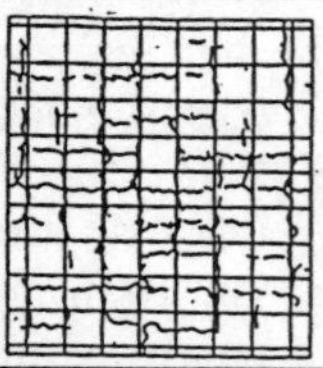	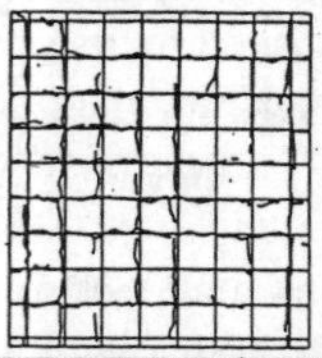	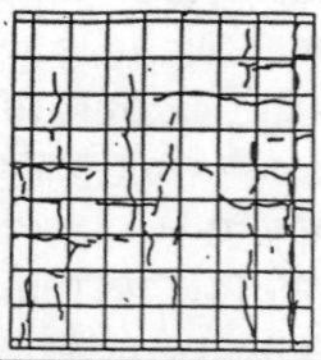
Electrode side	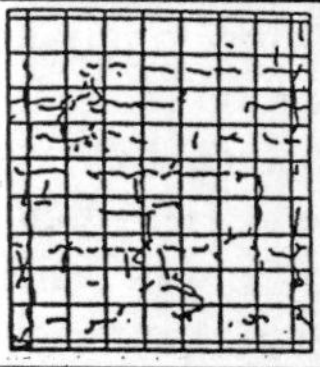	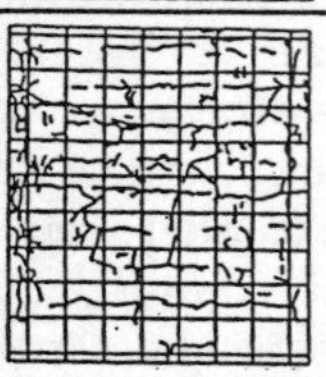	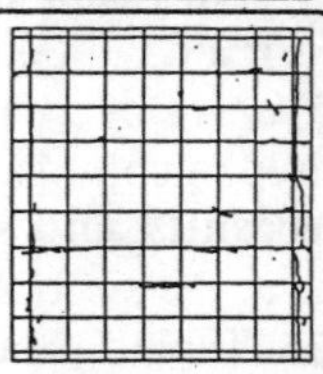	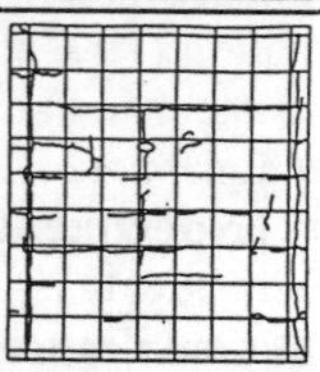

4 Measurements and loading arrangement

The instrumentation methods were designed to obtain the following data: (1) applied forces; (2) horizontal and vertical displacements; (3) rotation of the upper beam; (4) strains of the steel reinforcing bars in the wall panel.

A test setup is illustrated in Fig. 3. The vertical compression, which was kept constant during the experiment (compressive stress σ =1.96 MPa in the wall panel), was applied to the R/C structural wall by means of the servohydraulic actuator. The cyclic shearing force was statically applied to the test specimen of the R/C structural wall by means of one double acting hydraulic jack.

After the first crack occurred in the wall panel, the lateral force was applied in deflection increment levels equivalent to a interstory drift angle of the structural wall R=δ/h', where δ is the interstory drift and h' is the clear height of the structural wall. The loading pattern was cyclic with alternating displacement reversals. The peak drifts were increased stepwise with an incremental drift of 0.0025h' after two successive cycles at each displacement level.

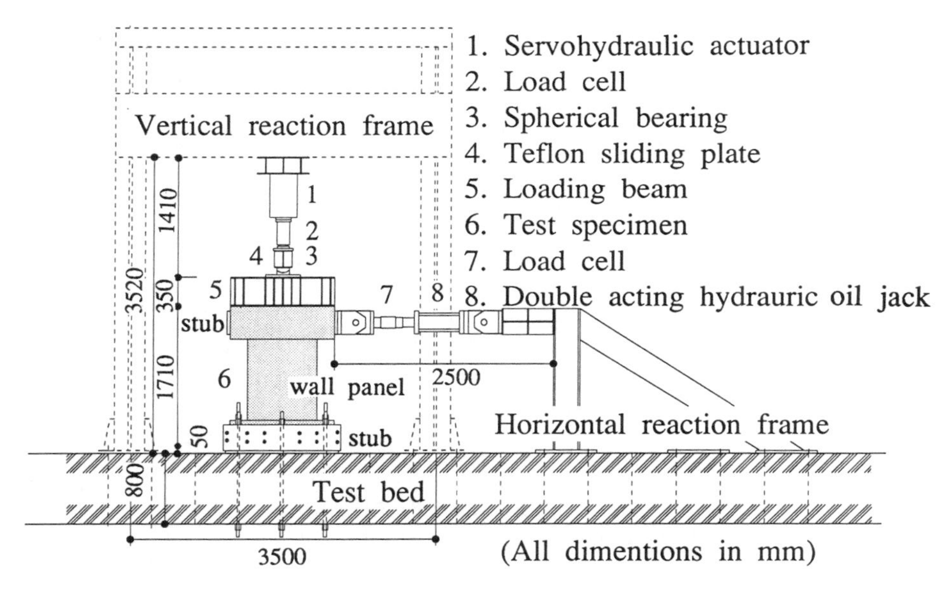

Fig. 3. Experimental test setup.

5 Experimental results and discussion

The crack patterns in the wall panels of R/C structural walls are illustrated at the second stage of each interstory drift angle R = 0.25, 0.5, 0.75 and 1% and at the final stage in Table 6. The Q-R relationships (hysteresis loops) and the mean vertical strain, ε_v

(=δ_v/h'), versus the interstory drift angle relationships of all specimens are shown in Table 7. The mean vertical strain ε_v is given by dividing a total of vertical displacements of the both sides of the wall panel by 2h' (h'= clear height of the structural wall). The skeleton curves for the Q-R curves in Table 7 are drawn in Fig. 4. Figure 5 shows an enlargement of the initial zone for the Q-R skeleton curves in Fig. 4. The initial stiffness for the specimens can be discussed according to Fig. 5. Figure 6 shows the comparison of the energy absorption capacity in all the specimens.

From these tables and figures, the behavior of the electrolytic corrosion damaged structural walls is shown to be different from the behavior of the normal test specimens without corrosion. The corrosion damaged structural walls exhibited a significant deterioration in ductility. However, the ultimate lateral loading capacities were almost the same, regardless of the extent of electrochemical corrosion damage. The corrosion

Table 6. Crack patterns in the wall panels at the second stage of each drift angle

	Drift angle / Specimen name	0.25%	0.50%	0.75%	1.00%	Final stage
Non corrosion	RCW-NN-1					R=1.50
	RCW-NN-2					R=1.25
Electrochemical corrosion	RCW-SCe-1 (923Ah)					R=1.25
	RCW-SCe-2 (1595Ah)					R=1.25
	RCW-SCe-3 (855Ah)					R=1.25
	RCW-SCe-4 (832Ah)					R=1.25

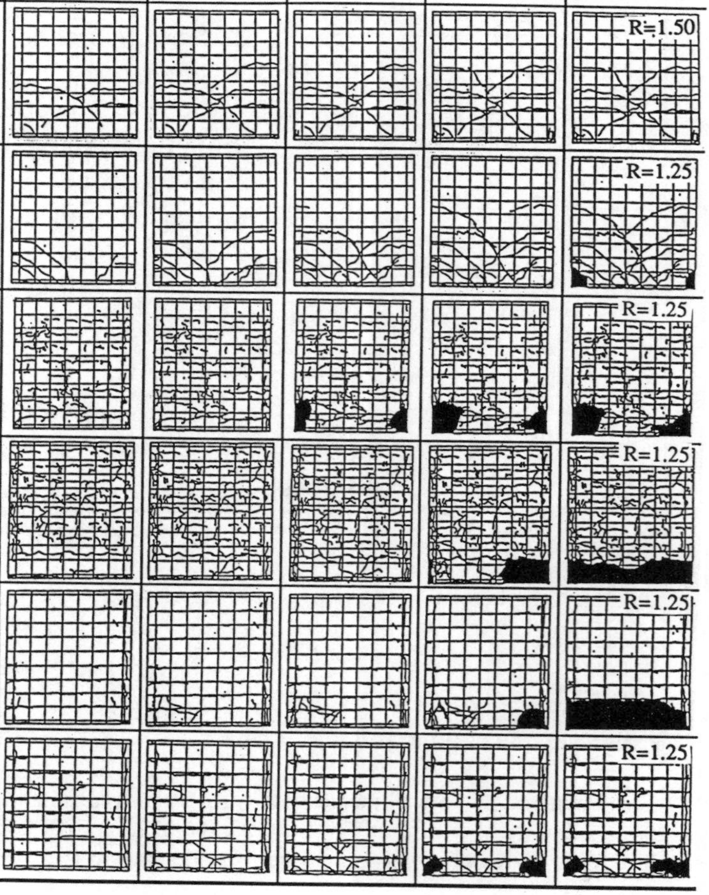

damaged specimens were higher in initial stiffness than non-corroded specimens (see Fig. 5). It may be considered as one of the reasons that the tension stiffening effect and bond strength increase when the longitudinal steel bars are corroded by an electrolytic corrosion testing method.

The failure modes of non-corroded specimens were of the flexural compressive failure type. On the other hand, the corrosion damaged cover concrete peeled off in the neighborhood of the bottom of the wall panel and buckling of the vertical reinforcing bars

Table 7. Experimental results on Q-R and ε_v- R relationships of specimens

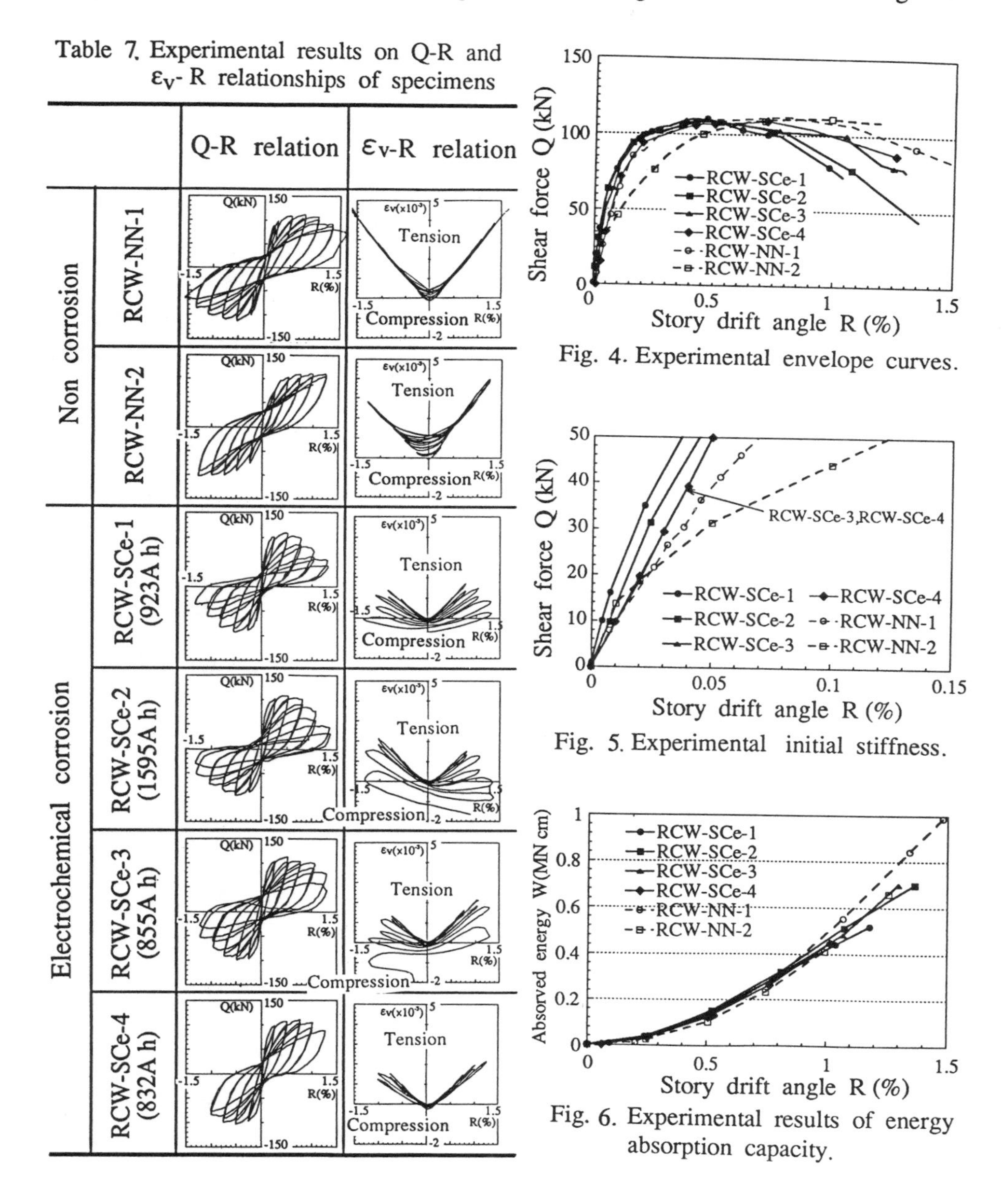

Fig. 4. Experimental envelope curves.

Fig. 5. Experimental initial stiffness.

Fig. 6. Experimental results of energy absorption capacity.

was observed in the final stage of the electrochemical corrosion damaged specimens. These test specimens, whose failure modes were similar to brittle shear failure, experienced a significant deterioration in ductility. This fact is recognized in Table 7, which shows the relation between the mean vertical strain ε_v and the interstory drift angle R. However, the deterioration of aseismatic behavior for the corrosion damaged specimen was not dominant as the product of the electric current and elapsed time became smaller (see Table 7 and Fig. 4).

6 Theoretical investigation

The calculations of the monotonic Q-R relationships of the R/C structural walls were carried out on the basis of the following assumptions and procedure.

(1) The transverse section of the wall panel is partitioned by a rectangular grid into a large number of small elemental areas. Each element is uniaxially stressed, and the section remains plane after deformation.

(2)The σ-ε curve for concrete is given by Umemura's exponential function, and it is illustrated in Fig. 7a. The tensile strength of the concrete is neglected. The σ-ε relation-

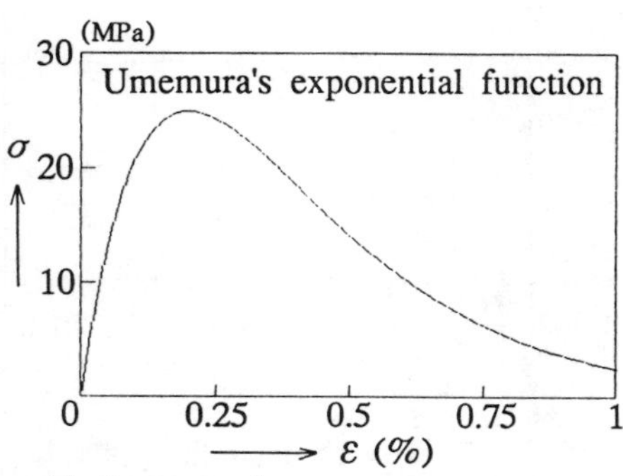

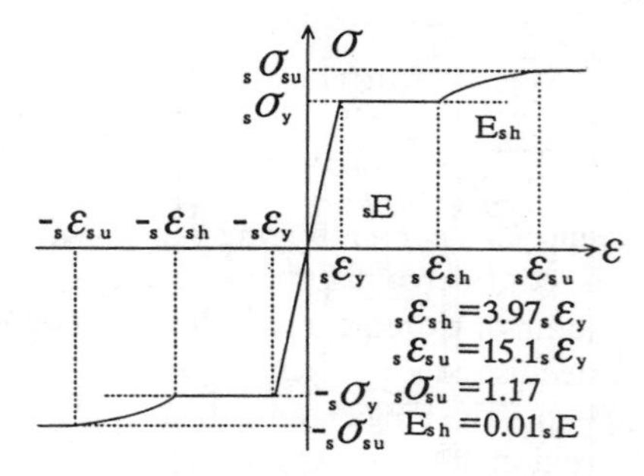

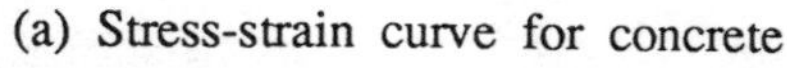
(a) Stress-strain curve for concrete (b) Stress-strain curve for steel

Fig. 7. Stress-strain curves for concrete and steel.

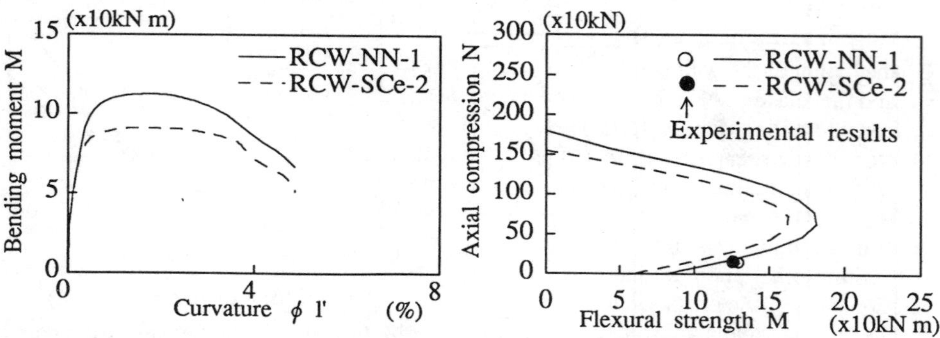

Fig. 8. Theoretical moment-curvature curves. Fig. 9. Theoretical interaction diagrams.

Table 8. Assumption of the curvature distribution along vertical axis

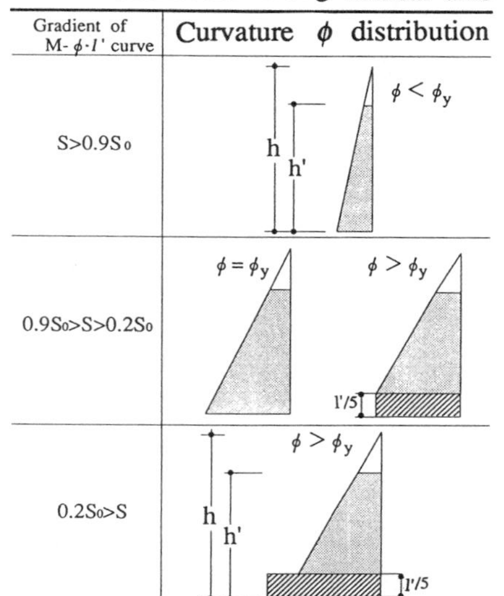

Gradient of M-ϕ-l' curve	Curvature ϕ distribution
S>0.9So	h, h', $\phi < \phi_y$
0.9So>S>0.2So	$\phi = \phi_y$; $\phi > \phi_y$, l'/5
0.2So>S	h, h', $\phi > \phi_y$, l'/5

h=height from wall panel bottom to center of the lateral load
l', h'=clear length and height of wall panel
S=gradient of M- ϕ l' curve, So=initial gradient of M- ϕ l' curve
ϕ =curvature of the structural wall
ϕ y=curvature at yield of the extreme tension reinforcing bar

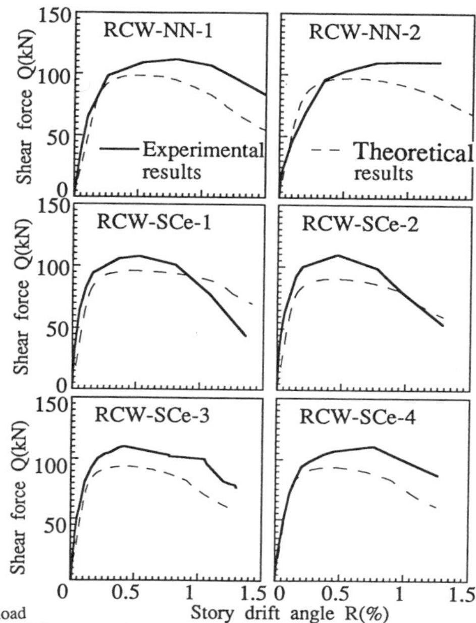

Fig. 10. Comparison between theoretical Q-R relations and experimental skeleton curves.

ship for steel reinforcing bars is perfectly elasto-plastic and the strain hardening region is considered as shown in Fig. 7b. As the width of the structural wall is large, the strain of extreme tension fiber is huge. The yield point stress of corrosion damaged steel is reduced according to Table 4.

(3) A plastic hinge with a length of (1/5)l' is assumed to be formed in the bottom end region of the structural wall (see Table 8). On referring to R/C column's plastic hinge length which is generally 1/3 of depth, its length of the R/C structural wall whose width corresponding to depth is large is assumed to be 1/5 of width l' in this paper. The tangent stiffness of the M-ϕ l′ curve in Fig.8 is defined to be S and the initial tangent stiffness as So. The tangent stiffness S and the curvature ϕ distribution along the height of the wall panel is assumed as shown in Table 8. The flexural deformation and the elastic shear distortion, except for the plastic hinge region, are taken into consideration. Moreover, the N-δ effect is considered.

(4) Non-elastic shear distortion and pulling out of reinforcing bars from the stub are not considered in this analysis because of their difficult estimations.

The moment-curvature relations are obtained by step-by-step application of the equilibrium equations. Iterative calculations are needed until an equilibrium condition between internal and external axial forces is satisfied. The bending moment-curvature relationships of the cross section of the R/C structural wall specimens are shown in Fig. 8. The compressive axial force-flexural strength interaction curves are illustrated for the cross section of R/C structural walls in Fig.9. Figure 10 shows the comparison between the theoretical results of monotonic Q-R relationship and experimental skeleton curves. The calculated results are almost similar to the tested ones.

7 Conclusions

The lateral loading capacity for the R/C structural walls is almost the same, regardless of corrosion damage. The initial stiffness of the corrosion damaged R/C structural walls is higher than that of the non-corroded structural walls. However, the deterioration of the ductility is remarkable as the corrosion damage increases.

Further investigations on the deterioration of aseismatic behavior of corrosion damaged reinforced concrete members are being planned on the basis of this preliminary research.

8 Acknowledgements

The author is grateful to thank Dr. Iraha, S., Assoc. Professor of Univ. of the Ryukyus, and Dr. Morinaga, S., Professor of Kyushu Tokai Univ. for their helpful guidances and useful suggestions. The author also wishes to thank Messrs. Matsunaga, T., and Nakayama, K., graduate students of Univ. of the Ryukyus, for their contributions in the experiment and preparation of this paper.

9 Reference

1. Yamakawa, T., Iraha, S., Morinaga, S. and Fujisaki, T. (1994) An experimental study on damage affecting a seismatic behavior of structural walls under chloride attack environment. Proceedings of International Conference on Corrosion and Corrosion Protection of Steel in Concrete, Sheffield, U.K., pp. 1320–1329.

150 SOFTENING OF REINFORCED CONCRETE FRAMES UNDER SEVERE LOADING CONDITIONS

K.S.P. DE SILVA and P.A. MENDIS
University of Melbourne, Australia

Abstract
Although properly detailed reinforced concrete structural elements possess remarkable ductile characteristics, there are a few inherent and some constitute deficiencies that do not warrant the use of elastic perfectly plastic idealisation for reinforced concrete. Concrete sections characteristically soften, that is the bending moment capacity decreases with increasing curvatures. This is more apparent when dealing with severe loading conditions where reinforced concrete structures are expected to go through elastic-plastic-softening phases. This paper examines the effect of member *softening* on the structural strength degradation, available displacement ductility, curvature ductility and current lateral load reduction criteria.
Keywords: Curvature ductility, displacement ductility, limited ductile moment resisting frames, over-strength, reinforced concrete, seismic loading, softening.

1 Introduction

The probability of a structure being subjected to severe loading conditions such as earthquakes, blasts and accidental loads during its design life is relatively low. However optimism is unwarranted due to the devastating nature of these events. The current design practice adopts a "dual" approach in mitigating normal and severe loading conditions. The philosophy governing the design under severe loading is prevention of collapse with significant structural damage whereas no structural damage is permitted under normal loading conditions. The energy imparted on to a structure under severe loading conditions needs to be efficiently absorbed or dissipated by the structural system within a relatively short duration. For example earthquakes, impulsive loads such as blasts and accidental loads can induce structural

Concrete Under Severe Conditions: Environment and loading (Volume Two) Edited by K. Sakai, N. Banthia and O.E. Gjørv. Published in 1995 by E & FN Spon. ISBN 0 419 19860 1

damage within 3-10 seconds during their application. The current design practice depends largely on the hysteretic energy dissipation capacity of the members in achieving the above requirement.

Therefore a good understanding and accurate modelling of full range behaviour of reinforced concrete structures is of great importance. There are a large number of experimental investigations available in the literature based on static, quasi-static and dynamic testing of single members, subessemblages and reduced scale models. These tests demonstrate that reinforced concrete sections with careful and appropriate reinforcement detailing possess remarkable ductile properties when subjected to full range destructive tests. A more comprehensive coverage of the factors affecting the energy dissipation capacity in terms of hysteretic energy is given in [1]. It has further emphasised that while some of these factors are inherent, others constitute deficiencies that can and should be avoided in the design stage. The inappropriate proportioning of members (strong beams and weak columns combination), insufficient transverse steel and inadequate joint detailing that aggravate bond failure are the commonly sighted deficiencies that need to be avoided in a well-engineered structure designed to severe loads. However, arguably, the said deficiencies are inherent in a large population of the building stock around the world except perhaps, the well-engineered building stock located in the high seismic regions. Many researchers around the world have observed the strength degrading characteristics of reinforced concrete possessing the said deficiencies when subjected to severe loading conditions [1,2,3]. This observation is more apparent in the low to medium ductile members, over-reinforced sections, sections with higher ratio of compression steel, high strength concrete members, axially loaded flexural members and beam column joints. In addition fatigue and load reversals adversely contribute to the rate of strength degrading nature.

This causes a serious concern over the current practice of using the design and response parameters based on elastic perfectly plastic idealisation for modelling of reinforced concrete members, especially in the areas of low seismicity where the major building stock is located. Fig. 1 illustrates the typical full range behaviour of low to medium ductile structural members under reversed loading. It must also be noted that due to the testing configuration particularly at large displacements, the P-δ contribution is partly responsible for the post yielding slope of the load deformation curve. However the proportion has not been clearly quantified in these experiments to differentiate the flexural softening, strength degrading due to bond failure and P-δ contribution. In fact the P-δ effect is present in real structures and more critical when softening occurs as this would enhance the overall structure softening slope. It is a fair approximation to assume that the combined influence is represented by a steeper slope in the analysis. However, if P-δ effect is neglected, and the total softening is attributed to flexural softening, which in effect improves the safety margin as the estimates will be conservative. The experimental observations suggest that the full range behaviour of a typical reinforced concrete hinge subjected to severe loading conditions is better characterised by the elastic-softening (strength degrading) rather than elastic-perfectly plastic idealisation (Fig. 2). "*Softening*" is the preferred term used through out this paper to describe the strength degrading character predominantly due to flexure.

This paper demonstrates the effect of the presence of softening hinges in a frame structure to its post yielding characteristics such as overall load displacement relationship and the local curvature ductility demands. This has a direct application on the plastic design, displacement ductility and the lateral load reduction approach in the seismic design of moment resisting frames. It is worth noting that most of the current design concepts and design parameters given in the design standards that deal with the said severe loading conditions are based on the elastic-perfectly plastic idealisation for reinforced concrete structures. This can be a costly misconception not only in financial terms but also in terms of human lives. It has been demonstrated that the presence of softening hinges in moment resisting frames amplifies the overall strength degrading nature of reinforced concrete frames that apparently accelerate to collapse under dynamic loading and P-δ effects [2]. In spite of this, a large portion of analytical studies, even at present, use the elastic perfectly plastic idealisation for post yield analysis of reinforced concrete frames. In addition it is found that if any one of such softening hinges reaches a critical softening value, the structure becomes unstable no matter how redundant the structure can still be [2,4,5]. This is quite contrary to the conventional plastic analysis techniques that assume the structure can sustain its load carrying capacity unless a mechanism is formed. This has a large impact on the dependable over-strength of redundant structures on which the current seismic lateral load reduction criterion is partly dependent for reserve strength. Therefore the displacement ductile capacity of a softening structure can be substantially different to that of an idealised elastic-perfectly plastic structure.

2 Softening

Fig. 1 illustrates the pseudo-dynamic behaviour of some typical reinforced concrete members designed to current building standards of Australia [2,3]. Empirical formulae to estimate the flexural softening slopes of reinforced concrete sections were suggested by various researchers [4,5]. These suggestions were based on the tests carried out on beams designed to the Australian and American building standards.

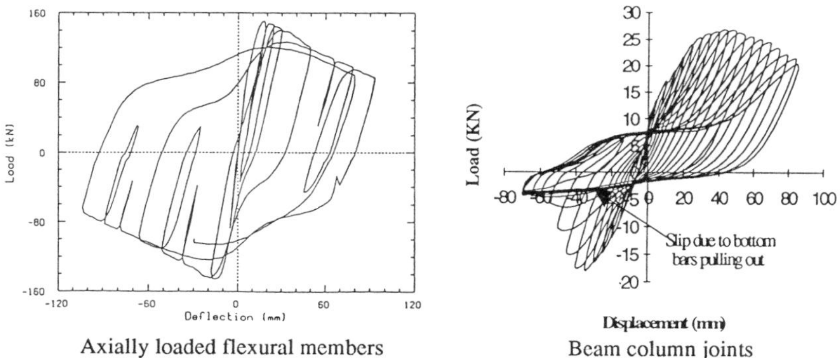

Axially loaded flexural members Beam column joints

Fig. 1. Typical strength degrading character of limited ductile members

The oscillators shown in Fig. 2 schematically illustrate the softening concept of a simple single degree of freedom structure. It is worth noting the reduction in elastic curvature due to the unloading branch of the elastic region whereas the increase of rotation via the softening branch of the hinge area. The displacement ductility of this system can be derived as given by Equation (1). The complete derivation is given in Reference [6].

$$\mu_\Delta = 1 + 3m(\gamma - 1)(1 - 0.5m) + \frac{\lambda}{a}\left[1 - (1 + a)(1 - m)^3\right] \tag{1}$$

Assuming the post yield displacement takes place via the hinge rotation only and neglecting the reduction of curvature due to the unloading of the elastic regions of the member, the curvature ductility can be expressed in terms of displacement ductility and softening hinge parameters as given by Equation (2).

$$\mu_\phi = \frac{1}{(1 - \lambda)}\left(\frac{\mu_\Delta - 1 - 3m(\gamma - 1)(1 - 0.5m)}{\left[1 - (1 + a)(1 - m)^3\right]} + \gamma\right) \tag{2}$$

where,

μ_Δ displacement ductility, Δ_u / Δ_y
μ_ϕ curvature ductility, ϕ_u / ϕ_y
m hinge length, (l_p)/ member length (L)
a softening slope of the hinge
γ plastic curvature, (ϕ_p) / yield curvature, (ϕ_y)
λ proportion of strength degradation

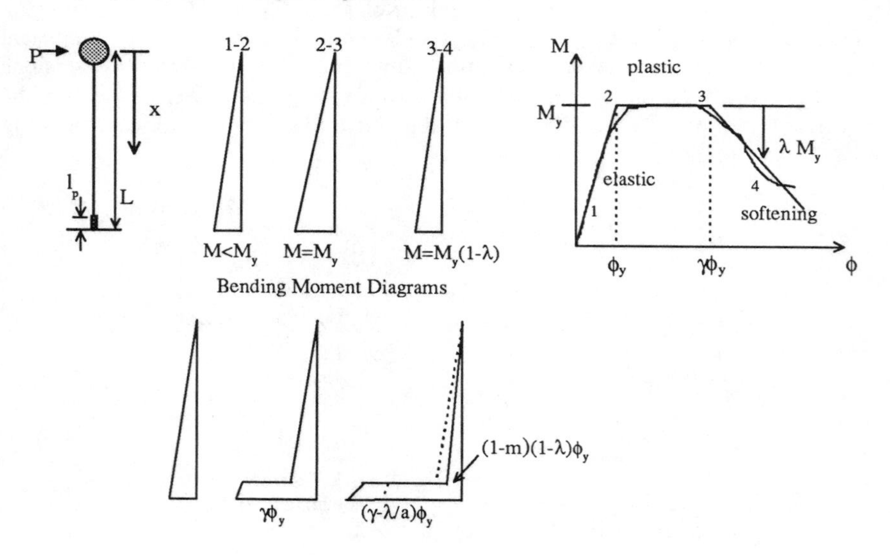

Fig. 2. A cantilever with a softening hinge

It is apparent from the above derivations that curvature ductility and displacement ductility relationships are not only dependent on hinge length ratio, m, as for an elastic perfectly plastic system [7] but also on the softening slope and greatly limited by the acceptable proportion of the strength degradation. The corresponding curvature ductility demands of an elastic perfectly plastic oscillator and elastic softening oscillator for a given displacement ductility are compared in Table 1. The values given in Table 1, for the elastic perfectly plastic (EPP) oscillators are taken from reference [7] and the values for elastic softening (ES) oscillator are calculated using Equations (1)&(2) given above. A 3% softening slope, 20% strength degrading limit and a displacement ductility of 4 are considered in this comparison.

Table 1. Comparison of curvature ductility demand of a cantilever analysed using EPP and ES idealisation for displacement ductility of 4.

$m=l_p/L$	0.05	0.1	0.15	0.20	0.25	0.30	0.35
μ_ϕ– EPP	21.5	11.5	8.2	6.6	5.6	4.9	4.5
μ_ϕ– ES	33.3	15.1	11.5	9.2	7.9	7.0	6.5

3 Structure softening and member softening

For a cantilever softening slope of the hinge and softening slope of the structure are the same. The relationship between local member softening and the overall structure softening of a redundant structure is somewhat difficult to ascertain and cannot be generalised due to its case dependency. However reasonable estimates can be made on a case by case basis. The relationship between the member softening slope, a, and the structure softening slope, B, for a sway frame with softening hinges at both ends is reported in Reference [2] and given in Equation (3). This formula is valid only when all the softening hinge parameters are identical and occur simultaneously.

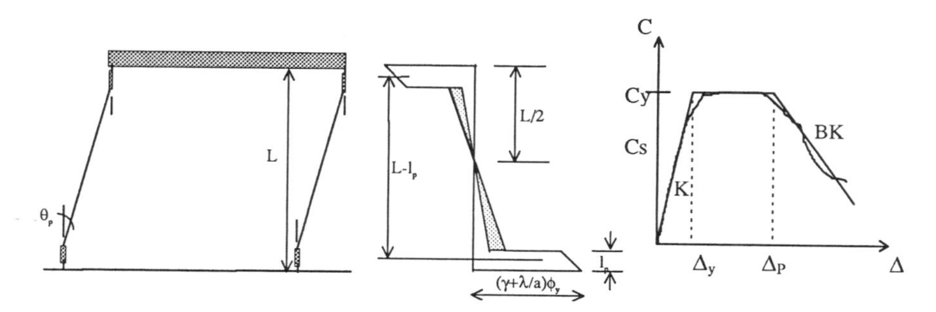

Fig. 3. Sway frame

$$B=\frac{1}{1-6em+12em^2-(6e+2)m^3} \quad (3)$$

where,

m - hinge length, (l_p)/ member length (L)

B - structure softening
e - 1-1/a

Fig. 4 shows the structure softening slopes corresponding to the typical member softening parameters of a sway frame. It is worth noting the significantly greater structure softening slopes opposed to local hinge softening slopes, at smaller hinge length ratio values.

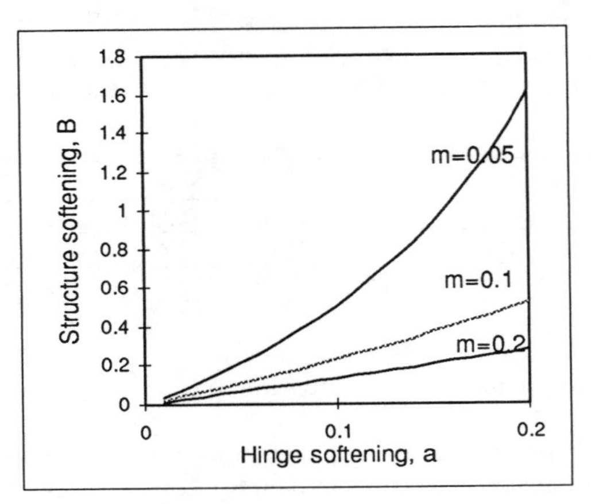

Fig. 4. Effect of local hinge softening on structure softening of a sway frame

The displacement and curvature ductility of the sway frame shown in Fig. 3 can be derived [6] similar to the cantilever as given in Equations (4)&(5). To simplify the derivation, identical softening hinge parameters and simultaneous hinge softening are assumed.

$$\mu_\Delta = 1 + 6m(\gamma - 1)(1 - m) + \frac{3\lambda}{a}[m(1 + a)(1 - m)(2 - m)] \tag{4}$$

$$\mu_\phi = \frac{1}{(1 - \lambda)}\left(\frac{\mu_\Delta - 1 - 6m(\gamma - 1)(1 - m)}{[3m(1 + a)(2 - m)(1 - m)]} + \gamma\right) \tag{5}$$

Table 2. Comparison of curvature ductility demand of a sway frame analysed using EPP and ES idealisation to achieve a displacement ductility of 4.

$m=l_y/L$	0.05	0.1	0.15	0.20	0.25	0.30	0.35
μ_ϕ– EPP	11.5	6.6	4.9	4.1	3.7	3.4	3.2
μ_ϕ– ES	13.1	7.1	5.1	4.3	3.7	3.4	3.2

As demonstrated above it is clear that on an analytical point of view, the curvature ductility demands for a given displacement ductility can be appreciably underestimated if elastic perfectly plastic idealisation is assumed especially for smaller hinge length ratios. Moreover the limited ductile structures with undesirable member proportioning tend to yield via column side sway failure mechanism under severe loading. It has been widely accepted that curvature ductility demands well in excess of available curvature ductility are resulted if columns yield prior to beams [7,8]. A study carried out by the authors revealed that existing and prospective limited ductile frames in Australia are vulnerable to this mode of failure subjected to earthquake forces [9]. Thus the elastic perfectly plastic idealisation without due consideration of softening tends to provide an upper bound collapse load in ultimate conditions.

4 Evaluation of over-strength, ductility and structure softening of redundant frames

A three storey frame given in Fig. 5 is used to demonstrate the effect of local softening on the above mentioned parameters. These parameters are especially investigated in this paper as they have a direct application in seismic resistive design, which is the most common severe loading condition the structural engineers deal with. A computer program was developed to investigate the optimum lateral load reduction of the softening reinforced concrete frames. A review of the current lateral load reduction concept in dealing with the seismic resistive design [10] and its applicability to low seismic regions are given elsewhere [11]. This program is especially designed to investigate the available strength of the typical gravity dominating softening structures in the low seismic regions.

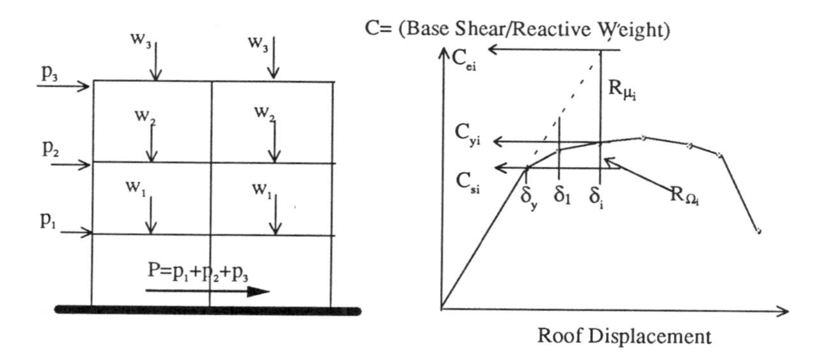

Fig. 5. Typical 3 storey frame analysed

The idea is to use the conventional gravity and wind design end product as a "seed" to evaluate the optimum lateral load reduction factor (similar to response modification factor used in the seismic design standards around the world). This program is attractive not only in the design of prospective structures but also in the assessment of existing structures that were not designed for severe loading conditions such as earthquakes. The program is capable of reporting the lateral load reductions, corresponding displacement ductility, curvature ductility, over-strength at each event (local yielding) for the full range behaviour of the structure up to collapse. It will "*flag*" all the events while keeping an eye on the hinges reaching critical softening slopes. The advantage of this simple, easy to use, analytical tool is that the designer is provided with sufficient information to make an engineering judgement to identify the optimum lateral load reductions permissible at each step of the post-yield deformation. The program is based on non-linear analysis of frames including elastic-plastic-softening phases.

Fig. 5 schematically illustrates the contributing factors to the lateral load reduction factor, R_f, which is defined as ratio of the elastic strength demand to the strength at the first yield.

$$R_f = \frac{C_{eu}}{C_s} = \frac{C_{eu}}{C_y}\frac{C_y}{C_s} = R_\mu R_\Omega \qquad (6)$$

Over-strength of a structural system is defined as the ratio between strength of the structure at first yield and the significant yield. In a redundant structure there can be a substantial amount of over-strength in-built due to a number of contributing factors such as redundancy, material safety factors, dominance of other design load combinations, over design sections and non structural members, etc. It is practically impossible to mathematically model all of these contributing factors. However if the post-elastic response is monitored event by event, the contribution of over-strength can be estimated for each subsequent event as shown in Fig. 5. It is clear that at each event, the proportion of over strength, R_Ω, and the proportion of strength compromised with the provision of ductility, R_μ, can be reasonably estimated. It must be noted that on one hand, as this procedure is essentially a non-linear static approach similar to the "push over analysis" the effect of damping cannot be taken into account. In fact, if damping had been taken into account it would have reduced the ductility dependant component. On the other hand the static approach does not take into account the dynamic amplification of the system opposed to a conventional dynamic analysis. To authors' knowledge generic dynamic amplification factors for non-linear structures have not been established. Also non-linear dynamic amplification factors are relatively small compared to linear dynamic amplification factors. Therefore it may be reasonable to assume that the effect of damping and the effect of non-linear dynamic amplification are likely to compensate each other, if not serving to improve the safety margin. The factors given in Equation (6) can be evaluated either based on equal displacement concept or on equal energy concept [7]. The equal energy approach is applicable to rigid structures (approximately, natural period < 1.5secs) whereas the equal displacement approach is applicable to more flexible structures (approximately natural periods > 1.5secs).

A three story frame with softening post yielding hinges is presented to demonstrate the developed procedure using equal displacement approach (natural period ~0.3 secs). It is our experience that the response of softening structures is highly case dependent. The configuration of the frame used in this example has two 8.4m bays and three 3.8m high stories. Vertical point loads of 50kN and inverted triangular lateral load distribution with p_1=0.15P, p_2=0.25P, p_3=0.6P are used. The yield moments of the members were set to 100kNm and elastic softening moment curvature with 3% softening slope is assumed. A hinge length ratio of m=0.1 is assumed. The results are presented in Figures 6(a)&6(b).

Figures 6(a)&6(b) demonstrate the effect of softening on the influential parameters mentioned before. It is worth noting that in this particular case, the load carrying capacity of the structure is improving well beyond the first yield of the structure to a maximum value and then experiences a sudden drop in the load carrying capacity that accelerates the structure to collapse. The other important observation is that there is a critical lateral load reduction that is largely case dependent. Operating within this critical lateral load reduction threshold is the most effective trade off between ductility and elastic strength of the softening structural systems. This is because, within this limit, the lateral load reduction is equally shared by the over-strength and the ductile capacity of the system. However if larger lateral load reductions are opted beyond this critical value, the sudden drop of the lateral load carrying capacity of the system imposes large, impulsive ductility demands. It should be borne in mind that by the time this situation arrises most of the reserve strength of the structural system is already exhausted and the hinges are approaching their critical softening slopes. Therefore it is quite illogical to assume that the state of the structure is capable of handling this additional ductility demand imposed due to the sudden loss of strength capacity. This is an observation quite contrary to the behaviour of elastic perfectly plastic systems and also unable to highlight if the elastic perfectly plastic idealisation is assumed. This observation disputes the use of generalised lateral load reduction criterion for limited ductile frames. It must also be noted that unlike in the plastic analysis, the sudden drop of the lateral load is not due to the formation of a mechanism but associated with some of the isolated hinges approaching critical softening slopes.

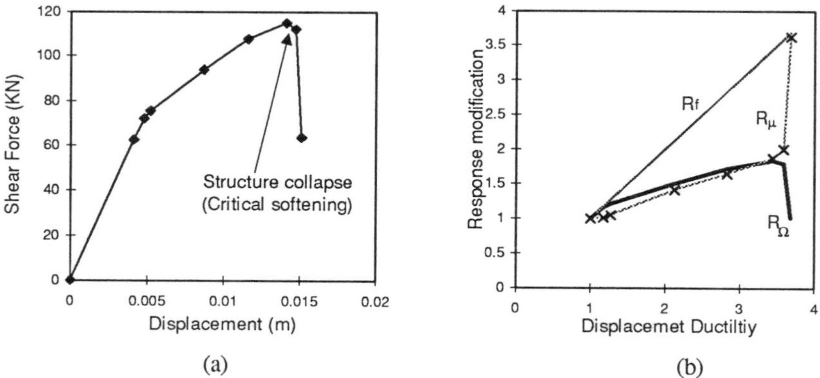

(a) (b)

Fig. 6. Three storey frame structure analysed using elastic softening idealisation

5 Conclusions

1. Elastic perfectly plastic idealisation of reinforced concrete members under severe loading conditions constitutes an upper bound approach to ultimate limit state design. This is especially the case in dealing with the low to medium ductile moment resisting frames with softening hinges.

2. The relationship between overall displacement ductility and local hinge curvature ductility of softening structures is substantially different to that of elastic perfectly plastic systems.
3. A more transparent procedure is suggested to assess the lateral load carrying capacity at each yielding event in the softening frames using non-linear, static, full-range (elastic-plastic-softening) analysis.
4. This procedure yields an optimum lateral load reduction level for each structure and highlights the threshold beyond which the lateral load reduction of softening structures is unwarranted.

6 References

1. Sozen, M.A. (1974) Hysteresis in structural elements, Applied mechanics in earthquake engineering, ASME, ASD8, 1974.
2. Sanjayan, G. (1988) Dynamic response of reinforced concrete structures with softening behaviour. Ph.D. Thesis, Monash University.
3. Covetti, J., Goldsworthy, G.J., and Mendis, P.A. (1993) Assessment of reinforced concrete exterior beam-column joints as specified in the new draft earthquake standard. *Thirteenth Australian conference on the mechanics of structures and materials,* University of Wollongong, Australia.
4. Mendis, P.A. (1986) Softening of reinforced concrete structures. Ph.D. Thesis, Monash University.
5. Darvall P.Lep. (1983) Some aspects of softening in flexural members. *Civil engineering research reports,* No. 3/1983, Monash University.
6. K.S.P. De Silva. Ph.D. Thesis, Seismic response of concrete frame structures in intra-plate regions, The University of Melbourne, Australia, 1995 (under preparation).
7. Park, R., and Paulay, T. (1975) *Reinforced concrete structures*, John Wiley Sons., Inc. USA.
8. Paulay, T., and Priestley, M.J.N. (1992) *Seismic design of reinforced concrete and masonry buildings,* John, Wiley & Sons, Inc., New York, pp. 1-94.
9. De Silva, K.S.P., Mendis, P., and Wilson, J.L. (1994) Performance of R/C frames designed in accordance with the new Australian earthquake loading standard; Accepted for publication in the IEAust transactions.
10. Uang, C.M. (1992) Seismic force reduction and displacement amplification factors. *Earthquake engineering, Tenth world conference,* Balkema, Rotterdam.
11. De Silva, K.S.P., Mendis, P.A., and Grayson, R. (1994) Ductility demand and response modification of reinforced concrete moment resisting frames in low seismic regions under Intra-plate type earthquakes, 3rd International Conf. on the Concrete Future, Malaysia, pp. 51-58.

PART TWENTY-FIVE

TIME DEPENDENCY AND TEMPERATURE EFFECT

151 EFFECT OF WATER CONTENT ON EXPANSION AND CREEP OF CONCRETE UNDER CONSTANT MULTI-AXIAL STRESS

S. KAWABE and T. OKAJIMA
Nagoya Institute of Technology, Nagoya, Japan

Abstract
The purpose of this study is to clarify the effect of water content on the thermal expansion of plain concrete under constant multi-axial stress. The thermal expansion of pre-heated concrete between 20°C and 100°C, under constant compressive stress including tri-axial loading in which maximum compressive stress is below one third of strength, can be concluded to be equal to the sum of elastic deformation caused by external forces and by the thermal expansion without external load. The thermal expansion of air dried and water saturated concrete is smaller than the thermal expansion of pre-heated concrete. The strain of concrete can be obtained as the sum of elastic deformation, free thermal expansion and creep strain. The creep strain of concrete at any temperature can be obtained by using the "Time Temperature Equivalence Principle."
Keywords: Concrete, creep, elevated temperature, multi–axial stress, time temperature equivalence principle, thermal expansion, water content.

1 Introduction

Concrete structures and its members are under constant multi-axial stress and exposed to various temperatures. It is important to understand concrete under multi-axial stress conditions and under various temperatures.

Information on the thermal expansion of concrete under constant multi-axial compressive stress is one of the important factors for the design and safety evaluation of structures.

This study has two main purposes. One is to clarify the thermal expansion of plain concrete under constant multi-axial compressive stress. The effect of wa-

Concrete Under Severe Conditions: Environment and loading (Volume Two) Edited by K. Sakai, N. Banthia and O.E. Gjørv. Published in 1995 by E & FN Spon. ISBN 0 419 19860 1

ter content on the thermal expansion of concrete is examined. The other is to predict the strain under multi-axial stress at the increasing temperature.

2 Experiment

2.1 Shape and size of specimen

The shape of the specimens was a 113mm cube with chamfered edges as shown in Fig.1. To ensure accuracy in the experiment, specially designed steel molds were made to cast the concrete.

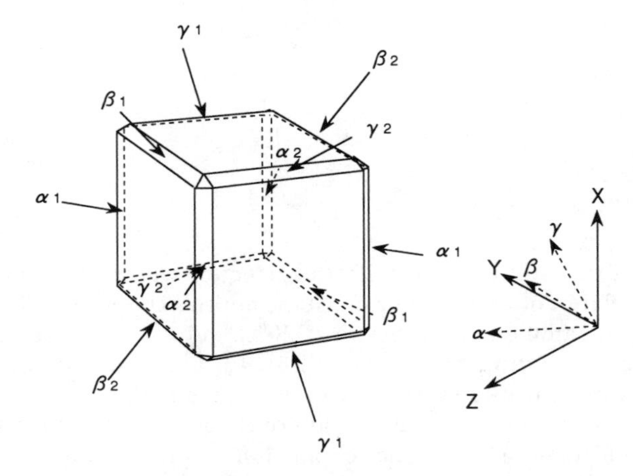

Fig. 1. The position of displacement measuring devices.

2.2 Mix proportion of concrete

The materials used in this study are ordinary portland cement, river sand, river gravel and air entraining agent. The physical properties of aggregates are shown in Table 1, and the mix proportion of concrete is shown in Table 2.

Table 1. Physical properties of aggregates

	Maximum size of aggregate (mm)	Water absorption (%)	Specific gravity (oven dry)	Fineness modulus (%)	Weight per unit volume (kg/m^3)
Fine aggregate	5	1.67	2.50	2.61	—
Coarse aggregate	20	1.25	2.58	7.05	1556

Table 2. Mix proportion of concrete

Slump (cm)	Air content (%)	Water cement ratio (%)	Maximam size of aggregate (mm)	Sand aggregate ratio (%)	Weight per unit volume of concrete (kg/m^3)				
					Water	Cement	Sand	Gravel	Admixture
15	4	57	20	42.7	174	305	725	1027	0.06

2.3 Curing

The specimens were demolded approximately 24 hours after casting. All specimens were cured in water for 12 weeks. Then the specimens were separated into three types of water conditions as follows.

The specimens of pre-heated condition were cured in air for one week. Then the specimens were pre-heated in an electric oven to 60°C for 6 hours and to 105°C for 3 days, so as to evaporate the free water of the specimens and also to reduce the moisture gradient.

The specimens of air dried condition were cured in air for one week.

The surface of the specimens of water saturated condition was wiped.

2.4 Loading apparatus

The testing apparatus is shown in Fig.2. The loading frame was designed with sufficient stiffness. The testing machine has four steel columns for each direction, and the diameter of each column is a 120mm. The three sets of loading frames can move independently of each other. At loading, the specimen is kept at the center by using spherical seating to avoid eccentricity . To reduce friction between the loading plates and the specimen, steel brush platens were used.

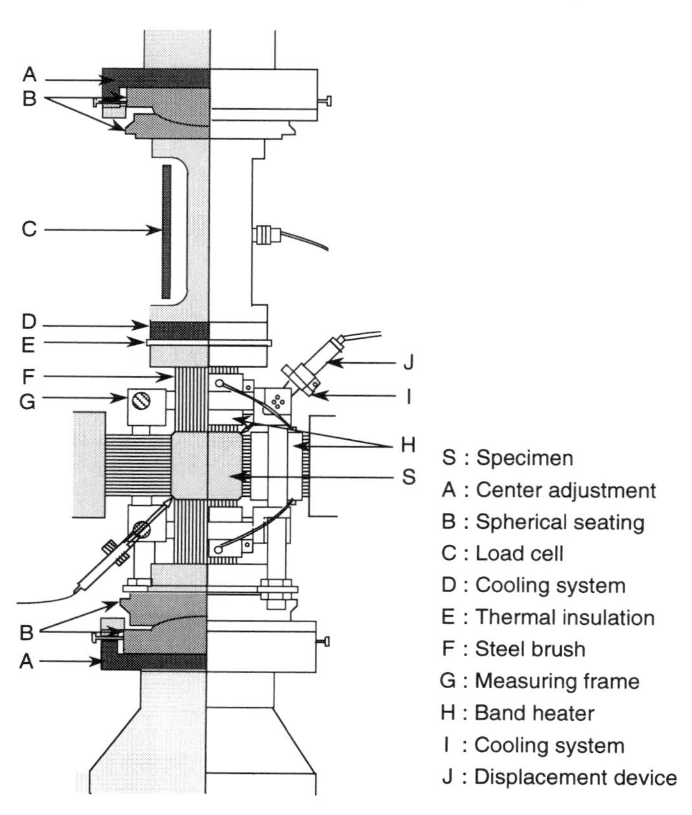

Fig. 2. Testing frame.

2.5 Strain measurement method

Three pairs of displacement measuring devices were set on the chamfered edges of the specimen in three directions, α,β and γ, as shown in Fig.1. The deformation in a specific direction was measured as the sum of the outputs from a pair of displacement devices placed in the diagonal and opposite edges in the same direction of the specimen. The strains, ε_X, ε_Y and ε_Z, in the X, Y and Z directions can be written in eq. (1), assuming that terms higher than second order are negligibly small.

$$\begin{pmatrix} \varepsilon_X \\ \varepsilon_Y \\ \varepsilon_Z \end{pmatrix} = \begin{pmatrix} -1 & 1 & 1 \\ 1 & -1 & 1 \\ 1 & 1 & -1 \end{pmatrix} \begin{pmatrix} \varepsilon_\alpha \\ \varepsilon_\beta \\ \varepsilon_\gamma \end{pmatrix} \quad (1)$$

2.6 Heating method

The specimen was heated with band heaters, which were installed on the outside of the individual steel brushes. Thermocouples (Cu-Co) embedded at three spots in the specimen were used to monitor the interior temperature of the specimen. The temperature was increased and decreased at a rate of about 10°C per hour.

2.7 Measurement of thermal expansion of concrete under constant multi-axtial stress

First, the specimens were loaded to the stress conditions shown Table 3.

Table 3. Types of loading

	X-axis	Y-axis	Z-axis
Free	-	-	-
Under uni-axial constant stress	#	-	-
Under bi-axial constant stress	#	#	-
Under tri-axial constant stress	#	#	#

: with constant stress of one third of strength - : without constant stress

Then the specimens were heated up to 100°C and cooled down to 30°C. The strains in the specimen were measured at every 10°C. A computer automatically controled the loads so that the stress was kept constant was during the test. Just before loading, the water content of the specimens was adjusted for the preheated, air dried and water saturated conditions. The specimens were not sealed.

2.8 Measurement of creep

Creep strain for 8 hours at temperatures of 20°C, 60°C and 100°C were measured.

3 Results

3.1 Strain under multi-axial constant stress

Figure 3 shows the strains due to the change of temperature without applied load. It can be seen that the greater the water content in the specimen, the smaller the strain increment with change in temperature. The strain increment of specimens was almost constant.

Figure 4 shows the relations between the total strain and the temperature in the X, Y and Z directions under uni-axial constant stress. Fig.5 and 6 show the relations under bi-axial loading and tri-axial loading respectively. The strains at room temperature indicate the instantaneous strains initiated by loading. As in Fig.3, Fig.4 shows that the higher the water content of the specimen the smaller the strain increment with changes in temperature. This effect is especially noted when the specimens have loads applied. The strain increment of the pre-heated specimens with load is almost the same as that without load. This implies that the strain increments of the air-dried and water saturated specimens with load were affected by creep. In the cooling process, the strain increments of the specimens under load were almost the same as those without load.

Table 4 shows the change in specimen weights before and after measurements. The higher the water content, the larger the change in weight.

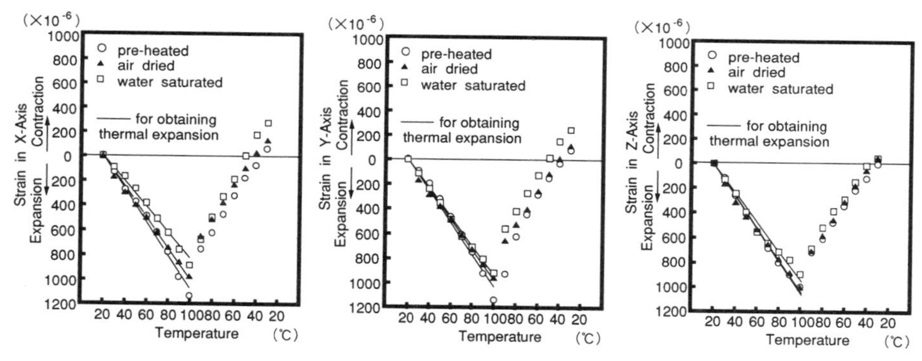

Fig. 3. The relations between the total strain and the temperature in the X, Y and Z directions (without loads).

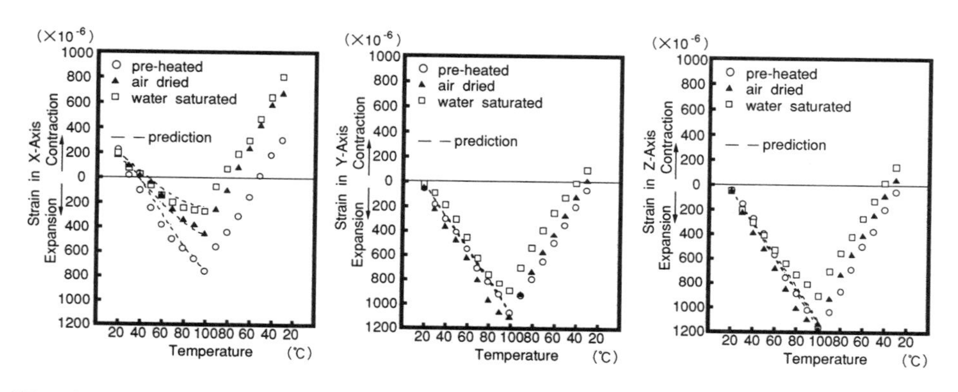

Fig. 4. The relations between the total strain and the temperature in the X, Y and Z directions (under uni-axial constant stress).

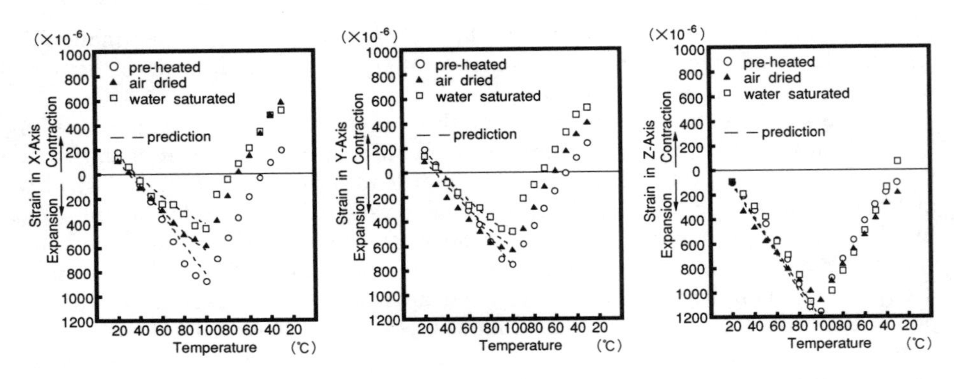

Fig. 5. The relations between the total strain and the temperature in the X, Y and Z directions (under bi-axial constant stress).

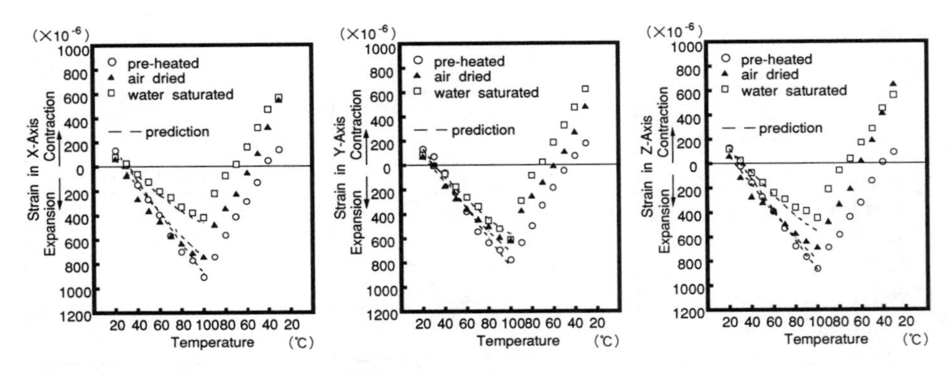

Fig. 6. The relations between the total strain and the temperature in the X, Y and Z directions (under tri-axial constant stress).

Table 4. Change of weight

Stress	Change of weight (g)		
(X-Y-Z)	pre-heated	air dried	water saturated
0- 0- 0	1.7	−45.0	−72.4
1/3- 0- 0	0.5	−61.2	−84.2
1/3-1/3- 0	1.4	−46.3	−77.8
1/3-1/3-1/3	1.9	−49.5	−74.8

3.2 Creep

Uni-axial creep strain was measured for 8 hours at temperatures of 20^{0}C, 60^{0}C and 100^{0}C as shown in Fig.7. The greater the specimen's water content and higher the temperature, the greater the creep.

Table 5 shows the change in specimens weights for after the 8 hours creep tests.

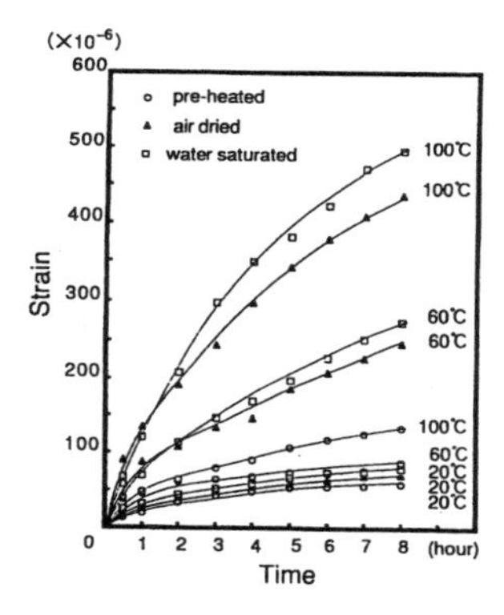

Fig. 7. Creep at constant temperature.

Table 5. Change of weight under creep test

Temperature (°C)	Change of weight (g)		
	pre-heated	air dried	water saturated
20	1.7	−2.8	−3.5
60	0.5	−25.5	−39.0
100	0.1	−81.3	−142.5

4 Predicted strain with external load and heating

4.1 Creep at any constant temperature

Creep at any constant temperature between 20°C and 100°C can be predicted by the "Time Temperature Equivalence Principle."

The original creep curves shown in Fig.7 were transformed into the curves shown in Fig.8, where ε is creep strain, T is absolute temperature and σ is loading stress.

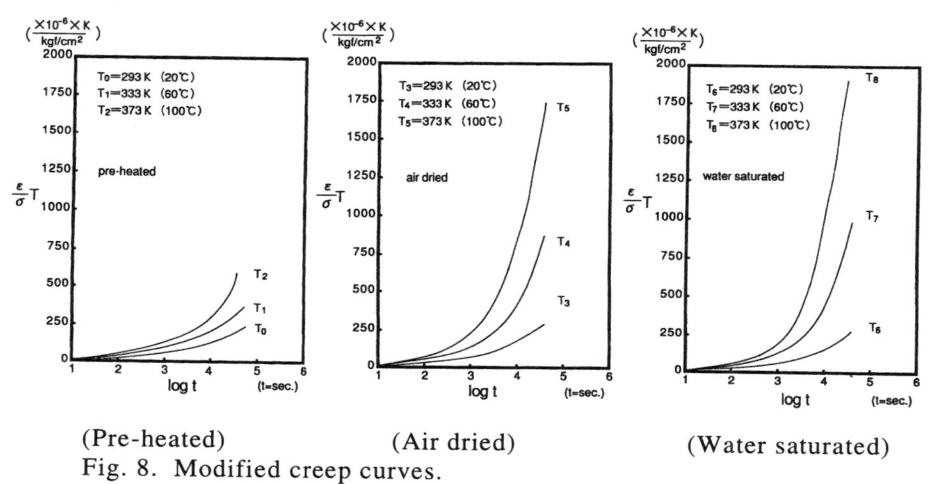

(Pre-heated) (Air dried) (Water saturated)

Fig. 8. Modified creep curves.

By shifting the curves T1 and T2 horizontally along the time axis and superimposing them on curve T0, a synthesis curve called as "master curve" can be obtained, as shown in Fig.9. The amount of time shift α_t can be represented as a function of the absolute temperature, as shown in Fig.10. Using the master curve and knowing the amount of time shift, the creep strain at any temperature between 20°C and 100°C can be obtained.

Figure 11 shows predicted creep curves under heating at 10°C intervals from 20°C to 100°C. Considering the fact that the temperature rises at the rate of 10°C per hour, creep strain in the heating process is predicted. Table 6 shows the prediction of creep (ε/σ) under uni-axial stress under heating.

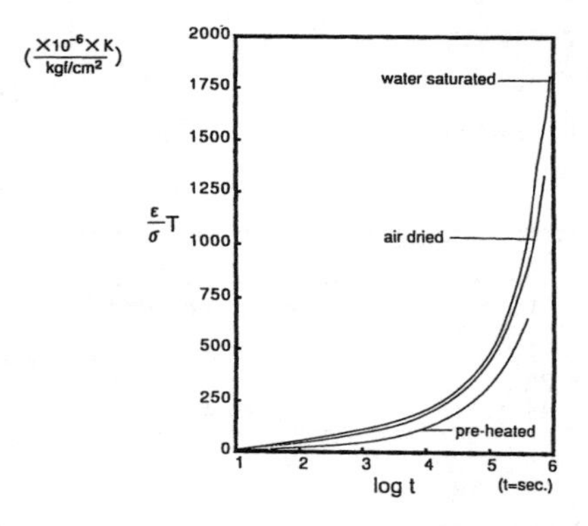

Fig. 9. Master curves.

Fig. 10. Relation of time shift and absolute temperature.

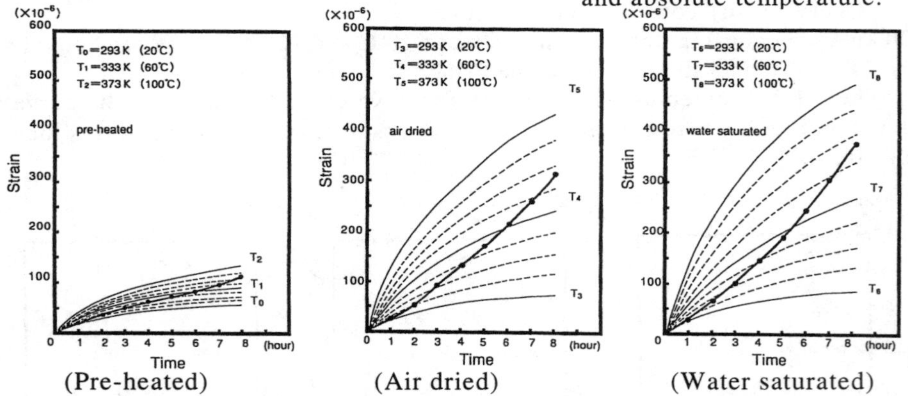

(Pre-heated) (Air dried) (Water saturated)

Fig. 11. Prediction of creep in the heating process.

Table 6. Prediction of creep (ε/σ) under uni-axial stress

Prediction of creep (ε/σ) $\left(\frac{\times 10^{-6}}{kgf/cm^2}\right)$	Temperature (°C)								
	20	30	40	50	60	70	80	90	100
pre-heated	0	0.22	0.41	0.47	0.69	0.76	0.89	1.07	1.33
air dried	0	0.40	0.67	1.01	1.45	2.00	2.62	3.40	4.35
water saturated	0	0.56	0.88	1.32	1.69	2.42	2.97	3.94	4.95

4.2 Predicted strain

The total strain of specimens under external load and heating can be obtained from a sum of three components as shown in eq. (2).

$$\varepsilon=\varepsilon^{e}+\varepsilon^{t}+\varepsilon^{c} \tag{2}$$

where ε= total strain, ε^{e}=elastic deformation caused by external forces, ε^{t}=thermal expansion without external load and ε^{c}= creep strain.

Supposing that creep from Poission's ratio is equal to the static one and using the principle of superposition, creep under multi-axial stress under increasing temperatures can be predicted. Predictions of strains are compared with experimental ones in Fig.s 4, 5 and 6. Predictions of strains are well fitted to the experimental ones.

Table 7 shows an example of the strain at 100⁰C under multi-axial stress.

Table 7. The strain at 100⁰C under multi-axial stress

		Water content								
		pre-heated			air dried			water saturated		
Strain in the each direction		X	Y	Z	X	Y	Z	X	Y	Z
Free expansion	experiment	1136	1134	1000	978	958	1003	882	914	897
	ε^{e}	1075	1031	1053	996	972	1037	825	934	948
Uni-axial stress	ε^{e}	-224	54	44	-182	48	47	-199	15	43
	ε^{t}	1075	1031	1053	996	972	1037	825	934	948
	ε^{c}	-106	23	23	-348	79	79	-396	58	58
	experiment	**756**	1077	1164	**448**	1110	1131	**266**	897	902
	$\varepsilon^{e}+\varepsilon^{t}+\varepsilon^{c}$	**745**	1108	1120	**466**	1099	1163	**230**	1007	1049
	$\varepsilon^{e}+\varepsilon^{t}$	**851**	1085	1097	**814**	1020	1084	**626**	949	991
Bi-axial stress	ε^{e}	-179	-188	103	-111	-95	98	-133	-136	190
	ε^{t}	1075	1031	1053	996	971	1038	825	934	948
	ε^{c}	-83	-83	47	-269	-269	159	-338	-338	115
	experiment	**879**	**750**	1170	**583**	**631**	1057	**442**	**483**	1222
	$\varepsilon^{e}+\varepsilon^{t}+\varepsilon^{c}$	**813**	**760**	1203	**616**	**607**	1295	**354**	**460**	1253
	$\varepsilon^{e}+\varepsilon^{t}$	**896**	**843**	1156	**885**	**876**	1136	**692**	**798**	1138
Tri-axial stress	ε^{e}	-131	-131	-124	-59	-69	-63	-80	-87	-120
	ε^{t}	1075	1031	1053	996	971	1038	825	934	948
	ε^{c}	-60	-60	-60	-181	-181	-181	-281	-281	-281
	experiment	**911**	**778**	**865**	**745**	**618**	**687**	**416**	**611**	**442**
	$\varepsilon^{e}+\varepsilon^{t}+\varepsilon^{c}$	**884**	**840**	**869**	**756**	**721**	**794**	**464**	**566**	**547**
	$\varepsilon^{e}+\varepsilon^{t}$	**944**	**900**	**929**	**937**	**902**	**975**	**745**	**847**	**828**

+: Expansion -: Contraction

5 Conclusion

1. It was found that the greater the specimen's water content, the smaller the strain increment with changes in the temperature, especially, when a load was applied to the specimen.
2. In the cooling process, the strain increment of specimens under load was almost the same as that without load.
3. The total strain of specimens under external load and heating can be obtained from the sum of three components as shown in eq. (2).

6 Acknowledgement

The authors wish to thank Professor Emeritus T. IWASHITA and the late Professor S. OHGISHI of Nagoya Institute of Technology.

7 References

1. Okajima, T. et al. (1984) The thermal expansion of plain concrete under constant biaxal stress. *Transactions of the Japan Concrete Institute*, pp. 293–300.
2. Ohgishi, S. (1971) On the rheological behavior of hardened concrete. *Transactions of the Japan Concrete Institute*, pp. 62–71.
3. Alfrey, T. and Gurnee, E.F. (1967) *Organic Polymers*, The Dow Chemical Company .
4. Schneider, U. (1993) Concrete creep at elevated temperatures. *Proceedings of the 5th International RILEM Symposium on Creep and Shrinkage of Concrete*, E&FN Spon, London, pp. 233–246.

152 CREEP AND SHRINKAGE OF ULTRA HIGH-PERFORMANCE STEEL FIBRE REINFORCED CONCRETE

A. LOUKILI
École Centrale de Nantes, Nantes, France
P. RICHARD
Direction Scientifique, Bouygues, St-Quentin en Yvelines, France

Abstract
This paper presents the results of an experimental investigation on the shrinkage and creep under uniaxial compressive loading of CRC (Compact Reinforced Concrete) at an early age. CRC is a new material with a very high compressive strength (140 to 180 MPa at 28 days). A very high autogenous shrinkage at an early age that actually stops at 8 days after batching was observed. This sudden break in shrinkage kinetics, as far as we know, has never been observed before. Drying shrinkage was very small, which confirms previous reported findings concerning silica fume concrete with a low water/cement ratio. The CRC creep was measured on five sealed specimens 2, 4, 7, 14 and 28 days after batching. The stress level on the specimens represented 20% of the concrete strength at the time of loading. The experimental results show significant age of loading effects on the creep strain.
Keywords : Creep, experimental study, high strength concrete, shrinkage.

1 Introduction

The long-term behaviour of normal strength concrete is relatively well explored. However, very few studies on the long-term characteristics of high strength concrete exist.

The present study focuses on the creep and shrinkage characteristics of ultra high strength concrete. An experimental programme was set up and numerous series of specimens were tested. Tests were performed on CRC which was a mixture of cement, silica fume, sand, water, dry superplasticizer and a 6% metallic fibres by volume. The water / binder ratio was 0.16. The creep and shrinkage tests' duration were 4 to 5 months.

Concrete Under Severe Conditions: Environment and loading (Volume Two) Edited by K. Sakai, N. Banthia and O.E. Gjørv. Published in 1995 by E & FN Spon. ISBN 0 419 19860 1

2 Experimental procedures

The composition of the steel fibre reinforced concrete is given in Table 1. The steel fibre had a diameter of 0.4 mm and a length of 12 mm.

Table 1. Steel fibre concrete mixture.

Ingredient	Binder	Silica Fume	Super-plastizer	Sand <0.25 mm	Sand 0.25-1 mm	Sand 1-4 mm	Steel fibres	Water
% by mass	33.3	1.05	0.01	6.04	12.3	24.57	17.26	5.42

Table 2 summarizes the mechanical properties of CRC measured on separate test specimens : the compressive strength f'_c, on 11x22 cm cylindrical specimens and the flexural strength f'_f on 4x4x16 cm prismatic specimens. The dynamic modulus of elasticity E_{dyn} was measured with the " *Grindosonic apparatus* " on 11x22 cm cylindrical specimens.

Table 2. Characteristics of the specimens for different ages.

Age of CRC after batching (days)	f'_c (MPa)	f'_f (Mpa)	E_{dyn} (Gpa)
1	83	17.5	50.2
2	95	-	51.0
3	105	-	52.2
4	115	25.0	52.3
14	124	-	52.7
28	140	35.0	56.4

All specimens were cast in the laboratory from a single batch. They were demoulded 18 hours after batching. Creep and autogenous shrinkage specimens were covered by a composite made of epoxy, an aluminium sheet and an aluminium tape. Subsequently, all specimens were stored in a climate chamber with a controlled environment of 20°C and 50% R.H.

3 Test apparatus

The test apparatus consisted of four shrinkage frames (Fig. 1) and five creep frames. All creep and shrinkage specimens were instrumented with three calibrated strain gauges arranged at 120° on the periphery of the specimens.

For all specimens, the measurments were carried out of 40 cm. The concrete temperature was recorded with individual thermal sensors embedded in the shrinkage specimens (Fig. 1).

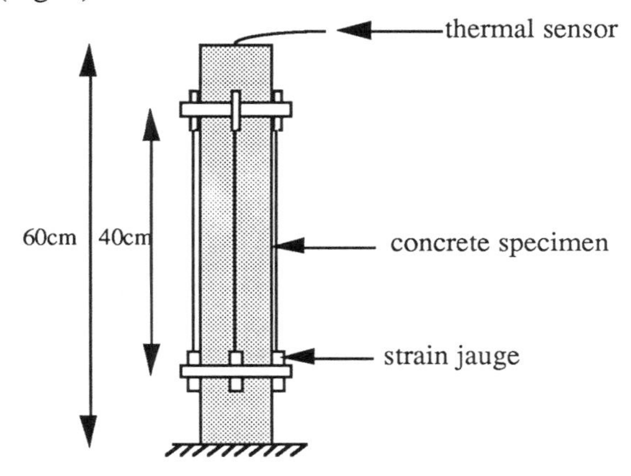

Fig. 1. Shrinkage frame

The creep frames were fitted out with hydraulic jacks connected to individual buffer tanks containing oil and nitrogen. This system guarantees a constant load during the test period, owing to the high compressibility of nitrogen. All data were logged on computer.

4 Test conditions

Three main conditions are respected for the completion of the creep and shrinkage tests :

1. constant load on the creep specimens.
2. constant temperature at 20 ± 1°C.
3. constant relative humidity at 50 ± 5%.

Loss of water in sealed specimens was reduced to a minimum. The combination of epoxy resin coat with an aluminium sheet and tape could be considered as sufficient. No weight loss was observed on control specimens.

5 Results

Measurements of shrinkage were started 22 hours after batching. All shrinkage curves were corrected for the deformation of thermal dilatation and contraction resulting from cement hydration.

5.1 Autogenous shrinkage

Autogenous shrinkage is the longitudinal strain in a specimen, without an external load, in the absence of any hygral exchange with the surrounding medium. The autogenous shrinkage was measured on two sealed specimens (R1 and R2) of 9 cm in diameter and 60 cm in height. Fig. 2 shows that the average autogenous shrinkage is approximately 500 microstrains. It should be noted that this value was already obtained after a 10-day testing period.

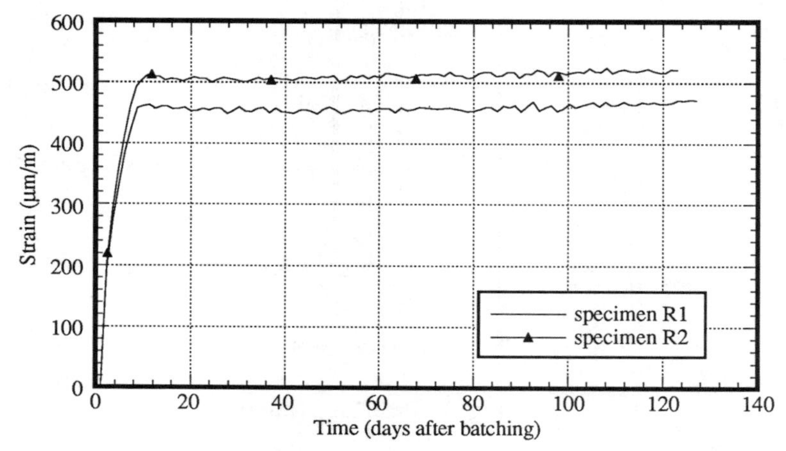

Fig. 2. Autogenous shrinkage of the 9x60 cm specimens.

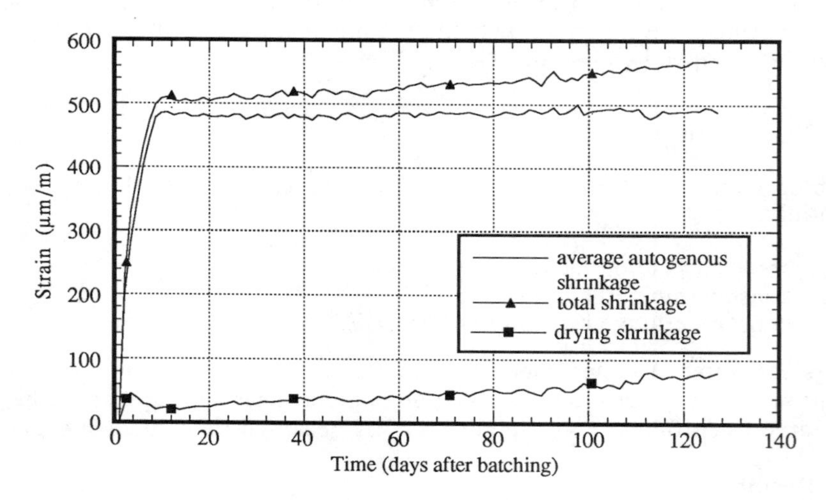

Fig. 3. Drying shrinkage of the 9x60 cm specimens.

5 .2 Drying shrinkage

Total shrinkage is the longitudinal strain in a specimen, in exchange with the surrounding medium, and drying shrinkage is the difference between total shrinkage and autogenous shrinkage. The total shrinkage was measured on two specimens of different diameters, 9 and 15 cm and 60 cm in height (R3 and R4). Fig. 3 shows that drying shrinkage remains very small compared to autogenous one. The total shrinkage of the two specimens R3 and R4 is illustrated in Fig. 4. The drying rate of the largest specimen was somewhat smaller than for the 9x60 cm specimen.

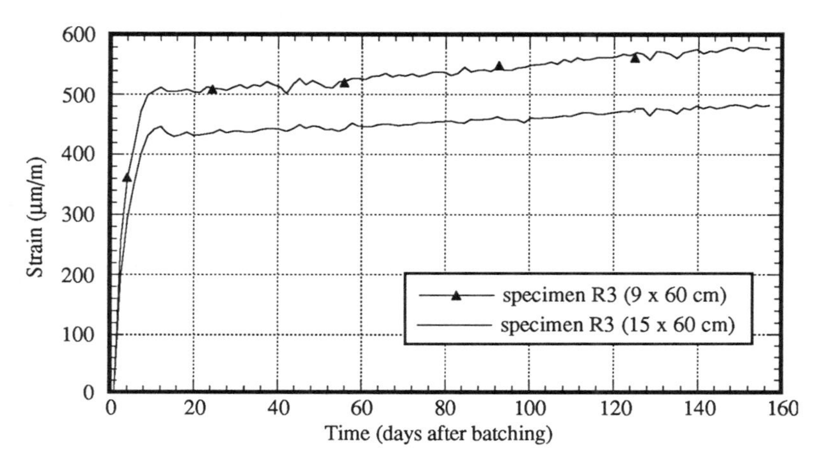

Fig. 4. Total shrinkage of the specimens R3 et R4.

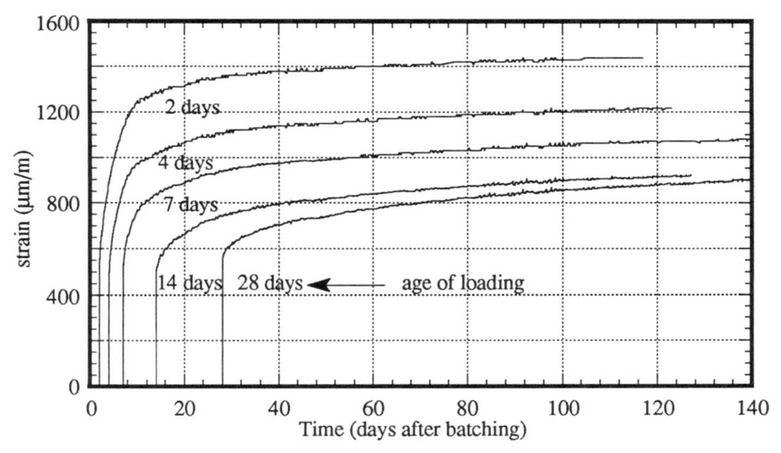

Fig. 5. Total deformation of CRC at different ages of loading.

5 .3 Creep

In this study, interest was focused on the basic creep, which is the longitudinal strain in a sealed specimen obtained under constant load minus the initial elastic strain and the shrinkage : $\varepsilon_c = \varepsilon_t - \varepsilon_i - \varepsilon_{sh}$, where ε_c is the creep strain, ε_t the total time-dependent deformation, ε_i the initial elastic strain at time of application of load t_o, and $\varepsilon_{sh;}$ the shrinkage strain since t_o.

The creep of CRC was measured on six sealed specimens loaded at 2, 4, 7, 14 and 28 days after batching. The load applied on the specimens corresponded to 20% of the CRC compressive strength at the time of loading. The total deformations measured on the five frames are given in Fig. 5.

The specific creep, J(t,to), is the term often used to compare the creep potential of concrete loaded at different ages. It is defined as the ratio of the creep strain to the applied stress : $\varepsilon_c(t) = \sigma\, J(t,t_o)$,
where σ is the total applied stress.

Fig. 6 represents the specific creep measured at different age of loading after a testing period of 120 days.

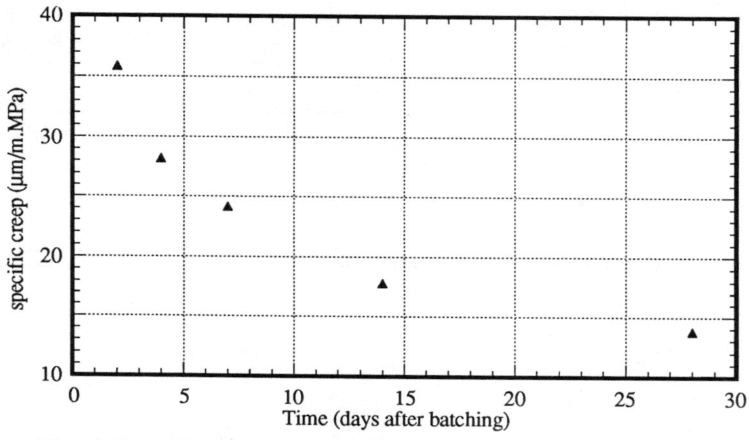

Fig. 6. Specific creep after 120 days loading

6 Discussion

A very high shrinkage was measured at an early age which actually stops 8 days after batching. This sudden break has, as far as we know, never been observed before. The cause that can be put forward was depletion of hydration water in the CRC. This phenomenon was confirmed by all shrinkage tests of our experimental investigation. Fig. 2 shows that the drying shrinkage was very small compared to the autogenous one because of the very low w/c ratio of the CRC. Most part of the water was used for the hydration of cement and only a small amount of water came out. The specific creep curve (Fig. 6.) illustrates the great importance of the age at loading on the creep strain. The specific creep of the specimen loaded the second day was more than twice that registered on the specimen loaded 28 days after batching.

7 Conclusions

The main conclusions concerning the studied steel fiber reinforced concrete under study can be expressed as follows :

1. a high shrinkage rate at an early age .
2. an autogenous shrinkage practically finished at 10 days after batching.
3. a minor drying shrinkage of CRC, while it can be equal to the autogenous shrinkage for the silica fume concrete ($60 < f'_{c28} < 100$ MPa) [1].
4. a creep strongly affected by the age of loading.

These results could be of great importance for practical applications of CRC and can allow the extrapolation of specimen results to structure level because there is a negligible scale effect.

8 References

1. Auperin, M., Richard, P., De Larrard, F. and Acker, P. (1989) Retrait et fluage de bétons à hautes performances. Influence de l'âge au chargement. *Annales de l'Institut technique du bâtiment et des travaux publics,* No. 474, pp. 49–75.
2. Le Roy, R. and De Larrard, F. (1993) Creep and shrinkage of high-performance concrete: the LCPC experience. Fifth International RILEM Symposium on Creep and Shrinkage of Concrete, Barcelona, pp. 499–505.
3. Loukili, A., Sieffert, J.G. and Richard, P. (1994) Réponses différées d'un mortier à très hautes performances renforcé de fibres d'acier. *Douzièmes rencontres universitaires de Génie Civil,* pp. 165–68.

153 THERMAL GRADIENTS IN CONCRETE ELEMENTS SUBJECTED TO COMPRESSIVE LOADS

A.K. AGGARWAL
PNG University of Technology, Lae, Papua New Guinea
N.D. VITHARANA
Department of Water Resources, NSW, Australia

Abstract
The paper reports the results of an experimental investigation carried out on reinforced concrete wall elements subjected to temperature variations. In the experimental programme, six concrete panels of varying thickness were exposed to uniform heat flow and temperature variations in them were monitored continuously by thermo-couples and data recorder both during the rise and drop of temperature. In three panels, axial compressive loads were also applied in addition to uniform heat and effect of load on the heat flow pattern was observed. The results indicate non-linearity of temperature gradient both during heating and cooling of test specimens. The time-temperature plots also show that the cooling gradients are much milder than those observed during the heating process.
Keywords: Concrete slabs, compressive loads, heat conductivity, thermal gradients.

1 Introduction

An accurate assessment of imposed loads is important for the design of safe and durable structures. Most structures are subjected to two kinds of loads (a) those arising from applied forces (dead and live loads) and (b) those generated by strains either on account of temperature variations within a member or due to their shrinkage. While the rules and procedures for the evaluation of dead and live load stresses on structures are well documented, little is known about strain induced stresses, their distribution and the procedure of incorporating them into design. In some structures, particularly those made of concrete, the strain induced stresses can be significantly large and their inaccurate assessment could result in overstressing of members. The mechanism of stresses generated by strain induced loads is not explicit as these depend on several factors like - stiffness of member, material properties and even on envoirnmental exposure conditions.

The importance of thermal stresses in bridge decks was realised by Reynold and Emanuel [1] in 1974, and in their conclusions, they stressed the need for a "continuous development of semi-empirical formulas for design". Conventionally, heat flow in a medium is a complex three-dimensional phenomenon and in most situations simplified assumptions have to be made to estimate thermal stresses. In

Concrete Under Severe Conditions: Environment and loading (Volume Two) Edited by K. Sakai, N. Banthia and O.E. Gjørv. Published in 1995 by E & FN Spon. ISBN 0 419 19860 1

1975, Hunt and Cooke [2] suggested a method for calculating thermal stresses in bridges, using a one-dimensional unsteady heat flow equation. Their theoretical stress predictions agreed reasonably well with the experimental values, thus suggesting that horizontal heat movements are generally small and can be ignored in most cases.

During the last two decades, several methods [3,4,5] have been suggested for the evaluation of temperature induced stresses in concrete but the research has mainly focussed on bridge decks subjected to temperature variations resulting from envoirnmental exposure. In recent years however, some research has been carried out to study the effects of imposed stresses on the chloride ion diffusion and water permeability in concrete members. It is well known that thermal properties of concrete are highly dependent on the mineralogical characteristics of the constituents, their composition and other factors such as moisture content. A literature review has indicated that little published material is available on the heat transmission characteristics of reinforced concrete elements under applied loads. Accordingly, an experimental program was initiated at the Papua New Guinea University of Technology to study the heat transfer properties of reinforced concrete elements both with and without the application of compressive loads. In the initial phase, reinforced concrete elements of varying wall thickness have been exposed to uniform heat flow, both with and without compressive loads. In a parallel study, a finite element model of the panel is being developed to determine temperature gradients theoretically. Unfortunately, at this stage experimental and theoretical temperatures cannot be compared because significant changes to the model are necessary particularly in view of the reinforcements in the panels. In the second phase, it is proposed to vary the concrete strength and reinforcement quantity in elements and determine their effect on heat transfer pattern.

2 Heat transfer in concrete

The heat transfer through an isotropic solid under a three-dimensional heat flow condition is governed by:

$$k \frac{\delta^2 T}{\delta x^2} + \frac{\delta^2 T}{\delta y^2} + \frac{\delta^2 T}{\delta z^2} = pc \frac{\delta T}{\delta t} \tag{1}$$

where k = is the coefficient of thermal conductivity of concrete, c = specific heat of concrete, p = mass density, T = temperature in the medium at a distance (x,y,z) at time 't'.

For walls in contact with air on one face and exposed to uniform heat flow conditions over the other, Eq. 1 can be simplified to a one-dimensional heat flow problem and can be represented as:

$$k \frac{d^2T}{d x^2} = \frac{dT}{dt} pc \tag{2}$$

where x is measured perpendicular to the wall surface.

It is generally difficult to obtain a closed-form solution to Eq. 2 particularly when the boundary conditions are not of a standard type [6]. The magnitude and the distribution of temperatures in a medium largely depend on the boundary conditions and therefore it is necessary to have an accurate assessment of boundary conditions, if any reasonable stress distribution has to be predicted.

Several methods for solving eq. 2 have been suggested but a technique which is most commonly used is based on the finite difference method [7]. With this method, problems with changes in thermal properties of materials and multi-layered systems can also be analysed. It may be noted that eq. 2 can also be replaced by a heat energy balance relation for each node lying along the thickness of the section and this can be used to analyse complex multi-layered heat flow problems.

2.1 Heat equilibrium at exposed surface

Under steady state condition, an equation of heat energy balance for an exposed surface can be written as:- (Ref Fig. 1)

$$q_s - q_c - q_x - q_r = 0.0 \qquad (3)$$

in which q_s = heat flux from an incident radiation, q_c = heat loss by convection and radiation to the atmosphere, q_x = heat flux conducted from surface into body, and q_r = heat used in raising the temperature of the surface. It should be noted that the term q_r has been ignored by some researchers [4] and under severe unsteady conditions, this could lead to significant errors in the near surface temperatures.

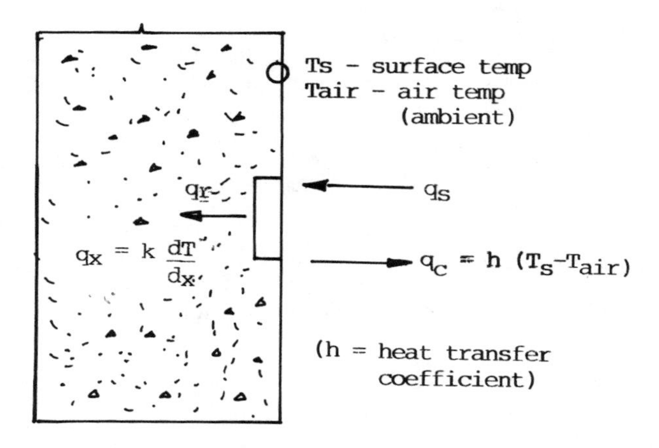

Fig 1. Heat energy equilibrium on an exposed surface.

3 Experimental program

3.1 Test unit parameters

To determine the temperature gradients which develop in reinforced concrete elements with uniform heat flow, a total of six panels were tested as a part of the experimental programme given in Table 1. All panels were made using 20 MPa nominal strength concrete and were dimensioned to size 700 mm x 900 mm. Two panels each of 100 mm, 150 mm and 200 mm thickness were cast with every panel having a single layer of reinforcement placed in both directions. All specimens were subjected to heat along one face while the other face was exposed to ambient conditions. Three specimens were subjected to thermal stresses arising from temperature variations only, while the other three specimens had externally applied axial compressive stress in addition to the thermal stresses. Two separate compressive stresses were applied to each test specimen and temperature variations along the cross-section were observed under increasing stress.

Table 1. Experimental programme

Specimen number	Panel thickness (mm)	Magnitude of applied compressive stress (MPa)
SL1	100	0.0
SL2	150	0.0
SL3	200	0.0
SL4	100	2.14 & 2.85
SL5	150	1.43 & 1.90
SL6	200	1.07 & 1.43

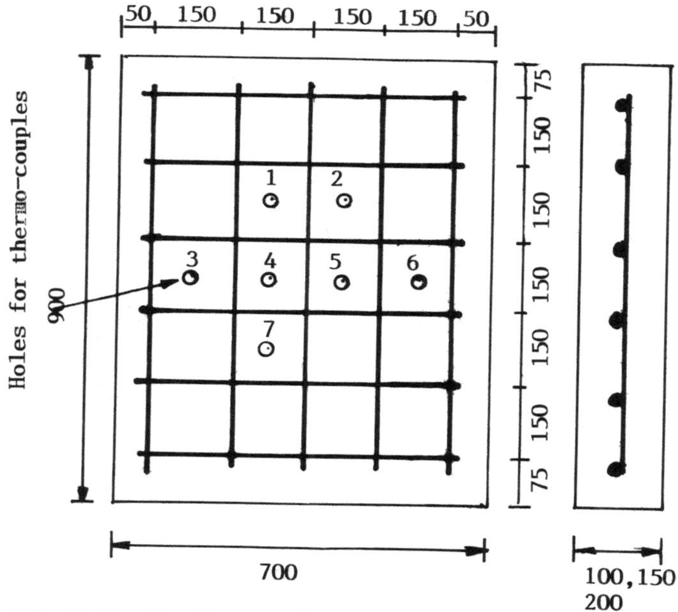

Fig. 2. Details of test specimen

Table 2. Depths of temperature measurement points

Slab thickness (mm)	Depth of thermo-couples (mm) (1)	(2)	(3)	(4)	(5)	(6)	(7)
100	11.6	16.5	57.1	46.0	56.7	42.9	65.6
150	84.6	56.2	99.6	74.5	97.5	75.4	115.2
200	19.0	42.0	56.0	85.0	119.0	167.0	163.0

In each panel, 12 mm diameter reinforcement was placed centrally in both directions according to the arrangement shown in Fig. 2. Seven holes of 6mm diameter and varying depth were cast to insert thermo-couples. The observation points for each panel are shown in Fig.2 while depths of temperature recording points (as measured from the heated surface) for each slab are given in Table 2. To avoid the loss of heat from the observation points, 6mm holes were plugged after inserting thermo-couples.

3.2 Testing arrangement

The heat to all specimens was applied through a 1.2 m long metal duct. A heat blower fan of 3000 watts capacity was placed at one end of the duct and the test specimen at the other. The test specimens were mounted vertically on a base frame. An arrangement used for testing specimens and applying compressive load is shown in a photograph - Fig. 3. The axis of heat flow was directed towards the centre of each panel and all edges between the metal duct and test specimens were sealed with epoxy. The heat was applied to the specimens till the temperature recorded by the thermo-couple nearest to the exposed surface showed an increase of 30 degrees centigrade approximately. On an average it took about 6 to 8 hours for the temperature to rise in different specimens and about 8 to 10 hours for it to drop.

Fig. 3. Typical concrete panel under test.

3.3 Material properties

While casting test panels, 3 standard size concrete cylinders were also made to determine the actual strength of concrete. An average 7 days and 28 days compressive strength was observed to be 17.5 MPa and 28.5 MPa respectively. The panels were cured for 28 days and were tested on an average 40 days after casting.

3.4 Loading of test specimens

Three test specimens were subjected to axial compressive loads using a Mohr & Fedderhaff loading machine. Each specimen was loaded to two discrete loads (150 kN and 200 kN) and the load was maintained both during heating and cooling of test panels.

3.5 Instrumentation
A data recorder coupled to a computer was used to measure temperature changes in the panels. The temperatures were recorded every minute from eight thermocouples. Of these, seven were inserted to varying depths in the test panels and the eighth was used to measure the temperature of the surface exposed to heat.

4 Measured Temperature Gradients

4.1 Panels without compressive load
The measured temperature distributions across the wall thickness for 100mm thick panel (SL1) for six different time intervals are shown in Fig. 4. The panel was heated for 5 hours and 45 minutes and during this period, the surface temperature increased from an ambient temperature of 23 degrees C to 66.2 degrees. The observations were recorded for 11 hours and at the end, an average temperature of 33 degrees was recorded over the entire cross-section of the panel.

It is observed from Fig. 4 that as the distance from the exposed surface increases, temperature drops, but the fall in temperature is not linear. Unfortunately, observations were recorded only to a depth of 65.6 mm. To have a complete temperature profile, it is prudent to record observations for the entire depth of the section.

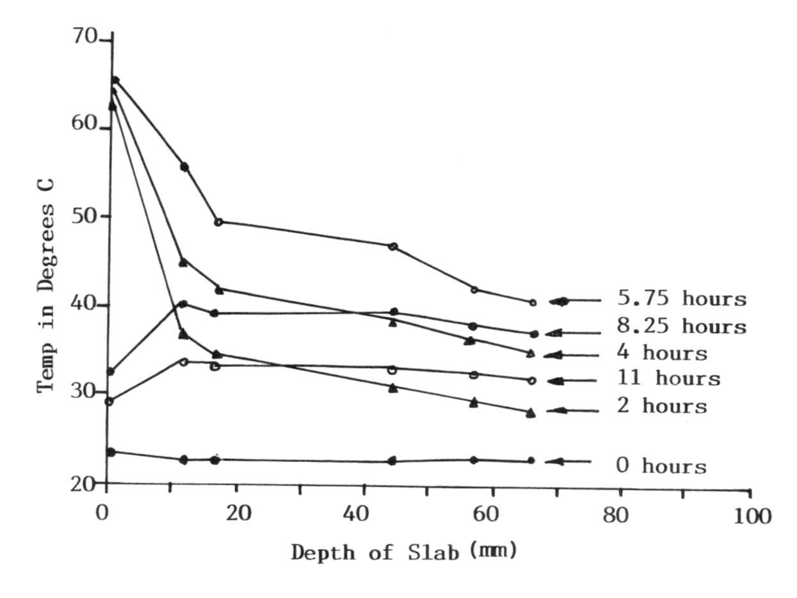

Fig. 4. Variation of temperature with wall thickness for panel SL1.

For test specimen SL2, the temperature distribution across the cross-section is seen to be different from the previous panel - (Fig. 5). It is observed from the plot that after a sharp drop of temperature from the exposed surface, the fall continues to an approximate depth of 75mm. Then the temperature rises again by a few degrees over a small depth of 25 mm, before gradually reducing to the atmospheric conditions at the other end. The most probable and obvious reason for this behaviour could be assigned to the position of the reinforcement in the panel. The steel reinforcing bars are good absorbers and radiators of heat and are therefore likely to increase temperatures locally. It should be noted that reinforcement in all panels was placed centrally and the local rise in temperature coincides with the location of steel.

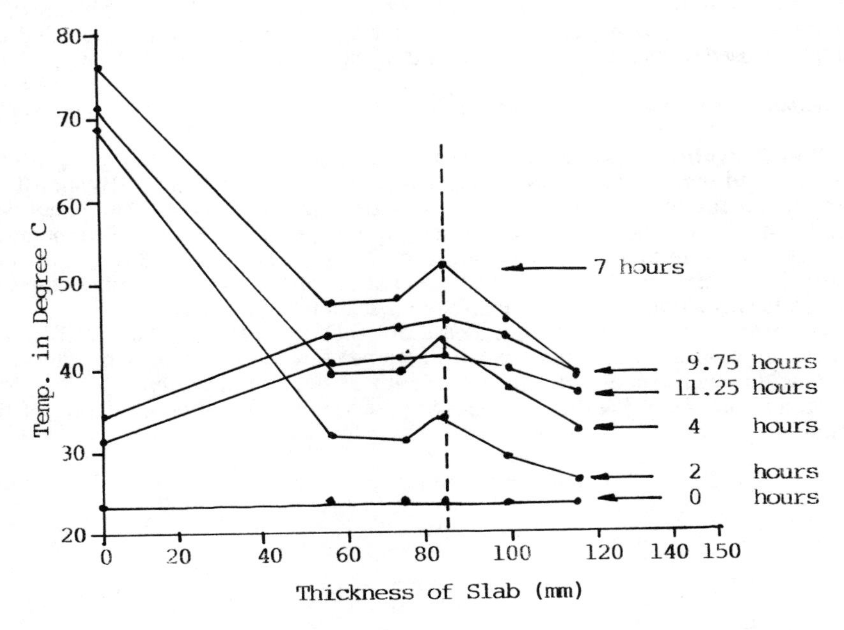

Fig. 5. Variation of temperature with wall thickness for panel SL2.

From the plots of temperature versus cross-sectional depth for specimen SL3 (Fig 6), the behaviour of the panel appeared almost similar to that observed for

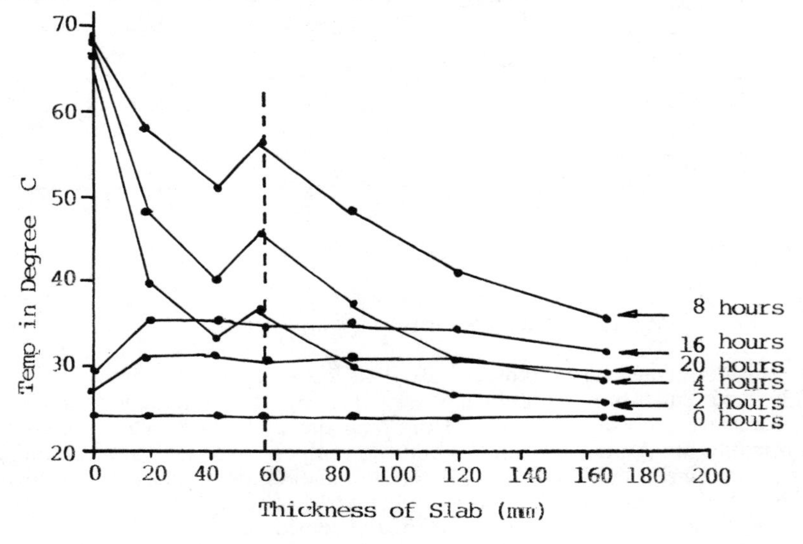

Fig. 6. Variation of temperature with wall thickness for SL3.

for specimen SL2, but the rise in temperature after an initial drop, did not coincide with the position of steel. The fall in temperature was observed only 40mm from the exposed surface as against an expected value around 100mm. As no reasonable explanation could be given to this behaviour, it was decided to repeat the experiment. However, on checking with a reinforcement locating meter, the bars were found to be approximately 65 mm from the heated surface and not 100mm as expected. It is likely that 75mm bar chairs may have been used for supporting the reinforcement.

Not unexpectedly, this specimen took much longer for the temperature to rise and drop. The panel was exposed to heat for just over 8 hours and it took nearly 14 hours for the temperature to drop. The variation of temperature along the depth of the panel for six different time intervals during heating and cooling operation are shown in Fig. 6.

4.2 Panels with axial compressive load

Three panels (SL4, SL5 and SL6) were subjected to axial compressive loads in addition to the uniform heat flow. Two different loads were applied to the panels on separate days. On the first day, a load of 150 kN was applied and after the specimens had cooled for 48 hours, they were reloaded to an increased compressive load of 200 kN. The average stresses resulting due to loads on different panels are given in Table 1.

To study the effect of compressive load on temperature gradients in concrete elements, a combined plot of specimens SL1 and SL4 were obtained (Fig 7).

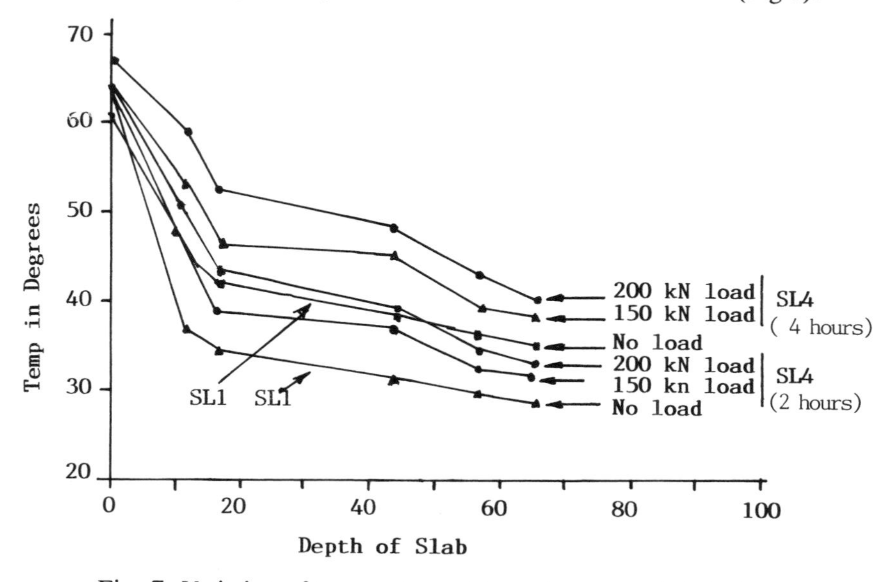

Fig. 7. Variation of temperature with thickness for SL1 & SL4 (Effect of compressive load on temperature gradient)

Both specimens were identical in construction and were subjected to similar heat exposure conditions. Two sets of plots are shown in Fig. 7 - (i) 2 hours and (ii) 4 hours after the heat was switched on. It is observed from the plots of specimen SL1 (no load cases), that the temperature gradient is almost linear after a cross-sectional depth of 20mm. However in the case of specimen SL4 (with compressive loads), plots are somewhat non-linear. Moreover in this specimen, a local increase in temperature is observed at 50 mm from the exposed surface - the location of the reinforcing bars. The temperature profile after 4 hours was almost similar to the 2 hours gradients, except that the temperatures in the former were higher.

Almost similar behaviour was observed when comparisons were drawn between specimens SL2 and SL5, and SL3 and SL6.

4.3 Time - temperature plots

Another objective of this experimental programme was to ascertain the variation in temperature with time, in panels with different wall thickness and with the application of compressive loads. As an indication, a plot of temperature versus time was obtained for specimen SL3 -Fig. 8.

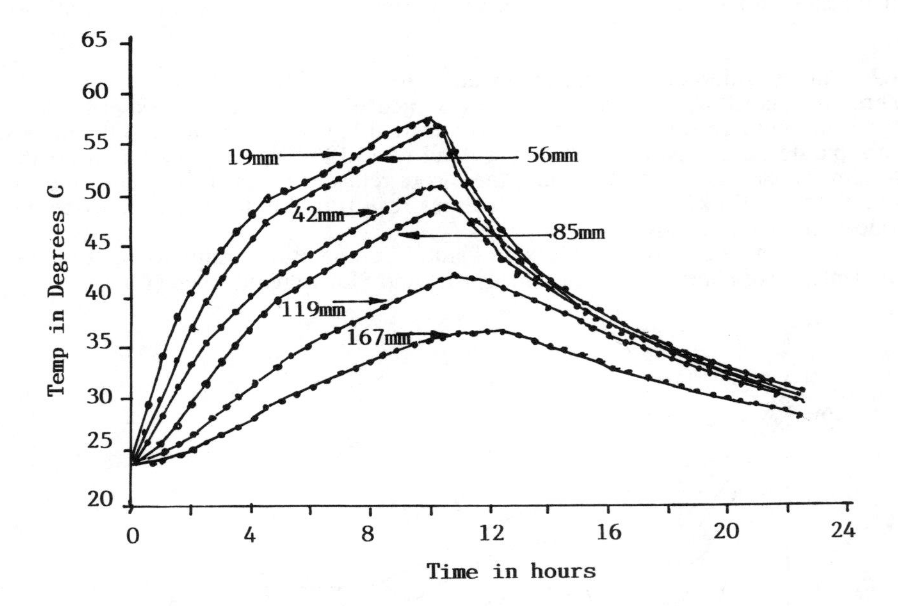

Fig. 8. Variation of temperature with time for specimen SL3.

It is observed from the plot, that the thermo-couple measuring the temperature at a depth of 19mm from the exposed surface recorded the highest temperature (56 degrees C) and the one at a distance of 167mm, records the lowest - 36 degress C. During the heating of the panel, temperatures recorded by six different thermo-couples indicate that there is a steady increase at all depths, but when the panel is cooled, exposed surface cools much faster than the inner core and all the plots tend to merge together soon after the heat is switched off. The plot also shows that the peak temperature was reached approximately 10 hours and 30 minutes after heating was commenced. Comparing the behaviour of specimens SL3 and SL6 (SL6 with 200 kN load), it is observed that the peak temperature of 56 degrees C in specimen SL6 was attained in 7 hours and 30 minutes only i.e. almost 3 hours earlier than specimen SL3. If the areas under the temperature -time plots for the two specimens are compared, they indicated that a lower volume of heat energy was necessary for specimen SL6. This behaviour could partly be explained in terms of the evaporation of moisture from the inner core of concrete, when the same panel was heated with a compressive load 150 kN. A detailed investigation is in progress to determine the cause of accelerated rise in temperature for panels with compressive loads.

5 Conclusions

From the two sets of experiments done on reinforced concrete panels, the following conclusions can be drawn:

1. The temperature gradients across the cross-section, are non-linear both during the heating and cooling of panels. The temperatures at the exposed surface are generally very high and are not true indicators of the gradients immediately below the surface.
2. The temperature differentials through the element thickness largely depend on their thickness and the results indicate that they could be as high as 30 degrees C.
3. In reinforced concrete, temperature variations across the cross-section depend on the position of reinforcement bars. A local increase in temperature is observed near steel bars.
4. Under compressive loads, the time required to raise the temperature of panels was observed to be lower than panels without load. The temperature-time plots also indicate the same behaviour, but the experimental evidence is small.

Finally, the conclusions drawn herein are from a small number of specimens and must be regarded as suggestive rather than definitive.

6 Acknowledgements

The financial support given by the PNG University of Technology for this project is gratefully acknowledged. Words of appreciation are also extended to the technical and secretarial staff in the Department of Civil Engineering.

7 References

1. Reynolds, J.C., and Emanuel, J.H. (1974) Thermal stresses and movements in bridges, *Journal of the Structural Division, ASCE,* Vol. 100, No. ST1, Proc. Paper 10275, pp. 63–78.
2. Hunt, B. and Cooke, N. (1975) Thermal calculations for bridge design, *Journal of the Structural Division, ASCE,* Vol. 101, No. ST9, Proc. Paper 11545, pp. 1163–81.
3. Priestley, N.J. (1978) Design of concrete bridges for temperature gradients, *Journal of the American Concrete Institute,* No. 75, pp. 209–17.
4. Thurston, S.J. (1978) *Thermal Stresses in Concrete Structures,* Ph.D. Thesis, University of Canterbury, New Zealand.
5. Vitharana, N.D. (1991) *Strain–induced Stressing of Concrete Storage Tanks,* Ph.D. Thesis, University of Canterbury, New Zealand.
6. Billington, N.S. (1952) *Thermal Properties of Buildings,* Cleaver–Hume Press Ltd., London.
7. Emerson, M. (1973) *The Calculation of the Distribution of Temperature in Bridges,* TRRL Report LR 561, Transport and Rod Research Laboratory, Crowthorne, Berkshire, U.K.

154 EARLY-AGE BEHAVIOUR OF CONCRETE SECTIONS UNDER STRAIN-INDUCED LOADINGS

N. VITHARANA
Department of Water Resources, Sydney, Australia
K. SAKAI
Hokkaido Development Bureau, Sapporo, Japan

Abstract
Concrete structures would be subjected to significant stress conditions due to the heat of hydration effects as well as autogeneous and drying shrinkage. While much attention is usually given to service and ultimate loading conditions, early-age distresses could result in severe adverse effects. The importance of rational evaluation of early-age stresses is not recognised in many current codes and design practices. The evaluation of early-age stresses itself is complex due to the interactive involvement of various age- and temperature-dependent parameters, in addition to ambient exposure conditions.

This paper presents the results of the theoretical part of an ongoing investigation on the early-age behaviour of concrete wall sections under heat of hydration effects and autogeneous shrinkage in particular. The paper critically reviews the validity of some of the conventional procedures adopted in practice in the prediction of early-age temperature and stress development, and also presents the significance of ambient conditions & creep on the predicted temperature stresses. It is shown that the cursory approaches could lead to inaccurate predictions of the early-age behaviour being either conservative or unconservative under different conditions.
Keywords: Autogeneous, concrete, cracking, creep, early-age, hydration, shrinkage, strain-induced, stresses.

1 Introduction

Concrete structures are subjected to strain-induced stresses, resulting from either temperature changes or moisture gain or loss (swelling and shrinkage) when the free strains are restrained. Strain-induced stresses are quite different from stresses caused by applied loadings as strain-induced stresses directly depend on the stiffness.

Concrete Under Severe Conditions: Environment and loading (Volume Two) Edited by K. Sakai, N. Banthia and O.E. Gjørv. Published in 1995 by E & FN Spon. ISBN 0 419 19860 1

Although much attention has been given to the design and detailing of structures for service and ultimate loading conditions, severe stress conditions could occur at early age due to heat of hydration and also autogeneous shrinkage possible with concretes of low water/cement ratios. Unless preventive measures are taken, strain-induced stresses could cause early-age cracking of concrete when they exceed the current concrete tensile strength which is a fraction of that of the well-hardened concrete.

Excessive cracking could lead to the corrosion of reinforcement steel and unsightly appearance. The other adverse effects would be the increased permeability, low abrasion resistance, liquid tightness and also structural aspects such as reduced strength and stiffness, stability and stress redistribution. The ignorance or irrational treatment of early-age stresses could incur heavy costs [eg, 1,2] requiring extensive rehabilitation work or even the total early replacement of the structure.

Enormous research has been carried out so far by researchers on the early-age temperature predictions [eg, 3,4]. The determination of temperature development itself would not however estimate the tendency to cracking which basically depends on the ratio of the induced stresses to the current tensile strength. It is also surprising to note that the recent sophisticated analysis models [5] have even neglected the matters of fundamental importance such as creep which is responsible for stress relaxation and reversal.

In this paper, some of the results of an investigation on the early-age behaviour of concrete structures, wall sections under heat of hydration effects in particular, are presented.

2 Heat of hydration

Hardening process (or hydration) of concrete is exothermic. The rate and the amount of heat produced would significantly modify the temperature and stress developments. The major factors affecting the rate and amount of heat generation are; unit cement content, cement type, fineness of cement, placing temperature, process temperature (the temperature within the concrete) etc. The process temperature in turn depends on hydration characteristics in addition to the heat transfer characteristics of the section and exposure conditions.

2.1 Hydration models

The heat of hydration models for adiabatic conditions are available [2,6], incorporating cement type and placing temperature. These models are widely used in conventional heat-transfer analysis and in estimating the total heat generating capacity of concrete. In adiabatic conditions, the heat already produced by the hydration accelerates the hydration process as no heat is lost to or gained from the outside. Adiabatic conditions would exist only in the interior of massive sections.

Being hydration is an thermally activated process, the process temperature would affect the rate of the reaction. The hydration process should therefore include the temperature/time history which varies within the section, depending on the hydration characteristics and also the heat transfer characteristics of the section. Every point will thus have a different temperature/time history and heat generating characteristics. This phenomenon would couple the heat of hydration and the heat transfer.

Rastrup [3] produced a heat generation model suitable for varying temperature environments using an equivalent time approach. This model was based on the heat generated by cement mortar samples under isothermal conditions. One draw-back with this model is considered to be the use of small samples unrepresentative of concrete. Several researchers (eg., Branco [5]) successfully used the Rastrup model to predict temperature rise of concrete structures. Accuracy can be improved by fine-tuning the model parameters, based on the measured temperatures in large concrete blocks.

Harada [4] has recently proposed a model primarily based on a relationship between the instantaneous activation energy and the accumulative heat, ie, without the time as a direct variable. The rather sensitive relationship was however derived from adiabatic conditions which have particular characteristics.

Hereinafter, the varying temperature environment model (Rastrup's) is referred to as the " non-adiabatic " model.

2.1.1 Adiabatic model

The adiabatic temperature rise of concrete for a given unit cement content S (kg/m^3) can be expressed;

$$T_{(t)} = T_{(\infty)} (1 - e^{-\sigma t}) \quad (1)$$

where $T_{(t)}$ is the temperature rise (℃) at time t (days), $T_{(\infty)}$ is the ultimate adiabatic temperature rise (℃) directly proportional to S, and σ is a thermal constant generally dependent only of the concrete placing temperature T_o for a given cement type. According to ACI[2], $T_{(\infty)}$ is independent of T_o and σ is independent of S. Typical values for $T_{(\infty)}$ and are given in JCI[6] for different cement types and compositions. According to JCI both $T_{(\infty)}$ and σ depend on both T_o and S, thus contradicting the conventional recommendations [2].

If the specific heat of concrete is c (J/kg/℃) and the mass density of concrete is p (kg/m^3), then the rate of heat generation $\dot{Q}$ for a given S can be obtained as ($J/m^3/s$);

$$\dot{Q} = c\, p\, T_{(\infty)}\, \sigma\, e^{-\sigma t} \quad (2)$$

2.1.2 Rastrup model

The heat of hydration at reaction time t (kJ/kg of cement) at a reference process temperature T_r is expressed as;

$$Q_{(t)} = A + B \exp\{-m[t_e]^{-n}\} \quad (3)$$

$$t_e = 2^{0.1(T - Tr)}\, dt \quad \text{or} \quad t_e = \Sigma\, 2^{0.1(T - Tr)}\, Dt$$

where A, B, m and n are constants dependent of the cement type and mix proportion. Therefore, the value of m obviously depends on T_r. The actual reaction time t is related to the equivalent time t_e, considering varying process temperature T obtained from the heat-transfer analysis of the section. The rate of heat generated per unit volume of concrete $\dot{Q}$ (kJ/m^3 of concrete/day) is then given by;

$$\dot{Q} = S\, [Q_{(te + dte)} - Q_{(te)}]/Dt \quad (4)$$

To illustrate the use of the Rastrup model for different temperature environments, adiabatic curves, Fig. 1, are generated for different concrete placing temperatures T_o. In this way, model parameters can also be slightly adjusted based on measured temperatures in large concrete blocks, if necessary. The parameters from Branco's tests [5] were used at a reference temperature T_r of 20 °C, and S was taken as 450 kg/m^3. The values of the parameters are; A=12.56, B = 328.7 (both in kJ/kg of cement), n = .42 and m = -1.029 (day$^{.42}$).

In developing the adiabatic curves to be used in the current study, the hydration is followed in time steps with the initial temperature equal to the concrete placing temperature T_o. The heat generated in each time step is converted to an equivalent temperature increase using c (equal to 1100 J/kg/°C) and add to the accumulative T before proceeding to the next time step. The Rastrup curves at isothermal conditions at T_o are also shown in Fig. 1. The adiabatic curves from JCI[6] are shown with the same $T_{(\infty)}$ for comparison. Figure 1 shows that significant differences exist between hydration characteristics under adiabatic and isothermal conditions and also between code recommended and actual adiabatic models.

3 Calculation of early-age temperature development

The temperature state depends on the heat of hydration characteristics as well as environmental exposure conditions such as solar radiation, ambient temperature etc. The environmental interaction would influence the overall temperature state of thin sections and near surface temperatures in thick sections.

3.1 Interior

The general one-dimensional heat transfer within the medium (interior) is described by;

$$k \frac{d^2T}{d^2x} + \dot{Q} = p\, c \frac{dT}{dt} \tag{5}$$

where k is the coefficient of thermal conductivity of concrete (J/m/s/ C), p is the mass density (kg/m^3), c is the specific heat (J/kg/°C), and T is the temperature at time t at point x.

The solution to Eq. 5 can be obtained from numerical techniques such as Finite difference or Finite element methods with known boundary conditions [7]. Finite difference technique was adopted in the current study.

3.2 Outside surface

By conservation of the heat energy, the equation for heat flux at the surface can be expressed as [7];

$$q_s - q_c - q_x - q_r = 0.0 \tag{6}$$

where q_s is the heat flux absorbed from the incident solar radiation, q_c is the net heat loss from convection and irradiation to the ambient, q_x is the heat flux conducted from surface to the body and q_r is the heat stored in raising the temperature of the surface layer.

Heat flux absorbed from the incident radiation is given by the product a_s S where a_s is the surface absorptivity factor and S is the incident radiation per unit surface area (W/m^2). It can be assumed that q_c is given by;

$$q_c = h\,(T_s - T_a) \tag{7}$$

where h is the surface heat transfer coefficient, T_s is the surface temperature and T_a is the ambient air temperature. The value of h largely depends on the speed of air movement past the wall surface [8]. The value of q_x is given by k dT/dx (where dT/dx is the temperature gradient at the surface).

In conventional guide-lines [eg, 6], the value of h is modified (h_m) to account for insulation and formwork, based on an implicit assumption of steady-state conditions, Eq. 8;

$$h_m = [1/h + \Sigma\,(dI/kI)]^{-1} \tag{8}$$

where dI and kI are the thickness and thermal conductivity of the insulation materials on the wall surface. To achieve steady-state conditions, the insulation should have a large thermal diffusivity and a smaller thickness. Therefore, when a relatively thick insulation is provided, a multi-layer analysis is recommended to avoid errors in the predicted temperatures, particularly under diurnal solar radiation.

4 Time and temperature-development of material properties

The development of material properties such as compressive strength f_c, tensile strength f_t, and Young's modulus of elasticity E_c, is strongly time (age)-dependent, particularly at early age. Due to the activation energy of hydration, the above properties are significantly temperature-dependent also.

In any rational attempt to estimate the tendency to cracking, the age and temperature dependency of material properties should be considered. The age-dependency of concrete compressive strength at a standard curing temperature of 20 °C is given by ACI[9];

$$f_c = t/(A + B\,t)\,f_{c(28)} = \beta\, f_{c(28)} \tag{9}$$

where A & B are material constants; typically A = 4.0, B = 0.857. It is generally known that the values of f_t and E_c are both proportional to the square root of f_c. Then;

$$f_t = \beta^{0.5}\, f_{t(28)} \tag{10}$$

$$E_c = \beta^{0.5}\, E_{c(28)} \tag{11}$$

The temperature-dependency of above properties can be incorporated by using an equivalent age t_e (similar to a maturity function) based on activation energy of hydration [10], Eq. 12;

$$t_e = \Sigma \exp\{-4000./(273+T) + 13.65\} Dt \tag{12}$$

where t_e is the equivalent age to be used in Eq. 9 and T is the temperature within the actual time period Dt. Equation 12 implies that the material properties within the concrete section would vary depending on the value of t_e.

5 Mechanism of restrained stress development and cracking

The cracking at early age could be caused either by internal restraining effects due to non-linear temperature distributions or external restraints such as base-fixity in concrete walls. Under fully-restrained conditions, the mechanism of early-age thermal cracking is primarily due to the increased E_c value during the cooling phase.

During temperature rise due to hydration, the concrete is in a plastic state with a very low E_c value. The resulting fully-restrained stresses are compressive, generally in the order of 1-2 MPa within the first day or two. Once the hydration retards after the peak period, the temperature starts to drop (or cooling) at a rate depending on the heat-transfer boundary conditions such as with or without formwork. The temperature decrease takes place under an increased E_c due to aging, and this results in a nett tensile stress. High early-age creep will have a pronounced effect on stress relaxation. If the induced tensile stress exceeds the current tensile strength, cracking would occur. The mechanism of early-age stress development is schematically shown in Fig. 2 comparing with the situation if E_c were constant in which case cracking does not occur.

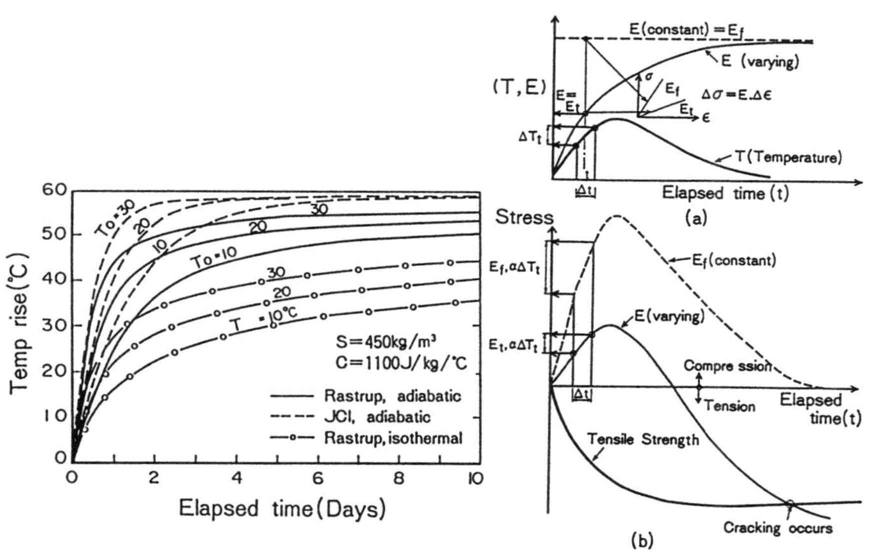

Fig. 1. Heat of hydration models.

Fig. 2. Early-age stress development.

(a) Temperature and Young's modulus

(b) Stress and strength

In the current study, an incremental analysis in time-steps was adopted, and linear elastic material behaviour was assumed. The incremental fully-restrained thermal stress $Df_{(t)}$ (compression negative) developed during time t and t+dt is given by;

$$Df_{(t)} = - E_{c(t+0.5dt)} \; \alpha \, [T_{(t+dt)} - T_{(t)}] \tag{13}$$

where T is the temperature and α is the coefficient of thermal expansion. Although Eq. 13 does not include the creep effect, this is described and incorporated in Section 6 below. The total stress at time t, $f_{(t)}$, is given by the algebraic sum of $Df_{(t)}$; $f_{(t)} = \Sigma \, Df_{(t)}$ up to time t. In calculating E_c, the equivalent time t_e from Eq. 12 should be used.

6 Creep and shrinkage effects

Shrinkage strains are analogous to temperature-induced strains. Concrete creeps under non-transient loadings, relaxing the strain-induced stresses. Creep characteristics depend on various parameters such as age at loading, composition, volume/surface ratio, moisture state, temperature etc. The available data on early-age creep is very limited and inconsistent. Even available limited data can not be used for a comprehensive study as other material properties for the same concrete are not usually available.

In the current study, the CEB[10] formulations intended for rather matured concrete are assumed to be valid from zero-age. The creep behaviour is complex even for matured concrete. With the available models [9,10], only mean creep behaviour of the section is possible based on average humidity, ie, point-wise creep considering moisture content distribution/history is not possible. The total strain $\varepsilon_{(t)}$ at age t caused by a constant stress σ applied at age $t = t_o$ is given by;

$$\varepsilon_{(t)} = [1 + \phi_{(t,to)}] \; \sigma / E_{c(to)} \tag{14}$$

where $E_{c(to)}$ is Young's modulus of elasticity at age $t = t_o$ and $\phi_{(t,to)}$ is the creep factor. Usually $\phi_{(t,to)}$ is expressed as; $\phi_{(t,to)} = \phi_o \, \beta_{c(t-to)}$ in which ϕ_o is the notional creep coefficient (ultimate creep value when time since loading $(t-t_o)$ approaches infinity) and β_c describes the time development of creep since loading at $t = t_o$. Values of ϕ_o and β_c can be obtained [10] incorporating various parameters such as temperature, humidity, member size, age at loading, concrete composition for both sealed and unsealed (basic and drying creep) conditions. Due to space limitations, those are not reproduced here, except β_c;

$$\beta_{c(t-to)} = \{(t-t_o)/(a+t-t_o)\}^{0.3} \tag{15}$$

where a is a constant depending on member size, relative humidity and current temperature. The incremental creep strain $d\varepsilon_{cr,to}$ occurring within time increment Dt due to a unit stress applied at age t_o is therefore given by;

$$d\varepsilon_{cr,to} = \phi_o \{ \beta_{c(t+dt-to)} - \beta_{c(t-to)} \} / E_{c(to)} \tag{16}$$

The effective age t_e should be used for t_o in estimating ϕ_o and $E_{c(to)}$ by using temperature/time history before loading (before age t_o) in Eq. 12. Consequently, every point in the section will have a separate creep function depending on its temperature/time history. The effect of temperature since loading (transient creep) can also be incorporated.

In the current study, the principal of superposition was used with a separate virgin creep function and characteristics assigned for the incremental stress developed at the beginning of each time-step. The total incremental creep strain due to each separate incremental stress is to be included in Eq. 13 resulting in;

$$Df_{(t)} = - E_{c(t+0.5dt)} \{ \alpha T_{(t+dt)} - \alpha T_{(t)} + D\varepsilon_{shr} + D\varepsilon_{cr} \} \tag{17}$$

$$D\varepsilon_{cr} = \Sigma\, Df_{(t)} \; d\varepsilon_{cr,to} \quad (\text{ from Eqs. 16 and 17 }) \tag{18}$$

where $D\varepsilon_{cr}$ is the algebraic sum of the creep strains from time-step 1 upto n, and $d\varepsilon_{shr}$ is the shrinkage or swelling strain within the current time step (note that the value of $d\varepsilon_{shr}$ is negative for a shrinkage, and $d\varepsilon_{cr,to}$ is positive for both compressive and tensile stress increments).

7 Influence of parameters on the behaviour of wall section

The hydration model parameters given in Section 2.1.2 were actual values obtained in field conditions [5] for normal portland cements. These parameters are used in the example calculations given below. The typical values assumed for material and thermal properties are; coefficient of thermal expansion of concrete $\alpha = 10*10^{-6}$ /°C, coefficient of thermal conductivity of concrete k = 2.30 J/m/s/°C, specific heat of concrete c = 1100 J/kg/°C, and mass density p = 2400 kg/m^3. The surface heat transfer coefficient h was assumed to be equal to 10 J/m^2/s/°C typical for still air conditions, and the unit cement content S was taken as 450 kg/m^3.

7.1 Environmental interaction

The importance of the accurate simulation of exposed surfaces to incorporate ambient effects such as solar radiation and also the effect of formwork or insulation, is shown herein.

The predicted early-age temperature distributions using a multi-layer and the modified h_m approach are shown in Fig. 3 for a 200 mm thick wall insulated by 30 mm thick wood formwork on both faces with $a_s = 0.75$. The wall is subjected to solar radiation and ambient temperature variations shown in Fig. 3b on east and west faces, typical for moderate climatic conditions [8]. Also shown are the predicted temperatures neglecting solar radiation and ambient temperature variations. As can be seen, the maximum error in the predicted temperature using h_m is about 20 °C higher, both at the edge and mid-thickness of the section. The magnitude of error however decreases with increasing wall thickness. The effect of ambient conditions is also quite significant over the heat of hydration, and could result in increased tendency to cracking following a rather cold period.

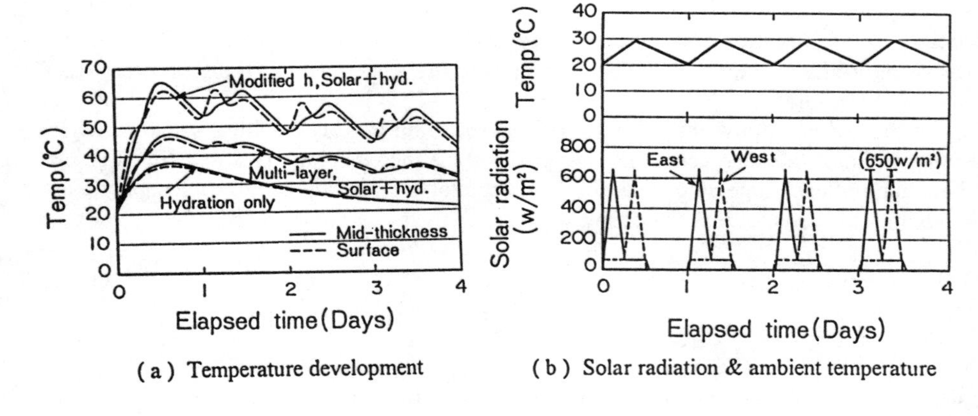

(a) Temperature development

(b) Solar radiation & ambient temperature

Fig. 3. Influence of ambient conditions on temperature (200 mm wall).

7.2 Use of adiabatic hydration models

Three wall sections of 200, 600 and 2000 mm thickness, which would cover the practical range of wall thicknesses, were analysed using adiabatic and non-adiabatic hydration models described in Section 2.1. This is to show the degree of error caused by the conventional use of adiabatic models in non-adiabatic environments. It is assumed that there is no insulation provided.

To maintain a consistency in the hydration parameters, the adiabatic model developed from the Rastrup model (Fig. 1) and the varying temperature Rastrup model (Eq. 3) were used. The other values assumed were; ambient temperature $T_a = 20$ ℃ (constant), concrete placing temperature $T_o = 20$ ℃, Young's modulus of elasticity $E_{c(28)} = 29{,}575$ MPa, and tensile strength $f_{t(28)} = 3.0$ MPa.

Figure 4a shows the predicted temperatures at the edge (S) and mid-thickness (M) of the wall for 200 and 2000 mm thick sections. The temperatures predicted using the adiabatic model are always higher with the maximum error occurring in the 200 mm wall. The maximum errors are about 4.2 and 3.3 ℃ for 200 and 2000 mm walls respectively. This is therefore a gross overestimate of the predicted temperature rise, eg, approximately 40 % for the 200 mm wall.

The average fully-restrained thermal stress (neglecting creep effects) is shown in Fig. 4b for 200 and 600 mm walls. As can be seen, the overestimate is quite significant with about 45 % and 25% for 200 and 600 mm walls respectively. Therefore, the use of adiabatic hydration models, as adopted in many conventional analysis procedures, could grossly overestimate the predicted early-age thermal stresses.

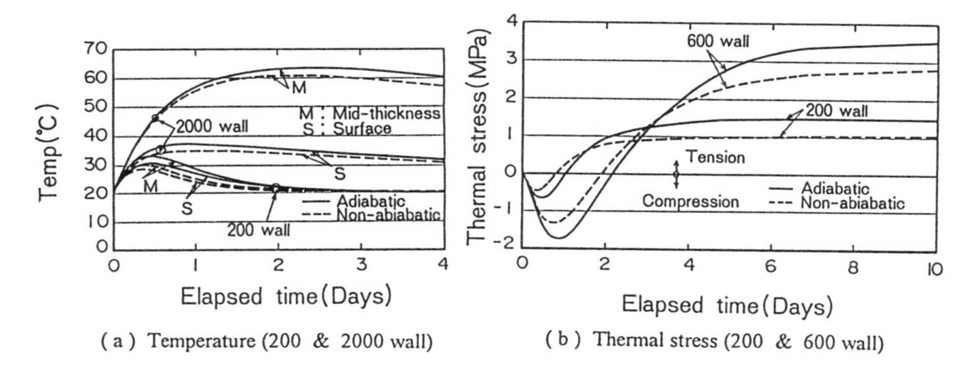

(a) Temperature (200 & 2000 wall)   (b) Thermal stress (200 & 600 wall)

Fig. 4. Use of adiabatic & non-adiabatic models.

7.3 Creep effects on early-age thermal stresses

Figure 5 shows the development of thermal stresses for 200 and 600 mm walls, using the non-adiabatic model (Eq. 3). Shown in Fig. 5 also are the average values of current tensile strength f_t and temperature across the wall thickness. The induced tensile stress is much higher in the 600 mm wall, and also the development of f_t is much rapid due to the activation effect of high temperatures. The creep effect on stress relaxation is significant with about 35% reduction at an age of 10 days. At the time of zero stress since hydration, the temperature rises ΔT is about 6 and 15 ℃ for 200 and 600 mm walls.

According to Fig. 5b, if creep effects are ignored, the ratio of induced tensile stress to f_t will be close to unity beyond 6 days showing that cracking would occur. Therefore, in any rational attempt to estimate the tendency to cracking, both creep and strength development should be considered.

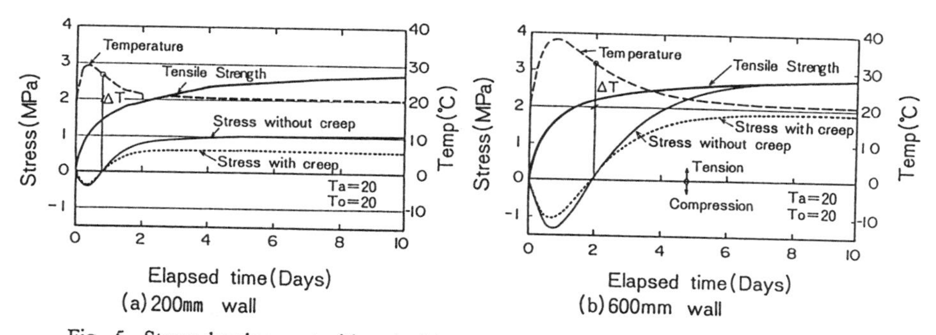

(a) 200mm wall   (b) 600mm wall

Fig. 5. Stress development with and without creep.

7.4 Autogeneous shrinkage stresses

It has been shown by a limited number of experiments [eg, 11] that the autogeneous shrinkage of concrete, possible with concretes of low water/cement ratios, could cause early-age cracking. This would act in combination with the heat of hydration effects increasing the tendency to cracking.

Figure 6a shows the development of free shrinkage with time for different w/c ratios. It was assumed that the development of autogeneous shrinkage is analogous to that of strength. The Fig. 6a model approximately fits onto the limited available data. For an accurate estimate, it would be however necessary to determine the shrinkage strains for a given concrete composition. Figure 6b shows the calculated pore humidity related to the Fig. 6a shrinkage strains [10], which can now be easily incorporated in a transient drying shrinkage analysis, if necessary. Figure 7 shows the rate of development of pore humidity and as can be seen, the rate is maximum at very early age.

Figure 8 shows the fully-restrained shrinkage stresses for a concrete composition of $f_{c(28)} = 50$ MPa and w/c = 0.28, with and without creep effects. As can be seen, this condition alone is severe enough to cause through-depth cracking in restrained members undergoing autogeneous shrinkage.

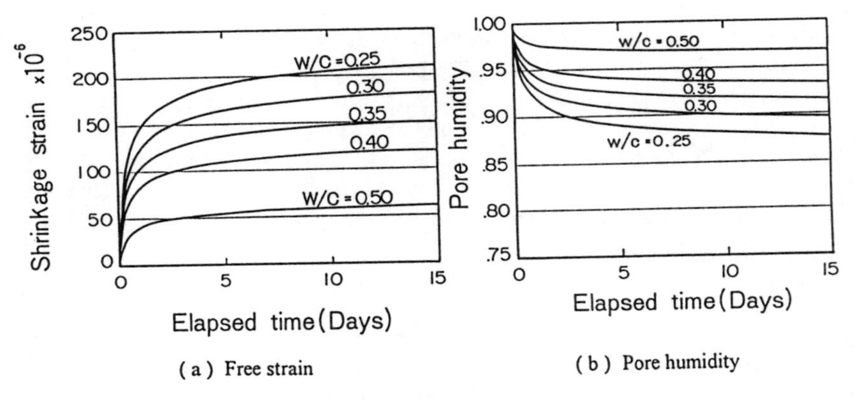

Fig. 6. Autogeneous shrinkage.

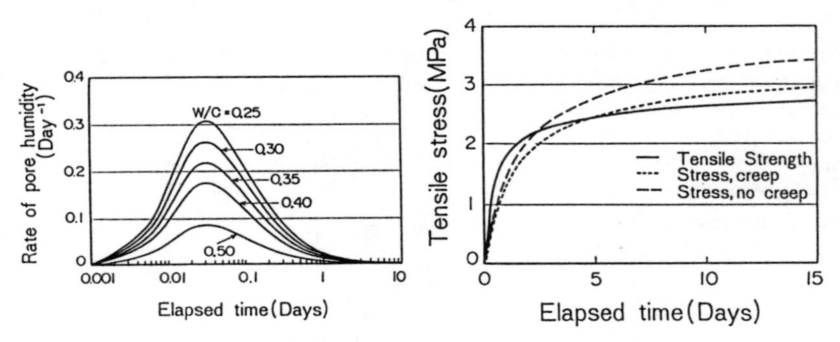

Fig. 7. Rate of pore humidity.

Fig. 8. Shrinkage-induced stresses.

8 Conclusion

The early-age tensile stresses are of significant magnitude in comparison with the current tensile strength. Rationally-based evaluation procedures should be adopted to arrive at durable, safe and economical designs, and these should include the environmental interaction, coupling of hydration and heat transfer mechanisms, creep effects and also the development of material properties. As shown by few example analyses, cursory approaches as adopted in some of the current design practices could lead to inaccurate predictions. The prediction of temperature itself is of not eventual importance as the cracking tendency depends on the ratio of induced stresses to the current tensile strength. The dependency of material and thermal properties on the time (age) and temperature should also be considered. The early-age creep effects would relax the early-age stresses remarkably in the order of 30 - 40%.

As the results described in this paper are based on the available material models with some extrapolated to the early age, an experimental evaluation of the early-age creep under tensile stresses in particular is urgently required. A thorough parametric study should then be undertaken, considering the local meteorological parameters, for the benefit of the designers.

9 Acknowledgment

The Japan Science and Technology Agency and the Australian Academy of Sciences are gratefully thanked for the visiting research fellowship awarded to the first author.

10 References

1. Holland, T.C., Krysa, A. et al (1991) Use of silica-fume concrete to repair abrasion-erosion damage in the Kinzua Dam stilling basin, *American Concrete institute, SP 91-40*, pp. 841-63.
2. ACI Committee 207 (1973) Effect of restraint, volume change and reinforcement on cracking of massive concrete, *ACI Journal*, pp. 445-69.
3. Rastrup, E. (1954) Heat of hydration in concrete, *Magazine of Concrete Research*, Vol. 6, No. 17. pp. 79-92.
4. Harada, S., Maekawa, K. et al. (1991) Nonlinear coupling analysis of heat conduction and temperature-dependent hydration of cement, *Concrete Library of JSCE*, No. 18. pp. 155-69.
5. Branco, F.A., Mendes, P.A. and Mirambell, E. (1992) Heat of hydration effects in concrete structures, *ACI Materials Journal*, Vol. 89, No. 2. pp. 139-45.
6. Japan Concrete Institute (1986) *Standard Specifications for Design and Construction of Concrete Structures, Part 2 (Construction)*.
7. Vitharana, N.D., Priestley, M.J.N. and Dean, J.A. (1991) *Strain-Induced Stressing of Concrete Storage Tanks*, PhD Thesis Report 91-8, University of Canterbury, New Zealand.
8. Billington, N.S. (1952) *Thermal Properties of Buildings*, Cleaver-Hume Press Ltd, London.
9. ACI Committee 209 (1982) Prediction of creep, shrinkage, and temperature effects in concrete structures, *Report No. ACI 209R-82*.
10. CEB-FIP. (1991) *Model Code for Concrete Structures*, Paris.
11. Bergner, H. (1993) Tests on high-strength concrete members under centrical restraint, *Darmstadt Concrete*, Germany, Vol. 8, pp. 11-8.

155 THERMAL BEHAVIOR OF COMPOSITE BEAMS UNDER EXTERNAL HEATING

I. KURODA, T. OHTA and S. HINO
Kyushu University, Fukuoka, Japan

Abstract

Heat conduction and thermal stress problems of steel plate-concrete composite beams as well as reinforced concrete (RC) beams are discussed from analytical and experimental points of view under external temperature loading.

In the analysis, an extended embedded model for reinforcement is presented for solving the thermal problems of the above mentioned composite structures. Another feature of the analysis is the consideration of the effect of thermal dependency upon heat conductivity and Young's modulus of concrete under high temperature.

External thermal loading tests have also been carried out on the composite and RC beams. The analytical results are compared with the experimental ones and their reasonable agreement is observed.

In addition, useful information about the thermal behavior of concrete beams reinforced by new materials such as CFRP and GFRP rods is presented herein. The effect of these materials and the usual steel on the heat conduction and thermal stress of the beam is discussed.

Keywords: Embedded model of reinforcements, finete element analysis, heat conduction, thermal stress.

1 Introduction

The heat conduction and thermal stress of concrete structures have mainly addressed to the subjects of mass concrete and nuclear reactor containment with in a low temperature range of 50 ~100°C (120 ~210°F). Most of studies did not consider the effect of the presence of steel in the concrete. But, it is necessary to clarify the exact heat conduction and thermal stress distribution

Concrete Under Severe Conditions: Environment and loading (Volume Two) Edited by K. Sakai, N. Banthia and O.E. Gjørv. Published in 1995 by E & FN Spon. ISBN 0 419 19860 1

with consideration for the effect of the presence of reinforcements in concrete under a high temperature condition because of the high heat conductivity of steel.

The present study deals with the subjects of heat conduction and thermal stress in steel plate-concrete composite beams and reinforced concrete (RC) beams under external heating. A thermal loading test on the above beams was carried out by using an electric heater up to 220°C (428°F). In the analysis, the embedded model[1] for reinforcement is applied. This model has been found that the cost of generating a finite element mesh can be reduced considerably. The analytical results are compared with the experimental values to confirm the validity of the presented analytical method for steel-concrete composite structures.

Subsequently, useful analytical information about the thermal behavior of concrete beams reinforced by new materials such as CFRP (Carbon Fiber Reinforced Plastics) or GFRP (Glass Fiber Reinforced Plastics) rods is presented. The effect of these new materials and the usual steel upon the heat conduction and thermal stress of the beams is discussed.

2 Experiment and analysis

2.1 External thermal loading test

A thermal loading test on steel plate-concrete composite beams and RC beams was carried out. These specimens are shown in Fig.1. The composite beam specimen (TSC

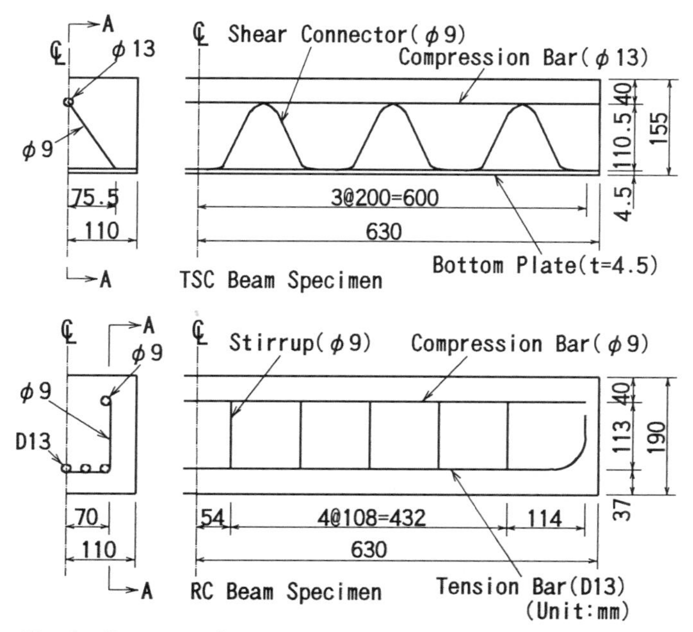

Fig.1 Beam specimens.

beam) consists of a bottom plate, a reinforcing bar, pyramidal shear connectors and concrete. The bottom plate, 1260 x 220 x 4.5mm, was made of steel with a minimum yield strength of 24kgf/mm² (240MPa) specified in Japanese Industrial Standards (JIS). Deformed and round bars with diameters of 13mm and 9mm were used for longitudinal reinforcing bars and stirrups. Concrete strength of the beams was 410kgf/cm² (40MPa) at the test.

The electric heating facility is shown schematically in Fig.2. As the specimen was placed on the heating facility upside down, the upper side of the beam was heated. The concrete surface was raised to 220°C in 6 hours (Fig.3). Here, the temperature curve (shown in Fig.3) was determined considering the capacity of the heating facility.

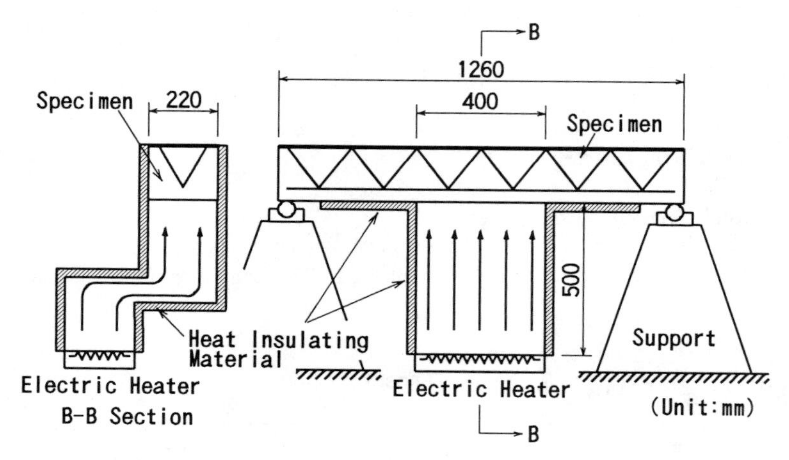

Fig.2 Heating facility.

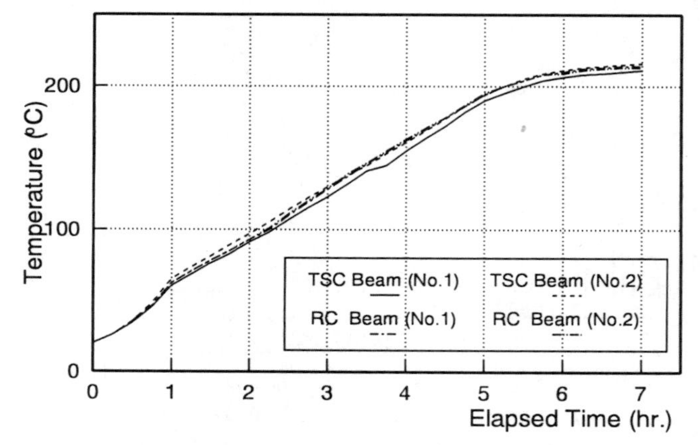

Fig.3 Applied temperature.

2.2 Heat conduction and thermal stress analysis

Heat conduction and thermal stress analysis is derived based on the finite element method. The Crank-Nicolson method is employed to deal with unsteady heat conduction. The material properties used in the analysis are shown in Table 1[2-3]. The heat conductivity and Young's modulus of concrete were values measured by experiments under high temperature. Figures 4a and 4b show these experimental values. Considering temperature dependency of the heat conductivity, the heat conduction analysis includes nonlinearity. Therefore, the direct iteration procedure is used.

In the analysis, the embedded model is applied to consider the effect of reinforcements such as reinforcing bars and shear connectors (Fig.5). This model allows the reinforcements to be located anywhere inside the concrete elements. In addition, the reinforcements do not necessarily have to connect to any nodes.

Table 1. Properties of concrete and steel

	Heat Conductivity kcal/mh°C	Heat Capacity kcal/m^3	Coefficient of Heat Transfer kcal/m^{2}h°C	Coefficient of Linear Expansion x10^{-6}/°C	Young's Modulus kgf/cm^2	Poisson's Ratio
Concrete	Fig.4	420	10	10	Fig.3	0.192
Steel	60-0.06T*	890	10	11.7	2.1x10^6	0.3

*T:Temperature

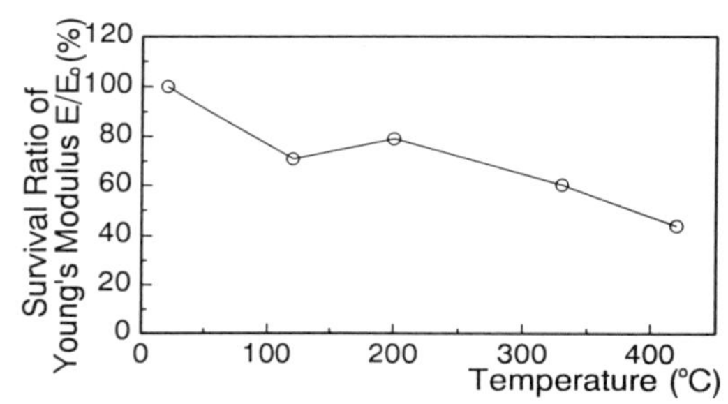

Fig.4a Relation between temperature and Young's modulus of concrete

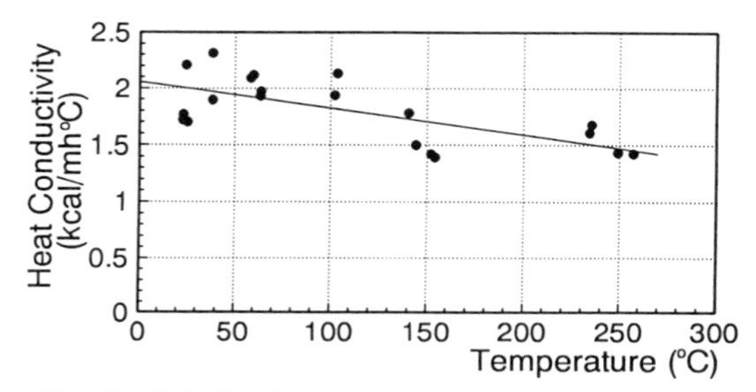

Fig.4b Relation between temperature and heat conductivity of concrete.

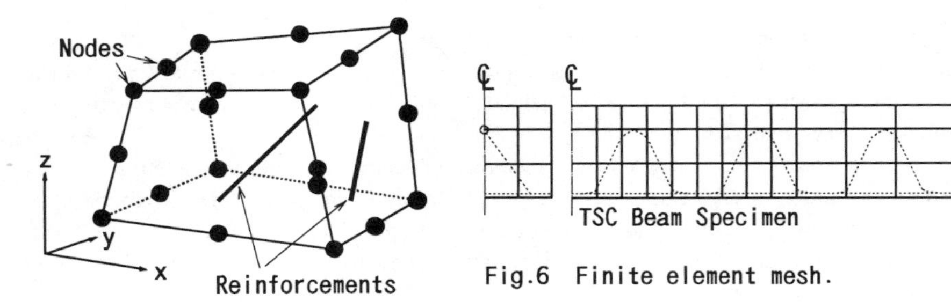

Fig.6 Finite element mesh.

Fig.5 Embedded model for reinforcement.

Therefore, the finite element mesh can be constructed regardless of the layout of the reinforcements. Accordingly, the embedded model helps to decrease the required number of nodes and elements. For example, Fig.6 shows the finite element mesh of TSC beam.

The heat conductivity matrix λ, the heat capacity matrix C and the stiffness matrix K of a composite element shown in Fig.5 are given by the summation of the contributions of concrete and reinforcements. Let the subscripts c and s denote the contributions of concrete and a reinforcement, respectively. Then these matrices of the composite element are given by

$$\lambda = \lambda_c + \Sigma \lambda_s \quad , \qquad C = C_c + \Sigma C_s \quad , \qquad K = K_c + \Sigma K_s \tag{1}$$

where Σ implies that multiple reinforcements may exist in an element.

3 Temperature and strain distribution in beams

Numerical values are compared with the experimental data to test the validity of the presented analytical method for tracing the heat conduction and thermal stress in composite structures.

Figures 7, 8 and 9 show temperature, axial strain and axial stress distributions, respectively, obtained in the test (the average of two specimens) and the numerical results in the cross section A-A in Fig.1. Shaded zones indicate the heated area. The figures beside the solid circles are the measured values at the points. The solid lines in the figures are the results of analysis.

The numerical results reflect the general trend of unsteady heat conduction; that is, the temperature gradient is very severe in the region near the heated zone and becomes moderate as the distance from the heated area increases. The experimental data are captured very well by the analysis, although some discrepancy between the experimental and numerical results is recognized. Thus, it can be seen that the heat conduction and thermal stress analysis procedure employed is adequate.

There is not enough space to comment on the results if the existance of

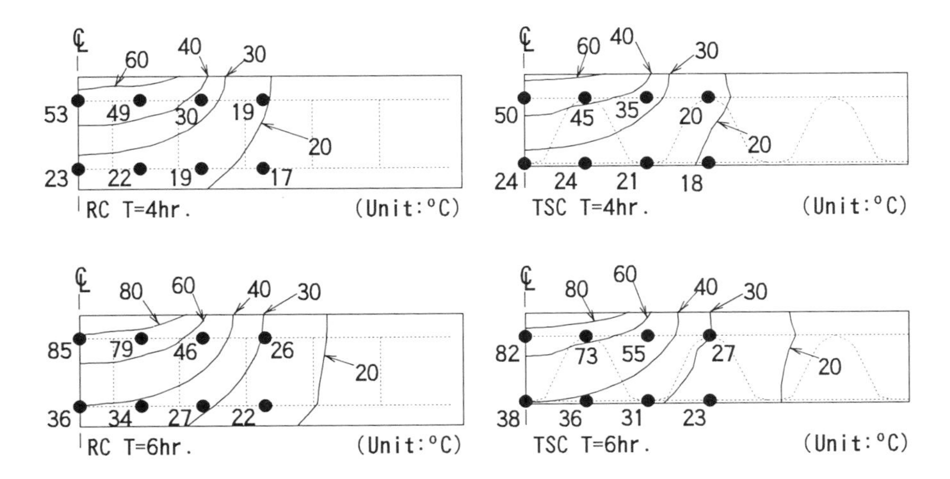

Fig.7 Temperature distribution (RC beam and TSC beam).

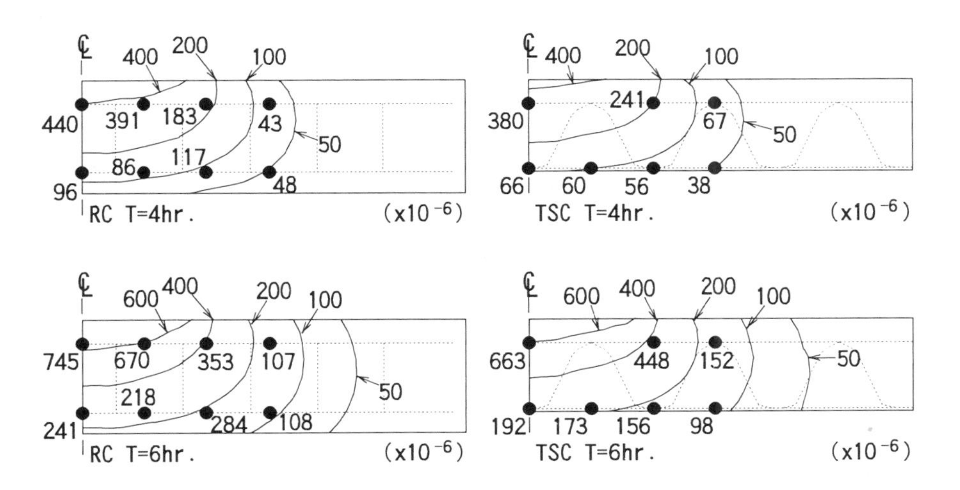

Fig.8 Axial strain distribution (RC beam and TSC beam).

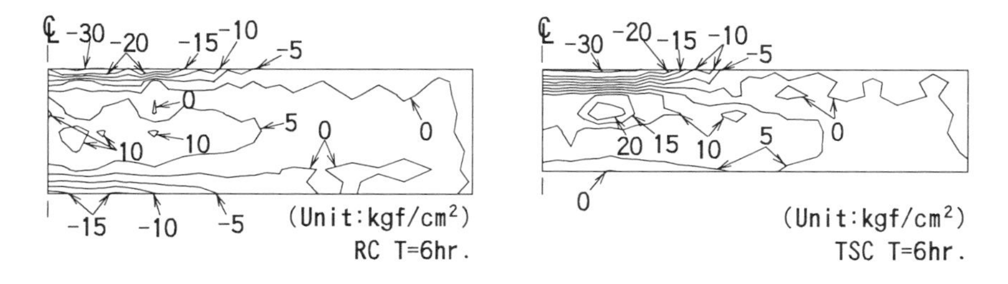

Fig.9 Axial stress distribution (RC beam and TSC beam).

reinforcement is neglected. The reader may refer to our paper[2]. In the paper, the effects of the quantity of reinforcement on heat conduction and thermal stress were explained in detail.

4 Thermal effect of FRP as reinforcement

Recently, some applications of new materials such as CFRP and GFRP for reinforcement in concrete structures have been investigated. But, there is no data on their effect upon the heat conduction and thermal stress of concrete structures. Therefore, an analytical simulation has been carried out to study this problem.

Two analytical models (CFRP beam and GFRP beam) have been used. These models have the same dimensions as the RC beam shown in Fig.1. CFRP or GFRP bars are substituted for reinforcing steel bars. That is, the cross-sectional areas of the carbon or glass fiber and steel bars are equal. The material properties of CFRP and GFRP are shown in Table 2[4].

Figures 10 and 11 show the analytical results of the temperature and axial stress distribution in the CFRP beam and the GFRP beam. Compared to the temperature distribution in the RC beam shown in Fig.7, the temperature gradients in the CFRP and the GFRP beams are more severe. Therefore, their maximum temperatures are higher. This result depends on the heat conductivity and the heat capacity of the reinforcement used. With respect to axial stress distributions, the CFRP beam has a remarkable characteristic. Namely, in the RC beam and the GFRP beam, tensile stresses are observed nearby the middle height of the beams. On the other hand, tensile stresses are absent in the CFRP beam. This characteristic must be caused by the negative coefficient of linear expansion of CFRP. Therefore, it can be said that the application of the CFRP makes it possible to control the tensile stress in concrete under external heating.

Table 2. Properties of carbon fiber and glass fiber

	Heat Conductivity kcal/mh°C	Heat Capacity kcal/m^3	Coefficient of Linear Expansion x10^{-6}/°C	Young's Modulus kgf/cm^2
Carbon Fiber	20.6	306	-0.3	2.96x10^6
Glass Fiber	0.89	477	4.9	7.76x10^5

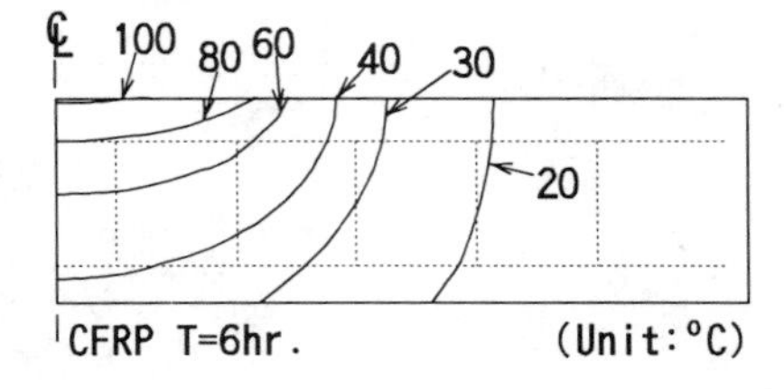

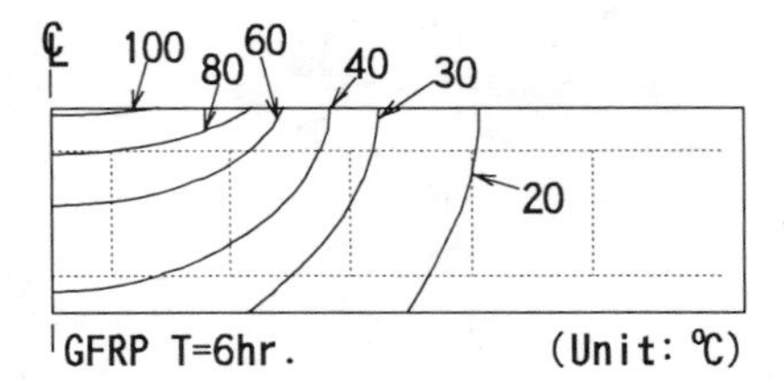

Fig.10 Temperature distribution (CFRP beam and GFRP beam).

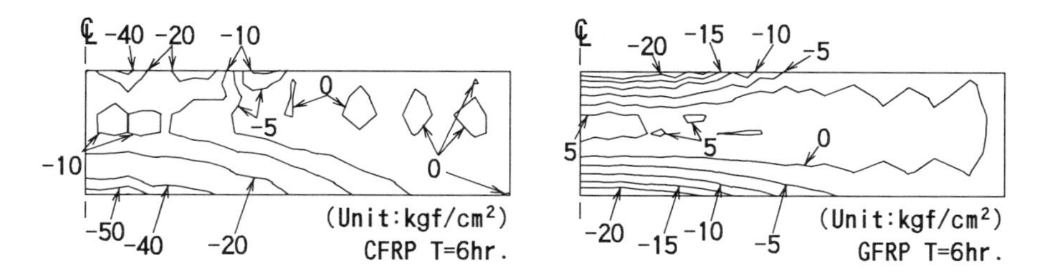

Fig.11 Axial stress distribution (CFRP beam and GFRP beam).

In addition, from the above results, it can be concluded that the effect of reinforcements such as the usual steel or FRP on the temperature and stress distributions in concrete structures cannot be neglected.

5 Concluding remarks

Numerical analysis as well as external heating tests were carried out in order to clarify the unsteady heat conduction and the thermal stress distribution in RC and composite structures. The measured temperature and strain distributions in the beams are modeled well by the analysis. Thus, it can be said that the analysis can give reasonable solutions to the unsteady heat conduction and thermal stress problems of beams subjected to external temperature loading.

In addition, it can be also concluded that the effect of reinforcements such as the usual steel or FRP upon these problems cannot be neglected.

6 References

1. ASCE Committee on Concrete and Masonry Structures (1981) *A State-of-the-art report on finite element analysis of reinforced concrete.* Task Committee on Finite Element Analysis of Reinforced Concrete Structures, ASCE Spec. Publ.
2. Ohta,T., Kuroda,I. and Hino,S. (1991) Heat Conduction and Thermal Stress in Reinforced Concrete Beams under High Temperature. *Transactions of Japan Concrete Institute,* Vol.13, pp.157-64(In English).
3. Tokuda,H. (1984) Thermal Properties of Concrete. *Concrete Journal,* Vol.22, No.3, pp.29-37(In Japanese).
4. Japan Society of Mechanical Eng. (1990) *Van Composite Materials.* Gihoudou-Shuppan, (In Japanese).

Unit convert table

SI unit	1.0MPa	1.0J
Non-SI unit	10.2kgf/cm^2	0.239cal

156 BOND BETWEEN A REINFORCEMENT BAR AND CONCRETE AT LOW TEMPERATURES

L. VANDEWALLE
Katholieke Universiteit Leuven, Leuven, Belgium

Abstract

The bond between a reinforcing steel bar and the surrounding concrete is of major importance and affects the behaviour (cracking, deformation, etc.) of the reinforced concrete composite. The bond stress(τ)-slip(δ) relationship is the common means to determine the steel-concrete interfacial mechanical properties.

The purpose of this investigation was to examine experimentally (beam test) as well as theoretically the bond between a deformed bar and concrete (-> τ-δ-course) at room temperature as well as at very low temperatures.

The experimental program consisted of 50 beam test specimens and 600 cubes. The following parameters were investigated: testing temperature (+20°C, -40°C, -80°C, -120°C and -165°C), storage conditions of the test specimens (fog room, air-dry and oven-dry) and age of the concrete (28 days and 90 days). From the experiments, it was found that the bond strength increased with a decreasing temperature and that so much the more as the free moisture content of concrete was higher. This was true for temperatures higher than -120°C, if the temperature, however, was lower than -120°C, the value of the bond strength decreased again.

Theoretically the τ-δ-relationship is approximated by the following expression:

$$\tau = \tau_u(1 - \mu e^{-\lambda\delta})$$

in which the bond strength τ_u is calculated as a function of the concrete cover on the rebar, the concrete quality and the temperature by means of the model of a thick-walled cylinder subjected to an internal pressure; μ and λ are constants.

Keywords: Bond, concrete, low temperatures, rebar, storage conditions.

Concrete Under Severe Conditions: Environment and loading (Volume Two) Edited by K. Sakai, N. Banthia and O.E. Gjørv. Published in 1995 by E & FN Spon. ISBN 0 419 19860 1

1 Introduction

The bond between a reinforcing steel bar and the surrounding concrete is of major importance and affects the behaviour (cracking, deformation, etc.) of the reinforced concrete composite. The bond stress-slip relationship is the common means to determine the steel-concrete interfacial mechanical properties.

The bond behaviour is mainly dependent on the profiling of the steel bar, the concrete cover on the bar and the concrete quality. The concrete properties, and consequently the bond stresses, are considerably influenced by temperature. In particular, the quantity of chemically unbound water in the concrete plays an important role at low temperatures.

With the financial support of the National Funds for Scientific Research in Belgium (NFWO) the possibility has been offered to carry out fundamental research (experiments-theoretical modelling) into the behaviour of these bond stresses both at room temperature and at low temperatures by means of a beam test.

2 Experiments

2.1 Survey of the program

The influence of the following parameters on the bond behaviour between a steel rebar and concrete has been examined:

- test temperature: +20°C, -40°C, -80°C, -120°C, -165°C;
- curing condition of the test specimen: fog room (T=+20°C,R.H.>95%), air-storage (T=+20°C,R.H.=60%), oven-storage (immediately after demolding, the specimen is stored for 14 days in a fog room and afterwards dried for 14 days at a temperature of +105°C);
- age of the concrete; 28 days, 90 days.

The concrete cover (d) on the rebar (diameter ϕ) was equal to 1.5ϕ for all specimens. A survey of the number of tested specimens is given in Table 1.

Table 1. Survey of the test program

Series	Age of the concrete	Curing condition	Test temperature				
			+20°C	-40°C	-80°C	-120°C	-165°C
I	28 days	fog room	3	3	3	3	3
II	90 days	fog room	2	3	2	2	2
III	28 days	air-storage	2	2	3	2	3
IV	28 days	oven-storage	2	2	3	2	3

2.2 Material properties

The concrete mixture (1m³) was composed of:

gravel 4-14	: 1300 kg
sand 0-5	: 550 kg
blast furnace slag cement	: 400 kg
water	: 165 l (W/C=0.413)

The mean compressive strength, mean splitting tensile strength respectively, of the concrete determined on cubes (side=150 mm) at different temperatures, are mentioned in Table 2.

Table 2. Mechanical properties of the concrete : mean compressive strength f_{cm} (mean splitting tensile strength $f_{ctm,sp}$) (N/mm²)

Series	Test temperature				
	+20°C	-40°C	-80°C	-120°C	-165°C
I	54.5(4.6)	93.8(8.3)	113.5(8.3)	134.4(8.6)	133.8(9.7)
II	61.5	93.8	110.4	114.4	117.2
III	47.2(3.8)	72.4(5.7)	91.6(7.3)	105.1(8.2)	111.8(8.1)
IV	68.3(4.8)	69.4(3.9)	84.9(6.1)	77.9(5.1)	89.2(5.7)

The reinforcement was Tempcore BE400 (ϕ = 25 mm). The rib-factor f_R (relative rib area) of the bar was at least equal to 0.065.

2.3 Test setup: beam test

The test setup is shown in Figure 1. The test specimen was composed of two prismatic blocks of reinforced concrete. These blocks were on the underside connected by means of the bar of which the bond was to be examined and on the upperside by means of a steel hinge. The embedment length between the concrete and the bar was three times the diameter of the bar. During application of the load, which was continued up to the total failure of the bond in both half-beams, the displacement of the bar relative to the concrete (δ) was measured at both ends of the bar. In this way, each test yielded two results, i.e. two τ-δ-curves; τ is the mean bond stress over

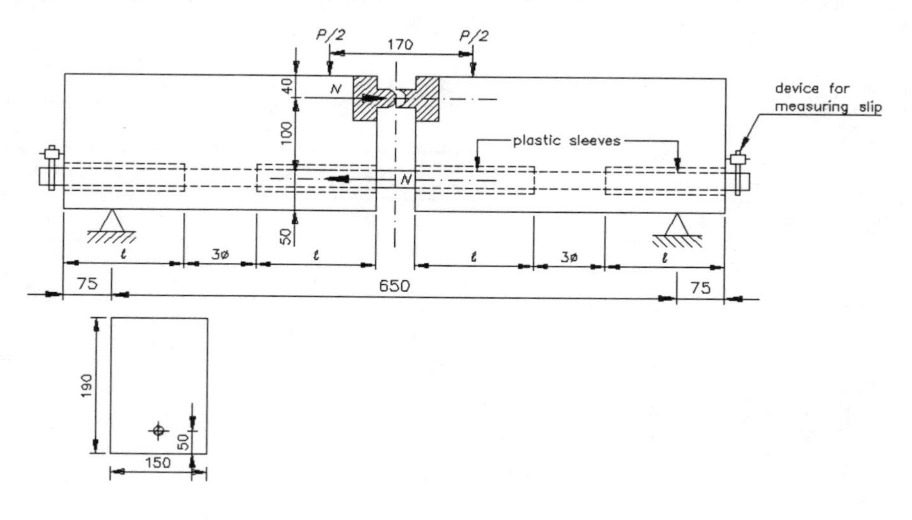

Fig. 1. Test setup.

the embedment length ($\tau = N/(3\pi\phi^2)$).

An advantage of the beam test is that it approximates well the situation of a beam in reality.

The tests were carried out in a temperature test room (CRYOSON type TRA-10) which was especially designed for the performance of mechanical tests at low temperatures. Cooling was achieved using liquid nitrogen.

2.4 Test results

The results of the tests, namely the bond stress-slip-curves, at the various temperatures are presented in Figures 2, 3, 4 and 5 for the curing condition "fog room-28 days", "fog room-90 days", "air-storage" and "oven-storage" respectively. Each curve in the diagrams is the middle curve of all experimentally measured τ-δ-curves at that specific temperature and curing condition. A presentation of all experimentally measured τ-δ-curves can be found in [1].

2.4.1 Influence of the curing condition

The influence of the curing condition of the specimen on the bond stress τ_x (=mean bond stress over the embedment length corresponding to a slip value "x") at a certain temperature was very significant. It was observed that the higher the free moisture content (m) of the specimen, as a result of its curing condition, the greater the bond stress τ_x became (see Figures 2 to 5). The average free moisture content of the concrete, corresponding to a particular curing condition, is noted in Table 3.

Table 3. Moisture content of the concrete as a function of its curing condition.

Curing condition	m (%)
fog room-28 days	4.38
fog room-90 days	4.21
air-storage	3.35
oven-storage	0

In general, the bond stress τ_x increased at lower temperatures. However, this increase in bond stress τ_x occurred only at temperatures down to -120°C. At lower temperatures, a slight decrease in bond stress τ_x was recorded. Around -115°C the ice changes of crystalline structure resulting in a volume decrease of about 20%. Because of this, the ice no longer completely fills the pore, and consequently, its absorbable force becomes smaller.

It was also observed for temperatures equal to and lower than -120°C that the slip changed abruptly. This was probably caused by the very brittle character of the concrete at those temperatures.

Splitting failure of the concrete cover on the rebar along the embedment length was found to be the failure mechanism for all test specimens.

2.4.2 Influence of the age of the concrete

To investigate the influence of the age of the concrete on the bond stress-slip-relationship, experiments were conducted at two different ages, namely 28 and 90

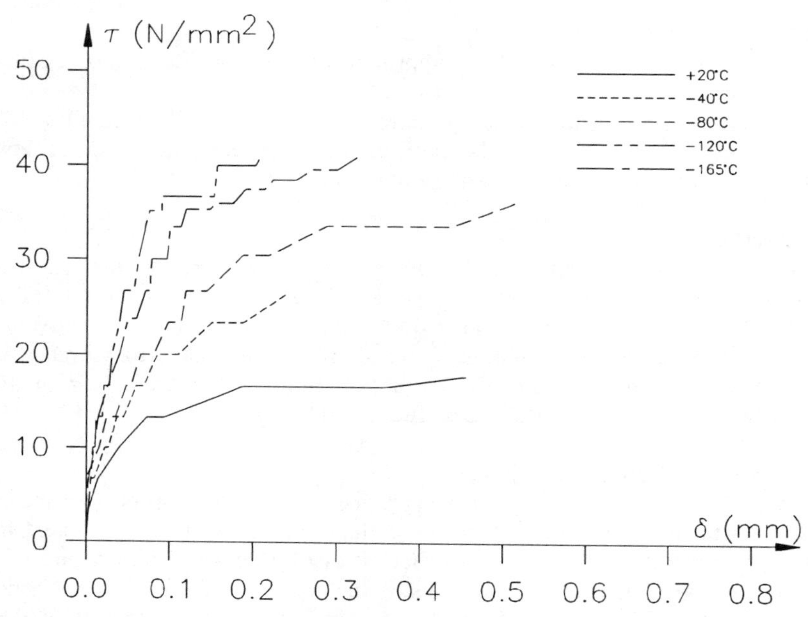

Fig. 2. τ-δ-curves for series I-specimens at various temperatures.

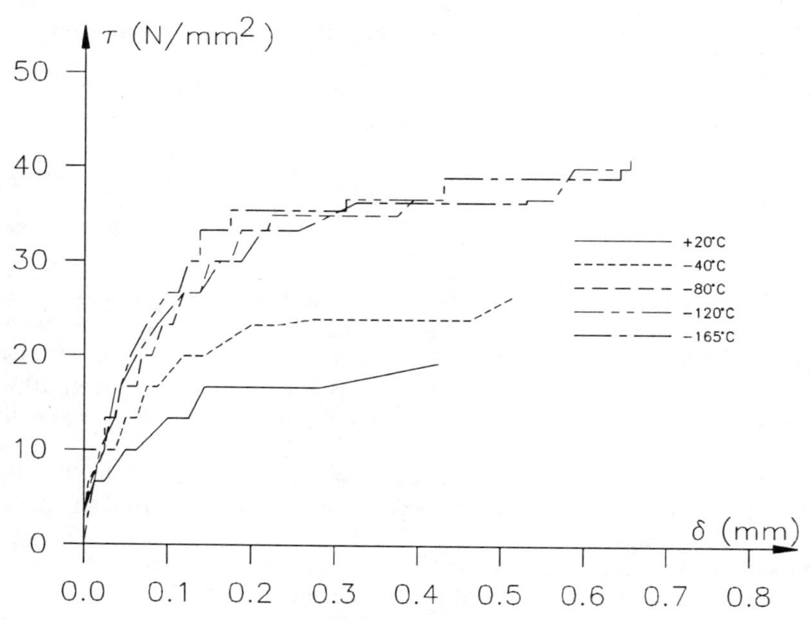

Fig. 3. τ-δ-curves for series II-specimens at various temperatures.

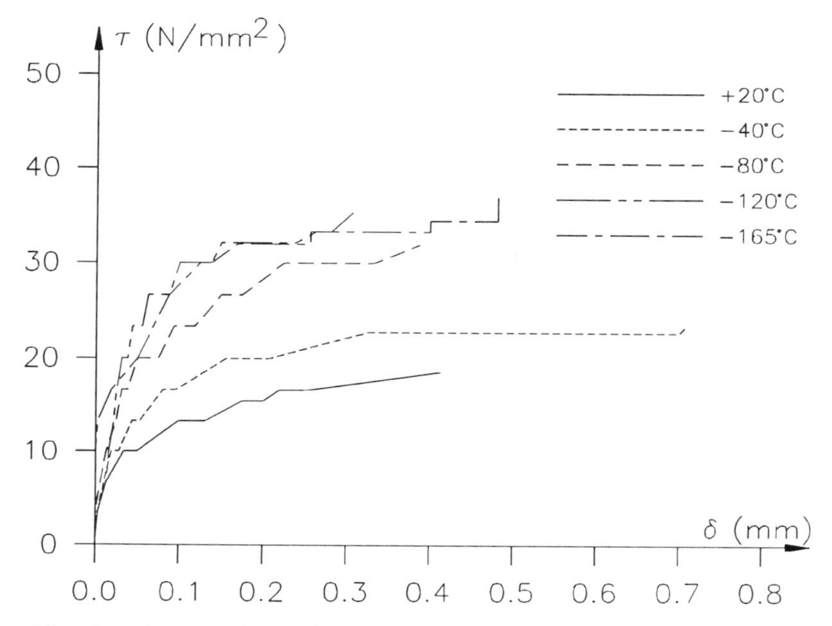

Fig. 4. τ-δ-curves for series III-specimens at various temperatures.

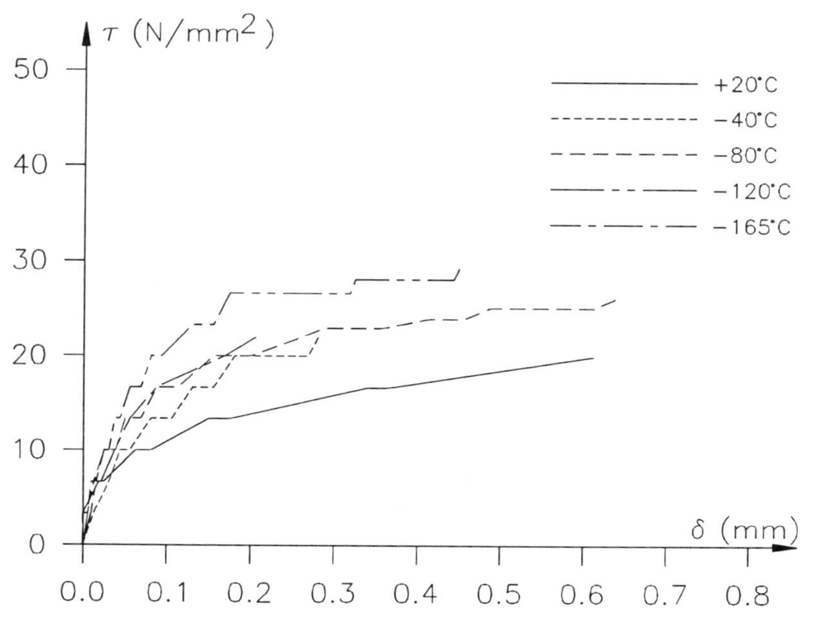

Fig. 5. τ-δ-curves for series IV-specimens at various temperatures.

days. Since the free moisture content of the concrete at both ages was almost the same (4.21% for 90 days old concrete and 4.35% for 28 days old concrete) a significant difference in the bond behaviour between the two series of tests was not found.

3 Theoretical modelling

3.1 Shape of the τ-δ-curve

A common expression for the analytical prediction of the bond stress-slip relationship is of the following form:

$$\tau = A\delta^B \tag{1}$$

in which the constants A and B are dependent on the concrete cover of the bar (d), the profiling of the bar surface (f_R), the position of the bar in the cross section, temperature, concrete quality, etc.

A similar τ-δ-relation has been derived by Rostasy [2], van der Veen [3], etc. Equation (1) has the merit that it leads to a closed analytical solution at the calculation of crack widths, etc. A demerit of it, however, is that for great values of δ, dependent on the constants A and B, the calculated bond stress is too great in comparison with bond stresses obtained by experiment. Therefore, this τ-δ-relation appears to be more suited for the calculation of structures under service load.

On the basis of the results of the beam tests, the τ-δ-relation was approximated at Leuven by the following expression:

$$\tau = \tau_u(1 - \mu e^{-\lambda\delta}) \qquad (\tau \text{ in N/mm}^2 - \delta \text{ in mm}) \tag{2}$$

with μ=0.78 and λ=9.78 mm^{-1}. The bond strength τ_u is a function of the concrete cover on the rebar, the bar diameter and the concrete quality (see 3.2).

Equation (2) has as a merit that it describes the τ-δ-course up to the bond failure. This τ-δ-relation may be used, in contrast with Equation (1), for both small and great values of δ. Consequently, by using this τ-δ-relation, calculations may be carried out for a structure under service load as well as in the ultimate limit state. Similarly, the cracking behaviour of a structural member may be followed up to its failure. An additional merit of Equation (2) is that it holds for temperatures from +20°C to -165°C. The temperature effect is completely taken into account by means of the bond strength τ_u.

Figures 6a and 6b show both the experimentally obtained τ-δ-curves at +20°C and -165°C, respectively, and the predicted τ-δ-relationship (bold line), in which τ_u has been taken equal to the average value of the ultimate bond strengths obtained during the corresponding experiments. From these figures, it follows that the τ-δ-relationship can be predicted with adequate accuracy using the proposed Equation (2).

3.2 Ultimate bond strength τ_u

Whilst the deformed bar is sliding in relation to the concrete, either the concrete around the bar is more or less crushed or a kind of plastic sliding threshold is

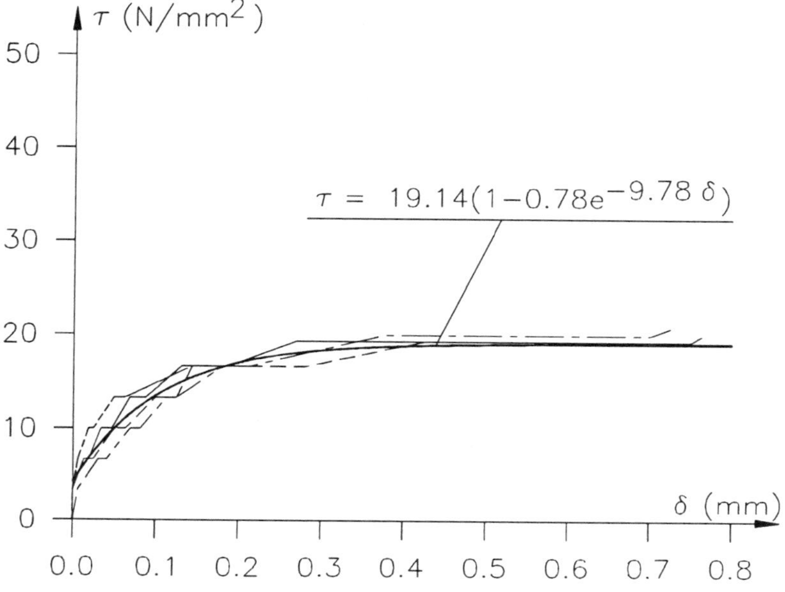

Fig. 6a. τ-δ-curves (test temperature = +20°C - fog room-90 days).

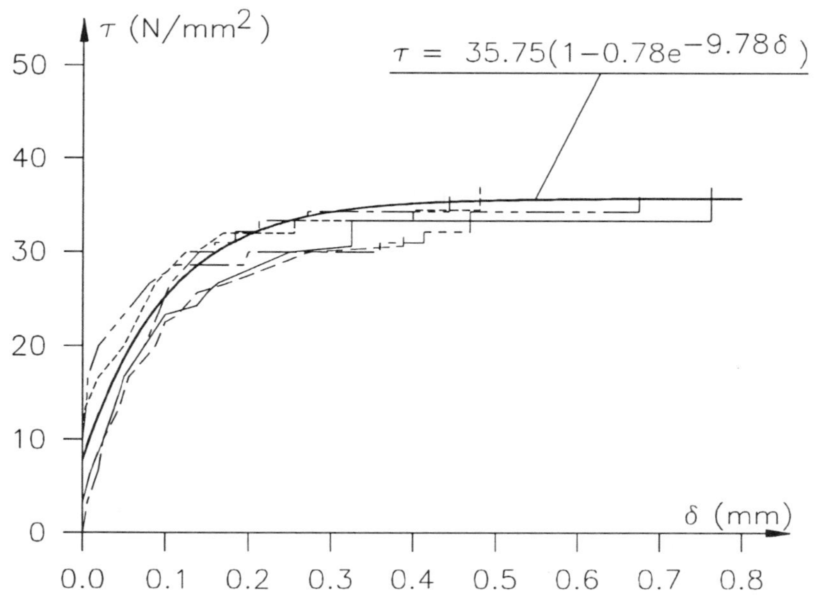

Fig. 6b. τ-δ-curves (test temperature = -165°C - air-storage).

attained in the concrete. In both suppositions, everything happens as if a radial compressive stress acts between the bar surface and the concrete. This radial compressive stress can be regarded as an inner hydraulic pressure "p" acting on a thick-walled concrete ring that approximately represents the effect of the surrounding concrete. The wall-thickness of the ring is determined by the smallest possible dimension that is given by the least thickness of the cover. Once the pressure p that causes failure of the concrete ring has been determined, τ_u can be calculated by using Coulomb's failure criterion if the failure mode of the specimen is "splitting of the concrete cover" [4]:

$$\frac{d}{\phi} \leq 3: \quad \frac{\tau_u}{f_c} = \frac{\sqrt{K}}{2} \left[1 + (1 - K)\, 0.353 \frac{\frac{\phi}{2} + d}{\frac{\phi}{2}}\right] \tag{3}$$

with K equal to f_{ct}/f_c; f_c (N/mm^2) and f_{ct} (N/mm^2) are the cylinder (150 x 300 mm) compressive strength and the uniaxial tensile strength of the concrete, respectively.

Rostasy and Schueurmann [2] performed a large number of pull-out tests. From these test results, it followed that the failure mode changes from splitting failure to shear failure for values of d/ϕ varying between 2.5 and 3.5. Therefore, for values of $d/\phi > 3$ (mean value), the bond strength, mostly as a result of shear failure, is equalized with τ_u corresponding to $d/\phi = 3$:

$$\frac{d}{\phi} > 3: \quad \frac{\tau_u}{f_c} = \frac{\sqrt{K}}{2} \left[1 + (1 - K)\, 2.473\right] \tag{4}$$

Mean values of τ_u at normal as well as at low temperatures can be obtained from Equations (3) and (4) because:

- the above-mentioned model and expressions hold for normal as well as for low temperatures;
- τ_u is expressed as a function of the concrete quality; a quantity which itself is dependent on the temperature.

In Figure 7 are presented the results of beam tests [1] and pull-out tests [2], [3]. The value d/ϕ is equal to 1.5 for the beam tests and 4.5 - 5 for the pull-out tests. All the mentioned tests were executed at temperatures varying between +20°C and -170°C. Besides these test results, also given in Figure 7 are the theoretically calculated values of τ_u/f_c for K=0.09. This curve can be considered as mean values for τ_u.

Since in reality the failure mode changes from splitting failure to shear failure for values of d/ϕ varying between 2.5 and 3.5, i.e. the boundary between those two failure modes does not correspond to "one" value of d/ϕ, a lower and upper limit are given for τ_u in Figure 7.

A comparison of the theoretically calculated τ_u/f_c with the test results shows that the correspondance between both those values is good for the pull-out tests of Rostasy. For the beam tests of Vandewalle and the pull-out tests of van der Veen, however, the theoretically calculated values can rather be considered as a lower limit.

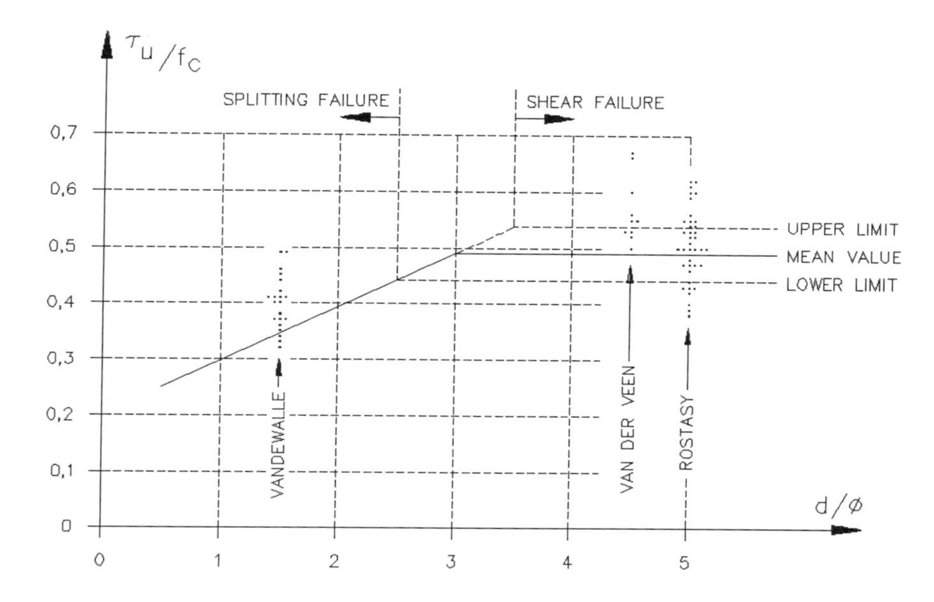

Fig. 7. τ_u/f_c as a function of d/ϕ.

4 Conclusions

From the experiments, it was shown that the bond stress τ_x increased when the temperature was lowered in the range from +20°C to -120°C; at lower temperatures, a slight decrease of the bond stress τ_x was observed. The increase of the bond stress τ_x was primarily a function of the free moisture content of the concrete.

The bond stress-slip relationships can be predicted analytically and a reasonable agreement with the test results at both normal and low temperatures was found.

5 References

1. Vandewalle, L.(1988) Hechting tussen wapening met verbeterde hechting en beton bij gewone en cryogene omstandigheden, Ph.D.thesis, K.U.Leuven, Leuven.
2. Rostasy, F.S. and Schueurmann, J.(1987) Verbundverhalten einbetonierten Betonrippenstahls bei extrem tiefer Temperatur, Deutscher Ausschuss für Stahlbeton, Heft 380, pp. 43-105.
3. van der Veen, C. (1990) Cryogenic bond stress-slip relationship, Ph.D.thesis, Delft University of Technology, Delft.
4. Vandewalle, L. (1992) Theoretical prediction of the ultimate bond strength between a reinforcement bar and concrete, Proceedings International Conference on Bond in Concrete : From Research to Practice, Riga (Latvia).

PART TWENTY-SIX
IMPACT AND FATIGUE

157 ANALYSIS OF SHOCK-TUBE TESTS ON CONCRETE SLABS

F. TOUTLEMONDE and P. ROSSI
Laboratoire Central des Ponts et Chaussées, Paris, France

Abstract

Tests on circular reinforced concrete slabs were carried out using a shock tube. The response of these hyperstatic structures to a quasi-instantaneous load step was derived from pressure, strain and displacement measurements. The natural frequency, an empiric damping factor and local compliance values were determined. Furthermore, a spectrum analysis with damping was performed and helped in defining an equivalent tangent rigidity, the significance of which could be checked by comparing simulations and experimental results. The paper emphasizes this modelling of the slab behaviour and its validity. An agreement of better than 30% has been found with this relatively simple approach. But further analysis is still required to account for permanent deformations, to predict the final failure mode, and to quantify in advance the evolution of the overall rigidity itself, related to the cracking process.

Keywords: Dynamics, modal analysis, reinforced concrete, shock-tube, slabs, structural response, tangent rigidity, tests.

1 Introduction

The French and European design codes lack for reliable models adapted to concrete structures in high strain rate dynamics. Thus, if one knows how to predict the response of structures in the case of accidental dynamic loadings, a possible improvement of cost and security is offered. But to check the predictions of computations, well controlled experiments have to be performed and properly analyzed. Therefore, tests on circular plain and reinforced concrete slabs were carried out using a shock tube [1]. The loading conditions were particularly representative of what may occur to a concrete structural element in the case of accidental impact or explosion.

In fact, accurate control of the loading history and of the boundary conditions was achieved. Material characteristics were obtained experimentally in tension and in compression, under a

Concrete Under Severe Conditions: Environment and loading (Volume Two) Edited by K. Sakai, N. Banthia and O.E. Gjørv. Published in 1995 by E & FN Spon. ISBN 0 419 19860 1

quasi-static regime and at higher rates using the Hopkinson Bars technique [2]. Namely, all this seemed essential for a reliable validation of computational methods. Concretes with various water-cement ratios and internal moisture conditions as well as regular and very high strength (VHS) reinforced concrete (RC) slabs were tested [3], to confirm, at a structural level, the influence of significant parameters pointed out at the material level [4]. This paper focuses on the evolution of the mechanical characteristics of reinforced concrete slabs (considered as model structures), when subjected to successive shocks with increasing load.

2 Characterization of the slab dynamic response

2.1 Loading process

The slabs at the end of the shock tube are hit by an aerian shock wave. The incident overpressure is controlled by the inflation pressure of the reservoir. This incident wave is reflected on the slab. Therefore, the "real" loading is due to the reflected wave. The duration of the loading is controlled by the length of the reservoir. It ranges from about 25 to 100 ms, which is more than 5 to 10 times the natural period of the slab and is long enough to reach a stationary regime. Therefore, transient effects can be neglected and the load can be considered as applied quasi-instantaneously. In fact, the maximum pressure is reached in less than a few tenths of microseconds, which is very small compared to the natural period of the slab (about 2,5 ms). Because of the length of the tube, the shock wave is quasi-plane and the pressure is applied over the whole available section. Thus, the slab is loaded uniformly over a diameter of 82 cm, which corresponds to the diameter of the supporting circle. After the load plateau, the pressure decreases slowly (about 10^6 Pa/s instead of 10^{10} Pa/s during the loading phase). Finally, the whole loading process can be characterized only by the maximum reflected overpressure, which is measured during the tests.

Because the displacements are measured in a local coordinate system, related to the support of the slab, it must be checked whether this system is galilean or not. If not, global inertia forces must be added to the loading. An accelerometer was placed on the support of the slab. Overall displacements and accelerations were not negligible, because during the shock the elasticity of the whole tube could be excited. Therefore, a correction had to be done, after eliminating high frequency noise on the accelerometer signal. However, this correction did not induce a significant change of the maximum load measured by the pressure sensor, and in a first approximation, the global shape of the loading can still be idealized as a Heaviside function.

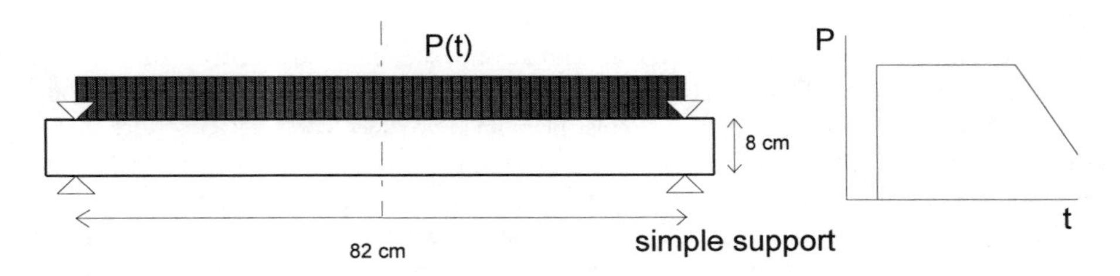

Fig. 1. Loading conditions of the slabs.

It was verified that the boundary conditions were very close to a simple support by comparing the experimental ratio support rotation / maximum deflection with its theoretical value. The best agreement was found with a zero-rotation restraint of the support. All these characteristics of the loading process are summed up in Fig. 1.

2.2 Deformations

As described in [1], the deformations of the slab were measured by strain gauges and displacement sensors. Gauges were glued on the inner and outer sides of the slab along a radius. It was thus possible to check the position of the middle plane. Displacement sensors allowed determining the deflections along a radius at 4 points (center, one-third, two-thirds radius, support) plus the rotation on support. Concerning the shape of these signals as time functions, they can be analyzed as a superimposition of damped vibrations and a quasi-static deformation. As presented on Fig. 2, six parameters can properly describe the curves: initial slope (velocity or strain rate), maximum deflection (or strain), mean stationary value, natural period of oscillations, damping factor, and final slope (giving the local compliance during "quasi-static" unloading).

This analysis was performed for all the measurement channels, and complete results have been synthetized in [5]. During the same shock, the empiric natural period is about the same for all the signals, corresponding to the structural mode I vibration. Therefore, only one mean value was determined. For the damping factor, defined as the logarithmic decrement of the vibrations, a mean value for the orthoradial strain signals seemed to be the most significant way to quantify it for each shot. The maximum or the stationary central deflection divided by the maximum applied pressure can quantify a global dynamic compliance, as well as the final slope of the pressure/deflection curve, the evolution of which may be related to the evolution of the natural period of the slab.

This evolution gives a global indication of the progressive mechanical degradation of a structure depending on the intensity of the shock it has been subjected to. Global indicators, such as the natural frequency, the damping factor and the global rigidity can still be significant after the first cracks have appeared, at least in the case of reinforced (regular or very high strength) concrete slabs (Fig. 3), because damage can first be regarded as diffuse and pseudo-symmetric at the level of the structure. For plain concrete slabs, however, cracking almost coincides with the onset of a block mechanism which these global indicators hardly account for.

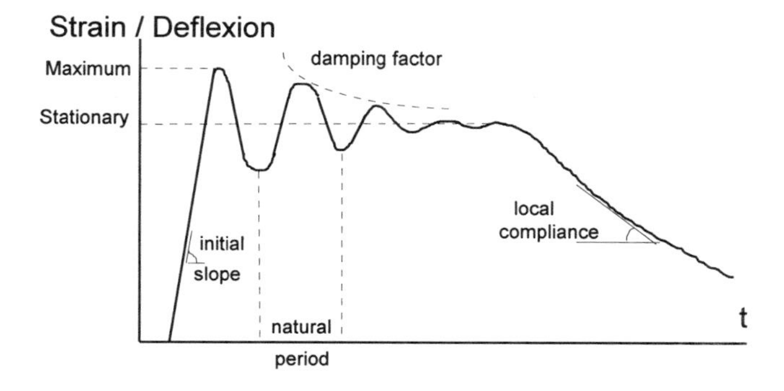

Fig. 2. Characteristic parameters of the time-deformations curves.

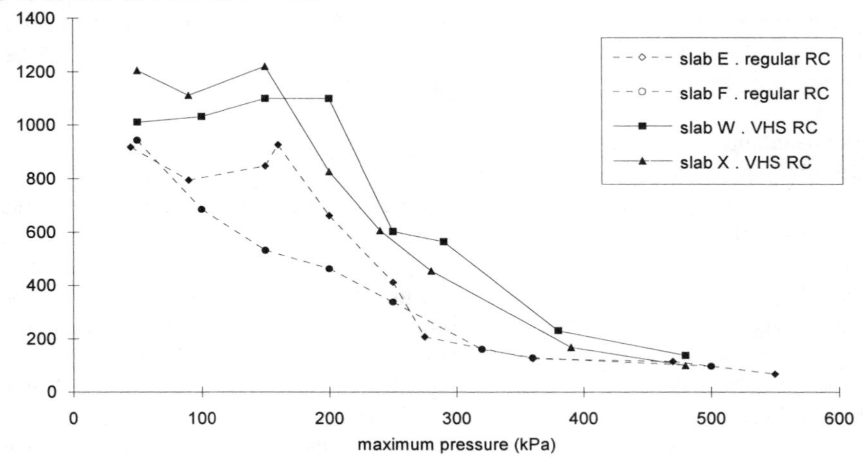

Fig. 3. Evolution of the rigidity of reinforced (regular or VHS) concrete slabs.

Even though the apparent rigidity, as shown on Fig. 3, is not intrinsic, it seems well correlated to macroscopic observations. The onset of visible cracks takes place for a 200 kPa overpressure in both types of concrete, which corresponds to the beginning of a progressive quasi-linear decrease of the rigidity. Beyond 400 kPa for regular concrete and 500 kPa for very high strength concrete, the mesh of fine cracks in the whole central part of the slab is fully developed, and most of the tensile rigidity is given by the reinforcement, which is the same square welded wire mesh (10 cm edge, dia. 7 mm with 2 cm concrete cover) in both cases [1]. Therefore the same residual rigidity is obtained in this ultimate phase. But permanent strains, crack openings and the final failure mode differ, which is hardly characterized with this simple indicator.

3 Elastic Analysis

3.1 Hypotheses and equations

A simplified elastic analysis of the problem was performed to check the validity of this approach for undamaged slabs and to specify the necessity of refined modelling.

Considering the radial symmetry of the problem and assuming that the Love-Kirchhoff hypothesis for thin plates applies in our case, the general equation to determine the deflection w as a function of time t and of the distance r to the axis reads:

$$\frac{\partial^4 w}{\partial r^4} + \frac{2}{r}\frac{\partial^3 w}{\partial r^3} - \frac{1}{r^2}\frac{\partial^2 w}{\partial r^2} + \frac{1}{r^3}\frac{\partial w}{\partial r} + \frac{\rho h}{D}\frac{\partial^2 w}{\partial t^2} = -\frac{p}{D}H(t) \tag{1}$$

p is the applied reflected overpressure (overall inertia is neglected), h the slab thickness, D the rigidity of the slab, the mass density and H(t) the Heaviside distribution. Solving this problem classically requires the separation of variables, i.e. to develop w as a superposition of eigenfunctions w_i, which verify the following homogeneous equations:

$$w_i = \psi_i(r).f_i(t),\quad \ddot{f}_i + f_i\, c_i^{\,2}\frac{D}{\rho h} = 0,\quad \nabla^4\psi_i - c_i^{\,2}\psi_i = 0,\quad \nabla\ cyl.\ gradient \tag{2}$$

It leads to a sine function for f_i and a Bessel's function for ψ_i. The boundary conditions allow the determination of discrete values c_i of c, because the determinant of the system expressing the nullity of the deflection and of the radial moment along the support radius R must be zero. This characteristic equation is the following:

$$2R\sqrt{c}\ \mathrm{I}_0\left(R\sqrt{c}\right).\mathrm{J}_0\left(R\sqrt{c}\right) = (1-\nu)\left[\mathrm{I}_1\left(R\sqrt{c}\right).\mathrm{J}_0\left(R\sqrt{c}\right) + \mathrm{I}_0\left(R\sqrt{c}\right).\mathrm{J}_1\left(R\sqrt{c}\right)\right] \tag{3}$$

With ν the Poisson's ratio of concrete approximately equal to 0.2, and using tabulated values of Bessel's functions [6], this leads to the first three eigenvalues:

$$R^2c_1 = 4.76 \qquad R^2c_2 = 29.59 \qquad R^2c_3 = 74.53 \tag{4}$$

These values are quite close to those reported in [7] with a Poisson's ratio equal to 0.3. To compute the coefficients of the modal distribution, an identification can be done with the static solution, which represents the stationary regime to which vibrations are superimposed in the shock response. The development has been limited to the first three eigenmodes. The final solution is thus the following (with R = 0.41 m):

$$w = -\frac{p}{D}\sum_{i=1}^{3}\frac{\varphi_i}{c_i^2}.\left[J_0\left(r\sqrt{c_i}\right) + \alpha_i.I_0\left(r\sqrt{c_i}\right)\right]\left(1-\cos\left(c_i\sqrt{\frac{D}{\rho h}}t\right)\right) \tag{5}$$

with
$$\frac{\varphi_1}{c_1^2} = 2.025\ 10^{-3}\ \mathrm{m}^4 \qquad \frac{\varphi_2}{c_2^2} = -2.415\ 10^{-5}\ \mathrm{m}^4 \qquad \frac{\varphi_3}{c_3^2} = 9.958\ 10^{-7}\ \mathrm{m}^4$$
$$\alpha_1 = -4.360\ 10^{-2} \qquad \alpha_2 = 6.778\ 10^{-4} \qquad \alpha_3 = 8.374\ 10^{-6} \tag{6}$$

A slightly more general solution of the equations would have admitted a complex value of c, which leads to a damping factor $\exp(-\beta_i t)$ in front of the cosine function. For sake of simplicity, and since only damping of the first mode can be measured, β will be taken in the following as the same for all modes (simultaneously damped by the same factor) and related to η, the observed logarithmic decrement on mode I, by the following formula :

$$\beta = \eta/T \qquad T = \textit{natural period of the slab} \tag{7}$$

3.2 Elastic simulation of the slabs behaviour

In fact, the previous analysis can be used to check the validity of a global analysis of the slabs as damped elastic structures, eventually with a reduced rigidity. First, as the concrete used in casting the slabs has been tested in compression and its Young's modulus measured, the theoretical rigidity can be evaluated as follows:

$$D = \frac{E.h^3}{12\left(1-\nu^2\right)} \tag{8}$$

The discrepancy between the elastic theoretical simulation using this D value and experimental results can be quantified by a quadratic relative deviation of the central deflection. This deviation can be computed considering the local compliance during the unloading phase of the shots. A quite good agreement has been obtained for the first shots, which is a good sign for the quality of boundary conditions and control of the material. The increase of the discrepancy (Fig. 4) expresses the progressive damage of the slab.

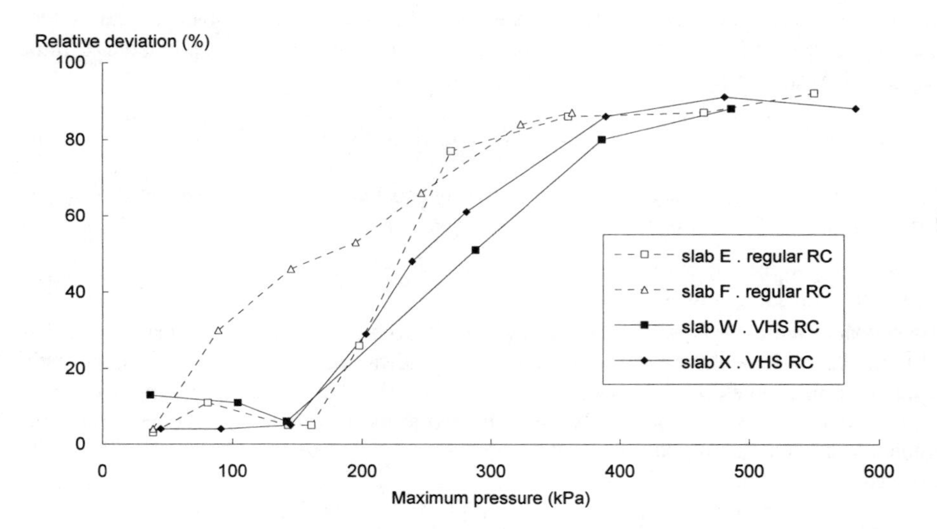

Fig. 4. Relative deviation between test and elastic simulation (central compliance).

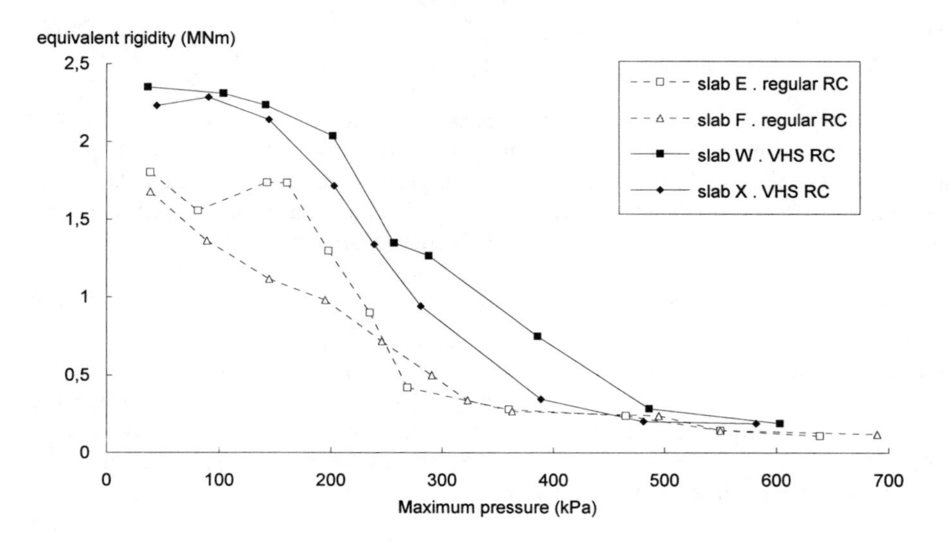

Fig. 5. Evolution of the fitted rigidity D with the loading parameter.

To account for this loss of rigidity, it has been attempted to fit an equivalent rigidity at each shot by minimizing the mean quadratic deviation for local compliances. Compliances are determined during the unloading phase of the shots. Since gauge signals are disturbed by the cracking process after the first shots, an average value is taken among the displacement signals. A deviation smaller than 15% for the central compliance is obtained by this process. The results of this determination are presented on Fig. 5. The evolution of this equivalent rigidity with the applied pressure is very close to the experimental decrease of the central rigidity (Fig. 3). A comparison between the experimental natural frequency of the slabs and the one determined using the fitted D can also be achieved and leads to an error of less than 15%, except for the shot where first significant cracks appear (while the frequency is measured during vibrations before cracks develop, D is determined in the unloading phase, which takes place afterwards).

4 Validity of the analysis

4.1 Damping of the vibrations

An elastic simulation of the slab response leads to undamped vibrations during all the time the pressure is applied. These oscillations have an amplitude twice as large as the static stationary solution. It clearly underestimates viscous and frictional dissipations. Thus, it was decided to integrate the measured damping factor (Fig. 6) in the simulation.

This evolution of the damping factor is hardly correlated to the cracking process, even if visible cracks (at least not too open ones) coincide with an increase in dissipation. The same damping factor defined according to Eq. (7) was used for all the modes because mode I is the most important component, especially for displacements. The correct prediction of the maximum deformed shape of the slab in the first oscillation after the shock can be considered as a test for this modelling (Table 1).

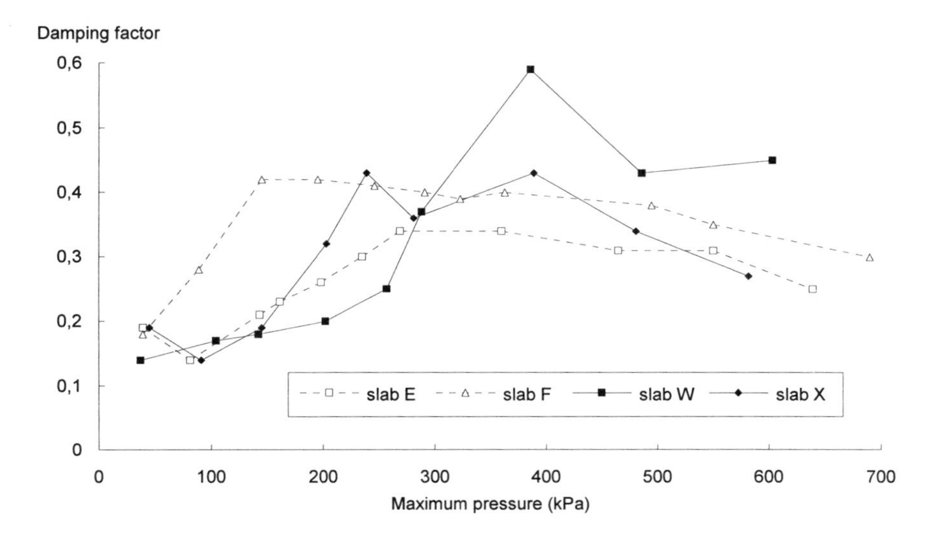

Fig. 6. Damping of the vibrations vs loading of the slabs.

Table 1. Maximum central deflections (computation and measure)

slab E			slab F			slab W			slab X		
p kPa	w_{exp} µm	w_{calc} µm	p kPa	w_{exp} µm	w_{calc} µm	p kPa	w_{exp} µm	w_{calc} µm	p kPa	w_{exp} µm	w_{calc} µm
30	120	79	34	108	85	34	83	58	39	92	74
70	203	192	77	251	233	92	192	165	78	139	147
126	422	299	129	549	450	128	292	233	130	281	247
143	425	336	176	698	689	183	429	361	184	463	419
177	622	548	246	1080	1190	253	704	686	235	618	617
235	946	930	323	2720	3330	285	842	797	281	947	1050
269	1520	2255	363	4010	4720	383	2040	1720	387	3240	3890
360	4430	4540	495	6560	7210	484	5410	5880	481	7530	8360
466	7780	6890	550	18100	13400	603	9470	10900	582	15300	11000
550	10800	13500	690	20100	20500						
639	21600	20900									

Except for one shot, all the simulations lead to an error smaller than 30%. For all the shots, the average deviation is 14.7%. More than the maximum deflection, the strains and displacements evolution can also be predicted (Fig. 7). A global visual comparison with the experimental curves (Fig. 8) can be done, and some quantified features as the initial central velocity can be compared to predictions. The computation leads to correct estimations (error smaller than 20%) only for uncracked slabs. It may be concluded that when cracks appear, an overall viscous-type damping factor, as it was used here, leads to overestimating the initial velocity and underestimating the damping of later vibrations, in neglecting part of the frictional dissipation. It suggests to account explicitly for the cracks mechanical consequences, namely permanent localized strains.

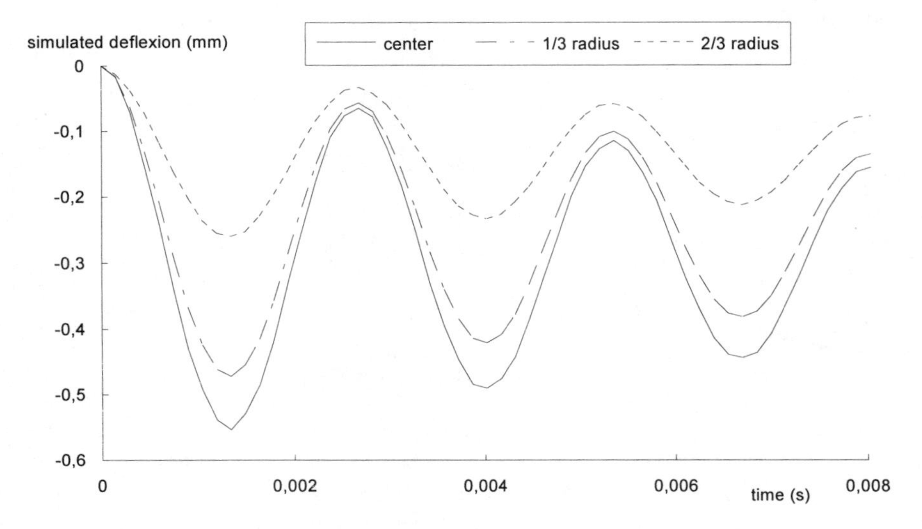

Fig. 7. Slab E. Loading pressure 198 kPa. Simulated deflections along a radius.

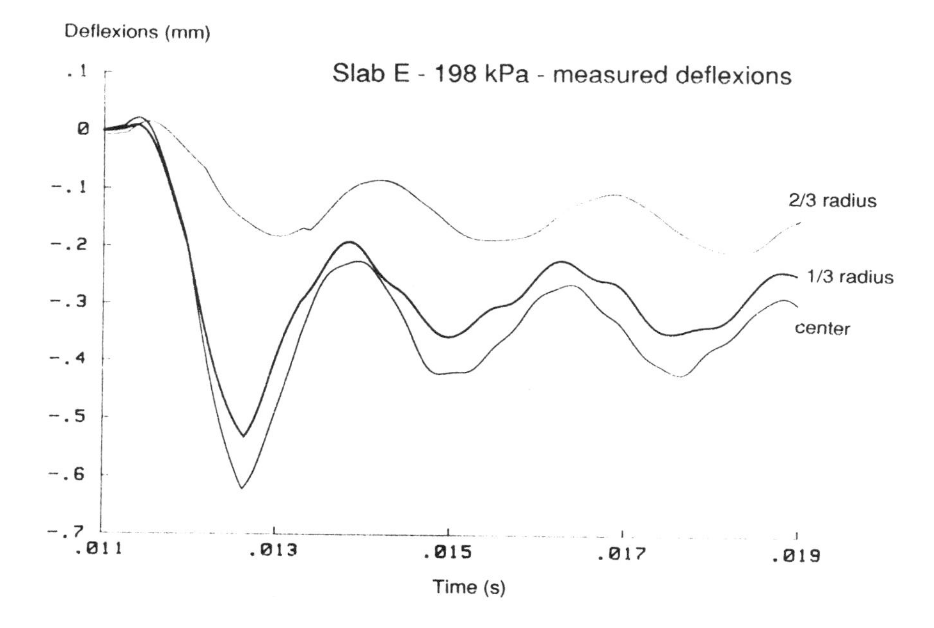

Fig. 8. Slab E. Loading pressure 198 kPa. Experimental deflections along a radius.

4.2 Permanent strains

The equivalent rigidity which has been determined accounts quite well for the elastic "tangent" properties of the slabs during the shots, as the correct prediction of compliances and natural frequency indicates. However, it cannot give an accurate description of permanent strains, which appear with the slab cracking and are related to the increasing discrepancy between the simulations and the experiments. Nevertheless, the difference between the measured stationary value of the deformation and the prediction which could be done with an elastic tangent modulus gives an estimation of the increment of irreversible strains, which can be interpreted as a sum of crack widths. These differences between the total deformations and recoverable strains have been computed and correlated with maximum crack openings observed in the center of the slabs (Table 2).

Permanent strains are related mainly to concrete cracks, slip between concrete and steel, and reinforcement yielding and can be roughly estimated at each shot. In the first step of the damage process, up to about 250 kPa, these permanent deflections are almost completely explained by the microcrack openings in the central part of the outer concrete side. Afterwards, the crack mesh develops, and the increment of permanent deflections may correspond to 3 or 4 crack openings along the radius, depending on the crack repartition due to the reinforcement. However, the failure mode actuated, that is, reinforcement yielding at the center of the slabs and bending failure with VHS concrete or slipping of the reinforcement and shear failure in the case of ordinary concrete slabs [4], is not accounted for in this simplified analysis and requires further investigation (and an explicit description of cracks).

Table 2. Increments of permanent deflections (estimated) and measured crack openings

slab E			slab F			slab W			slab X		
p kPa	w_{irr} µm	µm	p kPa	w_{irr} µm	µm	p kPa	w_{irr} µm	µm	p kPa	w_{irr} µm	µm
30	5		34	-20		34	3		39	-3	
70	-1		77	-1		92	-1		78	1	
126	23		129	51		128	-9		130	6	
143	33		176	84	30	183	53	30	184	86	60
177	45		246	28	30	253	115	60	235	70	120
235	115		323	208	150	285	91	100	281	95	120
269	55	30	363	357	150	383	881	150	387	650	200
360	806	n.m.	495	1140	600	484	1121	200	481	1489	600
466	1922	n.m.	550	9259	2500				582	8548	4000
550	2267	n.m.									

5 Conclusion

Analysing these slab tests has demonstrated that the experimental data are fully adapted to validate a material modelling of concrete, or a simplified approach to compute the structural response. Capabilities of a tangent spectrum analysis have been pointed out to explain maximum deflections (precision better than 30%) and interprete test results in determining permanent strains. But this analysis only accounts *a posteriori* for the observed cracks and their consecutive increase in equivalent compliance, in damping (friction) and in unrecoverable strains. Predicting permanent strains and the position of most critical cracks and knowing which failure mode is activated, are still challenges.

6 References

1. Toutlemonde, F., Boulay, C. and Gourraud, C. (1993) Shock-tube tests of concrete slabs. *Materials and Structures*, Vol. 26, pp. 38-42.
2. Rossi, P., van Mier, J.G.M., Toutlemonde, F., Le Maou, F. and Boulay, C. (1994) Effect of loading rate on the strength of concrete subjected to uniaxial tension. *Materials and Structures*, Vol. 27, pp. 260-264.
3. Toutlemonde, F., Rossi, P., Boulay, C., Gourraud, C. and Guédon, D. (1993) Comportement dynamique des bétons : essais de dalles au tube à choc. *Bulletin de Liaison des Laboratoires des Ponts et Chaussées*, No. 184. pp. 77-85 (in French).
4. Toutlemonde, F., Rossi, P. (1994) Shock-tested R.C. slabs : significant parameters, in *ASCE Structures Congress XII*, (ed. N.C. Baker and B.J. Goodno), pp. 227-232.
5. Toutlemonde, F. (1993) *Essais de dalles de béton au tube à choc. Rapport d'essais, données mode d'emploi*, Synthesis report, LCPC, Paris (in French).
6. Angot, A. (1965) *Compléments de mathématiques à l'usage des ingénieurs de l'électrotechnique et des télécommunications*, Ed. Revue d'Optique, Paris (in French).
7. Schaffar M. (1970) *Calcul de la déformation dynamique de plaques circulaires encastrées ou appuyées, uniformément chargées par un échelon de pression*, Institut franco-allemand de recherches de Saint-Louis, report 25/70 (in French).

158 DYNAMIC NONLINEAR RESPONSE ANALYSIS OF PLATES

H. ISHITANI and L.A.C. DIOGO
Escola Politécnica da Universidade de São Paulo, São Paulo, Brazil

Abstract
The aim of this paper is to present a procedure related to the finite element method that describes the nonlinear behavior of a plate under impact loads. In the development of this method, the von Mises' criterion of plastification and the membrane effect associated with a moderately large displacements were considered.
Keywords: Analysis, impact, nonlinear, plates.

1. Introduction

In a nonlinear analysis of a plate under impact loads, numerical methods have proved themselves a very interesting tool. The finite element method associated with processes of numerical integration of the dynamic equilibrium equations is a reliable path which was taken in this paper.

Through the process, some regions may undergo elasto-plastic strains during the loading and elastic strains during the unloading. In order to take this into account, we make use of the von Mises' criterion of plastification. In addition, even with small strains, we can have moderately large displacements, which corresponds to the well known membrane effect. In order to take this into account, we retain the nonlinear terms in the strain components.

An example of application is shown to illustrate the proposed formulation.

2. Finite element formulation

The vector $\underline{u}$ of the displacement components u, v, w in the x, y, z directions of a point of the middle plane characterized by the coordinates x, y is related with the nodal displacements $\underline{U}_e$ of the element by the shape function matrix $\underline{N}$, that is,

Concrete Under Severe Conditions: Environment and loading (Volume Two) Edited by K. Sakai, N. Banthia and O.E. Gjørv. Published in 1995 by E & FN Spon. ISBN 0 419 19860 1

$$\underline{u} = \begin{bmatrix} \underline{N}_c & \underline{0} \\ \underline{0} & \underline{N}_p \end{bmatrix} \begin{bmatrix} \underline{U}_{ce} \\ \underline{U}_{pe} \end{bmatrix} = \underline{N}\underline{U}_e \tag{1}$$

where $\underline{U}_{ce}$ is the displacement vector associated with the strain of the middle plane (which contains the displacements in the x, y directions) and $\underline{U}_{pe}$ is the displacement vector associated with the bending (which contains the displacement in the z direction and the rotations in the x, y directions).

The relation between the nodal displacements $\underline{U}_e$ of the element and the nodal displacements $\underline{U}$ of the structure can be indicated as

$$\underline{U}_e = \underline{T}_e \underline{U}. \tag{2}$$

The generalized strain vector $\underline{\varepsilon}$ is given by

$$\underline{\varepsilon} = \begin{bmatrix} \underline{\varepsilon}_c \\ \underline{\varepsilon}_p \end{bmatrix} = \begin{bmatrix} \underline{\varepsilon}_{oc} \\ \underline{\varepsilon}_{op} \end{bmatrix} + \begin{bmatrix} \underline{\varepsilon}_{1c} \\ \underline{0} \end{bmatrix} = \underline{\varepsilon}_o + \underline{\varepsilon}_1 \tag{3}$$

where $\underline{\varepsilon}_c$ contains the generalized strains of the middle plane (which is the sum of the linear term $\underline{\varepsilon}_{oc}$ due to the displacements $\underline{U}_{ce}$ and the nonlinear term $\underline{\varepsilon}_{1c}$ due to the displacements $\underline{U}_{pe}$) and $\underline{\varepsilon}_p$, the generalized strains associated with the bending (which is due to the displacements $\underline{U}_{pe}$).

Increments of the generalized strains correspond to increments of the nodal displacements, such that

$$d\underline{\varepsilon} = \underline{B}d\underline{U}_e = \underline{B}\underline{T}_e d\underline{U} \tag{4}$$

with

$$\underline{B} = \begin{bmatrix} \underline{B}_{oc} & \underline{0} \\ \underline{0} & \underline{B}_{op} \end{bmatrix} + \begin{bmatrix} \underline{0} & \underline{B}_{1p} \\ \underline{0} & \underline{0} \end{bmatrix} = \underline{B}_0 + \underline{B}_1 \tag{5}$$

where $\underline{B}_{oc}$ and $\underline{B}_{op}$ permit the obtaining of the linear terms $d\underline{\varepsilon}_{oc}$ and $d\underline{\varepsilon}_{op}$ of the increments $d\underline{\varepsilon}_c$ and $d\underline{\varepsilon}_p$ due to the increments $d\underline{U}_{ce}$ and $d\underline{U}_{pe}$, respectively, and $\underline{B}_{1p}$ permits the obtaining of the nonlinear term $d\underline{\varepsilon}_{1c}$ of the increment $d\underline{\varepsilon}_c$ due to the increment $d\underline{U}_{pe}$.

The generalized stress vector is given by

$$\underline{\sigma} = \begin{bmatrix} \underline{\sigma}_c \\ \underline{\sigma}_p \end{bmatrix} \tag{6}$$

where $\underline{\sigma}_c$ contains the generalized stresses associated with the strains of the middle plane (the membrane forces N_x, N_y, N_{xy}), and $\underline{\sigma}_p$, the generalized stresses associated with the bending (the bending and twisting moments M_x, M_y, M_{xy}).

Increments of the generalized stresses $d\underline{\sigma}$ correspond to increments of the generalized strains $d\underline{\varepsilon}$, such that

$$d\underline{\sigma} = \underline{D} d\underline{\varepsilon}. \tag{7}$$

In the regions of the plate undergoing only elastic strains or when there is unloading of the regions undergoing plastic strains,

$$\underline{D} = \underline{D}_E = \begin{bmatrix} \underline{D}_c & \underline{0} \\ \underline{0} & \underline{D}_p \end{bmatrix} \tag{8}$$

where $\underline{D}_c$ permits the obtaining of the increment $d\underline{\sigma}_c$ due to $d\underline{\varepsilon}_c$ and $\underline{D}_p$, the increment $d\underline{\sigma}_p$ due to $d\underline{\varepsilon}_p$. In the regions of the plate undergoing elasto-plastic strains,

$$\underline{D} = \underline{D}_{EP} = \left[\underline{D}_E - \frac{\underline{D}_E \underline{\phi} \underline{\phi}^T \underline{D}_E}{\underline{\phi}^T \underline{D}_E \underline{\phi}} \right] = \underline{D}_E - \underline{D}_P \tag{9}$$

where

$$\underline{\phi}^T = \left[\frac{\partial F}{\partial N_x} \quad \frac{\partial F}{\partial N_y} \quad \frac{\partial F}{\partial N_{xy}} \quad \frac{\partial F}{\partial M_x} \quad \frac{\partial F}{\partial M_y} \quad \frac{\partial F}{\partial M_{xy}} \right] \tag{10}$$

F being the yield surface corresponding to the generalization of the von Mises' criterion, given, according to [6], by

$$F = Q_N + Q_M + \frac{\sqrt{3}}{3} |Q_{NM}| - 1 \tag{11}$$

with

$$Q_N = \overline{N}_x^2 - \overline{N}_x \overline{N}_y + \overline{N}_y^2 + 3\overline{N}_{xy}^2 \tag{11.a}$$

$$Q_M = \overline{M}_x^2 - \overline{M}_x \overline{M}_y + \overline{M}_y^2 + 3\overline{M}_{xy}^2 \tag{11.b}$$

$$Q_{NM} = \overline{N}_x \overline{M}_x - \frac{1}{2} \overline{N}_x \overline{M}_y - \frac{1}{2} \overline{N}_y \overline{M}_x + \overline{N}_y \overline{M}_y + 3\overline{N}_{xy} \overline{M}_{xy} \tag{11.c}$$

and

$$\overline{N}_x = \frac{N_x}{N_e}, \ \ldots, \ \overline{M}_{xy} = \frac{M_{xy}}{M_e}. \tag{11.d}$$

In the above expressions,

$$N_e = \sigma_e h \quad e \quad M_e = \sigma_e \frac{h^2}{4} \tag{11.e}$$

where σ_e is the yielding stress and h the thickness of the plate.

The discretization by finite elements leads to the system of equilibrium equations

$$\underline{\psi} = \int \underline{B}^T \underline{\sigma} dS - \left(\underline{P} - \underline{M}\underline{\ddot{U}} - \underline{C}\underline{\dot{U}}\right) = \underline{R} - \left(\underline{P} - \underline{M}\underline{\ddot{U}} - \underline{C}\underline{\dot{U}}\right) = \underline{0} \tag{12}$$

where $\underline{R}$ is the generalized internal force vector, $\underline{P}$ the generalized external force vector, $\underline{M}$ the generalized mass matrix and $\underline{C}$ the generalized damp matrix.

From the definition of $\underline{\psi}$, we can write

$$d\underline{\psi} = \int \underline{B}^T d\underline{\sigma} dS + \int d\underline{B}^T \underline{\sigma} dS - \left(d\underline{P} - \underline{M}d\underline{\ddot{U}} - \underline{C}d\underline{\dot{U}}\right). \tag{13}$$

The contribution of each element to the first term of the above expression is given, considering the expressions (7) and (4), by

$$\int_{Se} \underline{B}^T d\underline{\sigma} dS = \int_{Se} \underline{B}^T \underline{D}_{EP} d\underline{\varepsilon} dS = \left(\int_{Se} \underline{B}^T \underline{D}_{EP} \underline{B} dS \right) d\underline{U}_e = \underline{k}^* d\underline{U}_e \tag{14}$$

and to the second term, according to (1), by

$$\int_{Se} d\underline{B}^T \underline{\sigma} dS = \underline{k}_\sigma d\underline{U}_e . \tag{15}$$

Expressions (14) and (15) enable us to write the expression (13) as

$$d\underline{\psi} = \underline{K}d\underline{U} - \left(d\underline{P} - \underline{M}d\underline{\ddot{U}} - \underline{C}d\underline{\dot{U}}\right) \tag{16}$$

or in the incremental form

$$\Delta\underline{\psi} = \underline{K}\Delta\underline{U} - \left(\Delta\underline{P} - \underline{M}\Delta\underline{\ddot{U}} - \underline{C}\Delta\underline{\dot{U}}\right). \tag{17}$$

When, at the the time $t + \Delta t$, we perform the iteration k, in which clearly $\Delta\underline{P} = \underline{0}$, we have

$$\begin{aligned} \underline{\psi}^k_{t+\Delta t} - \underline{\psi}^{k-1}_{t+\Delta t} = {} & \underline{K}^{k-1}_{t+\Delta t}\left(\underline{U}^k_{t+\Delta t} - \underline{U}^{k-1}_{t+\Delta t}\right) + \underline{M}\left(\underline{\ddot{U}}^k_{t+\Delta t} - \underline{\ddot{U}}^{k-1}_{t+\Delta t}\right) \\ & + \underline{C}\left(\underline{\dot{U}}^k_{t+\Delta t} - \underline{\dot{U}}^{k-1}_{t+\Delta t}\right) \end{aligned} \tag{18}$$

where

$$\underline{\psi}^k_{t+\Delta t} = \underline{0} \tag{18.a}$$

$$\underline{\psi}_{t+\Delta t}^{k-1} = \underline{R}_{t+\Delta t}^{k-1} - \underline{P}_{t+\Delta t} + \underline{M}\underline{\ddot{U}}_{t+\Delta t}^{k-1} + \underline{C}\underline{\dot{U}}_{t+\Delta t}^{k-1} \tag{18.b}$$

and, according to the trapezoidal rule,

$$\underline{U}_{t+\Delta t}^{k} = \underline{U}_{t} + \frac{1}{2}\left(\underline{\dot{U}}_{t} + \underline{\dot{U}}_{t+\Delta t}^{k-1}\right)\Delta t \tag{19.a}$$

$$\underline{\dot{U}}_{t+\Delta t}^{k} = \underline{\dot{U}}_{t} + \frac{1}{2}\left(\underline{\ddot{U}}_{t} + \underline{\ddot{U}}_{t+\Delta t}^{k-1}\right)\Delta t . \tag{19.b}$$

The acceleration $\underline{\ddot{U}}_{t+\Delta t}^{k}$ is obtained from the equilibrium equation (12). From expressions (18) and (19), we have

$$\begin{aligned}\left[\underline{K}_{t+\Delta t}^{k-1} + \frac{4}{\Delta t^2}\underline{M} + \frac{2}{\Delta t}\underline{C}\right]\Delta\underline{U}_{t+\Delta t}^{k} &= \underline{P}_{t+\Delta t} - \underline{R}_{t+\Delta t}^{k-1} \\ &-\underline{M}\left[\frac{4}{\Delta t^2}\left(\underline{U}_{t+\Delta t}^{k-1} - \underline{U}_{t}\right) - \frac{4}{\Delta t}\underline{\dot{U}}_{t} - \underline{\ddot{U}}_{t}\right] \\ &-\underline{C}\left[\frac{2}{\Delta t}\left(\underline{U}_{t+\Delta t}^{k-1} - \underline{U}_{t}\right) - \underline{\dot{U}}_{t}\right]\end{aligned} \tag{20}$$

Writing the above expression for $k = 1$ and noting that

$$\underline{U}_{t+\Delta t}^{o} = \underline{U}_{t} \;\; ; \;\; \underline{K}_{t+\Delta t}^{o} = \underline{K}_{t} \;\; ; \;\; \underline{R}_{t+\Delta t}^{o} = \underline{R}_{t} = \underline{P}_{t} - \underline{M}\underline{\ddot{U}}_{t} - \underline{C}\underline{\dot{U}}_{t}, \tag{21}$$

we obtain the expression corresponding to the increment Δt

$$\left[\underline{K}_{t} + \frac{4}{\Delta t^2}\underline{M} + \frac{2}{\Delta t}\underline{C}\right]\Delta\underline{U} = \Delta\underline{P}_{t} - \underline{M}\left[\frac{4}{\Delta t}\underline{\dot{U}}_{t} - 2\underline{\ddot{U}}_{t}\right] - \underline{C}\left[-2\underline{\dot{U}}_{t}\right]. \tag{22}$$

With the vector $\underline{U}_{t+\Delta t}^{k}$ we can determine the vector $\underline{\varepsilon}_{t+\Delta t}^{k}$ as well as the increment

$$\Delta\underline{\varepsilon}_{t+\Delta t}^{k} = \underline{\varepsilon}_{t+\Delta t}^{k} - \underline{\varepsilon}_{t+\Delta t}^{k-1} = \Delta\underline{\varepsilon}_{k} \tag{23}$$

and also the increment

$$\Delta\underline{\sigma}_{t+\Delta t}^{k} = \underline{\sigma}_{t+\Delta t}^{k} - \underline{\sigma}_{t+\Delta t}^{k-1} = \Delta\underline{\sigma}_{k} \tag{24}$$

where

$$\Delta\underline{\sigma}_{k} = \int_{0}^{\Delta\varepsilon_k} \underline{D}_{EP} d\underline{\varepsilon} = \int_{0}^{\Delta\varepsilon_k} \left(\underline{D}_{E} - \underline{D}_{P}\right) d\underline{\varepsilon} = \Delta\underline{\sigma}_{k}^{E} - \Delta\underline{\sigma}_{k}^{P} \tag{25}$$

with

$$\Delta\underline{\sigma}_k^E = \int_0^{\Delta\underline{\varepsilon}_k} \underline{D}_E d\underline{\varepsilon} = \underline{D}_E \int_0^{\Delta\underline{\varepsilon}_k} d\underline{\varepsilon} = \underline{D}_E \Delta\underline{\varepsilon}_k \tag{25.a}$$

$$\Delta\underline{\sigma}_k^P = \int_0^{\Delta\underline{\varepsilon}_k} \underline{D}_P d\underline{\varepsilon} = \int_{r\Delta\underline{\varepsilon}_k}^{\Delta\underline{\varepsilon}_k} \underline{D}_P d\underline{\varepsilon}. \tag{25.b}$$

In the above expression, r characterizes the point where, through the increment $\Delta\underline{\varepsilon}_k$, the yield surface was reached. A first estimate, r', of r can be obtained with the following linear interpolation:

$$r' = -\frac{F_o}{F_1 - F_o} \tag{26}$$

where $F_o = F(P_o) = F(\underline{\sigma}_{t+\Delta t}^{k-1}) < 0$ *and* $F_1 = F(P_1) = F(\underline{\sigma}_{t+\Delta t}^{k-1} + \Delta\underline{\sigma}_k^E) \geq 0.$

When $F_1 < 0$, we put $r = 1$, since there are no plastic strains; when $F_o = 0$ and $F_1 \geq 0$, we put $r = 0$, since there are plastic strains throughout the increment.

The above procedure, being an approximate one, determines the point P' with coordinates $\underline{\sigma}_{t+\Delta t}^{k-1} + r'\Delta\underline{\sigma}_k^E$, near the yield surface. The projection of this point P' on the yield surface through the direction P_oP_1 determines the point P corresponding to the beginning of the plastification. Considering the points P and P', we can write:

$$\Delta F = F - F' = 0 - F' = \left(\frac{\partial F}{\partial N_x}\right)' \Delta N_x + \ldots + \left(\frac{\partial F}{\partial M_{xy}}\right)' \Delta M_{xy} = \underline{\phi}'^T \Delta\underline{\sigma} \tag{27}$$

where $\underline{\phi}'$ is the value of $\underline{\phi}$ at the point P'.

Making $\Delta\underline{\sigma}$ proportional to $\Delta\underline{\sigma}_k^E$, that is,

$$\Delta\underline{\sigma} = \Delta r' \Delta\underline{\sigma}_k^E, \tag{28}$$

we have

$$\Delta r' = -\frac{F'}{\underline{\phi}'^T \Delta\underline{\sigma}_k^E}. \tag{29}$$

Now we can carry out numerically the integration of the expression (25.b) by dividing the interval $(1-r)\Delta\underline{\varepsilon}_k$ into p parts which correspond to $p + 1$ points on the yield surface. At a point P_j

$$\underline{\sigma}_{k,j} = \underline{\sigma}_{k,j-1} + \Delta\underline{\sigma}_{k,j}\,, \quad \text{where} \quad \Delta\underline{\sigma}_{k,j} = \Delta\underline{\sigma}_{k,j}^E - \Delta\underline{\sigma}_{k,j}^P \tag{30}$$

with

$$\Delta\underline{\sigma}^{E}_{k,j} = \underline{D}_{E}\left[\frac{1-r}{p}\Delta\underline{\varepsilon}_{k}\right] \quad and \quad \Delta\underline{\sigma}^{P}_{k,j} = \underline{D}_{P,j-1}\left[\frac{1-r}{p}\Delta\underline{\varepsilon}_{k}\right]. \tag{31}$$

The above procedure, being an approximate one, determines the point P_j', near the yield surface. The point P_j is determined by projecting the point P_j' on the yield surface through a normal to the surface at P_j'. Considering the points P_j and P_j', we can write:

$$\Delta F = F - F' = 0 - F' = \left(\frac{\partial F}{\partial N_x}\right)' \Delta N_x + \ldots + \left(\frac{\partial F}{\partial M_{xy}}\right)' \Delta M_{xy} = \underline{\phi}'^{T}\Delta\underline{\sigma} \tag{32}$$

where $\underline{\phi}'$ is the value of $\underline{\phi}$ at the point P_j'.

Making $\Delta\underline{\sigma}$ proportional to $\underline{\phi}'$, that is,

$$\Delta\underline{\sigma} = k\,\underline{\phi}', \tag{33}$$

we have

$$k = -\frac{F'}{\underline{\phi}'^{T}\,\underline{\phi}'}. \tag{34}$$

With the vector $\underline{\sigma}^{k}_{t+\Delta t}$, determined through the above procedure, we can determine the vector

$$\underline{\psi}^{k}_{t+\Delta t} = \underline{R}^{k}_{t+\Delta t} - \underline{P}_{t+\Delta t} + \underline{M}\,\underline{\ddot{U}}^{k}_{t+\Delta t} + \underline{C}\,\underline{\dot{U}}^{k}_{t+\Delta t} \tag{35}$$

and the residue

$$r^{k}_{t+\Delta t} = \frac{\left(\underline{\psi}^{k}_{t+\Delta t}\right)^{T}\left(\underline{\psi}^{k}_{t+\Delta t}\right)}{\underline{P}^{T}_{t+\Delta t}\,\underline{P}_{t+\Delta t}} \tag{36}$$

which is required to be less than a prescribed value (10^{-16} in this paper) before another increment is considered.

3. Example

With the previous formulation, we can determine the behavior of a simple supported square plate, showed in Fig. 1, under the uniform load ϕq, where $q = 0.010 kN / cm^2$. The material is such that: $E = 7000\, kN / cm^2$, $\nu = 0.35$, $f_y = 5\, kN / cm^2$ and $\rho = 2700\, kg / m^3$. We consider no damping present.

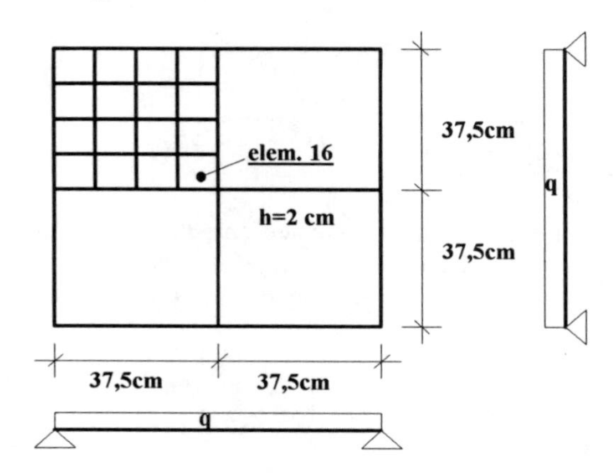

Fig. 1. Plate under uniform load.

Initially, we consider the loading to be applied statically and determine the load-deflection curves at the center of the plate showed in Fig. 2 and the bending moment curves for the element 16, Fig. 3.

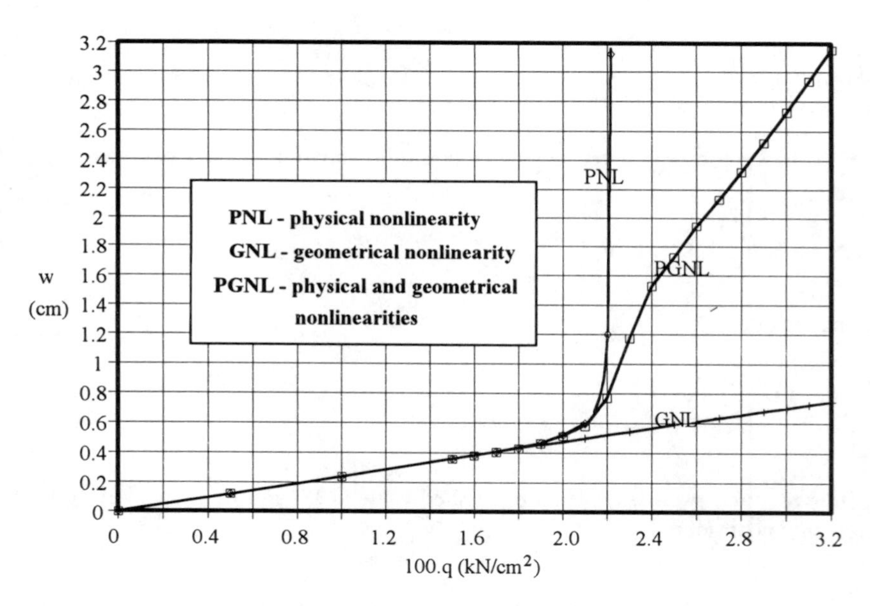

Fig. 2. Deflection at the center of the plate; static analysis.

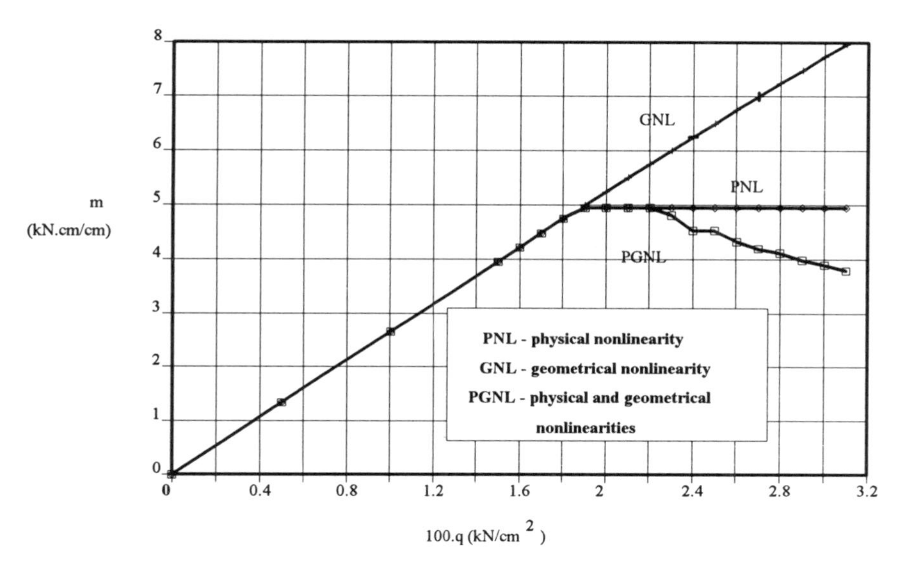

Fig. 3. Average bending moment at element 16; static analysis.

Then, we consider a dynamic analysis for an applied step-function load and obtain, with a time interval of *0.0003 sec*, the curves showed in Figs. 4 and 5.

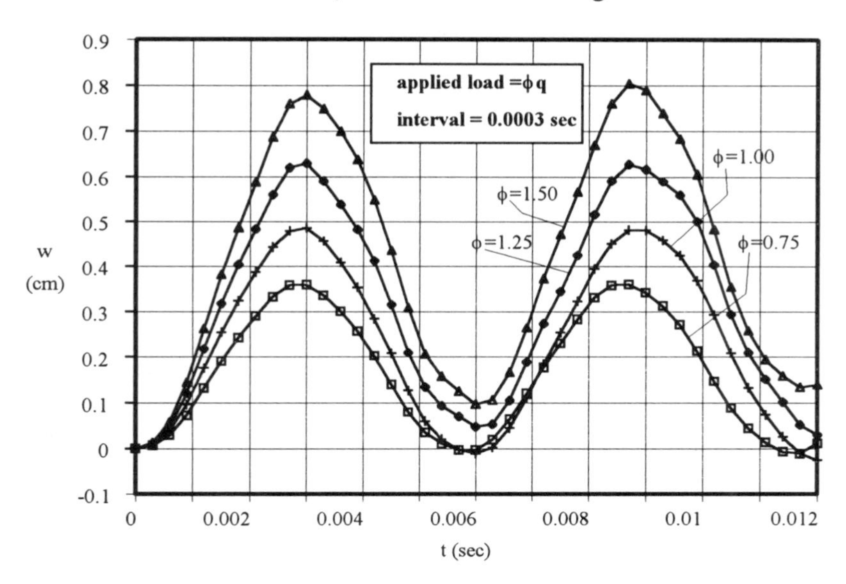

Fig. 4. Deflection at the center of the plate; dynamic analysis.

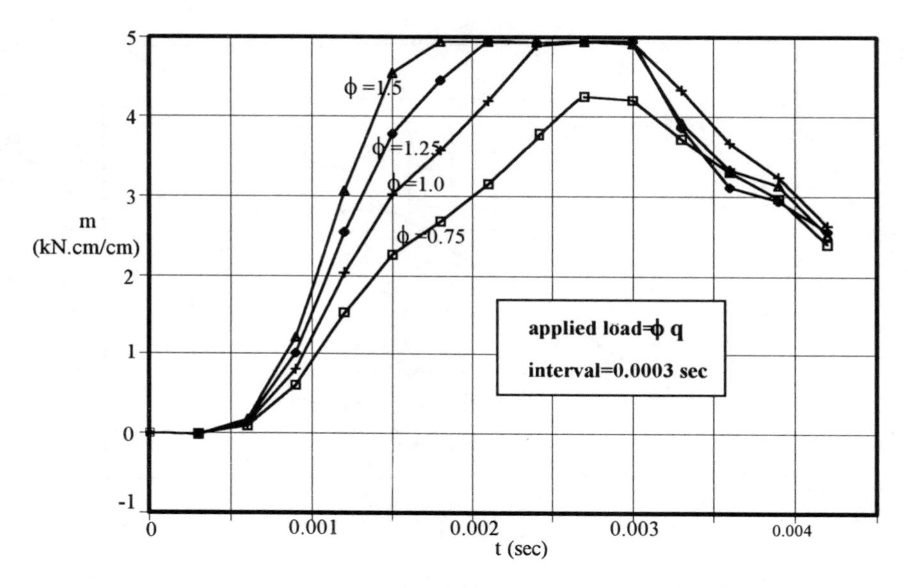

Fig. 5. Average bending moment at element 16; dynamic analysis.

4. Conclusion

It can be concluded from Fig. 2 that (1) when only physical nonlinearity is considered, the structure presents a well-defined load capacity and (2) when geometrical nonlinearity is also considered, there is stiffening of the structure due to the membrane effect. The more deformed is the structure as a result of the progressive plastification, the more important is this effect.

It can be concluded from Fig. 5 that the greater is the load applied, the greater the interval during which element 16 remains plastified.

5. References

1. Zienkiewics O. C. (1977) *The Finite Element Method*, McGraw-Hill.
2. Bathe K. J. (1982) *Finite Element Procedures in Engineering Analysis*, Prentice-Hall Inc., New Jersey, U.S.A.
3. Clough, R. W. and Penzien, J. (1975) *Dynamics of Sructures*, McGraw-Hill, Tokyo.
4. Owen, D. R. J. and Hinton E. (1980) *Finite Element in Plasticity: Theory and Practice*, Pineridge Press, Swansea, U. K.
5. Chen, W. F. (1982) *Plasticity in Reinforced Concrete*. McGraw-Hill.
6. Iliouchine, A. A. (1956) *Plasticité: Déformations Élastico-Plastiques*, Editions Eyrolles, Paris.

159 FULL SCALE IMPACT TEST OF PC ROCK-SHED WITH SHOCK ABSORBING SYSTEM

M. SATO and H. NISHI
Hokkaido Development Bureau, Sapporo, Japan
N. SUGATA and N. KISHI
Muroran Institute of Technology, Muroran, Japan

Abstract

In this paper, the dynamic behavior of a full scale Prestressed Concrete (PC) rock-shed with a shock absorbing system under an impact load generated by a 49 kN free falling weight is discussed. The PC rock-shed was composed of five T-shaped PC girders. The shock absorbing systems used here were a single layered sand cushion with 90 cm thickness and a three-layered absorbing system composed of a 50 cm thick sand layer (top), a 20 cm thick Reinforced Concrete (RC) core slab and a 100 cm thick Expanded Poly-Styrol layer (bottom). The influence of the stiffness of the PC rock-shed on the weight impact force, and the influence of the shock absorbing system and the lateral prestressing on the dynamic behavior of the PC rock-shed were investigated.

Keywords: impact behavior, load share ratio, PC rock-shed, shock absorbing system

1 Introduction

In Japan, there are many rock-sheds over mountainous and coastal roads with cliffs. The rock-sheds have been designed based on the design manual for anti-rock falling structures [1]. However, the design manual has specified only the maximum impact force equation and the construction procedure of the sand absorbing system. The evaluation method for sectional forces of rock-sheds has not been established yet. On the other hand, PC rock-sheds, one of the conventional multi-girder structures, are usually jointed by laterally prestressing the girders. Even though it is clear that they behave as a continuous structure, each girder is designed as an individual unit, ignoring the load share effect based on struc-

Concrete Under Severe Conditions: Environment and loading (Volume Two) Edited by K. Sakai, N. Banthia and O.E. Gjørv. Published in 1995 by E & FN Spon. ISBN 0 419 19860 1

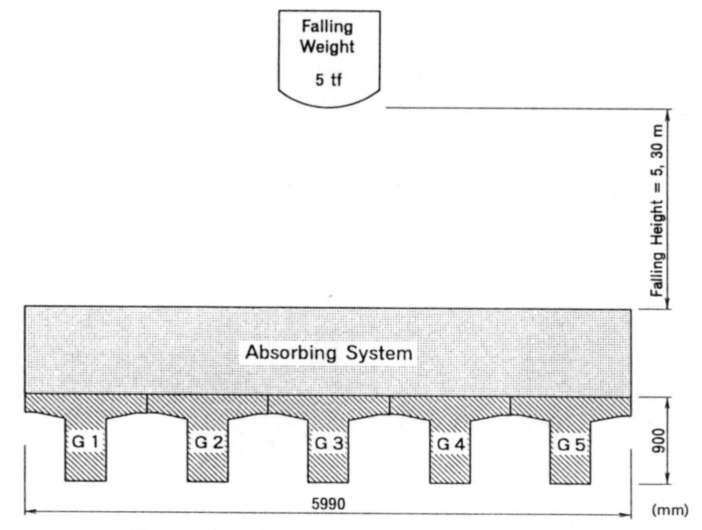

Fig. 1. Schematic diagram of impact test.

tural continuity. To have a rational design method, it is necessary to clarify the dynamic behavior and the load share ratio of each PC girder.

In this study, to establish the design method for PC rock-sheds, full scale impact tests of a PC multi-girder with two types of shock absorbing systems were executed. The impact force was generated by a 49 kN free falling weight. The shock absorbing systems applied in this study were: a single layered sand cushion with 90 cm thickness and a three-layered absorbing system composed of a 50 cm thick sand layer as the top layer, a 20 cm thick RC core slab and 100 cm thick Expanded Poly-Styrol (EPS) blocks as the bottom layer[2].

2 Outline of experiment

Impact tests were executed by using a PC multi-girder composed of five girders with two types of shock absorbing systems, as shown in Fig. 1, which is the main structure of a PC rock-shed and is used in the roof. The impact load was generated by a 49 kN free falling weight from a predetermined height. The cylindrical weight was 1 m in diameter and its spherical bottom was 17.5 cm high. The measured parameters were acceleration of the weight and strains in rebars. The former was used to estimate the impact force generated by the falling weight, and the latter were used to estimate the dynamic bending moments of each girder. Figure 2 shows the measuring points of strain in the rebars. All of the experimental cases considered here are listed in Table 1. In this study, in order to study the influence of lateral prestressing on the dynamic behavior of a PC multi-girder, the cases with/without prestressing were investigated. The impact tests, falling a weight onto the shock absorbing system mounted on an RC rigid foundation were also executed to study the effect of the stiffness of the PC rock-shed on the

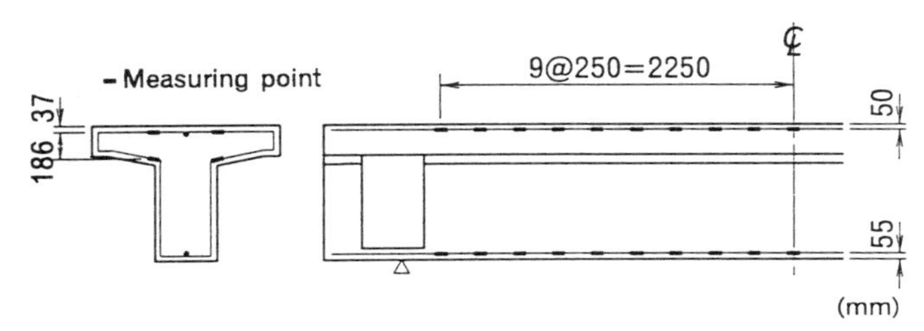

Fig. 2. Measuring points of strain in rebars.

Table 1. Experimental cases

Nominal name	Absorbing system	Falling height (m)	Falling point	Lateral prestress
S-5-2-P S-5-3-P S-5-4-P	Sand cushion	5	G2 G3 G4	Introduced
S-5-2-B S-5-3-B S-5-4-B	Sand cushion	5	G2 G3 G4	Released
D-30-2-P D-30-3-P D-30-4-P	3-layered absorbing system	30	G2 G3 G4	Introduced
D-30-2-B D-30-3-B D-30-4-B	3-layered absorbing system	30	G2 G3 G4	Released
S-5-S	Sand cushion	5	Mounted on	
D-30-S	3-layered absorbing system	30	an RC rigid foundation	

weight impact force. The falling heights of the weight were determined so that both types of absorbing systems behave elastically and produce almost the same bending moment of the girder.

2.1 The PC multi-girder

The PC multi-girder used for the experiments was composed of five T-shaped PC girders and laterally prestressed, as shown in Fig. 3. The dimensions of the PC girder were (a) height: 90 cm, (b) span length: 5 m and (c) upper flange width: 1.19 m. The 1 cm space between flanges was filled with shrinkage-compensating mortar. Each PC girder was designed based on the following conditions:

1. A single layered sand cushion (unit weight: 17.64 kN/m^3, Lamé's constant λ: 0.98 MPa) with 90 cm thickness is used.
2. A 29.4 kN weight rock falls on the sand cushion from a 10 m height.
3. An impact force of 1257 kN acts on the center of girder which is estimated based on the Hertz's contact theory[3]:

$$P = 33.41\lambda^{2/5}W^{2/3}H^{3/5} \tag{1}$$

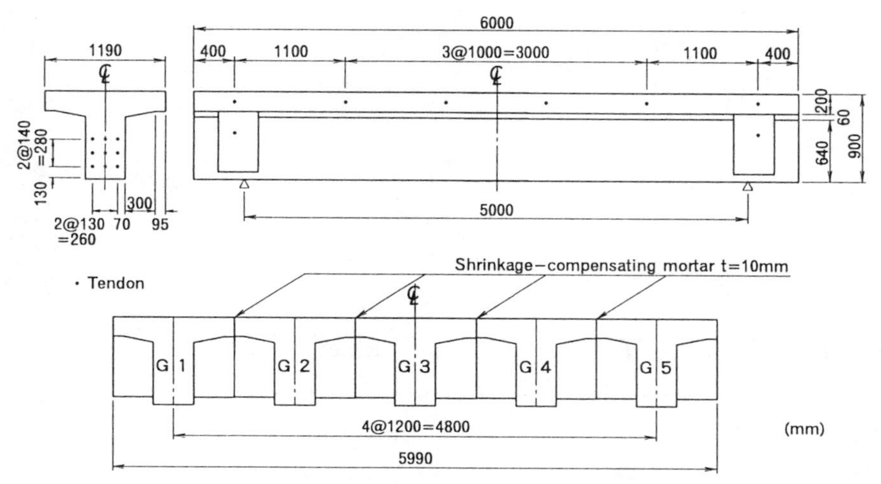

Fig. 3. The PC multi-girder.

Table 2. Material properties of tendon and rebar

	PC tendon	Rebar
Nominal name	SBPR 930/1080	SD295A
Diameter (mm)	26	–
Yield strength (MPa)	>930	>294
Tensile strength (MPa)	>1080	441–598

in which λ: Lamé's constant (MPa), W: weight of a falling object (kN) and H: falling height of a object (m).

4. Impact load is distributed on the upper surface of the multi-girder with an area of two times the thickness of the sand cushion.
5. Design strength, allowable tensile stress and Young's modulus of concrete are 73.5 MPa, 3.43 MPa and 34.5 GPa, respectively.

The PC girder was prestressed by using nine PC tendons with post-tensioning system. After introducing the prestress in the girders, the gap between a tendon and its duct was filled with mortar grout. The material properties of tendons and rebars used here are shown in Table 2. Each tendon was stretched up to 279.7 kN of the effective prestress (526.8 MPa), then the introduced effective prestress in the lower fiber became 15.0 MPa. Lateral prestressing of the multi-girder was introduced by PC tendons arranged with an interval of 1 m. After completion of prestressing, the gap between tendon and its duct was left ungrouted. The lateral surfaces of the girder flanges were greased to allow each girder to release perfectly from restraint when the lateral prestressing was set free. At the commencement of the experiment, the concrete had aged one year. The 28 day compressive strength and Young's modulus of concrete were 76.15 MPa and 37.24 GPa, respectively.

Table 3. Material properties of sand

Unit weight (kN/m^3)	Specific gravity	Uniformity coefficient	Coefficient of curvature
16.07	2.55	4.85	0.87

Table 4. Material properties of EPS

Unit weight (N/m^3)	Strength at 5 % strain (MPa)	Poisson's ratio	Max. of elastic strain (%)
196	0.108	0.05	1 or less

2.2 The sand cushion

The sand cushion used for these experiments was fine aggregate for concrete. The material properties of the sand are listed in Table 3. The sand was compacted with stamping at each 20 cm layer up to its 90 cm height. At the commencement of the experiment, the moisture content and the relative density of the sand were 6.0 % and 46.6 %, respectively.

2.3 The three-layered absorbing system

The three-layered absorbing system was composed of a 50 cm thick sand layer (top), a 20 cm thick RC core slab and 100 cm thick EPS blocks (bottom). The top layer used the same sand as the single layered sand cushion. The dimensions of the RC core slab were 6 m $\times$ 6 m $\times$ 20 cm. The RC slab was doubly reinforced with a rebar ratio of 1 %. The design strength of concrete was 20.58 MPa. At the commencement of the experiment, the age of concrete was 27 days and the compressive strength was 21.46 MPa. The material properties of the EPS used as the bottom layer are shown in Table 4. An EPS block (2 m $\times$ 1 m $\times$ 0.5 m) was used to form a layer.

3 Experimental results and considerations

3.1 Wave configuration of weight impact force

To study the wave configuration of weight impact force in the case using each absorbing system, the acceleration of a falling weight was measured. A weight impact force of a falling weight was evaluated by multiplying the mass of the weight with its measured acceleration. Figure 4 shows wave configurations of the weight impact force in the cases of S-5-3-P/B and D-30-3-P/B, which are compared the results for shock absorbing systems mounted on an RC rigid foundation. As for the results in the cases using the single layered sand cushion, as shown in Fig. 4 (a), although the durations of waves were different from each other, the wave configurations were almost. the same among them. The maximum impact force occurred about 30 msec after the wave was generated. After that, the impact force decreased to a value in between 1/4 to 1/3 of the maximum value and continued for 70 - 80 msec. The maximum weight impact forces of S-5-3-P, S-5-3-B and S-5-S were almost the same at 805.6 kN, 809.5 kN and 830.1 kN, respectively. As for the results in the cases using the three-layered absorbing system, as shown in

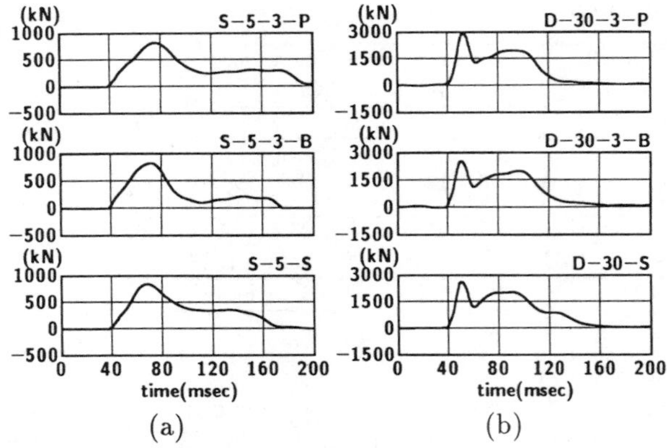

Fig. 4. Weight impact force.

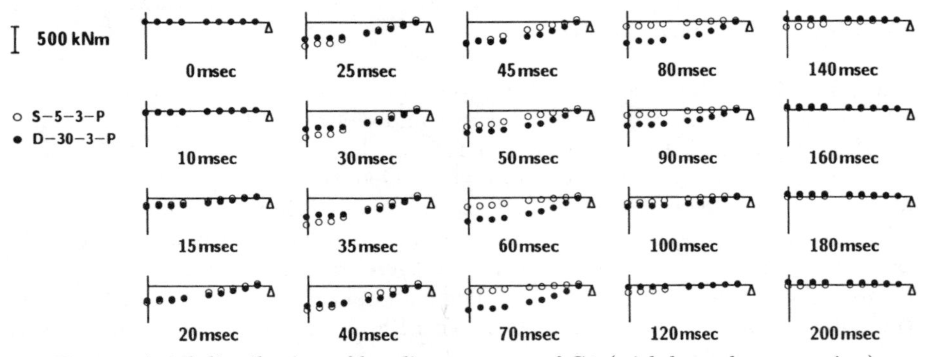

Fig. 5. Axial distribution of bending moment of G3 (with lateral prestressing).

Fig. 4 (b), the wave configuration of D-30-S 80 msec after wave generation was slightly different from the other cases. In all cases, the maximum weight impact force occurred about 10 msec after wave generation, and the second peak occurred about 50 msec later. The maximum impact forces of D-30-3-P, D-30-3-B and D-30-S were 2903 kN, 2474 kN and 2565 kN, respectively, and that of D-30-3-P was a little greater than the other cases. These results imply that the influence of the stiffness of the rock-shed on the weight impact force was small.

3.2 Bending moment of girder

The maximum strain in the lower rebars in all cases was 353 μ. From this result, it was concluded that no cracks occurred in the concrete; therefore, the girders behaved elastically. Thus, the bending moment of the girders could be calculated assuming the full effective sectional area.

Figure 5 shows the axial distributions of bending moment of G3 in the cases of S-5-3-P and D-30-3-P. The distribution of bending moment for S-5-3-P was almost linear, but that for D-30-3-P was somewhat parabolic. From these results, it was inferred that the impact force acted like a concentrated load when using

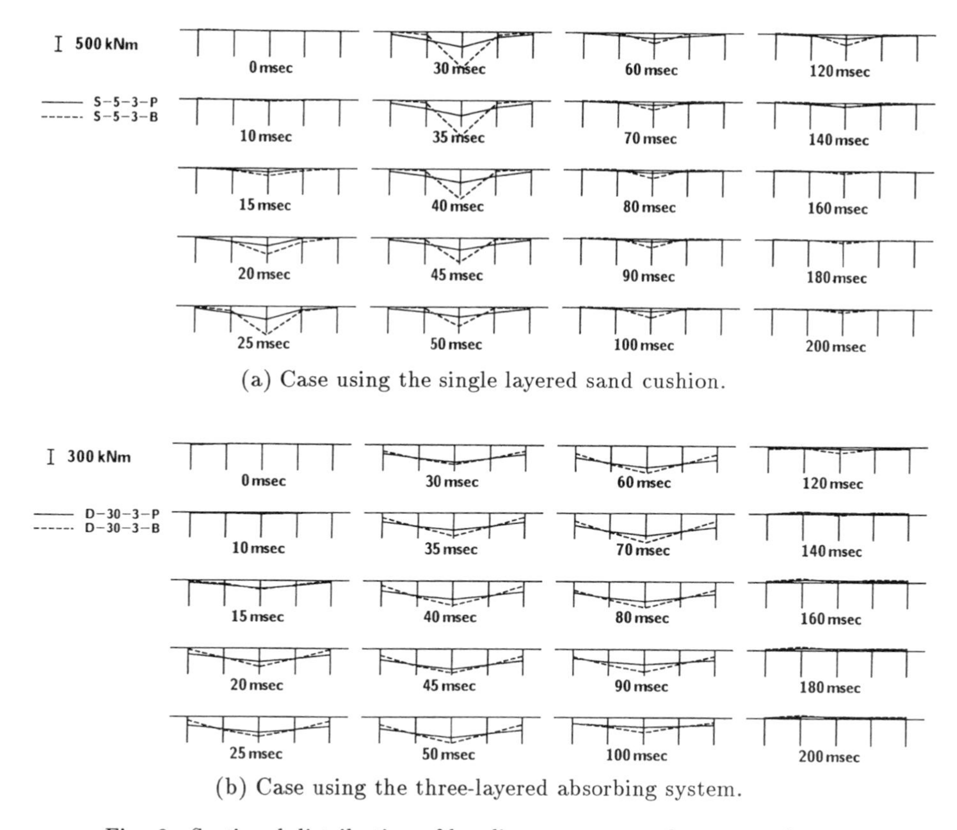

(a) Case using the single layered sand cushion.

(b) Case using the three-layered absorbing system.

Fig. 6. Sectional distribution of bending moment at the center of span.

the single layered sand cushion (S-5-3-P) and acted like a distributed load when using the three-layered absorbing system (D-30-3-P). In both cases, the bending moments were damped to a negligibly small value at 160 msec, and a negative loading state (rebound) did not occur.

Figure 6 shows the sectional distributions of the bending moment at the center of the span in the cases of S-5-3-P/B and D-30-3-P/B. With S-5-3-P, the bending moment of G3 was 5 times greater than those of G1 and G5, and the distribution was almost linear. However, in the case of S-5-3-B, the bending moment of G3 was the largest among the 5 girders, and those of the girders (G2 and G4) adjoining with G3 were almost zero. This means that the impact load was concentrated on the girder which was impacted by a free falling weight. This result, in the case without lateral prestressing, suggests that each girder behaved independently.

In the case of D-30-3-P, the bending moments were distributed with a small gradient; the bending moment of G3 was the largest among the 5 girders, and those of G1 and G5 were comparatively large. In the case of D-30-3-B, the distribution of bending moment was almost the same as that of S-5-3-P. Thus, the effect of lateral prestressing on the distribution of bending moment using the three-layered

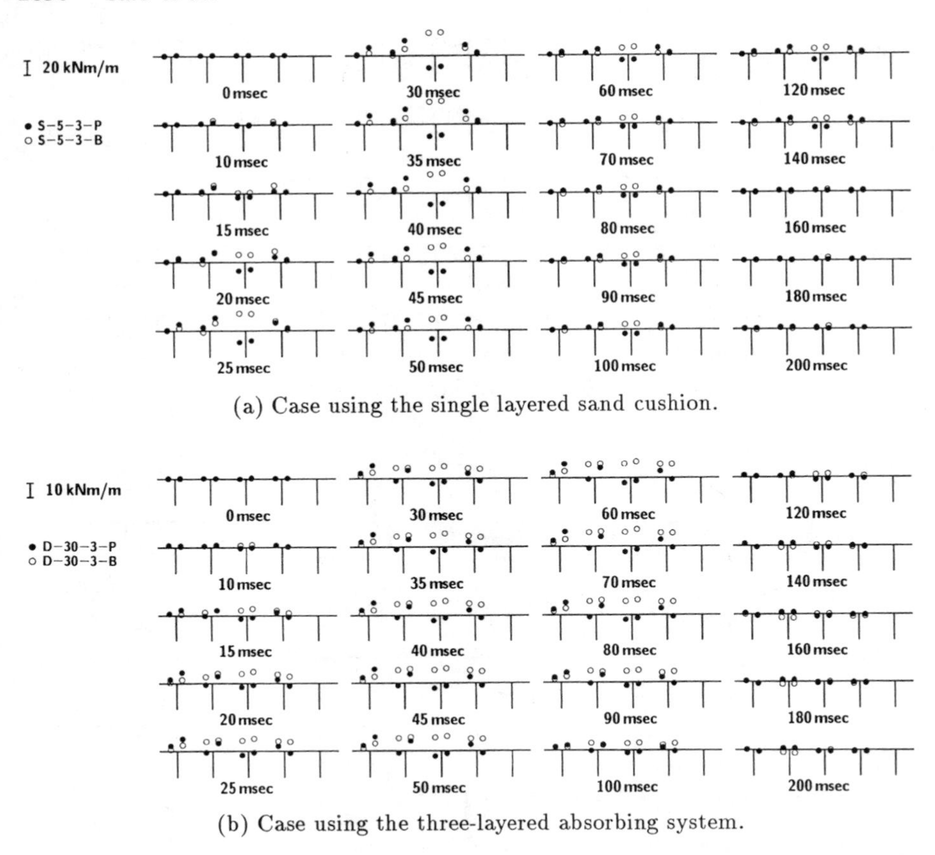

(a) Case using the single layered sand cushion.

(b) Case using the three-layered absorbing system.

Fig. 7. Bending moment of flange at the intersection of flange and web.

absorbing system would be less than that using the single layered sand cushion.

3.3 Bending moment of flange

Figure 7 shows the bending moments of the flanges at the intersection of the flange and the web in the cases of S-5-3-P/B and D-30-3-P/B. In this figure, a bending moment was plotted on the side of flange where the bending tension stress was generated. A positive and a negative moment were specified for the tension stress generating in the lower and the upper fiber of flange, respectively. With S-5-3-B, the negative bending moments occurred in the flanges of G3 and in the flanges of G2 and G4, attached to G3. This means that the impact load acted on G3 similarly to the design assumption. The maximum bending moment of 31.4 kNm/m was about 15 % greater than the design bending moment, 27.4 kNm/m. The negative flange moments of G2 and G4, attached to G3, may have been generated by the impact load applying onto those flanges which was dispersed by the sand cushion. Those maximum bending moments were one third less than the flange moment of G3. In the case of S-5-3-P, the distribution of the bending

Table 5. Load share ratio of the impacted girder

Nominal name	Load share ratio (%)	Nominal name	Load share ratio (%)
S-5-2-P	42.4	D-30-2-P	30.4
S-5-3-P	42.9	D-30-3-P	26.2
S-5-4-P	39.8	D-30-4-P	31.6
S-5-2-B	84.6	D-30-2-B	45.3
S-5-3-B	85.0	D-30-3-B	35.3
S-5-4-B	80.3	D-30-4-B	49.1

moment was similar to that of each girder rigidly connected; a positive bending moment occurred in the flange of G3. This means that when the multi-girder is laterally prestressed, the upper flange of each PC girder can not be designed by using the customary design method for the single girder and should be designed assuming alternative loads.

In the case of D-30-3-B, negative bending moments occurred in the flanges of all girders, and especially, the bending moments of G2, G3 and G4 were almost identical. The maximum value was about 9.8 kNm/m, similar to the result for the case of a uniform load ($q = 0.127$ MPa) acting statically on the flange. This implies that the impact load was uniformly distributed on the girders from G2 through G4. In the case of D-30-3-P, the bending moment of G3 was almost zero; however, those of the other girders (G1, G2, G4 and G5) were similar to those of S-5-3-P.

3.4 Load share ratio

The load share ratio of multi-girders is generally evaluated by using vertical displacements. However, in this paper, the load share ratio is evaluated by using bending moments at the center of girders because of the following two reasons: 1) the bending moment is calculated from the axial strain, 2) each girder behaves elastically, and then the relationship between the bending moment and the load acting on the girder becomes one-to-one. The load share ratio is evaluated by the ratio of the bending moment in each girder to the maximum total bending moment generated in time domain, which is obtained by summing up the each girder bending moment.

Table 5 shows the load share ratio of the girder subjected to weight falling. When the single layered sand cushion was used (S-5-), the load share ratios were between 40 % and 43 % with lateral prestressing and between 80 % and 85 % without lateral prestressing. The load share ratio for the cases without lateral prestressing was about twice than that for the cases with lateral prestressing. They were not affected by the position of the impacted girder and were almost the same. When the three-layered absorbing system was used (D-30-), the load share ratios were between 26 % and 32 % with lateral prestressing and between 35 % and 49 % without lateral prestressing. In both cases, the case of 3G loading with lateral prestressing (D-30-3-P) was the smallest one in all cases because G3 was placed at the center of the multi-girder and the loading condition was symmetrical.

4 Conclusions

To establish a rational design method for PC rock-sheds, full scale impact tests using the PC multi-girder were executed. In this study, two absorbing systems were applied: single layered sand cushion with 90 cm thickness and three-layered absorbing system. The impact tests were executed in the range of PC multi-girder behaving elastically. The results obtained from this study are summarized as follows:

1. The influence of the stiffness of rock-shed on the weight impact force is negligibly small.
2. The impact force when using the single layered sand cushion can be assumed as a concentrated load and that when using the three-layered absorbing system, as a distributed load.
3. When the multi-girder is laterally prestressed, the upper flange of each PC girder can not be designed by using the customary design method for the single girder and should be designed assuming alternative loads.
4. When the single layered sand cushion is used, the load share ratio of girders would be 45 % with lateral prestressing and 85 % without lateral prestressing.
5. When the three-layered absorbing system is used, the load share ratio of girders would be 30 % with lateral prestressing and 35 % without lateral prestressing.

5 Acknowledgments

We would like to thank Prof. Ken-ichi G. Matsuoka of the Muroran Institute of Technology and Dr. Osamu Nakano of the Hokkaido Development Bureau for their helpful suggestions and comments in executing this study. The PC multi-girder was fabricated by DPS Bridge Works Co., Ltd.

6 References

1. Japan Road Association (1983) *Design Manual for Anti-Rock Fall Structures* (in Japanese).
2. Kishi, N., Nakano, O., Mikami, H., Matsuoka, K. G. and Sugata, N. (1993) Field Test on Shock-Absorbing Effect of Three-Layered Absorbing System. *Proceedings of the 12th International Conference on Structural Mechanics in Reactor Technology*, Vol. JH13/6, pp.357–362.
3. Timoshenko, S. P. and Goodier, J. N. (1951) *Theory of Elasticity, 2nd edition*, McGraw-Hill, New York, pp.372–377.

160 SHOCK ABSORBING PERFORMANCE OF A THREE-LAYERED CUSHION SYSTEM USING RC CORE SLAB REINFORCED WITH AFRP RODS

H. MIKAMI and T. TAMURA
Mitsui Construction Co. Ltd, Tokyo, Japan
M. SATO
Hokkaido Development Bureau, Sapporo, Japan
N. KISHI
Muroran Institute of Technology, Muroran, Japan

Abstract
In this paper, the shock absorbing performance of a three-layered cushion system using an RC core slab reinforced with Aramid Fiber Reinforced Plastic (AFRP) rods is discussed on the basis of experimental results. This system consists of an RC core slab, sand (top) and Expanded Poly-Styrol (bottom). In order to study the influence of the properties of the RC core slab on the shock absorbing performance, both the material and the rebar ratio and the thickness of the RC core slab were taken as variables. Furthermore, these results were compared with the results when a single sand layer was used.

The results obtained from these experiments are as follows: (1) the transmitted impact stress of the three-layered cushion system was distributed more effectively than that of the single sand layer, (2) shock absorbing performances of the three-layered system were influenced by the properties of the RC core slab and (3) AFRP rods were useful for reinforcing the RC core slab in the three-layered cushion system.
Keywords: AFRP rods, cushion system, impact force, RC slab, shock absorbing performance, transmitted impact stress.

1 Introduction

In recent years, to ensure the greater safety of nuclear power plants, fuel tanks, rock-sheds and/or other important structures which are vulnerable to impact loads, extensive studies have been carried out. There are two aspects of investigating the impact behavior of a structure: (1) investigate the impact behavior of a structure, assuming that the impact loads are directly applied on the

Concrete Under Severe Conditions: Environment and loading (Volume Two) Edited by K. Sakai, N. Banthia and O.E. Gjørv. Published in 1995 by E & FN Spon. ISBN 0 419 19860 1

structure and (2) investigate a cushion system that attenuates the impact loads.

On the other hand, in order to ensure the further durability of concrete structures, extensive studies have been conducted on the applicability of continuous-fiber-reinforced plastic (FRP) rods as a concrete reinforcing material which has excellent corrosion resistance and high tensile strength[1-3].

With this in mind, the authors have conducted studies of impact behavior and impact resistance of RC beams, slabs and PC beams reinforced with FRP rods directly subjected to impact loads[4-6]. The test results indicated that the impact resistance of RC beams and slabs reinforced with braided AFRP rods coated with silica sand was superior to those of concrete members reinforced with deformed steel bars. Furthermore, the authors have studied the shock absorbing performances of a prototype three-layered cushion system, where the RC slab reinforced with AFRP rods was sandwiched between a sand layer (top) and an Expanded Poly-Styrol (EPS) layer (bottom)[7-8]. The test results suggested that the stiffness and bonding action of the reinforcing material in the RC core slab affected the shock absorbing performances of the three-layered cushion system under repeated impact loading.

In this paper, the shock absorbing performance of the three-layered cushion system using an RC core slab reinforced with AFRP rods is discussed based on the experimental results. The three-layered cushion system used here was composed of a 50 cm thick sand layer (top), a 20 or 30 cm thick RC core slab and a 50 cm thick EPS layer (bottom). From a mechanical point of view, the role of the RC core slab is to disperse the impact force coming through the top layer and to transfer it to the bottom layer. Ideally, the core material must remain stable and keep the load dispersion effect intact for a long period. This means that the reinforcing material of the core slab should have excellent corrosion resistance and shock absorbing ability. An AFRP rod has excellent corrosion resistance and high tensile strength, and its elastic elongation is greater than that of either steel bars or carbon fiber rods; thus, an AFRP rod would be an effective reinforcing material for RC core slab.

Taking this into consideration, the authors conducted studies on the shock absorbing performance of the three-layered cushion system by varying the thickness of the core slab, material and bond properties of rebar and rebar ratio in the core slab. Also, a 120 cm thick single sand layer was used to compare the shock absorbing performance. Throughout all experimental cases, the impact load was applied only once by a 29.4 kN steel weight free falling from a 30 m height onto the center of each cushion system.

2 Outline of experiment

2.1 Experimental methods and cases

Figure 1 illustrates the cross section and dimensions of the three-layered cushion system used for the experiments. The three-layered cushion system (400 cm × 400 cm with a 50 cm thick sand layer, a 20 and/or 30 cm thick RC core slab, and a 50 cm thick EPS layer, in order from the uppermost level)

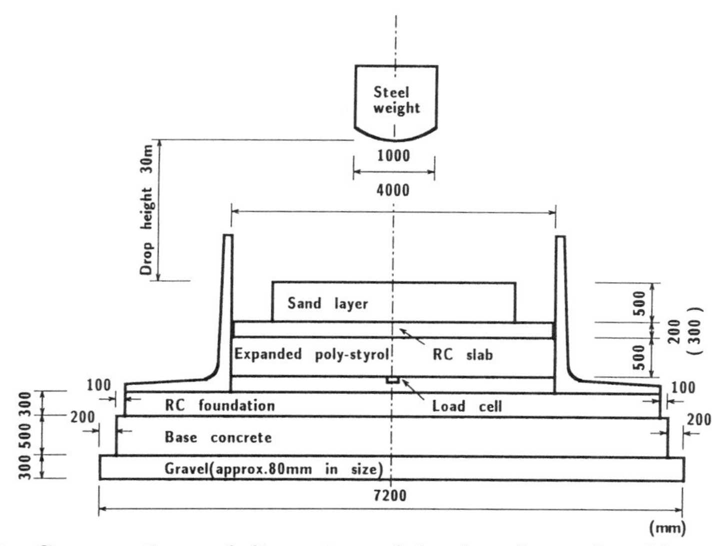

Fig. 1. Cross section and dimensions of the three-layered cushion system.

Table 1. Experimental cases

Core slab	Type of rebar	Thickness (cm)	Rebar ratio (%)
As-20-1.0	Sand surfaced rod (RA13S)	20	1.0
A-20-1.0	Non-sand surfaced rod (RA13)	20	1.0
As-20-0.5	Sand surfaced rod (RA9S)	20	0.5
As-30-1.0	Sand surfaced rod (RA15S)	30	1.0
D-20-1.0	Deformed steel bar (D13)	20	1.0
S-120	Single sand layer with 120 cm thickness		

was laid out on an RC foundation having the dimensions of 100 cm thick and 670 cm sq. The cushion system was subjected to impact loads by a 29.4 kN cylindrical steel weight free falling from a 30 m height. The diameter and height of the steel weight were 100 cm and 97.5 cm, respectively, with a spherical bottom of 17.5 cm height.

The experimental cases are listed in Table 1. In these experiments, five RC core slabs with various sectional properties for the three-layered cushion system were made and tested; the other one was the 120 cm thick single sand layer used for the comparative study. In denoting the test specimens, designations in core slab nomenclature refer to be reinforcing material, the thickness of RC core slab and the rebar ratio in sequence. For instance, As-20-1.0 stands for a 20cm thick RC core slab reinforced with 1% sand surfaced AFRP rods. The rebar ratio means the ratio of the cross sectional area of rebar to the effective sectional area of the slab.

Table 2. Material properties of rebars

Rebar denomination	RA13 RA13S	RA9S	RA15S	D13
Material	Aramid	Aramid	Aramid	SD30A
Nominal diameter (mm)	12.7	9.0	14.7	12.7
Nominal sectional area (mm^2)	127	63	170	127
Density (g/cm^3)	1.30	1.30	1.30	7.85
Yield strength (kN)	-	-	-	37.24
Tensile strength (kN)	188.16	94.08	250.88	55.86-75.46
Elastic modulus (GPa)	68.6	68.6	68.6	205.8
Elastic elongation (%)	2.0	2.0	2.0	0.2

2.2 Sectional properties of RC core slab

Table 2 shows the material properties of rebars. RA 13 is a braided AFRP rod made of mechanically braided aramid fibers which were impregnated in resin and finally hardened. RA9S and RA13S were formed by bonding NO.5 silica sand on its surface before the resin became hardened. All of the concrete slabs were doubly reinforced with rebars. The maximum size of coarse aggregate was 25 mm, and the design compressive strength of concrete was 20.58 MPa. At commencement of the experiment, the concrete had aged between nine to fifteen days. The strength ranged from 18.42 to 21.36 MPa, and elastic modulus, from 26.17 to 27.34 GPa.

2.3 Material properties of the top and bottom layers

The unit weight of the sand used for the top layer of the three-layered cushion system and the single sand layer was 15.68 kN/m^3. Its specific gravity and uniformity coefficient were 2.59 and 5.72, respectively. Sand was spread and compacted with stamping at each 20 cm layer until the predetermined thickness was formed. The unit weight of the EPS used for the bottom layer was 196 N/m^3, the compressive stress at a 5 % strain was 0.12 MPa and the poisson's ratio was 0.05. In the region where elastic behavior was observed in the EPS, the compressive strain was less than 2 %, with its stress being approximately two thirds of the compressive stress at a 5 % strain.

2.4 Measuring items

The acceleration of the steel weight was measured using 4 accelerometers to calculate the weight impact force generated when the steel weight collided with a cushion system. The impact stresses transmitted to the RC foundation were measured in one direction located along the centerline of the RC foundation using 39 load cells arranged with 10 cm intervals. The diameter and capacity of load cell were designed to be 32 mm and 10 MPa, respectively, and the maximum response frequency to be measured was 600 Hz. These load cells were manufactured in order to be able to measure the transmitted impact stresses. Outputs from the individual sensors were recorded by wide-band data recorders; then A/D conversion was made, and the data were finally processed using work stations.

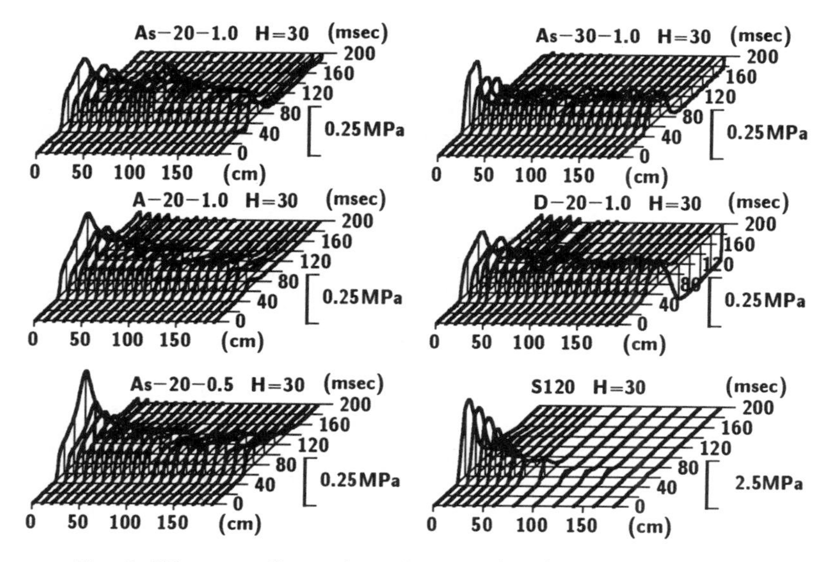

Fig. 2. Wave configuration of transmitted impact stresses.

3 Analysis of experimental results

3.1 Wave configuration of transmitted impact stress

Figure 2 shows the wave configurations of the transmitted impact stresses with respect to time measured by load cells from different experimental tests. In this figure, horizontal and vertical axes in the plane of paper show the location of load cells (cm) and transmitted impact stresses (MPa), respectively, and the axis perpendicular to the plane of the paper shows the duration time starting after A/D conversion (msec).

The origin of the figure is at the center of loading which is at the center of the RC foundation. The wave patterns on two sides of the center of loading were almost symmetrical to each other. Hence, this figure only shows the wave at the load cell located to one side of the center of loading between the center and the edge of RC foundation. The total sampling times in all test cases were 200 msec.

In order to investigate the effects of the stiffness of rebar on the transmitted impact stresses, a comparison of As-20-1.0 with D-20-1.0 was made. In the case of type As, the transmitted impact stresses at the loading point and at a distance of 80 to 90 cm from the loading point were relatively large, but stresses decreased smoothly from that point to the edge of RC foundation. In the case of type D, the transmitted impact stresses decreased smoothly from the loading point to the edge. Stress concentration around the loading points was not noticeable in either case, and the stress dispersibility was excellent. This suggests that the stiffness of the rebar did not markedly affect the dispersibility of the transmitted impact stresses. In the case of type D, there was a negative stress, i.e., uplift occurred at the edge of the RC foundation, but with the type As, this was less notable. This implies that the stiffness of the type D concrete slab was greater than that of the type As concrete slab.

In order to investigate the effects of the bond properties of rebar on the transmitted impact stresses, a comparison between As-20-1.0 and A-20-1.0 was made. In the case of the type A slab, the transmitted impact stress at the loading point was greater than that of the type As. This means that the dispersibility of the transmitted impact stress and shock absorbing ability of the type As slab was superior to that of the type A. This tendency was particularly remarkable in the portion within 50 cm of the center. This obviously suggests that the level of stress concentration at the loading point was affected by the bond properties of rebar.

In order to investigate the effects of the rebar ratio on the transmitted impact stresses, As-20-1.0 was compared with As-20-0.5. In case of the type -0.5 slab, the stress concentration beneath the loading point was more clearly notable than that of the type -1.0 slab, and the stress decreased hyperbolically from the loading point to the edge. Thus, the rebar ratio extremely affected the shock absorbing ability and the dispersibility of the transmitted impact force.

In order to investigate the effects of slab thickness, a comparison between As-20-1.0 and As-30-1.0 was made. In case of the type As-30 slab, the transmitted impact stress tended to concentrate in the area beneath the loading point, but the stress level was almost the same with that of the type As-20. Especially, the stress within a distance of 20 and 30 cm from the center was large, but the stress variation from that region to the edge was small, and the stress was distributed almost uniformly. Also, it was obvious that the duration of the response was extremely short. Furthermore, since the edge of the RC foundation sustained a negative stress, uplift might have occurred. This suggests that the stiffness of the RC core slab was greater in the type As-30 slab than in the type As-20 slab. Thus, slab thickness affected the dispersion property and the duration of the transmitted impact stress.

The shock absorbing features of the single sand layer were compared with those of the three-layered cushion system. In the case of the single sand layer, the transmitted impact stress was concentrated around the loading point, and the range of the stress dispersion became extremely limited to a radius of 80 cm from the center. The intensity of the transmitted impact stress beneath the loading point was approximately ten times greater than that using the three-layered cushion system. On the other hand, the duration of the transmitted impact stress in the single sand layer was shorter than in any of the three-layered cushion systems, and was three fourths of the type As-30 slab, which had the shortest in all of the three-layered cushion systems. These suggests that a larger sectional force in the structure was generated in the single sand layer than in the three-layered cushion system.

3.2 Wave configuration of impact force

In this paper, to evaluate the impact force generated by the weight dropped on the cushion system, the following two approaches were investigated: (1)

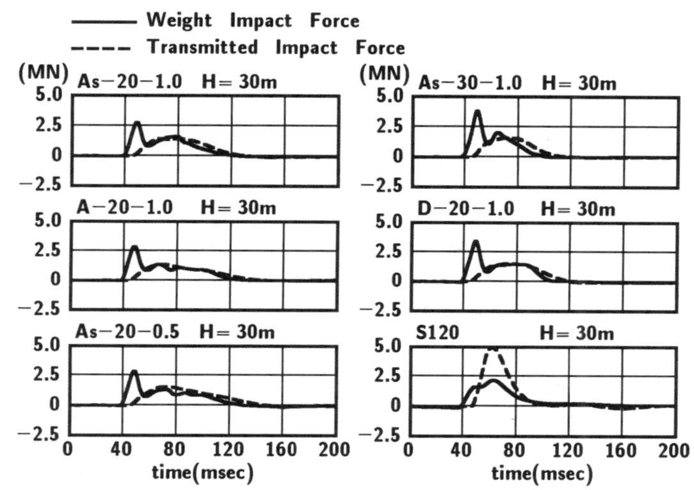

Fig. 3. Wave configurations of impact forces.

impact force obtained by multiplying the acceleration (G) of the steel weight by its weight (hereafter, referred to as the weight impact force) and (2) impact force obtained by summing up the transmitted impact stresses measured by the load cells (hereafter, referred to as the transmitted impact force). Assuming that the distribution of the transmitted impact stresses in two directions with respect to the origin were the same, the transmitted impact force was calculated by integrating the measured impact stresses on one side.

Figure 3 shows the wave configuration of the weight and transmitted impact forces with respect to time with solid and broken lines, respectively. In the case of the three-layered cushion systems, although durations of transmitted impact forces differed from each other, the maximum transmitted impact forces and their wave configurations were similar in each case. On the other hand, in the case of the weight impact force, although the second waveforms were somewhat different from each other, the first waveforms looked similar. In other words, the magnitudes of the weight and transmitted impact force and their durations were affected by the sectional properties of RC core slab. But the wave configurations of the three-layered cushion systems did not differ greatly from each other.

The wave configurations of both impact forces in the single sand layer and the three-layered cushion system were compared. In the case of the three- layered cushion system, the first waveform of the weight impact force indicated a clear peak and the second waveform behaved relatively smoothly. In the case of the single sand layer, the waveform corresponding to the first waveform of the three-layered cushion system was not observed. On the other hand, although the transmitted impact force in the single sand layer was greater than that in the three-layered cushion system, the duration was extremely short. Thus, the single sand layer and the three-layered cushion system sustained impact forces which were of considerably different magnitudes and durations.

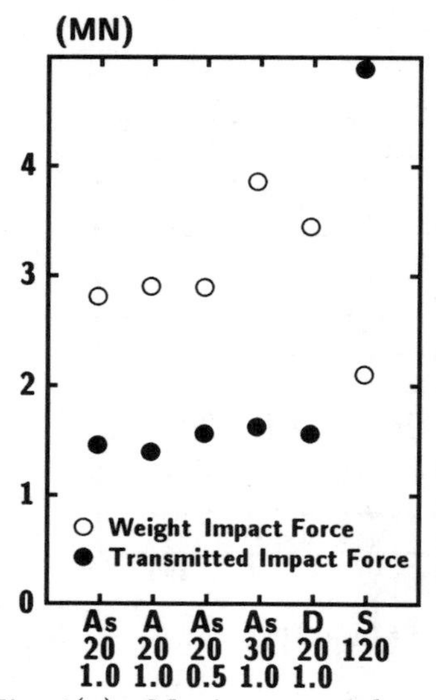

Fig. 4(a). Maximum weight and transmitted impact forces.

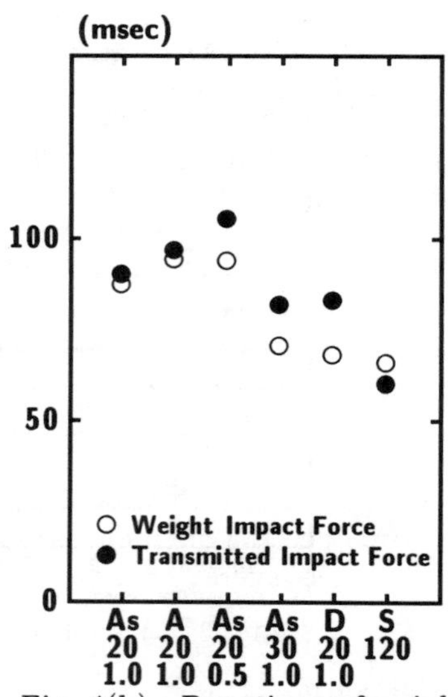

Fig. 4(b). Durations of weight and transmitted impact forces.

3.3 Maximum impact forces

Figure 4(a) shows the distribution of the maximum weight and transmitted impact forces. Figure 4(b) shows the durations of the weight and transmitted impact forces. From Figure 4(a), it is evident that the maximum weight impact forces became smaller in the following order: 30 cm thick RC slab, RC slab reinforced with deformed steel bars, 20 cm thick slab reinforced with AFRP rods and the single sand layer. In addition, regarding the cases with the 20 cm thick slab reinforced with AFRP rods, since there was no significant difference among them, the rebar ratio and the bond properties of rebars only slightly affected the maximum weight impact force. On the contrary, the slab thickness and the stiffness of the rebars greatly affected the magnitude of the maximum weight impact force. The greater the stiffness of the concrete slab was, the smaller the deformation of RC slab and the larger the maximum weight impact force were. In the case of the single sand layer, there was no RC slab, this causes the maximum weight impact force to be the smallest.

On the other hand, the maximum transmitted impact force generated in the single sand layer was 3 to 4 times greater than that in the three-layered cushion system. However, compared with the maximum weight impact force, the maximum transmitted impact force in the three-layered cushion system was not affected much by the sectional properties of RC core slab. This suggests that the shock absorbing performance of the three-layered cushion system was affected more by the shock absorbing ability of the EPS than the effects of the thickness of core slab and the stiffness of rebars. Thus, to reduce the maximum weight and transmitted impact forces, it would be effective to use a three-layered cushion system composed of a core RC slab with decreased thickness, using rebars with adequate quantity, lower stiffness and better bond properties.

Figure 4(b) shows the durations of both weight and transmitted impact forces. The duration of the weight impact force in the single sand layer was the shortest in all cases followed by the slab reinforced with deformed steel bars and the 30 cm thick slab. In the case of the three-layered cushion system with 20 cm thick slabs reinforced with AFRP rods, the duration of the weight impact force in A-20-1.0 was the longest one among the three cases, but the differences were small. On the other hand, the duration of the transmitted impact force in the single sand layer was the shortest followed by the 30 cm thick slab and the slab reinforced with deformed steel bars. Among all types of slabs, the longest duration of the transmitted impact force was observed in the case of the slab reinforced with the 0.5 % rebar ratio. Thus, the larger the stiffness of the slab was, the shorter the duration of the transmitted impact force. Therefore, in order to prolong the duration of impact force as well as to reduce the maximum transmitted impact force, it would be effective to use the three-layered cushion system.

4 Conclusion

The results obtained from these experiments could be summarized in relation to the different sectional properties of the RC core slabs in the three-layered cushion system on the features of impact forces, as follows:

1. The slab thickness, the rebar ratio and the bond property of rebars dominantly affected the wave configuration and dispersibility of the transmitted impact stress, but the stiffness of rebars did not affect them very much.
2. From the wave configurations of the weight and transmitted impact forces, it was evident that the durations and magnitudes of the impact forces varied with the sectional properties of the RC core slab, but the waveforms were not notably affected.
3. The maximum weight impact force was predominantly affected by the slab thickness and the stiffness of rebars, but the maximum transmitted impact force was not affected by it very much.
4. The duration of the impact force was affected by the slab thickness and the stiffness of rebars. In order to prolong the duration of the transmitted impact force, it would be effective to use a thin slab and to arrange an adequate quantity of rebars with lower stiffness and better bond property.
5. Comparing the shock absorbing ability in terms of the magnitude and the duration of the transmitted impact force, it was evident that the core slab reinforced with AFRP rods was more effective than the slab reinforced with deformed steel bars. With concrete slabs reinforced with 0.5% AFRP rods and 1 % deformed steel bars, the three-layered cushion system using the former slab had a shock absorbing ability equal to or even greater than that using the latter one. However, an increase of the slab thickness was not useful to ensure a further absorbing effect.

From the comparison of absorbing performances between the three-layered cushion system and the single sand layer, the following can be summarized:

6. The transmitted impact stress in the single sand layer concentrated around the loading point, and its intensity was about ten times greater than that of the three-layered cushion system.
7. The weight impact force generated in the case using the single sand layer was smaller than that in the case using the three-layered cushion system. The duration of the weight impact force in the case using the single sand layer was longer than in the case using the three-layered cushion systems. On the other hand, the transmitted impact force in the case using the single sand layer resulted in the reverse; the maximum transmitted impact force was 3 to 4 times greater than that in the case using the three-layered cushion system, and the duration of the transmitted impact force was half to two thirds of the case using the three-layered cushion system. Thus, the shock absorbing performance of the three-layered cushion system far excelled that of the single sand layer.

5 Acknowledgement

The authors sincerely thanks to Dr. Ken-ichi G. Matsuoka (Professor of Muroran Institute of Technology), Dr. Osamu Nakano (Vice Manager of Asahikawa Division of Hokkaido Development Bureau) and Mr. Masanori Takazawa (Graduate Student of Muroran Institute of Technology) who supported this investigation.

6 References

1. Mikami, H., Kato, M., Takeuchi, H. and Tamura,T. (1990) *Shear Behavior of Concrete Beams Reinforced with Braided High Strength Fiber Rods in Spiral Shape.* Proceedings of the FIP 11th International Congress on Prestressed Concrete, Vol. 2, pp.T44-T46.
2. Mikami, H., Kato, M., Tamura, T. and Ishibashi, K. (1990) *Fatigue Characteristics of Concrete Beams Reinforced with Braided AFRP Rods.* Transactions of the Japan Concrete Institute, VOL.12, pp.223-230.
3. Japan Society of Civil Engineers. (1993) *State-of-the-Art Report on Continuous Fiber Reinforcing Materials.* Concrete Engineering Series 3, in Japanese.
4. Mikami, H., Kishi, N., Matsuoka, K. and Nomachi, S. (1991) *Dynamic Behavior of Concrete Slabs Reinforced by Braided AFRP Rods under Impact Loads.* Proceedings of the 11th International Conference on Structural Mechanics in Reactor Technology, Vol. J, pp.45-50.
5. Nakajima, N., Kishi, N., Mikami, T., Matsuoka, K. and Nomachi, S. (1993) *Impact Resistance of RC Beams Reinforced with Steel Bars and Braided AFRP Rods.* Proceedings of the 12th International Conference on Structural Mechanics in Reactor Technology, Vol. J, pp.291-296.

6. Kishi, N., Matsuoka, K., Mikami, H. and Tamura,T. (1993) *Impact Resistance of PC Beams using Braided AFRP Rods as PC Tendons.* Proceedings of the FIP Symposium '93 Kyoto, Vol.2, pp.835-842.
7. Tamura, T., Mikami, H., Nakano, O. and Kishi, N. (1993) *Absorbing Capacity of Cushion System Using Concrete Slab Reinforced with AFRP Rods.* Fiber-Reinforced-Plastic Reinforcement for Concrete Structures, SP-138, ACI, pp.301-313.
8. Kishi, N., Nakano, O., Mikami, H., Matsuoka, K. and Nomachi,S. (1993) *Field Test on Shock-Absorbing Effect of Three-Layered Absorbing System.* Proceedings of the 12th International Conference on Structural Mechanics in Reactor Technology, Vol.J, pp.357-362.

161 FULL SCALE TEST ON IMPACT RESISTANCE OF PC GIRDER

M. SATO and H. NISHI
Hokkaido Development Bureau, Sapporo, Japan
O. NAKANO
Hokkaido Development Bureau, Ashahikawa, Japan
N. KISHI and K.G. MATSUOKA
Muroran Institute of Technology, Muroran, Japan

Abstract
The impact resistance of prestressed concrete girders frequently used in prestressed concrete rock-sheds were studied by executing full scale impact tests. The prestressed concrete girders were of 5 m span length, T shaped with 1.2 m wide upper flange and 0.9 m in height. The tests were performed by a 49 kN weight free falling onto the center of the prestressed concrete girder with a 1 m thick Expanded Poly-Styrol cushion. The falling height of the steel weight was varied from 1 m to 30 m.

The behavior and progress of damage of the girders against iterative and single loadings are discussed. The critical impact resistance of the prestressed concrete girder was also obtained.
Keywords: Impact resistance, PC girder, rock-shed.

1 Introduction

Prestressed Concrete (PC) rock-sheds, as well as Reinforced Concrete (RC) rock-sheds, are widely used in Japan. The use of PC rock-sheds is expected to increase in the future because they require comparatively less man power and a shorter construction time. The authors studied full scale impact tests of PC multi-girders using two types of absorbing systems to establish a rational design method of PC rock-sheds against falling rocks [1],[2]. On the other hand, in 1989, the colossal accident involving PC rock-sheds at the Echizen

Concrete Under Severe Conditions: Environment and loading (Volume Two) Edited by K. Sakai, N. Banthia and O.E. Gjørv. Published in 1995 by E & FN Spon. ISBN 0 419 19860 1

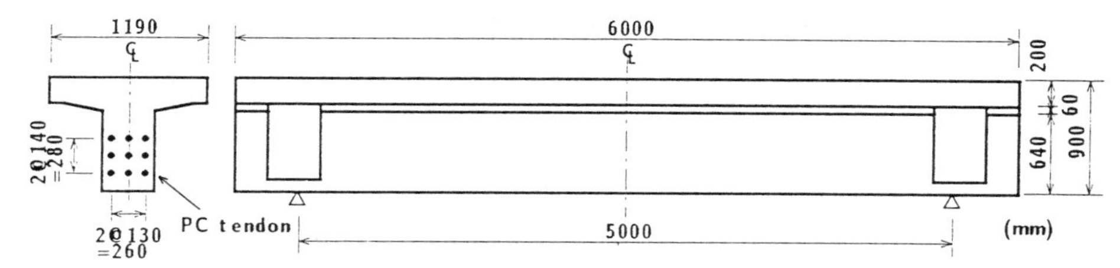

Fig. 1. Profile and dimensions of PC girder.

seashore in Fukui prefecture claimed more than 10 lives. After that, it was commonly recognized that the ultimate resistance of PC rock-sheds against impact loads should be studied. Matsuba et al. executed a full scale failure test of PC rock-sheds[3], and Sonoda et al. carried out the numerical analyses of PC rock-sheds in the range from small deformations through to complete failure by means of a distinct element method[4]. However, an evaluation method for the impact resistance of PC girder has not been fully established yet.

In this study, from these points of view, in order to evaluate the impact resistance of PC girders, full scale impact tests were executed. The impact loads were generated by a free falling 49 kN steel weight. Six PC girders were used for the experiments, one was used to study the damage progress of the PC girder under iterative loading and the others were used to study the impact behavior and the damage level with the steel weight dropped from heights of 10 m up to 30 m.

2 The PC girders

The PC girders were of 5 m span length, T-shaped with 1.19 m wide upper flange and 0.9 m in height (Fig. 1). They were designed as per the following conditions:

1. The PC girders were required to withstand an impact load caused by a 49 kN rock free falling from a height of 10 m.
2. A single layered sand cushion with 90 cm thickness was used as the absorbing system.
3. The impact force (P_{max} = 1257.3 kN) calculated on the basis of the Hertz's contact theory, assuming Lame's constant λ = 0.98 MPa, was loaded as the static force on the center of the PC girder.
4. Impact load was distributed on the upper surface of the girder with an area of two times the thickness of the sand cushion.

The design strength and the elastic modulus of concrete were assumed to be 73.5 MPa and 34.3 GPa, respectively. Prestress was introduced to the girder by a post-tensioning system with nine tendons (ϕ 26 mm, Grade B-1 SBPR 930/1080). After the introduction of prestresses, the gaps between tendon and concrete were grouted. Deformed steel bars (Grade SD295A) were used for reinforcement, stirrups and assemblage bars. The effective prestress and the stress in the PC girder at each loading step are listed in Table 1.

Table 1. Effective prestress and stress in girder

	Upper fiber	Upper rebar	Lower rebar	Lower fiber
Effective prestress (MPa)	1.71 (49.7 μ)	0.77 (22.6 μ)	-13.98 (-407.7 μ)	-15.00 (-437.4 μ)
Stress for dead load (MPa)	-0.84 (-24.6 μ)	-0.73 (-21.1 μ)	1.26 (36.9 μ)	1.40 (40.9 μ)
Total stress (MPA)	0.87 (25.1 μ)	0.04 (1.5 μ)	-12.72 (-370.8 μ)	-13.60 (-396.5 μ)
Stress by designed impact force (MPa)	-9.59 (-279.7 μ)	-8.17 (-238.2 μ)	14.35 (418.3 μ)	15.91 (463.7 μ)
Total stress (MPa)	-8.73 (-254.6 μ)	-8.12 (-236.7 μ)	1.63 (47.5 μ)	2.31 (67.2 μ)

( ) : Strain converted by using E_c = 34.3 GPa

Table 2. Experimental cases

Specimen	Falling height H (m)	Kinetic energy E (kNm)
PC-R	1, 2, 3, 5, 10, 15	49, 98, 147, 245, 735
PC-10	10	490
PC-15	15	735
PC-20	20	980
PC-25	25	1225
PC-30	30	1470

Yield stress and tensile strength of tendon were 931 MPa and 1078 MPa, respectively. It was assumed that the effective tensile stress of tendons was 526.8 MPa. The volume of stirrups was determined according to the Concrete Manual[5]. Stirrups (D13, Grade SD295A) were placed at spacings of 20 cm, 15 cm and 5 cm for the central part (2 m), the middle part (90 cm) and the outer part (remaining portion), respectively. The PC girders had been aged one year after construction. The concrete was mixed using high-early-strength Portland cement (water cement ratio: 0.335) and water-reducing agent. The compressive strength and elastic modulus of the concrete at four weeks were found to be 76.4 MPa and 37.24 GPa, respectively.

3 Outline of experiment

A total of six full scale PC girders were used as listed in Table 2. Specimen PC-R was used for the iterative loading tests in which the falling height of the steel weight was sequentially increased from 1 m up to 15 m. The PC-n specimens were used for the single loading tests, in which n equals to the falling height H (m) of the steel weight. The impact load was surcharged by a 49 kN

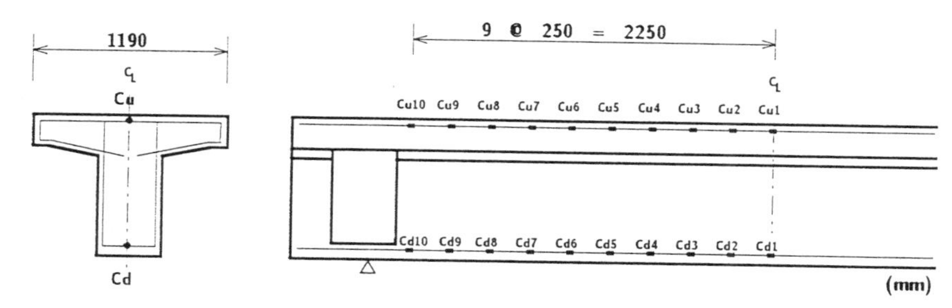

Fig. 2. Measuring points of strain in rebar.

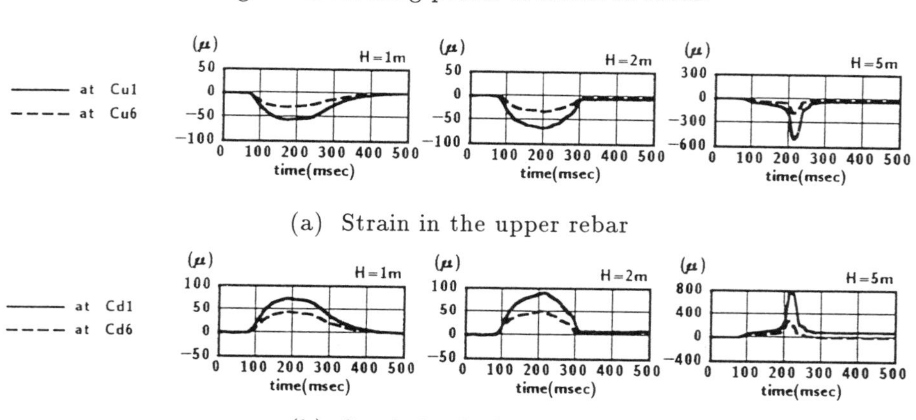

Fig. 3. Wave response of strains in the rebars in the case of PC-R.

steel weight free falling on the PC girders covered with 1 m high Expanded Poly-Styrol (EPS) blocks, which were used to keep the collision between PC girder and the steel weight soft. The unit weight of the EPS used here was 196 N /m ^{3}. The dimensions of the cylindrical steel weight were the following : diameter, 1 m; height, 97 cm and the height of spherical bottom, 17.5cm. The PC girder was mounted on steel bars, 50 mm in diameter, to get a simply supported condition.

The acceleration of the free-falling weight and the strain of rebars in the PC girder were measured. The measuring points of rebar strain are illustrated in Fig. 2. Other than the girder PC-R, it was impossible to measure rebar strain because of the excessive damage of girder.

4 Experimental results and discussions

4.1 Wave configuration of rebar strain for PC-R

The response wave of the strain in the upper and lower rebar at two sections ($l/2$ and $l/4$: l is the girder span) for different falling heights (H = 1,2 and 5 m) are shown in Fig. 3. Measurements of rebar strain became impossible in the case of H = 3 m due to problems in the measuring devices and in the cases of H = 10 and 15 m due to snapping of the sensor cables. Hence, response waves for these three falling heights (H = 3, 10 and 15 m) can not be reported here. It is shown in this figure that the PC girder behaved elastically as long as the falling height H did not exceed 2 m because in these cases, the maximum strain at $l/2$ as well as at $l/4$ did not exceed 100μ.

Assuming that the energy absorbed by the EPS was equal to the kinetic energy of the falling weight, the impact force and deformation of the EPS for the case of H = 2 m are estimated as 262.6 kN and 46.5 cm, respectively. Since this impact force is about 1/5th of the designed value, it seems that the EPS effectively acted as an absorbing material up to this level of dropping height. In the case of a 5 m falling height, the strains in the upper and lower rebars were abruptly increased and reached the maximum strain of 800μ after about 100 msec of wave generation. Since the strain in the lower rebar is obtained as -370 μ from Table 1 when dead load and the effective prestress are considered and obtained as 430 μ by summing up the strains of 800 μ and -370 μ, the cracks occurred in the lower part of the web in the mid span section of the girder. In this case, the required EPS deformation for perfectly absorbing the kinetic energy of a steel weight with 3 m free falling is 1.08 m. This means that the 1 m thick EPS did not perfectly absorb the kinetic energy input, and therefore, a severe impulsive load was surcharged onto the girder.

The duration of strain wave for the cases of 1 m, 2 m and 5 m falling height were 300 msec, 230 msec and 200 msec, respectively. It is noted that the duration of the strain wave became shorter with increased falling height. When EPS is adopted as an absorbing material, the fact that the duration of the impact load varies depending on the kinetic energy input from the weight dropping should be taken into consideration.

4.2 Distribution of bending moment of PC-R

Distributions of bending moment of the girder PC-R with respect to time for different falling heights (H = 1, 2 and 5 m) are shown in Fig. 4. Bending moment was calculated using the measured rebar strain assuming 1) plane conservation, 2) maximum effective tensile strain of concrete = 300 μ and 3) no axial component of the force. Since the girder behaved elastically as long as falling height was 2 m, bending moment could be calculated considering no cracks occurred in the concrete. For the case of H = 1 m, the distribution of bending moment curve took a shape similar to that of a simply supported beam with a static and partially distributed load at the mid span. But for the cases of H = 2 m and 5 m, the distributions of bending moment became almost linear, closely resembling that of a simply supported beam with a static and concentrated load at the mid span.

The maximum bending moments for H = 1 m and 2 m were about 265.6 kNm and 322.4 kNm, respectively. Assuming that the concentrated load was statically surcharged at the mid span, the impact loads calculated backward from the bending moments for H = 1 m and H = 2 m were 211.7 kN and 258.7 kN, respectively. On the other hand, transmitted impact forces were estimated as 229.3 kN and 262.6 kN, respectively, assuming that the energy absorbed by the EPS was equal to the kinetic energy of the falling weight[6].

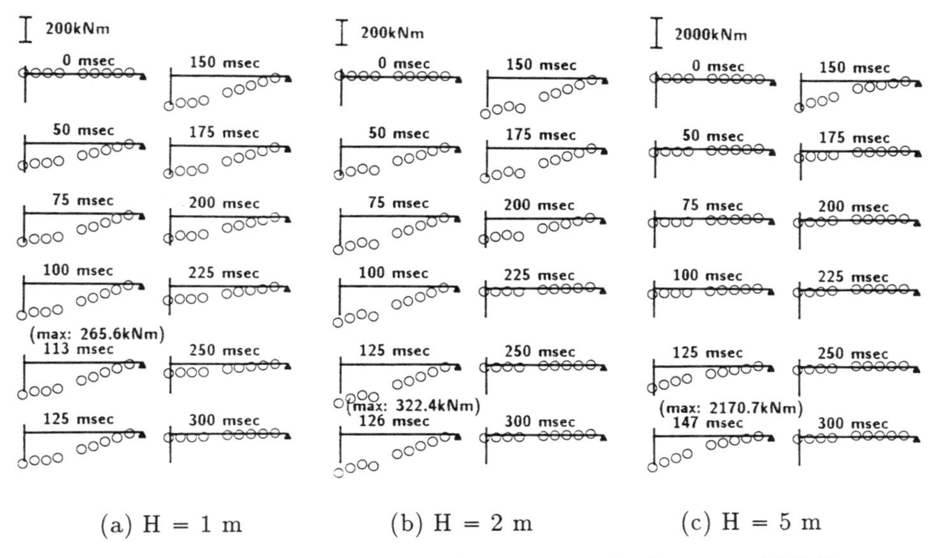

Fig. 4. Axial distribution of bending moment in the case of PC-R.

It is observed that the experimental value is similar to the theoretical value. Therefore, it can also be confirmed that the impact energy was effectively absorbed by the EPS up to H = 2 m.

The maximum bending moment for H = 5 m was found to be about 2170.7 kNm. In this case, transmitted impact force was estimated as 1736.6 kN by calculating backward as mentioned above. Clearly, this experimental value is more than six times the theoretical value. It is clear that the impact energy was not effectively absorbed by the EPS due to its insufficient thickness.

4.3 Final crack pattern and deformation of girder

Final crack patterns on both sides of the girders are shown in Fig. 5. In the case of the lowest falling height (PC-10), it is clear that even cracks have appeared, but the concrete was not torn neither in the flange part nor in the web part. Cracks at both the flange and the web beneath the loading area developed in a vertical direction, directly while diagonal cracks with 45° inclination developed in the area between the edge of loading area and the support. This suggests that bending cracks developed in the proximity of the loading area and shearing cracks developed in the remaining area. Diagonal cracks, in the area between the edge of loading area and the support, were 40 ~50 cm long. This supports the idea that an arch action was formed in the area, and the impact force was transmitted to the support by this action. The cracks in the lateral side of flange might have been generated by rebounding of the girder due to the prestress which was introduced by the PC tendons allocated in the lower part of the girder.

In all cases of PC-15 ~ 30, each girder suffered substantial damage and the concrete was torn off in some areas of the web and flange of the girder. The

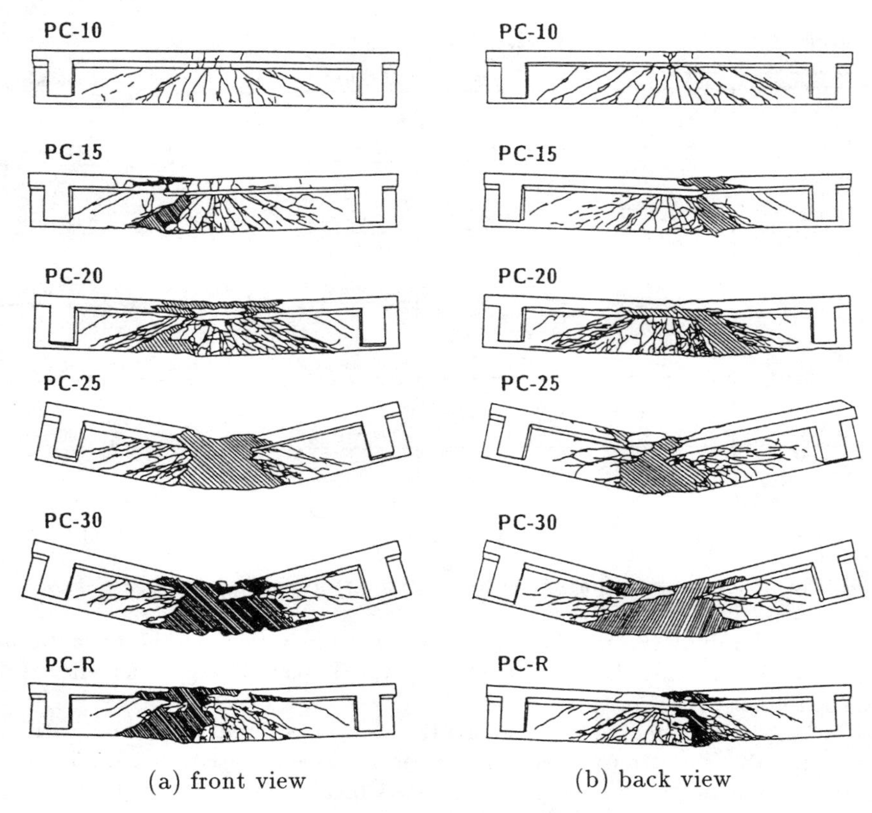

(a) front view (b) back view

Fig. 5. Crack pattern of girder after final dropping.

area of torn concrete became wider according to the drop height of the steel weight. In the girders of PC-15 and PC-20, even though the density of cracks generated beneath the loading area became large according to the dropping height of the steel weight, their crack patterns were similar to those of PC-10. Only the concrete in the transition area between the bending and shearing actions in the girders of PC-15 and PC-20 were torn off. This concrete tearing action might have been generated from the following reasons: 1) the compressive force acting on the arch increased in accordance with the impact load increment, 2) the transition area became weak and 3) finally, the bending area beneath the loading area was crushed out. The crack pattern of girder PC-R, which was subjected to iterative loadings, showed an intermediate pattern between those of PC-10 and PC-15. This means that there was little difference between iterative and single loading on the damage of girders.

On the other hand, in the cases of PC-25 and PC-30, the area of concrete collapse covered an area equal to the sectional area of the steel weight. Cracks developed at an inclination of 45°, and the arch action no longer existed as it had in the cases of PC-10 and PC-15. It is considered that the girders were snapped by the concrete crushing at the loading area and rebars and PC

Table 3. Residual deformation and break angle

Specimen	Residual deformation Δ (cm)	Break angle $2\Delta/l$
PC-R	9.3	0.037
PC-10	-0.2	-0.001
PC-15	6.4	0.026
PC-20	10.2	0.041
PC-25	48.4	0.194
PC-30	63.0	0.252

tendons yielding.

Residual deformation Δ and the average final break angle of the PC girders are listed in Table 3. The table shows that the break angles for PC-10, PC-15 and PC-20 were less than 0.05 radian, while the break angles for PC-25 and PC-30 were almost equal and more than 0.2 radian: more than four times larger than those of the former. Based on the final crack pattern, the residual deformation and break angle, the critical impact load for the PC girders used in this study may be represented by PC-20, the case in which a 49 kN heavy weight was allowed to fall from a height of 20 m.

5 Conclusions

To study the impact resistance of PC girders which form a component of PC rock-sheds, a full scale experimental investigation was conducted. The results obtained from these experiments are as follows:

1. The PC girder studied behaved elastically up to a falling height H = 2 m.
2. In the case of H = 5 m, the impact load converted into a static load was observed to be 1736.6 kN, exceeding the design load (1257.3 kN). Cracks were developed at the lower part of the PC girder.
3. The crack patterns of PC-R (maximum dropping height of 15 m) were of intermediate nature between those of PC-15 and PC-20. The difference between iterative and single loading on the damage of girder was not significant.
4. In the cases of PC-10, PC-15 and PC-20, arch action was formed between the edge of loading area and the supports, and crack patterns were seen to be similar among them.
5. In the cases of PC-25 and PC-30, the girders suffered a severe damage, causing the concrete under the loading area to be crushed into chips. As long as the falling height did not exceed 20 m, the break angle of the girder remained below 0.05 radian, while it quadrupled for a falling height greater than 20 m. From residual deformation and break angle considerations, a 49 kN weight falling from a height of 20 m can be regarded as the critical state for the PC girder used in this experiment.

6 Acknowledgement

The authors sincerely thank Mr. Masanori Takazawa (Graduate Student of Muroran Institute of Technology) who assisted in this investigation.

7 References

1. Nakano, O., Kishi, N., Sugata, N. and Satake, T. (1993) *Dynamic Behavior of PC Rock-Shed with 3-Layered Shock Absorbing System under Impact Load*, Proceedings of Annual Congress of JSCE Hokkaido Branch, 49, pp.145-150, in Japanese.
2. Sugata, N., Sato, M., Nishi, H. and Kishi, N. (1994) *Experimental Study on Impact Behavior of Prototype PC Rock-Shed Composed of Cushion*, Proceedings of JCI, 16, 2, pp.949-954, in Japanese.
3. Matsuba, Y., Goto, Y., Sato, A. et al. (1993) *On Impact Failure Tests of Prototype PC Rock-Shed under Rock Falling (1), (2)*, Proceedings of 2nd Symposium on Impact Problems, JSCE, pp.241-253, in Japanese.
4. Sonoda, Y., Sato, H. and Ishikawa, N. (1993) *Impact Response Analysis of PC Rock-Shed*, Proceedings of 2nd Symposium on Impact Problems, JSCE, pp.264-269, in Japanese.
5. *Concrete Manual*, JSCE, (1991), in Japanese.
6. Yoshida, H., Matsuba, Y., Hohki, K. and Kubota, T. (1991) *An Experimental Study on Shock Absorbing Effect of Expanded Poly-Styrol against Falling Rocks*, Proceedings of JSCE, 427/VI-14, pp.143-152, in Japanese.

162 BEHAVIOR OF REINFORCED CONCRETE BEAMS AT HIGH STRAIN RATES

S.M. KULKARNI and S.P. SHAH
Northwestern University, Evanston, USA

Abstract
This paper presents the results of tests conducted at Northwestern University on reinforced concrete beams at rates of straining typically observed during earthquakes. These tests were conducted using a digitally controlled, closed-loop servo-hydraulic testing machine. Six pairs of singly reinforced beams (without shear reinforcement) were tested under displacement control. From each pair, one beam was tested at a "static" rate (piston velocity = 0.00071 cm/sec) and the other at a "high" rate (piston velocity = 38 cm/sec). Peak load and energy absorption capacity were found to increase with the rate of straining. Crack pattern did not change significantly with the rate. Deflected shape of the beams did not change significantly with the rate of straining. For two of the pairs tested, final failure mode shifted from a shear type of failure at the static rate to a flexural type of failure at the high rate.
Keywords: Ductility, dynamic loads, earthquakes, flexural strength, impact, reinforced concrete beams, shear failure, strain rate, structural testing.

1 Introduction

In recent years, there has been considerable interest in the behavior of concrete subjected to high strain rates. Design of structures to withstand the severity of events such as earthquakes, impacts, and blasts requires knowledge of behavior at high strain rates, both at the material and the structural level. Recent research activity has provided some information regarding the significant effect of high rate of straining on plain concrete, reinforcing steel, and the bond between them. Reviews of some of the recent work done can be found in references [1], [2] and [3]. It is it is well known that strength and stiffness of plain concrete under compression increase with the rate of straining. Tensile strength of concrete is known to increase with the rate. For reinforcing bars, yield strength, yield strain, and strain at the beginning of strain hardening are known to increase with the rate. Considerable disagreement still exists among researchers regarding influence of strain rate on some of the key properties (for example, strain at maximum stress for plain concrete in compression). Part of the reason for this is the fact that different researchers used different testing equipment. Also measurement techniques and analyses of results (especially incorporation of inertial effects) were different in the various studies.

To obtain high strain rates, devices which induce impact [4][5][6][7] or

Concrete Under Severe Conditions: Environment and loading (Volume Two) Edited by K. Sakai, N. Banthia and O.E. Gjørv. Published in 1995 by E & FN Spon. ISBN 0 419 19860 1

explosion [8][9] have been used in the past. Several studies on plain concrete and reinforced concrete structural elements subjected to impact and impulsive loading are now available. Using such devices, relatively high strain rates (~ 10 /sec) can be generated. Another advantage is that testing using these devices can closely simulate a natural impact or blast event. Severe inertial forces generally associated with such devices, however, often complicate the analysis and interpretation of test results. Also, response of test specimens may contain reversal of displacements, implying that straining of the material may not be monotonic.

Influence of nature of the loading mechanism used on the response of beams, especially considering the inertial effects, was demonstrated by Kulkarni and Shah [10]. They showed that for a given duration of load, if a test is conducted under displacement control using a closed-loop testing machine (instead of load control), inertial effects are significantly reduced, permitting a more truly monotonic straining at a high rate. (See also reference [9].) Displacement controlled tests are thus quite suitable for examination of effects of monotonic, high-rate straining on structural elements such as beams. The strain rates available in this type of testing are generally limited to about 1 /sec. Thus, such testing is ideal for examining the effect of strain rates generally observed during earthquakes.

Available information on the effect of earthquake type of strain rates on reinforced concrete beams is quite limited. Bertero et al. [11] conducted some such tests and found that for predominantly flexural beams, the high rate of straining produced an increase in moment capacity at first yield. The failure mode for such beams did not change with the rate. However, possibility of a transition in the mode of failure from flexural to a more brittle, shear type of failure was suggested by them. They recommended further investigation on shear-critical beams to examine this possibility. Wakabayshi et al. [12] examined the effect of strain rate on steel and reinforced concrete beams. For both these types of beams, their theoretical prediction of load-deflection behavior was not sufficient to quantitatively describe the observed experimental behavior. The possible reasons for this were not discussed. Mutsuyoshi and Machida [13] conducted a relatively systematic investigation on several beams with different a/d ratios and reinforcement percentages. Of the beams tested, four pairs exhibited a transition in the mode of failure from flexure to shear as the rate of straining was increased.

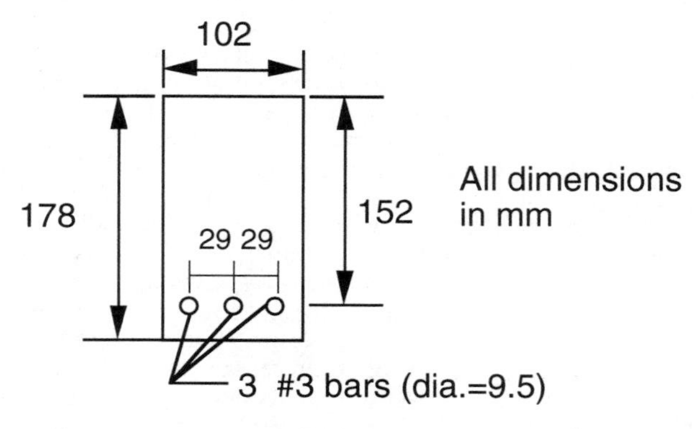

Fig. 1. Cross-section of the specimens used.

With the aim of obtaining further understanding of the behavior of reinforced concrete beams under earthquake type of strain rates, a systematic experimental investigation was started at Northwestern University. The focus was on shear-critical beams. This paper presents highlights of the results obtained so far. A relatively detailed analysis of the inertial effects involved and additional considerations regarding closed-loop testing are discussed in [10]. It was shown from theoretical analysis and some experimental observation that for the highest rate of strain available at Northwestern University, the test beams behaved in a quasi-static manner.

Fig. 2. Picture of the test setup (with a trial specimen).

2 Experiments

2.1 Specimen preparation

All specimens had the cross-section shown in Fig. 1. No shear reinforcement was provided. Three bars of nominal diameter 9.5 mm and made of Grade 60 steel (guaranteed *minimum* yield strength of 414 MPa) were used. Reinforcement ratio was 1.38%. Type I ordinary portland cement, sand graded #2, and pea gravel of maximum size 9.5 mm were used. Mix-proportion of concrete used was 1 : 2.85 : 2.89 : 0.5 (Cement : Sand : Gravel : Water). Specimens were cast in wooden molds in two layers, and compacted by hand rodding. The casting procedure closely followed ASTM recommendations C192-90a. Specimens were demolded after 24 hours and moist cured until test time.

From each batch of concrete, a pair of beam specimens and several companion cylinders (10.2 cm X 20.3 cm) were cast. One beam in each pair was later tested at the "static" rate, and the other, at the highest available rate. The companion cylinders were tested along with the beam specimens to determine the compressive strength.

2.2 Test equipment

A specially designed, digitally controlled, closed-loop servo-hydraulic test system was recently purchased by the Center for Advanced Cement-Based Materials at Northwestern University for conducting high-rate tests on medium-to-large scale structures. Fig. 2 is a photograph of the machine. It has an axial capacity of 1 MN. The machine is equipped with a hydraulic pump that supplies 1.262×10^{-3} m^3/sec of oil at 20.7 MPa, and a close-coupled accumulator package rated at 7.572×10^{-3} m^3. Depending on the type of specimen being tested, the machine can achieve a piston velocity of about 38 cm/sec over a piston travel of about 3 cm. The machine is controlled by a Pentium processor-based personal computer. A more detailed description of the test system and considerations required for its operation are given in [10].

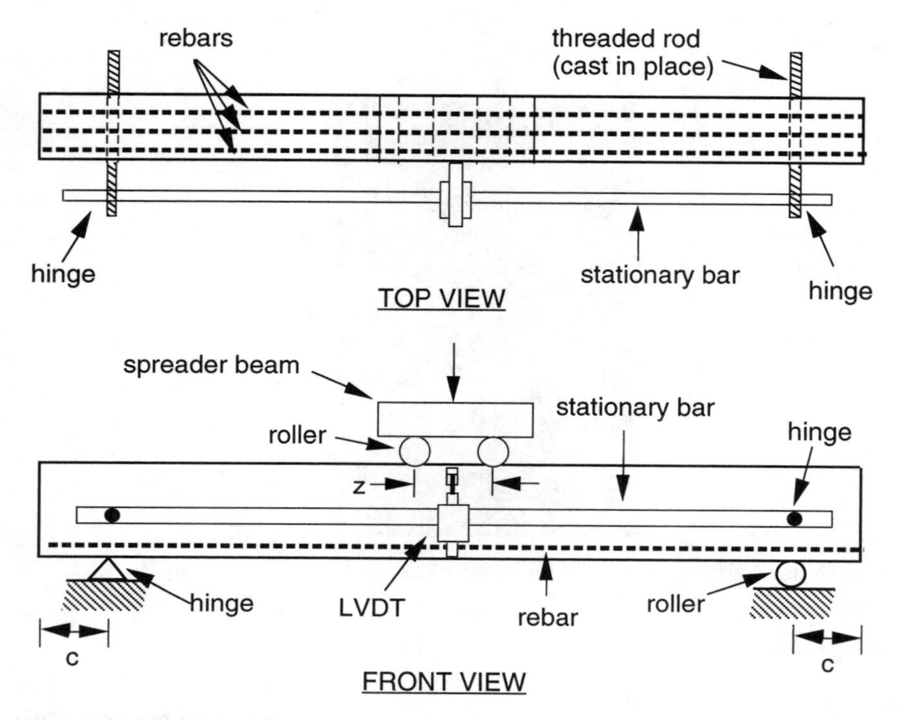

Fig. 3. Sketch of test setup.

2.3 Test configuration

All specimens were simply supported and were tested in a four-point bend configuration as shown in Fig. 3. The fixtures for support points and the loading points were designed to insure absence of any twisting moment. Deflections were measured with reference to a stationary aluminum bar as shown in Fig. 3. This avoids the influence of machine deformation on the deflection measurements. Central deflections were measured by an LVDT (Linear Variable Differential Transformer). All tests were conducted using the actuator stroke control. However, for the setup used in this investigation, this control was shown by Kulkarni and Shah [10] to be virtually identical to central deflection control. Data acquisition for static tests was done through the controlling computer at a rate of about 1 Hz. For the high-rate tests, one or more digital oscilloscopes were used. Acquisition rate was about 100 kHz.

2.4 Test program

2.4.1 Beam testing

The beam test program is shown in Table 1. A total of 6 pairs of beams were tested. From each pair, one beam was tested at a piston velocity of 0.00071 cm/sec, and the other at the highest available piston velocity of about 38 cm/sec.

Table 1. Properties of beam specimens and some results

Specimen designation	a/d	Dist. bet. load pts. (z) (mm)	Over-hang (c) (mm)	Conc. comp. strength (MPa)	Piston vel. (cm/sec)	Max. moment (kN-m)	Max. shear force (kN)	Failure mode
B3SE03_S	4.5	305	229	45.02	0.00071	15.79	23.02	Shear
B3SE03_H	4.5	305	229	45.02	38	17.71	25.82	Flexure
B4JL25_S	5.5	152	102	41.51	0.00071	16.03	19.13	Flexure
B4JL25_H	5.5	152	102	41.51	38	18.01	21.48	Flexure
B3OC25_S	5.0	152	229	46.20	0.00071	15.80	20.73	Flexure
B3OC25_H	5.0	152	229	46.20	38	17.58	23.06	Flexure
B3DE03_S	4.5	152	229	42.96	0.00071	15.85	23.11	Flexure
B3DE03_H	4.5	152	229	42.96	38	18.14	26.44	Flexure
B3NO15_S	4.0	152	229	42.75	0.00071	13.82	22.66	Shear
B3NO15_H	4.0	152	229	42.75	38	18.17	29.80	Flexure
B3NO30_S	3.5	152	229	45.23	0.00071	12.93	24.24	Shear
B3NO30_H	3.5	152	229	45.23	38	14.59	27.36	Shear

2.4.2 Cylinder testing

For each batch of concrete, three companion cylinders (10.2 cm X 20.3 cm) were tested along with the beams, according to ASTM specifications C39-86, to obtain the compressive strength. The average values obtained are reported in Table 1. Cylinders were capped with sulfur capping compound prior to testing.

2.4.3 Rebar testing

The static yield strength was obtained by testing tensile coupons. Approximate strain rate in these tensile tests was 0.00003 /sec. An average value of yield stength of 518 MPa was obtained from three specimens tested. This value is applicable for all the pairs except the pair B4JL25_S and B4JL25_H. The steel used for this pair was also grade 60, but was from a different batch. For this steel, no measurement of yield strength was done, however, a value of 445 MPa was obtained from the data supplied by the manufacturer.

2.4.4 Deflected shape and strain measurements

Two of the pairs of beams tested were instrumented with more sensors in order to obtain more information on the overall mechanics of the beams at the two rates and therefore facilitate a better understanding of the phenomena observed and to aid future modeling. Strain gages were attached to rebars in the constant moment zone for specimens B4JL25_S and B4JL25_H. All strain gages were purchased from the Measurements Group. These were self-temperature compensated and had a gage length of 6.4 mm. Provision against moisture and abrasion was provided in the form of coatings recommended by the manufacturer.

When significant inertial forces are associated with deformation of a beam in a short duration test, the deflected shape of the beam, the magnitude and distribution of shear and moment along the span, and the moment-to-shear ratio are different compared to those in a static test. Such differences have been considered by Hughes and Beeby [4] as a possible reason for a different mode of failure under impact loading. Cracking behavior is somewhat different at a high rate ([7][13]), implying a different cracked moment of inertia. The effect of this on the deflected shape is not well known yet.

In the light of these facts, it was decided to measure the deflections at different locations along the span of a typical beam and observe any differences when strain rate changed from static to high. Deflections were measured for the beam pair B3SE03_S and B3SE03_H at five equispaced points in a half of the span.

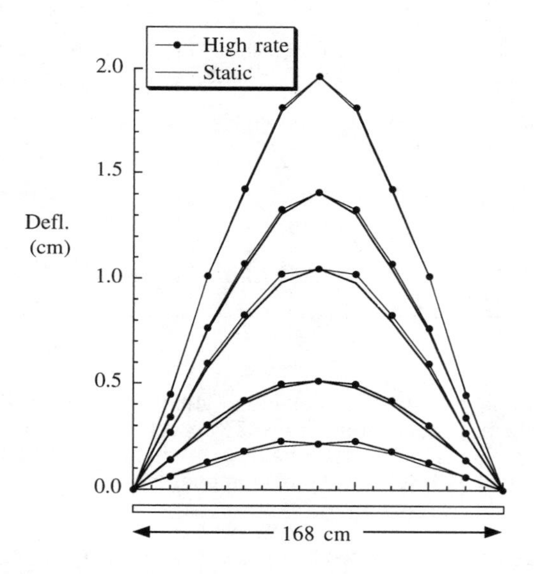

Fig. 4. Deflected shape for beams B3SE03_S (static rate) and B3SE03_H (high rate).

3 Results and discussion

3.1 Deflected shapes

Deflected shapes at the two rates are plotted in Fig. 4. The shapes are not significantly different at the two rates.

3.2 Steel strains

Generally speaking, the setup for strain measurement worked well. However, signals from the strain gages were found to contain unacceptable features, such as sudden jumps, after a strain of about 0.7%, depending on the test. Since the yield strain of steel was about 0.28% the main feature of yielding could still be observed. From strain data, average strain rates were computed to be 0.3 /sec at the high rate and 3 x 10^{-6} /sec at the static rate. The ratio of the two strain rates (~ 100000) is higher than the ratio of the corresponding piston velocities (~54000). This implies that, for a given central deflection, steel strain is higher at a higher rate. Fig. 5(a) shows that this is indeed true. The reasons behind this are not clear. Similar observation was made by Mutsuyoshi and Machida [13], however, no explanation was attempted.

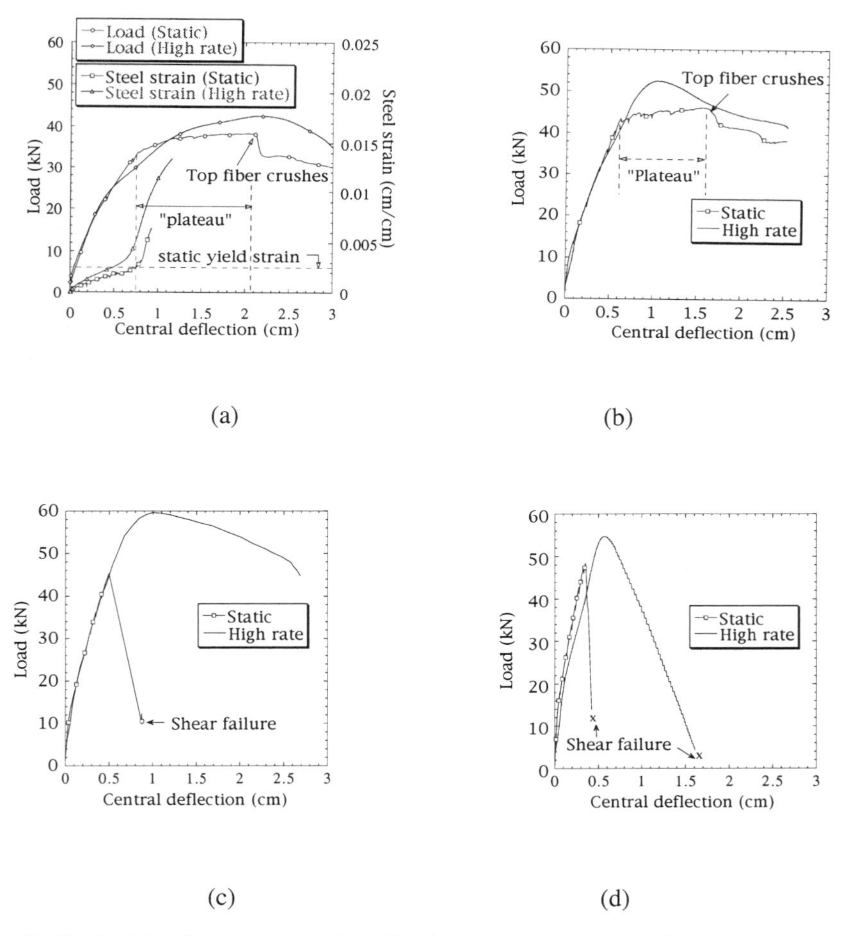

Fig. 5. Typical load versus central deflection diagrams: (a) specimens B4JL25_S and B4JL25_H (a/d =5.5) (b) specimens B3DE03_S and B3DE03_H (a/d = 4.5) (c) specimens B3NO15_S and B3NO15_H (a/d = 4.0) (d) specimens B3NO30_S and B3NO30_H (a/d = 3.5).

3.3 Crack patterns

Fig. 6 shows the crack patterns for specimens B3SE03_S and B3SE03_H. Crack pattern does not seem to change significantly with the rate. However, average crack spacing appears slightly higher at the high rate.

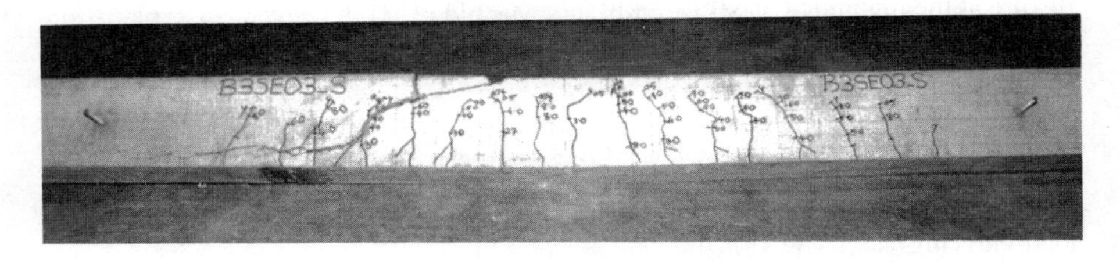

(a) Static test on specimen B3SE03_S (shear failure)

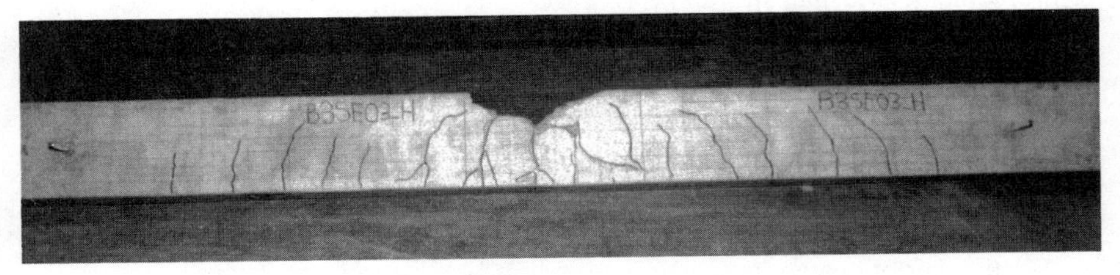

(b) High-rate test on specimen B3SE03_H (flexural failure)

Fig. 6. Pictures of failed specimens.

3.4 Load-deflection curves and failure modes

3.4.1 General

All the beams tested were relatively close to the shear failure regime. Therefore, final failure sometimes could not easily be classified as shear or flexural failure. To aid this decision, the failure pattern, load-deflection curve, and maximum moment obtained were studied carefully. Flexural failure is characterized by crushing of concrete in the compression zone between the load-points. After the peak load, the load-deflection curve of a beam failing in flexure shows a relatively slow reduction of load with an increasing deflection. Shear failure is characterized by a predominant shear crack in one half of the span. After the peak load, the load-deflection curve of a beam failing in shear shows a very sharp decrease in the load. The maximum moments sustained by all the beams failing in flexure at a given rate showed very good consistency, irrespective of the a/d ratio. (This is expected since the maximum moment sustained by a beam failing in flexure is a sectional property and is independent of the a/d ratio.) This is not true for beams failing in shear (Table 1). Maximum moment considerations can therefore give an additional check regarding the type of failure.

3.4.2 Load-deflection curves and energy absorption

Load-deflection curve for a beam that failed in flexure (Figs. 5(a) and 5(b)) at the static rate shows a distinct yield "plateau," that is, a region of very small (positive) slope compared to the initial slope of the curve. The maximum moment is associated with visual observation of crushing of the top concrete in the constant moment zone. Tests were however continued beyond this point and the ensuing responses are shown in Figs. 5(a) and 5(b). The load-deflection curve for a beam that failed in flexure at the high rate (Figs. 5(a), 5(b), and 5(c)) does not show a "plateau." At the present time, the reason for this is not clear.

For beams that failed in shear at the static rate (Figs. 5(c) and 5(d)), the load dropped to (almost) zero rapidly after the peak. Since the data was acquired at a rate of about 1 Hz for static tests, no data points were obtained after the peak. Beam B3NO30_H failed in shear at the high rate. The associated high speed of data acquisition permitted obtaining the post-peak data (Fig. 5(d)).

From Table 1, it is apparent that for all the pairs tested, there is an increase in load-carrying capacity with increase in the rate. This is in agreement with the findings of virtually every other researcher. Increase in load-carrying capacity for beams failing in flexure is generally attributed to the increase in yield strength of reinforcement. Mechanism of shear failure of beams without shear reinforcement has not been completely understood even at static rates. Understanding of rate effect on shear failure is therefore much more difficult. Mutsuyoshi and Machida [13] attributed the increase in shear strength to the increase in tensile and compressive strength of plain concrete.

For the pairs of beams for which flexural failure was obtained at both the rates, the absorbed energy at a given central deflection increased with the rate, due to the higher loads sustained at the high rate (See Figs. 5(a) and 5(b)). Mutsuyoshi and Machida [13], Banthia [6], and Ragan [7] observed similar increases.

For the pair of beams for which shear failure was observed at both the rates (Fig. 5(d)), the total deflection at failure and the total energy absorbed increased remarkably with the rate. Mutsuyoshi and Machida and Ragan observed similar increases.

For two of the pairs tested (B3SE03_S and B3SE03_H; B3NO15_S and B3NO15_H) the failure mode changed from shear at the static rate to flexure at the high rate. Energy absorbed at a given deflection also increased with the rate (See Fig. 5(c)). This is a transition towards a ductile type of failure and is quite the opposite of the transition observed by Mutsuyoshi and Machida and Ragan. These researchers observed that with an increase in the rate, the failure mode for some of their specimens shifted from flexure to shear. However, these two observations are not necessarily contradictory since these experiments were not conducted under identical conditions. In particular, the properties of steel and concrete used, reinforcement used (presence or absence of shear reinforcement, percentage of longitudinal reinforcement, presence or absence of top reinforcement), and the nature of loading (impact versus quasi-static) were different in different studies.

At present, quantitative understanding of the influence of most these factors on rate dependence of structural failure is not sufficient to be able to explain this apparent contradiction easily. Some understanding of the possible influence of the nature of loading on the mode of failure can be obtained from the work of Hughes and Beeby [4]. As noted by them, under direct impact conditions, the deflected shape is different from that at a static rate due to the presence of inertial forces. The altered deflected shape at the high rate is responsible for higher maximum shear and lower maximum moment compared to those at a static rate. (The reader is also referred to the work of Keenan [8].) This is at least partly responsible for the observed tendency in impact tests towards a transition from flexural failure at a static rate to shear failure at the impact rate. For high-rate quasi-static tests such as the ones reported in this paper, deflected shape is essentially independent of the rate. Therefore, the above-mentioned tendency is not applicable. Quantitative understanding of influence of other factors on final structural failure should be the subject of further investigation.

4 Conclusions

A high-rate, quasi-static setup was used to examine the effect of rate of (monotonic) straining on reinforced concrete beams. Measurement of deflection revealed that the profile of deflected shape of the beam did not change with the rate. Crack patterns were similar at the two rates used. Peak load was found to increase with the rate. Area under the load versus central deflection diagram (up to a given value of central deflection) increased with the rate. For two of the pairs tested, the mode of failure changed from shear at the static rate to flexure at the high rate.

5 Acknowledgments

This study is part of a joint project between Northwestern University and Pennsylvania State University funded by the National Science Foundation through grant no. MSS-402-1992. The authors also gratefully appreciate financial support from the NSF Center for Advanced Cement-Based Materials headquartered at Northwestern University.

6 References

1. Fu, H.C., Erki, M.A., and Seckin, M. (1991) Review of effects of loading rate on reinforced concrete. *Journal of Structural Engineering, ASCE*, Vol. 117, No. 12. pp. 3660-3679.
2. Fu, H.C., Seckin, M., and Erki, M.A. (1991) A review of the effects of loading rate on concrete in compression. *Journal of Structural Engineering, ASCE*, Vol. 117, No. 12. pp. 3645-3659.
3. Bischoff, P.H. and Perry, S.H. (1991) Compressive behavior of concrete at high strain rates. *Materiaux et constructions,* Vol. 24. pp. 425-450.
4. Hughes, G. and Beeby, A.W. (1982) Investigation of the effect of impact loading on concrete beams. *The Structural Engineer,* Vol. 60B, No. 3. pp. 45-52.
5. Masuya, H., Kajikawa, Y., and Shibata, Y. (1991) Experimental study on behavior of reinforced concrete beams subjected to impact load. *Transactions of the Japan Concrete Institute*, Vol. 13. pp. 379-386.
6. Banthia, N. (1987) *Impact resistance of concrete*. Ph.D. Thesis, Department of Civil Engineering, University of British Columbia, Vancouver, B.C., Canada.
7. Ragan, S. (1994) *Rate of loading effects on reinforced concrete beams.* M.S. Thesis, Department of Civil Engineering, Pennsylvania State University, University Park, PA, U.S.A.
8. Keenan, W. A. (1965) Dynamic shear strength of reinforced concrete beams Part I, Technical report R395, U.S. naval civil engineering laboratory, Port Hueneme, CA, U.S.A.
9. Penzien, J. and Hansen, R.J. (1954) Static and dynamic elastic behavior of reinforced concrete beams. *Journal of the American Concrete Institute*, Vol. 25, No. 7. pp. 545-567.

10. Kulkarni, S.M. and Shah, S.P.(1994) Some aspects of closed-loop controlled testing of reinforced concrete beams at high rates, in *New experimental techniques for evaluating concrete material and structural performance,* (Ed. D.J. Stevens and M.A. Issa), ACI Special Publication SP-143, Detroit, Michigan. pp. 123-143.
11. Bertero, V.V., Rea, D., Mahin, S., and Atalay, M.B. (1973) Rate of loading effects on uncracked and repaired reinforced concrete members. *Proceedings, Fifth World Conference on Earthquake Engrg.*, Rome, Italy, Vol. 2. pp. 1461-1470.
12. Wakabayashi, M., Nakamura, T., Yoshida, N., Iwai, S., and Watanabe, Y. (1980) Dynamic loading effects on the structural performance of concrete and steel materials and beams. *Proceedings, Seventh World Conference on Earthquake Engineering*, Istanbul, Turkey, Vol. 6, No. 3. pp. 271-278.
13. Mutsuyoshi, H. and Machida, A. (1984) Properties and failure of reinforced concrete members subjected to dynamic loading. *Transactions of the Japan Concrete Institute*, Vol. 6. pp. 521-528.

163 FRACTURE OF FIBER REINFORCED CONCRETE UNDER IMPACT

N. BANTHIA and S. MINDESS
University of British Columbia, Vancouver, Canada
J.-F. TROTTIER
Technical University of Nova Scotia, Halifax, Canada

Abstract
An adequate resistance of concrete to impact loads is essential in many civil engineering applications. Unfortunately, given the complexities involved in testing concrete under impact and in analyzing the data, our understanding of the resistance of concrete to impact loads is quite limited. This paper describes the construction of a simple impact machine capable of conducting impact tests on concrete in the uni-axial tensile mode. Some impact data for normal strength, medium strength, high strength and fiber reinforced concrete is presented.
Keywords: Impact, fiber reinforced concrete, tension, strength, toughness.

1 Introduction

Resistance of concrete to impact and impulsively applied loads is of concern in many structures [1]. Several technique have been developed in the past to test concrete under impact which include the *Drop Weight Machines* [2,3,4,5], *Swinging Pendulum Machines* [6,7,8], and the *Split Hopkinson Pressure Bar Test* [9,10]. Although plain as well as fiber reinforced concrete have been reported as stress-rate sensitive under all modes of loading [11-15], there is little agreement in the published literature, data appears to be machine dependent and the exact mechanisms responsible for the apparent stress-rate sensitivity are not understood. Clearly, continued research is needed in this area to strengthen our understanding of the stress-rate sensitivity of concrete.

Concrete Under Severe Conditions: Environment and loading (Volume Two) Edited by K. Sakai, N. Banthia and O.E. Gjørv. Published in 1995 by E & FN Spon. ISBN 0 419 19860 1

2 Experimental

2.1 The impact machine

The machine used is similar in principle to a smaller version described previously [16]. The setup is schematically shown in Figure 1. In principle, the specimen bridges two supports, A and B, with support B mounted on rollers (called the *trolley*) and support A being fixed. Support B is struck by the swinging pendulum on impact points located on either side of the specimen and in the same plane as the specimen. Support A being fixed, this causes tensile loading in the specimen.

The 67 kg impact hammer carries two dynamic load cells (222.4 kN capacity) mounted on either side that record the contact load vs. time pulse during the impact. Under an impact, the specimen fractures and the trolley (mass = 20.53 kg) travels toward the shock absorbers. On its way, the trolley passes through two photocell assemblies 37 mm apart where its post-event velocity is recorded. The other instrumentation provided includes accelerometers (± 500 g, 9.98 mV/g) mounted both on the trolley and the hammer and a photocell assembly to record the hammer velocity during approach. The photocell assembly recording hammer velocity supplies the triggering pulse for the data acquisition system which then records data at a sampling rate of 1.2 MHz.

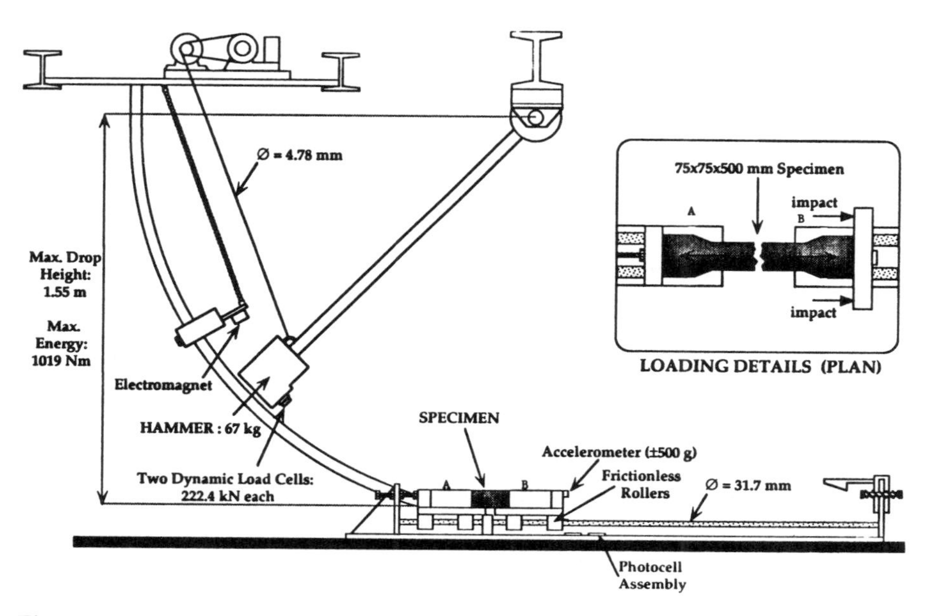

Fig. 1. Impact Test Setup

2.2 Data analysis

The objective of the impact testing in the present study was to obtain the strength and fracture energy values for the various cementitious systems. Considering the free body diagram of the *trolley* itself, the horizontal force equilibrium (ignoring damping) may be written as:

$$P_t(t) = P_i(t) + P_s(t) \tag{1}$$

The hammer load $P_t(t)$ in the above equation is the sum of the loads recorded by the two load cells mounted on either side of the hammer and $P_s(t)$ is the specimen load. $P_i(t)$ is the inertial load [2] given rise to by the accelerations in the system. If $a_t(t)$ is the trolley acceleration and m_t and m_s are the masses of the trolley and the specimen, respectively, Eqn. 1 can be written as:

$$P_t(t) = a_t \left[m_t + \frac{m_s}{2} \right] + \sigma_s(t) A_s \tag{2}$$

$\sigma_s(t)$ and A_s in the above Eqn. are the specimen stress and the cross-sectional area, respectively. With trolley accelerations $a_t(t)$ recorded by the accelerometer, Eqn. 2 can be solved for $\sigma_s(t)$, the peak value of which could then be taken as the tensile strength of the composite under impact.

Using the impulse-momentum principle [6,7,8], the energy lost by the pendulum (E_h) during its contact with the trolley can be written as:

$$E_h = \frac{1}{2} m_p v_i^2 - \frac{1}{2} m_p \left[v_i - \frac{1}{m_p} \int P(t)dt \right]^2 \tag{3}$$

where,

m_p = mass of the pendulum

$\int P(t)dt$ = contact impulse

v_i = initial velocity of pendulum

= $\sqrt{2gh}$

g = earth's gravitational acceleration

h = height of hammer drop

If one can ignore the frictional and other losses of energy, the energy lost by the pendulum can be regarded as the sum of the energies consumed by the specimen during fracture (E_s) and that gained by the trolley as post-fracture kinetic energy (E_t). In other words,

$$E_h = E_t + E_s \tag{4}$$

Further, if v_t is the post-fracture velocity of the trolley as recorded by the base-mounted photocell assemblies, then,

$$E_t = \frac{1}{2}\left[m_t + \frac{1}{2}m_s\right] v_t^2 \qquad (5)$$

Using Eqns. 3, 4 and 5 and solving for E_s,

$$E_s = \frac{1}{2}m_p\left[v_i^2 - \left[v_i - \frac{1}{m_p}\int P(t)dt\right]^2\right] - \frac{1}{2}\left[m_t + \frac{m_s}{2}\right] v_t^2 \qquad (6)$$

With all quantities on the right hand side known, the fracture energy consumed by the specimen can be calculated. For tests carried out with no specimen in the system, the recorded tup loads and the calculated inertial loads matched as seen in Figure 2 (see Eqn. 1). However, when a specimen was placed in the system, it was observed that the tup load vs. time trace had two distinct peaks, with the first peak fully accounted for by inertia and the second peak representing the true specimen loading with a negligible inertial load. Clearly, in the early part of an impact, the applied load is utilized only in accelerating the system with no load applied to the specimen yet. It is only after the trolley gains enough momentum that the specimen experiences the tensile load which manifests itself as the second peak in the recorded impulse. After the first peak, the accelerations died down and the inertial forces could be ignored. The stress-rate generated in a test ($\dot{\sigma}$) was estimated from the "second hump" as illustrated in Figure 2.

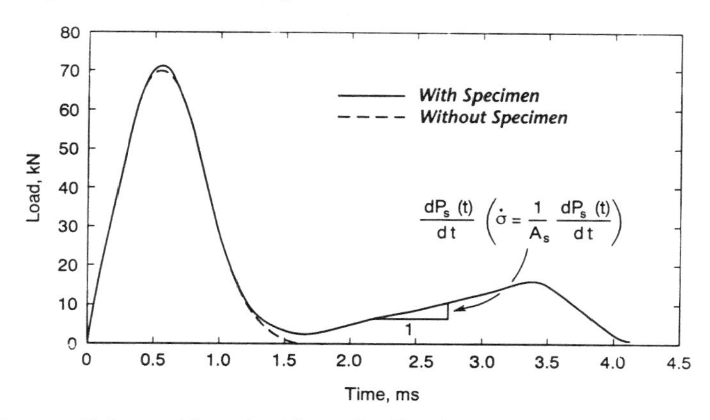

Fig. 2. Contact Pulses with and without the Specimen

2.3 Impact and companion static testing

Specimens with tapered ends (75 mm x 75 mm critical section; Figure 1) were cast with normal strength, mid-strength and high strength concretes with mix details (cement:water:sand:coarse aggregate) as follows:Normal Strength (Mix I): 1:0.45:2.14:2.61; Mid-Strength (Mix II): 1:0.35:1.90:2.32; High Strength (Mix III): 1:0.30:1.72:2.27. Another series of test specimens were cast with fiber reinforced concrete, where the above three matrices were reinforced with 40 kg/m^3 of four types of deformed steel fibers with details given below:

Fiber F1: Hooked-ends, circular cross-section 0.8 mm diameter, 60 mm long, TS = 1115 MPa, fiber mass = 263 mg.

Fiber F2: Crimped along the length, circular cross-section, 1.0 mm diameter, 60 mm long, TS = 1037 MPa, fiber mass = 420 mg.

Fiber F3: Crimped along the length, crescent cross-section 2.3 mm x 0.55 mm, 52 mm long, TS = 1050 MPa, fiber mass = 393 mg.

Fiber F3: Twin-cone, circular cross-section 1.0 mm diameter, 62 mm long, TS = 1198 MPa, fiber mass = 403 mg.

Conventional static properties for the plain mixes are given below. These properties were not found to change significantly due to the presence of fibers.

Normal Strength Matrix: f'_c = 40 MPa; E_c = 39 GPa; *MOR* = 5.2 MPa

Mid-Strength Matrix: f'_c = 52 MPa; E_c = 38 GPa; *MOR* = 7.2 MPa

High Strength Matrix: f'_c = 85 MPa; E_c = 46 GPa; *MOR* = 10.2 MPa.

Impact tests were performed twenty eight days after casting with a hammer approach velocity of 2.94 m/s (hammer incident energy = 291 N-m). This produced an average stress-rate of 3500 MPa/s and an average strain rate of 0.08 sec^{-1}. Six specimens were tested in each category.

Companion static tests were carried out using the same setup but with a hydraulic jack ($\dot{\sigma} \approx 0.04$ *MPa/s*) with a dynamic to static stress-rate ratio of 87500. The low stiffness of the loading system, however, led to unstable fractures and hence unreliable estimations of fracture energies. Consequently, beams (100 mm x 100 mm x 350 mm) were tested in four-point bending as per ASTM C1018 and fracture energies to a central deflection of 2 mm were noted [17].

3 Results and discussion

3.1 General observations

Some representative contact load-time plots are shown in Figure 3 for plain concrete specimens with different compressive strengths. The first peak in these curves, being entirely due to inertia, is nearly the same regardless of specimen type; it is only in the second peak that the differences between the various concretes emerge. An increase in the maximum load attained in the second peak with an increase in the concrete compressive strength may be noted.

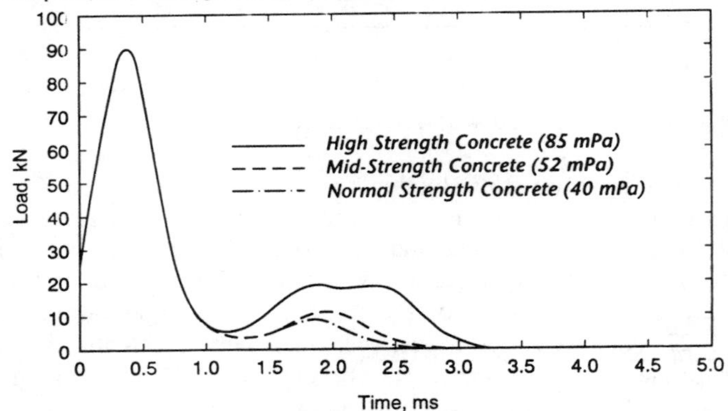

Fig. 3. Contact Pulses for Concretes with Different Compressive Strengths

In Figures 4-7, the load-time curves for steel fiber reinforced concrete with various fiber geometries and the mid-strength matrix are shown. Notice an increases in the area under the curve { ∫ Pdt} due to the presence of fibers which meant (Eqn. 3) that the pendulum lost a greater amount of energy during impact indicating a tougher material. Detailed results are given in Table 1.

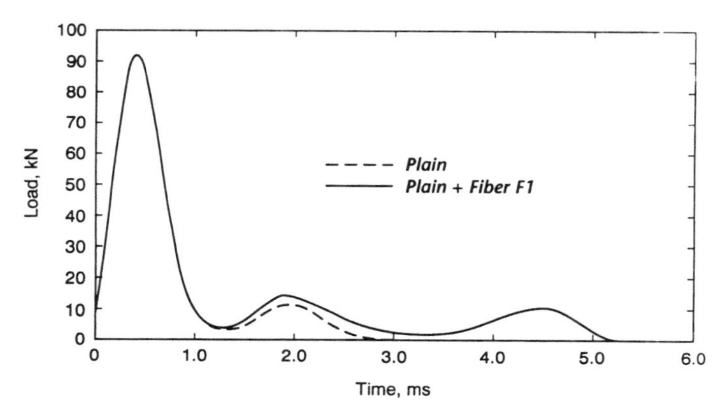

Figure 4. Contact Pulses for Plain and Fiber Reinforced Concrete (Fiber F1)

3.2 Tensile strengths

An increase in the tensile strength occurred under impact when compared to static loading for both plain and fiber reinforced concrete. The increase in strength under impact became more pronounced with an increase in the compressive strength of the matrix and also with fiber reinforcement as noted from the impact/static strength ratios in Table 1.

For an ideally brittle material, the strength (σ_c) in tension can be expressed as a function of stress-rate ($\dot{\sigma}$) using the principles of LEFM [18]:

$$\ln\sigma_c = \frac{1}{N+1}\ln B\dot{\sigma} + \frac{1}{N+1}\ln\left(\sigma_i^{N-2} - \sigma_f^{N-2}\right) \tag{7}$$

where, N is a material constant (slope of the stress intensity factor, K_I, vs. crack velocity, V, plot on a logarithmic scale), subscripts i and f refer to initial and final conditions, respectively, and B is given by [13,18]:

$$B = \frac{2K_{IC}^{2-N}(N+1)}{A\ Y^2\ (N-2)} \tag{8}$$

where A and Y are constants. In Table 1, the values of constant N obtained on the basis of Eqn. (7) are given. Notice that N varies between wide limits and is, in general, higher than normally reported, particularly for the plain, unreinforced matrices. In other words, the sensitivity of concrete to stress-rate as observed in these tests is less pronounced than that observed by others [15,19,20]. This is surprising given that concrete is known to be far more sensitive to stress-rate in uni-axial tension than in any other mode. The following are the probable causes:

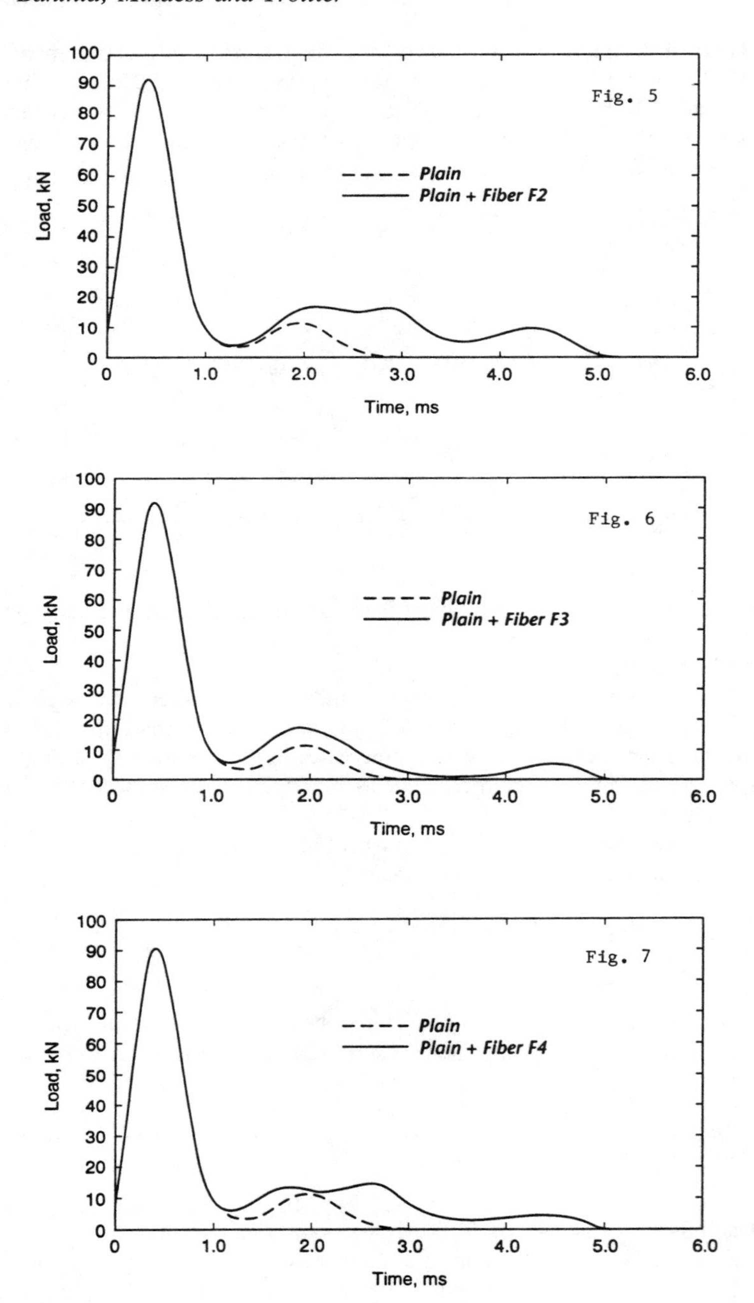

Figures 5, 6 and 7. Contact Pulses for Fiber Reinforced Concretes with Fibers F2, F3 and F4, respectively, and their Plain Counterparts (Mid-Strength Matrix)

Table 1. Static and impact data

			Tensile strength			Fracture energy					
			Static (MPa)	Impact (MPa)	Ratio impact/static	N**	Static***	Impact	Ratio impact/static	Ratio FRC/plain Static	Ratio FRC/plain Impact
	Plain		2.07	1.95	0.95	-	0.44	67	152	-	-
Normal strength (f'_c = 40 MPa)	Fiber reinforced	F1	2.19	3.09	1.41	32	37.80	161	4.3	85.9	2.4
		F2	2.20	2.47	1.12	100	24.08	106	4.4	54.2	1.6
		F3	*	*	*	*	18.69	*	*	42.5	*
		F4	1.81	2.96	1.63	22	32.79	173	5.2	74.5	2.6
	Plain		2.30	2.56	1.11	109	0.44	85	193	-	-
Mid-strength (f'_c = 52 MPa)	Fiber reinforced	F1	2.40	2.81	1.17	72	34.10	166	4.9	77.5	1.9
		F2	2.37	2.71	1.14	86	21.48	120	5.6	48.8	1.4
		F3	2.31	2.95	1.28	45	17.57	129	7.3	39.9	1.5
		F4	2.51	3.03	1.20	61	26.68	165	6.2	60.6	1.9
	Plain		2.39	3.01	1.26	48	0.77	99	128	-	-
High strength (f'_c = 85 MPa)	Fiber reinforced	F1	2.54	3.45	1.36	36	23.61	169	7.1	30.6	1.7
		F2	2.55	2.81	1.10	118	20.39	108	5.3	26.5	1.1
		F3	2.21	2.92	1.32	40	16.91	116	6.9	21.9	1.2
		F4	2.32	3.16	1.36	36	33.18	117	3.5	43.1	1.2

* Unsuccessful test. ** $\frac{1}{1+N}$ is the slope of log σ_c vs. log $\dot{\sigma}$ curve. *** from 4-point flexure to 2 mm deflection.

The Estimated Stress Rate: The imposed stress-rates under impact were calculated from the slopes of the contact load vs. time pulses in rising part of the "second hump" (see Figure 2) in the load-time plots and then averaged. This gives only an approximate stress-rate.

Eccentricity in Loading: The load eccentricity in rapidly applied impact loads (where the specimen did not get much time to align itself in the direction of the load) was greater than that in slow static tests. This may have led to decreased apparent strengths under impact loading and hence lower impact/static strength ratios.

Lack of a Linear Response: As pointed out by Mindess [13], the assumption of a linear elastic fracture response assumed in Eqn. (7) is not entirely valid. Concrete is not ideally brittle and the σ-ϵ response for both concrete and its fiber reinforced composites is far from linear.

Stress-Rate Vs. Strain Rate: Assuming the material is linearly elastic, the imposed stress-rate ($\dot{\sigma}$) in a test can be related to the imposed strain-rate ($\dot{\epsilon}$) through a simple Equation:

$$\dot{\sigma} = E\,\dot{\epsilon} \tag{9}$$

where E is the elastic modulus. Eqn. (9) implies that for a given applied stress-rate a stiffer material would be subjected to a lower strain-rate. Which means that if the failure is governed by a limiting strain rather than a limiting stress value, the data must be normalized and different materials must be compared only on an equal strain-rate basis. This is, however, not attempted here given the lack of appropriate values of dynamic moduli for the materials tested.

The above points have beem dealt with in details elsewhere [21].

3.3 Fracture energies

In Table 1, notice the higher fracture energy values under impact loading as compared to static loading. This improvement, however, appears to be far more pronounced for the plain, unreinforced matrix than for the fiber reinforced composites. The observed increases in the fracture energy absorption due to fiber reinforcement are well anticipated and in tune with most published data. The improvement in the fracture energy absorption under impact do not appear to be dependent upon the static compressive strength of the matrix or the geometry of the fiber.

The effectiveness of fibers in improving the toughness (Table 1) appears to be decreasing with an increase in the static strength of the matrix both under static as well as impact loading. One may relate this to the dependence of the fiber pull-out resistance on the strength of the matrix; at higher matrix strengths, more instances of fiber fractures (rather than complete pull-out) and matrix splitting have been reported to occur [22].

The very high absorption of energy under impact for plain concrete, also reported previously [15] using a very different impact machine, is puzzling. Significant amount of additional research is needed before the true influence of machine characteristics on the measured fracture energy values can be understood.

4 Conclusions

1. Meaningful material properties under impact loading can be obtained by using the proposed test technique.
2. Both plain and fiber reinforced cement-based materials are stronger as well as tougher under impact loading as compared to static loading.
3. Fiber reinforcement is significantly effective in improving the fracture energy absorption under impact. The improvements are, however, not as pronounced as those under static conditions and are less pronounced for high strength concrete.

5 Acknowledgements

The continued support of the Natural Sciences and Engineering Research Council of Canada is gratefully acknowledged.

6 References

1. Struck, W. and Voggenreiter, W. (1975) Examples of Impact and Impulsive Loading in the Field of Civil Engineering, Mat. & Struc. 8(44).
2. Banthia, N. et al (1989) Impact Testing of Concrete Using a Drop Weight Impact Machine, Expt. Mech., 29(2), pp. 63-69.
3. Naaman, A. and Gopalaratnam, V. (1983) Impact Properties of Steel Fiber Reinforced Concrete in Bending, Int. J. of Cem. Compo. and Light Weight Conc. 3(1), pp. 2-12.
4. Suaris, W. and Shah, S.P. (1983) Properties of Concrete and Fiber Reinforced Concrete Subjected to Impact Loading, American Society of Civil Engineers, J. of the Struc. Div. 109, ST7, July, pp. 1717-1741.
5. Gokoz, U. and Naaman, A.E. (1983) Effect of Strain Rate on the Pull-out Behaviour of Fibers in Mortar, Int. J. of Cem. Compo. and Light Weight Conc. 3 (3), pp. 187-202.
6. Hibbert, A.P. (1979) Impact Resistance of Fiber Concrete, Ph.D. Thesis, University of Surrey (UK).
7. Gopalaratnam, V., Shah, S.P and John, R. (1984) A Modified Instrumented Charpy Tests for Cement Based Composites, Expt. Mech. 24(2), pp. 102-111.
8. Banthia , N. and Ohama, Y. (1989) Dynamic Tensile Fracture of Carbon Fiber Reinforced Cements, Proc. Int. Conf. on Recent Developments in Fiber Reinforced Cements and Concretes, Cardiff, UK, pp. 251-260.
9. Zielinski, A.J. (1982) Fracture of Concrete and Mortar Under Uniaxial Loading, Ph.D. Thesis, Delft University of Technology (The Netherlands).
10. Malvern, L.E. et al (1986) Dynamic Compressive Strength of Cementitious Materials, in Cement-Based Composites: Strain Rate Effects on Fracture, MRS Symp. Proc. 64, pp. 119-138.
11. Sierakowski, R.L. (1985) Dynamic Effect in Concrete Materials, in Application of Fracture Mechanics to Cementitious Composites (Ed. S.P. Shah), Martinus

Nijhoff Publishers, Dordrecht, pp. 535-557.
12. Reinhardt, H.W. (1986) Strain Rate Effects on the Tensile Strength of Concrete as Predicted by Thermodynamic and Fracture Mechanics Models. in Cement-Based Composites: Strain Rate Effects on Fracture, MRS Symp. Proc., 64, pp. 1-14.
13. Mindess, S. (1985) Rate of Loading Effects on the Fracture of Cementitious Materials, In Application of Fracture Mechanics to Cementitious Composites (Shah, S.P. ed.), Martinus Nijhoff Publishers, Dordrecht, pp. 617-636.
14. Gopalaratnam, V. and Shah, S.P. (1986) Properties of Steel Fiber Reinforced Concrete Subjected to Impact Loading, J. of Amer. Conc. Inst. 83 (1), pp. 117-126.
15. Banthia, N., Mindess, S. and Bentur, A. (1987) Impact Behavior of Concrete Beams, Mat. and Struc. 20 (119), pp. 293- 302.
16. Banthia, N. et al (1994) Fiber Reinforced Cement-Based Composites under Tensile Impact, Advanced Cement Based Materials, 1, pp. 131-141.
17. Banthia, N. and Trottier, J.-F. (1995) Toughness Characterization in Steel Fiber Reinforced Concrete, Parts 1 and 2, ACI Mat. J. in press.
18. Nadeau, J.S. et al (1982) An Explanation of the Rate-of-Loading and Duration-of-Load Effects in Wood in Terms of Fracture Mechanics, J. of Mat. Sci. 17, pp. 2831-2840.
19. Ross, C.A. (1985) Fracture of Concrete at High Strain-Rates, in Toughening Mechanisms in Quasi-Brittle Materials (Ed. S.P. Shah), NATO ASI Series, Kluwer Academic Publishers, 1991, pp. 577-596.
20. Zielinski, A., and Reinhardt, H.W. (1982) Stress-Strain Behavior of Concrete and Mortar at High Rates of Tensile Loading, Cem. and Conc. Res. 12, pp. 309-319.
21. Banthia N. et al (1995) Impact Resistance of Fiber Reinforced Concrete, ACI Materials Journal (submitted).
22. Banthia, N. (1990) A Study of Some Factors Affecting the Fiber-Matrix Bond in Steel Fiber Reinforced Concrete, Cdn. J. of Civ. Engg. 17(4), pp. 610-620.

164 ELASTIC MODULUS VARIATION OF CONCRETE SUBJECTED TO LOW NUMBER CYCLIC LOADINGS

M. SAICHI and H. SHINOHE
Tohoku Institute of Technology, Sendai, Japan

Abstract
An experimental investigation on the behavior of plain concrete subjected to low number cyclic loadings is described. The test results showed that even a limited number of cyclic loadings degrades the elastic modulus to values of 5 to 22 percent lower than that obtained from monotonically loaded reference concrete specimens.

Experimental results indicated that the elastic modulus was more sensitive to small changes in the loading history than was the compressive strength. The stress-strain diagrams of test specimens subjected to cyclic loading were compared with those of monotonically loaded specimens, and the elastic modulus deterioration was correlated to the shape or gradient of the ascending branch of the curves.

These findings might be applicable for the assessment of the quality or condition of concrete in existing structures.
Keywords: Assessment, compressive strength, concrete, cyclic load, elastic modulus, fatigue, quality, stress-strain curve

1 Introduction

The mechanical properties of concrete cylinders under compression are used to design concrete structural members. Previously, most investigators studying concrete behavior have primarily attached importance to determining the strength variation. On the other hand, analyzing ordinary concrete structures and determining their probable deformation require knowledge of the elastic constants, namely, the elastic modulus and Poisson's Ratio of concrete under compression. In particular, the elastic modulus of concrete under compression is a basic material property in computing deflections and other displacements needed when analyzing concrete structures. Since deformation properties provide indirect

Concrete Under Severe Conditions: Environment and loading (Volume Two) Edited by K. Sakai, N. Banthia and O.E. Gjørv. Published in 1995 by E & FN Spon. ISBN 0 419 19860 1

information concerning the internal structure as well as the failure mechanism of concrete, knowledge of the elastic modulus of concrete is desirable if the behavior of concrete structures is to be interpreted properly. Such knowledge may only rarely be crucial, but it is helpful for understanding the material itself and in-situ performance.

While elastic modulus is mainly affected by the type of aggregate, concrete strength, age, curing history and loading rate, cyclic loading also seems to be important[1][2]. The elastic modulus of concrete is conventionally taken as constant; however, this assumption does not explain the behavior of concrete structures under cyclic loading. It is hoped in this way, to further understand the response of elastic modulus of concrete subjected to variable load histories.

The investigation reported in this paper aims at the comparison of sensitivity for damaging between compressive strength and elastic modulus, and also aims at the prediction of pre-loading effects due to a small number of cyclic loadings on the elastic modulus. The effects of these cyclic loads may be divided into a primary simulation for the mechanical damaging. Although these cyclic loadings may considered to be over simplification, such loadings are essential for a understanding of low cycle fatigue behavior or damage. For example, T.M.Chrisp et al.[3] have used the five cycles of loading for a new approach as a possible solution to simulate the internal damage caused by A.A.R. and there have been numerous proposals to define damaging by means of various cyclic loadings e.g.[4][5]. The information presented in this paper may conceivably be applicable when evaluating the deterioration of reinforced concrete members subjected to a specified load history.

2 Mix proportions and curing

Ordinary portland cement was used to prepare concrete. The coarse aggregate was river gravel with a maximum size of 20 mm, and the fine aggregate was river sand. The concrete mix proportion is given in Table 1. No chemical admixtures or no mineral additions were applied to the concrete mix. Standard $\phi 100 \times 200$ mm steel-cylinder molds were used in accordance with JIS-A-1132, and molding was done according to the specifications of JIS-A-1138.

Approximately 48 hours after casting, specimens were removed from the steel molds and transferred to a standard moist curing room where curing was continued until the day before testing at approximately 20°C in accordance with JIS-A-1132. Specimens were tested at an age of 28 days. To prevent stress concentration during testing, one surface of the specimen was polished with carborundum sand while the concrete was still green. The flatness of specimens was checked, and the tolerance was found to be within 0.02mm.

Table 1. Concrete mix proportion.

		Quantities (kg/m^3)			
Water/Cement Ratio	Slump (cm)	Water	Cement	Sand	Gravel
0.63	18	207	327	745	1004

3 Testing procedure

All the experimental tests described in this paper were conducted on concrete cylinders whose dimensions were $\phi 100 \times 200$ mm. Simultaneous with the compressive loading, the longitudinal deformation was measured with a compressometer equipped with two linear variable differential transformer transducers (LVDTs). These unbonded sensing devices were provided for measuring the average deformation of two diametrically opposite gage lines. The specimens were loaded without any polymeric packing layers.

Table 2 shows the specimen designation and the magnitude of the repeated stress. Before testing, three cylinders were loaded up to the strength. Based on their average strength, five stress levels were selected and six specimens were cyclically loaded under each level. The stress/strength ratios for the cyclic loading were 0.25, 0.33, 0.50, 0.67 and 0.75, approximately equal to the values of 6.5, 9.0, 12.5, 16.5 and 19.0 MPa, respectively. The rate of loading was maintained manually at about 0.2 to 0.3 MPa/sec. Each cylinder was loaded until it reached the stress described in Table 2, and then the load was released at the same rate.

After unloading, the load was applied again for a total of five cycles, and then the load was applied until the ultimate stress was reached without any interruptions in loading. The output from the device was fed through an amplifier to an X-Y recorder. Thus, the load-displacement curve was drawn automatically during the test. The applied load and the developed displacement were converted into stress and strain, respectively. A typical stress-strain curve is shown in Figure 1 as an example.

The static elastic modulus on each ascending branch was calculated from the stress-strain diagram. The chord modulus, referred to as "elastic modulus" in this paper, is calculated from the following relation:

$$Ec = \frac{S_2 - S_1}{\epsilon_2 - \epsilon_1} \qquad (1)$$

where S_2 is the stress corresponding to 33.3 percent (one-third) of the compressive stress, ϵ_2 is the longitudinal strain produced by S_2 , ϵ_1 is a constant strain of 50×10^{-6} with eliminates the residual strain from respective cycles, and S_1 is the stress corresponding to the strain of ϵ_1 .

Table 2. Specimen identification.

Designation	Stress/Strength ratio for cyclic load	Upper load for cyclic load	Upper Stress for cyclic load
Pnon	-	-	-
P1/4	0.25	50 kN	6.5MPa
P1/3	0.33	70	9.0
P1/2	0.50	100	12.5
P2/3	0.67	130	16.5
P3/4	0.75	150	19.0

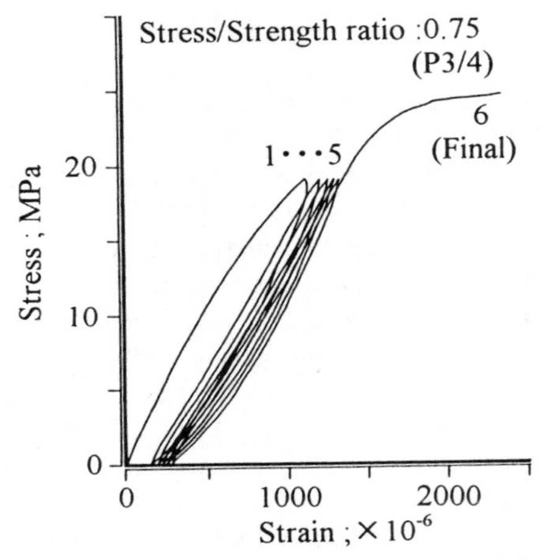

Fig.1 Typical stress-strain diagram for 5 times cyclically loaded concrete specimen.

4 Results and Discussion

The compressive strength of the reference cylinders at 28 days, the concrete specimens without any cyclic loading, ranged between 24.9 MPa and 26.6 MPa with a mean strength of 25.3 MPa, while the elastic modulus varied from 20.8 GPa to 23.9 GPa. The strength and elastic modulus of specimens subjected to cyclic loading five times, namely sixth cycle, are shown in Tables 3 and 4 and illustrated for each specimen in Figures 2 and 3.

Figure 2 shows the compressive strength as a function of the stress/strength ratio of the cyclic loading level. The strength does not show a marked increase or decrease over the entire range of the stress/strength ratios. The value of strength was, thus, found to be unaffected by low number of cyclic loadings.

Figure 3 presents the elastic modulus as a function of the stress/strength ratio of the cyclic loading level. All plot families belonging to each stress/strength ratio exhibited continuous deterioration of elastic modulus as the stress/strength ratio was increased. In the case of a high stress/strength ratio, e.g., three-fourths of the ultimate stress, the deterioration was found to be in the order of 20 percent for the type of concrete used in this investigation. By comparing Figure 2 with Figure 3, the following observations are made:

(1) The elastic modulus decreased remarkably as the stress/strength ratio was increased, while the strength variation remained practically unchanged.

(2) Strength was less sensitive to the loading history applied to the concrete than was the elastic modulus.

Figure 4 presents the elastic modulus as a function of the number of loading cycles for all stress/strength families, excluding the result of "0.25"(P1/4) loading. At "0.33"(P1/3) loading, between the first and fifth cycles, the elastic modulus appeared to be reasonably constant (Fig.4(a)). For higher magnitudes of cyclic load, at least higher than one-half ("0.5"loading, P1/2), there tended to be a significant variation of values. At "0.5"(P1/2) loading, the total data could be approximated by a straight line, and the variations among

Table 3. Summary of compressive strength at different levels of cyclic loading.

		Reference Specimen	Cyclically loaded specimens				
Designation of specimen		Pnon	P1/4	P1/3	P1/2	P2/3	P3/4
Avereage	MPa	25.3	26.0	25.6	25.0	25.5	26.0
S.D	MPa	0.9	1.1	0.5	1.0	1.2	1.3
C.V	%	3.6	4.4	2.1	4.1	4.9	5.1

Table 4. Summary of elastic modulus at different levels of cyclic loading.

		Reference Specimen	Cyclically loaded specimens				
Designation of specime		Pnon	P1/4	P1/3	P1/2	P2/3	P3/4
Avereage	GPa	22.2	23.1	22.4	21.0	18.9	17.4
S.D	GPa	1.1	0.9	0.7	0.7	0.6	1.5
C.V	%	4.8	4.0	2.9	3.4	3.3	8.7

※ S.D: Standard deviation C.V: Coefficient of variation

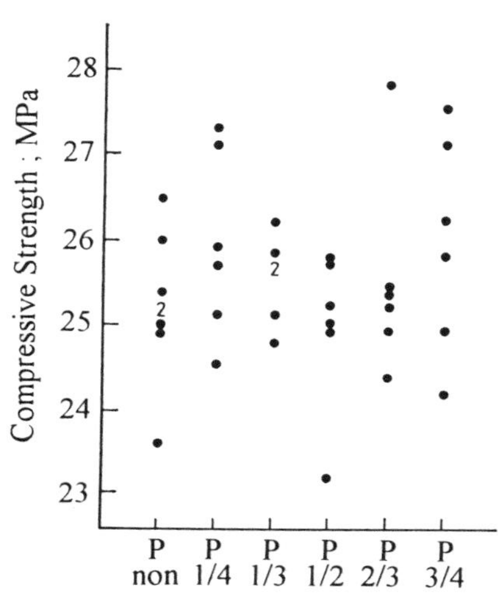

Fig.2 The variation in compressive strength of concrete specimens subjected to cyclic loading 5 times.

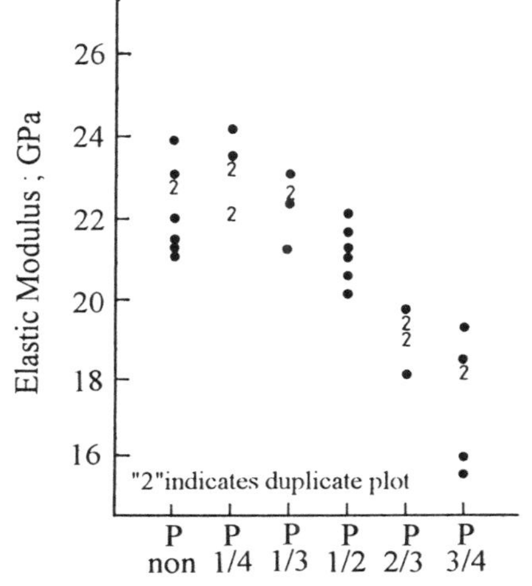

Fig.3 The variation in elastic modulus of concrete specimens subjected to cyclic loading 5 times.

the datum were relatively small as apparent from Fig.4(b). The deterioration of the elastic modulus for the "0.67"(P2/3) and "0.75"(P3/4) stress/strength ratios were non-linear and could be represented by the second or third power functions of the number of cycles. Increasing the stress/strength ratio from "0.5"(P1/2) to "0.75"(P3/4) resulted in a 17 percent decrease of elastic modulus. At high values of the stress/strength ratio, the elastic modulus deterioration was large, and the effect of the magnitude of the repeated stress was more significant than that of the number of cycles.

Figure 4 also presents a compilation of the deterioration of elastic modulus as the number of cycles was increased. It is apparent that the elastic modulus in each loading history deteriorated, except at the low stress/strength ratio of "0.33"(P1/3), where the elastic modulus remained almost constant. It follows that degradation in the elastic modulus depended primarily on the level of the cyclic stress. The onset of elastic modulus degradation corresponded to the level at which the stress/strength ratio was equivalent

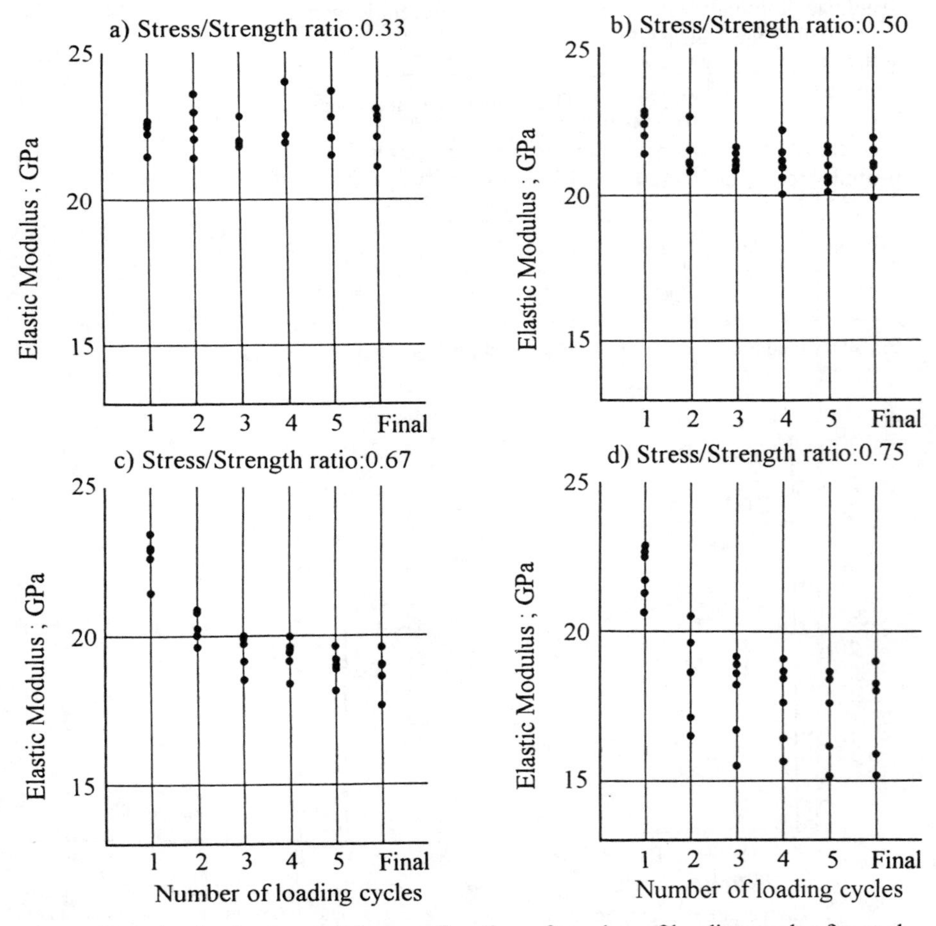

Fig. 4 Variation in elastic modulus as a function of number of loading cycles for each stress/strength ratio.

to "0.5"(P1/2).

The significance of the stress-strain diagram is that elastic modulus is calculated at the point on the ascending branch of this diagram, i.e., the secant modulus, or the chord modulus. Since the cyclic loading was found to affect the elastic modulus deterioration, it is possible to say that the stress-strain diagram found under a cyclic loading might be different from that found under a monotonical loading. When the total stress-strain data obtained from families in each type of loading are combined, a general comparison with other families can be made. Hence, the averaged line of ascending branch of the stress-strain curve is considered to be a convenient parameter for comparing the deformation characteristics of different groups. The averaged sixth ascending branch, reflecting the influence of five cyclic loadings with the five levels of cyclic stress on the stress-strain diagram, is shown in Figure 5. For the purpose of comparison, the average curves calculated from monotonically loaded reference specimens ("Pnon" family) are also plotted in Figures 5(a) to (e).

As shown in each diagram, there is a distinct difference between the cyclically loaded stress-strain average curve and the monotonically loaded curve. A significant change exists in the shape of the cyclically loaded stress-strain curve is that the ascending branch becomes more increasingly concave towards the strain axis than does the monotonically loaded curve. S.Popovics[6] notes that, in relation to this change, the magnitude of the repeated stress greatly influences the stress-strain diagram of concrete.

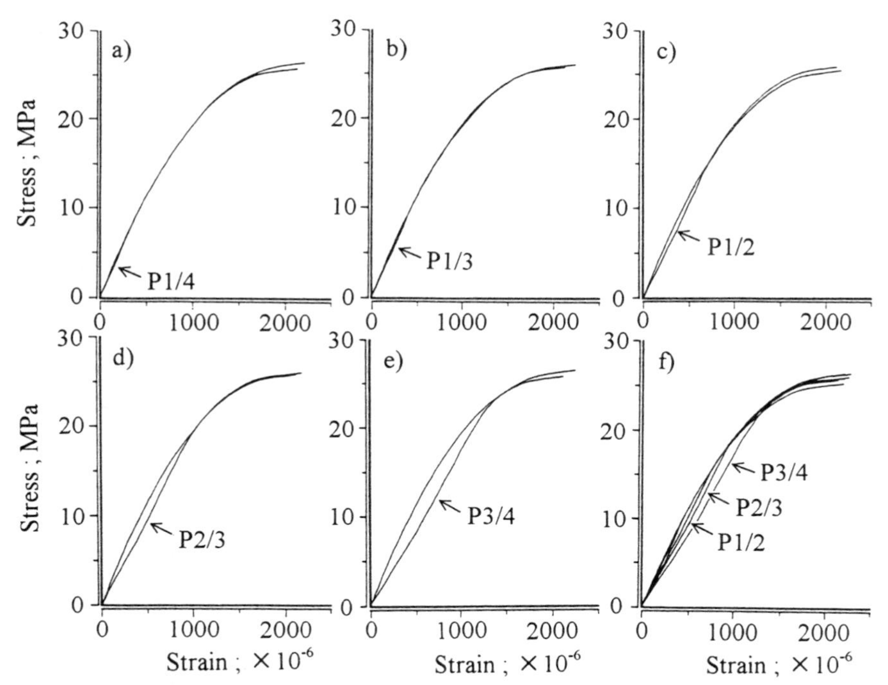

Fig.5 Comparison of the ascending portion of stress-strain curves resulting from different cyclic stress levels with the reference unloaded concrete : for the following stress/strength ratios: a)0.25, b)0.33, c)0.50, d)0.67, e)0.75 and f)compilation.

In fact, this concave behavior occurred primarily in the higher stress/strength ratio range. The gradient of the stress-strain diagram at lower levels of cyclic maximum stresses from "0.25"(P1/4) to "0.33"(P1/3) usually remained approximately unchanged. At the higher levels, however, the corresponding shape became concave to the strain axis, and the gradient decreased in comparison with the initial gradient; consequently, the elastic modulus showed deterioration.

Figure 5(f) presents a compilation of the average stress-strain curves, revealing that the degree of curvature of the various curves may differ considerably. Significant consequences of cyclic loading are evident, especially larger magnitudes of repeated stress; the influence of cyclic stresses above at least one half of the complessive strength could be very important at low numbers of cyclic loading.

Cyclic loading results in progressive crack propagation[7] which depends both on the rate and on the pattern of cyclic loading. Thomas T.C.Hsu et al.[8], using a microscope, reported findings obtained by direct observation of plain concrete. They concluded that, above approximately 30 percent of the ultimate load, the bond cracks at the interface between coarse aggregates and mortar begin to increase in length, width, and number; hence, the stress-strain diagram begins to deviate from a straight line. A similar observation has also been reported by other investigators[9][10][11].

The stress-strain diagram of concrete subjected to cyclic loading deviates gradually from the initial curve primarily due to the progressive propagation of internal cracking in the specimen. In other words, the shape or gradient of the stress-strain curve of concrete is determined by the process of micro-cracking[12][13]. Hence, taking into account that the ascending branch of the stress-strain diagram is concaved under a small number of cyclic loadings of high stress level, the propagation of micro-cracks appears to be responsible for the deterioration of the elastic modulus.

5 Conclusions

Based on the results of the investigation reported in this paper, and in view of experimental variability in the evaluation of elastic modulus, the following conclusions are made:

Generally, the measured strength of concrete specimens showed essentially the same value for all conditions of cyclic loading. The elastic modulus deterioration of concrete previously subjected to cyclic loading is very clear, which suggests that the elastic modulus is more sensitive to the loading history than is the compressive strength. The elastic modulus of concrete subjected to five loadings can be 5 to 22 percent lower than that obtained from the monotonically loaded concrete. The rate of elastic modulus deterioration depends more significantly on the magnitude of repeated stress rather than on the number of repetitions.

Acknowledgement

The authors would like to thank the students in the Department of Architecture of the Tohoku Institute of Technology for their assistance in the experiment.

References

1. Ornum, J.L.V. (1903) The fatigue of cement products, *Transactions ASCE*, Vol.50, pp.443-445.
2. Slate, F.O. and Mayers, B.L. (1968) Determination of plain concrete. *Fifth international Symposium on the Chemistry of Cement,* Cement Association of Japan, Tokyo, Paper No. 3-33, pp. 142-151.
3. Popovics, S. (1973) A review of stress-strain relationships for concrete, *Journal of the ACI*, Vol.67, pp.243-248.
4. Chrisp, T.M., Waldron, P. and Wood, J.G.M. (1993) Development of a non-destructive test to quantify damage in deteriorated concrete, *Magazine of Concrete Research*, Vol.45, No.165, pp.247-256.
5. Grzybowski, M. and Meyer, C. (1993) Damage accumulation in concrete with and without fiber reinforcement, *ACI Materials Journal*, Vol.90, No.6, pp.594-604.
6. Suaris, W. and Fernand, V. (1987) Ultrasonic pulse attenuation as a measure of damage growth during cyclic loading of concrete, *ACI Materials Journal*, Vol.84, No.3, pp.185-193.
7. Shah, S.P. and Chandra, S. (1970) Fracture of concrete subjected to cyclic and sustained loading, *Journal of the ACI*, Vol.67, pp.816-825.
8. Hsu, T.T.C., Slate, F.O., Stutman, G.M. and Winter, G. (1963) Microcracking of plain concrete and the shape of the stress-strain curve, *Journal of the ACI*, Vol.60, pp.209-224.
9. Shah, S.P. and Sanker, R. (1987) Internal cracking and strain-softening response of concrete under uniaxial compression, ACI Materials Journal, No.84-M22, pp.200-212.
10. Sturman,G.M., Shah, S.P. and Winter, G. (1965) Effect of flexural strain gradients on microcracking and stress-strain behavior of concrete, *Journal of the ACI*, Vol.62, pp.805-822.
11. Khalifa, S. (1974) Effect of sustained load on the load-deformation curve of concrete, *First Australian Conference on Engineering Materials*, Sydney, pp.191-209.
12. Popovics, S. (1969) The fracture mechanism in concrete:How much do we know ?, *Proceedings ASCE*, Vol.95, EM3, pp.531-544.
13. Sriravindrarajah, R and Swamy, R.N. (1989) Load effects on fracture of concrete, *Materials and Structures*, Vol.22, pp.15-22.

165 ANALYTICAL EVALUATION FOR THE FATIGUE STRENGTH OF STEEL–CONCRETE SANDWICH BEAMS WITHOUT SHEAR REINFORCEMENT

M. ZAHRAN, T. UEDA and Y. KAKUTA
Hokkaido University, Sapporo, Japan

Abstract

This study predicts the fatigue strength of steel–concrete sandwich beams using the finite element method. The analysis was done by using a nonlinear finite element method computer program (WCOMR). The analysis procedure used in this study is based upon reducing the concrete strength and stiffness with increasing the number of loading cycles (N), or increasing the stress range (R). The sandwich beams were analyzed for different external load ranges. The fatigue life for the ultimate failure of the sandwich beams was investigated. The output load–deflection curves are presented. The S–N relationship for the fatigue failure of the sandwich beams is also presented. These analytical results are compared with experimental data.
Keywords: fatigue strength, finite element analysis, steel–concrete sandwich beam

1 Introduction

Recently, the applications of composite structures have become increasingly popular. A new type of composite structures, which is the steel–concrete sandwich structure, has been greatly developed to fulfill complicated structural requirements. As shown in Fig. 1, the steel–concrete sandwich member is composed of core concrete, steel skin plates, shear reinforcing steel plates, and shear connectors to transfer the shear between the concrete and the steel skin plates. The sandwich member has proved its high load carrying capacity and high ductility in bending and shear. The sandwich member has also proved good constructibility since the cost of formwork and the period of construction can be considerably saved. The sandwich members have many practical applications in various structures such as tunnels, underground walls, bridge decks, marine structures, etc.

Concrete Under Severe Conditions: Environment and loading (Volume Two) Edited by K. Sakai, N. Banthia and O.E. Gjørv. Published in 1995 by E & FN Spon. ISBN 0 419 19860 1

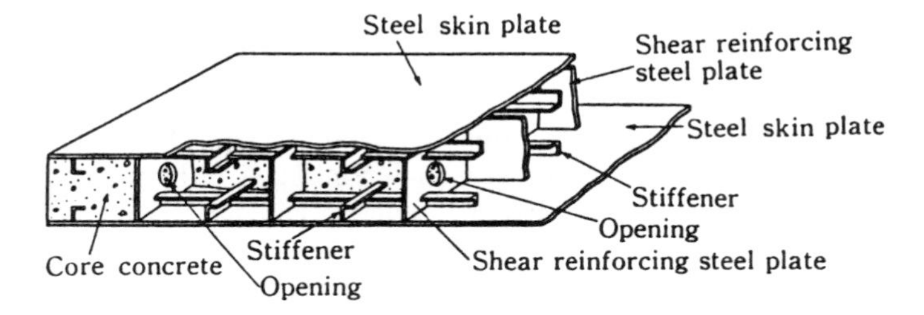

Fig. 1. Steel–concrete sandwich member.

Marine structures are subjected to wave forces repeatedly. Bridge decks are also subjected to traffic loads repeatedly. These repetitive loads may result in progressive cracking and sometimes fatigue failure if the load amplitude is sufficiently large. Therefore, the fatigue endurance of the steel–concrete sandwich members has to be adequately investigated. This study presents an attempt to predict analytically the fatigue strength of the steel–concrete sandwich beams by using the finite element method.

2 Description of the investigated sandwich beam

The steel–concrete sandwich beam investigated in this study is shown in Fig. 2, which has a span length of 2.65 m and a cross section of (250×400) mm. The shear span to effective depth ratio (a/d) is equal to 3.0. The thickness of the upper and lower steel plates is 16 mm. The compressive strength of concrete (f_c) is 25 MPa. The tensile strength of concrete (f_t) is 2.5 MPa. The yield strength of the steel plates is equal to 400 MPa. The sandwich beam is not provided with web reinforcement. Shear connectors (steel angles 40×4 mm) are provided at the interface between the concrete and the steel plates. The steel angles are welded to the steel plates. The beam was designed in such a way that the flexural capacity is much higher than the shear capacity in order to avoid fatigue fracture of the tensile steel plate, which was indicated in a previous study[1].

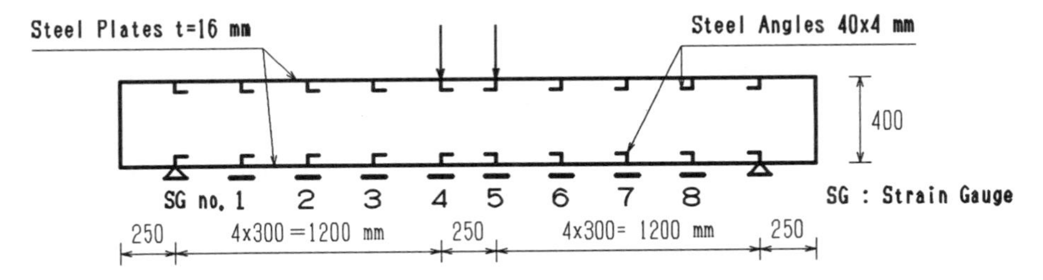

Fig. 2. Geometry and loading configuration of the sandwich beam.

3 Finite element analysis

3.1 Finite element idealization

A nonlinear finite element method computer program (WCOMR)[2] was used to analyze the steel–concrete sandwich beam shown in Fig. 2. The finite element mesh of the beam is shown in Fig. 3. For the concrete and steel elements, eight–node quadratic elements are used. The constitutive models for the concrete and steel elements are given in references [2] and [3]. Bond elements are provided to simulate the shear connectors and the interface between the concrete and the lower steel plate. A linear bond stress–slip relationship was adopted as a constitutive law for the bond elements[3]. Enforced displacements are given at the loading point as shown in Fig.3.

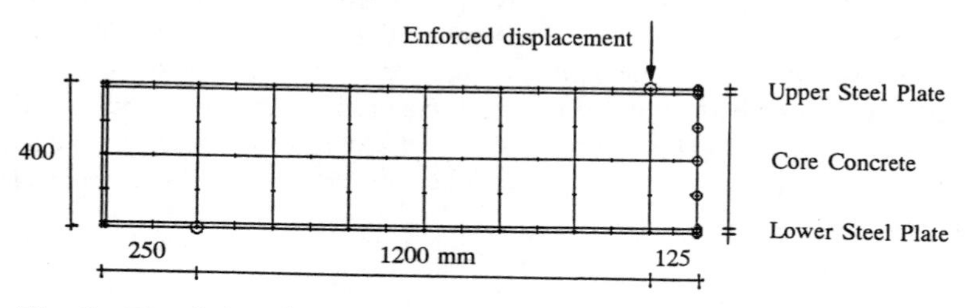

Fig. 3. The finite element mesh.

3.2 Analytical models for fatigue

An experimental S–N relationship is adopted as a constitutive law for concrete under fatigue loading. This S–N relationship is given by the following equation,

$$(F_{max}/F_u) = 1.0 - 0.0685\ (1 - R) \times \text{Log } N \tag{1}$$

where: F_u = the static strength
$R = F_{min}/F_{max}$ ($0 \le R \le 1.0$)
F_{min} , F_{max} = the minimum and maximum stress, respectively
N = the number of loading cycles

This S–N relationship was proposed by Tepfers[4][5] to predict the tensile and compressive fatigue strengths of plain concrete. It is noted that this S–N relationship accounts for the influence of the number of loading cycles (N) as well as the influence of the stress range (R).

A number of previous studies [6][7] have indicated that the secant modulus of elasticity of concrete (E_s) is reduced during the fatigue life. Therefore, the linear relationship shown in Fig. 4 is adopted in this study to model the stiffness degradation of concrete under the fatigue loading. This linear relationship is given by the following equation [7],

$$R_N = 299 - 2.99\ (E_f / E_s) \qquad ((E_f/E_s) \text{ in percentage}) \tag{2}$$

where: E_s = the modulus of elasticity of concrete under static loading
E_f = the reduced modulus of elasticity of concrete under the fatigue loading
R_N = the percentage consumed of the fatigue life

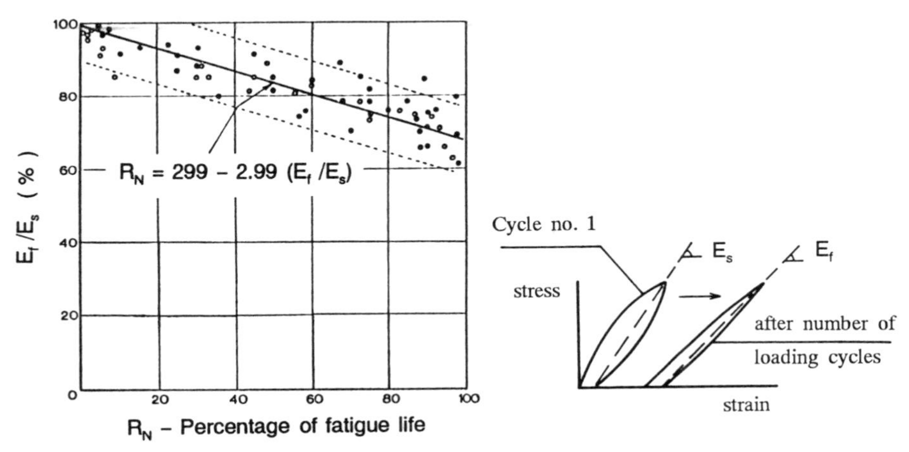

Fig. 4. Decrease of modulus of elasticity under repeated loading [7].

3.3 Analysis procedure

The analysis procedure used in this study is illustrated in Fig. 5. At first, a static loading cycle (OAB) is applied. The maximum principal compressive and tensile stresses at every concrete gauss point (σ_{cmax},σ_{tmax}) are stored at point A. Similarly, the minimum principal compressive and tensile stresses at every concrete gauss point (σ_{cmin},σ_{tmin}) are stored at point B. Then, these stored principal stresses are used to calculate the mean and the deviatoric stresses as shown in Fig. 6 [2], hence:

$$\sigma_{m,max} = \sqrt{2}/2\ (\sigma_{tmax} + \sigma_{cmax}) \tag{3}$$
$$\tau_{d,max} = \sqrt{2}/2\ (\sigma_{tmax} - \sigma_{cmax}) \tag{4}$$

similarly:

$$\sigma_{m,min} = \sqrt{2}/2\ (\sigma_{tmin} + \sigma_{cmin}) \tag{5}$$
$$\tau_{d,min} = \sqrt{2}/2\ (\sigma_{tmin} - \sigma_{cmin}) \tag{6}$$

Thereafter, the maximum and minimum equivalent stresses (S_{max},S_{min}) are calculated as shown in Fig. 7 [2], hence:

$$S_{max} = [(\ a\ \sigma_{m,max}\ )^2 + (\ b\ \tau_{d,max}\ )^2]^{0.5} \leq 1.0 \tag{7}$$

$$S_{min} = [(\ a\ \sigma_{m,min}\ )^2 + (\ b\ \tau_{d,min}\ )^2]^{0.5} \leq 1.0 \tag{8}$$

where: $a = 0.6/f_c$, $b = 1.3/f_c$

These equivalent stresses (S_{max},S_{min}) indicate the level of applied stresses at any gauss point under the plane stress condition. Then, substituting (F_{max}=S_{max}),(F_{min}=S_{min}), and (F_u=1.0) in Eq.(1), the number of loading cycles (N_C) could be calculated. The number of cycles (N_C) is defined as the number of loading cycles required to induce compression fatigue failure at this concrete gauss point.

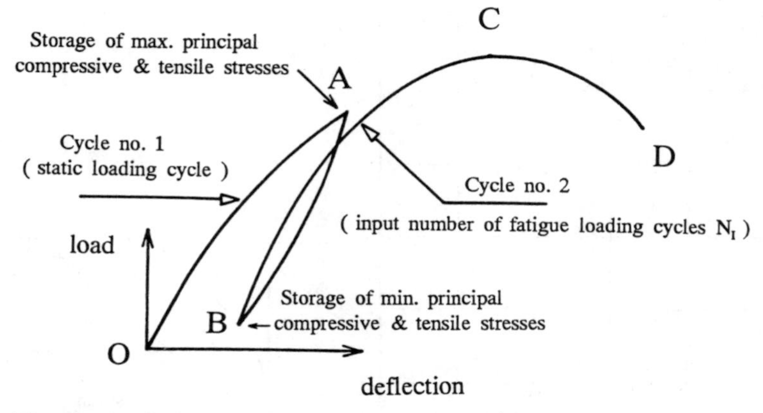

Fig. 5. Analysis procedure.

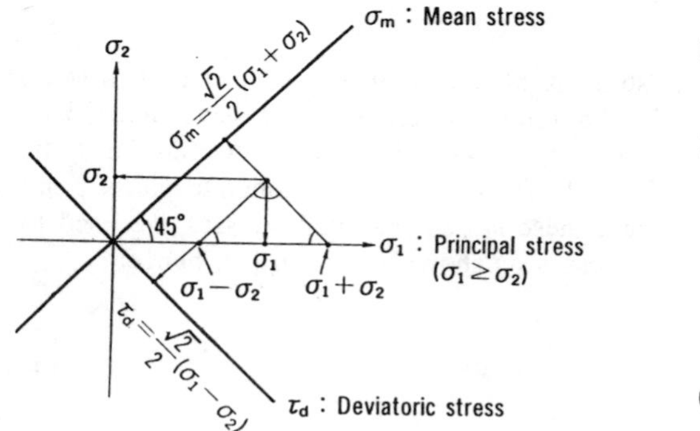

Fig. 6. Mean and deviatoric stress coordinates[2].

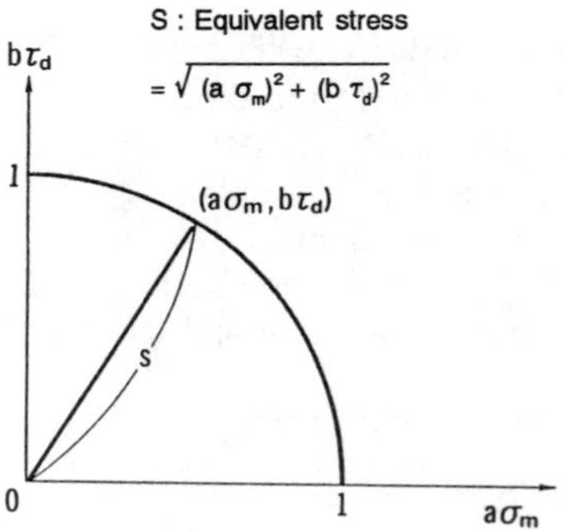

Fig. 7. Definition of equivalent stress[2].

The stored principal stresses are also used together with the cracking criterion of concrete as shown in Figs. 8(a) and 8(b)[2]. In Fig. 8(a), the point (σ_{tmin},σ_{cmin}) is plotted, and hence the ratio (R_{min}) is calculated as,

$$R_{min} = \sigma_{tmin} / f_{t1} \leq 1.0 \tag{9}$$

Similarly, in Fig. 8(b), the point (σ_{tmax},σ_{cmax}) is plotted, and hence the ratio (R_{max}) is calculated as,

$$R_{max} = \sigma_{tmax} / f_{t2} \leq 1.0 \tag{10}$$

Then, substituting (F_{max}=R_{max}), (F_{min}=R_{min}), and (F_u=1.0) in Eq.(1), the number of loading cycles (N_T) could be calculated. The number of cycles (N_T) is defined as the number of loading cycles required to induce tensile fatigue failure at this concrete gauss point.

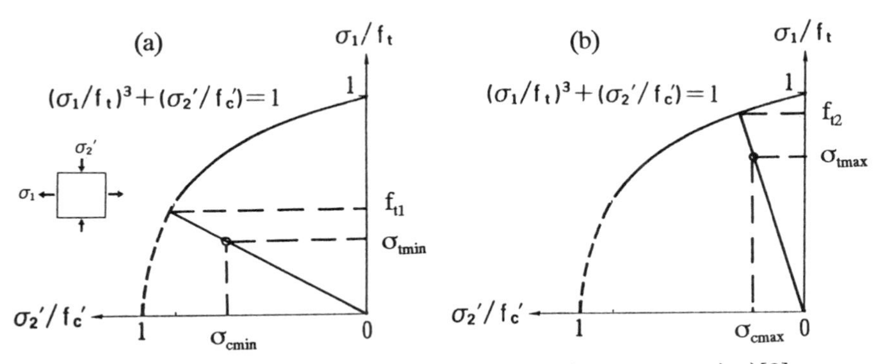

Fig. 8. Cracking criterion for concrete (biaxial tension–compression)[2].

Thereafter, a second loading cycle (BCD) ia applied with an input number of fatigue loading cycles (N_I)(see Fig. 5). The input number of cycles (N_I) is compared with the calculated ones (N_C and N_T) :

– if $N_I \geq N_C$, the concrete gauss point is considered to fail in compression, and therefore, the compressive strength of this gauss point is reduced to [$f_c \times S_{max}$].

– if $N_I \geq N_T$, the concrete gauss point is considered to fail in tension, and therefore, the tensile strength of this gauss point is reduced to [$f_t \times R_{max}$]. Also, the stiffness of concrete in tension is reduced to 66% of the initial stiffness. The ratio 66% is the minimum value given by Eq.(2) when (R_N = 100%).

– if $N_I < N_T$, the concrete gauss point is considered not to have failed yet. In this case, the stiffness of concrete in tension is reduced according to Eq.(2), in which the percentage of fatigue life (R_N) is given by,

$$R_N = (N_I / N_T) \times 100 \tag{11}$$

Therefore, during the loading cycle (BCD), the strength and the stiffness of the concrete gauss points are reduced, which in turn results in reducing the overall stiffness of the sandwich beam. Finally, the sandwich beam is considered to fail due to the fatigue loading (N_I cycles) if the peak load of the second cycle (point C) is approximately equal to the peak load of the first cycle (point A). In this case, the input number of cycles (N_I) is considered to be equal to the fatigue life of the sandwich beam.

4 Outline of the experimental work

An experimental work was carried out for the steel–concrete sandwich beam shown in Fig. 2. Tests were carried out for four specimens. Specimen no.1 was tested under static monotonic loading, while the other three specimens were tested under fatigue loading. In the fatigue tests, the specimens were loaded statically during the first hundred cycles and then loaded dynamically with 240 cycles per minute until failure. The minimum fatigue load (P_{min}) was kept constant at 20 kN, while the maximum fatigue load (P_{max}) was chosen to be 130 kN for specimen no.2, 160 kN for specimen no.3, and 190 kN for specimen no.4. For specimen no.2, fatigue failure did not occur until 2×10^6 cycles, and therefore, the maximum load was increased to 220 kN. The mid–span deflections, the strains in the steel plates, and the crack growth were measured with increasing the number of loading cycles (N)(i.e., 1, 10, 100, 10^3, 10^4, 10^5 cycles,...,etc.).

5 Analytical and experimental results

5.1 Static monotonic loading

At first, the sandwich beam was tested under static monotonic loading. The sandwich beam was also analyzed under static monotonic loading by using the finite element method. The experimental as well as the analytical load–deflection curves are shown in Fig. 9. It is observed that the load increases with high stiffness until about 230 kN. At this load, main diagonal cracking occurs. This could be illustrated by the crack patterns in Fig. 10 (i.e.,P_{cr}= 230 kN). Then, the load–deflection curve increases further with decreased stiffness until the ultimate failure load (P_u). The experimental ultimate failure load ($P_{u,exp}$) was equal to 318 kN, while the analytical ultimate failure load ($P_{u,FEM}$) was equal to 340 kN (see Fig. 9).

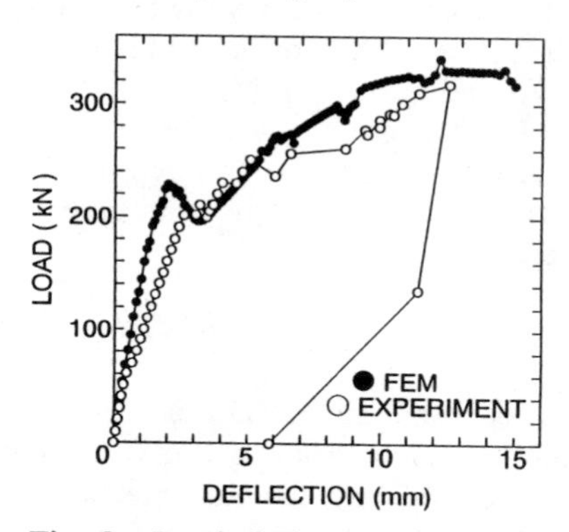

Fig. 9. Load–deflection curves (static loading).

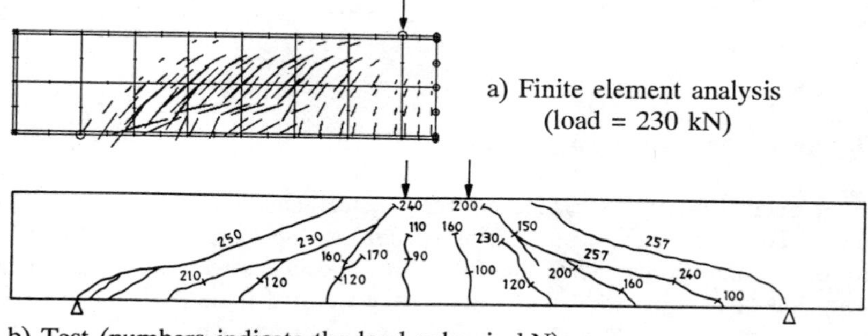

b) Test (numbers indicate the load value in kN)

Fig. 10. Crack patterns under static monotonic loading.

5.2 Fatigue loading

Four fatigue tests were carried out. The maximum fatigue load (P_{max}) was chosen to be 130, 160, 190, and 220 kN, which is 40.9%, 50.3%, 59.8%, and 69.2%, respectively, of the experimental static strength of the beam ($P_{u,exp}$= 318 kN). The results of the fatigue tests are summarized in Table 1. The crack patterns of the sandwich beam in fatigue tests no.2 and no.4 are shown in Fig. 11. The numbers written on the crack patterns indicate the number of fatigue loading cycles (N). In fatigue test no.1 (P_{max}= 130 kN), fatigue failure did not occur until $2{\times}10^6$ cycles. In fatigue tests no.2 (P_{max}= 160 kN) and no.3 (P_{max}= 190 kN), fatigue failure was caused by fracture of the tensile steel plate at the supporting point. The cross section of the steel plate at the fractured point is shown at top of Fig. 11, in which the white area indicates the fatigue crack, while the hatched area indicates the part that finally fractured in tension after the growing fatigue crack weakened the plate. In fatigue test no.4 (P_{max}= 220 kN), fatigue failure occurred due to crushing of the concrete between diagonal cracks in the vicinity of the support. These fatigue failure modes are shown in Fig. 11. More details about the experimental results can be found in reference [8].

Table 1. Results of the fatigue tests.

Specimen No.	Fatigue test no.	P_{max} (kN)	P_{min} (kN)	P_{max} / $P_{u,exp}$ (%)	Fatigue life (cycles)	Failure mode
2	1	130	20	40.9	2,000,000	NF[1)]
3	2	160	20	50.3	624,221	FS[2)]
4	3	190	20	59.8	120,081	FS[2)]
2	4	220	20	69.2	79,948	WC[3)]

1) NF : no fatigue failure until 2,000,000 cycles
2) FS : fracture of the lower steel plate at the supporting point (see Fig. 11)
3) WC : diagonal web crushing (crushing of concrete between diagonal cracks in the vicinity of the support)(see Fig. 11)

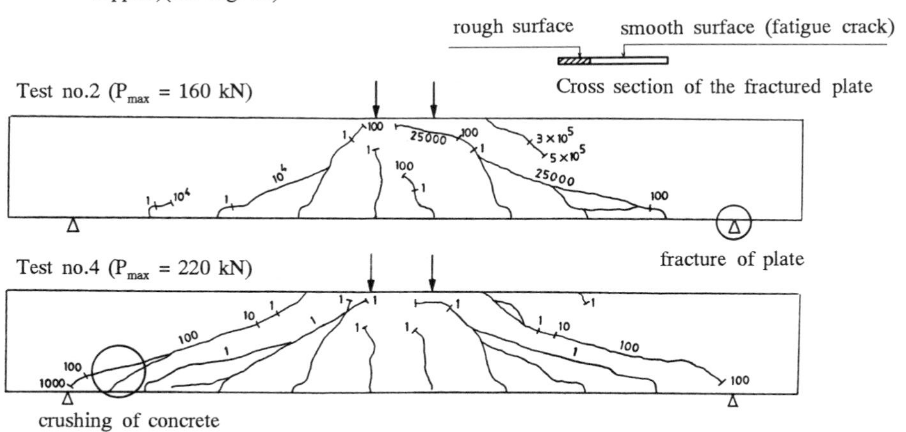

Fig. 11. Crack patterns under fatigue loading.

The sandwich beam was analyzed for different external load ranges. The minimum fatigue load (P_{min}) was kept constant at 20 kN. The maximum fatigue load (P_{max}) was chosen to be 240, 270, and 300 kN, which is 70.6%, 79.4%, and 88.2%, respectively, of the analytical static strength of the beam ($P_{u,FEM}$= 340 kN). The analytical results are summarized in Table 2. The output load–deflection curves for beams no.2 and no.3 are shown in Fig. 12. It should be noticed that the fatigue failure mode predicted by the finite element method is the crushing of concrete failure mode (see Fig. 11). More modifications for the computer program (WCOMR) are needed in order to predict the fracture of steel failure mode. The experimental and the analytical S–N relationships for the sandwich beam are illustrated in Fig. 13.

Table 2. Results of the fatigue analysis.

Beam No.	P_{max} (kN)	P_{min} (kN)	$P_{max}/P_{u,FEM}$ (%)	Fatigue life (cycles)	Failure mode
1	240	20	70.6	20,000	CC[1)]
2	270	20	79.4	1000	CC[1)]
3	300	20	88.2	50	CC[1)]

1) Crushing of core concrete

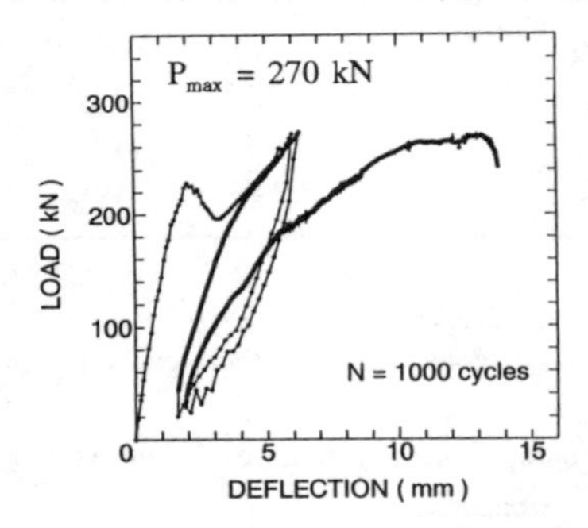

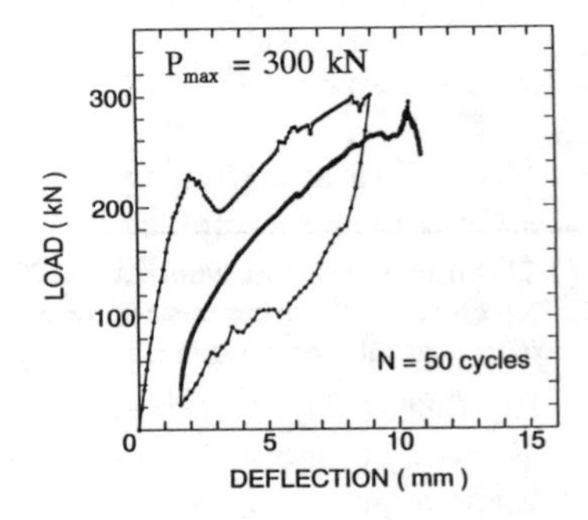

Fig. 12. Analytical load–deflection curves.

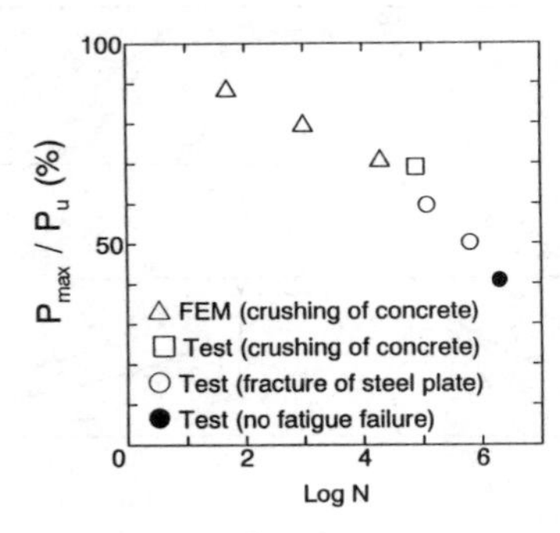

Fig. 13. S–N relationships for the sandwich beam.

6 Conclusions

Based on the present experimental work and analytical study using the finite element method, the following conclusions can be drawn :

1. For a large maximum fatigue load (P_{max}= 69.2% of the static strength), the failure mode is crushing of concrete between diagonal cracks, but for smaller loads (P_{max}= 50.3% and 59.8% of the static strength), the failure mode is fracture of the tensile steel plate at the supporting point.
2. Under the fatigue loading, the fatigue strength of the beam with the crushing of concrete failure could be predicted by using the finite element method in which the compressive strength, the tensile strength, and the stiffness of concrete are reduced with increasing the number of cycles (N) or increasing the stress range (R).

In order to propose design equations for the fatigue strength of steel–concrete sandwich beams, much research is still needed in this field. These future studies should cover the case of sandwich beams without shear reinforcement as well as the case with different types of shear reinforcement. At the present time, experimental and analytical studies are being carried out by the authors to estimate the fatigue strength of sandwich beams provided with shear reinforcement. The results of these investigations will be presented in subsequent papers.

7 References

1. Yokota, H. and Kiyomiya, O. (1989) Fatigue Behaviours of Steel–Concrete Hybrid Beams, Transactions of the JCI, Vol.11, pp.455–462.
2. Okamura, H. and Maekawa, K. (1991) Nonlinear Analysis and Constitutive Models of Reinforced Concrete, Gihodo Shuppan, Tokyo.
3. Pantaratorn, N. (1991) Finite Element Analysis on Shear Resisting Mechanism of RC Beams, Dissertation Submitted to the University of Tokyo, pp.74–102.
4. Tepfers, R. and Kutti, T. (1979) Fatigue Strength of Plain , Ordinary , and Lightweight Concrete, ACI Journal, Vol.76, No.5, pp.635–652.
5. Tepfers, R. (1979) Tensile Fatigue Strength of Plain Concrete, ACI Journal, Vol.76, No.8, pp.919–933.
6. Holmen, J.O. (1982) Fatigue of Concrete Structures, ACI, SP–75, Detroit, pp.71–110.
7. Bennett, E.W. and Raju, N.K. (1969) Cumulative Fatigue Damage of Plain Concrete in Compression, The Proceedings of the Southampton 1969 Civil Engineering Materials Conference, Part 2, pp.1089–1102.
8. Zahran, M. (1994) Shear–Fatigue Behavior of Steel–Concrete Sandwich Beams Without Web Reinforcement, Proceedings of the JCI Conference, Vol.16, pp.1217–1222.

166 INVESTIGATION ON THE REDUCTION OF FATIGUE STRENGTH OF SUBMERGED CONCRETE

S. OZAKI, N. SUGATA and K. MUKAIDA
Muroran Institute of Technology, Muroran, Japan

Abstract
The compressive fatigue strength of concrete submerged in water is significantly lower than that of dry concrete in air. In this experimental investigation, compressive fatigue tests on concrete cylindrical specimens were performed both in submerged and in air-dry conditions. From the results of strain measurements under repeated loading, it was observed that the energy loss at the first cycle of loading is small, and there is a rapid increase in bulk strain immediately prior to sudden failure under the submerged condition. However, for dry concrete in air, the increase in strain is gradual, and cracks propagate slowly until failure.

The pH level of the water in the fatigue test tank was observed to increase with repeated loading; the amount of calcium in the water oozed out of the submerged specimens was analyzed.

From the results of pore size distribution measurements, the disappearance of micro pores in the surface layer of the failed specimens under the submerged fatigue test and the increase of micro pores in the air-dry fatigue test specimens were verified.
Keywords: bulk strain, compression test, fatigue strength, micro pore, pH, submerged concrete

1 Introduction

The fatigue strength of concrete immersed in water or in a wet condition is substantially lower than that of dry concrete in air. The fatigue strength loss due to factors affecting static strength was conventionally considered to be equivalent to the static strength loss; the fatigue strength of concrete was said to be normally 60

Concrete Under Severe Conditions: Environment and loading (Volume Two) Edited by K. Sakai, N. Banthia and O.E. Gjørv. Published in 1995 by E & FN Spon. ISBN 0 419 19860 1

Table 1. Mix proportions of concrete

Water-cement ratio (%)	Fine aggregate ratio (%)	Unit content (kg/m^3)				
		Water	Cement	Fine agg.	Coarse agg.	AE agent
52	43	165	318	796	1068	0.095

% of its static strength. As a result of fatigue tests in water conducted about 20 years ago, however, the fatigue strength was found to be reduced to nearly 30 % of the static strength, indicating that fatigue strength is more strongly affected by water within concrete than the static strength [1]. Thus, in relation to the ratio of maximum stress to static strength S_1, ratio of minimum stress to static strength S_2 and fatigue life N_f, expressed by the equation $\log N_f = K(1 - S_1)/(1 - S_2)$, given in the Standard Specification for Design and Construction of Concrete Structures by JSCE, 10 is specified as the value of K for concrete saturated with water, instead of 17 for normal concrete [2].

Such fatigue strength loss in water has been attributed to the pumping action [3] or the wedging action [4] of water. While the authors recognize that such actions of water cause concrete failure, they consider that the fatigue strength loss results from weakened bonds at the interfaces between aggregate and cement paste due to the presence of water; thus, they have continued to investigate fatigue strength improvement by filling the voids at the paste-aggregate interfaces [5]. The bond loss, they considered, occurs because the paste-aggregate interfaces, where calcium hydroxide exists in large quantities, can become weak points in concrete saturated with water. In fact, when cyclic loads are applied to concrete in water, the water in the fatigue test tank becomes turbid. Measurement of the pH level in the water suggests that the turbidity is caused by calcium hydroxide dissolved from within the concrete.

In this study, the authors intend to elucidate the effects of water on the fatigue strength of concrete from a microscopic standpoint by analyzing the ions dissolved into water from concrete specimens and measuring the pore diameter distribution in the mortar phase of the specimens. The measured items also include those the authors have conventionally adopted, i.e., measurement of the pH levels of tank water and the strain of the specimens.

2 Materials, proportioning, and specimens

Ordinary portland cement concrete was used in this study. The specific gravity and fineness of the ordinary portland cement were 3.15 and 3200 cm^2/g, respectively. The fine aggregate was sea sand with a specific gravity of 2.77 and a fineness modulus of 2.56. The coarse aggregate was crushed stone with a maximum size of 20 mm and a specific gravity of 2.69. An air-entraining agent was used as a chemical admixture. The concrete was proportioned as given in Table 1.

Three 50-liter batches were mixed with a forced-action mixer, producing 27 cylindrical specimens 10 cm in diameter and 20 cm in length from each batch for a total of 81 specimens. The slump of each batch was 14.0, 14.5, and 15.5 cm,

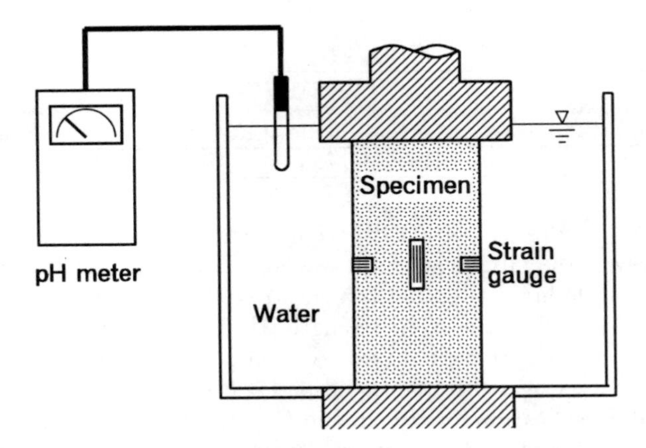

Fig. 1. Water tank for fatigue testing and pH measurement.

and the air content, 8.0, 8.0, and 9.0 %, respectively.

The specimens made from the first and second batches were water-cured for approximately 7 months at 20°C in a curing tank until fatigue testing in water. Those from the third batch were water-cured at 20°C for 2 months and then were air-dried indoors for approximately 8 months until fatigue testing in air.

The static strength of the specimens from each batch was 28.2, 27.4, and 26.1 MPa at 28 days. The reference static strengths of the specimens in water were 34.6 and 33.7 MPa, and that of the specimens in air was 33.1 MPa, at the time of the fatigue test.

3 Fatigue and related test procedures

The fatigue test in water was conducted by removing the specimens from the curing water tank and placing them in a transparent fatigue test tank (Fig. 1) 31 cm in inside diameter and 29 cm in height. Test loads were applied by a 30-ton hydraulic servo fatigue tester for structures.

Cyclic sinusoidal loads of 2–9 Hz were applied, with the minimum stress ratio (S_2) being 5 % of the static strength and the maximum stress ratio (S_1) being 65, 60, 55, 50, 45, and 40 % in water and 75, 70, and 65 % in air. These cyclic loads were applied continuously until failure, and the numbers of cycles were recorded.

Bases for strain gauges were made in advance with an epoxy resin adhesive on each specimen, on which four strain gauges (2 in longitudinal and 2 in transverse directions) were affixed with an instant adhesive. After three-core gauges were connected, they were waterproofed with an elastomeric adhesive. The specimens were then stored in water until testing.

A measuring system consisting of a DC amplifier (signal conditioner), an A/D converter, and a personal computer was used to observe and to record the strain.

A specified amount of the tank water was sampled for atomic absorption analysis, in order to analyze the ingredients dissolved from the specimens as the cyclic

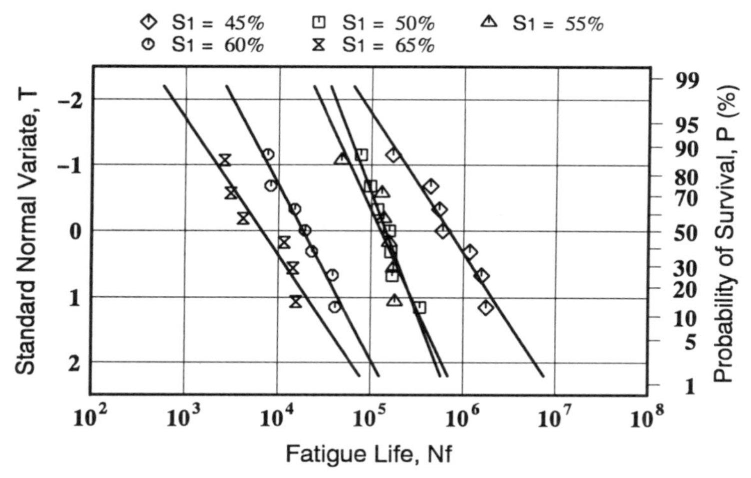

Fig. 2. P(T)-N diagram of concrete by fatigue test in water.

loading proceeded.

After the fatigue test was completed, the specimens were cleft into 4 layers from the surface to the center. Mortar phase was taken from each layer to measure the pore diameter distribution using a mercury porosimeter. The pore diameter distribution of the mortar phase of reference concrete specimens that did not undergo fatigue testing and any loading was also measured for comparison.

4 Results and discussion

4.1 Fatigue strength

The expected values of the probability of survival, P, against the fatigue life of the specimens were obtained by $P = 1 - r/(n+1)$ where r is the rank of the specimen and n is the total number of specimens for each maximum stress ratio. On the assumption that these values would follow a logarithmic normal distribution, Figs. 2 and 3 were obtained by linear regressions. In addition to the probability of survival, P, the values of the standard normal variate, T, which are convenient for expressing the regression line, are indicated on the ordinate axes. The S-N diagram (not shown here) is obtained by a regression of the plotted points of the numbers of cycles corresponding to $P = 50$ % ($T = 0$) on these regression lines.

By utilizing the relationship of the modified Goodman diagram and converting it to the equation of full unreversed compression with the minimum stress ratio, $S_2 = 0$ %, the S-N diagram is as shown in Fig. 4. On the other hand, if the S-N diagram is expressed as a regression equation that passes $S_1 = 100$ % when $N_f = 1$, so as to satisfy the static standard strength specified by the Standard Specification by JSCE, it is as shown in Table 2.

These results agree with past results in that the differences between the fatigue strengths in water and in air are distinguished.

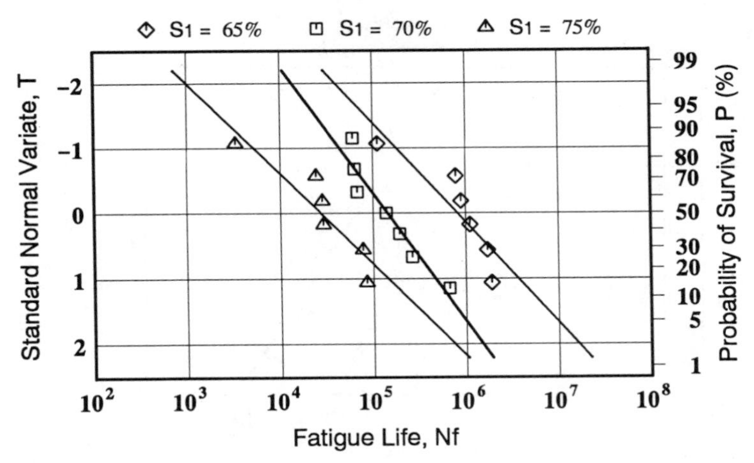

Fig. 3. P(T)-N diagram of concrete by fatigue test in air.

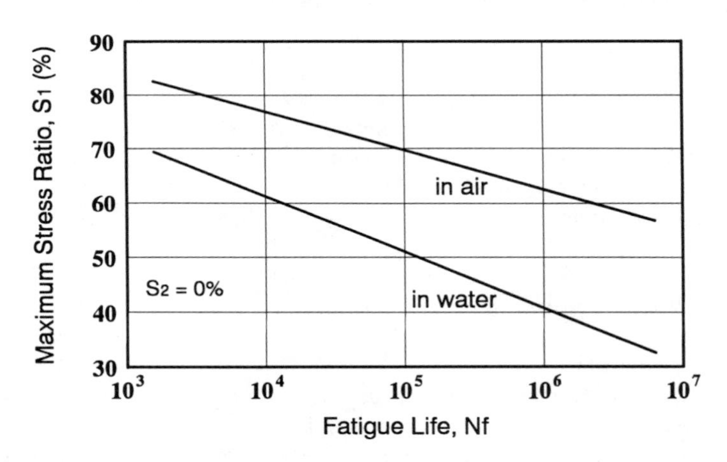

Fig. 4. S-N diagram of full unreversed compressive fatigue ($S_2 = 0$).

Table 2. Regression equations of S-N diagram by JSCE method

Test conditions	Regression equation	Fatigue str. ratio at 2 million cycles (%)	K value
In water	$(S_1 - S_2)/(1 - S_2) = 1 - 0.098 \log N_f$	38	10.2
In air	$(S_1 - S_2)/(1 - S_2) = 1 - 0.061 \log N_f$	61	16.4

4.2 Strain analysis results

Figures 5 and 6 show the results of the bulk strain incorporating the longitudinal compressive strain and the transverse tensile strain. As seen from these figures, the strain behaviors in water and in air greatly differ. In the bulk strain in water, the concrete maintains a constant compression, and when the strain shifts towards the expansion side, it rapidly leads to failure.

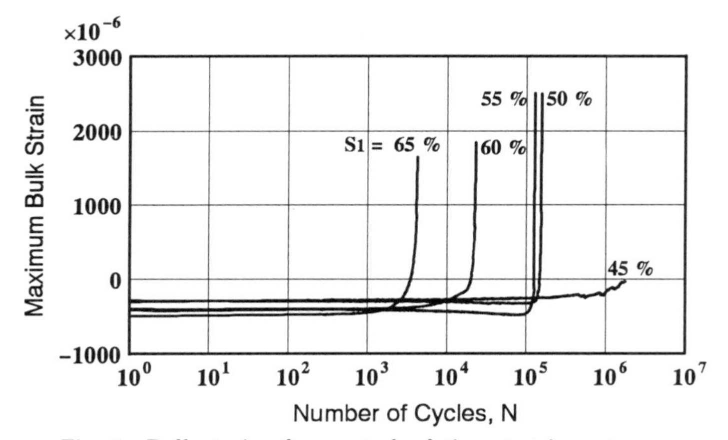

Fig. 5. Bulk strain of concrete by fatigue test in water.

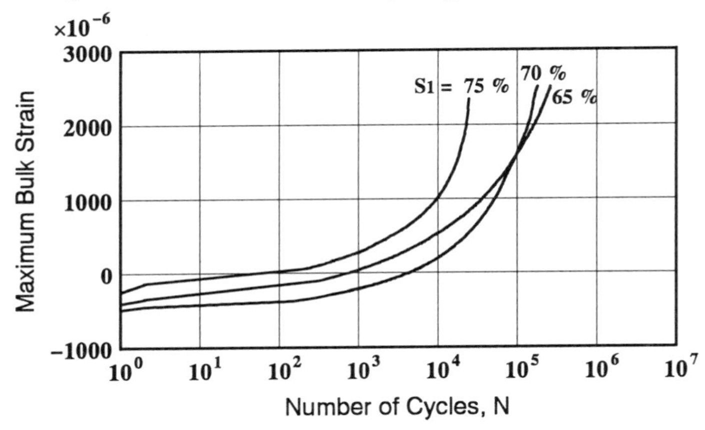

Fig. 6. Bulk strain of concrete by fatigue test in air.

Such a behavior specific to concrete undergoing fatigue loading in water agrees with past results, indicating that crack initiation leads to a rapid increase and growth of cracks due to the water action. This shows the strong adverse effects of the water action, when compared with the specimens in air with no effect from water action. The bulk strain of the specimens in air begins to increase at an early loading cycle, continuing to increase slowly, even after the shift towards the expansion side, and finally leads to failure. The bulk strain in water tends to show a turn towards the compression side before shifting towards the expansion side and leads to failure. This is inferred by the decrease in pore volume of the specimens. This decrease is also suggested by the pore diameter distribution on the surfaces of the concrete specimens failed in water which mentioned later in this paper, as well as by the reduction in chloride content in the concrete surfaces during the separately conducted fatigue tests in seawater.

Meanwhile, the stress-strain curve of concrete takes different courses during loading and unloading, producing a hysteresis loop. Figures 7 and 8 show examples

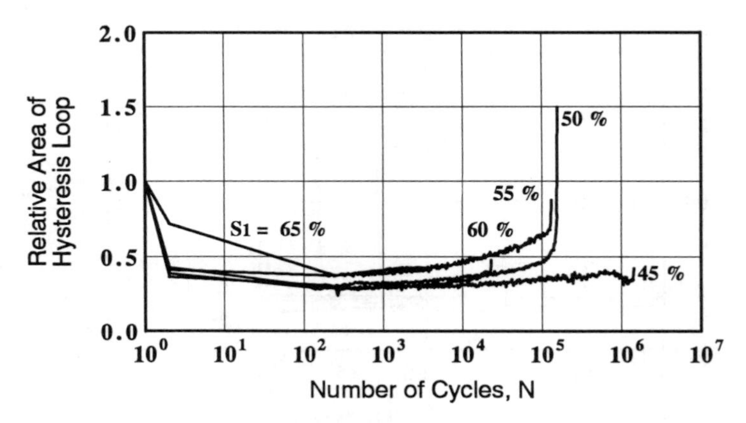

Fig. 7. Relative hysteresis loop area by fatigue test in water.

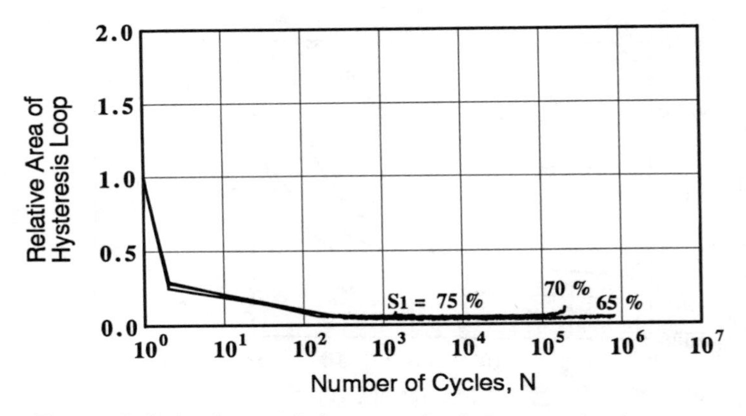

Fig. 8. Relative hysteresis loop area by fatigue test in air.

of vertical compression of the relationship between the number of cycles (N) and the ratio (An/A_1) of the area of the hysteresis loop at N cycles (An) to the area of the loop at the first cycle (A_1). By assuming the area of a hysteresis loop to be an irreparable energy loss, the relationship between the energy loss at the first cycle (W_1) and the fatigue life (N_f) is shown in Fig. 9. As seen from these results, the energy loss of ordinary portland cement concrete in water at the first cycle is as small as 0.2-1.5 kJ/m^3, but the losses thereafter remain as large as 40 %, as shown in Fig. 7, and then again increase to failure. On the other hand, the energy loss of concrete in air at the first cycle is as large as 4-8 kJ/m^3, indicating that a great amount of energy is consumed by crack development during the first cycle. However, the losses thereafter become marginal, as shown in Fig. 8. The losses then maintain a constant level, and the fatigue damage gradually increases until failure. Thus, the energy losses are considered to be consumed by the viscous friction in the liquid phase and the dislocation in the solid phase in submerged concrete under low stresses that cause little crack development.

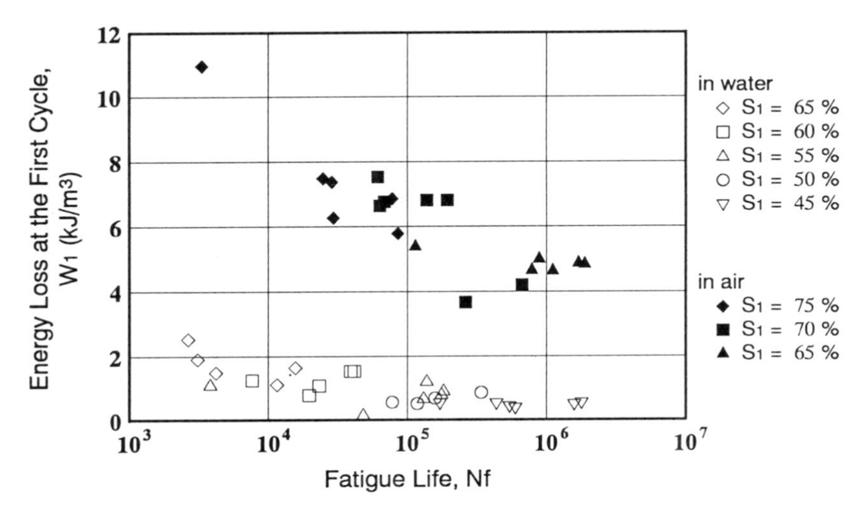

Fig. 9. Relationship between the energy loss at the first cycle and the fatigue life.

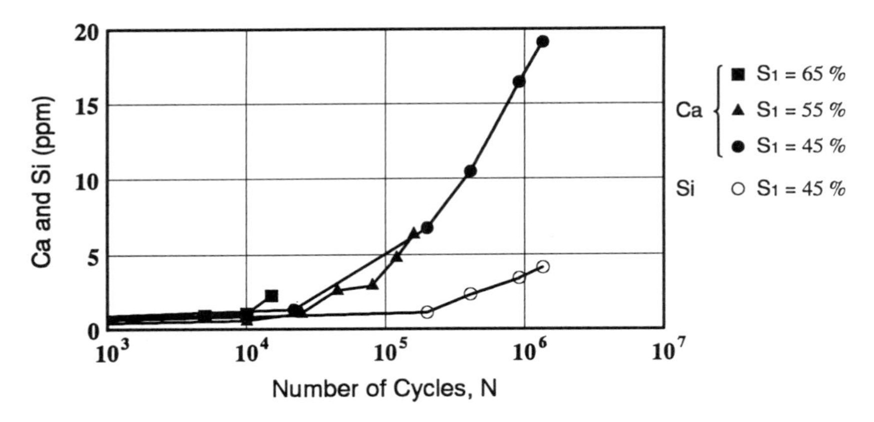

Fig. 10. Changes in Ca and Si concentration during fatigue test in water.

4.3 Ions dissolved into water during fatigue test

The results of the atomic absorption spectrometry dissolution concentration analysis of the Ca and Si ions produced by calcium hydroxide dissolution in the water tank during fatigue testing on concrete specimens in water are shown in Fig. 10. All of these values increase slowly in the beginning and rapidly later.

The absence of Si with the upper stress ratio, S_1, at 65 % and 55 % in Fig. 10, indicates that there is no dissolution of calcium silicate hydrate gel, and that the increase in the Ca concentration is attributed solely to the dissolution of calcium hydroxide. On the other hand, when S_1 is 45 %, Si is detected. Though the dissolution of pulverized CSH gel is conceivable, the detected Si concentration is very low. When comparing the period in which the Ca concentrations increase with the crack initiation period in which the bulk strain increases, the dissolution

Table 3. Changes in pore diameter distribution in percentages

	Surface layer of specimen			Core part of specimen		
	Peak-1 3–6nm	Peak-2 6–100nm	Peak-3 0.1–20μm	Peak-1 3–6nm	Peak-2 6–100nm	Peak-3 0.1–20μm
Before fatigue test	6.5	48.1	45.4	9.2	59.2	31.6
After test in water	0.0	57.9	42.1	9.2	61.8	29.0
Before fatigue test	5.1	40.5	54.4	6.6	55.2	38.2
After test in air	10.0	61.4	28.6	11.8	55.3	32.9

is found to occur before the crack initiation in all cases. It is therefore probable that the Ca dissolution is due to the damage by the dislocation of defects at paste-aggregate interfaces that have existed from the beginning

4.4 Pore diameter distribution in mortar

As a result of analysis by mercury porosimetry, the pore diameter distribution of mortar taken from concrete specimens was found to have 3 peaks in the ranges of 3nm–6nm, 6nm–0.1μm, and 0.1μm–20μm. Unless microcracks exist, these peaks are considered to correspond to the gel voids, capillary voids, and voids at paste-aggregate interfaces, respectively.

The values are expressed in percentages in Table 3, in order to distinguish the changes in the pore volumes before fatigue test and after the fatigue failure in the core area and in the surface layer of the specimens.

In the case of the fatigue test in water, Peak-1 disappears from the surface after the fatigue testing. This may be because the Peak-1 pores are enlarged due to the dissolution of calcium hydroxide or CSH gel, and the resulting pores can be classified as Peak-2. Since no such changes as found in the surface layer are observed in the core area, the dissolution of calcium is found to be limited to the surface layer. No appreciable changes are observed after the fatigue test in water, except the loss of Peak-1 in the surface layer. This suggests that micro-cracking did not disperse all over the specimen of submerged concrete during the test.

In the case of the fatigue test in air, the number of Peak-1 pores on the surface and in the core doubled, and Peak-2 pores on the surface also increased. This indicates the occurrence of microcracks. Though it has been considered that crack occurrence would lead to an increase in the total pore volume, there were no actual changes. This may be because the fine grains produced by microcracking filled large Peak-3 pores in the mortar.

5 Conclusion

Fatigue tests on ordinary portland cement concrete were conducted in water, in order to investigate the changes in its microstructure and fatigue properties. The following were found:

1. In spite of the small energy loss at the first cycle, the fatigue life of concrete in water is short, and the process towards failure is quick, which is specific to the fatigue properties of concrete in water. Such behavior is quite different from

that of concrete in air, in which the strain gradually increases starting from early cycles, and the cracking slowly develops to failure.

2. As the loading cycles proceeded, calcium increasing was found to have dissolved into the tank water by the analysis of the water. In the fatigue tests that lasted to large numbers of cycles, the dissolution of marginal amounts of silicon was confirmed as well.
3. As a result of the measurement of the pore diameter distribution of mortar taken from concrete specimens broken in the fatigue test in water, it was confirmed that the micropores near the surfaces disappeared. This is considered to be due to the dissolution of CSH gel and agrees with the results of 2. above.
4. An increase in the number of micropores was observed in the fatigue test in air. This is because no dissolution occurred due to the absence of water and because the microcracks increased as the cyclic loading proceeded. This agrees with the results of 1. above.

As an inference, the damage due to dislocation occurs gradually in water, because the friction resistance is not effective, and it is considered to occur at paste-aggregate interfaces. The damage at paste-aggregate interfaces is considered to turn into cracks, quickly proceeding to failure under water pressure.

6 References

1. Ozaki, S. and Shimura, M. (1980) Compressive Fatigue Strength of Concrete in Water. *Proceedings of the 35th Annual Conference of the Japan Society of Civil Engineers*, Vol. 5, pp.293–294.
2. JSCE (1986) *Standard Specification for Design and Construction of Concrete Structures, Part 1 (Design)*. pp.26–27.
3. Waagaard, K. (1982) Fatigue Strength Evaluation of Offshore Concrete Structures. ACI SP-75, pp.373–397.
4. Muguruma, H. and Watanabe, F. (1984) On the Low-cycle Compressive Fatigue Behaviour of Concrete Under Submerged Condition. *Proceedings of the 27th Japan Congress on Materials Research*, pp.219–224.
5. Suzuki, T. and Ozaki, S. (1985) Improvement of Compressive Fatigue Strength of Concrete in Water. *Proceedings of the 40th Annual Conference of the Japan Society of Civil Engineers*, Vol. 5, pp.241–242.

PART TWENTY-SEVEN
FLEXURE AND SHEAR

167 FLEXURAL BEHAVIOURS OF RC BEAMS HAVING A VERTICAL CONSTRUCTION JOINT

Y. TSUJI and C. HASHIMOTO
Gunma University, Gunma, Japan
T. MORIWAKI
Nihon Kasei Corporation, Saitama, Japan

Abstract
Construction joints have to be executed carefully. Generally, at the time of execution of construction joints, fresh concrete is placed after having made a green cut for joints. While the old concrete has already cured, cement paste, cement mortar or epoxy resin is placed just before fresh concrete is placed. In this paper, we turned our attention to polymer-modified mortar as a joint material in order to have sufficient open-time, i.e., the time between placement of polymer-modified mortar on the joint and placement of fresh concrete[1]. The flexuarl behavior of reinforced concrete beam having a vertical construction joint was investigated in order to ascertain the performance of polymer-modified mortar as a joint material[2]. As a result of tests, it became clear that the polymer-modified mortar is effective as a vertical construction joint material, allowing an open-time as long as two weeks.
Keywords: Flexural behavior, polymer-modified mortar, reinforced concrete beam, vertical construction joint.

1 Introduction

When forming construction joints between new and old concrete, it has been mandatory for laitance to be removed and the old concrete surface to be moistened prior to placing new concrete, while cement paste, mortar, and epoxy resin have been used as joining materials. Particularly, in the case of vertical construction joints, it is prescribed in "Construction Volume" of the Japan Society of Civil Engineers Standard Specifications for Concrete that one of these construction joint materials be used in executing the work. However, when any one of these construction joint materials has been coated on the construction joint surface of old concrete, it is necessary for new concrete to be placed immediately, and this comprises a

Concrete Under Severe Conditions: Environment and loading (Volume Two) Edited by K. Sakai, N. Banthia and O.E. Gjørv. Published in 1995 by E & FN Spon. ISBN 0 419 19860 1

constraint on reinforcing bar workmen and formwork erectors in carrying out the work. Performances of construction joints are the most sensitive to the durability of concrete structures in the severe conditions.

The study here is of using polymer cement mortar added to a conventional material to make reinforced concrete (RC) beams having vertical construction joints. Flexural behaviors using each kind of construction joint material were compared with beams not having construction joints, and the results of the examinations are reported.

2 Outline of experiments

Concrete mix proportions were made to have unit water content constant at 164 kg/m^3, and water-cement ratios were of the two levels of 0.65 and 0.45. The cement used was an ordinary portland cement, while the aggregates were river sand and river gravel from the Watarase River in Gunma Prefecture. The water-cement ratio of cement paste used as construction joint material was 0.30, while the water-binder ratio of polymer cement mortar was 0.28.

Specimens were reinforced concrete beams, 10 cm in width, 20 cm in height and 110 cm in length, as shown in Fig.1. Effective depth was constant at 17 cm, and loading was done by third-point loading of a 90 cm span.

Each beam had two D10 bars as tension reinforcement, and was made to have a construction joint at the middle.

Along with the RC beams, prism specimens, 10 cm in width, 10 cm in height and 40 cm in length with construction joints at the middle cross sections were made. Flexural strength tests were conducted by third-point loading of a 30 cm span with the construction joint at the middle of the span.

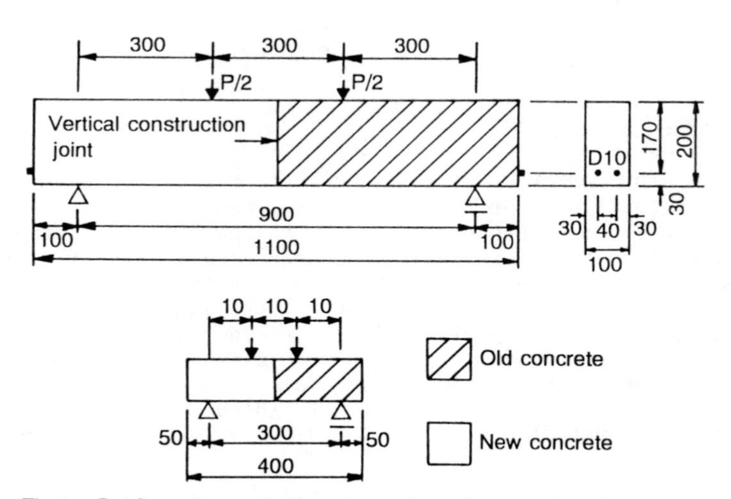

Fig.1. Configuration and dimensions of specimen and loading method.

When old concretes had reached ages of 24 hours and 48 hours, respectively, the construction joint surfaces were roughened by wire brush, and construction joint material was coated on after curing under a moist cloth until an age of 14-days. When cement paste and epoxy resin were used, new concrete was placed immediately, while when polymer cement mortar was used, the time from application until the placing of new concrete (hereinafter referred to as "open time") was varied from none at all to 3, 7, and 14 days. And, strength tests were performed after moist-cloth curing of the new concrete until an age of 28-days.

3 Construction joint strength of concrete

The flexural strengths of prism specimens having vertical construction joints are shown in Fig.2. The flexural strengths of specimens with polymer cement mortar coated often become higher with open times extended to 14 days, but it appears the influence of open time is not very great. And when cement paste or epoxy resin is used, the construction joint strengths are generally higher in comparison with polymer cement mortar, but there are also cases when the strengths are lower.

As for the flexural strength of concrete itself, it is higher the lower the water-cement ratio. Conversely, construction joint strength often becomes slightly lower with lower water-cement ratio, even when treatments of joint surfaces are the same. This is thought to be due to the construction joint being a relatively weak point when concrete strength is increased.

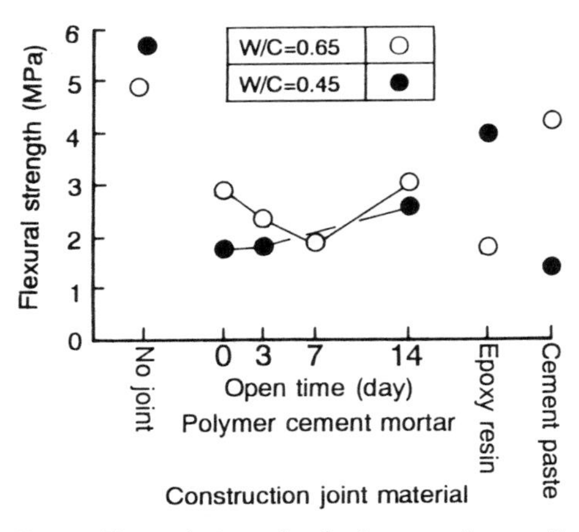

Fig.2. Flexural strength of prism specimen with construction joint.

4 Strain of tension fiber

The value measured by a wire strain gauge attached, straddling the construction joint, to the tension fiber of a reinforced concrete beam having a vertical construction joint is called the strain of tenion fiber, and the relationship of this strain with bending moment is shown in Fig.3.

When the water-cement ratio of concrete is 0.65, as shown in Fig.3(a), reinforced concrete beams joined immediately after and 7 days after coating with polymer cement mortar have larger strains of tension fiber from a small load stage compared with others. This is thought to be due to minute cracks formed along the construction joint prior to loading tests because of the dead weight of the specimen. However, a beam joined after an open time of 14 days, even with the same stripping and curing methods, did not have flexural cracks occur from the construction joint but from the new concrete side away from the strain gauge. Beams for which epoxy resin and cement paste were used had flexural cracks occur with bending moments of the same degree.

Cases with a water-cement ratio of 0.45 are shown in Fig.3(b). Flexural cracking moments of beams joined immediately after coating with epoxy resin and polymer cement mortar are less than those of beams provided with other construction joint treatments, but when compared with the case of concrete with a water-cement ratio of 0.65, the differences according to treatment methods of construction joints are not prominent.

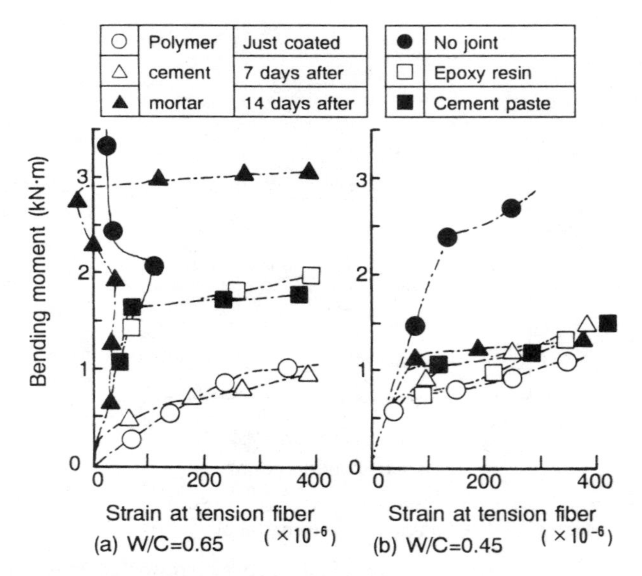

Fig.3. Strain at tension fiber.

With reinforced concrete beams having vertical construction joints, similarly to construction joint strengths of prism specimens, when high-strength concrete with a water-cement ratio of 0.45 was used instead of concrete with a water-cement ratio of 0.65, it was seen that flexural cracking moments were conversely decreased for beams joined 14 days after coating with polymer cement mortar, epoxy resin, and cement paste.

5 Shifting of neutral axis of construction joint portion

When performing loading tests, pi gauges were arranged at the sides of beams straddling construction joints, as shown in Fig.4. Measurements were also made regarding location of the neutral axis using this arrangement.

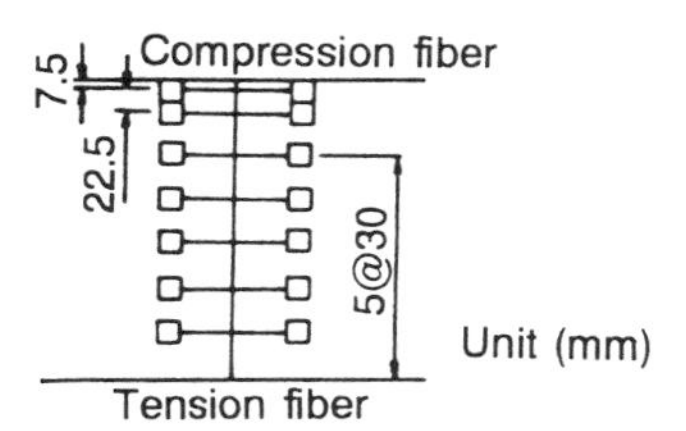

Fig.4. Arrangement of pi gauges.

The relationships between the location of the neutral axes and bending moments in the case of a concrete water-cement ratio of 0.65 are shown in Fig.5(a). For beams coated of polymer cement mortar, the neutral axes of beams with no open time and beams with a 7 day open time are close to the location of the neutral axis according to elasticity calculations ignoring the tensile force of concrete from the stage when the acting bending moment is small. In contrast, beams joined 14 days after coating with polymer cement had their neutral axes gradually approaching the compression fibers, and indicated the same shifting as beams without construction joints. From this, it may be concluded that with a beam joined 14 days after polymer cement application, the tensile force borne by the concrete side is gradually lost, similarly to beams without construction joints.

Beams using epoxy resin also had gradually rising neutral axes. However, with beams using cement paste, neutral axes rose abruptly and, after this, converged to the calculated value. From this, it may be concluded that construction joints using cement paste showed brittle failure and flexural cracking which developed at once to the top part of the cross section.

The relationship between the locations of neutral axes and bending moments in the case of a water-cement ratio of 0.45 are shown in Fig.5(b). In spite of differences in construction joint materials, neutral axes gradually converged on the calculated value, around a bending moment of 6.0 kN•m which is similar to the case with a water-cement ratio of 0.65.

Beams joined 14 days after application of polymer cement mortar, differing from cases with a water-cement ratio of 0.65, showed neutral axes converging from a stage when the acting bending moment was still small. Neutral axes of beams joined 7 days after coating with polymer cement mortar were lower than those of beams using epoxy resin and cement paste. It may be considered from this that beams joined after an open time of 7 days have lower rate of loss of tensile force shared than when using epoxy resin or cement paste.

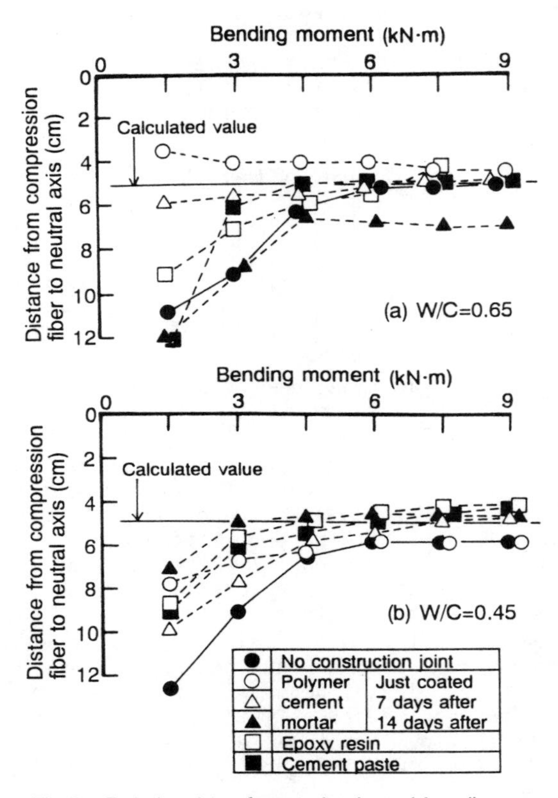

Fig.5. Relationship of neutral axis and bending moment.

6 Strain of tension reinforcement

The relationships between tension reinforcement strains and open times of RC beams joined using polymer cement mortar when bending moments were 1.0 kN·m, 4.0 kN·m, and 8.0 kN·m are shown in Fig.6. Excepting RC beams joined with open times of 0, 3, and 7 days, the strains of reinforcing bars at the construction joint location at a bending moment of 1.0 kN·m before flexural cracking occurs are about the same as for RC beams with no construction joints.

However, after flexural cracking has occured, when bending moment is 4.0 kN•m, flexural cracks become predominant at the construction joint section so that strains of reinforcement at the construction joint sections of RC beams with vertical construction joints become larger than the strains of reinforcement of RC beams without construction joints. However, with bending moments as large as 8.0 kN•m, the difference in strains of reinforcement in RC beams with and without construction joints tends to become slightly smaller. This is considered to be due to numerous flexural cracks occurring at locations other than construction joints as bending moment becomes larger.

Although a slight amount of difference can be seen in strains depending on the types of construction joint materials, there are hardly any differences in the increases of tension reinforcement strains accompanying increases in bending moment.

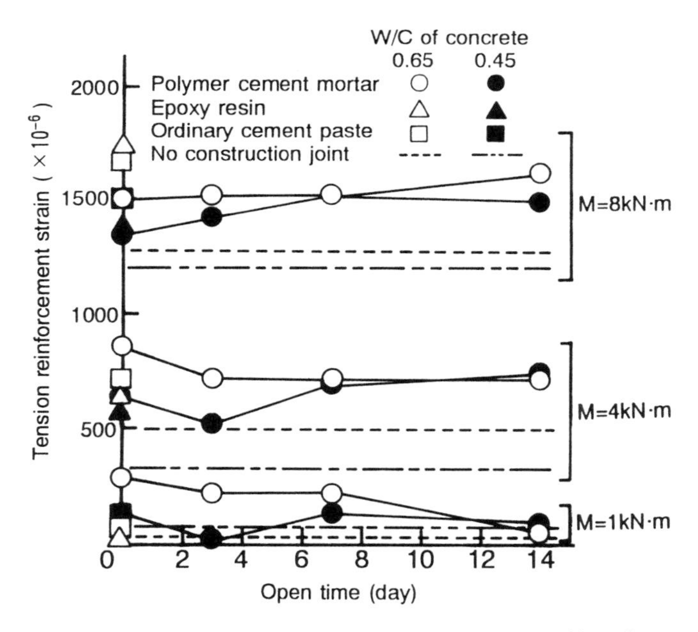

Fig.6. Relationship of tension reinforcement strain and bending moment.

7 Flexural crack width

The bending moments when flexural cracks in RC beams have reached 0.1 mm are shown in Fig.7. The bending moments at which flexural crack widths of RC beams having construction joints become 0.1 mm are smaller 20 to 30 % than those of RC beams without construction joints.

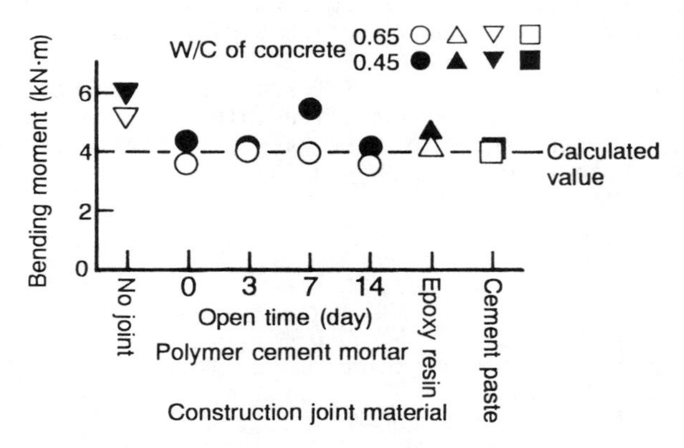

Fig.7. Bending moment when flexural crack width reaches 0.1 mm.

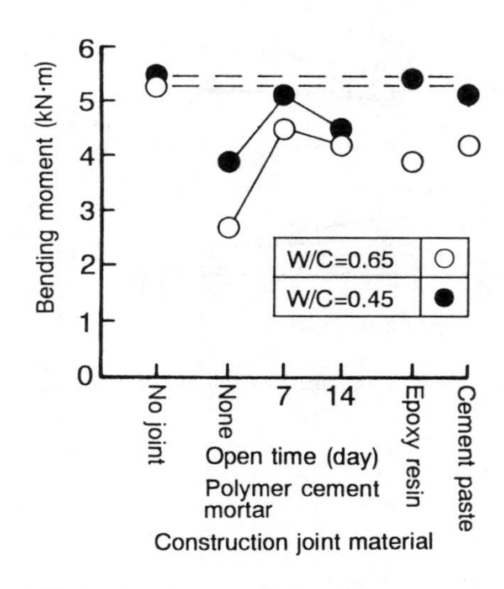

Fig.8. Bending moment when deflection of beam becomes 0.5 mm.

8 Deflection of beam

The bending moments when beam deflections reached 0.5 mm are shown in Fig.8. The bending moments of concrete specimen with high water-cement ratios were small when there were construction joints. However, when the water-cement ratio became 0.45, there were beams with construction joints in which the bending moments were of the same degree as beams without construction joints. They were beams which had used epoxy resin or cement paste 7 days after application of polymer cement mortar. Among beams on which polymer cement mortar had been used, the bending moments were largest for those with an open time of 7 days, irrespective of the water-cement ratio of concrete.

9 Failure bending moment

The bending moments at failure are shown in Fig.9. Beams with construction joints which failed in flexural tension had bending moment values about the same as those without construction joints. This is thought to have occurred because the cross-sectional specifications of reinforcing bars and others were the same as for beams without construction joints, and because construction joints which bear compressive force possessed adequate strengths.

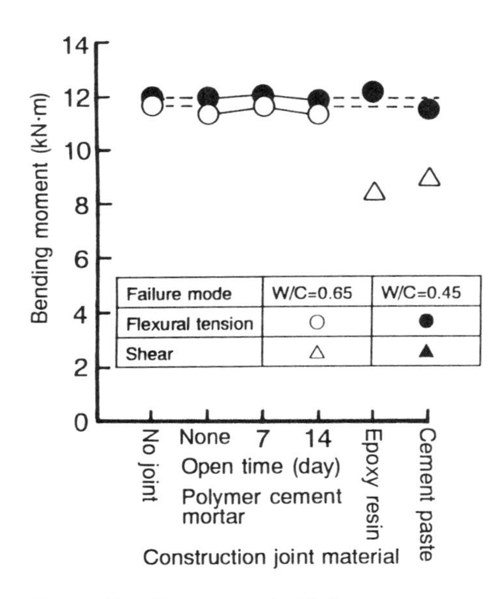

Fig.9. Bending moment at failure.

There were beams which failed in shear at loads smaller than at the time of flexural tension failure, and a difference occurred in types of failure. The reason for this is unknown at present, but since diagonal cracks had occurred in other beams also, that a slight torsion was produced between old and new concrete when making a construction joint and the development of diagonal cracking that was predominant are thought to be factors.

10 Conclusions

Reinforced concrete beams having a vertical construction joint were made using various construction joint materials, and flexural tests of those beams were conducted. The following conclusions may be drawn from these tests:

1. The flexural cracking moment of a beam having a vertical construction joint, in the case of an application of polymer cement mortar, is often larger the longer the open time. The tendency for construction joint strength decline of prism specimens having vertical construction joints to become greater using high-strength concrete was recognized in the flexural cracking moment of the beam also.
2. Strain of tension reinforcement at the construction joint, flexural crack width, and deflection become larger than in RC beams without construction joints due to flexural cracks being predominant in vertical construction joints. It was possible for such decline in quality to be remedied by appropriate selection of construction joint material and method of work execution.
3. Flexural cracking moments were of approximately the same degree irrespective of whether or not there were construction joints and the methods of treating construction joints.

11 References

1. Tsuji,Y.,Furusawa,M.,Hasegawa,M.and Moriwaki,T.(1989)Constructing methods of construction joints by using polymer-cement. Proc. of JCI,Vol.11,No.1.pp.721-726.(in Japanese)
2. Tsuji,Y.,Tanaka,K.,Furusawa,M. and Moriwaki,T.(1990) Mechanical behaviors of RC beam having vertical or horizontal construction joint. Proc. of JCI,Vol.12,No.2.pp.209-214.(in Japanese)

168 FLEXURAL AND SHEAR STRENGTH OF RC BEAMS STRENGTHENED WITH EXTERNAL PRESTRESSING CABLE

E. KONDO and M. SANO
Sho-Bond Corporation, Omiya, Japan
H. MUTSUYOSHI and H. TAKAHASHI
Saitama University, Urawa, Japan

Abstract
In the current study, the effectiveness of an external cable system for the strengthening of concrete bridges is discussed. Experiments were conducted with the objective of obtaining basic data on the flexural and shear strengthening effects. From this, strengthening effects against flexure and shear were confirmed empirically. As for the flexural properties, a comparative study was conducted between the experimental results and analytical prediction based on the technique proposed in a past study; while for the shear, the experimental results were compared with the computed results obtained in compliance with the specifications of the Japan Society of Civil Engineers.
Keywords: External prestress, flexural strength, partial prestressing, shear strength.

1 Introduction

Recently, design live load in Specifications for highway bridges, Japan has been revised from 20 to 25 tonf (from 196 kN to 245 kN). Furthermore, deterioration of RC and PC bridges due to corrosion of reinforcement inside concrete has become a big problem. As a result, the necessity to strengthen and repair existing RC or PC bridges has been raised. An externally prestressing method is one of appropriate method to strengthen and repair concrete structures [1]. In addition, together with enhancing the flexural strength, shear strength could also be improved by external cable system.
However, a rational design method remains not to be established for

Concrete Under Severe Conditions: Environment and loading (Volume Two) Edited by K. Sakai, N. Banthia and O.E. Gjørv. Published in 1995 by E & FN Spon. ISBN 0 419 19860 1

strengthening existing structures.

Generally, RC or PC members are designed to fail primarily in flexure. However, in the case of strengthening, the increase in the flexure and shear capacities may not be the same. As such, there is a possibility that the failure mode may change from flexure to shear. Accordingly, with the objective of obtaining basic data necessary to establish a design method, prestress was introduced by external cables to RC beam specimens, and the effects of strengthening on flexure and shear were investigated both experimentally and analytically in this study.

2 Outline of experiment

2.1 Test specimens and experimental variables

The specimens consisted of 2.7 m span RC beams having a T−shaped section as shown in Fig. 1. D6 stirrups were used as shear reinforcement at a spacing of 18 cm or 4.5 cm, while four D19 were used for flexural reinforcement. SWPR7B prestressing strand wires were used as the external cables, which were bent up at an angle about 7 degrees at the deviators (positioned at the points of loading) and anchored at the ends of the specimen. The average compressive strength of the concrete was 37 MPa. The mechanical properties of materials used are shown in Table 1.

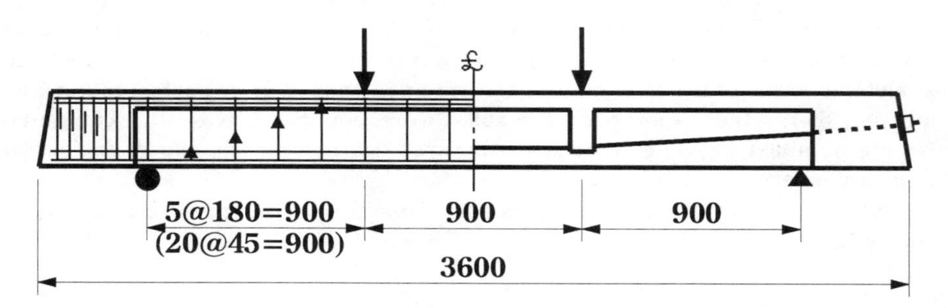

(a) Side view

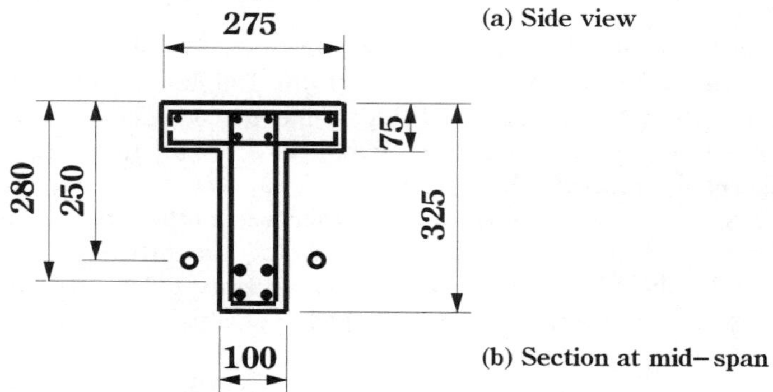

(b) Section at mid−span

Fig. 1. Detail of test specimens (mm).

Table 1. Mechanical properties of materials

Type	Diameter (mm)	Sectional area (mm^2)	Yield strength (N/mm^2)	Tensile strength (N/mm^2)	Young's modulus (N/mm^2)
D 6	6.53	31.67	392	559	179000
D10	9.53	71.33	441	588	195000
D19	19.1	286.5	431	579	189000
Prestressing strand wire	12.7	98.7	1579	1853	196000

Table 2. Experimental variables

Specimen No.	Type of beam	Pitch of stirrup (mm)	Tension reinforcement ratio (%)	Shear reinforcement ratio (%)	Prestressing force (kN)
No. 1	RC	45	1.49	1.41	---
No. 2	RC	180	1.49	0.35	---
No. 3	PPC	45	1.49	1.41	137
No. 4	PPC	180	1.49	0.35	137
No. 5	RC → PPC	180	1.49	0.35	137

The experimental variables are shown in Table 2. Specimens No. 1 and No. 2 were RC beams without external cables. To cause shear failure in No. 2, the stirrups were spaced at larger intervals. Specimens No. 3 and No. 4 were PPC (partially prestressed concrete) beams, with the same variables as No. 1 and No. 2, except the introduction of a prestressing force. In addition, No. 5 was modeled to simulate the behavior of an existing structure. This was achieved by causing bending and shear cracks in the specimen at RC state, and later, prestressed by the external cable system. In this specimen, the deviators were post-attached with epoxy resin and bolts to prevent cracking of the deviator in the RC state. The maximum possible prestressing force was applied to the external cables, without exceeding the allowable stress in tension at the top fiber. This value was found to be 137 kN in total; thus, 68.5 kN force was applied to each cable.

2.2 Loading method and observed measurements

Static two point symmetrical loading by a hydraulic jack was adopted as the loading method, with distance between the loading points set at 90 cm. During loading, displacements at mid-span and at loading points were measured using displacement transducers. Strains in the stirrups and main reinforcement at various locations and concrete strain at the mid-span were measured by strain gauges. The tension in the external cables were monitored by load cells installed at both ends of the cables. The propagation of cracks were checked visually, and crack widths were measured by π -gauges at critical locations.

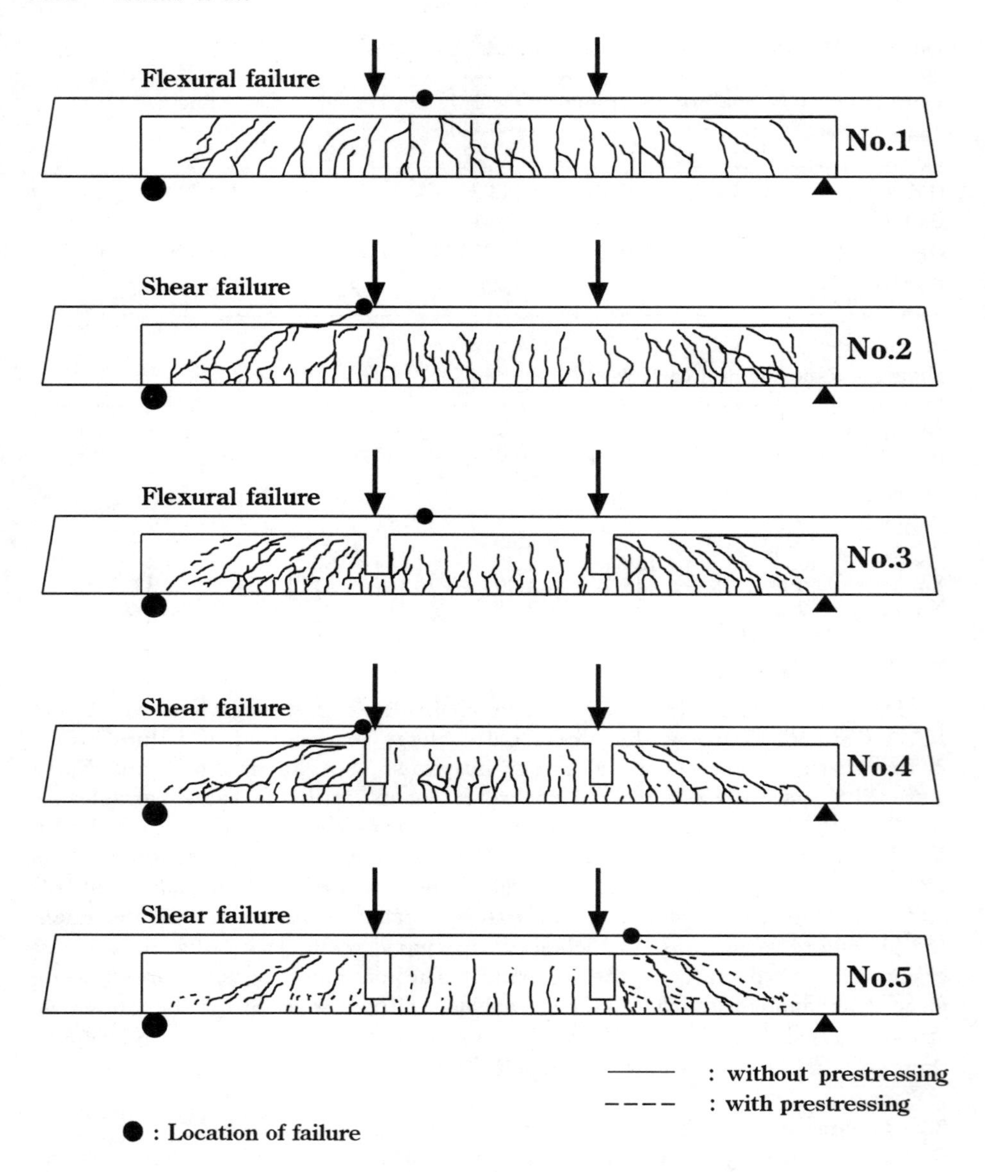

Fig. 2. Crack pattern and location of failure.

3 Test results and discussions

3.1 Mode of failure

The location of failure and crack patterns are shown in Fig. 2. In all the specimens, except in No. 1, flexural cracks invariably developed in the vicinity of the mid-span, and shear cracks appeared later in the web within the shear span. As for specimen No. 1, after the development of flexural cracks in the vicinity of the mid-span, additional flexural cracks appeared in the shear span. As the load increased, these flexural cracks developed into shear cracks as they further propagated. These shear cracks of No.1 did not develop into the flange, and the maximum crack width was about 0.15 mm. After the yielding of the main reinforcement, the crushing of concrete occurred closer to the loading point within the uniform moment region, and collapse took place. As for No. 2 and No. 5, immediately following the yielding of the main reinforcement, the shear cracks extended into the flange, and failure occurred at the neighborhood of the loading point by crushing of concrete. Meanwhile, shear cracks about 0.15 mm wide developed in No. 3, but did not progress as far as to the flange. After the main reinforcement yielded, the concrete crushed closer to the loading point within the uniform moment region, and the beam collapsed. In case of No. 4, shear cracks propagated into the flange before the main reinforcement yielded, and crushing of concrete in the neighborhood of the loading point occurred which led to failure.

3.2 Flexural strengthening effect gained by external prestressing

The experimental results are given in Table 3, while the variation of load with displacement is illustrated in Fig. 3. By comparing RC specimen No. 1 with PPC specimen No. 3, the influence of prestressing by external cables on the flexural strength could be seen. As seen from Table 3, the load at which the first visible flexural crack occurred showed about an 86% increase in PPC specimen No. 3, while the ultimate load of No.3 showed an increase of a 36% over No.1.

With reference to Fig. 3, under the same load of 127 kN (i.e., when the main reinforcement reached the allowable stress), the displacement was observed to be 6.2 mm in the RC specimen and 3.6 mm in the PPC specimen, showing a reduction of about a 42%. This indicates that displacement restraining effect could be obtained by the introduction of prestress. From these observations, the effectiveness of external prestressing for flexural strengthening was verified.

Table 3. Experimental results

Specimen No.	Type of beam	Cracking load		0.2mm [1] cracking load	Load at [2] yielding of stirrup	Ultimate load
		Flexure (kN)	Shear (kN)	(kN)	(kN)	(kN)
No. 1	RC	37	---	---	245	255
No. 2	RC	39	78	98	152	260
No. 3	PPC	69	127	---	245	348
No. 4	PPC	74	132	147	147	294
No. 5	RC → PPC	83 [3]	88 [3]	152	196	363

1) Load at 0.2 mm crack width.
2) Load at about 2200 μ strain of stirrup.
3) Load at crack reopening of PPC state.

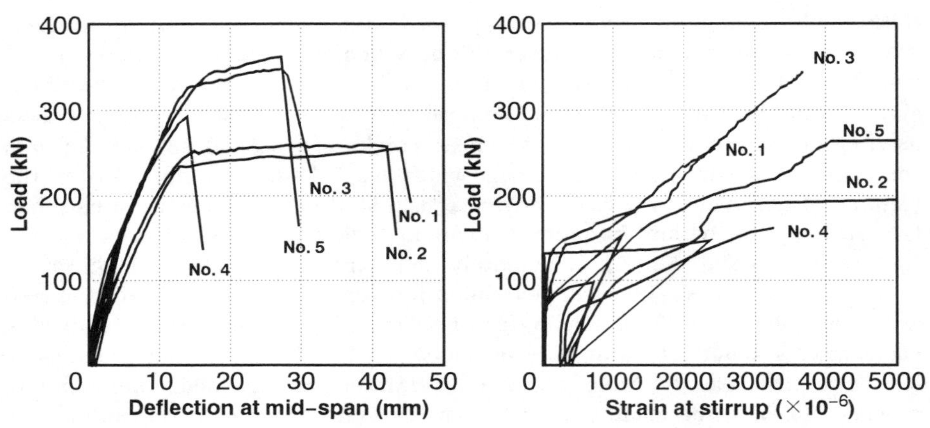

Fig.3. Load–deflection curves. Fig.4. Load–strain relationship at stirrup.

3.3 Shear strengthening effect gained by external prestressing

The variation of load with stirrup strain at the same location is shown in Fig. 4. In this case, comparing RC specimen No. 2 with PPC specimen No. 4, the influence of prestressing by external cable on the shear strength could be observed. As seen from Table 3, the load at which the visible shear crack occurred showed an increase of about a 68% for PPC specimen over that for RC one. Further more, the load when the shear crack width was 0.2 mm showed approximately a 50% increase for PPC specimen No. 4. Likewise, the ultimate load in No. 4 showed an increase of about a 13% over No. 2. With reference to the crack pattern in Fig. 2, the inclination of the shear cracks in No. 4 is smaller than that of No. 2, which seems to suggest that the prestressing by external cables has influenced the direction of principal stress. This may be one of the causes for the increase in loads mentioned above. Though the loads at the yielding of stirrups in No. 2 and No. 4 were practically the same, the process of yielding was different. As shown in Fig. 4, in specimen No. 4, the strain in the stirrup in the vicinity of the shear crack, which was approximately 10 μ just before the shear crack occurred, suddenly increased to 1,600 μ immediately after the shear crack appeared. This may be due to the fact that the load at which the shear crack developed in No. 4 was much greater than that of No. 2 and that the rapid transfer of high loads from concrete to stirrups at cracking, led to the sudden yielding of stirrups in No. 4. However, in No. 2, since the load was small at the occurrence of shear cracking, the load transferred from concrete was borne by the stirrups without yielding. Later, with the increase of load, the stirrups yielded gradually in No. 2. But this phenomena did not occur in specimen No. 3, where the stirrups were spaced at a closer intervals.

3.4 Influence of cracks on Shear strengthening effect

In actual practice, the existing structure may have already been cracked when the strengthening work is executed. To investigate such behaviors, the results of RC specimen No. 2 shall be compared with those of No. 5, where the specimen was prestressed by external cables after cracks were produced in its RC state. With reference to Table 3, shear cracks in No. 5 started to reopen when the load was 88 kN. At 152 kN loading, the shear crack width was 0.2 mm. Compared with No. 2, where this load was only 98 kN, about a 55% increase was achieved by strengthening. In terms of maximum strength, specimen No. 5 showed an increase of about a 40% over that of No. 2, which suggests that sufficient shear strengthening effects could be achieved by prestressing with external cables, even in situations where cracks already exist.

Next, comparing the results of No. 5 with No.4 (a specimen in which prestress was introduced in its crack−free state), as seen from Table 2, the loads were practically the same when the shear crack width was 0.2 mm. However, in terms of maximum strength, No. 5 showed an approximately a 23% increase over No. 4. As mentioned earlier, in terms of failure mode, No. 4 underwent shear failure before the main reinforcement yielded; while in No.5, the shear failure took place after the main reinforcement had yielded. In addition, rapid increasing of strain did not occur following the development of shear cracks in No. 5, as observed in No. 4.

One of the factors attributed to this difference may that, in No. 5, the prestressing was introduced after the beam was in the cracked state. Due to the conversion of the structure from RC to PPC by prestressing, the principal stress direction had been changed. As such, the direction of propagation of shear cracks was changed compared to the inclination of the previously produced cracks in the RC state. This is believed to have obstructed the propagation of shear cracks in the specimen after the PPC state was produced; thus, this resulted in the yielding of the main reinforcement. This suggests that the condition of the beam at the introduction of prestress may affect the shear strength, depending on whether the beam is cracked or not. This mechanism was not clarified within the scope of the present study. It is suggested that further study should be conducted in this regard.

4 Comparison of experimental and analytical results

4.1 Flexural strength

A simple method for the calculation of flexural strength is not to be established when external cables are used. In this study, flexural strength was calculated by the method proposed in the past studies [2] [3], where an analytical technique which gives consideration to compatible conditions for displacements and changes in the cable positions is used. Fig. 5 shows the variation of load with mid−span displacement for specimen No. 3. There is good agreement in the behavior of the structure, especially, in the predicted and observed yield loads. The slight difference in the analytical and experimental curve, generally after the development of shear cracks, may be attributed to the fact that the influence of shear was not incorporated in the above calculations.

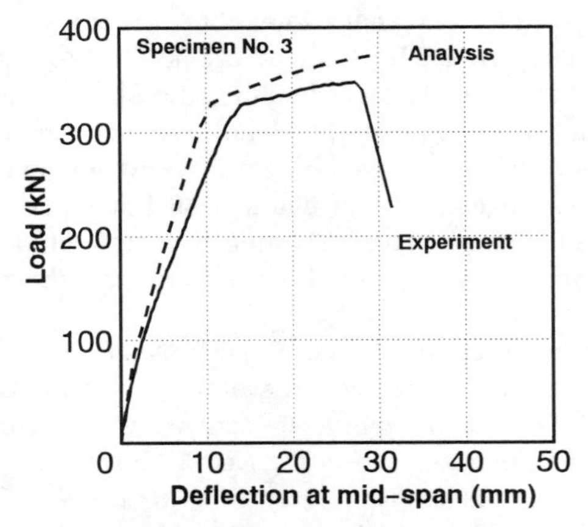

Fig. 5 Comparison of load–deflection curves.

4.2 Shear strength

The shear strength was estimated in accordance with the shear strength formula given in the specifications [3] and compared with the experimental values.

$$V = V_c + V_s + V_p \tag{1}$$

$$V_c = 0.9(f_c)^{1/3} \cdot \beta_d \cdot \beta_p \cdot (1 + 2M_0 / M_u)\, b_w \cdot d \tag{2}$$

$$\beta_d = (100/d)^{1/4}, \quad \text{if } \beta_d > 1.5 \text{ then } \beta_d = 1.5$$

$$\beta_p = (100 p_w)^{1/3}, \quad \text{if } \beta_p > 1.5 \text{ then } \beta_p = 1.5$$

$$V_s = (A_w \cdot f_{wy} \cdot d / 1.15) / s \tag{3}$$

$$V_p = P \sin \alpha \tag{4}$$

where;

- V : Shear strength(kgf)
- f_c : Compressive strength of concrete(kgf/cm^2)
- d : Effective depth(cm)
- p_w : Tension bar ratio
- M_0 : Decompression moment(kgf·cm)
- M_u : Bending failure moment at mid–span section(kgf·cm)
- b_w : Web width(cm)
- A_w : Sectional area of stirrup(cm^2)
- f_{wy} : Yield strength of stirrup(kgf/cm^2)
- s : Stirrup spacing(cm)
- P : Introduced prestress(kgf)
- α : Bent–up angle of external cables($^\circ$)

Table 4. Summary of test results in comparison to calculated results

Specimen No.	Type of beam	Cracking load at shear			Ultimate load		
		Exp. (kN)	Cal. (kN)	Exp./Cal.	Exp. (kN)	Cal. (kN)	Exp./Cal.
No. 1	RC	---	---	---	255	272 [1)]	0.94
No. 2	RC	78	73	1.07	260	141	1.84
No. 3	PPC	127	107	1.19	348	373 [1)]	0.93
No. 4	PPC	132	106	1.24	294	207	1.42
No. 5	RC → PPC	---	---	---	363	208	1.75

1) Due to the observed failure modes of specimen No. 1 and No. 3 as flexure, the calculated value is based on flexural calculation[2] [3].

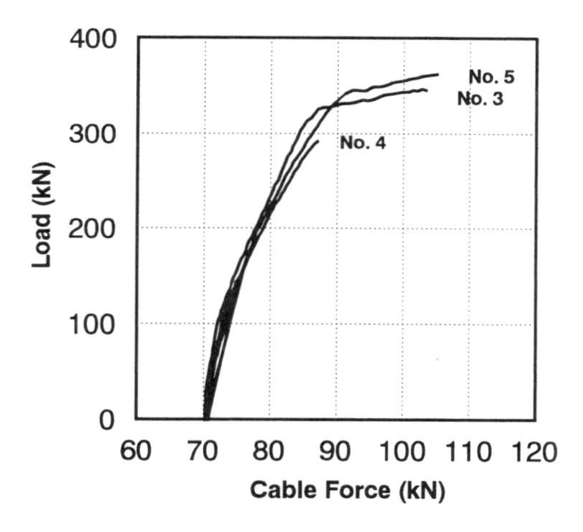

Fig. 6 Variation of load with cable force.

A comparison of the experimental values with the calculated ones are given in Table 4, while Fig. 6 shows the variation of load with cable force. The calculated load at which the shear cracks appear (based on equation (2)), shows practical agreement with experimental results. As for the shear strength, the experimental values were found to exceed the calculated values.

This difference may be due to the following assumptions made in equation (1): a) the web is to resist the total shear neglecting the flange area, b) the inclination of cracks is taken as 45 degrees and c) the increase in the external cable force as the load increases is not taken into account. These points call for further study.

5 Conclusions

The following may be concluded based on the experimental and analytical results from this study.

1. As a result of prestressing, there was an increase in the cracking load by about an 86% for the flexural case and a 68% for the shear.
2. The flexural and shear strength showed an increase of approximately a 36% and a 13%, respectively, due to the influence of prestressing.
3. Strengthening effects may be obtained, even in a specimen previously cracked in the RC state, by introducing external prest ressing.
4. There is good agreement between the experimental and analytical results in the case of flexure.
5. In the case of shear, the cracking load results practically agreed with the calculated values. The ultimate load, however, was found to be far greater than the calculated values, meaning that the experimental values were on the safer side.

From the present study, we were able to obtain basic data related to the flexural and shear strengthening of existing structures. The results of this study are encouraging. However, further study is recommended to establish a suitable design method for such applications.

6 Acknowledgment

The authors wish to thank Mr. K. Tsuchida and Mr. T.Yamaguchi (former graduate students of Saitama University) for their assistance during the course of the experiment and analysis as well as graduate student Mr. T.Aravinthan for his helpful suggestions in preparing the manuscript.

7 References

1. Sano, M., Tokumitsu, S., Maruyama, K. and Mutsuyoshi, H. (1993) Retrofit of damaged PC beams by external prestressing cables. *FIP symposium*, Vol. 2, pp. 923–926. (in English)
2. Matupayont, S., Mutsuyoshi, H., Tsuchida, K. and Machida, A. (1994) Loss of tendon's eccentricity in external prestressed concrete beam. *Proceedings of the JCI*, Vol. 16, No. 2, pp. 1033–1038. (in English)
3. Mutsuyoshi, H. and Machida, A. (1993) Behavior of prestressed concrete beams using FRP as external cable. *ACI SP–138*, pp. 651–670. (in English)
4. Japan Society of Civil Engineers. (1991) *Standard Specification for Design and Construction of Concrete Structures (Design)*. JSCE, Tokyo.(in Japanese)

Conversion factors

1 MPa = 10.1972 kgf / cm^2
1 N / mm^2 = 10.1972 kgf / cm^2
1 kN = 101.972 kgf

169 LOCAL FAILURE MODE AND STRENGTH OF CABLE ANCHOR BY ONE-THIRD SCALE MODEL TESTS

M. SAKURADA and S. KAMIYAMA
Hokkaido Development Bureau, Obihiro, Japan
S. HANADA and M. INOUE
Hokkaido Engineering Consultants, Sapporo, Japan
T. TAKEDA
Kajima Technical Research Institute, Tokyo, Japan

Abstract

Tokachi Bridge is a 3–span continuous prestressed concrete cable–stayed bridge with a bridge length of 501m and a center span of 251m. While the cable anchors play an important role in ensuring the safety of cable–stayed bridges as a whole, their design methods with regard to local failure due to concentrated loads are not yet fully established. The tests were conducted on a 1/3–scale model of the Tokachi Bridge cable anchors to investigate the local failure mode and to confirm their strengths. In addition, the actual stress of the cable anchors on the bridge under construction was measured by strain gauges, and the adequacy of the design was confirmed by comparing the test results with measured values.

Keywords: Cable anchor, cable–stayed bridge, failure mode, model test, strength

1 Introduction

The Tokachi Bridge, which is now being constructed on the Tokachi River is a 3–span continuous prestressed concrete cable–stayed bridge with a total length of 501m and a center span of 251m. The superstructure is a 32.8m, four–cell box girder. The cables are fixed to the central part of the main girder section, because they are arranged as a semi–harp system with single central cable plane. The general arrangemant of the bridge is shown in Figure 1.

In cases of the prestressed concrete cable–stayed bridge arranged along a single plane, the cable tensile forces generally concentrate on the cable anchors at the central part of the main girder section. Because the local failure of the cable anchors has a great influence on the safety of the bridge in its entirety, these cable anchors must be comfirmed to have sufficient loading capacity. However, the cable anchors

Concrete Under Severe Conditions: Environment and loading (Volume Two) Edited by K. Sakai, N. Banthia and O.E. Gjørv. Published in 1995 by E & FN Spon. ISBN 0 419 19860 1

have complicated structures and the mechanism of their behavior is not well understood and so reliable design methods are not yet established. At present, the arrangement of steels is determined using FEM analysis.

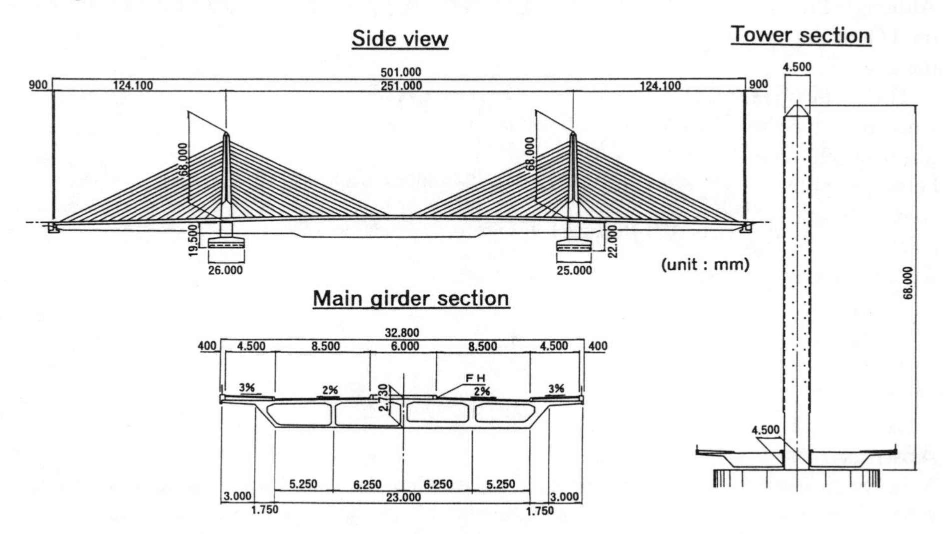

Fig. 1. General arrangements.

Large–scale model tests were conducted before the construction of large cable–stayed bridge such as the Aomori Bay Bridge and the Usui Bridge for confirmation of their behavior and strengths and the results were used in design of the bridges.

The main girders of these bridges were 3–cell box girder sections and the cables were anchored to the intersection of the upper slab and the cross beam of the central box. Besides, becouse the cable was arranged with an angle of approximately 60 ° , the cable tension had large vertical component. As a result, those bridges showed failure modes, such as the punching shear of the anchors. In the case of Tokachi Bridge, however, two cables are anchored to the cross beam, sandwiching the middle web, and the cables were arranged with an angle of 45°C. Therefore the cable tension had large horizontal component as well as vertical component. As a result, the stress transmission and the stress distribution are different from those at the Aomori Bay Bridge and the Usui Bridge, and it was judged that past test results could not be used in evaluating the strength of the cable anchors of this bridge against local failure.

Hence, load tests using large–scale model were conducted aiming at the following:

(1) Investigate the mode and mechanism of the failure of cable anchors.
(2) Develop a method to estimate the strength of the cable anchors.
(3) Confirm that the strength of the cable anchors of the actual bridge is over the specified tensile strength of the stay cable (1.0 Pu, Pu: specified tensile strength of the stay cable).

In addition, the stress of the cable anchers on the bridge under construction was measured by strain gauges.

2 Specimen and the test method

2.1 Design of specimen

Although full–scale testing of anchorage is more appropriate, the test was conducted on 1/3–scale model, considering the available laboratory equipment and the cost for the tests.

The tests were concerned the cable anchors of the cable with the shortest length since they take the largest vertical component of cable tension when subjected to the ultimate factored load during earthquakes. As is shown in Figure 2, the interval between cables of this bridge is narrow over the side spans, and wide over the center span, owing to the arrangement of cables at the tower. In the tests, the side span was selected as the object of the tests, as it was considered to be under greater influence from punching shear.

To simplify the construction process of the specimen, the 2 central cells under much influence from punching failure were strictly reduced to the scale of 1/3, and each of the outside cells, being under little influence, was modeled by a beam whose bending and axial rigidities are the same as those of 4–cellular boxes combined. Thus, the 4–cellular box girder was modeled by a 2–cellular box girder section, as shown in Figure 3. In the design of the specimen, FEM analysis was conducted in order to confirm that the stress and its distribution near the anchors of the specimen would be the same as those of the actual bridge when $(1/3)^2$ of the tension force during the ultimate factored load of the actual bridge was applied to the cables of the specimen.

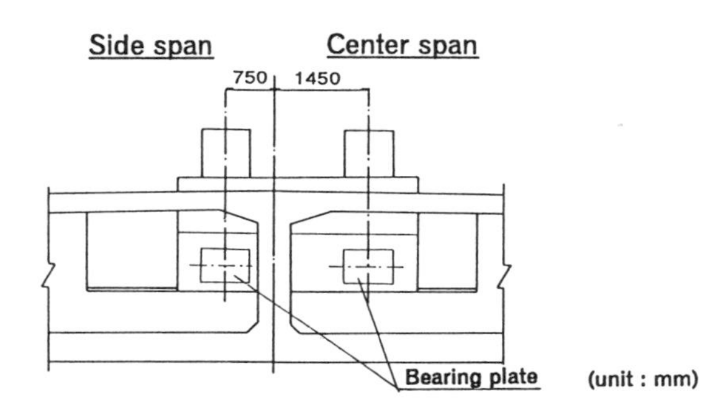

Fig. 2. Type of the anchors.

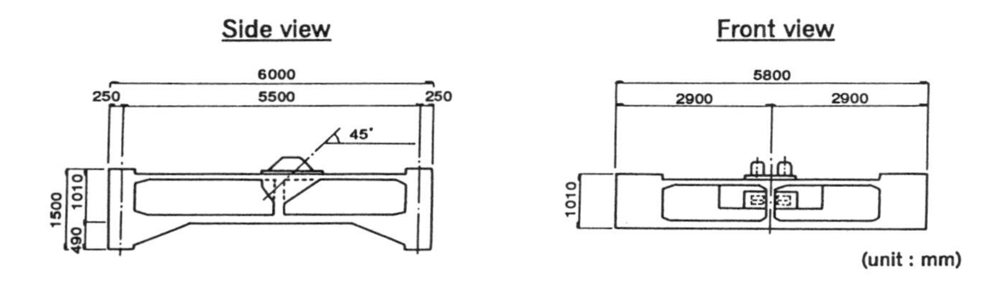

Fig. 3. Structure and size of the specimen.

2.2 Materials

The maximum aggregate size in the main girder of the actual bridge was 20 mm, and the specified concrete compressive strength was 400 kgf/cm ^{2} . In the test specimen, the compressive strength was kept the same, but the maximum aggregate size was reduced to 5 mm. The steel ratio in the test section was accorded with that of the actual bridge. The size of reinforcing bars was accorded with the reduced scale of the model, using items corresponding to SD345, such as D6 (D19 in the actual bridge) and D10 (D25 in the actual bridge). The prestressing steel with the same material characteristics and with a diameter corresponding to the reduced scale was used.

The use of 2–F.K.K. Freyssinet stay cables (61 T15.2 mm) was planned for the actual bridge. However, it was necessary to confirm how much safety the cable anchors have against the specified tensile strength of the cable members. Therefore, 8 prestressing steel bars (ϕ 36) were used as cable members in the specimen so that approximately 3 times the specified tensile strength could be loaded.

2.3 Loading method

The cable tension was applied by 2 jacks containing a central threaded rod, which reacted against a portal frame. The axial force of the main beam was applied by 3 jacks from the reaction block installed in the axial direction of the specimen. The cable tension and the axial force were applied until the ultimate factored load state, so that, at the position of cable anchors, the stress caused by the bending moment and the axial force might be equal to that in the actual bridge. After reaching the state of the ultimate factored load, the cable tension was increased to the point of failure, keeping the axial force constant.

The tests were conducted for three days. On the first day, the specimen was unloaded after reaching the ultimate factored load. On the second day, it was loaded to a cable tension of 760 tf (2.11 Pu) and then unloaded. On the third day, the same loading hysteresis as that on the second day was carried out until failure.

3 Test results

3.1 Failure process

The changes in the specimen condition according to the amount of cable tension until the moment of failure are described below.

3.1.1. The first day

(1) 230tf (0.64 Pu: during the application of the service load): A bending crack occurred in the upper slab of the main beam.

(2) 240tf (0.67 Pu): A crack occurred in the cross beam and the middle web around the anchors.

(3) 281tf (0.78 Pu: during the application of the ultimate factored load): With the expansion of the crack, the rate of deformation increased more than that observed in elastic range.

3.1.2. The second day

(1) 300tf (0.83 Pu): A crack occurred in the cross beam near the upper slab.

(2) 600tf (1.67 Pu): The crack in the middle web on the side of the bearing plate increasely grew. Another crack was found in the middle web, also on the side of the tower. After this time, a rotational displacement of the whole cable anchor around the axis of the tip of anchor block on the side of the tower became notable, as shown in Figure 4. The overall rigidity also began to decrease.

(3) 760tf (2.11 Pu: the maximum load): The vertical steel near the anchors of the cross beam nearly yielded, and the vertical steel of the middle web on the side of the bearing plate yielded. The load scarcely rose, and only the deformation tended to increase. (The condition of specimen at the time of the maximum load is shown in Figure 4.)

3.1.3. The third day

(1) The residual deformation of the second day was made the initial value of the third day, and the specimen showed the behavior that was progressing foward the maximum point on the second day.

(2) 700tf (1.94 Pu): While the crack at the side of the anchors on the tower side and the bearing plate began to be notable, the load scarcely rose, and only the deformation increased.

(3) 748tf (2.08 Pu): Bending failure of specimen occurred, resulting from the compression failure of the lower slab.

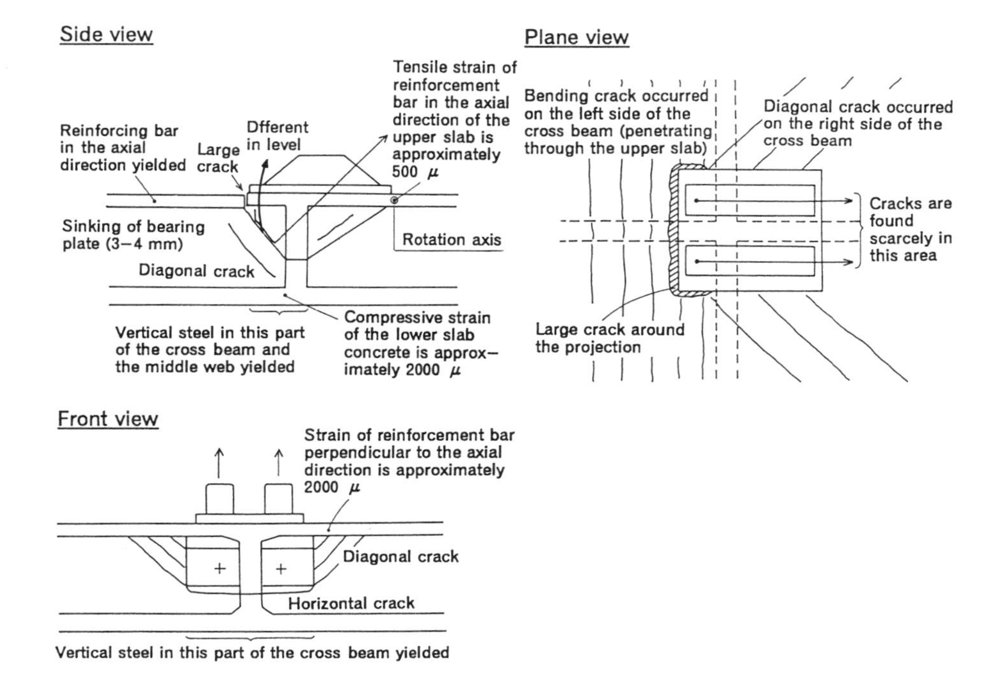

Fig. 4. Condition of the specimen during the maximum load.

Figure 5 shows the relation between the cable tension and the displacement of the point where the cross beam and the middle web intersect.

Figure 6 shows the displacement distribution at each load stage.

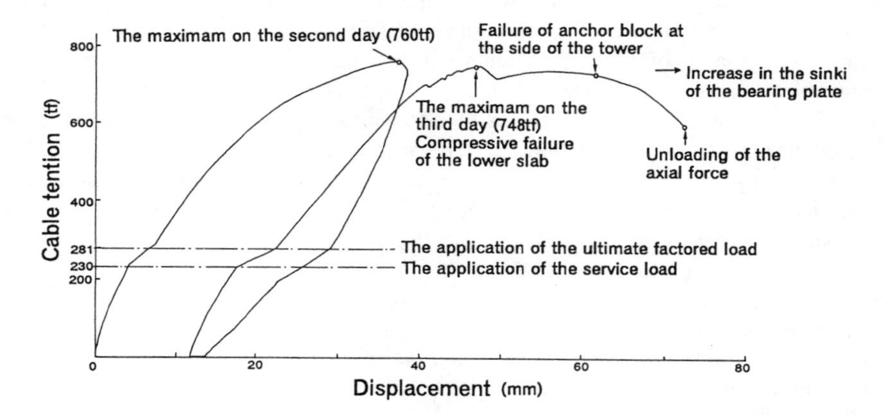

Fig. 5. Load–displacement relation.

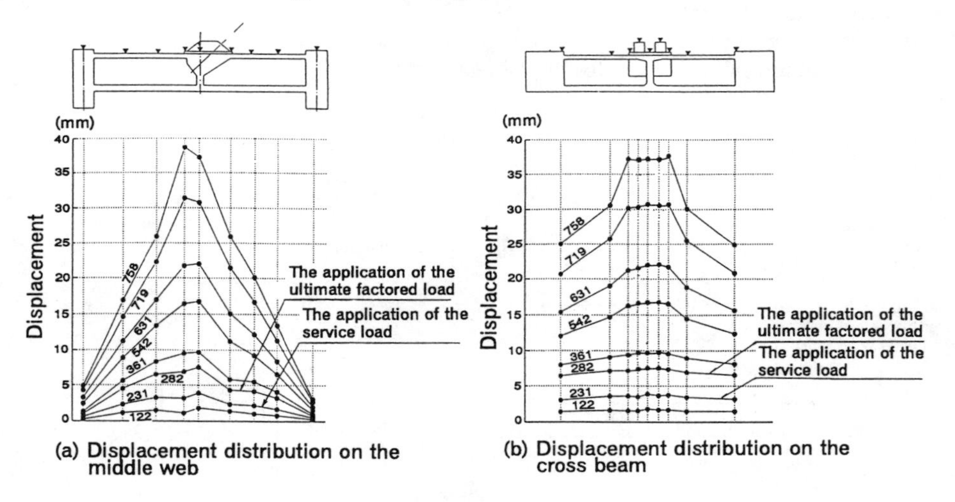

Fig. 6. Vertical displacement of specimen.

3.2 Failure mode of the specimen

The ultimate failure mode of the specimen was considered to be bending failure caused by the compressive failure of concrete of the lower slab, which occurred when the cable tension was 748 tf. When the loading continued, however, the load did not rise and only the deformation of the specimen increased; the exfoliation of the concrete of the anchors and middle web on the side of the bearing plate and the sinking of the bearing plate were verified. Judging from the time of this occurrence, the bearing failure of concrete near the bearing plate and the bending failure of the specimen seemed to have occurred almost simultaneously.

3.3 Local failure and the strength near the anchors

Judging from the strain distributions (as shown in Figure 8) of the vertical steel and the stirrup of the cross beam and the middle web and the crack pattern of concrete, the failure near the cable blocks appeared to be upward punching with shear rotating on the axis of the junction between the block and the upper slab on the side of the tower. From the failure mode in which most of the major steel, which was considered to contribute to the strength, yielded at the maximum load, the strength was estimated to be nearly equal to the maximum load.

The failure mode in past tests of cable anchors (at the Aomori Bay Bridge and the Usui Bridge) whole cable anchors being punched upwards, and a failure mode such as anchors rotating seen in this test could not be observed then. This appears to be due to the fact that the cable angle in the past tests were approximately 60°, with a large vertical component of cable tension, while the cable angle in this test was 45°, allowing a greater influence of the horizontal component. It also appears that the existence of the middle web provided the strength which enabled the block to resist the reaction force acting on the rotation axis.

In estimating the strength in the failure mode described above, an equation of equilibrium was conceived, using the model shown in Figure 7 and considering the forces acting on the following elements.

(1) Vertical steel of the middle web
(2) Stirrup of the middle web
(3) Reinforcing bar of the middle web in the axial direction of the bridge
(4) Vertical steel of the cross beam
(5) Stirrup of the cross beam
(6) Reinforcing bar of the upper slab in the axial direction of the bridge
(7) Shear transfer at the cracks
(8) Cable tension

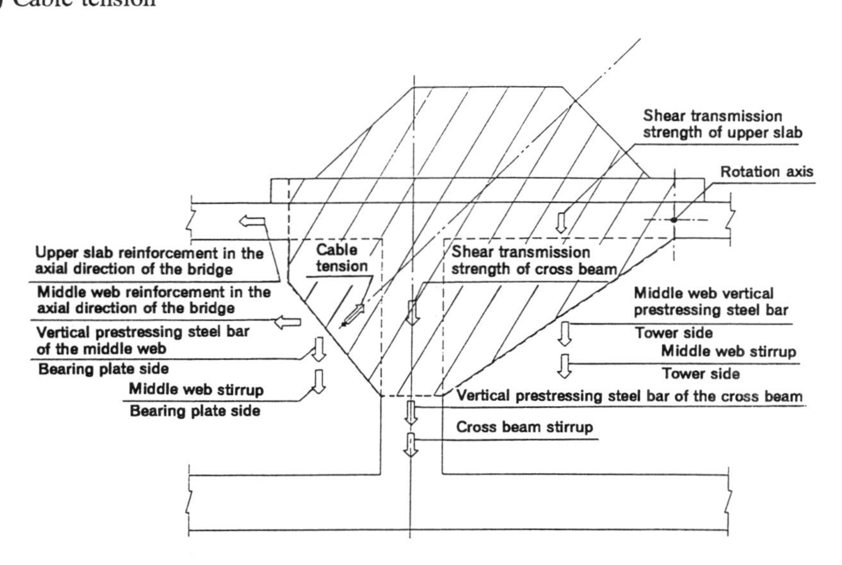

Fig. 7. Model for estimating failure strength.

Among these, the stress distribution models of the vertical steel and the stirrup of the middle web and the cross beam were presumed as shown in Figure 8, based on the measured strain distribution. The reinforcing bar in the axial direction of the middle web on the side of the tower was not taken into account, as its influence was considered to be negligible. The shear transfer at the cracks of the cross beam and the upper slab was evaluated as the shear transmission strength at the junction between the anchors and the cross beam and the junction between the block and the upper slab. According to a past study [4], the shear transmission strength of mortar is approximately 60 % of concrete, owing to the influence of the aggregate diameter. Based on this, 60 % of the value calculated by the equation of the shear strength specified in the Standard Specifications for Design and Construction of Concrete Structures (Japan Society of Civil Engineers) was used in the evaluation of the shear strength in this test.

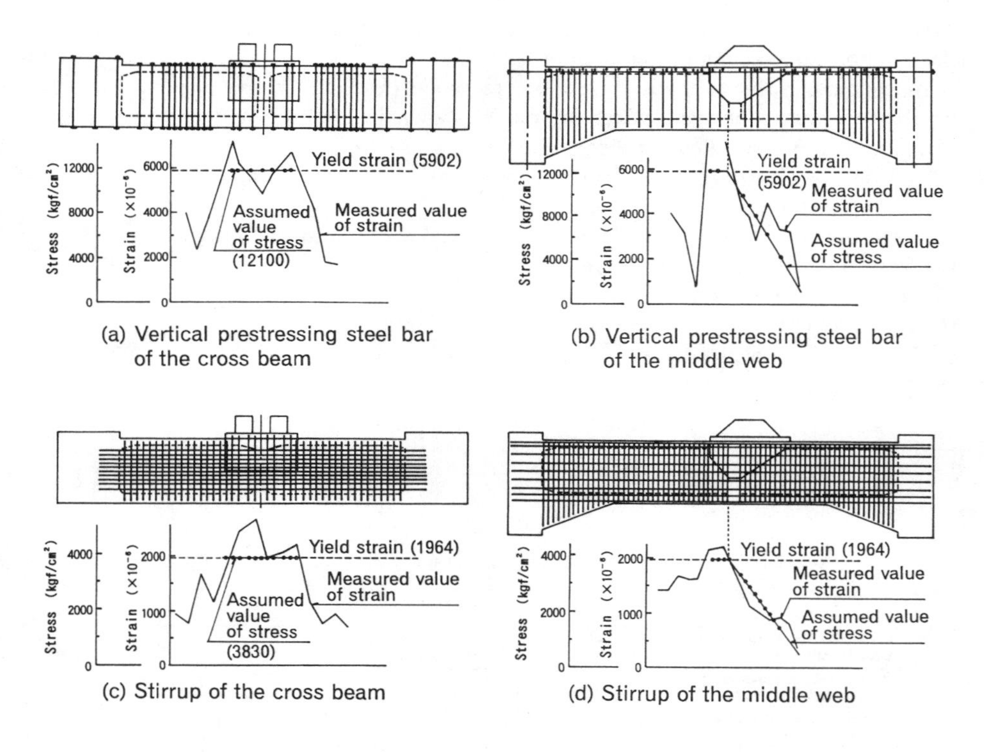

(a) Vertical prestressing steel bar of the cross beam

(b) Vertical prestressing steel bar of the middle web

(c) Stirrup of the cross beam

(d) Stirrup of the middle web

Fig. 8. Stress and strain distribution of prestressing steel bar and stirrup.

Table 1 shows the numerical results of the horizontal force, the vertical force, and the moment at the rotation center, based on the above assumption. From the table, the cable tension, in equilibrium with the strength of the cable anchors, was calculated to be 730 tf, which was 2.03 times as much as the specified cable tensile strength. It was thus confirmed that the numerical results were an accurate estimation of the test results (2.11 Pu).

Table 1. Numerical results

			Horizontal force (tf) (ratio)	Vertical force (tf) (ratio)	Moment (tf•m) (ratio)
Steel	Upper slab reinforcement in the axial direction of the bridge (D6 SD345)		64.8 (92)	0.0 (0)	0.0 (0)
	Vertical prestressing steel bar of the cross beam (ϕ 11 SBPR930/1080)		0.0 (0)	161.0 (35)	112.7 (40)
	Stirrup of the cross beam (D10 SD345)		0.0 (0)	60.1 (13)	42.1 (15)
	Vertical prestressing steel bar of the middle web (ϕ 11 SBPR930/1080)	Bearing plate side	0.0 (0)	23.0 (5)	22.1 (8)
		Tower side	0.0 (0)	35.2 (8)	13.0 (5)
	Stirrup of the middle web (D10 SD345)	Bearing plate side	0.0 (0)	21.9 (5)	19.4 (7)
		Tower side	0.0 (0)	29.8 (6)	11.5 (4)
	Middle web reinforcement in the axial direction of the bridge (D6 SD345)		6.0 (8)	0.0 (0)	-1.7 (-1)
Shear transmission strength		Cross beam	0.0 (0)	65.5 (14)	45.8 (16)
		Upper slab	0.0 (0)	63.6 (14)	19.6 (7)
Total			70.8(100)	460.0(100)	284.5(100)

4 Actual stress

The strain of the cable anchor at the Tokachi Bridge under construction is being measured continuously. Figure 9 shows the location of strain gauges, and Figure 10 shows the relation between the cable tension and the strain.

And the adequacy of the design is being confirmed by comparing the test results with measured values.

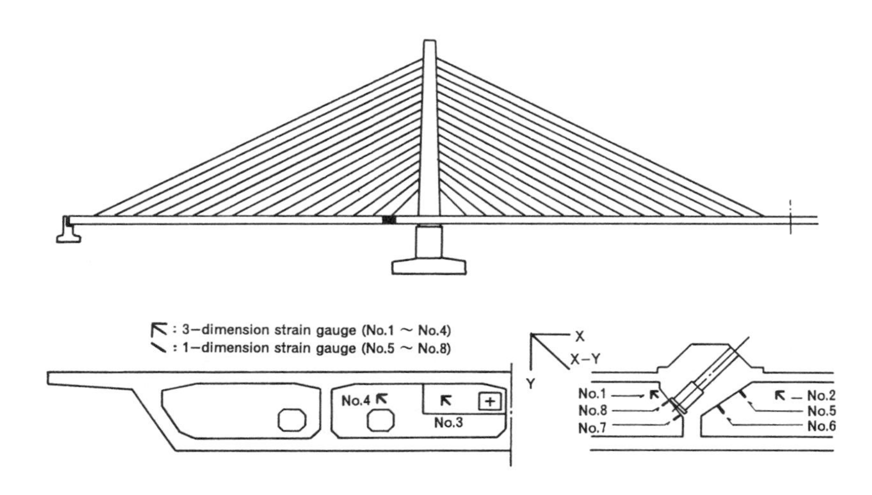

Fig. 9. Locations of the strain gauge.

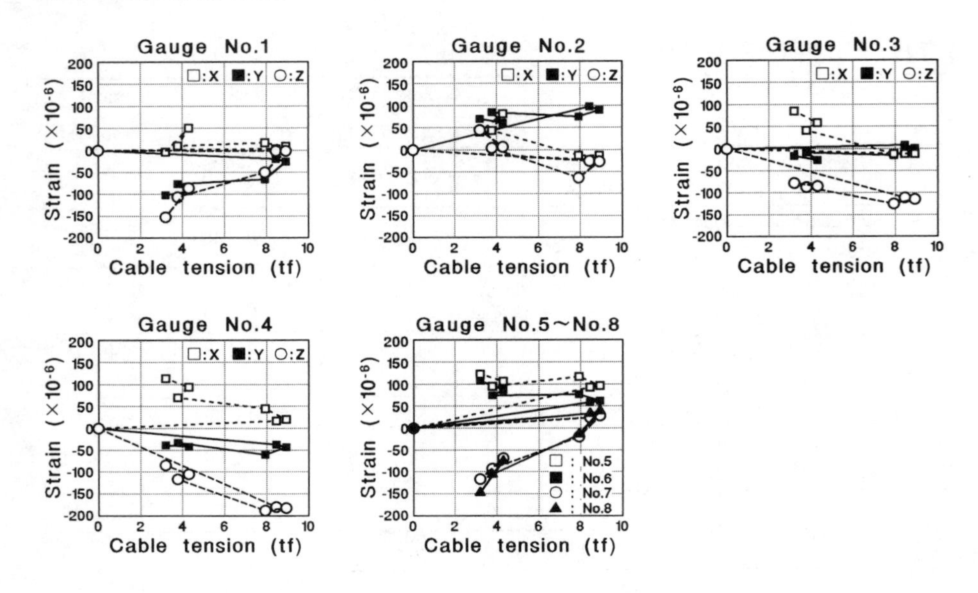

Fig. 10. Relation between the cable tension and the strain.

5 Conclusions

As a result of the test, the maximum strength of the anchors was found to be approximately 2.1 times greater than the specified cable tensile strength. The local failure mode of anchors appeared to have been the following: whole anchors were punched upwards while rotating around the rotation axis as shown in Figure 4. The calculated value of the anchor strength, taking into consideration this failure mode, showed a good agreement with the test results. When the anchor strength of the actual bridge was estimated in this method, it was confirmed that it was higher than the specified tensile strength of the cables.

6 References

1. Japan Road Association (1990), *Specification for Highway Bridges and Commentaries, Ⅲ Concrete Bridge*
2. Japan Society of Civil Engineers (1991), *Standard Specifications for Design and Construction of Concrete Structures*
3. Li, B. and Maekawa, K. (1988), Stress transfer constitutive equation for cracked plane of concrete based on the contact plane density function, *Concrete Journal*, Vol.26, No.1
4. Yamada, K. (1982), A study on the design of RC containment for nuclear power stations with particular reference to rationaligation of RC shell elements against shear forces, *Special Report of the Technical Research Institute of Maeda Construction Co.,Ltd.*, No.22–1

170 THREE DIMENSIONAL SHEAR FAILURE OF RC COLUMNS

T. ICHINOSE
Nagoya Institute of Technology, Nagoya, Japan
K. HANYA
Nippon Steel Corporation, Futtsu, Japan

Abstract
The effects of sub-ties and spacing of shear reinforcement on shear strength of reinforced concrete beams are discussed; tests were conducted on four specimens having an identical shear reinforcement ratio but different detailings. The observed strength of a specimen with smaller spacing and with sub-ties was 25% higher than that with larger spacing and without sub-tie. After the tests, the specimens were sawed in transverse directions; V-shaped crack patterns were observed in the specimen with smaller spacing and without sub-ties, indicating three dimensional shear failure.

Analyses were performed assuming rigid-plasticity of steel and concrete, utilizing the upper bound theorem of plasticity. Shear failure was assumed to occur three-dimensionally, sliding at V-shaped cracks connecting the corners of shear reinforcement. The analyses quantitatively explained the observed effects of detailing on shear strength. The analyses also explained the observed difference in shear failure.
Keywords: beam, crack, detailing, plasticity, reinforced concrete, shear reinforcement, shear strength, spacing, strain, sub-ties

1 Introduction

Application of plastic theory to reinforced concrete members has been successful in predicting their shear strengths [1]. The theory has been accepted in design codes such as in Europe [2] and Japan [3]. These codes assume that shear reinforcement is distributed densely enough, but the definition of "denseness" has not been made clear. Also, the codes assume two dimensional stress field; but if a member section is wide enough and does not have sub-ties, the stress might not be uniform in the direction of member width. Experiments by Kobayashi et al. [4] showed that shear strength is not very sensitive to

Concrete Under Severe Conditions: Environment and loading (Volume Two) Edited by K. Sakai, N. Banthia and O.E. Gjørv. Published in 1995 by E & FN Spon. ISBN 0 419 19860 1

the spacing of shear reinforcement. On the contrary, analyses by the authors [5] showed that the effect of spacing on shear strength is large when shear reinforcement ratio is large. This paper investigates the effects of sub-ties and the spacing of shear reinforcement on shear strength of reinforced concrete beams experimentally and analytically.

2 Experiment

2.1 Loading apparatus and specimens

The loading and measuring apparatus is shown in Fig. 1. The two jacks were controlled to achieve antisymmetric deformation in the shaded region in Fig. 1. Rotation angles of the two ends of the specimens were monitored by the two displacement transducers.

The compressive strength of concrete was 24.8 MPa. The yield strength of shear reinforcement was as high as 1397 MPa to enhance the effect of sub-ties and spacing. The yield strength of longitudinal reinforcement was as high as 1020 MPa to prevent flexural failure.

Details of the specimens are shown in Table 1 and Figs. 2 and 3. The first portion of the specimens' names, In or Out, means that supplementary ties were placed in the *inner* part of the section or on the perimeter (*outer* part). The second portion, 50 or 150, indicates the spacing of shear reinforcement. In specimens In150 and Out150, spiral bars were used to make the shear reinforcement ratio the same as those of the 50 series, p_w = 1.71 %. Specimen In50t had lateral tie-bars with anchor plates at their ends to examine the effects of lateral constraint.

2.2 Test results

The relationships observed between shear force and deflection are shown in Fig. 4. The strength of In50 was 25% higher than that of Out150 though the shear reinforcement ratio was the same. Specimen In50t showed a response almost identical to that of In50,

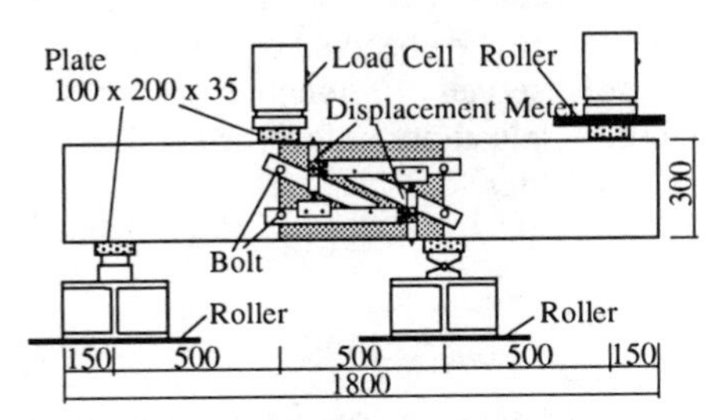

Fig. 1 Loading and measuring apparatus

Table 1 List of specimens

Specimen	Pitch (mm)	Inner Tie	Age (days)
In 50	50	2	35
Out 50		0	35
In 150	150	2	34
Out 150		0	33
In 50t	50	2 (Tie-Bar)	36

Shear reinforcement ratio p_w = 1.71%

Table 2 Stirrup stresses

Specimen	Stirrup Stress (MPa)	
	at 95% Strength	at 100% Strength
In 50	225	301
Out 50	191	225
In 150	149	195
Out 150	83	148

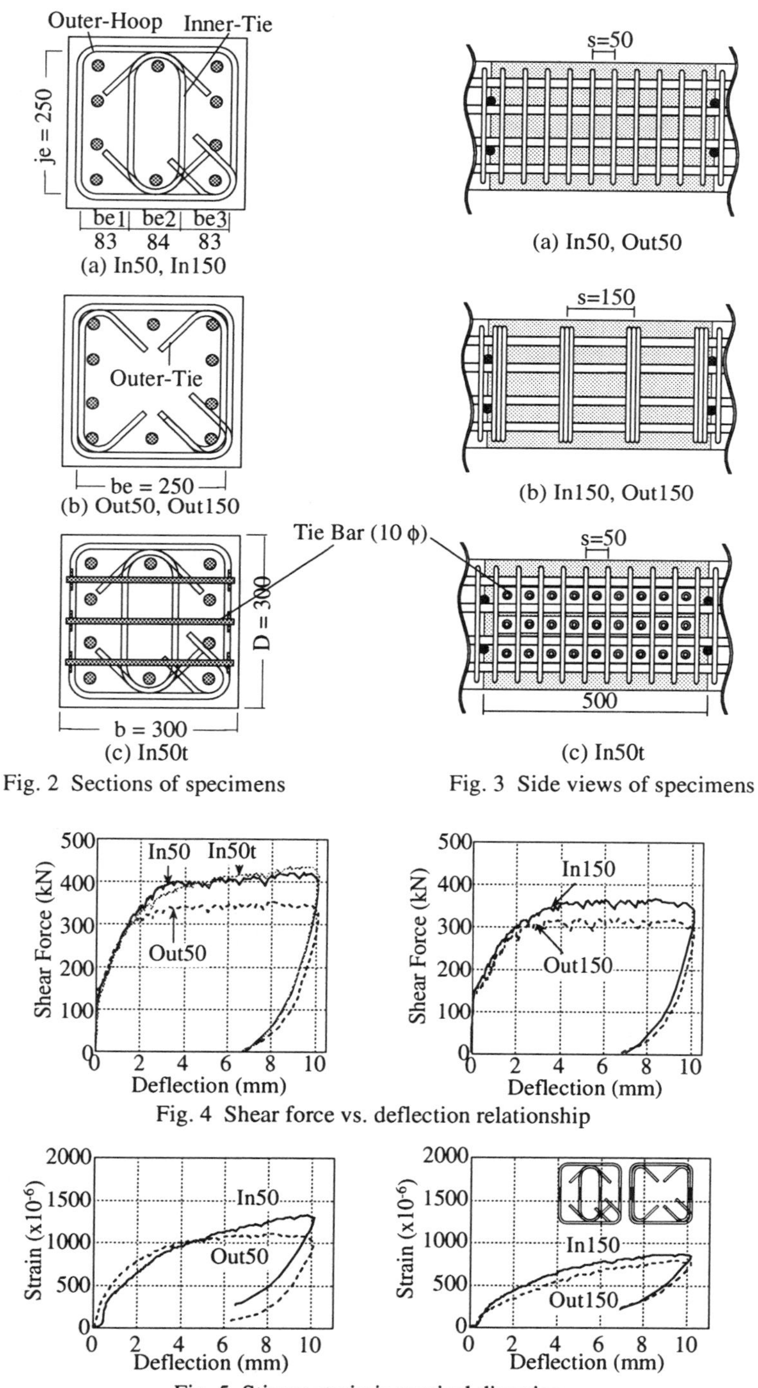

Fig. 2 Sections of specimens

Fig. 3 Side views of specimens

Fig. 4 Shear force vs. deflection relationship

Fig. 5 Stirrup strain in vertical direction

which means that the effect of tie-bars was small.

Figure 5 shows the relationships between stirrup strain and deflection. The strain of specimen In50 was about twice that of Out150. Table 2 shows the stresses in the stirrups when the shear forces of the specimens reached 95% and 100% of their full strengths. These are used in the analysis presented below.

The vertical axis on the left of Fig. 6 shows the curvature of stirrups measured with strain gages on the upper and lower faces of the stirrup. Assuming that the curvature was distributed uniformly, the uplift at the center of stirrup was calculated and shown in Fig. 6. Uplift was large in the Out-series specimens, which implies three dimensional shear failure explained later.

3 Analyses

3.1 Assumptions

1) Longitudinal reinforcement is sufficiently strong and rigid in the longitudinal direction.
2) Dowel action of longitudinal reinforcement is sufficiently small.
3) Shear reinforcement is rigid-plastic.
4) Concrete is rigid-plastic having a yield locus as shown in Fig. 7, where s_e is effective strength calculated as follows. (See Ichinose et al. [5] for derivation.)

$$\sigma_B < 20 \text{ MPa} \quad \sigma_e = 0.85\,\sigma_B \tag{1}$$

$$\sigma_B \geq 20 \text{ MPa} \quad \sigma_e = 0.4\,\sigma_B + 9 \text{ (MPa)} \tag{2}$$

where σ_B is compressive strength of concrete.Tensile strength of concrete is negligible.

5) Concrete is effective inside the shaded region in Fig. 8, which is confined by the outer stirrup. [Statically, the reason for ignoring the cover concrete is that the compressive field of the truss action inside a section is as shown in Fig. 9. Kinematically, the inclination of the failure line in the cover η' is smaller than that inside the confined region η, as shown in Fig. 10. Thus the internal work by the upper and lower cover is negligible. Similar phenomena will occur also in the right and left sides of the section.]
6) Concrete fails at the failure plane shown in Fig. 11(a) if the section does not have sub-

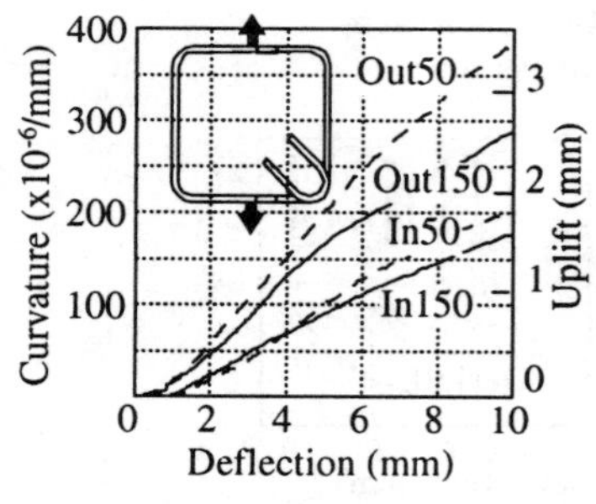

Fig. 6 Curvature and uplift of stirrup

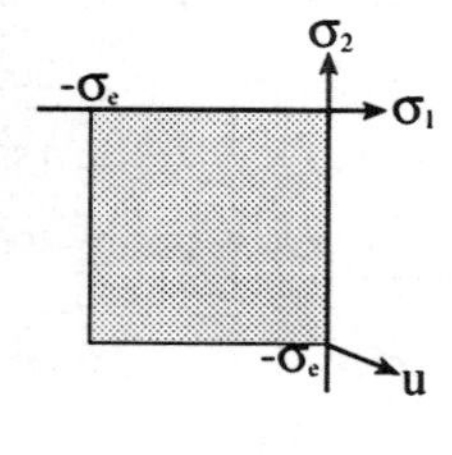

Fig. 7 Yield locus of concrete

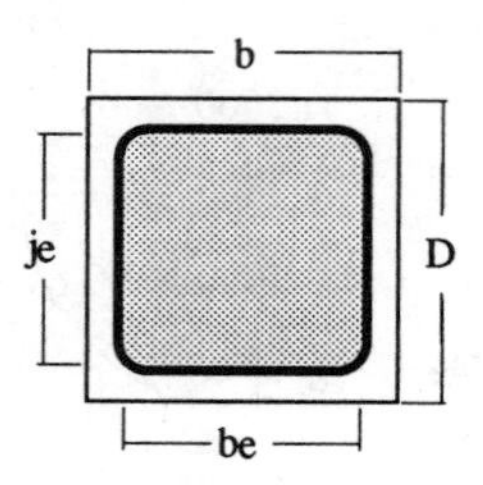

Fig. 8 Effective concrete section

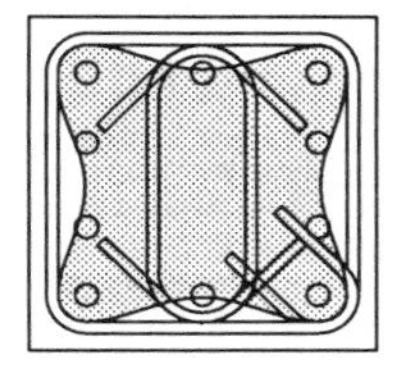

Fig. 9 Compressive field due to truss action

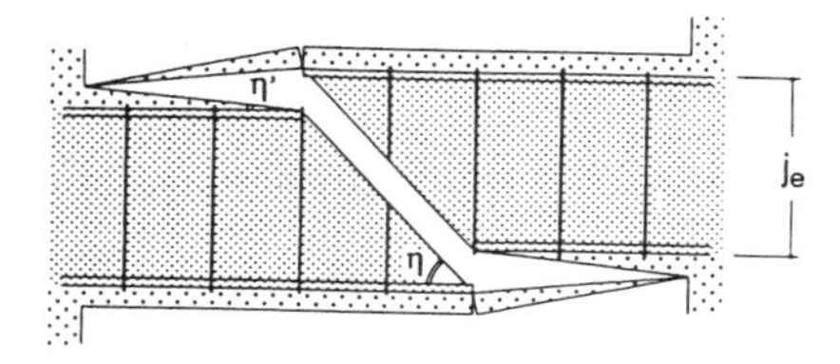

Fig. 10 Failure modes with cover concrete

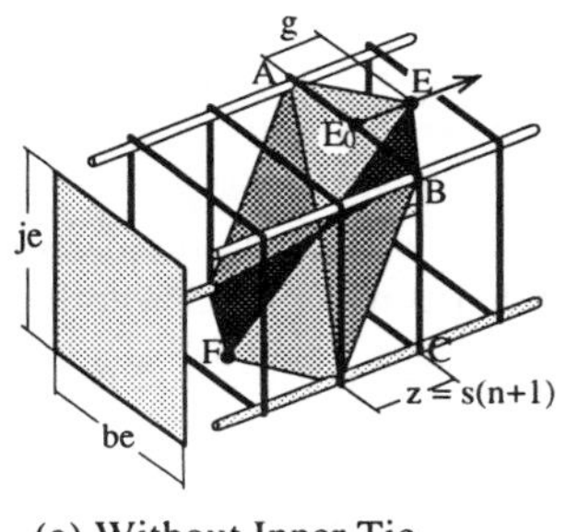

(a) Without Inner Tie

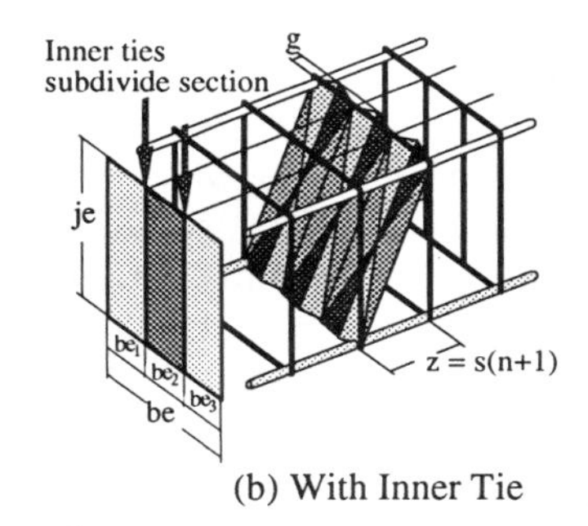

(b) With Inner Tie

Fig. 11 Assumed failure plane

ties. If it does, failure plane would be as Fig. 11(b). [The depth of the valley, *g* in Fig. 11, is calculated according to the upper bound theorem.]

3.2 Shear strength

Since the longitudinal reinforcement is sufficiently strong and rigid, the relative displacement occurs vertically as shown in Fig. 10. The work equation is as follows:

$$W_e = W_c + W_s \tag{3}$$

where W_e is the external work done by the shear force; W_c and W_s are the internal work done by concrete and shear reinforcement, respectively.

The internal work done in the minimal failure plane, *dA* in Fig. 12, is as follows:

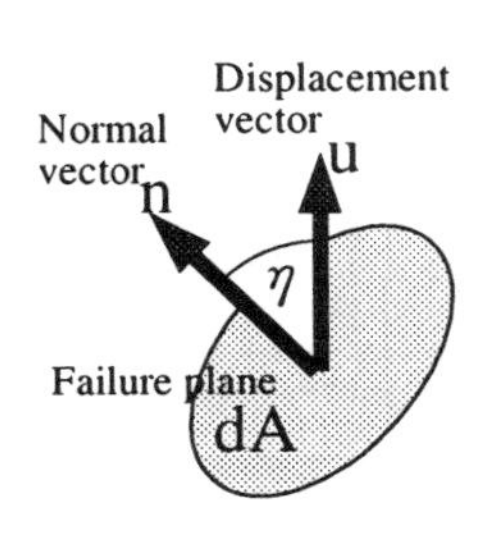

Fig. 12 Minimal failure plane

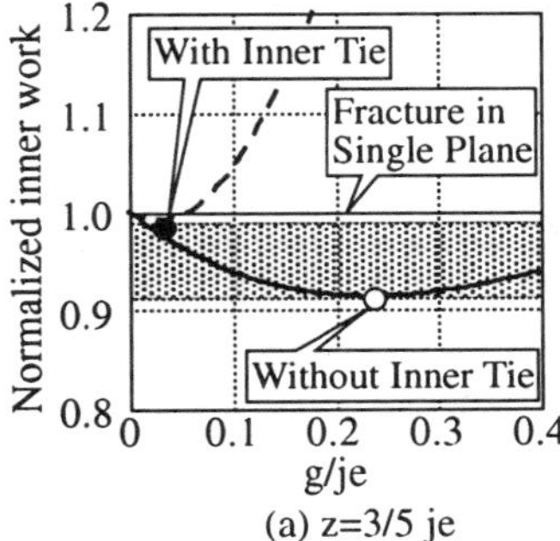

(a) z=3/5 je

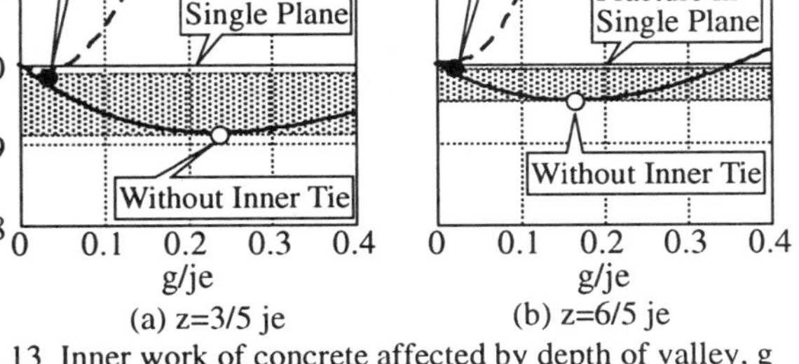

(b) z=6/5 je

Fig. 13 Inner work of concrete affected by depth of valley, g

$$dW_c = \frac{(1 - \cos \eta_i)\sigma_e dA}{2} u \tag{4}$$

where η_i is the angle between the displacement vector u and the normal vector n.

Integrating this over the whole failure plane, we have W_c. This W_c is a function of the depth of the valley, g, as shown in Fig. 13.

3.3 Results

The analytically obtained relationship between normalized shear force vs. stirrup stress is shown in Fig. 14, where *n* is the number of stirrups crossing a failure plane (e.g., in the case of Fig. 11, *n* = 0). Theoretically, *n* must be zero, since the stirrup used in the specimens was strong enough. The white circles in Fig. 14 show the predicted strengths and the stirrup stresses of the specimens. The black circles show the predicted strengths corresponding to the stirrup stresses at 95% of the strengths in Table 2, which will be called 'quasi-theoretical results' hereafter. The white squares show the observed full strengths and the corresponding stirrup stresses. The black squares show 95% of the observed strengths and the corresponding stirrup stresses. Analytical results generally agreed with observations. However, the analysis slightly overestimated the strength of Out 50 and underestimated that of In 150. In other words, the analysis overestimated the effects of stirrup spacing and underestimated the effects of sub-ties. These must be solved in future research.

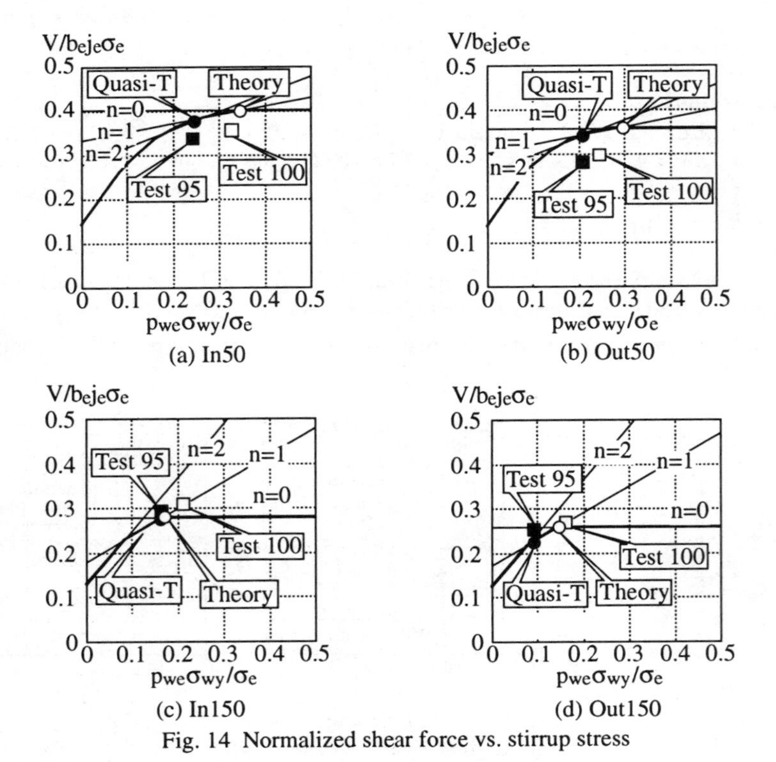

Fig. 14 Normalized shear force vs. stirrup stress

The theoretical failure planes of the theoretical results are shown in Fig. 15. After the tests, the specimens were sawed in longitudinal and transverse directions. The theoretical failure planes in the sawed sections are shown by the broken lines in Figs. 16 through 19, whereas the quasi-theoretical failure planes are shown by the solid lines. The observed cracks in the sawed sections as well as the cracks observed on the surface are shown in Figs. 20 through 23. The observed cracks would almost correspond to fracture planes, if not completely. We should note that the inclination of cracks would vary according to non-linearity of the specimens: the theoretical results would represent the fracture at ultimate states whereas the quasi-theoretical results would represent the fracture at 95% strength. Thus, the observed cracks should be compared with both the theoretical and quasi-theoretical failure planes.

Through such comparison, we can note the following.

1) The inclinations of the observed cracks inside the specimens are generally steeper than those on the surface.
2) The cracks are more dense in the specimens of 50 series than in those of 150 series. This corresponds with the calculated results that the specimens of 50 series have more fracture planes than those of 150 series. In Fig. 22 (c) of In 150, we can observe cracks connecting stirrups like the analysis, Fig. 18 (b)
3) In Fig. 21 (d) of Out 50, we can observe the V-shaped cracks in the upper half of the section like the analysis, Fig. 17 (c).
4) The observed inclination of the cracks in Fig. 21 (b) of Out 50 along the central main bars is steeper than those in Fig. 21 (c) along the outer main bars as they are in the analysis, Figs. 17 (a) and (b).

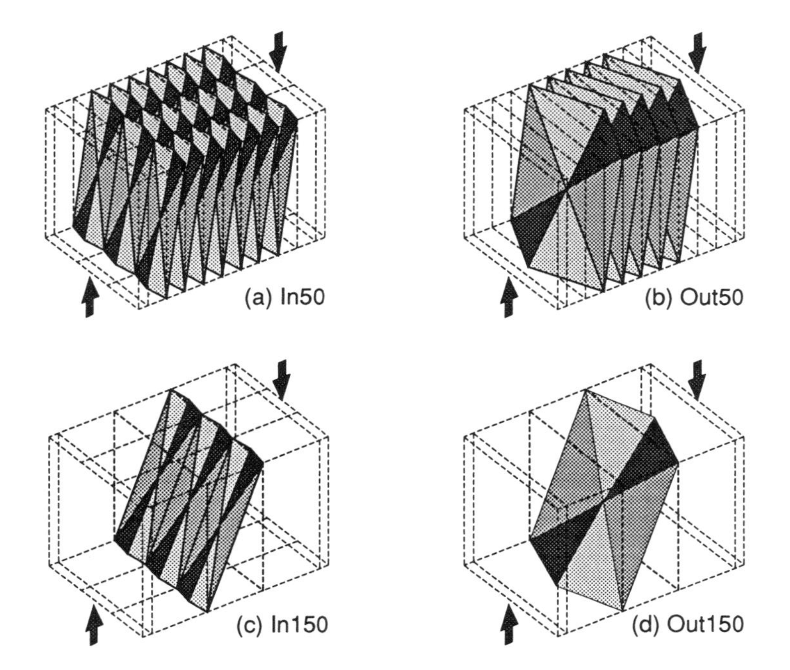

Fig. 15 Theoretical failure planes

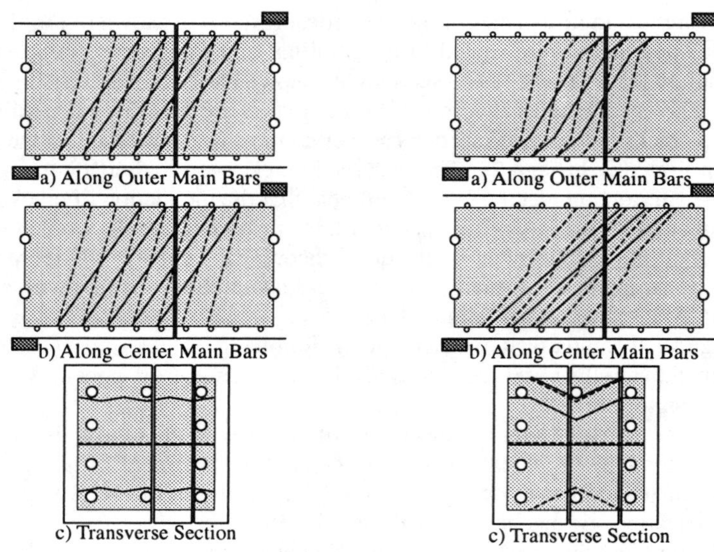

Fig. 16 Calculated failure section of In 50

Fig. 17 Calculated failure section of Out 50

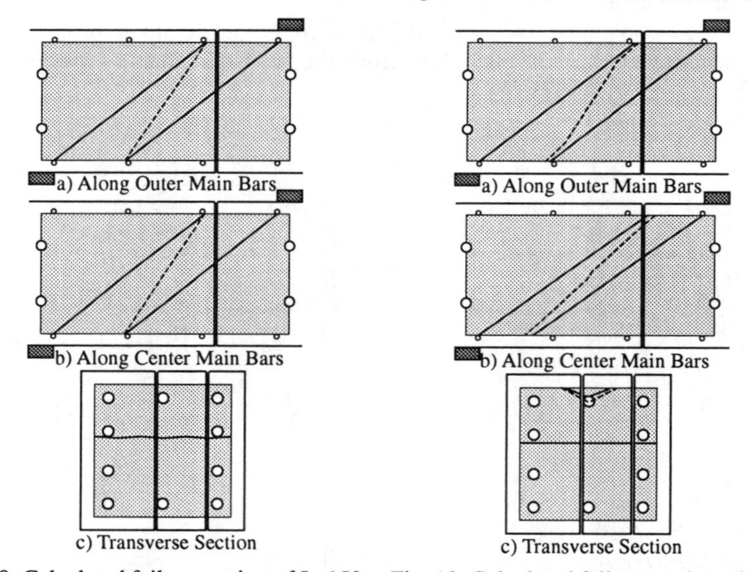

Fig. 18 Calculated failure section of In 150

Fig. 19 Calculated failure section of Out 150

4 Conclusions

Spacing of shear reinforcement and sub-ties affect shear strength of RC member as much as 25 %. We may explain these effects using the three dimensional fracture planes as shown in Fig. 15 and the upper bound theorem of the limit analysis. The observed crack patterns inside the specimens were similar to the calculated fracture planes.

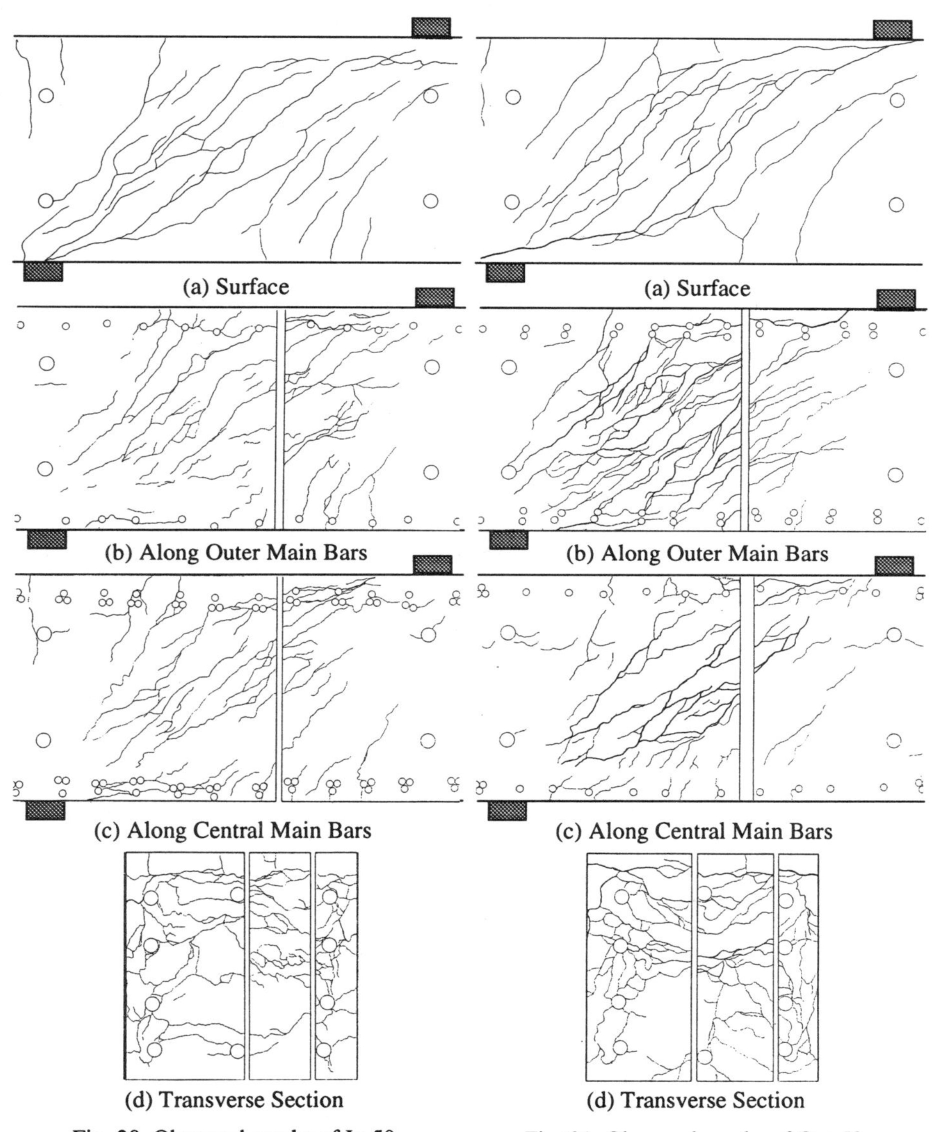

Fig. 20 Observed cracks of In 50

Fig. 21 Observed cracks of Out 50

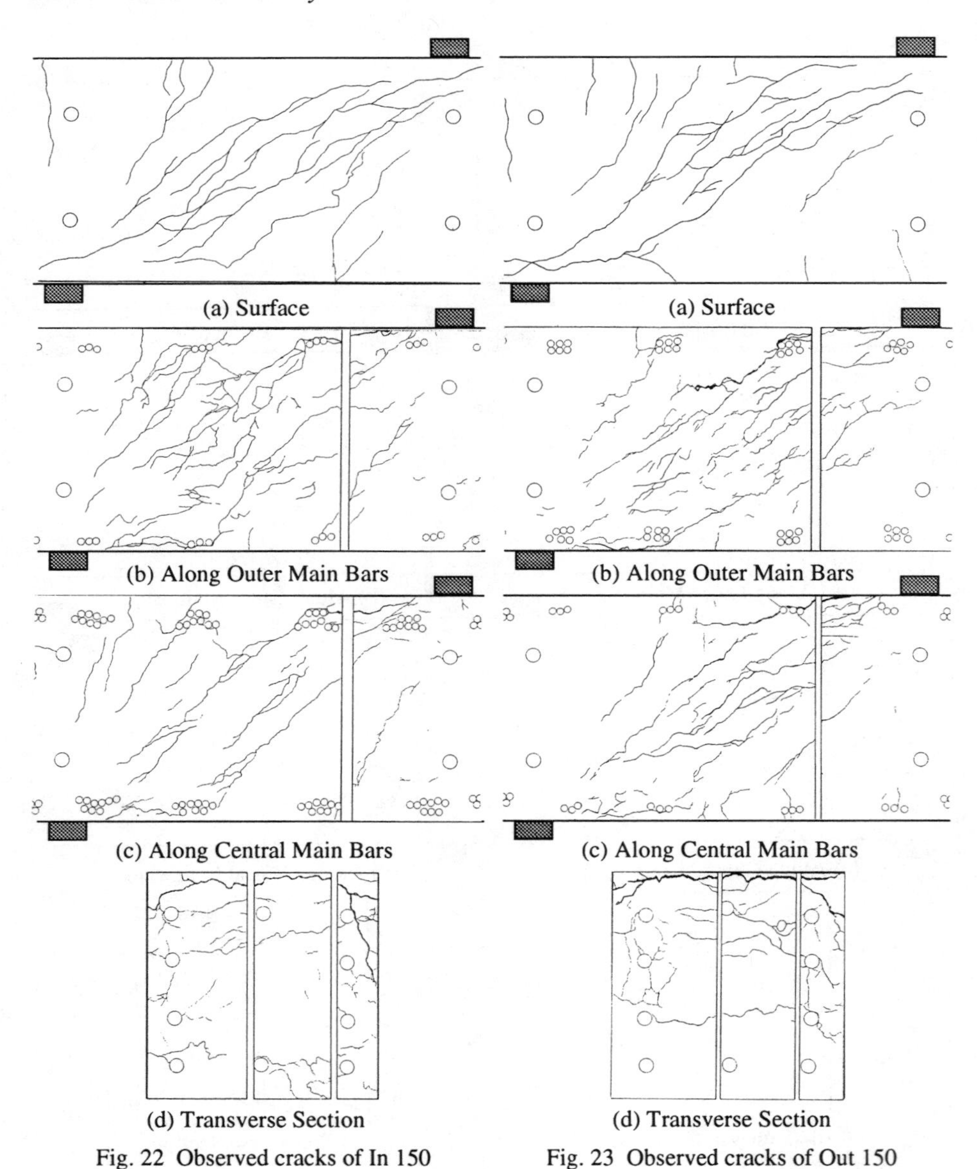

Fig. 22 Observed cracks of In 150

Fig. 23 Observed cracks of Out 150

5 References

1. Nielsen, M. P. (1984). *Limit analysis and concrete plasticity*, Prentice Hall, USA.
2. Comite Euro-International du Beton. (1978) *CEB-FIP model code*, Lausanne, Switzerland.
3. Architectural Institute of Japan. (1990) *Design guidelines for earthquake resistant reinforced concrete buildings based on ultimate strength concept*, Tokyo, Japan. (in

Japanese)
4. Kobayashi, K. et al. (1991). Shear behavior of RC beams with concentrated stirrup arrangement, *Proc. of the Japan Concrete Institute,* Vol. 13, No. 2, pp. 181-184 (in Japanese)
5. Ichinose, T. and S. Yokoo, (1992). Effect of sparseness of shear reinforcement on shear strength of R/C beam, *J. of Struct. and Constr. Engng,* Architectural Institute of Japan, No. 437, pp. 97-103 (in Japanese)

171 MECHANICAL BEHAVIOR OF CYLINDER LAP-JOINTS

K. SASAKI, M. INUZUKA and T. UESHIMA
Hokkaido Institute of Technology, Sapporo, Japan
M. MIYAJIMA
Iwata Construction Co. Ltd, Sapporo, Japan

Abstract
Concrete work in cold regions is often subjected to subzero temperatures. Precast concrete can be a solution to avoid a risk of freezing fresh concrete in winter. The connection of units is the key to success in this procedure. This paper introduces a system of joints which is composed of a splice and a constricting steel cylinder wrapped with a heat–generating wire. The use of sulfur mortar marks a departure from conventional lap–joints. The mortar can be workable to flow over certain temperatures. The cylinder can hold a device which can heat it. During construction, the mortar is pressed into the cylinder heated by electricity. When the mortar is cooled short while after turning off electricity, the joint can be ready for loading. The cylinder plays a vital role in both constricting the lap–joints and heating the limited space. Therefore, the joint is named the cylinder lap–joint. In this paper the connecting procedure in practice is illustrated. Then the experimental data concerning the mechanical behavior of the joints are discussed, in relation to the fracture mechanism. The mortar is stressed by three dimensional compressions with the bond strength increased, which derives the characteristic stress distribution on the steel bars. Also, the ability to control the joint strength by adjusting the heating is discussed. There are many unknown mechanical factors under severe conditions. If the structure can maintain minimum safety in the most adverse conditions, it may be beneficial to develop demountability in precast concrete structures, which will require the capacity to disconnect members by removing joints.
Keywords: Cylinder, lap–joints, design of experiments, heating wire, sulfur mortar, shear strength, tensile strength, three dimensional compression.

1 Introduction

The objective of this research is to obtain information on precast concrete joints at freezing temperatures; the connection of these units is a vital question in construction. If cement mortar in conventional joints freezes, the joints will fail to function. In order to avoid such damage in conventional concrete work, a large scale heating system has been employed. Heating in this manner, however, is accompanied

Concrete Under Severe Conditions: Environment and loading (Volume Two) Edited by K. Sakai, N. Banthia and O.E. Gjørv. Published in 1995 by E & FN Spon. ISBN 0 419 19860 1

with a cost increase and instability of concrete hardening. Therefore the systematic heating of specific joint locations joint can be a solution for this problem in setting winter concrete.

The cylinder lap–joints were originally used for connecting precast members with conventional cement mortar [1]. One advantage is the simple procedure at the construction site. Mechanically, an advantage is related to the use of the cylinder embedded in the precast unit to cover the lap joint. The lap–joint, when subjected to a tensile force, experiences an increased volume of mortar around the joint. The cylinder , restricting the volume change of mortar within it, will increase the mortar bond strength by creating an increased three dimensional compression. This preparation of precast units and their connection at the construction site is different from that of conventional precast concrete. The necessary parts such as steel cylinders, however, are easily available. This method can be combined with grouting materials which change physical properties at different temperatures. Sulfur mortar is commonly available and easy to handle. The combination of the cylinder and sulfur mortar may make a new joint method with unique characteristics as follows:

1. Temperature control in joints becomes extremely easy since the amount of calories required is limited for related materials.
2. Strength of joints becomes controllable, making it possible to either connect or disconnect units at the discretion of engineers.

These points permit concrete construction in winter when temperatures often are sub–zero, since the joints can develop strength regardless of the ambient temperatures. It is also possible to disconnect the units by reheating the joints. Thus, the adjustment of joints becomes possible when it is necessary. In the winter, the ground level can change due to frost heave. Further benefit of the adjustable joints is that the over design of structures is avoidable. If the damage is not serious and the repair costs are assessable, a design anticipating such damage may reduce the construction cost. Particularly in severe climate conditions, repairable structures can be more economical since the loading range can be magnified on assumptions. If systematic repair work proves to be practicable, the joints may provide with concrete structures an application different from the conventional ones. Thus, the present joints need information for connection and disconnection.

2 Outline of cylinder–lap joints with heating device

The lap–joints consist of two steel bars in the joint and the cylinder to cover the lap. The cross section of the side view of the joint is shown in Fig.1. The steel bars belong to different parts: the M (male) part and the F (female) part. The left steel bar is projected toward F from M. The cylinder, wrapped with nickel chrome heating wire, is embedded in F, with two openings left for inserting the sulfur mortar, as shown with the arrows. The M and F parts are connected to form a lap–joint in the cylinder, as shown in Fig.2. Since the space allows free movement, the two parts must be held in the proper position until the connection is completed. Sulfur mortar is inserted into the space only after the position of both has been fixed. Figure 3 shows the cross section of the cylinder when it is filled with sulfur mortar [2].

3 Strength of sulfur mortar

Sulfur powder (through #20 mesh) was mixed with a mixture of natural river sand and the standard one of fine grains. Ingredients and the compaction are closely related since the hardening shrinkage is considerable. Segregation of materials is dependent on temperatures. The experimental results of the strengths in the Michaelis test are shown in Figs 4 and 5. The workability of the mortar was improved with increased amounts of the standard sand. Thus, three mixtures were tested for the mortar.

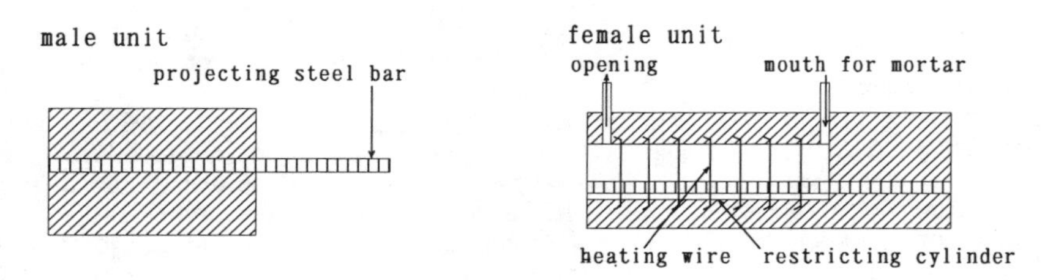

Fig.1 Cross section of units to be connected

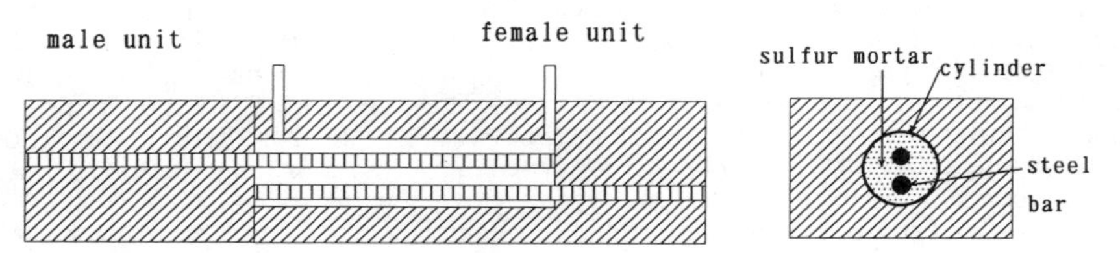

Fig.2 Connected units.

Fig.3 Lateral cross section of unit.

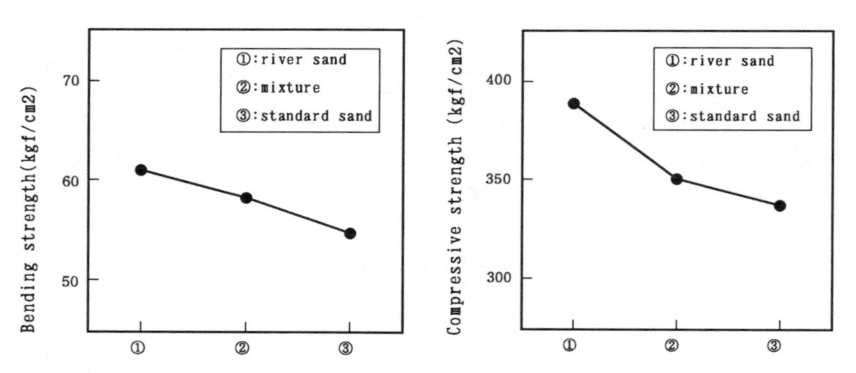

Fig.4 Bending strength.

Fig.5 Compressive strength.

4 Experiments of connecting strength

4.1 Design of experiments

The factorial design method is applied to the experiment of three factors ; kind of aggregate, diameter of the restricting cylinders and lap–lengths are chosen for the experimental factors. Experimental conditions are shown in Table 1. At level 1 in factor A the steel cylinder was with an inner diameter of 44 mm and with thickness of 2.3 mm. At level 3, it was 2.8 mm. The lap–length is given in order to cause the extraction of steel bars avoiding their fracture.

The design of experiments is shown in Table 2 for 27 specimens with experimental results. " No restriction " indicates a lap–joint specimen without the cover cylinder.

4.2 Experimental procedures

Mortar was mixed in a steel pan which was placed on an electrical heater. Each specimen was subjected to a tensile test, pulling out one steel bar while the other held

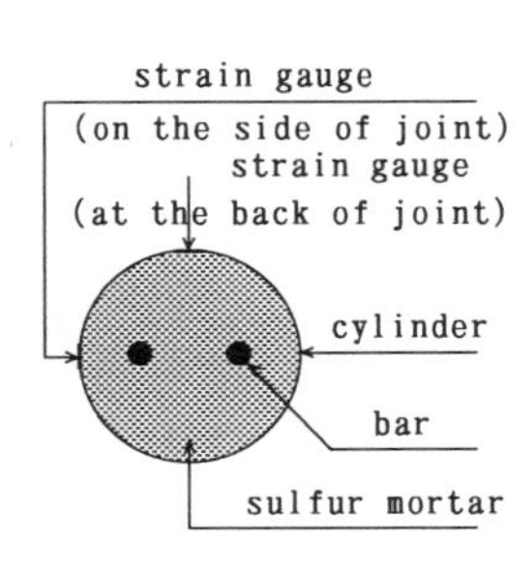

Fig.6 Positions of strain gauges

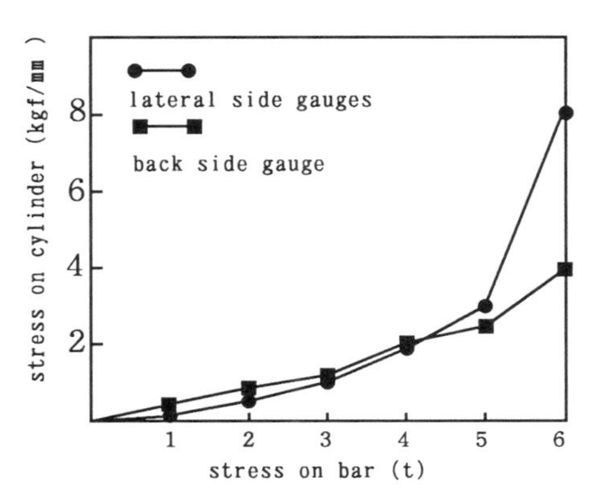

Fig.7 Relationship between bar stress and cylinder stress

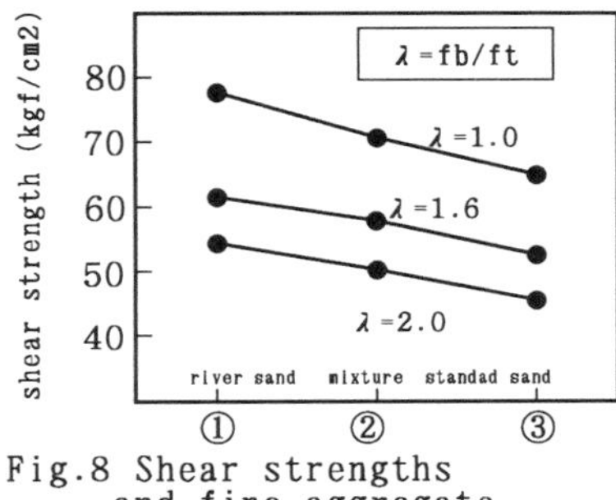

Fig.8 Shear strengths and fine aggregate

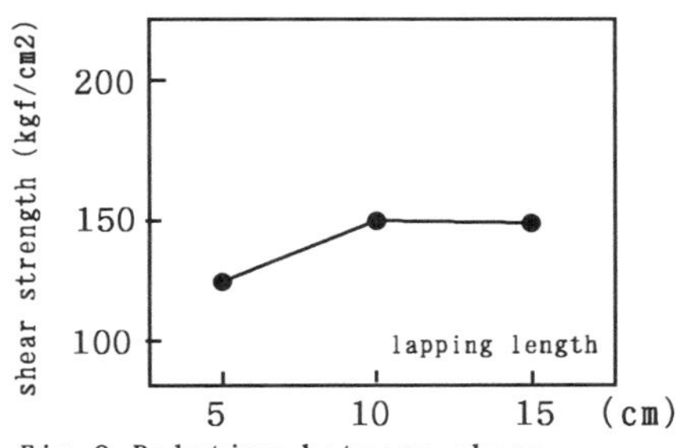

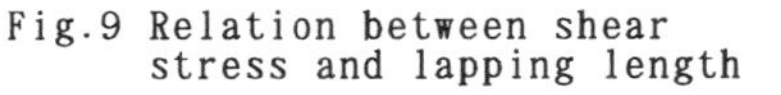

Fig.9 Relation between shear stress and lapping length

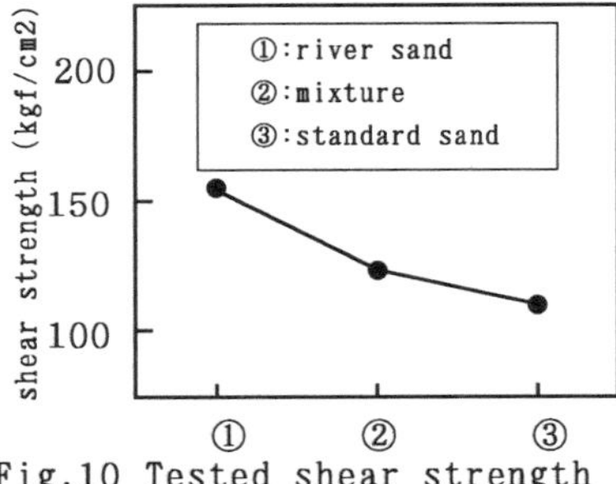

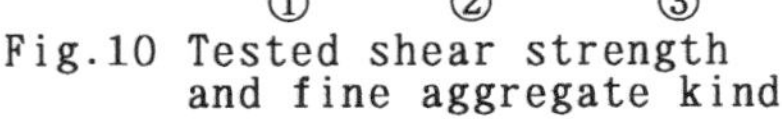

Fig.10 Tested shear strength and fine aggregate kind

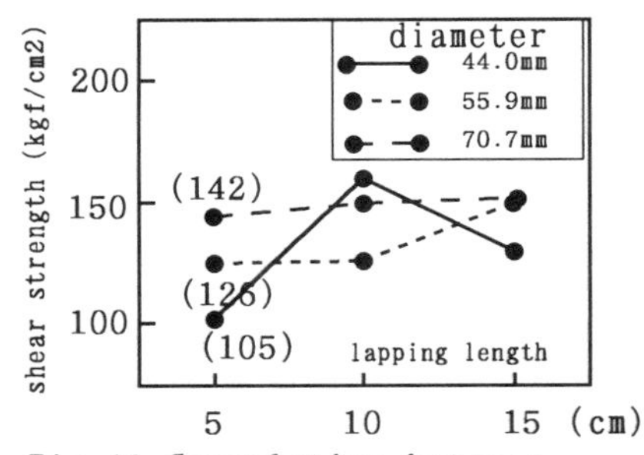

Fig.11 Correlation between lapping length and cylinder diameter

stationary. Loading was the same as that used in the tensile test of steel bars. D indicates the diameter of the bar. Steel bars made of SD30 were all D16 in diameter. The lapping length L was equal to the cylinder length. The joint strength at the maximum load P attained in the test was converted to the tensile stress σs and the seeming shear stress τ of the bar. The cross-sectional area of the D16 bar was assumed to be As. Since the area was slightly smaller than the actual one, the calculated stress σs can be more than the real one. When σs was about 60 kg/sq.mm, either fracture or extraction of steel bars took place.

Another area that was approximated was the cylindrical hole surface which was formed when the steel bar was extracted. This area A was also larger than what was actually formed, so stress τ is less than the real one.

$$\sigma s = P/As \quad \text{where } As = \pi D \ /4$$

$$\tau = P/A \quad \text{where } A = \pi DL$$

4.3 Stress on the restricting cylinder

Obviously, the tensile stress born by the steel bar was multiplied up to 28 times when the restricting cylinder was used, relative to that without a restricting cylinder. As the stress condition is complex, the increase is assumed to be due to the Mohr's fracture criterion. When there was no cylinder over the joint, the principal stress in the wrapping mortar seemed to easily exceed the fracture stress. Besides the cylinder restriction prevents transformation of fractures. Even the shortest cylinder used in the test series seemed to cause three dimensional compression and, thus enhance the bond strength. A distribution analysis on the seeming shear τ is shown in Table 3.

The stresses due to the hoop tension in the cylinder were measured with strain gauges about specimen No.15 in Table 2. The positions of the gauges are shown in Fig.6. The stresses increased as the tensile force increased, as shown in Fig.7.

4.4 Experimental results of joint strength

Since experimental data on three dimensional tests on sulfur mortar are available, the shear strengths assumed are those of cement mortar which has a λ from 1.6 to 2. The λ is the rate between the bending and the compressive strengths. Thus, the assumed shear strength without normal stress is given as follows:

$$\tau = 1/2 \ \mathrm{sqrt}(Fc \ Ft)$$

where Fc is compressive strength and Ft is tensile strength.
Thus the shear strength may be assumed different values of the rate λ, as shown in Fig.8. The rate λ is used for assuming the tensile strength from Fc and Fb experimented.

The analysis of statistical distribution is shown in Table 3. The three factors, lapping length, the kind of sand and the overlap interwoven between AxB, are found statistically significant for the shear increase. The effect of lapping length is shown in Fig.9. The effect of the kind of sand is shown in Fig.10. These three graphs show the considerable increase of shear strength, which proved to be due to the restriction of steel cylinder. Considerable shrinkage takes place when the sulfur mortar hardens; therefore, the three dimensional compression can be distributed in a rather complex manner. Figure.11 shows that an appropriate combination can increase the shear strength. Preliminary experiments showed that joints with certain lapping lengths will break. The bearing stress of the steel bar increased with increasing lapping length, as shown in Fig.12. It is obvious that there is no three dimensional compression when there is no lapping part or the lapping length is zero. Then the shear strength should be as shown in Fig.8. Since the steel stress can not increase when the lapping length exceeds a certain length, the simple average of shear strength should decrease

Table 1. Factors and levels in joints tests

factor levels	1	2	3
A. specimen diameter	44.0	55.9	70.7
b. lapping length	5	10	15
c. fine aggregate	river sand	mixture	standard sand

Table 2. Design of experiments and experimental results

No	A	B	C	restriction		no restriction
				tensile stress (kgf/cm^2)	shear stress (kgf/cm^2)	tensile stress (kgf/cm^2)
1	1	1	1	20	158	1.8
2	1	1	2	13	105	1.6
3	1	1	3	6	51	1.5
4	1	2	1	45	179	1.8
5	1	2	2	40	159	1.6
6	1	2	3	34	134	1.5
7	1	3	1	54	141	2.9
8	1	3	2	51	135	3.4
9	1	3	3	45	119	1.6
10	2	1	1	18	142	2.3
11	2	1	2	16	124	1.6
12	2	1	3	14	111	1.2
13	2	2	1	39	154	4.3
14	2	2	2	30	117	3.2
15	2	2	3	29	114	2.7
16	2	3	1	60	157	5.9
17	2	3	2	59	155	5.5
18	2	3	3	49	128	5.0
19	3	1	1	20	157	4.1
20	3	1	2	18	140	3.3
21	3	1	3	16	129	2.5
22	3	2	1	39	156	5.9
23	3	2	2	39	153	5.8
24	3	2	3	35	139	4.2
25	3	3	1	58	152	7.6
26	3	3	2	57	149	5.6
27	3	3	3	52	136	4.2

Table 3. Distribution analysis on shear strength

factor	sq. sum	deg. of freedom	mean sq. sum	F ratio
A. cylinder diameter	1082	2	541	3.16
B. lapping length	2239	2	1120	6.55 *
C. fine aggregate	6240	2	3120	18.24 **
correlated factor A×B	2787	4	697	4.07 *
correlated factor A×C	1130	4	283	1.65
correlated factor B×C	971	4	243	1.42
error	1368	8	171	

** 1% level of significance * 5% level of significance

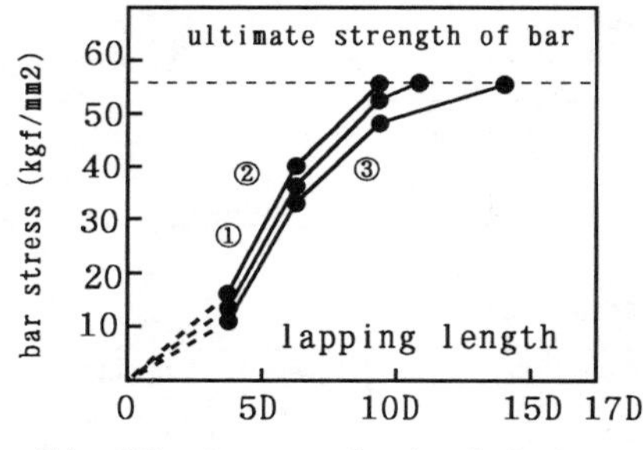

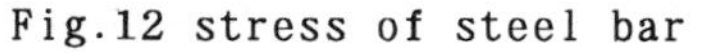
Fig.12 stress of steel bar

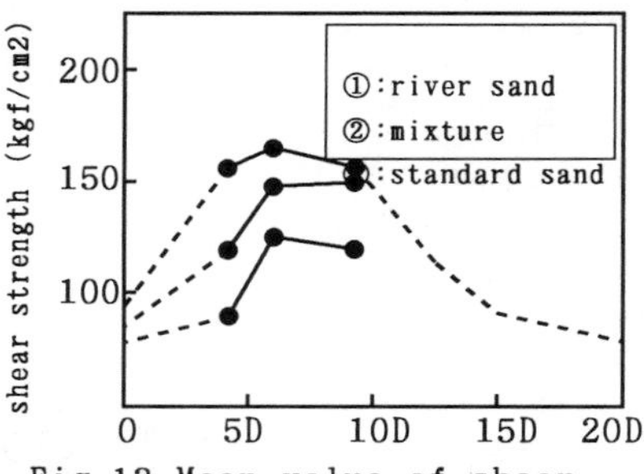

Fig.13 Mean value of shear strength of sulfur mortar

over the length. This relationship is found in Fig.13. Consequently, the joint shows the strength increase due to the reinforced sulfur mortar in a relatively short cylinder. When the joint used mortar with standard sand and its lapping length was 20 times of diameter, the steel bars were able to deform more than the original bar bearing the stress near the ultimate. This joint showed a unique behavior, as shown in Fig.14. The joint can be used for structures in very severe conditions in which larger deformation is allowable when the loading scale exceeds the predicted level.

4.5 Demounting

Demounting, itself, is an interesting topic in concrete engineering [3]. When the joints are to be demounted, steel bars are pulled out at a force which ranges widely. The force required depends on many factors. When joints are at normal ambient temperatures, the steel bar will be subjected to a high stress. The demounting force can be reduced as the volume of mortar, heated over 119 ℃, increases. It is possible to molten all the mortar by heating for days. There may be a demand to demount in a limited period. In order to reduce the demounting load to a minimum at which the self weight of a unit can be barely sustained, a heating system ought to be effective. A system combining sulfur mortar with cement mortar can reduce the heating time required for demounting. When the sulfur volume is minimized for the work, heating time can be reduced as shown in Fig.15.

5 Case studies

Heating joints were applied to a wide culvert on peat. The culvert, which consisted of 17 precast concrete units, was to be completed in midwinter. Low temperatures were severe for concreting. The peat foundation was severe in that there were many unpredictable factors in designing by the conventional philosophy. Incidentally, the culvert had steel grating ceilings for snow to pass through. Over twelve years have passed without any traffic problems since it was completed.
Work procedures may be traced in Fig. 16 which is reproduced by courtesy of the producer of the units, Sanko Kenzai Co, Ltd., Sapporo, Japan.

6 Conclusions

1. A systematic heating system permits two members to be connected with thermoplastic materials in winter. Heating and cooling can be used as the signal for controlling physical properties in order to carry out structural work. Particularly when a non-freezing grouting mortar, such as sulfur mortar, is available for joints, connecting procedure can be simple and reliable at low

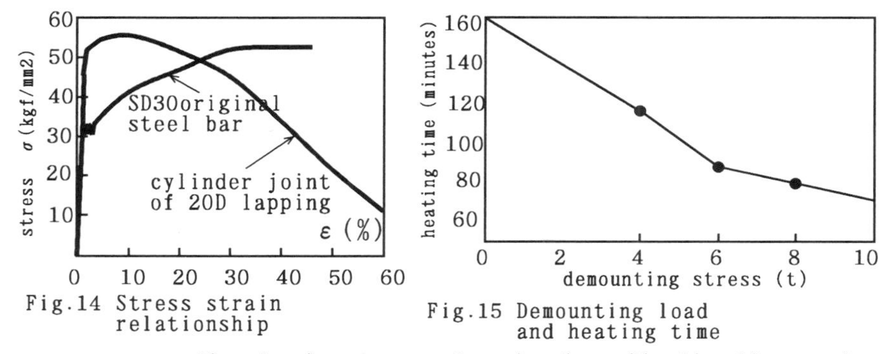

Fig.14 Stress strain relationship

Fig.15 Demounting load and heating time

temperatures. Since heating costs can be reduced considerably, this procedure may prove to be practicable in freezing conditions.

2. Restricting cylinders can play various roles in precast concrete joints. Their restriction of the mortar around steel bars enhances three dimensional compression to raise the shear strength. Besides, a certain safety level of the structure will be retained, because the joint allows a larger deformation than that of the original steel bars. If fractures are not accompanied with critical damage, a concrete design taking such fractures into account may contribute to reductions in construction costs in severe conditions. More benefits may result from the introduction of systematic demountability for repairs in severe conditions.

Fig.16 Wrapping work of heating wire

7 References

1. Inuzuka, M. and Sasaki, K. (1984) New Method of Jointing Members, *Prc. for the Seminar on Tall Structures and Use of Prestressed Concrete in Hydraulic Structures, held in Srinagar,* ING/IABSE, Vol.2, pp.163–72.
2. Inuzuka, M. and Sasaki, K. (1988) Mechanical Behavior of Demountable Splice of Sulphur Mortar , Journal of Structural Engineering, Vol.35 A, pp.1279–90.
3. Reinhart, H. W. et al (editors) (1985) Demountable Concrete Structures, *Prc. of symp. in Rotterdam,* Delft University Press.

Author index

Volume One, pages 1-870; Volume Two, pages 871-1774.

Subject index

Volume One, pages 1-870; Volume Two, pages 871-1774.

This index is compiled from the keywords assigned to the papers, edited and extended as appropriate. The page references are to the first page of the relevant paper.